IMAGING IN DERMATOLOGY

IMAGING IN DERMATOLOGY

Edited by

MICHAEL R. HAMBLIN
PINAR AVCI
GAURAV K. GUPTA

AMSTERDAM • BOSTON • HEIDELBERG • LONDON
NEW YORK • OXFORD • PARIS • SAN DIEGO
SAN FRANCISCO • SINGAPORE • SYDNEY • TOKYO
Academic Press is an imprint of Elsevier

Academic Press is an imprint of Elsevier
125 London Wall, London EC2Y 5AS, United Kingdom
525 B Street, Suite 1800, San Diego, CA 92101-4495, United States
50 Hampshire Street, 5th Floor, Cambridge, MA 02139, United States
The Boulevard, Langford Lane, Kidlington, Oxford OX5 1GB, United Kingdom

Library of Congress Cataloging-in-Publication Data
A catalog record for this book is available from the Library of Congress

British Library Cataloguing-in-Publication Data
A catalogue record for this book is available from the British Library

ISBN: 978-0-12-802838-4

Publisher: Mica Haley
Editorial Project Manager: Lisa Eppich
Production Project Manager: Edward Taylor
Designer: Mark Rogers

Typeset by TNQ Books and Journals

Transferred to Digital Printing in 2017

Dedication

To the love of my life, my beautiful wife Angela, to whom I have been devoted for 36 years.

Michael R. Hamblin

To Ari, Afsin, Atul, Thao, Thomo, Theo, Yair and Zehra whose advice, encouragement and support have been genuine and precious.

Pinar Avci

To my parents, Dr. Ram P Gupta and Sudha Gupta, and my best friend and loving wife, Dr. Tanupriya Agrawal.

Gaurav K. Gupta

Contents

38. From Image to Information: Image Processing in Dermatology and Cutaneous Biology

M.O. VISSCHER, S.A. BURKES, R. RANDALL WICKETT, K.P. EATON

List of Contributors

J. Aguirre Institute of Biological and Medical Imaging, Helmholtz Zentrum München, Neuherberg, Germany

I. Alarcon Ramon y Cajal Hospital, Madrid, Spain

A. Alavi Hospital of the University of Pennsylvania, Philadelphia, PA, United States

Z. Apalla Aristotle University, Thessaloniki, Greece

G. Argenziano Second University of Naples, Naples, Italy

P. Avci Harvard Medical School & Massachusetts General Hospital, Boston, MA, United States

V. Beylergil Metrohealth Campus of Case Western University, Cleveland, OH, United States

M. Bonmarin Zurich University of Applied Sciences, Winterthur, Switzerland

A.C. Bourgeois The University of Tennessee Graduate School of Medicine, Knoxville, TN, United States

Y.C. Bradley The University of Tennessee Graduate School of Medicine, Knoxville, TN, United States

S.A. Burkes University of Cincinnati, Cincinnati, OH, United States

C.H. Camp, Jr. National Institute of Standards and Technology, Gaithersburg, MD, United States

S. Campos Hospital de Santo António dos Capuchos, Centro Hospitalar de Lisboa Central, Lisboa, Portugal

G.C. Casazza University of Utah School of Medicine, Salt Lake City, Utah, United States

F. Castagnetti Arcispedale Santa Maria Nuova-IRCCS, Reggio Emilia, Italy

G. Charles-Edwards Guy's and St Thomas' NHS Foundation Trust, London, United Kingdom; King's College London, London, United Kingdom

K. Chen Nanyang Technological University, Singapore

R. Cicchi National Institute of Optics, National Research Council (INO-CNR), Sesto Fiorentino, Italy; European Laboratory for Non-linear Spectroscopy (LENS), Sesto Fiorentino, Italy

A. Doronin University of Otago, Dunedin, New Zealand

K.P. Eaton University of Cincinnati, Cincinnati, OH, United States

B.M. Erovic Medical University of Vienna, Vienna, Austria

C.L. Evans Harvard Medical School & Massachusetts General Hospital, Boston, MA, United States

S. Fardin Hospital of the University of Pennsylvania, Philadelphia, PA, United States

D.L. Farkas Spectral Molecular Imaging Inc., Beverly Hills, CA, United States; University of Southern California, Los Angeles, CA, United States

X. Feng University of Massachusetts, Lowell, MA, United States

S. Gardini Arcispedale Santa Maria Nuova-IRCCS, Reggio Emilia, Italy

S. Gholami Hospital of the University of Pennsylvania, Philadelphia, PA, United States

K.E. Göbel University Medical Center Freiburg, Freiburg im Breisgau, Germany; University of Freiburg, Freiburg im Breisgau, Germany

V. Goh Guy's and St Thomas' NHS Foundation Trust, London, United Kingdom; King's College London, London, United Kingdom

S. González Ramon y Cajal Hospital, Madrid, Spain; Universidad de Alcalá, Madrid, Spain; Instituto Ramón y Cajal de Investigación Sanitaria (IRYCIS), Madrid, Spain; Memorial Sloan-Kettering Cancer Center, New York, NY, United States

N. Griffin Guy's and St Thomas' NHS Foundation Trust, London, United Kingdom

G. Guan University of Dundee, Dundee, Scotland, United Kingdom; University of Washington, Seattle, WA, United States

G.K. Gupta Harvard Medical School & Massachusetts General Hospital, Boston, MA, United States

M.R. Hamblin Harvard Medical School & Massachusetts General Hospital, Boston, MA, United States

J. Hegyi Inštitút klinickej a experimentálnej dermatovenerológie, Bratislava, Slovakia

V. Hegyi Inštitút klinickej a experimentálnej dermatovenerológie, Bratislava, Slovakia

H.W. Higgins II Brown University, Providence, RI, United States

D. Ho University of California at Davis, Sacramento, CA, United States; Sacramento VA Medical Center, Mather, CA, United States

Z. Huang University of Dundee, Dundee, Scotland, United Kingdom

D. Ioannides Aristotle University, Thessaloniki, Greece

S.L. Jacques Oregon Health & Science University Portland, OR, United States

J. Jagdeo University of California at Davis, Sacramento, CA, United States; Sacramento VA Medical Center, Mather, CA, United States; State University of New York Downstate Medical Center, Brooklyn, NY, United States

N. Jain Seth G S Medical College and KEM Hospital, Mumbai, India

G.B.E. Jemec Zealand University Hospital, Roskilde, Denmark

L. Kadletz Medical University of Vienna, Vienna, Austria

D. Kapsokalyvas European Laboratory for Non-linear Spectroscopy (LENS), Sesto Fiorentino, Italy

U. Khopkar Seth G S Medical College and KEM Hospital, Mumbai, India

E. Kraeva Sacramento VA Medical Center, Mather, CA, United States

A. Lallas Aristotle University, Thessaloniki, Greece

E. Lazaridou Aristotle University, Thessaloniki, Greece

T.K. Lee BC Cancer Agency, Vancouver, BC, Canada; Vancouver Coastal Health Research Institute and University of British Columbia, Vancouver, BC, Canada

F.A. Le Gal Geneva University Hospital, Geneva, Switzerland

A. Lencastre Hospital de Santo António dos Capuchos, Centro Hospitalar de Lisboa Central, Lisboa, Portugal

R.M. Levenson UC Davis Medical Center, Sacramento, CA, United States

C. Li University of Dundee, Dundee, Scotland, United Kingdom; University of Washington, Seattle, WA, United States

Q. Liu Nanyang Technological University, Singapore

C. Longo Arcispedale Santa Maria Nuova-IRCCS, Reggio Emilia, Italy

H. Lui University of British Columbia and Vancouver Coastal Health Research Institute, Vancouver, BC, Canada; BC Cancer Agency, Vancouver, BC, Canada

N. MacKinnon Spectral Molecular Imaging Inc., Beverly Hills, CA, United States

A. Mamalis University of California at Davis, Sacramento, CA, United States; Sacramento VA Medical Center, Mather, CA, United States

I. Markhvida Vancouver Coastal Health Research Institute and University of British Columbia, Vancouver, BC, Canada

D.I. McLean University of British Columbia and Vancouver Coastal Health Research Institute, Vancouver, BC, Canada; BC Cancer Agency, Vancouver, BC, Canada

I. Meglinski University of Otago, Dunedin, New Zealand; University of Oulu, Oulu, Finland

M.M. Monroe University of Utah School of Medicine, Salt Lake City, Utah, United States

A.J. Moy University of Texas at Austin, Austin, TX, United States

V.A. Neel Massachusetts General Hospital, Boston, MA, United States

S.F. Nemec Medical University of Vienna, Vienna, Austria

V. Ntziachristos Institute of Biological and Medical Imaging, Helmholtz Zentrum München, Neuherberg, Germany

M. Omar Institute of Biological and Medical Imaging, Helmholtz Zentrum München, Neuherberg, Germany

Y.H. Ong Nanyang Technological University, Singapore

J.R. Osborne Memorial Sloan Kettering Cancer Center, New York, NY, United States; Weill-Cornell Medical College, New York, NY, United States

Alexandru D.P. Papoiu Therapeutics Inc., San Diego, CA, United States

A.S. Pasciak The University of Tennessee Graduate School of Medicine, Knoxville, TN, United States; The University of Tennessee Medical Center, Knoxville, TN, United States

F.S. Pavone National Institute of Optics, National Research Council (INO-CNR), Sesto Fiorentino, Italy; European Laboratory for Non-linear Spectroscopy (LENS), Sesto Fiorentino, Italy

G. Pellacani University of Modena and Reggio Emilia, Modena, Italy

J.A. Perez Memorial Sloan Kettering Cancer Center, New York, NY, United States

B. Peters University Hospital Antwerp, Edegem, Belgium; AZ Sint-Maarten, Mechelen-Duffel, Belgium

B.W. Petersen Brown University, Providence, RI, United States

S. Piana Arcispedale Santa Maria Nuova-IRCCS, Reggio Emilia, Italy

M. Ragazzi Arcispedale Santa Maria Nuova-IRCCS, Reggio Emilia, Italy

R. Randall Wickett University of Cincinnati, Cincinnati, OH, United States

A.H. Rook Hospital of the University of Pennsylvania, Philadelphia, PA, United States

C. Rowlands Massachusetts Institute of Technology

M. Schwarz Institute of Biological and Medical Imaging, Helmholtz Zentrum München, Neuherberg, Germany

A. Sidoroff Medical University of Innsbruck, Innsbruck, Austria

P.T.C. So Massachusetts Institute of Technology; Laser Biomedical Research Center; Singapore-MIT Alliance for Science and Technology (SMART) Center

L. Tchvialeva Vancouver Coastal Health Research Institute and University of British Columbia, Vancouver, BC, Canada

L. Themstrup Zealand University Hospital, Roskilde, Denmark

J.W. Tunnell University of Texas at Austin, Austin, TX, United States

F.M. Vanhoenacker University Hospital Antwerp, Edegem, Belgium; AZ Sint-Maarten, Mechelen-Duffel, Belgium; University Hospital Ghent, Ghent, Belgium

F. Vasefi Spectral Molecular Imaging Inc., Beverly Hills, CA, United States

M.O. Visscher University of Cincinnati, Cincinnati, OH, United States

H. Wang Harvard Medical School & Massachusetts General Hospital, Boston, MA, United States

L.V. Wang Washington University in St. Louis, St. Louis, Missouri, United States

R. Wang University of Washington, Seattle, WA, United States

T.J. Werner Hospital of the University of Pennsylvania, Philadelphia, PA, United States

O. Westerland Guy's and St Thomas' NHS Foundation Trust, London, United Kingdom

X. Wortsman University of Chile, Santiago, Chile

A.N. Yaroslavsky University of Massachusetts, Lowell, MA, United States; Massachusetts General Hospital, Boston, MA, United States

E. Yew Singapore-MIT Alliance for Science and Technology (SMART) Center

C. Yuen Nanyang Technological University, Singapore

H. Zeng University of British Columbia and Vancouver Coastal Health Research Institute, Vancouver, BC, Canada; BC Cancer Agency, Vancouver, BC, Canada

J. Zhao University of British Columbia and Vancouver Coastal Health Research Institute, Vancouver, BC, Canada; BC Cancer Agency, Vancouver, BC, Canada

Y. Zhou Washington University in St. Louis, St. Louis, Missouri, United States

CHAPTER

1

Introduction to Imaging in Dermatology

M.R. Hamblin, P. Avci, G.K. Gupta

Harvard Medical School & Massachusetts General Hospital, Boston, MA, United States

OUTLINE

Dermatology is one of the most important medical specialties. The prevalence of skin diseases exceeds that of obesity, hypertension, and cancer added together [1]. Skin disease accounts for 12.4% of primary care visits in the United States, and it is estimated that one out of three people in the United States has a skin disease at any given time. Although the number of dermatologists has grown more quickly than the US population (increasing from 1.9 to 3.2 per 100,000 persons between 1970 and 2010, respectively) [2], there is still considered to be an overall shortage of dermatologists [3].

The origins of dermatology have always relied heavily on visual observation of the skin. In *The Canon of Medicine* by Avicenna (who was also known as Ibn Sina) written in Persia in 1025, treatments were described for a variety of skin conditions, including skin cancer [4]. In 1572, Girolamo Mercuriale published in Venice, Italy, *De morbis cutaneis (On the Diseases of the Skin)*, considered to be first scientific work to be devoted to dermatology [5]. Daniel Turner, who was born in London, received the first doctoral degree from the College of the Academy of Yale in Connecticut in the American colonies (later to become the United States) [6]. Interestingly, Turner also published a book of the same name *De morbis cutaneis*, subtitled *A Treatise of Diseases Incident to the Skin* in 1712. This work was the first book published in English devoted to dermatology [7]. In 1777 Anne-Charles Lorry (1726–1783) published the 700 page *Tractatus de morbis cutaneis* and for the first time referred to the skin as an organ [8]. Jean-Louis Alibert (1768–1837) was a pioneer of dermatology in France. Alibert was personal physician to Louis XVIII and Charles X, and was appointed Professor of Materia Medica and Therapeutics in Paris in 1823 [9]. Alibert introduced a classification system for skin diseases that became known as the "Tree of Dermatoses." He produced the first illustrated atlas of dermatology called *Descriptions des maladies de la peau: observées à l'Hôpital Saint-Louis, et exposition des meilleures méthodes suivies pour leur traitement* in 1806 [10]. The Vienna School of Dermatology was founded by Ferdinand von Hebra in the mid-19th century [11] and became the one of leading academic centers for the study of dermatology.

In the last half of the 19th century and going on into the 20th century, dermatopathology assumed an increasingly important role in dermatology, and microscopic examination of biopsies became the gold standard for diagnosis of a wide range of dermatological conditions. However, now that we are well into the 21st century, this status quo may be beginning to change. One of the main reasons for this change is the almost unbelievable rise in the use of noninvasive imaging in (almost) all branches of medicine.

The origin of this explosive growth in medical imaging can be traced back to the discovery of X-rays. Wilhelm Conrad Röntgen (1845–1923) discovered this highly penetrating form of radiation in 1895, and called them *X-rays* (signifying an unknown quantity), although many others referred to them as "Röntgen rays." Röntgen was awarded the Nobel Prize for Physics in

Imaging in Dermatology
http://dx.doi.org/10.1016/B978-0-12-802838-4.00001-7

1901. It did not take long before the new science of radiology was put to practical use. In the Greco-Turkish war of 1897, battlefield radiographic imaging was used to detect bullets in injured soldiers. The radiographs were produced by an apparatus manufactured by the London company Miller and Woods, that was then shipped to Piraeus in 15 crates and powered by electricity from accumulators (a forerunner of lead-sulfuric acid batteries) [12].

In the same year as Röntgen's discovery, Henri Becquerel (1852–1908, Professor of Physics at Muséum National d'Histoire Naturelle in Paris) was studying phosphorescent uranium salts. He initially thought that the penetrating radiation he found was phosphorescence emission from the salt produced by exposure to bright sunlight, but soon realized that the radiation came from the uranium itself without any external excitation. This discovery of radioactivity earned him the Nobel Prize in Physics in 1903 in conjunction with Marie Curie and her husband Pierre Curie [13].

For the first half of the 20th century radiographs remained the only widely employed imaging modality. In the 1950s nuclear medicine emerged as a medical specialty after radionuclides (radioactive isotopes) were first produced for medical use by the Oak Ridge National Laboratory in Tennessee. The development of the rectilinear scanner and the gamma scintillation camera helped establish nuclear medicine as a fully developed medical imaging specialty [14].

Conventional tomography (rotating the X-ray tube and the film synchronously in opposite directions) had been described in a patent issued in 1922 to the French dermatologist Andre Bocage (1892–1953) [15]. However, it was not until development of sufficient computing power in the 1970s that radiographic computed tomography (CT) emerged as the one of the dominant imaging modalities with its ability to image soft tissue as well as bone. The first CT scan took place on a patient with a suspected frontal lobe brain tumor in 1971 at Atkinson Morley's Hospital, in London, England. The patient was scanned with a prototype scanner developed by Godfrey Hounsfield and his team at EMI Central Research Laboratories in Hayes [16].

Professor Isidor I. Rabi (1898–1988), while working in the Pupin Physics Laboratory in Columbia University, New York City, in 1937 observed the quantum phenomenon dubbed *nuclear magnetic resonance (NMR)*. He discovered that atomic nuclei (particularly hydrogen atoms) will absorb and emit radio waves when exposed to a sufficiently strong magnetic field. He received the Nobel Prize in Physics in 1944 for this work [17]. Raymond Vahan Damadian (born on March 16, 1936) is an American of Armenian origin, credited with being the inventor of the principle of the magnetic resonance imaging (MRI) device [18]. His research into sodium and potassium in living cells led him to his first experiments with NMR, which caused him to first propose the MR body scanner in 1969. Damadian discovered that tumors and normal tissue could be distinguished in vivo by NMR because of the longer relaxation times in tumors, both T1 (spin-lattice relaxation) or T2 (spin–spin relaxation) [19]. Damadian was the first to perform a full body scan of a human being in 1977 to diagnose cancer. However Damadian's point scanning approach called *field focused NMR (FONAR)* was time-consuming, and it was the rival device of Paul C. Lauterbur and Sir Peter Mansfield, which was based on field gradients and was able to provide linear localization, that eventually succeeded. In a controversial decision, the Nobel Committee awarded the Nobel Prize in Physiology or Medicine of 2003 to Lauterbur and Mansfield only, whereas Damadian was excluded [20]. It was remarked upon by commentators that the Nobel citation was able to include up to three recipients.

Positron emission tomography (PET) is a nuclear medicine imaging technique that produces a three-dimensional image of active processes occurring in the body. The system detects pairs of gamma rays emitted when a positron emitted from a particular type of radionuclide isotope decomposes. The PET isotope is introduced into the body as a tracer by tagging it to a biologically active molecule. Three-dimensional images of tracer concentration within the body are then constructed by computer analysis. In modern PET-CT scanners, three-dimensional imaging is enabled with the aid of a concurrent CT radiography scan performed on the patient in the same machine [21]. In 1953, Sweet and Brownell reported the use of positron-emitting isotopes to localize brain tumors [22], The use of 2-fluoro-2-deoxy-D-glucose (^{18}F) as a glucose-analog tracer was introduced by a collaborative group consisting of Martin Reivich, David Kuhl, and Abass Alavi at the Hospital of the University of Pennsylvania and Alfred Wolf at Brookhaven National Laboratory [23]. The compound was first administered to two normal human volunteers by Alavi in August 1976 at the University of Pennsylvania. Brain images obtained with an ordinary (non-PET) nuclear scanner demonstrated the concentration of ^{18}F-fluorodeoxyglucose (FDG) in that organ [24]. The PET-CT scanner was developed by Dr. David Townsend, Dr. Ronald Nutt, et al. [25] and was named by *Time Magazine* as the medical invention of the year in 2000.

Ultrasound as used for diagnostic imaging is called *ultrasonography*. English-born physicist John Wild (1914–2009) first used ultrasound to assess the thickness of bowel tissue as early as 1949 [26]; he has been described as the "father of medical ultrasound" [27]. Professor Ian Donald et al. at the Glasgow Royal Maternity Hospital were the first to use ultrasound to diagnose live volunteer patients with abdominal masses [28].

Donald and Dr. James Willocks then refined their techniques to obstetrical applications including fetal head measurement to assess the size and growth of the fetus [29].

Optical imaging covers such a large field that it is difficult to decide what was the first medical application. Could it be said that the introduction of spectacles in 1270, in Florence, Italy was the first use of optical imaging in medicine? Or the introduction in 1590 of the compound microscope by the father and son team of Hans and Zacharias Janssen in the Netherlands? It is more likely that optical imaging (as understood by the general scientific community) and the application of biomedical optics to diagnose various diseases, came to prominence with the discovery of optical coherence tomography (OCT) in the 1990s [30], although there had been a variety of fluorescence and other simple optical imaging techniques being sporadically explored for many years earlier. Now there are many sophisticated optical imaging methodologies being studied and explored for diagnosis, such as in vivo confocal microscopy [31], optical frequency domain imaging [32], diffuse optical imaging [33], fluorescence tomography [34], Brillouin microscopy [35], Cerenkov imaging [36], polarization sensitive techniques [37], photoacoustic techniques [38], and so on.

The present volume attempts to gather together information on the use of a variety of medical imaging technologies applied to the general area of dermatology.

This text book has been divided into eight broad sections. The first section describes simple optical imaging modalities including dermoscopy, trichoscopy, and onychoscopy that are routinely used in clinical practice. In Chapter 2, Sidoroff discusses the current function and role of clinical photography in dermatology. In Chapter 3, Lallas et al. provide an overview of the basic dermoscopic findings seen in melanocytic and nonmelanocytic tumors, as well as the inflammatory and infectious skin diseases. In Chapter 4, Khopkar and Jain highlight the diagnostic features of noncicatricial and cicatricial alopecias and genetic hair shaft disorders, as well as psoriasis and seborrheic dermatitis. In Chapter 5, Lencastre and Campos describe dermatoscopic findings in tumors of the nail apparatus as well as bacterial and fungal nail infections and inflammatory nail diseases. Themstrup and Jemec, in Chapter 6, review applications of OCT in nonmelanoma skin cancer, with an emphasis on basal cell carcinoma and actinic keratosis. In Chapter 7, Mamalis et al. focus on a specific application of OCT, assessment of skin fibrosis. In Chapter 8, Lee at al. discusse the utilization of interference and polarization techniques for evaluation of skin roughness, which aids in differentiating melanoma from other benign skin lesions, such as seborrheic keratoses. In Chapter 9, Hegyi and Hegyi cover the use of fluorescence in the detection and localization of poorly demarcated skin lesions. In Chapter 10, Longon et al. describe the clinical applications of the novel imaging technique of ex vivo fluorescence confocal microscopy (FCM). In Chapter 11, Wang and Evans describe coherent Raman scattering, microscopy which provides not only the morphological/structural information of the skin, but also the chemical and molecular information. Zhao et al., in Chapter 12, present a rapid real-time Raman system and an imaging-guided confocal Raman system, both of which can be utilized for in vivo skin evaluation. Chen and associates, in Chapter 13, introduce the concept of surface-enhanced Raman spectroscopy and discuss how nontoxic nanoscale substrates and a variety of strategies can bring the substrates and target molecules together for intradermal measurements. In Chapter 14, Camp gives an overview of broadband coherent anti-Stokes Raman scattering microspectroscopy. In Chapter 15, Alarcon et al. describe reflectance confocal microscopy, which enables the analysis of the skin horizontally with a nearly histological resolution. Hyperspectral and multispectral imaging in dermatology is described by Vasefi et al. in Chapter 16. Hyperspectral imaging generates a three-dimensional data cube that contains absorption, reflectance, or fluorescence spectrum data for each image pixel. Moy and Tunnell, in Chapter 17, cover diffuse reflectance spectroscopy and its applications in dermatology, including skin cancer, port wine stain, erythema, sunscreen evaluation, and burns. Moreover, future directions combining diffuse reflectance with other optical methods are also presented. In Chapter 18, Ho et al. provide a broad description of spectral imaging in vivo and ex vivo skin specimens. So et al. discusse the uses of multiphoton imaging to study skin immunoresponse, aging, and regeneration in Chapter 19. In Chapter 20, Cicchi et al. mainly describe use of two-photon microscopy, second-harmonic generation microscopy, and their combination for differentiation of epidermal layers and characterization of the skin dermis. In Chapter 21, Jacques highlights the principles of confocal reflectance and polarized light imaging. Yaroslavsky et al. give an overview of polarization optical imaging of skin pathology and aging in Chapter 22. Huang et al.define mechanical characterization of skin using surface acoustic waves, a novel combination of phase-sensitive OCT technology with a simple mechanical impulse surface wave stimulation in Chapter 23. In Chapter 24, Zhou and Wang discuss use of photoacoustic tomography of both primary and metastatic melanomas. Wortsman summarizes use of ultrasound in detection of common skin, nail, and hair diseases in Chapter 25. Raster scan optoacoustic mesoscopy, a high-resolution optical imaging technique that can penetrate several millimeters in tissues is described by Schwarz et al. in Chapter 26.

Chapter 27 by Petersen and Higgins explicates the use of total body photography and serial digital dermoscopy in dermatology. Papoiu introduces utilization of functional MRI in detection of brain processing of itch in Chapter 28. In Chapter 29, Gobel provides an overview of MRI of skin. Westerland et al. discusse a specific application of MRI in the management of anogenital hidradenitis suppurativa in Chapter 30. Bonmarin and Gal, in Chapter 31, review the current technologies and applications of thermal imaging in dermatology. Bourgeois et al. describe the use of PET combined with CT in staging, imaging, and surveillance of cutaneous melanoma in Chapter 32. In Chapter 33, Beylergil et al. introduce the concept of molecular imaging in Merkel cell carcinoma. On the other hand, Lorenz et al. describe the use of other imaging modalities, such as ultrasound, CT, MRI, and lymphoscintigraphy in Merkel cell carcinoma in Chapter 34. In Chapter 35, Fardin et al. describe the use of FDG−PET-CT in cutaneous lymphoma. A general overview of imaging in cutaneous squamous cell carcinoma of the head and neck is given by Casazza and Monroe in Chapter 36. Peters and Vanhoenacker outline the imaging patterns of metastatic melanoma in Chapter 37. In the final chapter, Chapter 38, Visscher et al. discuss most up-to-date technologies, emerging methods, and unmet needs of image processing in dermatology.

References

[1] Bickers DR, et al. The burden of skin diseases 2004: a joint project of the American Academy of Dermatology Association and the Society for Investigative Dermatology. J Am Acad Dermatol 2006;55(3):490−500.

[2] Yoo JY, Rigel DS. Trends in dermatology: geographic density of US dermatologists. Arch Dermatol 2010;146(7):779.

[3] Kimball AB, Resneck Jr JS. The US dermatology workforce: a specialty remains in shortage. J Am Acad Dermatol 2008;59(5):741−5.

[4] Abdel-Halim RE. The role of Ibn Sina (Avicenna)'s medical poem in the transmission of medical knowledge to medieval Europe. Urol Ann 2014;6(1):1−12.

[5] Siraisi NG. History, antiquarianism, and medicine: the case of Girolamo Mercuriale. J Hist Ideas 2003;64(2):231−51.

[6] Editorial. Daniel Turner (1667−1740): dermatologist, surgeon, physician. JAMA 1970;213(5):863−4.

[7] Loewenthal LJ. Daniel Turner and "De morbis cutaneis". Arch Dermatol 1962;85:517−23.

[8] Everett MA. Anne-Charles Lorry: the first French dermatologist. Int J Dermatol 1979;18(9):762−4.

[9] Karamanou M, et al. Baron Jean-Louis Alibert (1768−1837) and the first description of mycosis fungoides. J BUON 2014;19(2): 585−8.

[10] Alibert JL. Description des maladies de la peau: observées à l'Hôpital Saint-Louis, et exposition des meilleures méthodes suivies pour leur traitement. Bruxelles, Belgium: Wahlen; 1825.

[11] Finnerud CW. Ferdinand von Hebra and the Vienna school of dermatology. AMA Arch Derm Syphilol 1952;66(2):223−32.

[12] Ramoutsaki IA, Giannacos EN, Livadas GN. Birth of battlefield radiology: Greco-Turkish war of 1897. Radiographics 2001;21(1): 263−6.

[13] Dutreix J, Dutreix A. Henri Becquerel (1852−1908). Med Phys 1995;22(11 Pt 2):1869−75.

[14] Williams LE. Anniversary paper: nuclear medicine: fifty years and still counting. Med Phys 2008;35(7):3020−9.

[15] Editorial. Andr'e Bocage (1892−1953): French tomographer. JAMA 1965;193:233.

[16] Webb S. Historical experiments predating commercially available computed tomography. Br J Radiol 1992;65(777):835−7.

[17] Shampo MA, Kyle RA, Steensma DP. Isidor Rabi: 1944 Nobel laureate in physics. Mayo Clin Proc 2012;87(2):e11.

[18] Prasad A. The (amorphous) anatomy of an invention: the case of magnetic resonance imaging (MRI). Soc Stud Sci 2007;37(4): 533−60.

[19] Damadian R. Tumor detection by nuclear magnetic resonance. Science 1971;171(3976):1151−3.

[20] Dreizen P. The Nobel prize for MRI: a wonderful discovery and a sad controversy. Lancet 2004;363(9402):78.

[21] Bailey DL, et al. Positron emission tomography: Basic sciences. Secaucus, NJ: Springer-Verlag; 2005.

[22] Sweet WH, Brownell GL. Localization of brain tumors with positron emitters. Nucleonics 1953;11:40−5.

[23] Gallagher BM, et al. Radiopharmaceuticals XXVII: ^{18}F-labeled 2-deoxy-2-fluoro-d-glucose as a radiopharmaceutical for measuring regional myocardial glucose metabolism in vivo: tissue distribution and imaging studies in animals. J Nucl Med 1977;18(10): 990−6.

[24] Reivich M, et al. Measurement of local cerebral glucose metabolism in man with ^{18}F-2-fluoro-2-deoxy-d-glucose. Acta Neurol Scand Suppl 1977;64:190−1.

[25] Beyer T, et al. A combined PET/CT scanner for clinical oncology. J Nucl Med 2000;41(8):1369−79.

[26] Wild JJ, Neal D. Use of high-frequency ultrasonic waves for detecting changes of texture in living tissues. Lancet 1951; 1(6656):655−7.

[27] Brady T. Wild ideas: medical researcher and inventor John Julian Wild led the field of ultrasound medicine. Minn Med 2010;93(3): 19−21.

[28] Donald I, Macvicar J, Brown TG. Investigation of abdominal masses by pulsed ultrasound. Lancet 1958;1(7032):1188−95.

[29] Willocks J, et al. Foetal cephalometry by ultrasound. J Obstet Gynaecol Br Commonw 1964;71:11−20.

[30] Huang D, et al. Optical coherence tomography. Science 1991; 254(5035):1178−81.

[31] Villringer A, et al. Confocal laser microscopy to study microcirculation on the rat brain surface in vivo. Brain Res 1989;504(1): 159−60.

[32] Pogue BW, et al. Initial assessment of a simple system for frequency domain diffuse optical tomography. Phys Med Biol 1995;40(10):1709−29.

[33] Pogue B, et al. Comparison of imaging geometries for diffuse optical tomography of tissue. Opt Express 1999;4(8):270−86.

[34] Graves EE, et al. A submillimeter resolution fluorescence molecular imaging system for small animal imaging. Med Phys 2003; 30(5):901−11.

[35] Scarcelli G, Yun SH. In vivo Brillouin optical microscopy of the human eye. Opt Express 2012;20(8):9197−202.

[36] Li C, Mitchell GS, Cherry SR. Cerenkov luminescence tomography for small-animal imaging. Opt Lett 2010;35(7):1109−11.

[37] de Boer JF, et al. Two-dimensional birefringence imaging in biological tissue by polarization-sensitive optical coherence tomography. Opt Lett 1997;22(12):934−6.

[38] Li C, Wang LV. Photoacoustic tomography and sensing in biomedicine. Phys Med Biol 2009;54(19):R59−97.

CHAPTER

2

The Role of Clinical Photography in Dermatology

A. Sidoroff

Medical University of Innsbruck, Innsbruck, Austria

OUTLINE

INTRODUCTION

Clinical dermatology depends very much on optical impressions. Although palpation, odor, and a patient's given history may enhance diagnostic decisions, clinicians rely heavily on what they actually see when making an initial assessment of skin lesions. Verbal descriptions, accurate as they attempt to be, cannot come close to and will never replace visual perception. In the days before photography was invented, there were only three possible ways to depict and communicate such visual perceptions: drawings by a skilled illustrator; moulages (three-dimensional wax reproductions of disease-affected body parts); or actually seeing the disease of a patient on-site. The main objective was learning and teaching; visualization was a major part of that medical education. A sea change was unleashed at the beginning of the 19th century [1] through the launch of Joseph Nicéphore Niépce's lithography and Louis Daguerre's daguerreotype, but it took several decades before photography became accessible to the general public. Around 1900, photographic pictures were used in scientific medical publications (Fig. 2.1).

Ever-increasing opportunities of taking clinical photographs made a deep impact on the transfer of optical information, and not only in dermatology. It was above all else the accessibility of the motive, ie, skin, which predisposed dermatology to benefit inordinately highly from this technique. Pictures of skin diseases could not only be printed but also projected onto screens at clinical conferences and lectures at teaching institutions. The development and availability of computers and the resulting opportunity to digitize clinical pictures (at the beginning by scanning analog photographs or

Imaging in Dermatology
http://dx.doi.org/10.1016/B978-0-12-802838-4.00002-9

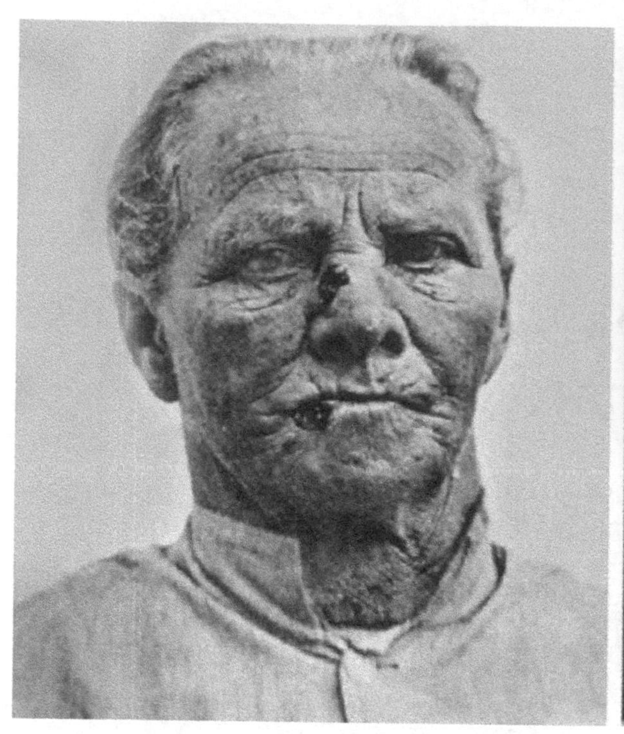
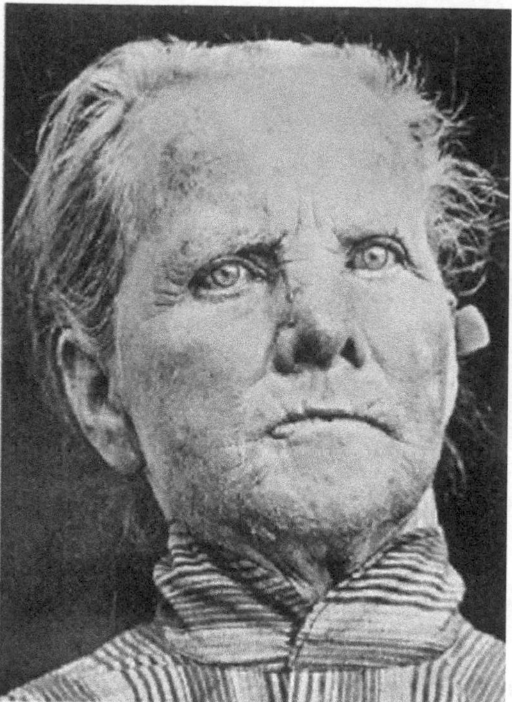

FIGURE 2.1 Early documentation of a photodynamic treatment of NMSC around 1900.

slides) was the next step. The development of digital photography then became the major breakthrough in heightening the status of photography in modern clinical dermatology [2]. Pairing affordable digital cameras and computers gives every dermatologist a way to take and store pictures of patients digitally in an extremely simple process. The aim of this chapter is to discuss the implications of this technical status quo in the context of dermatology [3–5].

THE EQUIPMENT

Today a broad spectrum of digital cameras is available on the market. Ranging from expensive high-resolution camera systems to integrated cameras in mobile phones, every demand can be met. The choices range from camera systems costing thousands of US dollars to photographic equipment nearly everybody already owns and has readily available, for example, as a feature of a mobile phone [6,7].

The relevant question, of course, is the purpose of the picture. Nonetheless, even cameras in high-end smart phones used under optimal conditions suffice to meet the basic resolution requirements of reproduction in printed media, eg, medical journals. For projections at talks or lectures, resolution requirements are even lower. The need for large-scale high-resolution prints is rather limited and not part of routine clinical use.

Given the extremely rapid development in the sector of photographic devices, only a few basic questions can be posed to aid in selecting the right equipment:

- Does the camera have the capability to take overview pictures as well as close-ups? Many simple "point and shoot" devices have a limited capacity in the macro range, which means that close-up pictures are often ruled out. One cannot rely on the manufacturers' nearest-distance specifications. They are based on measurements only in optimal light conditions.
- Is the autofocus system [8] able to deal with low-contrast pictures? Especially in close-up situations the contrast between normal skin and pathologically influenced skin could be very low. Cheaper camera systems might have difficulties bringing the region of interest into focus.
- Is an integrated flash sufficient? Frontal flashes have the disadvantage of forfeiting the three-dimensional aspect of a lesion, ie, it could get lost. (Additional lighting from the side is often essential to make this third dimension visible, eg, urticarial lesions, granuloma anulare.)
- How much influence can be exercised on the resulting picture? Most cheap and easy-to-use cameras have automated algorithms for standard situations but very limited possibilities to influence the ultimate pictures. Taking clinical pictures in dermatology is not a standard situation. The ability to optimize the

camera's behavior to a suit dermatologist's actual needs demands such possibilities as modification of focus, white balance, and labeling of the picture. One must be aware, though, that changing settings is a time-consuming process that needs a lot of expertise.

- How standardized do pictures need to be? Especially before-and-after pictures have to be taken under the exactly same light conditions for camera settings to be meaningful. Lighting from a different angle or a slight shift in the automatic white balance can make a useful comparison of two pictures impossible. This also holds true for follow-up pictures of pigmented lesions (if not documented with special mole-documentation devices). In such cases, cameras that automatically regulate their parameters are of little use. It is clear that not only camera parameters but also positioning of the patient has to be standardized to make pictures comparable [9].
- How should the connection between the camera system and the existing computer system be set up? Ordinarily, cameras store pictures on memory cards and the data are then transferred to a computer. This can be a cumbersome procedure. There are different levels to facilitate this, from integration of the import function into the clinical documentation software to complex (and expensive) systems with full control of the camera from the utilized software. The main issue in this context is that full integration is usually restricted to a limited number of cameras.
- Can data security or authenticity be verified? As mentioned, many camera systems regulate their illumination and focusing parameters themselves. Because these algorithms are based on general consumer demands, the resulting picture might not reproduce the desired result in a dermatological setting. Postprocessing might be imperative to give an account to the typical color of, for example, a heliotrope erythema in dermatomyositis. On the other hand, the picture has been "manipulated." Some systems emphasize the point that manipulation of pictures is not possible. But this necessitates that the original picture is perfect the moment it is taken, which is rarely the case.
- What is the quality of illumination? From all the points mentioned, it is clear that light conditions are as important as the camera system itself, especially when pictures need to be compared or a third dimension needs to be documented. Integrated flashes, ring flashes around the objective, external flashlights, natural light, or artificial light in a room all deliver different results.

In conclusion, it can be said that there is no optimal camera device for dermatologists, because optical documentation needs are quite different and no general standard has been established thus far. The choice of one's camera system is a highly individual decision and depends on the purpose of the pictures, technical skills of the physician, and costs.

THE "ART" OF TAKING DERMATOLOGICAL CLINICAL PICTURES

Poor quality pictures are not usually a result of inadequate technology or insufficient options of camera equipment, but of the unskilled use of technology. As with every skill, there is a learning curve, which cannot be skipped. Low-quality pictures are usually caused by insufficient care on the part of the photographer. In particular, illumination and selection of the details to focus on are significant components of the visual information a picture ultimately provides. Just *point and shoot* (as propagated in other photographic situations) might not lead to ideal results in imaging skin diseases. For example, as elaborated, if one wants to show the infiltration/elevation of a lesion, side lighting is important. A frontally integrated camera flash will never be able to highlight this morphological feature. Or a close-up view of one lesion will not show the distribution pattern of a rash or its potential change over time. Although it has become an everyday routine, dermatological education has not honed this skill in its educational curriculum; moreover, few articles and courses deal with the topic. In most educational institutions there is much room for improvement in teaching dermatologists how to take qualitative, informative clinical pictures. It is important to get to know the camera being used and how to use it. It requires just a few hours to learn the essentials. The rest comes with accumulated experience.

WHY PHOTOGRAPHS?

According to estimates, dermatology has to deal with 3000 diagnoses [at least in Europe, where autoimmune diseases, malignancies, allergology, sexually transmitted diseases (STDs), and other pathologies are part of this specialty]. As mentioned, dermatologists have developed a special terminology to verbally describe the morphology of skin diseases. *Primary lesions, secondary lesions, distribution pattern,* and, in particular, use of color terms can, when properly used, describe the clinical appearance in such a way that a diagnosis can often be made. On the other hand, it is clear that verbal descriptions depend heavily on the language used. Furthermore, it takes a lot of time to compose these descriptions. The old saying that a picture is worth more than a thousand words holds true to a certain extent. Good clinical pictures can replace a host of descriptive sentences and, in addition, are more objective, ie, not susceptible to the describer's bias. Time is precious. It takes but a few minutes to snap some pictures and store them to a patient's chart, whereas the time needed to elucidate a skin disease in words may require more than an hour. However, the obvious timesaving advantage of taking photographs has one drawback, which is usually underestimated; the necessity to verbally describe what one sees makes it imperative to analyze the patient thoroughly. This has long been the cornerstone of gaining diagnostic skills. On the one hand, one does not have to be a dermatologist to take a picture; taking photographs does not need the analytical competence involved in verbal description. However, the better documentation becomes through photographs, the less attention is paid to what can really be seen. It is a vicious cycle; photographs all too often replace instead of complement clinical observations.

PHOTOGRAPHS AS AN EDUCATIONAL TOOL

There is no doubt that clinical pictures are extremely helpful for educational purposes. Dermatology is a specialty in which visual clinical experience plays a very important, in fact an unparalleled, role. Once one has consciously seen a rare disease, the linked synapses will remember it and call it into mind when it is seen another time. Pictures are only a pale runner-up compared with the real thing; the distance separating the two modes of perception is larger than ordinarily assumed. Associative memory is a very complex procedure. It means that apart from the "typical lesion" seen in a photograph in a textbook, dealing directly with a patient offers a whole range of details that may not have anything to do with the disease itself. The face of the patient, the situation, the patient's name, and the local setting are all apparently unimportant pieces of information but can turn out to be very helpful when recalling a disease. A picture in a textbook or on a piece of paper lacks these subliminary "add-ons." Most of the time a photograph is not given the same attention as a patient, especially when it is only part of the flood of information in a textbook. Nonetheless, as an addition to the correlating plain text, the correct picture can be extremely helpful in giving an impression about a disease. Even then one has to keep aware that years might go by before a dermatologist sees a certain skin disease in the real world that he or she has once seen in a textbook or lecture, if ever at all.

PHOTOGRAPHS AS A DIAGNOSTIC TOOL

A typical constellation in daily clinical routine is to examine a (rare) skin disease and not be able to make a diagnosis. The plethora of cutaneous diseases and the fact that they: (1) often have atypical presentations, and (2) may change their appearance over time make it impossible to keep all the possibilities in mind. A good clinician will therefore come up with a variety of differential diagnoses. In such a case, locating a suitable, comparable photograph in the literature or a photo database can be very helpful. Although several books about differential diagnoses in dermatology are available, it still can be a tedious and cumbersome task to find an appropriate picture. The basic idea of creating dermatological atlases has been picked up by many institutions (academic and commercial). Whereas it may be easy to find pictures of a certain disease, it still remains a challenge to find pictures via descriptive search terms.

This is where the skill of verbal description and the use of a correct, standard vocabulary come back into play. Unfortunately, up to now even the challenge of compiling a list and/or code of skin diseases has not been satisfactorily accomplished. All the same, one can say that good reference pictures are a helpful tool in finding the correct diagnosis.

THREE ADDITIONAL ASPECTS THAT NEED TO BE CONSIDERED IN THIS CONTEXT

First, with the new technology it is relatively easy to transmit pictures via electronic media. However, data security and confidentiality have to be guaranteed. Once this condition is met, getting a second opinion in a challenging case becomes extremely easy. Moreover, different approaches to teledermatology can spare patients unnecessary trips from remote places to specialized centers if the transferred information is sufficient [10,11].

The second point concerns the dynamics of skin diseases. It is a common scenario that by the time a patient can see a dermatologist, the clinical picture has changed in such a way that a clear diagnosis is no longer possible. More and more patients try to overcome this problem by taking pictures themselves during the acute phase of the disease. Although to our knowledge, this method has not been formally evaluated and statistics are not available, personal experience suggests that this approach is very helpful, particularly for diseases where clinical pictures are subject to short-term changes or the time lag between the acute phase and the appointment at the doctor is longer than desired.

Third, the availability of a clinical picture can be very useful for clinicopathological correlation in countries or institutions where histological slides are still seen by a clinically experienced dermatopathologist. Although the usual approach is to first look at the histopathological slide unbiased, the possibility of having a look at the clinical picture can help to confirm or question the result of a skin biopsy, especially when interaction between clinician and histopathologist is possible.

PHOTOGRAPHY AS A TOOL FOR DOCUMENTATION

Documentation has become one of the most time-consuming parts of medical routine. There are many nonmedical reasons for extensive documentation, but from a medical point of view, clinical pictures can be very helpful when it is a question of viewing the evolution of a skin disease or when different doctors are caring for a patient, as is routinely the case in larger hospitals.

PHOTOGRAPHY AS A TOOL IN RESEARCH

Dermatology has a long history. The names of many diseases (eg, mycosis fungoides, a cutaneous T-cell lymphoma that has no relation to fungi) and their classification go far back in history. Not wishing to get involved in a discussion about splitters and lumpers, it is nonetheless clear that new insights underscore the necessity of reevaluating many clinical patterns [12]. Diseases that look similar often have utterly different causes; disparate-appearing morphologies are commonly variants of the selfsame disease. A lot of work has to be put into transferring basic research findings into clinical practice. One approach is to look at clinical pictures and pinpoint subtle differences in them. But one still has to find the lowest common denominator to subordinate a variety of clinical appearances to a superordinate disease. Doing this, particularly for rare diseases, often takes longer than today's timetable for publication allows. Many institutions have a large collection of clinical pictures (including digital and printed photographs and slides). Retrospective screening of these archives in the context of the disease in question can provide new insight to an evaluation of morphological clinical findings. But again, the result of this type of research can only be as good as the written documentation attached to it.

PHOTOGRAPHS FROM THE PATIENT'S PERSPECTIVE

From everyday life we know that many healthy people do not like to have their pictures taken. This discomfort obviously gets worse in a clinic when a disease is present as a visible flaw [13,14]. The main strategy thus far has been to explain to the patient that clinical pictures are a routine part of dermatological documentation (eg, by comparing it with a radiograph in other disciplines). It can often be helpful to place emphasis on the fact that skin diseases are a dynamic process and imaging is the best way to document their stages and changes. Making lesions visible, for example, if they are located on a part of the body where they cannot be seen by the patient, or using techniques like fluorescence diagnosis to make subclinical lesions visible, can also be a good tool to convince patients about the need for treatment. It goes without saying, discomfort for patients while taking pictures should be reduced as much as possible so as not to stigmatize them even more than the disease does. An example is covering a patient's genital area when its documentation is not part of the purpose of the photograph. Another strategy is to avoid taking pictures of the face. Although from the doctor's point of view, this approach has disadvantages because doctors commonly recognize patients sooner from their faces than from their names and makes allocation and relocation of pictures easier. Sometimes it is important for the diagnosis and documentation of a disease that certain parts of the body are not affected. This should be communicated to the patient when photographs are taken of obviously healthy parts of the skin. For patients, the camera system used also seems to matter. In a study on 300 patients, 97.7% preferred a hospital-owned camera device over the use of a physician's camera or smartphone [15].

PHOTOGRAPHS FROM A LEGAL POINT OF VIEW

As can be concluded from the previously mentioned considerations, from a medical point of view there is no doubt that today's possibilities of digital imaging are of great advantage in education and clinical practice. However (allowing for differences from country to country), a series of legal aspects also needs to be considered [16]. The most obvious one is that the identity of the patient has to be protected on pictures seen by others apart from the treating physicians. It hardly needs be said that barring or pixeling the eyes by digital image processing often deteriorates the quality of information intrinsic to the clinical appearance of the disease. Second, more and more journals and publishers make written consent of a patient a prerequisite for publishing one or more of his or her pictures. This clearly makes sense in cases where publication could lead to a disadvantage for the patient. In many cases, however, it is just a tool to protect publishers from legal risks. Such an overly defensive attitude often leads to good, informative pictures not being used in textbooks or publications simply because they were taken long before it was common practice to ask patients for their consent to publish pictures in which their identity is not recognizable. Although a weighing of interests is essential in publishing such pictures, the overly strict rules of many journals make use of many highly informative pictures nearly impossible. A typical example from the author's own experience is pictures of toxic epidermal necrolysis. In a rare disease like this (1–2 cases per million inhabitants per year) with a mortality of over 30%, the bureaucratic action to get written consent to publish clinical pictures is utterly disproportionate to the patient's actual physical and psychological condition.

The sticking point here is the use of clinical pictures in malpractice lawsuits, a phenomenon that has reached an almost unbearable level, especially in the United States. Once a definite diagnosis is known, it is often easy to recognize features of a disease on documented photographs, even if they are not typical. Yet the information on a diagnosis, which relies on doctors highly specialized in the disease, can lead to the conclusion that "it would have been possible to make a certain diagnosis earlier." In this context, this a posteriori judgment (ie, when the diagnosis is known) is submitted to a significant bias. If physicians at the same level of expertise (general practitioner, specialist, expert) and blind to the diagnosis were the ones to evaluate a clinical photograph, the assessment of a physician's malpractice would be much more objective. One must be aware that the many faces of cutaneous diseases and the constant flow of new publications supersede the amount of information that any single person can cope with. As a result of this, one has to be exceedingly careful about judging whether a doctor should have recognized a disease (earlier) on the basis of clinical pictures at a point in time when the diagnosis is already known. That said, in clear cases of misdiagnosis, pictures taken at a relevant moment can often help a patient achieve his or her rights.

On the other hand, because clinical picture-taking in most countries is neither compulsory nor reimbursed, it is no wonder that this method is not utilized as often as it might be.

PHOTOGRAPHY AS AN INFORMATIVE TOOL FOR THE GENERAL POPULATION

With the availability and reproducibility of clinical photographs, health and prevention campaigns now have a viable tool to inform the general public and indicate what people need to watch out for. Such pictures can be used in leaflets, posters, newspapers, or television clips. The reverse side of the coin is the enormous availability of images on the Internet. There are many doubtless informative sites with serious and high-quality content. However, one can also find thousands of websites with inadequate or downright incorrect information. That includes sites published by people through mistaken personal perceptions, as well as sites that pursue commercial interests. The amount of misinformation is overwhelming. Unfortunately, all these pictures and unfiltered data give Internet users the false impression that they can be their own doctors, not realizing that even if a posted clinical picture resembles their own perception of their disease, the associated information might be completely wrong. In search of solutions for their problems, patients often resort to clutching at the straws the Internet has to offer, paying little attention to the seriousness of the source of information. The most commonly used tools in this context are carefully selected before-and-after pictures. Without knowledge of the correct background, they are one of the best means of misleading and manipulating patients' expectations.

CONCLUSIONS

Today's technical possibilities of taking high-quality pictures of skin disease through a very simple act has significantly changed the way dermatological disorders can be documented for purposes of education, research, teledermatology, and patient documentation. Nevertheless, it is up to the physician/photographer to optimally employ this tool. To put it simply, the task of taking good clinical photographs depends not so much on the quality of the equipment as on the photographer's skills.

Digital photography and computer technology provide us the possibility of transferring and storing visual information better and more easily than ever before. But one has to be aware of the danger that for most cases imaging alone is not sufficient. For dermatologists, not only documentation but a thorough analysis of what one sees is a major part of diagnosis and experience.

In addition, clinical pictures can be helpful in less obvious constellations, such as interdisciplinary communication, patient guidance, prevention and/or early detection, and information designed for the general public. The caveat remains: dermatological photographs always have to be seen in their clinical or general context.

Last but not least, the importance of clinical photography in dermatology needs to be reflected to a greater extent in educational curricula and reimbursement policies. It needs to become an integral part of everyday dermatology.

References

[1] Neuse WH, Neumann NJ, Lehmann P, Jansen T, Plewig G. The history of photography in dermatology: milestones from the roots to the 20th century. Arch Dermatol 1996;132(12):1492–8.
[2] Levy JL, Trelles MA, Levy A, Besson R. Photography in dermatology: comparison between slides and digital imaging. J Cosmet Dermatol 2003;2(3–4):131–4.
[3] Kaliyadan F, Manoj J, Venkitakrishnan S, Dharmaratnam AD. Basic digital photography in dermatology. Indian J Dermatol Venereol Leprol 2008;74(5):532–6.
[4] Ratner D, Thomas CO, Bickers D. The uses of digital photography in dermatology. J Am Acad Dermatol 1999;41(5 Pt 1):749–56.
[5] Witmer WK, Lebovitz PJ. Clinical photography in the dermatology practice. Semin Cutan Med Surg 2012;31(3):191–9.
[6] Ashique KT, Kaliyadan F, Aurangabadkar SJ. Clinical photography in dermatology using smartphones: an overview. Indian Dermatol Online J 2015;6(3):158–63.
[7] Helm TN, Wirth PB, Helm KF. Inexpensive digital photography in clinical dermatology and dermatologic surgery. Cutis 2000;65(2):103–6.
[8] Taheri A, Yentzer BA, Feldman SR. Focusing and depth of field in photography: application in dermatology practice. Skin Res Technol 2013;19(4):394–7.
[9] Halpern AC, Marghoob AA, Bialoglow TW, Witmer W, Slue W. Standardized positioning of patients (poses) for whole body cutaneous photography. J Am Acad Dermatol 2003;49(4):593–8.
[10] Kaliyadan F. Digital photography for patient counseling in dermatology – a study. J Eur Acad Dermatol Venereol 2008;22(11):1356–8.
[11] Leggett P, Gilliland AE, Cupples ME, McGlade K, Corbett R, Stevenson M, et al. A randomized controlled trial using instant photography to diagnose and manage dermatology referrals. Fam Pract 2004;21(1):54–6.
[12] Katugampola R, Lake A. The role of photography in dermatology research. J Vis Commun Med 2012;35(1):5–10.
[13] Hacard F, Maruani A, Delaplace M, Caille A, Machet L, Lorette G, et al. Patients' acceptance of medical photography in a French adult and paediatric dermatology department: a questionnaire survey. Br J Dermatol 2013;169(2):298–305.
[14] Leger MC, Wu T, Haimovic A, Kaplan R, Sanchez M, Cohen D, et al. Patient perspectives on medical photography in dermatology. Dermatol Surg 2014;40(9):1028–37.
[15] Hsieh C, Yun D, Bhatia AC, Hsu JT, Ruiz de Luzuriaga AM. Patient perception on the usage of smartphones for medical photography and for reference in dermatology. Dermatol Surg 2015;41(1):149–54.
[16] Kunde L, McMeniman E, Parker M. Clinical photography in dermatology: ethical and medico-legal considerations in the age of digital and smartphone technology. Australas J Dermatol 2013;54(3):192–7.

CHAPTER

3

Dermoscopy

A. Lallas, Z. Apalla, E. Lazaridou, D. Ioannides

Aristotle University, Thessaloniki, Greece

OUTLINE

INTRODUCTION

Dermoscopy is a noninvasive imaging technique that enables the visualization of submacroscopical structures invisible to the naked eye. The handheld dermatoscope is an inexpensive, easy to use device using either nonpolarized or polarized light and providing a ×10 magnification of the examined skin lesions [1,2].

Imaging in Dermatology
http://dx.doi.org/10.1016/B978-0-12-802838-4.00003-0

Dermoscopy Is an Integral Part of Clinical Examination

Although representing an imaging technique, dermoscopy should not be regarded as a second-level examination to be applied only in clinically preselected lesions. Instead, the dermatoscope should be considered as the dermatologist's stethoscope, because it is easy to carry and use, and it provides diagnostic information that cannot be otherwise acquired [3]. Overall, dermoscopy is not more (or less) important than any other part of the clinical examination. In contrast, the findings of the dermoscopic examination should always be integrated with all the information acquired from the macroscopic inspection, palpation, or patient's history, and interpreted within the context of a given patient. Opponents of the method argue that dermoscopy is time consuming, but it has been shown that, with experience, it adds only a little extra time to the clinical consultation [4]. The applicability of the method has been significantly enhanced by the introduction of new-generation dermatoscopes, which, by using polarized light and not requiring immersion fluid, allow a rapid screening of multiple lesions.

Initially, dermoscopy has been mainly used for the evaluation of melanocytic skin tumors, with research efforts focusing mainly on identification of dermoscopic characteristics of nevi and melanoma. With time, continuously gathering evidence established the value of dermoscopy in improving melanoma detection, and the technique gained global appreciation for assessment of melanocytic tumors [5]. Meanwhile, the dermoscopic patterns of several nonmelanocytic pigmented and nonpigmented tumors were described [6,7]. More recently, dermoscopy has been shown to be useful also for the assessment of infectious and inflammatory dermatoses [8]. Overall, the dermatoscope is now regarded as the dermatologist's stethoscope, providing to a clinician experienced in the technique useful additional information on the morphology of skin lesions or eruptions.

DERMOSCOPY OF MELANOCYTIC SKIN TUMORS

Dermoscopy as an in vivo, noninvasive technique enables the visualization of diagnostic features of pigmented skin lesions, which are not seen with the naked eye [1,2]. The value of dermoscopy in significantly improving the discrimination between melanoma and nevi has been confirmed by several meta-analyses [5,9]. In everyday practice, dermoscopy is considered a first-level clinical tool that facilitates the evaluation of pigmented skin lesions by allowing the recognition of early melanoma signs, prompting clinicians to check clinically banal-looking lesions and digitally monitor their patients [10]. At the same time, dermoscopy helps clinicians minimize the unnecessary excisions of nevi that might clinically look worrisome, by revealing their characteristic dermoscopic architecture.

Dermoscopy of pigmented skin lesions is based on various analytic approaches or algorithms, which take into consideration established specific dermoscopic features that create different patterns. These dermoscopic patterns represent the backbone for the morphologic diagnosis of nevi and melanoma.

A two-step procedure has been suggested as the optimal approach to evaluate a pigmented skin lesion [11]. The first step aims to differentiate melanocytic from nonmelanocytic tumors, assessing the presence or absence of predefined structures that are associated with melanocytic lesions.

The dermoscopic criteria considered to be suggestive of a melanocytic tumor are:

1. pigment network (reticular pattern)
2. globules (globular pattern)
3. streaks (starburst pattern)
4. homogeneous blue pigmentation (homogeneous pattern)
5. parallel pattern (for acral lesions).

If the lesion is judged as melanocytic, it enters the second analytic step, which aims to distinguish nevi from melanoma. This step is mainly based on the so-called "pattern analysis," namely the assessment of the global dermoscopic morphology (pattern), as well as the presence of local features. Alternatively, several algorithms attempting to quantify the presence of dermoscopic criteria have been suggested, including the ABCD rule, Menzies' scoring method, the 7-point and the revised 7-point checklist, and the 3-point checklist [12–14] (Table 3.1).

Pattern Analysis

Pattern analysis is regarded as the classic dermoscopic method for evaluating pigmented skin lesions [15,16]. This method includes the assessment of the symmetry of the lesion, the presence of one or more colors, the global dermoscopic appearance of the lesion according to predefined patterns, and the presence of local features. The global pattern results from predominant features occupying large areas of the lesion. A global pattern usually consists of one (usually) or two (less often) predominant features. In the presence of more than two predominant features, the pattern is classified as *multicomponent*. Instead, local features can be

TABLE 3.1 Algorithms for Evaluation of Melanocytic Lesions

ABCD rule
Asymmetry in 0, 1 ή 2 axes, regarding contour, color and structures – Score 0-2
Border abruptly interrupted at the periphery in 0-8 segments – Score 0-8
Colour : Presence of up to 6 colors (white, red, light brown, dark brown, blue-gray, black) – Score 1-6
Dermoscopic structures: presence of pigment network, dots, globules, streaks, structureless homogeneous areas – Score 1-5
Total dermoscopic score (TDS): (A score × 1.3) + (B score × 0.1) + (C score × 0.5) + (D score × 0.5) = TDS
Interpretation of TDS: < 4.75 Benign melanocytic lesion, 4.75–5.45 suspicious lesion, > 5.45 Melanoma

Menzies method	
Negative features	**Positive features**
Symmetry of dermoscopic structures	Blue white veil
Presence of a single color	Multiple brown dots
	Pseudopods
	Radial streaks
	Scar like depigmentation
	Peripheral dots/globules
	Multiple (5 or 6) colors
	Multiple blue-gray dots
	Broadened network
Interpretation: A diagnosis of melanoma is made when both negative features are absent and 1 or more of the 9 positive features are present	

7 point checklist	
Dermoscopic features	**Score**
Major criteria	
1. Atypical pigment network	2
2. Blue-white veil	2
3. Atypical vascular pattern	2
Minor criteria	
4. Irregular streaks	1
5. Irregular dots and globules	1
6. Irregular blotches	1
7. Regression structures	1
Interpretation: Total score ≥ 3: melanoma, total score < 3: nonmelanoma	

3 point checklist	
1. Asymmetry of colors and/ or structures	
2. Atypical pigment network	Presence of more than one criterion ⟶ excision
3. Blue-white structures	
For nonexperts in dermoscopy–It aims in the recognition of suspicious lesions	

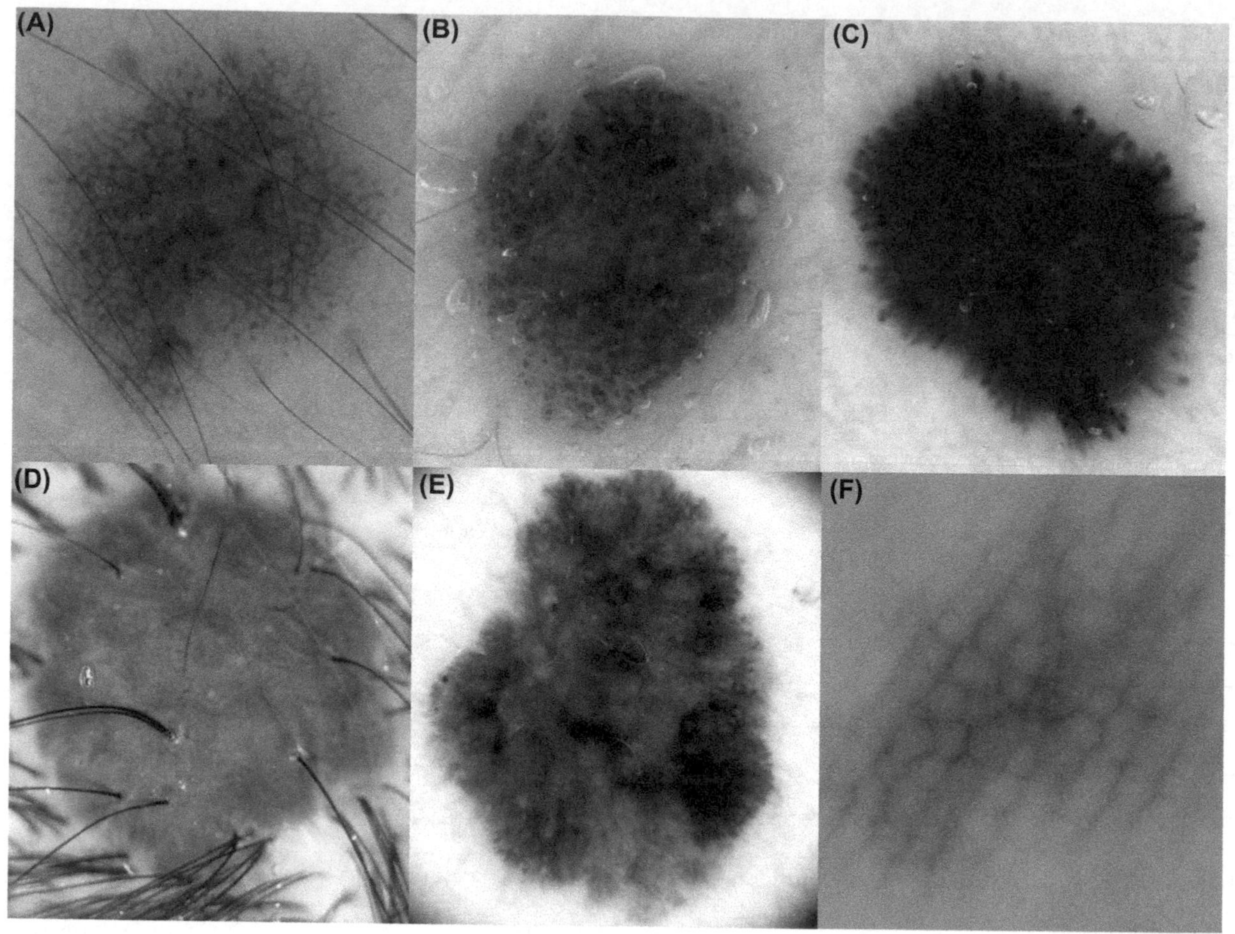

FIGURE 3.1 Global dermoscopic patterns. (A) Reticular (nevus). (B) Globular (nevus). (C) Starburst (Spitz nevus). (D) Homogeneous (blue nevus). (E) Multicomponent (melanoma). (F) Parallel furrow pattern (acral nevus).

recognized as single or grouped characteristics, and several of them can coexist in the same lesion.

There are five basic global patterns, including reticular (resulting from pigment network), globular (resulting from multiple globules), starburst (resulting from peripheral streaks or pseudopods), homogenous (resulting from structureless pigmentation), and multicomponent (resulting from the combination of more than two of the above patterns). The first four patterns can be seen in both nevi and melanoma, whereas the multicomponent pattern is directly suggestive of melanoma (see Fig. 3.1). Two different patterns can be combined in nevi (eg, globular and reticular, homogenous and reticular), but this combination also follows some kind of structured architecture (eg, globular in the center and reticular at the periphery). If a lesion exhibits one of these four patterns, further assessment will be based on the overall symmetry, colors, and the presence of local features, so-called "melanoma-specific criteria."

In general, nevi are characterized by symmetry of structures and display one or two colors. In contrast, melanoma exhibits architectural disorder and often more than two colors.

Melanoma-Specific Criteria

Atypical pigment network: brown-black network with irregular meshes and irregularly distributed lines of different thickness (high specificity for the diagnosis of melanoma).

Irregular dots and/or globules: brown-black or gray, dots and globules of different size, irregularly distributed within the lesion.

Irregular blotches: black, brown, or gray areas with irregular shape and/or distribution.

Irregular streaks and/or pseudopods: radial lines irregularly distributed at the periphery of the lesion (streaks), sometimes with a bulbous projection at their peripheral ending (pseudopods).

Regression structures: seen in the flat area of the lesion and may exhibit as either white scarlike areas corresponding to fibrosis or blue-gray areas (peppering) corresponding to melanophages.

Blue-white veil: seen in the elevated part of the lesion; blue-gray or blue-white, diffuse, irregular pigmentation.

Irregular vascular structures: polymorphous vessels, coexistence of dotted, linear, or hairpin vessels in the same lesion.

Dermoscopy may also offer a preoperative assessment of the Breslow thickness and sentinel lymph node positivity [17,18]. An atypical pigment network is usually found in thin melanomas with a Breslow thickness of <0.76 mm, whereas atypical vascular patterns, radial streaming, and blue-white areas are usually seen in deeper lesions with a Breslow thickness of >0.75 mm [19].

Pigmented melanocytic lesions in certain locations (face, palms, soles, and nails) exhibit unique dermoscopic features because of the specific skin anatomy.

Facial lesions are characterized by a pseudonetwork, which consists of structureless pigmentation interrupted by the numerous, and often enlarged, follicular openings [20]. A pseudonetwork can be found both in melanocytic and nonmelanocytic lesions and, effectively, the first step of the two-step algorithm does not work in facial lesions. Nevi on the face are characterized by brown color, symmetric perifollicular pigmentation, and regular borders. Furthermore, nevi on the face of elderly individuals are usually dermal, papillomatous, and minimally pigmented, whereas flat pigmented nevi are exceedingly rare on the face of elderly individuals. Subsequently, facial melanoma [lentigo maligna (LM) type] does not have to be differentiated from nevi, but mainly from other pigmented flat tumors, including pigmented actinic keratosis (AK) and solar lentigo (SL) [20]. The early dermoscopic criteria of LM include gray color, asymmetric perifollicular pigmentation, and granular and rhomboidal structures [21,22]. In contrast, SL/early seborrheic keratosis (SK) rarely displays gray color under dermoscopy, unless undergoing regression, forming the so-called "lichen planus-like keratosis (LPLK)" [20].

Melanocytic lesions on the palms and soles exhibit unique dermoscopic patterns because of the peculiar anatomy of the acral skin. Nevi show pigmentation along the furrows, whereas melanoma shows pigmentation along the ridges [23].

Benign dermoscopic patterns in acral lesions include:

1. the parallel furrow pattern (see Fig. 3.1F): pigmentation along the furrows of the skin markings (the most common dermoscopic pattern in acral melanocytic nevi)
2. the lattice-like pattern: pigmentation along and across the furrows
3. the fibrillar pattern: fine fibrillar pigmentation perpendicular to the furrows
4. the globular pattern
5. the homogeneous pattern
6. the reticular pattern.

Malignant dermoscopic patterns are:

1. the parallel ridge pattern: pigmentation located on the ridges of the skin markings (acral melanoma in situ or early invasive melanoma)
2. diffuse pigmentation
3. the multicomponent pattern.

Nail Pigmentation

The most common clinical presentation of melanocytic lesions of the nail plate is melanonychia striata. Dermoscopy improves the assessment of pigmented nail bands, allowing early melanoma detection and reducing the number of diagnostic surgical interventions. Subungual nevi are characterized by brown longitudinal parallel lines with regular pattern. Subungual melanoma exhibits irregular multicolor longitudinal bands and the micro-Hutchinson sign, namely a clinically invisible but dermoscopically evident pigmentation of the proximal nail fold [24].

Digital dermoscopy is a useful strategy in the management of patients at risk of melanoma and in monitoring melanocytic lesions. Short- or long-term dermoscopic follow-up observation with special equipment helps the clinician recognize thin melanomas with subtle changes and minimize unnecessary excisions of benign nevi [25,26].

DERMOSCOPY OF COMMON BENIGN NONMELANOCYTIC SKIN TUMORS

A broad spectrum of heterogeneous cutaneous neoplasms is described under the umbrella of *benign nonmelanocytic skin tumors*. Among them, the most common are the vascular tumors, including pyogenic granuloma (PG), hemangioma and angiokeratoma, SK, and dermatofibroma. Dermoscopy is considerably helpful in the diagnosis of the latter group of skin neoplasms [27].

Seborrheic Keratosis

SK is the most common benign epidermal tumor in the elderly. Sites of predilection are the trunk, face, scalp, and extremities. Common clinical variants of SK are the acanthotic, reticulated, and verrucous types, whereas uncommon variants include the clonal and irritated types, LPLK, melanoacanthoma, and stucco keratosis. The dermoscopic picture of an SK significantly depends on its clinical type. However, some dermoscopic features, when present, are considered pathognomonic

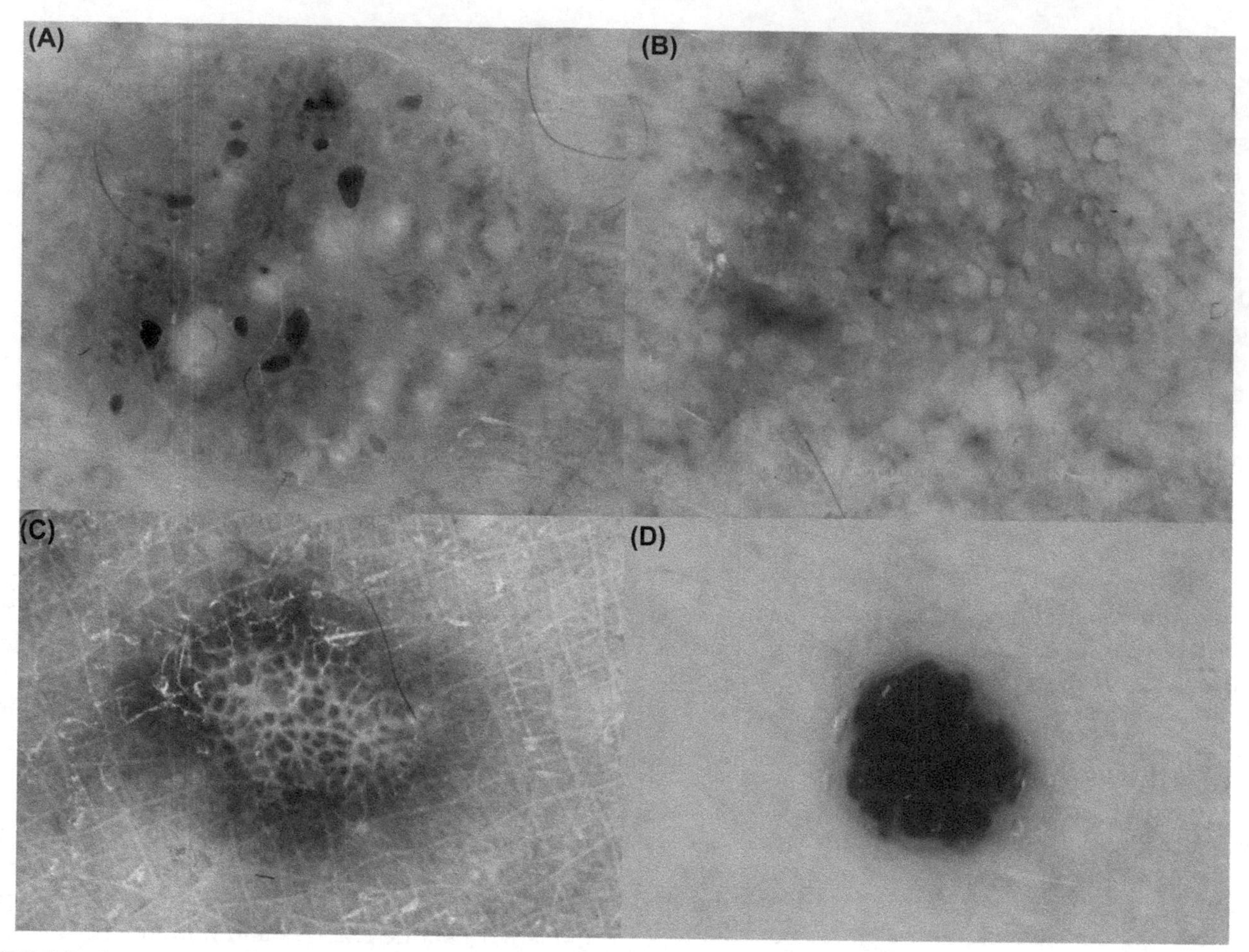

FIGURE 3.2 (A) Multiple milia-like cysts and comedo-like openings are typical for seborrheic keratosis (SK). (B) Solar lentigo (SL), dermoscopically displaying a fine network and sharply demarcated borders. (C) Dermatofibroma is dermoscopically characterized by a peripheral delicate network, whereas the center might display a white network, as in this case, or a whitish structureless area. (D) Angioma is dermoscopically typified by the characteristic well-demarcated red globules (lacunae).

for the diagnosis. Milia-like cysts and comedo-like openings are the classic dermoscopic structures found in an SK (Fig. 3.2A). Milia-like cysts are differently sized, roundish, white or white-yellowish structures corresponding to intraepidermal horn globules. Milia-like cysts are mostly present in acanthotic SK. Comedo-like openings are roundish, ovoid, or even irregularly shaped, sharply defined structures with coloration ranging from brown-yellow to brown-black. Irregularly shaped comedo-like openings are also called *irregular crypts*. Histopathologically they correspond to keratin plugs that fill dilated follicular openings. Additional features that improve the diagnostic accuracy, such as the fingerprint and fat finger signs, the "moth-eaten" border, a "brainlike" appearance (fissures and ridges), and the delicate pigment network have been described [28]. The latter dermoscopic finding is highly representative of SL (see Fig. 3.2B). Ink-spot lentigo, a distinct form of SL, is dermoscopically characterized by a special reticular pattern forming a sharply in-focus, black, broken-up network in the absence of any additional features. This pattern is virtually diagnostic of ink-spot lentigo [29].

Furthermore, in pale-skinned patients, we can easily recognize the specific vascular pattern of SK, mainly characterized by the presence of hairpin and dotted vessels. Nonpolarized dermoscopy is recommended for the examination of SK, because it highlights the presence of milia-like cysts and comedo-like openings. LPLK, which represents SK in regression, displays a distinguished dermoscopic pattern, mainly consisting of gray granules, corresponding to melanophages in histology [30]. In terms of differential diagnosis, the most difficult to diagnose variants are the clonal and the melanoacanthoma type, because they may closely mimic melanoma, both clinically and dermoscopically [31–33].

Dermatofibroma

Dermatofibroma, or fibrous histiocytoma, is a common cutaneous benign neoplasm mostly affecting young and middle-aged adults, with a female predominance.

Clinically, dermatofibromas present in palpation as single or multiple firm and hard papules, or nodules, usually characterized by color variability ranging from light yellowish to dark brown, or purple-red. They can develop anywhere on the body, with a predilection for the lower extremities. In the majority of cases, the diagnosis is set on a clinical basis; however, dermoscopy can be useful in challenging lesions, where differentiation from other benign or malignant tumors is difficult. The prototype of a dermatofibroma in dermoscopy consists of a white scarlike patch in the center and a fine pigment network at the periphery of the lesion. Homogenous pigmentation and a white network are other common dermoscopic features of dermatofibroma (see Fig. 3.2C) [34,35]. The previously mentioned dermoscopic structures may combine together to form 10 different dermoscopic patterns, as described by Zaballos et al.:

Pattern 1 pigment network located throughout the lesion
Pattern 2 delicate pigment network at the periphery and central white scarlike patch
Pattern 3 delicate pigment network at the periphery and central white network
Pattern 4 delicate pigment network at the periphery and central homogeneous pigmentation
Pattern 5 white network throughout the lesion
Pattern 6 homogeneous pigmentation throughout the lesion
Pattern 7 total scarlike patch and a variant with multiple white scarlike patches regularly distributed
Pattern 8 peripheral homogeneous pigmentation and central white scarlike patch
Pattern 9 peripheral homogeneous pigmentation and central white network
Pattern 10 atypical pattern that consists of the presence of atypical pigment network, atypical scarlike patch or white network, atypical homogeneous pigmentation, or irregular distribution of these structures [35].

Aneurysmal dermatofibroma is a relatively rare form of histiocytoma representing less than 2% of total cases [36]. The latter entity and the atypical dermatofibroma share many clinical and dermoscopic similarities with other skin tumors, especially malignant melanoma and Kaposi sarcoma, which can make differentiation problematic [36,37].

Vascular Tumors

Dermoscopy improves the diagnostic accuracy in the clinical evaluation of vascular lesions such as hemangioma, angiokeratoma, and PG. The dermoscopic hallmarks of the vascular lesions are the red, blue, or black lacunae and the red-bluish or red-black homogenous areas (see Fig. 3.2D). Lacunae are well-circumscribed, roundish, or ovoid areas with a reddish, red-bluish, or dark-red to black coloration. Histopathologically they correspond to dilated vascular spaces situated in the upper dermis [38,39]. A rare variant of hemangioma is the targetoid hemosiderotic hemangioma, which may clinically be worrisome; however, the presence of the characteristic lacunae in the central elevated part of the lesion in dermoscopy is indicative of the benign nature of the lesion [40]. Venous lake, also known as *phlebectases*, is a solitary, soft, compressible, dark-blue to violaceous papule commonly involving sun-exposed areas, with a predilection for the lip vermilion, face, and ears. Lesions are common among the elderly. Homogenous blue is the dermoscopic hallmark of the lesion and can be particularly useful in the differentiation of melanoma [41]. Variations on the theme of red, blue, or black lacunae may be occasionally observed in subungual and subcorneal hematomas [24,42,43].

Regarding PG, the most commonly seen dermoscopic features include red homogeneous areas, the white collaret, "white rail lines" that intersect the lesion, and ulceration. Even though the latter dermoscopic criteria may be suggestive of a PG, it is important to underline that amelanotic melanoma represents a major potential diagnostic pitfall. Therefore histopathological confirmation is highly recommended for all lesions with a clinical dermoscopic or differential diagnosis of PG [44,45].

Angiokeratoma is a rare malformation of the vascular network of the upper dermis that clinically presents as single or multiple dark red to black papules, nodules, or plaques. Lacunae, whitish veil, erythema, and hemorrhagic crusts represent the main dermoscopic structures observed in angiokeratoma. Combination of the features results in three distinct dermoscopic patterns. Pattern 1 is composed of dark lacunae and a whitish veil; pattern 2 consists of dark lacunae, whitish veil, and peripheral erythema; and pattern 3 is characterized by dark lacunae, whitish veil, and presence of hemorrhagic crusts [41].

DERMOSCOPY OF COMMON MALIGNANT NONMELANOCYTIC TUMORS

Basal Cell Carcinoma

Basal cell carcinoma (BCC) is the most common form of skin cancer. It mostly affects the sun-exposed body sites, especially the head and neck area. There are five clinicopathological types of BCC (namely, nodular, infiltrative, micronodular, morpheaform, and superficial), each of which has a distinct biological behavior [46]. The variability of BCC in dermoscopy is a result

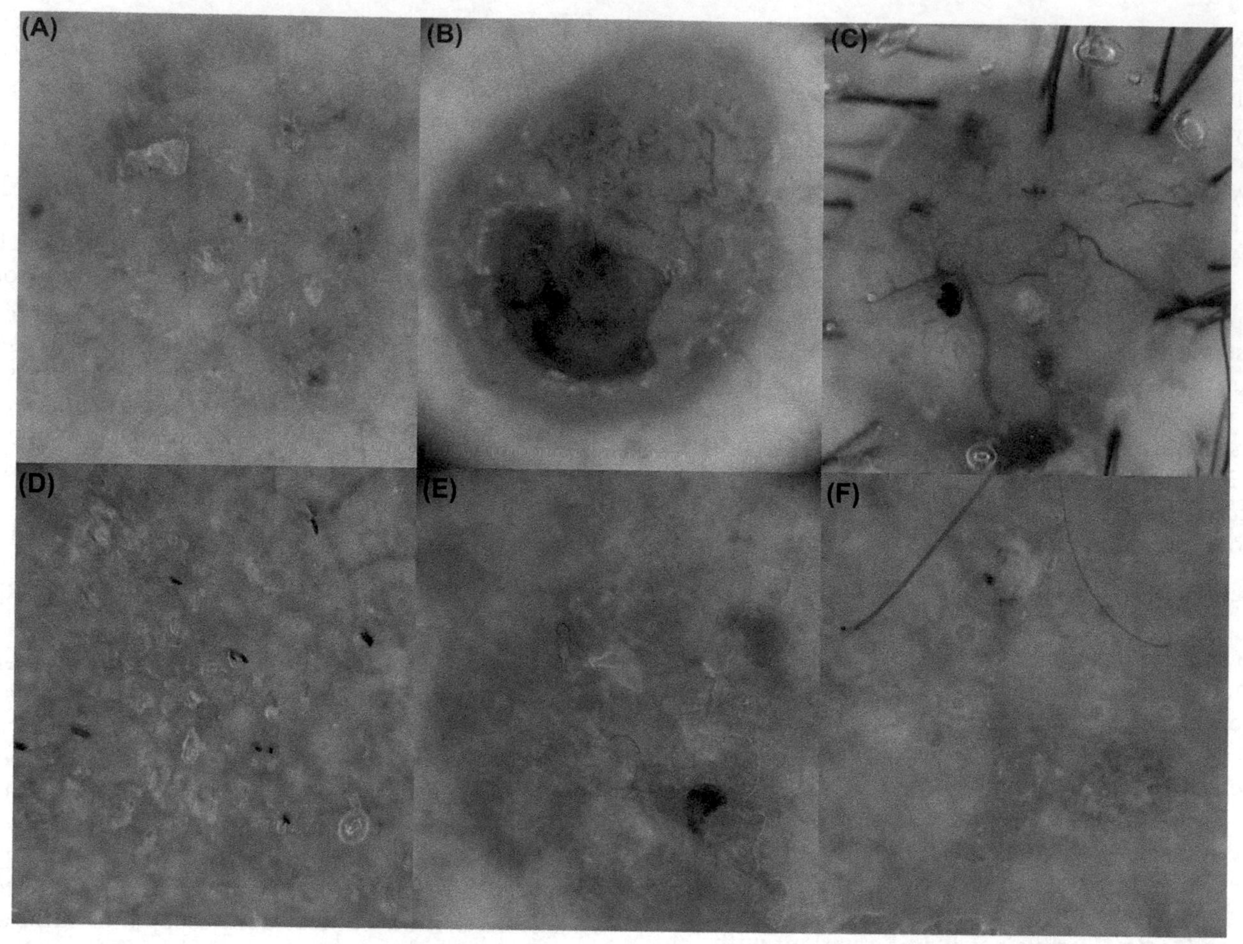

FIGURE 3.3 (A) Dermoscopy of superficial basal cell carcinoma (BCC) usually reveals short, fine telangiectasia; multiple small erosions; and brown-colored pigmented structures. (B) and (C) Nodular BCC displays large arborizing vessels, large ulcerations, and, if pigmented, blue-gray ovoid nests. (D) The "strawberry" pattern of actinic keratosis, consisting of a reddish color interrupted by the white to yellowish follicular openings, which may be filled with keratin plugs. (E) Dotted and glomerular vessels combined with yellow scales comprise the typical dermoscopic pattern of Bowen disease. (F) The white circles surrounding follicular openings represent the most specific dermoscopic criterion of well-differentiated squamous cell carcinoma (SCC).

of different combinations of the dermoscopic features, depending on various factors, including clinicopathological subtype, location, gender, age, and pigmentary trait (Fig. 3.3A–C) [47].

Arborizing vessels are the dermoscopic hallmark of nodular BCC, but can also be seen in all the other subtypes. They represent the supportive neovasculature of the tumor nests and they are large in diameter, branching irregularly into fine terminal capillaries. Their color is bright red, and these vessels are perfectly in focus in the images because of their location on the surface of the tumor (just below the epidermis) [47].

Another vascular dermoscopic structure typically characterizing the superficial BCC (sBCC) is superficial fine telangiectasia (SFT). SFTs appear in focus as short, fine, linear vessels with very few branches [48].

Apart from the vessels, pigmented structures are very representative of a BCC.

Blue-gray ovoid nests are sharply demarcated, usually confluent, ovoid or elongated configurations that histopathologically correspond to large, well-defined tumor nests with pigment aggregates, invading the dermis. Blue-gray ovoid nests represent a stereotypical feature of nodular, pigmented BCC, but they can be also seen in all subtypes except the superficial type BCC [49,50].

Multiple blue-gray dots and globules are numerous, loosely arranged, roundish well-defined structures, which are smaller than the nests. Their histopathological correlation is small, roundish tumor nests with central pigmentation, localized to the papillary or reticular dermis. Blue-gray dots and globules can be observed in all BCC subtypes [49,50].

Maple leaf-like areas represent a highly specific BCC feature and can be seen in all subtypes, but more commonly in sBCC. Maple leaf-like areas are translucent brown to gray-blue peripheral bulbous extensions that closely mimic the shape of the leaves of a maple tree. In histopathology, they correlate to multifocal tumor nests containing pigment aggregates connected to each other by lobular extensions. They are usually found in the epidermis or, less often, in the upper dermis [49,50].

Spoke-wheel areas are a variation of maple leaf-like areas. These are well-defined radial projections, usually brown but sometimes blue or gray, connecting to each other via a darker central axis. Their histopathological correlation is tumor nests arising and connected to the epidermis, characterized by fingerlike projections and centrally located pigmentation. Spoke-wheel areas are highly specific for BCC and can be seen in all subtypes, but they are more common in the superficial type [46–50].

In-focus dots is a term used to describe small, loosely arranged, and well-circumscribed gray dots, which are sharply in focus. Histopathologically, they correspond to melanophages or free pigment deposition in the papillary and reticular dermis [46–50].

Concentric structures are irregularly shaped roundish structures with various colors (blue, gray, brown, black) and a darker central area. They possibly represent variations or precursors of the spoke-wheel areas. They are more common in sBCC [46–50].

It is well known that BCC is a fragile neoplasm that bleeds easily. Minimal trauma results in ulceration, which under dermoscopy may be seen as one or more large structureless areas of red to black-red color. At the sites of ulceration there is a loss of the epidermis, usually covered by hematogenous crusts in histology. Ulceration is mostly seen in nodular BCC. Similarly, multiple small erosions are often seen as small brown-red to brown-yellow crusts. They correlate to thin crusts overlying superficial loss of the epidermis. They are typical features of sBCC [46–50].

Shiny, white/red structureless areas have been reported as a dermoscopic feature of sBCC and may correlate to fibrotic tumoral stroma [47,48].

Another interesting recently described dermoscopic finding in BCC is the presence of chrysalis structures or short white streaks. They can be seen only under polarized light as orthogonal short and thick crossing white lines. They may be attributed to the presence of collagenous stroma and fibrosis in the dermis [51].

Keratinocyte Skin Cancer

Historically, keratinocyte tumors were subdivided into premalignant or precursor lesions (AK), tumors of intermediate biological nature [Bowen disease (BD)], and highly malignant ones [invasive squamous cell carcinoma (SCC)] [52]. However, AK and BD are now classified as in situ SCC, whereas keratinocyte skin cancer is considered to represent an apparent continuum of malignant neoplasms in different progression stages, with AK on one edge and poorly differentiated SCC on the other [46].

Actinic Keratosis

Also known as *solar keratosis*, AK represents the most common carcinoma (in situ) in humans [53]. The incidence of AK is significantly higher in individuals with skin types I–III and in regions with a sunny climate.

The reported risk of an individual AK to progress to invasive SCC varies from 0.1% to 20% [54]. However, patients with multiple AKs have a 5-year cumulative probability of 14% to develop SCC, either within the AK or de novo, highlighting the need of regular follow-up examinations [54].

AKs typically present as erythematous hyperkeratotic macules, papules, or plaques on chronically sun-exposed areas such as the bald scalp, ears, face, forearms, and dorsum of the hands [46].

According to a recently introduced clinical classification scheme, grade I AKs are slightly palpable (better felt than seen), grade II includes AKs of moderate thickness (easily felt and seen), and grade III AKs are clinically obvious, very thick, and usually hyperkeratotic [55]. It has been suggested that these clinical grades of AK also dermoscopically correspond to three different patterns. Grade I AKs display a red pseudonetwork and white scales, and grade II lesions are typified by the so-called "strawberry" pattern, consisting of an erythematous background interrupted by white to yellow enlarged follicular openings with or without keratin plugs (see Fig. 3.3D). In grade III AKs, the dense hyperkeratosis, seen as a white-yellow structureless area, often impedes the visualization of the follicular openings, which are typically filled with keratotic plugs [7]. The diagnostic sensitivity and specificity of dermoscopy in the diagnosis of nonpigmented AK has been reported to reach 98% and 95%, respectively [56].

Less often, AK may be slightly or heavily pigmented [pigmented AK (PAK)], clinically presenting as a red-brownish or even brown macules. In such cases, it has to be discriminated from SL and early LM [20]. When located on the face, dermoscopy of PAK typically reveals a pseudonetwork consisting of a diffuse brown pigmentation interrupted by nonpigmented follicular openings, histopathologically corresponding to pigmented keratinocytes along the flattened dermoepidermal junction of the facial skin. The latter dermoscopic pattern can be also seen in SL and LM and, accordingly, the differential diagnosis of a pigmented facial macule relies on the detection of additional specific criteria [20].

PAK is known to occasionally display several of the dermoscopic criteria of LM, such as asymmetrically pigmented follicular openings, rhomboidal structures, and gray dots or globules, rendering the discrimination between these two entities highly troublesome. Dermoscopic features highly suggestive of PAK are superficial scales, keratin plugs, rosettes, and white circles. Furthermore, a potentially useful clue is that in contrast to LM, the pigmentation in PAK does not have the tendency to obscure the visualization of the follicular openings. Notably, the discrimination between the two entities may even be histopathologically difficult when it is not clear whether the pigmented atypical cells in the basal layer are keratinocytes or melanocytes [20]. The dermoscopic recognition of SL (which is considered a type of early SK) is usually feasible based on the absence of gray color and the detection of light brown fingerprint areas, yellow opaque areas, milia-like cysts, moth-eaten border, and a sharp demarcation [20].

Bowen Disease (Intraepidermal Carcinoma)

BD is defined as an SCC in situ with full epidermal thickness dysplasia that has the potential for significant lateral spread before invasion [57].

BD may progress to invasive SCC in 3–20% of cases. Notably, SCC developing on preexisting BD is associated with an unfavorable prognosis and a high rate of regional or distant metastasis [58].

Typically BD presents as an asymptomatic, slowly enlarging, erythematous, well-demarcated, scaly patch or plaque. This unspecific clinical presentation often results in a delayed diagnosis, sometimes complicating the management of the tumor.

The typical dermoscopic pattern of BD consists of dotted and/or glomerular vessels, white to yellowish surface scales, and a red-yellowish background color (see Fig. 3.3E) [59]. Glomerular (or coiled) vessels represent a variation of dotted vessels, which are larger in size and characterized by tortuous capillaries, reminiscent of the histological appearance of the glomerular apparatus of the kidney. Both dotted and glomerular vessels often appear within the same lesion and are usually distributed in small, densely packed clusters or groups [46,59].

The characteristic vessel morphology and distribution seen in BD is particularly useful in distinguishing the disease from clinically similar skin tumors and inflammatory skin diseases [60]. Discrimination from sBCC is usually straightforward based on the vessels' morphology, which is dotted/glomerular in BD and linear in sBCC. The differential diagnosis between BD and psoriasis is problematic in cases where the latter manifests with one or few plaques distributed on nontypical sites, which can be misinterpreted as BD. Conversely, BD developing in psoriatic patients, especially those undergoing phototherapy, might be easily overlooked among the plethora of psoriatic lesions. The dermoscopic discrimination among the two entities might also be difficult, because psoriasis typically displays regularly distributed dotted vessels and white scales, closely resembling the features seen in BD [61]. However, some useful clues do exist and include: (1) the diameter of the vascular structures, which is typically larger in BD (glomerular vessels vs. red dots); (2) the distribution of the vascular structures, which is almost always regular (symmetric) in psoriasis and most often clustered in BD; and (3) the presence of yellow-colored scales, which minimizes the possibility of psoriasis, although scales are very common in BD [8].

It has been shown that the simultaneous presence of a clustered vascular pattern and glomerular vessels is associated with a diagnostic probability of 98% for BD, compared with psoriasis and sBCC [60].

A characteristic linear arrangement of glomerular or dotted vessels has been described in pigmented BD [62]. Although not very common (present in approximately 10% of cases), the latter finding has been suggested to represent a highly specific feature of pigmented BD, allowing its discrimination from melanoma in particular [62].

In addition to the vascular criteria, dermoscopy of pigmented BD has been shown to reveal two main patterns: a brown structureless pattern and a mixed pattern combining a hypopigmented structureless eccentric area and small brown/black dots. The dots in pigmented BD are arranged either in a patchy distribution or in peripheral lines, the latter representing a highly specific arrangement [62].

Squamous Cell Carcinoma

Cutaneous SCC is the second most common skin cancer [63]. The majority (70%) of SCCs develop on the head and neck, with an additional 15% arising on the upper extremities.

Clinically, SCC usually presents as an indurated hyperkeratotic nodule with or without ulceration. Less often, SCC lacks signs of keratinization and manifests as an ulcer. The presence of AKs is usually evident on the neighboring and surrounding skin surface [46].

The dermoscopic pattern of SCC depends on the grade of histopathological differentiation [46]. Specifically, well differentiated tumors exhibit a white predominant color, resulting from one or more of the following structures: white structureless areas, white circles (surrounding the follicular openings), white halos (surrounding vessels), and white amorphous masses of keratin (see Fig. 3.3F) [7,46,64]. White structureless areas represent the most common but less specific feature. In contrast, white circles represent the most specific feature of SCC when compared with other common nonpigmented skin tumors [64]. White halos and amorphous white keratin masses are indicative of a keratinizing tumor, but cannot predict a specific diagnosis. Vascular structures may be dermoscopically seen in well-differentiated SCC, usually as linear irregular or hairpin vessels of large diameter [7]. However, the quantity of vascular structures is usually low in well-differentiated SCC, with white structures typically predominating. A specific combination of central keratin masses surrounded by hairpin or linear irregular vessels distributed at the periphery of the tumor has been suggested to typify keratoacanthoma [7,64].

In contrast, poorly differentiated SCC is clinically typified by a flat appearance and dermoscopically by a red predominant color, attributed to the absence of scaling and the presence of bleeding and/or dense vascularity [65]. Vessel quantity is significantly correlated to the differentiation grade of SCC, because tumors displaying vessels in more than 50% of the lesion surface have a 30- to 120-fold increased possibility of being poorly differentiated. Vessels caliber also represents a significant predictor of differentiation grade, with a small caliber associated with poor differentiation [65].

DERMOSCOPY IN GENERAL DERMATOLOGY

Continuously gathering evidence suggests that in addition to its usefulness for the evaluation of skin tumors, dermoscopy is also helpful for the assessment of nontumoral lesions [8,66]. The latter is based on the observation that apart from pigmentation structures formed by melanin deposition, dermoscopy may also reveal vascular alterations, color variegation, follicle disturbances, and other features invisible to the unaided eye. The dermoscopic patterns of several inflammatory and infectious skin diseases have been described. It has been suggested that four parameters should be assessed when applying dermoscopy in the realm of inflammatory and infectious diseases: (1) morphological vascular patterns, (2) arrangement of vascular structures, (3) colors, and (4) follicular abnormalities, although other specific features (clues) should also be evaluated [8]. In Table 3.2, the dermoscopic characteristics of several inflammatory skin diseases are presented.

Papulosquamous Skin Diseases

Psoriasis

Dotted vessels represent the most common dermoscopic feature of psoriasis, and are typically present in every psoriatic plaque. Effectively, detection of any other morphological type of vessel should raise doubts about the diagnosis of psoriasis.

However, red dots do not represent a specific finding, because they can be found in several other inflammatory diseases. Instead, their uniform or regular distribution within the lesion represents the dermoscopic hallmark of psoriasis, being particularly useful in differential diagnosis. Another less common but equally specific vessel arrangement pattern for psoriasis is the so-called "red globular rings." Other types of vessel distribution are extremely rare in psoriasis.

Light red background color and white superficial scales represent two common additional dermoscopic criteria of psoriasis. The scale color is of particular value for differentiating the disease from dermatitis, which typically displays yellow scales [61].

Psoriatic lesions located on specific body sites exhibit the same pattern, with variations in the degree of scaling. For example, in psoriatic balanitis and inverse psoriasis lesions that lack scaling, the typical vascular pattern of regularly distributed red dots is prominent under dermoscopic examination. Conversely, in scalp or palmoplantar psoriasis, the thick hyperkeratotic plaque surface does not allow visualization of the underlying vascular structures, which are highlighted after removal of the scales [67].

Dermatitis

Typically, dermatitis dermoscopically exhibits red dots in a patchy distribution and fine, diffuse, yellowish scales [61]. Morphologically, no difference exists between the vessels of dermatitis and those of psoriasis. However, the vessels are not symmetrically distributed but are usually aggregated or clustered in some sites of the lesion and absent in others, forming an overall asymmetrical "patchy" pattern. Most importantly, dermatitis lesions typically display yellowish scales, which are particularly useful clues for the recognition of the disease [61]. Notably, yellow scale color can be dermoscopically detected not only in cases of acute dermatitis, but also in long-standing lesions. Although the dermoscopic pattern of each disease subtype has not been separately investigated, several case studies including contact dermatitis, nummular eczema, generalized dermatitis, chronic dermatitis, seborrheic dermatitis, and other subtypes report on similar dermoscopic findings (as described) [61,68,69].

Lichen Planus

White crossing streaks (Wickham striae) are considered the dermoscopic hallmark of lichen planus, being a constant finding in almost all types of lesions associated with the disease [61,70]. Vessels of mixed morphology (dotted and linear), usually distributed at the periphery of the lesion, represent additional dermoscopic findings of the disease [61].

Pityriasis Rosea

A yellowish background color and peripheral whitish scales (collarette) are the most important dermoscopic features of pityriasis rosea. Dotted vessels can be also

TABLE 3.2 Dermoscopic Criteria of Inflammatory Skin Diseases

Disease	Dermoscopic criteria
Darier disease	Pseudocomedones, erythema, dotted/linear vessels
Dermatitis	Dotted vessels with patchy distribution, yellow crusts/scales
Discoid lupus erythematosus	Early lesions: perifollicular whitish halo, follicular plugging, and white scales
	Late lesions: telangiectasias, pigmentation structures, and whitish structureless areas
Erythema multiforme	Linear vessels peripherally, bluish patches in the center
Granuloma annulare	Dotted, linear, or dotted/linear vessels; white, red, or yellow background
Granuloma faciale	Dilated follicular openings, perifollicular whitish haloes, pigmentation structures, follicular keratotic plugs, elongated or linear branching vessels
Henoch–Schönlein purpura	Irregularly shaped red patches with blurred borders
Lichen planus	Wickham striae, peripheral dotted/linear vessels
Lichen sclerosus	Genital lesions: white-yellowish structureless areas, linear vessels
	Extragenital lesions: white/yellowish structureless areas, yellowish keratotic plugs (pseudocomedones)
Livedo reticularis	Linear vessels with a regular distribution
Mastocytosis	Light-brown blot, pigment network, reticular vascular pattern, or yellow-orange blot
Morphea	Whitish fibrotic beams, linear vessels
Mycosis fungoides	Short linear vessels, orange-yellowish areas, spermatozoa-like structures
Necrobiosis lipoidica	Prominent network of linear arborizing vessels and a yellow background color
Pigmented purpuric dermatoses	Purpuric dots or globules, orange-brown background
Pityriasis rosea	Yellowish background, peripheral white scales, dotted vessels with patchy distribution
Pityriasis rubra pilaris	Yellowish areas, dotted and linear vessels with patchy or peripheral distribution
Porokeratosis	White-yellowish or brownish peripheral annular structure; in the center, brownish pigmentation, dotted/linear vessels, or structureless whitish areas
Psoriasis	Dotted vessels with regular distribution, white scales
Rosacea	Erythematotelangiectatic type: polygonal vessels
	Papulopustular type: follicular plugs, follicular pustules, polygonal vessels
Sarcoidosis	Orange-yellowish globules or areas, linear vessels
Sweet syndrome	Structureless bluish patches
Urticaria	Network of linear vessels surrounding avascular areas
Urticarial vasculitis	Purpuric dots or globules, orange-brown background

found in several lesions, but they are arranged in an irregular or patchy pattern, unlike the characteristic regular distribution of psoriasis [61].

Granulomatous Skin Diseases

Dermal granulomas dermoscopically project as translucent orange-yellowish patches or structureless areas. These structures, often associated with linear vessels, are highly indicative of a granulomatous skin disease, including sarcoidosis, lupus vulgaris, and granulomatous rosacea. However, dermoscopy is insufficient to differentiate among these entities [71].

Discoid Lupus Erythematosus

Perifollicular whitish halo, follicular plugging, and white scales represent the predominant features of early lesions of discoid lupus erythematosus (DLE), whereas telangiectasias, pigmented structures, and whitish structureless areas characterize longer-standing lesions [72]. By highlighting the characteristic follicular disturbances of DLE or the yellowish patches of lupus pernio (cutaneous sarcoidosis) and lupus vulgaris, dermoscopy might significantly facilitate this particularly difficult differential diagnosis [71].

Rosacea

A characteristic dermoscopic vascular pattern of polygonal vessels has been described in erythematotelangiectatic rosacea (ER) [71]. Intense vasodilation, which is well known to represent a major pathophysiological alteration of the disease, results in a characteristic morphological pattern of dermoscopic vascular polygons. Telangiectasias may also be detected on chronically sun-damaged, atrophic facial skin, but they usually lack the characteristic arrangement in polygons. Additional dermoscopic findings of ER include follicular plugs, white scales, features related to the presence of *Demodex*, namely "*Demodex* tails," and whitish amorphic follicular material [73]. In papulopustular rosacea, dermoscopy might highlight clinically nonvisible pustules, providing a useful diagnostic clue.

Lichen Sclerosus and Morphea

White-yellowish structureless areas represent the predominant dermoscopic feature of genital and extragenital lichen sclerosus. Genital lesions often also display linear vessels, whereas early extragenital lesions commonly exhibit keratotic plugs and may be surrounded by an erythematous halo, which represents a marker of disease activity. In contrast, linear vessels within a lilac ring and "fibrotic beams" (correlating histopathologically with dermal sclerosis) have been reported to characterize morphea [74].

Urticaria and Urticarial Vasculitis

A red, reticular network of linear vessels has been described to dermoscopically characterize common urticaria [75]. In contrast, urticarial vasculitis displays purpuric dots or globules, suggestive of the underlying vasculitis, on an orange-brown background [76].

Pigmented Purpuric Dermatoses (Capillaritis)

Five distinct entities are traditionally described under the term *pigmented purpuric dermatoses (PPD)*: Schamberg disease, Majocchi purpura, eczematoid purpura of Doucas and Kapetanakis, lichen aureus, and pigmented purpuric lichenoid dermatitis of Gougerot–Blum. The typical dermoscopic pattern of PPDs consists of purpuric dots or globules and orange-brown areas [77].

Mastocytosis

Four dermoscopic patterns characterize cutaneous mastocytosis: light-brown blot, pigment network, reticular vascular pattern, and yellow-orange blot [78,79]. Notably, an association has been suggested between the dermoscopic pattern and the disease subtype. In detail, light-brown blot and pigment network were associated with maculopapular mastocytosis, yellow orange blot with solitary mastocytoma, and a reticular vascular pattern was detected in all cases of telangiectasia macularis eruptiva perstans [79].

Mycosis Fungoides

Short linear vessels and orange-yellowish areas represent the most common dermoscopic findings of early mycosis fungoides (MF), whereas dotted vessels might also be present [69]. A peculiar vascular structure consisting of a dotted and a linear component (spermatozoon-like structure) can be found in half of MF cases. The dermoscopic pattern of MF might be particularly useful for its discrimination from dermatitis. Specifically, dermatitis typically displays only dotted vessels, with the exception of lesions under long-term treatment with topical steroids, which might result in atrophy and telangiectasia [69].

Dermoscopy of Infectious Skin Diseases

Specific dermoscopic patterns have been described for several infectious skin diseases, including those of viral, fungal, and parasitic origin [80]. Of note, use of the new-generation dermatoscopes that do not require direct contact to the skin minimizes the risk of transfection. Table 3.3 summarizes the main dermoscopic findings in several infectious skin diseases, and the most common of them are also described in the following sections.

Scabies

The typical dermoscopic pattern of scabies consists of small, dark brown, triangular structures located at the end of whitish curved or wavy lines, giving an appearance reminiscent of a delta-wing jet with a contrail. Microscopically, the brown triangle corresponds to the pigmented anterior part of the mite, whereas the burrow of the mite correlates dermoscopically to the contrail feature [81].

The diagnostic accuracy of dermoscopy has been reported to be at least equal to traditional ex vivo microscopic examination, but requiring less time, cost, and experience [81,82].

Mycoses

Tinea nigra is dermoscopically typified by a reticulated pattern consisting of superficial fine, wispy, light brown strands or pigmented spicules [83]. In tinea capitis, dermoscopy typically reveals comma hairs, broken dystrophic hairs, or corkscrew or convoluted hairs.

TABLE 3.3 Dermoscopic Criteria of Infectious Skin Diseases

Disease	Dermoscopic criteria
Cutaneous lavra migrans	Translucent brownish structureless areas in a segmental arrangement
Human papillomavirus infections	Common warts: multiple densely packed papillae with a central red dot or loop, surrounded by a whitish halo; hemorrhages (small red to black dots or streaks) may also be present
	Plantar warts: prominent hemorrhages within a well-defined, yellowish papilliform surface in which skin lines are interrupted
	Plane warts: regularly distributed red dots, light brown to yellow background
	Genital warts: mosaic pattern (early/flat lesions), fingerlike and knoblike pattern (raised/papillomatous lesions), unspecific pattern
Leishmaniasis	Orange-yellowish globules or areas, linear vessels, erythema, follicular plugs, hyperkeratosis, central ulceration
Lupus vulgaris	Orange-yellowish globules or areas, linear vessels
Molluscum contagiosum	Central pore or umbilication, white to yellow amorphous structures, peripheral linear or branching vessels ("red corona")
Pediculosis	The lice itself, ovoid brownish structures (nits with vital nymphs), ovoid translucent structures (empty nits)
Scabies	"Jet with contrail" structure
Spider leg spines	Small black spines
Tick bites	Visualization of the anterior legs protruding from the skin surface, brown to gray translucent "shield"
Tinea nigra	Reticulated pattern, consisting of superficial fine, wispy, light brown strands or pigmented spicules
Tungiasis	White to light brown color, targetoid brownish ring surrounding a black central pore

Molluscum Contagiosum

Dermoscopy is particularly useful for the diagnosis of molluscum contagiosum by revealing a characteristic pattern consisting of a central umbilication in conjunction with polylobular white to yellow amorphous structures and surrounded by linear or branched vessels [84].

Pediculosis

Dermoscopy allows a rapid and reliable diagnosis of pediculosis by revealing the lice itself or the nits fixed to the hair shaft [85]. Nits containing vital nymphs dermoscopically display ovoid brown structures, whereas the empty nits are translucent and typically show a plane and fissured free ending. Additionally, dermoscopy enables the discrimination between nits and the so-called "pseudo-nits," such as hair casts and debris of hair spray or gel. The latter are not firmly attached to the hair shaft and appear dermoscopically as amorphous, whitish structures [86].

CONCLUSION

Dermoscopy has gained an irreplaceable role in the evaluation of skin tumors, because it significantly improves the performance of clinicians. Furthermore, with novel dermoscopic patterns of several skin diseases continuously coming to light, the dermatoscope gradually acquires a role similar to the pathologist's stethoscope.

References

[1] Malvehy J, Puig S, Argenziano G, Marghoob AA, Soyer HP, International Dermoscopy Society Board Members. Dermoscopy report: proposal for standardization. Results of a consensus meeting of the International Dermoscopy Society. J Am Acad Dermatol 2007;57(1):84–95.

[2] Menzies SW, Zalaudek I. Why perform dermoscopy? The evidence for its role in the routine management of pigmented skin lesions. Arch Dermatol 2006;142(9):1211–2.

[3] Lallas A, Zalaudek I, Apalla Z, Longo C, Moscarella E, Piana S, et al. Management rules to detect melanoma. Dermatology 2013;226(1):52–60.

[4] Zalaudek I, Kittler H, Marghoob AA, Balato A, Blum A, Dalle S, et al. Time required for a complete skin examination with and without dermoscopy: a prospective, randomized multicenter study. Arch Dermatol 2008;144(4):509–13.

[5] Kittler H, Pehamberger H, Wolff K, Binder M. Diagnostic accuracy of dermoscopy. Lancet Oncol 2002;3(3):159–65.

[6] Menzies SW, Westerhoff K, Rabinovitz H, Kopf AW, McCarthy WH, Katz B. Surface microscopy of pigmented basal cell carcinoma. Arch Dermatol 2000;136(8):1012–6.

[7] Zalaudek I, Giacomel J, Schmid K, Bondino S, Rosendahl C, Cavicchini S, et al. Dermatoscopy of facial actinic keratosis,

intraepidermal carcinoma and invasive squamous cell carcinoma: a progression model. J Am Acad Dermatol 2012;66(4):589–97.
[8] Lallas A, Giacomel J, Argenziano G, García-García B, González-Fernández D, Zalaudek I, et al. Dermoscopy in general dermatology: practical tips for the clinician. Br J Dermatol 2014;170(3): 514–26.
[9] Vestergaard ME, Macaskill P, Holt PE, Menzies SW. Dermoscopy compared with naked eye examination for the diagnosis of primary melanoma: a meta-analysis of studies performed in a clinical setting. Br J Dermatol 2008;159(3):669–76.
[10] Argenziano G, Albertini G, Castagnetti F, De Pace B, Di Lernia V, Longo C, et al. Early diagnosis of melanoma: what is the impact of dermoscopy? Dermatol Ther 2012;25(5):403–9.
[11] Argenziano G, Soyer HP, Chimenti S, Talamini R, Corona R, Sera F, et al. Dermoscopy of pigmented skin lesions: results of a consensus meeting via the Internet. J Am Acad Dermatol 2003; 48(5):679–93.
[12] Johr RH. Dermoscopy: alternative melanocytic algorithms, the ABCD rule of dermatoscopy, Menzies' scoring method, and 7-point checklist. Clin Dermatol 2002;20(3):240–7.
[13] Argenziano G, Catricalà C, Ardigo M, Buccini P, De Simone P, Eibenschutz L, et al. Seven-point checklist of dermoscopy revisited. Br J Dermatol 2011;164(4):785–90.
[14] Soyer HP, Argenziano G, Zalaudek I, Corona R, Sera F, Talamini R, et al. Three-point checklist of dermoscopy: a new screening method for early detection of melanoma. Dermatology 2004;208(1):27–31.
[15] Pehamberger H, Steiner A, Wolff K. In vivo epiluminescence microscopy of pigmented skin lesions. I. Pattern analysis of pigmented skin lesions. J Am Acad Dermatol 1987;17(4):571–83.
[16] Braun RP, Rabinovitz HS, Oliviero M, Kopf AW, Saurat JH. Pattern analysis: a two-step procedure for the dermoscopic diagnosis of melanoma. Clin Dermatol 2002;20(3):236–9.
[17] De Giorgi V, Carli P. Dermoscopy and preoperative evaluation of melanoma thickness. Clin Dermatol 2002;20(3):305–8.
[18] Gonzalez-Alvarez T, Carrera C, Bennassar A, Vilalta A, Rull R, Alos L, et al. Dermoscopy structures as predictors of sentinel lymph node positivity in cutaneous melanoma. Br J Dermatol 2015;172(5):1269–77.
[19] Argenziano G, Fabbrocini G, Carli P, De Giorgi V, Delfino M. Clinical and dermatoscopic criteria for the preoperative evaluation of cutaneous melanoma thickness. J Am Acad Dermatol 1999;40(1): 61–8.
[20] Lallas A, Argenziano G, Moscarella E, Longo C, Simonetti V, Zalaudek I. Diagnosis and management of facial pigmented macules. Clin Dermatol 2014;32(1):94–100.
[21] Stolz W, Schiffner R, Burgdorf WH. Dermatoscopy for facial pigmented skin lesions. Clin Dermatol 2002;20(3):276–8.
[22] Pralong P, Bathelier E, Dalle S, Poulalhon N, Debarbieux S, Thomas L. Dermoscopy of lentigo maligna melanoma: report of 125 cases. Br J Dermatol 2012;167(2):280–7.
[23] Saida T, Koga H, Uhara H. Key points in dermoscopic differentiation between early acral melanoma and acral nevus. J Dermatol 2011;38(1):25–34.
[24] Braun RP, Baran R, Le Gal FA, Dalle S, Ronger S, Pandolfi R, et al. Diagnosis and management of nail pigmentations. J Am Acad Dermatol 2007;56(5):835–47.
[25] Kittler H, Binder M. Risks and benefits of sequential imaging of melanocytic skin lesions in patients with multiple atypical nevi. Arch Dermatol 2001;137(12):1590–5.
[26] Fikrle T, Pizinger K, Szakos H, Panznerova P, Divisova B, Pavel S. Digital dermatoscopic follow-up of 1027 melanocytic lesions in 121 patients at risk of malignant melanoma. J Eur Acad Dermatol Venereol 2013;27(2):180–6.
[27] Marghoob AA, Usatine RP, Jaimes N. Dermoscopy for the family physician. Am Fam Physician 2013;88(7):441–50.
[28] Braun RP, Rabinovitz HS, Krischer J, Kreusch J, Oliviero M, Naldi L, et al. Dermoscopy of pigmented seborrheic keratosis: a morphological study. Arch Dermatol 2002;138(12):1556–60.
[29] Bottoni U, Nisticò S, Amoruso GF, Schipani G, Arcidiacono V, Scali E, et al. Ink spot lentigo: singular clinical features in a case series of patients. Int J Immunopathol Pharmacol 2013;26(4): 953–5.
[30] Zaballos P, Blazquez S, Puig S, Salsench E, Rodero J, Vives JM, et al. Dermoscopic pattern of intermediate stage in seborrhoeic keratosis regressing to lichenoid keratosis: report of 24 cases. Br J Dermatol 2007;157(2):266–72.
[31] Longo C, Zalaudek I, Moscarella E, Lallas A, Piana S, Pellacani G, et al. Clonal seborrheic keratosis: dermoscopic and confocal microscopy characterization. J Eur Acad Dermatol Venereol 2014; 28(10):1397–400.
[32] Rossiello L, Zalaudek I, Ferrara G, Docimo G, Giorgio CM, Argenziano G. Melanoacanthoma simulating pigmented spitz nevus: an unusual dermoscopy pitfall. Dermatol Surg 2006; 32(5):735–7.
[33] Shankar V, Nandi J, Ghosh K, Ghosh S. Giant melanoacanthoma mimicking malignant melanoma. Indian J Dermatol 2011;56(1): 79–81.
[34] Espasandín-Arias M, Moscarella E, Mota-Buçard A, Moreno-Moreno C, Lallas A, Longo C, et al. The dermoscopic variability of dermatofibromas. J Am Acad Dermatol 2015;72(Suppl. 1):S22–4.
[35] Zaballos P, Puig S, Llambrich A, Malvehy J. Dermoscopy of dermatofibromas: a prospective morphological study of 412 cases. Arch Dermatol 2008;144(1):75–83.
[36] Zaballos P, Llambrich A, Ara M, Olazarán Z, Malvehy J, Puig S. Dermoscopic findings of haemosiderotic and aneurysmal dermatofibroma: report of six patients. Br J Dermatol 2006;154(2): 244–50.
[37] Morariu SH, Suciu M, Vartolomei MD, Badea MA, Cotoi OS. Aneurysmal dermatofibroma mimicking both clinical and dermoscopic malignant melanoma and Kaposi's sarcoma. Rom J Morphol Embryol 2014;55(Suppl. 3):1221–4.
[38] Oiso N, Kawada A. The dermoscopic features in infantile hemangioma. Pediatr Dermatol 2011;28(5):591–3.
[39] Haliasos EC, Kerner M, Jaimes N, Zalaudek I, Malvehy J, Lanschuetzer CM, et al. Dermoscopy for the pediatric dermatologist. II. Dermoscopy of genetic syndromes with cutaneous manifestations and pediatric vascular lesions. Pediatr Dermatol 2013; 30(2):172–81.
[40] Piccolo V, Russo T, Mascolo M, Staibano S, Baroni A. Dermoscopic misdiagnosis of melanoma in a patient with targetoid hemosiderotic hemangioma. J Am Acad Dermatol 2014;71(5):179–81.
[41] Zaballos P, Daufí C, Puig S, Argenziano G, Moreno-Ramírez D, Cabo H, et al. Dermoscopy of solitary angiokeratomas: a morphological study. Arch Dermatol 2007;143(3):318–25.
[42] Oztas MO. Clinical and dermoscopic progression of subungual hematomas. Int Surg 2010;95(3):239–41.
[43] Massa AF, Dalle S, Thomas L. Subcorneal hematoma. Ann Dermatol Venereol 2013;140(1):63–4.
[44] Zaballos P, Carulla M, Ozdemir F, Zalaudek I, Bañuls J, Llambrich A, et al. Dermoscopy of pyogenic granuloma: a morphological study. Br J Dermatol 2010;163(6):1229–37.
[45] Zaballos P, Rodero J, Serrano P, Cuellar F, Guionnet N, Vives JM. Pyogenic granuloma clinically and dermoscopically mimicking pigmented melanoma. Dermatol Online J 2009;15(10):10.
[46] Lallas A, Argenziano G, Zendri E, Moscarella E, Longo C, Grenzi L, et al. Update on non-melanoma skin cancer and the value of dermoscopy in its diagnosis and treatment monitoring. Expert Rev Anticancer Ther 2013;13(5):541–58.
[47] Lallas A, Apalla Z, Argenziano G, Longo C, Moscarella E, Specchio F, et al. The dermatoscopic universe of basal cell carcinoma. Dermatol Pract Concept 2014;4(3):11–24.

[48] Giacomel J, Zalaudek I. Dermoscopy of superficial basal cell carcinoma. Dermatol Surg 2005;31(12):1710–3.

[49] Lallas A, Argenziano G, Kyrgidis A, Apalla Z, Moscarella E, Longo C, et al. Dermoscopy uncovers clinically undetectable pigmentation in basal cell carcinoma. Br J Dermatol 2014;170(1):192–5.

[50] Lallas A, Tzellos T, Kyrgidis A, Apalla Z, Zalaudek I, Karatolias A, et al. Accuracy of dermoscopic criteria for discriminating superficial from other subtypes of basal cell carcinoma. J Am Acad Dermatol 2014;70(2):303–11.

[51] Balagula Y, Braun RP, Rabinovitz HS, Dusza SW, Scope A, Liebman TN, et al. The significance of crystalline/chrysalis structures in the diagnosis of melanocytic and nonmelanocytic lesions. J Am Acad Dermatol 2012;67(2):194.e1-8.

[52] MacKie RM, Quinn A. Non-melanoma skin cancer and other epidermal skin tumours. In: Burns T, Breathnach SM, Cox NH, Griffiths CEM, editors. Rook's textbook of dermatology. 7th ed. Oxford: Blackwell Publishing; 2004. p. 36.1–36.39.

[53] Memon AA, Tomenson JA, Bothwell J, Friedmann PS. Prevalence of solar damage and actinic keratosis in a Merseyside population. Br J Dermatol 2000;142(6):1154–9.

[54] Glogau RG. The risk of progression to invasive disease. J Am Acad Dermatol 2000;42(1 Pt 2):23–4.

[55] Rowert-Huber J, Patel MJ, Forschner T, Ulrich C, Eberle J, Kerl H, et al. Actinic keratosis is an early in situ squamous cell carcinoma: a proposal for reclassification. Br J Dermatol 2007;156(Suppl. 3):8–12.

[56] Huerta-Brogeras M, Olmos O, Borbujo J, Hernández-Núñez A, Castaño E, Romero-Maté A, et al. Validation of dermoscopy as a real-time noninvasive diagnostic imaging technique for actinic keratosis. Arch Dermatol 2012;148(10):1159–64.

[57] Lee M-M, Wick MM. Bowen's disease. CA Cancer J Clin 1990; 40(10):237–42.

[58] Kossard S, Rosen R. Cutaneous Bowen's disease: an analysis of 1001 cases according to age, sex, and site. J Am Acad Dermatol 1992;27(3):406–10.

[59] Zalaudek I, Argenziano G, Leinweber B, Citarella L, Hofmann-Wellenhof R, Malvehy J, et al. Dermoscopy of Bowen's disease. Br J Dermatol 2004;150(6):1112–6.

[60] Pan Y, Chamberlain AJ, Bailey M, Chong AH, Haskett M, Kelly JW. Dermatoscopy aids in the diagnosis of the solitary red scaly patch or plaque-features distinguishing superficial basal cell carcinoma, intraepidermal carcinoma, and psoriasis. J Am Acad Dermatol 2008;59(2):268–74.

[61] Lallas A, Kyrgidis A, Tzellos TG, Apalla Z, Karakyriou E, Karatolias A, et al. Accuracy of dermoscopic criteria for the diagnosis of psoriasis, dermatitis, lichen planus and pityriasis rosea. Br J Dermatol 2012;166(6):1198–205.

[62] Cameron A, Rosendahl C, Tschandl P, Riedl E, Kittler H. Dermatoscopy of pigmented Bowen's disease. J Am Acad Dermatol 2010;62(4):597–604.

[63] Alam M, Ratner D. Cutaneous squamous cell carcinoma. N Engl J Med 2001;344(13):975–83.

[64] Rosendahl C, Cameron A, Argenziano G, Zalaudek I, Tschandl P, Kittler H. Dermoscopy of squamous cell carcinoma and keratoacanthoma. Arch Dermatol 2012;148(12):1386–92.

[65] Lallas A, Pyne J, Kyrgidis A, Andreani S, Argenziano G, Cavaller A, et al. The clinical and dermoscopic features of invasive cutaneous squamous cell carcinoma depend on the histopathological grade of differentiation. Br J Dermatol 2015;172(5):1308–15.

[66] Lallas A, Zalaudek I, Argenziano G, Longo C, Moscarella E, Di Lernia V, et al. Dermoscopy in general dermatology. Dermatol Clin 2013;31(4):679–94.

[67] Lallas A, Apalla Z, Argenziano G, Sotiriou E, Di Lernia V, Moscarella E, et al. Dermoscopic pattern of psoriatic lesions on specific body sites. Dermatology 2014;228(3):250–4.

[68] Vazquez-Lopez F, Kreusch J, Marghoob AA. Dermoscopic semiology: further insights into vascular features by screening a large spectrum of nontumoral skin lesions. Br J Dermatol 2004;150(2): 226–31.

[69] Lallas A, Apalla Z, Lefaki I, Tzellos T, Karatolias A, Sotiriou E, et al. Dermoscopy of early stage mycosis fungoides. J Eur Acad Dermatol Venereol 2013;27(5):617–21.

[70] Vázquez-López F, Manjón-Haces JA, Maldonado-Seral C, Raya-Aguado C, Pérez-Oliva N, Marghoob AA. Dermoscopic features of plaque psoriasis and lichen planus: new observations. Dermatology 2003;207(2):151–6.

[71] Lallas A, Argenziano G, Apalla Z, Gourhant JY, Zaballos P, Di Lernia V, et al. Dermoscopic patterns of common facial inflammatory skin diseases. J Eur Acad Dermatol Venereol 2014;28(5): 609–14.

[72] Lallas A, Apalla Z, Lefaki I, Sotiriou E, Lazaridou E, Ioannides D, et al. Dermoscopy of discoid lupus erythematosus. Br J Dermatol 2012;168(2):284–8.

[73] Segal R, Mimouni D, Feuerman H, Pagovitz O, David M. Dermoscopy as a diagnostic tool in demodicidosis. Int J Dermatol 2010; 49(9):1018–23.

[74] Shim W-H, Jwa SW, Song M, Kim HS, Ko HC, Kim MB, et al. Diagnostic usefulness of dermatoscopy in differentiating lichen sclerous et atrophicus from morphea. J Am Acad Dermatol 2012; 66(4):690–1.

[75] Vázquez-López F, Fueyo A, Sánchez-Martín J, Pérez-Oliva N. Dermoscopy for the screening of common urticaria and urticaria vasculitis. Arch Dermatol 2008;144(4):568.

[76] Vazquez-Lopez F, Maldonado-Seral C, Soler-Sánchez T, Perez-Oliva N, Marghoob AA. Surface microscopy for discriminating between common urticaria and urticarial vasculitis. Rheumatology (Oxford) 2003;42(9):1079–82.

[77] Zaballos P, Puig S, Malvehy J. Dermoscopy of pigmented purpuric dermatoses (lichen aureus): a useful tool for clinical diagnosis. Arch Dermatol 2004;140(10):1290–1.

[78] Akay BN, Kittler H, Sanli H, Harmankaya K, Anadolu R. Dermatoscopic findings of cutaneous mastocytosis. Dermatology 2009; 218(3):226–30.

[79] Vano-Galvan S, Alvarez-Twose I, De las Heras E, Morgado JM, Matito A, Sánchez-Muñoz L, et al. Dermoscopic features of skin lesions in patients with mastocytosis. Arch Dermatol 2011; 147(8):932–40.

[80] Zalaudek I, Giacomel J, Cabo H, Di Stefani A, Ferrara G, Hofmann-Wellenhof R, et al. Entodermoscopy: a new tool for diagnosing skin infections and infestations. Dermatology 2008; 216(1):14–23.

[81] Walter B, Heukelbach J, Fengler G, Worth C, Hengge U, Feldmeier H. Comparison of dermoscopy, skin scraping, and the adhesive tape test for the diagnosis of scabies in a resource-poor setting. Arch Dermatol 2011;147(4):468–73.

[82] Park JH, Kim CW, Kim SS. The diagnostic accuracy of dermoscopy for scabies. Ann Dermatol 2012;24(2):194.

[83] Piliouras P, Allison S, Rosendahl C, Buettner PG, Weedon D. Dermoscopy improves diagnosis of tinea nigra: a study of 50 cases. Australas J Dermatol 2011;52(3):191–4.

[84] Zaballos P, Ara M, Puig S, Malvehy J. Dermoscopy of molluscum contagiosum: a useful tool for clinical diagnosis in adulthood. J Eur Acad Dermatol Venereol 2006;20(4):482–3.

[85] Di Stefani A, Hofmann-Wellenhof R, Zalaudek I. Dermoscopy for diagnosis and treatment monitoring of pediculosis capitis. J Am Acad Dermatol 2006;54(5):909–11.

[86] Zalaudek I, Argenziano G. Dermoscopy of nits and pseudonits. N Engl J Med 2012;367(18):1741.

CHAPTER

4

Trichoscopy: The Dermatologist's Third Eye

U. Khopkar, N. Jain

Seth G S Medical College and KEM Hospital, Mumbai, India

OUTLINE

INTRODUCTION

Dermoscopy/dermatoscopy, or epiluminescence microscopy, is the examination of skin lesions with a dermatoscope (digital magnifier, typically ×10). In 2006, Lidia Rudnicka and Malgorzata Olszewska coined the term *trichoscopy* for the dermoscopy of the hair and scalp [1].

When introduced, dermatoscopy was extensively used for early and noninvasive diagnosis of melanoma. However, in the later years, multiple other significant uses of a dermatoscope in evaluating tumoral/nontumoral skin conditions and hair and scalp disorders has been reported.

Trichoscopy has recently evolved as a simple, noninvasive, and relatively inexpensive technique to evaluate hair and scalp. The other advantages include being able to inspect a larger area in less time and the fact that it can be easily mastered if one has a keen eye.

Most trichoscopes available come with an in-built software that makes record-keeping easy with good quality digital images. Comparison of the pretreatment and post-treatment images helps in guiding the course of therapy and evaluating the treatment results. It has a high patient satisfaction quotient and obviates the need for biopsy or choice of the best site for biopsy when one is indicated. Trichoscopy is also being used to

Imaging in Dermatology
http://dx.doi.org/10.1016/B978-0-12-802838-4.00004-2

calculate the follicular density in the donor area before follicular unit transplantation. Thus the trichoscope has become a must-have gadget in a dermatologist's arsenal.

TECHNICAL CONSIDERATIONS

A dermoscope can be of contact or noncontact type. Most of the manual hand-held dermoscopes are contact dermoscopes that require an interface solution, such as oil or alcohol. Pigment patterns are best visualized through a contact dermoscope.

Videodermoscopes are noncontact dermoscopes that usually have three modes: white light, ultraviolet light, and polarized light (PL). The interfollicular patterns, which relate to vascular structures and pigmentation, are visualized only with a polarizing light source or a polarizing filter. Vascular patterns are best seen through a videodermoscope because direct contact can result in blanching [2,3]. Videodermoscopes are in-motion dermoscopes and have higher patient satisfaction because the doctor and patient can simultaneously view the videographic images on the monitor and record the selected images for comparison during subsequent follow-up visits.

For scalp examination, a manual dermoscope (×10 magnification) or a videodermoscope with lenses ranging from ×20 to ×1000 magnification can be used [2,3] (Fig. 4.1A–C).

DERMOSCOPIC FEATURES

Systematic trichoscopic evaluation mandates evaluating the hair and scalp for

1. Follicular signs
   a. Yellow dots
   b. White dots
   c. Black dots
2. Hair shaft characteristics
3. Interfollicular patterns
   a. Vascular patterns
   b. Pigment patterns

What does normal (physiologic) hair and scalp look like? Knowing the normal is a must to differentiate it from the abnormal. Trichoscopic findings may vary with variations in skin color and racial types. Dermoscopy of normal healthy scalp shows follicular units containing two to four terminal hairs and one or two vellus hairs (Fig. 4.2).

In darker races, a prominent brown homogenous honeycomb pigment network is seen over the scalp, which is accentuated over sun-exposed areas [4,5]. Vascular patterns are easily visualized in the fair skin population using a videodermoscope with a polarizing filter. However, in darkly pigmented skin, the heavy pigment prevents visualization of the underlying dermal vasculature.

FIGURE 4.2 Dermoscopy of normal scalp with follicular units bearing two to four terminal hairs and a uniform pigment network in the background.

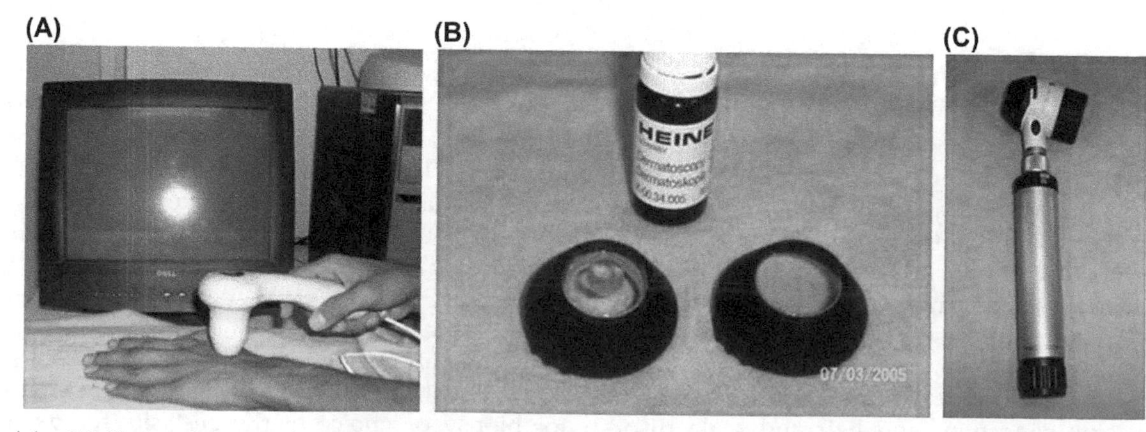

FIGURE 4.1 (A) Various lenses and interface solution for contact dermoscopy. (B) Manual Heine delta 20 hand-held dermoscope. (C) A standard videodermoscope.

Follicular Signs

Follicular signs observed on trichoscopy in various conditions can be correlated with the pathological changes occurring in the surface and subsurface structures. Perifollicular and interfollicular inflammation may result in alteration in vascular patterns, pigmentary changes, and scaling. Hair shaft affection results in changes in hair shaft diameter and breakage resulting in black dots. Common follicular signs described are discussed here.

Yellow Dots

Yellow dots is the term used to describe the follicular infundibulum, which is clogged with degenerating keratinocytes and excess sebum [2–6]. These are usually round and are best seen under PL. In lighter skin shades, they appear yellow, whereas in brown/darker skin types they appear pale against the pigmented background. Yellow dots are seen in alopecia areata (AA), androgenetic alopecia (AGA), and alopecia incognito Fig. 4.3.

In AA, yellow dots are the most common and most sensitive finding. They represent keratinous debris, which is not cleared from the infundibulum because of the presence of dystrophic/broken hairs. Yellow dots are usually associated with other findings of AA, such as black dots and cadaverized hair [2,6]. Yellow dots may have a hair strand within them or may even be empty.

In AGA, pearly white to yellowish dots are seen prominently over areas with sparse terminal hairs (ie, the frontoparietal and temporal areas). Distended follicles with hypertrophied sebaceous glands account for this finding. These are seen in advanced stages of AGA (Fig. 4.4).

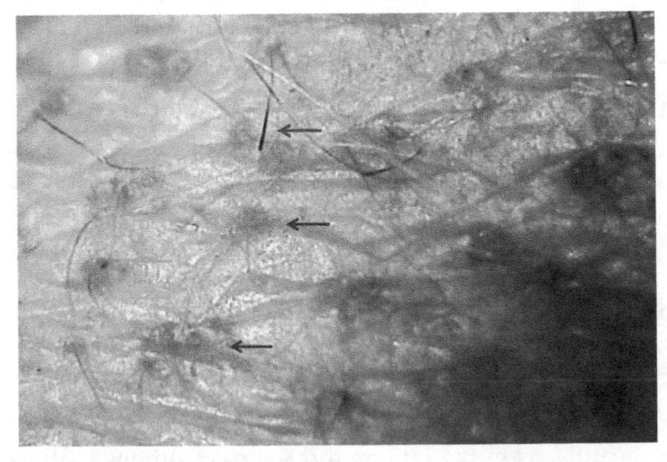

FIGURE 4.3 Follicular yellow dots in a case of alopecia areata (*blue arrows*). A single exclamation mark hair (*red arrow*) can be seen as well along with multiple vellus hairs. Black dots can also be observed (*green arrow*).

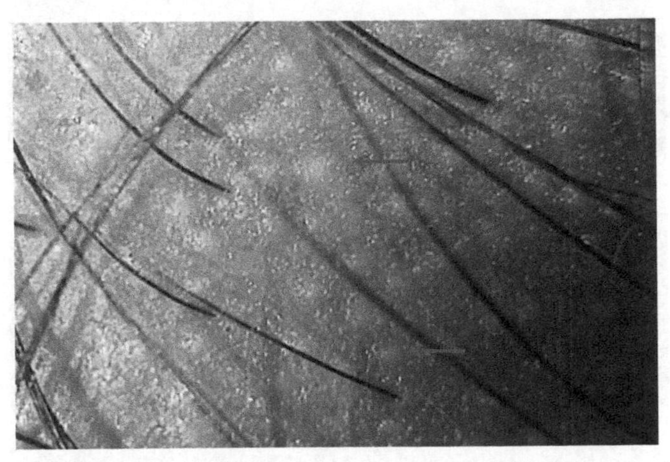

FIGURE 4.4 Advanced stage of androgenetic alopecia. All of the follicles show a single hair within. Multiple yellow to pearly white dots can also be observed (*red arrows*).

In alopecia incognita, they may be seen even in the areas with terminal hairs [2].

"Three-dimensional" soap bubbles like yellow dots have been described in dissecting cellultis of scalp [7,8].

White Dots

Follicles destroyed because of scarring are visualized as pale white dots in conditions that spare the interfollicular areas [2,3,7,9]. White dots represent fibrous tracts of scarred follicles seen in cicatricial alopecias, such as lichen planopilaris (LPP) and folliculitis decalvans [9]. White dots were reported by Kossard and Zagarella in cicatricial alopecia [10]. They are best seen in dark skin types and over tanned, sun-exposed areas where the honeycomb pattern pigment background provides a good contrast (Fig. 4.5).

Conditions causing complete scarring with affliction of the interfollicular areas cause a break in the scalp pigment network; hence the white dots are not visualized. Eccrine duct openings may look similar but can be differentiated because they are well-defined, rounded structures that are numerous and are seen over the diseased and the normal scalp. Sometimes active sweat secretion may be seen emanating from them [11,12].

Black Dots

Black dots represent broken/fractured dystrophic hairs and the remnants of exclamation mark hairs that break at the proximal end [2,3,7,13]. Black dots may be seen in AA, tinea capitis (noninflammatory black dot variant more commonly), dissecting cellulitis, and trichotillomania.

Black dots (cadaverized hairs) have been described most commonly in AA and are a finding reportedly associated with disease activity. They have been described to be the most specific markers by Inui et al. [14] (Fig. 4.6).

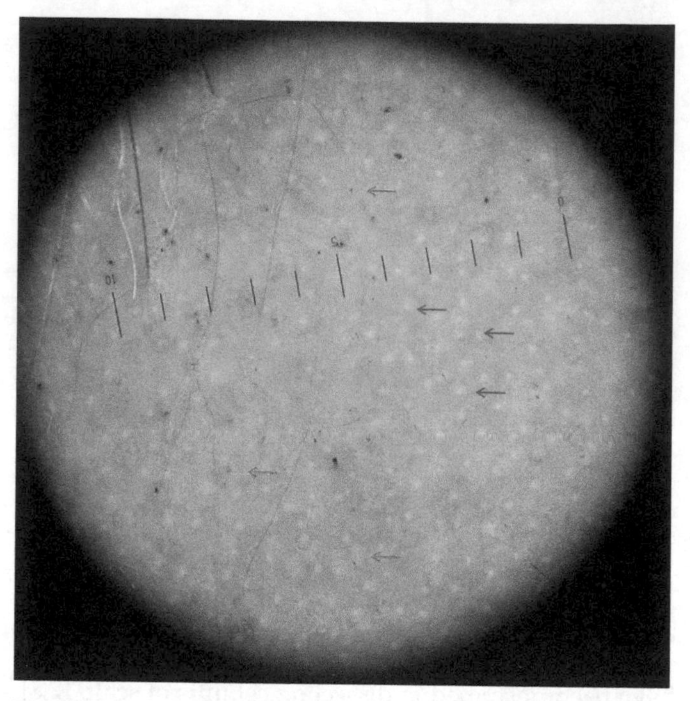

FIGURE 4.5 A case of cicatricial alopecia with multiple white dots that represent fibrosed follicles (*blue arrows*) and some white dots have broken hair shafts within. A prominent feature is the eccrine duct openings (*red arrows*). In addition, a break in the honeycomb pigment network can be seen at the sites of scarring.

In AA, black dots are usually present within yellow dots. These are dystrophic hairs that fracture before emergence from the scalp [3,7,14,15].

Hair Shaft Characteristics

Trichoscopy of the hair shaft is useful to diagnose genetic and acquired hair shaft defects. Variation in thickness and pigment characteristics can be readily picked up. In addition, hair shaft changes in other conditions, such as AGA, telogen effluvium, and alopecia incognito help in confirming the diagnosis and monitoring therapy response.

Normal hair shafts are usually terminal with up to 10% being vellus hair shafts. The terminal hairs have uniform distribution of pigment and are thicker. They may be medullated or nonmedullated. Vellus hairs are thin, short, hypopigmented, and nonmedullated.

The hair shaft should be evaluated from the proximal to the distal end to look for alteration in thickness, pigmentation, length, presence of fractures, nodes, twists, and casts. Specific hair signs associated with different conditions have been described.

Evaluation of hair shaft thickness can be done for evaluating treatment response in conditions, such as AGA, which is characterized by progressive miniaturization of the hair follicle and an increased proportion of vellus hairs. Videodermoscopes with an in-built software that helps in detailed evaluation can be used for this purpose.

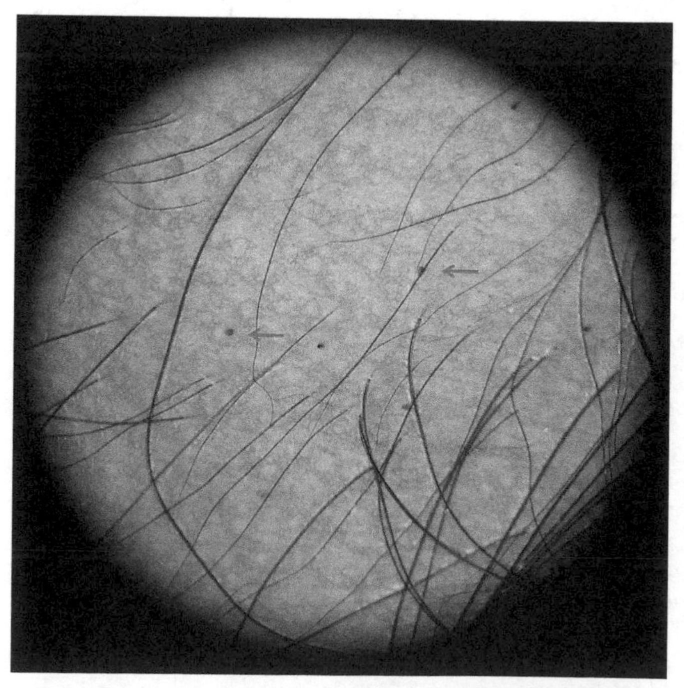

FIGURE 4.6 Multiple black dots (*blue arrows*) in a case of alopecia areata. A homogenous honeycomb pigment network can be seen in the background.

Interfollicular Patterns

Pathological changes in the interfollicular areas can be observed as changes in the pigment pattern and changes in the normal vascular pattern. Primary and secondary cicatricial alopecias and nonalopecic scalp conditions, such as psoriasis and seborrheic dermatitis can be diagnosed.

Vascular Patterns

Various vascular patterns associated with unaffected and affected scalp have been best described by Tosti and Duque-Estrada [2,3]. In pigmented skin types, the vascular patterns are generally obfuscated by the overlying prominent pigment network. Vascular patterns are best observed through a videodermoscope (noncontact) using the polarizing filter. The characteristic patterns described are as follows:

- *Interfollicular simple red loops*: These are seen in normal healthy scalp [2,3]. Hair and scalp conditions that do not affect the epidermis also show this pattern of capillary arrangement. This capillary pattern appears as multiple regularly spaced hairpin-like structures.

Absent loops indicate epidermal atrophy; thus they are not seen in advanced cases of discoid lupus erythematosus (DLE).

- *Interfollicular twisted loops*: These appear as twisted coils and are best seen with the probe placed tangentially to the scalp surface [2]. Conditions characterized by acanthosis, such as psoriasis and folliculitis decalvans, show this pattern [2–5]. Some cases of inflammatory seborrheic dermatitis also reveal this capillary pattern. The number, visibility, and tightness of the coiling correlate with disease severity. The presence of twisted capillaries on dermoscopy correlates with the histological finding of tortuous interpapillary loops in psoriasis.
- *Arborizing red lines*: These are seen as lines that underlie the loops in normal and affected scalp in all conditions [2,3]. They are of wider caliber. These are best seen at higher magnifications and are believed to represent the subpapillary plexus.

Pigment Pattern

At higher magnification, a diffuse homogenous, honeycomb pigment network is classically seen in normal scalp and is more pronounced in individuals with darker skin shades (Fig. 4.6) [2,3]. Bald areas and areas with sparse hair, as seen in men with advanced AGA, have a darker pigment network that corresponds to tanning due to excessive sun exposure. Even in normal scalp the extent of the pigment varies over the scalp depending upon the sun exposure and the density of hair.

This pattern is characterized by grid (irregular lines) and holes. The lines are hyperchromic and represent melanotic rete ridges, whereas holes are the hypochromic suprapapillary epidermis [2].

Variation in the continuity of the pigment pattern is generally seen in cicatricial alopecias, which affect the interfollicular epidermis. In conditions, such as LPP, where the interfollicular epidermis is spared, pigmentary changes are seen in the perifollicular areas.

TRICHOSCOPY FINDINGS IN COMMON HAIR AND SCALP CONDITIONS

Androgenetic Alopecia

Look for the following:

1. Hair shaft diameter diversity
2. Increase in proportion of vellus hairs with reduction of terminal-to-vellus hair ratio
3. Predominance of follicles with single hair
4. Peripilar brown halo
5. Yellow dots
6. Accentuated pigment pattern

AGA is the most common cause of hair loss across the world, affecting men and women alike. The diagnosis of AGA is mostly clinical. Confusion arises in cases occurring in very young individuals, fast progression, slow/nonresponse to therapy, and association with systemic conditions or where multiple conditions may co-exist. In such circumstances, trichoscopy helps to differentiate between AGA and conditions, such as telogen effluvium, AA, frontal fibrosing alopecia (FFA), and alopecia incognito.

The pathogenesis of AGA involves progressive miniaturization of hair follicles, which results in a progressive reduction in the hair shaft diameter. The corresponding trichoscopic finding is an increase in the proportion of vellus hairs. The earliest diagnostic feature of AGA has been described as variation in hair shaft diameter involving more than 20% of hair shafts [2,4,16].

Progressive miniaturization of the follicle also leads to a reduction in the number of terminal hairs per follicle. On trichoscopy, this is seen as a predominance of follicles bearing single hairs (normal follicles have two to four terminal hairs) over affected areas. A similar finding can also be seen in telogen effluvium and anagen effluvium. In patients with AGA, comparative analysis with the unaffected areas shows the disparity. Several commercially available software help to calculate the terminal-to-vellus hair ratio, monitor hair shaft thickness, grade the severity, and monitor treatment response at a later date.

Another significant finding in early cases of AGA is the peripilar sign. Mild perifollicular inflammation seen in early stages gives rise to a subtle brown halo that is usually missed on clinical examination (Fig. 4.7).

Yellow dots, described earlier, are seen as pearly white to yellowish rounded structures (papules) in

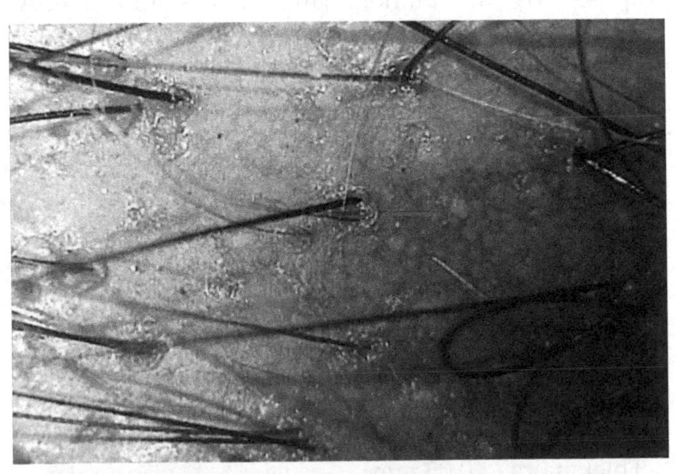

FIGURE 4.7 Early stage of androgenetic alopecia with increased number of vellus hair shafts and peripilar brown halo. Some follicles show single hair.

advanced cases, more commonly over the temporoparietal areas. These represent hypertrophied sebaceous glands [13,15]. The sebaceous gland activity is intact even in a miniaturized terminal follicle and in fact the gland may be hypertrophied because of increased end-organ sensitivity to circulating androgens (Fig. 4.4). Advanced cases may show a prominent honeycomb pigment pattern over the bald areas and the presence of yellow dots.

Most findings in pattern baldness are similar in both sexes. In female AGA (FAGA), focal areas of baldness (atrichia) are more commonly seen [13].

Kibar et al. evaluated the trichoscopic findings and their relations with disease severity in AGA in males and females and found no significant relation between trichoscopic findings and severity in male AGA and FAGA. In addition, this study described multiple other findings, such as brown dots, white dots, multihair follicular units, and hidden hair [17].

Trichoscopic criteria for diagnosing FAGA have been devised by Rakowska et al. based on a trichoscopy study of 131 females [16]. The study was a comparative analysis of the frontal and occipital areas visualized in patients of chronic telogen effluvium (39), FAGA (59), and healthy controls (33). In every patient the frontal, occipital, and right and left temporal areas were visualized, each for five images: one at 20-fold magnification and four images at 70-fold magnification.

The criteria evaluated were as follows: (1) number of vellus hairs; (2) hair thickness [percentage of thin (<0.03 mm), medium (0.03–0.05 mm), and thick (>0.05 mm) hairs]; (3) percentage of pilosebaceous units with one, two, and three hairs at 20-fold magnification; (4) number of yellow dots; and (5) percentage of perifollicular hyperpigmentation at 20-fold magnification. A trichoscopy record scheme in tabular format has been proposed by the authors [16]. To diagnose FAGA, comparative analysis of the frontal and occipital area is important. Major and minor criteria have been devised. The presence of two major or one major and two minor criteria diagnoses FAGA with a 98% specificity [16].

The criteria are as follows:

Major criteria

1. Total number of yellow dots in four fields of vision (FAGA criteria more than four yellow dots in frontal area)
2. Mean hair thickness in millimeters [1, thin hairs (<0.03 mm); 2, medium hairs (0.03–0.05 mm); FAGA criteria—lower hair thickness in frontal area and >10% thin hairs in frontal area]
3. Thick hairs (>0.05 mm)

Minor criteria

1. Percentage of units in one field of vision at 20-fold magnification (single hair, two-hair units, three-hair units; FAGA criteria—ratio of single-hair units, frontal area:occipital area >2:1)
2. Total number of vellus hairs in four fields of vision at 70-fold magnification (FAGA criteria—frontal area: occipital area >1.5:1)
3. Percentage hair follicles with perifollicular discoloration at 20-fold magnification (FAGA criteria—frontal area: occipital area >3:1)

Alopecia Areata

Clinical features are as follows:

1. Black dots
2. Yellow dots
3. Tapering hairs/exclamation hairs, cadaverized hairs
4. Short regrowing vellus hairs

AA is one of the most common forms of autoimmune, patchy, noncicatricial alopecias in all age groups affecting scalp and nonscalp sites. Although clinical diagnosis is simple, trichoscopy helps in differentiating it from other patchy alopecias, especially in pediatric patients.

The pathogenesis in AA is an abrupt arrest of hair cycle and formation of dystrophic hairs [2,18]. Black dots and yellow dots have been described above in detail. Trichoscopy findings may vary depending upon severity and disease duration.

Tapering hairs (exclamation mark hair) are short dystrophic hair strands with a narrow proximal end. It is a marker of active disease and can be seen at the periphery of a patch [7,13,15,19] (Fig. 4.8A and B). Exclamation mark hairs may also be seen in trichotillomania [7].

Yellow dots are seen in all of the stages of the disease and correlate well with disease severity [7,15] (Figs. 4.3 and 4.8A). Population variation in occurrence of yellow dots (may not be easily seen in dark-skinned individuals) can be attributed to variations in skin color and cleansing habits [2,3]. Yellow dots usually contain fractured dystrophic hair (cadaverized hair, black dots), short vellus hairs, or telogen hairs.

Various studies conducted in different parts of the world differ in their conclusions over the sensitivity and specificity of individual markers [4–6,14,19]. A recent study from Egypt concludes that in pediatric AA cases, black dots are the most common finding and can be a sensitive marker if associated with other findings of AA [19] (Fig. 4.8C).

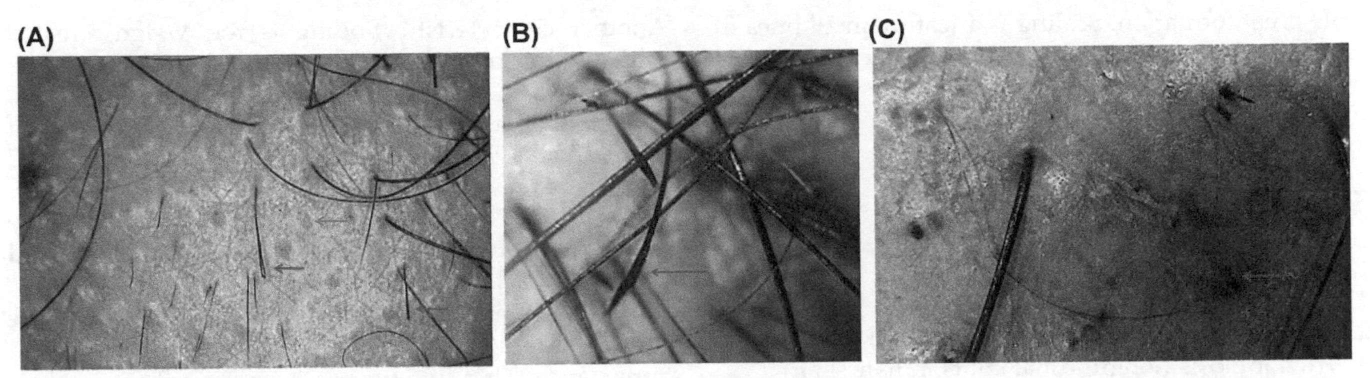

FIGURE 4.8 (A) A case of alopecia areata with multiple exclamation mark hairs (*red arrows*) yellow dots (*blue arrows*) and (B) regrowing short vellus hairs (*red arrow*). (C) Multiple black dots can be seen in a case of alopecia areata (*red arrows*).

Acute progressive cases are characterized by exclamation mark hairs and black dots. In chronic cases, the dystrophic hairs may be shed; thus the follicles may appear empty. Regrowing vellus hairs may also be seen. Short vellus hairs are a sensitive marker of hair regrowth. These regrowing hairs may be coiled as a pigtail as described by Rudnicka et al, [7].

The coudability sign, which represents terminal hair kinking toward the proximal end when pushed perpendicular to the scalp surface, can be seen in active disease [20].

Trichoscopic examination of epilated hair can be done by placing the hair against a light background and visualizing the roots at higher magnification (dermoscopic trichogram). Active disease is characterized by dystrophic hairs with fractured roots and telogen hairs [2,3].

Tinea Capitis

Clinical features are as follows:

1. Comma hair, corkscrew hairs
2. Black dots
3. Scales
4. Short broken hairs
5. Blotchy pigmentation
6. Erythema
7. Pustules, follicular scale crust

Tinea capitis is a fungal infection of the scalp, and it is the single greatest cause of alopecia in children. Trichoscopy helps in differentiation from other patchy alopecias, such as AA and trichotillomania. The black dot variant may especially cause confusion, and culture is considered to be the gold standard for diagnosis. Trichoscopy is convenient because it gives instant confirmation of the diagnosis. Standard sterile measures should be observed, or a separating transparent film can alternatively be used.

Features vary between inflammatory and noninflammatory variants. Comma-shaped hair stubs, which are slightly curved, and fractured hair shafts are a specific feature [7,13,15,21,22]. Hughes et al. has described that comma-shaped hairs and corkscrew hairs are a feature of zoophilic infection [19,23].

Numerous black dots, which are the remnants of broken hairs/dystrophic hairs, may be seen in the black dot variant (noninflammatory tinea). These black dots represent breakage of hair shafts infested by fungal hyphae. Unlike in AA, black dots in tinea are numerous and are not associated with yellow dots and tapering hairs (Fig. 4.9).

Other features, such as broken hairs, damaged hairs, and zigzag hairs have been described [7,19]. Short broken hairs are most common but a nonspecific feature. The zigzag hairs, corkscrew hair seems to be a variation of the comma hair, manifesting in Black patients [19,23].

Inflammatory tinea capitis is characterized by blotchy pigmentation, scaling, erythema, pustules, and follicular

FIGURE 4.9 Tinea capitis with multiple black dots, scaling, and blotchy pigment pattern.

scale crust formation. Scaling is a feature in all tinea infections. Videodermoscopes with additional ultraviolet light mode help in demonstrating fluorescence caused by fungi.

Trichotillomania

Clinical features are as follows:

1. Broken hairs with variable length
2. Coiled hairs, hook hairs (question mark hair)
3. Trichoptilosis (longitudinal splits in hair shaft)
4. Flame hairs, V-sign
5. Tulip hairs, hair powder
6. Regrowing pigtail hairs

Trichotillomania is a psychocutaneous disorder, characterized by an impulse to pull out hair. It is commonly seen in children and young adults, with a significant female preponderance of 70–93% [24].

Clinically, trichotillomania is a type of patchy hair loss with patches mostly over the easy-access areas of the scalp-like vertex [25,26]. An extensive tonsure pattern may be seen at times [27]. Because of the similar age group of affliction and patchy nature, conditions, such as AA and tinea capitis need to be differentiated. Findings, such as black dots, yellow dots, and exclamation mark hairs may cause confusion because of their nonspecificity.

Classically, trichoscopy reveals broken hair shafts of variable length with longitudinal splitting/fraying (trichoptilosis) [24,26,27] (Fig. 4.10A). Some fractured hairs may be coiled because of the excessive pulling force applied, and its distal part may contract and coil [3]. Partial coiling may give the hook hair [24]/question mark hair appearance [28] (Fig. 4.10B).

The pulling tractional force and subsequent fracturing may also give the "flame hairs" sign. Flame hairs are short proximal hair stubs that look twisted/wavy and thinned out, left behind following anagen hair breakage [24,26]. They are seen in active disease. Another characteristic finding is the V-sign, created when two or more hairs that originate from one follicular unit break at same level. Rakowska et al. have reported that the V-sign was observed in 57% of trichotillomania cases [24,26]. Diagonally fractured hair shafts may have tulip-flower–shaped distal hyperpigmentation. This finding is called tulip hairs. These hairs are short with tulip-flower–shaped, darker distal ends [24,26,27].

In severe mechanical trauma, hair shafts may be completely damaged, giving a shattered/sprinkled appearance. This finding has been described as "hair powder" by Rakowska et al. [24,26,27].

Other well-known trichoscopic findings of trichotillomania includes short vellus hairs, yellow dots, black dots, and exclamation mark hairs [3,19,24,26,27]. Short vellus hairs may sometimes be seen but differ from AA because they are never white. Furthermore, yellow dots are generally less numerous than in AA [19]. Other less common findings include perifollicular erythema, pigmentation, and hemorrhages [21]. Coexistence of other conditions should be looked for.

Telogen Effluvium

Clinical features are as follows:

1. Predominance of follicles with single hair
2. Upright regrowing hairs
3. Decreased hair density and empty follicles

Trichoscopy is of limited use in diagnosing telogen effluvium because no specific features have been described. On trichoscopy, telogen effluvium is a diagnosis of exclusion [7,15,29]. However, a predominance of hair follicles with a single hair and upright regrowing hairs have been described as a common finding [7,29]. In one case report yellow dots and short vellus hairs have been described. It needs to be differentiated from AGA and AA. Telogen effluvium can be easily differentiated from AGA because of the absence of hair shaft diameter

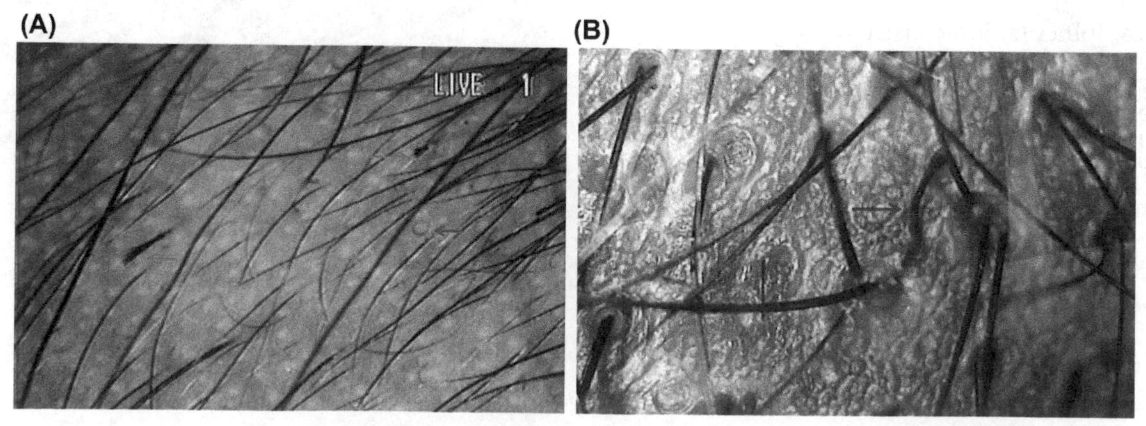

FIGURE 4.10 Trichotillomania: Longitudinal splitting/fraying of hair shafts (*blue arrows*) and coiled hairs (*red arrow*) can be seen. (A) Hair shafts of variable length are a feature. (B) Hairs with partial coiling, called question mark/hook hairs, can also be visualized Fig. 4b.

variation and peripilar halo. It affects the entire scalp, unlike AGA.

Lichen Planopilaris

Clinical features are as follows:

1. Sparing of interfollicular epidermis
2. Peripilar casts
3. Target pattern blue-gray dots
4. White dots

LPP is the most common cause of cicatricial alopecia of the scalp and is occasionally also seen on other body areas. On histopathology, active LPP is characterized by perifollicular interface dermatitis and pigmentary incontinence. These features are recognized on surface dermoscopy as perifollicular scales, which form tubular casts and may extend up to a few millimeters above the skin surface [3,7,13,15]. Elongated, concentrically arranged blood vessels can also be observed [7,8,30] (Fig. 4.11A).

Pigment incontinence is characterized by presence of blue-gray dots in a target pattern around the follicles [3,9,13,15] (Fig. 4.11B). The hair pull test reveals anagen roots with thickened hair sheaths [2,15,30,31].

The fibrotic stage is characterized by multiple white dots, which represent the scarred follicles, replaced by fibrous tracts [3,8,9,15].

In brown skin types, the spared interfollicular epidermis shows an intact, homogenous, honeycomb pigment pattern and some spared, terminal-hair–bearing follicles.

Discoid Lupus Erythematosus

Clinical features are as follows:

1. Loss of follicular ostia
2. Arborizing/branching capillaries
3. Hyperkeratotic follicular plugs
4. Blue-gray dots in speckled pattern
5. White dots

DLE is an uncommon cause of cicatricial alopecia. It presents as single or multiple, well-defined, erythematous scaly plaques over the prominent sun-exposed areas of the scalp. It can easily be confused with other patchy alopecias, especially LPP.

Dermoscopy of a DLE plaque shows atrophic skin with complete loss of follicular ostia. Arborizing telangiectasia and scaling are prominent features [8,9,15,30] (Fig. 4.12).

Histological features of DLE, such as follicular keratotic plugs, interface dermatitis, and pigment incontinence, can be seen as hyperkeratotic plugging (more so over the margins of the plaque) and a blue-gray pigment pattern on surface dermoscopy.

Inactive DLE plaques are characterized by white dots, reminiscent of follicles that are replaced by fibrous tracts. Whitish to milky red areas caused by fibrosis of the interfollicular epidermis and loss of follicular ostia are also seen [8,9,30].

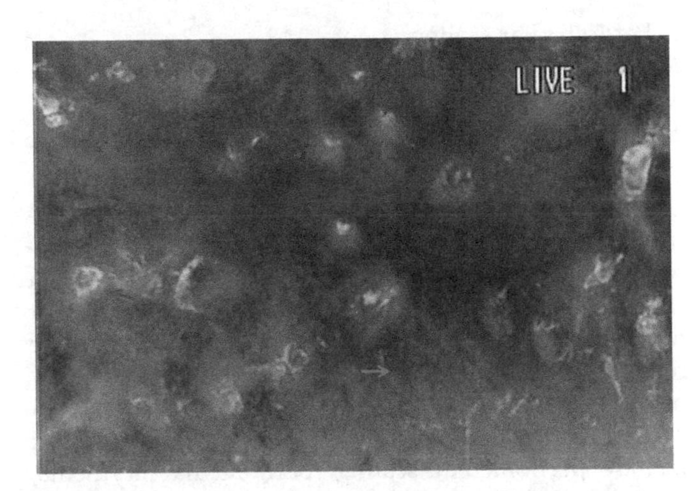

FIGURE 4.12 Discoid lupus erythematosus: Characteristic keratotic plugs (*red arrow*) and arborizing telangiectasia (*blue arrow*) can be seen. In addition, follicular paucity, atrophy, and scaling are visible findings.

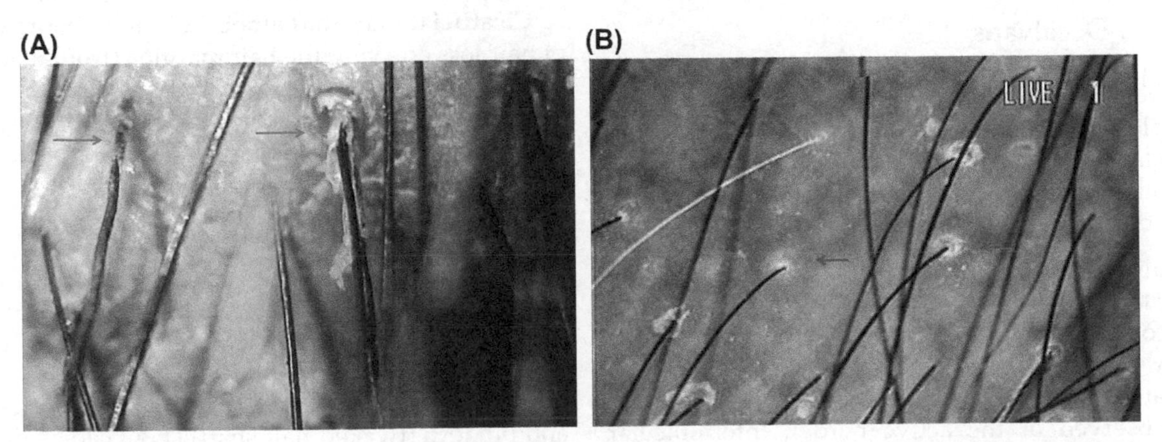

FIGURE 4.11 (A) A case of lichen planopilaris. Tubular hair casts of thick perifollicular scales caused by perifollicular inflammation can be seen. (B) Perifollicular inflammation with pigment incontinence (*red arrows*) gives the target blue-gray dot appearance.

Large yellow dots have been described [7,8]. Radial, thin, arborizing vessels emerging from the dot are considered characteristic for DLE. This feature is sometimes referred to as "red spider in yellow dot" [7,8].

Differentiation between LPP and DLE may be required in certain cases because features may overlap. The most important differentiating feature is the sparing of the interfollicular epidermis in LPP. Thus the honeycomb pigment pattern is a feature of LPP but is not seen in DLE. In addition, the blue-gray pigment seen in LPP is distributed in a "target pattern" around the follicles because of follicular pigment incontinence, whereas in DLE it is distributed in a "speckled" fashion because the pigment incontinence also affects the interfollicular areas. In addition, in LPP some follicles may be spared whereas in DLE all of the follicles in the plaque are affected. White dots are a feature seen in both conditions [9,15,30].

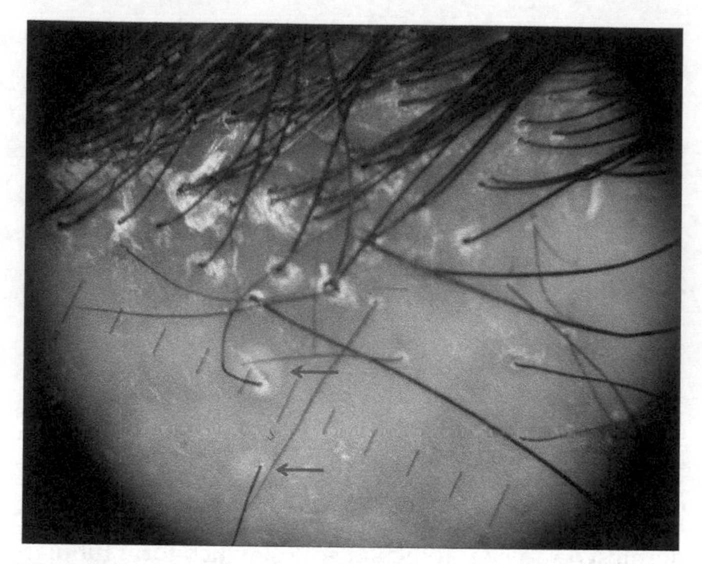

FIGURE 4.13 Folliculitis decalvans: Characteristic follicular scaling (*blue arrow*) and follicular pustules (*red arrow*) at the periphery of the patch are visible. Inactive scarred area can be seen as atrophic, shiny patch with complete follicular paucity. A break in the honeycomb pigment pattern is evident.

Frontal Fibrosing Alopecia

Clinical features are as follows:

1. Perifollicular scaling
2. Perifollicular erythema
3. Loss of follicular ostia
4. Branching capillaries

FFA is a rare cause of cicatricial alopecia seen in postmenopausal women. It is usually patchy and affects the frontotemporal areas. Patients present with frontotemporal hair recession and eyebrow loss [32,33]. It can sometimes be misdiagnosed as AA or AGA because of the distribution pattern because it mostly causes hairline recession and is patchy. FFA is considered to be a variety of LPP [32].

Trichoscopy findings in FFA that have been described include perifollicular scaling, perifollicular erythema, and loss of follicular ostia [7–9,32]. Branching/arborizing vessels [9] as well as predominance of single-hair–bearing follicles has been described [7].

Folliculitis Decalvans

Clinical features are as follows:

1. Multiple hairs emerging from single follicular ostium
2. Follicular scaling and follicular pustules
3. Interfollicular twisted capillary loops
4. White dots

Folliculitis decalvans is characterized by the presence of multiple upright hairs (>5) emerging from a single ostium, corresponding to the classic clinical picture [7]. Follicular scaling is seen, and it represents follicular inflammation (Fig. 4.13). In addition, follicular pustules can be observed at the active border. Interfollicular twisted/coiled capillary loops may be present around affected follicles as well as white dots [7,34]. These vascular patterns may not be well appreciated in the darker pigmented skin types even with polarizing light source. The inactive scarred areas are seen as pinkish-white patches with absent follicular openings.

Other Cicatricial Alopecias

Dissecting cellulitis is characterized by yellow dots, appearing as three-dimensional structures imposed over dystrophic hairs [7,15]. Advanced cases are difficult to differentiate from other scarring alopecias because of their fibrosed patchy appearance with absent follicular ostia.

Pseudopelade of Brocq is a diagnosis of exclusion because nonspecific features are usually seen. Scarred hypopigmented areas with follicular paucity and few dystrophic hairs are the most common features [7,15].

Cicatricial marginal alopecia is an uncommon cause of hair loss affecting the hair margins (frontal, temporal, and occipital). Dermoscopic findings include low hair density, loss of follicular ostia, and thinning of the remaining hair shafts [35].

Hair Shaft Disorders

Hair shaft characteristics can be seen easily at higher magnifications through a videodermoscope. Hair shaft disorders, such as monilethrix (hair shaft beading) [36–38], trichorrhexis nodosa (brush-like fractures) [38], trichorrhexis invaginata (hair shaft nodes) [1,36], and pili torti (twisted hair shaft) [1,36] can be diagnosed conveniently without the help of hair mounts and microscopes.

Monilethrix

This congenital hair shaft disorder is characterized by regularly spaced elliptical nodes and internodes (intermittent constrictions of hair shafts). The term *regularly bended ribbon sign* has been coined for this finding [37,38] (Fig. 4.14A and B).

Trichorrhexis Nodosa

Trichorrhexis nodosa is characterized by nodes located along the hair shafts. These nodes represent multiple longitudinal splits of the hair shaft, which on higher magnification are seen as brush-like ends (Fig. 4.15A and B). These nodes are fragile, causing hair shaft breakage. Trichoscopy shows the nodes clearly and detailed brush-like fibers at higher magnification [7,39].

Trichorrhexis Invaginata

Trichorrhexis invaginata, or bamboo hair, on trichoscopy is seen as hair shaft telescoping into itself. At lower magnification, this is seen as multiple nodes along the hair shaft. The nodes are weak areas and tend to easily fracture. The fractured proximal end appears cupped. This finding is called "golf-tee hairs" [7,39–41].

Pili Torti

Pili torti can be genetic or acquired in origin. Dermoscopy reveals flattened and irregularly twisted hair shafts.

Pili Annulati

Trichoscopy in pili annulati is characterized by alternating dark and light bands. The lighter bands are shorter than the darker area [7,36].

Other Scalp Conditions

Differentiating Scalp Psoriasis From Seborrheic Dermatitis

Localized scalp psoriasis sparing other body areas may be easily confused with seborrheic dermatitis. Differentiation on dermoscopy can be done by looking for

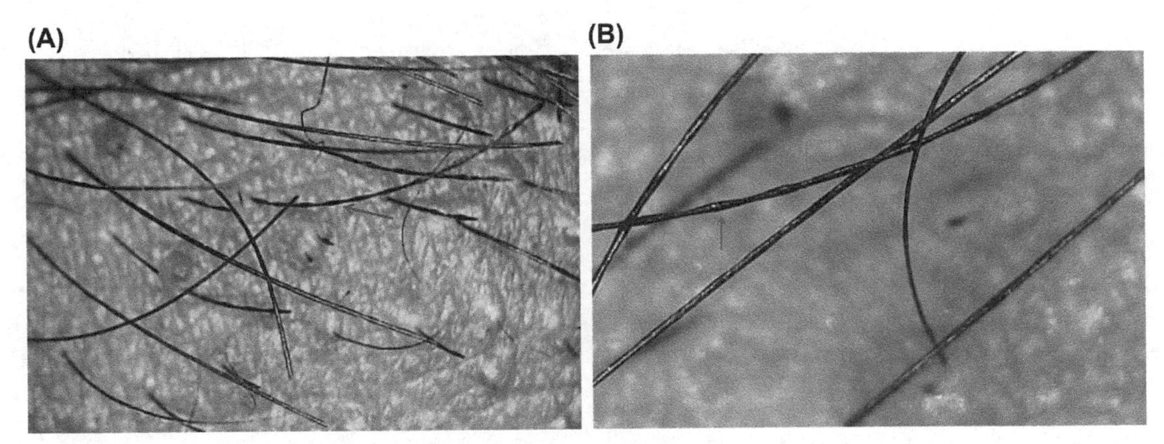

FIGURE 4.14 Monilethrix: (A) Hair shafts broken at the weaker internodes can be seen (*red arrow*). (B) Beaded hair shafts with regularly spaced nodes (*red arrow*) and internodes (regularly bended ribbon sign; *blue arrow*).

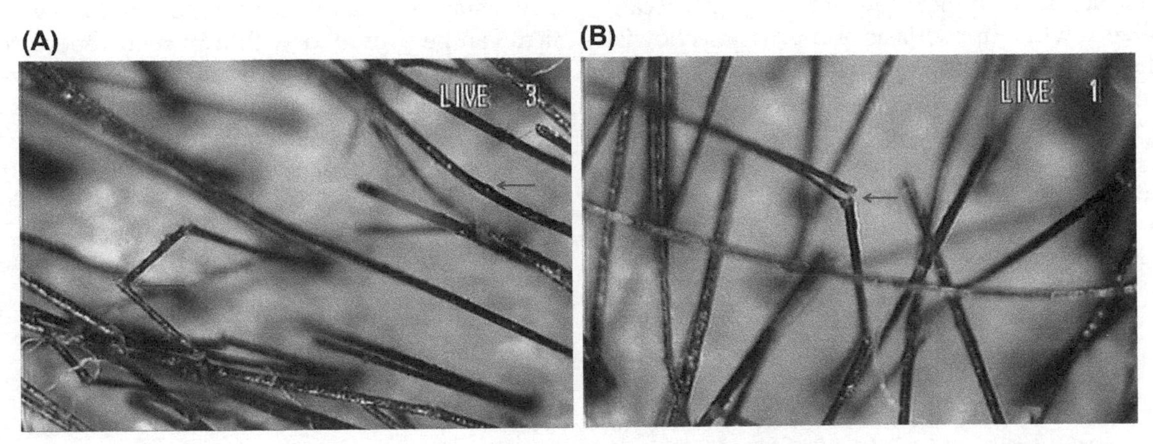

FIGURE 4.15 Trichorrhexis nodosa: Multiple nodes along individual hair shafts that represent brush-like longitudinal splitting of the shafts.

TABLE 4.1 Algorithmic Approach for Diagnosis of Cicatricial and Noncicatricial Alopecia

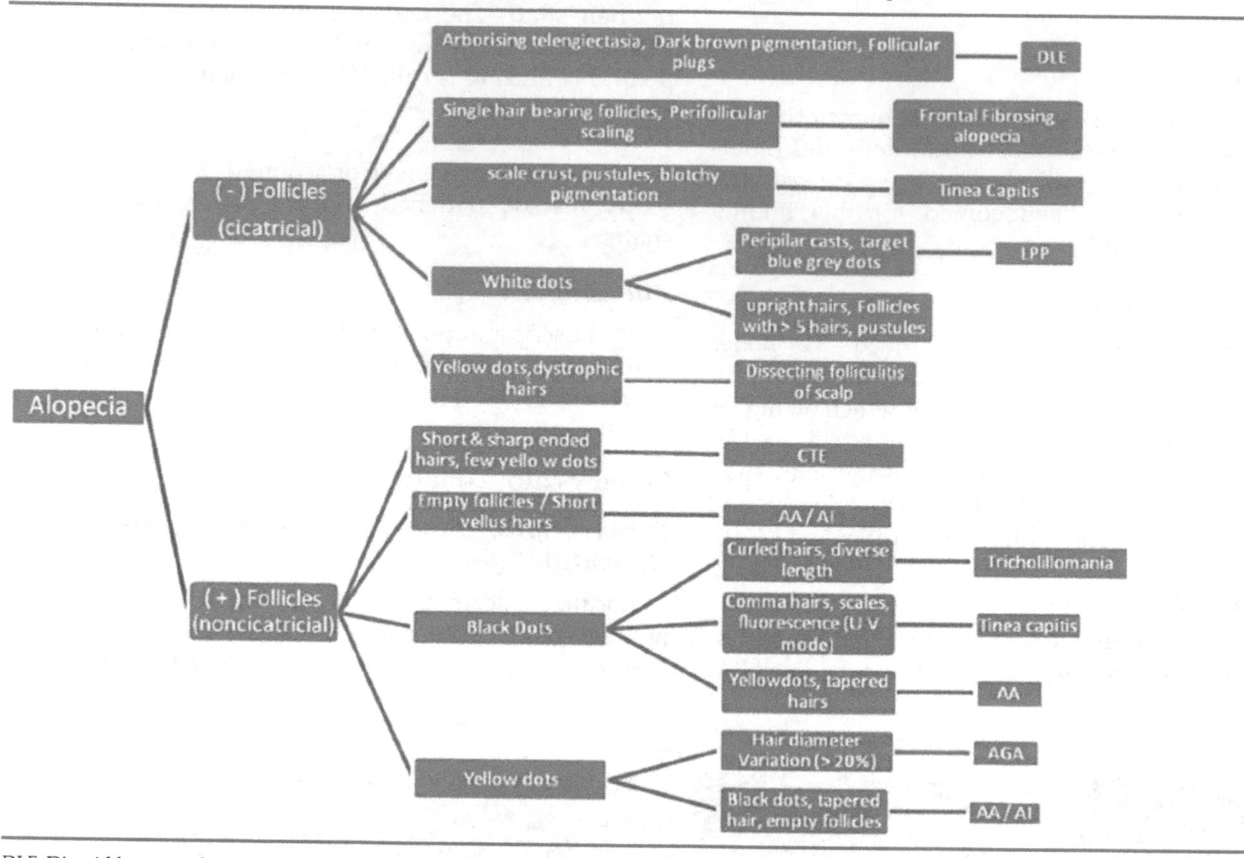

DLE, Discoid lupus erythematosus; *LPP*, lichen planopilaris; *CTE*, chronic telogen effluvium; *AA*, alopecia areata; *AI*, alopecia incognito; *AGA*, androgenetic alopecia; *UV*, ultraviolet.
From Kharkar V. Overview of trichoscopy. In: Khopkar U, editor. Dermoscopy and trichoscopy in diseases of the brown skin. 1st ed. New Delhi: Jaypee; 2012. p. 169–81.

the vascular patterns and types of scales. Psoriasis is characterized by an extensive array of red dots, globules, and glomerular vessels [2,7,42]. The red dots appear as twisted capillary loops on higher magnification. These vascular findings correspond to the dilated capillaries seen in the dermal papillae on histopathology. The number of twisted loops correlates with disease severity [2].

Seborrheic dermatitis is characterized by thin arborizing capillaries and atypical red vessels [7]. In scales are dry, silvery white in psoriasis and greasy, yellowish in seborrheic dermatitis [7]. A study by Kim et al. found no significant difference in the frequency and characteristics of the scales in both of the conditions on dermoscopy. It was concluded that vascular patterns are more valuable for differentiation [42]. An algorithmic approach to trichoscopy aided diagnosis of alopecia [13] (Table 4.1).

CONCLUSION

In this chapter, we have attempted to cover the trichoscopic evaluation and findings in the most common hair and scalp conditions; however, much remains to be seen and explored. It hereby needs to be mentioned that most of the dermoscopic studies conducted across the world pertain predominantly to tumoral and some nontumoral conditions affecting the skin, mainly melanoma. Trichoscopy has recently been in vogue because it is simple to perform and gives gratifying results in terms of quick and easy diagnosis. Moreover, easy retrieval of data at a later stage is possible for comparative analysis. The patient can be provided with a printed report, thus escalating the standards of consultation, follow-up, and overall patient satisfaction. We intend to continue our exploration and hope to come up with trichoscopic findings in a wider spectrum of diseases in the future.

Abbreviations

AA Alopecia areata
AGA Andrgenetic alopecia
DLE Discoid lupus erythematosus
FAGA Female androgenetic alopecia
FFA Frontal fibrosing alopecia
LPP Lichen planopilaris
PL Polarized light

References

[1] Rudnicka L, Olszewska M, Rakowska A, Kowalska-Oledzka E, Slowinska M. Trichoscopy: a new method for diagnosing hair loss. J Drugs Dermatol 2008;7:651–4.

[2] Tosti A, Ross EK. Patterns of scalp and hair disease revealed by videodermoscopy. In: Tosti A, editor. Dermoscopy of hair and scalp disorders. 1st ed. London: Informa Healthcare; 2007. p. 1–14.

[3] Tosti A, Duque-Estrada B. Dermoscopy in hair disorders. J Egypt Womens Dermatol Soc 2010;7:1–4.

[4] Lacarrubba F, Dall'Oglio F, Rita Nasca M, Micali G. Videodermatoscopy enhances diagnostic capability in some forms of hair loss. Am J Clin Dermatol 2004;5:205–8.

[5] Ross EK, Vincenzi C, Tosti A. Videodermoscopy in the evaluation of hair and scalp disorders. J Am Acad Dermatol 2006;55: 799–806.

[6] Mane M, Nath AK, Thappa DM. Utility of dermoscopy in alopecia areata. Indian J Dermatol 2011;56:407–11.

[7] Rudnicka L, Olszewska M, Rakowska A, Slowinska M. Trichoscopy update 2011. J Dermatol Case Rep 2011;5:82–8.

[8] Rakowska A, Slowinska M, Kowalska-Oledzka E, Warszawik O, Czuwara J, Olszewska M, et al. Trichoscopy of cicatricial alopecia. J Drugs Dermatol 2012;11:753–8.

[9] Duque-Estrada B, Tamler C, Sodré CT, Barcaui CB, Pereira FB. Dermoscopy patterns of cicatricial alopecia resulting from discoid lupus erythematosus and lichen planopilaris. An Bras Dermatol 2010;85:179–83.

[10] Kossard S, Zagarella S. Spotted cicatricial alopecia in dark skin. A dermoscopic clue to fibrous tracts. Australas J Dermatol 1993;34: 49–51.

[11] de Moura LH, Duque-Estrada B, Abraham LS, Barcaui CB, Sodre CT. Dermoscopy findings of alopecia areata in an African–American patient. J Dermatol Case Rep 2008;2:52–4.

[12] Abraham LS, Piñeiro-Maceira J, Duque-Estrada B, Barcaui CB, Sodré CT. Pinpoint white dots in the scalp: dermoscopic and histopathologic correlation. J Am Acad Dermatol 2010;63: 721–2.

[13] Kharkar V. Overview of trichoscopy. In: Khopkar U, editor. Dermoscopy and trichoscopy in diseases of the brown skin. 1st ed. New Delhi: Jaypee; 2012. p. 169–81.

[14] Inui S, Nakajima T, Nakagawa K, Itami S. Clinical significance of dermoscopy in alopecia areata: analysis of 300 cases. Int J Dermatol 2008;47:688–93.

[15] Jain N, Doshi B, Khopkar U. Trichoscopy in alopecias: diagnosis simplified. Int J Trichol 2013;5:170–8.

[16] Rakowska A, Slowinska M, Kowalska-Oledzka E, Olszewska M, Rudnicka L. Dermoscopy in female androgenic alopecia: method standardization and diagnostic criteria. Int J Trichology 2009;1: 123–30.

[17] Kibar M, Aktan S, Bilgin M. Scalp dermatoscopic findings in androgenetic alopecia and their relations with disease severity. Ann Dermatol 2014;4:478–84.

[18] Paus R, Olsen EA, Messenger AG. Hair growth disorders. In: Wolff K, Goldsmith LA, Katz SI, Gilchrist BA, Paller AS, Lefell DJ, editors. Fitzpatrick's dermatology in general medicine. 7th ed. USA: McGraw Hill; 2008. p. 753–77.

[19] El-Taweel A-E, El-Esawy F, Abdel-Salam O. Different trichoscopic features of tinea capitis and alopecia areata in pediatric patients. Dermatol Res Pract 2014. 848763.

[20] Shuster S. 'Coudability': a new physical sign of alopecia areata. Br J Dermatol 1984;111:629.

[21] Thakkar V, Haldar S. Trichoscopy of patchy alopecia. In: Khopkar U, editor. Dermoscopy and trichoscopy in diseases of the brown skin. 1st ed. New Delhi: Jaypee; 2012. p. 182–201.

[22] Slowinska M, Rudnicka L, Schwartz RA, Kowalska-Oledzka E, Rakowska A, Sicinska J, et al. Comma hairs: a dermatoscopic marker for tinea capitis: a rapid diagnostic method. J Am Acad Dermatol 2008;59:S77–9.

[23] Hughes R, Chiaverini C, Bahadoran P, Lacour JP. Corkscrew hair: a new dermoscopic sign for diagnosis of tinea capitis in black children. Arch Dermatol 2011;147:355–6.

[24] Yorulmaz A, Artuz F, Erden O. A case of trichotillomania with recently defined trichoscopic findings. Int J Trichol 2014;6:77–9.

[25] Sah DE, Koo J, Price VH. Trichotillomania Dermatol Ther 2008;21: 13–21.

[26] Rakowska A, Slowinska M, Olszewska M, Rudnicka L. New trichoscopy findings in trichotillomania: flame hairs, V-sign, hook hairs, hair powder, tulip hairs. Acta Derm Venereol 2014;94:303–6.

[27] Kumar B, Verma S, Raphael V, Khonglah Y. Extensive tonsure pattern trichotillomania-trichoscopy and histopathology aid to the diagnosis. Int J Trichol 2013;5:196–8.

[28] Rudnicka L, Ozewska M, Rakowska A. Trichotillomania and traction alopecia. In: Rudnicka L, Olszewska M, Rakowska A, editors. Atlas of trichoscopy: dermoscopy in hair and scalp disease. 2nd ed. London: Springer-Verlag; 2012. p. 257–78.

[29] Inui S. Trichoscopy: a new frontier for the diagnosis of hair diseases. Expert Rev Dermatol 2012;7:1–8.

[30] Ankad BS, Beergouder SL, Moodalgiri VM. Lichen planopilaris versus discoid lupus erythematosus: a trichoscopic perspective. Int J Trichol 2013;5:204–7.

[31] Kang H, Alzolibani AA, Otberg N, Shapiro J. Lichen planopilaris. Dermatol Ther 2008;21:249–56.

[32] Inui S, Nakajima T, Shono F, Itami S. Dermoscopic findings in frontal fibrosing alopecia: report of four cases. Int J Dermatol 2008;47:796–9.

[33] Rubegni P, Mandato F, Fimiani M. Frontal fibrosing alopecia: role of dermoscopy in differential diagnosis. Case Rep Dermatol 2010; 2:40–5.

[34] Baroni A, Romano F. Tufted hair folliculitis in a patient affected by pachydermoperiostosis: case report and videodermoscopic features. Skinmed 2011;9:186–8.

[35] Goldberg LJ. Cicatricial marginal alopecia: is it all traction? Br J Dermatol 2009;160:62–8.

[36] Rakowska A, Slowinska M, Kowalska-Oledzka E, Rudnicka L. Trichoscopy in genetic hair shaft abnormalities. J Dermatol Case Rep 2008;2:14–20.

[37] Rakowska A, Slowinska M, Czuwara J, Olszewska M, Rudnicka L. Dermoscopy as a tool for rapid diagnosis of monilethrix. J Drugs Dermatol 2007;6:222–4.

[38] Jain N, Khopkar U. Monilethrix in pattern distribution in siblings: diagnosis by trichoscopy. Int J Trichology 2010;2:56–9.

[39] Kharkar V, Gutte R, Thakkar V, Khopkar U. Trichorrhexis nodosa with nail dystrophy: diagnosis by dermoscopy. Int J Trichology 2011;3:105–6.

[40] de Berker DA, Paige DG, Ferguson DJ, Dawber RP. Golf tee hairs in Netherton disease. Pediatr Dermatol 1995;12:7–11.

[41] Bittencourt MJS, Mendes AD, Moure ERD, Deprá MM, Pies OTC, Mello ALP. Trichoscopy as a diagnostic tool in trichorrhexis invaginata and Netherton syndrome. An Bras Dermatol 2015;90: 114–6.

[42] Kim GW, Jung HJ, Ko HC, Kim MB, Lee WJ, Lee SJ, et al. Dermoscopy can be useful in differentiating scalp psoriasis from seborrhoeic dermatitis. Br J Dermatol 2011;164:652–6.

CHAPTER

5

Dermatoscopic Correlates of Nail Apparatus Disease

A Look at the Nail From Another Scope, the Dermatoscope

S. Campos, A. Lencastre

Hospital de Santo António dos Capuchos, Centro Hospitalar de Lisboa Central, Lisboa, Portugal

OUTLINE

INTRODUCTION

The nail organ is an integral component of the digital tip. It is a highly versatile tool that protects the fingertip, contributes to tactile sensation by acting as a counterforce to the fingertip pad, and aids in peripheral thermoregulation via glomus bodies in the nail bed and matrix [1]. The nail apparatus includes the nail plate, the matrix (sterile and germinal), the proximal and lateral nail folds, the eponychium, and the hyponychium (Fig. 5.1). The nail plate emerges from the proximal nail fold and is bordered on either side by the lateral nail folds [2]. The skin proximal to the nail that covers the nail fold is the eponychium. The tissue distal to the eponychium in contact with the nail represents the cuticle. Extending from proximal to distal on the nail is a half-moon shaped white arc known as the lunula (see Fig. 5.1). The lunula is the distal extent of the germinal matrix. This characteristic white

Imaging in Dermatology
http://dx.doi.org/10.1016/B978-0-12-802838-4.00005-4

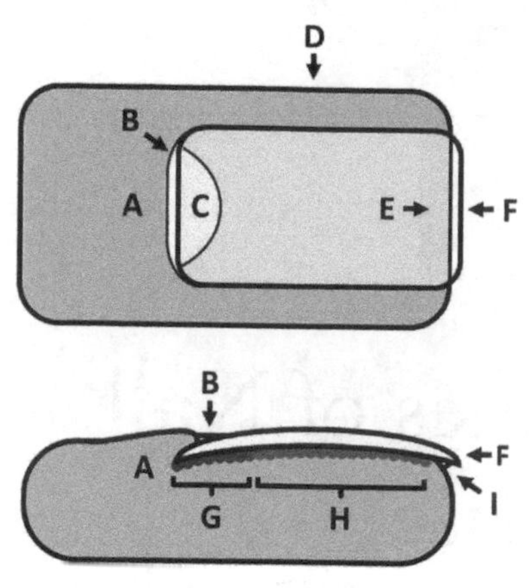

FIGURE 5.1 Nail anatomy. (A) Proximal nail fold; (B) dorsal cuticle; (C) lunula; (D) lateral nail fold; (E) onychodermal band; (F) free edge of nail plate; (G) germinal matrix; (H) sterile matrix (or nail bed); (I) hyponychium (with ventral cuticle).

color is caused by the persistence of nail cell nuclei in the germinal matrix. Distal to this location, the nail adheres to the sterile matrix (ie, nail bed), traverses the onychodermal band (its last point of attachment) before ending freely detached in a white colored segment. The hyponychium underlies this free margin of the nail plate. It is located after the onychodermal band proximally and ends with the distal nail groove. This region is susceptible to contamination from environmental interactions. Between the nail free edges the hyponychium is a keratin plug that acts as a mechanical barrier to protect against infectious inoculation, the ventral cuticle (solehorn). Also found within this keratin plug are polymorphonuclear leukocytes and lymphocytes contributing an immunological barrier to the mechanical one established by the keratin plug [3].

Dermatoscopy is a noninvasive method that helps narrow the differential diagnosis of pigmented lesions of the skin and can assist in the recognition of early melanoma. Recently, dermatoscopy has been described to be useful in the evaluation and diagnosis of numerous nail diseases, and can be used to evaluate various parts of the nail apparatus including the nail plate surface and free edge, the nail matrix, the nail bed, the periungual folds, and the hyponychium [4–6].

The evaluation of nailfold capillaroscopic abnormalities is widely utilized in the diagnosis of several types of connective tissue diseases (CTDs) and dermatoscopy has been considered equivalent to capillaroscopy in the identification of simple changes [7–9].

METHODS IN NAIL DERMATOSCOPY

Two types of dermatoscopes are available: those that use nonpolarized light with liquid immersion and those that use polarized light. A recent study investigated the differences between contact and noncontact polarized light dermatoscopy and contact nonpolarized dermatoscopy (standard dermatoscopy). The authors concluded that although the techniques did not provide equivalent images, they were complementary. Polarized light images seem to offer a better view of structures located deeper in the skin, whereas immersion-contact nonpolarized dermatoscopy allows for the visualization of more superficial structures. Also noncontact polarized dermatoscopy allows much better visualization of blood vessels [10]. Depending on the site of the nail apparatus to be evaluated, it is necessary to adapt the dermatoscopy device, type of light (polarized vs nonpolarized), and immersion solution (Table 5.1).

It is necessary to vary the focus of the device to correctly evaluate nail pigmentation and to appreciate thin lines withina nail pigment band [6]. When evaluating melanonychia, it is important to keep in mind that nail dermatoscopy does not provide direct analysis of the nail matrix or nail bed. These are the anatomical origins of the pigmentation [5]. Thus the distribution of the pigment in the nail plate may not accurately represent the characteristics of the underlying lesion [11]. The direct visualization of the pigment production site is only possible with intraoperative nail matrix and bed dermatoscopy. This procedure involves the reflection of the proximal nail fold and/or partial/complete

TABLE 5.1 Type of dermoscopy device, light, and immersion fluid to use depending on the site of the nail apparatus to be evaluated

Nail site/technique	Dermoscopy device	Type of light	Immersion fluid (if needed)
Nail plate	Videodermatoscope, hand-held device	Contact nonpolarized	Ultrasound gel
Free edge of the nail	Videodermatoscope, hand-held device	Contact nonpolarized	Alcohol
Hyponychium (skin, capillaries)	Videodermatoscope preferred	Contact nonpolarized	Alcohol
Perionychium	Videodermatoscope preferred	Contact nonpolarized preferred, or polarized	Alcohol
Capillaries of the proximal nail fold	Videodermatoscope, hand-held device	Polarized	Oil, ultrasound gel
Intraoperative dermoscopy[a]	Hand-held device	Polarized	No fluid

[a] *Before or after the procedure.*

avulsion of the nail plate. To avoid contact with potentially contaminated blood and tissue, it is better to use a polarized light dermatoscope [4,5,12,13]. Immediate ex vivo dermatoscopy after collection of the nail specimen can also be performed [14].

All pictures accompanying this text were taken with a hand-held device. This way, the reader can have a sense of the effortlessness accuracy with which one can evaluate the nail apparatus.

DERMATOSCOPY OF NAIL PIGMENT CHANGES (CHROMONYCHIA)

Melanonychia

Onychoscopy is one method to evaluate nail-pigmented lesions. It does not replace the clinical history or physical examination. Histopathology remains the gold standard for the definitive diagnosis of melanonychia. Looking for the Hutchinson sign and the application of the *ABCDEF* acronym for nail pigmented lesions (patient *A*ge, *B*and color and breadth, history of *C*hange, *D*igit involved, *E*xtension to eponychium, and *F*amily or personal history) are also important tools in the evaluation of melanonychia, particularly in the assessment of differential diagnoses [15,16].

Melanonychia, a brown or black pigmentation of the nail plate usually caused by the presence of melanin, commonly appears as a longitudinal band [longitudinal melanonychia (LM)]. LM mostly starts from the matrix and extends to the tip of the nail plate. Less commonly, the pigmentation can involve the whole nail plate (total melanonychia) (Fig. 5.2) or present as a transverse band (transverse melanonychia). Total melanonychia and transverse melanonychia are much rarer occurrences [4]. LM is a diagnostic challenge, because it may occur with both benign lesions and nail malignancies [5].

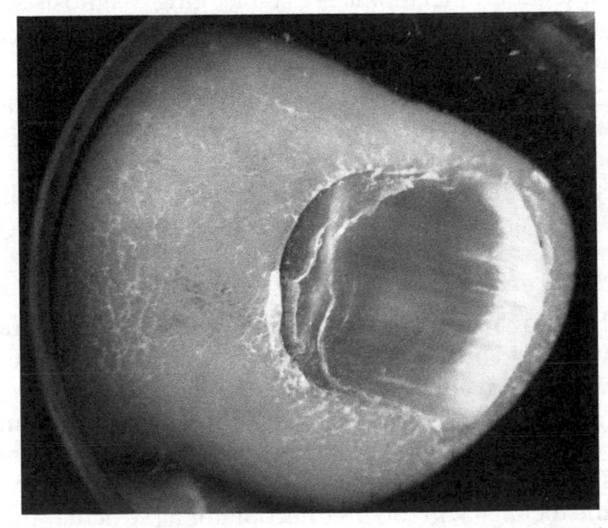

FIGURE 5.2 Total melanonychia caused by subungual hemorrhage.

One can assess the matrical site of nail plate pigment production with dermatoscopy of the nail plate free edge. If the pigment origin lies on the proximal nail matrix, the pigment is located on the upper portion of the free edge, whereas in the case of distal matrix involvement, the lower portion will be pigmented (as is more commonly observed) [4,17]. When in doubt, a nail clipping and Fontana-Mason staining may be performed [6,17].

When considering a brown pigment band, the physician should distinguish melanic from nonmelanic nail pigmentation, where pigmentation caused by hemorrhage is the main differential diagnosis. If the pigment is melanin, then it is mandatory to evaluate whether this deposition is secondary to melanocyte activation or melanocyte proliferation and, in case of proliferation, it must be determined whether the lesion is benign (lentigo, nevus) or malignant [4,18]. To differentiate melanonychia resulting from melanocytic activation from that caused by melanocytic proliferation, Ronger et al. proposed some dermatoscopic criteria. These are still controversial because they are not standardized enough to make such distinction, especially when the melanonychia is associated with periungual pigmentation [18]. These criteria are the following (Fig. 5.3) [6,18,19]:

1. Black or dark red round globules (often with a filamentous distal end) represent blood.
2. The color of the band can be used as a marker to differentiate between activation and proliferation. Gray bands represent melanocytic activation; these

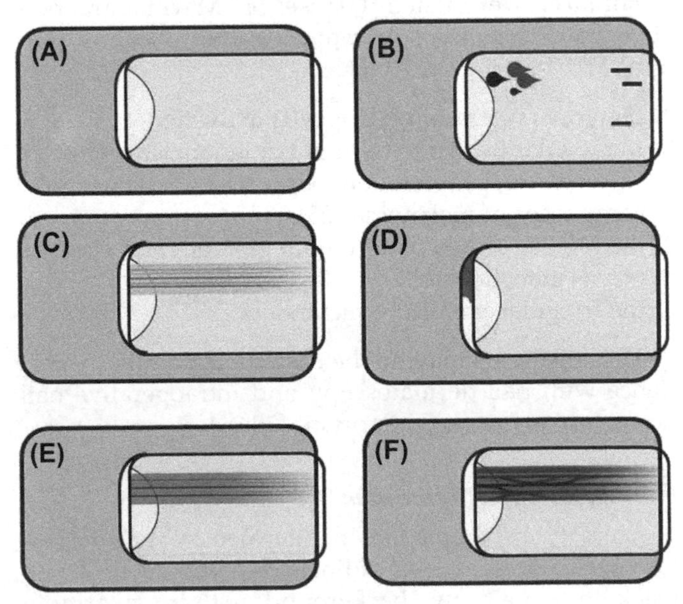

FIGURE 5.3 Melanonychia patterns. (A) Normal nail. (B) Blood spots (proximal) and splinter hemorrhages (distal). (C) Gray band with parallel lines. (D) Micro-Hutchinson sign. (E) Brown band with parallel lines. (F) Brown band with criss-crossing lines of variable width and hue.

are benign and require no histopathological evaluation. Brown-black bands represent melanocytic proliferation. These may have: [20]:

a. A regular pattern, suggestive of benign proliferation, composed of individual brown-black lines with similar thickness and color that are regularly spaced and parallel [21].
b. An irregular pattern, suggestive of malignant proliferation. The brown-black lines lose parallelism, are irregularly spaced, and vary in width and color (not only in color shades) [21]. Some argue that these criteria are not valid in children, where benign lesions often have an irregular pattern [4], and are also often difficult to applyeven in adults [22].

3. The micro-Hutchinson sign can best be visualized by dermatoscopy (not clinically). This pigmentation of the cuticle is indicative of malignancy [4]. Special care must be made not to confuse the appearance of nail plate pigmentation under the relatively translucent cuticle (pseudo-Hutchinson sign), with pigmentation of the cuticle and/or proximal nail fold.
4. Melanin granular inclusions are more adequately seen under high magnification (they have a diameter of less than 0.1 mm) and are present in case of melanocyte proliferation [6].

These criteria were reassessed and reorganized in a severity-based classification of LM by Sawada et al. [23]. An automated calculated index, based on color variegation of the LM assessed by dermatoscopy, has also been developed [24].

Hirata et al. evaluated 100 cases of LM with intraoperative nail dermatoscopy reporting the following patterns [5]:

1. the gray pattern of melanocytic activation;
2. the regular brown pattern of benign melanocytic hyperplasia/proliferation, observable, however, in some cases of melanoma [25];
3. the regular brown pattern with globules and blotches of melanocytic nevi;
4. the irregular pattern of melanoma.

These patterns may not be easy to distinguish. Experience with nail dermatoscopy and intraoperative nail dermatoscopy is very important [22,26].

Nail Apparatus Melanoma

An early melanoma may be revealed by such dermatoscopic features as longitudinal criss-crossing brown to black lines in a brown background, with irregularity in color, spacing, and thickness (Fig. 5.4) [4,19]. A pitfall, however, is that subungual melanoma may present a regular pattern or a thin band of light brown pigment (<3 mm) [6,22]. Nevi in children, on the other hand, tend to present with brown black bands with an irregular pattern [4].

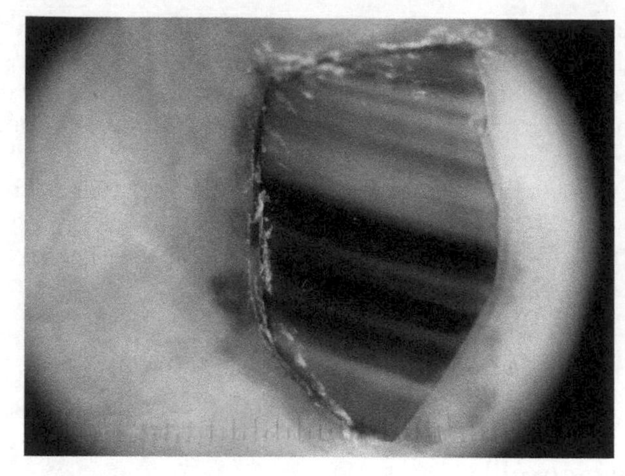

FIGURE 5.4 Nail apparatus melanoma. Dermoscopy shows irregular bands, micro-Hutchinson sign, and pigment on the perionychium favoring the diagnosis of melanoma. *Image kindly ceded by Trakatelli, CM. Second Dermatology Clinic of Aristotle University, Papageorgiou Hospital, Thessaloniki, Greece.*

Abrupt stoppage of the lines and the micro-Hutchinson sign also favor the diagnosis of melanoma (see Fig. 5.4) [19,27,28]. As observed clinically, rapidly growing melanomas may present a triangular pigment band where the proximal end (facing toward the proximal nail fold) is wider than the distal portion [6]. Another potential pitfall is that blood spots may be observed in cases of melanoma. Their presence does not rule out the diagnosis and should not be attributed to hemorrhage, especially when other suspicious dermatoscopic signs are present [19].

A few authors have mentioned erosion or microscopic grooves of the nail plate surface as signs of melanoma, although statistically convincing evidence is lacking [19,21].

Amelanotic melanoma is a challenging diagnosis. In a later stage, when the nail is partially or totally destroyed by an apparently nonpigmented reddish tumor, traces of pigmentation may be detected with the dermatoscope, establishing a differential diagnosis with pyogenic granuloma and other nonmelanocytic nail tumors [4,21]. A polymorphic vascular pattern combined with milky red areas and red spots may also be observed [25].

Melanocytic Nevus

The brown background associated with regular parallel lines of identical color, spacing, and width suggests a benign lesion, either a nevus or lentigo [27].

Nail matrix nevi may be congenital or acquired and are often seen in children and young adults. In general, nail matrix nevi occur more often on the fingernails than on the toenails with no predilection for a particular digit [4,6]. Clinically they appear as one or more longitudinal pigmented bands varying in size from a few millimeters to the whole nail width and in color from light brown to black [4]. Dermoscopically, nevi of the nail matrix show brown longitudinal parallel lines with regular spacing and thickness (Fig. 5.5) [6]. Difficult cases, however, exist especially in congenital nevi of the nail apparatus (Fig. 5.6). Usually they are heavily pigmented with irregular lines and can clinically simulate melanoma [6,21]. Also, a clinical Hutchinson sign, LM with a broader proximal end, darkening and spreading of the band, and thinning and grooving of the nail plate are not uncommon features of congenital nevi in children [4,11]. For these reasons, onychoscopy is not useful for the evaluation of melanonychia in childhood. Other limitations in the assessment of melanonychia with the dermatoscope include a thick nail and total melanonychia (see Fig. 5.2), which make proper visualization difficult. The Hutchinson sign, especially on the hyponychium, is not always evidence of subungual melanoma because it can be seen in both melanoma and melanocytic nevus. However, there is a wide difference in dermatoscopic features between the two. A linear brushy pattern is suggestive of nevus, whereas an irregular ridge pattern or haphazard surface pigmentation suggests melanoma [29]. These features are analogous to those that are found when evaluating pigmented acral skin lesions. Thus this dermoscopic differentiation is useful in assessing Hutchinson sign, allowing the differentiation between benign and malignant conditions as well as all other characteristics that are observable through nail dermatoscopy [11].

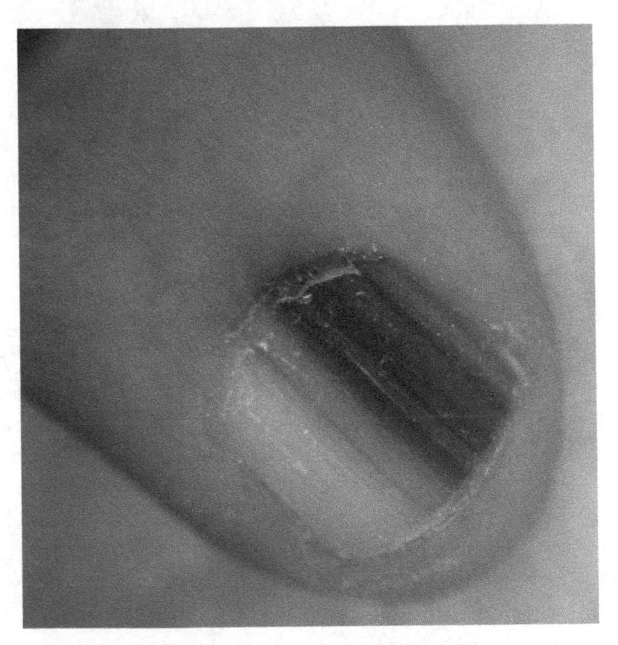

FIGURE 5.5 Nail matrix nevus. Dermoscopy shows brown, longitudinal parallel lines with regular spacing and thickness.

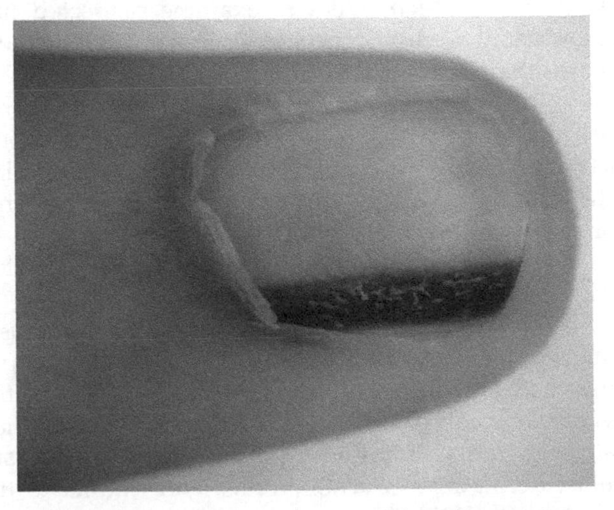

FIGURE 5.6 Congenital nevus of the nail apparatus.

Blue nevi are characterized by a stable, hazy blue coloration seen through the nail plate [21,26].

Nonmelanocytic Nail Tumors

Melanonychia caused by melanocyte activation can also be caused by benign (epidermal inclusion cyst, onychomatrichoma) and malignant [Bowen's disease (BD)] tumors that either lie within or close to the nail apparatus (Fig. 5.7) [18,27,30].

Melanocyte Activation

There are multiple causes of melanocyte activation (Fig. 5.8). Clinically, the diagnosis may be promptly suggested by the presence of pale bands and by the involvement of several nails [11]. The pattern of benign melanonychia caused by melanocyte activation (traumatic, ethnic-type, inflammatory nail diseases like lichen planus and psoriasis, or drug-induced pigmentation)

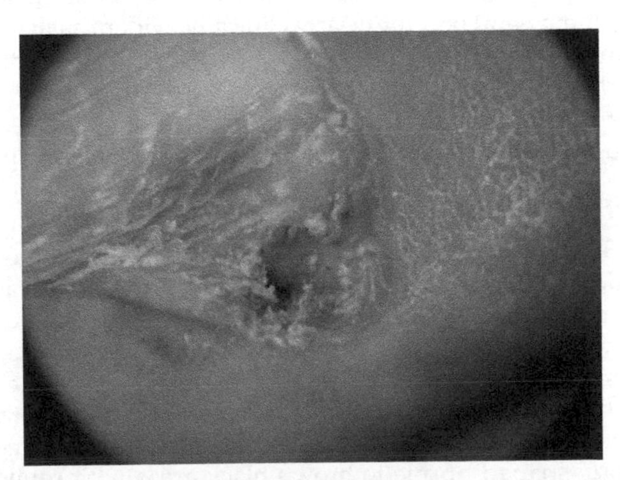

FIGURE 5.7 Squamous cell carcinoma (SCC) in a 60-year-old woman. Dermoscopy shows longitudinal melanonychia (LM). The association between clinical and dermoscopic changes was very important to establish the diagnosis in this patient.

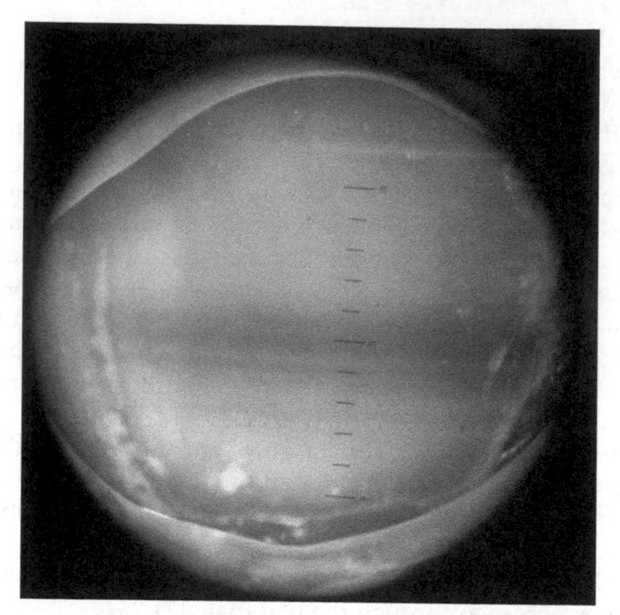

FIGURE 5.8 Longitudinal melanonychia (LM) of the toenail caused by nail matrix lentigo.

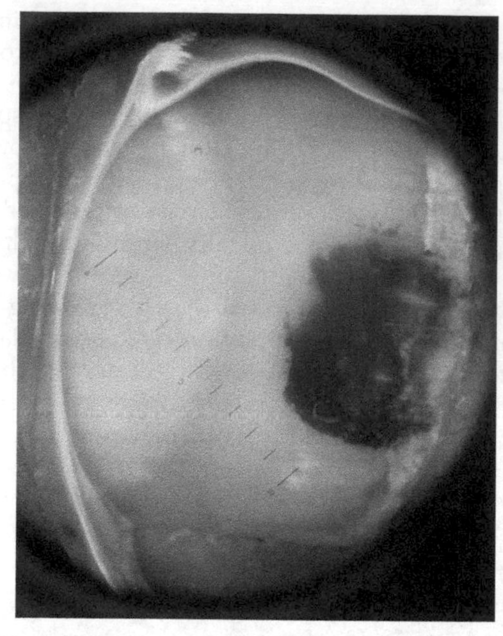

FIGURE 5.9 Subungual hemorrhage. Dermoscopy shows an irregularly shaped, large, *reddish-black* area with round, *dark red spots* at the periphery. A rounded proximal border and a streaked "filamentous" distal end are discernable.

exhibits a homogeneous gray discoloration of the background with thin, longitudinal gray lines. Because of the light scattering Tyndall effect, the color may vary, depending on the thickness of the nail and melanin location within the nail plate [26,28,31]. When pigmentation is caused by repetitive trauma (onychotillomania, onychophagia, frictional melanonychia of the fourth and fifth toenails) dermatoscopy also shows tiny dark red to brown spots representing blood extravasation [4,11,21]. In ethnic pigmentation, the grayish longitudinal lines are often multiple and polydactylic [21].

Melanocytes can also be activated by inflammatory skin diseases that affect the nail unit such as psoriasis, Hallopeau acrodermatitis, lichen planus, and chronic paronychia [1]. The pigmentation can be subtle at the beginning and become clearly clinically evident as the inflammatory process progresses. It is usually associated with nail scarring and abnormalities of the surface of the nail plate. The clinical and dermatoscopic aspects caused by inflammatory skin diseases are not specific [6].

Subungual Hemorrhage

The presence of subungual blood usually appears as a reddish to reddish-black pigment seen through the nail plate, depending on the age of the bleed. Subungual hemorrhage may pose as an important differential diagnosis of nail melanoma when a longitudinal placement of the hemorrhage is observed. Dermatoscopically, the blood spot pattern is nonetheless characterized by irregularly shaped purple to brown-black areas with round, dark red spots at the periphery and a streaked "filamentous" distal end (Fig. 5.9) [4,6,11,19,21]. Small reddish to reddish-black globules along the proximal and lateral margins of the pigment can also be seen. The pigment is homogenous and no melanin granules can be observed. An isolated report coined the term *pseudopods* at the distal end of a nail hemorrhage, which may represent an equivalent of the streaked distal end described previously [32]. It is important to remember that a subungual hematoma can, on rare occasions, be caused by an episode of hemorrhage or caused by neovascularization within a tumor and thus the presence of subungual blood should not be used to rule out the diagnosis of melanoma [6,19]. Furthermore, a history of trauma to the digit has been mentioned in the context of nail unit melanoma, adding another potential confounding factor [33,34]. It may be prudent to examine suspected nail hemorrhaging 3–4 months later to assess its movement toward the distal end and proximal clearance [21]. Hence any subungual hemorrhage that does not grow out with the nail or that recurs at the same place requires special attention and further examination, including radiological workup (eg, by imaging of the distal phalange to exclude exostosis). If the diagnosis cannot be determined by radiological studies, then the lesion should be biopsied [6].

On the other hand, splinter hemorrhages are a consequence of blood extravasation from the nail bed capillaries with longitudinal arrangement and from the successive incorporation of blood in the ventral nail plate (Fig. 5.10). This finding can appear in psoriasis, contact dermatitis, onychomycosis, and trauma to the nail apparatus [35].

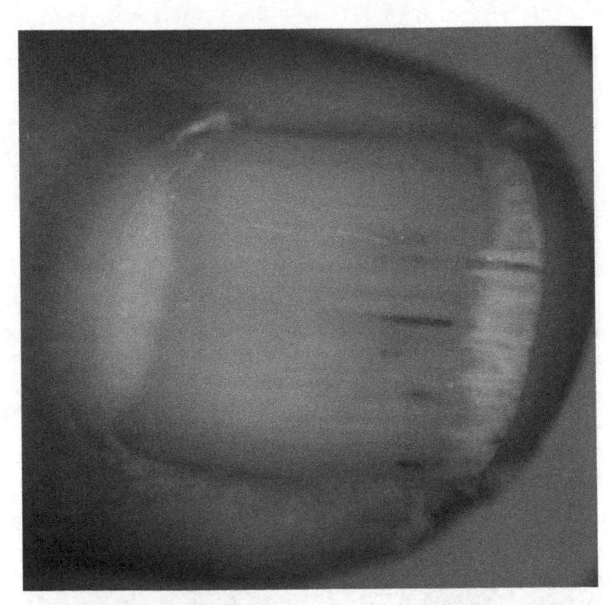

FIGURE 5.10 Splinter hemorrhages with the characteristic longitudinal arrangement of blood extravasation.

Fungal and Bacterial Nail Infection

Some bacteria and fungi may be responsible for brown-black discoloration of the nail plate [11]. Fungal melanonychia, as a result of fungal infection, is a relatively rare nail disorder [36], and the melanonychia may be caused by melanocyte activation but also by direct melanin production by the fungi (Fig. 5.11) [4]. Absence of granular inclusions and the presence of other suggestive signs (jagged edge with spikes and longitudinal striae), should elicit the cause [6,37].

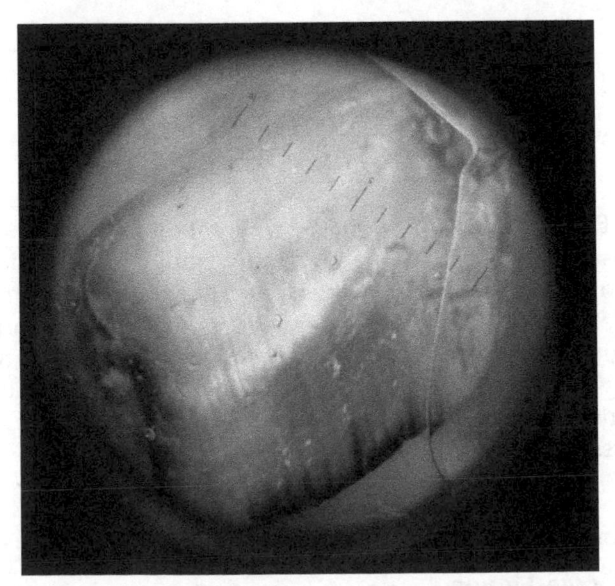

FIGURE 5.11 Fungal melanonychia. Dermoscopic image with longitudinal brown-black discoloration of the nail plate caused by onychomycosis.

Dermatoscopy in Other Causes of Chromonychia

The causes of nail dyschromia are well known. There are, however, a few reports of the use of dermatoscopy in leukonychia, erythronychia, and chloronychia, and isolated reports on the association between orange-brown chromonychia and Kawasaki disease in children [38]. The data described in the literature, on some of these topics, are discussed in the following sections.

Leukonychia

Opaque white discoloration of the nail or leukonychia is the most common form of nail dyschromia. Leukonychia has been classified into three different types: true leukonychia, apparent leukonychia, and pseudoleukonychia. Dermatoscopy provides optimal evaluation of leukonychia and allows easy distinction between true leukonychia and pseudoleukonychia. True leukonychia is whitening of the nail plate itself and is further separated into total and subtotal, or partial, the latter occurring as punctate, transverse, and longitudinal. The total and subtotal forms of true leukonychia are more commonly associated with a systemic disorder, whereas the transverse and punctate forms are more commonly the result of localized trauma [18,39]. Dermatoscopy has highlighted the presence of longitudinal leukonychia associated with tumors of the nail apparatus, particularly onychopapilloma and onychomatricoma [40,41]. Onychoscopy of thumbnail dystrophy in three patients with Hailey–Hailey disease has revealed multiple parallel longitudinal white lines of variable width [42].

Pseudoleukonychia is an alteration of the nail bed with a normal nail plate that results in a nail unit that is clinically white and can be associated to onychomycosis (see below) or keratin deposits due to nail varnish.

Apparent leukonychia is a white-appearing nail plate with a variety of causes, such as onycholysis, wherein the whiteness is not from either the nail plate or bed [43].

Erythronychia

Longitudinal erythronychia is a linear red band on the nail plate that usually originates at the proximal nail fold and extends to the free edge of the nail plate. It appears as a linear band, pink to red in color, and the longitudinal streak usually ranges in width from less than 1 to 3 mm; however, wider bands have been observed. Associated morphological findings that can be present at the distal tip of the nail with longitudinal erythronychia include fragility, onycholysis, splinter hemorrhaging, subungual keratosis, thinning, or a V-shaped nick [44].

For purposes of evaluation and management, longitudinal erythronychia can be divided into two groups, depending on the extent of nail involvement (a single nail or multiple nails). Both groups, localized longitudinal erythronychia (LLE) and polydactylous longitudinal erythronychia (PLE), are associated with a limited differential diagnosis. When limited to one digit (LLE), the etiology is usually neoplastic, with onychopapilloma being the most common diagnosis. However, glomus tumor, BD, melanoma in situ, and basal cell carcinoma have all been reported in this presentation. When identified on multiple digits (PLE), there is usually an underlying regional or systemic cause. The most common include lichen planus and Darier disease (Fig. 5.12). Other important, less common causes include systemic amyloidosis, graft-versus-host disease, and hemiplegia. In certain circumstances, biopsy is indicated. A partial nail plate avulsion coupled with a longitudinal biopsy of matrix and bed is the recommended technique for diagnosis in most circumstances [45,46].

Chloronychia

Green nails, also known as *chloronychia* or *green nail syndrome*, are characterized by green discoloration of the nail plate (greenish-yellow, greenish-brown, greenish-black) (Fig. 5.13). The cause is an infection, mostly from *Pseudomonas aeruginosa* (Fig. 5.14). The clinical appearance consists in a typical triad: green discoloration of the nail plate associated with proximal chronic paronychia and distal-lateral onycholysis. Exposure to moist environment, microtrauma, and associated nail diseases such as psoriasis may promote infection by *Pseudomonas* [47,48].

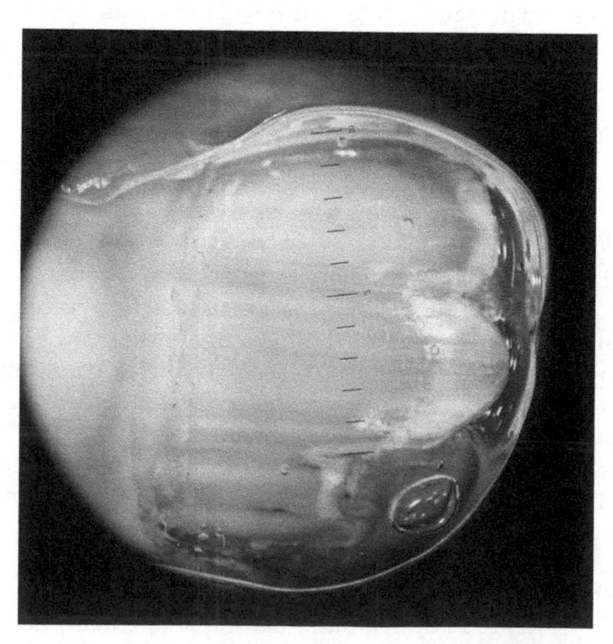

FIGURE 5.12 Erythronychia in a patient with Darier disease.

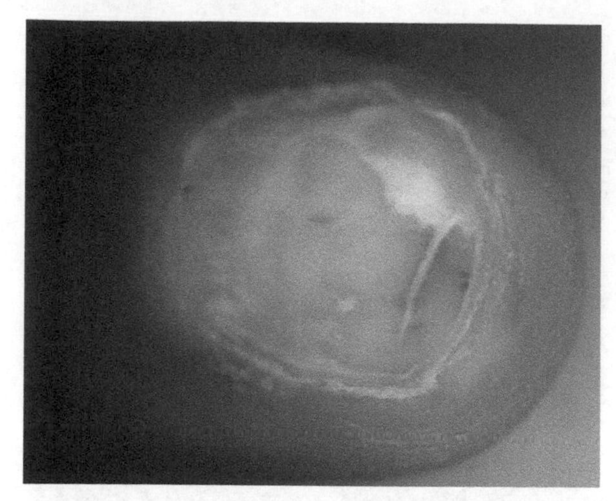

FIGURE 5.13 Secondary infection of a subungual hematoma.

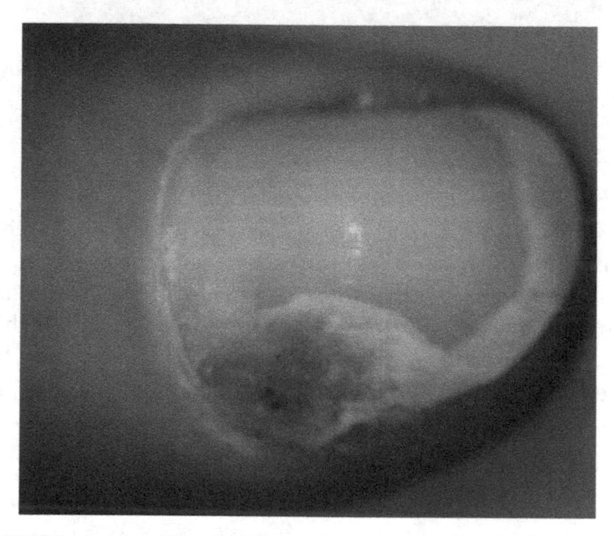

FIGURE 5.14 Chloronychia caused by *Pseudomonas aeruginosa* infection.

DERMATOSCOPY OF OTHER NAIL CHANGES

Beyond the scope of nail discoloration and the diagnosis of melanocytic activation versus hyperplasia/neoplasia, dermatoscopy has found a place in the evaluation of other nail tumors as well as nail diseases and signs as common as onychomycosis, nail psoriasis, and onycholysis. It can be particularly helpful when typical clinical features of these diseases are either discrete or missing.

Nonmelanocytic Tumors of the Nail Apparatus

Glomus Tumor

This benign vascular tumor of habitual fingernail location is identified clinically and with dermatoscopy

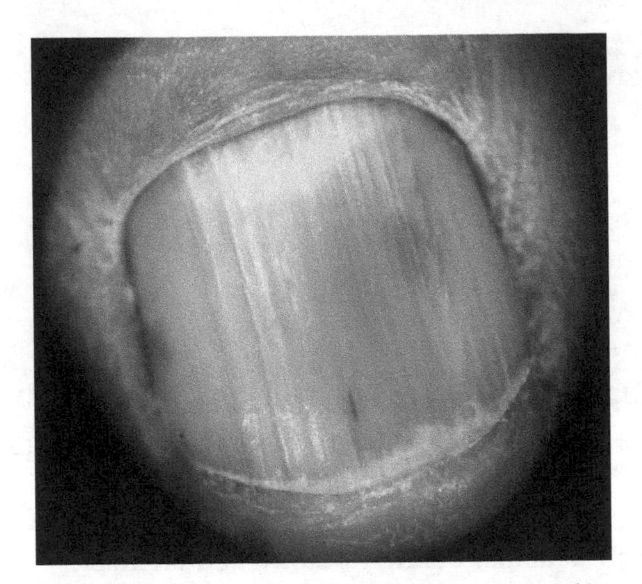

FIGURE 5.15 Glomus tumor. Dermatoscopy reveals a blue-red homogeneous patch beneath the nail plate. Notice how the blue patch partially covers the lower limit of the lunula.

as a blue-red homogeneous patch beneath the nail plate with or without visible blood vessels (Fig. 5.15) [49]. It has been reported that in the intraoperative setting, nail bed/matrix dermatoscopy may facilitate estimation of tumor margins and postexcisional assessment for remaining tumor [45].

Onychopapilloma

This benign nail bed or nail matrix tumor will usually present itself as a solitary, distal linear segment of longitudinal erythronychia, ending above a hyperkeratotic papule located immediately beneath the adjacent free edge (Fig. 5.16) [49]. These signs are distinctly observed with dermatoscopy and the previously mentioned linear band may be white [40] or brown-black [30], aside from red, in color, and start anywhere from the proximal nail fold onward.

Onychopapilloma may present with LM and yet conserve some of the more classic features described in this section [30,50].

Onychomatricoma

This benign nail matrix tumor will grow within the nail plate, increasing its thickness and convexity with a yellow to brownish tinge and multiple splinter hemorrhaging [51]. Longitudinal leukonychia and splinter hemorrhages are observed on the nail plate, whereas dermatoscopy of the free edge will aid further in the detection of the pathognomonic woodworm channel-like structures (Fig. 5.17) [51–53].

As stated for onychopapilloma, onychomatricoma may present with LM and at the same time maintain the more classic features previously mentioned [50,54,55]. LM has been described for this tumor, from clinical examination; yet to our knowledge, very few dermatoscopic descriptions have been published [56]. Pigmented onychomatricoma should therefore also be included in the differential of nail unit melanoma.

Viral Warts

To date, no specific studies have been published regarding dermatoscopic findings for periungual or subungual warts. Previous studies concerning plantar warts, [57,58] however, suggest that identifying characteristic thrombosed capillaries may be aided by this technique, especially in the case of subclinical lesions (Fig. 5.18).

Bowen's Disease and Squamous Cell Carcinoma

A plethora of nail plate and periungual changes may all be either clinically or dermatoscopically appreciated

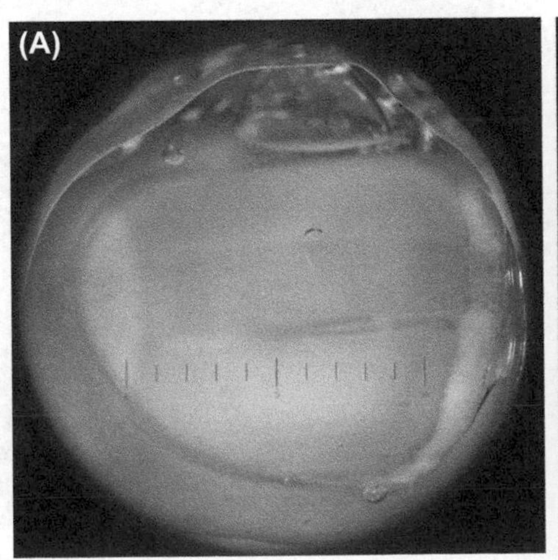

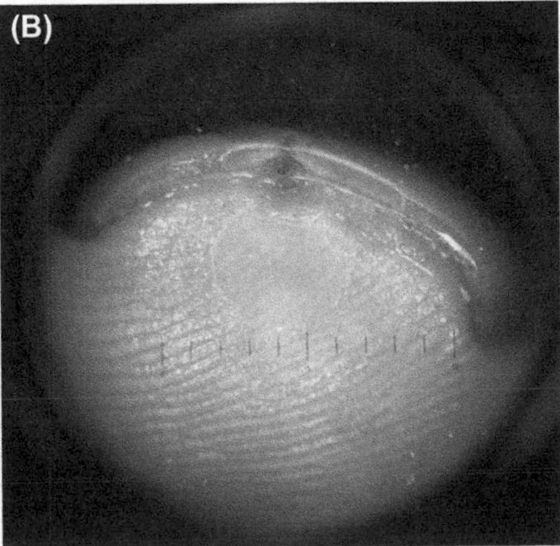

FIGURE 5.16 (A and B) Onychopapilloma. A solitary, distal linear segment of longitudinal erythronychia ending above a hyperkeratotic papule located immediately beneath the adjacent free edge is the typical presentation of onychopapilloma.

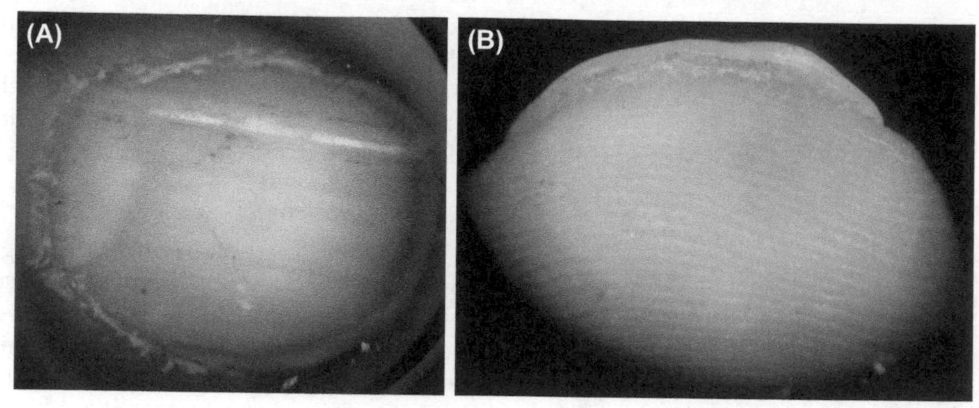

FIGURE 5.17 Onychomatrichoma. (A) Longitudinal xanthonychia and splinter hemorrhages are observed on the nail plate. (B) Dermatoscopy of the free edge showing the pathognomonic woodworm channel-like structures.

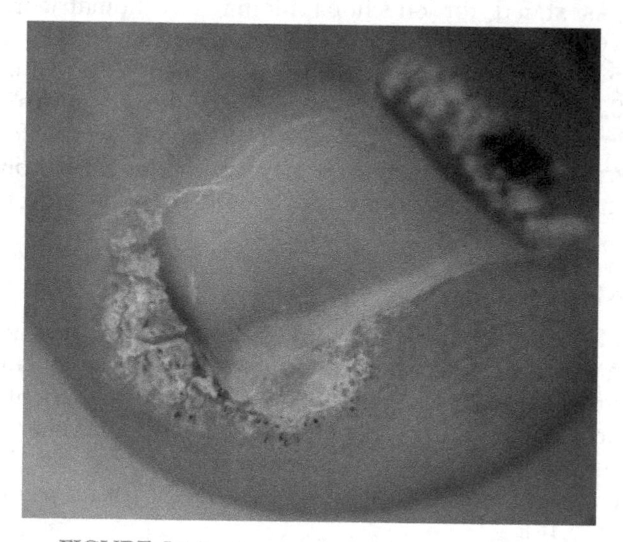

FIGURE 5.18 Dermoscopy of a periungual wart.

in the case of these conditions, including erythema, scaling, melanonychia, erythronychia, onycholysis, periungual verrucous growths, ulceration, and vascular changes [59,60]. Only very sporadic reports have addressed their onychoscopic features, so we speak from experience and on the basis of published data.

Although LM has long been described as a potential feature of BD of the nail bed or the periungual area [61], its specific dermatoscopic attributes remain to be described. Also, the LM of BD has been recorded to clinically bear close resemblance to melanoma [18,62,63]. Associated onycholysis and/or a warty growth may nevertheless be more suggestive of BD [50].

In advanced cases of squamous cell carcinoma (SCC), considerable damage to the nail will result in ulceration and oozing (see Fig. 5.7). In these cases the typical dermatoscopic patterns of SCC located on other sites may develop, namely brown points along imaginary lines (in the case of pigmented BD) or in glomerular or polymorphous vascular patterns for invasive SCC [49].

Nail Trauma, Foreign Bodies, and Artifacts

Acute or chronic trauma to the nail or surrounding tissues may give rise to subungual hematoma, splinter hemorrhages (see Subungual Hemorrhage) or onycholysis (see Onycholysis) [37,64–66]. Accurate observation of piercing injuries of the nail and external causes of artificial pigmentation can also be visualized via dermatoscopy (Fig. 5.19) [49].

Chronic exposure to occupational or environmental stressors such as water, chemicals, or antiseptics, the frequent use of nail polish, artificial nails, or nail polish removers can, depending on the case, alter the production of a healthy nail or interfere with the nail plate water content. The consequences are complaints of nail

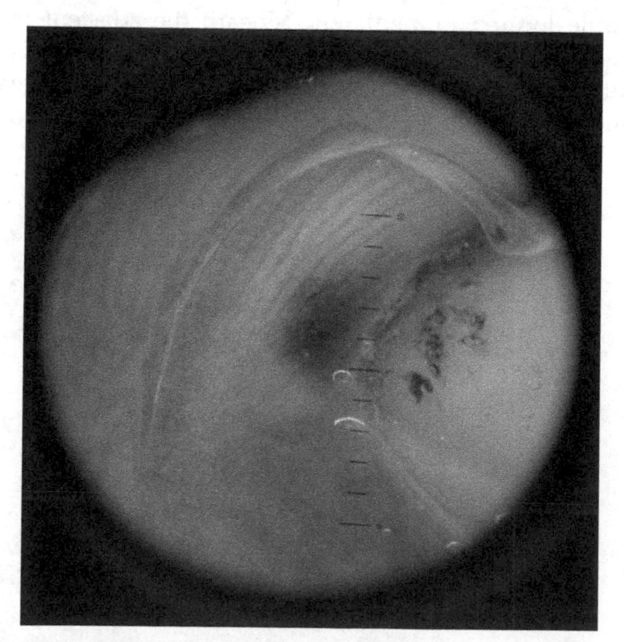

FIGURE 5.19 Exogenous pigmentation of the nail caused by purple varnish. Dermoscopy was done to observe the nearby nevus of the lateral nail fold.

brittleness, fragility, and splitting. Flaking nail layers in lamellar onychoschizia will stand out as transverse leukonychia, whereas longitudinal fissures, grooves, and distal splitting in onychorrhexis may appear slightly pigmented because of the presence of exogenous material [67].

Onychomycosis

The way in which fungi invade the nail plate and/or bed produces different clinical subtypes of onychomycosis and will evidently yield corresponding dermatoscopic pictures in white superficial onychomycosis, proximal subungual onychomycosis, distal subungual onychomycosis, endonix onychomycosis, and total dystrophic onychomycosis.

Irregularly distributed, white to yellow patches of different shapes and sizes on the nail plate surface is the main finding of superficial white onychomycosis (Fig. 5.20) [67]. Subclinical cognates are probably easier to detect with dermatoscopy.

Two findings have been reported to be both specific and sensitive for the diagnosis of distal subungual onychomycosis: a jagged edge with spikes and longitudinal striae [37,68]. When evaluating the proximal area of onycholysis of onychomycosis, sharp whitish longitudinal indentations called *spikes*, directed toward the proximal nail fold, can be observed (Fig. 5.21). Spikes correspond to the progressing edge of the dermatophyte infection, going along the nail bed longitudinal ridges. Longitudinal striation of different colors ranging from white to yellow, orange, or brown can be observed as well (see Fig. 5.11).

The detached nail plate may acquire different chromonychia (matte homogeneous white, yellow, orange, brown, or gray-black), subungual debris may be observable underneath the distal free edge and subungual hemorrhages may be present [37,69]. These are not yet proven to be specific for onychomycosis but may otherwise accompany the clinical picture.

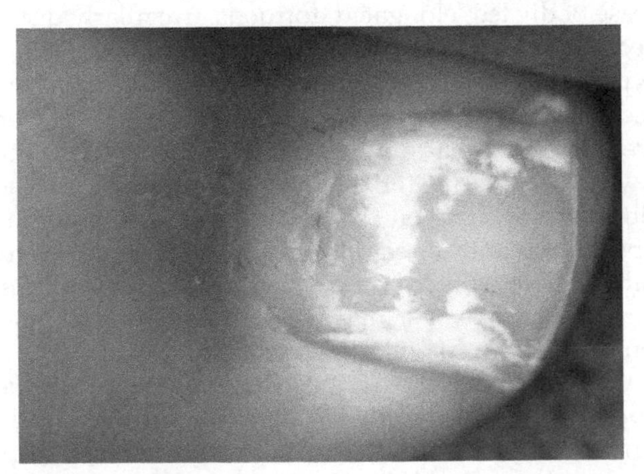

FIGURE 5.20 Superficial white onychomycosis. Irregularly distributed, *white to yellow patches* of different shapes and sizes on the nail plate surface.

FIGURE 5.21 Distal subungual onychomycosis. Dermoscopic image with *jagged edge with spikes* and *longitudinal striae*.

Fungal LM may be caused by the fungi *Scytalidium dimidiatum* or *Trichophyton rubrum* var. *nigricans*. A longitudinal brown to black band may even simulate melanoma [6,36]. However, other accompanying signs may less likely do so: scattered dots and globules caused by hemorrhage, or an isolated band or a few bands of distally wider LM [36,37,70]. In a review of 20 nails with fungal melanonychia, the following dermatoscopic criteria were summarized: (1) multicolored pigmentation (yellow, brown, gray, black, or red), (2) matte black pigment (lines, homogeneous areas, coarse granules, and/or pigment clumps), (3) black reverse triangles, (4) superficial transverse striation, and (5) blurred appearance of pigmentation under dermatoscopy [71]. Notably, the pigment, usually affecting the toes, may be accompanied by other signs of onychomycosis [50]. In the presence of fungal melanonychia, avulsion of the nail for matrix and nail bed dermatoscopy may render impossible the observation of any dermatoscopic structure [12].

Dermatoscopy can also accentuate the presence of dermatofitoma in difficult-to-treat onychomycoses. This mass of fungal and organic debris, enclosed within the nail plate above and nail bed below, appears as a patch of xanthonychia that does not touch the nail free edge (Fig. 5.22).

As disease progresses toward the stage of total dystrophic onychomycosis, some of these signs may change or disappear [69].

In summary, in the appropriate clinical setting, the presence of a jagged edge with spikes and longitudinal striae mandate the collection of a nail specimen for mycology, when readily available.

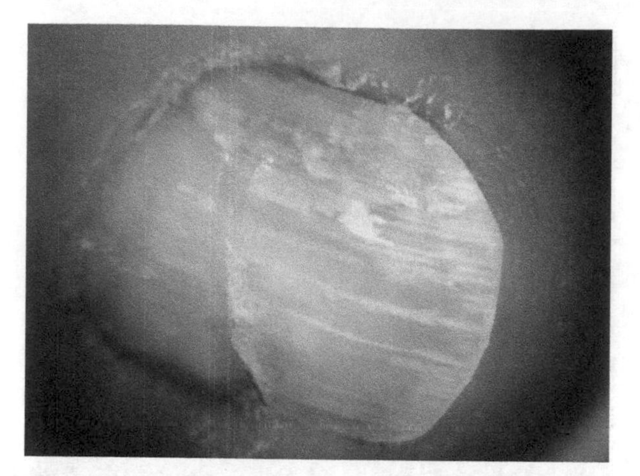

FIGURE 5.22 Dermatofitoma appears as a patch of xanthonychia that does not touch the nail free edge.

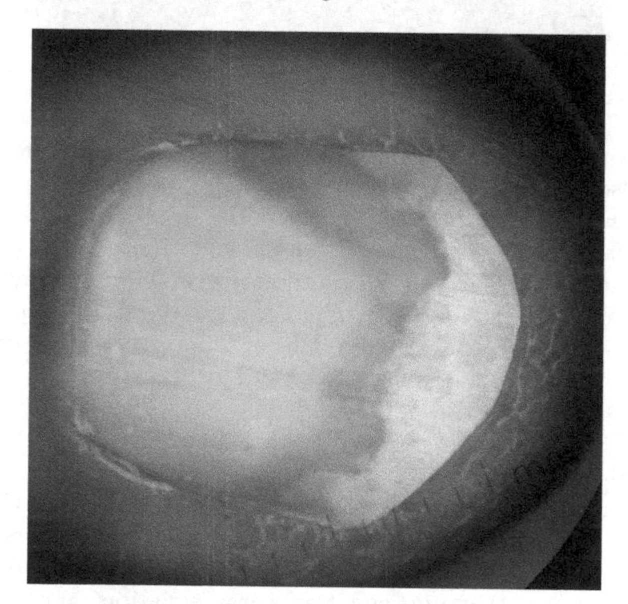

FIGURE 5.23 Onycholysis, salmon patch and oil spot.

Onycholysis

In contrast to onychomycosis, a linear edge confines the proximal area of onycholysis attributed to other causes (eg, inflammatory nail disease, traumatic onycholysis, or idiopathic onycholysis) (Fig. 5.23) [37]. This sign, along with absence of other diagnostic signs that indicate onychomycosis (eg, spikes), excludes the need for mycology. A nail afflicted by trauma may, in addition (and in adjacency) to onycholysis, present with subungual hematoma and/or splinter hemorrhage [37,64].

Inflammatory Nail Disease (Psoriasis, Eczema, and Lichen Planus)

Psoriasis

The main differentials for nail psoriasis are onychomycosis, idiopathic onycholysis (or onycholysis of other causes), and nail bed lichen planus.

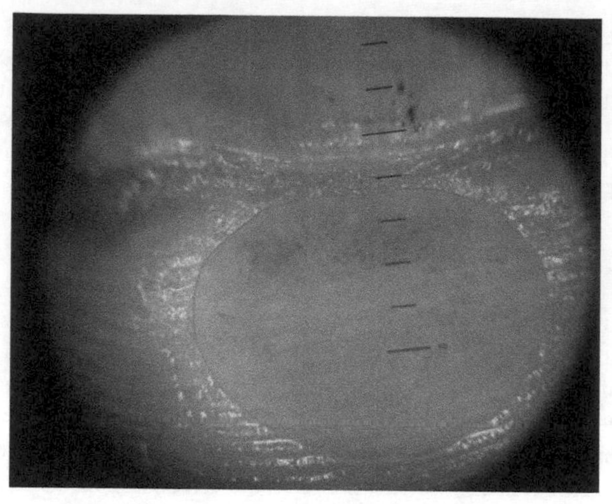

FIGURE 5.24 Dermoscopy in psoriasis reveals regular *red dots* of the hyponychium.

In order to rule out other diseases, the following signs of localized nail psoriasis can be readily observed with the use of a dermatoscope: pitting, nail crumbling, nail bed hyperkeratosis, oil spot, salmon patch, and splinter hemorrhages [35,67,72]. Psoriasis can afflict the nail bed, nail matrix, or both, and produce these different signs in accordance [73].

Irregular punctate depressions with a bright, whitish, circular border correspond to nail plate pits, a finding that has also been reported in patients with atopic dermatitis [49]. Subclinical salmon patches may be more easily rendered visible with the use of a dermatoscope (see Fig. 5.23). They are red or orange patches with irregular shape or size and an important finding is their presence in adjacency to an area of onycholysis [35]. Other than irregular pitting, salmon patch, and oil spots (see Fig. 5.23), the remaining signs are less specific for nail psoriasis.

Dermatoscopy of the hyponychium reveals an increase of dilated, elongated, tortuous, irregularly distributed hyponychial capillaries (Fig. 5.24) [35,74]. Hyponychial capillary numbers may also correlate with disease severity and response to treatment [74]. It is therefore interesting that dermatoscopy of the hyponychium may even be useful for diagnosis of mild nail psoriasis with discrete signs on the nail plate. It is the authors' experience that these capillaries may be observed as regular red dots with hand-held dermatoscopy. Their clear visualization, however, may best be aided by a magnification of at least ×40 with the videodermatoscope [74].

Yadav et al. have described dilated globose nail bed capillaries longitudinally arranged along the onychodermal band, surrounded by a prominent halo, in patients with psoriasis [72]. We have also seen this sign in a few patients, but further studies are required to characterize their relation with psoriasis.

Lichen Planus

Dermatoscopy may aid the visualization of nail lichen planus signs. Trachyonychia, longitudinal streaking, pitting, chromonychia (namely PLE), and nail plate fragmentation are observed in more than half of cases, whereas splinter hemorrhages, onycholysis, and subungual keratosis are less common (Figs. 5.25 and 5.26) [74]. A handful of reports have addressed the dermatoscopic changes of nail lichen planus [74,75]; however, the role of dermatoscopy in the diagnosis of nail lichen planus nonetheless requires further studies. Although some reports suggest that dermatoscopy can also aid in the monitoring of treatment response [64], again, further studies are required.

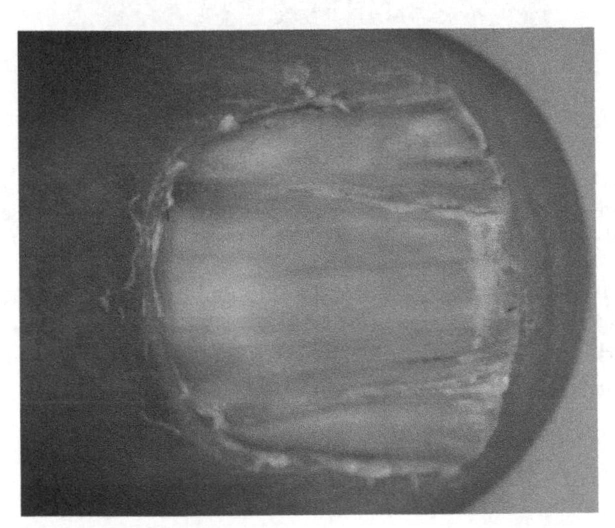

FIGURE 5.25 Nail lichen planus with longitudinal erythronychia and longitudinal streaking.

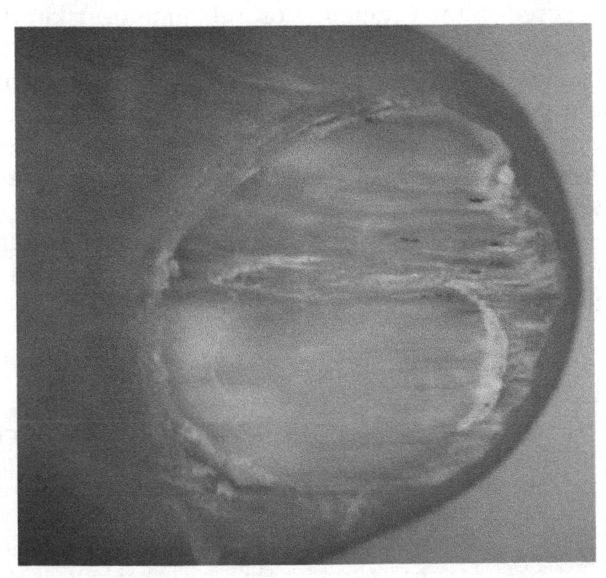

FIGURE 5.26 Nail lichen planus with splinter hemorrhages and onychorrhexis.

As previously stated, nail inflammatory disorders, specifically psoriasis and lichen planus, may be causes of LM [50].

In conclusion, dermatoscopy is a practical clinical adjuvant to the diagnosis and management of nail psoriasis and possibly lichen planus as well as eczema.

Connective Tissue Disease and Nail Fold Capillaroscopy

Standard microscope capillaroscopy has found its place in the diagnosis of several types of CTD, monitoring their progression and response to therapy.

Dermatoscopy has been considered equivalent to capillaroscopy with regard to identification of simple changes, although the latter may provide more detailed information and thus providing higher sensitivity changes [7,8,76]. Several studies have already established the use of dermatoscopy in the evaluation of CTD [7,8,77–79], and the detection of capillary changes of the proximal nail fold is yet another technique that may conveniently be performed with the handy portable dermatoscope [8,39,80]. Polarized or nonpolarized epiluminescence methods have been reported to be equivalent [81]. Videodermatoscopy has been reported to be useful in the distinction of patients with primary Raynaud phenomenon (ie, normal capillary pattern) from those with an associated CTD (who may exhibit capillary changes) [77].

In a healthy individual, a normal proximal nail fold capillary plexus will demonstrate an even, regular distribution of thin U-shaped vessel loops (Fig. 5.27). In contrast, reflecting the course of CTD, progressive

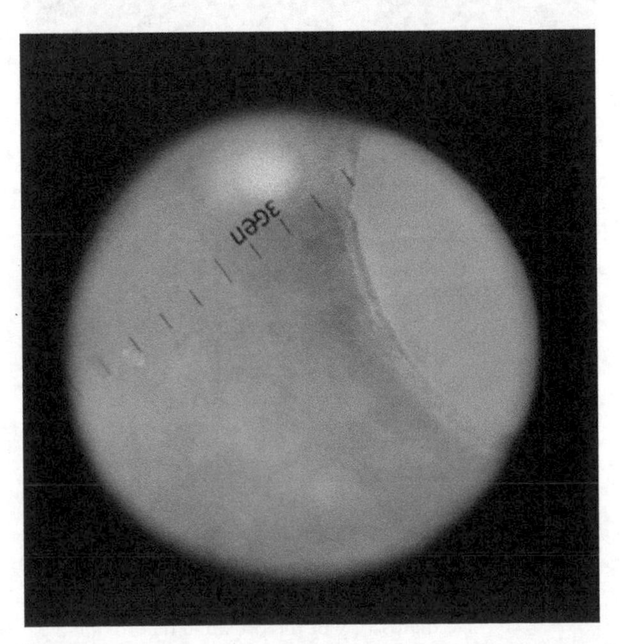

FIGURE 5.27 Dermatoscopy of the proximal nail fold in a healthy patient.

autoimmune vascular injury will lead to several changes in vasculature. Therefore correlation between capillary changes and disease activity can be assumed. Furthermore, even early local microvascular lesions can be either screened or monitored by dermatoscopy of the proximal nail fold capillaries [7,81,82].

The capillary pathological changes produced have been described in a pattern conventionally called *scleroderma–dermatomyositis* or *scleroderma pattern.* Classically, systemic sclerosis is responsible for such a pattern of changes. However, dermatomyositis, mixed CTD, undifferentiated CTD, and overlap syndromes also show similar alterations.

Generally, a scleroderma pattern will be present if two or more fingers demonstrate at least two of the following signs: capillary loss with irregular capillary distribution, enlargement of capillary loops, capillary shape changes (as part of an autoregulatory response), or areas of hemorrhage (from vessel wall damage) [78,83]. The morphological changes in the capillaries occur because of local tissue anoxia, as part of an autoregulatory response, and specific capillary changes include loss of at least two contiguous capillaries defining an avascular area. To compensate, new vascularization acquires tortuous, branching, and/or anastomosed shapes, and/or becomes either continuously/homogenously enlarged (termed *megacapillary*) or irregularly/restrictedly enlarged (termed *microaneurism*). Hemorrhage is viewed as dotted, linear, or blotted (Figs. 5.28 and 5.29) [84].

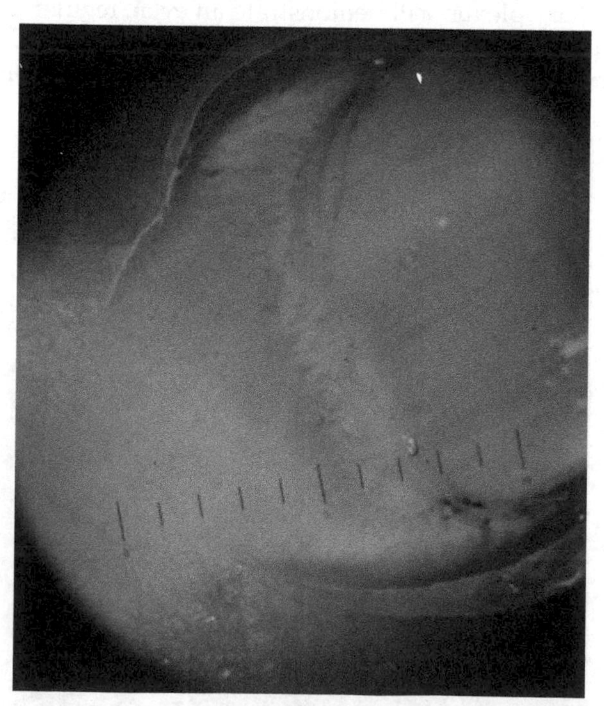

FIGURE 5.28 Dermatoscopy of the proximal nail fold in a patient with scleroderma. Dermoscopy reveals a scleroderma pattern with megacapillary, avascular area, and hemorrhage.

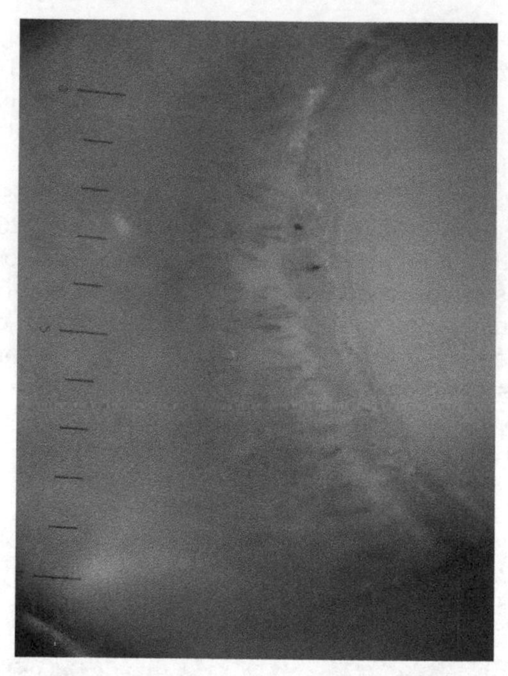

FIGURE 5.29 Dermatoscopy of the proximal nail fold in a patient with dermatomyositis. Dermoscopy reveals a scleroderma pattern with megacapillary, avascular area, and hemorrhage.

In accordance with published findings, videodermatoscopy or hand-held devices can be safely recommended for use in clinical screening, assessment, and follow-up observation of patients with CTD.

CONCLUSION

The widening uses of dermatoscopy have long reached beyond the evaluation of melanocytic skin lesions. Aside from other skin cancers, dermatoscopy is being used in the evaluation of inflammatory skin disease, hair disease, and nail diseases. In the latter case, it is a useful partner for an increasing number of nail disease experts for nail disease evaluation, treatment, and research. Specific training in this area may be in order for dermatology residents, whereas practicing dermatologists should become familiar with the most common and most important signs described, and now and then look at the nail through the scope of their own dermatoscopy device.

References

[1] Haneke E. Surgical anatomy of the nail apparatus. Dermatol Clin 2006;24(3):291–6.

[2] Fleckman P, Allan C. Surgical anatomy of the nail unit. Dermatol Surg 2001;27(3):257–60.

[3] de Berker D, Baran R. Science of the nail apparatus. In: Baran R, de Berker D, Holzberg M, Thomas L, editors. Baran & Dawber's diseases of the nails and their management. NY: Wiley-Blackwell; 2012. p. 1–50.

[4] Tosti A, Piraccini B, de Farias D. Dealing with melanonychia. Semin Cutan Med Surg 2009;28(1):49–54.

[5] Hirata S, Yamada S, Enokihara M, Di Chiacchio N, Ameida F, Enokihara M, et al. Patterns of nail matrix and bed of longitudinal melanonychia by intraoperative dermatoscopy. J Am Acad Dermatol 2011;65(2):297–303.

[6] Braun R, Baran R, Le Gal F, Dalle S, Ronger S, Pandolfi R, et al. Diagnosis and management of nail pigmentations. J Am Acad Dermatol 2007;56(5):835–47.

[7] Dogan S, Akdogan A, Atakan N. Nailfold capillaroscopy in systemic sclerosis: is there any difference between videocapillaroscopy and dermatoscopy? Skin Res Technol 2013;19(4):446–9.

[8] Baron M, Bell M, Bookman A, Buchignagni M, Dunne J, Hudson M, et al. Office capillaroscopy in systemic sclerosis. Clin Rheumatol 2007;26(8):1268–74.

[9] Micalli G, Lacarruba F, Massimino D, Schwartz R. Dermatoscopy: alternative uses in daily clinical practice. J Am Acad Dermatol 2011;64(6):1135–46.

[10] Braun R, Oliviero M, Kolm I, French L, Marghoob A, Rabinovitz H. Dermoscopy: what's new? Clin Dermatol 2009;27(1):26–34.

[11] Lencastre A, Lamas A, Sa D, Tosti A. Onychoscopy. Clin Dermatol 2013;31(5):587–93.

[12] Hirata S, Yamada S, Almeida F, Tomomori-Yamashita J, Enokihara M, Paschoal F, et al. Dermoscopy of the nail bed and matrix to assess melanonychia striata. J Am Acad Dermatol 2005;53(5):884–6.

[13] Hirata S, Almeida F, Enokihara M, Yamada S, Rosa I, Enokihara M, et al. Dermoscopic examination of the nail bed and matrix. Int J Dermatol 2006;45(1):28–30.

[14] Pinto-Gouveia M, Coutinho I, Vieira R, Gonçalo M, Cardoso J, Figueredo A. Immediate ex-vivo dermoscopy of a nail bed biopsy specimen: a useful procedure for melanonychia. J Eur Acad Dermatol Venereol 2014. http://dx.doi.org/10.1111/jdv.12783 [Epub ahead of print].

[15] Levit E, Kage M, Scher R, et al. The ABC rule for clinical detection of subungual melanoma. J Am Acad Derm 2000;42(2 Pt 1):269–74.

[16] Braun R, Baran R, Saurat J, Thomas L. Surgical pearl: dermoscopy of the free edge of the nail to determine the level of nail plate pigmentation and the location of its probable origin in the proximal or distal nail matrix. J Am Acad Dermatol 2006;55(3):512–3.

[17] Tosti A, Argenziano G. Dermoscopy allows better management of nail pigmentation. Arch Dermatol 2002;138(10):1369–70.

[18] Ronger S, Touzet S, Ligeron C, Balme B, Viallard A, Barrut D, et al. Dermoscopic examination of nail pigmentation. Arch Dermatol 2002;138(10):1327–33.

[19] Husain S, Scher R, Silvers D, Ackerman A. Melanotic macule of the nail unit and its clinicopathologic spectrum. J Am Acad Dermatol 2006;54(4):664–7.

[20] Thomas L, Dalle S. Dermoscopy provides useful information for the management of melanonychia striata. Dermatol Ther 2007; 20(1):3–10.

[21] Di Chiacchio N, Hirata S, Enokihara M, Michalany N, Fabbrocini G, Tosti A. Dermatologists' accuracy in early diagnosis of melanoma of the nail matrix. Arch Dermatol 2010;146(4):382–7.

[22] Sawada M, Yokota K, Matsumoto T, Shibata S, Yasue S, Sakakibara A, et al. Proposed classification of longitudinal melanonychia based on clinical and dermoscopic criteria. Int J Dermatol 2014;53(5):581–5.

[23] Koga H, Yoshikawa S, Sekiguchi A, Fujii J, Saida T, Sota T. Automated evaluation system of dermoscopic images of longitudinal melanonychia: proposition of a discrimination index for detecting early nail apparatus melanoma. J Dermatol 2014;41(10):867–71.

[24] Phan A, Dalle S, Touzet S, Ronger-Savlé S, Balme B, Thomas L. Dermoscopic features of acral lentiginous melanoma in a large series of 110 cases in white population. Br J Dermatol 2010;162(4): 765–71.

[25] Koga H, Saida T, Uhara H. Key point in dermoscopic differentiation between early nail apparatus melanoma and benign longitudinal melanonychia. J Dermatol 2011;38(1):45–52.

[26] John R, Izakovic J. Dermatoscopy/ELM for the evaluation of nail-apparatus pigmentation. Dermatol Surg 2011;27(3):315–22.

[27] Kawabata Y, Ohara K, Hino H, Tamaki K. Two kinds of Hutchinson's sign, benign and malignant. J Am Acad Dermatol 2001;44(2): 305–7.

[28] Di Chiacchio N, Farias D, Piraccini B, Hirata S, Richert B, Zaiac M, et al. Consensus on melanonychia nail plate dermoscopy. An Bras Dermatol March–April 2013;88(2):309–13.

[29] Miteva M, Fanti P, Romanelli P, Zaiac M, Tosti A. Onychopapilloma presenting as longitudinal melanonychia. J Am Acad Dermatol 2012;66(6):242–3.

[30] Lambiase M, Gardner T, Altman C, Albertini J. Bowen disease of the nail bed presenting as longitudinal melanonychia: detection of human papillomavirus type 56 DNA. Cutis 2003;72(4):305–9.

[31] Haas N, Henz B. Pitfall in pigmentation: pseudopods in the nail plate. Dermatol Surg 2002;28(10):966–7.

[32] Grazzini M, Rossari S, Gori A, Corciova S, Guerriero G, Lotti T, et al. Subungueal pigmented lesions: warning for dermoscopic melanoma diagnosis. Eur J Dermatol 2011;21(2):286–7.

[33] Phan A, Touzet S, Dalle S, Ronger-Savle S, Balme B, Thomas L. Acral lentiginous melanoma: a clinicoprognostic study of 126 cases. Br J Dermatol 2006;155(3):561–9.

[34] Farias D, Tosti A, Chiacchio N, Hirata S. Dermoscopy in nail psoriasis. An Bras Dermatol 2010;85(1):101–3.

[35] Finch J, Arenas R, Baran R. Fungal melanonychia. J Am Acad Dermatol 2012;66(5):830–41.

[36] Piraccini B, Balestri R, Starace M, Rech G. Nail digital dermoscopy (onychoscopy) in the diagnosis of onychomycosis. J Eur Acad Dermatol Venereol 2013;27(4):509–13. http://dx.doi.org/10.1111/j.1468-3083.2011.04323.x.

[37] Tessarotto L, Rubin G, Bonadies L, Valerio E, Cutrone M. Orange-brown chromonychia and Kawasaki disease: a possible novel association? Pediatr Dermatol 2015;32(3):e104–5. http://dx.doi.org/10.1111/pde.12529.

[38] Bergman R, Sharony L, Schapira D, Nahir M, Balbir-Gurman A. The handheld dermatoscope as a nail-fold capillaroscopic instrument. Arch Dermatol 2003;139(8):1027–30.

[39] Criscione V, Telang G, Jellinek N. Onychopapilloma presenting as longitudinal leukonychia. J Am Acad Dermatol 2010;63(3): 541–2.

[40] Piraccini B, Antonucci A, Rech G, Starace M, Misciali C, Tosti A. Onychomatricoma: first description in a child. Pediatr Dermatol 2007;24(1):46–8.

[41] Bel B, Jeudy G, Vabres P. Dermoscopy of longitudinal leukonychia in Hailey-Hailey disease. Arch Dermatol 2010;146(10):1204.

[42] Grossman M, Scher R. Leukonychia: review and classification. Int J Dermatol 1990;29(8):535–40.

[43] Cohen P. Longitudinal erythronychia, individual or multiple linear red bands of the nail plate: a review of clinical features and associated conditions. Am J Clin Dermatol 2011;12(4):217–31.

[44] Maehara Lde S, Ohe E, Enokihara M, Michalany N, Yamada S, Hirata S. Diagnosis of glomus tumor by nail bed and matrix dermoscopy. An Bras Dermatol 2010;85(2):236–8.

[45] Jellinek N. Longitudinal erythronychia: suggestions for evaluation and management. J Am Acad Dermatol 2011;64(1):167.

[46] Maes M, Richert B, Brasssinne M. Green nail syndrome or chloronychia. Rev Med Liege 2002;57(4):233–5.

[47] Chiriac A, Brzenzinski P, Foia L, Marincu I. Chloronychia: green nail syndrome caused by *Pseudomonas aeruginosa* in elderly persons. Clin Interv Aging 2015;10:265–7.

[48] Haenssle H, Brehmer F, Zalaudek I, Hofmann-Wellenhof R, Kreush J, Stolz W, et al. Dermoscopy of nails. Der Hautarzt 2014;65(4):301–11. http://dx.doi.org/10.1007/s00105-013-2707-x.

[49] Piraccini B, Dika E, Fanti P. Tips for diagnosis and treatment of nail pigmentation with practical algorithm. Dermatol Clin 2015; 33(2):185–95.
[50] Richert B, André J. L'onychomatricome. Ann Dermatol Venereol 2011;138(1):71–4.
[51] Di Chiacchio N, Tavares G, Padoveze E, Bet D, Di Chiacchio N. Onychomatricoma. Surg Cosmet Dermatol 2013;5(1):10–4.
[52] Tavares G, Chiacchio N, Chiacchio N, Souza M. Onychomatricoma: a tumor unknown to dermatologists. An Bras Dermatol 2015;90(2):265–7.
[53] Fayol J, Baran R, Perrin C, Labrousse F. Onychomatricoma with misleading features. Acta Derm Venereol 2000;80(5):370–2.
[54] Wynes J, Wanat K, Huen A, Mlodzienski A, Rubin A. Pigmented onychomatricoma: a rare pigmented nail unit tumor presenting as longitudinal melanonychia that has potential for misdiagnosis as melanoma. J Foot Ankle Surg 2015;54(4):723–5.
[55] Wanat K, Reid E, Rubin A. Onychocytic matricoma: a new, important nail-unit tumor mistaken for a foreign body. JAMA Dermatol 2014;150(3):335–7.
[56] Bae J, Kang H, Kim H, Park Y. Differential diagnosis of plantar wart from corn, callus and healed wart with the aid of dermoscopy. Br J Dermatol 2009;160(1):220–2.
[57] Lee D, Park J, Lee J, Yang J, Lee E. The use of dermoscopy for the diagnosis of plantar wart. J Eur Acad Dermatol Venereol 2009; 23(6):726–7.
[58] Thomas L, Zook E, Haneke E, Drapé J, Baran R. Tumors of the nail apparatus and adjacent tissues. In: Baran R, de Berker D, Holzberg M, Thomas L, editors. Diseases of the nails and their management. Oxford: John Wiley & Sons, Ltd; 2012. p. 657–61.
[59] Lecerf P, Richert B, Theunis A, André J. A retrospective study of squamous cell carcinoma of the nail unit diagnosed in a Belgian general hospital over a 15-year period. J Am Acad Dermatol 2013;69(2):253–61.
[60] Baran R, Simon C. Longitudinal melanonychia: a symptom of Bowen's disease. J Am Acad Dermatol 1988;18(6):1359–60.
[61] Sau P, McMarlin S, Sperling L, Katz R. Bowen's disease of the nail bed and periungual area: a clinicopathologic analysis of seven cases. Arch Dermatol 1994;130(2):204–9.
[62] Sass U, Andre J, Stene J, Noel J. Longitudinal melanonychia revealing an intraepidermal carcinoma of the nail apparatus: detection of integrated *HPV16* DNA. J Am Acad Dermatol 1998; 39(3):490–3.
[63] Piraccini B, Bruni F, Starace M. Dermoscopy of non-skin cancer nail disorders. Dermatol Ther 2012;25(6):594–602.
[64] Haenssle H, Blum A, Hofmann-Wellenhof R, Kreusch J, Stolz W, Argenziano G, et al. When all you have is a dermatoscope: start looking at the nails. Dermatol Pract Concept 2014;4(4):11–20.
[65] Bakos R, Bakos L. Use of dermoscopy to visualize punctate hemorrhages and onycholysis in "playstation thumb". Arch Dermatol 2006;142(12):1664–5.
[66] Tosti A, Piraccini B, de Farias D. Nail diseases. In: Dermatoscopy in clinical practice: beyond pigmented lesions. London: Informa Healthcare Ltd; 2010.
[67] Jesús-Silva M, Fernández-Martínez R, Roldán-Marín R, Arenas R. Dermoscopic patterns in patients with a clinical diagnosis of onychomycosis-results of a prospective study including data of potassium hydroxide (KOH) and culture examination. Dermatol Pract Concept 2015;5(2):39–44. http://dx.doi.org/10.5826/dpc.0502a05 [eCollection 2015].
[68] De Crignis G, Rezende P, Leverone A. Dermatoscopy of onychomycosis. Int J Dermatol 2014;53(2):e97–9. http://dx.doi.org/10.1111/ijd.12104.
[69] Haneke E. Pigmentations of the nails. Pigment Disord 2014;1(5): 1–11. http://dx.doi.org/10.4172/2376-0427.1000136.
[70] Kilinc Karaarslan I, Acar A, Aytimur D, Akalin T, Ozdemir F. Dermoscopic features in fungal melanonychia. Clin Exp Dermatol 2015;40(3):271–8.
[71] Yadav T, Khopkar U. Dermoscopy to detect signs of subclinical nail involvement in chronic plaque psoriasis: a study of 68 patients. Indian J Dermatol 2015;60(3):272–5.
[72] Baran R. How to diagnose and treat psoriasis of the nails. Presse Med 2014;43(11):1251–9.
[73] Iorizzo M, Dahdah M, Vicenzi C, Tosti A. Videodermoscopy of the hyponychium in nail bed psoriasis. J Am Acad Dermatol 2008; 58(4):714–5.
[74] Nakamura R, Broce A, Palencia D, Ortiz N, Leverone A. Dermatoscopy of nail lichen planus. Int J Dermatol June 2013;52(6):684–7.
[75] Nakamura R, Costa M. Dermatoscopic findings in the most frequent onychopathies: descriptive analysis of 500 cases. Int J Dermatol 2012;51(4):483–5.
[76] Hughes M, Moore T, O'Leary N, Tracey A, Ennis H, Dinsdale G, et al. A study comparing videocapillaroscopy and dermoscopy in the assessment of nailfold capillaries in patients with systemic sclerosis-spectrum disorders. Rheumatology (Oxford) 2015;54(8): 1435–42.
[77] Beltrán E, Toll A, Pros A, Carbonell J, Pujol R. Assessment of nailfold capillaroscopy by ×30 digital epiluminescence (dermoscopy) in patients with Raynaud phenomenon. Br J Dermatol 2007;156(5): 892–8.
[78] Hasegawa M. Dermoscopy findings of nail fold capillaries in connective tissue diseases. J Dermatol 2011;38(1):66–70. http://dx.doi.org/10.1111/j.1346-8138.2010.01092.x.
[79] Ohtsuka T. Dermoscopic detection of nail fold capillary abnormality in patients with systemic sclerosis. J Dermatol 2012;39(4): 331–5.
[80] Bauersachs R. The poor man's capillary microscope. a novel technique for the assessment of capillary morphology. Ann Rheum Dis 1997;56(7):435–7.
[81] Mazzotti N, Bredemeier M, Brenol C, Xavier R, Cestari T. Assessment of nailfold capillaroscopy in systemic sclerosis by different optical magnification methods. Clin Exp Dermatol March 2014; 39(2):135–41.
[82] Muroi E, Hara T, Yanaba K, Ogawa F, Yoshizaki A, Takenaka M, et al. A portable dermatoscope for easy, rapid examination of periungual nailfold capillary changes in patients with systemic sclerosis. Rheumatol Int 2011;31(12):1601–6.
[83] Maricq H. Wide-field capillary microscopy. Arthritis Rheum 1981; 24(9):1159–65.
[84] Gallucci F, Russo R, Buono R, Acampora R, Madrid E, Uomo G. Indications and results of videocapillaroscopy in clinical practice. Adv Med Sci 2008;53(2):149–57.

CHAPTER

6

Optical Coherence Tomography for Skin Cancer and Actinic Keratosis

L. Themstrup, G.B.E. Jemec

Zealand University Hospital, Roskilde, Denmark

OUTLINE

INTRODUCTION

Common keratinocyte skin cancers such as basal cell carcinoma (BCC) and squamous cell carcinoma (SCC) constitute the overwhelming majority of nonmelanocytic skin cancers (NMSCs), and are the most common cancer forms in fair-skinned people. The incidence and lifetime risk of developing skin cancer has been steadily increasing over time, and are reaching epidemic proportions [1–3]. Ultraviolet (UV) light irradiation from the environment is a well-known class I carcinogen and societal trends promoting UV exposure are thought to play a major role in creating the NMSC epidemic [4,5]. These trends include aesthetic preferences for tanning, increased leisure time, and more affordable opportunities for acquiring a tan either through tanning salons or through low-cost trips to sunny holiday destinations [6].

The concept of field cancerization is well established and particularly relevant to skin cancer. The irradiation of large areas of skin may cause focal mutations at different points in time, leading to the development of dysplastic or neoplastic changes [7]. These subsequently may or may not develop into clinically apparent skin cancers. Taken all together, an average skin cancer patient will present a spectrum of changes at any given point in time, with some patches of cells containing UV-induced mutations, some spots of subclinical dysplasia, and some areas of clinical dysplasia as well as the lesions of skin cancer that brought the patient to the attention of the dermatologist. Therefore it is rational to change the therapy offered from being exclusively targeted at clinically apparent skin cancer to a therapy that also target subclinical lesions [8].

In order to achieve this goal, a number of topical medical treatments became available. These include topical application of antimetabolites (5-flouuracil cream), stimulants of the immune system (imiquimod and diclofenac), photosensitizers (aminolevulinic acid), and empirically identified agents in which the exact mechanism remains to be described (ingenol mebutate) [9]. These treatments share the ability to treat subclinical lesions, dysplastic lesions, and most superficial neoplastic

Imaging in Dermatology
http://dx.doi.org/10.1016/B978-0-12-802838-4.00006-6

lesions effectively with the added benefit of a good cosmetic result. The cosmetic aspect cannot be underestimated, because these tumors are most often located on the face and other visible areas of the body. With the exception of photodynamic therapy using photosensitizers, the treatments are self-administered by the patient after the initial diagnostic and planning consultation, and may therefore help ease the burden of treating skin cancer. Because the topical treatments are approved for dysplastic lesions and thin BCCs only, diagnosis procedures should be performed before choosing one of these treatments. Diagnostic biopsies are always possible, but have to be directed at a visible lesion and often leave scars.

The management of this combination of a skin cancer epidemic and the availability of nonsurgical treatment options therefore can benefit from the introduction of a noninvasive diagnostic method that is able to help diagnose and delineate lesions, identify subclinical lesions, and accurately assess their thickness. One such method is optical coherence tomography (OCT) [10].

OPTICAL COHERENCE TOMOGRAPHY

OCT is a macrooptical imaging modality using light–tissue interaction for generating images. In its simplest form, the OCT system consists of a light source, a beam splitter, and a detector, also known as the *Michelson interferometer*. OCT measures reflected or backscattered light from tissue by correlating it with light that has traveled a known reference path [11]. From the resulting interference signal, one can derive the reflectivity profile along the beam axis. Repeating these measurements across three dimensions allows high-resolution two- and three-dimensional mapping of the tissue [12,13]. OCT was initially introduced in the clinical field of ophthalmology and the first in vivo OCT image was presented in 1993 [11]. In 1997 OCT was introduced in dermatology and since then several manufactures of OCT systems have made OCT imaging for dermatological purposes commercially available (eg, Telesto (Thorlabs Inc., Newton, New Jersey, United States), VivoSight (Michelson Diagnostics Ltd, Maidstone, Kent, UK), and SKINTELL (Agfa HealthCare, Mortsel, Belgium)). The OCT systems can create in vivo cross-sectional and en face images of skin with an axial resolution of 3–5 μm and a lateral resolution of 3–8 μm, depending on the system. Usually authors distinguish between conventional OCT (higher penetration, lower resolution) and high-definition OCT (HD-OCT) with lower penetration and higher resolution. Generally light can penetrate deeper into tissue at longer wavelengths, although it depends on the scattering and absorbing properties of the tissue. For OCT systems, the center wavelength of the light source has to be such that it allows maximal imaging depth in the highly scattering skin tissue. Because longer wavelengths come at the expense of lower resolution, most OCT systems operate at wavelengths of around 900–1300 nm (near-infrared spectrum), making it possible to reach skin imaging depths of up to 2 mm (highly dependent on the system and the image resolution) and thus placing OCT in the imaging gap between reflective confocal microscopy (RCM) and high-frequency ultrasound (HFUS) [11,14].

One of the advantages of OCT compared with other skin imaging modalities is that it produces real-time, dynamic images of the skin in a very short time; usually the imaging itself takes less than 30 seconds. It also has a rather large field of view, ranging from 1.8 × 1.5 mm to 10 × 10 mm, depending on the system used. The possibility to evaluate OCT images in a cross-sectional view makes it easier to compare it with histology sections [15] but in contrast to biopsies and histological evaluations, OCT is noninvasive, allowing the tissue to be inspected in vivo without inducing trauma to the skin and so the skin morphology stays unaltered during the examination. This also means that the exact same skin area can be evaluated repeatedly over time, which is especially useful in follow-up evaluations [14].

During the last decade OCT has advanced from being an interesting scientific tool in the laboratory to being a useful bedside tool for supplementing the clinical diagnosis and also for treatment monitoring. These advances are the result of great technical development of the OCT systems that in turn has led to higher image quality, better design, smaller hand-held probes, and also more compact and moveable systems that fit better in the clinic. The biggest potential for OCT in dermatology has thus far been in the diagnosing, delineating, and treatment monitoring of NMSC, especially BCCs. Pigmented lesions, on the other hand, continue to pose great challenges in OCT imaging, and in the diagnosis of malignant melanoma (MM) OCT is not as accurate as dermoscopy or RCM. Another limitation of OCT is the restricted imaging depth of maximum 2 mm that usually makes it inadequate in the evaluation of skin changes stretching beyond the reticular dermis. The limited tissue penetration depth of OCT is of particular concern when it comes to the evaluation and delineation of nonsuperficial malignant skin tumors.

THE SQUAMOUS SPECTRUM

In contrast to BCC and melanoma, the spectrum of SCC covers the full range from local patches of dysplastic cells to actinic keratosis (AK) and

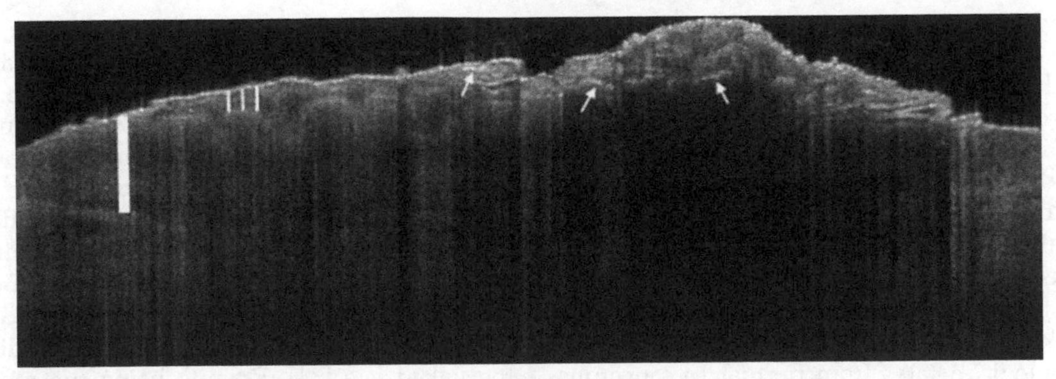

FIGURE 6.1 (Image measurements 6 × 2 mm) Conventional cross-sectional optical coherence tomography (OCT) image acquired by the VivoSight OCT scanner (Michelson Diagnostics, Kent, UK). Actinic keratosis (AK) located on the trunk showing disruption of normal layering and thickening of epidermis *(thick white bar)*. A wide hyperkeratotic area is demarcated at the surface *(thin white bars)* and white streaks are seen in epidermis *(white arrows)*.

full-blown SCC. The histological similarity over the range of these different lesions is great, and some authors have argued that all the lesions may be seen as SCC, albeit with differing potentials for morbidity [16]. These tumors can clinically be described as hyperkeratotic and erythematous to varying degrees. At one end of the clinical spectrum, the mildest form of AK is more often felt as a surface irregularity and hyperkeratosis than seen, whereas the full-blown SCC may present as an indurated hyperkeratotic ulcerating tumor easily felt and seen. Therefore these tumors pose a diagnostic challenge and lead to numerous biopsies. The clinical diagnostic accuracy of dysplastic lesions (AK) versus cancer (SCC) depends on the setting, the background, and the training of the observer, but rarely range above 50% [17–22]. The clinical diagnostic accuracy for intraepithelial changes such as Bowen disease is generally lower.

In addition to the clinical uncertainty regarding the diagnosis of SCC versus AK, there is histological uncertainty in some cases either because of the passing of information and specimens to and from the pathologist or, more rarely, because of discrepancies in interpreting the histopathology [23,24]. Histologically these tumors (AK and SCC) are characterized by a disorderly arrangement of dysplastic keratinocytes with large, hyperchromatic nuclei. Mitosis and apoptotic or dyskeratotic cells are seen throughout the thickened epidermis. Involvement of hair follicles is commonly observed, and the tumors are often associated with a dense chronic inflammatory infiltrate of a basophil-degenerated dermis. The invasion by the tumor into the dermis through the basement membrane is the diagnostic hallmark of SCC. The leading edge of the invasion is usually associated with a stromal reaction and most easily recognizable in poorly differentiated tumors. Histologically, tumor differentiation and perineural invasion are important prognostic factors for SCCs [25].

Optical Coherence Tomography Imaging of Actinic Keratosis

The hyperkeratotic nature of AK has negative effects on the image quality in OCT because of additional superficial reflective particles. Images may therefore be blurred in parts, limiting sensitivity and specificity in a clinical setting. Compared with normal skin, AK presents as a thickened epidermis, often with alternating hypertrophy and atrophy [26,27]. The increased thickness of AK is associated with disease severity, and thinning occurs after appropriate treatment [27,28]. Disrupted layering is seen, and usually the epidermis contains hyperreflective steaks/dots, possibly indicating inflammation or acantholysis (Fig. 6.1). Occasionally ulcerations may be seen.

HD-OCT images can be acquired in greater detail because of a higher lateral resolution. The HD-OCT images are displayed in en face view and do, in some ways, resemble images obtained with reflectance confocal microscopy. In AK, the normal honeycomb pattern of the cells in the basal, spinous, and granular layers becomes irregular with cells varying in size, shape, and reflectivity. Furthermore, the superficial parts of the adnexae can be visualized. Follicular infundibulum appears as a central dark hole surrounded by a perifollicular hyporeflective band. This band is encircled by a thin, dark ring. In AK, adnexal involvement can be seen as a disappearance of these concentric rings [27].

Imaging of Squamous Cell Carcinoma

There is a gradual transition in OCT as well as in histology between AK and SCC, with the previously mentioned features becoming progressively more prominent. Larger studies are lacking, but a study of Boone et al. suggests that loss of a clearly outlined DEJ, pronounced variation between hypertrophic/atrophic epidermis, and intralesional acantholysis (seen as dots) are indicative of SCC over AK. In HD-OCT particularly the degree of honeycomb atypia and absence of concentric rings surrounding adnexal ostia are important [29]. A pilot study investigating the use of OCT in Mohs surgery indicated good correlation between the histology and the OCT results [30].

Similar qualitative morphological features have previously been used to study the use of OCT for cancers of the uterine cervix [31]. An early study focused on the DEJ, the epithelial thickness, and the reflectivity of the tissue in a range of pathological states from normal to cancerous. On the mucosa, a sharp DEJ was associated with normal tissue, whereas increasing stromal reflectivity, swelling of the epidermal and stromal layers, and loss of an identifiable DEJ were associated with increasing levels of dysplasia and malignancy. OCT has also been tested for noninvasive diagnosis of vulvar intraepithelial neoplasia (a precancerous condition that can progress to vulvar SCC). When compared with normal tissue, differences both in thickness and the attenuation coefficient of the tissue containing vulvar intraepithelial neoplasia were identified by OCT [32].

The current data suggest that OCT may be able to identify morphological features of the spectrum from AK to SCC that would enable its use in diagnosis and management of SCC and its differential diagnosis, but additional large-scale prospective studies are necessary to confirm the preliminary results and establish the role of functional measures available with this technology.

BASAL CELL CARCINOMA

BCC is the most prevalent cancer in the Caucasian population. Clinically and histologically, BCCs can be divided into three most common subtypes: nodular BCC, superficial BCC, and infiltrative BCC (including morpheic and sclerosing BCC). Treatment decisions are based on the subtype.

BCC lesions most often occur on the head and neck; however, superficial BCCs differ from other subtypes of BCC by occurring more commonly on the trunk and in younger patients [33]. Clinically, nodular BCC typically presents as a smooth, semitranslucent, pinkish papule or nodule with pearly borders. Central ulceration and arborizing telangiectasias may be visible to the naked eye. Nodular BCCs are usually clinically distinct, but differential diagnoses may include other tumors, such as dermal nevus, sebaceous hyperplasia, epidermoid cyst, compound nevi, SCC, and amelanotic melanoma [34]. Nonpigmented superficial BCC presents clinically as a flat erythematous macule or patch with a slightly raised outer edge and a shiny surface with or without scale. Other characteristics can include the presence of multiple small ulcerations or blood crusts on the surface of the lesion. The clinical diagnosis of superficial BCC can be challenging because it resembles many other pink lesions of the skin, including Bowen disease, AK, pink lichenoid keratosis, tinea corporis, or inflammatory lesions like eczema and psoriasis [34]. Infiltrative BCC is the least common of the mentioned subtypes and clinically presents as a whitish, scarlike plaque with ill-defined edges. Infiltrative BCC easily resembles other lesions with atrophic and benign appearances, thereby making it challenging to diagnose with the naked eye [34]. Despite being the most accessible and used assessment, the diagnostic accuracy for the clinical examination of BCC is often not known. In a review of the literature, the overall sensitivity for clinical diagnosis of NMSC was found to be 56–90%, and specificity 75–90%, with highest values for BCC diagnosis [35]. Another study found the naked-eye sensitivity for diagnosing malignant neoplasms (including melanoma and NMSC) to be 70% [36].

The histological features of BCC include basaloid cells with scant cytoplasm (higher nucleus-to-cytoplasm ratio) and elongated hyperchromatic nuclei. A peripheral cell layer in which the nuclei form a palisade surrounds tumor masses, and peritumoral clefting and mucinous alteration of the surrounding stroma can be seen [37]. Each of the BCC subtypes has their own histological characteristics, including large tumor nodules in the dermis (nodular BCC), tumor nests growing multifocally from the epidermis (superficial BCC), and angulated narrow tumor nests growing in an infiltrative manner at the leading edge of the tumor (infiltrative BCC). The different histopathological subtypes of BCC are associated with different results and prognoses and are not reliably distinguishable by punch biopsy [24]. The infiltrative BCC subtype has poorly defined borders, making it difficult to delineate and it is generally considered to be the most aggressive subtype of BCC. Poor prognostic histological factors include infiltrative rather than expansile tumor growth, dense fibrous stroma, and loss of peripheral palisading [37,38].

The imaging of BCC is one of the prime examples of indications for OCT. Promising results have already been demonstrated in delineating these tumors, and also the diagnostic criteria for BCC have been established [11,35,39–42]. In normal skin, OCT can reliably identify the distinct layers of the skin usually down to the deep reticular dermis (depending on the imaged skin region and the OCT system used) and the DEJ appears as an intact narrow hyporeflective line (Fig. 6.2).

In BCC lesions, loss of normal skin architecture is an overall finding. The specific OCT characteristics of BCC include alteration of the DEJ and dark ovoid areas in the dermis (basal cell nests) surrounded by a white halo (stroma), sometimes referred to as a *honeycomb pattern.*

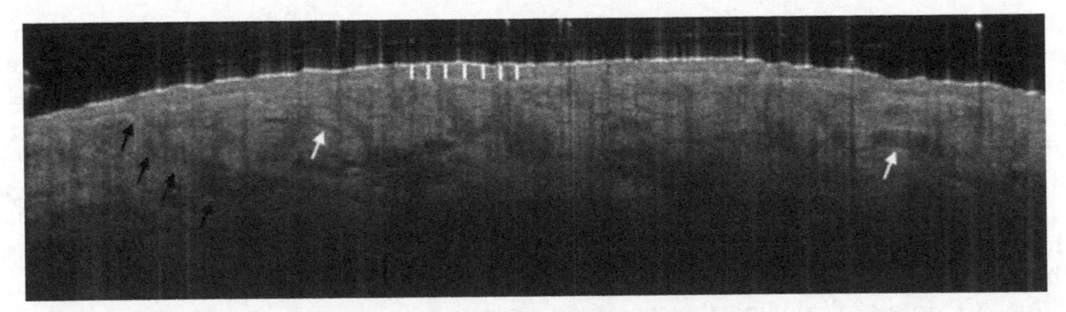

FIGURE 6.2 (Image measurements: 6 × 1.3 mm) Skin imaging depth 0.9 mm. Conventional cross-sectional optical coherence tomography (OCT) image acquired by the VivoSight OCT scanner (Michelson Diagnostics, Kent, UK). Normal skin located on the chin. The layering is intact and the dermo-epidermal junction (DEJ) is seen as an unbroken, fine, hyperreflective line at the interface between the epidermis *(white bars)* and dermis. The hyporeflective diagonal line corresponds to a hair follicle and the hair shaft protruding through the epidermis *(black arrows)*. Vessels are marked by *white arrows*.

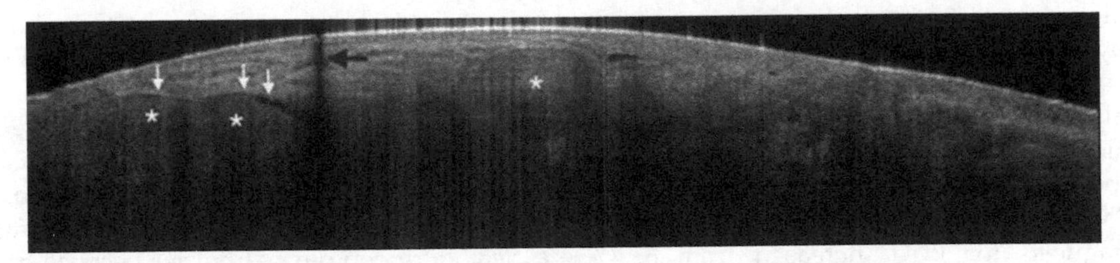

FIGURE 6.3 (Image measurements: 6 × 1.4 mm) Skin imaging depth 0.9 mm. Conventional cross-sectional optical coherence tomography (OCT) image acquired by the VivoSight OCT scanner (Michelson Diagnostics, Kent, UK). Nodular basal cell carcinoma located on the cheek showing disruption of normal layering and hyporeflective oval structures *(white asterisks)* corresponding to tumor islands. An unreflective line is seen bordering the tumor islands *(white arrows)* corresponding to mucinous clefting/peripheral palisading often recognized in histology sections. The *black arrow* marks a hair casting shadow.

Cellular palisading/peritumoral clefting at the margins of basal cell nests is often seen as a low-intensity OCT signal at the periphery of the cell nests (Fig. 6.3). This is also seen in en face view (Fig. 6.4).

Secondary features include absence of normal hair follicles and glands and altered dermal capillaries directed toward the basaloid cell islands [42]. Several studies have shown the correlation of OCT morphology with histology for several types of skin tumors, although earlier studies found it difficult to differentiate between BCC subtypes [39,41,43–46]. HD-OCT has an increased lateral resolution (at the cost of penetration depth) and may permit differentiation of BCC subtypes [47]. The features described in HD-OCT can be visualized only superficially in the lesion and include the combination of distinct lobular organization, a dominant vascular pattern in the papillary plexus, and the presence/absence of a stretching effect on the stroma.

As mentioned earlier, OCT provides relatively high accuracy in distinguishing lesions from normal skin, which is of great importance in identifying tumor borders. In differentiating normal skin from NMSC lesions, a sensitivity of 79–94% and specificity of 85–96% was found for OCT [48]. Looking specifically at the diagnostic accuracy of OCT in identifying BCC, recent extensive studies have investigated this aspect [49,50]. By

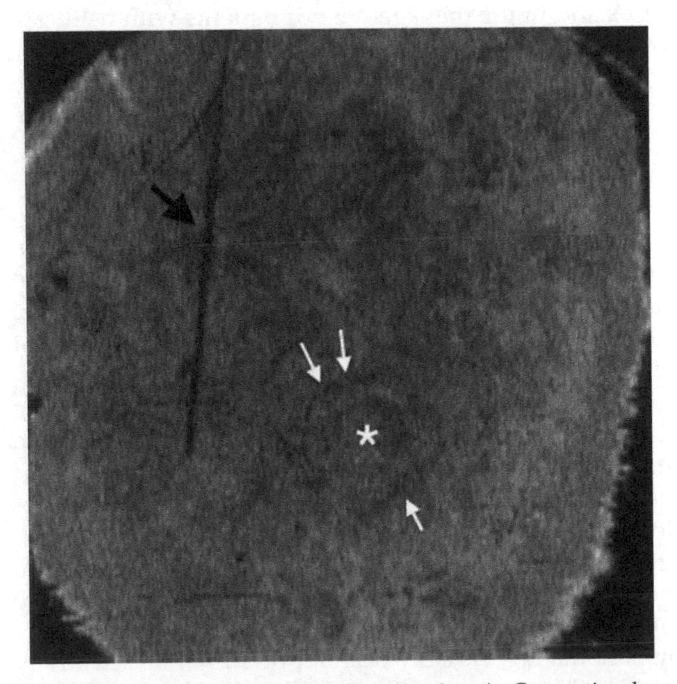

FIGURE 6.4 (Image measurements 6 × 6 mm). Conventional en face optical coherence tomography (OCT) image acquired by the VivoSight OCT scanner (Michelson Diagnostics, Kent, UK). The image shows the same nodular basal cell carcinoma (BCC) lesion as in Fig. 6.1 but in an en face view. A hyporeflective structure *(white asterisk)* bordered by an unreflective rim *(white arrows)* corresponds to a BCC tumor island and mucinous clefting. The *black arrow* marks a hair casting shadow.

using a specific scoring system (Berlin score) based on five predetermined diagnostic OCT criteria, Wahrlich et al. found that the sensitivity and specificity for multi-beam OCT amounted to 96.6% (95% CI 80.4–99.8) and 75.2% (95% CI 52.5–90.9) when evaluated by a dermatology specialist familiar with OCT. Meaning that 88% of all BCC diagnoses based on OCT were correctly classified, confirmed by histopathology [49]. In a multicenter study by Ulrich and Mayer et al., the diagnostic value of OCT for BCC in a typical clinical setting was investigated. One hundred and fifty-five patients with 235 nonpigmented pink lesions suspicious for BCC underwent clinical assessment, dermoscopy, OCT, and biopsy/histological examination, with the diagnosis recorded at each stage. The results showed that sensitivity was high for all three techniques, increasing from 90.0% by clinical examination only to 95.7% with the addition of OCT. However, there was a marked and statistically significant increase in specificity from 28.6% by clinical assessment to 75.3% with the addition of OCT. The positive predictive value and negative predictive value were greatest for OCT, and overall the accuracy of diagnosis for BCC increased from 65.8% (clinical examination alone) to 87.4% with the addition of OCT [50]. The authors of this study emphasize that OCT should not be used as a replacement for clinical examination and dermoscopy, but that OCT is best utilized as an adjunct noninvasive tool especially in difficult cases and in the management of patients with field cancerization or a large number of suspicious skin tumors.

PIGMENTED NEVI AND MELANOMA

Pigmented nevi are the most common melanocytic tumors and appear clinically as a papule or macule, tan-brown, uniformly pigmented, and usually small (6 mm or less). Caucasian people averagely have 20–30 nevi. Compared with benign nevi, skin melanomas present clinical warning signs such as changes in color of pigmented lesion, enlargement of an existing mole, itching or pain in an existing mole, and irregular borders. Skin melanomas most commonly appear on the head and neck, back, and lower extremities. Clinically and histologically, melanomas can be divided into four most common subtypes: acral lentiginous, lentigo maligna, nodular, and superficial spreading. Diagnostic accuracy of naked eye examination for the diagnosis of primary melanoma has variably been reported with a sensitivity of 43–100% and a specificity of 71–94% [51]. Trained use of dermoscopy has been shown in meta-analyses to improve the diagnostic accuracy of melanoma unequivocally compared with naked eye examination and to reduce excision rates of benign melanocytic lesions in clinical trials [52]. Histologically benign pigmented nevi are composed of nevomelanocytic nests uniform in size and are classified according to the histological location of the nests. The architecture of the rete is regular and without bridging or fusion [53]. No single histological feature is pathognomonic for melanoma, but many characteristic features exist. Cytological atypia is almost always present, with enlarged cells containing large, pleomorphic, hyperchromic nuclei with prominent nucleoli, and numerous mitotic figures often are noted.

Thus far, only few studies have investigated OCT imaging of pigmented nevi and melanomas. The reason is that melanin is a strong scatterer of light; images of pigmented lesions like MM and nevi have proven more difficult to obtain by techniques based on the penetration of light [14]. Conventional OCT with an imaging resolution of 5–7 μm cannot visualize cellular features and therefore the image analysis of melanocytic lesions must rely on distinct morphological changes in the tissue. In benign nevi, an intact DEJ and acantholysis is often seen in OCT images. In comparison, MMs often show marked architectural disarray and rarely display a clear dermoepidermal border because of the infiltrative tumor growth [14,54]. The most characteristic feature seen in MM with conventional OCT is large, vertical, icicle-shaped structures that are not observed in the benign nevi; however, despite this finding, conventional OCT cannot demonstrate enough clear-cut differences between malignant and benign pigmented lesions to be used as a diagnostic tool for MM [54,55]. The same goes for the rare cases of amelanotic melanoma that clinically can present as a pink patch, plaque, or nodule and may resemble BCC. OCT criteria for amelanotic melanoma are largely unknown and further studies are required to determine whether differentiation between amelanotic melanoma and BCC is possible [50]. In superficial melanomas, not exceeding 2 mm in depth, conventional OCT may be able to assess the vertical tumor size. Compared with HFUS, OCT seems to be more exact as far as thickness determination of thin melanocytic skin lesions is concerned [56]. However, the opposite seems to be the case when HFUS thickness measurements of melanocytic lesions are compared with an OCT system using the wavelength of 930 nm [57]. For future developments of conventional OCT for imaging of melanocytic lesions, it may be feasible to combine OCT with functional methods to retrieve quantitative data on, eg, the attenuation coefficient or blood flow [58].

Compared with conventional OCT, HD-OCT has a high enough resolution to image the tissue on a cellular level but with a limited penetration depth complementary to that of RCM. The high resolution of HD-OCT allows detailed imaging of pathological changes in melanocytic lesions, and studies have suggested that HD-OCT can provide morphological imaging that can discriminate architectural patterns and cytological features of pigmented lesions and cells in the epidermis and superficial dermis [59,60]. With regard to diagnostic accuracy of HD-OCT in MM, a recent multicenter study employing one blinded investigator assessed the diagnostic performance of HD-OCT in the differentiation of benign melanocytic skin lesions and MM [61]. The study included 93 histopathologically proven melanocytic skin lesions, of which 27 were MMs. The sensitivity of HD-OCT was found to be 74.1% (95% CI 53.7–88.8%) and the specificity was 92.4% (95% CI 83.2–97.5%). The positive predictive value was 80% and the negative predictive value was 89.7%. The performance of HD-OCT was dependent on tumor thickness and the presence of borderline lesions, indicated by high false negative rates in very thin melanomas and high false positive rates in dysplastic nevi. In conclusion, the performance in diagnosing MM by OCT is still inferior to other competing techniques such as RCM, although recent studies have provided some encouragement [61]. To optimize early diagnosis of MM, OCT may develop into a valuable adjunct tool; however, further technical development and more extensive studies of OCT are required before this can be achieved.

OTHER SKIN TUMORS

Lymphoma

The main focus of the limited literature on OCT imaging of other skin cancers has been on cutaneous lymphomas. The recognition of these cancers is often delayed and dependent on multiple biopsies. Noninvasive diagnostics may therefore benefit patients, both as a primary diagnostic tool as well as a tool to direct biopsies in order to achieve the highest likelihood of a correct diagnosis.

A body of literature is accumulating in the field of ophthalmology, suggesting that specific locations within the complex anatomy of the eye and lesional patterns may indicate the presence of lymphoma [62,63]. In skin, the observations are far fewer but indicate that OCT can identify cellular accumulations corresponding to those seen in histological studies and where the likelihood of a positive biopsy is therefore greater [64,65]. The primary diagnostic accuracy of OCT in lymphoma patients has not been addressed.

Hemangiomas

Due to their high content of erythrocytes hemangiomas are most often recognizable on simple clinical examination, but may occasionally be a differential diagnosis to MM. OCT has been used to diagnose the presence of hemangiomas in the eye [66,67], but no large-scale systematic studies have been performed on skin OCT imaging. Limited reports suggest that vascular tumors appear as characteristic structures, which may allow positive identification of hemangiomas [68–70].

CONCLUSION

The development of skin imaging technologies is challenging not only because of the scattering and absorption properties of the skin but also because the ease and tradition for clinical inspection of the skin is a barrier to implementing new imaging techniques in the clinic. The increased use of noninvasive treatments, especially in skin cancer and the widespread request for close monitoring of patients, however, means that there can be great benefit from the introduction of noninvasive diagnostic methods. OCT fills the imaging gap between RCM and HFUS and provides high-resolution images and reasonable penetration depth combined with ease of use. In OCT imaging of BCC, the diagnostic criteria have already been established and recent extensive studies have shown good diagnostic accuracy. The limited use of OCT in pigmented lesions and the restricted imaging depth still pose challenges in OCT imaging, but it is speculated that the continued technological development can propel the method to a greater level of use in a variety of dermatological diseases.

Acknowledgments

The project has received funding from the European Union's ICT Policy Support Programme as part of the Competitiveness and Innovation Framework Programme. It reflects only the author's views and the European Union is not liable for any use that might be made of information contained therein.

References

[1] Perera E, Gnaneswaran N, Staines C, Win AK, Sinclair R. Incidence and prevalence of non-melanoma skin cancer in Australia: a systematic review. Australas J Dermatol 2015;56(4):258–67.

[2] Ahmad AS, Ormiston-Smith N, Sasieni PD. Trends in the lifetime risk of developing cancer in Great Britain: comparison of risk for those born from 1930 to 1960. Br J Cancer 2015;112(5):943–7.

[3] Rogers HW, Weinstock MA, Harris AR, et al. Incidence estimate of nonmelanoma skin cancer in the United States, 2006. Arch Dermatol 2010;146(3):283–7.

[4] monographs.iarc.fr/eng/classification. [last accessed 01.05.15].

[5] Chang C, Murzaku EC, Penn L, et al. More skin, more sun, more tan, more melanoma. Am J Public Health 2014;104(11):e92–9.

[6] Mogensen M, Jemec GB. The potential carcinogenic risk of tanning beds: clinical guidelines and patient safety advice. Cancer Manag Res 2010;2:277–82.

[7] Vanharanta S, Massague J. Field cancerization: something new under the sun. Cell 2012;149(6):1179–81.

[8] Philipp-Dormston WG. Field cancerization: from molecular basis to selective field-directed management of actinic keratosis. Curr Probl Dermatol 2015;46:115–21.

[9] Gupta AK, Paquet M. Network meta-analysis of the outcome 'participant complete clearance' in nonimmunosuppressed participants of eight interventions for actinic keratosis: a follow-up on a Cochrane review. Br J Dermatol 2013;169(2):250–9.
[10] Mogensen M, Thrane L, Jorgensen TM, Andersen PE, Jemec GB. OCT imaging of skin cancer and other dermatological diseases. J Biophotonics 2009;2(6–7):442–51.
[11] Gambichler T, Jaedicke V, Terras S. Optical coherence tomography in dermatology: technical and clinical aspects. Arch Dermatol Res 2011;303(7):457–73.
[12] Vakoc BJ, Fukumura D, Jain RK, Bouma BE. Cancer imaging by optical coherence tomography: preclinical progress and clinical potential. Nat Rev Cancer 2012;12(5):363–8.
[13] Dasgeb B, Kainerstorfer J, Mehregan D, Van Vreede A, Gandjbakhche A. An introduction to primary skin imaging. Int J Dermatol 2013;52(11):1319–30.
[14] Sattler E, Kastle R, Welzel J. Optical coherence tomography in dermatology. J Biomed Opt 2013;18(6):061224.
[15] Drakaki E, Vergou T, Dessinioti C, Stratigos AJ, Salavastru C, Antoniou C. Spectroscopic methods for the photodiagnosis of nonmelanoma skin cancer. J Biomed Opt 2013;18(6):061221.
[16] Ackerman AB, Mones JM. Solar (actinic) keratosis is squamous cell carcinoma. Br J Dermatol 2006;155(1):9–22.
[17] Green A, Leslie D, Weedon D. Diagnosis of skin cancer in the general population: clinical accuracy in the Nambour survey. Med J Aust 1988;148(9):447–50.
[18] Cooper SM, Wojnarowska F. The accuracy of clinical diagnosis of suspected premalignant and malignant skin lesions in renal transplant recipients. Clin Exp Dermatol 2002;27(6):436–8.
[19] Hallock GG, Lutz DA. A prospective study of the accuracy of the surgeon's diagnosis and significance of positive margins in nonmelanoma skin cancers. Plast Reconstr Surg 2001;107(4):942–7.
[20] Hillson TR, Harvey JT, Hurwitz JJ, Liu E, Oestreicher JH, Pashby RC. Sensitivity and specificity of the diagnosis of periocular lesions by oculoplastic surgeons. Can J Ophthalmol [J canadien d'ophtalmologie] 1998;33(7):377–83.
[21] Matteucci P, Pinder R, Magdum A, Stanley P. Accuracy in skin lesion diagnosis and the exclusion of malignancy. J Plast Reconstr Aesthet Surg 2011;64(11):1460–5.
[22] Zemelman V, Valenzuela CY, Fich F, Roa J, Honeyman J. Assessment of clinical diagnostic accuracy for skin cancer. Revista Med de Chile 2003;131(12):1421–7.
[23] Comfere NI, Sokumbi O, Montori VM, et al. Provider-to-provider communication in dermatology and implications of missing clinical information in skin biopsy requisition forms: a systematic review. Int J Dermatol 2014;53(5):549–57.
[24] Trotter MJ, Bruecks AK. Interpretation of skin biopsies by general pathologists: diagnostic discrepancy rate measured by blinded review. Arch Pathol Lab Med 2003;127(11):1489–92.
[25] LeBoeuf NR, Schmults CD. Update on the management of high-risk squamous cell carcinoma. Semin Cutan Med Surg 2011; 30(1):26–34.
[26] Banzhaf CA, Themstrup L, Ring HC, Mogensen M, Jemec GB. Optical coherence tomography imaging of non-melanoma skin cancer undergoing imiquimod therapy. Skin Res Technol 2014;20(2): 170–6.
[27] Boone MA, Norrenberg S, Jemec GB, Del Marmol V. Imaging actinic keratosis by high-definition optical coherence tomography. Histomorphologic correlation: a pilot study. Exp Dermatol 2013; 22(2):93–7.
[28] Schmitz L, Bierhoff E, Dirschka T. Optical coherence tomography imaging of erythroplasia of Queyrat and treatment with imiquimod 5% cream: a case report. Dermatology 2014;228(1):24–6.
[29] Boone MA, Marneffe A, Suppa M, et al. High-definition optical coherence tomography algorithm for the discrimination of actinic keratosis from normal skin and from squamous cell carcinoma. J Eur Acad Dermatol Venereol 2015;29(8):1606–15.
[30] Durkin JR, Fine JL, Sam H, Pugliano-Mauro M, Ho J. Imaging of Mohs micrographic surgery sections using full-field optical coherence tomography: a pilot study. Dermatol Surg 2014;40(3):266–74.
[31] Escobar PF, Belinson JL, White A, et al. Diagnostic efficacy of optical coherence tomography in the management of preinvasive and invasive cancer of uterine cervix and vulva. Int J Gynecol Cancer 2004;14(3):470–4.
[32] Wessels R, de Bruin DM, Faber DJ, et al. Optical coherence tomography in vulvar intraepithelial neoplasia. J Biomed Opt 2012; 17(11):116022.
[33] McCormack CJ, Kelly JW, Dorevitch AP. Differences in age and body site distribution of the histological subtypes of basal cell carcinoma: a possible indicator of differing causes. Arch Dermatol 1997;133(5):593–6.
[34] Giacomel J, Zalaudek I. Pink lesions. Dermatol Clin 2013;31(4): 649–78. ix.
[35] Mogensen M, Jemec GB. Diagnosis of nonmelanoma skin cancer/ keratinocyte carcinoma: a review of diagnostic accuracy of nonmelanoma skin cancer diagnostic tests and technologies. Dermatol Surg 2007;33(10):1158–74.
[36] Rosendahl C, Tschandl P, Cameron A, Kittler H. Diagnostic accuracy of dermatoscopy for melanocytic and nonmelanocytic pigmented lesions. J Am Acad Dermatol 2011;64(6):1068–73.
[37] Nakayama M, Tabuchi K, Nakamura Y, Hara A. Basal cell carcinoma of the head and neck. J Skin Cancer 2011;2011, 496910.
[38] Strutton GM. Pathological variants of basal cell carcinoma. Australas J Dermatol 1997;38(Suppl. 1):S31–5.
[39] Gambichler T, Orlikov A, Vasa R, et al. In vivo optical coherence tomography of basal cell carcinoma. J Dermatol Sci 2007;45(3): 167–73.
[40] Olmedo JM, Warschaw KE, Schmitt JM, Swanson DL. Optical coherence tomography for the characterization of basal cell carcinoma in vivo: a pilot study. J Am Acad Dermatol 2006;55(3): 408–12.
[41] Coleman AJ, Richardson TJ, Orchard G, Uddin A, Choi MJ, Lacy KE. Histological correlates of optical coherence tomography in non-melanoma skin cancer. Skin Res Technol 2013;19(1):10–9.
[42] Hussain AA, Themstrup L, Jemec GB. Optical coherence tomography in the diagnosis of basal cell carcinoma. Arch Dermatol Res 2015;307(1):1–10.
[43] Forsea AM, Carstea EM, Ghervase L, Giurcaneanu C, Pavelescu G. Clinical application of optical coherence tomography for the imaging of non-melanocytic cutaneous tumors: a pilot multi-modal study. J Med Life 2010;3(4):381–9.
[44] Bechara FG, Gambichler T, Stucker M, et al. Histomorphologic correlation with routine histology and optical coherence tomography. Skin Res Technol 2004;10(3):169–73.
[45] Boone MA, Norrenberg S, Jemec GB, Del Marmol V. Imaging of basal cell carcinoma by high-definition optical coherence tomography. Histomorphological correlation: a pilot study. Br J Dermatol 2012;167(4):856–64.
[46] Gambichler T, Plura I, Kampilafkos P, et al. Histopathological correlates of basal cell carcinoma in the slice and en face imaging modes of high-definition optical coherence tomography. Br J Dermatol 2014;170(6):1358–61.
[47] Boone MA, Suppa M, Pellacani G, et al. High-definition optical coherence tomography algorithm for discrimination of basal cell carcinoma from clinical BCC imitators and differentiation between common subtypes. J Eur Acad Dermatol Venereol 2015; 29(9):1771–80.
[48] Mogensen M, Joergensen TM, Nurnberg BM, et al. Assessment of optical coherence tomography imaging in the diagnosis of nonmelanoma skin cancer and benign lesions versus normal skin:

observer-blinded evaluation by dermatologists and pathologists. Dermatol Surg 2009;35(6):965–72.
[49] Wahrlich C, Alawi SA, Batz S, Fluhr JW, Lademann J, Ulrich M. Assessment of a scoring system for basal cell carcinoma with multi-beam optical coherence tomography. J Eur Acad Dermatol Venereol 2015;29(8):1562–9.
[50] Ulrich M, Maier T, Kurzen H, et al. The sensitivity and specificity of optical coherence tomography for the assisted diagnosis of non-pigmented basal cell carcinoma: an observational study. Br J Dermatol 2015;173(2):428–35.
[51] Vestergaard ME, Macaskill P, Holt PE, Menzies SW. Dermoscopy compared with naked eye examination for the diagnosis of primary melanoma: a meta-analysis of studies performed in a clinical setting. Br J Dermatol 2008;159(3):669–76.
[52] Menzies SW. Evidence-based dermoscopy. Dermatol Clin 2013; 31(4):521–4. vii.
[53] Chamlin SL, Williams ML. Pigmented lesions in adolescents. Adolesc Med 2001;12(2):195–212. v.
[54] Gambichler T, Regeniter P, Bechara FG, et al. Characterization of benign and malignant melanocytic skin lesions using optical coherence tomography in vivo. J Am Acad Dermatol 2007;57(4): 629–37.
[55] Kardynal A, Olszewska M. Modern non-invasive diagnostic techniques in the detection of early cutaneous melanoma. J Dermatol Case Rep 2014;8(1):1–8.
[56] Hinz T, Ehler LK, Voth H, et al. Assessment of tumor thickness in melanocytic skin lesions: comparison of optical coherence tomography, 20-MHz ultrasound and histopathology. Dermatology 2011;223(2):161–8.
[57] Meyer N, Lauwers-Cances V, Lourari S, et al. High-frequency ultrasonography but not 930-nm optical coherence tomography reliably evaluates melanoma thickness in vivo: a prospective validation study. Br J Dermatol 2014;171(4):799–805.
[58] Wessels R, de Bruin DM, Relyveld GN, et al. Functional optical coherence tomography of pigmented lesions. J Eur Acad Dermatol Venereol 2015;29(4):738–44.
[59] Boone MA, Norrenberg S, Jemec GB, Del Marmol V. High-definition optical coherence tomography imaging of melanocytic lesions: a pilot study. Arch Dermatol Res 2014;306(1):11–26.
[60] Picard A, Tsilika K, Long-Mira E, et al. Use of high-definition optical coherent tomography (HD-OCT) for imaging of melanoma. Br J Dermatol 2013;169(4):950–2.
[61] Gambichler T, Schmid-Wendtner MH, Plura I, et al. A multicentre pilot study investigating high-definition optical coherence tomography in the differentiation of cutaneous melanoma and melanocytic naevi. J Eur Acad Dermatol Venereol 2015;29(3): 537–41.
[62] Baryla J, Allen LH, Kwan K, Ong M, Sheidow T. Choroidal lymphoma with orbital and optic nerve extension: case and review of literature. Can J Ophthalmol 2012;47(1):79–81.
[63] Shields CL, Pellegrini M, Ferenczy SR, Shields JA. Enhanced depth imaging optical coherence tomography of intraocular tumors: from placid to seasick to rock and rolling topography. The 2013 Francesco Orzalesi Lecture. Retina 2014;34 (8):1495–512.
[64] Christian Ring H, Hansen I, Stamp M, Jemec GB. Imaging cutaneous T-cell lymphoma with optical coherence tomography. Case Rep Dermatol 2012;4(2):139–43.
[65] Ring HC, Hussain AA, Jemec GB, Gniadecki R, Gjerdrum LM, Mogensen M. Imaging of cutaneous T-cell lymphomas by optical coherence tomography: a case series study. J Eur Acad Dermatol Venereol 2015.
[66] Heimann H, Jmor F, Damato B. Imaging of retinal and choroidal vascular tumours. Eye (Lond) 2013;27(2):208–16.
[67] Qin XJ, Huang C, Lai K. Retinal vein occlusion in retinal racemose hemangioma: a case report and literature review of ocular complications in this rare retinal vascular disorder. BMC Ophthalmol 2014;14:101.
[68] Liu G, Jia W, Nelson JS, Chen Z. In vivo, high-resolution, three-dimensional imaging of port wine stain microvasculature in human skin. Lasers Surg Med 2013;45(10):628–32.
[69] Zhao S, Gu Y, Xue P, et al. Imaging port wine stains by fiber optical coherence tomography. J Biomed Opt 2010;15(3):036020.
[70] Zhou Y, Yin D, Xue P, et al. Imaging of skin microvessels with optical coherence tomography: potential uses in port wine stains. Exp Ther Med 2012;4(6):1017–21.

CHAPTER

7

Optical Coherence Tomography Imaging of Skin Scarring and Fibrosis

A. Mamalis[1,2], D. Ho[1,2], J. Jagdeo[1,2,3]

[1]University of California at Davis, Sacramento, CA, United States; [2]Sacramento VA Medical Center, Mather, CA, United States; [3]State University of New York Downstate Medical Center, Brooklyn, NY, United States

OUTLINE

INTRODUCTION

Optical coherence tomography (OCT) is a real-time imaging device that is altering clinical diagnosis in dermatology. OCT allows real-time two-dimensional imaging of tissue through the use of interferometry [1–6]. OCT was initially used in ophthalmology to record precise measurements of the eye and has since become a standard in the clinical management of several eye conditions [7]. Since introduction to medicine, OCT has gained recognition for its applications in dermatology and various other medical fields [2,7–12]. OCT allows dermatologists to visualize the epidermis, dermis, skin appendages, and superficial blood vessels [2]. This allows dermatologists to noninvasively diagnose certain skin conditions and longitudinally evaluate their response to therapy in lieu of invasive skin biopsies.

Collagen proteins are a key component of the skin that are imaged by OCT. Collagen proteins are the most abundant extracellular matrix components of the skin because they contribute approximately 80% of the dry weight of the dermis [13]. Collagen's inherent orientation, organization, and reflective properties allow OCT to detect and visualize skin collagen [14]. Diseases such as keloids, hypertrophic scars, chronic graft-versus-host disease, and systemic sclerosis are fibrotic skin conditions characterized by their increased skin collagen levels. Skin fibrosis is challenging to evaluate and is often managed based upon the clinician's subjective evaluation of disease burden and response to treatment. We foresee OCT imaging becoming a standard in the evaluation and management of diseases associated with skin fibrosis.

Many other noninvasive imaging methods have been studied for their ability to evaluate skin. However, imaging methods generally have an inverse relationship between penetration depth and resolution (Table 7.1) [5]. Fig. 7.1 illustrates the different penetration depths of various imaging methods. High-frequency ultrasound (US) is an imaging modality possessing a good

Imaging in Dermatology
http://dx.doi.org/10.1016/B978-0-12-802838-4.00007-8

TABLE 7.1 Maximal Penetration Depth and Resolution of Noninvasive Imaging Techniques

Imaging modality	Maximum penetration depth	Maximum resolution
Confocal microscopy	0.2 mm [1]	0.5–1 μm [1]
High-definition optical coherence tomography	0.57 mm [38,39]	<3 μm [38,39]
Optical coherence tomography	2 mm [1]	4–10 μm [1]
High-frequency ultrasound	15 mm [15]	300 μm [15]
Computed tomography	Total body penetration [1]	100 μm [1]
Magnetic resonance imaging	Total body penetration [1]	100 μm [1]

penetration depth of approximately 15 mm at the expense of lower resolution 300 μm; this lower resolution limits the ability of high-frequency US to evaluate fine tissue variations within the skin [15,16]. Computed tomography (CT) and magnetic resonance imaging are technologies that allow excellent penetration depth; however, their resolution of 100 μm limits their ability to evaluate structural changes within the skin and cost makes them impractical in clinical dermatology [1,16]. Confocal laser microscopy has a superior resolution of 1 μm, but its penetration depth of 0.2 mm and time requirement limit its utility in evaluating skin collagen alterations [1,5]. However, OCT balances a penetration depth of 2 mm with a resolution between 4 and 10 μm, providing an optimal imaging window to assess collagen in real time. Thus, OCT allows imaging of deeper structures than confocal microscopy while providing a greater resolution than US [17].

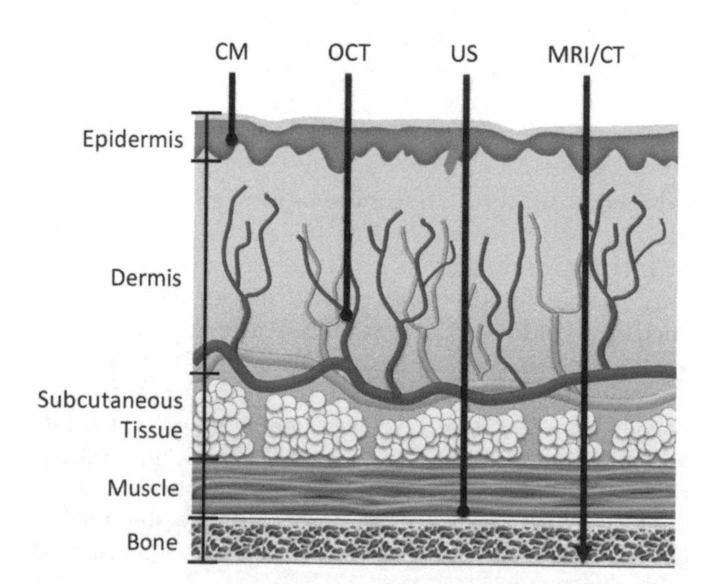

FIGURE 7.1 Diagram of the penetration depth of various imaging modalities. Computed tomography (CT), magnetic resonance imaging (MRI), and high-frequency ultrasound (US) have good depth of penetration at the expense of reduced resolution. Confocal microscopy (CM) has good resolution with a low depth of penetration. Optical coherence tomography (OCT) has an optimal balance of penetration depth and resolution that facilitates imaging the dermis. This makes OCT an excellent modality for imaging the increased dermal collagen that characterizes fibrotic skin disorders.

The concepts of OCT imaging are similar to US imaging. However, whereas US uses sound waves to generate images, OCT uses light or laser to generate images of the skin. OCT imaging uses interferometry to capture and record the desired reflected light. OCT systems then use this reflected light to generate two-dimensional images and some specialized systems can generate composite three-dimensional images [17–21]. Thus, OCT can provide cross-sectional images of skin that are structurally similar to images gathered from skin biopsy and histology [21]. However, although OCT can provide a gross structural evaluation of tissue and collagen content, the cellular information provided by OCT is limited.

Early OCT technology was based on a time-domain interferometry technique that utilized a moving reference arm to measure the time it took light to be reflected from the sample [5]. More recent OCT systems utilize a frequency-domain (FD) technique, which takes advantage of a static reference mirror and allows the entire depth of the tissue to be analyzed simultaneously and provides increased sensitivity and imaging speed [5]. Swept-source OCT (SS-OCT) is based on this FD technique and can provide structural information on skin collagen based on the density of collagen bundles, and it is inversely related to the amount of extracellular fluid [22]. Thus, using SS-OCT, fibrotic areas of skin that possess high collagen content can be easily distinguished from areas with lower collagen content.

Not only can OCT assess the gross structural features of skin, but it can also be used to obtain quantitative data on skin collagen levels [23]. Polarization-sensitive OCT (PS-OCT) is a technique that measures and displays OCT images based on the polarization state of the reflected light. The rate that polarization changes is directly related to the overall collagen content of the skin [24]. Thus, tissues with a large collagen burden as seen in skin fibrosis may lead to a rapid change in the polarization state of light leading to high phase retardation rates, whereas tissues with normal or decreased levels of collagen have comparatively lower rates of polarization and low phase retardation rates [24]. Other components of the extracellular matrix can also contribute to alterations in the polarization state; however, because skin collagen is the primary extracellular component and major contributor of skin anisotropy, PS-OCT imaging can provide valuable information on collagen when evaluating the burden of skin fibrosis.

OCT has several strengths that have the potential to alter clinical practice and dermatological research. Specifically, OCT is a significant improvement to the clinical evaluation of patients with skin fibrosis because it is a real-time, noninvasive, and safe imaging method. In contrast to biopsy, this real-time imaging gives OCT the advantage of time efficiency by allowing clinicians to examine skin sites in less than 10 s [25]. OCT also demonstrates excellent inter-rater reliability that may lead to more standardized methods of clinically diagnosing and evaluating skin fibrosis [25]. In addition, OCT requires minimal training to operate and allows users to collect skin images that can then be saved, stored, and forwarded [25]. We foresee OCT significantly altering how we manage and research skin fibrosis in the future. Using the visual and quantitative information provided by OCT images, physicians will be able to digitally record and standardize clinical evaluations, strengthen clinical trials data, track individual response to therapy, and improve telemedical care of skin fibrosis patients.

The purpose of this chapter is to review the available clinical evidence on OCT imaging of normal and fibrotic skin and its potential for diagnosis and evaluation of diseases that feature skin fibrosis. A detailed list of included studies is presented in Table 7.2.

OPTICAL COHERENCE TOMOGRAPHY IMAGING OF NORMAL SKIN COLLAGEN

Collagen is the key extracellular matrix component of the dermis. Skin collagen content normally varies by anatomical location, with areas such as the back possessing increased levels of collagen compared with areas such as the face [26]. Researchers have sought to characterize the variations in collagen levels of normal skin at different anatomical sites [4,27,28]. One study measured the skin birefringence in OCT images collected at the temple, dorsal hand, and lower back of normal healthy volunteers [26]. The investigators were able to conclude that OCT measurements demonstrated increased phase retardation rates of skin from the lower back, intermediate values from the dorsal hand, and the lowest values from the temple [26]. These findings were expected and correlated well with the known differences in collagen content at these skin sites. This study demonstrates that OCT is capable of distinguishing the baseline tissue variations in collagen that are physiological; thus, it supports that OCT likely possesses the sensitivity necessary to distinguish the much larger variations seen in skin fibrosis. Additional research is necessary to validate normal baseline values of OCT-measured skin collagen content for clinical use. These values would facilitate the diagnosis and evaluation of skin diseases characterized by variations in skin collagen content.

One study used OCT imaging to examine normal skin of the forehead, ear lobe, nose, cheek, chin, neck, chest, hands, arms, and calf in healthy patients [4]. They concluded that OCT was capable of visualizing and distinguishing the epidermis, papillary dermis, and reticular dermis at all of these sites [4]. Alterations of these areas are key characteristics of many fibrotic skin diseases, and this study demonstrates that OCT has the resolution necessary to visualize these key locations associated with skin fibrosis pathology [29].

Most OCT studies focus on two-dimensional OCT images; however, three-dimensional images are possible and have been utilized in studies investigating normal skin [1,25,30]. We foresee three-dimensional OCT imaging gaining further utilization in the future as clinicians utilize it to understand the impact of collagen fiber orientation and the total burden of fibrosis. In the future, OCT has the potential to become a standard measure for assessing skin fibrosis and the response to therapy. Further research is necessary to establish normal and pathological ranges for OCT-measured collagen burden. These standardized measurements will empower clinicians to make objective evaluations and may even allow earlier diagnosis and treatment of fibrotic skin diseases before clinical signs are apparent.

OPTICAL COHERENCE TOMOGRAPHY IMAGING OF SKIN FIBROSIS AS A FEATURE OF SYSTEMIC SCLEROSIS

The skin manifestation of systemic fibrosis, also known as scleroderma, is a fibrotic skin disease that is caused by an immune-mediated increase in collagen deposition in the skin and other organ systems [29,31,32]. In systemic sclerosis, skin fibrosis is a key prognostic factor and is often a primary measure in clinical trials. Unfortunately, there are no validated imaging biomarkers to quantify the dermal collagen levels [25]. Most systemic sclerosis trials use a clinical scoring system for skin fibrosis, often the modified Rodnan skin score (MRSS), as their primary measure [31]. The MRSS requires the palpation and measurement of skin

TABLE 7.2 Summary of Studies Investigating the Use of Optical Coherence Tomography to Evaluate Skin Fibrosis

Authors	Study aim	Population characteristics	Findings	Limitations
NORMAL SKIN				
Mogensen et al. [4]	PS-OCT imaging to describe normal skin at various body sites	Healthy volunteers aged 0.5–59 years ($n = 20$)	PS-OCT could identify birefringent differences among the epidermis, papillary dermis, and reticular dermis.	Most patients older despite large age range.
Pierce et al. [26]	PS-OCT imaging to describe normal skin at various body sites	Healthy volunteers aged 24–35 years ($n = 5$)	Mean phase retardation rate was highest at back skin and lowest at temple skin.	Small sample size.
Pircher et al. [27]	PS-OCT imaging to describe normal skin at various body sites	Healthy volunteers	Three-dimensional PS-OCT provided improved contrast and provided orientation of birefringent skin structures.	Fingertip and hand were only regions investigated.
Yasuno et al. [28]	To investigate normal skin birefringence using PS-OCT	Healthy volunteers	PS-OCT could measure differences in human skin birefringence.	Limited patient information.
FIBROSIS				
Abignano et al. [25]	Evaluate the use of SS-OCT as an imaging biomarker of skin fibrosis	Systemic sclerosis patients ($n = 21$)	Fibrotic skin had decreased optical density in the papillary dermis that correlated with MRSS score.	Only 2 of 21 patients underwent biopsy for histological comparison.
Gong et al. [35]	To present a method to characterize dermal scar tissue by measurement of the attenuation coefficient using OCT	Burn patients with hypertrophic and normotrophic scar ($n = 6$); examined various scar sites	On average, the dermal attenuation coefficient is 36% lower in scarred skin compared with normal skin. An algorithm was developed to remove significant artifact from the calculation of the attenuation coefficient by masking blood vessels because of the high light-scattering properties of red blood cells.	Small sample size.
Kunzi-Rapp et al. [37]	Evaluation of collagen biosynthesis after scar treatment with the Er:YAG laser	Post-traumatic and acne scar patients aged 12–39 years ($n = 12$)	OCT was capable of imaging collagen production after Er:YAG laser treatment.	OCT was not primary objective of study but rather used in assessment.
Liew et al. [34]	OCT evaluation of vascularity in hypertrophic scars	Hypertrophic scar patients with mean age of 32 years ($n = 8$)	Increase in mean density of vasculature in hypertrophic scar tissues (38%) when compared with normal, unscarred skin (22%); proliferation of larger vessels.	Small sample size.
Pierce et al. [18]	To demonstrate the capability of OCT in detecting features of skin fibrosis	Fibrotic scar patient ($n = 1$)	OCT measured significant polarization differences between normal skin and the fibrotic site.	Fibrosis examination restricted to the hand; $n = 1$.
Ring et al. [36]	To explore the feasibility of OCT imaging for in vivo assessment and monitoring of collagen deposition disorders	Patients with ordinary scars, hypertrophic scars, keloids, lichen sclerosis et atrophicus, and localized or systemic scleroderma ($n = 33$); examined various scar sites	OCT identified important hallmark characteristics for each lesion type.	Limited sample size. Lack specificity, sensitivity, and positive predictive value of aforementioned hallmark characteristics from OCT imaging compared with histology.

PS-OCT, polarization-sensitive optical coherence tomography; *SS-OCT*, swept-source optical coherence tomography; *OCT*, optical coherence tomography; *MRSS*, modified Rodnan skin score.

thickness at various anatomical sites and requires a high level of skill to minimize interobserver variability [25,31]. OCT imaging may provide researchers and clinicians an imaging biomarker for skin fibrosis that has improved interobserver variability and reproducibility compared with the MRSS.

Studies are beginning to investigate OCT as a quantitative imaging biomarker for assessing systemic sclerosis [25]. A study looking at the use of OCT to evaluate systemic sclerosis utilized SS-OCT to image the hands and forearms of 21 patients with system sclerosis [25]. The investigators found that OCT imaging quantification of collagen burden correlated well with histological evaluation and the patients' calculated MRSS [25]. Furthermore, the authors demonstrated that visualization of the dermoepidermal junction decreased as the burden of skin fibrosis increased. This correlation between OCT imaging, histology, and clinical MRSS score demonstrates the tremendous potential of OCT in evaluating and monitoring response to therapy in systemic sclerosis.

Further research is needed to validate and characterize the ability of OCT to assess systemic sclerosis and other diseases characterized by skin fibrosis. In addition, research into preclinical skin changes in patients with scleroderma may lead to earlier diagnosis and better patient outcomes. In the future, we believe that OCT will be utilized to aid clinical assessment, measure response to therapy, and quantify collagen burden in clinical trials on systemic sclerosis.

OPTICAL COHERENCE TOMOGRAPHY IMAGING OF FIBROTIC SCARS

OCT has also shown promise in its ability to assess hypertrophic scars, a fibrotic disorder that results from cutaneous insults such as a cut or burn. Hypertrophic scars result when cutaneous injuries lead to an excess production of dermal collagen synthesis [33,34]. Several studies have investigated OCT's ability to evaluate the fibrotic changes resulting from hypertrophic scars, burn scars, and other scars.

Early case reports have demonstrated that OCT can be utilized to easily differentiate scar tissue from normal tissue [18]. More recent studies have investigated methods of further characterizing scar tissue using OCT. One such study developed an algorithm to characterize dermal scar tissue by measurement of the attenuation coefficient using OCT [35]. The attenuation coefficient reflects the rate at which the OCT signal decreases with depth in the tissue. It has been hypothesized that changes in the tissue microstructure (such as collagen fibers and vasculature) between scarred and normal skin may result in differences in this attenuation coefficient. Background vasculature is a significant artifact for calculation of the attenuation coefficient because of the large light-scattering properties of red blood cells, and the proposed algorithm successfully masked the background vasculature [35]. The results from six patients with hypertrophic and normotrophic burn scars demonstrated on average 36% lower attenuation coefficients for scarred skin compared with healthy skin [35]. This led to improved structural imaging of the underlying scar tissue and a characteristic birefringence pattern of scar tissue [35]. Using this algorithm with masking of background vasculature, OCT has the potential to serve as an objective assessment tool to study scars in greater detail.

A different study identified hallmark characteristics from OCT imaging with hypertrophic scars, keloids, and scleroderma [36]. Hypertrophic scars revealed characteristic white longitudinal dermal streaks due to increased deposition of compacted collagen [36]. Keloids demonstrated disarray of signal-rich streaks that correspond to the high density of fibroblasts and highly organized and unidirectional collagen fibrils on histology [36]. Scleroderma displayed more cohesive backscattering in the dermis, which may indicate a difference in density of collagen or other dermal structures. OCT allows serial imaging of the same wound over time without the need for invasive biopsy that may disrupt the wound-healing process, and it requires further research with randomized trials to evaluate the sensitivity, specificity, and positive predictive values with OCT as compared with histology findings from biopsy. Another study investigated the use of OCT in evaluating response to therapy after treatment of 2940-nm Erbium:YAG laser for facial rejuvenation, post-traumatic scars, and acne scars [37]. The authors report that OCT was able to visualize post-treatment dermal remodeling and newly synthesized collagen [37]. It is likely that OCT can also be used to evaluate treatment response in other ablative and nonablative lasers. Given that OCT provides real-time structural imaging of the epidermis and dermis, it is likely that OCT can also be used to determine appropriate ablative laser settings that are individualized to the specific patient and lead to improved patient outcomes.

In addition to collagen, local vasculature contributes to the pathogenesis of skin fibrosis. One clinical study utilized OCT to characterize vessels present in eight patients' hypertrophic scars and found that vessel density was significantly increased in hypertrophic scars compared with normal skin, and large vessels (>100 μm) present in hypertrophic scars were completely absent in normal scars and unscarred skin [34]. This study illustrates that OCT may identify additional imaging biomarkers that can be utilized in the diagnosis and management of hypertrophic scars.

Although studies directly investigating keloid scars and other types of scars are limited, we believe that

OCT will also prove to have utility in the identification and evaluation of these scar types. The future use of OCT imaging as a quantitative imaging biomarker of fibrosis is promising, and OCT may lead to improved patient outcomes through noninvasive clinical and research evaluations while foregoing the need for biopsies that may worsen scars.

THE FUTURE OF OPTICAL COHERENCE TOMOGRAPHY

There is a lack of randomized clinical trials evaluating the efficacy of OCT compared with other methods of evaluating skin fibrosis and collagen content. However, clinical studies have consistently demonstrated that there are visual and measurable differences between normal and fibrotic skin. We believe that OCT will become widely used in many diseases characterized by skin fibrosis such as chronic graft-versus-host disease, morphea, or lichen sclerosis et atrophicus. Additional studies are required to evaluate the correlation between OCT imaging and histology of skin fibrosis—the current gold standard in the diagnosis and evaluation of many of the aforementioned diseases.

OCT does possess some technical limitations than may be addressed in the future. Penetration depth is a significant limitation that may be overcome by combining OCT with other compatible acousto-optical imaging systems. An OCT imaging system offering multiple wavelengths for different depths of penetration and improved algorithms for identifying background scatter may result in improved image quality.

Thus far OCT has proven to be useful in the evaluation of skin fibrosis. However, although clinicians can capture rapid qualitative images of fibrotic regions of skin, quantitative visual assessment of collagen content currently requires third-party applications or algorithms. More advanced native OCT imaging software with real-time quantification features would enhance OCT's real-world clinical utility.

OCT has the potential to redefine how fibrotic skin diseases are evaluated in the teledermatology and research setting. Using remote or store-and-forward features available to OCT, remote and/or unbiased blinded evaluations of skin fibrosis can be made. Furthermore, OCT's provision of visual and quantitative data on skin collagen content may be the solution to teledermatology's tactile limitation.

One current barrier for widespread use and adoption of OCT is the cost and portability of available OCT devices. Unfortunately, there is both an element of limited competition with the OCT market and a high technology cost. In the near future, innovations within the telecommunication sector will hopefully trickle down to OCT technology and lead to reduced cost and reductions in the size of current OCT devices. It is likely that as costs continue to decline, adoption by dermatologists and researchers will continue to increase. Despite current limitations, OCT imaging for the assessment of skin fibrosis is still in its formative stages and is poised to make tremendous strides that parallel the development and refinement of US over the past 30 years.

CONCLUSION

OCT is an imaging modality that uses light or laser to produce an image of the skin that approximates histological architecture. OCT imaging provides a noninvasive real-time approach to the diagnosis and evaluation of several skin diseases ranging from systemic sclerosis to hypertrophic scars. OCT images can also provide a quantitative assessment of collagen levels within skin. This technology is poised to transform the way dermatologists manage skin fibrosis by allowing a longitudinal repeatable measure of collagen content. Using OCT collagen assessment as a noninvasive imaging biomarker for skin fibrosis will not only improve clinical dermatology but also facilitate research that may lead to new insights into the physiology, pathology, and management of skin fibrosis.

Abbreviations

CT Computed tomography
FD Frequency domain
MRI Magnetic resonance imaging
MRSS Modified Rodnan skin score
OCT Optical coherence tomography
PS-OCT Polarization-sensitive optical coherence tomography
SS-OCT Swept-source optical coherence tomography
TD Time domain
US Ultrasound

References

[1] Dalimier E, Salomon D. Full-field optical coherence tomography: a new technology for 3D high-resolution skin imaging. Dermatology 2012;224(1):84–92.
[2] Gambichler T, Jaedicke V, Terras S. Optical coherence tomography in dermatology: technical and clinical aspects. Arch Dermatol Res 2011;303(7):457–73.
[3] Matcher SJ. Practical aspects of OCT imaging in tissue engineering. Methods Mol Biol 2011;695:261–80.
[4] Mogensen M, Morsy HA, Thrane L, Jemec GB. Morphology and epidermal thickness of normal skin imaged by optical coherence tomography. Dermatology 2008;217(1):14–20.
[5] Sattler E, Kastle R, Welzel J. Optical coherence tomography in dermatology. J Biomed Opt 2013;18(6):061224.
[6] Welzel J. Optical coherence tomography in dermatology: a review. Skin Res Technol 2001;7(1):1–9.
[7] Fujimoto JG. Optical coherence tomography for ultrahigh resolution in vivo imaging. Nat Biotechnol 2003;21(11):1361–7.

[8] Cahill RA, Mortensen NJ. Intraoperative augmented reality for laparoscopic colorectal surgery by intraoperative near-infrared fluorescence imaging and optical coherence tomography. Minerva Chir 2010;65(4):451–62.

[9] Chu CR, Izzo NJ, Irrgang JJ, Ferretti M, Studer RK. Clinical diagnosis of potentially treatable early articular cartilage degeneration using optical coherence tomography. J Biomed Opt 2007;12(5):051703.

[10] Unterhuber A, Povazay B, Bizheva K, Hermann B, Sattmann H, Stingl A, et al. Advances in broad bandwidth light sources for ultrahigh resolution optical coherence tomography. Phys Med Biol 2004;49(7):1235–46.

[11] Lamirel C, Newman N, Biousse V. The use of optical coherence tomography in neurology. Rev Neurol Dis 2009;6(4):E105–20.

[12] Tearney GJ, Brezinski ME, Bouma BE, Boppart SA, Pitris C, Southern JF, et al. In vivo endoscopic optical biopsy with optical coherence tomography. Science 1997;276(5321):2037–9.

[13] Krieg T, Aumailley M, Koch M, Chu M, Uitto J. Collagens, Elastic fibers, and other extracellular matrix proteins of the dermis. In: Fitzpatrick's dermatology in general medicine. 8th ed. New York: McGraw-Hill; 2012.

[14] Liu B, Vercollone C, Brezinski ME. Towards improved collagen assessment: polarization-sensitive optical coherence tomography with tailored reference arm polarization. Int J Biomed Imaging 2012;2012:892680.

[15] Crisan M, Crisan D, Sannino G, Lupsor M, Badea R, Amzica F. Ultrasonographic staging of cutaneous malignant tumors: an ultrasonographic depth index. Arch Dermatol Res 2013;305(4):305–13.

[16] Han JH, Kang JU, Song CG. Polarization sensitive subcutaneous and muscular imaging based on common path optical coherence tomography using near infrared source. J Med Syst 2011;35(4):521–6.

[17] Mogensen M, Thrane L, Joergensen TM, Andersen PE, Jemec GB. Optical coherence tomography for imaging of skin and skin diseases. Semin Cutan Med Surg 2009;28(3):196–202.

[18] Pierce MC, Strasswimmer J, Park BH, Cense B, de Boer JF. Advances in optical coherence tomography imaging for dermatology. J Invest Dermatol 2004;123(3):458–63.

[19] Gladkova ND, Petrova GA, Nikulin NK, Radenska-Lopovok SG, Snopova LB, Chumakov YP, et al. In vivo optical coherence tomography imaging of human skin: norm and pathology. Skin Res Technol 2000;6(1):6–16.

[20] Tadrous PJ. Methods for imaging the structure and function of living tissues and cells: 3. Confocal microscopy and microradiology. J Pathol 2000;191(4):345–54.

[21] Wessels R, De Bruin DM, Faber DJ, Van Leeuwen TG, Van Beurden M, Ruers TJ. Optical biopsy of epithelial cancers by optical coherence tomography (OCT). Lasers Med Sci 2013. http://dx.doi.org/10.1007/s10103-013-1291-8.

[22] Phillips KG, Wang Y, Levitz D, Choudhury N, Swanzey E, Lagowski J, et al. Dermal reflectivity determined by optical coherence tomography is an indicator of epidermal hyperplasia and dermal edema within inflamed skin. J Biomed Opt 2011;16(4):040503.

[23] Sakai S, Yamanari M, Lim Y, Nakagawa N, Yasuno Y. In vivo evaluation of human skin anisotropy by polarization-sensitive optical coherence tomography. Biomed Opt Express 2011;2(9):2623–31.

[24] Pierce MC, Sheridan RL, Hyle Park B, Cense B, de Boer JF. Collagen denaturation can be quantified in burned human skin using polarization-sensitive optical coherence tomography. Burns 2004;30(6):511–7.

[25] Abignano G, Aydin SZ, Castillo-Gallego C, Liakouli V, Woods D, Meekings A, et al. Virtual skin biopsy by optical coherence tomography: the first quantitative imaging biomarker for scleroderma. Ann Rheum Dis 2013. http://dx.doi.org/10.1136/annrheumdis-2012-202682.

[26] Pierce MC, Strasswimmer J, Hyle Park B, Cense B, De Boer JF. Birefringence measurements in human skin using polarization-sensitive optical coherence tomography. J Biomed Opt 2004;9(2):287–91.

[27] Pircher M, Goetzinger E, Leitgeb R, Hitzenberger C. Three dimensional polarization sensitive OCT of human skin in vivo. Opt Express 2004;12(14):3236–44.

[28] Yasuno Y, Makita S, Sutoh Y, Itoh M, Yatagai T. Birefringence imaging of human skin by polarization-sensitive spectral interferometric optical coherence tomography. Opt Lett 2002;27(20):1803–5.

[29] Jimenez SA, Derk CT. Following the molecular pathways toward an understanding of the pathogenesis of systemic sclerosis. Ann Intern Med 2004;140(1):37–50.

[30] Pan Y, Farkas DL. Noninvasive imaging of living human skin with dual-wavelength optical coherence tomography in two and three dimensions. J Biomed Opt 1998;3(4):446–55.

[31] Gabrielli A, Avvedimento EV, Krieg T. Scleroderma. New Engl J Med 2009;360(19):1989–2003.

[32] Varga J, Abraham D. Systemic sclerosis: a prototypic multisystem fibrotic disorder. J Clin Invest 2007;117(3):557–67.

[33]] Oliveira GV, Chinkes D, Mitchell C, Oliveras G, Hawkins HK, Herndon DN. Objective assessment of burn scar vascularity, erythema, pliability, thickness, and planimetry. Dermatol Surg 2005;31(1):48–58.

[34] Liew YM, McLaughlin RA, Gong P, Wood FM, Sampson DD. In vivo assessment of human burn scars through automated quantification of vascularity using optical coherence tomography. J Biomed Opt 2013;18(6):061213.

[35] Gong P, Chin L, Es'haghian S, Liew YM, Wood FM, Sampson DD, et al. Imaging of skin birefringence for human scar assessment using polarization-sensitive optical coherence tomography aided by vascular masking. J Biomed Opt 2014;19(12):126014.

[36] Ring HC, Mogensen M, Hussain AA, Steadman N, Banzhaf C, Themstrup L, et al. Imaging of collagen deposition disorders using optical coherence tomography. J Eur Acad Dermatol Venereol 2014. http://dx.doi.org/10.1111/jdv.12708.

[37] Kunzi-Rapp K, Dierickx CC, Cambier B, Drosner M. Minimally invasive skin rejuvenation with Erbium: YAG laser used in thermal mode. Lasers Surg Med 2006;38(10):899–907.

[38] Boone M, Norrenberg S, Jemec G, Del Marmol V. High-definition optical coherence tomography: adapted algorithmic method for pattern analysis of inflammatory skin diseases: a pilot study. Arch Dermatol Res 2013;305(4):283–97.

[39] Boone MA, Norrenberg S, Jemec GB, Del Marmol V. High-definition optical coherence tomography imaging of melanocytic lesions: a pilot study. Arch Dermatol Res 2013. http://dx.doi.org/10.1007/s00403-013-1387-9.

C H A P T E R

8

Polarization Speckles and Skin Applications

T.K. Lee[1,2], L. Tchvialeva[2], I. Markhvida[2], H. Zeng[1,2], H. Lui[1,2], A. Doronin[3], I. Meglinski[3,4], D.I. McLean[1,2]

[1]BC Cancer Agency, Vancouver, BC, Canada; [2]Vancouver Coastal Health Research Institute and University of British Columbia, Vancouver, BC, Canada; [3]University of Otago, Dunedin, New Zealand; [4]University of Oulu, Oulu, Finland

O U T L I N E

INTRODUCTION

Skin cancer is a worldwide health problem. Just the United States alone has more than 3 million new cases each year [1]. The annual incidence rate is so high, equal to or greater than all other cancers combined, that many cancer registries are reluctant to ascertain, code, and recode these cancer cases [2]. Among the different types of skin cancer, malignant melanoma (MM) is the most well studied (and feared) because it accounts for most skin cancer deaths. Australia and New Zealand have the highest melanoma incidence rates in the world, more than 35 cases per 100,000 population [3]. Most European countries have a rate of approximately 10 per 100,000, but the northern European countries, such as Norway and Denmark reach a rate of approximately 15 per 100,000 [3].

The melanoma incidence rate (per 100,000 population) has been growing steadily and rapidly. For instance, within the past 30 years, the Canadian melanoma rate has been increasing sharply from 4.1 for males and 5.2 for females in 1973 [4] to 15.1 and 12.2 for males and females, respectively, in 2013 [5]. A similar trend was also found in the United Kingdom, where melanoma, a relatively rare cancer in the 1990s with only approximately 5900 cases annually and ranked as the 13th and 11th most common cancer for males and females, respectively [6], jumped to the 6th most common cancers in both males and females with more than 13,384 new cases in 2011 [7]. Both countries have reported a heavy financial burden to their health-care systems because of the many annual cases, social impact, and related treatment costs. It was estimated that in 2004 alone the total economic burden was $443 million in Canada. Among this, $30 million was direct medical-related cost and $414 million was indirect cost related to mortality and morbidity [8]. A recent UK study showed that the related health-care cost imposed on the National Health Services was £22 million in 2008, but the projected cost will reach £37 million in 2020 [9].

Melanoma is life threatening. The standard treatment for localized melanomas is surgery, and it has not been

Imaging in Dermatology
http://dx.doi.org/10.1016/B978-0-12-802838-4.00008-X

changed for many years. Removing melanoma at an early stage has an excellent prognosis with a greater than 90% 10-year survival rate. However, the rate drops to as low as 15% for patients with an advanced disease [10]. Hence, early detection is the key to combat the disease.

Although melanoma manifests itself on the skin surface and can be readily seen by patients and physicians, diagnosing the disease by naked-eye examination is challenging because many benign skin lesions, such as melanocytic nevi (also known as pigmented skin lesions and moles) and seborrheic keratoses (SKs) may resemble melanomas [11,12]. The reported diagnostic sensitivity ranges from approximately 60–90% [11,13,14]. It has been observed that dermatologists perform much better than nondermatologists [15] because the current diagnosis method relies primarily on visual inspection for lesional-specific morphologic features, such as color, shape, border, configuration, distribution, elevation, and texture. Although classification rules can be applied to visual diagnosis [16], the overall clinical approach is subjective and qualitative, with a critical dependence on training and experience. In terms of specificity, a study of primary care practitioners reported that more than 80% of skin biopsies performed by nondermatologists proved to be benign [17]. Thus diagnostic tools are especially needed for primary care providers, the gatekeepers of the medical system, to increase the detection sensitivity so that melanomas can be detected, treated, or referred in the earliest possible time, and, at the same time, to improve the specificity and reduce the number of excisions of benign lesions.

Over the past 20 years several noninvasive techniques for measuring the skin's physical properties have been developed to extend the accuracy of expert visual assessment. For example, fluorescence spectroscopy detects and analyzes the presence and distribution of fluorophores. Raman spectroscopy provides fingerprinting of the molecules within the skin. Dermoscopy, multispectral imaging, optical coherence tomography, confocal microscopy, and multiphoton microscopy generate two-dimensional (2D) high-resolution images showing structures in the epidermis and dermis. All of these developments demonstrate the potential for noninvasive "optical biopsy," particularly for differentiating benign from malignant lesions. However, many of these devices are very expensive; in addition, the 2D imaging methods usually require experts to interpret complex 2D data, which without intelligent image analysis programs are not suitable for primary care providers.

In this study, we propose a novel approach, analyzing the coherence and polarization properties of light. Light coherence and polarization, the manifestation of the wave nature of light, have a characteristic scale of approximately the wavelength of light that corresponds to the dimensions of tissue microstructure. Therefore coherence and polarization are often used as sensitive tools in biomedical optics [18,19]. The unified theory of light reveals the mutual connection between the degree of coherence and polarization of light [20]. Backscattered coherent laser light forms an interference pattern called a speckle pattern. The light maintains its coherence and polarization within a speckle grain [21]. However, the spatial distribution of polarization properties of light are called polarization speckles [22] and are not uniform but have a statistical distribution related to the tissue morphology [23]. A recently developed theoretical description of polarization speckles has been limited by the special cases of fully coherent and polarized light [24]. Practically speaking, backscattered light from the skin light is partially depolarized. Because of the lack of theoretical understanding of polarization speckle phenomena, experimental studies in this area are of great interest. In this paper, we report some dermatological application of polarization speckle. We develop a method to measure skin roughness in vivo and suggest polarization metrics for lesion diagnostics. We believe the new approach will improve diagnostic accuracy in dermatology and the resultant instrument will be easy to operate. This paper will present the current development of the technique.

Coherence

Coherence is one of the most important characteristics of light. It is less well known than intensity and color because coherence cannot be directly perceived by the human eye. Coherence is responsible for all interference phenomena, especially when we consider coherent laser light. For example, when a laser beam (coherent light) is backscattered by a rough surface we observe the well-known phenomenon called a speckle pattern. The interference (speckle) pattern is manifested as bright and dark spots. The areas with high and low intensities correspond to the addition and subtraction of the individual backscattered waves. It was known that the coherence of the light source is equal to the contrast C of the speckle pattern, which is defined as the ratio of the standard deviation of light intensity σ_I over the average intensity $<I>$ [25]:

$$C = \frac{\sigma_I}{<I>}. \tag{8.1}$$

The range of contrast C is from 0 (no coherence) to 1 (full coherence).

Speckle patterns encode information about the light source, the targeted object, and the geometry of the image formation process [25]. The speckle theory was established several decades ago [26]. However, the theoretical formulations were not developed into practical instruments in the early years because of technical limitations. Recent advances in light sources and devices for registration of images have now revived interest in speckle techniques. The relationship between speckle contrast and surface roughness has been well studied [27–36]. Industrial applications, such as the detection of microtextures of a pavement surface have been reported [37]. Speckle contrast has also been used for estimating tissue optical properties. The absorption and scattering coefficients of diffusing media can be derived by measuring contrast as a function of either tissue thickness or the coherence length of the light source [38]. Temporal averaging of contrasts for dynamic biospeckle has been used to monitor tissue conditions [39], visualize internal inhomogeneities [40–43], measure viscoelasticity of atherosclerotic plaques [44], and evaluate blood flow [45–47].

Polarization

Polarization is another important property of light. In physics, polarization is described as the orientation of the oscillation of the electric field perpendicular to the traveling path of the light. When the electric fields oscillate in the same orientation, these rays have the same polarization. Polarization can be classified as linear, circular, and elliptical, depending on the sweeping pattern of the oscillation. However, these classifications cannot be applied to depolarized light where the oscillations are stochastically changing direction.

When the oscillation orientation is altered for a portion of the rays, either a new polarization state or depolarization results. If the altered rays maintain coherence with the initial light, then the resulting light remains polarized but the polarization state is changed. If the altered rays lose coherence, the resulting light became partially depolarized. For example, when polarized light enters a scattering medium, light is scattered and depolarized. The amount of depolarization has been shown to be a function of the average distance traveled by the light in the media [48]. Thus light penetrating less deeply below a surface retains more of its original polarization, whereas deeply penetrating light eventually becomes fully depolarized with the rays oscillating in random orientations. The amount of light retaining its original polarization state can be quantified by a measure known as the degree of polarization.

Changes in depolarization and polarization states can also be caused by birefringence. Certain media, such as calcite crystals, liquid crystal displays, and the skin proteins keratin and collagen split the incident light into two paths according to its polarization state and the anisotropic refractive index of the media. Thus a double refraction effect is generated.

Polarization techniques have been actively investigated in medical applications, including skin. On the basis of the amount of depolarization and penetration depth, the so-called "polarization filtering" or "polarization gating" allows a specified layer of superficial biological tissue to be viewed and examined [49–51]. Dermoscopy is the well-known example of the generation of an internal view of a skin lesion by manipulating a polarizing filter [52]. Steven Jacques reported that polarization filtering could be used to determine skin cancer margins not visible to naked-eye examinations [51,53]. The technique was also used to generate a Raman spectrum of only the superficial skin layer [54]. The techniques mentioned here provide an internal view of the skin based on the polarization principle. Techniques based on numeric analyses of polarization measurements have also been pursued for biomedical applications. Studies of phantoms [55,56] and biological tissues [51,57] demonstrated that changes of the degree of polarization depend on the scatterers' sizes and concentration. Collagen and keratin alter polarization of propagated light via birefringence, optical activity, and de-attenuation [58]. Thinly sliced samples of biotissues, such as spleen, heart, and kidney were tested and demonstrated that the altered polarization signals could be analyzed quantitatively and could identify the tissue properties [59]. Exposing pig skin to radiation [60] and internal pathological changes caused by skin diseases [61] also modify the polarization state of backscattered light. Kim et al. [62] demonstrated that the degree of linear polarization is highest in the regions of melanocytic lesions that contained the most pigment. A recent in vitro study of cancerous and healthy lung tissues embedded in paraffin wax showed that the cancerous and normal tissues could be distinguished using circular polarized light [63]. A laparoscopic version of the polarimetry imaging system for endometrial lesions has also been developed [64]. Commonly, the state-of-the-art polarimetric techniques characterize tissues by a complicated mathematical formulation describing how polarized illumination interacts with the media. Determination of all of the components in the mathematical formulation requires a lengthy step of manipulating polarizers in the azimuth angle scanning [58,60,61]. A recent study avoided moving parts by using a complex setup with sequential illumination from multiple light sources [65,66]. Despite a great diagnostic potential in these techniques, the multistep requirement in data acquisition poses a challenge of applying the techniques in vivo in a clinical setting.

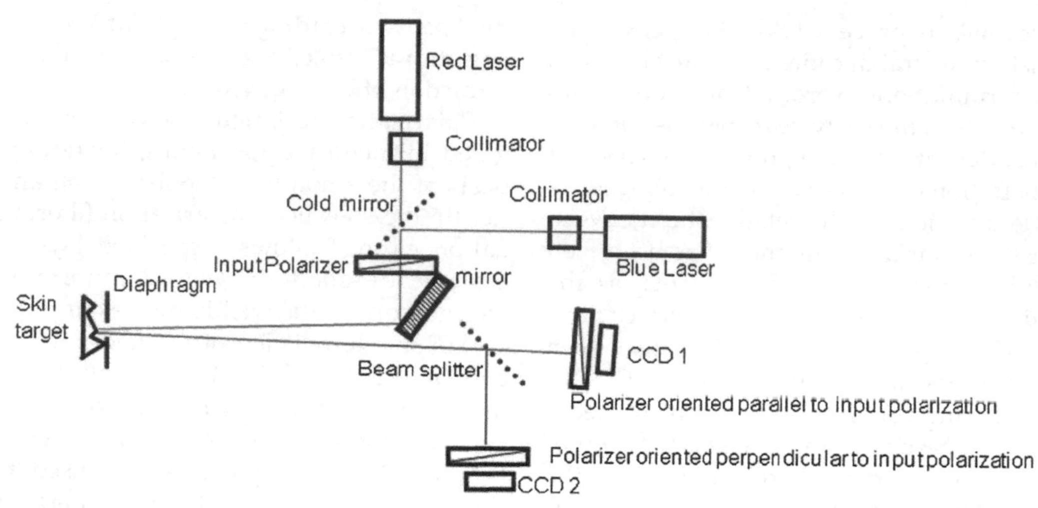

FIGURE 8.1 Schematic of the polarization speckle prototype. *Reprinted with permission from Tchvialeva L, Dhadwal G, Lui H, Kalia S, Zeng H, McLean DI, et al. Polarization speckle imaging as a potential technique for in vivo skin cancer detection. J Biomed Opt June 2013;18(6):061211.*

Polarization Speckle

Polarization is often expressed in Stokes vector formalism. A Stokes vector, $S = [S_0, S_1, S_2, S_3]$, has four components, and each of them can be calculated as

$$\begin{aligned} S_0 &= I_0 + I_{90} \\ S_1 &= I_0 - I_{90} \\ S_2 &= I_{45} - I_{135} \\ S_3 &= I_{LHC} - I_{RHC}, \end{aligned} \tag{8.2}$$

where I_0, I_{90}, I_{45}, and I_{135} are the intensities passing through a linear polarizer with a 0, 90, 45, and 135 degrees orientation. I_{LHC} and I_{RHC} are the left- and right-circular components, respectively. Conceptually, Stokes vectors are easy to measure. The four Stokes components directly relate to the polarization ellipse parameters, light intensity, and degree of polarization. It was shown theoretically that Stokes vectors obey the wave equation [22]. It means that for coherent light the spatial distribution of any of Stokes components and their combinations will reveal the speckle structure, which was called polarization speckles [22,24]. Compared with the classic intensity laser speckles, which possess the properties of random amplitude, random phases, and uniform polarization, polarization speckles manifest random amplitude, random phases, and spatially random polarization. In other words, polarization speckle could encode more information than traditional laser speckles.

We are interested in developing a new, noninvasive detection technique for melanoma-based polarization speckles. In particular, the speckle patterns of the coherence light can be used to quantify the skin surface roughness, an important property for differentiating MM and SK, a benign and pigmented skin condition. These two types of skin lesions often have a similar appearance that causes confusion and misdiagnosis. However, their surface profile is the key diagnostic factor: melanoma tends to have a smooth surface whereas SK is a much rougher lesion. Technically, we calculate the contrast C of S_1 of a speckle pattern, which can be considered as the second statistical moment over the first moment of S_1. In addition, the new device can also use the mathematical moments of the spatial distribution of $S_1(x, y)/S_0(x, y)$ to analyze the morphology of the skin tissues at the cellular level.

A prototype device has been constructed, and its schematic is shown in Fig. 8.1.

The device consists of two polychromatic diode lasers, which can generate polarization speckles. One of the lasers is a red laser ($\lambda = 663$ nm, coherence length is ~300 μm) and the other one is a blue laser ($\lambda = 407$ nm, coherence length is ~200 μm). These two lasers operate independently and sequentially. The reason for using two lasers is to test the dependence on the wavelength of the light sources. Two charge-coupled device cameras without an objective lens were used to simultaneously capture two speckle images. Both cameras are equipped with a polarizer; the orientation of the polarizers is placed in such a way that one of them is parallel (0-degree) and another one is perpendicular (90-degree) to the polarization of the illumination. To enhance the surface reflection and, at the same time, separate it from the volume backscattering, we used three filtering methods: spatial, polarization, and spectral. Spatial filtering relies on the property that weakly scattered light emerges at positions close to the illuminating spot [67]. Therefore the superficial signal can be enhanced by limiting the emerging light. Using an opaque diaphragm centered with the incident beam allows the weakly scattered light to be collected. Polarization filtering is based on the polarization-maintaining property of weakly scattered light. When linearly polarized light illuminates a scattering medium, the weakly scattered light emerging from the superficial region maintains its original polarization orientation whereas the multiply scattered light emerging from deeper regions is randomly polarized [68]. Subtracting the cross-polarized pattern from the co-polarized pattern suppresses volume scattering. Spectral filtering [50] is based on the spectral dependence of skin attenuation coefficients [69]. Shorter wavelengths are attenuated more heavily by biological tissue and yield a higher output of weakly scattered light than longer wavelengths. Thus the superficial depth for blue light is expected to be shallower than that for red light; therefore we used a blue laser for skin roughness measurements [70]. When one of the lasers is turned on, its light is directed to the mouth of the device after passing through a linear polarizer, and the backscattering light will be filtered spatially at the opening before the light is split and directed to the two cameras. Because scanning is not involved, acquisition time is extremely fast, within milliseconds. The device can be used to quantify the surface roughness by examining the speckle pattern of the enhanced surface reflection and to analyze the depolarization of the volume backscattering by examining the spatial distribution of the depolarization. Details of these analyses are presented in the next two sections.

SKIN SURFACE ROUGHNESS QUANTIFICATION

The current de facto method for measuring skin roughness is a two-step approach of first making a silicone skin replica and then mechanically or optically scanning the imprint offline. The main drawbacks of this indirect approach are that the replica-making process is time-consuming and prone to artifacts and distortions. On the basis of optical triangulation with a diode laser [71] and fringe projection of white light [72–74], a few in vivo methods have then been developed. However, they are mainly used to measure deep wrinkles and do not have the sensitivity to detect the fine differences in roughness between different body parts and skin conditions. Other potential applications for this device are for aging studies and the management of other rough lesions, such as warts, actinic keratoses, and psoriasis.

Taking advantage of the fact that speckles encode surface roughness information, we adapted the technique for skin applications. Specifically, we developed the theoretical formulation that relates the speckle contrast C of S_1 with the root mean square roughness Rq of the skin [75,76], and we selected the appropriate light source suitable for detecting skin roughness in the range of 20–100 μm [77]. Because the regular reflection is rather weak, it was enhanced and the volume backscattering was suppressed by applying spatial and polarization filtering [78]. Finally, the geometry and the setup of the device was carefully analyzed to ensure that the speckle pattern can be used for surface roughness assessment [79].

To validate our technique, we conducted a clinical study, systematically measuring 24 body sites from 72 adult volunteers recruited from the Skin Care Centre, Vancouver General Hospital. The speckle images produced by the blue light were processed and the skin surface roughness was computed. (The blue light was more suitable for surface roughness assessment.) The results were analyzed statistically and showed that males have a rougher skin surface than females. When the body sites were categorized into minimally, intermittently, and maximally sun exposed, the data revealed that the roughness of maximally sun-exposed sites was significantly higher than that of the other two categories (manuscript is under preparation). We currently are working to further validate the technique and apply it to the roughness evaluation of abnormal skin.

MELANOMA SCREENING

The device can also be used to analyze the polarimetric properties of the volume backscattering. We are particularly interested in the spatial distribution of the depolarization ratio D, which is defined as the difference between the parallel- and the perpendicular-polarized intensities, denoted by I_{parallel} and $I_{\text{perpedicular}}$, respectively, normalized by the sum of the two intensities as shown in Eq. (8.3) [80]:

$$D = \frac{I_{\text{parallel}} - I_{\text{prependicular}}}{I_{\text{parallel}} + I_{\text{prependicular}}} = \frac{S_1}{S_0}. \tag{8.3}$$

The term *depolarization* instead of *degree of linear polarization* is used to emphasize that we are interested in detecting the amount of depolarization, or the difference of polarization between the incidence light and the backscattered signals. Because our two cameras had a small misalignment due to a small tilt used to avoid the specular reflection, we corrected the alignment by applying a rigid registration computer program to the two polarization speckle images, and then we were able to generate a pixel-by-pixel depolarization map (image) $D(x, y)$, defined as

$$D(x,y) = \frac{I_{\text{parallel}}(x,y) - I_{\text{prependicular}}(x,y)}{I_{\text{parallel}}(x,y) + I_{\text{prependicular}}(x,y)} = \frac{S_1(x,y)}{S_0(x,y)}. \tag{8.4}$$

The distribution shape of the map $D(x, y)$ with N number of pixels is characterized by using the first to the fourth order of mathematical moments, denoted by M, σ, A, and E, respectively, which can be loosely described as the mean, variance, skewness, and kurtosis, respectively, of the distribution. The four mathematical moments are defined as

$$M = \frac{1}{N}\sum_{1}^{N} |D(x,y)|,$$

$$\sigma = \frac{1}{N}\sum_{1}^{N} D(x,y)^2,$$

$$A = \frac{1}{\sigma^{3/2}}\left(\frac{1}{N}\sum_{1}^{N} D(x,y)^3\right),$$

$$E = \frac{1}{\sigma^2}\left(\frac{1}{N}\sum_{1}^{N} D(x,y)^4\right). \tag{8.5}$$

A clinical study was conducted at the Skin Care Centre, Vancouver General Hospital [80]. Two-hundred fourteen skin patients, including 25 MMs, 11 squamous cell carcinomas (SCCs), 31 basal cell carcinomas (BCCs), 76 nevi, and 71 SKs, were recruited after a written informed consent. Their lesions were imaged and analyzed using depolarization-ratio maps. MMs had the highest first- and second-order moments but the lowest third- and fourth-order moments than that of the other lesions for both lasers. A Kruskal–Wallis statistical test confirmed a significant difference among all lesion types for all four moments ($p = .0001$). Dunn's multiple comparison tests identified pairs with significant difference (Table 8.1). The blue light was good at separating MM from SCC, BCC, and SK, whereas the red light could differentiate SK from MM, nevus, and BCC. It appears that the fourth moment is the most promising diagnostic predictor for separating MM from other lesion types under the blue laser and for separating SK under the red laser. MM and SK have been well studied because they often resemble and are confused with each other by patients and physicians. Many studies attempted to differentiate them. As shown in Fig. 8.2, we compared the receiver operating characteristic curve of

TABLE 8.1 Statistical Significant Pairs Detected by the Dunn's Multiple Comparison Test for the Blue and Red Light

BLUE LASER			
MM vs. SK***	MM vs. SK***	MM vs. SK***	MM. SK***
MM vs. SCC***	MM vs. SCC***	MM vs. SCC**	MM vs. SCC***
		MM vs. BCC*	MM vs. BCC*
			Nevus vs. SK*
			Nevus vs. SCC*
RED LASER			
SK vs. MM***	SK vs. MM***	SK vs. MM***	SK vs. MM***
SK vs. Nevus***	SK vs. Nevus***	SK vs. Nevus***	SK vs. Nevus***
SK vs. BCC*		SK vs. BCC**	SK vs. BCC**
		SCC vs. MM*	SCC vs. MM*

BCC, basal cell carcinoma; *MM*, malignant melanoma; *SCC*, squamous cell carcinoma; *SK*, seborrheic keratosis.
*$p < .05$; **$p < .001$; ***$p < .0001$.

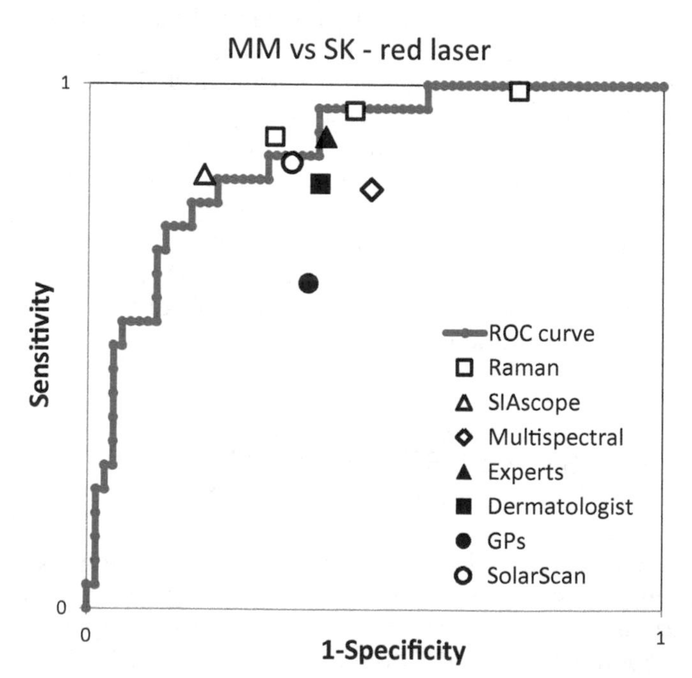

FIGURE 8.2 Comparing the separation of malignant melanoma and seborrheic keratosis for our method (receiver operating characteristic analysis using the fourth-order moment E of the polarization speckle patterns) with literature values. Published sensitivity and specificity values from studies with Raman (*open square*) [81], SIAscope (*open triangle*) [82], multispectral imaging (*open diamond*) [83], SolarScan (*open circle*) [84], and all three groups of physicians [84] [dermatologists specialized in melanoma (*closed triangle*), general dermatologists (*closed square*), and general practitioners (*closed circle*)] are also included. *Reprinted with permission from Tchvialeva L, Dhadwal G, Lui H, Kalia S, Zeng H, McLean DI, et al. Polarization speckle imaging as a potential technique for in vivo skin cancer detection. J Biomed Opt June 2013;18(6): 061211.*

the fourth-order moment on the red laser data with the reported literature values from the studies of Raman [81], SIAscope [82], a multispectral device [83], SolarScan [84], and physicians (melanoma experts, general dermatologists, and general practitioners) [84] and found that our method had a comparable performance as the Raman study [81] and melanoma experts [84]; on the other hand, our technique outperformed a multispectral device [83], general practitioners, and general dermatologists [84].

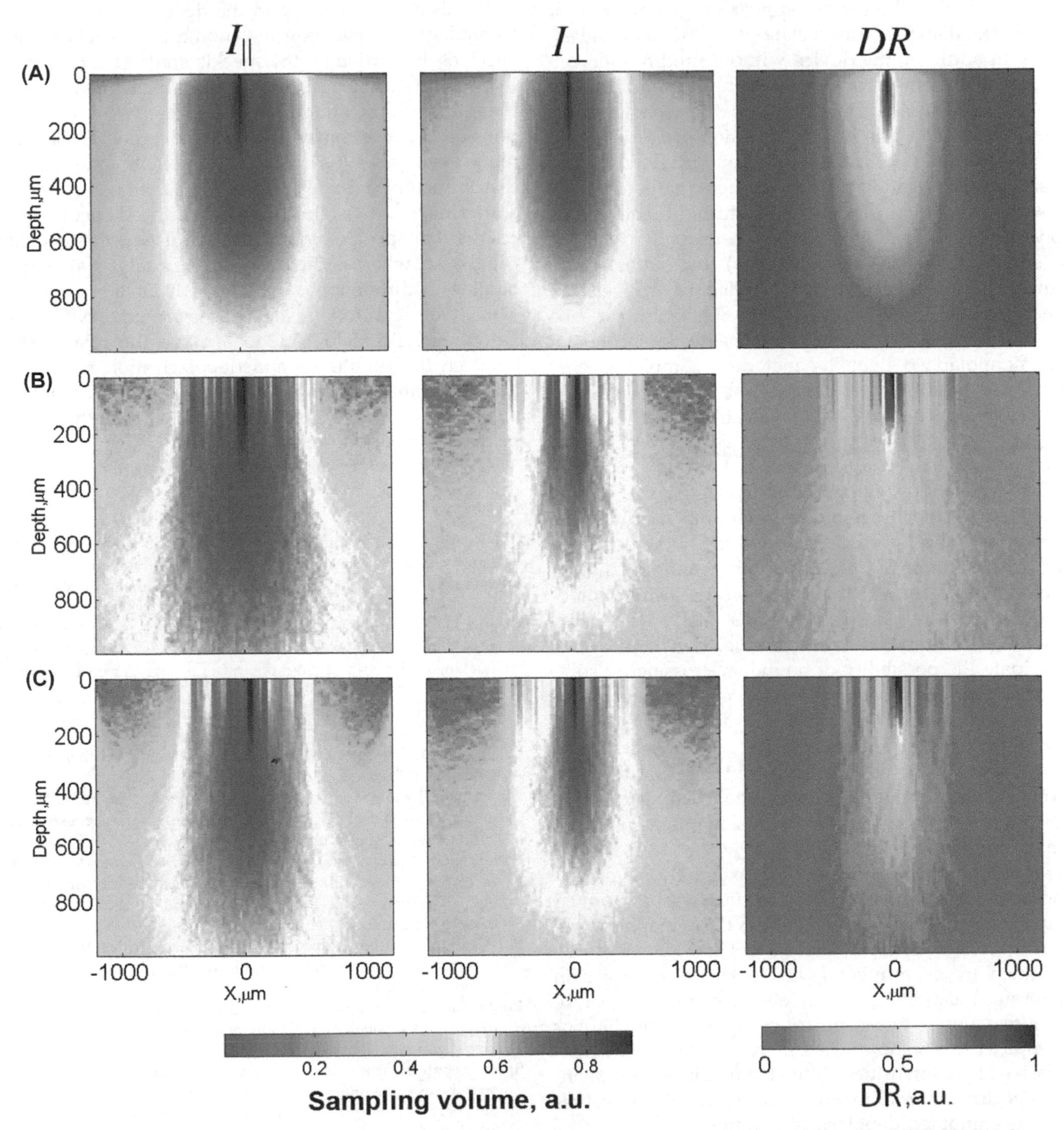

FIGURE 8.3 Spatial distribution of the co- and cross-polarized light intensity by the penetration depth (in micrometers) within the phantoms and their depolarization ratios, computed for various surface root mean square roughness values: (A) 2.5 μm, (B) 34.4 μm, (C) 65.8 μm. *Reprinted with permission from Tchvialeva L, Markhvida I, Lee TK, Doronin A, Meglinski I. Depolarization of light by rough surface of scattering phantoms. Proc SPIE 2013;8592:859217. San Francisco.*

ELECTRIC FIELD MONTE CARLO SIMULATION

The clinical results were encouraging and generated much interest. Using two polarized speckle images, our prototype seems to be able to detect, with reasonable accuracy, differences between skin cancers and benign lesions and that the results appear to be similar to or even better than the current state-of-art methods and experts. In addition, the device without modification can analyze skin surface roughness. To understand the process in detail and further develop the technique, we started a Monte Carlo (MC) simulation project.

We are interested in the effect that surface roughness has on the propagation of polarized light in turbid media. However, radiation propagation in turbid media can be solved analytically only for simple (perturbation method) or special (diffusion approximation) cases; other problems require the help of MC simulations. We selected the Electric Field MC approach [85,86], which expends the traditional scalar MC methods mainly for incoherent and nonpolarized light, by including complex electric fields and tracking co- and cross-polarization states. On the basis of Jones formalism, the Electric Field MC model accounts for the wave properties of light, temporal coherence, polarization, phase change, and a rough boundary [87–89]. The model assigns an initial weight to a photon package and launches it into a semi-infinite medium with a rough interface. The interaction of the photon package and the medium is tracked by the random photon-package step size, which is based on the distribution of scattering events. The Henyey–Greenstein phase function [90] is used to describe scattering. The "hop" events anticipate the possibility of boundary crossing by splitting the photon package into reflected and transmitted parts. Polarization of the photon packages are propagated along the trajectories, and the co- and cross-polarized components of the electric field are tracked at the scattering events. A model was developed to simulate the depolarization ratios using two rough-skin phantoms under the illumination of a red diode laser ($\lambda = 663$ nm, 5 mW). Typical simulations launched approximately 10^8 photon packages. A 10^{11}-package simulation was performed using parallel programming on NVIDIA graphics processing units in approximately 6 h. The MC model was validated by experimental data obtained from a skin phantom study [89,91]. The model can be used to analyze the interaction between polarized light and media. For example, Fig. 8.3 shows the difference of co- and cross-polarized light intensities within the media of three phantoms of different root mean square *Rq* roughness values and the computed depolarization ratios.

The MC simulation results showed that the co-polarized light has a wide and deeper penetration than the cross-polarized light. On the other hand, skin with a higher *Rq* roughness value subdues the light intensities, and the depolarization ratios.

CONCLUSION

In this chapter, we report the development of a new technology for detecting melanoma using polarization speckles. In particular, the speckle contrast can be used to measure skin surface roughness, which is a diagnostic parameter for melanoma and SK. These two skin conditions often cause confusion and misdiagnosis. The technique can also be used to evaluate the spatial distribution of light depolarization due to volume backscattering. The two clinical studies showed encouraging results that speckle contrast can be used to detect skin roughness, whereas the polarimetric analysis appear to be able to differentiate melanoma from other skin lesions. We began to investigate a new type of MC simulation, Electric Field MC, to analyze the relationship between the roughness and depolarization ratio. This line of research aims to develop a portable and rapid device that is suitable for busy clinics, especially for the primary care provider.

Glossary

Coherence A fundamental property of light. Coherent light beams have constant phase difference and the same frequency.

Depolarization of light The loss of initial polarization after light reflection/propagation from/inside media.

Electric Field Monte Carlo simulation A special type of Monte Carlo simulation that allows the simulation of complex electric fields and tracking of the co- and cross-polarization states for coherence and polarized light.

Graphics processing units A special computer process unit allowing fast parallel processing.

Laser speckle The interference pattern generated by coherent light when it is reflected by a rough surface.

Mathematical moments The first to fourth orders of mathematical moments are loosely described as the mean, variance, skewness, and kurtosis, respectively, of a distribution.

Polarization A property of light, polarization is the orientation of the oscillation of the electric field perpendicular to the traveling path of the light. When the electric fields oscillate in the same orientation, these rays have the same polarization.

Polarization speckles A special type of laser speckles, polarization speckles manifest random amplitude, random phases, and spatially random polarization.

Root mean square roughness A common method to quantify surface roughness by calculating the square root of the average of a set of the roughness height squares.

Speckle contrast The ratio of the standard deviation of light intensity over the average intensity. Speckle contrast is used to characterize the coherence of light. A contrast of 0 implies no coherence, and contrast of 1 implies full coherence.

Stokes vector A vector of four elements that is used to represent the polarization properties of light.

List of Acronyms and Abbreviations

2D Two-dimensional
A Third mathematical moment
BCC Basal cell carcinoma
C Contrast
CCD Charge-coupled device
D Depolarization
E Fourth mathematical moment
GPU Graphics Processing Unit
I Intensity
M First mathematical moment
MC Monte Carlo
MM Malignant melanoma
mW Milliwatts
N Number of pixels
nm Nanometer
***p* value** Statistical significant level of a test
ROC Receiver operating characteristic
Rq Root-mean-square roughness
$S = [S_0, S_1, S_2, S_3]$ A Stokes vector of four elements S_0, S_1, S_2, and S_3
SCC Squamous cell carcinoma
SK Seborrheic keratosis
x *x* coordinate
y *y* coordinate
λ Wavelength
σ Standard deviation
μm Micrometer

Acknowledgments

This work was supported in part by grants from the Canadian Dermatology Foundation, the Canadian Institutes of Health Research, the Natural Sciences and Engineering Research Council of Canada, the VGH & UBC Hospital Foundation, and the University of British Columbia Faculty of Medicine.

References

[1] American Cancer Society. Cancer facts and figures 2015. Atlanta: American Cancer Society; 2015.
[2] Gallagher RP, Lee T. Assessing incidence rates and secular trends in nonmelanocytic skin cancer: which method is best? J Cutan Med Surg 1998;3(1):35–9.
[3] Forman D, Bray F, Brewster DH, Gombe Mbalawa C, Kohler B, Piñeros M, et al. Cancer incidence in five continents (electronic version), vol. X. Lyon: IARC; 2013. Available from: http://ci5.iarc.fr.
[4] National Cancer Institute of Canada. Canadian cancer statistics 2002. Toronto, Canada; 2002.
[5] Canadian Cancer Society's Advisory Committee on Cancer Statistics. Canadian cancer statistics 2013. Toronto, Canada: Canadian Cancer Society; 2013.
[6] Gavin A, Walsh P. Melanoma of skin. Smps No 68: cancer atlas of the UK and Ireland. Office for National Statisics; 2005.
[7] Cancer Research UK. Skin cancer incidence statistics 2015. February 7, 2015. Available from: http://www.cancerresearchuk.org/cancer-info/cancerstats/types/skin/incidence/uk-skin-cancer-incidence-statistics.
[8] Canadian Partnership Against Cancer. The economic burden of skin cancer in Canada: current and projected. February 26, 2010.
[9] Vallejo-Torres L, Morris S, Kinge JM, Poirier V, Verne J. Measuring current and future cost of skin cancer in England. J Public Health (Oxf) March 2014;36(1):140–8.
[10] Balch CM, Gershenwald JE, Soong SJ, Thompson JF, Atkins MB, Byrd DR, et al. Final version of 2009 AJCC melanoma staging and classification. J Clin Oncol December 20, 2009;27(36): 6199–206. PubMed PMID: 19917835.
[11] Witheiler DD, Cockerell CJ. Sensitivity of diagnosis of malignant melanoma: a clinicopathologic study with a critical assessment of biopsy techniques. Exp Dermatol November 1992;1(4):170–5.
[12] Izikson L, Sober AJ, Mihm Jr MC, Zembowicz A. Prevalence of melanoma clinically resembling seborrheic keratosis: analysis of 9204 cases. Arch Dermatol December 2002;138(12):1562–6.
[13] Grin CM, Kopf A, Welkovich B, Bart R, Levenstein M. Accuracy in the clinical diagnosis of malignant melanoma. Arch Dermatol 1990;126:763–6.
[14] Mayer J. Systematic review of the diagnostic accuracy of dermatoscopy in detecting malignant melanoma. Med J Aust August 18, 1997;167(4):206–10. PubMed PMID: 9293268.
[15] Cassileth BR, Clark Jr WH, Lush EJ, Frederick BE, Thompson CJ, Walsh WP. How well do physicians recognize melanoma and other problem lesions. J Am Acad Dermatol 1986;14:555–60.
[16] Rapini R. Clinical and pathologic differential diagnosis. In: Bolognia JL, Jorizzo JL, Rapini RP, editors. Dermatology. London: Mosby; 2003.
[17] Jones TP, Boiko PE, Piepkorn MW. Skin biopsy indications in primary care practice: a population-based study. J Am Board Fam Pract November–December 1996;9(6):397–404.
[18] Ghosh N, Vitkin IA. Tissue polarimetry: concepts, challenges, applications, and outlook. J Biomed Opt November 2011;16(11): 110801.
[19] Boas DA, Dunn AK. Laser speckle contrast imaging in biomedical optics. J Biomed Opt January–February 2010;15(1):011109. PubMed PMID: 20210435.
[20] Wolf E. Introduction to the theory of coherence and polarization of light. Cambridge: Cambridge University Press; 2007.
[21] Elies P, LeJeune B, LeRoyBrehonnet F, Cariou J, Lotrian J. Experimental investigation of the speckle polarization for a polished aluminium sample. J Phys D Appl Phys January 7, 1997;30(1): 29–39.
[22] Takeda M, Wang W, Hanson SG. Polarization speckles and generalized Stokes vector wave: a review. Proc SPIE 2010:73870V.
[23] Angelsky OV, Ushenko AG, Ushenko YA, Ushenko YG, Tomka YY, Pishak VP. Polarization-correlation mapping of biological tissue coherent images. J Biomed Opt 2006;10(6):064025.
[24] Wang W, Hanson SG, Takeda M. Statistics of polarization speckle: theory versus experiment. Proc SPIE 2009:738803.
[25] Goodman JW. Speckle phenomena in optics: theory and application. Greenwood Village, Colorado: Roberts and Company Publishers; 2006.
[26] Briers J. Surface roughness evaluation. In: Sirohi RS, editor. Speckle metrology. CRC Press; 1993.
[27] Goodman JW. Statistical properties of laser speckle patterns. In: Dainty JC, editor. Laser speckle and related phenomena. Springer series topics in applied physics, vol. 9. Heidelberg: Springer-Verlag; 1975. p. 9–75.
[28] Pedersen HM. On the contrast of polychromatic speckle patterns and its dependence on surface roughness. J Mod Opt 1975;22(1): 15–24.
[29] Pedersen HM. Second-order statistics of light diffracted from gaussian, rough surfaces with applications to the roughness dependence of speckles. J Mod Opt 1975;22(6):523–35.
[30] Iwai T, Takai N, Asakura T. Space-time correlation function of the dynamic polychromatic laser speckle. J Mod Opt 1983;30:759–76.
[31] Parry G. Speckle patterns in partially coherent light. In: Dainty JC, editor. Laser speckle and related phenomena. Springer series topics in applied physics, vol. 9. Berlin; New York: Springer-Verlag; 1984. p. 77–122.

[32] McKechnie TS. Image-plane speckle in partially coherent illumination. Opt Quan Electronics 1976;8(1):61–7.
[33] Hu Y-Q. Dependence of polychromatic-speckle-pattern contrast on imaging and illumination directions. Appl Opt 1994;33(13):2707–14.
[34] George N, Jain A. Space and wavelength dependence of speckle intensity. Appl Phys 1974;4(3):201–12.
[35] Huntley JM. Simple model for image-plane polychromatic speckle contrast. Appl Opt 1999;38:2212–5.
[36] Rodrigues CMP, Pinto JL. Contrast of polychromatic speckle patterns and its dependence to surface height distribution. Opt Eng 2003;42(6):1699–703.
[37] Hun C, Caussignac J-M, Bruynooghe MM. Speckle techniques for pavement surface analysis. In: Gastinger K, Lokberg OJ, Winther S, editors. Proc SPIE; 2003. p. 261–6.
[38] McKinney JD, Webster KJ, Webb KJ, Weiner AM. Characterization and imaging in optically scattering media by use of laser speckle and a variable-coherence source. Opt Lett 2000;25:4–6.
[39] Zimnyakov DA, Agafonov DN, Sviridov AP. Speckle-contrast monitoring of tissue thermal modification. Appl Opt 2002; 41(28):5989–96.
[40] Tearney GJ, Bouma BE. Atherosclerotic plaque characterization by spatial and temporal speckle pattern analysis. Opt Lett 2002;27(7): 533–5.
[41] Li J, Ku G, Wang LV. Ultrasound-modulated optical tomography of biological tissue by use of contrast of laser speckles. Appl Opt 2002;41(28):6030–5.
[42] Yu P, Mustata M, Turek JJ, French PMW, Melloch MR, Nolte DD. Holographic optical coherence imaging of tumor spheroids. Appl Phys Lett 2003;83(3):575–7.
[43] Nothdurft R, Yao G. Imaging obscured subsurface inhomogeneity using laser speckle. Opt Express 2005;13(25):10034–9.
[44] Nadkarni SK, Bouma BE, Yelin D, Gulati A, Tearney GJ. Laser speckle imaging of atherosclerotic plaques through optical fiber bundles. J Biomed Opt 2008 ;13(5):054016.
[45] Tanin LV, Dick SC, Alexandrov SA, Loiko MM, Kumeisha AA, Markhvida IV, et al., editors. Laser specklometer for determining the biomechanical parameters of skeletal muscles and the microhaemodynamics of human skin. Laser optics '95: biomedical applications of lasers. Proc. SPIE, 2769; 1996. p. 94–100.
[46] Cheng H, Luo Q, Wang Z, Gong H, Chen S, Liang W, et al. Efficient characterization of regional mesenteric blood flow by use of laser speckle imaging. Appl Opt 2003;42(28):5759–64.
[47] Zakharov P, Volker AC, Wyss MT, Haiss F, Calcinaghi N, Zunzunegui C, et al. Dynamic laser speckle imaging of cerebral blood flow. Opt Express August 3, 2009;17(16):13904–17.
[48] Zimnyakov DA, Sinichkin YP, Tuchin VV. Polarization reflectance spectroscopy of biological tissues: diagnostic applications. Radiophysics Quan Electronics 2004;47(10–11):860–75.
[49] Morgan SP, Stockford I. Surface-reflection elimination in polarization imaging of superficial tissue. Opt Lett 2003;28(2):114–6.
[50] Demos SG, Radousky HB, Alfano RR. Deep subsurface imaging in tissues using spectral and polarization filtering. Opt Express 2000; 7(1):23–8.
[51] Jacques SL, Ramella-Roman JC, Lee K. Imaging skin pathology with polarized light. J Biomed Opt July 2002;7(3):329–40.
[52] Braun RP, Rabinovitz HS, Oliviero M, Kopf AW, Saurat J-H. Dermoscopy of pigmented skin lesions. J Am Acad Dermatol 2005;52: 109–21.
[53] Jacques S. Video imaging with polarized light finds skin cancer margins not visible to dermatologists: Steven Jacques. 1998 [cited July 2000]. Available from: http://omlc.bme.ogi.edu/news/feb98/polarization/index.html.
[54] Smith ZJ, Berger AJ. Surface-sensitive polarized Raman spectroscopy of biological tissue. Opt Lett June 1, 2005;30(11):1363–5.
[55] Ghosh N, Patel HS, Gupta PK. Depolarization of light in tissue phantoms - effect of a distribution in the size of scatterers. Opt Express 2003;11(18):2198–205.
[56] Sankaran V, Joseph T, Walsh J, Maitland DJ. Polarized light propagation through tissue phantoms containing densely packed scatterers. Opt Lett 2000;25(4):239–41.
[57] Sankaran V, Joseph T, Walsh J, Maitland DJ. Comparative study of polarized light propagation in biologic tissues. J Biomed Opt 2002; 7(3):300–6.
[58] Wood MFG, Ghosh N, Moriyama EH, Wilson BC, Vitkin IA. Proof-of-principle demonstration of a Mueller matrix decomposition method for polarized light tissue characterization in vivo. J Biomed Opt 2009;14:014029.
[59] Angelsky OV, Ushenko AG, Ushenko YA, Ushenko YG, Tomka YY, Pishak VP. Polarization-correlation mapping of biological tissue coherent images. J Biomed Opt 2005;10(6):064025.
[60] Boulvert F, Boulbry B, Le Brun G, Le Jeune B, Rivet S, Cariou J. Analysis of the depolarizing properties of irradiated pig skin. J Opt A Pure Appl Opt 2005;7(1):21–8.
[61] Angelsky OV, Ushenko AG, Ushenko YA, Ushenko YG. Polarization singularities of the object field of skin surface. Jphys D Appl Phys 2006;39(16):3547–58.
[62] Kim J, John R, Wu PJ, Martini MC, Walsh Jr JT. In vivo characterization of human pigmented lesions by degree of linear polarization image maps using incident linearly polarized light. Lasers Surg Med 2010;42(1):76–85.
[63] Kunnen B, Macdonald C, Doronin A, Jacques S, Eccles M, Meglinski I. Application of circularly polarized light for non-invasive diagnosis of cancerous tissues and turbid tissue-like scattering media. J biophotonics October 18, 2014;9999(9999).
[64] Detection of endometrial lesions by degree of linear polarization maps. In: Kim J, Fazleabas A, Walsh JT, editors. SPIE Photonics West; 2010; January 24, 2010. San Francisco.
[65] Lemaillet P, Ramella-Roman JC. Hemispherical Stokes polarimeter for early cancer diagnosis. Proc SPIE 2011;7883:788304.
[66] Ghassemi P, Lemaillet P, Germer TA, Shupp JW, Venna SS, Boisvert ME, et al. Out-of-plane Stokes imaging polarimeter for early skin cancer diagnosis. J Biomed Opt July 2012;17(7):076014.
[67] Phillips K, Xu M, Gayen S, Alfano R. Time-resolved ring structure of circularly polarized beams backscattered from forward scattering media. Opt Express 2005;13(20):7954–69.
[68] Stockford IM, Morgan SP, Chang PC, Walker JG. Analysis of the spatial distribution of polarized light backscattered from layered scattering media. J Biomed Opt July 2002;7(3):313–20.
[69] Salomatina E, Jiang B, Novak J, Yaroslavsky AN. Optical properties of normal and cancerous human skin in the visible and near-infrared spectral range. J Biomed Opt 2006 ;11(6):064026.
[70] Tchvialeva L, Zeng H, Lui H, McLean DI, Lee TK. Comparing in vivo Skin surface roughness measurement using laser speckle imaging with red and blue wavelengths. In: The 3rd world congress of noninvasive skin imaging; May 7–10, 2008; Seoul, Korea; 2008.
[71] Potorac A, Toma I, Mignot J. In vivo skin relief measurement using a new optical profilometer. Skin Res Technol 1996;2:64–9.
[72] Jaspers S, Hopermann H, Sauermann G, Hoppe U, Lunderstadt R, Ennen J. Rapid in vivo measurement of the topography of human skin by active image triangulation using a digital micromirror device. Skin Res Technol 1999;5:195–207.
[73] Piche E, Hafner HM, Hoffmann J, Junger MFOITS. (fast optical in vivo topometry of human skin): new approaches to 3-D surface structures of human skin. Biomed Tech (Berl) November 2000; 45(11):317–22.
[74] Levy JL, Servant JJ, Jouve E. Botulinum toxin A: a 9-month clinical and 3D in vivo profilometric crow's feet wrinkle formation study. J Cosmet Laser Ther May 2004;6(1):16–20.

[75] Markhvida I, Tchvialeva L, Lee TK, Zeng H. Influence of geometry on polychromatic speckle contrast. J Opt Soc Am A Opt Image Sci Vis January 2007;24(1):93–7.
[76] Tchvialeva L, Lee TK, Markhvida I, McLean DI, Lui H, Zeng H. Using a zone model to incorporate the influence of geometry on polychromatic speckle contrast. Opt Eng 2008;47(7):074201.
[77] Tchvialeva L, Markhvida I, Zeng H, McLean DI, Lui H, Lee TK. Surface roughness measurement by speckle contrast under the illumination of light with arbitrary spectral profile. Opt Lasers Eng 2010;48:774–8.
[78] Tchvialeva L, Zeng H, Markhvida I, Dhadwal G, McLean L, McLean DI, et al. Optical discrimination of surface reflection from volume backscattering in speckle contrast for skin roughness measurements. Proc SPIE 2009;7161:71610I.
[79] Tchvialeva L, Markhvida I, Lee TK. Error analysis for polychromatic speckle contrast measurements. Opt Lasers Eng December 2011;49(12):1397–401.
[80] Tchvialeva L, Dhadwal G, Lui H, Kalia S, Zeng H, McLean DI, et al. Polarization speckle imaging as a potential technique for in vivo skin cancer detection. J Biomed Opt June 2013;18(6):061211.
[81] Lui H, Zhao J, McLean D, Zeng H. Real-time Raman spectroscopy for in vivo skin cancer diagnosis. Cancer Res May 15, 2012;72(10): 2491–500.
[82] Moncrieff M, Cotton S, Claridge E, Hall P. Spectrophotometric intracutaneous analysis: a new technique for imaging pigmented skin lesions. Br J Dermatol 2002;146(3):448–57.
[83] Farina B, Bartoli C, Bono A, Colombo A, Lualdi M, Tragni G, et al. Multispectral imaging approach in the diagnosis of cutaneous melanoma: potentiality and limits. Phys Med Biol May 2000; 45(5):1243–54.
[84] Menzies SW, Bischof L, Talbot H, Gutenev A, Avramidis M, Wong L, et al. The performance of SolarScan: an automated dermoscopy image analysis instrument for the diagnosis of primary melanoma. Arch Dermatol November 2005;141(11): 1388–96.
[85] Doronin A, Meglinski I. Online object oriented Monte Carlo computational tool for the needs of biomedical optics. Biomed Opt Express September 1, 2011;2(9):2461–9.
[86] Doronin A, Meglinski I. Peer-to-peer Monte Carlo simulation of photon migration in topical applications of biomedical optics. J Biomed Opt September 2012;17(9):090504.
[87] Doronin A, Macdonald C, Meglinski I. Propagation of coherent polarized light in turbid highly scattering medium. J Biomed Opt 2014;19(2):025005.
[88] Doronin A, Radosevich AJ, Backman V, Meglinski I. Two electric field Monte Carlo models of coherent backscattering of polarized light. J Opt Soc Am A Opt Image Sci Vis November 1, 2014;31(11): 2394–400.
[89] Deleted in review.
[90] Henyey LG, Greenstein JL. Diffuse radiation in the galaxy. Astrophys J January 1941;93(1):70–83.
[91] Tchvialeva L, Markhvida I, Lee TK, Doronin A, Meglinski I. Depolarization of light by rough surface of scattering phantoms. Proc SPIE 2013;8592:859217. San Francisco.

CHAPTER

9

New Developments in Fluorescence Diagnostics

J. Hegyi, V. Hegyi

Inštitút klinickej a experimentálnej dermatovenerológie, Bratislava, Slovakia

OUTLINE

INTRODUCTION

Each medical discipline will win recognition from the population usually only on the basis of successful therapy. Patients are rarely interested in etiology and pathogenesis of the disease and if so, then only to know how to avoid a relapse. This is even more relevant in dermatology than in other specializations, since all blemishes and symptoms of the disease are clearly visible.

This visual background of dermatology has understandably led in the past to attempts to capture images of the surface and the depths of the skin. Dermatology is a visual science requiring a picture or a view , ranging from simple curiosity to forensic use. Some of these first attempts are clearly visible in some renaissance paintings and sculptures. The first documentation of disease was by using drawings, diagrams, and, much later, photography, although it was not anywhere near the quality that we know today. Also, wax sculptures were used in order to preserve the size, shape, and structure of lesions. Over time, the standard became photography and has become an important part of any dermatological practice. In the last decade, significant advances have been achieved in the direct viewing of the skin. The dermatologist in clinical practice is no longer restricted to examination by eyes, palpation, and biopsy. New methods of photography were developed using other parts of the spectrum. The advent of digital photography allowed us to analyze images at a later date using different approaches and filters. These new methods of imaging the skin allow us to see what was hidden from us in the past and give us the possibility to achieve better therapeutic results, earlier diagnosis, and excellent follow-up care of patients. However, most importantly, imaging methods finally give the dermatologist analyzable data that can be quantified. Although many of these new techniques have been applied so far in research only, some of them have become standard tools in clinical use. Typical examples include fluorescence analysis and diagnosis, confocal laser microscopy [1], high frequency ultrasound, and optical coherence tomography [2].

FLUORESCENCE HISTORY

Fluorescence is an optical phenomenon in which the molecular absorption of energy in the form of photons triggers the emission of fluorescent photons with

Imaging in Dermatology
http://dx.doi.org/10.1016/B978-0-12-802838-4.00009-1

a longer wavelength. Usually, the photon is absorbed in the ultraviolet (UV) spectrum, and radiation is emitted in the visible spectrum. Fluorescence, unlike phosphorescence, is only short-lived and disappears after the excitation ceases [3]. The history of photoactive agents dates back to H. Scherer, who first produced hematoporphyrin in 1841 [4]. However, until 1867 the fluorescent properties of hematoporphyrin were not known [6]. The characteristic red fluorescence was first observed in 1924 by Policard in rat sarcoma using a Wood lamp [7]. This lamp was constructed by Robert W. Wood in 1903 and was used in dermatology for the detection of fungal infections of the hair [8]. A Wood lamp emits long-wave UV light (also called *black light*) in the range of 320–400 nm with a peak at 365 nm. When tissue absorbs Wood light UV, fluorescence is emitted with a longer wavelength, usually in the range of visible light. This can be further highlighted using special filters to eliminate the remaining UV radiation or autofluorescence, which can lead to greater contrast between normal and diseased skin. This approach is the basis for modern fluorescence detection and its use in clinical practice [9]. The first localization of malignant tumors by fluorescence detection dates back to 1942, when Auler and Banzer administered porphyrins topically [10]. In 1948 Figgis and Weiland used protoporphyrins (hematoporphyrin, coproporphyrin) and observed fluorescence in mouse models with experimentally induced and transplanted tumors [11]. It was also reported that some fluorescence could be observed in healthy tissue as well, such as in lymph nodes, the omentum, fetal tissue, placental tissue, and healing wounds. In humans, the first attempts to detect tumors via fluorescence date back to the 1950s. In 1955, Rassmussan and Taxdal described the typical red fluorescence after intravenous application of hematoporphyrin hydrochloride to patients with various types of malignant and benign lesions. It was observed that the fluorescence in tumors increased proportionally with the applied dose of porphyrin. The authors even managed to locate a breast carcinoma through intact skin and an adenocarcinoma through the abdominal wall [12].

Currently the main use of fluorescence diagnosis (FD) in practice is the exact detection and localization of poorly demarcated neoplasms. An ever-increasing number of patients with nonmelanocytic skin cancer (NMSC) have led to the development of new noninvasive, diagnostic, and therapeutic methods, and one of the most common detection procedures for NMSC has become FD. Generally if an intense red fluorescence is visible, it is very likely that a tumor or tumor recurrence is present. This fact has been confirmed in many cases by histological analysis. Although FD is a very sensitive method, it lacks specificity. However, several new imaging devices and methods have recently become available, which compensate for this drawback.

FLUORESCENCE PRINCIPLES

FD can either be performed by point measurements or by imaging geometry. In point monitoring mode, the entire fluorescence spectrum is obtained from one small tissue spot. In imaging mode, a larger area is monitored, but usually the fluorescence is detected with a much lower spectral resolution. FD on tissue can either use the differences in autofluorescence (the natural fluorescence of the tissue itself) between normal and diseased tissue, or the fluorescence from a reporter or marker molecule that is specifically accumulated in the tumors and that has very strong fluorescence.

When the fluorescence of porphyrins in the tissue is being detected, the most widely used substances are aminolevulinic acid (ALA) and the methyl ester of ALA (MAL). Although ALA/MAL is taken up by both affected and healthy skin; differences in metabolism and vascularization are the key factors in different fluorescence intensities. ALA is a precursor of heme and enters the metabolic process, which results in a light-sensitive product [protoporphyrin IX (PpIX)] before getting further transformed into heme that is then incorporated into hemoglobin. This process occurs in a healthy cell, but because neoplastic cells contain lower amounts of ferrochelatase, and when the feedback inhibition is bypassed by administration of ALA, incomplete conversion occurs and an intermediate product of this cascade PpIX lingers in the abnormal cells. PpIX has the ability to emit fluorescence after blue or UVA light irradiation [5]. The advantage of MAL over ALA is the fact that the cell uptake is much higher and occurs more rapidly, which leads to a roughly 30% increase in efficiency of the ALA methyl ester compared with unmodified ALA.

FLUORESCENCE CAPTURE AND DETECTION

When dealing with noninvasive imaging, digital photography is an important issue. Even though this technology is now considered standard in medical documentation, it still holds many untapped resources for the clinician. Modern digital cameras allow the capture of images at increasingly higher image resolution. Additionally, various special modes can be applied using clinical photography. Hereby, it is possible to image fluorescence with a greater sensitivity than our eyes allow

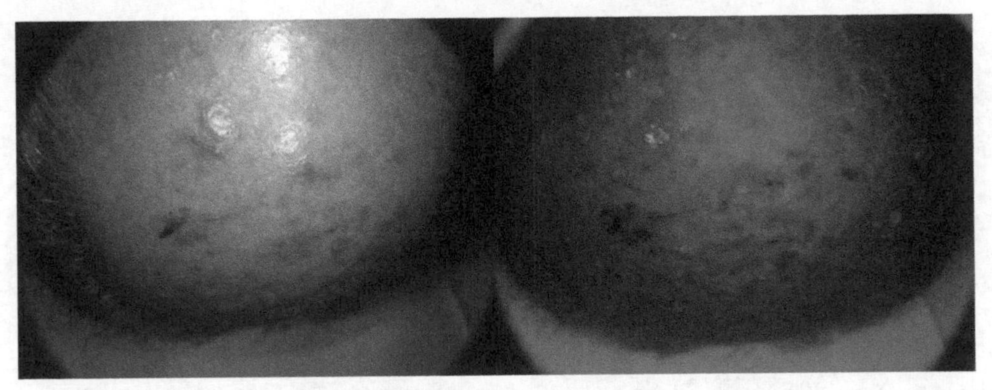

FIGURE 9.1 Aminolevulinic acid (ALA)-mediated fluorescence of actinically altered skin imaged with a commercially available digital camera (3.8 Mpix). The light source used was a Wood lamp. *Photo by author.*

(Fig. 9.1). One of the drawbacks of this technology is the necessity to employ high ISO (International Organization for Standardization) or DIN (Deutsches Institut für Normung (German Institute for Standardization)) values, which lead to long exposure times and thus to the risk of blurred images. To avoid this drawback, static positioning of both the device and the target is necessary. Additionally, the light source can be attached directly to the camera and thus ensure constant illumination. One such device incorporating this type of design is the Curalux fluorescence detection device (Saalmann GmbH, Herford, Germany). The device also incorporates a specific cut-off filter that eliminates reflected excitation light. The obtained images are therefore not oversaturated in the blue channel and far better contrast is achieved. The advantage of the Curalux is its versatility, easy manipulation, and very good image quality. This setup also requires only one static point in order to eliminate blur (Figs. 9.2 and 9.3).

FIGURE 9.2 A portable device for fluorescence imaging incorporating a digital camera, ultraviolet (UV) light source in the form of an LED light, and a power source (Saalmann GmbH, Herford, Germany).

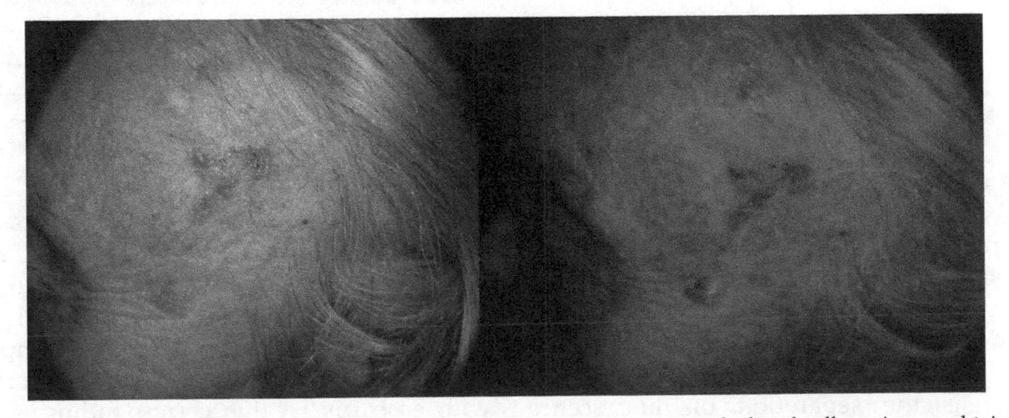

FIGURE 9.3 Methyl ester of aminolevulinic acid (MAL)-mediated fluorescence imaging of a basal cell carcinoma obtained with a portable device for fluorescence imaging incorporating a digital camera, ultraviolet (UV) light source in the form of an LED light, and a power source. *Photo by author.*

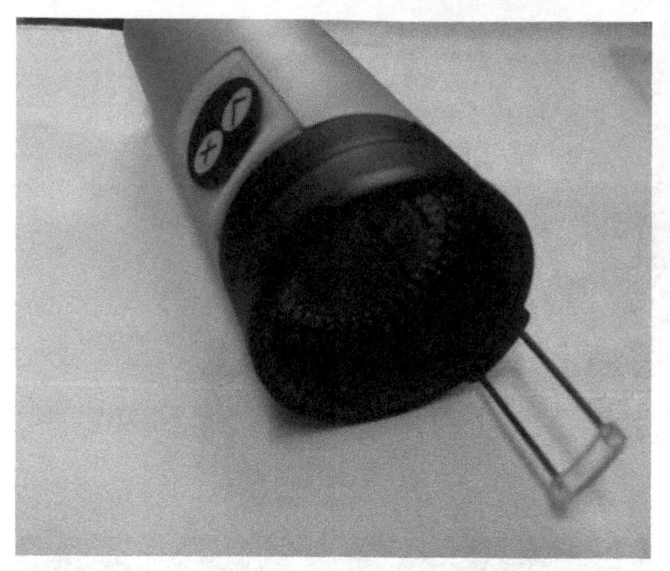

FIGURE 9.4 Handpiece of a dedicated fluorescence imaging device with a stabilizing pod and LED light sources. *Photo by author.*

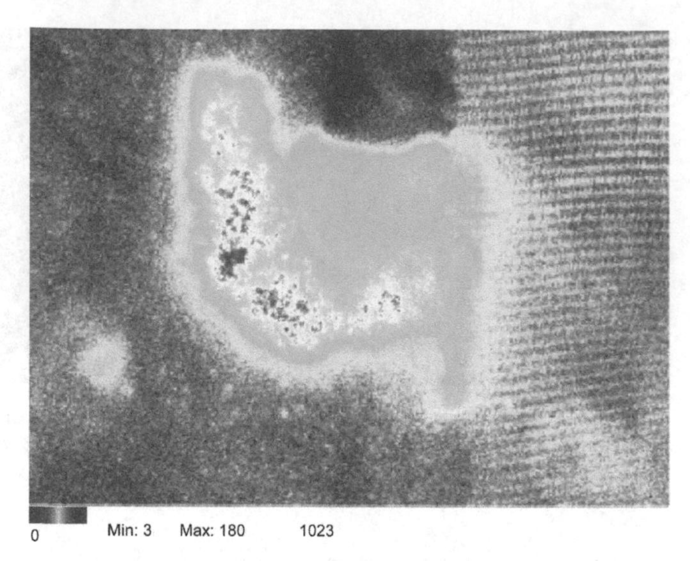

FIGURE 9.5 Color coding of a basal cell carcinoma fluorescence image using look-up tables (LUTs). The *red* color represents the strongest intensity of fluorescence, and *green* the weakest. *Photo by author.*

A different approach to acquiring quality images is the use of dedicated fluorescence detection devices. These usually contain an array of cameras connected to a detector with integrated imaging software and a handpiece designed specifically for fluorescence capture. Usually different light sources and lens systems are incorporated and may be customizable. The most common sources include visible light illumination and UV light. Switching between these light sources allows both normal images and fluorescence images to be obtained. These can then be merged together in the form of overlays that provide additional data on fluorescence localization. A disadvantage is that the devices are usually bulky and require much space. The fact that a handpiece is used helps to offset the mass and the weight of the device but means that a stabilizing unit has to be connected to ensure sharp images (Fig. 9.4).

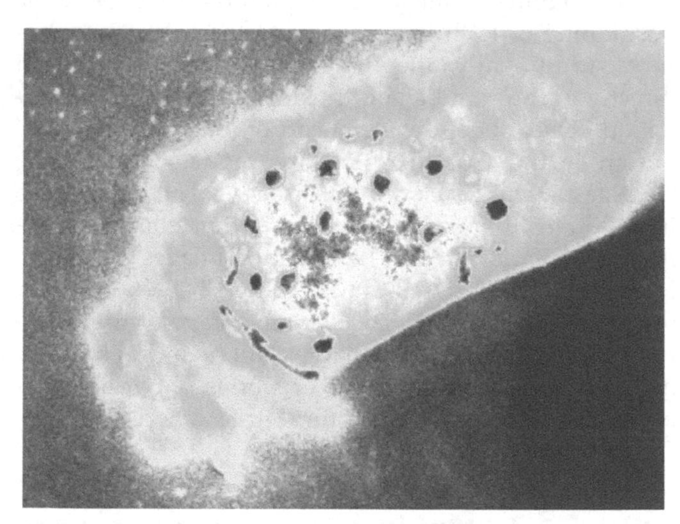

FIGURE 9.6 Color coding of a basal cell carcinoma fluorescence image using look-up tables (LUTs). The *red* color represents the strongest intensity of fluorescence, and *green* the weakest. *Blue dots* represent the expected clinical border of the lesion. *Photo by author.*

IMAGE ENHANCEMENT

Once images are obtained, further software analysis of the images can comprise increasing the sharpness of the image, measuring the intensity of the signal and applying specific colors to different intensities, so-called "look-up tables (LUTs)." This modern digital approach provides a much better and precise evaluation of the pictures [13] (Figs. 9.5 and 9.6). A helpful and viable approach is the separation of fluorescence and autofluorescence, which allows isolation of the desired spectrum. A similar effect can be achieved by placing specific cut-off filters in front of the camera lens when obtaining images. It is also possible to use charged-coupled device (CCD) cameras for imaging. Depending on the desired wavelength, a CCD camera with different filters is used most commonly. Hereby, the intensity can be detected over a two-dimensional area. The CCD camera is connected to a computer for image software analysis. The data consists of grayscale images with each pixel value corresponding to the measured intensity value (Fig. 9.7). This can be converted to a predefined color scale for better distinction of different intensities [14]. Ideally a threshold value can be defined to differentiate between neoplastic and normal tissue.

FIGURE 9.7 Grayscale charge-coupled device (CCD) image of a basal cell carcinoma lesion in the medial canthus. The image can be further amplified using software modeling, allowing sharper demarcation and contrast. *Photo by author.*

AUTOFLUORESCENCE

Autofluorescence is the tissue-endogenous fluorescence caused by several different fluorophores. These include collagen and elastin as components of the connective tissue, tryptophan as a component of most proteins, and nicotinamide adenine dinucleotide (NAD), a coenzyme found in all living cells. The fluorescence lifetimes vary between the mentioned chromophores, and this fact forms the basis for the differentiation between separate tissue types. Collagen and melanin, for instance, have very short lifetimes (0.2–0.4 ns) and flavin, on the other side of the spectrum, has a long lifetime (3.5–5.2 ns). The spectra depend on the excitation wavelength and are also influenced by the optical properties of the tissue, in which the fluorophores are not uniformly distributed. Strong absorbers, such as hemoglobin, can reabsorb fluorescent light emitted at certain wavelengths and thus change the fluorescence spectrum.

Normal and diseased tissue can be distinguished because of differences in structure and metabolism. Metabolic fluorophores include reduced NAD (NADH)–NAD+ and flavins, and structural fluorophores are collagen and elastin. Whereas the emission spectrum of autofluorescence is in the green spectrum, the emission spectrum of ALA/MAL is in the red range.

The advantage of the use of autofluorescence in cancer demarcation is that there is no need to apply exogenous substances, which may be time consuming and potentially harmful [15–18].

ALTERNATIVE DETECTION METHODS

The second most widely used method for fluorescence detection is spectroscopy. The analysis of skin by means of spectrophotometry is a relatively new concept. The concept is scanning and analyzing reflected light from the skin after exposure to an activating light source. Fluorescence spectroscopy is a very sensitive technique for qualitative as well as quantitative measurements of tissue constituents. Traditionally fluorescence emission spectra have been used almost exclusively, although there are many other properties of fluorescence emission that can be used for diagnostics, eg, differences in peak excitation wavelength, the fluorescence lifetime, and the polarization of the fluorescence [19,20]. The most common spectrophotometers use light sources in the UV and visible regions of the spectrum, and some of these instruments also operate in the near-infrared region. The spectrophotometer quantitatively compares the light emitted from the lesion and compares it with a reference. Both emitted and reflected light are guided to the spectrometer, which transforms the light into a "rainbow" of wavelengths, and the excitation and emission can be easily separated. The intensity of the reflected light is measured by the spectrometer with a photodiode or other light sensor, and the spectrum is then compared with the reference. The method takes advantage of chromophores, which are located in the skin and which are substances that absorb and reflect light of different wavelengths. Various components of skin produce different spectral emission and these are then evaluated with appropriate computer software reconstructs. The attained data is then visualized by curves showing exact values to corresponding wavelengths. The method may not be as fast and easy to use as conventional digital or CCD imaging but, when coupled with a raster scanning device, gives more precise values that can overcome the inherent flaws of FD. Among many advantages of FD, the most important one is that it can easily be used for in vivo investigations [21–25].

By using optical fibers, skin but also internal organs, such as lung or cervix, can be accessed. Furthermore, fluorescence spectroscopy may be performed by utilizing other imaging techniques, making fluorescence investigations less time consuming, especially when screening larger areas [26].

Fluorescence and autofluorescence detection has become a mainstream diagnostic tool in dermatology. It has come a long way and evolved into a method with viable uses in detection of aberrant tissues. The fact that images can be digitized and analyzed under different conditions offers an even more powerful tool to the examiner and opens new alternatives in tumor detection.

References

[1] Rajadhyaksha M. Confocal microscopy of skin cancers: translational advances toward clinical utility. Conf Proc IEEE Eng Med Biol Soc 2009;1:3231–3.

[2] Mogensen M, Thrane L, Joergensen TM, Andersen PE, Jemec GB. Optical coherence tomography for imaging of skin and skin diseases. Semin Cutan Med Surg 2009;28:196–202.

[3] Gregorie Jr HB, Horger EO, Ward JL, Green JF, Richards T, Robertson Jr HC, et al. Hematoporphyrin derivative fluorescence in malignant neoplasms. Ann Surg 1968;167:829.

[4] Figge FHJ, Weiland GS. The affinity of neoplastic embryonic and traumatized tissue for porphyrins and metalloporphyrins. Anat Rec 1948;100:659.

[5] Henderson BW, Dougherty TJ. How does photodynamic therapy work? Photochem Photobiol 1992;55:145–57.

[6] Scherer H. Chemisch-physiologische Untersuchungen. Ann Chem Pharm 1841;40:1.

[7] Policard A. Etudes sur les aspects offerts par des tumeurs experimentales examinées a la lumière de Wood. C R Soc Biol 1924;91: 1423–8.

[8] Wood RW. Secret communications concerning light rays. J Physiol 1919. In: Asawanonda P, Charles TR, editors. Wood's light in dermatology. Int J Dermatol, 38; 1999. p. 801–7.

[9] Kennedy JC, Pottier RH. Endogenous protoporphyrin IX: a clinically useful photosensitizer for photodynamic therapy. J Photochem Photobiol (B) 1992;14:275–92.

[10] Auler H, Banzer G. Untersuchungen über die Rolle der Porphyrine bei geschwulstkranken Menschen und Tieren. Z Krebsforsch 1942;53:65–8.

[11] Figge FHJ, Weiland GS, Manganiello LOJ. Cancer detection and therapy: affinity of neoplastic, embryonic and traumatized regenerating tissues for porphyrins and metalloporphyrins. Proc Soc Exp Biol Med 1948;68:640–1.

[12] Rassmussan-Taxdal DS, Ward GE, Figge FHJ. Fluorescence of human lymphatic and cancer tissues following high doses of intravenous hematoporphyrin. Cancer 1955;8:78–81.

[13] Andersson-Engels S, Canti G, Cubeddu R, Eker C, Klinteberg C, Pifferi A, et al. Preliminary evaluation of two fluorescence imaging methods for the detection and the delineation of basal cell carcinomas of the skin. Lasers Surg Med 2000;26:76–82.

[14] Han X, Lui H, McLean DI, Zeng H. Near-infrared autofluorescence imaging of cutaneous melanins and human skin in vivo. J Biomed Opt 2009;14:024017.

[15] Fritsch C, Becker-Wegerich PM, Menke H, Ruzicka T, Goerz G, Olbrisch RR. Successful surgery of multiple recurrent basal cell carcinomas guided by photodynamic diagnosis. Aesthet Plast Surg 1997;21:437–9.

[16] Wennberg AM, Gudmundson F, Stenquist B, Ternesten A, Molne L, Rosen A, et al. In vivo detection of basal cell carcinoma using imaging spectroscopy. Acta Derm Venereol 1999;79:54–61.

[17] Fischer F, Dickson EF, Pottier RH, Wieland H. An affordable, portable fluorescence imaging device for skin lesion detection using a dual wavelength approach for image contrast enhancement and aminolevulinic acid-induced protoporphyrin IX. I. Design, spectral and spatial characteristics. Lasers Med Sci 2001;16(3):199–206.

[18] Tope WD, Ross EV, Kollias N, Martin A, Gillies R, Anderson RR. Protoporphyrin IX fluorescence induced in basal cell carcinoma by oral delta-aminolevulinic acid. Photochem Photobiol 1998;67: 249–55.

[19] Yavari N. Optical spectroscopy for tissue diagnostics and treatment control [Doctoral thesis]. Department of Physics and Technology, University of Bergen; April 2006.

[20] Sandberg C. Aspects of fluorescence diagnostics and photodynamic therapy in non-melanoma skin cancer [Doctoral thesis]. Department of Dermatology and Venereology, Sahlgrenska University Hospital, Institute of Clinical Sciences, Sahlgrenska Academy University; 2009.

[21] Lohmann W, Nilles M, Bodeker RH. In situ differentiation between nevi and malignant melanomas by fluorescence measurements. Naturwissenschaften 1991;78:456–7.

[22] Sterenborg HJCM, Motamedi M, Wagner Jr RF, Duvic M, Thomsen S, Jacques SL. In vivo fluorescence spectroscopy and imaging of human skin tumours. Lasers Med Sci 1994;9:191–201.

[23] Schomacker KT, Frisoli JK, Compton CC, Flotte TJ, Richter JM, Nishioka NS, et al. Ultraviolet laser-induced fluorescence of colonic tissue: basic biology and diagnostic potential. Lasers Surg Med 1992;12:63–78.

[24] Papazoglou TG. Malignancies and atherosclerotic plaque diagnosis: is laser induced fluorescence spectroscopy the ultimate solution? J Photochem Photobiol B 1995;28:3–11.

[25] Fritsch C, Ruzicka T. Fluorescence diagnosis and photodynamic therapy of skin diseases handbook and atlas. 1st ed. Springer; 2003.

[26] Sapozhnikova VV, Shakhova NM, Kamensky VA, Petrova SA, Snopova LB, Kuranov RV. Capabilities of fluorescence spectroscopy using 5-ALA and optical coherence tomography for diagnosis of neoplastic processes in the uterine cervix and vulva. Laser Phys 2005;12:1664–73.

C H A P T E R

10

Ex Vivo Fluorescence Microscopy: Clinical Applications in Dermatology and Surgical Pathology

C. Longo[1], *S. Gardini*[1], *S. Piana*[1], *F. Castagnetti*[1], *G. Argenziano*[2], *G. Pellacani*[3], *M. Ragazzi*[1]

[1]**Arcispedale Santa Maria Nuova-IRCCS, Reggio Emilia, Italy;** [2]**Second University of Naples, Naples, Italy;** [3]**University of Modena and Reggio Emilia, Modena, Italy**

O U T L I N E

INTRODUCTION

Precise surgical removal of tumor with minimal damage to the surrounding healthy tissue is achieved with a series of excisions that are guided by the histology of frozen sections. Examples include micrographic Mohs surgery (MMS) for the removal of epithelial skin cancer; excisions of oral mucosal lesions, thyroid nodules, parathyroid glands, and bone during head-and-neck surgery; and needle core biopsies and lumpectomies of breast cancer and many other tissues. MMS typically is time-consuming. In fact, frozen histology sections are prepared of each sample removed during surgery, and then the examination for the presence and location of tumor in the histology sections of each excision guides the subsequent excision. The preparation of frozen histology sections is labor intensive and slow, meaning that MMS usually lasts from 2 to several hours. Approximately 1 million MMSs are performed per year in the United States, with total treatment costs of approximately $1 billion [1]. The costs for conventional frozen histology are 15%, which amounts to $150 million per year.

Ex vivo fluorescence confocal microscopy (FCM) offers an attractive alternative to frozen histology because nuclear morphology may be imaged in real time and directly in freshly excised tissue [1–22]. It has been

Imaging in Dermatology
http://dx.doi.org/10.1016/B978-0-12-802838-4.00010-8

applied firstly on skin cancer and in particular on basal cell carcinoma (BCC) and later on visceral tumors. Gray-scale images may be stitched together to create mosaics to display large areas of tissue, as necessary for surgical pathology. The large field of view of 12 × 12 mm corresponds approximately to a view with 2× magnification that is routinely used by Mohs surgeons when examining frozen histology with a standard light microscope. Mosaics were created in less than 9–10 min whereas preparation of frozen histology requires 20–45 min per excision. Thus large-area mosaicking may offer a means for rapid examination of BCCs and other tumors directly in fresh excisions while minimizing labor and costs. Herein we summarize the main applications of this new and intriguing technique that may evolve into an adjunct or an entirely new alternative to frozen histology.

CONTRAST AGENTS

In the past, acetic acid was used as a contrast agent to "aceto-whiten" or brighten nuclei in reflectance imaging [2]. However, the strongly scattering normal dermis that surrounds the lesion also appeared bright. For this reason, small, tiny strands and sparsely nucleated tumors as seen in micronodular, infiltrative, and sclerosing/infiltrative BCCs were not visualized. The detection power and therefore the perceived contrast and visual detectability of a tumor depends on its size relative to the surrounding background. A nucleus in a BCC tumor is typically seen as 100 pixels in an individual image but subsequently appears as only one pixel in a mosaic. This is due to scaling down of full-resolution mosaics by 10× to match the pixilation and resolution to that of a 2× view of histology. To overcome this issue, fluorescence imaging was explored. In fact, in fluorescence, when using a contrast agent that specifically stains nuclei, very little light may be collected from the surrounding dermis. Detection of both large and small BCC tumors and also squamous cell carcinomas (SCCs) using fluorescence from methylene blue and toluidine blue was reported. However, acridine orange is the most used agent in the clinical setting because it offers the best contrast. Acridine orange differentially stains nuclear DNA and cytoplasmic RNA in endothelial cells. Acridine orange has a quantum yield of 75% when bound to DNA and an extinction coefficient of approximately 53,000 L/mole-cm. Using an analytical model for detectability in a fluorescence confocal microscope, the nuclear contrast was estimated to drop from 105 to 103 relative to a significantly darkened dermis. This implies that even small and tiny tumors of BCCs or other carcinomas can be seen in mosaics.

INSTRUMENT AND IMAGE ACQUISITION

Confocal mosaics are acquired using a newer version of an ex vivo fluorescence confocal microscope (Vivascope 2500, Mavig, Munich, Germany; FCC Class 2 A, EC). The laser illumination wavelength used is 488 nm; the illumination laser power is automatically set in the microscope and the depth manually adjusted to image the surface. Imaging is with a 30×, 0.9 numerical aperture water immersion lens, which provides optical sectioning of approximately 1.5 μm and resolution of approximately 0.4 μm at the 488-nm wavelength. Acridine orange was chosen as the contrast agent. Acridine orange provides a strong contrast between nucleus and dermis because it specifically stains the DNA and RNA of nucleated cells. With FCM, only weak fluorescence is collected from the dermis and subcutaneous fat, thereby increasing the contrast of epithelial cells up to 1000-fold, including the epidermis, adnexal structures, and tumor cells.

Specimens are freshly excised lesions from MMS. Instruments and procedure have been described elsewhere [23]. In brief, the procedure can be summarized in five steps as follows [19,22]:

Step 1: Specimens are freshly excised tissue of Mohs surgery. A drawing with the schematic representation of the anatomic area (ie, face) is provided along with the orientation of the specimen that is usually done by using silk suture at one or more poles of the tissue. In brief, all margins are numbered using the o'clock procedure and labeled as A, B, C, D, and E.

Step 2: The tissue of 2- to 3-mm thickness is immersed in a 0.6 mM solution of acridine orange dye for 10–20 s, and then the excess dye is removed by wiping with paper.

Step 3: The tissue is sandwiched between two glass slides that are fixed with modeling clay and then positioned onto the platform of the inverted microscope. The button "Scan" is set up. The imaging acquisition can be done as "A/C" (automatic contrast) or set up manually by using the best fluorescent contrast of the tissue.

Step 4: For each skin specimen (one layer/margin), a two-dimensional sequence of images is acquired and the images stitched together into a mosaic that displays the entire area of the tissue. The size of the mosaic ranges from 7 × 7 to 12 × 12 mm.

Step 5: After FCM imaging, the "sandwich" can be easily broken and the specimen processed for frozen sections with complete preservation of tissue characteristics.

Current technical effort is focused on increasing the speed and efficiency of imaging with high-speed strip mosaicism confocal microscopy. A digital staining approach is also being developed to transform grayscale (black and white) contrast of mosaics to purple and pink to simulate the appearance of standard histopathology [13,14,18].

APPLICATION OF EX VIVO FLUORESCENCE MICROSCOPY IN DERMATOLOGY

Basal Cell Carcinoma

BCCs constitute approximately 70–80% of all skin cancer, and BCC is the most common malignancy among the human population. Standard surgery is generally considered the treatment of choice. Even with surgery, there are several tumor characteristics that are associated with higher recurrence rates. These include tumors located around the eyes, nose, lips, and ears; morphoeic, infiltrative, micronodular, and basosquamous histopathological subtypes; BCCs with ill-defined margins; recurrent lesions; incompletely excised lesions; and those with perineural or perivascular involvement. BCCs with these characteristics are preferably treated with MMS, which allows for complete examination of all tissue margins, minimizing the risk of recurrence and avoiding unnecessary removal of healthy tissue.

FCM has been applied in the setting of MMS, and a definite set of classic and new histological criteria has been adapted to interpret the FCM grayscale images [19,22]. The FCM criteria include the presence of the following:

1. *Fluorescence*: The presence of fluorescence was determined when bright-white images were seen on the screen. Fluorescence corresponds to nucleated cells stained with acridine orange. As an area of higher fluorescence compared with a darker background, fluorescence is typically seen as forming structures/aggregates.
2. *Tumor demarcation*: Tumor shape was divided into two categories: ill-defined when a line could not be clearly drawn to separate the tumor from the surrounding tissue and well demarcated when a line could be drawn.
3. *Nuclear crowding*: Nuclear crowding was determined when the nuclear density was higher than that of the surrounding epidermis and adnexal structures.
4. *Peripheral palisading*: Palisading is described as peripheral polarized and aligned fluorescent ellipses, being the counterpart of the so-called criteria in formalin-fixed hematoxylin and eosin stained slides; it corresponds to the prominent tendency of the outermost row of basal cells to be arranged in a parallel-polarized way.
5. *Clefting*: A hypofluorescent space surrounding basaloid islands; it partially outlines the islands in the micronodular and infiltrative BCC subtype whereas it is more evident in superficial and nodular tumors.
6. *Nuclear pleomorphism*: Nuclear pleomorphism is a deviation from the normal round or oval nuclear outline present in normal keratinocytes.
7. *Enlarged nucleus-to-cytoplasm ratio*: BCC nests are seen as crowded masses of elongated heterogeneous spots of fluorescence (prominent nuclei) with poor or absent cytoplasm.
8. *Stroma*: Tumoral stroma is the modified dermis surrounding the BCC mass. When viewed in FCM mosaics, the stroma is seen as a more densely nucleated dermis as fluorescent dots within a black background.

Considering the different BCC subtypes in FCM images, they reveal specific morphologic aspects [22]. In detail, superficial BCCs are typified by the presence of a proliferation of atypical basaloid cells that form an axis parallel to the epidermal surface and demonstrate slit-like retraction of the palisaded basal cells from the subjacent stroma (cleft-like spaces). Nodular BCCs reveal small to large nodules with peripheral palisading (Fig. 10.1) and clefting whereas the micronodular subtype shows monotonous small, round, and well-defined islands with roughly the same shape and contour. Infiltrative BCCs are the most challenging tumor because they appear as columns and cords of basaloid cells one to two cells thick with sharp angulation, enmeshed in a densely collagenized stroma; palisading and clefting are not always present.

A study conducted on 80 BCCs using FCM criteria demonstrated an overall sensitivity and specificity of detecting residual BCC of 88% and 99%, respectively [21]. Moreover, the new technique reduced by almost two-thirds the time invested when compared with the processing of a frozen section.

Several pitfalls may occur when evaluating FCM images [22]. In particular, using FCM, tiny and angulated cords and strands of fluorescent cells that are typically found in infiltrative BCCs can be more difficult to recognize. Furthermore, it can be difficult to distinguish the

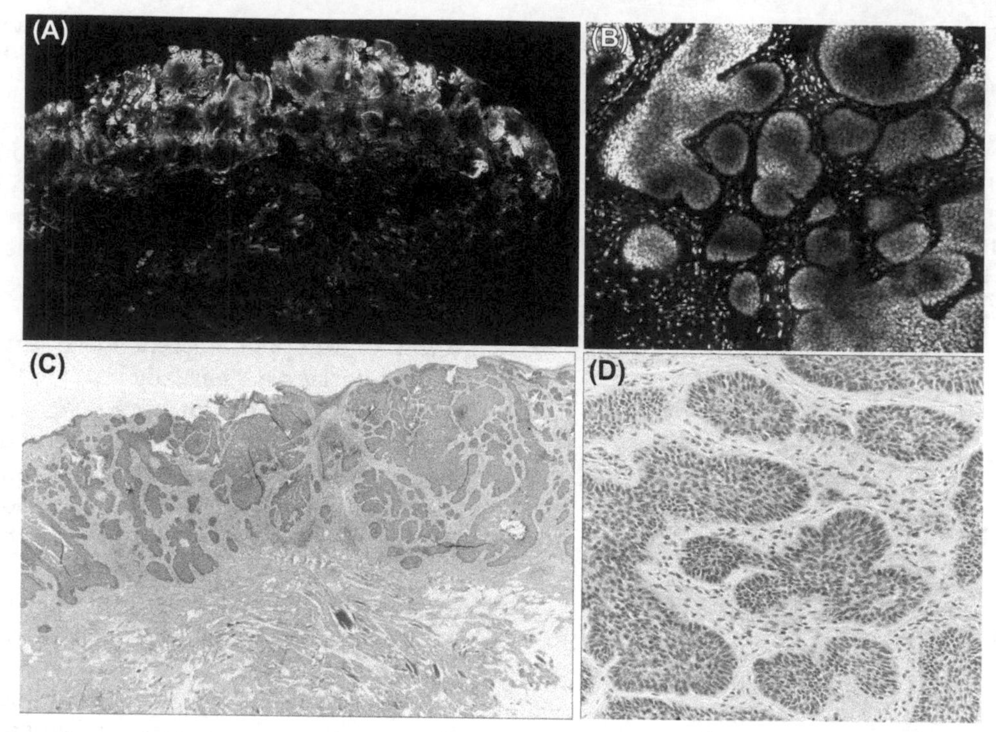

FIGURE 10.1 Nodular basal cell carcinoma. (A) Highly fluorescent, well-demarcated tumor proliferation that at high magnification (B) appears as compact rounded tumor islands. (C) Low and (D) high magnification histological images of the corresponding fluorescence confocal microscopy pictures.

infiltrative cords from the surrounding stroma, although the latter showed no tendency to cluster. Another possible pitfall is due to the presence of several sebaceous glands that may be confused with BCC islands. However, the former showed no palisading, less fluorescence, and the presence of a centrally located nucleus compared with the tumors.

Squamous Cell Carcinoma

A few preliminary reports have described the feasibility of FCM for SCC diagnosis and margin assessment. Longo et al. [22] defined the FCM criteria to grade SCC tumors. The pilot study demonstrated that the presence of a well-defined tumor silhouette, numerous keratin pearls, keratin formation, and scarce nuclear pleomorphism on FCM images were correlated with the diagnosis of well-differentiated SCC. Conversely, an ill-defined tumor silhouette, paucity or absence of keratin pearls, and marked nuclear pleomorphism was observed in poorly differentiated tumors. SCCs that were moderately differentiated revealed an intermediate pattern of growth with the presence of keratin formation (Fig. 10.2).

APPLICATION OF EX VIVO FLUORESCENCE MICROSCOPY IN GENERAL SURGERY

Studies on ex vivo fluorescence microscopy in general pathology have been less common than those focused on dermatology and MMS, but they have shown promising results. The first studies on confocal microscopy in general pathology started in 2000 and were mainly focused on endoscopic and gastroenterological pathology. These studies were conducted in the reflectance mode, based on the native differences in refractive indices between subcellular structures within the tissues and without using fluorescence dyes. In this setting, the present authors analyzed untreated mucosal specimens from the esophagus, stomach, and colon to explore the potential use of this fast and nondestructive imaging technique as an alternative to conventional histology [23]. Pilot studies on liver biopsies and pancreas from the rat were also performed [24,25].

Later on, other studies were conducted exclusively using reflectance-mode confocal microscopy. They were mainly focused on the parathyroid gland and oral pathology as well as tissues from head-and-neck surgery, knee cartilage biopsies, and stereotactic breast biopsies [26–33].

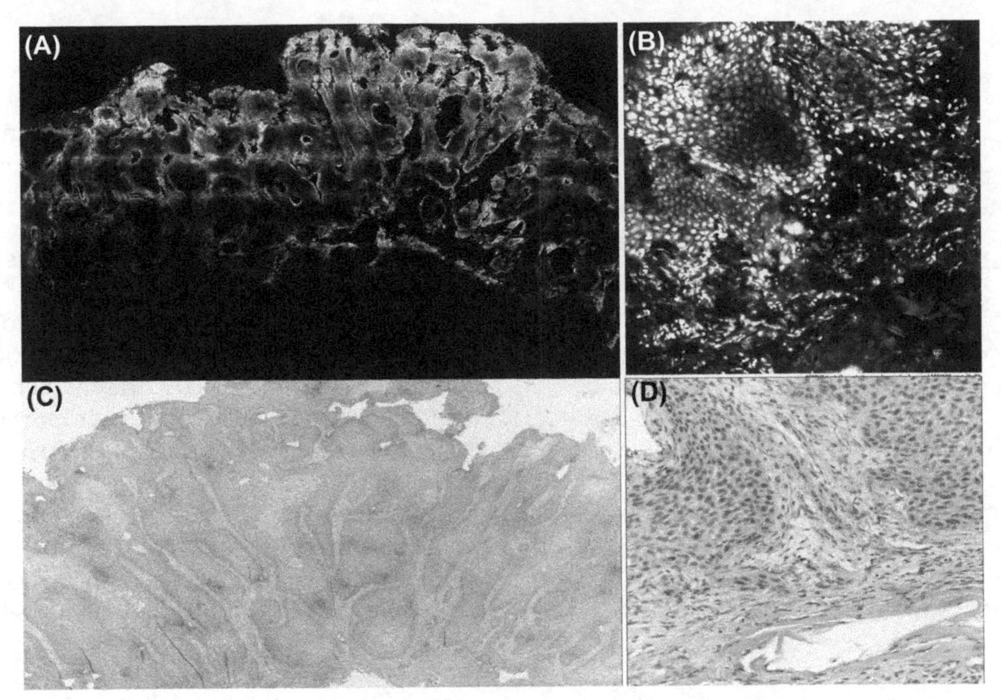

FIGURE 10.2 Moderately differentiated squamous cell carcinoma. (A) Fluorescence confocal microscopy (FCM) mosaic displays a well-defined tumor. (B) Highly fluorescent keratinocytes with low pleomorphism. (C) Histological picture revealing an excellent correspondence with FCM image as seen in panel A. (D) Histology shows a moderately differentiated squamous cell carcinoma.

In 2005 Carlson and colleagues described the possibility to assess and grade cervical dysplasia in 25 cervical biopsies using an inverted Leica confocal laser scanning fluorescence microscope based on autofluorescence [34]. This was mainly concerned with detecting cytokeratins and the nicotinamide adenine dinucleotide/flavin adenine dinucleotide contained in the mitochondria of the epithelial cells. The authors described a normal cervical epithelium characterized by a basal layer of proliferative cells with cytoplasmic fluorescence due to high mitochondrial activity and a thick intermediate and superficial region composed of mature cells with only peripheral fluorescence caused by cytokeratins. They observed that with cancer progression the cytoplasmic fluorescence tended to become prominent and to progressively involve the entire epithelial thickness [34].

In 2013 the first three studies taking advantages of fluorescent dyes were almost simultaneously published [35–37]. El Hallani et al. investigated the accuracy of confocal microscopy to detect oral high-grade dysplasia lesions in 31 surgical specimens of oral mucosa. After testing five clinically used fluorescent contrast agents in vitro in cell culture, they decided to use acriflavine hydrochloride. Confocal fluorescence microscopy was performed on a bench-top Carl Zeiss Axio Imager Z1 equipped with a custom laser-scanning confocal attachment [35].

Ragazzi and coauthors recently conducted a study on 35 fresh tissue samples from different surgical specimens including breast, thyroid, colon, and lymph node tissues [36]. They used the confocal microscope Vivascope 2500 and acridine orange as a fluorescent contrast agent. They described the feasibility of the technique to provide similar histological information to that offered by conventional histological slides stained with hematoxylin and eosin to discern neoplastic from normal tissue. They also proposed possible applications and the main limitations of the device [36].

Dobbs et al. described 70 breast tissue samples, including surgical specimen and core needle biopsies, using the same device reported above but with proflavine as the fluorescent agent. These authors analyzed the possibility to also distinguish (in addition to normal and neoplastic tissues) benign lesions, such as mild hyperplasia, chronic inflammation, fibrocystic changes, and fibrosis [37].

In a very recent paper Dobbs et al. used confocal fluorescence microscopy to evaluate tumor cellularity in 25 breast core needle biopsies of inflammatory breast carcinoma [38]. The data reported in the following subsections regarding potential application of ex vivo fluorescence microscopy in general surgery are mainly based on the results of these experimental studies.

Oral Mucosa

FCM has been applied on oral mucosa to study dysplasia of the squamous epithelium and SCC. It has been shown that it could be a useful adjunct to detect precancerous lesions that are at high risk of cancer progression, obtained by direct biopsy, to overcome the problem of an inadequate sample that could delay appropriate surgical treatment and delineate excision margins during surgical procedures.

Breast Cancer

Confocal fluorescence microscopy has been demonstrated to easily distinguish invasive carcinoma from normal ducts, and from reactive processes such as fibrosis and inflammation, on samples obtained with the Mammotome sponge (Figs. 10.3 and 10.4). Moreover FCM allows the visualization of adipose tissue as accurately as in conventional histology while avoiding frozen-section-related artifacts. Similar good results were obtained by analyzing fine needle core biopsies and surgical specimens. Therefore it could be used for real-time evaluation of the appropriateness of a core biopsy; to verify the presence of a neoplastic lesion, in and can immediately repeat the procedure. It could be also applied as an alternative to frozen sections in the assessment of margin status, in which the presence of artifacts due to adipose tissue could compromise the correct evaluation, and in the study of small tumors (<1 cm) in which frozen sections are not indicated because it is important to preserve sufficient tissue to eventually perform biological predictive markers.

An interesting improvement in confocal microscopy applications could be provided by the use of nanoparticles conjugated to antibodies. Bickford and coworkers applied silica-gold nanoshells as potential

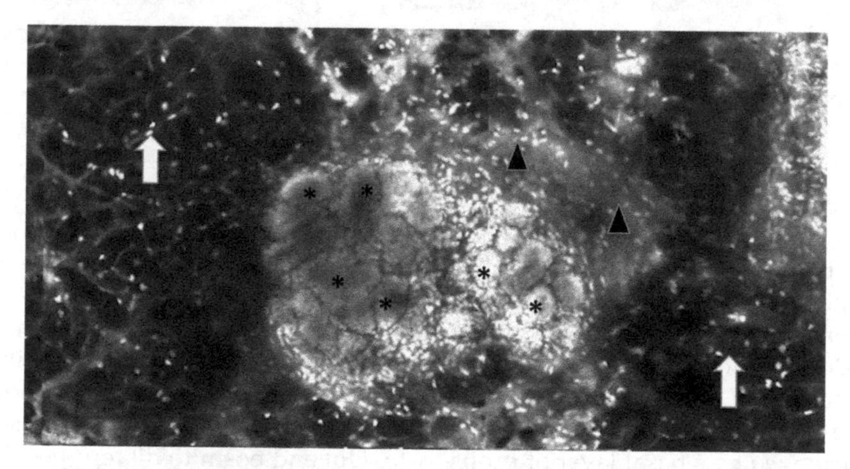

FIGURE 10.3 Normal breast. Normal breast parenchyma is mainly composed of well-circumscribed lobular structures made of small acini (*asterisks*) surrounded by loose fibrous stroma (*triangles*) on a black-gray background, which corresponds to fibro-fatty breast tissue (*arrows*).

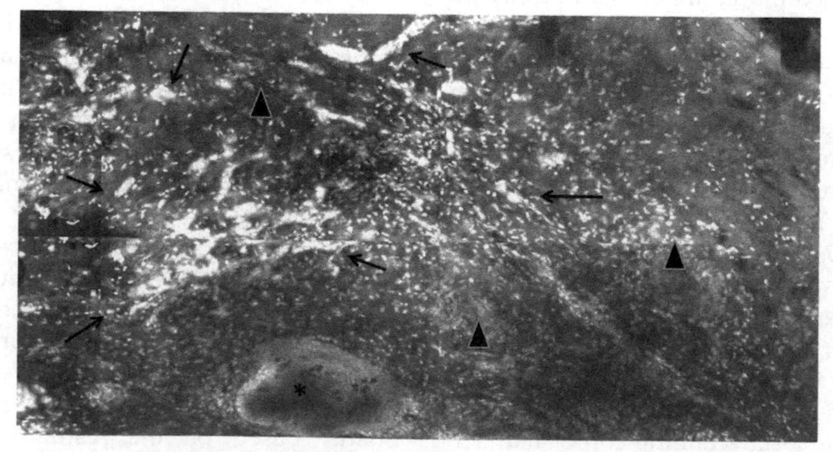

FIGURE 10.4 Breast carcinoma. Cancer is easily recognized because normal breast parenchyma becomes distorted and highly fluorescent: invasive duct carcinoma corresponds to elongated bright bands and cords (*arrows*) haphazardly distributed and infiltrating a dense gray desmoplastic stroma (*triangles*). More cohesive large nest represents in situ component (*asterisk*).

intraoperative molecular probes for assessment of Her2-overexpression with encouraging results [39,40].

Colon Cancer

FCM imaging permits easy identification of the border between tumor and normal mucosa in colorectal resection. This suggests that it could be used to assess tumor margins in gastrointestinal surgical pathology. In addition, the evaluation of samples obtained from endoscopic mucosectomy to detect tumor extent and margins during endoscopic resections could represent an interesting field of application.

Thyroid Cancer

Using FCM images, the papillary architecture is well recognizable from follicular architecture and clearly distinguishable in normal follicles. Therefore FCM could be applied for intraoperative evaluation of lesions with suspected cytology, especially for those with a small size (less than 1 cm) in which frozen section is not indicated. It could be also a useful tool for the intraoperative study of calcified or cystic tumors in which frozen section analysis is inapplicable. Because the confocal technique cannot evaluate nuclear detail, it is not possible to identify follicular variants of papillary carcinoma, a limit that should also be taken into consideration in frozen section analysis.

Other Tumors

As suggested by Ragazzi and coworkers, the possibility to apply immunohistochemistry could implement further applications of this approach (eg, for the assessment of lymph node metastases) [36]. High fluorescence due to the high proportion of small lymphocytes currently limits the possibility to clearly evaluate the lymph node parenchyma and to discriminate metastases. This limitation particularly applies when the deposits are small or when the neoplastic cells have a similar shape and size to lymphocytes, and when they present a diffuse distribution within the lymphoid parenchyma, such as in lobular carcinoma of the breast.

CONCLUSIONS

Ex vivo fluorescence microscopy in general surgery represents a useful tool for real-time evaluation of freshly excised tissue. It allows to completely preserve the integrity of the specimen, not only for conventional histopathological diagnosis but also for ancillary studies, including immunohistochemistry and molecular testing. These characteristics encourage its applications as an alternative to frozen sections and to assess the adequacy of the tissue sample (eg, evaluation of tumor cellularity in breast core needle biopsies, evaluation of kidney biopsies to confirm presence of glomeruli, cellular adequacy for molecular studies in non-small lung cancer small biopsies, evaluation for tissue banking).

Larger studies are needed to better define the new semiology of FCM for the different tissues visualized by this device, and prospective clinical trials need to be performed to establish its diagnostic accuracy and reliability.

List of Acronyms and Abbreviations

BCC Basal cell carcinoma
FCM Fluorescence confocal microscopy
MMS Micrographic Mohs surgery
SCC Squamous cell carcinoma

References

[1] Bialy L, Whalen J, Veledar E, Lafreniere D, Spiro J, Chartier T, et al. Mohs micrographic surgery versus traditional surgical excision: a cost comparison analysis. Arch Dermatol 2004;140:736–42.

[2] Rajadhyaksha M, Menaker G, Flotte TJ, Dwyer PJ, Gonzalez S. Rapid confocal examination of nonmelanoma cancers in skin excisions to potentially guide Mohs micrographic surgery. J Invest Dermatol 2001;117:1137–43.

[3] Rajadhyaksha M, Menaker G, Flotte T, Dwyer PJ, González S. Confocal examination of nonmelanoma cancers in thick skin excisions to potentially guide Mohs micrographic surgery without frozen histopathology. J Invest Dermatol 2001;117:1137–43.

[4] Gauthier P, Ngo H, Azar K, Allaire A, Comeau L, Maari C, et al. Mohs surgery – a new approach with a mould and glass discs: review of the literature and comparative study. J Otolaryngol 2006; 35(5):292–304.

[5] Chung VQ, Dwyer PJ, Nehal KS, Rajadhyaksha M, Menaker GM, Charles C, et al. Use of ex vivo confocal scanning laser microscopy during Mohs surgery for nonmelanoma skin cancers. Dermatol Surg 2004;30:1470–8.

[6] Patel YG, Nehal KS, Aranda I, Li Y, Halpern AC, Rajadhyaksha M. Confocal reflectance mosaicing of basal cell carcinomas in Mohs surgical skin excisions. J Biomed Opt 2007;12(3):034027.

[7] Gareau DS, Patel YG, Li Y, Aranda I, Halpern AC, Nehal KS, et al. Confocal mosaicing microscopy in skin excisions: a demonstration of rapid surgical pathology. J Microsc 2009;233(1):149–59.

[8] Schüle D, Breuninger H, Schippert W, Dietz K, Moehrle M. Confocal laser scanning microscopy in micrographic surgery (three-dimensional histology) of basal cell carcinomas. Br J Dermatol 2009;161(3):698–700.

[9] Ziefle S, Schüle D, Breuninger H, Schippert W, Moehrle M. Confocal laser scanning microscopy vs 3-Dimensional histologic imaging in basal cell carcinoma. Arch Dermatol 2010;146(8): 843–7.

[10] Kaeb S, Landthaler M, Hohenleutner U. Confocal laser scanning microscopy—evaluation of native tissue sections in micrographic surgery. Lasers Med Sci 2009;24(5):819–23.

[11] Yaroslavsky AN, Barbosa J, Neel V, DiMarzio C, Anderson RR. Combining multispectral polarized light imaging and confocal

microscopy for localization of nonmelanoma skin cancer. J Biomed Opt 2005;10(1):14011.
[12] Al-Arashi MY, Salomatina E, Yaroslavsky AN. Multimodal confocal microscopy for diagnosing nonmelanoma skin cancers. Lasers Surg Med 2007;39(9):696–705.
[13] Gareau DS, Li Y, Huang B, Eastman Z, Nehal KS, Rajadhyaksha MJ. Confocal mosaicing microscopy in Mohs skin excisions: feasibility of rapid surgical pathology. J Biomed Opt 2008;13(5):054001.
[14] Gareau DS, Karen JK, Dusza SW, Tudisco M, Nehal KS, Rajadhyaksha M. Sensitivity and specificity for detecting basal cell carcinomas in Mohs excisions with confocal fluorescence mosaicing microscopy. J Biomed Opt 2009;14(3):034012.
[15] Karen JK, Gareau DS, Dusza SW, Tudisco M, Rajadhyaksha M, Nehal KS. Detection of basal cell carcinomas in Mohs excisions with fluorescence confocal mosaicing microscopy. Br J Dermatol 2009;160(6):1242–50.
[16] Bennàssar A, Vilalta A, Carrera C, Puig S, Malvehy J. Rapid diagnosis of two facial papules using ex vivo fluorescence confocal microscopy: toward a rapid bedside pathology. Dermatol Surg 2012; 38(9):1548–51.
[17] Abeytunge S, Li Y, Larson B, Toledo-Crow R, Rajadhyaksha M. Rapid confocal imaging of large areas of excised tissue with strip mosaicing. J Biomed Opt 2011;16(5):050504.
[18] Abeytunge S, Li Y, Larson B, Peterson G, Seltzer E, Toledo-Crow R, et al. Confocal microscopy with strip mosaicing for rapid imaging over large areas of excised tissue. J Biomed Opt 2013; 18(6):61227.
[19] Longo C, Ragazzi M, Castagnetti F, Gardini S, et al. Inserting ex vivo fluorescence confocal microscopy perioperatively in Mohs micrographic surgery expedites bedside assessment of excision margins in recurrent basal cell carcinoma. Dermatology 2013; 227(1):89–92.
[20] Bennàssar A, Carrera C, Puig S, Vilalta A, Malvehy J. Fast evaluation of 69 basal cell carcinomas with ex vivo fluorescence confocal microscopy: criteria description, histopathological correlation, and interobserver agreement. JAMA Dermatol 2013;149(7): 839–47.
[21] Bennàssar A, Vilata A, Puig S, Malvehy J. Ex vivo fluorescence confocal microscopy for fast evaluation of tumor margins during Mohs surgery. Br J Dermatol 2014;170(2):360–5.
[22] Longo C, Ragazzi M, Gardini S, Piana S, Moscarella E, Lallas A, Raucci M, Argenziano G, Pellacani G. Ex vivo fluorescence confocal microscopy in conjunction with Mohs micrographic surgery for cutaneous squamous cell carcinoma. J Am Acad Dermatol 2015;73(2):321–2.
[23] Inoue H, Igari T, Nishikage T, Ami K, Yoshida T, Iwai T. A novel method of virtual histopathology using laser-scanning confocal microscopy in-vitro with untreated fresh specimens from the gastrointestinal mucosa. Endoscopy 2000;32(6):439–43.
[24] Keck T, Campo-Ruiz V, Warshaw AL, Anderson RR, Fernandez-del Castillo C, Gonzalez S. Evaluation of morphology and microcirculation of the pancreas by ex vivo and in vivo reflectance confocal microscopy. Pancreatology 2001;1(1):48–57.
[25] Campo-Ruiz V, Ochoa ER, Lauwers GY, Gonzalez S. Evaluation of hepatic histology by near-infrared confocal microscopy: a pilot study. Hum Pathol 2002;33(10):975–82.
[26] White WM, Tearney GJ, Pilch BZ, Fabian RL, Anderson RR, Gaz RD. A novel, noninvasive imaging technique for intraoperative assessment of parathyroid glands: confocal reflectance microscopy. Surgery 2000;128(6):1088–100.
[27] Clark AL, Gillenwater AM, Collier TG, Alizadeh-Naderi R, El-Naggar AK, Richards-Kortum RR. Confocal microscopy for real-time detection of oral cavity neoplasia. Clin Cancer Res Off J Am Assoc Cancer Res 2003;9(13):4714–21.
[28] White WM, Baldassano M, Rajadhyaksha M, Gonzalez S, Tearney GJ, Anderson RR, et al. Confocal reflectance imaging of head and neck surgical specimens. A comparison with histologic analysis. Arch Otolaryngol-Head Neck Surg 2004;130(8):923–8.
[29] Campo-Ruiz V, Patel D, Anderson RR, Delgado-Baeza E, Gonzalez S. Evaluation of human knee meniscus biopsies with near-infrared, reflectance confocal microscopy. A pilot study. Int J Exp Pathol 2005;86(5):297–307.
[30] Tilli MT, Cabrera MC, Parrish AR, Torre KM, Sidawy MK, Gallagher AL, et al. Real-time imaging and characterization of human breast tissue by reflectance confocal microscopy. J Biomed Opt 2007;12(5):051901.
[31] Yoshida S, Tanaka S, Hirata M, Mouri R, Kaneko I, Oka S, et al. Optical biopsy of GI lesions by reflectance-type laser-scanning confocal microscopy. Gastrointest Endosc 2007;66(1):144–9.
[32] Schiffhauer LM, Boger JN, Bonfiglio TA, Zavislan JM, Zuley M, Fox CA. Confocal microscopy of unfixed breast needle core biopsies: a comparison to fixed and stained sections. BMC Cancer 2009;9:265.
[33] Anuthama K, Sherlin HJ, Anuja N, Ramani P, Premkumar P, Chandrasekar T. Characterization of different tissue changes in normal, betel chewers, potentially malignant lesions, conditions and oral squamous cell carcinoma using reflectance confocal microscopy: correlation with routine histopathology. Oral Oncol 2010;46(4):232–48.
[34] Carlson K, Pavlova I, Collier T, Descour M, Follen M, Richards-Kortum R. Confocal microscopy: imaging cervical precancerous lesions. Gynecol Oncol 2005;99(3 Suppl. 1):S84–8.
[35] El Hallani S, Poh CF, Macaulay CE, Follen M, Guillaud M, Lane P. Ex vivo confocal imaging with contrast agents for the detection of oral potentially malignant lesions. Oral Oncol 2013;49(6):582–90.
[36] Ragazzi M, Piana S, Longo C, Castagnetti F, Foroni M, Ferrari G, et al. Fluorescence confocal microscopy for pathologists. Mod Pathol 2014;27(3):460–71.
[37] Dobbs JL, Ding H, Benveniste AP, Kuerer HM, Krishnamurthy S, Yang W, et al. Feasibility of confocal fluorescence microscopy for real-time evaluation of neoplasia in fresh human breast tissue. J Biomed Opt 2013;18(10):106016.
[38] Dobbs J, Krishnamurthy S, Kyrish M, Benveniste AP, Yang W, Richards-Kortum R. Confocal fluorescence microscopy for rapid evaluation of invasive tumor cellularity of inflammatory breast carcinoma core needle biopsies. Breast Cancer Res Treat 2015; 149(1):303–10.
[39] Bickford LR, Agollah G, Drezek R, Yu TK. Silica-gold nanoshells as potential intraoperative molecular probes for HER2-overexpression in ex vivo breast tissue using near-infrared reflectance confocal microscopy. Breast Cancer Res Treat 2010;120(3): 547–55.
[40] Bickford LR, Langsner RJ, Chang J, Kennedy LC, Agollah GD, Drezek R. Rapid stereomicroscopic imaging of HER2 overexpression in ex vivo breast tissue using topically applied silica-based gold nanoshells. J Oncol 2012;2012:291898.

CHAPTER

11

Coherent Raman Scattering Microscopy in Dermatological Imaging

H. Wang, C.L. Evans

Harvard Medical School & Massachusetts General Hospital, Boston, MA, United States

OUTLINE

INTRODUCTION

Skin Structure

Skin is one of the largest organs of the human body, consisting of a cellular epidermis layer and a dermis of connective tissue. The epidermis can be divided into four sublayers: (1) the stratum corneum is the most outside layer that contains nonviable keratinocytes and mainly functions as a permeability barrier; (2) the stratum granulosum contains granular cells that promote cross-linking of keratin; (3) the stratum spinosum, consisting of spinous cells, is the place where the keratinization process begins; and (4) the stratum basale provides germinal cells and is necessary for epidermal regeneration. Keratinocytes in the epidermis proceed from their germinative state to finally differentiated cells filled with keratin, and this important cell differentiation process is called *keratinization*. The main components of the dermis are collagen and elastic fibers, as well as skin appendages such as hair follicles, sweat glands, and sebaceous glands (SGs). In between the epidermis and the dermis, the dermal–epidermal junction (DEJ) can be found, which provides both mechanical support for the epidermis and a physical barrier that prevents exchange of cells and large molecules between the two layers.

Human skin serves as an excellent physical barrier against external chemicals and microbes, and has a

Imaging in Dermatology
http://dx.doi.org/10.1016/B978-0-12-802838-4.00011-X

variety of homeostasis functions in response to external stimuli via the temperature and pressure receptors. A very unique type of cell within skin, the melanocyte, can produce melanin and protect the body against ultraviolet (UV) radiation in sunlight. Skin appendages such as SGs secrete a lipid-rich compound known as *sebum* that acts to waterproof the skin and hairs, preventing water loss. Skin provides a means to regulate body temperature. For example, blood vessels in the dermis will dilate to increase blood flow to the surface of the body when there is an increase of temperature in the environment. The secretion of sweat is another form of cooling mechanism in humans [1].

Optical Techniques Used in Dermatological Imaging

Because our eyes are sensitive to changes in skin color and texture, clinical assessment in dermatology has traditionally depended on visual inspection. A number of noninvasive approaches have been developed over the years to aid in a more objective evaluation of the skin.

Wood Lamp

The invention of the Wood lamp dates back to 1903, when fluorescent light was proposed to help visualize the skin [2]. Clinically, Wood lamp with a UV light source was first used to induce fluorescence to detect fungal infection of hair [3]. This technique has also been used to visualize skin conditions such as porphyrin disorders [4], cutaneous infections [5], and pigmented disorders [6]. However, fluorescence generated from other exogenous components such as soap residue and topical medications/creams may interfere with tissue autofluorescence under Wood lamp examination.

Dermoscopy

Dermoscopy is essentially a form of skin surface microscopy, which provides detailed visualization of skin structures. Dermoscopy was first used by dermatologists in the 1950s to assess pigmented skin conditions [7]. Today, applications of dermoscopy in dermatology include the evaluation of inflammatory and infectious diseases [8], autoimmune diseases [9], and nonpigmented skin tumors [10], as well as the monitoring of treatment response [8,11,12]. Compared with pure visual inspection, the diagnostic accuracy of dermoscopy is 5–30% greater, although this does depend on the experience of the dermatologist as well as the type of lesion under inspection [13]. Limitations of dermoscopy include low resolution and lack of optical sectioning capabilities, resulting in a limited diagnosis accuracy.

Reflectance Confocal Microscopy

Confocal scanning laser microscopy (CSLM), which allows for the imaging of thick samples layer by layer, [14] has thrived in dermatology applications. Reflectance confocal microscopy (RCM), a branch of CSLM, exploits the endogenous differences in refractive indices as sources of image contrast, and therefore no exogenous labeling is required [15]. Because it delivers high-resolution images at fast imaging speeds hundreds of micrometers deep in the skin, RCM is useful for many in vivo dermatological applications. Confocal images of normal skin can reveal cellular details in the epidermis, collagen, and elastic fibers in the dermis, as well as circulating blood cells in the dermal capillaries [16]. Because of the high image contrast provided by melanin, RCM has been used to study benign and malignant melanocytic lesions [17]. Other applications of RCM include diagnosis of basal cell carcinoma (BCC) [18], allergic contact dermatitis [19], and actinic keratosis [20] as well as the mapping of tumor margins to guide Mohs surgery [21]. RCM has also been implemented in combination with confocal Raman spectroscopy to differentiate benign and malignant skin cells [22].

Optical Coherence Tomography

Optical coherence tomography (OCT) is another noninvasive technique that provides high-resolution cross-sectional images of the microstructure of living tissue. OCT was first introduced to dermatology in 1997, and has been employed in both laboratory and clinical skin research [23]. Under OCT imaging, the epidermis, the upper dermis, blood vessels, and skin appendages can be well visualized [24]. OCT can also be used to differentiate diseased skin states. For example, parakeratosis can be seen when imaging psoriasis using OCT [25]. Inflammatory skin conditions such as contact dermatitis have also been visualized using OCT [23,25–27]. Blood vessels of inflamed skin have been imaged with a subbranch of OCT known as *optical microangiography* [28,29]. More applications of OCT in skin research include evaluation of blistering diseases [30–32], onychomycosis [33], skin cancers, and precancers [34–36]. OCT has also been applied to study therapeutic effects such as effects of moisturizers [37] and laser thermal therapy [38]. However, the contrast in standard OCT arises from the scattering of light and does not readily provide molecular specificity.

Multiphoton Microscopy

Multiphoton microscopy (MPM) techniques using near-infrared (NIR) wavelength as the excitation were developed to achieve considerable larger imaging depth while providing comparable image resolution to CSLM and inherent optical sectioning capability. MPM requires an ultrafast (typically femtosecond pulse duration) laser source in order to achieve the extremely high photon density at the focal plane needed to excite

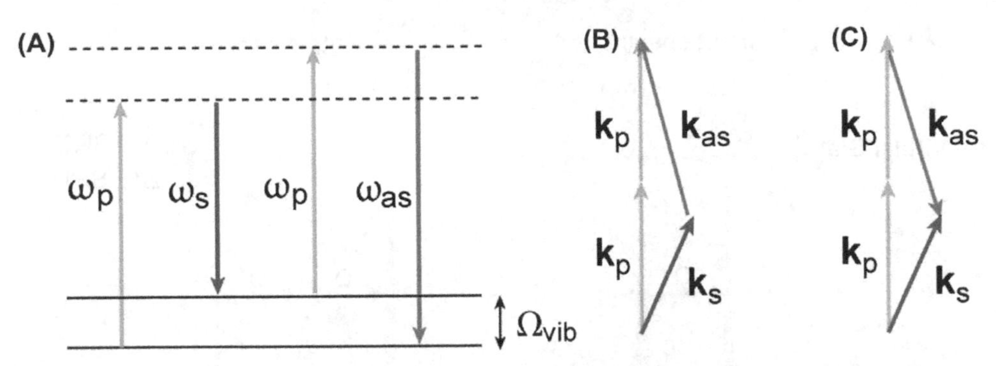

FIGURE 11.1 (A) Diagram of the coherent anti-Stokes Raman scattering *(CARS)* process. When the difference between the pump and Stokes frequencies ($\omega_p - \omega_s$) matches the molecular vibrational frequency, Ω_{vib}, the anti-Stokes signal is generated at a frequency $\omega_{as} = 2\omega_p - \omega_s$. (B) Phase-matching condition for forward-generated CARS. (C) Phase-matching condition for backward- (epi-) generated CARS. k is known as the wave vector, and is given by $k = 2\pi/\lambda$. Here, k_p, k_s, and k_{as} represent the pump, Stokes, and anti-Stokes wave vectors, respectively. *Adapted from Evans CL, Xie XS. Coherent anti-Stokes Raman scattering microscopy: chemical imaging for biology and medicine. Annu Rev Anal Chem 2008;1:883–909, with permission.*

two-photon absorption-based fluorescence. Endogenous two-photon excitable fluorophores inside tissues include nicotinamide adenine dinucleotide (phosphate) [NAD(P)H], flavin, and elastin. Ultrafast light sources can also be used to generate image contrast that comes via second harmonic generation (SHG), which allows for visualizing noncentrosymmetric molecules such as collagens inside the skin.

The number of applications of MPM in the field of dermatology has increased dramatically since its introduction. For example, two-photon fluorescence (TPF) and SHG images of in vivo human skin can be acquired where both cellular and fiber structures can be clearly visualized [39,40]. Because TPF and SHG can image elastic fibers and collagen fibers, respectively, MPM has also been used to quantify the degree of skin photoaging based on the ratio of elastic fiber to collagen fiber (aging index of the dermis) [41]. More applications in skin research include studying drug delivery [42], nonmelanoma skin cancers [43,44], and melanoma itself [45]. A great advance toward clinical application is the development of rapid-scan multiphoton microscopic systems, which allow video-rate imaging speeds and therefore can compensate for involuntary body movement to prevent image distortion [46]. More recently, a commercial in vivo MPM instrument (DermaInspect, JenLab, Jena, Germany) has been developed that integrates an ultrafast laser, optic components, scanning units, and electric circuits into one miniaturized portable unit. A scanning probe head on an articulated arm enables the measurement of skin at different body sites. Other than direct imaging of healthy or diseased skin, MPM can also be utilized to perform two-photon absorption-based photothermolysis. Because of the highly selective absorption within the tissue, this could potentially become a highly targeted therapy. With both the imaging and treatment capabilities, two-photon absorption can be used to treat and monitor the treatment response simultaneously [47].

Coherent Raman Scattering

Label-free chemical contrast is highly desirable in the field of biomedical imaging, especially in clinical dermatological imaging. Because different molecules contain chemical structures that have unique vibrational frequencies, vibrational spectroscopy and microscopy techniques can provide intrinsic and specific chemical contrast. Among these vibrational techniques, Raman spectroscopy has been extensively utilized in biomedical applications such as tumor detection [48]. However, one major limitation of Raman microscopy is its inherent long acquisition time caused by the extremely weak spontaneous Raman effect. Coherent Raman scattering (CRS), including coherent anti-Stokes Raman scattering (CARS), and stimulated Raman scattering (SRS), can provide a coherent enhancement to the weak Raman signal through nonlinear interactions, which is more suitable for fast image acquisition speeds required for in vivo biological imaging. Moreover, CRS makes use of nonlinear processes that have similar "virtual biopsy" sectioning capabilities as MPM.

When the frequency difference between two pulsed laser beams, $\Delta\omega = \omega_1 - \omega_2$ (beat frequency), matches the frequency of a given molecular vibration, the two input fields can provide a coherent driving force to all vibrationally resonant molecules in the focal spot, enabling the generation of a strong anti-Stokes Raman signal at $\omega_{as} = 2\omega_p - \omega_s$, where ω_{as}, ω_s, and ω_p represent the frequencies of the anti-Stokes, Stokes, and pump signals. An energy diagram illustrating the CARS process can be found in Fig. 11.1 [49]. When $\omega_p - \omega_s$ is tuned to the frequency of molecular vibration, the anti-Stokes signal is significantly enhanced; however, when ω_p and ω_s are far from electronic resonance, an electronic nonresonant background will still be observed. CARS signals can propagate in both forward and backward (epi) directions. For thick skin tissues, the image contrast in CARS microscopy is mainly derived from backscattering of the far stronger forward-propagating CARS signals.

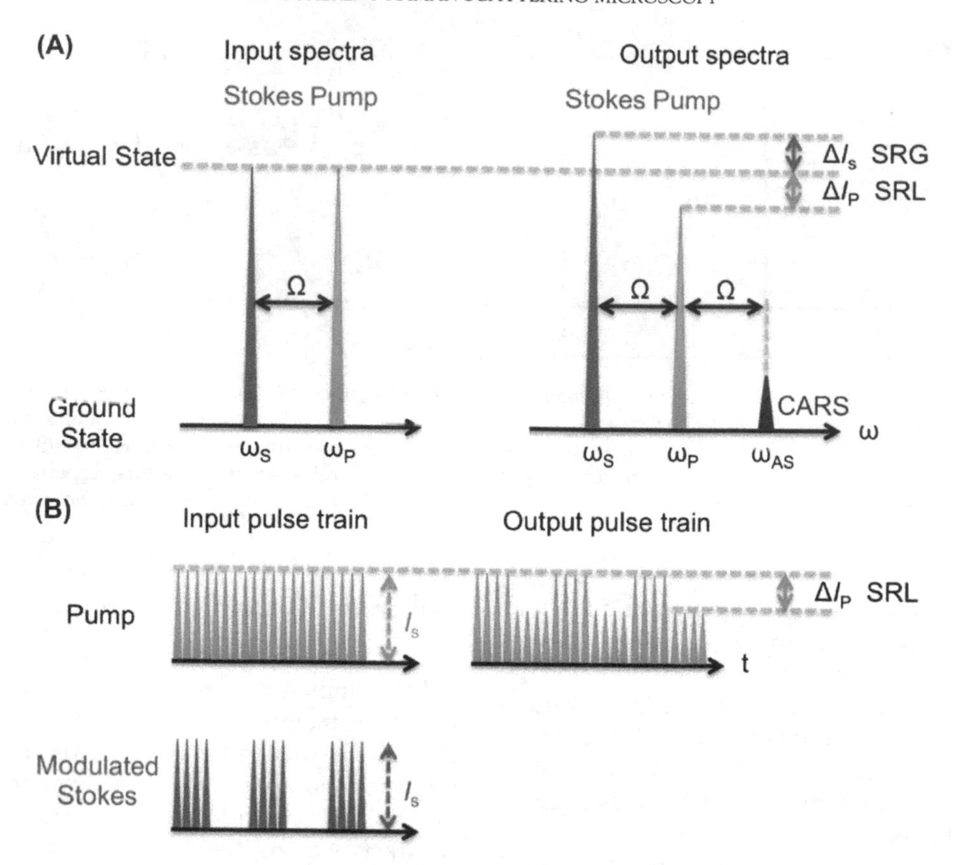

FIGURE 11.2 (A) Input and output spectra of stimulated Raman scattering (SRS). SRS leads to an intensity increase in the Stokes beam [stimulated Raman gain (*SRG*)] and an intensity decrease in the pump beam [stimulated Raman loss (*SRL*)]. Also shown (not to scale) is the coherent anti-Stokes Raman scattering (*CARS*) signal generated at the anti-Stokes frequency (ω_{as}). (B) SRL detection scheme. The Stokes beam is modulated at high frequency (MHz), at which the resulting amplitude modulation of the pump beam caused by SRL can be detected. *Adapted from Freudiger, Christian Wm et al. Label-free biomedical imaging with high sensitivity by stimulated Raman scattering microscopy. Science 2008; 322(5909): 1857–1861, with permission.*

SRS also belongs to the family of CRS, which requires two ultrafast laser beams [50]. When the frequency difference of the two laser beams at $\Delta\omega = \omega_1 - \omega_2$ matches the vibrational frequency of a molecule Ω, there will be an increase (ΔIs) in the intensity of the Stokes beam (I_S) and a decrease (ΔIp) in the intensity of the pump beam (I_p). The changes in the intensity ΔI_S and ΔI_P are called stimulated Raman gain (SRG) and stimulated Raman loss (SRL), respectively. The occurrence of SRG and SRL requires strict frequency matching; therefore SRS does not contain a nonresonant background. Similar to CARS, the detected epi signal from tissue is derived from backscattering of the forward-going signals. The input and output spectra, as well as the detection scheme, of SRS can be found in Fig. 11.2.

Applications of Coherent Raman Techniques in Biomedical Research

Coherent Raman techniques, including both CARS and SRS microscopy, have been extensively used in biomedical imaging. For example, SRS was demonstrated to be capable of noninvasively imaging living cells and brain tissue by Freudiger and colleagues in 2008 [50]. The distribution of Ω_3 fatty acids and saturated lipids in cells has also been differentiated [50]. Moreover, neuron bundles in corpus callosum of mouse brain have been successfully imaged using SRS, where individual neurons can be clearly visualized [50]. CARS in conjunction with TPF microscopy was able to image both corneal epithelial and endothelia cells based on strong lipid CARS signal from plasma membrane in intact mouse cornea [51]. CARS microscopy has also been used to image plaque lesions in atherosclerotic mice, and an increase in overall lipid content was observed in the plaque lesions of the Western diet-fed mice [52]. To understand the link between obesity and breast cancer, Le et al. have pointed out that mammary glands of obese rats contain increased levels of adipocytes with increased size of lipid droplets [52]. Moreover, CARS together with SHG microscopy has also been used to image mammary stroma [53].

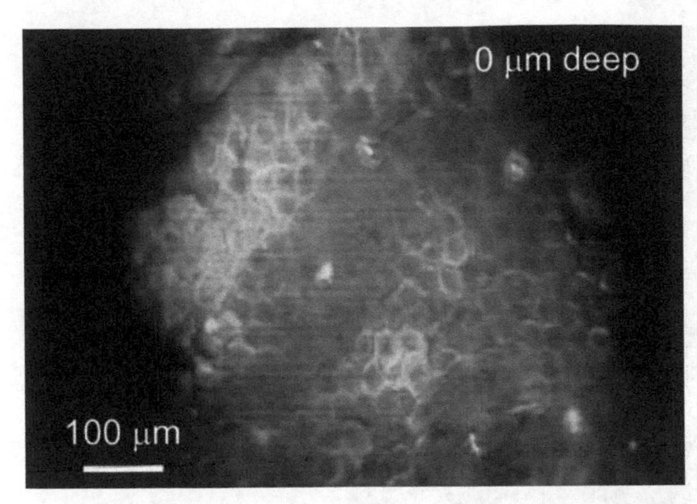

FIGURE 11.3 Coherent anti-Stokes Raman scattering (CARS) image of a hairless mouse ear. The Raman shift is set at 2845 cm^{-1} (ω_p = 816.8 nm) to address the lipid methylene (CH_2) symmetric stretch vibration. The frames are averaged for 2 s. Note the stratum corneum with bright signals from the lamellar lipid intercellular space that surrounds the polygonal corneocytes. Bright punctuated dots are ducts of sebaceous glands. *Adapted from Evans CL, Potma EO, Puoris'haag M, Côté D, Lin CP, Xie XS. Chemical imaging of tissue in vivo with video-rate coherent anti-Stokes Raman scattering microscopy. Proc Natl Acad Sci U S A 2005;102:16807–12, with permission.*

IMAGING SKIN STRUCTURES USING COHERENT RAMAN SCATTERING

Normal Skin

Because skin is easily accessible, skin structures as well as skin appendages have been extensively studied using CRS. CARS microscopy was the first to be demonstrated as a live-skin imaging technique with chemical specificity in mice [54]. For example, when tuned into the saturated lipid, methylene (CH_2) vibrational mode at 2845 cm^{-1}, a lipid-rich "mortar" could be easily identified surrounding each polygonal-shaped keratinocyte (Fig. 11.3). CARS could be used to visualize specific lipid structures at depths throughout murine ear skin. Subcutaneous fat at depths beyond 120 μm could be well visualized, illustrating the capability of CARS to be used in highly scattering skin tissues [55]. Likewise, SRS has also been demonstrated for skin imaging. By tuning into the methyl group (CH_3) stretching vibrations of proteins (2950 cm^{-1}), keratinocytes with different sizes in the stratum corneum and epidermis can be well visualized [50].

Apart from cellular structures inside the skin, skin appendages such as hair follicles, sweat glands, and SGs can also be visualized using CRS. Because of its easy access, human hair has become one of the most important models for investigating protein structures and functions [56–58]. Ultrabroadband multiplex CARS microspectroscopy has been used to study hairs with or without chemical/mechanical treatments. The acquired CARS images showed inhomogeneous chemicalcompositions and protein secondary structures in the hair. Using polarization-sensitive measurement techniques, treatment-induced changes of protein secondary structures could be clearly observed [59].

SGs are present at the base of every hair and secret the lipid-rich sebum that serves as a natural barrier of the skin, preventing water loss and keeping the skin moisturized. Because SG physiology plays an important role in the development of acne vulgaris, the ability to noninvasively visualize SGs may potentially aid in building better therapies for acne. Because the main chemical composition of SG is lipid, CARS imaging can be carried out at 2845 cm^{-1} to specifically image SGs in mouse skin. It was found that CARS spectra of SGs and adipocyte showed some level of difference as a result of their different chemical lipid composition; murine SGs were found to have greater saturated-to-unsaturated fat content than adipocytes [54]. Because CARS microscopy can be performed at video rate (30 frames per second) and can be used for long-term imaging without introducing perturbations, sebum production, cell proliferation, and cell migration may be investigated using CARS. Indeed, Jung et al. have demonstrated the capability of CARS microscopy to reveal dynamic changes in SGs during the holocrine lipid secretion process, as well as in response to cold treatment [60]. The migration of each individual sebocyte was visualized and tracked with CARS microscopy (Fig. 11.4).

CARS and SRS have also been used to investigate different properties of creams, shampoos, and other cosmetic products. Cosmetic companies have shown great interest in better understanding how their products interact with human skin/hair. Some common parameters include the penetration depth, time of penetration, and the mechanisms of change in the skin. CRS techniques are ideal to reveal these fundamental questions and may provide a new way to evaluate skin and hair products. CARS microscopy has been used to determine the distribution and concentration of selected compounds in human hair. For example, Zimmerley and colleagues have utilized CARS microscopy to measure both water and externally applied D-glycine (deuterated glycine) in the cortical region of human hair [61]. Ratiometric CARS contrast was then generated to perform quantitative analysis that aided in understanding the chemical and physical mechanisms that underlie hair-care products.

In addition to SGs, skin appendages such as sweat glands can also be studied with CRS. CARS together

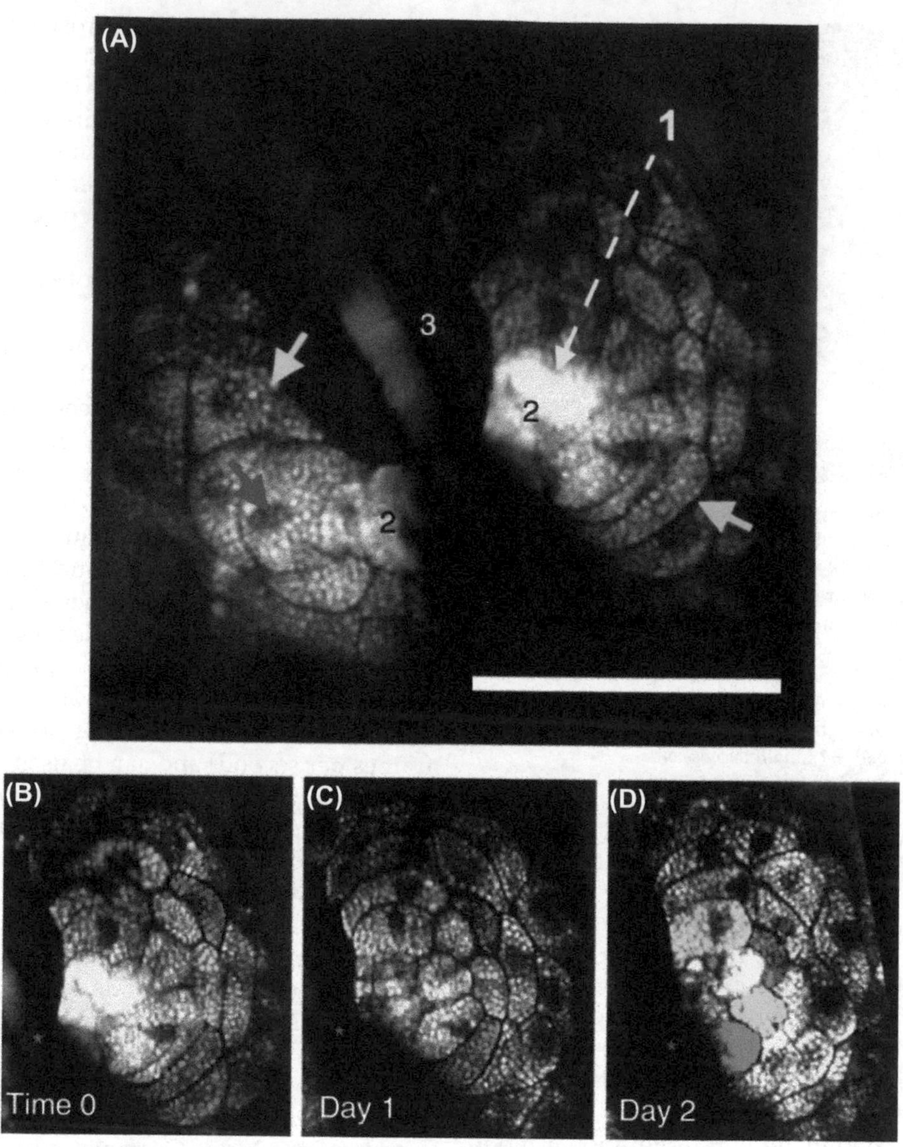

FIGURE 11.4 Coherent anti-Stokes Raman scattering (CARS) imaging of normal sebaceous glands. (A) A sebaceous gland showing intracellular lipid granules *(blue arrow)*, nuclei *(red arrow)*, and cell membranes *(green arrow)*. 1, CARS signal intensity increases as sebocytes approach the gland duct, corresponding to lipid accumulation as sebocytes mature. 2, Sebocytes near the duct show the highest CARS signal and lose cellular structures corresponding to cell death and lipid content release. 3, Hair shafts are coated with secreted lipid. (B), (C), and (D) Sebocyte migration in the same sebaceous gland shown in three consecutive days. Five sebocytes are marked in separate colors to facilitate identification. As the sebocytes migrate to the gland duct (*), there is a loss in intracellular structures and a concomitant increase in CARS signal. Scale = 50 μm. *Adapted from Jung Y, Tam J, Jalian HR, Anderson RR, Evans CL. Longitudinal, 3D in vivo imaging of sebaceous glands by coherent anti-Stokes Raman scattering microscopy: normal function and response to cryotherapy. J Invest Dermatol 2014, with permission.*

with TPF microscopy have been used to image a section (18 μm below the skin surface) of an eccrine sweat gland [62]. The main contrast for CARS imaging was lipid, whereas NAD(P)H was used as the main cellular fluorophore excited by TPF microscopy.

Using noninvasive imaging methods such as CRS microscopy, baseline parameters can be derived to characterize normal skin. These parameters can then be further referenced to guide researchers and physicians in skin disease diagnosis and evaluation.

Diseased Skin

The diagnosis of skin diseases has historically been largely dependent on subjective, operator-dependent clinical evaluations of symptoms. The rise of recent spectroscopy and imaging tools promises to introduce more quantitative and objective evaluation methods for the evaluation and diagnostics of skin diseases. Imaging tools such as CRS microscopy have the potential to capture data from skin that, alongside automated routines, computer vision tools, and the expertise of dermatologists, may provide improved and timely disease diagnostics. For example, psoriasis is a very common skin condition that is usually chronic and characterized by skin lesions with red, scaly patches. The causes of psoriasis are not fully understood, but it has been associated with an increased risk of certain cancers, cardiovascular disease, and other autoimmune disorders. CRS techniques in combination with TPF microscopy have recently been applied to reveal differences on the cellular level between healthy and psoriasis-affected skin lesions where a reduced CH_2 lipid signal captured by CARS was found in the intercellular areas of the psoriasis-affected regions compared with normal skin [63]. This optical signal offers a potential route for initial diagnosis as well as for monitoring psoriasis plaques.

Another potential application is in BCC, the most common of skin cancers. Despite the fact that these cancers are superficial in the skin, the routine identification of BCC tumor margins can still be challenging. In a recent study, a large BCC section (9.5 × 6.35 mm) was studied with CARS and MPM [64]. When the CARS system was tuned to interrogate the 2850 cm^{-1} lipid band, the authors found that strong CARS signals were found at the topmost layer of epidermis within the cancerous area. This suggested that the CARS lipid imaging contrast within the dermis could be an important indicator for assessing the invasiveness of a BCC tumor. The mosaic CARS images obtained (Fig. 11.5) were in good agreement with the gold standard of hematoxylin and eosin (H&E) histology, where areas shaded in gray marks the diagnosis made by pathologists. Moreover, SHG was combined with CARS imaging to identify differences in collagen content in the dermis. Along with these two imaging modalities, TPF was also incorporated and used to reveal the distribution of autofluorescence in tissue. As each of the three techniques provides complementary information, their combination provided a more complete set of information that is comparable to that of conventional H&E and offers potentially high clinical diagnostic value.

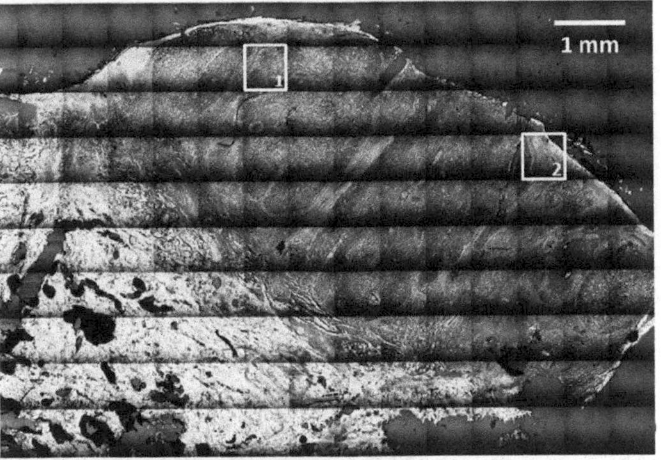

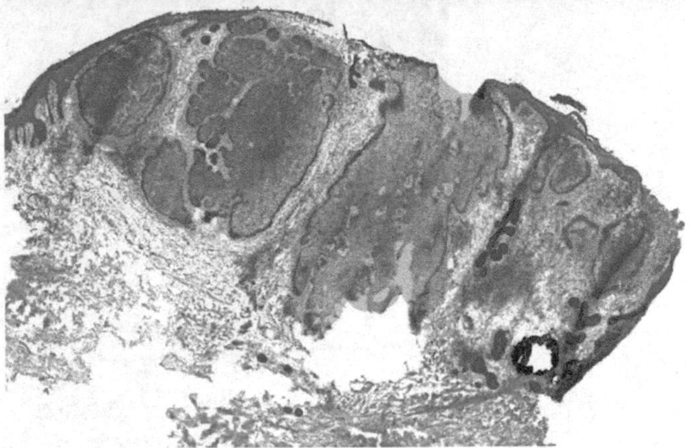

FIGURE 11.5 Coherent anti-Stokes Raman scattering (CARS) image *(top)* in comparison to a hematoxylin and eosin (H&E) stained parallel section *(bottom)* with the tumor areas marked *(shading)*. *Adapted from Vogler N, Heuke S, Akimov D, Latka I, Kluschke F, Röwert-Huber H-J, et al. Discrimination of skin diseases using the multimodal imaging approach. In: Spie Photonics Europe; 2012, pp. 842710–842710-8, with permission.*

Squamous cell carcinoma (SCC) is another form of nonmelanoma skin cancer. Clinically, SCC in situ appears as multiple scaly, reddish macules or papules on sun-exposed sites, whereas invasive SCC usually manifests as fast-growing papules or nodules with most lesions showing scales, keratin crusts, and ulcerations around the center of the lesion [65–67]. Skin sections from five SCC patients have been investigated by multimodal nonlinear microscopy including CARS, TPF, and SHG to identify optical features of SCC [68]. CARS was tuned to 2850 cm^{-1} to detect the CH_2-stretching vibrations of lipids, whereas TPF was detected at 435–485 nm to visualize NADH. Images acquired from one of the representative SCC sections are shown in Fig. 11.6. Marked keratinization infiltrating into the reticular dermis can be visualized in the CARS image. Moreover, aggregates of keratin within the SCC tumor nest show high lipid CARS signals. Tumor islands with enlarged cell nuclei were found as well. These morphological features corresponded well with histological findings.

Keloids are a scar subset usually caused by an overgrowth of granulation tissue (collagen type III), which is gradually replaced by collagen type I. Even though keloid scars are benign, they may cause severe pain and itching. A multimodal system including CARS, TPF, and SHG imaging capabilities was employed in a case study comparing normal skin and keloid tissue [69]. Human skin samples (three normal and three keloid) excised during surgery were cut into 20-μm thick slices without further preparation. Lipid CARS images

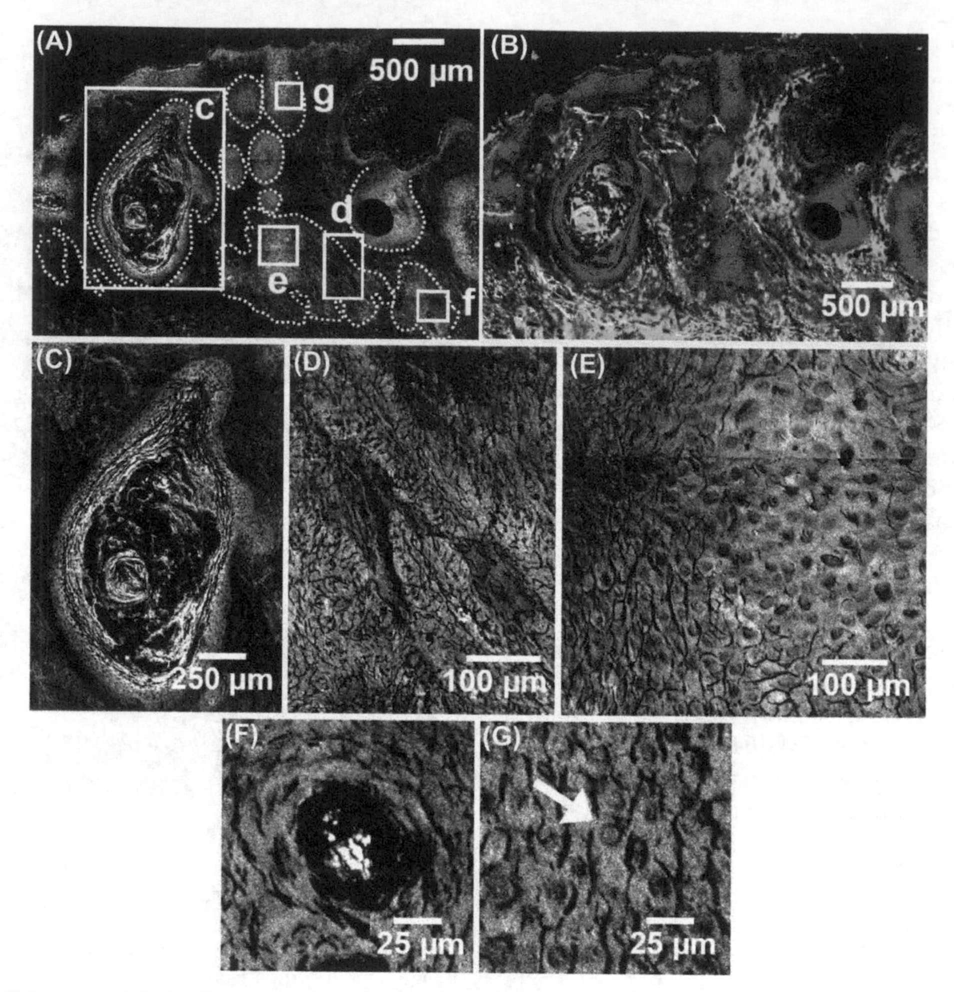

FIGURE 11.6 (A) Coherent anti-Stokes Raman scattering (CARS). (B) Image section illustrating the tumorous region, which is located by the *dotted lines* in (A). (C) Keratinizing tumor. (D) Squamous cell carcinoma (SCC) tumor nest. The tumorous cells can be discerned from the embedding dermal tissue. (E) Tumor cells with pleomorphic nuclei. (F) Keratin pearl. (G) SCC cells. The *white arrow* points towards a nucleus appearing dark in CARS. Compared with noncancerous tissue, the tumorous tissue possesses an increased cell density, larger nuclei, and therefore an elevated nuclear cytoplasm ratio. *Adapted from Heuke S, Vogler N, Meyer T, Akimov D, Kluschke F, Röwert-Huber H-J., et al. Detection and discrimination of non-melanoma skin cancer by multimodal imaging. In: Healthcare; 2013, pp. 64–83, with permission.*

demonstrated the ability of CARS to visualize individual cells with strong signals in the stratum corneum of the keloid tissue, potentially offering a noninvasive imaging biomarker for keloids. However, it should be noted that for in vivo keloid imaging, even with all three modalities, the penetration depth was limited to approximately 200 μm. Therefore tools such as OCT might be required to overcome this limitation.

IMAGING DRUG PENETRATION AND CHEMICAL DIFFUSION USING COHERENT RAMAN SCATTERING

In addition to imaging skin structures, CRS microscopy has been found useful in the imaging and evaluation of drug penetration and chemical diffusion inside the skin. Although the barrier function of human skin is essential for homeostasis, the low permeability of skin makes the transdermal delivery of drugs challenging [70]. Traditionally, tape stripping has been used to remove the stratum corneum layer by layer, and the extracted layers can be analyzed chemically to estimate the depth penetration of certain compounds. However, the results based on this approach may not be accurate because the tape stripping technique can be perturbative. Another set of methods for tracking drug penetration includes fluorescence imaging, but it is challenging to perform fluorescence imaging for small molecule drugs because the addition of tags/labels can alter the molecule and its transport. For example, a fluorophore may be the same size or bigger than a given drug molecule. Therefore in order to accurately track drugs, a technique is needed that can visualize the drug molecule via intrinsic contrast. Coherent Raman imaging matches this need for many

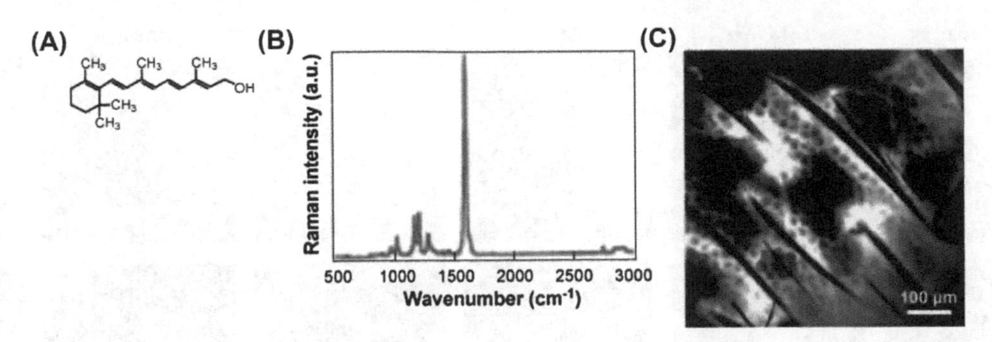

FIGURE 11.7 (A) Structure of retinol. (B) Raman spectrum of retinol, showing the strong characteristic peak at 1594 cm^{-1} arising from the conjugated polyene structure. (C) 620 μm × 620 μm image of mouse ear skin treated with a 10% retinol in myritol 318 solution. *a.u.*, arbitrary unit; *CH_3*, methyl group. *Adapted from Evans CL, Xie XS. Coherent anti-Stokes Raman scattering microscopy: chemical imaging for biology and medicine. Annu Rev Anal Chem 2008;1:883–909, with permission.*

molecules and can provide molecular visualization on the microscale to reveal diffusion, transport, and uptake of drugs.

Retinol and Retinoic Acid

Retinols are a class of vitamin A derivatives that have many effects in skin [71]. Retinoic acid (RA), a metabolite of retinol, has been actively developed by the pharmaceutical industry for many years to induce a wide number of effects, including reduction of acne [72] and stimulation of collagen synthesis [73] as well as reduction of psoriasis [74]. SRS has been demonstrated to have the capability to track the penetration of topically retinols into mouse skin [50]. In order to image RA, an SRS system was tuned to a highly specific vibrational band unique to RA at 1570 cm^{-1}. Using SRS, the hydrophobic molecule RA was observed to penetrate skin via the lipid-rich intercellular space between keratinocytes in the epidermis. CARS microscopy also revealed that retinol can penetrate via hair shaft into SG, and localizes in the tissue region immediately surrounding the hair shaft and its associated SGs (Fig. 11.7) [49].

Nonsteroidal Anti-inflammatory Drugs

Nonsteroidal anti-inflammatory drugs(NSAIDs) such as ketoprofen and ibuprofen are often used to treat skin rashes, dermatitis, and actinic keratosis, aside from their general use as pain killers. The structure of ketoprofen gives rise to a strong vibrational resonance at the aromatic CH-bond stretching frequency of 1599 cm^{-1} that can be used for visualization. Although the structure of ibuprofen does not give rise to a strong and unique vibrational frequency, it is possible to synthesize the molecule in a deuterated form, where hydrogens are replaced with their heavier isotope deuterium. Because carbon-deuterium (CD) bonds do not appear in abundance in nature, this deuterium labeling approach enables the specific visualization of ibuprofen at the CD vibration band at 2120 cm^{-1}. Three-dimensional SRS images were taken at different time points targeting lipid, ibuprofen, ketoprofen, and propylene glycol (PG) in both deuterated and normal forms. A set of SRS images showing the penetration of PG (deuterated) and ketoprofen across the stratum corneum can be found in Fig. 11.8. SRS images have shown that these compounds penetrate via the intercellular lipids of stratum corneum as well as through hair shafts [75]. The authors also observed a formation of drug crystals at the tissue surface after topically applying a solution of ibuprofen (deuterated) in PG (normal).

Oils

Many cosmetics contain oils and other lipid-rich compounds, which have long been thought to penetrate into skin through the lipid domains of the stratum corneum. In a mouse experiment, video-rate CARS microscopy was tuned to the CH_2 lipid band and used to monitor the change in skin lipid content after the application of baby oil (Johnson & Johnson) [54]. Three-dimensional video-rate imaging with CARS was able to capture the slow penetration of the oil between the cells of the stratum corneum and the eventual diffusion of the oil into the deeper layers of the dermis over the course of 20 min (Fig. 11.9).

CHALLENGES AND NEW ADVANCES OF COHERENT RAMAN TECHNIQUES IN SKIN IMAGING

As optical imaging technologies, coherent Raman imaging face similar challenges as related techniques

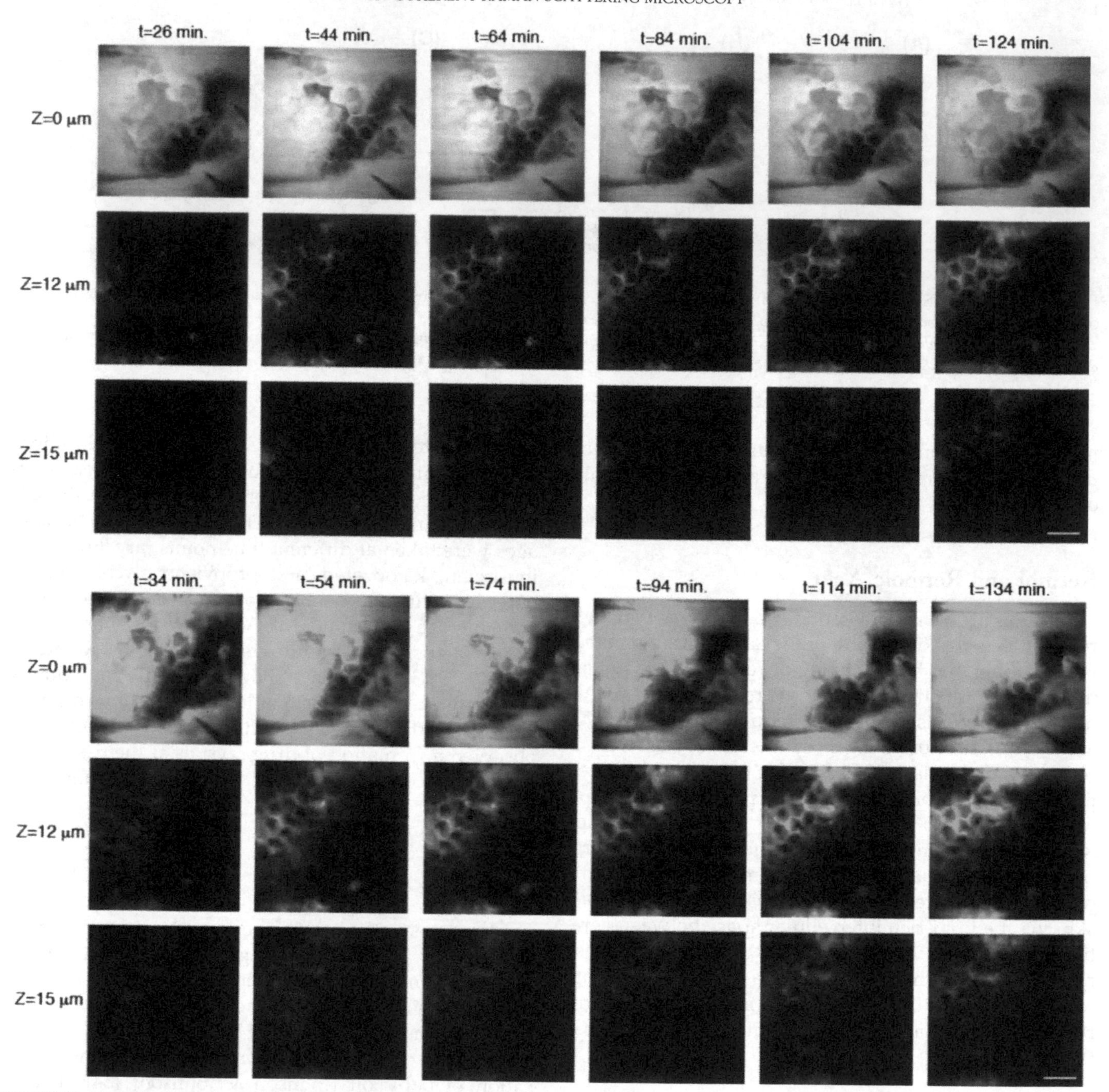

FIGURE 11.8 Imaging the penetration of deuterated propylene glycol (PG) *(upper panel)* and ketoprofen *(lower panel)* across the stratum corneum. Images acquired at the depths indicated down the left-hand side of the figure, and times indicated along the top show the penetration of cosolvent and drug into the tissue using stimulated Raman scattering (SRS) contrast at 2120 and 1599 cm^{-1}, respectively. Scale = 50 μm. *Adapted from Saar BG, Contreras-Rojas LR, Xie XS, Guy RH. Imaging drug delivery to skin with stimulated Raman scattering microscopy, Mol Pharm 2011;8: 969–75, with permission.*

such as multiphoton and confocal microscopies. These include limitations in imaging depth and the need to capture greater sources of contrast from samples. Coherent Raman imaging techniques in particular face an additional barrier to translation because of their current complexity, which must be overcome for these promising tools to make inroads in everyday use.

Imaging Depth

Imaging deep within skin is a challenge to all optical techniques because of the high degree of turbidity of the skin. For efficient imaging, excitation light must first penetrate to the focal plane, then efficiently generate emission, and then the emitted light must reach a

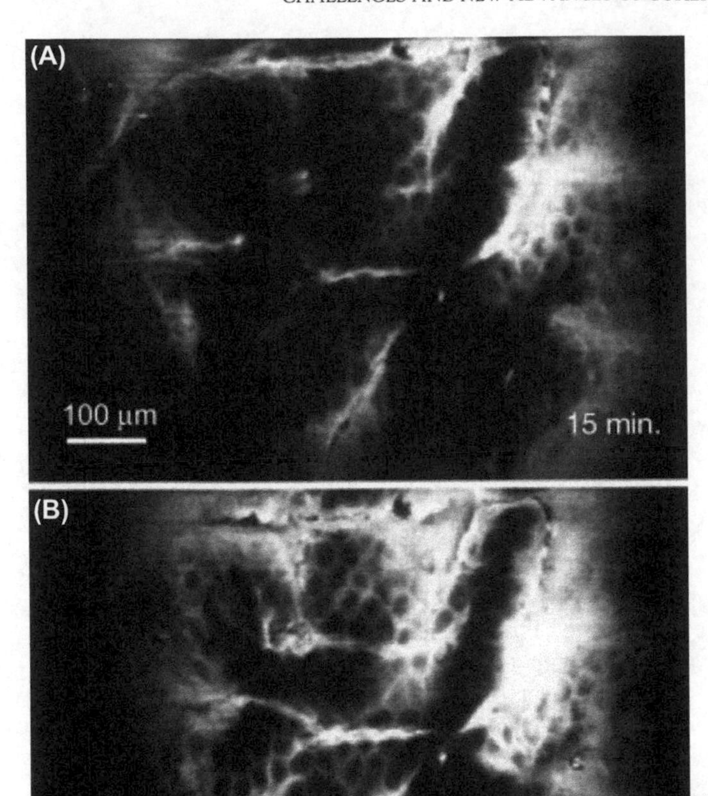

FIGURE 11.9 Diffusion of mineral oil through mouse epidermis. (A) Externally applied mineral oil penetrates the stratum corneum through the lipid clefts between corneocytes. Image was taken 20 μm below the surface 15 min after application of the oil. Raman shift is set to 2845 cm^{-1}, yielding a bright signal from the oil. (B) The same area is shown 5 min later. Brighter signal indicates a higher oil concentration caused by time-dependent diffusion, which can be clearly seen during the 5-min time window. *Adapted from Evans CL, Potma EO, Puoris' haag M, Côté D, Lin CP, Xie XS. Chemical imaging of tissue in vivo with video-rate coherent anti-Stokes Raman scattering microscopy. Proc Natl Acad Sci U S A 2005;102:16807–12, with permission.*

detector. Multiphoton techniques can particularly suffer in turbid environments, because the quality of the focal spot is critical to the nonlinear generation process. Scattering can both attenuate the focusing beam as well as broaden the focal spot size; both mechanisms act to reduce the peak power per volume, and thus the efficiency of the nonlinear process. Scattering also can affect coherent Raman imaging in another negative way; the pump and Stokes colors can experience differing degrees of scattering, leading to reduced focal overlap.

One way of overcoming this challenge is to simply use longer wavelengths of light, which will experience less scattering and therefore reach deeper layers in the skin. In CARS or SRS, molecular imaging contrast arises when the difference between two colors matches a molecular vibration; therefore there is no one "ideal" wavelength range that is required. Although NIR light is optimal over visible or UV light to avoid the photodamage and the generation of nonresonant signals in CARS microscopy, any pair of wavelengths can be chosen as long as they have the requisite energy difference. The improved penetration depth of longer wavelength pump and Stokes were demonstrated by Evans et al., who showed that 900/1200 nm pump-Stokes pairs could achieve improve penetration over 700/900 nm beams in brain tissue [76]. However, it should be noted that this improvement comes at a resolution cost. Longer wavelengths of light naturally have larger excitation volumes and thus reduced imaging resolution. Thus the choice of wavelengths depends on the application, the depth required, and the minimum resolution that can be tolerated.

Adaptive optics techniques can also act to improve the imaging depth of coherent Raman imaging. Borrowed from astronomy, adaptive optics uses a deformable mirror to alter the wavefront of the focused beam to compensate for optical aberrations within a sample [77]. Although this cannot overcome scattering, adaptive optics techniques can vastly improve imaging depth. The shape of the mirror can be practically determined in two ways. First, one can adjust the shape of the mirror according to its Zernike modes (or polynomials), which are sets of orthogonal disk shapes. The intensity of the coherent Raman signal is generally monitored while algorithms alter the mirror shape according to Zernike modes, with the optimal set of modes selected through an iterative process. Alternatively, each element in the mirror can be independently adjusted using an iterative algorithm, such as a random search or genetic algorithm, to find the optimal mirror shape. This latter approach was utilized by Wright et al. [78] with CARS microscopy, where the imaging depth in muscle tissue was pushed beyond 700 μm. Though adaptive optics approaches require a degree of know-how and can be complex, these methods can offer substantial improvements in imaging depth.

Whereas the former two approaches make alterations to the imaging system, it is possible to alter the tissue itself to improve imaging depth. Tissue clearing is such a process, where a chemical is introduced into the skin that acts to match the local index of refraction and reduce scattering. In tissue clearing, a chemical such as dimethyl sulfoxide(DMSO) or glycerin is added to tissue that has an index of refraction that lies between that of water and lipids. As it diffuses through the tissue, the chemical reduces the local index of refraction mismatch, which in turn acts to lessen the degree of light scattering. This process can be highly effective; when glycerin is applied to the skin through disrupted stratum corneum, the tissue becomes relatively transparent to the bare eye, allowing for improved microscopy imaging depth. Tissue clearing can improve the depth of coherent Raman imaging, although great care is needed when applying this technique. As shown by Evans et al. [54], the strong CARS signal visible from tissue is dominated by forward-generated CARS light backscattered by tissue immediately beyond the focal spot. SRS signal detected in the epi direction is similarly relayed in this manner. Thus if the tissue volume of interest is entirely cleared, very little CARS or SRS signal will likely be detected. Tissue clearing can be effective if the region of interest lies near the edge of the diffusing index matching front; in this way there is still tissue present to backscatter the forward-propagating signal. Whereas this may not always be practical, the tissue clearing mechanism does enable substantially improved imaging depth.

Improved Contrast

In the most common CARS or SRS set-up, a laser system is tuned to a particular vibrational mode and images are gathered at that single vibrational band. This stands in contrast to Raman microspectroscopy, where an entire spectrum rich in chemical information is collected at a tissue location. "Hyperspectral" imaging attempts to bridge these two extremes so that spectral and spatial information are collected together. This approach, which has been successfully implemented in white light and fluorescence imaging, separates spectral contributions within the image to improve image contrast and feature classification, as well as increase the amount of obtainable information. In fluorescence microscopy, for example, collecting spectra at each pixel allows, in postprocessing, spectral unmixing of overlapping fluorescence spectra. Whereas spectral overlap might limit fluorescence imaging to three or perhaps four fluorophores, hyperspectral fluorescence microscopy can visualize and quantify eight or more in a given image.

Hyperspectral CARS and SRS operate in a similar way: either through microspectroscopy, where a complete spectrum is obtained at each pixel [79]; or via serial scanning, where images are sequentially collected at each Raman shift to construct a spectrum [80,81]. This allows for the collection of far greater information than standard, single-Raman shift coherent Raman imaging, including the ability to measure the concentrations of species, determine the ratio between species, and simultaneously visualize multiple interacting molecules. With many of the advantages of traditional Raman spectroscopy, the only downside to this approach is image acquisition rate. In most samples, simply too few photons for CARS are generated and too minimal Raman gain/loss occurs at biologically unperturbative incident laser powers to carry out hyperspectral imaging at video rate, with most hyperspectral data sets collected over seconds or minutes.

Making Coherent Raman Imaging Accessible

Although coherent Raman imaging brings many advantages for skin imaging, one of the current challenges is that it remains relatively inaccessible to most dermatological researchers. This is the result of several factors, the most important being the complex laser set-ups that are required for high-contrast, high-resolution imaging. CARS and SRS require two colors of short-pulsed light whose femtosecond or picosecond pulses must arrive at exactly the same location at the same time. Furthermore, at least one color must be tunable over a fairly large wavelength range. Few available laser systems meet these requirements, and those that do tend to require stringent laboratory conditions such as dry atmosphere and vibration isolation. Whereas all major advances in coherent Raman imaging have so far been made with these laser table-based systems, future applications of CARS and SRS require portable, flexible, and more forgiving instruments for routine use. Laser table-based systems can also be quite expensive, requiring hundreds of thousands of dollars to purchase and assemble. It should be noted that at least one company offers a portable CARS imaging system based on current technology, but the price and complexity of this tool may limit its general applicability.

Several advances now point the way toward cart-mounted, turn-key, and cost-effective coherent Raman imaging tools. Several groups have developed fiber-based methods to generate the Stokes beam for CARS and SRS imaging from the pump laser, which removes complexity and the high cost of a second light source [82]. New optical parametric oscillator light sources can generate both the pump and the Stokes beams simultaneously and colinearly, removing the need for complex beam lines and temporal overlap setups [83]. The largest advance, however, is appearing in the form of fiber lasers, which are far more stable than their free-space precursors and offer considerably improved value. Unlike traditional lasers, fiber lasers can be manipulated, bumped, and even dropped without losing alignment and functionality—key features required for cart-mounted, clinic-ready imaging tools. Xie and colleagues have already demonstrated CARS imaging systems based on these fiber light sources

[84,85], with new startup companies promising rollable coherent Raman imaging systems in the next several years. The only major limitation of fiber lasers is their tunability, a weakness that is the focus of numerous research teams.

These new imaging systems will make CRS far more accessible to both researchers and clinicians alike. Simple, low-cost, and easy-to-use lasers will bring CARS and SRS to more laboratories, especially if they can be offered as plug-in modules to existing microscopy equipment. Similar to how fluorescence lifetime imaging microscopy is now available for most commercial microscopes, CARS and SRS could be readily introduced to microscopes through the addition of a laser system, light path, and control software. In a similar way, fiber-based CARS and SRS systems can be readily developed for clinical use. Cart-mounted and combined with either hand-held or C-arm mounted scan heads, CRS imaging systems could closely resemble the VivaScope confocal reflectance systems currently in clinical use. When combined with other modalities such as fluorescence and reflectance, such platforms will greatly aid clinical research.

CONCLUSIONS

As a noninvasive, chemical-specific imaging technology, CRS microscopy is primed to make a major impact in dermatology. With the ability to visualize specific molecules without the need for exogenous contrast agents, CRS is ideal for clinical imaging. New imaging systems under development will be of extreme importance in bridging the gap between research and practice, especially those that offer tunable, simple operation. The major onus now rests on current CARS and SRS researchers, who must continue to reach out to our dermatology colleagues to advance the science and application of these exciting toolkits. Research focused on visualizing the uptake of drugs, the alterations of skin structure, and the characterization of skin diseases is needed to drive forward the in vivo and clinical relevance of CRS. For example, the recent use of SRS to generate H&E-like contrast in brain tissue without the use of stains or labels [85] would find great utility in dermatology. Another important avenue is to engage pharmaceutical and cosmeceutical laboratories and companies in these next steps as partners in the CRS field. Many of the current problems faced by these companies, from drug uptake to the myriad effects of actives, can be addressed and quantified using CRS microscopy.

The rapid development and wide range of applications make CRS microscopy an exciting and promising technology. This tool kit will undergo major advances in the coming years, supported by collaborations between researchers, technology companies, pharmaceutical and cosmeceutical companies, and clinicians. By overcoming barriers (physical, practical, and economical) CRS is poised to have a bright future in dermatology.

List of Acronyms and Abbreviations

BCC Basal cell carcinoma
CARS Coherent anti-Stokes Raman scattering
CSLM Confocal scanning laser microscopy
DEJ Dermal epidermal junction
DMSO Dimethyl sulfoxide
MPM Multiphoton microscopy
NSAID Nonsteroidal antiinflammatory drug
OCT Optical coherence tomography
PG Propylene glycol
RA Retinoic acid
RCM Reflectance confocal microscopy
SCC Squamous cell carcinoma
SG Sebaceous gland
SHG Second harmonic generation
SRS Stimulated Raman scattering
TPF Two-photon fluorescence
UV Ultraviolet

References

[1] Kurosumi K, Shibasaki S, Ito T. Cytology of the secretion in mammalian sweat glands. Int Rev Cytol 1984;87:253.

[2] Wood R. Secret communications concerning light rays. J Physiol 1919;5e(p. t IX).

[3] Margarot J, Deveze P. Aspect de quelques dermatoses lumiere ultraparaviolette: note preliminaire. Bull Soc Sci Med Biol Montpellier 1925;6:375–8.

[4] Halprin KM. Diagnosis with Wood's light. II. The porphyrias. JAMA 1967;200:460.

[5] Jilson O. Wood's light: an incredibly important diagnostic tool. Cutis 1981;28:620, 623.

[6] Sanchez NP, Pathak MA, Sato S, Fitzpatrick TB, Sanchez JL, Mihm MC. Melasma: a clinical, light microscopic, ultrastructural, and immunofluorescence study. J Am Acad Dermatol 1981;4:698–710.

[7] Braun RP, Rabinovitz HS, Oliviero M, Kopf AW, Saurat J-H. Dermoscopy of pigmented skin lesions. J Am Acad Dermatol 2005;52:109–21.

[8] Vázquez López F, Kreusch J, Marghoob A. Dermoscopic semiology: further insights into vascular features by screening a large spectrum of nontumoral skin lesions. Br J Dermatol 2004;150:226–31.

[9] Bergman R, Sharony L, Schapira D, Nahir MA, Balbir-Gurman A. The handheld dermatoscope as a nail-fold capillaroscopic instrument. Arch Dermatol 2003;139:1027–30.

[10] Argenziano G, Zalaudek I, Corona R, Sera F, Cicale L, Petrillo G, et al. Vascular structures in skin tumors: a dermoscopy study. Arch Dermatol 2004;140:1485–9.

[11] Vázquez-López F, Manjón-Haces JA, Vázquez-López AC, Pérez-Oliva N. The hand-held dermatoscope improves the clinical evaluation of port-wine stains. J Am Acad Dermatol 2003;48:984–5.

[12] Bianchi L, Orlandi A, Campione E, Angeloni C, Costanzo A, Spagnoli L, et al. Topical treatment of basal cell carcinoma with tazarotene: a clinicopathological study on a large series of cases. Br J Dermatol 2004;151:148–56.

[13] Binder M, Puespoeck-Schwarz M, Steiner A, Kittler H, Muellner M, Wolff K, et al. Epiluminescence microscopy of small pigmented skin lesions: short-term formal training improves the diagnostic performance of dermatologists. J Am Acad Dermatol 1997;36:197–202.

[14] Minsky M. Microscopy apparatus. Google Patents, editor; 1961.

[15] Rajadhyaksha M, Grossman M, Esterowitz D, Webb RH, Anderson RR. In vivo confocal scanning laser microscopy of human skin: melanin provides strong contrast. J Invest Dermatol 1995;104:946–52.

[16] Rajadhyaksha M, González S, Zavislan JM, Anderson RR, Webb RH. In vivo confocal scanning laser microscopy of human skin. II. Advances in instrumentation and comparison with histology. J Invest Dermatol 1999;113:293–303.

[17] Langley RG, Rajadhyaksha M, Dwyer PJ, Sober AJ, Flotte TJ, Anderson RR. Confocal scanning laser microscopy of benign and malignant melanocytic skin lesions in vivo. J Am Acad Dermatol 2001;45:365–76.

[18] González S, Tannous Z. Real-time, in vivo confocal reflectance microscopy of basal cell carcinoma. J Am Acad Dermatol 2002;47:869–74.

[19] Astner S, Gonzalez E, Cheung A, Rius-Diaz F, González S. Pilot study on the sensitivity and specificity of in vivo reflectance confocal microscopy in the diagnosis of allergic contact dermatitis. J Am Acad Dermatol 2005;53:986–92.

[20] Aghassi D, Anderson RR, González S. Confocal laser microscopic imaging of actinic keratoses in vivo: a preliminary report. J Am Acad Dermatol 2000;43:42–8.

[21] Rajadhyaksha M, Menaker G, Flotte T, Dwyer PJ, González S. Confocal examination of nonmelanoma cancers in thick skin excisions to potentially guide Mohs micrographic surgery without frozen histopathology. J Invest Dermatol 2001;117:1137–43.

[22] Wang H, Tsai TH, Zhao J, Lee A, Lo BKK, Yu M, et al. Differentiation of HaCaT cell and melanocyte from their malignant counterparts using microRaman spectroscopy guided by confocal imaging. Photodermatol Photoimmunol Photomed 2012;28:147–52.

[23] Welzel J, Lankenau E, Birngruber R, Engelhardt R. Optical coherence tomography of the human skin. J Am Acad Dermatol 1997;37:958–63.

[24] Gambichler T, Jaedicke V, Terras S. Optical coherence tomography in dermatology: technical and clinical aspects. Arch Dermatol Res 2011;303:457–73.

[25] Welzel J, Bruhns M, Wolff HH. Optical coherence tomography in contact dermatitis and psoriasis. Arch Dermatol Res 2003;295:50–5.

[26] Izatt JA, Kulkarni MD, Yazdanfar S, Barton JK, Welch AJ. In vivo bidirectional color Doppler flow imaging of picoliter blood volumes using optical coherence tomography. Opt Lett 1997;22:1439–41.

[27] Welzel J. Optical coherence tomography in dermatology: a review. Skin Res Technol 2001;7:1–9.

[28] Wang H, Baran U, Wang RK. In vivo blood flow imaging of inflammatory human skin induced by tape stripping using optical microangiography. J Biophotonics 2015;8:265–72.

[29] Wang RK, Jacques SL, Ma Z, Hurst S, Hanson SR, Gruber A. Three dimensional optical angiography. Opt Express 2007;15:4083–97.

[30] Gladkova ND, Petrova G, Nikulin N, Radenska Lopovok S, Snopova L, Chumakov YP, et al. In vivo optical coherence tomography imaging of human skin: norm and pathology. Skin Res Technol 2000;6:6–16.

[31] Mogensen M, Morsy HA, Nurnberg BM, Jemec GB. Optical coherence tomography imaging of bullous diseases. J Eur Acad Dermatol Venereol 2008;22:1458–64.

[32] Mogensen M, Thrane L, Jørgensen TM, Andersen PE, Jemec GB. OCT imaging of skin cancer and other dermatological diseases. J Biophotonics 2009;2:442–51.

[33] Abuzahra F, Spöler F, Först M, Brans R, Erdmann S, Merk HF, et al. Pilot study: optical coherence tomography as a non-invasive diagnostic perspective for real time visualisation of onychomycosis. Mycoses 2010;53:334–9.

[34] Jørgensen TM, Tycho A, Mogensen M, Bjerring P, Jemec GB. Machine learning classification of non-melanoma skin cancers from image features obtained by optical coherence tomography. Skin Res Technol 2008;14:364–9.

[35] Strasswimmer J, Pierce MC, Park BH, Neel V, de Boer JF. Polarization-sensitive optical coherence tomography of invasive basal cell carcinoma. J Biomed Opt 2004;9:292–8.

[36] Gambichler T, Regeniter P, Bechara FG, Orlikov A, Vasa R, Moussa G, et al. Characterization of benign and malignant melanocytic skin lesions using optical coherence tomography in vivo. J Am Acad Dermatol 2007;57:629–37.

[37] Sand M, Gambichler T, Moussa G, Bechara F, Sand D, Altmeyer P, et al. Evaluation of the epidermal refractive index measured by optical coherence tomography. Skin Res Technol 2006;12:114–8.

[38] Yang VX, Pekar J, Lo SS, Gordon ML, Wilson BC, Vitkin IA. Optical coherence and Doppler tomography for monitoring tissue changes induced by laser thermal therapy: an in vivo feasibility study. Rev Sci Instrum 2003;74:437–40.

[39] Masters BR, So P, Gratton E. Multiphoton excitation fluorescence microscopy and spectroscopy of in vivo human skin. Biophysical J 1997;72:2405.

[40] Lee A, Wang H, Yu Y, Tang S, Zhao J, Lui H, et al. In vivo video rate multiphoton microscopy imaging of human skin. Opt Lett 2011;36:2865–7.

[41] Lin S-J, Jee S-H, Chan J-Y, Wu R-J, Lo W, Tan H-Y, et al. Monitoring photoaging by use of multiphoton fluorescence and second harmonic generation microscopy. In: Biomedical optics; 2006. p. 607803–607803-7.

[42] König K, Ehlers A, Stracke F, Riemann I. In vivo drug screening in human skin using femtosecond laser multiphoton tomography. Skin Pharmacol Physiol 2006;19:78–88.

[43] Paoli J, Smedh M, Wennberg A-M, Ericson MB. Multiphoton laser scanning microscopy on non-melanoma skin cancer: morphologic features for future non-invasive diagnostics. J Invest Dermatol 2008;128:1248–55.

[44] Lin S-J, Jee S-H, Kuo C-J, Wu Jr R, Lin W-C, Chen J-S, et al. Discrimination of basal cell carcinoma from normal dermal stroma by quantitative multiphoton imaging. Opt Lett 2006;31:2756–8.

[45] Dimitrow E, Ziemer M, Koehler MJ, Norgauer J, König K, Elsner P, et al. Sensitivity and specificity of multiphoton laser tomography for in vivo and ex vivo diagnosis of malignant melanoma. J Invest Dermatol 2009;129:1752–8.

[46] Wang H, Lee A, Frehlick Z, Lui H, McLean DI, Tang S, et al. Perfectly registered multiphoton and reflectance confocal video rate imaging of in vivo human skin. J Biophotonics 2013;6:305–9.

[47] Wang H, Zandi S, Lee A, Zhao J, Lui H, McLean DI, et al. Imaging directed photothermolysis through two photon absorption demonstrated on mouse skin: a potential novel tool for highly targeted skin treatment. J Biophotonics 2014;7:534–41.

[48] Wang H, Huang N, Zhao J, Lui H, Korbelik M, Zeng H. Depth-resolved in vivo micro-Raman spectroscopy of a murine skin tumor model reveals cancer-specific spectral biomarkers. J Raman Spectrosc 2011;42:160–6.
[49] Evans CL, Xie XS. Coherent anti-Stokes Raman scattering microscopy: chemical imaging for biology and medicine. Annu Rev Anal Chem 2008;1:883–909.
[50] Freudiger CW, Min W, Saar BG, Lu S, Holtom GR, He C, et al. Label-free biomedical imaging with high sensitivity by stimulated Raman scattering microscopy. Science 2008;322:1857–61.
[51] Ammar DA, Lei TC, Kahook MY, Masihzadeh O. Imaging the intact mouse cornea using coherent anti-Stokes Raman scattering (CARS). Invest Ophthalmol Vis Sci 2013;54:5258–65.
[52] Lim RS, Kratzer A, Barry NP, Miyazaki-Anzai S, Miyazaki M, Mantulin WW, et al. Multimodal CARS microscopy determination of the impact of diet on macrophage infiltration and lipid accumulation on plaque formation in ApoE-deficient mice. J Lipid Res 2010;51:1729–37.
[53] Le TT, Rehrer CW, Huff TB, Nichols MB, Camarillo IG, Cheng J-X. Nonlinear optical imaging to evaluate the impact of obesity on mammary gland and tumor stroma. Mol Imaging 2007;6:205.
[54] Evans CL, Potma EO, Puoris' haag M, Côté D, Lin CP, Xie XS. Chemical imaging of tissue in vivo with video-rate coherent anti-Stokes Raman scattering microscopy. Proc Natl Acad Sci U S A 2005;102:16807–12.
[55] Djaker N, Lenne P-F, Marguet D, Colonna A, Hadjur C, Rigneault H. Coherent anti-Stokes Raman scattering microscopy (CARS): instrumentation and applications. Nucl Instrum Methods Phys Res Sect A Accel Spectrom Detect Assoc Equip 2007;571:177–81.
[56] Robbins CR. Chemical and physical behavior of human hair, vol. 4. Springer; 2002.
[57] Hearle J. A critical review of the structural mechanics of wool and hair fibres. Int J Biol Macromol 2000;27:123–38.
[58] Feughelman M. Natural protein fibers. J Appl Polym Sci 2002;83: 489–507.
[59] Bito K, Okuno M, Kano H, Tokuhara S, Naito S, Masukawa Y, et al. Protein secondary structure imaging with ultrabroadband multiplex coherent anti-Stokes Raman scattering (CARS) microspectroscopy. The J Phys Chem B 2012;116:1452–7.
[60] Jung Y, Tam J, Jalian HR, Anderson RR, Evans CL. Longitudinal, 3D in vivo imaging of sebaceous glands by coherent anti-Stokes Raman scattering microscopy: normal function and response to cryotherapy. J Invest Dermatol 2014.
[61] Zimmerley M, Lin C-Y, Oertel DC, Marsh JM, Ward JL, Potma EO. Quantitative detection of chemical compounds in human hair with coherent anti-Stokes Raman scattering microscopy. J Biomed Opt 2009;14:044019–044019-7.
[62] Breunig H, Weinigel M, Kellner-Höfer M, Bückle R, Darvin M, Lademann J, et al. Combining multiphoton and CARS microscopy for skin imaging. In: Spie BiOS; 2013. p. 85880N–85880N-7.
[63] Breunig HG, Bückle R, Kellner Höfer M, Weinigel M, Lademann J, Sterry W, et al. Combined in vivo multiphoton and CARS imaging of healthy and disease affected human skin. Microsc Res Tech 2012;75:492–8.
[64] Vogler N, Heuke S, Akimov D, Latka I, Kluschke F, Röwert-Huber H-J, et al. Discrimination of skin diseases using the multimodal imaging approach. In: Spie Photonics Europe; 2012. p. 842710–842710-8.
[65] Berardesca E, Maibach H, Wilhelm K. Non invasive diagnostic techniques in clinical dermatology. Springer Science & Business Media; 2013.
[66] Zalaudek I, Kreusch J, Giacomel J, Ferrara G, Catricalà C, Argenziano G. How to diagnose nonpigmented skin tumors: a review of vascular structures seen with dermoscopy. I. Melanocytic skin tumors. J Am Acad Dermatol 2010;63:361–74.
[67] Zalaudek I, Giacomel J, Schmid K, Bondino S, Rosendahl C, Cavicchini S, et al. Dermatoscopy of facial actinic keratosis, intraepidermal carcinoma, and invasive squamous cell carcinoma: a progression model. J Am Acad Dermatol 2012;66:589–97.
[68] Heuke S, Vogler N, Meyer T, Akimov D, Kluschke F, Röwert-Huber H-J, et al. Detection and discrimination of non-melanoma skin cancer by multimodal imaging. In: Healthcare; 2013. p. 64–83.
[69] Vogler N, Medyukhina A, Latka I, Kemper S, Böhm M, Dietzek B, et al. Towards multimodal nonlinear optical tomography: experimental methodology. Laser Phys Lett 2011;8:617.
[70] William A. Transdermal and topical drug delivery from theory to clinical practice. London: Pharmaceutical Press; 2003. p. 37–84.
[71] Varani J, Warner RL, Gharaee-Kermani M, Phan SH, Kang S, Chung J, et al. Vitamin a antagonizes decreased cell growth and elevated collagen-degrading matrix metalloproteinases and stimulates collagen accumulation in naturally aged human skin1. J Invest Dermatol 2000;114:480–6.
[72] Shalita AA, Weiss J, Chalker D, Ellis C, Greenspan A, Katz H, et al. A comparison of the efficacy and safety of adapalene gel 0.1% and tretinoin gel 0.025% in the treatment of acne vulgaris: a multicenter trial. J Am Acad Dermatol 1996;34:482–5.
[73] Schwartz E, Cruickshank FA, Mezick JA, Kligman LH. Topical all-trans retinoic acid stimulates collagen synthesis in vivo. J Invest Dermatol 1991;96:975–8.
[74] Fredriksson T, Pettersson U. Severe psoriasis: oral therapy with a new retinoid. Dermatology 1978;157:238–44.
[75] Saar BG, Contreras-Rojas LR, Xie XS, Guy RH. Imaging drug delivery to skin with stimulated Raman scattering microscopy. Mol Pharm 2011;8:969–75.
[76] Evans CL, Xu X, Kesari S, Xie XS, Wong ST, Young GS. Chemically-selective imaging of brain structures with CARS microscopy. Opt Express 2007;15:12076–87.
[77] Booth MJ. Adaptive optics in microscopy. Philos Trans R Soc Lond A Math Phys Eng Sci 2007;365:2829–43.
[78] Wright A, Poland S, Girkin J, Freudiger C, Evans C, Xie X. Adaptive optics for enhanced signal in CARS microscopy. Opt Express 2007;15:18209–19.
[79] Parekh SH, Lee YJ, Aamer KA, Cicerone MT. Label-free cellular imaging by broadband coherent anti-Stokes Raman scattering microscopy. Biophysical J 2010;99:2695–704.
[80] Zhang D, Wang P, Slipchenko MN, Ben-Amotz D, Weiner AM, Cheng J-X. Quantitative vibrational imaging by hyperspectral stimulated Raman scattering microscopy and multivariate curve resolution analysis. Anal Chem 2012;85:98–106.
[81] Brideau C, Poon K, Stys P. Broadly tunable high-energy spectrally focused CARS microscopy with chemical specificity and high resolution for biological samples. In: Spie BiOS; 2013. p. 85880E–85915E.
[82] Murugkar S, Brideau C, Ridsdale A, Naji M, Stys PK, Anis H. Coherent anti-Stokes Raman scattering microscopy using photonic crystal fiber with two closely lying zero dispersion wavelengths. Opt Express 2007;15:14028–37.
[83] Ganikhanov F, Carrasco S, Sunney Xie X, Katz M, Seitz W, Kopf D. Broadly tunable dual-wavelength light source for coherent anti-Stokes Raman scattering microscopy. Opt Lett 2006;31:1292–4.
[84] Wang K, Freudiger CW, Lee JH, Saar BG, Xie XS, Xu C. Synchronized time-lens source for coherent Raman scattering microscopy. Opt Express 2010;18:24019–24.
[85] Freudiger CW, Pfannl R, Orringer DA, Saar BG, Ji M, Zeng Q, et al. Multicolored stain-free histopathology with coherent Raman imaging. Lab Invest 2012;92:1492–502.

CHAPTER

12

Rapid Real-Time Raman Spectroscopy and Imaging-Guided Confocal Raman Spectroscopy for In Vivo Skin Evaluation and Diagnosis

J. Zhao[1,2], *H. Lui*[1,2], *D.I. McLean*[1,2], *H. Zeng*[1,2]

[1]University of British Columbia and Vancouver Coastal Health Research Institute, Vancouver, BC, Canada; [2]BC Cancer Agency, Vancouver, BC, Canada

OUTLINE

Imaging in Dermatology
http://dx.doi.org/10.1016/B978-0-12-802838-4.00012-1

INTRODUCTION

Raman Effect

Raman is an inelastic scattering effect in which the scattered light is changed in frequency after interaction of the excitation light with biomolecules. It was first discovered experimentally by the Indian scientist C. V. Raman in 1928 and has been named after him ever since [1]. The Raman spectrum measures the vibrational modes of molecules, and it is highly dependent on the structure and conformation of biochemical constituents at the molecular level.

As an example, the Raman spectrum of cholesterol powder under 785-nm excitation is presented in Fig. 12.1. The location of a peak, measured in Raman shift (in cm^{-1}), is a measure of the frequency change of the Raman signal from the excitation light and is assigned to a specific molecule. For example, the 1445-cm^{-1} peak is assigned to CH_2 and CH_3 molecules in the bending mode, and the 1650-cm^{-1} peak is assigned to C=C bond in the stretching mode. The peak intensity (sometimes the area under the peak) is a measure of the quantity of the molecule in the sample. Therefore, by measuring the changes in the Raman spectrum, one can quantify the difference in molecular composition. These differences provide diagnostic information for different diseases, including those in skin. Different from infrared absorption, the Raman signal is relatively insensitive to the water content in tissue; therefore it is of particular interest for in vivo applications.

Raman Spectroscopy Studies in Skin

Raman spectroscopy has been used to study dysplasia and cancer in various tissue types, such as skin [2–5], lung [6–8], breast [9–12], stomach [13–16], colon [17–20], cervix [21–23], prostate [24,25], and oral

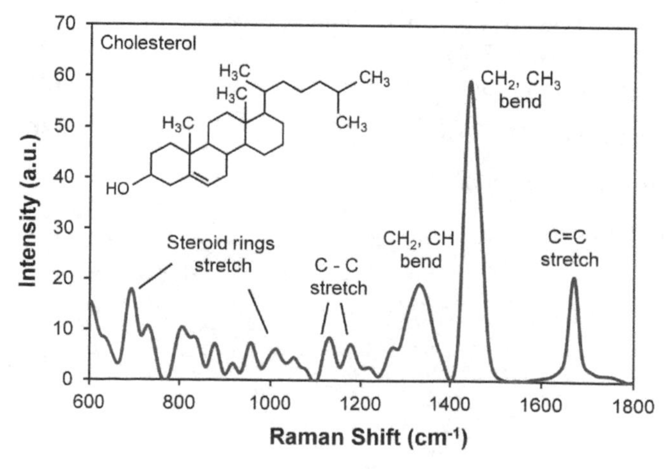

FIGURE 12.1 Raman spectrum of cholesterol powder under 785-nm excitation.

cancers [26]. Raman spectroscopy was first introduced into skin investigation by Edwards et al. in 1992 [27,28]. They measured several skin samples ex vivo using a Fourier transform Raman (FT-Raman) system and characterized the major Raman peaks for skin samples. It took approximately 30 min to acquire a single Raman spectrum using the FT-Raman system. Later on, Puppels et al. developed a confocal microscopic Raman system and measured the Raman properties of different skin layers in vitro and in vivo [29–32]. They measured normal skin and basal cell carcinoma (BCC) ex vivo with a 850-nm laser (12 nodular BCC and 3 superficial BCC) and achieved a sensitivity of 100% and specificity of 93% [33]. They also measured BCC and normal skin in the high-frequency region with a 720-nm laser (19 BCC and 9 normal) and found that Raman spectroscopy could differentiate BCC from normal tissue with high accuracy [34]. It took from 10 s to 8 min to acquire a single spectrum in their studies. We developed a state-of-the-art rapid Raman system for in vivo skin studies under 785-nm excitation, which reduced the integration time to 1–5 s, making it possible for in vivo clinical measurements [3,5]. We measured the Raman properties of human melanin in vivo [35] and found that combining Raman signals with the corresponding fluorescence background may improve tumor diagnosis [36]. Diagnosis of melanoma and nonmelanoma skin cancer ex vivo using FT-Raman spectroscopy was reported by Gniadecka et al. [37–41]. In a study, they measured 223 skin samples including 22 melanomas, 41 pigmented nevi, 48 BCC, 23 seborrheic keratosis (SK), and 89 normal skin ex vivo using an FT-Raman system with an integration time of approximately 7 min for each spectrum at 1064-nm excitation and achieved a sensitivity of 85% and specificity of 99% using a neural network classification model [37]. Mahadevan-Jansen et al. developed a confocal microscopic Raman system for ex vivo and in vivo skin cancer diagnosis [42,43]. It took approximately 30 s for confocal microscopic Raman spectral acquisition. They demonstrated that confocal microscopic Raman spectroscopy could be used for skin cancer diagnosis with high diagnostic performance. However, their in vivo study was limited to a small number of cases [9 BCC, 4 squamous cell carcinoma (SCC), 8 inflamed scars, and 21 normal]. We reported the first large-scale clinical study of Raman spectroscopy for in vivo skin cancer diagnosis [2]. It was the first study to demonstrate that Raman spectroscopy could be used for in vivo clinical skin cancer diagnosis with comparable diagnostic performance to that of clinicians and other diagnostic aids [2].

Raman spectroscopy can also be used for other skin-related studies, such as measuring the carotenoid level in the skin [44], monitoring cutaneous topical delivery of drugs [45–47], monitoring microphotothermolysis

[48], and objectively evaluating the effects of cosmetic products [49,50]. Huang et al. [51] have systematically studied the Raman properties of mouse skin and other organs in the fingerprint region and high-frequency region. The accumulated literature has demonstrated that Raman spectroscopy is a viable technique for skin science and for skin cancer diagnosis.

Raman Spectroscopy for In Vivo Applications

The traditional FT-Raman system requires long integration times to acquire a single spectrum, which prevents it from being used in in vivo applications. The use of Raman spectroscopy for in vivo detection in a clinical setting depends on the feasibility of measuring adequate Raman spectra in a few seconds or less using fiber-optic probes. In this chapter, we present the implementation of a rapid real-time Raman system for in vivo skin cancer diagnosis and an imaging-guided confocal Raman system for in vivo skin assessment. We also present the modeling of in vivo skin Raman spectra using Monte Carlo (MC) simulation. In this chapter, we will focus particularly on in vivo applications of Raman spectroscopy in skin studies.

RAPID REAL-TIME RAMAN SYSTEM

Instrumentation

The prototype of the rapid real-time Raman system for in vivo skin cancer diagnosis is shown schematically in Fig. 12.2. It consists of a diode laser, a custom-made Raman probe, a spectrometer, and a fiber bundle for delivery of the laser and capture of the signal. The laser is a wavelength-stabilized diode laser at 785 nm, which has been found to be the optimal wavelength for skin Raman measurement. It has several advantages over other wavelengths, (1) including that the light at this wavelength penetrates deeper into the tissue than shorter wavelengths, (2) the fluorescence background of tissue at this excitation wavelength is relatively weaker than that at shorter excitation wavelengths, (3) the Raman quantum yield of the tissue at this excitation wavelength is higher than that of longer excitation wavelengths, and (4) the wavelength of Raman signal under 785-nm excitation falls within the sensitivity range of silicon-based charge-coupled device (CCD) cameras.

The laser beam is delivered to the Raman probe by a 200-μm core-diameter single fiber. The raw signal, which is composed of Raman signal and tissue fluorescence background, is collected by the Raman probe and transmitted to the spectrograph through a customized fiber bundle. The fiber bundle is composed of 58 100-μm core-diameter low-OH fibers for higher near-infrared (NIR) transmission. The distal end of the fiber bundle that is connected to the Raman probe is packed into a 1.3-mm diameter circular area. The proximal end of the fiber bundle that is connected to the spectrograph is designed such that the fiber tips are aligned along a parabolic arc that is in an inverse orientation to the image aberration of the transmissive spectrograph. The center fiber is used for calibration so that the image of the fibers is symmetrical along the centerline of the CCD detectors.

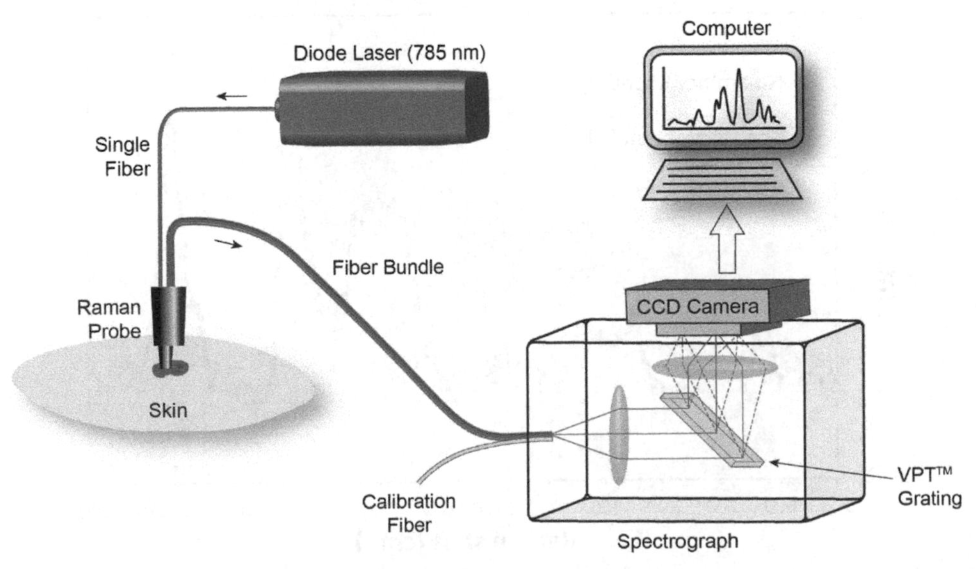

FIGURE 12.2 Schematic configuration of the rapid real-time Raman system for in vivo skin cancer diagnosis. *CCD*, charge-coupled device. *Adapted from Lui H, Zhao J, McLean D, Zeng H. Real-time Raman spectroscopy for in vivo skin cancer diagnosis. Cancer Res 2012;72(10):2491–500 with permission from the American Association for Cancer Research.*

With this fiber arrangement, the image aberration is fully corrected, allowing us to achieve full-chip vertical hardware binning. The signal-to-noise ratio of the rapid real-time Raman system is improved by 16 times compared with that using software binning [5]. The CCD has a 16-bit dynamic range and is liquid nitrogen-cooled to −120°C. The spectral resolution of the system is 8 cm^{-1}.

System Calibration and Fluorescence Background Removal

Several calibration procedures are needed before reliable measurement of the spectrum can be achieved, including wavelength calibration and intensity calibration [3]. Wavelength calibration is needed to establish the correlation between the CCD pixels and wavelengths (or Raman shifts). In calibration of the rapid real-time Raman system, we choose 10 major peaks from two known light sources for wavelength calibration (Hg−Ar and Kr lamps). A fifth-order polynomial fitting is used to correlate the CCD pixel positions with the wavelengths.

The intensity calibration is needed to correct the optical transfer function and spectral response of the CCD camera, which varies from system to system. It is performed using a National Institute of Standards and Technology (NIST)-traceable tungsten calibration lamp or NIST-certified intensity calibration disk (ie, SRM 2241). The ratios of the known spectrum to the measured spectrum of the lamp or disk lead to the intensity calibration factors. The intensity-calibrated spectrum is then obtained from the measured spectrum by multiplying by the intensity calibration factors.

A dedicated software package is implemented for rapid real-time Raman measurement, which includes dark-noise subtraction, intensity calibration, and fluorescence background removal. We proposed a Vancouver Raman algorithm that is found to be optimal for skin fluorescence background removal [52].

Special Lighting for In Vivo Clinical Raman Measurement

The Raman signal is conventionally acquired in a dark environment. Darkness is inconvenient for the operator and the patient in a clinical setting. We proposed a method to implement real-time in vivo Raman measurement with a light-emitting diode (LED) illumination source, which is based on multiple filtering of the illumination and collection optics [53]. The proposed multiple-filtering system consists of a broad band-pass filter, a long-pass filter, and a narrow band-pass laser-line filter. The broad band-pass filter, placed in front of the LED light source, allows the visible light to pass through and blocks the NIR light that is longer than 700 nm. The long-pass filter, placed in the Raman signal collection path, blocks the LED illumination and laser light for the Raman measurement. The narrow band-pass laser-line filter, placed in the Raman excitation path, suppresses the laser side bands and emissions from within the fiber. The Raman spectra of palm skin obtained with the specially designed LED illumination source switched on and switched off are shown in Fig. 12.3 and show no statistically significant difference ($p = .4591$). This study demonstrated that clinical Raman spectra can be measured in an

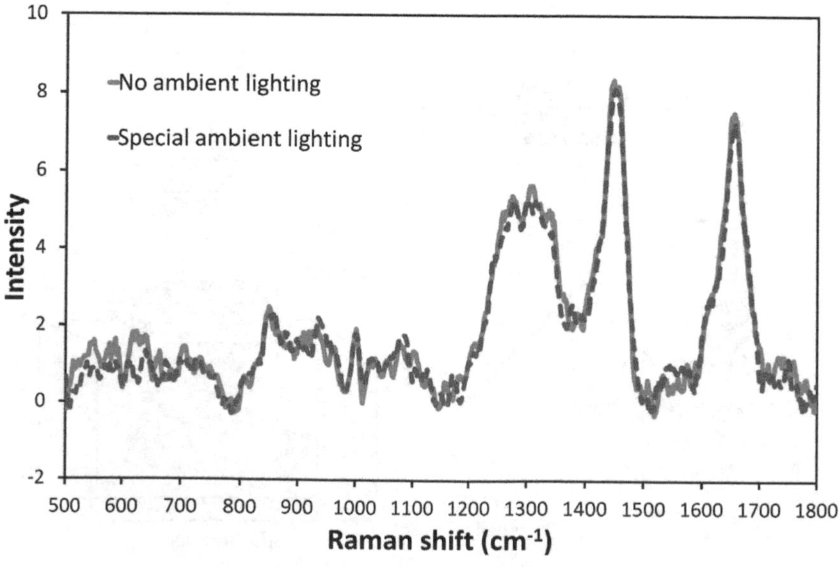

FIGURE 12.3 Raman spectra of palm skin with and without the specially designed illumination light source. The difference between the two spectra is statistically insignificant ($p = .4591$). *Adapted from Zhao J, Short M, Braun T, Lui H, McLean D, Zeng H. Clinical Raman measurements under special ambient lighting illumination. J Biomed Opt 2014;19(11):111609 with permission from SPIE.*

illuminated environment by the multiple-filtered light source without interference.

RAPID REAL-TIME RAMAN SYSTEM FOR IN VIVO SKIN CANCER DIAGNOSIS

Patient Recruitment

This study was approved by the Clinical Research Ethics Board of the University of British Columbia (Vancouver, BC, Canada; Protocol C96-0499).The rapid real-time Raman spectrometer system is used for skin cancer diagnosis in vivo. In this chapter, we report analysis of 645 cases in total. The first set of 518 cases in the previous study was acquired between January 2003 and May 2011, hereafter referred to as "previous cohort" [2]. The second set of 127 completely independent new cases was measured between June 2011 and May 2014, hereafter referred to as "new cohort" [54]. When the two sets of patients are consolidated according to pathology, it is referred to as the "consolidated cohort." The detailed distribution of the lesions, including diagnostic subtypes, is provided in Table 12.1. The lesion types include malignant and premalignant lesions that require treatment, including malignant melanoma (MM), SCC, BCC, and actinic keratosis (AK), as well as benign conditions that can visually mimic skin cancer, including SK, atypical nevi (AN), junctional nevi (JN), compound nevi (CN), intradermal nevi (IN), and blue nevi (BN). For the purposes of this study each individual lesion was considered an experimental unit for analysis.

TABLE 12.1 Summary of Lesions Evaluated by Raman Spectroscopy

Final lesion diagnosis	Previous cohort[a]	New cohort[b]	Consolidated cohort[c]
Lentigo maligna	20	0	20
Lentigo maligna melanoma	8	2	10
Superficial spreading melanoma	14	7	21
Other types of melanoma	2	0	2
BCC superficial	28	6	34
BCC nodular	73	24	97
BCC other	8	5	13
SCC in situ	18	5	23
SCC invasive	28	13	41
Other types of SCC	1	0	1
Actinic keratosis	32	15	47
Atypical nevus	57	16	73
Blue nevus	13	0	13
Compound nevus	30	2	32
Intradermal nevus	38	7	45
Junctional nevus	34	1	35
Seborrheic keratosis	114	24	138
Total	518	127	645

BCC, basal cell carcinoma; SCC, squamous cell carcinoma.
[a]*Refers to the validated cases measured between January 2003 and May 2011 for the analysis [2].*
[b]*Refers to the validated cases measured between June 2011 and May 2014 for an independent validation.*
[c]*The combination of the previous cohort and the new cohort according to diagnosis.*

Three Clinical Diagnostic Tasks

The diagnostic performance of the rapid real-time Raman system for classifying skin lesions was tested according to three tasks based on clinical relevance: (1) to discriminate cancerous and precancerous lesions that require treatment (MM, BCC, SCC, and AK) from benign conditions (AN, BN, CN, IN, JN, and SK); (2) to discriminate melanoma (all forms, MM) from benign pigmented skin lesions (AN, BN, CN, IN, JN, and SK); and (3) to discriminate melanoma (all forms, MM) from SK, which can be confused with melanoma because of similarities in appearance.

Multivariate Analysis: Principal Component Generalized Discriminant Analysis and Partial Least Squares

Multivariate analysis including principal component generalized discriminant analysis (PC-GDA) and partial least squares (PLS) were each used separately for lesion classification according to three clinical diagnostic tasks. A diagram of the PC-GDA is shown in Fig. 12.4. It starts from randomly dividing the set of spectra into training spectra and test spectra. The training spectra are then scaled and normalized by subtracting the mean and dividing by the standard deviation before applying the PC-GDA. The principal component (PC) factors and the PC loadings of the training spectra are calculated. A general discrimination model is developed based on the PC factors derived from the training spectra, where the classifications are known a priori. The discrimination model is then used to classify the test spectra: the test spectra are scaled and normalized using the mean and standard deviation obtained from the training spectra; the PC factors of the test spectra are calculated based on the PC loadings of the training spectra; the test spectra are classified based on the discrimination model developed from the training spectra; and posterior probabilities (the probability of a spectrum to be classified into skin cancer based on the discrimination

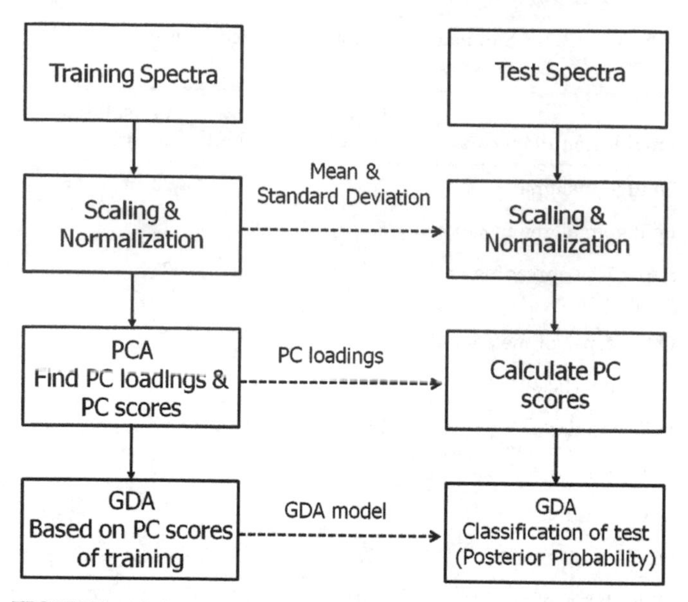

FIGURE 12.4 Diagram of principal component and general discrimination analysis. For leave-one-out cross-validation algorithm, successive single spectra were left out as "test spectra," with the remaining spectra being used for "training spectra." *GDA*, general discrimination analysis; *PC*, principal component; *PCA*, principal component analysis.

model) of the test spectra for skin cancer are then obtained. All multivariate data analyses in this chapter are implemented based on "leave-one-out cross-validation" (LOO-CV; unless otherwise stated), in which successive single spectra were left out for "testing," with the remaining spectra being used for "training." The procedures of PLS analysis based on LOO-CV are similar to those of PC-GDA.

Receiver Operating Characteristic Curves and Diagnostic Parameters

The receiver operating characteristic (ROC) curve [ie, sensitivity vs. (1 − specificity)] is calculated from the posterior probabilities previously derived and represents the diagnostic performance. With good discrimination between two groups the ROC curve moves toward the left and top boundaries of the graph, whereas poor discrimination yields a curve that approaches the diagonal line. The area under the ROC curve (AUC) is calculated using the trapezoidal rule [55]. The 95% confidence interval (CI) and the significance of the AUCs are performed in a standard fashion [56–58]. The comparison between two ROC curves is calculated from *z*-score by (Graphpad, La Jolla, California, USA)

$$z = |AUC_1 - AUC_2|/SQRT \times \left(SE_1^2 + SE_2^2 - 2 \times \gamma \times SE_1 \times SE_2\right),$$

where AUC_1 and AUC_2 are the areas under the two ROC cures, SE_1 and SE_2 are the standard errors of the two ROC curves, and γ is the correlation between the two ROC curves [56]. For two unpaired sets, $\gamma = 0$. In Excel, a two-tailed p value is calculated as $p = 2 \times [1 - \text{NORMSDIST}(z)]$. All of the ROC analyses are based on nonparametric techniques and are performed separately for the PC-GDA and PLS analyses and for each of the three clinical diagnostic tasks.

Other diagnostic parameters include sensitivity, specificity, positive predictive value (PPV), negative predictive value (NPV), biopsy ratio, and accuracy [59]. The definition of these diagnostic parameters can be found from Table 12.2.

Raman Spectra of Different Pathologies

The mean Raman spectra for different skin pathologies are shown in Fig. 12.5. All spectra are normalized to their respective areas under the curve (AUC) before being averaged in aggregate according to diagnosis. Overall the skin lesions all appear to share similar Raman peaks and bands. The strongest Raman peak is located around 1445 cm^{-1}, which represents the CH_2

TABLE 12.2 Summary of the Optimal Diagnostic Waveband, the Area Under the Receiver Operating Characteristic Curve for Principal Component Generalized Discriminant Analysis, and Partial Least Squares Analysis

Diagnosis	Waveband (cm^{-1})	AUC	
		PCA-GDA (95% CI)	PLS (95% CI)
Skin cancers + actinic keratosis vs. benign lesions	500–1800	0.879 (0.829–0.929)	0.896 (0.846–0.946)
Melanoma vs. benign pigmented lesions	1055–1800	0.823 (0.731–0.915)	0.827 (0.735–0.929)
Melanoma vs. seborrheic keratosis	1055–1800	0.898 (0.797–0.999)	0.894 (0.793–0.995)

AUC, area under the receiver operating characteristic curve; *CI*, confidence interval; *PC-GDA*, principal component generalized discriminant analysis; *PLS*, partial lease squares.
Adapted from Lui H, Zhao J, McLean D, Zeng H. Real-time Raman spectroscopy for in vivo skin cancer diagnosis. Cancer Res 2012;72(10):2491–500 with permission from the American Association for Cancer Research.

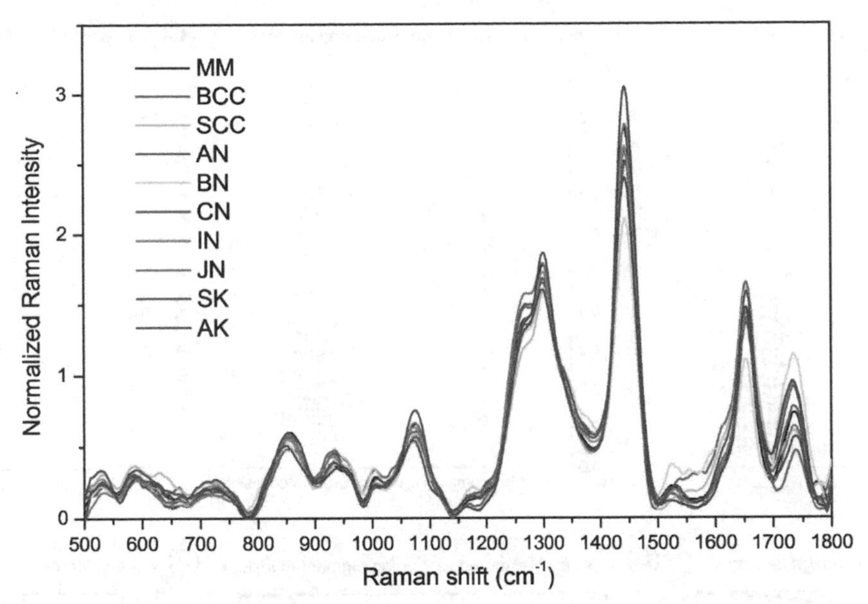

FIGURE 12.5 Mean Raman spectra by diagnosis. All lesion spectra were normalized to their areas under the receiver operating characteristic curve (by multiplication of data numbers in the range) before averaging by diagnosis. *AK*, actinic keratosis; *AN*, atypical nevus; *BCC*, basal cell carcinoma; *BN*, blue nevus; *CN*, compound nevus; *IN*, intradermal nevus; *JN*, junctional nevus; *MM*, malignant melanoma; *SCC*, squamous cell carcinoma; *SK*, seborrheic keratosis. *Adapted from Lui H, Zhao J, McLean D, Zeng H. Real-time Raman spectroscopy for in vivo skin cancer diagnosis. Cancer Res 2012;72(10):2491–500 with permission from the American Association for Cancer Research.*

and CH_3 bonds of lipids and proteins in the bending vibrational mode. Other major Raman bands are centered at 855, 936, 1002, 1271, 1302, 1655, and 1745 cm^{-1}. The two peaks at 855 and 936 cm^{-1}, assigned to C–C stretching mode, are signatures of collagen type I. The region from 1000 to 1150 cm^{-1} contains information about aliphatic hydrocarbon chains. The phenylalanine peak at 1002 cm^{-1} is observed in all of the pathologies. The peak at 1265 cm^{-1} is assigned to amide III, and the peak at 1304 cm^{-1} is assigned to a twisting deformation of the CH_2 methylene groups of intracellular lipid acyls. The band around 1655 cm^{-1} is attributed to keratin protein vibrational modes involving amide I bonds. The band around 1742 cm^{-1} is assigned to the C=O stretching mode of the lipid ester, but it may have traces of contribution of melanin in vivo [35]. Note that there are no distinctive Raman peaks or bands that can be uniquely assigned to specific skin pathology by visual inspection alone. Therefore statistical methods are used to extract the diagnostic information from these complex and data-rich Raman spectra.

Raman Spectroscopy Distinguishes Skin Cancers From Benign Lesions: Previous Cohort

We have found that the diagnostic performance of Raman spectra is spectral band dependent. The higher spectral range from 1055 to 1800 cm^{-1} is optimal for diagnostic task (ii) differentiating melanomas from nonmelanoma pigmented lesions, and diagnostic task (iii) differentiating melanomas from SKs; the full spectral range from 500 to 1800 cm^{-1} is optimal for diagnostic task (i) separating skin cancers and precancers from benign skin lesions. The PLS approach yields results similar to those for PC-GDA. There are no statistical differences between the PLS and PC-GDA methodologies for these three clinical diagnostic tasks ($p = .0644$, .7494, and .1646, respectively).

The posterior probabilities and the ROC curves for these three clinical diagnostic tasks based on PC-GDA LOO-CV analysis are shown in Fig. 12.6. When Raman spectroscopy is used to distinguish cancerous and precancerous skin lesions ($n = 232$) from benign skin lesions ($n = 286$), the ROC AUC is 0.879 (95% CI: 0.829–0.929, PC-GDA) and statistically significant ($p < .001$). Fig. 12.6A shows the posterior probability for each lesion to be classified as a skin cancer or precancer. From the distribution of posterior probabilities, the ROC curve with 95% CIs is generated and shown in Fig. 12.6D. At a sensitivity of 90%, the overall specificity is more than 64%, with a PPV of 67% and NPV of 89%. The estimated biopsy ratio at a sensitivity of 90% is approximately 0.49:1.

When only lesions with pigment are considered, Raman spectroscopy can separate MM ($n = 44$) from nonmelanoma pigmented skin lesions (AN, BN, CN, IN, JN, SK, $n = 286$) with an ROC AUC of 0.823 (95% CI: 0.731–0.915, PC-GDA) and is statistically significant ($p < .001$) based on PC-GDA LOO-CV analysis.

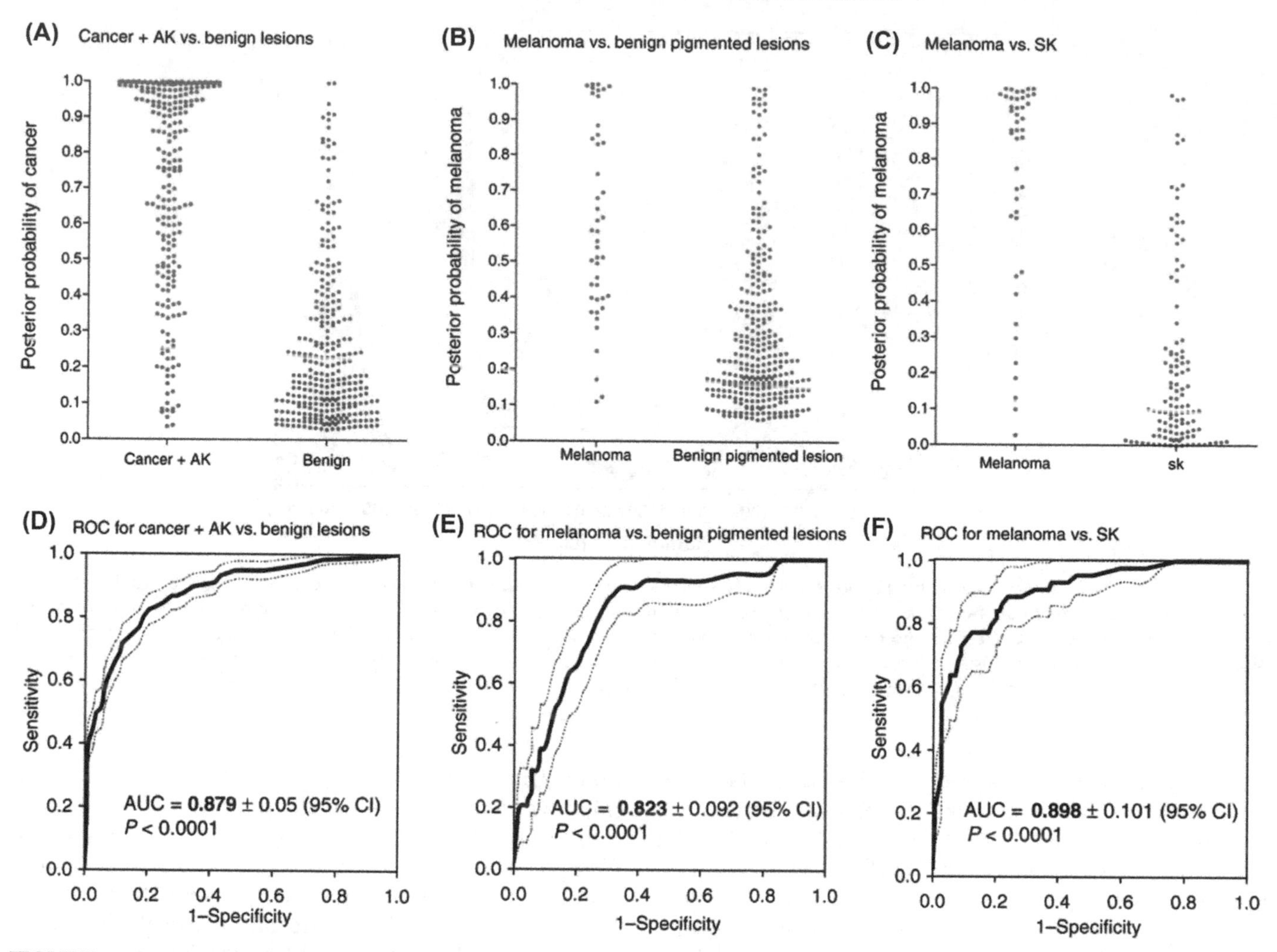

FIGURE 12.6 Lesion classification by Raman spectroscopy based on principal component generalized discriminant analysis (previous cohort). Posterior probabilities for discriminating (A) skin cancers and precancers (including malignant melanoma, basal cell carcinoma, squamous cell carcinoma, AK, $n = 232$) from benign skin disorders (including AN, BN, CN, IN, JN, and SK, $n = 286$; (B) melanoma ($n = 44$) from benign pigmented skin diseases (including AN, BN, CN, IN, JN, SK, $n = 286$; and (C) melanoma ($n = 44$) from SK ($n = 114$). (D–F) The corresponding ROC curves and 95% CIs are derived from the respective posterior probabilities, and all AUCs are significant ($p < .0001$). *AK*, actinic keratosis; *AN*, atypical nevi; *AUC*, area under the ROC curve; *BN*, blue nevi; *CI*, confidence interval; *CN*, compound nevi; *IN*, intradermal nevi; *JN*, junctional nevi; *ROC*, receiver operating characteristic; *SK*, seborrheic keratosis. *Adapted from Lui H, Zhao J, McLean D, Zeng H. Real-time Raman spectroscopy for in vivo skin cancer diagnosis. Cancer Res 2012;72(10):2491–500 with permission from the American Association for Cancer Research.*

Fig. 12.6B shows the posterior probability for each lesion to be classified as melanoma. From the distribution of posterior probabilities, the ROC curve with 95% CIs for discriminating MM from nonmelanoma pigmented skin lesions is generated and shown in Fig. 12.6E. Our results show the biopsy ratio from 5.6:1 to 2.3:1 for sensitivities corresponding to 99–90% and specificities from 15% to 68%, respectively.

Fig. 12.6C and F, shows the posterior probabilities and the ROC curves with 95% CIs for cases of melanoma or SK to be classified by Raman spectroscopy as melanoma based on PC-GDA LOO-CV analysis. The AUC for the corresponding ROC curve is 0.894 (95% CI: 0.793–0.995) and is statistically significant ($p < .001$). The biopsy ratio ranges from 1.96:1 to 0.92:1 for sensitivities ranging from 99% to 90% and specificities from 25% to 68%.

The performance and other diagnostic parameters of rapid real-time Raman spectroscopy for these three diagnostic tasks are summarized in Tables 12.2 and 12.3, including the AUCs of the ROC curves as well as the sensitivity, specificity, PPV, NPV, and biopsy ratio. The diagnostic parameters of PLS are similar to those of PC-GDA and are listed in Tables 12.2 and 12.3.

Comparison With Previous Clinical and Experimental Studies: Previous Cohort

For clinical studies, Cohen et al. studied 1250 patients and showed that the biopsy ratios ranged from 135:1 to

TABLE 12.3 Summary of Raman Spectroscopy Diagnostic Parameters From Receiver Operating Characteristic Curves According to Various Levels of Sensitivity

		PCA-GDA				PLS			
Diagnosis	Sensitivity (95% CI)	Specificity (95% CI)	PPV[a]	NPV[b]	Biopsy ratio[c]	Specificity (95% CI)	PPV[a]	NPV[b]	Biopsy ratio[c]
Cancers + AK vs. benign lesions	0.99 (0.98–1.00)	0.17 (0.13–0.21)	0.49	0.95	1.03:1	0.24 (0.19–0.29)	0.51	0.97	0.95:1
	0.95 (0.92–0.99)	0.41 (0.35–0.48)	0.57	0.91	0.77:1	0.52 (0.48–0.58)	0.62	0.93	0.62:1
	0.90 (0.86–0.94)	0.64 (0.58–0.70)	0.67	0.89	0.49:1	0.66 (0.61–0.71)	0.68	0.89	0.47:1
Melanoma vs. benign pigmented lesions	0.99 (0.96–1.00)	0.15 (0.11–0.19)	0.15	0.99	5.58:1	0.14 (0.10–0.18)	0.15	0.99	5.65:1
	0.95 (0.89–1.00)	0.38 (0.32–0.44)	0.19	0.98	4.24:1	0.44 (0.38–0.50)	0.21	0.98	3.83:1
	0.90 (0.81–0.99)	0.68 (0.63–0.73)	0.30	0.98	2.31:1	0.63 (0.57–0.69)	0.27	0.98	2.67:1
Melanoma vs. seborrheic keratosis	0.99 (0.96–1.00)	0.25 (0.17–0.33)	0.34	0.98	1.96:1	0.46 (0.37–0.55)	0.41	0.99	1.41:1
	0.95 (0.89–1.00)	0.54 (0.45–0.63)	0.44	0.97	1.25:1	0.52 (0.43–0.61)	0.43	0.96	1.31:1
	0.90 (0.81–0.99)	0.68 (0.59–0.77)	0.52	0.95	0.92:1	0.66 (0.57–0.75)	0.51	0.94	0.98:1

AK, actinic keratosis; *CI*, confidence interval; *NPV*, negative predictive value; *PC-GDA*, principal component generalized discriminant analysis; *PLS*, partial lease squares; *PPV*, positive predictive value.
[a]*Ratio of true positives to the total of true positives and false positives.*
[b]*Ratio of true negatives to the total of true negatives and false negatives.*
[c]*Number of biopsied false positives for each biopsied true positive.*
Adapted from Lui H, Zhao J, McLean D, Zeng H. Real-time Raman spectroscopy for in vivo skin cancer diagnosis. Cancer Res 2012;72(10):2491–500 with permission from the American Association for Cancer Research.

576:1 for patients with/without a personal history of melanoma [60]. From a retrospective study of 4741 pigmented skin lesions evaluated by 468 general practitioners, English et al. [61] found that the biopsy ratio ranged from 58:1 to 21:1 for new versus experienced general practitioners, respectively. Heal et al. [62] found that the clinical diagnosis of skin cancers and precancers was associated with a sensitivity of 63.9% for BCC, 41.1% for SCC, and 33.8% for MM. The PPVs for their study were 72.7% for BCC, 49.4% for SCC, and 33.3% for MM, respectively. Millier et al. [63] found that the accuracy to rule out melanoma was only approximately 64% for the average dermatologist. In their study, 128 of the 354 melanomas were not recognized by dermatologists using conventional microscopy. MacKenzie-Wood et al. [64] found that the diagnostic accuracy ranged from 50% for clinicians with less than 5 years clinical experience to 72.1% with more experienced clinicians, similar to the data of Lindelof et al. [65]. Menzies et al. [66] reported that the sensitivities and specificities were approximately 90%, 81%, 85%, and 62% and 59%, 60%, 36%, and 63% for experts, dermatologists, trainees, and general practitioners, respectively. From Fig. 12.6 and Table 12.3, we found that Raman spectroscopy had comparable or superior diagnostic performance in comparison with previous clinical studies.

Skin cancer diagnosis appears to be substantially improved with the use of newer technologies, such as dermoscopy, surface microscopy, and multispectral imaging. Robinson et al. [67] reported a biopsy ratio of 47:1 to rule out melanoma with digital epiluminescence microscopy. Banky et al. [68] reported a biopsy ratio as low as 3:1 for patients at high risk of melanoma using a combination of baseline images and dermoscopy, but at the expense of a relatively low sensitivity of only 72%. Binder et al. [69] investigated 120 lesions (including 39 MM) using dermoscopy alone and found that depending on the sample size and selection of lesions, the sensitivity and specificity varied from 93% to 38% and from 84% to 50%, respectively. Menzies et al. [66] studied 2430 lesions (382 MM) using a dermoscopy-based automated diagnostic called SolarScan and found a sensitivity of 85% and specificity of 65%. Farina et al. [70] studied 237 pigmented lesions using multispectral imaging (67 MM and 170 non-MM), and found the AUC of the ROC curve was 0.779 with a sensitivity of 80% and specificity of 51%. Moncrieff et al. [71] studied 348 pigmented lesions (52 MM) using the SIAscope, a narrow-band spectral imaging technique, and found the sensitivity of 82.7% and specificity of 80.1%. Westerhoff et al. [72] found that the accuracy of melanoma diagnosis could be improved from 63% to 76% with the aid of surface microscopy. Monheit et al. [73] found that the estimated biopsy ratio ranged from 7.6:1 to 10.8:1 for identifying melanoma with or without borderline lesions using multispectral imaging techniques (ie, MelaFind). At a sensitivity of 98.3%, the average specificity was 9.9%, superior to that of clinicians at 3.7% [73]. Wells et al. [74] studied 23 melanomas and 24 pigmented benign lesions using the multispectral imaging-based MelaFind and found that the sensitivity and specificity were approximately 96% and 8%, similar to what was found by Monheit et al. [73]. Our studies support the use of Raman spectroscopy for guiding skin cancer diagnosis at different levels of clinical interest. For all three diagnostic tasks, the specificity of the Raman approach is greater than 15% at a sensitivity of 99% and is indeed higher than other studies. Overall, Raman spectroscopy appeared to be more effective in comparison with clinicians and other technical diagnostic aids.

Independent Validation: New Cohort

As a translational study, we further evaluated the performance of the previously derived algorithm for in vivo skin cancer diagnosis using completely independent new measurements. In this section we report the classification of the new cohort of patients based on the discrimination model generated from the previous cohort of patients [54]. PC-GDA and PLS were applied to validate previous analyses. Fig. 12.7A shows the posterior probability for each lesion to be classified as a skin cancer or precancer by PC-GDA for the new cohort of skin Raman spectra ($n = 127$) based on the previous cohort for training ($n = 518$). Fig. 12.7B shows the ROC curve (solid line) with 95% CIs (dashed lines) obtained from the distribution of posterior probabilities in Fig. 12.7A. The ROC AUC for the new cohort was 0.861 (95% CI: 0.796–0.927; PC-GDA), comparable to the previous results in which the ROC AUC was 0.879 (95% CI: 0.829–0.929; PC-GDA, LOO-CV). The specificity ranged from 24% to 54% for sensitivities ranging

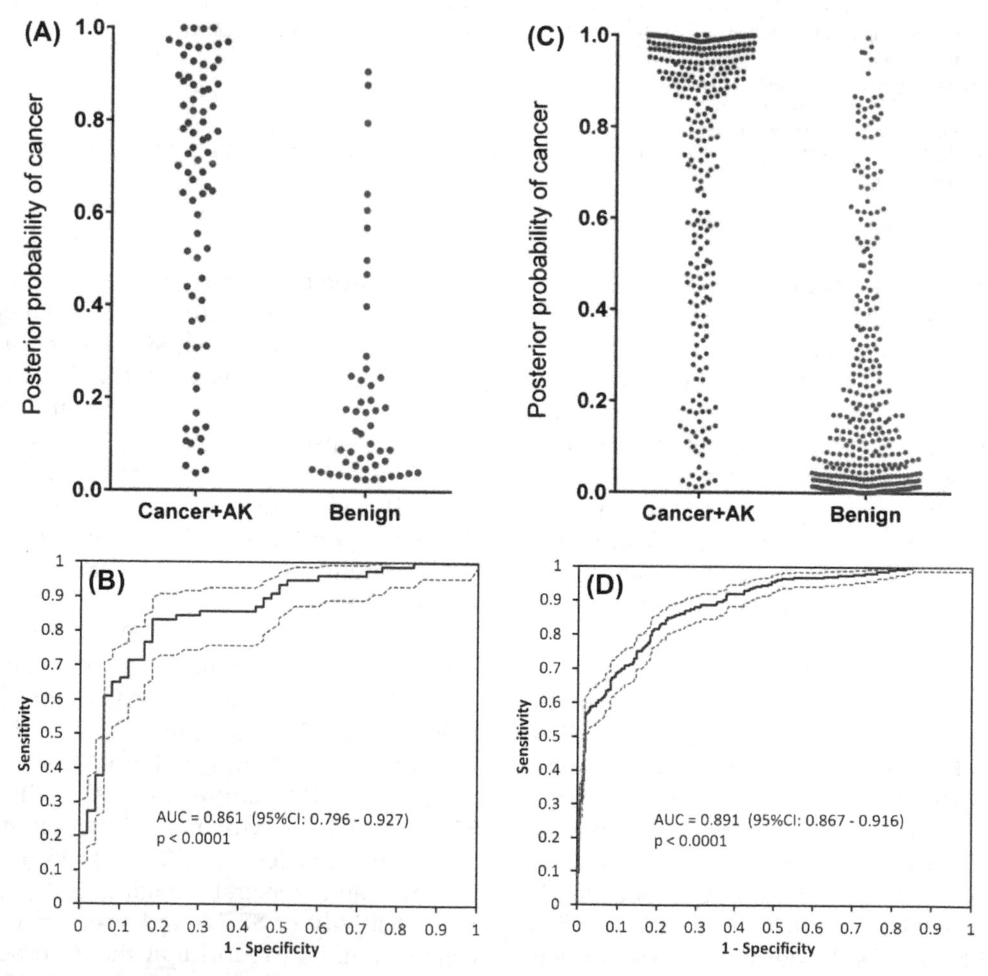

FIGURE 12.7 Lesion classification by Raman spectroscopy based on PC-GDA for new cohort and consolidated cohort. (A and C) Posterior probabilities for discriminating skin cancers and precancers (including malignant melanoma, basal cell carcinoma, squamous cell carcinoma, AK) from benign skin disorders (including atypical, blue, compound, intradermal, and junctional nevi and seborrheic keratosis). (A): Previous cohort for training ($n = 232$ cancer vs. $n = 286$ benign) and new cohort for testing ($n = 77$ cancer vs. $n = 50$ benign), (C): consolidated cohort based on PC-GDA with leave-one-out cross-validation ($n = 309$ cancer vs. $n = 336$ benign). (B and D) The corresponding receiver operating characteristic curves (*solid lines*) and 95% CIs (*dashed lines*), derived from the respective posterior probabilities in panels (A and C). All AUCs are statistically significant ($p < .0001$). *AK*, actinic keratosis; *AUC*, area under the receiver operating characteristic curve; *CI*, confidence interval; *PC-GDA*, principal component generalized discriminant analysis.

TABLE 12.4 Performance of Raman Spectroscopy for Discriminating Skin Cancers From Benign Skin Lesions for Different Cross-Validations

Dataset	PLS (95% CI)	PC-GDA (95% CI)
Previous cohort, LOO-CV	0.896 (0.846–0.946)	0.879 (0.829–0.929)
Previous cohort training new cohort testing	0.889 (0.834–0.944)	0.861 (0.796–0.927)
Consolidated cohort LOO-CV	0.894 (0.870–0.918)	0.891 (0.867–0.916)

CI, confidence interval; *LOO-CV*, leave-one-out cross-validation; *PC-GDA*, principal component with general discriminate analysis; *PLS*, partial least squares.

from 99% to 90%, similar to the results of the previous cohort, in which the specificity ranged from 17% to 64% for the same sensitivities.

Independent Validation: Consolidated Cohort

Fig. 12.7C shows the posterior probability for each lesion to be classified as a skin cancer or precancer by PC-GDA for the consolidated cohort ($n = 645$) based on PC-GDA LOO-CV analysis. Fig. 12.7D shows the ROC curve (solid line) with 95% CIs (dashed lines) obtained from the distribution of posterior probabilities in Fig. 12.7C. The ROC AUC for the consolidated cohort was 0.894 (95% CI: 0.870–0.918; PC-GDA, LOO-CV), similar to the previous cohort in which the ROC AUC was 0.879 (95% CI: 0.829–0.929; PC-GDA, LOO-CV) and the new cohort in which the ROC AUC was 0.861 (95% CI: 0.796–0.927; PC-GDA). Note that the consolidated cohort has the narrowest CI. The specificities ranged from 25% to 65% for sensitivities from 99% to 90%, comparable to the previous cohort and the new cohort.

The results for PLS analysis are similar to those of PC-GDA. The diagnostic performance of rapid real-time Raman spectroscopy for the previous cohort, the new cohort, and the consolidated cohort are summarized in Tables 12.4 and 12.5. The independent study validated all of the previous findings. The previous diagnostic model could be applied directly to new patients with comparable diagnostic performance. However, a diagnostic model based on more cases could be even more reliable with a narrower CI.

MODELING OF IN VIVO SKIN RAMAN SPECTRA

The Raman spectrum of in vivo skin is altered by the absorption and scattering of skin tissue. It is important to have an accurate theoretical model for complete understanding of in vivo skin Raman measurements. MC simulation has been found to be an effective theoretical tool to study fluorescence and Raman in multilayered biological tissues [75–79]. Feld et al. [80] utilized MC simulations to elucidate the relationship between Raman and diffuse reflectance for finite and semi-infinite biological samples. Everall et al. [81] applied MC simulations to study time-resolved Raman scattering under isotropic and forward scattering conditions. Wilson et al. [82] built a layered MC model to

TABLE 12.5 Summary of Raman Spectroscopy Diagnostic Specificities Derived From Receiver Operating Characteristic Curve According to Various Levels of Sensitivity for Discriminating Skin Cancers and Precancers From Benign Skin Lesions Based on Partial Least Squares and Principal Component With General Discriminant Analysis for Independent Validation

	Specificity level (95% CI)		
Sensitivity level (95% CI)	**Previous cohort LOO-CV**	**Previous cohort training new cohort testing**	**Consolidated cohort LOO-CV**
PLS ANALYSIS			
0.99 (0.98–1.00)	0.24 (0.19–0.29)	0.30 (0.18–0.45)	0.24 (0.19–0.28)
0.95 (0.92–0.99)	0.52 (0.48–0.58)	0.46 (0.32–0.61)	0.54 (0.48–0.59)
0.90 (0.86–0.94)	0.66 (0.61–0.71)	0.58 (0.43–0.72)	0.67 (0.62–0.72)
PC-GDA			
0.99 (0.98–1.00)	0.17 (0.13–0.21)	0.24 (0.13–0.38)	0.24 (0.20–0.29)
0.95 (0.92–0.99)	0.41 (0.35–0.48)	0.48 (0.34–0.63)	0.54 (0.48–0.59)
0.90 (0.86–0.94)	0.64 (0.58–0.70)	0.54 (0.39–0.68)	0.65 (0.60–0.70))

CI, confidence interval; *LOO-CV*, leave-one-out cross-validation; *PC-GDA*, principal component generalized discriminant analysis; *PLS*, partial least squares.

understand the effect of overlying tissue layers on the detected Raman signal. Matousek et al. [83] and Keller et al. [84] designed an MC model to investigate the effects of tissue and probe geometry. Reble et al. [85] designed an MC model and calculated the depth-dependent sensitivity or sampling volume on a two-layer skin model. Mo et al. [86] designed an MC model for an endoscopic fiber-optic probe. Here in this section we propose an eight-layer skin model for inhomogeneous in vivo human skin tissue. We are the first to use experimentally measured intrinsic Raman spectra of different tissue layers in an MC modeling. We will present the details of the MC simulation and its application for modeling in vivo skin Raman spectra.

An Eight-Layer Normal Human Skin Model

We have published a systematic study on the modeling of light propagation and light–tissue interaction in normal human skin [77–79]. Here we present an eight-layer skin model in the NIR wavelength range for in vivo human skin Raman spectra. Table 12.6 outlines the parameters of the eight-layer skin model, including thickness (d), refractive index (n), and the optical transport parameters (absorption coefficient, μ_a; scattering coefficient, μ_s; scattering anisotropy, g) at 840 nm for each skin layer. Some of the parameters are compiled from Salomatina et al. [87] and Meglinski et al. [88].

The refractive index of the ambient medium above the eight-layer tissue is 1.0, which is assumed to be air, and the ambient medium below the eight-layer tissue is 1.37, which is assumed to be muscle. In this work, 785 nm is the wavelength of the laser light used for Raman excitation. To investigate the skin Raman properties, we compiled the parameters of each skin layer at 18 wavelengths ranging from 838 to 914 nm, corresponding to $800 - 1800\ \text{cm}^{-1}$. Details of other considerations can be found in Wang et al. [89].

Procedures of the Monte Carlo Simulation

MC simulation of in vivo human skin Raman spectra was implemented with codes modified from Wang and Jacques et al. [90,91] according to the eight-layered skin model. Similar to MC simulation of human skin fluorescence [78], the MC simulation of in vivo human skin Raman spectrum consists of the following steps:

1. Calculate the excitation light distribution inside of the model skin, $\Phi(\lambda_{ex}, r, z, \theta)$. It is in units of W/cm^2, λ_{ex} is the excitation wavelength whereas r, z, and θ represent local positions in the cylindrical coordinates.
2. Calculate escape functions for different emission wavelengths and at different depths, $E(\lambda_{em}, r, z, \theta)$, where λ_{em} is the emission wavelength. The MC code was modified to simulate the light propagation process for an isotropic Raman point source buried at depth z inside of the tissue. After integrating with respect to r and θ in the cylindrical coordinate, the escape function can be simplified as a function of emission wavelength and depth, $E(\lambda_{em}, z)$.
3. Calculate Raman detection efficiency, $\eta(\lambda_{em}, z)$, as a function of emission wavelength for different skin layers. The Raman detection efficiency is defined as the likelihood of obtaining a Raman signal from a specific skin layer, which illustrates how the intrinsic Raman spectra are distorted by the tissue reabsorption and scattering during the escape process. It is an integral of the product of the

TABLE 12.6 Parameters for an Eight-Layer Skin Optical Model, Including Thickness (d) of Each Skin Layer, Depths and Intervals in the Monte Carlo Model, and the Transport Parameters (μ_a, μ_s, g, n) at 840 nm

Layer	d (μm)	n	μ_s (cm^{-1})	μ_a (cm^{-1})	G	Depths and intervals in model
Air	–	1.0	–	–	–	–
Stratum corneum	10	1.45	176.125	0.7405	0.8	5 depths at 2 μm
Epidermis	80	1.4	176.125	1.3	0.8	8 depths at 10 μm
Papillary dermis	100	1.4	106.25	1.05	0.8	10 depths at 10 μm
Upper blood plexus	80	1.39	145.625	1.427	0.818	8 depths at 10 μm
Reticular dermis	1500	1.4	106.25	1.05	0.8	30 depths at 50 μm
Deep blood plexus	70	1.34	460.625	4.443	0.962	7 depths at 10 μm
Lower dermis	160	1.4	106.25	1.05	0.8	8 depths at 20 μm
Subcutaneous fat	3000	1.46	97.125	0.975	0.8	30 depths at 100 μm
Muscle	–	1.37	–	–	–	–

excitation light distribution inside of the tissue and the Raman escape function,

$$\eta_{\text{layer}z_1 \to z_2}(\lambda) = \int_{z_1}^{z_2} \Phi(z)E(\lambda, z)dz$$

where $z_2 - z_1 = d$, the thickness of skin layers.

4. Calculate the simulated Raman spectra, $R(\lambda_{ex}, \lambda_{em}, r)$. The reconstructed in vivo Raman spectrum of skin tissue is a linear combination of the product of the intrinsic spectrum and the Raman detection efficiency of all of the excited molecules. The intrinsic Raman spectrum of any skin layer, $\beta(\lambda_{ex}, \lambda_{em}, z)$, can be obtained from excised human skin tissue [92]. The simulated Raman spectrum is computed by the convolution of the excitation light distribution, the intrinsic Raman spectrum, and the Raman escape function:

$$R(\lambda_{ex}, \lambda_{em}, r) = \int_0^D \int_0^{2\pi} \int_0^{\infty} \Phi(\lambda_{ex}, r', z', \theta)\beta(\lambda_{ex}, \lambda_{em}, z') \times E\left(\lambda_{em}, \sqrt{r^2 + r'^2 - 2rr' \cos\theta'}, z'\right) r' dr' d\theta' dz',$$

where D is the total thickness of the model skin.

In our MC simulation, the excitation distribution and the escape function were simulated at 18 emission wavelengths and 106 depths in the skin. The wavelength covered from 840 to 910 nm at an interval of 5 nm (corresponding to 800–1800 cm^{-1} under 785 nm excitation). Another three wavelengths at 838, 858, and 914 nm were used in the simulation corresponding to 800, 1080, and 1800 cm^{-1}, respectively. The 106 depth levels in the simulations were listed in Table 12.6. In total we performed 1908 simulations to reconstruct in vivo human skin Raman spectra. In each simulation, 1,000,000 photons were launched.

Light Distribution in the Skin

Light distribution in the skin is shown in Fig. 12.8, assuming the excitation beam is infinitely wide on the skin at normal incidence. Because of back-scattering, photons pile up near the air–tissue interface, reaching as high as 3.7 times the incident fluence in the stratum corneum layer. The light fluence decreases monotonically with depth. For a skin model listed in Table 12.6, light can penetrate deep into the subcutaneous fat layer, but the fluence drops to approximately 2% of the initial fluence at the stratum corneum.

Raman Escape Function in the Skin

Once a Raman photon is generated, it has to escape out of the skin to be observed. Assuming it can be considered as an isotropic source buried in the skin

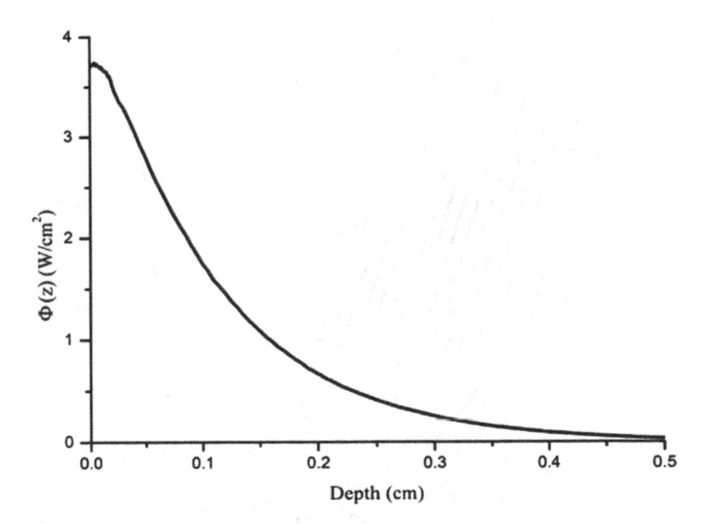

FIGURE 12.8 Distribution of the excitation laser light (785 nm) in skin tissue from the Monte Carlo simulation, assuming that an infinite wide beam illuminates the tissue at normal incidence with incident power density of 1 W/cm^2. *Reproduced with permission from Wang S, Zhao J, Lui H, He Q, Bai J, Zeng H. Monte Carlo simulation of in vivo Raman spectral measurements of human skin with a multi-layered tissue optical model. J Biophotonics 2014;7(9):703–12, copyright Wiley-VCH Verlag GmbH & Co. KGaA.*

tissue, the simulated escape function, $E(\lambda_{em}, z)$, in the 838–914-nm wavelength range, is shown in Fig. 12.9, plotted as a function of depth within the tissue at a specific wavelength. It shows that the escape function has a minimal dependence on emission wavelength, λ_{em}, but is heavily dependent on the depth in the tissue. It decreases with tissue depth, indicating that it is more difficult for Raman photons in the deeper layer to escape out of the tissue surface. However, because of changes in refractive index at the boundaries of different tissue layers, a slight increase in the subcutaneous fat layer at a depth of 0.28 cm is noticed. The refractive index of the subcutaneous fat layer is higher than both its upper and lower layers, and thus some form of internal reflection may occur, therefore Raman photons are accumulated in this layer. A similar effect is also noticed for the reticular dermis layer around 0.035 cm in depth, which also has a higher refractive index than the neighboring layers.

Raman Detection Efficiency

Raman detection efficiency, $\eta(\lambda_{em}, z)$, is a measure of the likelihood to obtain a Raman photon from a specific skin layer, which describes the distortion of the intrinsic Raman signal by tissue reabsorption and scattering. The simulated Raman detection efficiency from our model is shown in Fig. 12.10. The stratum corneum and the subcutaneous fat layers have lower Raman detection efficiency, indicating that stratum corneum and subcutaneous fat layers have less contribution to the

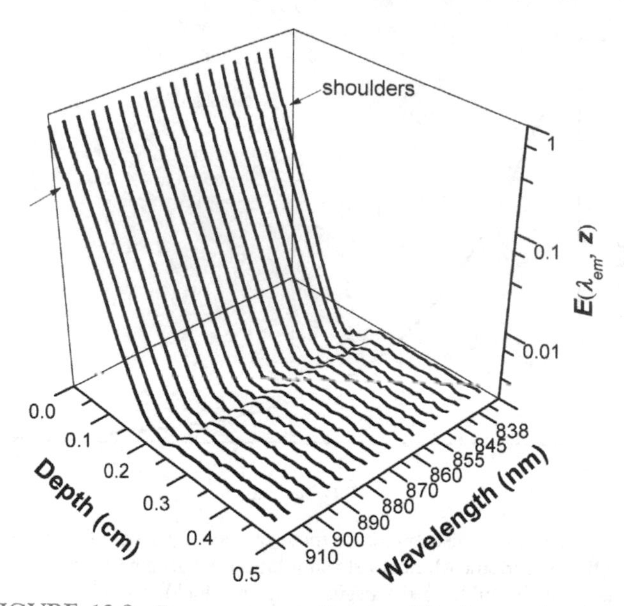

FIGURE 12.9 Raman escape efficiency, $E(\lambda, z)$, as a function of depth within the tissue at specific wavelength. The results were calculated by Monte Carlo simulation according to the parameters in Table 12.6. *Reproduced with permission from Wang S, Zhao J, Lui H, He Q, Bai J, Zeng H. Monte Carlo simulation of in vivo Raman spectral measurements of human skin with a multi-layered tissue optical model. J Biophotonics 2014;7(9):703–12, copyright Wiley-VCH Verlag GmbH & Co. KGaA.*

observed in vivo Raman spectrum. Most Raman photons are generated in the epidermis and dermis layers. The detection efficiencies of the stratum corneum layer, the subcutaneous fat layer, and the epidermis layer are minimally dependent on the emission wavelength. The detection efficiency of the dermis layer increases with emission wavelength, which may be related to the decreasing absorption of the dermis layer.

Simulated Raman Spectra of In Vivo Skin

As shown in section "Procedures of the Monte Carlo Simulation", the simulated Raman spectrum is a convolution of the excitation light distribution, the intrinsic Raman spectrum, and the Raman escape function. The intrinsic Raman spectrum, $\beta(\lambda_{ex}, \lambda_{em}, z)$, in this study was obtained from ex vivo experimental studies [92]. Fig. 12.11 shows the simulated Raman spectrum compared with in vivo skin Raman spectra, which are normalized to their respective AUC. In general, the simulated Raman spectrum matches reasonably well with the in vivo skin Raman spectra. In particular, the five major Raman peaks reconstructed by this MC simulation around 855, 935, 1265, 1445, and 1745 cm^{-1} match very well with the in vivo spectra. Differences between the reconstructed and the in vivo Raman spectra are also seen around the 900-, 1080-, and 1650-cm^{-1} bands, probably because the intrinsic Raman signal obtained from the microscopic measurement is different from the in vivo skin because they are obtained from different subjects and using different Raman systems. Nevertheless, we can assess the contribution of each skin layer to the measured in vivo Raman spectrum. For example, using the strongest Raman peak at 1445 cm^{-1}, we estimated that the contributions from the stratum corneum, epidermis, dermis, and the subcutaneous fat layer to

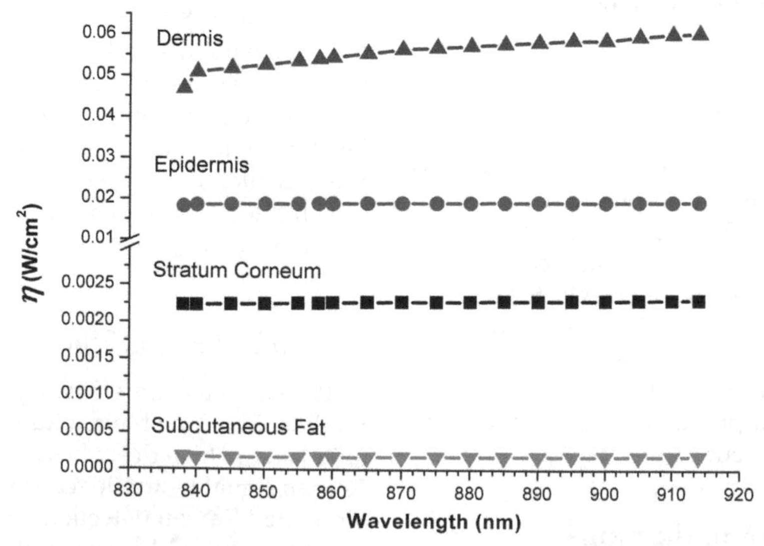

FIGURE 12.10 Monte Carlo simulated Raman detection efficiency, η, as a function of wavelength for the stratum corneum, epidermis, dermis, and subcutaneous fat layer. *Reproduced with permission from Wang S, Zhao J, Lui H, He Q, Bai J, Zeng H. Monte Carlo simulation of in vivo Raman spectral measurements of human skin with a multi-layered tissue optical model. J Biophotonics 2014;7(9):703–12, copyright Wiley-VCH Verlag GmbH & Co. KGaA.*

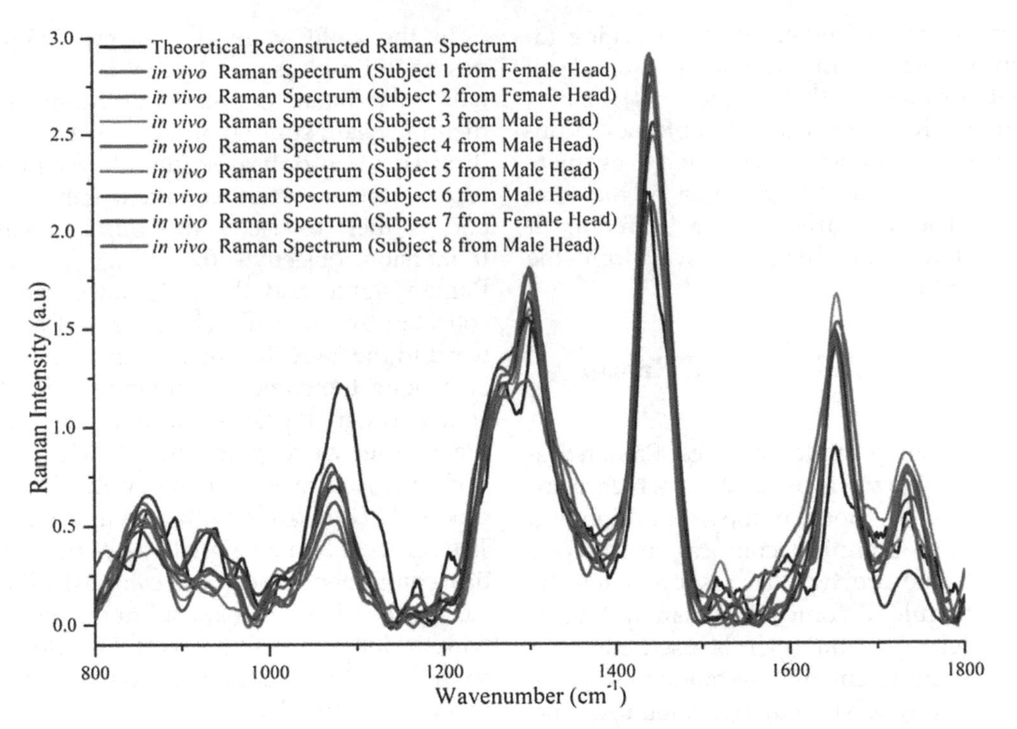

FIGURE 12.11 Comparison of the reconstructed facial skin Raman spectrum by Monte Carlo simulation with eight in vivo Raman spectra of facial skin tissue. *Reproduced with permission from Wang S, Zhao J, Lui H, He Q, Bai J, Zeng H. Monte Carlo simulation of in vivo Raman spectral measurements of human skin with a multi-layered tissue optical model. J Biophotonics 2014;7(9):703–12, copyright Wiley-VCH Verlag GmbH & Co. KGaA.*

the in vivo Raman spectrum are approximately 1%, 28%, 70%, and 1%, respectively.

In summary, an eight-layer optical model was proposed to represent skin tissue. The Raman properties of normal human skin were studied by MC simulation. The reconstructed Raman spectrum was reasonably in good agreement with the measured in vivo skin Raman spectra. These results provide unique insights about the contribution of the microscopic inhomogeneous Raman scatter to the in vivo Raman spectra. The absorption and scattering coefficients vary for different skin types or skin conditions. Therefore, further studies are needed to investigate the Raman properties of different skin types of normal skin, benign pigmented skin, and malignant skin tissues.

IMAGING-GUIDED CONFOCAL RAMAN SPECTROSCOPY FOR IN VIVO SKIN ASSESSMENT

Confocal Raman spectroscopy, which combines confocal microscopy and Raman spectroscopy, is mainly used for ex vivo Raman measurements because of the long integration times, but it has been applied to study several in vivo human tissues, such as estimation of stratum corneum thickness [93], monitoring drug penetration inside of the skin [94], or diagnosis of skin cancers [32,42,43]. However, there are several limitations in the current confocal Raman system. For example, the confocal Raman spectrum may contain information from the out-of-target area because of movement of the subject during the long integration time, the region of interest (ROI) for Raman measurement and for the confocal imaging are often not co-registered because the laser sources for confocal imaging and the Raman spectroscopy are usually different, and the Raman spectrum is often off of the target because there is no real-time monitoring. In this section we will present a novel, state-of-the-art design of the multimodality imaging-guided confocal Raman system for in vivo skin assessment [95]. In the design we not only incorporated laser scanning reflectance confocal microscopy, which provides morphological information based on refractive index variations, but we also incorporated other imaging modalities, including two-photon fluorescence (TPF) and second harmonic generation (SHG), which provide complementary information for skin tissue diagnosis. There are several advantages of this state-of-the-art, imaging-guided confocal Raman system. (1) Raman spectroscopy and reflectance confocal microscopy share the same laser source; therefore the reflectance confocal imaging and confocal Raman measurements are co-registered. (2) The three imaging modes, including laser scanning confocal microscopy, TPF, and SHG, are co-registered; therefore confocal Raman spectra can be measured under guidance of either mode of imaging. (3) The ROI for confocal Raman measurements can be well defined without turning off the scanners. (4) Finally, because the confocal Raman spectrum is measured under real-time imaging guidance, it is feasible to evaluate the contribution of the out-of-target area to the total Raman spectrum. The measurement can be redone if the target was found to have moved away from the Raman measurement ROI.

Multimode Imaging-Guided Confocal Raman Instrumentation

The multimode imaging-guided confocal Raman system is schematically shown in Fig. 12.12, which combines confocal Raman spectroscopy, multiphoton microscopy, and laser scanning confocal microscopy into one system. There are two laser sources for the multimode imaging-guided confocal Raman system. A continuous-wave (cw) 785-nm laser is used for both confocal Raman measurement and cw-mode reflectance confocal microscopy (cwRCM) imaging. A femtosecond (fs) Ti:Sapphire laser is used for multiphoton microscopy, including TPF imaging, SHG imaging, and fs-mode reflectance confocal microscopy (fsRCM) imaging.

In the cwRCM mode the cw 785-nm laser beam passes through several lenses, a band-pass filter, a half-wave plate, a polarizing beam splitter (PBS), a dichroic beam splitter, and a quarter-wave plate, and is directed to an optical scanning system, which consists of a resonance scanner and a galvanometer scanner, and is then focused onto human skin by a water-immersion objective (60×, NA = 1.0). The confocal Raman signal and the reflectance confocal signal are collected by the same objective and are directed backward to the cwRCM detector and the Raman spectrometer along the same illumination path. The reflectance confocal signal passes through the scanning system, the quarter-wave plate, the dichroic mirror, and the PBS and is directed to the cwRCM avalanche photodiode (APD) module with a 30-μm pinhole in the front. The confocal Raman signal passes through the same optical components and is directed to the Raman spectrometer by the dichroic beam splitter, which in combination with the long-pass filter rejects the reflected laser light. A 50-μm fiber that is connected to the Raman spectrometer works as a pinhole.

In the fsRCM mode the fs laser beam passes through a half-wave plate and a PBS (which is used to control the fs laser power), a dispersion correction system (which

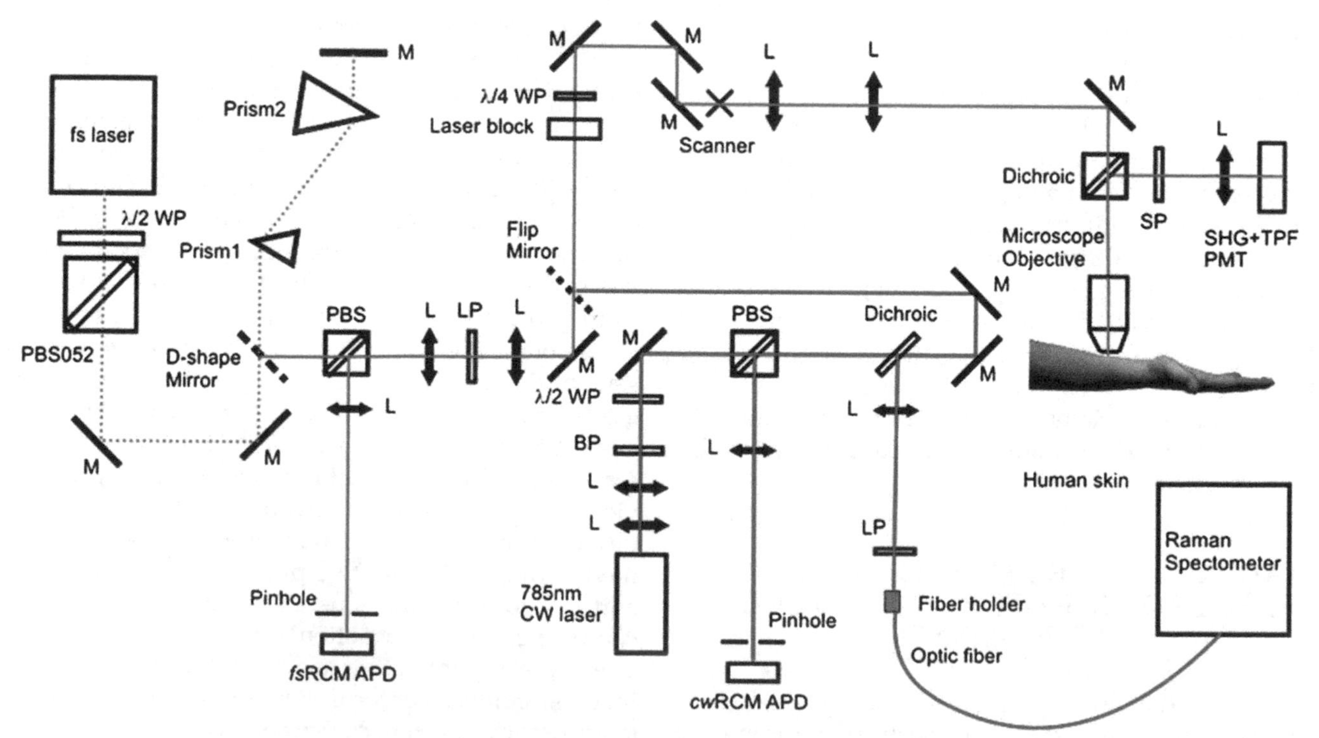

FIGURE 12.12 System diagram of in vivo multimodal confocal Raman spectroscopy, reflectance confocal, and multiphoton imaging. *APD*, avalanche photodiode; *BP*, band-pass filter; *cw*, continuous wave; *cwRCM*, cw reflectance confocal microscopy; *fs*, femtosecond; *fsRCM*, fs reflectance confocal microscopy; *L*, lens; *LP*, long-pass filter; *M*, mirror; *PBS*, polarizing beam splitter; *PMT*, photo multiplier tube; *SHG*, second harmonic generation; *SP*, short-pass filter; *TPF*, two-photon generation; *WP*, wave plate. *Adapted from Wang H, Lee AMD, Lui H, McLean DI, Zeng H. A method for accurate in vivo micro-Raman spectroscopic measurements under guidance of advanced microscopy imaging. Sci Rep 2013;3:1890 with permission from Nature Publishing Group.*

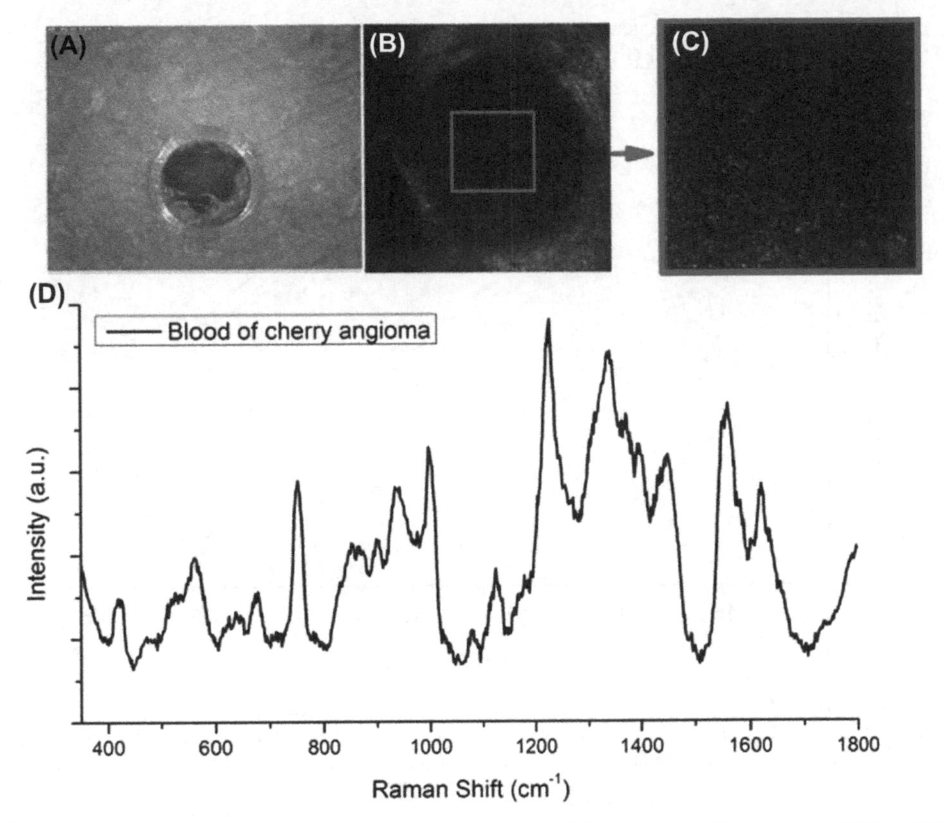

FIGURE 12.13 (A) Dermoscopy image of the cherry angioma lesion. (B) cwRCM image of the blood vessel, FOV = 300 × 300 μm. (C) cwRCM image of the blood vessel of the region of interest from the area of the *red square* in panel (B), FOV = 100 × 100 μm. (D) Confocal Raman spectrum of the blood vessel measured from the area marked as *red squares* in panels (B and C). The exposure time is 20 s. *cwRCM*, continuous-wave reflectance confocal microscopy; *FOV*, field of view. *Adapted from Wang H, Lee AMD, Lui H, McLean DI, Zeng H. A method for accurate in vivo micro-Raman spectroscopic measurements under guidance of advanced microscopy imaging. Sci Rep 2013;3:1890 with permission from Nature Publishing Group.*

consists of Prism1, Prism2, and a mirror), a PBS, and a long-pass filter, and is combined into the same optical path as the cwRCM mode from the quarter-wave plate. The fs laser beam is then directed to the same optical scanning system and objective for imaging. The TPF and SHG signals, as well as the fsRCM reflectance confocal signals, are collected by the same objective and are delivered to the respective imaging detectors. A photomultiplier tube, which is placed right behind the objective, is used to generate integrated SHG and TPF images. The SHG and TPF signals are reflected by the dichroic beam splitter behind the objective and filtered by the short-pass filter. The fsRCM signal is collected by the objective; passes through the scanning system, the quarter-wave plate, and the long-pass filter; and is reflected by the PBS to the fsRCM APD with a 50-μm pinhole in the front, as shown in Fig. 12.12. A flip mirror is used to switch between the cwRCM and fsRCM mode, which is placed right before the quarter-wave plate. The laser power on the target skin is 27 mW for the cw laser and 40 mW for the fs-laser. The imaging field of view (FOV) can be varied from 10 × 10 μm to 300 × 300 μm.

Imaging-Guided Confocal Raman Spectroscopy of Blood In Vivo

In this section we demonstrate the application of the multimode imaging-guided confocal Raman system to measure human blood in vivo. The target is a cherry angioma lesion on the upper arm of an Asian male volunteer. To find the ROI, the fsRCM and SHG + TPF modes were first switched on to locate a blood vessel inside of the cherry angioma lesion. The cwRCM imaging mode was then switched on for confocal Raman spectral measurement under the cwRCM imaging guidance. The FOV was adjusted so that only the ROI was imaged and measured. The dermoscopy image, the cwRCM, and the confocal Raman spectra of the blood vessel are shown in Fig. 12.13. Confocal Raman measurement that is out of the ROI because of movement can be easily rejected from real-time monitoring. Strong Raman peaks from hemoglobin (752 cm^{-1}), glucose (1123 and 1343 cm^{-1}), and protein (940 and 1665 cm^{-1}) can be easily identified from the Raman spectrum. Other microstructures, such as cellular structures can also be measured.

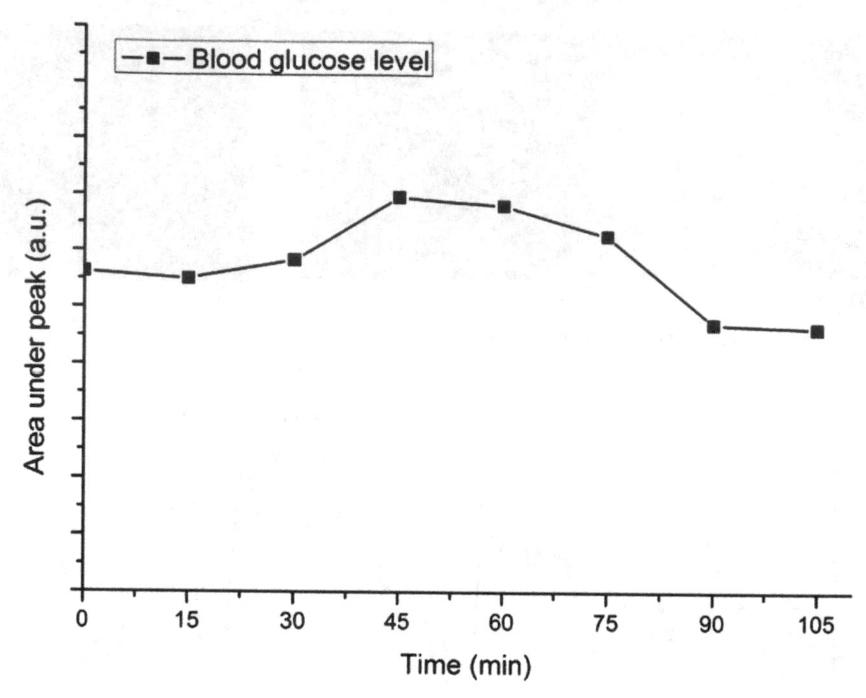

FIGURE 12.14 In vivo monitoring of changes of blood glucose level of a cherry angioma lesion on the upper arm skin of a volunteer using confocal Raman spectroscopy guided with reflectance confocal imaging. *Adapted from Wang H, Lee AMD, Lui H, McLean DI, Zeng H. A method for accurate in vivo micro-Raman spectroscopic measurements under guidance of advanced microscopy imaging. Sci Rep 2013;3:1890 with permission from Nature Publishing Group.*

Monitoring Glucose Level In Vivo

As shown in section "Imaging-Guided Confocal Raman Spectroscopy of Blood In Vivo", the Raman peaks around 1124 and 1343 cm^{-1} are assigned to glucose, which can be easily identified from the confocal Raman spectra. Therefore we designed an experiment to monitor glucose levels over time in vivo. The volunteer was asked to drink a standard glucose liquid, and the imaging-guided confocal Raman system was used to measure the Raman spectra every 15 min for 2 h. An in vivo confocal Raman spectrum was taken as a baseline reference before the experiment. The ratio of the area under the glucose peak at 1124 cm^{-1} to the area under the protein peak at 1450 cm^{-1} was used as a measure of the glucose level. The protein peak at 1450 cm^{-1} was considered as a stable Raman signal in the blood not affected by glucose concentration. The results are shown in Fig. 12.14. It was found that blood glucose increased with time and reached the highest level ($\sim$10% higher than reference concentration) approximately 45–60 min after taking glucose liquid.

It is interesting to note that the glucose signal is very clear in the confocal Raman spectra with greatly reduced interferences from other tissue components. Although Raman spectroscopy has been tested for noninvasive monitoring of glucose level in vivo [96,97], our system provides the feasibility of microscale monitoring of glucose level in vivo. Under confocal imaging guidance, the glucose measurement can be guaranteed from only within the blood vessel, not elsewhere in the tissue. This improves the diagnostic accuracy of blood glucose level. Furthermore, the quality of the Raman spectrum is significantly improved compared with previous publications, probably due to the confocal geometry, where blood vessels can be precisely targeted under confocal imaging guidance. Multimode imaging-guided confocal Raman spectroscopy can also be applied to quantify other microscale localized biochemical information, such as the concentration of oxyhemoglobin and deoxyhemoglobin.

SUMMARY

In summary, we present a rapid, real-time Raman system for in vivo skin cancer diagnosis and an imaging-guided confocal Raman system for microscopic in vivo Raman measurement. We also present an MC model to simulate in vivo skin Raman spectra. Because Raman spectroscopy provides diagnostic information at the molecular level, it provides unique opportunities compared with other techniques. We expect that Raman spectroscopy will become widely accepted in dermatology and skin science and eventually become a standard clinical tool for diagnosis purposes.

Acknowledgments

This project was supported by the Canadian Institutes of Health Research, the Canadian Cancer Society, the Canadian Dermatology Foundation, the VGH and UBC Hospital Foundation in it for Life Fund, and the BC Hydro Employees Community Service Fund. We gratefully acknowledge Sunil Kalia, Wei Zhang, Hequn Wang, and Shuang Wang for their assistance and providing some of the materials.

References

[1] Raman CV. A change of wave-length in light scattering. Nature 1928;121:619.

[2] Lui H, Zhao J, McLean D, Zeng H. Real-time Raman spectroscopy for in vivo skin cancer diagnosis. Cancer Res 2012;72(10):2491–500.

[3] Zhao J, Lui H, McLean DI, Zeng H. Integrated real-time Raman system for clinical in vivo skin analysis. Skin Res Technol 2008; 14(4):484–92.

[4] Zeng H, Zhao J, Short M, McLean DI, Lam S, McWilliams A, et al. Raman spectroscopy for in vivo tissue analysis and diagnosis, from instrument development to clinical applications. J Innovation Opt Health Sci 2008;1(1):95–106.

[5] Huang Z, Zeng H, Hamzavi I, McLean DI, Lui H. Rapid near-infrared Raman spectroscopy system for real-time in vivo skin measurements. Opt Lett 2001;26(22):1782–4.

[6] Huang Z, McWilliams A, Lui H, McLean DI, Lam S, Zeng H. Near-infrared Raman spectroscopy for optical diagnosis of lung cancer. Int J Cancer 2003;107(6):1047–52.

[7] Short MA, Lam S, McWilliams A, Zhao J, Lui H, Zeng H. Development and preliminary results of an endoscopic Raman probe for potential in vivo diagnosis of lung cancers. Opt Lett 2008;33(7): 711–3.

[8] Short MA, Lam S, McWilliams AM, Ionescu DN, Zeng H. Using laser Raman spectroscopy to reduce false positives of autofluorescence bronchoscopies: a pilot study. J Thorac Oncol 2011;6(7):1206–14.

[9] Barman I, Dingari NC, Saha A, McGee S, Galindo LH, Liu W, et al. Application of Raman spectroscopy to identify microcalcifications and underlying breast lesions at stereotactic core needle biopsy. Cancer Res 2013;73(11):3206–15.

[10] Frank CJ, Mccreery RL, Redd DCB. Raman-spectroscopy of normal and diseased human breast tissues. Anal Chem 1995; 67(5):777–83.

[11] Haka AS, Volynskaya Z, Gardecki JA, Nazemi J, Shenk R, Wang N, et al. Diagnosing breast cancer using Raman spectroscopy: prospective analysis. J Biomed Opt 2009;14(5).

[12] Keller MD, Vargis E, de Matos Granja N, Wilson RH, Mycek M-A, Kelley MC, et al. Development of a spatially offset Raman spectroscopy probe for breast tumor surgical margin evaluation. J Biomed Opt 2011;16(7):077006.

[13] Teh SK, Zheng W, Ho KY, Teh M, Yeoh KG, Huang Z. Diagnostic potential of near infrared Raman spectroscopy in the stomach: differentiating dysplasia from normal tissue. Br J Cancer 2008; 98:457–65.

[14] Teh SK, Zheng W, Ho KY, Teh M, Yeoh KG, Huang Z. Near-infrared Raman spectroscopy for optical diagnosis in the stomach: identification of Helicobacter-pylori infection and intestinal metaplasia. Int J Cancer 2010;126(8):1920–7.

[15] Teh SK, Zheng W, Ho KY, Teh M, Yeoh KG, Huang Z. Near-infrared Raman spectroscopy for early diagnosis and typing of adenocarcinoma in the stomach. Br J Surg 2010;97(4):550–7.

[16] Feng S, Chen R, Lin J, Pan J, Wu Y, Li Y, et al. Gastric cancer detection based on blood plasma surface-enhanced Raman spectroscopy excited by polarized laser light. Biosens Bioelectron 2011; 26(7):3167–74.

[17] Short MA, Tai IT, Owen D, Zeng H. Using high frequency Raman spectra for colonic neoplasia detection. Opt Express 2013;21(4): 5025–34.

[18] Widjaja E, Zheng W, Huang Z. Classification of colonic tissue using near-infrared Raman spectroscopy and support vector machines. Int J Oncol 2008;32:653–62.

[19] Andrade PO, Bitar RA, Yassoyama K, Martinho H, Santo AME, Bruno PM, et al. Study of normal colorectal tissue by FT-Raman spectroscopy. Anal Bioanal Chem 2007;387(5):1643–8.

[20] Chowdary MVP, Kumar KK, Thakur K, Anand A, Kurien J, Krishna CM, et al. Discrimination of normal and malignant mucosal tissues of the colon by Raman spectroscopy. Photomed Laser Surg 2007;25(4):269–74.

[21] Duraipandian S, Zheng W, Ng J, Low JJH, Ilancheran A, Huang Z. Near-infrared-excited confocal Raman spectroscopy advances in vivo diagnosis of cervical precancer. J Biomed Opt 2013;18(6): 67007.

[22] Mahadevan-Jansen A, Mitchell MF, Ramanujam N, Malpica A, Thomsen S, Utzinger U, et al. Near-infrared Raman spectroscopy for in vitro detection of cervical precancers. Photochem Photobiol 1998;68(1):123–32.

[23] Utzinger U, Heintzelman DL, Mahadevan-Jansen A, Malpica A, Follen M, Richards-Kortum R. Near-infrared Raman spectroscopy for in vivo detection of cervical precancers. Appl Spectrosc 2001; 55:955–9.

[24] Patel II, Martin FL. Discrimination of zone-specific spectral signatures in normal human prostate using Raman spectroscopy. Analyst 2010;135(12):3060–9.

[25] Crow P, Molckovsky A, Sone N, Uff J, Wilson B, Song LMW. Assessment of fiberoptic near-infrared Raman spectroscopy for diagnosis of bladder and prostate cancer. Urology 2005;65(6): 1126–30.

[26] Guze K, Pawluk HC, Short M, Zeng H, Lorch J, Norris C, et al. Pilot study: Raman spectroscopy in differentiating premalignant and malignant oral lesions from normal mucosa and benign lesions in humans. Head & Neck 2014;37(4):511–7.

[27] Barry BW, Edwards HGM, Williams AC. Fourier transform Raman and infrared vibrational study of human skin: assignment of spectral bands. J Raman Spectrosc 1992;23:641–5.

[28] Williams AC, Edwards HGM, Barry BW. Fourier transform Raman spectroscopy: a novel application for examining human stratum corneum. Int J Pharm 1992;81:R11–4.

[29] Caspers PJ, Jacobsen ADT, Lucassen GW, Wolthuis R, Bruining HA, Puppels GJ. Raman microspectroscopy of human skin. Spectrosc Biol Mol Mod Trends 1997:453–4.

[30] Caspers PJ, Lucassen GW, Wolthuis R, Bruining HA, Puppels GJ. In vitro and in vivo Raman spectroscopy of human skin. Biospectroscopy 1998;4:S31–9.

[31] Bakker Schut TC, Witjes MJ, Sterenborg HJ, Speelman OC, Roodenburg JL, Marple ET, et al. In vivo detection of dysplastic tissue by Raman spectroscopy. Anal Chem 2000;72(24):6010–8.

[32] Caspers PJ, Lucassen GW, Carter EA, Bruining HA, Puppels GJ. In vivo confocal Raman microspectroscopy of the skin: noninvasive determination of molecular concentration profiles. J Invest Dermatol 2001;116(3):434–42.

[33] Nijssen A, Schut TCB, Heule F, Caspers PJ, Hayes DP, Neumann MHA, et al. Discriminating basal cell carcinoma from its surrounding tissue by Raman spectroscopy. J Invest Dermatol 2002;119:64–9.

[34] Nijssen A, Maquelin K, Santos LF, Caspers PJ, Bakker Schut TC, den Hollander JC, et al. Discriminating basal cell carcinoma from perilesional skin using high wave-number Raman spectroscopy. J Biomed Opt 2007;12(3):034004.

[35] Huang Z, Lui H, Chen XK, Alajlan A, McLean DI, Zeng H. Raman spectroscopy of in vivo cutaneous melanin. J Biomed Opt 2004; 9(6):1198–205.

[36] Huang Z, Lui H, McLean DI, K M, Zeng H. Raman spectroscopy in combination with background near-infrared autofluorescence enhances the in vivo assessment of malignant tissues. Photochem Phobiol 2005;81(5):1219–26.

[37] Gniadecka M, Philipsen PA, Sigurdsson S, Wessel S, Nielsen OF, Christensen DH, et al. Melanoma diagnosis by Raman spectroscopy and neural networks: structure alterations in proteins and lipids in intact cancer tissue. J Invest Dermatol 2004;122:443–9.

[38] Gniadecka M, Wulf HC, Mortensen NN, Nielsen OF, Christensen DH. Diagnosis of basal cell carcinoma by Raman spectroscopy. J Raman Spectrosc 1997;28:125–9.

[39] Gniadecka M, Nielsen OF, Wulf HC. Water content and structure in malignant and benign skin tumours. J Mol Struct 2003;661–662: 405–10.

[40] Sigurdsson S, Philipsen PA, Hansen LK, Larsen J, Gniadecka M, Wulf HC. Detection of skin cancer by classification of Raman spectra. IEEE Trans Biomed Eng 2004;51(10):1784–93.

[41] Philipsen PA, Knudsen L, Gniadecka M, Ravnbak MH, Wulf HC. Diagnosis of malignant melanoma and basal cell carcinoma by in vivo NIR-FT Raman spectroscopy is independent of skin pigmentation. Photochem Photobiol Sci 2013;12(5):770–6.

[42] Lieber CA, Majumder SK, Ellis DL, Billheimer D, Mahadevan-Jansen A. In vivo nonmelanoma skin cancer diagnosis using Raman microspectroscopy. Lasers Surg Med 2008;40(7): 461–7.

[43] Lieber CA, Majumder SK, Billheimer D, Ellis DL, Mahadevan-Jansen A. Raman microspectroscopy for skin cancer detection in vitro. J Biomed Opt 2008;13:024013.

[44] Darvin ME, Sandhagen C, Koecher W, Sterry W, Lademann J, Meinke MC. Comparison of two methods for noninvasive determination of carotenoids in human and animal skin: Raman spectroscopy versus reflection spectroscopy. J Biophotonics 2012; 5(7):550–8.

[45] Lawson EE, Williams AC, Edwards HGM, Barry BW. The interactions between penetration enhancers and human skin assessed by FT-Raman spectroscopy. Fourier Transform Spectrosc 1998;430: 306–7.

[46] Melot M, Pudney PDA, Williamson A-M, Caspers PJ, Van Der Pol A, Puppels GJ. Studying the effectiveness of penetration enhancers to deliver retinol through the stratum cornum by in vivo confocal Raman spectroscopy. J Control Release 2009; 138(1):32–9.

[47] Bonnist EYM, Gorce JP, Mackay C, Pendlington RU, Pudney PDA. Measuring the penetration of a skin sensitizer and its delivery vehicles simultaneously with confocal Raman spectroscopy. Skin Pharmacol Physiol 2011;24(5):274–83.

[48]] Huang Y, Zhao J, Lui H, McLean D, Zeng H. Monitoring changes of skin Raman spectra induced by ultrafast laser irradiation: a porcine skin model study. In: Optics in the life sciences; 2015. Vancouver: Optical Society of America; April 12, 2015. JT3A.16.

[49] Chrit L, Bastien P, Biatry B, Simonnet JT, Potter A, Minondo AM, et al. In vitro and in vivo confocal Raman study of human skin hydration: assessment of a new moisturizing agent, pMPC. Biopolymers 2007;85(4):359–69.

[50] Tfayli A, Guillard E, Manfait M, Baillet-Guffroy A. Molecular interactions of penetration enhancers within ceramides organization: a Raman spectroscopy approach. Analyst 2012;137(21):5002–10.

[51] Huang N, Short M, Zhao J, Wang H, Lui H, Korbelik M, et al. Full range characterization of the Raman spectra of organs in a murine model. Opt Express 2011;19(23):22892–909.

[52] Zhao J, Lui H, McLean DI, Zeng H. Automated autofluorescence background subtraction algorithm for biomedical Raman spectroscopy. Appl Spectrosc 2007;61(11):1225–32.

[53] Zhao J, Short M, Braun T, Lui H, McLean D, Zeng H. Clinical Raman measurements under special ambient lighting illumination. J Biomed Opt 2014;19(11):111609.

[54] Zhao J, Lui H, Kalia S, Zeng H. Real-time Raman spectroscopy for automatic in vivo skin cancer detection: an independent validation. Anal Bioanal Chem 2015;407(27):8373–9.

[55] Hanley JA, McNeil BJ. The meaning and use of the area under a receiver operating characteristic (ROC) curve. Radiology 1982; 143(1):29–36.

[56] Hanley JA, McNeil BJ. A method of comparing the areas under receiver operating characteristic curves derived from the same cases. Radiology 1983;148(3):839–43.

[57] Metz CE, Herman BA, Shen JH. Maximum likelihood estimation of receiver operating characteristic (ROC) curves from continuously-distributed data. Stat Med 1998;17(9):1033–53.

[58] Hajian-Tilaki KO, Hanley JA, Joseph L, Collet J-P. A comparison of parametric and nonparametric approaches to ROC analysis of quantitative diagnostic tests. Med Decis Making 1997;17(1): 94–102.

[59] Akobeng AK. Understanding diagnostic tests 1: sensitivity, specificity and predictive values. Acta Paediatr 2007;96(3):338–41.

[60] Cohen M, Cohen BJ, Shotkin JD, Morrison PT. Surgical prophylaxis of malignant melanoma. Ann Surg 1991;213(4):308.

[61] English DR, Del Mar C, Burton RC. Factors influencing the number needed to excise: excision rates of pigmented lesions by general practitioners. Med J Aust 2004;180(1):16–9.

[62] Heal CF, Raasch BA, Buettner P, Weedon D. Accuracy of clinical diagnosis of skin lesions. Br J Dermatol 2008;159(3):661–8.

[63] Miller M, Ackerman AB. How accurate are dermatologists in the diagnosis of melanoma? Degree of accuracy and implications. Arch Dermatol 1992;128(4):559–60.

[64] MacKenzie-Wood AR, Milton GW, Launey JW. Melanoma: accuracy of clinical diagnosis. Australas J Dermatol 1998;39(1):31–3.

[65] Lindelöf B, Hedblad M-A. Accuracy in the clinical diagnosis and pattern of malignant melanoma at a dermatological clinic. The J Dermatol 1994;21(7):461–4.

[66] Menzies SW, Bischof L, Talbot H, Gutenev A, Avramidis M, Wong L, et al. The performance of SolarScan: an automated dermoscopy image analysis instrument for the diagnosis of primary melanoma. Arch Dermatol 2005;141(11):1388–96.

[67] Robinson JK, Nickoloff BJ. Digital epiluminescence microscopy monitoring of high-risk patients. Arch Dermatol 2004;140(1): 49–56.

[68] Banky JP, Kelly JW, English DR, Yeatman JM, Dowling JP. Incidence of new and changed nevi and melanomas detected using baseline images and dermoscopy in patients at high risk for melanoma. Arch Dermatol 2005;141(8):998–1006.

[69] Binder M, Kittler H, Seeber A, Steiner A, Pehamberger H, Wolff K. Epiluminescence microscopy-based classification of pigmented skin lesions using computerized image analysis and an artificial neural network. Melanoma Res 1998;8(3):261–6.

[70] Farina B, Bartoli C, Bono A, Colombo A, Lualdi M, Tragni G, et al. Multispectral imaging approach in the diagnosis of cutaneous melanoma: potentiality and limits. Phys Med Biol 2000;45(5):1243.

[71] Moncrieff M, Cotton S, Claridge E, Hall P. Spectrophotometric intracutaneous analysis: a new technique for imaging pigmented skin lesions. Br J Dermatol 2002;146(3):448–57.

[72] Westerhoff K, McCarthy W, Menzies S. Increase in the sensitivity for melanoma diagnosis by primary care physicians using skin surface microscopy. Br J Dermatol 2000;143(5):1016–20.

[73] Monheit G, Cognetta AB, Ferris L, Rabinovitz H, Gross K, Martini M, et al. The performance of MelaFind: a prospective multicenter study. Arch Dermatol 2011;147(2):188–94.

[74] Wells R, Gutkowicz-Krusin D, Veledar E, Toledano A, Chen SC. Comparison of diagnostic and management sensitivity to melanoma between dermatologists and MelaFind: a pilot study. Arch Dermatol 2012;148(9):1083–4.

[75] Pavlova I, Weber CR, Schwarz RA, Williams M, El-Naggar A, Gillenwater A, et al. Monte Carlo model to describe depth

selective fluorescence spectra of epithelial tissue: applications for diagnosis of oral precancer. J Biomed Opt 2008;13(6):064012–3.
[76] Pavlova I, Weber CR, Schwarz RA, Williams MD, Gillenwater AM, Richards-Kortum R. Fluorescence spectroscopy of oral tissue: Monte Carlo modeling with site-specific tissue properties. J Biomed Opt 2009;14(1):014009–10.
[77] Zeng H, MacAulay C, McLean DI, Palcic B. Reconstruction of in vivo skin autofluorescence spectrum from microscopic properties by Monte Carlo simulation. J Photochem Photobiol B 1997; 38(2–3):234–40.
[78] Wang S, Zhao J, Lui H, He Q, Zeng H. Monte Carlo simulation of near infrared autofluorescence measurements of in vivo skin. J Photochem Photobiol B 2011;105(3):183–9.
[79] Chen R, Huang Z, Lui H, Hamzavi I, McLean DI, Xie S, et al. Monte Carlo simulation of cutaneous reflectance and fluorescence measurements–the effect of melanin contents and localization. J Photochem Photobiol B 2007;86(3):219–26.
[80] Shih W-C, Bechtel KL, Feld MS. Intrinsic Raman spectroscopy for quantitative biological spectroscopy part I: theory and simulations. Opt Express 2008;16(17):12726–36.
[81] Everall N, Hahn T, Matousek P, Parker AW, Towrie M. Photon migration in Raman spectroscopy. Appl Spectrosc 2004;58(5):591–7.
[82] Wilson RH, Dooley KA, Morris MD, Mycek M-A. Monte Carlo modeling of photon transport in buried bone tissue layer for quantitative Raman spectroscopy. In: SPIE BiOS: biomedical optics; 2009. International Society for Optics and Photonics; 2009. p. 716604–10.
[83] Matousek P, Morris M, Everall N, Clark I, Towrie M, Draper E, et al. Numerical simulations of subsurface probing in diffusely scattering media using spatially offset Raman spectroscopy. Appl Spectrosc 2005;59(12):1485–92.
[84] Keller MD, Wilson RH, Mycek M-A, Mahadevan-Jansen A. Monte Carlo model of spatially offset Raman spectroscopy for breast tumor margin analysis. Appl Spectrosc 2010;64(6):607–14.
[85] Reble C, Gersonde I, Lieber CA, Helfmann J. Influence of tissue absorption and scattering on the depth dependent sensitivity of Raman fiber probes investigated by Monte Carlo simulations. Biomed Opt Express 2011;2(3):520–33.
[86] Mo J, Zheng W, Huang Z. Fiber-optic Raman probe couples ball lens for depth-selected Raman measurements of epithelial tissue. Biomed Opt Express 2010;1(1):17–30.
[87] Salomatina E, Jiang B, Novak J, Yaroslavsky AN. Optical properties of normal and cancerous human skin in the visible and near-infrared spectral range. J Biomed Opt 2006;11(6):064026–9.
[88] Meglinski IV, Matcher SJ. Quantitative assessment of skin layers absorption and skin reflectance spectra simulation in the visible and near-infrared spectral regions. Physiol Meas 2002;23(4):741.
[89] Wang S, Zhao J, Lui H, He Q, Bai J, Zeng H. Monte Carlo simulation of in vivo Raman spectral measurements of human skin with a multi-layered tissue optical model. J Biophotonics 2014;7(9): 703–12.
[90] Wang L, Jacques SL, Zheng L. MCML—Monte Carlo modeling of light transport in multi-layered tissues. Comput Methods Programs Biomed 1995;47(2):131–46.
[91] Wang L, Jacques SL, Zheng L. CONV—convolution for responses to a finite diameter photon beam incident on multi-layered tissues. Comput Methods Programs Biomed 1997;54(3):141–50.
[92] Wang S, Zhao J, Lui H, He Q, Zeng H. A modular Raman micro-spectroscopy system for biological tissue analysis. J Spectrosc 2010;24(6):577–83.
[93] Egawa M, Hirao T, Takahashi M. In vivo estimation of stratum corneum thickness from water concentration profiles obtained with Raman spectroscopy. Acta Derm Venereol 2007;87(1):4–8.
[94] Caspers PJ, Williams AC, Carter EA, Edwards HGM, Barry BW, Bruining HA, et al. Monitoring the penetration enhancer dimethyl sulfoxide in human stratum corneum in vivo by confocal Raman spectroscopy. Pharm Res 2002;19(10):1577–80.
[95] Wang H, Lee AMD, Lui H, McLean DI, Zeng H. A method for accurate in vivo micro-Raman spectroscopic measurements under guidance of advanced microscopy imaging. Sci Rep 2013;3:1890.
[96] Berger AJ, Itzkan I, Feld MS. Feasibility of measuring blood glucose concentration by near-infrared Raman spectroscopy. Spectrochim Acta A 1997;53A(2):287–92.
[97] Enejder AMK, Scecina TG, Oh J, Hunter M, Shih W, Sasic S, et al. Raman spectroscopy for noninvasive glucose measurements. J Biomed Opt 2005;10(3):031114.

CHAPTER

13

Surface-Enhanced Raman Spectroscopy for Intradermal Measurements

K. Chen, Y.H. Ong, C. Yuen, Q. Liu

Nanyang Technological University, Singapore

OUTLINE

INTRODUCTION

Intradermal Measurements and Relevant Techniques

The skin is the largest organ of the body. There are four layers of the skin: stratum corneum (SC), epidermis, dermis, and subcutis, which all have different optical characteristics. Therefore the skin can be considered a complex, variable, and multilayered optical medium.

Evaluation of physical appearance and analysis of intradermal biochemical components are two common techniques used in clinical diagnosis for a wide range of skin-related diseases, such as eczema, psoriasis, cutaneous tuberculosis, and melanoma, as well as non-skin-related diseases, such as diabetes and AIDS [1]. Because the skin is easily accessible, most dermatological conditions are initially examined by visual inspection, either by naked eyes or aided by a dermoscope [2]. However, the diagnostic accuracy and precision by visual inspection varies significantly among studies because of variation in doctors' experience and the type of assessment algorithm used [3,4]. Because the visual inspection of skin structures is not always reliable, any skin disorder diagnosis has to be confirmed by histopathological examination [5]. Histopathological examination involves the analysis of intradermal biochemical components in biopsy specimens or body fluid, which contain a wealth of substances predictive of clinical outcome. However,

Imaging in Dermatology
http://dx.doi.org/10.1016/B978-0-12-802838-4.00013-3

sample preparation in histopathological examination involves invasive procedures such as the surgical removal of tissue specimens and syringe collection of body fluid, which are labor-intensive, time-consuming, expensive, and painful to patients.

To assist the clinical diagnosis of dermatological conditions, a broad variety of diagnostic technologies capable of noninvasively performing in vivo intradermal measurements are being explored. *Intradermal measurements* refer to the detection and characterization of cutaneous biochemical components including the contents of both blood and interstitial fluid that circulates and bathes the skin layers. It should be noted that histopathological examination, which involves examination of ex vivo skin specimens from surgically removed biopsies, is the current gold standard for disease confirmation and tissue characterization. All the different diagnostic methods in dermatology discussed in the following sections are aimed at matching the clinical diagnostic accuracy achieved by histopathological examination but without the need of specimen excision or body fluid collection.

Optical techniques have been widely explored for intradermal measurements to extract clinically important information from skin tissues noninvasively. Optical coherence tomography (OCT) is an optical imaging modality analogous to ultrasound in principle, which uses light to achieve three-dimensional images of skin structures with high spatial resolution [6]. OCT provides cross-sectional images of skin tissue up to a few millimeters deep, in situ and in real time. It has been widely used to study various morphological features of healthy skin such as assessing the thicknesses of the SC [7] and epidermis [8] and visualizing skin microstructures including blood capillaries and skin appendages [9]. OCT is also capable of visualizing and monitoring pathological skin changes in various skin disorders and dermatological conditions that can be correlated with routine histological findings [10,11]. However, typical OCT images do not contain information about the biochemical makeup of the skin that is vital for the identification and classification of tissue neoplasms and skin cancer.

Confocal laser scanning microscopy (CLSM) represents another emerging medical imaging technique for noninvasive intradermal measurements capable of providing fine details about tissue histomorphology from a desired depth with submicrometer resolution [12,13]. CLSM can be performed in either reflectance or fluorescence mode. Reflectance CLSM generates image contrast based on the variation in the refractive indices of tissue microstructures, whereas fluorescence CLSM generates images based on the contrast of fluorescence signals from either exogenous fluorescence dyes or endogenous fluorophores. Despite the excellent optical sectioning capability of CLSM in generating optical-sectioned images of skin tissue with the highest resolution of all optical techniques used in dermatology, the technique is restricted by its limited penetration depth and slow acquisition speed. Moreover, highly experienced experts are required to interpret CLSM images for histomorphological analysis of the skin [12].

Diffuse reflectance spectroscopy is also being explored as a simple and fast tool for intradermal measurements, because diffuse reflectance measurements are sensitive to the alteration in optical properties of skin tissues [14]. Diffuse reflectance spectra are correlated with numerous clinically important parameters including blood oxygen saturation, hemoglobin concentration, and tissue hydration [15]. It has been investigated to assess the viability of skin flaps [15,16], differentiate various types of tissues ex vivo [17] and in vivo [18], and identify benign and malignant tissues in situ during surgery [19]. However, diffuse reflectance spectroscopy has no optical sectioning capability and thus it needs to be integrated with confocal imaging or OCT to be more suitable for layer-specific intradermal measurements. Moreover, its data interpretation requires complicated light transport models and extensive assumptions.

Autofluorescence spectroscopy and imaging is another attractive noninvasive analytic tool that is widely employed for intradermal measurements. Skin tissues contain numerous endogenous fluorophores such as flavins, collagen, protoporphyrin IX, and tryptophan, which fluoresce upon excitation by ultraviolet visible light [20]. Skin malignancy alters the concentrations of these fluorophore molecules, leading to changes in skin autofluorescence intensity, which can be used to provide clinically important biochemical information useful for the identification of various skin diseases and demarcation of skin cancer from normal tissue [21,22], the differentiation of cutaneous pigmented lesions [23], the assessment of skin aging [24], and the monitoring of cutaneous wound healing [25]. However, this technique suffers from the fact that broad and overlapping fluorescence emission peaks of intrinsic fluorophore molecules are difficult to distinguish from each other, hindering the multiplex detection capability of this technique.

Raman Spectroscopy

Spontaneous Raman Spectroscopy

Raman spectroscopy, a vibrational spectroscopic technique, has been widely explored as an analytical tool in the field of medical diagnostics for its high chemical specificity, noninvasiveness, and label-free nature. This technique complements other optical modalities that provide structural and morphological information by providing additional biochemical information about tissue samples down to the molecular level [26]. This information is diagnostically important because pathological changes in skin tissue result in alterations in molecular composition and functional groups as well as variation in molecular bonding, which can be detected by Raman spectroscopy in terms of minute shifts in Raman peak location, relative intensity, and peak width in the detected spectrum [27]. However, the application of Raman spectroscopy as an in vivo intradermal analytical tool has been so far limited because of the low Raman-scattering cross-sections of cutaneous Raman-active biomolecules. Minute alterations in a spontaneous Raman spectrum are usually masked by larger variation in the stronger autofluorescence background, and are attenuated by tissue absorption and scattering. The difficulty in effectively extracting these minute Raman changes against the background of strong autofluorescence has significantly limited the advancement of Raman spectroscopy in dermatology as well as other clinical applications in vivo.

Surface-Enhanced Raman Spectroscopy

Surface-enhanced Raman spectroscopy (SERS) is a Raman spectroscopic technique with high molecular selectivity and surface sensitivity, in which Raman scattering from Raman-active analyte molecules is strongly enhanced when these molecules are adsorbed on or are in close proximity to a metallic surface. Since its early discovery by Fleischmann et al. [28] in 1974, a huge amount of work has been carried out to understand and characterize the mechanism of this phenomenon, leading to the rapid and dramatic growth of SERS techniques over the past few decades in a broad stream of disciplines, including physics, chemistry, material science, nanoscience, and medical fields including dermatology. A typical SERS enhancement factor (EF) is on the order of 10^4–10^6 and an EF of 10^8–10^{14} has been observed for surface-enhanced resonance Raman scattering [29], making SERS a powerful tool for sensing molecules in a trace amount, previously undetectable by spontaneous Raman scattering spectroscopy. In the following sections, we will briefly discuss the mechanism of signal enhancement in SERS and review the potential of SERS as an intradermal measurement tool.

Mechanisms of Surface-Enhanced Raman Spectroscopy

SERS enhancement can be explained by two underlying mechanisms: (1) electromagnetic (EM) field enhancement through the localization of optical fields in metallic nanostructures, and (2) chemical enhancement (CE) as a result of the increase of the Raman cross-section when the molecule is in contact with metal nanostructures [30]. The amplification or EF of SERS is the result of these two major mechanisms, with a major contribution from the EM mechanism [31]. These two mechanisms arise because the intensity of Raman scattering is directly proportional to the square of the induced dipole moment, which is the product of the Raman polarizability and the magnitude of the incident EM field.

Electromagnetic Enhancement

When the incident light (an EM wave) interacts with a metal surface, localized surface plasmons are excited, whereby the magnitude of the EM fields at the surface are enhanced. Localized surface plasmon resonance (LSPR) occurs when the frequency of the incident light matches the natural frequency of surface electrons. The field enhancement is the greatest during LSPR when the collective oscillation of these valence electrons in the metal surface is in resonance with the frequency of incident light. A rough surface is required to create a component of the plasmon that oscillates perpendicular to the surface plane for scattering to occur [32]. The EM enhancement is therefore dependent on the presence of roughness features in the metal surface. Rough metal surfaces and nanoparticles of different shapes, such as nanorod [33], nanocrescent [34], nanotriangle [35], nanoflower [36], nanocube [37], or nanopyramid [38], have been fabricated to provide areas or hotspots on which these localized collective oscillations can occur.

The intensity of the Raman-scattered radiation is proportional to the square of the magnitude of the EM field incident on a test molecule, ie:

$$I_R \propto E^2 \tag{13.1}$$

where I_R is the intensity of the Raman field and E is the overall EM field coupling with the analyte (summation of the EM field on the analyte in the absence of all roughness features and the field emitted from the particulate metal roughness feature) [39,40]. SERS amplification is extremely strong because field enhancement occurs twice. First, LSPR amplifies the intensity of incident light that excites test molecules on the metal surface. Then the Raman-scattered light experiences enhancement by the same mechanism, augmenting the scattered light. Thus the total enhancement experienced by each test molecule is approximately the product of these two causes of field enhancements:

$$\text{Enhancement Factor (EF)} \approx |E(\omega)|^2 |E(\omega')|^2 \tag{13.2}$$

where $E(\omega)$ represents the field EF experienced by the incident light at a frequency of ω and $E(\omega')$ represents the field EF experienced by the Raman signal at a Stokes-shifted frequency of ω'. Therefore the EM SERS enhancement shows a fourth-power dependency on the EM field amplitude, $|E(\omega')|^4$, when $\omega = \omega'$ [41,42]. This EF experienced by the test molecule is the strongest when it is in direct contact with the metal surface. For molecules at a distance, d, away from the surface of a metal sphere, the EF deteriorates by a coefficient of $[r/(r+d)]^{12}$, where r is the radius of the spherical surface [30,43]. This shows that the average distance of metallic surfaces/nanoparticles introduced to specific cutaneous molecules of interest will directly affect the overall EF of intradermal SERS signals. Therefore the methods of introduction and fabrication of these metallic surfaces/nanoparticles are critical in defining the effectiveness of SERS as an intradermal analytic tool.

Chemical Enhancement

CE provides one or two orders of magnitude enhancement in the intensity of Raman scattering. Despite the fact that the detail of this mechanism is less clear than EM enhancement, it is evident that CE is the result of the change in the electronic structure of an analyte molecule when absorbed onto a metal surface that enhances its Raman polarizability [30,31]. When an analyte molecule is adsorbed onto a metal surface, the interaction between the analyte molecule and metal surface may create a pathway of electronic coupling, allowing for electron transfer between the orbital of the molecule and that of the metal. The electronic coupling between the metal and the molecule establishes a new equilibrium for the relaxation of the molecule, increasing its Raman-scattering cross-section [44,45]. Apart from the formation of charge-transfer intermediates, CE has also been described in several pathways of metal–adsorbate interaction, including metal–molecule electron tunneling [46], intensity borrowing [47], and formation of metal–molecular surface states [48]. Despite the difference in their enhancement mechanisms, both EM enhancement and CE require a very small distance between the metallic surface and the analyte molecule. Therefore the method of fabricating these metallic surfaces/nanoparticles and introducing them into the skin to get as close as possible to specific cutaneous components are the main consideration to the success of SERS in dermatology.

Surface-Enhanced Raman Spectroscopy in Biomedical Applications

In spontaneous Raman spectroscopy, Raman signals from the bulk tissue are excited as the excitation light illuminates and propagates through the skin. Accurately identifying and separating the Raman signal originated from targeted cutaneous components is challenging, and this problem is complicated further by the generation of an autofluorescence background, which is often several orders of magnitude stronger than the Raman signals arising from the same tissue components. SERS is superior to spontaneous Raman spectroscopy for intradermal measurements because of its high sensitivity and molecular selectivity. The SERS effect amplifies the intensity of Raman signals originating from molecules in close proximity to metal surfaces so strongly that they overwhelm the bulk signal, and rival the intensity of the fluorescence background. The strong enhancement of SERS can reduce incident laser power required and shorten data acquisition time that would reduce potential skin damage that might be caused during intradermal measurements.

Advances in fabrication of metal nanostructures has driven SERS to become a powerful analytical tool in a large number of biological and medical diagnostic applications, so that it can sense multiple biomolecules at the attomolar concentration [49] or even at the single-molecule level [50]. The sharp SERS spectra with high molecular specificity, and the capability of SERS in simultaneously detecting multiple biomolecules at low concentrations, make SERS better in multiplex detection than fluorescence methods, which suffer from spectral overlapping among the broad emission spectra arising from multiple target fluorophores, and strong background autofluorescence. In addition to the multiplexing capacity, Raman signals are immune to photobleaching that can be problematic in fluorescence, which makes the precise quantification of target molecule concentration possible. Visible and near-infrared (NIR) light is often used for Raman excitation to reduce the autofluorescence background of biological tissue. Because NIR light can penetrate deep into a biological tissue and reach the vascularized dermis layer with its rich blood content, SERS detection has rapidly moved toward in vivo biomedical applications, particularly for intradermal measurements, in a hope to reduce the needs of invasive tissue biopsy and blood sampling.

Because SERS selectively amplifies Raman signals from molecules close to metal surfaces, metal surfaces or nanoparticles have to be introduced into the skin to enhance SERS signals from cutaneous biomolecules. In the next section, we will review the current state-of-art techniques for intradermal SERS measurements, in which the design considerations in each method will be discussed.

SURVEY OF INTRADERMAL SURFACE-ENHANCED RAMAN SPECTROSCOPY APPROACHES

In order to achieve strong enhancement in SERS measurements, metallic nanoparticles have to be introduced in close proximity to the biomolecules. The most straightforward method to achieve this goal is to diffuse tiny nanoparticles into tissue specimens or to mix body fluid samples with nanoparticle suspension. Falamas et al. [51] immersed ex vivo skin samples in a suspension of silver (Ag) colloids and demonstrated the SERS detection of nucleic acids. Hsu et al. [52] demonstrated the quantitative detection of lactic acid at a physiological concentration from the mixture of human blood serum and Ag colloids. Yin et al. [53] identified characteristic bands in the SERS spectra of blood plasma mixed with Ag colloids that can be used to differentiate between healthy skin and squamous cell carcinoma in mouse models.

Despite strong enhancement in Raman signals from biomolecules as shown, these studies have been limited to ex vivo measurements that require invasive tissue biopsy or biological fluid sampling. Driven by the promising results of SERS in detecting biomolecules at physiological concentrations, numerous approaches have been investigated to introduce these metallic surfaces or nanoparticles to the close proximity of biomolecules in vivo for minimally invasive SERS measurements from tissues or tissue phantoms. To facilitate discussion, we have grouped these approaches into four categories based on the characteristics of SERS substrates: (1) injection of SERS-active nanoparticles, (2) implantation of SERS-active structures, (3) administration of SERS-active acupuncture needles, and (4) application of SERS-active microneedles.

Injection of Surface-Enhanced Raman Spectroscopy—Active Nanoparticles

SERS-active nanoparticles have been used in SERS applications in vivo for disease-related tissue imaging. These SERS-active nanoparticles, also known as *nanotags*, are usually functionalized with various antibodies and receptor moieties to actively target them to specific biomarkers for in vivo multiplex detection. SERS nanotags are fabricated by chemisorption of strong Raman-active molecules, also known as reporter molecules (RMs), on gold (Au) nanoparticles, before being encapsulated in a polyethylene glycol [54,55], silica [56], or bovine serum albumin (BSA) [57] shell. In addition to providing physical robustness, signal stability, and protection in the in vivo biochemical environment, the shells also serve as "handles" for bioconjugation [58]. These SERS nanotags are superior to fluorescence-based nanoparticles, such as quantum dots, for multiplex detection because of their sharp and narrow spectral fingerprints. Besides that, SERS nanotags generate strong and stable signals in the NIR spectral region, are not susceptible to photobleaching, and have low cytotoxicity owing to the usage of inert Au nanoparticles [58].

SERS nanotags have been successfully used for the detection of cancer biomarkers [57,59]. Samanta et al. [57] discovered and synthesized a novel compound, CyNAMLA-381, as a highly sensitive NIR SERS-active reporter that could be used as an ultrasensitive SERS probe by conjugating it with an antibody for the detection and discrimination of human epidermal growth factor 2—positive cancer cells. Wang et al. [59] developed a specific and sensitive method using the bioconjugation of epidermal growth factor and Au nanoparticles, tagged with QSY RMs, to rapidly detect circulating tumor cells in peripheral blood from patients with different stages of squamous cell carcinoma in the head and neck. Qian et al. [55] demonstrated in vivo SERS imaging, where they successfully targeted in vivo cancerous tumors using Au nanoparticles conjugated with single-chain variable fragment antibodies and introduced into a mouse model via tail vein injection. In another study by Maiti et al. [60], in vivo SERS multiplexing capability was demonstrated when they injected three different BSA-encapsulated nanotags constructed with NIR-active RMs (CyNAMLA-381, Cy7LA, and Cy7.5LA) through the tail vein of a living mouse and studied their kinetics and localization in the tumor and liver of the mouse over 8 days. Kang et al. [61] demonstrated the capability of deep-tissue SERS imaging by effectively detecting signals from SERS nanotags constructed with Au/Ag hollow shells and aromatic RMs, which were passively injected into porcine tissue up to 8-mm depth. These studies showed that the antibody-functionalized nanoparticles were able to travel through the blood circulation systems of the animal models to reach and tag specific targets, and that SERS signals can be detected from as deep as 8 mm into the skin.

Although none of the discussed studies investigate directly the intradermal measurements of a particular analyte in the skin, the methods involved are equally applicable to the skin. In all of the examples mentioned, the measured signals originated from SERS RMs. It should be noted that these signals do not provide direct biochemical information about targeted biomarkers. To explore the full capability of SERS for intradermal measurements, a second excitation laser could be employed at a frequency where the intrinsic Raman signals from the targeted biomarkers dominate those from the nanotags [62].

Another consideration when employing SERS nanotags for in vivo imaging is studying long-term possible toxicity of metal nanoparticles in the body. A study carried out by Asharani et al. [63] suggested that Ag nanoparticles (AgNPs) have the potential to cause health risk and ecotoxicity in a concentration-dependent manner, in which phenotypic deformities, altered physiological function and increased mortality rates in zebrafish embryos were observed when treated with AgNPs. Bar-llan et al. [64] have systematically assessed the toxicity of Ag and Au nanoparticles of different sizes in zebrafish embryos and found that Ag and Au nanoparticles exhibited significantly different toxicity profiles. The study showed that AgNPs were toxic, whereas Au nanoparticles were inert in all tested sizes, suggesting that Au is a better candidate for in vivo SERS intradermal measurements because it is less likely to induce any hazardous effect to the host body. Recently, the biocompatibility and biodistribution of Au nanoparticles was studied by Wang et al. [65], in which they microinjected Au nanotags directly into zebrafish embryos at the one-cell stage and assessed their distribution in various cell types and tissues in the developing embryo. The normal morphology and gene expression of zebrafish embryos at various developmental stages after SERS Au nanoparticle injection suggested that the Au nanoparticles were not toxic and were suitable to be used for the monitoring of in vivo physiological and pathological processes. Similar studies in the context of intradermal measurements need to be performed in more advanced animal models in order to predict the safety of SERS nanotags in humans.

Surface-Enhanced Raman Spectroscopy—Active Structures

Besides the nanoparticles and nanotags used in SERS measurements, complicated structures such as nanowires [66], biopatches [67], and film structures [68] have been developed, which could facilitate intradermal SERS measurements. They have been studied for intradermal measurements in an animal model or skin-mimicking phantoms.

To enable easy penetration into a tissue, Han et al. [66] demonstrated an NIR probe using silicon nanowires (SiNWs) for SERS measurements that achieved high-resolution intracellular pH detection. AgNPs with uniform size were coated on the surface of SiNW to create a structure of AgNPs@SiNW. Then AgNPs@SiNW was immersed in a pH-sensitive *p*-mercaptobenzoic acid (pMBA) solution (1×10^{-5} M). Raman spectra were acquired from the pMBA absorbed on AgNPs@SiNW. To evaluate the performance of AgNPs@SiNW, the SERS probe was loaded onto the tip of a glass pipet and was then inserted into culture media to measure the pH value. The intensity variation of pMBA Raman bands near 1395 cm^{-1}, which was measured using two different excitation lasers (633 and 785 nm) with a power of 0.5 mW and an acquisition time of 30 s, showed good correlation with the pH change from 4.0 to 9.0. Furthermore, there was no significant difference between the spectra recorded from freshly prepared pMBA-functionalized AgNPs@SiNW and the same material stored in deionized water for 7 days. This result indicated that AgNPs@SiNW had considerable stability in an aqueous environment so as to achieve continuous monitoring in situ. All these properties of AgNPs@SiNW suggest its great potential as a sensor for pH sensing or other biochemical measurements in the skin, although the strength and biocompatibility of the nanowires need to be further evaluated before applied in vivo.

To eliminate the need of injecting nanoparticles into human body, Park et al. [67] reported a nanoplasmonic biopatch for in vivo SERS measurements. The biopatch consisted of nanoparticles embedded inside an agarose hydrogel. The nanoparticles and agarose mixture was recast into a polydimethylsiloxane model to form a 1-mm thick biopatch. The entire structure was both highly biocompatible and had no Raman signal. When attached to the skin surface, this biopatch will absorb biomolecules from the body. In this work, 10-mM rhodamine 6G (R6G) was spotted on the top surface of the biopatch to mimic biomolecules being released from the skin and subsequent diffusion into the biopatch. A 633-nm laser with a power of 2.5 mW was used for excitation. SERS spectra were acquired within 10 seconds after spotting, at a depth of 100 μm below the top surface. The Raman signal decreased after 10 seconds because R6G diffused to a broader region and thus the concentration at focal area decreased. The concentration near the spotting area was larger but the scattering and absorption of the biopatch affected the Raman signal. Therefore the Raman signal not only depended on the concentration of AgNPs and analyte but also was related to the focal depth and diffusion time. Moreover, biomolecules including γ-aminobutyric acid (200 mM), which is a neurotransmitter, and β-amyloid (1.2 mM), which is a biomarker of Alzheimer disease, were tested in the biopatch and showed good Raman performance. This nanoplasmonic biopatch can facilitate highly sensitive label-free detection because of the low Raman background and high biocompatibility, which could be useful for monitoring the skin or other tissues. Nevertheless, this approach requires an invasive pathway for allowing intradermal biomolecules to diffuse to the surface and get absorbed by the biopatch. The optimal method to create an invasive pathway on the skin needs to be further demonstrated and evaluated.

An implantable approach has been demonstrated by Stuart et al. [68], who implanted a substrate with "silver—film over nanosphere" (AgFON) into a rat for in vivo glucose measurements. A layer of orderly arranged 390-nm latex polystyrene nanospheres was coated onto a circular copper mesh with a diameter of 18 mm. Then an Ag film with 200-nm thickness was deposited on top of the layer of the nanospheres to form the AgFON substrate. This AgFON substrate was placed subcutaneously in a rat at a location externally accessible through a glass window fixed at the midline along the back of the rat. In the SERS measurement, 785-nm laser light was focused obliquely through the window onto the substrate and the emitted signal was collected along the direction perpendicular to the window surface. The set-up was able to evaluate the glucose concentration in the rat with a root mean square error of 2.97 mM, when the excitation power was 50 mW and the exposure time was 2 minutes. This strategy of implanting the SERS substrate inside the body would be suitable for long-term intradermal measurements in humans after the long-term impact of such implantation on biological tissues had been thoroughly investigated.

Surface-Enhanced Raman Spectroscopy—Active Acupuncture Needles

The methods of nanoparticle injection or nanostructure implantation introduce SERS substrates to target molecules in the skin, both of which pose challenges in biocompatibility and health risk. The alternative way is to take body fluid such as blood out to SERS substrates. However, drawing blood or interstitial fluid from human skin requires the use of invasive tools such as hypodermic needles, which induces pain and the risk of infection. To minimize the invasion and discomfort during body fluid extraction [69], acupuncture needles have been explored to take body fluid out of body, which is directly applicable to intradermal measurements.

Dong et al. [70] have measured the depth profile of SERS spectra in agarose phantoms using a 0.2-mm acupuncture needle coated by Au nanoshells with inner and outer radii of 55 nm and 80 nm, respectively. This SERS-active acupuncture needle was inserted as deep as 7 mm into the investigation site for analyte absorption, after which the needle was retracted for SERS measurements. A 785-nm laser with an excitation power of 60 mW was sequentially focused on a series of locations starting at 1 mm from the needle tip and ending at the needle root with a step size of 1 mm to take SERS measurements. With an acquisition time of 10 seconds at each location, the acquired SERS signal could detect 6-mercaptopurine, which is a drug for treating acute lymphoblastic leukemia, at a concentration above 1×10^{-6} M in anticoagulated blood. This strategy has a unique advantage of obtaining the depth profile of SERS at the needle insertion site by correlating the depth beneath the skin surface to the distance from the tip of the acupuncture needle at which the SERS measurement was performed.

The same group [71] further designed a glucose-responsive multifunctional acupuncture needle by incorporating glucose oxidase and 4-mercaptobezoic acid (4-MBA), into their SERS-active acupuncture needle via microporous polystyrene coating. Glucose oxidase converted glucose to gluconolactone and gluconic acid, which changed the pH value. Then the change of pH was reported by the Raman signal of 4-MBA. The Au nanoshells coated on the needle enhanced the Raman signal of 4-MBA for the quantification of glucose concentration. The SERS results measured from glucose solutions were compared with those from agarose phantoms with the same glucose concentrations, which suggests that the insertion process in phantom measurements did not affect the measurement accuracy. Furthermore, the glucose-responsive needle was utilized to measure glucose concentrations at the center of the tendon and ear vein of a rabbit before and after glucose injection into the bloodstream of the rabbit. The similarity in the trends between SERS reading and glucose concentration measured by a glucose meter suggests the excellent promise of this technique for intradermal measurements in human skin. One potential problem is that the possible metal aggregation in human body caused by unstable coating on needles left behind in the skin can induce a health risk, which needs to be further investigated.

Surface-Enhanced Raman Spectroscopy—Active Microneedles

Compared with conventional hypodermic needles and acupuncture needles that are long (thus requiring precise control on the penetration depth and likely causing pain), the major advantage of a microneedle is that it can penetrate the SC to reach body fluid, but is short enough to avoid reaching nerve endings, causing nearly no pain [72,73]. Consequently, a microneedle patch can be applied onto the skin conveniently without the concern of going too deep. For example, the microneedle has been shown convenient for intradermal vaccination [74].

Yuen and Liu [75] constructed an Ag-coated stainless steel microneedle patch for SERS detection of glucose that was buried inside an agarose skin-mimicking phantom. A 785-nm laser was focused into the bottom phantom layer that contained glucose via the lumen created by the microneedle insertion. The emitted Raman signals of glucose traveled through the lumen in an opposite direction and reached a spectrometer for spectral acquisition. In this study, the SERS effect augmented the Raman signal from glucose molecules in close proximity to AgNPs coated on the microneedles. Moreover, microneedle insertion created a thin channel of propagation for the excitation and emission light. Without the microneedle, the Raman light would undergo significant attenuation because of optical absorption or scattering in superficial layers. Because the introduction of SERS-active nanoparticles or structures into the body requires skillful injection or implantation, the easy administration and less pain [76] using the microneedles could make it useful in low-resource regions where experienced clinicians are not available. However, one significant disadvantage with stainless steel microneedles used in this study is the high cost [72,77], which may limit its potential use.

To overcome the disadvantage of high cost associated with the stainless steel microneedles and prevent the reuse of metal microneedles, Yuen and Liu [78] have developed an Ag-coated agarose needle/microneedle for skin measurements, which was demonstrated for intradermal measurements in a two-layered skin phantom. In this method, a 1-mL pipette tip was used as the mold for the agarose needle and an acupuncture needle was placed at the center of the pipette tip to create a lumen during the solidification of agarose gel. To achieve SERS measurements, the agarose needle was coated by an Ag layer inside the lumen with the Tollen method. Then the needle was cut by a blade to create a sharp tip. The lumen of the agarose needle made in this manner facilitated SERS measurements, as illustrated in Fig. 13.1A. The performance of the agarose needle was evaluated in a two-layered skin-mimicking phantom in the same way as in their previous paper [75]. The results demonstrated that glucose in the bottom layer of the skin phantom, which mimicked the dermis layer, could be detected through the agarose needle by using a 785-nm laser with a power of 5 mW and

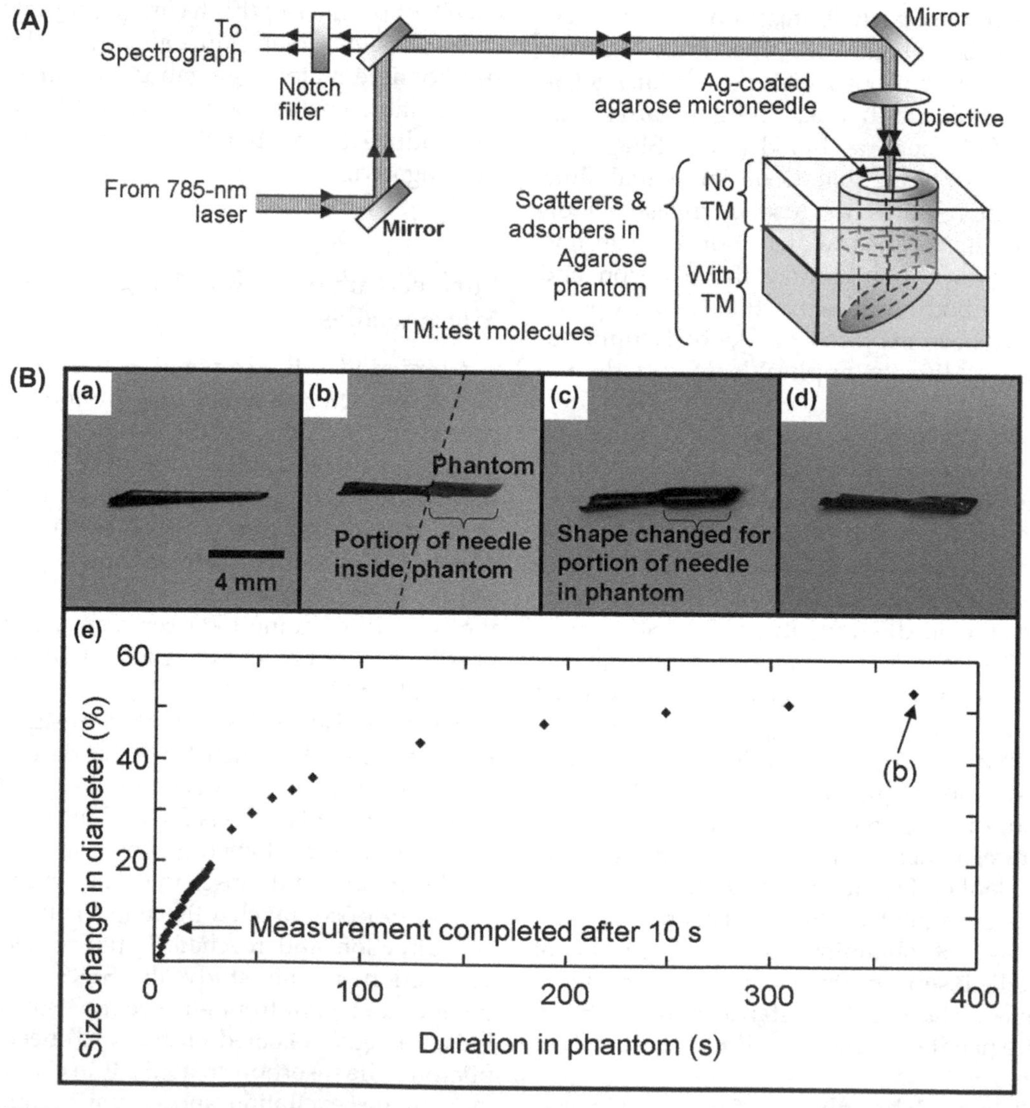

FIGURE 13.1 (A) Schematic of the Raman configuration for intradermal measurements in a skin-mimicking phantom using an agarose needle. (B) Size change of agarose needle after insertion in a phantom. The size of the agarose needle increased by 53.9% in 6 minutes after insertion inside the phantom. *Ag*, Silver. *Adapted from Yuen C, Liu Q. Hollow agarose microneedle with silver coating for intradermal surface-enhanced Raman measurements: a skin-mimicking phantom study. J Biomed Opt 2015; 20(6):061102, with written permission from the publisher.*

an exposure time of 10 seconds. Moreover, it was observed that the shape of the needle changed significantly after insertion in the phantom for a couple of minutes, as shown in Fig. 13.1B. This shape-changing phenomenon would ensure one-time use and prevent potential contamination caused by reuse. All these results suggested the high potential of this low-cost agarose needle for intradermal measurements in vivo.

Comparison Across All Methods

To facilitate the selection of an appropriate method for a particular intradermal application, Table 13.1 compares SERS methods in terms of sensing depth, sensitivity or accuracy, test model, and major advantage(s) and disadvantage(s). Because these data were obtained under various experimental configurations by different groups, the table is only intended to serve for the initial evaluation of these SERS methods.

Various nanoparticles or nanotags for injection allow flexibility in the surface modification of nanoparticles, which provides a variety of choices for optimal enhancement and characterization of biomolecules. However, light scattering and absorption of skin tissues can induce significant attenuation of light and result in low SERS signals and significant interference from superficial skin layers. Moreover, the potential toxicity of metal nanoparticles in the human body for this approach needs to be further evaluated. Nanostructure implantation has the potential to be used for continuous intradermal measurements and it has been tested in an animal model. Nevertheless it also faces an issue of biocompatibility. The biopatch approach is highly biocompatible to skin, but it is difficult to detect molecules deep in the skin in vivo. The AgNPs@SiNW approach is capable of detecting SERS from one single cell but the strength of the nanowire may be a problem for penetration into a deep skin layer.

Compared with the previously discussed approaches, the needle-based methods carry a much lower risk of leaving nanoparticles behind in the skin, because the needles stay in the skin only temporarily for sampling or SERS measurements and are then withdrawn. Among these methods, the SERS-active acupuncture needle approach achieves the deepest penetration in tissue phantoms. An acupuncture needle can reach a depth up to 10 mm, which would be effective for subcutaneous detection [79]. The sampling time in the acupuncture needle approach can be quite short to minimize the discomfort to patients. Although the depth profile in SERS has been demonstrated, the influence of blood present at small depths on blood sampled from large depths needs to be investigated to minimize cross-contamination.

For minimal invasiveness and convenient administration, Ag-coated microneedles would be an excellent choice. The microneedle can be made out of agarose, which is biocompatible, and the tip becomes blunt after use to reduce sharp waste and reuse risk. Another advantage of this approach that has not been explored in the literature is that the microneedle patch with an array of tips can be used for intradermal measurements in a large tissue area. This can be useful in applications requiring the estimation of concentration gradient of analytes. The penetration depth that can be achieved by the agarose needle may be limited in the skin compared with other methods because of the mechanical strength of the needle but it could increase as the fabrication procedure is improved. It should be highlighted that all the needle-based approaches can be combined with the aforesaid nanoparticles or nanotags to further improve signal enhancement.

FUTURE TRENDS OF INTRADERMAL SURFACE-ENHANCED RAMAN SPECTROSCOPY DEVELOPMENT

SERS has demonstrated great potential in intradermal measurements. Although the methods discussed have made much progress, there are several challenges to be addressed before this technology could be applied clinically.

First, continuous intradermal SERS measurement continues to be a challenge. Although the AgFON film has been implanted in an animal model and demonstrated for glucose measurements, it requires a glass window to expose the deep region of the skin. This would not be feasible to be used in humans. However, it is worth trying the implantation approach without the glass window to enable continuous monitoring in an animal model and eventually in the human body. A modern surgery technique could create a very small cut on the skin for superficial implantation [80], which can be useful in the long-term monitoring of multiple intradermal biochemical components.

Second, for those SERS techniques involving the injection of nanoparticles or implantation of SERS substrate without a glass window, the penetration of the laser light and the detection of Raman signals from a large target depth can be a challenge. The manipulation of light delivery and detection configuration can help achieve a larger sensing depth. For example, spatially offset Raman spectroscopy [81], in which the source and detector fibers are placed apart, has shown a sensing depth of 25 mm in muscle tissue. The cone-shell light delivery and detection geometry [82] has been demonstrated to minimize the contribution from the superficial layer of a two-layered tissue phantom in deep fluorescence measurements, which can be easily transferred to a Raman measurement system.

TABLE 13.1 Comparison of the Current Methods Suitable for Intradermal SERS Measurements in Terms of Sensing Depth, Advantages and Disadvantages, Test Models, and Reported Sensitivity/Accuracy

Category of methods	Sensing depth range	Advantages for intradermal measurements	Disadvantages for intradermal measurements	Tested models	Sensitivity/accuracy
Injection of nanoparticles [51,57,60,61,79]	Up to 8 mm [61]	**1.** Minimally invasive compared with biopsy **2.** Capable of characterizing multiple biomolecules	**1.** Nanoparticle aggregation inside human body may cause biohazard **2.** Raman measurements are highly affected by light scattering and absorption properties of skin tissues in deep measurements	Animal models [51,57,60,61,79]	Not available
SERS-active structures [66–68]	Subcutaneous measurements feasible for AgFON [68]	**1.** Capable of continuous real time monitoring (for the AgNPs@SiNW and AgFON approaches) **2.** Noninvasive (for the biopatch approach) or minimally invasive (for the AgNPs@ SiNW approach)	**1.** The approaches of AgNPs@ SiNW and biopatch have not been tested in an animal model or human study **2.** The approaches of AgFON film and biopatch require invasive treatment when measuring biomolecules inside the skin **3.** AgNPs@SiNW may be fragile when the nanowire is long	Phantoms for AgNPs@ SiNW [66] and biopatch approaches [67]; animal model for AgFON film approach [68]	AgNPs@SiNW: capable of detecting a pH change of 0.5 in the range from 4.0 to 9.0 [66] Biopatch: tested for 10 mM R6G, 200 mM γ-aminobutyric acid and 1.2 mM β-amyloid [67] AgFON: tested for glucose at 38.9 mM [68]
SERS-active acupuncture needle [70,71]	Up to 7 mm [70]	**1.** Capable of reaching deep layers **2.** Light attenuation caused by tissue scattering and absorption is circumvented by ex vivo measurements **3.** Limited pain and low risk of infection compared with hypodermic needles **4.** The measurement of SERS depth profile is possible	**1.** Acupuncture needles can still cause pain and discomfort **2.** Metal needles face the risks of sharp waste injury and potential contamination caused by reuse	Phantom and animal models [70,71]	1×10^{-6} M for 6-mercaptopurine [70] and 2.8 mM for glucose [71]
SERS-active microneedle [75,78]	0.7 mm [75]	**1.** Nearly no pain and minimally invasive **2.** Nonreusability and size changing after use prevent sharp injury and risk of cross-contamination (for agarose needles) **3.** Cost-effective (agarose needles)	**1.** Limited working depth **2.** High cost and risk of reuse (for stainless steel microneedles) **3.** The strength of agarose needle needs to be tested systematically on human skin	Skin-mimicking phantoms [75,78]	Silver coated stainless steel microneedle: 1×10^{-6} M for R6G and RMSE of 3.1 mM in glucose measurements in a range from 0 to 140 mM [75] Silver-coated agarose needle: 1×10^{-6} M for crystal violet and RMSE of 5.1 mM in glucose measurements in a range from 0 to 140 mM [78]

AgFON, silver—film over nanosphere; *AgNP*, silver nanoparticle; *R6G*, rhodamine 6G; *RMSE*, root mean square error, *SERS*, surface-enhanced Raman spectroscopy; *SiNW*, silicon nanowire.

Third, the focus of intradermal SERS is mostly limited to biomolecule detection, such as glucose, sequentially at individual spots. The biomolecular information at a single spot does not provide the entire view of the skin status that is important in applications such as skin lesion boundary assessment. One solution to overcome this limitation is to achieve SERS imaging in human skin. SERS imaging can be implemented either in the form of wide-field Raman imaging or with the help of a microlens array, each of which has its own pros and cons. Whereas the approach of wide-field Raman imaging [83] offers high spatial resolution but is limited in the overall area of examination, the approach using the microlens array [84,85] provides greater excitation power density at each focal spot but there is a large separation between each pair of adjacent focal spots. The SERS imaging technique [86,87] can be used to evaluate the boundary of skin lesions to facilitate the diagnosis or guide the surgery.

Fourth, in vivo human measurements have not yet been achieved. All experiments reported in this chapter were performed on either tissue phantoms or animal models. To fill this gap, some important issues such as the potential toxicity of nanoparticles or structures and laser safety in humans need to be fully addressed, for which more systematic studies on nanotoxicology are needed. There are a handful number of nanoparticles that have been approved for specific applications [88]. Investigation of these nanoparticles rather than others for intradermal measurements may facilitate the future approval.

Fifth, time-consuming and labor-intensive sample preparation has limited the efficiency of ex vivo SERS measurements. Most sample preparation steps could be implemented in microfluidic chips [89]. Sample extraction and enrichment for body fluid extracted from human skin can be integrated in the chips for SERS measurements [90]. This technology cannot only reduce labor and time for sample preparation but also improve the reproducibility of results.

Finally, the equipment for typical Raman or SERS measurements is relatively expensive compared with fluorescence or reflectance measurements. Significant effort has been made in the optics community to develop cost-effective Raman spectrometer [91]. This is especially practical for SERS measurements, given that the signal intensity in SERS measurements is comparable to that of fluorescence but requires a higher spectral resolution. It has been shown feasible to reconstruct Raman spectra with a high spectral resolution from Raman measurements with a low spectral resolution [92]. Therefore it is possible to combine a cost-effective spectrometer whose spectral resolution is low and the method of spectral reconstruction for SERS measurements. This approach can be extended to SERS imaging, for which narrow-band Raman measurements can be used to further enhance SERS signals [93]. Additionally, progress has been achieved in cost-effective Raman substrates [94]. All these new avenues will help speed up the transfer of SERS technology from the benchtop to beside in the near future.

CONCLUSION

SERS enabled by noble metal nanoparticles or nanostructures has demonstrated many advantages suitable for biological sample measurements, such as high sensitivity, large penetration depth, precise molecular characterization, and multiplexing capability. Various strategies, including direct injection, implantation, biopatches, acupuncture needles and microneedles, have been explored to bring the nanoparticles/nanostructures to close proximity with the test molecules in skin phantoms or animal models for SERS measurements, as reviewed in this chapter. These results have demonstrated the excellent potential of SERS for intradermal measurements in human skin.

Although the development of SERS for intradermal measurements has made significant progress, there are several challenges to be overcome before this technology could be applied clinically because of the following issues: the biocompatibility of metal nanoparticles, slow point measurements, lack of means for continuous measurements, labor-intensive sample preparation, and high cost of Raman equipment. Most of these issues can be addressed by the existing complementary methods, but these methods need to be specifically developed for intradermal measurements. With the continuous advances in Raman instrumentation, nontoxic nanoscale substrates and a variety of strategies for bringing the substrates and target molecules together, it can be anticipated that SERS will eventually play an important role in intradermal measurements.

List of Acronyms and Abbreviations

4-MBA 4-Mercaptobezoic acid
AgFON Silver—film over nanosphere
AgNP Silver nanoparticle
CE Chemical enhancement
CLSM Confocal laser scanning microscopy
EM Electromagnetic
LSPR Localized surface plasmon resonance
NIR Near-infrared
OCT Optical coherence tomography
pMBA *p*-Mercaptobenzoic acid
R6G Rhodamine 6G
RM Reporter molecule
SERS Surface-enhanced Raman spectroscopy
SiNW Silicon nanowire

References

[1] Habif TP, Chapman MS, Campbell JL, Dinulos JGH, Zug KA. Skin disease: diagnosis and treatment. Elsevier Health Sciences; 2011. p. 360–81.
[2] Soyer HP, Argenziano G, Hofmann-Wellenhof R, Zalaudek I. Dermoscopy: the essentials. Elsevier Health Sciences; 2011. p. 3–25.
[3] Vestergaard ME, Macaskill P, Holt PE, Menzies SW. Dermoscopy compared with naked eye examination for the diagnosis of primary melanoma: a meta-analysis of studies performed in a clinical setting. Br J Dermatol 2008;159(3):669–76.
[4] Binder M, PuespoeckSchwarz M, Steiner A, et al. Epiluminescence microscopy of small pigmented skin lesions: short-term formal training improves the diagnostic performance of dermatologists. J Am Acad Dermatol 1997;36(2):197–202.
[5] Orchard G, Nation B. Histopathology. Oxford University Press; 2011. p. 6–29.
[6] Drexler W, Fujimoto JG. Optical coherence tomography: technology and applications. Springer Science & Business Media; 2008. p. 73–113.
[7] Fruhstorfer H, Abel U, Garthe CD, Knuttel A. Thickness of the stratum corneum of the volar fingertips. Clin Anat 2000;13(6): 429–33.
[8] Weissman J, Hancewicz T, Kaplan P. Optical coherence tomography of skin for measurement of epidermal thickness by shapelet-based image analysis. Opt Express 2004;12(23):5760–9.
[9] Welzel J, Lankenau E, Birngruber R, Engelhardt R. Optical coherence tomography of the human skin. J Am Acad Dermatol 1997; 37(6):958–63.
[10] Bechara FG, Gambichler T, Stucker M, et al. Histomorphologic correlation with routine histology and optical coherence tomography. Skin Res Technol 2004;10(3):169–73.
[11] Gambichler T, Hyun J, Moussa G, et al. Optical coherence tomography of cutaneous lupus erythematosus correlates with histopathology. Lupus 2007;16(1):35–8.
[12] Ulrich M, Lange-Asschenfeldt S. In vivo confocal microscopy in dermatology: from research to clinical application. J Biomed Opt 2013;18(6):061212.
[13] Pawley J. Handbook of biological confocal microscopy. Springer Science & Business Media; 2010. p. 141–50.
[14] Török P, Kao FJ. Optical imaging and microscopy: techniques and advanced systems. Springer Science & Business Media; 2003. p. 3–19.
[15] Payette JR, Kohlenberg E, Leonardi L, et al. Assessment of skin flaps using optically based methods for measuring blood flow and oxygenation. Plast Reconstr Surg 2005;115(2):539–46.
[16] Zhu C, Chen S, Chui CH-K, Tan B-K, Liu Q. Early prediction of skin viability using visible diffuse reflectance spectroscopy and autofluorescence spectroscopy. Plast Reconstr Surg 2014;134(2): 240E–7E.
[17] Stelzle F, Tangermann-Gerk K, Adler W, et al. Diffuse reflectance spectroscopy for optical soft tissue differentiation as remote feedback control for tissue-specific laser surgery. Lasers Surg Med 2010;42(4):319–25.
[18] Stelzle F, Adler W, Zam A, et al. In vivo optical tissue differentiation by diffuse reflectance spectroscopy: preliminary results for tissue-specific laser surgery. Surg Innov 2012;19(4):385–93.
[19] Bensalah K, Peswani D, Tuncel A, et al. Optical reflectance spectroscopy to differentiate benign from malignant renal tumors at surgery. Urology 2009;73(1):178–81.
[20] Sauer M, Hofkens J, Enderlein J. Handbook of fluorescence spectroscopy and imaging: from ensemble to single molecules. John Wiley & Sons; 2010. p. 90–115.
[21] Stender RNI, Wulf HC. Can autofluorescence demarcate basal cell carcinoma from normal skin? A comparison with protoporphyrin IX fluorescence. Acta Derm Venereol 2001;81(4):246–9.
[22] Brancaleon L, Durkin AJ, Tu JH, Menaker G, Fallon JD, Kollias N. In vivo fluorescence spectroscopy of nonmelanoma skin cancer. Photochem Photobiol 2001;73(2):178–83.
[23] Borisova EG, Troyanova PP, Avramov LA. Fluorescence spectroscopy for early detection and differentiation of cutaneous pigmented lesions. Optoelectron Adv Mat Rapid Commun 2007; 1(8):388–93.
[24] Koetsier M, Nur E, Han C, et al. Skin color independent assessment of aging using skin autofluorescence. Opt Express 2010; 18(14):14416–29.
[25] Mokry M, Gal P, Vidinsky B, et al. In vivo monitoring the changes of interstitial pH and FAD/NADH ratio by fluorescence spectroscopy in healing skin wounds. Photochem Photobiol 2006;82(3): 793–7.
[26] Ghomi M. Applications of Raman spectroscopy to biology: from basic studies to disease diagnosis. IOS Press; 2012. p. 1–31.
[27] Talari ACS, Movasaghi Z, Rehman S, Rehman IU. Raman spectroscopy of biological tissues. Appl Spectrosc Rev 2015;50(1):46–111.
[28] Fleischmann M, Hendra PJ, McQuilla AJ. Raman-spectra of pyridine adsorbed at a silver electrode. Chem Phys Lett 1974;26(2): 163–6.
[29] Kneipp K, Kneipp H, Itzkan I, Dasari RR, Feld MS. Ultrasensitive chemical analysis by Raman spectroscopy. Chem Rev 1999;99(10): 2957–76.
[30] Yuen C, Zheng W, Huang Z. Surface-enhanced Raman scattering: principles, nanostructures, fabrications, and biomedical applications. J Innov Opt Health Sci 2008;1(2):267–84.
[31] Kambhampati P, Child CM, Foster MC, Campion A. On the chemical mechanism of surface enhanced Raman scattering: experiment and theory. J Chem Phys 1998;108(12):5013–26.
[32] Nima ZA, Biswas A, Bayer IS, et al. Applications of surface-enhanced Raman scattering in advanced bio-medical technologies and diagnostics. Drug Metab Rev 2014;46(2):155–75.
[33] Chaney SB, Shanmukh S, Dluhy RA, Zhao YP. Aligned silver nanorod arrays produce high sensitivity surface-enhanced Raman spectroscopy substrates. Appl Phys Lett 2005;87(3):31908–10.
[34] Li K, Clime L, Cui B, Veres T. Surface enhanced Raman scattering on long-range ordered noble-metal nanocrescent arrays. Nanotechnology 2008;19(14):145305.
[35] Chandran SP, Chaudhary M, Pasricha R, Ahmad A, Sastry M. Synthesis of gold nanotriangles and silver nanoparticles using aloe vera plant extract. Biotechnol Prog 2006;22(2):577–83.
[36] Xie J, Zhang Q, Lee JY, Wang DIC. The synthesis of SERS-active gold nanoflower tags for in vivo applications. ACS Nano 2008; 2(12):2473–80.
[37] Camargo PHC, Rycenga M, Au L, Xia Y. Isolating and probing the hot spot formed between two silver nanocubes. Angew Chem Int Ed 2009;48(12):2180–4.
[38] Lin T-H, Linn NC, Tarajano L, Jiang B, Jiang P. Electrochemical SERS at periodic metallic nanopyramid arrays. J Phys Chem C 2009;113(4):1367–72.
[39] Moskovits M. Surface-enhanced spectroscopy. Rev Mod Phys 1985;57(3):783–826.
[40] Garrell RL. Surface-enhanced Raman-spectroscopy. Anal Chem 1989;61(6):401A–11A.
[41] Le Ru EC, Etchegoin PG. Rigorous justification of the $|E|^4$ enhancement factor in surface enhanced Raman spectroscopy. Chem Phys Lett 2006;423(1–3):63–6.
[42] Cialla D, Maerz A, Boehme R, et al. Surface-enhanced Raman spectroscopy (SERS): progress and trends. Anal Bioanal Chem 2012;403(1):27–54.
[43] McCall SL, Platzman PM, Wolff PA. Surface enhanced Raman-scattering. Phys Lett A 1980;77(5):381–3.
[44] Otto A, Billmann J, Eickmans J, Erturk U, Pettenkofer C. The adatom model of SERS (surface enhanced Raman-scattering): the present status. Surf Sci 1984;138(2–3):319–38.

[45] Zhao LL, Jensen L, Schatz GC. Pyridine-Ag-20 cluster: a model system for studying surface-enhanced Raman scattering. J Am Chem Soc 2006;128(9):2911–9.
[46] Persson BNJ. On the theory of surface-enhanced Raman-scattering. Chem Phys Lett 1981;82(3):561–5.
[47] Lombardi JR, Birke RL, Lu TH, Xu J. Charge-transfer theory of surface enhanced Raman spectroscopy: Herzberg-Teller contributions. J Chem Phys 1986;84(8):4174–80.
[48] Alexson DA, Badescu SC, Glembocki OJ, Prokes SM, Rendell RW. Metal-adsorbate hybridized electronic states and their impact on surface enhanced Raman scattering. Chem Phys Lett 2009; 477(1–3):144–9.
[49] Kwon MJ, Lee J, Wark AW, Lee HJ. Nanoparticle-enhanced surface plasmon resonance detection of proteins at attomolar concentrations: comparing different nanoparticle shapes and sizes. Anal Chem 2012;84(3):1702–7.
[50] Li L, Hutter T, Steiner U, Mahajan S. Single molecule SERS and detection of biomolecules with a single gold nanoparticle on a mirror junction. Analyst 2013;138(16):4574–8.
[51] Falamas A, Dehelean C, Pinzaru SC. Raman and SERS characterization of normal pathological skin. Stud Univ Babes-Bol Chem 2011;56(4):89–96.
[52] Hsu P-H, Chiang HK. Surface-enhanced Raman spectroscopy for quantitative measurement of lactic acid at physiological concentration in human serum. J Raman Spectrosc 2010;41(12): 1610–4.
[53] Yin WZ, Guo ZY, Zhuang ZF, Liu SH, Xiong K, Chen SJ. Application of surface-enhanced Raman in skin cancer by plasma. Laser Phys 2012;22(5):996–1001.
[54] Maiti KK, Dinish US, Fu CY, et al. Development of biocompatible SERS nanotag with increased stability by chemisorption of reporter molecule for in vivo cancer detection. Biosens Bioelectron 2010;26(2):398–403.
[55] Qian XM, Peng XH, Ansari DO, et al. In vivo tumor targeting and spectroscopic detection with surface-enhanced Raman nanoparticle tags. Nat Biotechnol 2008;26(1):83–90.
[56] Kuestner B, Gellner M, Schuetz M, et al. SERS labels for red laser excitation: silica-encapsulated SAMs on tunable gold/silver nanoshells. Angew Chem Int Ed 2009;48(11):1950–3.
[57] Samanta A, Maiti KK, Soh K-S, et al. Ultrasensitive near-infrared Raman reporters for SERS-based in vivo cancer detection. Angew Chem Int Ed 2011;50(27):6089–92.
[58] Dinish US, Balasundaram G, Chang Y-T, Olivo M. Actively targeted in vivo multiplex detection of intrinsic cancer biomarkers using biocompatible SERS nanotags. Sci Rep 2014;4:4075.
[59] Wang X, Qian XM, Beitler JJ, et al. Detection of circulating tumor cells in human peripheral blood using surface-enhanced Raman scattering nanoparticles. Cancer Res 2011;71(5):1526–32.
[60] Maiti KK, Dinish US, Samanta A, et al. Multiplex targeted in vivo cancer detection using sensitive near-infrared SERS nanotags. Nano Today 2012;7(2):85–93.
[61] Kang H, Jeong S, Park Y, et al. Near-infrared SERS nanoprobes with plasmonic Au/Ag hollow-shell assemblies for in vivo multiplex detection. Adv Funct Mater 2013;23(30):3719–27.
[62] McAughtrie S, Faulds K, Graham D. Surface enhanced Raman spectroscopy (SERS): potential applications for disease detection and treatment. J Photochem Photobiol C Photochem Rev 2014; 21:40–53.
[63] Asharani PV, Wu YL, Gong ZY, Valiyaveettil S. Toxicity of silver nanoparticles in zebrafish models. Nanotechnology 2008;19(25): 255102.
[64] Bar-Ilan O, Albrecht RM, Fako VE, Furgeson DY. Toxicity assessments of multisized gold and silver nanoparticles in zebrafish embryos. Small 2009;5(16):1897–910.
[65] Wang Y, Seebald JL, Szeto DP, Irudayaraj J. Biocompatibility and biodistribution of surface-enhanced Raman scattering nanoprobes in zebrafish embryos: in vivo and multiplex imaging. ACS Nano 2010;4(7):4039–53.
[66] Han XM, Wang H, Ou XM, Zhang XH. Silicon nanowire-based surface-enhanced Raman spectroscopy endoscope for intracellular pH detection. ACS Appl Mater Inter 2013;5(12):5811–4.
[67] Park SG, Ahn MS, Oh YJ, Kang M, Jeong Y, Jeong KH. Nanoplasmonic biopatch for in vivo surface enhanced Raman spectroscopy. Biochip J 2014;8(4):289–94.
[68] Stuart DA, Yuen JM, Shah N, et al. In vivo glucose measurement by surface-enhanced Raman spectroscopy. Anal Chem 2006; 78(20):7211–5.
[69] Vuckovic D, de Lannoy I, Gien B, et al. In vivo solid-phase microextraction: capturing the elusive portion of metabolome. Angew Chem 2011;123(23):5456–60.
[70] Dong J, Chen Q, Rong C, Li D, Rao Y. Minimally invasive surface-enhanced Raman scattering detection with depth profiles based on a surface-enhanced Raman scattering-active acupuncture needle. Anal Chem 2011;83(16):6191–5.
[71] Dong J, Tao Q, Guo MD, Yan TY, Qian WP. Glucose-responsive multifunctional acupuncture needle: a universal SERS detection strategy of small biomolecules in vivo. Anal Methods 2012;4(11): 3879–83.
[72] Allen MG, Prausnitz MR, McAllister DV, Cros FPM. Microneedle devices and methods of manufacture and use thereof. US Patent No. US6334856 B1. 2002.
[73] Henry S, McAllister DV, Allen MG, Prausnitz MR. Microfabricated microneedles: a novel approach to transdermal drug delivery. J Pharm Sci 1998;87(8):922–5.
[74] Van Damme P, Oosterhuis-Kafeja F, Van der Wielen M, Almagor Y, Sharon O, Levin Y. Safety and efficacy of a novel microneedle device for dose sparing intradermal influenza vaccination in healthy adults. Vaccine 2009;27(3):454–9.
[75] Yuen C, Liu Q. Towards in vivo intradermal surface enhanced Raman scattering (SERS) measurements: silver coated microneedle based SERS probe. J Biophotonics 2014;7(9):683–9.
[76] Kim Y, Park J, Prausnitz MR. Microneedles for drug and vaccine delivery. Adv Drug Deliv Rev 2012;64(14):1547–68.
[77] Sherman FF, Yuzhakov VV, Gartstein V, Owens GD. Apparatus and method for manufacturing an intracutaneous microneedle array. US Patent No. US6312612 B1. 2001.
[78] Yuen C, Liu Q. Hollow agarose microneedle with silver coating for intradermal surface-enhanced Raman measurements: a skin-mimicking phantom study. J Biomed Opt 2015;20(6):061102.
[79] Habif TP. Clinical dermatology: a color guide to diagnosis and therapy. Mosby; 2004. p. 23–55.
[80] Böcker D, Fruhstorfer H. Cutting device for skin for obtaining small blood samples in almost pain-free manner. US Patent No. US6210421 B1. 2001.
[81] Stone N, Faulds K, Graham D, Matousek P. Prospects of deep Raman spectroscopy for noninvasive detection of conjugated surface enhanced resonance Raman scattering nanoparticles buried within 25 mm of mammalian tissue. Anal Chem 2010;82(10): 3969–73.
[82] Ong YH, Liu Q. Fast depth-sensitive fluorescence measurements in turbid media using cone shell configuration. J Biomed Opt 2013;18(11):110503.
[83] Schlücker S, Schaeberle MD, Huffman SW, Levin IW. Raman microspectroscopy: a comparison of point, line, and wide-field imaging methodologies. Anal Chem 2003;75(16):4312–8.
[84] Fujita K, Nakamura O, Kaneko T, Kawata S, Oyamada M, Takamatsu T. Real-time imaging of two-photon–induced fluorescence with a microlens-array scanner and a regenerative amplifier. J Microsc 1999;194(2–3):528–31.
[85] Zhu C, Ong YH, Liu Q. Multifocal noncontact color imaging for depth-sensitive fluorescence measurements of epithelial cancer. Opt Lett 2014;39(11):3250–3.

[86] Mallia RJ, McVeigh PZ, Veilleux I, Wilson BC. Filter-based method for background removal in high-sensitivity wide-field-surface-enhanced Raman scattering imaging in vivo. J Biomed Opt 2012; 17(7):0760171–5.

[87] McVeigh PZ, Mallia RJ, Veilleux I, Wilson BC. Widefield quantitative multiplex surface enhanced Raman scattering imaging in vivo. J Biomed Opt 2013;18(4):046011.

[88] Dreaden EC, Alkilany AM, Huang X, Murphy CJ, El-Sayed MA. The golden age: gold nanoparticles for biomedicine. Chem Soc Rev 2012;41(7):2740–79.

[89] Chen L, Choo J. Recent advances in surface-enhanced Raman scattering detection technology for microfluidic chips. Electrophoresis 2008;29(9):1815–28.

[90] März A, Mönch B, Rösch P, Kiehntopf M, Henkel T, Popp J. Detection of thiopurine methyltransferase activity in lysed red blood cells by means of lab-on-a-chip surface enhanced Raman spectroscopy (LOC-SERS). Anal Bioanal Chem 2011;400(9):2755–61.

[91] Malinen J, Rissanen A, Saari H, et al. Advances in miniature spectrometer and sensor development. In: Next-generation spectroscopic technologies VII. Proc. SPIE 9101; 2014. 91010C.

[92] Chen S, Lin X, Yuen C, Padmanabhan S, Beuerman RW, Liu Q. Recovery of Raman spectra with low signal-to-noise ratio using Wiener estimation. Opt Express 2014;22(10):12102–14.

[93] Chen S, Ong YH, Liu Q. Fast reconstruction of Raman spectra from narrow-band measurements based on Wiener estimation. J Raman Spectrosc 2013;44(6):875–81.

[94] Han Y-A, Ju J, Yoon Y, Kim S-M. Fabrication of cost-effective surface enhanced Raman spectroscopy substrate using glancing angle deposition for the detection of urea in body fluid. J Nanosci Nanotechnol 2014;14(5):3797–9.

CHAPTER

14

Broadband Coherent Anti-Stokes Raman Scattering

C.H. Camp, Jr.

National Institute of Standards and Technology, Gaithersburg, MD, United States

OUTLINE

INTRODUCTION

Molecules vibrate. Within each molecule individual atoms are constantly in motion. In fact, the motion between bonded atoms and small groups of linked atoms is oscillatory. The exact frequency (energy) of the vibrations is determined by such factors as the atoms involved, molecular orientation, and temperature. In addition, these vibrations occur in "modes" with descriptive names such as "stretching," "twisting," "breathing," and "rocking." Of significant consequence is that the particular vibrational modes between particular sets of atoms occur at predictable frequencies (or over small frequency ranges); thus molecules/chemicals can be characterized by their vibrational fingerprint. Furthermore, with an imaging technique that could interrogate these vibrational modes, one could visualize the molecular content within materials, cells, and tissues, extracting cell state and functional information that could, for example, be utilized to diagnose pathologies before morphological features are even present. Broadband coherent anti-Stokes Raman scattering (BCARS) microspectroscopy is one such method of interrogating these vibrational fingerprints with success in imaging cells, tissues, pharmaceutical tablets, and polymers. However, for all of the proposed promise and more than a decade in development, only recently has BCARS truly demonstrated high-quality imaging and spectroscopy with an unparalleled combination of speed, spectral clarity, and range of addressable vibrational frequencies. In this chapter an overview of BCARS technology and its theoretical underpinnings will be provided. In addition, demonstrative examples of tissue imaging will be presented.

BACKGROUND

Analytical methods of measuring molecular vibrations are not new, with commercial systems available

Imaging in Dermatology
http://dx.doi.org/10.1016/B978-0-12-802838-4.00014-5

for more than 50 years. The two classic methods of vibrational spectroscopy are infrared (IR) absorption spectroscopy and Raman scattering spectroscopy. At room temperature, most molecular vibrational modes occur at frequencies within the IR region of the electromagnetic spectrum. If a sample is illuminated with IR light, photons at the particular frequencies (wavelengths) corresponding to an existing vibrational mode may be absorbed. By comparing the incident spectrum with the spectrum transmitted through a sample, one may determine the vibrational spectrum—thus the basis of IR absorption spectroscopy. The absorption of IR radiation was first explored in the late 1800s and early 1900s [1,2], IR microscopy was commercialized in the 1950s, and today modern methods such as Fourier transform IR spectroscopy are still in common use.

An alternative method of exploring the vibrational structure of molecules is through Raman scattering spectroscopy, which was first explored in the 1920s [3–6]. The Raman effect (or scattering), named for the pioneering Indian physicist and Nobel laureate Sir Chandrasekhara Venkata (C.V.) Raman, is an inelastic scattering event in which incident photons interact with molecular vibrations, gaining or losing the amount of energy of the vibrational mode (see Fig. 14.1A). Thus in Raman scattering spectroscopy one observes light at new wavelengths from those incident on the sample.

These classic techniques, although in use today and commercially available, have significant technical challenges that limit their ubiquity. For example, IR absorption spectroscopy provides coarse spatial resolution in the range of multiple micrometers because of the diffraction limit of IR light. In addition, as an absorptive technique this method requires a transmissive geometry unless the sample is highly scattering (ie, there is significant back-reflection), and it does not provide a direct route to three-dimensional imaging. Moreover, the IR absorption of water often necessitates sample dehydration for the highest sensitivity of cellular constituents. On the other hand, Raman spectroscopy, as an emissive technique, can provide three-dimensional images in a transmissive or reflective geometry and may be performed at visible wavelengths; thus spatial resolution can be comparable to that of other standard light microscopies. However, one major challenge is autofluorescence, which may obscure the Raman spectrum. In addition, Raman scattering is a rare phenomenon, affecting approximately one in every tens of millions of incident photons; thus a single spectrum may require tens of milliseconds to seconds to acquire. Even a small 100×100 pixel image at 100-ms dwell times would require more than 15 min.

Although Raman microspectroscopy systems are continually improving, the inherent improbability of the scattering events and the limitations on the amount of incident illumination that is tolerable to cells and tissues significantly restrict the ultimate possibility of a high-throughput imaging platform. An alternative mechanism of inelastic scatter from vibration modes is through coherent Raman scattering (CRS). The aforementioned Raman effect/scattering is a spontaneous process in which minute quantum thermodynamic fluctuations dictate which vibrational modes are accessible in a stochastic fashion and thus the rarity of a scattering event occurring [7–9]. In CRS, pulsed lasers are utilized to actively excite molecular vibrations; thus enhancing the probability of another incident photon inelastically scattering [10]. In addition, the efficiency of collecting the scattered photons is dramatically enhanced as emitted photons are in a directed beam [10,11] whereas in traditional Raman scattering the emission is typically isotropic; thus even with high numerical aperture (NA) lenses the collection efficiency may be significantly less than 20%. Finally, CRS is a nonlinear process in which the intensity of the scattered light is quadratically related to the analyte concentration and cubically related to the incident light intensity (spontaneous Raman scattering is linear in both concentration and incident intensity).

The most prevalent CRS mechanism utilized is coherent anti-Stokes Raman scattering (CARS). The CARS mechanism, as described in Fig. 14.1B, occurs when a "pump" photon and a "Stokes" photon beat at a frequency that corresponds to that of a molecular vibration, which stimulates the vibrational mode [11–15]. A "probe" photon inelastically scatters off of the excited mode, gaining the energy of that mode. In both CRS and spontaneous Raman scattering, the probe photon is capable of losing or gaining the energy of the vibrational mode—"Stokes" and "anti-Stokes" scattering, respectively. In the spontaneous Raman scattering process at physiological temperatures, the probability of Stokes Raman scattering is orders of magnitude more likely than anti-Stokes scattering; thus the term Raman spectroscopy almost always implies Stokes scattering rather than anti-Stokes. However, in CRS Stokes and anti-Stokes scattering can both be readily achieved. The advantage to anti-Stokes scattering is that the gain of energy in the scattered photon equates to a blue-shift in wavelength; thus there is no overlap with autofluorescence, which is red-shifted.

The CARS mechanism was first described by Maker and Terhune [16] of the Ford Motor Company in 1965 (ironically, the abbreviation *CARS* did not originate until nearly a decade later) [17]. Although there was significant interest in CARS spectroscopy at the time, the first CARS microscope was not described until 1982 [18]. The CARS microscope, which is often shortened to just "CARS" (there is some ambiguity in terminology

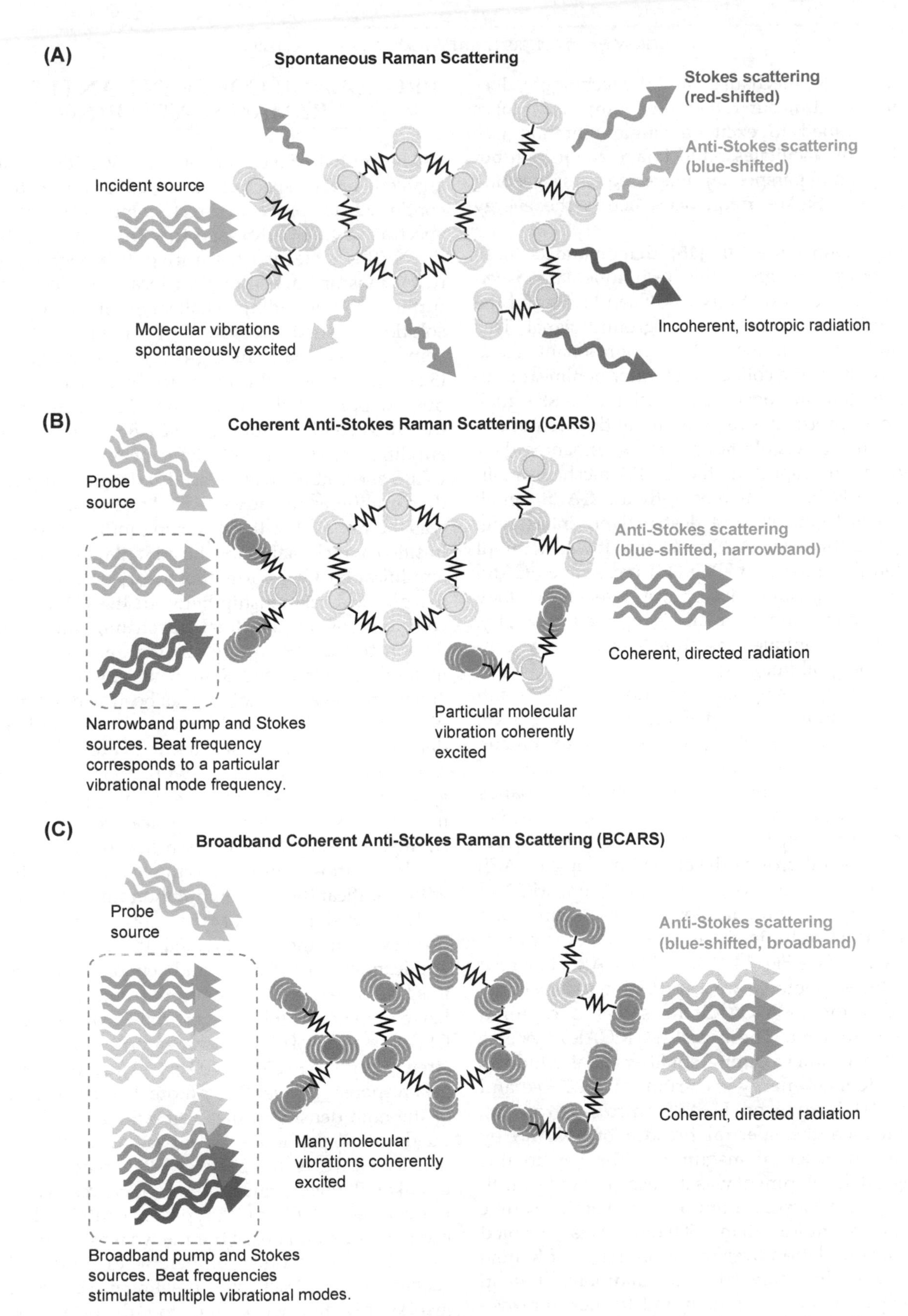

FIGURE 14.1 (A) In Raman scattering, incident photons interact with spontaneously excited molecular vibrations, inelastically scattering with increased or decreased energy (frequency). (B) In coherent anti-Stokes Raman scattering (CARS), excitation sources (pump and Stokes) are tuned so that their frequency difference corresponds to the frequency of a particular molecular vibration, thus stimulating the vibrational mode. Probe photons inelastically scatter off of the excited vibration with higher energy (frequency). (C) In broadband coherent anti-Stokes Raman scattering (BCARS), similar to CARS, lasers coherently excite vibrational modes. However, in BCARS broadband sources are used to simultaneously excite multiple molecular vibrations.

between the CARS mechanism and the technique), does not acquire spectra, but rather the pump and Stokes sources are tuned to excite a single vibration (see Fig. 14.1B). This facilitates rapid image acquisition but recording a full hyperspectral image requires sweeping the pump and Stokes frequencies across the energy spectrum.

Although Duncan et al. [18] demonstrated high-speed, label-free imaging, this technique languished because of limited availability of pulsed lasers and the overwhelming existence of a background signal. This background, later termed the "nonresonant background" (NRB), is the collective of other nonlinear optical processes that are cogenerated with the CARS signal. Because these processes are predominantly electronic in nature, the use of visible lasers greatly enhances their generation with respect to the CARS mechanism. In addition, the NRB is coherent with the CARS signal; thus it is constructively and destructively interfering, distorting the effective CARS spectrum. It was not until 1999, when Zumbusch and Xie [13] reexamined CARS using newly available Ti:Sapphire laser technology with IR radiation and a greatly simplified geometry, that the CARS technique took off as a microscopy technique for biological imaging.

Although progressing as a technology, CARS still faced two fundamental challenges: (1) the NRB, although reduced by IR lasers, still distorted the Raman spectral landscape and (2) the limited tunability of lasers limited the range of Raman vibrational bands that were practically accessible. To address the second challenge and to transform CARS into a spectroscopic techniques, several groups developed multiplex CARS (MCARS) microspectroscopy [19,20]. With MCARS techniques the pump and/or Stokes source are broadband; thus multiple vibrational bands are simultaneously excited (see Fig. 14.1C). Unlike CARS, in which the anti-Stokes photons are captured with a single-element detector, in MCARS the signal is captured with a spectrometer. These earliest MCARS systems were limited to interrogating small spectral windows within the Raman energy spectrum. In 2004 Kee and Cicerone [21] introduced BCARS microspectroscopy, which extended the spectral breadth of MCARS by more than an order of magnitude. The key to this technological development was the use of a nonlinear fiber to generate a "supercontinuum" (SC) Stokes source that spanned more than 600 nm. This enabled vibrational stimulation over the entire range of Raman energies typically analyzed with traditional Raman spectroscopy. This development and further improvements in nonlinear fibers for SC generation facilitated hyperspectral imaging of materials [22,23]; cells [24–28]; and more recently, tissues [29–31].

BROADBAND COHERENT ANTI-STOKES RAMAN SCATTERING

The initial developments in BCARS technology enabled broad spectral interrogation, but the NRB continued to prevent direct comparison of Raman spectra with BCARS spectra. In addition, because the NRB is molecularly sensitive (to a degree), spectral distortions are dependent on local sample conditions, further compounding challenges in analysis. One solution pursued across the CARS and MCARS/BCARS communities was reducing the generation of the NRB [32–37]. However, this approach had an acknowledged but unappreciated ramification. Although disruptive, the NRB actually amplifies the CARS signal (heterodyne amplification) [38]. By attacking the NRB generation, the CARS spectral intensity was also dramatically reduced, thus nullifying any advantage. An alternative approach was the removal of the distorting influence of the NRB in silico, which maintains the advantages of heterodyne amplification. On information theory grounds and using the physical relationship between the CARS and NRB contributions to the total spectrum, Vartiainen [39] demonstrated the utility of the maximum entropy method to extract the Raman vibrational signal. This "phase retrieval" technique has been widely applied to biomolecules [40,41] and biological imaging [26,29,31], demonstrating CARS spectra that are comparable in shape to those measured with spontaneous Raman spectroscopy. Later an alternative phase retrieval method was proposed on physical grounds using the Kramers–Kronig (KK) relation that obtains equivalent results to those with maximum entropy method but with significantly less computation time [42,43].

The development of the optical hardware and processing methods expanded the scope of possible applications of BCARS and enabled hyperspectral imaging in less than an hour, but one significant challenge remained: the "fingerprint" Raman region. In CRS imaging of biological samples, the CH-/OH-stretch region (~2700–3400 cm^{-1}) presents the strongest response, frequently by more than 10 times, because of the high density of these vibrational modes and the quadratic dependence of signal strength on mode density. However, there are few Raman peaks. The aptly monikered fingerprint region, which resides at lower frequencies (<1800 cm^{-1}), typically contains the highest number of Raman peaks but the weakest response, with the NRB intensity often comparable to or larger than the Raman peaks. Practically, the fingerprint region was just simply too weak and too distorted for much practical use in imaging. However, Camp et al. [30] recently developed a new system architecture that significantly enhanced the sensitivity of BCARS to the fingerprint region by a factor of approximately 100. This new BCARS approach uses a unique combination of lasers that disproportionately stimulates vibrational modes within the fingerprint region using "impulsive" stimulation (also known as "three-color" excitation). In addition, this system design does not require using higher power lasers—it is just more efficient.

THEORY

Vibrational Stimulation and Scattering

Although a complete description of the theory is outside of the scope of this work, a brief summary will facilitate understanding the differences in performance between the various BCARS methods. Classically, the CARS mechanism is described by the beating of pump (E_p) and Stokes (E_S) electric fields with a material described by a third-order nonlinear susceptibility tensor, $\chi^{(3)}$. A probe field, E_{pr}, inelastically scatters off of the excited molecular vibration at a new frequency. If the center frequency of the pump, Stokes, and probe fields are ω_{p0}, ω_{S0}, and ω_{pr0}, respectively, and a vibrational mode (Ω) exists at $\omega_{p0} - \omega_{S0}$, the scattered radiation is at a frequency $\omega_{p0} - \omega_{S0} + \omega_{pr0}$. The total generated signal, I, in the frequency-domain (ω), may be described mathematically as [12]

$$I(\omega) \propto \left| \iiint \chi^{(3)}(\omega; \omega_p, -\omega_S, \omega_{pr}) E_p(\omega_p) E_S^*(\omega_S) E_{pr}(\omega_{pr}) \delta(\omega - \omega_p + \omega_S - \omega_{pr}) d\omega_p d\omega_S d\omega_{pr} \right|^2, \quad (14.1)$$

where ω_p, ω_S, and ω_{pr} are frequency spaces and δ is a Dirac delta function that maintains conservation of energy. This equation may be written in a mathematically identical but more tractable form [30]:

$$I(\omega) \propto \left| \left\{ [E_S(\omega) \star E_p(\omega)] \chi^{(3)}(\omega) \right\} * E_{pr}(\omega) \right|^2, \quad (14.2)$$

where "★" and "∗" are the cross-correlation and convolution operations, respectively. Eq. (14.2) highlights some important properties of CARS systems:

1. The spectral resolution is determined by the bandwidth of the probe source.
2. The stimulation bandwidth is determined by the bandwidth of the cross-correlation of the Stokes and pump sources.
3. The absolute frequency of the CARS spectrum is relative to the probe frequency.

With regards to property 1, in CARS microscopies, the pump, Stokes, and probe sources emanate from picosecond lasers with bandwidths frequently in the range of 1–2 cm^{-1}, which is less than or comparable to most Raman lineshapes. In BCARS microspectroscopies, the pump and/or Stokes sources are SC sources with bandwidths frequently in excess of 3000 cm^{-1}, but the probe source is narrowband ($\leq$10 cm^{-1}); thus CARS microscopies and BCARS microspectroscopies have similar spectral resolution, but the region of interrogation is dramatically different (property 2). Property 3 states that all recorded Raman spectra will be relative to the probe frequency (often described as the "Raman shift"). As an inelastic scattering process, the probe photons will change energy relative to their initial energy (frequency). For example, a Raman peak at 3000 cm^{-1} will be measured 3000 cm^{-1} away from the probe center frequency, regardless of the actual wavelength of the probe. The energy (frequency) indifference to wavelength motivates the use of a frequency unit (cm^{-1}, "wavenumber") rather than wavelength in Raman spectroscopy.

Beyond spectral resolution and stimulative bandwidth, it is also pertinent to analyze signal spectral intensity. For simplicity, we will ignore the nonlinear susceptibility and assume all laser sources have real Gaussian fields in the form of $E(\omega) = E_0 \exp[i(\omega - \omega_0)^2/2\sigma^2]$, where E_0 is the field amplitude, ω_0 is the center frequency, and σ is a bandwidth parameter that is related to the full-width half-maximum (FWHM) as FWHM $= 2(2\ln 2)^{1/2}\sigma$. For this field, the average power, P, is proportional to $\langle |E|^2 \rangle = |E_0|^2 I \sqrt{\pi}\sigma$. Under these conditions Eq. (14.2) may be rewritten as

$$I(\omega) \propto \left| 2\pi \frac{E_{S0} E_{p0} E_{pr0} \sigma_S \sigma_p \sigma_{pr}}{\sqrt{\sigma_S^2 + \sigma_p^2 + \sigma_{pr}^2}} \exp\left\{ \frac{-(\omega - \omega_{p0} + \omega_{S0} - \omega_{pr0})^2}{2\left(\sigma_S^2 + \sigma_p^2 + \sigma_{pr}^2\right)} \right\} \right|^2 \quad (14.3)$$

$$\propto 4\sqrt{\pi} \frac{P_S P_p P_{pr} \sigma_S \sigma_p \sigma_{pr}}{\sigma_S^2 + \sigma_p^2 + \sigma_{pr}^2} \exp\left\{ \frac{-(\omega - \omega_{p0} + \omega_{S0} - \omega_{pr0})^2}{\left(\sigma_S^2 + \sigma_p^2 + \sigma_{pr}^2\right)} \right\}, \quad (14.4)$$

where E_{p0}, E_{S0}, and E_{pr0} are the spectral field intensities of the pump, Stokes, and probe sources, respectively; σ_p, σ_S, and σ_{pr} are the bandwidth parameters; and ω_{p0}, ω_{S0}, and ω_{pr0} are the center frequencies. From Eq. (14.4) one can also analyze the total (integrated) signal intensity:

$$\int I(\omega)d\omega \propto 4\pi \frac{P_S P_p P_{pr} \sigma_S \sigma_p \sigma_{pr}}{\sqrt{\sigma_S^2 + \sigma_p^2 + \sigma_{pr}^2}}. \quad (14.5)$$

Using Eqs. (14.4) and (14.5) we can elucidate important characteristics of CARS microscopy and different BCARS implementations. In the most common BCARS microspectroscopies (and CARS microscopies), the pump and probe sources are degenerate ($P_p = P_{pr} = P_{p,pr}$; $\sigma_p = \sigma_{pr} = \sigma_{p,pr}$). In the past this has been referred to as "two-color" stimulation, but for clarity we will refer to it as "interpulse" stimulation (see Fig. 14.2A). Under this condition the spectral intensity and total intensity are described as

$$I(\omega) \propto 4\sqrt{\pi} \frac{P_S P_{p,pr}^2 \sigma_S \sigma_{p,pr}^2}{\sigma_S^2 + 2\sigma_{p,pr}^2} \exp\left\{ \frac{-(\omega - 2\omega_{p0,pr0} + \omega_{S0})^2}{\left(\sigma_S^2 + 2\sigma_{p,pr}^2\right)} \right\}, \quad (14.6)$$

$$\int I(\omega)d\omega \propto 4\pi \frac{P_S P_{p,pr}^2 \sigma_S \sigma_{p,pr}^2}{\sqrt{\sigma_S^2 + 2\sigma_{p,pr}^2}}. \quad (14.7)$$

From Eq. (14.6) one can see that the generated signal is Gaussian centered at $\omega = 2\omega_{p0,pr0} - \omega_{S0}$ (or a Raman shift of $\omega_{p0} - \omega_{s0}$) with an $\text{FWHM} = 2\left[\left(\sigma_S^2 + 2\sigma_{p,pr}^2\right)\ln 2\right]^{1/2}$. From Eq. (14.7) we can compare the total generated signal from CARS ($\sigma_S = \sigma_{p,pr}$) microscopies and BCARS microspectroscopies with interpulse stimulation ($\sigma_S \gg \sigma_{p,pr}$):

$$\text{CARS microscopy}: \quad \int I(\omega)d\omega \propto 4\pi \frac{P_S P_{p,pr}^2 \sigma_{p,pr,S}^2}{\sqrt{3}}. \quad (14.8)$$

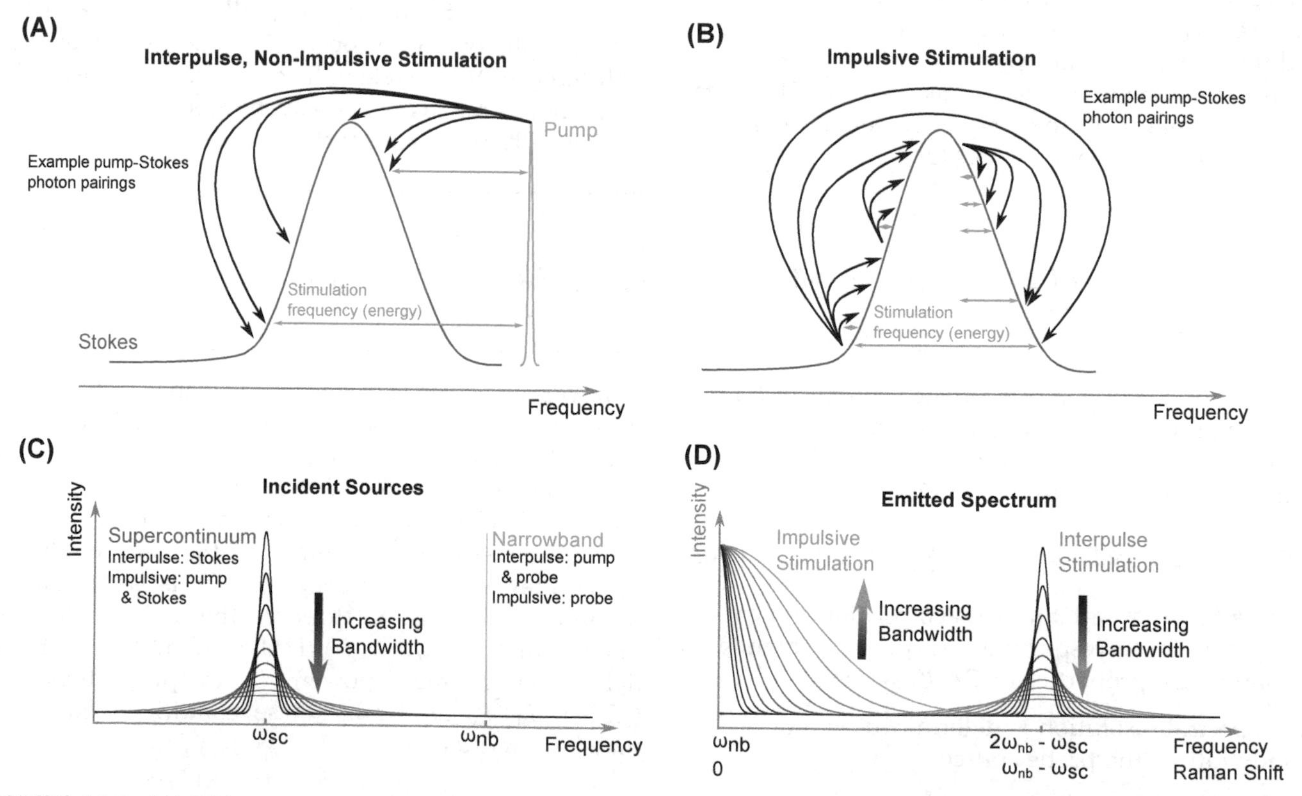

FIGURE 14.2 (A) With interpulse stimulation, pump and Stokes photons emanate from separate sources. *Black arrows* depict possible frequency differences between the pump and Stokes source. *Blue arrows* indicate the energy (frequency) of some of these pairings. No two pairings have the same frequency. (B) In impulsive stimulation the pump and Stokes source emanates from the same source. The pump–Stokes photon pairings are from different permutations across the spectrum. There are many photon pairs with the same stimulation frequency—maximum for the smallest frequencies (see *blue arrows* showing example pairings). (C) As the supercontinuum (SC) source bandwidth is increased, and average power is maintained, the peak spectral intensity of the SC drops. (D) As the SC bandwidth increases, the total coherent anti-Stokes Raman scattering (CARS) signal generated by interpulse stimulations remains constant, but the energy is spread over a broader spectrum (peak intensity drops). With intrapulse stimulation, the peak CARS intensity at 0 cm^{-1} remains fixed and the rest of the spectrum increases with increasing bandwidth. The total CARS signal generated with impulsive stimulation increases with increased SC bandwidth.

Interpulse stimulation BCARS :

$$\int I(\omega)d\omega \propto 4\pi P_S P_{p,pr}^2 \sigma_{p,pr}^2. \tag{14.9}$$

Comparing Eqs. (14.8) and (14.9), we see that the total signal generated is similar in magnitude. However, in BCARS the total signal is divided over a large bandwidth; thus the intensity at any given frequency is much smaller than if probed by CARS microscopy. This fundamentally explains how CARS microscopy can probe a single vibrational band in hundreds of nanoseconds to microseconds but BCARS microspectroscopy has traditionally required tens of milliseconds to acquire a full spectrum. In addition, this inherent weakness in BCARS has precluded significant acquisition of fingerprint region spectra.

An alternative stimulation mechanism is "impulsive" or "intrapulse" stimulation (historically, termed "three-color"). This stimulation paradigm has one small difference with large ramifications: the pump and Stokes sources are degenerate (see Fig. 14.2B). Under this condition, the total generated signal is (from Eq. (14.7) and assuming $\sigma_{p,S} \gg \sigma_{pr}$):

Impulsive stimulation BCARS :

$$\int I(\omega)d\omega \propto 4\pi \frac{P_{p,S}^2 P_{pr} \sigma_{p,S}^2 \sigma_{pr}}{\sqrt{2\sigma_{p,S}^2 + \sigma_{pr}^2}} \approx 4\pi \frac{P_{p,S}^2 P_{pr} \sigma_{p,S} \sigma_{pr}}{\sqrt{2}}. \tag{14.10}$$

Comparing Eqs. (14.9) and (14.10), we see that for the same incident power, impulsive stimulation generates approximately $0.71\sigma_{p,S}/\sigma_{pr}$ times more signal, which can be a factor of 100 or more.

Although impulsive stimulation in BCARS can generate much more signal, there is another benefit that is just as important: it disproportionately probes the lowest energy levels. Using the same derivation as used for Eq. (14.6) and noting that the crosscorrelation in Eq. (14.2) is now an auto-correlation [30]:

Impulsive stimulation BCARS :

$$I(\omega) \propto 4\sqrt{\pi}\, \frac{P_{p,S}^2 P_{pr} \sigma_{p,S}^2 \sigma_{pr}}{2\sigma_{p,S}^2 + \sigma_{pr}^2} \exp\left\{ \frac{-(\omega - \omega_{pr0})^2}{\left(2\sigma_{p,S}^2 + \sigma_{pr}^2\right)} \right\}. \tag{14.11}$$

With impulsive stimulation, the center frequency is at $\omega - \omega_{pr0}$, which corresponds to a Raman shift of 0 cm^{-1}. With impulsive stimulation the strongest signal is generated at the smallest energies, which corresponds to the dense-and-weak fingerprint Raman region.

Finally, there is one more important difference between interpulse and impulsive stimulation: the signal intensity with increasing Stokes bandwidth. With interpulse stimulation the total CARS signal generated remains constant with increasing Stoke source bandwidth and fixed average power (as shown in Fig. 14.2C and D, and described in Eq. (14.9)). Thus the spectral intensity at any particular frequency will diminish. In contrast, the total signal will rise with increasing Stokes bandwidth using impulsive stimulation (see Fig. 14.2C and D and Eqs. (14.10) and (14.11)). Interestingly, the maximum intensity at 0 cm^{-1} will remain approximately constant. Therefore impulsive stimulation can be viewed as more efficient than interpulse stimulation.

The Nonlinear Susceptibility, Nonresonant Background, and Raman Spectrum Extraction

As previously described, the CARS signal is generated along with an NRB. The NRB is so ubiquitous that in theoretical presentations the CARS and NRB mechanisms are not separate but rather both presented collectively as the CARS mechanism.

The material response is described by a third-order nonlinear susceptibility tensor, $\chi^{(3)}$ (in this work, we will disregard the tensor nature). Classically, the nonlinear susceptibility is split into two components: a Raman component that describes the chemically resonant response, χ_R, and a chemically nonresonant component, χ_{NR}, that will generate the NRB. Furthermore, the Raman component is approximated by a summation of complex Lorentzian lineshapes that are akin to damped harmonic oscillators:

$$\begin{aligned} \chi^{(3)}(\omega) &= \chi_R(\omega) + \chi_{NR}(\omega) \\ &= \chi_{NR}(\omega) + \sum_m \frac{A_m}{\Omega_m - \omega - i\Gamma_m}, \end{aligned} \tag{14.12}$$

where A_m, Ω_m, and Γ_m are the amplitude factor, center frequency, and half-linewidth of the mth Raman vibrational component. Fig. 14.3A shows a simulated nonlinear susceptibility with a real nonresonant component, which is relatively accurate away from electronic resonances. To a first-order approximation, spontaneous Raman spectroscopy measures the imaginary portion of χ_R. On the other hand, CARS methods (see Eq. (14.2)) measure a signal proportional to $|\chi|^2$. This represents a coherent mixing of χ_R and χ_{NR}, which mathematically is described as:

$$I(\omega) \propto |\chi|^2 = |\chi_R|^2 + |\chi_{NR}|^2 + 2\mathrm{Re}\{\chi_R \chi_{NR}^*\}, \tag{14.13}$$

where Re denotes the real part. The net effect is a distorted spectrum as demonstrated in Fig. 14.3B. Although the NRB is vibrationally nonresonant, it is still chemically sensitive to a certain degree; thus it cannot be simply removed, and raw BCARS spectra cannot be directly compared with spontaneous Raman spectra.

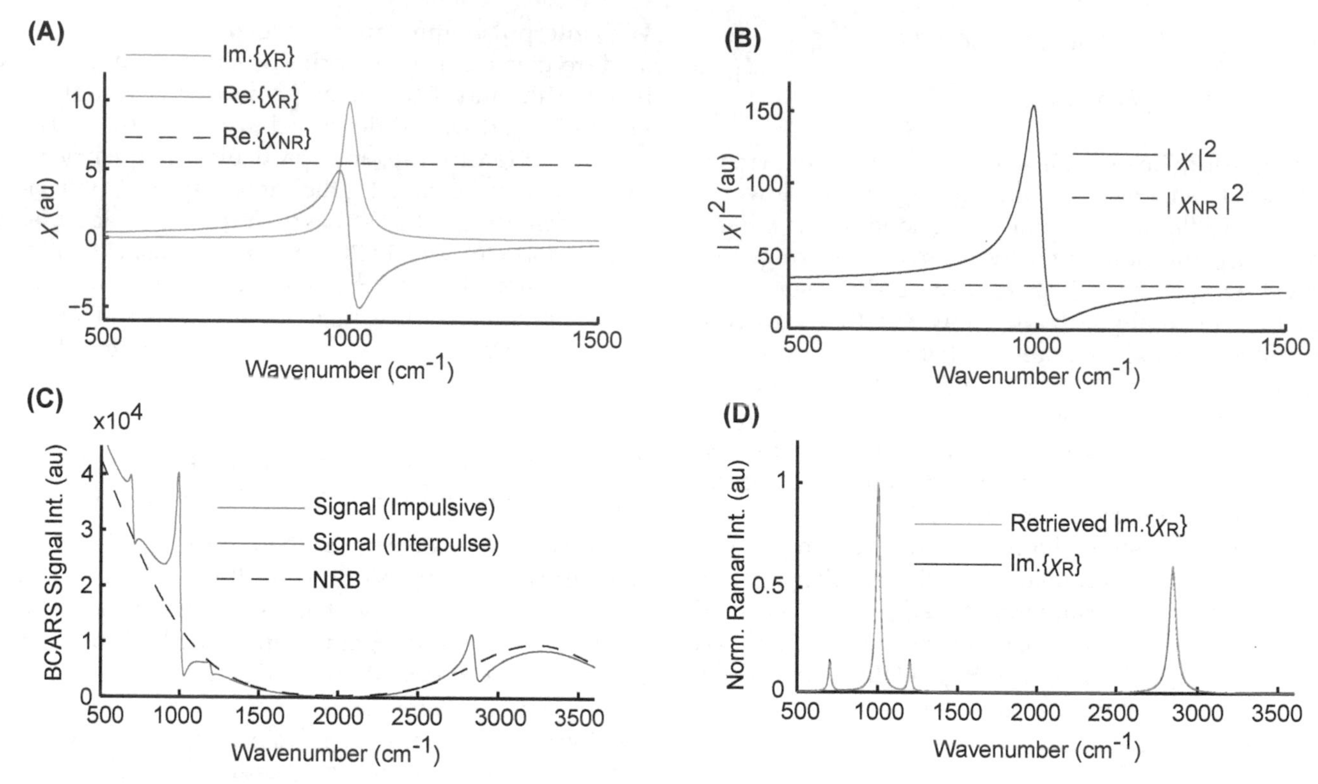

FIGURE 14.3 (A) Real (Re) and imaginary (Im) parts of the chemically resonant and nonresonant components of the nonlinear susceptibility. The nonresonant component is assumed to be real away from electronic resonances, which is typically the case with infrared excitation. (B) The measured broadband coherent anti-Stokes Raman scattering (BCARS) spectrum is proportional to the modulus of the nonlinear susceptibility squared. The coherent mixing of the resonant and nonresonant components leads to spectral distortions that prevent direct comparison of spontaneous Raman spectra and BCARS spectra. (C) Simulated BCARS spectrum of four Raman peaks showing significant peak distortion due to the nonresonant background. (D) Retrieved Raman spectrum using the Kramers–Kronig relation and the ideal Raman spectrum.

As previously mentioned, the phase retrieval techniques calculate the real and imaginary parts of the nonlinear susceptibility. The two most prominent methods are the maximum entropy method [39,40] and by using the KK relation [42]. Both of these methods provide equivalent results, although the KK relation is computationally more expedient [43]. Fig. 14.3C shows a simulated BCARS spectrum with an excitation profile similar to that of the system in Ref. [30]. Because of coherent mixing with the NRB, the Raman signature is distorted to the point of unusable. Fig. 14.3D shows the retrieved Raman spectrum in comparison to Im $\{\chi_R\}$, the ideal result.

A final significant aspect to all CARS-based methods is the nonlinear spectral intensity with respect to analyte concentration. χ_R is linearly proportional to concentration; thus spontaneous Raman spectra are linear with concentration and CARS spectra are quadratically proportional. The net effect is that for high-concentration species CARS spectra are intense but rapidly diminish with decreasing concentration. With impulsive stimulation BCARS demonstrated sensitivities in the tens of millimoles per liter range (within an approximately femtoliter volume) [30]. Although this does not necessarily represent the ultimate detection limit, it does indicate the current status of the technology and indicates that CARS-based imaging methods are most applicable to dense biomolecules, such as structural proteins and lipids bodies.

EXPERIMENTAL RESULTS: TISSUE IMAGING

BCARS has only recently matured to the point of practical application to tissue imaging. On the basis of the success of Raman spectroscopy for dermatological applications [44–50], BCARS appears poised to make significant contributions in upcoming years. In this section, BCARS images of histological sections of murine tissue will be presented, specifically that of the vaginal lining and a pancreatic artery.

System Architecture

The BCARS system is built upon two coseeded fiber lasers: one providing an ~3.4-ps narrowband source at 771 nm and the other a 16-fs SC that spans from

~900 to 1350 nm (both sources 40-MHz repetition rate). This laser design, in particular, acquires spectra with impulsive excitation from ~470 to 2000 cm^{-1} and interpulse stimulation from ~2000 to 3600 cm^{-1}. The lasers are temporally overlapped using a delay line and collinearly combined using a dichroic beamsplitter. The sources are routed into an inverted microscope and focused onto the sample using a water-immersion objective lens (NA = 1.2). The scattered light (elastic and inelastic) is collected in the transmissive direction with a long working distance objective (NA = 0.7) and spectrally filtered with shortpass dichroics. The anti-Stokes signal is focused onto the front slit of a spectrograph and the spectra recorded on a cooled charge-coupled device camera with 3.5-ms integration time. The sample is mounted on a piezoelectric driven stage that allows for raster scanning of the sample over a 200 × 200 μm area.

The acquired hyperspectral images are processed using in-house developed software. The basic workflow is as follows: dark signal removal, denoising via singular value decomposition, extraction of Raman-like spectra via the KK relation, and finally baseline detrending. The finer details of the system architecture and processing methodology are described in Ref. [30].

Histological Imaging

Fig. 14.4 shows BCARS images and spectra acquired from a histological slice of murine vaginal tissue. Similar to the human epidermis, the murine vagina is composed of a keratinized stratified squamous epithelium [51]. The longitudinal sectioning of the folded tissue surface allows multiple epithelial surfaces to come into contact around the lumen. Fig. 14.4A shows a pseudocolor BCARS image highlighting collagen in red, DNA/RNA in orange, and phenylalanine (Phe) in blue. In particular, collagen is identified by the 1250 cm^{-1} amide III peak [52,53]. The nuclei are highlighted using the 785 cm^{-1} peak that emanates from the phosphodiester-stretch backbone vibration and the pyrimidine ring-breathing mode found in cytosine, thymine, and uracil [54–56]. Phe content is highlighted using the ring-breathing 1004 cm^{-1} peak [55]. The enhanced Phe content, as shown in Fig. 14.4B, clearly delineates the epithelial layer from the fibrous lamina propria. The shape and position of the nuclei (also shown independently in Fig. 14.4C), relative to the cornified surface and the lamina propria, identify the epithelial layers. Single-pixel spectra from different histological features are shown in Fig. 14.4E. In addition, an image highlighting the CH_2-symmetric stretch, which is particularly prominent in lipids, is shown in Fig. 14.4D.

From the spectra in Fig. 14.4E one can appreciate the depth and complexity of the Raman signatures. The presented images used single peaks to assign molecular content; however, this is not always possible and limits chemical specificity. In the next example, several peaks and spectral shoulders are utilized to distinguish structural proteins with highly similar spectra.

Fig. 14.5 shows colorized BCARS imagery of a pancreatic artery. Fig. 14.5A presents in grayscale the total BCARS signal and highlights the nuclei in blue (785 cm^{-1}) and lipids in red (2857 cm^{-1}). From this image the wrinkled inner surface of the artery is exposed, as is the surrounding smooth muscle [57]. In addition, one can identify several endothelial cells defining the tunica intima. Fig. 14.5B shows a BCARS image highlighting nuclei in orange, collagen in blue, smooth muscle in red, and the elastic lamina in pink. Smooth muscle was identified by using the spectral shoulder at 1342 cm^{-1}, which is likely due to actin and/or myosin [58,59], and removing the contributions from nearby peaks by subtracting the linear interpolation from 1312 to 1353 cm^{-1}. The collagen was highlighted using 855 cm^{-1}. Both collagen and the elastic lamina contain the 855 cm^{-1} peak, which may be due solely to elastin [52] or there may also be a collagen component [59]. However, the elastic lamina contains a peak at 907 cm^{-1} (C–C–N stretch of elastin [52]) and a prominent peak at 1106 cm^{-1}, which is attributed to the desmosine and isodesmosine in elastin [52,60]. The 1106 cm^{-1} peak overlaps significantly with several other peaks, such as the 1093 cm^{-1} peak (phosphate symmetric stretch of nucleic acids [54]); thus subtraction of these contributions is necessary. Fig. 14.4C shows the elastic lamina in isolation, defining the tunica media in between. For comparison, Fig. 14.4E shows the second harmonic generation (SHG) of collagen fibrils [61,62] and two-photon-excited (auto-) fluorescence for species sensitive to light greater than 900 nm [62]. Because of their quasi-crystalline noncentrosymmetric structure, collagen fibrils strongly generate SHG. Comparing the SHG image with the BCARS image shows significant similarity. The differences may be attributable to SHG only produced by fibers with a particular orientation relative to the incident laser polarization or fiber degradation [61,63]. However, Raman scattering does not cease from these effects, but rather the shape of particular peaks is modulated [53,64]. Fig. 14.4D shows single-pixel BCARS spectra of the three connective tissues in this image. These spectra clearly demonstrate the challenge of distinguishing between individual proteins because the spectra are dense and have significant similarity.

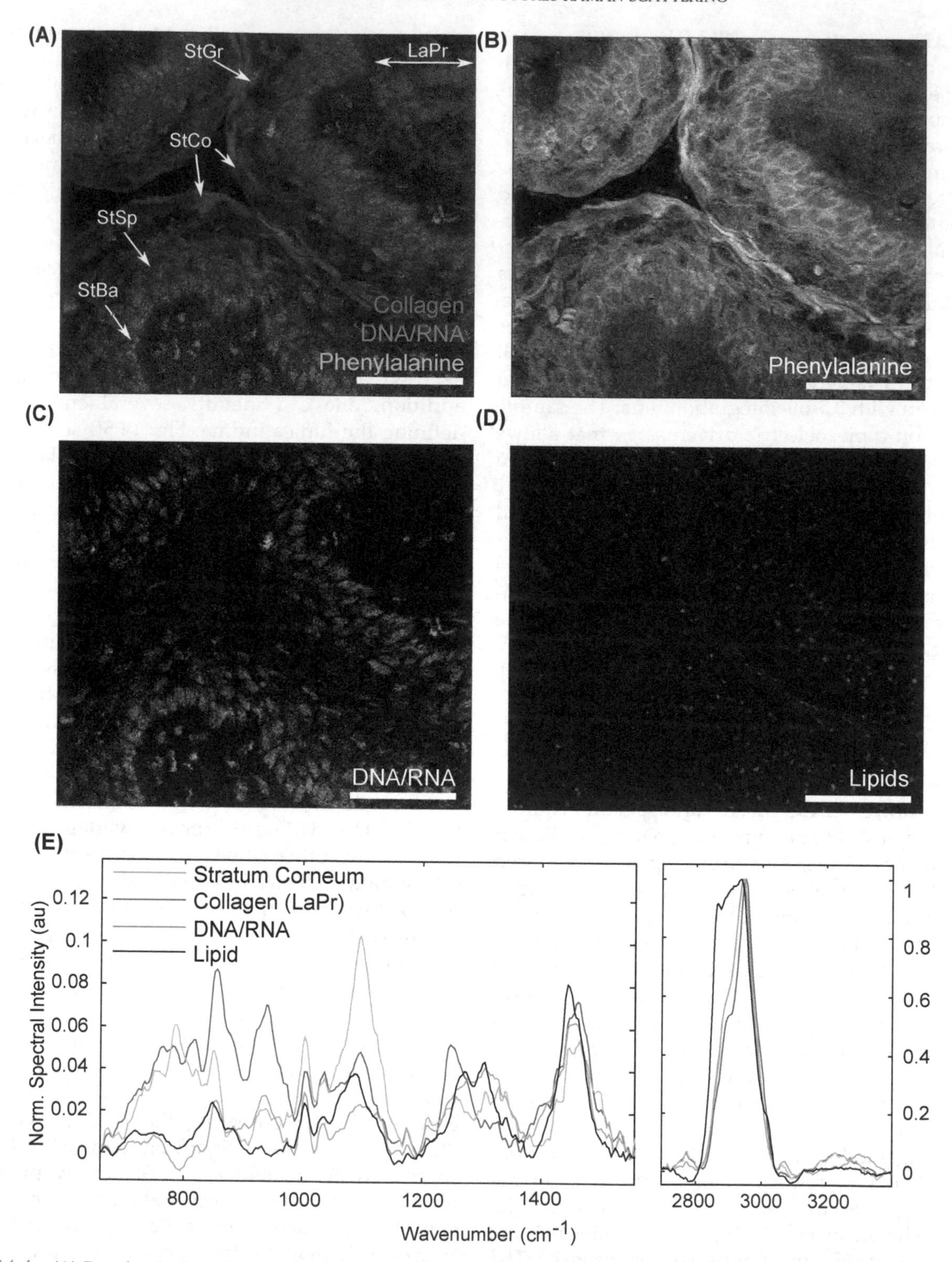

FIGURE 14.4 (A) Pseudocolor broadband coherent anti-Stokes Raman scattering (BCARS) image of murine vaginal tissue. *Arrows* point to pertinent epithelial structures. *StCo, Stratum corneum; StGr, stratum granulosum; StSp, stratum spinosum;* StBa, *stratum basale;* LaPr, *lamina propria.* (B) Grayscale BCARS image of phenylalanine content. Note the highest content in the cornified layers and the weakest in the collagenous regions. (C) Grayscale BCARS image of DNA/RNA. (D) Grayscale BCARS image highlighting lipids. (E) Single-pixel BCARS spectra. Scale bars are 50 μm.

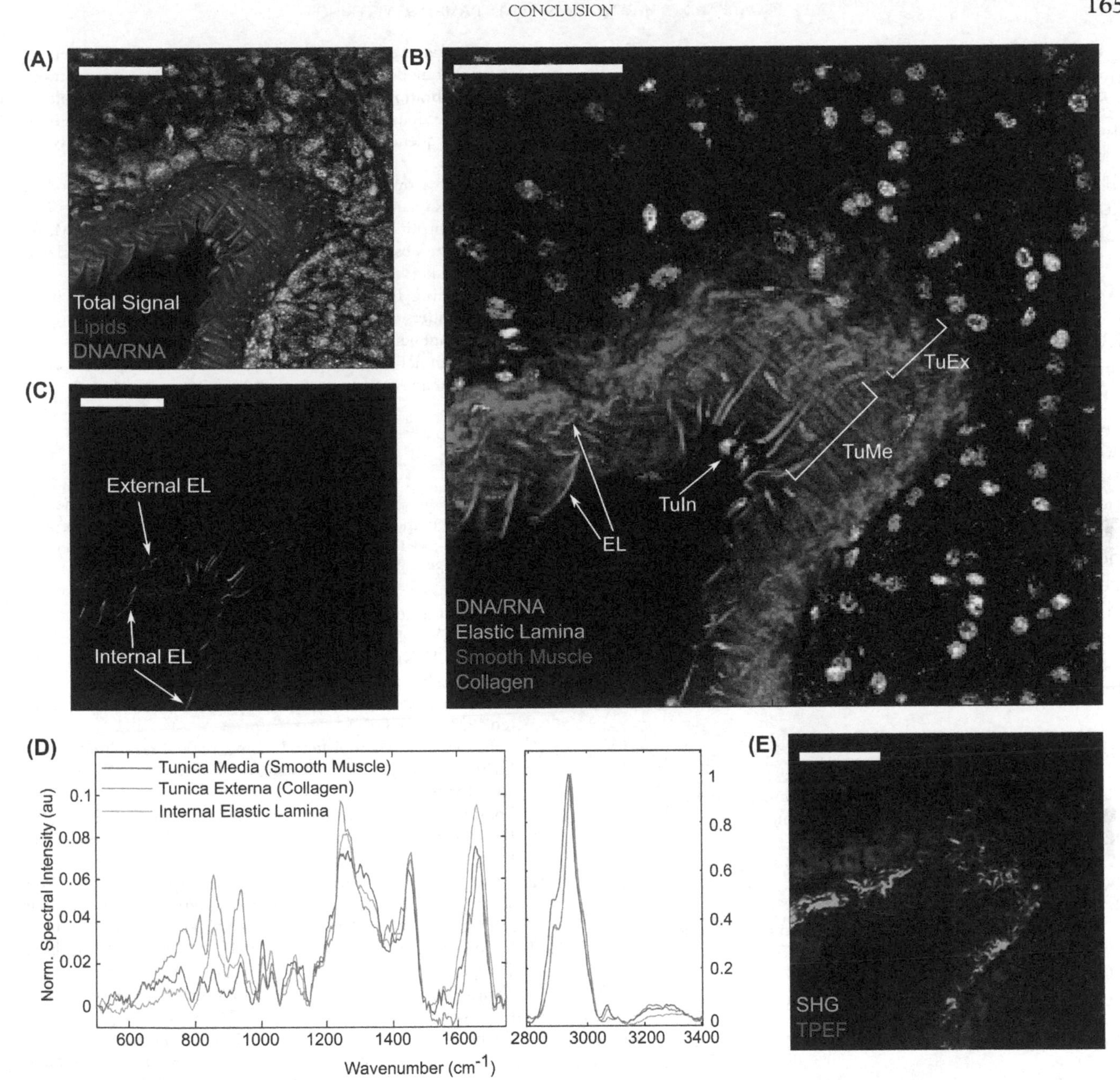

FIGURE 14.5 (A) Pseudocolor broadband coherent anti-Stokes Raman scattering (BCARS) image of a murine pancreatic artery. (B) Pseudocolor BCARS image highlighting DNA/RNA, the elastic lamina, smooth muscle, and collagen. *Arrows* identify pertinent histological features of the arterial wall. *EL*, Elastic lamina; *TuIn*, *tunica intima*; *TuMe*, *tunica media*; *TuEx*, *tunica externa*. (C) Grayscale BCARS image highlighting the elastic lamina (1106 cm^{-1}). (D) Single-pixel BCARS spectra of structural proteins. (E) Pseudocolor image of second harmonic generation (SHG, collagen) and two-photon excited fluorescence (autofluorescence). Scale bars are 50 μm.

CONCLUSION

For nearly a century, vibrational spectroscopy has provided a unique glimpse into the chemical world, detailing molecular composition and state in their native environment. Capturing this rich information within cells and tissues could fundamentally alter our understanding of biology and disease. Despite the significant promise, the classic techniques, Raman scattering and IR absorption spectroscopies, do not offer the speed and spatial resolution together necessary for high-throughput microscopy, significantly hampering their widespread adoption. Coherent Raman methods offer vibrationally sensitive imaging with greatly enhanced speed. Recent developments in BCARS microspectroscopy facilitates new applications and frontiers for vibrational analysis.

BCARS microspectroscopy provides a unique platform for dermatological analysis that currently is being applied to tissue resections. With further development in laser sources, detectors, and fiber probes, in vivo imaging could provide a powerful tool for evaluation of suspicious lesions and surgical margin determination. IR and Raman techniques have been used for in vivo spectroscopy of dermatological tissues and pathologies [44,49,65–67] (as well as in other tissue types [68–72]). However, the vast majority of these studies provided only single spectra from points of interest and those with imaging capability provided limited spatial quality. Coherent Raman methods have demonstrated high-resolution imagery but with limited chemical specificity [73–77]. These earlier developments lay the foundation for porting BCARS to in vivo imaging. Although BCARS is typically performed in a transmissive geometry, recent imaging experiments of pharmaceutical tablets [78] provide evidence that a reflection geometry is possible.

In this work, the contextual background, basic theory, and experimental demonstration of BCARS microspectroscopy have been presented. Images of murine vagina and pancreas serve to highlight the complexity of the spectral content and the depth of information available. This reinvigorated technology could provide the chemical and spatial clarity to enhance our understanding of the chemical markers of disease and facilitate clinical diagnoses.

List of Abbreviations

BCARS Broadband coherent anti-Stokes Raman scattering
CARS Coherent anti-Stokes Raman scattering
CRS Coherent Raman scattering
FWHM Full-width at half-maximum
IR Infrared
KK Kramers–Kronig relation
MCARS Multiplex coherent anti-Stokes Raman scattering
NA Numerical aperture
NRB Nonresonant background
Phe Phenylalanine
SC Supercontinuum
SHG Second harmonic generation

Glossary

Broadband coherent anti-Stokes Raman scattering A coherent Raman scattering method that utilizes broadband sources to simultaneously excite multiple Raman vibrations. Full spectra collected at each image pixel.

Coherent anti-Stokes Raman scattering (mechanism) A nonlinear optical scattering mechanism in which a "pump" and "Stokes" photon coherently excite a vibrational mode. A probe photon inelastically scatters from the excited mode, gaining energy equal to that of the vibrational energy.

Coherent anti-Stokes Raman scattering (method) Microscopy technique in which narrowband laser sources are tuned to excite a single vibrational mode (frequency) via the coherent anti-Stokes Raman scattering mechanism. Images are captured to record the intensity at the tuned-to frequency. To capture multiple frequencies or to effectively record a spectrum, the laser frequencies are tuned between acquisitions.

Coherent Raman scattering Class of mechanisms in which molecular vibrational modes are actively driven by incident laser sources.

Molecular vibrational modes/vibrational modes Bonded atoms or small collections of bonded atoms vibrate within a molecule at certain frequencies and certain motion (eg, rocking, twisting, stretching).

Raman microscopy/microspectroscopy Microscopy method that collects a Raman spectrum at each image pixel.

Infrared absorption spectroscopy/infrared spectroscopy Measurement of the absorbance of infrared light on a sample, detailing the molecular vibrational modes.

Raman scattering Inelastic scattering mechanism in which an incident photon scatters off of molecular vibration, gaining or losing an energy amount equal to that of the vibrational mode.

Raman vibrational modes Molecular vibrational modes that can take part in Raman scattering.

References

[1] de Wiveleslie Abney W, Festing ER. On the influence of the atomic grouping in the molecules of organic bodies on their absorption in the infra-red region of the spectrum. Philos Trans R Soc Lond 1881;172:887–918.
[2] Coblentz WW. Investigations of infrared spectra Part I. Washington, DC: Carnegie Institution of Washington; 1905.
[3] Raman CV. A new radiation. Indian J Phys 1928;398:368–76.
[4] Raman CV, Krishnan KS. A new type of secondary radiation. Nature 1928;121:501–2.
[5] Smekal A. Zur Quantentheorie der Dispersion. Naturwissenschaften 1923;11(43):873–5.
[6] Landsberg G, Mandelstam L. Eine neue Erscheinung bei der Lichtzerstreuung in Krystallen. Naturwissenschaften 1928; 16(28):557–8.
[7] Long DA. Raman spectroscopy. New York: McGraw-Hill; 1977.
[8] Konigstein JA. Introduction to the theory of the Raman effect. Dordrecht, Holland: D. Reidel; 1972.
[9] Stevenson TL, Vo-Dinh T. Signal expressions in Raman spectroscopy. In: Laserna JJ, editor. Modern techniques in Raman spectroscopy. New York: John Wiley & Sons; 1996. p. 1–40.
[10] Petrov GI, Arora R, Yakovlev VV, Wang X, Sokolov AV, Scully MO. Comparison of coherent and spontaneous Raman microspectroscopies for noninvasive detection of single bacterial endospores. Proc Natl Acad Sci USA 2007;104:7776–9.
[11] Cheng J-X, Volkmer A, Xie XS. Theoretical and experimental characterization of coherent anti-Stokes Raman scattering microscopy. J Opt Soc Am B 2002;19(6):1363–75.
[12] Gomez JS. Coherent raman spectroscopy. In: Laserna JJ, editor. Modern techniques in Raman spectroscopy. Chichester: John Wiley & Sons; 1996. p. 305–42.
[13] Zumbusch A, Holtom G, Xie XS. Three-dimensional vibrational imaging by coherent anti-Stokes Raman scattering. Phys Rev Lett 1999;82(20):4142–5.
[14] Potma EO, Mukamel S. Theory of coherent Raman scattering. In: Cheng J-X, Xie XS, editors. Coherent Raman scattering microscopy. Boca Raton: CRC; 2013. p. 3–42.
[15] Mukamel S. Principles of nonlinear optical spectroscopy. New York: Oxford University; 1995.
[16] Maker P, Terhune R. Study of optical effects due to an induced polarization third order in the electric field strength. Phys Rev 1965;137(3):A801–18.
[17] Begley RF, Harvey AB, Byer RL. Coherent anti-Stokes Raman spectroscopy. Appl Phys Lett 1974;25(7):387–90.
[18] Duncan MD, Reintjes J, Manuccia TJ. Scanning coherent anti-Stokes Raman microscope. Opt Lett 1982;7(8):350–2.

[19] Cheng J-X, Volkmer A, Book LD, Xie XS. Multiplex coherent anti-Stokes Raman scattering microspectroscopy and study of lipid vesicles. J Phys Chem B 2002;106(34):8493–8.
[20] Müller M, Schins JM. Imaging the thermodynamic state of lipid membranes with multiplex CARS microscopy. J Phys Chem B 2002;106(14):3715–23.
[21] Kee TW, Cicerone MT. Simple approach to one-laser, broadband coherent anti-Stokes Raman scattering microscopy. Opt Lett 2004;29(23):2701–3.
[22] von Vacano B, Meyer L, Motzkus M. Rapid polymer blend imaging with quantitative broadband multiplex CARS microscopy. J Raman Spectrosc 2007;38:916–26.
[23] Lee YJ, Moon D, Migler KB, Cicerone MT. Quantitative image analysis of broadband CARS hyperspectral images of polymer blends. Anal Chem 2011;83(7):2733–9.
[24] Lee YJ, Vega SL, Patel PJ, Aamer KA, Moghe PV, Cicerone MT. Quantitative, label-free characterization of stem cell differentiation at the single-cell level by broadband coherent anti-Stokes Raman scattering microscopy. Tissue Eng Part C Methods 2014; 20(7):562–9.
[25] Kano H, Hamaguchi H-O. Vibrationally resonant imaging of a single living cell by supercontinuum-based multiplex coherent anti-Stokes Raman scattering microspectroscopy. Opt Express 2005;13(4):1322–7.
[26] Rinia HA, Burger KNJ, Bonn M, Müller M. Quantitative label-free imaging of lipid composition and packing of individual cellular lipid droplets using multiplex CARS microscopy. Biophys J 2008;95(10):4908–14.
[27] Parekh SH, Lee YJ, Aamer KA, Cicerone MT. Label-free cellular imaging by broadband coherent anti-Stokes Raman scattering microscopy. Biophys J (Biophysical Society) 2010;99(8):2695–704.
[28] Afonso PV, Janka-Junttila M, Lee YJ, McCann CP, Oliver CM, Aamer KA, et al. LTB4 is a signal-relay molecule during neutrophil chemotaxis. Dev Cell 2012;22(5):1079–91.
[29] Pohling C, Buckup T, Pagenstecher A, Motzkus M. Chemoselective imaging of mouse brain tissue via multiplex CARS microscopy. Biomed Opt Express 2011;2(8):2110–6.
[30] Camp Jr CH, Lee YJ, Heddleston JM, Hartshorn CM, Walker ARH, Rich JN, et al. High-speed coherent Raman fingerprint imaging of biological tissues. Nat Photon (Nature Publishing Group) 2014;8(8):627–34.
[31] Billecke N, Rago G, Bosma M, Eijkel G, Gemmink A, Leproux P, et al. Chemical imaging of lipid droplets in muscle tissues using hyperspectral coherent Raman microscopy. Histochem Cell Biol 2014;141(3):263–73.
[32] Volkmer A, Book LD, Xie XS. Time-resolved coherent anti-Stokes Raman scattering microscopy: imaging based on Raman free induction decay. Appl Phys Lett 2002;80(9):1505–7.
[33] Cui M, Joffre M, Skodack J, Ogilvie JP. Interferometric Fourier transform coherent anti-Stokes Raman scattering. Opt Express 2006;14(18):8448–58.
[34] Dudovich N, Oron D, Silberberg Y. Single-pulse coherently controlled nonlinear Raman spectroscopy and microscopy. Nature 2002;418(6897):512–4.
[35] Cheng J-X, Book LD, Xie XS. Polarization coherent anti-Stokes Raman scattering microscopy. Opt Lett 2001;26(17):1341–3.
[36] Garbacik ET, Korterik JP, Otto C, Mukamel S, Herek JL, Offerhaus HL. Background-free nonlinear microspectroscopy with vibrational molecular interferometry. Phys Rev Lett 2011; 107(25):253902.
[37] Evans CL, Xie XS. Coherent anti-Stokes Raman scattering microscopy: chemical imaging for biology and medicine. Annu Rev Anal Chem 2008;1:883–909.
[38] Müller M, Zumbusch A. Coherent anti-Stokes Raman scattering microscopy. ChemPhysChem 2007;8(15):2156–70.
[39] Vartiainen EM. Phase retrieval approach for coherent anti-Stokes Raman scattering spectrum analysis. J Opt Soc Am B 1992;9(8): 1209–14.
[40] Vartiainen EM, Rinia HA, Müller M, Bonn M. Direct extraction of Raman line-shapes from congested CARS spectra. Opt Express 2006;14(8):3622–30.
[41] Rinia HA, Bonn M, Vartiainen EM, Schaffer CB, Müller M. Spectroscopic analysis of the oxygenation state of hemoglobin using coherent anti-Stokes Raman scattering. J Biomed Opt 2006;11(5):050502.
[42] Liu Y, Lee YJ, Cicerone MT. Broadband CARS spectral phase retrieval using a time-domain Kramers–Kronig transform. Opt Lett 2009;34(9):1363–5.
[43] Cicerone MT, Aamer KA, Lee YJ, Vartiainen E. Maximum entropy and time-domain Kramers–Kronig phase retrieval approaches are functionally equivalent for CARS microspectroscopy. J Raman Spectrosc 2012;43(5):637–43.
[44] Lieber CA, Majumder SK, Ellis DL, Billheimer DD, Mahadevan-Jansen A. In vivo nonmelanoma skin cancer diagnosis using Raman microspectroscopy. Lasers Surg Med 2008;40(7):461–7.
[45] Gniadecka M, Philipsen PA, Sigurdsson S, Wessel S, Nielsen OF, Christensen DH, et al. Melanoma diagnosis by Raman spectroscopy and neural networks: structure alterations in proteins and lipids in intact cancer tissue. J Invest Dermatol 2004;122(2): 443–9.
[46] Gniadecka M, Wulf HC, Mortensen NN, Nielsen OF, Christensen DH. Diagnosis of basal cell carcinoma by Raman spectroscopy. J Raman Spectrosc 1997;28:125–9.
[47] Nijssen A, Bakker Schut TC, Heule F, Caspers PJ, Hayes DP, Neumann MH, et al. Discriminating basal cell carcinoma from its surrounding tissue by Raman spectroscopy. J Invest Dermatol 2002;119(1):64–9.
[48] Hata TR, Scholz TA, Ermakov IV, McClane RW, Khachik F, Gellermann W, et al. Non-invasive Raman spectroscopic detection of carotenoids in human skin. J Invest Dermatol 2000;115(3): 441–8.
[49] Lui H, Zhao J, McLean D, Zeng H. Real-time Raman spectroscopy for in vivo skin cancer diagnosis. Cancer Res 2012;72(10): 2491–500.
[50] Sigurdsson S, Philipsen PA, Hansen LK, Larsen J, Gniadecka M, Wulf HC. Detection of skin cancer by classification of Raman spectra. IEEE Trans Biomed Eng 2004;51(10):1784–93.
[51] Rendi MH, Muehlenbachs A, Garcia RL, Boyd KL. Female reproductive system. In: Treutling PM, Dintzis SM, editors. Comparative anatomy and histology: a mouse and human atlas. St. Louis: Academic Press; 2011. p. 253–84.
[52] Frushour BG, Koenig JL. Raman scattering of collagen, gelatin, and elastin. Biopolymers 1975;14(2):379–91.
[53] Bonifacio A, Sergo V. Effects of sample orientation in Raman microspectroscopy of collagen fibers and their impact on the interpretation of the amide III band. Vib Spectrosc (Elsevier B.V.) 2010;53(2):314–7.
[54] Deng H, Bloomfield VA, Benevides JM, Thomas GJ. Dependence of the Raman signature of genomic B-DNA on nucleotide base sequence. Biopolymers 1999;50(6):656–66.
[55] De Gelder J, De Gussem K, Vandenabeele P, Moens L. Reference database of Raman spectra of biological molecules. J Raman Spectrosc 2007:1133–47.
[56] Erfurth SC, Kiser EJ, Peticolas WL. Determination of the backbone structure of nucleic acids and nucleic acid oligomers by laser Raman scattering. Proc Natl Acad Sci USA 1972;69(4): 938–41.
[57] Kempf SC, Hortsch M, MacCallum D. Don MacCallum's Michigan histology. Ann Arbor: University of Michigan Technology Transfer; 2013.

[58] Romer TJ, Brennan JF, Fitzmaurice M, Feldstein ML, Deinum G, Myles JL, et al. Histopathology of human coronary atherosclerosis by quantifying its chemical composition with Raman spectroscopy. Circulation 1998;97(9):878–85.

[59] Buschman HP, Deinum G, Motz JT, Fitzmaurice M, Kramer JR, Van Der Laarse A, et al. Raman microspectroscopy of human coronary atherosclerosis: biochemical assessment of cellular and extracellular morphologic structures in situ. Cardiovasc Pathol 2001;10(2):69–82.

[60] Manoharan R, Wang Y, Feld MS. Histochemical analysis of biological tissues using Raman spectroscopy. Spectrochim Acta Part A Mol Biomol Spectrosc 1996;52(2):215–49.

[61] Williams RM, Zipfel WR, Webb WW. Interpreting second-harmonic generation images of collagen I fibrils. Biophys J (Elsevier) 2005;88(2):1377–86.

[62] Zipfel WR, Williams RM, Christie R, Nikitin AY, Hyman BT, Webb WW. Live tissue intrinsic emission microscopy using multiphoton-excited native fluorescence and second harmonic generation. Proc Natl Acad Sci USA 2003;100(12):7075–80.

[63] Sun Y, Chen W-L, Lin S-J, Jee S-H, Chen Y-F, Lin L-C, et al. Investigating mechanisms of collagen thermal denaturation by high resolution second-harmonic generation imaging. Biophys J (Elsevier) 2006;91(7):2620–5.

[64] Dong R, Yan X, Pang X, Liu S. Temperature-dependent Raman spectra of collagen and DNA. Spectrochim Acta A 2004;60(3):557–61.

[65] McIntosh LM, Jackson M, Mantsch HH, Mansfield JR, Crowson AN, Toole JWP. Near-infrared spectroscopy for dermatological applications. Vib Spectrosc 2002;28:53–8.

[66] Caspers PJ, Lucassen GW, Carter EA, Bruining HA, Puppels GJ. In vivo confocal Raman microspectroscopy of the skin: noninvasive determination of molecular concentration profiles. J Invest Dermatol 2001;116:434–42.

[67] Greve TM, Kamp S, Jemec GBE. Disease quantification in dermatology: in vivo near-infrared spectroscopy measures correlate strongly with the clinical assessment of psoriasis severity. J Biomed Opt 2013;18:037006.

[68] Bergholt MS, Zheng W, Lin K, Ho KY, Teh M, Yeoh KG, et al. Raman endoscopy for in vivo differentiation between benign and malignant ulcers in the stomach. Analyst 2010;135(12):3162–8.

[69] Shim MG, Wong Kee Song L-M, Marcon NE, Wilson BC. In vivo near-infrared Raman spectroscopy: demonstration of feasibility during clinical gastrointestinal endoscopy. Photochem Photobiol 2000;72(1):146–50.

[70] Krafft C, Kirsch M, Beleites C, Schackert G, Salzer R. Methodology for fiber-optic Raman mapping and FTIR imaging of metastases in mouse brains. Anal Bioanal Chem 2007;389(4):1133–42.

[71] Motz JT, Gandhi SJ, Scepanovic OR, Haka AS, Kramer JR, Dasari RR, et al. Real-time Raman system for in vivo disease diagnosis. J Biomed Opt 2005;10(3):031113.

[72] Matthäus C, Dochow S, Bergner G, Lattermann A, Romeike BFM, Marple ET, et al. In vivo characterization of atherosclerotic plaque depositions by Raman-probe spectroscopy and in vitro coherent anti-Stokes Raman scattering microscopic imaging on a rabbit model. Anal Chem 2012;84(18):7845–51.

[73] Evans CL, Potma EO, Puoris'haag M, Côté D, Lin CP, Xie XS. Chemical imaging of tissue in vivo with video-rate coherent anti-Stokes Raman scattering microscopy. Proc Natl Acad Sci USA 2005;102(46):16807–12.

[74] Ji M, Orringer DA, Freudiger CW, Ramkissoon S, Liu X, Lau D, et al. Rapid, label-free detection of brain tumors with stimulated Raman scattering microscopy. Sci Transl Med 2013;5(201):1–10.

[75] Fu Y, Wang H, Huff TB, Shi R, Cheng J-X. Coherent anti-Stokes Raman scattering imaging of myelin degradation reveals a calcium-dependent pathway in lyso-PtdCho-induced demyelination. J Neurosci Res 2007;85(13):2870–81.

[76] Bélanger E, Henry FP, Vallée R, Randolph MA, Kochevar IE, Winograd JM, et al. In vivo evaluation of demyelination and remyelination in a nerve crush injury model. Biomed Opt Express 2011;2(9):2698–708.

[77] Jung Y, Tam J, Jalian HR, Anderson RR, Evans CL. Longitudinal, 3D in vivo imaging of sebaceous glands by coherent anti-Stokes Raman scattering microscopy: normal function and response to cryotherapy. J Invest Dermatol (Nature Publishing Group) 2014;135(1):1–6.

[78] Hartshorn CM, Lee YJ, Camp CH, Liu Z, Heddleston J, Canfield N, et al. Multicomponent chemical imaging of pharmaceutical solid dosage forms with broadband CARS microscopy. Anal Chem 2013;85(17):8102–11.

CHAPTER

15

In Vivo Reflectance Confocal Microscopy in Dermatology

I. Alarcon[1], *C. Longo*[2], *S. González*[1,3,4,5]

[1]Ramon y Cajal Hospital, Madrid, Spain; [2]Arcispedale Santa Maria Nuova-IRCCS, Reggio Emilia, Italy; [3]Universidad de Alcalá, Madrid, Spain; [4]Instituto Ramón y Cajal de Investigación Sanitaria (IRYCIS), Madrid, Spain; [5]Memorial Sloan-Kettering Cancer Center, New York, NY, United States

OUTLINE

Imaging in Dermatology
http://dx.doi.org/10.1016/B978-0-12-802838-4.00015-7

INTRODUCTION

The clinical diagnostic accuracy of the diagnosis of skin lesions has been reported to be 24–64% [1], being low even for experienced dermatologists who diagnose correctly approximately 75–80% of skin tumors [2]. Hence the final diagnosis of skin conditions ultimately relies on histopathology analysis, which is a painful invasive method that leaves a scar and might require several biopsies taken before the diagnosis can be finally achieved.

In the 1990s dermoscopy enabled visualizing new subsurface features to differentiate between malignant and benign lesions [3], allowing the connection between clinical evaluation and histopathology analysis by enabling the visualization of morphological features, which are not discernible by the unaided eye examination. This resulted in an improvement in diagnostic sensitivity of 10–30% [4]. However, dermoscopy still relies heavily on the subjective judgement of the observer, and it may actually lower the diagnostic accuracy in the hands of less trained dermatologists [5]. Despite this, dermoscopy is the fastest growing method to image skin for diagnosis. In comparison with other specialties, dermatologists have been slow to adopt advanced technological diagnostic aids.

In the past decade new computer-based technologies that enable noninvasive, in vivo diagnosis have been developed as a response to the need to improve the diagnostic accuracy and sensitivity for skin conditions and to optimize lesion selection for biopsy and pathology review. These techniques include ultrasound, reflectance confocal microscopy (RCM), magnetic resonance imaging, optical coherence tomography, multispectral imaging, and multiphoton microscopy. Although dermoscopy has probably reached the method's inherent potential diagnostic accuracy because of its lack of cellular-level resolution, further improvements could be expected by in vivo RCM [6].

RCM is a novel noninvasive technique that allows the evaluation of the skin with high, quasi-histological resolution and good contrast. Among the newer techniques, it is the only one with a cellular resolution necessary for histological analysis that could offer an alternative to biopsy [2]. In this chapter we will provide an updated literature review and a critical appraisal of the advantages and limitations of the applications of RCM in dermatology.

THE TECHNOLOGY: REFLECTANCE CONFOCAL LASER MICROSCOPE

In 1995 Rajadhyaksha et al. reported the first RCM for real-time in vivo imaging of the human skin [7]. Several years before, this noninvasive methodology was described as a very promising tool for morphometrical studies of living human skin at the cellular level [8]. The technology was based on the tandem scanning reflected light microscope invented by Petran and Hadravsky. It provided real-time vision in the confocal mode. In 1993 Corcuff et al. explored human skin in vivo, at the cellular level, to a depth of 150 μm. Two different anatomical sites were investigated: the back of the hand and the volar aspect of the forearm. The stratification of the corneal layers, the nuclei of the living keratinocytes through the whole epidermis, and the capillary loops within the superficial dermis were used as the reference points, allowing thickness measurements of the stratum corneum and epidermis with an accuracy of 1 μm. Furthermore, it performed nondestructive optical sectioning of biological material to a depth that depends on the transparency of the tissue, adding a fourth dimension (time) to the study of living specimens [9].

The device uses a diode laser at 830 nm with a power lower than 35 mW at tissue level. A 30× water-immersion objective lens of numerical aperture 0.9 is used with either water (refractive index, 1.33) or gel (refractive index, 1.3335) as an immersion medium. It captures images with a spatial resolution of 0.5–1.0 μm in the lateral dimension and 4–5 μm in the axial dimension [10].

In RCM, the laser beam is focused by the objective lens into a single spot within the skin and backscattered by the tissue through the optical system onto a pinhole placed in front of the detector. When the beam is moved in one direction (scanning), a line of reflected signals is generated (x-axis) and when the beam is moved in the other direction (y-axis), a complete area can be scanned leading to an en face image of the tissue comparable to an "optical slice." By moving that plane into or out of the tissue (z-axis, parallel to the beam direction), a stack of images can be generated representing the optical image of a tissue volume, enabling the confocal microscope to visualize at a slice in the sample. The detected signal intensity is digitalized using a gray scale [11].

For the living tissue use, there are three commercially available microscopes: the VivaScope 1500, VivaScope 1500 Multilaser, and the VivaScope 3000 (Caliber I.D., Rochester, NY). VivaScope 3000 is a compact handheld device that simplifies examining anatomically inaccessible skin regions such as the nasolabial fold and the free edge of the eyelid.

NORMAL SKIN UNDER REFLECTANCE CONFOCAL MICROSCOPY

Under RCM, the individual layers of the epidermis and the dermis can be identified by the morphological appearance of the cells, the depth of the section from the stratum corneum, and the visualization of distinctive structures such as dermal papillae [12]. The confocal images are based on the detectability of the presence of contrast, being melanin, the strongest endogenous contrast present in the human skin [7]. Other sources of contrast are provided by cytoplasmic organelles, keratin, nuclear chromatin, and the dermis itself.

The stratum corneum is visualized in optical sections as the most superficial layer where large (10–30 μm), very bright, anucleated, and polygonal corneocytes are observed forming "islands" separated by skin folds, which appear very dark. The stratum granulosum is located at a depth of approximately 15–20 μm below the stratum corneum and is made of two to four layers of corneocytes with a diameter of 25–35 μm arranged in a cohesive pattern. The morphology of granular cells is characterized by the presence of centrally located dark, oval nuclei and bright grainy cytoplasm consistent with its content in organelles, keratohyalin, and melanosomes. At an approximate depth of 20–100 μm below the stratum corneum, smaller (15–25 μm), tightly packed corneocytes with distinct borders; dark nuclei; and clear, homogeneous cytoplasm are located forming the stratum spinosum [13]. RCM reveals that the stratum granulosum and spinosum cells are arranged in a characteristic "honeycomb pattern" [14]. The basal layer is observed at an average depth of 50–100 μm beneath the stratum corneum and is composed of 7- to 12-μm diameter cells. The melanin aggregates forming bright disks above the nuclei make the basal cells appear highly refractive [13]. The melanocytes are visualized as bright, round, oval and spindle cells or dendritic cells, located in the dermoepidermal junction (DEJ).

The dermal papillae are visualized as dark, round, or elliptical structures with capillary loops at their center displaying clearly visible blood flow and surrounded by bright rings of the basal layer cells. In the papillary dermis, collagen fibers (1–5 μm) and bundles (5–25 μm) form a network at an average depth of 100–300 μm [10]. Eccrine sweat ducts can be visualized in optical sections as bright, centrally hollow structures, penetrating out through the epidermis and dermis. Highly refractive hair shafts emerge from hair follicles associated with pilosebaceous units and visualized as centrally hollow structures with elliptical, elongated cells at the circumference [15].

REFLECTANCE CONFOCAL MICROSCOPY FOR PIGMENTED TUMORS

Early detection of malignant melanoma (MM) is one of the most challenging problems in clinical dermatology. Hence, the research into noninvasive methods that improve the clinical diagnostic accuracy for cutaneous melanoma has been prompt. RCM has demonstrated that it is capable of identifying distinct patterns and cytological features of benign and malignant pigmented skin lesions in vivo.

Acquired Melanocytic Nevi

RCM enables the in vivo characterization of different aspects of melanocytic nevi. The presence of a preserved architecture of the suprabasal layers characterized by a regular honeycombed and cobblestone pattern is related to benign melanocytic lesions. The edged papillae are a feature present in melanocytic nevi and are visualized as dermal papillae surrounded by a rim of bright cells, appearing as bright rings contrasting with the dark background. The junctional nests could be seen as clusters or thickenings. The junctional clusters are seen as compact cellular aggregates bulging within the dermal papillae connected with the basal cell layer. The junctional thickenings are enlargements of the interpapillary space formed by aggregated cells. Clusters and thickenings are characteristic of benign lesions. Melanocytic nevi show at the level of the upper dermis bright cells with centered nuclei aggregated in dense homogeneous nests [10,16,17].

RCM also reveals cell clusters that correspond to dermoscopic pigment globules and to well-circumscribed nests that histopathologically are composed of large epithelioid cohesive cells with fine dusty melanin, located in the lower epidermis or in the upper dermis.

In junctional nevi the nevus cells localize only at the DEJ; in compound nevi they also localize in the upper dermis. Dermal nevi are characterized by the presence of nevomelanocyte nests within the papillary and reticular dermis [18] (Fig. 15.1).

Congenital Nevi

In a study aimed to determine the utility of RCM in the in vivo evaluation of congenital melanocytic nevi that are suggestive of having developed melanoma, the RCM evaluation was able to characterize the morphological features of the cells including regular shape, comparatively small cell size, and monomorphic cellular architecture. This correlated well with the findings of routine histopathological analysis. These findings illustrated the potential application of RCM as a noninvasive screening tool in the assessment of congenital melanocytic nevi that undergo clinical changes suggestive of MM. The depth of penetration of the near-infrared light represents a limitation to the ability of RCM for fully assessing deep dermal cells of congenital nevi, failing to identify cellular structures within the reticular dermis. This excludes important information from the diagnostic process. However, this limitation is mitigated by the fact that most melanomas arising from these smaller congenital nevi appear to originate at the DEJ [19].

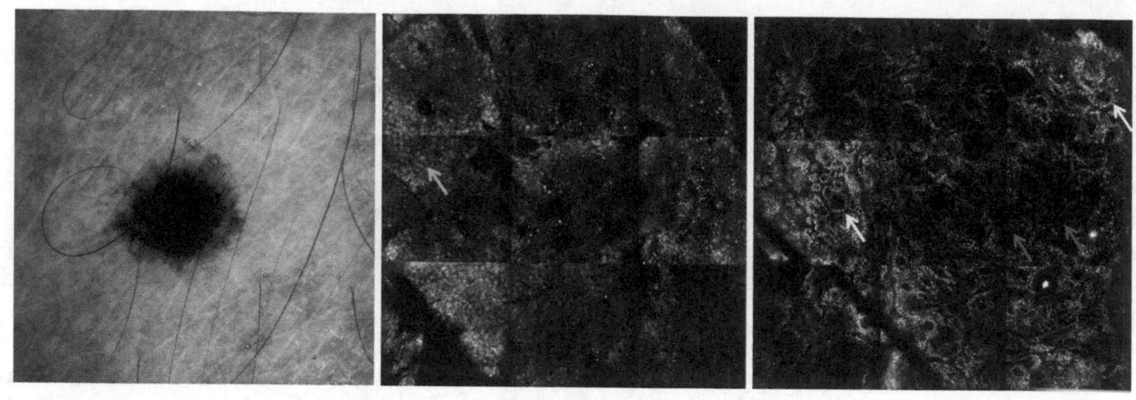

FIGURE 15.1 **Nevus.** A symmetric lesion with pigmented network and a structureless area is seen by dermoscopy. The image in the center shows a cobblestone pattern which corresponds to pigmented melanocytes at the level of epidermis (*blue arrow*). The image on the right shows edged papillae (*yellow arrows*) and nevomelanocyte nests as junctional thickenings (*red arrows*).

Dysplastic Nevi

The pathogenetic model for stepwise development of MM having dysplastic nevi as precursors is still controversial, and precise correlation between clinical and histopathological features is required. Dermoscopy is not able to identify dysplastic nevi because it cannot evaluate cytology [20]. The cellular resolution provided by RCM allows correlating dermoscopy and histopathology. A study aimed to determine whether specific histological features in dysplastic nevi have reliable correlates in confocal microscopy and to develop an in vivo microscopic grading system was published in 2012. The authors evaluated 60 melanocytic lesions with equivocal dermatoscopic aspects, corresponding to 19 nondysplastic nevi, 27 dysplastic nevi, and 14 melanomas.

The lesions were analyzed by confocal microscopy and histopathology using the Duke grading criteria. All of the architectural and cytological features of the Duke grading score had significant RCM correlates. Under RCM, dysplastic nevi were characterized by a ringed pattern associated with a meshwork pattern in a large proportion of cases, along with atypical junctional cells in the center of the lesion, and irregular junctional nests with short interconnections. A simplified algorithm was developed to distinguish dysplastic nevi from MM and nondysplastic nevi. The contemporary presence of cytological atypia and of atypical junctional nests (irregular, with short interconnections, and/or with nonhomogeneous cellularity) was suggestive of histological dysplasia, whereas a widespread pagetoid infiltration, widespread cytological atypia at the junction, and non-edged papillae suggested melanoma diagnosis [21]. The possibility to differentiate dysplastic nevi from MM in vivo, discriminating lesions that can be followed up safely from those that should be removed promptly, led to an appropriate management decision [21].

Spitz Nevi

Spitz nevi can clinically present either in the classical (reddish pink) or the pigmented (brownish black) variant. Under dermoscopy none of these show dermoscopic patterns clearly distinguishable from melanoma. Even histopathologically, a clear-cut differentiation between benign and malignant spitzoid neoplasms is often difficult [22]. The largest series published of Spitz nevi evaluated with RCM [23] showed that some histological aspect characteristics for Spitz nevus diagnosis were correlated with confocal features, comprising some that can be useful for atypical Spitz nevus classification. The most striking features for differentiating Spitz nevi from MMs were the presence of sharp border cutoff, junctional nests, and melanophages. No correlates were found for other aspects, such as Kamino bodies, hyperkeratosis, acanthosis, mitoses, and maturation with depth. The impossibility of exploring deeper aspects hampered an accurate distinction from MMs.

Melanoma

Concerns about the increasing incidence of melanoma in Caucasian populations have focused on the well-documented association between early excision and the reduction of mortality [24]. Cutaneous MM represents one of the greatest challenges in early or preventive detection. In early stages of tumor development, surgical excision is almost always curative, whereas delayed recognition increases the risk of tumor growth and death from metastatic disease [25,26]. Dermoscopy represents to date the clinical standard in routine skin tumor screening [27]. A meta-analysis has shown that the diagnostic accuracy for melanoma was significantly higher with than without dermoscopy with an improvement of 49%. The authors concluded that dermoscopy improves the diagnostic accuracy for melanoma in comparison with inspection by the unaided eye, but only for experienced examiners [28].

RCM represents a breakthrough in melanoma diagnosis and knowledge on the biology of melanocytic lesions [18]. Because melanosomes and melanin are strong sources of endogenous contrast, melanocytic cells are particularly evident by means of this technique [7,10]. Several studies have identified RCM features of melanocytic lesions [16,29,30], suggesting that further improvement in diagnostic accuracy for MM could be achieved [6,25,26], especially when this technique is used in combination with dermoscopy [18,31].

Superficial Spreading Type

A study by Pellacani and colleagues [32] aimed to evaluate the frequency of confocal features in benign and malignant melanocytic lesions and their diagnostic significance for melanoma identification where a second end point was to identify the most relevant features for melanoma diagnosis. The authors included 102 consecutive melanocytic lesions [37 MMs, 49 acquired nevi (21 junctional, 27 compound, and 1 intradermal), and 16 epithelioid and/or spindle cell nevi (3 junctional Spitz, 8 compound Spitz, and 5 Reed)]. After describing the RCM features of every type of lesion included, six criteria were identified as independently correlated with a diagnosis of melanoma: two major criteria were the presence of cytological atypia (mild or marked) and of nonedged papillae at the basal layer and four minor criteria including the presence of roundish cells in superficial layers spreading upward in a pagetoid fashion, pagetoid cells widespread throughout the lesion, cerebriform clusters in the papillary dermis, and nucleated cells within dermal papilla. The study concluded that RCM was useful for the in vivo characterization of equivocal melanocytic lesions at clinical and dermoscopic evaluation.

In 2007 Pellacani et al. developed an algorithm aimed to define the impact of RCM features that distinguish melanomas and nevi with standard statistical methods [17]. They defined the following RCM features for the diagnosis of melanoma:

Superficial (granulosum/spinosum) layers: The presence of a marked epidermal disarray was more frequently observed in melanomas, although it was also present in one-third of neviConversely, a homogeneous epidermis characterized by regular honeycombed and/or cobblestone pattern was strongly related to benign lesions. Pagetoid infiltration of roundish cells was reported in 78% of melanomas and 19% of nevi, whereas the observation of dendritic pagetoid cells, although significant, had a relatively lower odds ratio for melanomas. More than three cells per image and the presence of cells larger than 20 μm were predominantly observed in melanomas. Pleomorphism and widespread diffusion throughout the lesion of pagetoid cells were specific but not sensitive markers of malignancy.

Dermoepidermal junction: The presence of nonedged papillae was observed in 90% of melanomas and in 41% of nevi, whereas edged papillae were predominantly present in nevi. Mild to marked atypia was observed in 73% and 27% of melanomas and nevi, respectively. Cells distributed in sheet-like structures, disrupting the papillary architecture of the basal layer, were highly specific, but with low sensitivity for melanoma. On the other hand, junctional nests, both clusters and thickenings, were characteristic of benign lesions.

Upper dermis: Immediately below the dermal epidermal junction, nested cell aggregates were visible in more than half of the lesions. Whereas regular dense nests were significantly more represented in nevi, the presence of atypical nests (eg, the nonhomogeneous nests, sparse cell nests, and/or cerebriform nests) correlated with malignancy (53% of melanomas), although they were also observed in 26% of nevi. Within the dermal papilla, almost half of the melanomas showed large nucleated cells, compared with 13% of nevi. No difference in the frequency of plump bright cells (bright small cells, and/or hyper-reflecting spots), and broadened reticulated and/or large bundles of fibers was reported within the two groups. When comparing melanomas equal or thinner than 1 mm with thicker ones, epidermal disarray, cells in sheet-like structures, cerebriform nests, and nucleated cells within dermal papilla were significantly associated with thick melanomas [17].

Lentigo Maligna

Lentigo maligna (LM) is an early form of melanoma with an incidence that has increased during the past 2 decades [33]. Clinically, it is a pigmented lesion that occurs on sun-exposed skin, particularly the head and neck areas of elderly patients [34]. Untreated, LM is associated with a 5–50% risk of progressing to LM melanoma, the invasive counterpart, which is also increasing in individuals older than 45 years [35].

Several studies demonstrated that RCM can be used to differentiate LM from other pigmentations of the face [36,37]. It can also assist in defining the peripheral margins of LM [38], even on amelanotic tumors [39].

In 2010 Guitera et al. [40] identified six features that were independently correlated with diagnosis of LM by means of discriminant analysis, corresponding to, in order of relevance, nonedged papillae, pagetoid round and large cells, nucleated cells in a dermal papilla, three or more atypical cells at the DEJ in five $0.5 \times 0.5\ mm^2$ images, follicular localization of pagetoid cells and/or atypical junctional cells, and the single

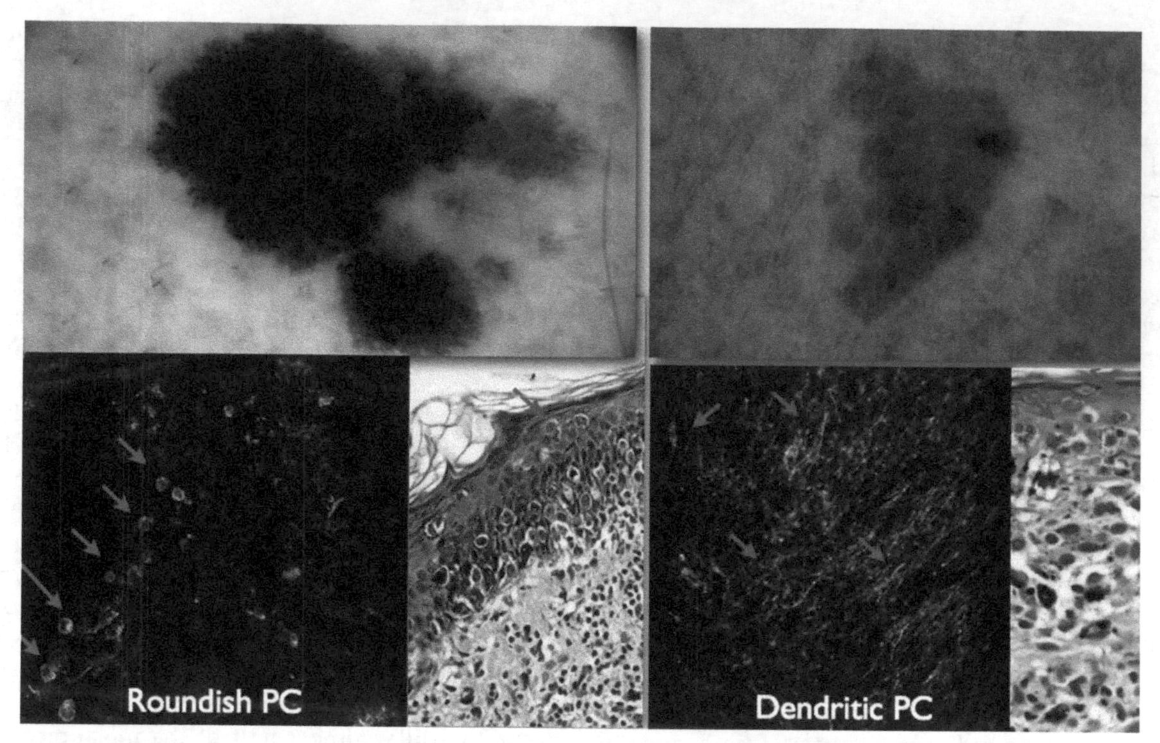

FIGURE 15.2 **Melanoma.** Distinct melanoma types are shown here: On the left, a superficial spreading melanoma with multiple brown dots, pseudopods, radial streaming, scar-like depigmentation, and irregular and broad network. Under reflectance confocal microscopy (RCM) we observe epidermal disarray and roundish pagetoid cells (*red arrows*) corresponding to the pagetoid spread under histopathology. On the right, a melanoma appearing on chronically sun-damaged skin; under RCM we can observe dendritic pagetoid cells (*red arrows*) that are also visible on histology.

negative (benign) feature of a broadened honeycomb pattern of the epidermis.

In a study performed to determine whether in vivo RCM mapping of difficult LM cases might determine the most efficient route of patient care and management, the authors compared the differences in the margin of LM as established by RCM versus dermoscopy versus histopathological analysis. On the basis of the information obtained using this technique, the authors showed that they could potentially assist multidisciplinary medical teams in their management of difficult and challenging cases [41] (Fig. 15.2).

Nodular Melanoma

Under RCM, nodular melanoma (NM) exhibits several differential features at RCM compared with superficial spreading MM. These differences are often correlated with dermoscopic and histopathological findings. Within the epidermis, NM lacks characteristic features of melanoma such as epidermal disarrangement and pagetoid spreading. Instead, they usually show a honeycomb pattern or a peculiar broadened pattern consisting of polygonal cells with black nuclei and bright, thick borders. At the DEJ, the typical papillary architecture is not visible in the nodules, corresponding to the epidermal flattening caused by massive proliferation of malignant cells in the dermis. Hyporefractive nests, called "cerebriform nests," are observed in NM, correlating with deep tumor infiltration. Plump cells are present and correlate with dermal macrophages, usually associated with a moderate degree of inflammation [42].

Featureless and Amelantoic Melanoma

Small-diameter melanomas represent a distinct clinical-pathological entity because they do not follow the ABCD rule and they do not always display enough histopathological criteria that help to distinguish melanoma from atypical nevi [43]. The RCM findings demonstrated that small melanomas frequently reveal specific dermoscopic and confocal features that help to differentiate them from nevi [43].

The diagnosis of amelanotic melanoma is challenging because the classical clinical and dermoscopic features of pigmented melanoma are usually not present. Therefore RCM is a valuable tool in the diagnosis of partially and completely amelanotic tumors [44–46] (Table 15.1).

Diagnostic Scoring Systems for Melanoma

A two-step algorithm for the in vivo diagnosis of skin tumors by RCM, including melanocytic and nonmelanocytic lesions, was developed in 2009 [47]. The first step of the method consists of differentiating melanocytic from nonmelanocytic lesions. Four single confocal features helped to identify melanocytic lesions: the

TABLE 15.1 Reflectance Confocal Microscopy Algorithms for the Diagnosis of Melanoma and Lentigo Maligna

Reflectance confocal microscopy score malignant melanoma score [17]	Reflectance confocal microscopy score lentigo maligna score [40]
Two major features, each scored 2 points	*Two major features, each scored 2 points*
Nonedged papillae Cellular atypia at dermoepidermal junction	Nonedged papillae Pagetoid cells round and >20
Four minor features, each scored 1 point	*Three minor features, each scored 1 point*
Roundish pagetoid cells Widespread pagetoid infiltration Cerebriform nests Nucleated cells within the papilla	More than three atypical cells at the junction in five images Follicular localization of pagetoid cells and/or atypical junctional cells Nucleated cells within the papilla One minor negative feature, score −1 point (broadened honeycomb pattern)
Sensitivity of 91.9% and a specificity of 69% with a threshold ≥3	*Sensitivity of 85% and a specificity of 76% with a threshold ≥2*

Adapted from Guitera P, Pellacani G, Crotty KA, Scolyer RA, Li LX, Bassoli S, et al. The impact of in vivo reflectance confocal microscopy on the diagnostic accuracy of lentigo maligna and equivocal pigmented and nonpigmented macules of the face. J Invest Dermatol 2010;130(8):2080–91.

presence of the cobblestone pattern of epidermal layers, pagetoid spread, mesh appearance of the DEJ, and dermal nests. In the second step, within melanocytic lesions, the presence of roundish suprabasal cells and atypical nucleated cells in the dermis was associated with melanoma, and the presence of edged papillae and typical basal cells was associated with nevi.

For LM, a simplified algorithm termed "LM score" was developed [40]. This new LM score algorithm produced a sensitivity of 93% and a specificity of 61% for a score ≥1. However, the threshold of ≥2 produced the highest diagnostic accuracy with a sensitivity of 85% and a specificity of 76% [odds ratio 18.6 (95% confidence interval: 9.3–37.1)]. The addition of RCM analysis to dermoscopy has reduced unnecessary excisions with a high diagnostic accuracy and could lessen the economic impact associated with the management of skin cancer [48].

REFLECTANCE CONFOCAL MICROSCOPY FOR EPITHELIAL TUMORS

Nonmelanoma skin cancer (NMSC) is the most common type of cancer in Caucasian populations [49]. Basal cell carcinoma (BCC) and squamous cell carcinoma (SCC) show an increasing incidence rate worldwide, but stable or even decreasing mortality rates. The rising incidence rates of NMSC are probably caused by a combination of increased exposure to ultraviolet (UV) or sunlight, increased outdoor activities, changes in clothing style, increased longevity, ozone depletion, genetics, and in some cases immune suppression. An intensive UV exposure in childhood and adolescence was causative for the development of BCC whereas for the etiology of SCC a chronic UV exposure in the earlier decades was accused.

Basal Cell Carcinoma

The RCM features of BCC have been described, with the presence of tumor islands and cords considered as the distinctive criteria of the tumor. In 2002 González et al. defined for the first time the in vivo histological features of BCC using RCM [50]. At the epidermal level, RCM imaging shows alteration in architecture of the normally regular honeycombed pattern of the epidermis.

At the DEJ extending downward or disconnected from the epidermis, distinct aggregates of tightly packed cells, not seen in normal papillary dermis, appeared upon RCM observation. The nuclei of the cells at the periphery of the tumor are arranged parallel to each other ("palisading"). These densely packed cells form trabeculae or cordlike structures and nodules and are surrounded by nonrefractile peritumoral spaces. These were usually regarded as an artifact resulting from tumor retraction occurring during routine tissue processing for the preparation of light-microscopical section. Correlation between RCM images and histopathological samples showed that the diameter of the hyporefractile areas on RCM as well as the thickness of the peritumoral mucin deposits correspond to the peritumoral cleft-like spaces seen in BCC [51].

The stroma displays low refraction whereas tumor cells show higher refractivity. The nucleocytoplasmic ratio is high, and prominent nucleoli may be present. Abundant blood vessels demonstrating prominent tortuosity were seen, as well as prominent, predominantly mononuclear inflammatory infiltrate admixed or in close apposition with BCC cells. Leukocyte rolling along the endothelium and increased blood flow may also be seen in the vicinity of the tumor. Between BCC cells, bright round cells corresponding to leucocytes are occasionally seen [50] (Fig. 15.3).

Diagnostic Scoring Systems for Basal Cell Carcinoma

In 2004 Nori et al. [52] developed an RCM-based diagnostic algorithm for the diagnosis of BCC. RCM

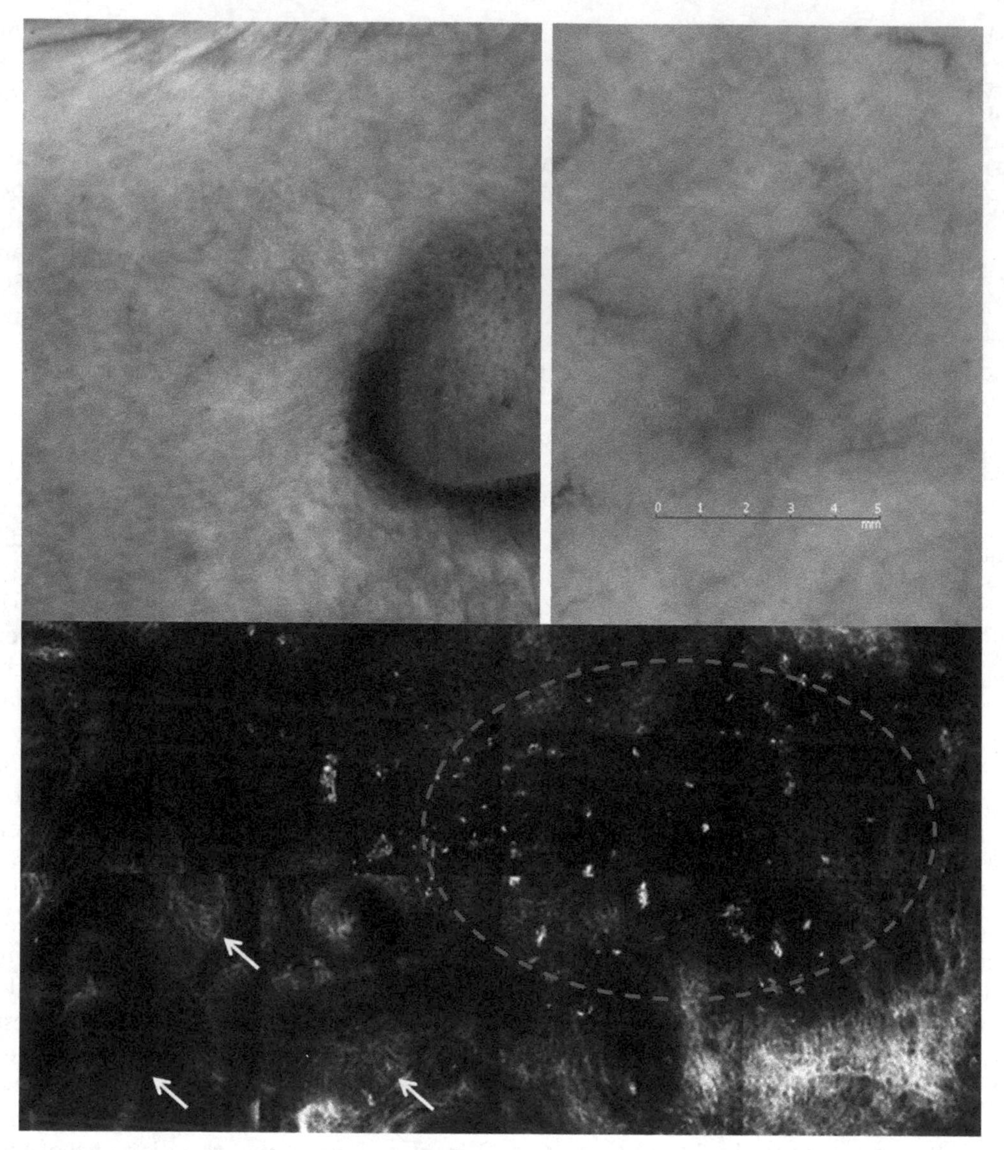

FIGURE 15.3 **Basal cell carcinoma.** Upper images: The presence of a plaque showing white-red structureless areas and featureless areas is observed. Lower image: Basaloid islands (*yellow arrows*) showing a peripheral palisading. Presence of hyporefractile areas (*red circle*) corresponding to basaloid islands devoid of pigment with inflammatory infiltrate.

images from four institutions and 152 skin lesions representing various benign and malignant diagnoses were evaluated. The lesions were examined clinically, with biopsies recorded for all of the 83 BCCs detected. On the basis of a previous study, a set of five histologically correlated confocal imaging criteria for diagnosing BCC was established. These included the presence of epidermal pleomorphism, the presence of elongated, monomorphic nuclei in the basal layer of the epidermis, the orientation of the tumor cell nuclei along the same axis, prominent inflammatory cell infiltrate, and increase of capillary vessels in the upper layers of the dermis. Blinded retrospective analysis of the images from the 152 lesions was performed by a single novice reviewer to determine the sensitivity and specificity of these five RCM criteria for diagnosing BCC.

The presence of two or more criteria is 100% sensitive for the diagnosis of BCC, the presence of four out of five of these criteria allows for the diagnosis of BCC with specificity equal to 95.7% and a sensitivity of 82.9%. Fulfilling five of these criteria increases the specificity of the diagnosis to 100% while reducing the sensitivity to 48.8%.These results were found to have little variability across study sites and across BCC subtypes.

In 2009 another study defined a two-step method for diagnosis: (1) melanocytic lesions were first distinguished from nonmelanocytic lesions and (2) criteria were then applied to differentiate melanomas from nevi. The authors described RCM features of BCC based on the analysis of 26 cases. Moreover, typical clinical cases of BCCs were not excised; hence the diagnosis of each case was not pathologically confirmed [47].

TABLE 15.2 Two-Step Reflectance Confocal Microscopy Method for the Diagnosis of Skin Tumors

Five positive features	Three negative features
• Polarized elongated features in the superficial layer was the most powerful feature. It has also been reported to be correlated with polarized and elongated nuclei of basal cells [52]. • Linear telangiectasia-like horizontal vessels. • Basaloid cord and nodules have also been well described in nodular BCCs, as well as circumscribed nests of hyporeflective cells tightly packed in a palisading way. • Epidermal shadow, defined as a large featureless area with blurred borders disrupting the normal epidermis and corresponding to the horizontal clefting (due to hyporeflective stroma) [51]. • Convoluted glomerular-like vessels.	• Disarray of the epidermis, is more specific of MMs, and the honeycomb pattern was recognized in more than 90% of the BCCs of this series. • Nonvisible papillae, meaning that BCC structures altered the normal junction organization. • Cerebriform nests, very specific of MM.

The diagnostic accuracy of this BCC algorithm was 100% sensitivity, 88.5% specificity.
BCC, *basal cell carcinoma;* MM, *malignant melanoma*.
Adapted from Guitera P, Menzies SW, Longo C, Cesinaro AM, Scolyer RA, Pellacani G. In vivo confocal microscopy for diagnosis of melanoma and basal cell carcinoma using a two-step method:analysis of 710 consecutive clinically equivocal cases. J Invest Dermatol 2012;132(10):2386–2394.

A two-step RCM method for the diagnosis of skin tumors based on a large series of melanocytic and nonmelanocytic lesions was published in 2012 [53]. The authors analyzed the RCM characteristics of lesions from patients treated at two specialized skin cancer clinics to define a model for accurately diagnosing BCCs and MMs and to compare it with previously published methods of confocal diagnosis. First, they characterized the features of BCCs because the diagnosis of BCCs is more straightforward than that of melanocytic lesions. The method for diagnosing BCC is based on eight independently significant features (Table 15.2).

Classification of Basal Cell Carcinoma Subtypes With Reflectance Confocal Microscopy

According to the different subtypes of BCC, the current guidelines aim to aid in the selection of the most appropriate approach for the management of this epithelial tumor [54]. For superficial BCC (sBCC) tumors, nonsurgical therapy is considered the first option. For nodular BCC (nBCC) surgical excision is the standard criterion; for infiltrative BCC (iBCC) the Mohs micrographic surgery represents the optimal choice to avoid incomplete excision or recurrence [55].

A recent study aimed to provide the key confocal features of different subtypes of BCC and investigated whether RCM, as an adjunct to dermoscopy, could enhance the accurate subtype classification. The description of the confocal features led to the identification of specific criteria associated with different BCC subtypes [56].

BCC subtype revealed the presence of cords connected to the epidermis. The streaming of the epidermis, assessed as the most significant parameter for BCC diagnosis [52,53], was found to be associated with a 50% increased odds of sBCC, but this result was not statistically significant. However, the interobserver agreement for the interpretation of this criterion was poor, highlighting the limitation of this parameter [56].

nBCC was primarily typified by the presence of large tumor nests, although small ones were also detected, corresponding possibly to the micronodular histopathological subtype. Clefting was more frequently seen in nBCC than in sBCC and iBCC. Finally, the vascularization was increased in all subtypes, in nBCC the vessel caliber being larger compared with sBCC and iBCC [56].

iBCC showed a characteristic RCM feature, which was the presence of dark silhouettes and abundant bright compact collagen. Dark silhouettes were particularly useful for the discrimination of iBCC from other subtypes. Furthermore, in the absence of small or big tumor islands and cords connected to the epidermis, iBCC was the most common diagnosis. For this criterion, the interobserver agreement was good, but because dark silhouettes appear hyporefractive, their recognition requires extensive training with RCM [56].

Actinic Keratoses

The global RCM features describing the architectural pattern and cytological features of actinic keratoses (AK) have been described elsewhere [57–59]. The clinical applicability of in vivo RCM for the diagnosis of AK in correlation with routine histology has been also investigated [60].

On the basis of previously described RCM parameters [57], a study defined the RCM features of AK as follows: stratum corneum disruption, parakeratosis, individual corneocytes, exocytosis, and architectural disarray at the level of granular and spinous layers. From these features, those most consistently found by RCM included the presence of superficial disruption, architectural disarray, and cellular pleomorphism at the level of the spinous and granular layer. The latter two are the best predictors of AK according to a sensitivity and specificity analysis [60].

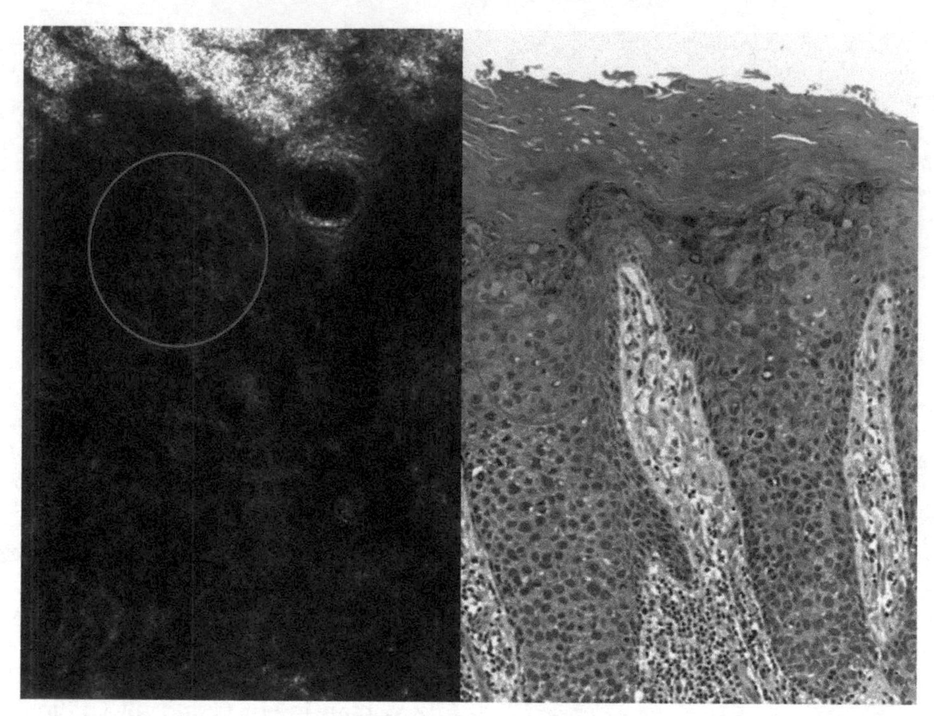

FIGURE 15.4 **Bowen disease.** Left image: Atypical honeycomb in the epidermis with large round cells (*red circle*) with a dark center and a bright rim corresponding to different degrees of dyskeratosis observed under routine histopathology (right image).

Actinic cheilitis represents the equivalent of AK on the lip. The use of RCM for this entity has been evaluated [61]. RCM was able to correctly identify several criteria for diagnosis of actinic cheilitis, including the presence of atypia at the stratum spinosum and granulosum with atypical honeycomb pattern. Marked inflammation represents a potential diagnostic pitfall and should be evaluated with caution.

Squamous Cell Carcinoma

The diagnosis of AK is mainly made upon clinical evaluation. In contrast, if a lesion is suspected to be an SCC, a biopsy will be collected to seek histological confirmation. This makes SCC a difficult disease to identify early. Although most SCCs display only limited dermoscopic features, namely scale and dotted or glomerular vessels, a wider range of diagnostic descriptors can be observed with RCM [59].

Several reproducible RCM features of SCC were described by Rishpon et al. [59] AK manifest similar features to a lesser extent. These findings are defined as follows and represent an important step toward the use of RCM in the diagnosis of SCC at the bedside.

- *Scale:* Variably refractile, amorphous material
- *Polygonal nucleated cells at the stratum corneum:* Sharply delineated cells with refractive thin outline surrounding a dark nucleus
- *Atypical honeycomb*
- *Disarranged epidermal pattern:* Severe architectural disarray of the spinous-granular layer in which the honeycomb pattern is no longer visible
- *Round nucleated cells at the spinous-granular layer:* Sharply delineated cells with refractive outline surrounding a dark nucleus that correspond to atypical keratinocytes or dyskeratotic cells
- *Round blood vessels traversing the dermal papilla*

These RCM features described for SCC do not determine whether RCM has the ability to distinguish between SCC and AK. A recent study published by Peppelman et al. with the objective to determine in vivo RCM features that are specific for making a distinction between AK and SCC revealed that the combination of architectural disarray in the stratum granulosum, architectural disarray in the spinous layer, and nest-like structures in the dermis can be used to distinguish SCC from AK [62] (Fig. 15.4).

Seborrheic Keratosis

Under RCM, seborrheic keratosis shows a preserved honeycomb pattern of the epidermis and densely packed, round to polymorphous, well-circumscribed dermal papillae at the DEJ. Both features are suggestive of benign neoplasms. The confocal features associated with the diagnosis of seborrheic keratosis include epidermal projections and keratin-filled invaginations at the lesion surface, corneal pseudocysts at epidermal layers, and melanophages and dilated round and linear blood vessels in the papillary dermis. These features are present despite their variable clinical and dermoscopic appearances [63].

REFLECTANCE CONFOCAL MICROSCOPY FOR INFLAMMATORY LESIONS

Psoriasis

RCM examination allows the identification of the main histological features of psoriatic lesions and revealed a high degree of correlation between RCM findings and routine histology. In 1999 González et al. characterized the histological features of psoriasis in vivo in lesional and nonlesional skin by RCM. In psoriatic lesions, nucleated corneocytes and collections of infiltrating inflammatory cells were clearly seen. Morphometric parameters such as epidermal height, length of papillary dermis, and the count of dermal papillae were also easily quantified. In the upper dermis, dilated capillary loops were present [64]. This was the first study providing evidence on how RCM was a suitable technique for histologically evaluating psoriasis in vivo.

Ardigo et al. investigated the in vivo RCM features of plaque psoriasis and analyzed their correlation with histopathological findings. They evaluated psoriatic lesions from 36 patients with an established diagnosis of plaque psoriasis. Subsequently, a 4-mm punch biopsy specimen of the same imaged areas was taken for histopathological examination. Normal skin from similar topographical areas of 12 healthy volunteers was evaluated as control samples. The results showed that several RCM features of psoriasis correlate well with histopathological evaluation. In more than 90% of the cases RCM revealed hyperkeratosis, parakeratosis, reduced or absent granular layer, papillomatosis, and dilated blood vessels. Acanthosis was observed in psoriasis cases, with thickness ranging from 75 to 300 μm, compared with normal skin, which ranged from 60 to 90 μm. The diameter of the dermal papillae was also enlarged (>100 μm) compared with what was observed in normal skin (>80 μm). These results led to the conclusion that RCM is useful for microscopic evaluation of plaque psoriasis features and offers a good correlation with histopathological findings [65].

Further description of the cellular features of psoriatic skin was performed by Wolberink et al. in [66]. In this study the morphological and cell biological features of plaque psoriasis and noninvolved skin were evaluated and quantified using RCM.

Correlation to hematoxylin and eosin (HE)-stained transverse and en face tissue sections were assessed, as well as to immunohistochemistry staining with filaggrin and CD3 antibody. Overall the results showed that RCM exhibits high correlation with histology, confirming the earlier results of González and Ardigo et al. [64,65] and identifying new features. The pathognomonic Munro microabscess could be also visualized, and the quantification and evaluation of cell biological and histological features of plaque psoriasis with RCM correlated highly to evaluation in HE-, CD3-, and filaggrin-stained histology [66].

More recently, the dynamics of neutrophil migration in active psoriatic lesions by RCM imaging has been evaluated. The active lesions showed a cyclic pattern of neutrophil migration, consisting of squirting papillae, transepidermal migration, accumulation in the stratum spinosum, accumulation in the stratum corneum, and degeneration of the abscesses. This study established the dynamics and time phasing of neutrophil migration in psoriatic lesions in vivo and demonstrated that RCM might distinguish between active or chronic psoriatic areas, which might contribute to new insights into the pathogenesis of psoriasis [67].

Allergic and Irritant Contact Dermatitis

Acute irritant contact dermatitis (ICD) and allergic contact dermatitis (ACD) may be difficult to distinguish by clinical or histological assessment. RCM has been used to noninvasively examine ACD and ICD in vivo and to describe the dynamics of the cutaneous changes in the two forms of acute contact dermatitis [68–73].

By correlating RCM results with clinical findings and histology [73,74], RCM has proved to be a useful be tool in the distinction of ACD and ICD reactions. Epidermal disruption can be easily visualized using RCM, and the dynamics of the RCM findings correlate well with the clinical scores. The degrees of epidermal spongiosis, vesicle formation, and exocytosis are less distinctive and less specific in the differentiation of ICD versus ACD. Intraepidermal vesicle formation was typically present more prominently in ACD and necrosis more commonly found in ICD.

Another RCM parameter of interest is the evolution of epidermal thickness. The marked increase in irritant reactions can only be partially explained by the presence of spongiosis because increased epidermal thickness is not evident in allergic reactions with similar degrees of spongiosis. For ACD, focal parakeratosis, hyperkeratosis, and significantly increased epidermal thickness have previously been described in late phases during follow-up. RCM allows repeatedly measuring the epidermal depth in vivo and following the evolution of epidermal thickness over time [72]. In ICD, early changes seen within the stratum corneum are a more pronounced disruption and demarcation and a focal parakeratosis interestingly seen as bright nuclei within the stratum corneum.

RCM was also used to evaluate whether susceptibility to ICD differs between black and white skin. This technique was able to track early pathophysiological events revealing differences between black and white skin during the development of ICD. For instance, participants with white skin had more severe clinical reactions than those with black skin, and RCM was able to reveal microscopic changes even without clinical evidence of irritation [74].

REFLECTANCE CONFOCAL MICROSCOPY FOR MONITORING TREATMENT RESPONSE OF SKIN TUMORS

Several reports have shown that RCM is useful in the monitoring of treatment response over time.

Reflectance Confocal Microscopy for Monitoring the Response of Basal Cell Carcinoma to Therapy

The nonsurgical treatment of BCC mainly includes photodynamic therapy (PDT), imiquimod, and oral hedgehog inhibitors. PDT is a well-established treatment option for BCC with a clearance rate of 87% and 53% for sBCC and nBCC, respectively. RCM has been utilized to monitor PDT efficacy in sporadic and Gorlin–Goltz patients [75], and it was considered as the optimal modality to detect BCC persistence.

Longo et al. analyzed a series of BCCs at different time points to obtain a precise analysis of biological changes after the treatment and to assess PDT efficacy in a long-term follow-up. Although all cases of BCC were correctly identified by dermoscopy and RCM at the baseline, only RCM allowed the detection of persisting tumors, which were judged clinically and dermoscopically as complete response to the treatment [76]. RCM is not only a valuable practical tool to monitor tumor response of BCC to PDT, but it also plays a role in the early assessment of the subclinical recurrences and in the better management of the lesions that may benefit from a different treatment option [77,78].

When RCM was used to establish the need for surgical intervention after imiquimod treatment for BCC, the results obtained by RCM correlated well with the histological analysis of tumor response to therapy, suggesting that confocal microscopy may help determine the need for surgery [79].

Vismodegib is an oral hedgehog inhibitor that has shown significant efficacy in the treatment of BCC. The evaluation of tumor regression by using RCM detects changes seen only in traditional histopathology by allowing the visualization of the response noninvasively and identifying characteristic signs of BCC response to vismodegib as the formation of pseudocysts and fibrosis [80].

Reflectance Confocal Microscopy for Monitoring the Response of Actinic Keratoses to Therapy

A study has evaluated RCM for monitoring the response to 5-aminolevulinic acid PDT in organ transplant recipients that display AK. In this study, RCM proved to be an excellent tool to monitor the response to PDT. This application of RCM is particularly important for immunocompromised individuals who have a high prevalence of AK [81].

The response of AK to other topical treatments such as imiquimod 5% and diclofenac 3% has been evaluated [82,83]. During topical therapy, the use of RCM allowed the noninvasive visualization and monitoring of sequential pharmacodynamic changes over time. Furthermore, RCM aided in the detection of subclinical AK in the setting of field cancerization, potentially increasing diagnostic accuracy compared with clinical evaluation alone.

Reflectance Confocal Microscopy for Monitoring the Response of Lentigo Maligna to Therapy

The off-label use of topical imiquimod in LM has been proposed as an alternative treatment to surgery and an adjunctive modality after surgical excision. Alarcon et al. evaluated the use of in vivo RCM to accurately monitor the response of LM to nonsurgical treatment with topical imiquimod [84]. They found that clinical evaluation overestimates the tumor clearance after imiquimod therapy (85%), with a high number of false-negative results, whereas dermoscopy underestimates the response (60%), but it identifies fewer false-negative results than clinical examination. RCM identified 70% of patients as responders without false-negative results, and when compared with histopathology, there was no statistically significant difference between this technique and histopathology in evaluating the response to imiquimod [85].

REFLECTANCE CONFOCAL MICROSCOPY FOR INFESTATIONS AND INFECTIONS

In vivo visualization of cutaneous organisms with RCM at the bedside could provide an alternative to scrapings, microscopy, cultures, and biopsies [86].

Onychomycosis and Dermatomycosis

RCM was evaluated for the diagnosis and treatment monitoring of toenail onychomycosis in a prospective study [87]. The aim of this study was to evaluate the accuracy of RCM for the diagnosis of onychomycosis compared with standard mycological tests. Under RCM, the fungal infection was observed as filamentous and/or roundish structures in the nail plate, corresponding respectively to septate hyphae and/or arthroconidia. The sensitivity and specificity of RCM for the diagnosis of onychomycosis was 52.9% and 90.2%, respectively.

The evaluation of tinea corporis with RCM seems to be useful for microscopic evaluation of mycelium features and has scientific value in the study of superficial cutaneous fungal infections [88,89].

Syphilis

Lesions suggestive of secondary syphilis were evaluated to assess the usefulness of in vivo RCM to detect *Treponema pallidum*. Elongated, small, bright particles with a spiral shape intermingled with the keratinocytes were seen under RCM and corresponded with immunohistochemical findings that revealed several spirochetes infiltrating the epidermis [90].

Pityriasis Folliculorum

RCM has been compared to standardized skin surface biopsy (SSSB) for measuring the density of Demodex mites. It was suggested to be a fast, direct, and noninvasive method for Demodex-associated diseases and superior to SSSB for Demodex mite detection [91–93].

Leishmaniasis

In vivo microscopic features of cutaneous leishmaniasis have been characterized by RCM. Some features were apparent on dermoscopy and RCM, including linear and comma-shaped vessels corresponding to local angiogenesis, amorphous material, and follicular plugging. Other features were strongly correlated with histopathology, including the presence of multinucleated giant cells and mixed inflammatory infiltrate. Despite the limitation in depth penetration, some multinucleated giant cells were observed, often located in the reticular dermis. This illustrates the ability of RCM to identify the cellular morphology of the immunological response to intracellular organisms and suggests that RCM could be a complementary diagnostic tool in these commonly self-healing lesions that may avoid unnecessary excisions [94].

Virus

Cutaneous herpesvirus infections have been imaged using RCM for immediate noninvasive bedside detection. RCM revealed the presence of pleomorphic ballooned keratinocytes and multinucleated giant cells in a loose aggregate of keratinocytes, inflammatory cells, and debris. These findings correlated well with those of routine histology [95,96].

RCM has proved to be useful in the evaluation of the human papilloma virus infection. The images obtained in genital warts relate well to the dermoscopy and histological findings. The papilomatosis and vascular patterns were similar to the dermoscopy and histological exam. It was also possible to identify the epidermic architecture variation in some areas, presenting koilocytosis and dendritic structures that were only identifiable by histopathological exam [97,98].

REFLECTANCE CONFOCAL MICROSCOPY APPLICATION IN COSMETICS

In vivo RCM offers the opportunity to determine in vivo the kinetics of the skin after the application of topical products [99] and to assess the influence of aging on epidermal turnover.

Skin Aging

Skin aging is a complex biological process that is traditionally classified as intrinsic and extrinsic aging. Several clinical scores and instrumental devices have been applied to obtain a precise assessment of skin aging. Among them, RCM has emerged as a new technique capable of assessing cytoarchitectural changes with a nearly histopathological resolution.

RCM has showed that young skin is characterized by regular polygonal keratinocytes and thin reticulated collagen fibers whereas aged skin is characterized by loss of small skin furrows, irregularity of the epidermal honeycomb pattern with more irregularly shaped keratinocytes, irregularly distributed (mottled) pigmented keratinocytes/melanocytes, irregularity of the papillary rings and/or effacement of the rete ridges, and loss of thin collagen fibers and presence of collagen clods corresponding to elastosis [100–104].

Longo et al. developed a score to quantify the skin aging-related signs with RCM. Combining the previously identified confocal features, three different semiquantitative scores were calculated: epidermal disarray score (irregular honeycombed pattern + epidermal thickness + furrow pattern), epidermal hyperplasia score (mottled pigmentation + extent of polycyclic papillary + epidermal thickness), and collagen score (curled fibers, 2 for huddles of collagen, 1 for coarse collagen structures, and 0 for thin reticulated collagen). This could allow to quantify the aging signs in vivo [105].

Monitoring Cosmetic Treatment

A study revealed that the morphological changes at different time points after the application of topical products in the skin surface can be analyzed by RCM. The presence of higher interkeratinocytes' brightness is observed when a moisturizer is applied, but not for the control area. This RCM finding could be linked to keratinocyte membrane protein exposure and/or substance release in the interkeratinocytic space [99].

The effect of topically applied vitamin C on the epidermal turnover (in terms of papillary index) was investigated with RCM. Density of functional entities present in the dermal papillae can be evaluated more accurately when compared with conventional histology partly because of avoidance of artifacts that occur after fixation. After the treatment with vitamin C, the smaller size of granular layer cells in the treated areas could be interpreted as a sign for higher proliferative activity of the epidermis. Another implication of the increase in papillae was that new blood vessels were formed during the treatment with vitamin C. The newly formed blood vessels showed a normal anatomical structure under RCM, and they were integrated in a healthy vascular architecture. Moreover no pathological changes in the vasculature were observed [101].

The assessment of a superficial chemical peel combined with a multimodal, hydroquinone-free skin brightener using RCM showed that in vivo RCM images at baseline and at postpeel correlated well with the improvements observed by the investigator [106].

The epidermal changes and collagen remodeling induced by laser skin rejuvenation have been evaluated by in vivo confocal microscopy, illustrating that RCM can be an essential adjunct for clinicians to measure the effects of laser treatment and possibly to gain new insights into the development of side effects [107,108].

CONCLUSIONS

In this chapter we have reviewed the rapidly increasing applications of RCM in the clinical and research arenas of dermatology. In vivo RCM is a true means of "optical biopsy." The use of RCM offers many opportunities to improve the speed and accuracy of the diagnosis in several skin conditions, including differentiating between benign and malignant skin lesions, monitoring response to medical or surgical treatments, and pathophysiological study of inflammatory processes including infections and infestations, among others.

Having the penetration depth to the upper dermis as a limitation, the quasi-histological imaging achieved by in vivo RCM has demonstrated to significantly improve the sensitivity and specificity in comparison with other noninvasive techniques. Although a training process is necessary to master this technique, teledermatological platforms through skill development and exchange of experiences can prove a valuable support for the beginners.

Potential clinical applications of RCM are limited by fundamental principles of light interactions with human tissue and by the current level of technology development.

Because optical images can be recorded remotely—in near real time without need for tissue removal—and image analysis can be automated, we believe that RCM affords many important advantages over traditional techniques, including the potential to reduce the number of unnecessary biopsies or quickly assess whether a tumor is responding to a treatment.

The detectability of contrast agents for confocal reflectance imaging of skin and microcirculation is currently limited to imaging at the cellular, nuclear, and general architectural levels because of the lack of microstructure-specific contrast [109]. Development of specific contrast agents can help to dramatically improve visualization of subcellular details. The trend toward new techniques that fully exploit the superior performance of RCM microscopes is expected to accelerate, which will quickly establish the confocal microscope as a routine laboratory tool for the dermatologist.

List of Abbreviations

ACD Allergic contact dermatitis
AK Actinic keratoses
BCC Basal cell carcinoma

DEJ Dermoepidermal junction
HE Hematoxylin and eosin
iBCC Infiltrative basal cell carcinoma
ICD Irritant contact dermatitis
LM Lentigo maligna
MM Malignant melanoma
nBCC Nodular basal cell carcinoma
NM Nodular melanoma
NMSC Nonmelanoma skin cancer
PDT Photodynamic therapy
RCM Reflectance confocal microscopy
sBCC Superficial basal cell carcinoma
SCC Squamous cell carcinoma
SSSB Standardized skin surface biopsy
UV Ultraviolet

Acknowledgments

The work was partially supported by Ministerio de Economía y Competitividad(MINECO, FIS PI12/01253) and Comunidad de Madrid (CAM, S2010/BMD-2359), Spain.

Conflict Statement: S.G. serves as a consultant for Caliber ID, manufacturer of reflectance confocal microscopes.

References

[1] Masood A, Al-Jumaily AA. Computer aided diagnostic support system for skin cancer: a review of techniques and algorithms. Int J Biomed Imag 2013;2013:323268.

[2] Calzavara-Pinton P, Longo C, Venturini M, Sala R, Pellacani G. Reflectance confocal microscopy for in vivo skin imaging. Photochem Photobiol 2008;84(6):1421–30.

[3] Rigel DS, Russak J, Friedman R. The evolution of melanoma diagnosis: 25 years beyond the ABCDs. CA Cancer J Clin 2010; 60(5):301–16.

[4] Mayer J. Systematic review of the diagnostic accuracy of dermatoscopy in detecting malignant melanoma. Med J Aust 1997; 167(4):206–10.

[5] Piccolo D, Ferrari A, Peris K, Diadone R, Ruggeri B, Chimenti S. Dermoscopic diagnosis by a trained clinician vs a clinician with minimal dermoscopy training vs computer-aided diagnosis of 341 pigmented skin lesions: a comparative study. Br J Dermatol 2002;147(3):481–6.

[6] Gerger A, Hofmann-Wellenhof R, Samonigg H, Smolle J. In vivo confocal laser scanning microscopy in the diagnosis of melanocytic skin tumours. Br J Dermatol 2009;160(3):475–81.

[7] Rajadhyaksha M, Grossman M, Esterowitz D, Webb RH, Anderson RR. In vivo confocal scanning laser microscopy of human skin: melanin provides strong contrast. J Invest Dermatol 1995;104(6):946–52.

[8] Corcuff P, Bertrand C, Leveque JL. Morphometry of human epidermis in vivo by real-time confocal microscopy. Arch Dermatol Res 1993;285(8):475–81.

[9] Corcuff P, Lévêque JL. In vivo vision of the human skin with the tandem scanning microscope. Dermatology 1993;186(1):50–4.

[10] Busam KJ, Charles C, Lee G, Halpern AC. Morphologic features of melanocytes, pigmented keratinocytes, and melanophages by in vivo confocal scanning laser microscopy. Mod Pathol 2001; 14(9):862–8.

[11] Sheppard CJ, Wilson T. Depth of field in the scanning microscope. Opt Lett 1978;3(3):115.

[12] Branzan AL, Landthaler M, Szeimies RM. In vivo confocal scanning laser microscopy in dermatology. Lasers Med Sci 2007;22(2): 73–82.

[13] Rajadhyaksha M, González S, Zavislan JM, Anderson RR, Webb RH. In vivo confocal scanning laser microscopy of human skin II: advances in instrumentation and comparison with histology. J Invest Dermatol 1999;113(3):293–303.

[14] González S, Swindells K, Rajadhyaksha M, Torres A. Changing paradigms in dermatology: confocal microscopy in clinical and surgical dermatology. Clin Dermatol 2003;21(5):359–69.

[15] Rajadhyaksha M, Anderson RR, Webb RH. Video-rate confocal scanning laser microscope for imaging human tissues in vivo. Appl Opt 1999;38(10):2105–15.

[16] Langley RG, Walsh N, Sutherland AE, Propperova I, Delaney L, Morris SF, et al. The diagnostic accuracy of in vivo confocal scanning laser microscopy compared to dermoscopy of benign and malignant melanocytic lesions: a prospective study. Dermatology 2007;215(4):365–72.

[17] Pellacani G, Guitera P, Longo C, Avramidis M, Seidenari S, Menzies S. The impact of in vivo reflectance confocal microscopy for the diagnostic accuracy of melanoma and equivocal melanocytic lesions. J Invest Dermatol 2007;127(12):2759–65.

[18] Pellacani G, Cesinaro AM, Seidenari S. In vivo assessment of melanocytic nests in nevi and melanomas by reflectance confocal microscopy. Mod Pathol 2005;18(4):469–74.

[19] Marghoob AA, Charles CA, Busam KJ, Rajadhyaksha M, Lee G, Clark-Loeser L, Halpern AC. In vivo confocal scanning laser microscopy of a series of congenital melanocytic nevi suggestive of having developed malignant melanoma. Arch Dermatol 2005; 141(11):1401–12.

[20] Soyer HP, Kenet RO, Wolf IH, Kenet BJ, Cerroni L. Clinicopathological correlation of pigmented skin lesions using dermoscopy. Eur J Dermatol 2000;10(1):22–8.

[21] Pellacani G, Farnetani F, Gonzalez S, Longo C, Cesinaro AM, Casari A, et al. In vivo confocal microscopy for detection and grading of dysplastic nevi: a pilot study. J Am Acad Dermatol 2012;66(3):e109–21.

[22] Ferrara G, Gianotti R, Cavicchini S, Salviato T, Zalaudek I, Argenziano G. Spitz nevus, Spitz tumor, and spitzoid melanoma: a comprehensive clinicopathologic overview. Dermatol Clin 2013;31(4):589–98.

[23] Pellacani G, Longo C, Ferrara G, Cesinaro AM, Bassoli S, Guitera P, et al. Spitz nevi: in vivo confocal microscopic features, dermatoscopic aspects, histopathologic correlates, and diagnostic significance. J Am Acad Dermatol 2009;60(2):236–47.

[24] Balch CM, Gershenwald JE, Soong SJ, Thompson JF, Atkins MB, Byrd DR, et al. Final version of 2009 AJCC melanoma staging and classification. J Clin Oncol 2009;27(36):6199–206.

[25] Gerger A, Koller S, Kern T, Massone C, Steiger K, Richtig E, et al. Diagnostic applicability of in vivo confocal laser scanning microscopy in melanocytic skin tumors. J Invest Dermatol 2005;124(3): 493–8.

[26] Gerger A, Hofmann-Wellenhof R, Langsenlehner U, Richtig E, Koller S, Weger W, et al. In vivo confocal laser scanning microscopy of melanocytic skin tumours: diagnostic applicability using unselected tumour images. Br J Dermatol 2008;158(2): 329–33.

[27] Pehamberger H, Steiner A, Wolff K. In vivo epiluminescence microscopy of pigmented skin lesions. I. Pattern analysis of pigmented skin lesions. J Am Acad Dermatol 1987;17(4): 571–83.

[28] Kittler H, Pehamberger H, Wolff K, Binder M. Diagnostic accuracy of dermoscopy. Lancet Oncol 2002;3(3):159–65.

[29] Pellacani G, Cesinaro AM, Seidenari S. Reflectance-mode confocal microscopy for the in vivo characterization of pagetoid melanocytosis in melanomas and nevi. J Invest Dermatol 2005; 125(3):532–7.

[30] Busam KJ, Charles C, Lohmann CM, Marghoob A, Goldgeier M, Halpern AC. Detection of intra epidermal malignant melanoma

in vivo by confocal scanning laser microscopy. Melanoma Res 2002;12(4):349–55.

[31] Pellacani G, Cesinaro AM, Longo C, Grana C, Seidenari S. Microscopic in vivo description of cellular architecture of dermoscopic pigment network in nevi and melanomas. Arch Dermatol 2005; 141(2):147–54.

[32] Pellacani G, Cesinaro AM, Seidenari S. Reflectance-mode confocal microscopy of pigmented skin lesions—improvement in melanoma diagnostic specificity. J Am Acad Dermatol 2005; 53(6):979–85.

[33] Swetter SM, Boldrick JC, Jung SY, Egbert BM, Harvell JD. Increasing incidence of lentigo maligna melanoma subtypes: northern California and national trends 1990–2000. J Invest Dermatol 2005;125(4):685–91.

[34] Cohen LM. Lentigo maligna and lentigo maligna melanoma. J Am Acad Dermatol December 1995;33(6):923–36. quiz 937–940.

[35] Erickson C, Miller SJ. Treatment options in melanoma in situ: topical and radiation therapy, excision and Mohs surgery. Int J Dermatol 2010;49(5):482–91.

[36] Tannous ZS, Mihm MC, Flotte TJ, González S. In vivo examination of lentigo maligna and malignant melanoma in situ, lentigo maligna type by near-infraredreflectance confocal microscopy: comparison of in vivo confocal images with histologic sections. J Am Acad Dermatol 2002;46(2):260–3.

[37] Langley RG, Burton E, Walsh N, Propperova I, Murray SJ. In vivo confocal scanning laser microscopy of benign lentigines: comparison to conventional histology and in vivo characteristics of lentigo maligna. J Am Acad Dermatol 2006;55(1):88–97.

[38] Chen CS, Elias M, Busam K, Rajadhyaksha M, Marghoob AA. Multimodal in vivo optical imaging, including confocal microscopy, facilitates presurgical margin mapping for clinically complex lentigo maligna melanoma. Br J Dermatol 2005;153(5): 1031–6.

[39] Curiel-Lewandrowski C, Williams CM, Swindells KJ, Tahan SR, Astner S, Frankenthaler RA, et al. Use of in vivo confocal microscopy in malignant melanoma: an aid in diagnosis and assessment of surgical and nonsurgical therapeutic approaches. Arch Dermatol 2004;140(9):1127–32.

[40] Guitera P, Pellacani G, Crotty KA, Scolyer RA, Li LX, Bassoli S, et al. The impact of in vivo reflectance confocal microscopy on the diagnostic accuracy of lentigo maligna and equivocal pigmented and nonpigmented macules of the face. J Invest Dermatol 2010;130(8):2080–91.

[41] Guitera P, Moloney FJ, Menzies SW, Stretch JR, Quinn MJ, Hong A, et al. Improving management and patient care in lentigo maligna by mapping with in vivo confocal microscopy. JAMA Dermatol 2013;149(6):692–8.

[42] Segura S, Pellacani G, Puig S, Longo C, Bassoli S, Guitera P, et al. In vivo microscopic features of nodular melanomas: dermoscopy, confocal microscopy, and histopathologic correlates. Arch Dermatol 2008;144(10):1311–20.

[43] Pupelli G, Longo C, Veneziano L, Cesinaro AM, Ferrara G, Piana S, et al. Small-diameter melanocytic lesions: morphological analysis by means of in vivo confocal microscopy. Br J Dermatol 2013;168(5):1027–33.

[44] Busam KJ, Hester K, Charles C, Sachs DL, Antonescu CR, Gonzalez S, et al. Detection of clinically amelanotic malignant melanoma and assessment of its margins by in vivo confocal scanning laser microscopy. Arch Dermatol 2001;137(7): 923–9.

[45] Curchin C, Wurm E, Jagirdar K, Sturm R, Soyer P. Dermoscopy, reflectance confocal microscopy and histopathology of an amelanotic melanoma from an individual heterozygous for MC1R and tyrosinase variant alleles. Australas J Dermatol 2012;53(4): 291–4.

[46] Maier T, Sattler EC, Braun-Falco M, Korting HC, Ruzicka T, Berking C. Reflectance confocal microscopy in the diagnosis of partially and completely amelanotic melanoma: report on seven cases. J Eur Acad Dermatol Venereol 2013;27(1):e42–52.

[47] Segura S, Puig S, Carrera C, Palou J, Malvehy J. Development of a two-step method for the diagnosis of melanoma by reflectance confocal microscopy. J Am Acad Dermatol 2009; 61(2):216–29.

[48] Alarcon I, Carrera C, Palou J, Alos L, Malvehy J, Puig S. Impact of in vivo reflectance confocal microscopy on the number needed to treat melanoma in doubtful lesions. Br J Dermatol 2014;170(4): 802–8.

[49] Leiter U, Eigentler T, Garbe C. Epidemiology of skin cancer. Adv Exp Med Biol 2014;810:120–40.

[50] González S, Tannous Z. Real-time, in vivo confocal reflectance microscopy of basal cell carcinoma. J Am Acad Dermatol 2002; 47(6):869–74.

[51] Ulrich M, Roewert-Huber J, González S, Rius-Diaz F, Stockfleth E, Kanitakis J. Peritumoral clefting in basal cell carcinoma: correlation of in vivo reflectance confocal microscopy and routine histology. J Cutan Pathol 2011;38(2):190–5.

[52] Nori S, Rius-Díaz F, Cuevas J, Goldgeier M, Jaen P, Torres A, et al. Sensitivity and specificity of reflectance-mode confocal microscopy for in vivo diagnosis of basal cell carcinoma: a multicenter study. J Am Acad Dermatol 2004;51(6):923–30.

[53] Guitera P, Menzies SW, Longo C, Cesinaro AM, Scolyer RA, Pellacani G. In vivo confocal microscopy for diagnosis of melanoma and basal cell carcinoma using a two-step method:analysis of 710 consecutive clinically equivocal cases. J Invest Dermatol 2012;132(10):2386–94.

[54] Telfer NR, Colver GB, Morton CA. British Association of Dermatologists. Guidelines for the management of basal cell carcinoma. Br J Dermatol 2008;159(1):35–48.

[55] Sterry W. European dermatology Forum guideline committee. Guidelines: the management of basal cell carcinoma. Eur J Dermatol 2006;16(5):467–75.

[56] Longo C, Lallas A, Kyrgidis A, Rabinovitz H, Moscarella E, Ciardo S, et al. Classifying distinct basal cell carcinoma subtype by means of dermatoscopy and reflectance confocal microscopy. J Am Acad Dermatol 2014;71(4):716–24. e1.

[57] Aghassi D, Anderson RR, González S. Confocal laser microscopic imaging of actinic keratoses in vivo: a preliminary report. J Am Acad Dermatol 2000;43(1 Pt 1):42–8.

[58] Horn M, Gerger A, Ahlgrimm-Siess V, Weger W, Koller S, Kerl H, et al. Discrimination of actinic keratoses from normal skin with reflectance mode confocal microscopy. Dermatol Surg 2008; 34(5):620–5.

[59] Rishpon A, Kim N, Scope A, Porges L, Oliviero MC, Braun RP, et al. Reflectance confocal microscopy criteria for squamous cell carcinomas and actinic keratoses. Arch Dermatol 2009; 145(7):766–72.

[60] Ulrich M, Maltusch A, Rius-Diaz F, Röwert-Huber J, González S, Sterry W, et al. Clinical applicability of in vivo reflectance confocal microscopy for the diagnosis of actinic keratoses. Dermatol Surg 2008;34(5):610–9.

[61] Ulrich M, González S, Lange-Asschenfeldt B, Roewert-Huber J, Sterry W, Stockfleth E, et al. Non-invasive diagnosis and monitoring of actinic cheilitis with reflectance confocal microscopy. J Eur Acad Dermatol Venereol 2011;25(3):276–84.

[62] Peppelman M, Nguyen KP, Hoogedoorn L, van Erp PE, Gerritsen MJ. Reflectance confocal microscopy: non-invasive distinction between actinic keratosis and squamous cell carcinoma. J Eur Acad Dermatol Venereol October 30, 2014. http://dx.doi.org/10.1111/jdv.12806 [Epub ahead of print].

[63] Ahlgrimm-Siess V, Cao T, Oliviero M, Laimer M, Hofmann-Wellenhof R, Rabinovitz HS, et al. Seborrheic

keratosis: reflectance confocal microscopy features and correlation with dermoscopy. J Am Acad Dermatol 2013;69(1): 120–6.

[64] González S, Rajadhyaksha M, Rubinstein G, Anderson RR. Characterization of psoriasis in vivo by reflectance confocal microscopy. J Med 1999;30(5–6):337–56.

[65] Ardigo M, Cota C, Berardesca E, González S. Concordance between in vivo reflectance confocal microscopy and histology in the evaluation of plaque psoriasis. J Eur Acad Dermatol Venereol 2009;23(6):660–7.

[66] Wolberink EA, van Erp PE, Teussink MM, van de Kerkhof PC, Gerritsen MJ. Cellular features of psoriatic skin: imaging and quantification using in vivo reflectance confocal microscopy. Cytometry B Clin Cytom 2011;80(3):141–9.

[67] Wolberink EA, Peppelman M, van de Kerkhof PC, van Erp PE, Gerritsen MJ. Establishing the dynamics of neutrophil accumulation in vivo by reflectance confocal microscopy. Exp Dermatol 2014;23(3):184–8.

[68] González S, González E, White WM, Rajadhyaksha M, Anderson RR. Allergic contact dermatitis: correlation of in vivo confocal imaging to routine histology. J Am Acad Dermatol 1999;40(5 Pt 1):708–13.

[69] Sakanashi EN, Matsumura M, Kikuchi K, Ikeda M, Miura H. A comparative study of allergic contact dermatitis by patch test versus reflectance confocal laser microscopy, with nickel and cobalt. Eur J Dermatol 2010;20(6):705–11.

[70] Astner S, Gonzalez E, Cheung A, Rius-Diaz F, González S. Pilot study on the sensitivity and specificity of in vivo reflectance confocal microscopy in the diagnosis of allergic contact dermatitis. J Am Acad Dermatol 2005;53(6):986–92.

[71] Astner S, González S, Gonzalez E. Noninvasive evaluation of allergic and irritant contact dermatitis by in vivo reflectance confocal microscopy. Dermatitis 2006;17(4):182–91.

[72] Astner S, González E, Cheung AC, Rius-Díaz F, Doukas AG, William F, et al. Non-invasive evaluation of the kinetics of allergic and irritant contact dermatitis. J Invest Dermatol 2005; 124(2):351–9.

[73] Swindells K, Burnett N, Rius-Diaz F, González E, Mihm MC, González S. Reflectance confocal microscopy may differentiate acute allergic and irritant contact dermatitis in vivo. J Am Acad Dermatol 2004;50(2):220–8.

[74] Hicks SP, Swindells KJ, Middelkamp-Hup MA, Sifakis MA, González E, González S. Confocal histopathology of irritant contact dermatitis in vivo and the impact of skin color (black vs white). J Am Acad Dermatol 2003;48(5):727–34.

[75] Segura S, Puig S, Carrera C, Lecha M, Borges V, Malvehy J. Noninvasive management of non-melanoma skin cancer in patients with cancer predisposition genodermatosis: a role for confocal microscopy and photodynamic therapy. J Eur Acad Dermatol Venereol 2011;25(7):819–27.

[76] Longo C, Casari A, Pepe P, Moscarella E, Zalaudek I, Argenziano G, et al. Confocal microscopy insights into the treatment and cellular immune response of Basal cell carcinoma tophotodynamic therapy. Dermatology 2012;225(3):264–70.

[77] Goldgeier M, Fox CA, Zavislan JM, Harris D, Gonzalez S. Noninvasive imaging, treatment, and microscopic confirmation of clearance of basal cell carcinoma. Dermatol Surg 2003;29(3): 205–10.

[78] Venturini M, Sala R, Gonzàlez S, Calzavara-Pinton PG. Reflectance confocal microscopy allows in vivo real-time noninvasive assessment of the outcome of methyl aminolaevulinate photodynamic therapy of basal cell carcinoma. Br J Dermatol 2013;168(1): 99–105.

[79] Torres A, Niemeyer A, Berkes B, Marra D, Schanbacher C, González S, et al. 5% imiquimod cream and reflectance-mode confocal microscopy as adjunct modalities to Mohs micrographic surgery for treatment of basal cell carcinoma. Dermatol Surg 2004;30(12 Pt 1):1462–9.

[80] Maier T, Kulichova D, Ruzicka T, Berking C. Noninvasive monitoring of basal cell carcinomas treated with systemic hedgehog inhibitors: pseudocysts as a sign of tumor regression. J Am Acad Dermatol 2014;71(4):725–30.

[81] Astner S, Swindells K, González S, Stockfleth E, Lademann J. Confocal microscopy: innovative diagnostic tools for monitoring of noninvasive therapy in cutaneous malignancies. Drug Discov Today Dis Mech 2008;5(1):e81–91.

[82] Ulrich M, Krueger-Corcoran D, Roewert-Huber J, Sterry W, Stockfleth E, Astner S. Reflectance confocal microscopy for noninvasive monitoring of therapy and detection of subclinical actinic keratoses. Dermatology 2010;220(1):15–24.

[83] Malvehy J, Roldán-Marín R, Iglesias-García P, Díaz A, Puig S. Monitoring treatment of field cancerisation with 3% diclofenac sodium 2.5% hyaluronic acid by reflectance confocal microscopy: a histologic correlation. Acta Derm Venereol 2015;95(1): 45–50.

[84] Nadiminti H, Scope A, Marghoob AA, Busam K, Nehal KS. Use of reflectance confocal microscopy to monitor response of lentigo maligna to nonsurgical treatment. Dermatol Surg 2010;36(2): 177–84.

[85] Alarcon I, Carrera C, Alos L, Palou J, Malvehy J, Puig S. In vivo reflectance confocal microscopy to monitor the response of lentigo maligna to imiquimod. J Am Acad Dermatol 2014;71(1): 49–55.

[86] Slutsky JB, Rabinovitz H, Grichnik JM, Marghoob AA. Reflectance confocal microscopic features of dermatophytes, scabies, and demodex. Arch Dermatol 2011;147(8):1008.

[87] Pharaon M, Gari-Toussaint M, Khemis A, Zorzi K, Petit L, Martel P, et al. Diagnosis and treatment monitoring of toenail onychomycosis by reflectance confocal microscopy: prospective cohort study in 58 patients. J Am Acad Dermatol 2014;71(1): 56–61.

[88] Hui D, Xue-cheng S, Ai-e X. Evaluation of reflectance confocal microscopy in dermatophytosis. Mycoses 2013;56(2):130–3.

[89] Cinotti E, Perrot JL, Labeille B, Moragues A, Raberin H, Flori P. Cambazard F groupe imagerie cutanée non invasive de la Société française de dermatologie. Tinea corporis diagnosed by reflectance confocal microscopy. Ann Dermatol Venereol 2014;141(2): 150–2.

[90] Venturini M, Sala R, Semenza D, Santoro A, Facchetti F, Calzavara-Pinton P. Reflectance confocal microscopy for the in vivo detection of *Treponema pallidum* in skin lesions of secondary syphilis. J Am Acad Dermatol 2009;60(4):639–42.

[91] Turgut Erdemir A, Gurel MS, Koku Aksu AE, Bilgin Karahalli F, Incel P, Kutlu Haytoğlu NS, et al. Reflectance confocal microscopy vs standardized skin surface biopsy for measuring the density of Demodexmites. Skin Res Technol 2014;20(4):435–9.

[92] Yuan C, Wang XM, Guichard A, Lihoreau T, Sophie MM, Lamia K, et al. Comparison of reflectance confocal microscopy and standardized skin surface biopsy for three different lesions in a pityriasis folliculorum patient. Br J Dermatol October 31, 2014. http://dx.doi.org/10.1111/bjd.13506 [Epub ahead of print].

[93] Veasey J, Framil V, Ribeiro A, Lellis R. Reflectance confocal microscopy use in one case of Pityriasis folliculorum: a Demodex folliculorum analysis and comparison to other diagnostic methods. Int J Dermatol 2014;53(4):e254–7.

[94] Alarcon I, Carrera C, Puig S, Malvehy J. In vivo confocal Microsc features Cutan leishmaniasis. Dermatology 2014; 228(2):121–4.

[95] Goldgeier M, Alessi C, Muhlbauer JE. Immediate noninvasive diagnosis of herpesvirus by confocal scanning laser microscopy. J Am Acad Dermatol 2002;46(5):783–5.

[96] Debarbieux S, Depaepe L, Poulalhon N, Dalle S, Balme B, Thomas L. Reflectance confocal microscopy characteristics of eight cases of pustular eruptions and histopathological correlations. Skin Res Technol 2013;19(1):e444–52.

[97] Veasey JV, Framil VM, Nadal SR, Marta AC, Lellis RF. Genital warts: comparing clinical findings to dermatoscopic aspects, in vivo reflectance confocal features and histopathologic exam. An Bras Dermatol 2014;89(1):137–40.

[98] González S, Gilaberte-Calzada Y. In vivo reflectance-mode confocal microscopy in clinical dermatology and cosmetology. Int J Cosmet Sci 2008;30(1):1–17.

[99] Manfredini M, Mazzaglia G, Ciardo S, Simonazzi S, Farnetani F, Longo C, et al. Does skin hydration influence keratinocyte biology? in vivo evaluation of microscopic skin changes induced by moisturizers by means of reflectance confocal microscopy. Skin Res Technol 2013;19(3):299–307.

[100] Wurm EM, Longo C, Curchin C, Soyer HP, Prow TW, Pellacani G. In vivo assessment of chronological ageing and photoageing in forearm skin using reflectance confocal microscopy. Br J Dermatol 2012;167(2):270–9.

[101] Sauermann K, Jaspers S, Koop U, Wenck H. Topically applied vitamin C increases the density of dermal papillae in aged human skin. BMC Dermatol 2004;4(1):13.

[102] Longo C, Casari A, Beretti F, Cesinaro AM, Pellacani G. Skin aging: in vivo microscopic assessment of epidermal and dermal changes by means of confocal microscopy. J Am Acad Dermatol 2013;68(3):e73–82.

[103] Haytoglu NS, Gurel MS, Erdemir A, Falay T, Dolgun A, Haytoglu TG. Assessment of skin photoaging with reflectance confocal microscopy. Skin Res Technol 2014;20(3):363–72.

[104] Kawasaki K, Yamanishi K, Yamada H. Age-related morphometric changes of inner structures of the skin assessed by in vivo reflectance confocal microscopy. Int J Dermatol 2015; 54(3):295–301.

[105] Longo C, Casari A, De Pace B, Simonazzi S, Mazzaglia G, Pellacani G. Proposal for an in vivo histopathologic scoring system for skin aging by means of confocal microscopy. Skin Res Technol 2013;19(1):e167–73.

[106] Goberdhan LT, Mehta RC, Aguilar C, Makino ET, Colvan L. Assessment of a superficial chemical peel combined with a multimodal, hydroquinone-free skin brightener using in vivo reflectance confocal microscopy. J Drugs Dermatol 2013;2(3):S38–41.

[107] Longo C, Galimberti M, De Pace B, Pellacani C, Bencini PL. Laser skin rejuvenation: epidermal changes and collagen remodeling evaluated by in vivo confocal microscopy. Lasers Med Sci 2013; 28(3):769–76.

[108] Shin MK, Kim MJ, Baek JH, Yoo MA, Koh JS, Lee SJ, et al. Analysis of the temporal change in biophysical parameters after fractional laser treatments using reflectance confocal microscopy. Skin Res Technol February 2013;19(1):e515–20. http://dx.doi.org/10.1111/srt.12003 [Epub September 7, 2012].

[109] Rajadhyaksha M, Gonzalez S, Zavislan JM. Detectability of contrast agents for confocal reflectance imaging of skin and microcirculation. J Biomed Opt 2004;9(2):323–31.

CHAPTER

16

Hyperspectral and Multispectral Imaging in Dermatology

F. Vasefi[1], *N. MacKinnon*[1], *D.L. Farkas*[1,2]

[1]Spectral Molecular Imaging Inc., Beverly Hills, CA, United States; [2]University of Southern California, Los Angeles, CA, United States

OUTLINE

INTRODUCTION

Hyperspectral imaging (HSI) is an emerging field in which the advantages of optical spectroscopy as an analytical tool are combined with two-dimensional object visualization obtained by optical imaging [70]. In HSI, each pixel of the image contains spectral information, which is added as a third dimension of values to the two-dimensional spatial image, generating a three-dimensional data cube, sometimes referred to as *hypercube data* or as an *image cube* [34]. A simple, well-known example of a three-dimensional data cube is the common RGB color image, where each pixel has red, green, and blue color. Hyperspectral data cubes can contain absorption, reflectance, or fluorescence spectrum data for each image pixel [37]. It is assumed that HSI data is spectrally sampled at more than 20 equally distributed wavelengths [35]. The spectral range in hyperspectral data can extend beyond the visible range (ultraviolet, infrared). *Multispectral imaging (MSI)* is a term that should probably be reserved for imaging that simultaneously uses two or more different spectroscopy methods in the imaging mode (eg, wavelength and fluorescence lifetime). The result of imaging with a couple of color bands/filters should not be termed *spectral imaging,* much less *multispectral,* but unfortunately

Imaging in Dermatology
http://dx.doi.org/10.1016/B978-0-12-802838-4.00016-9

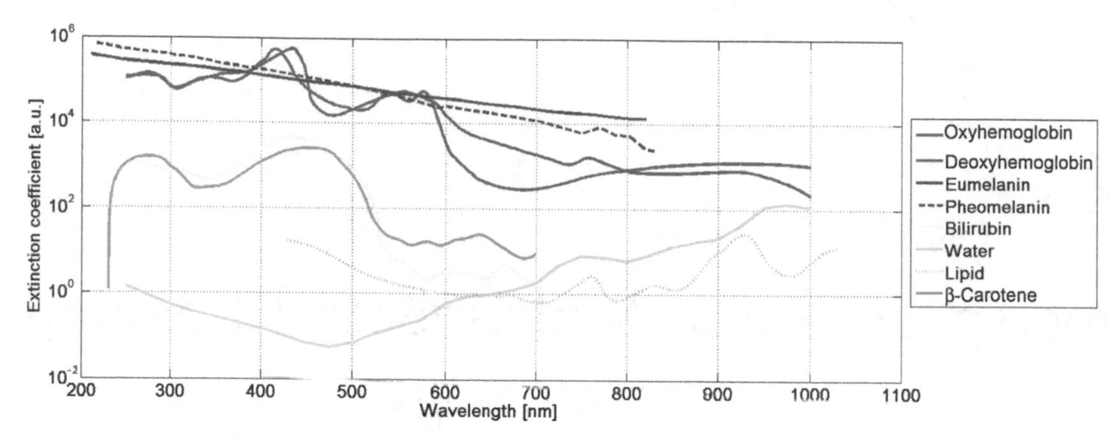

FIGURE 16.1 Absorption spectra of skin constituents. *a.u.*, Arbitrary unit.

these terms have been somewhat misused, especially in the biological literature. With these reservations, in the following review MSI systems typically record images at fewer than 20 wavelengths and can include noncontiguous as well as wide and narrow spectral bands.

HSI applications in industry (machine vision [62]) and remote sensing (including satellite reconnaissance [33]) are relatively well-known because the technique was originally defined by Goetz in the late 1980s [18]. However, within the past decade, a surge of interest in HSI technology has been seen in life sciences with applications in fields as diverse as agriculture [36,38], food quality and safety [19], pharmaceuticals [40], and particularly in healthcare.

The surface of the body is an excellent area for deployment of optical research methods, and HSI technology is being applied in ever more applications in dermatology for noninvasively targeting cancer detection, skin oxygenation mapping for diabetic ulcers [71], spectral unmixing of fluorescently labeled antigens, and more. In this chapter, we will introduce the tissue optics principles used for hyperspectral skin imaging. Different spectral imaging technologies for a variety of skin imaging applications will be described.

TISSUE OPTICS PRINCIPLES

Information about skin physiology, morphology, and composition can be obtained noninvasively by optical imaging methods. When light interacts with tissue, it is usually altered in some way before being remitted and detected by an image sensor or a point probe. Photons can be scattered because of the refractive index fluctuations on a microscopic level [55] by collagen fibers or by membranes, have their polarization altered after multiple scattering events, or be absorbed by molecules such as hemoglobin or melanin. Skin has a compound layered structure comprising dermal and epidermal layers. Epidermis thickness ranges between 30 and 150 μm [2]. Melanin is found in the epidermis in the form of red/yellow pheomelanin and/or brown/black eumelanin that absorbs very broadly in the visible spectrum with higher values for shorter wavelengths. Human skin color is characterized by variable concentrations of eumelanin, where the volume fraction of the epidermis occupied by melanosomes varies and can range from 1.3% [light Caucasian skin (type I)] to 43% [very dark in African skin (type VI)] [24]. The dermis thickness can vary from 0.6 μm to 3 mm [78] and consists of the papillary dermis and reticular dermis. The layers are composed of connective tissues, blood vessels, and nerves. Blood concentration in the dermis varies from 0.2% to 7% in volume fraction [24]. In the blood cells there are several natural chromophores, primarily hemoglobin, which absorbs blue and green light and gives blood its reddish color. The hemoglobin oxygenation varies from 47% in veins to 90–95% hemoglobin oxygenation in arteries [3]. The total hemoglobin concentration in blood is between 134 and 173 g/L [80]. Hemoglobin is found mainly in its oxyhemoglobin and deoxyhemoglobin forms in the microvascular network of the dermis, normally 50–500 μm below the skin surface [84]. Other chromophores present can include bilirubin and β-carotene that when found in the dermis contribute to the yellowish or olive tint of human skin. Fig. 16.1 displays the absorption spectra of major skin constituents that have a distinctive spectral absorption signature.

The contribution of melanin, epidermal thickness, and hemoglobin variations are evident in Fig. 16.2, which shows their effect on skin spectral absorption. The skin absorptions were calculated using the Biophysically-Based Spectral Model of Light Interaction with Human Skin (BioSpec) modeling software developed by researchers at University of Waterloo, Canada. Figs. 16.2A and B show the skin absorption spectra variation resulting from changing the percentage of epidermis occupied by melanosomes from 1% to 5%

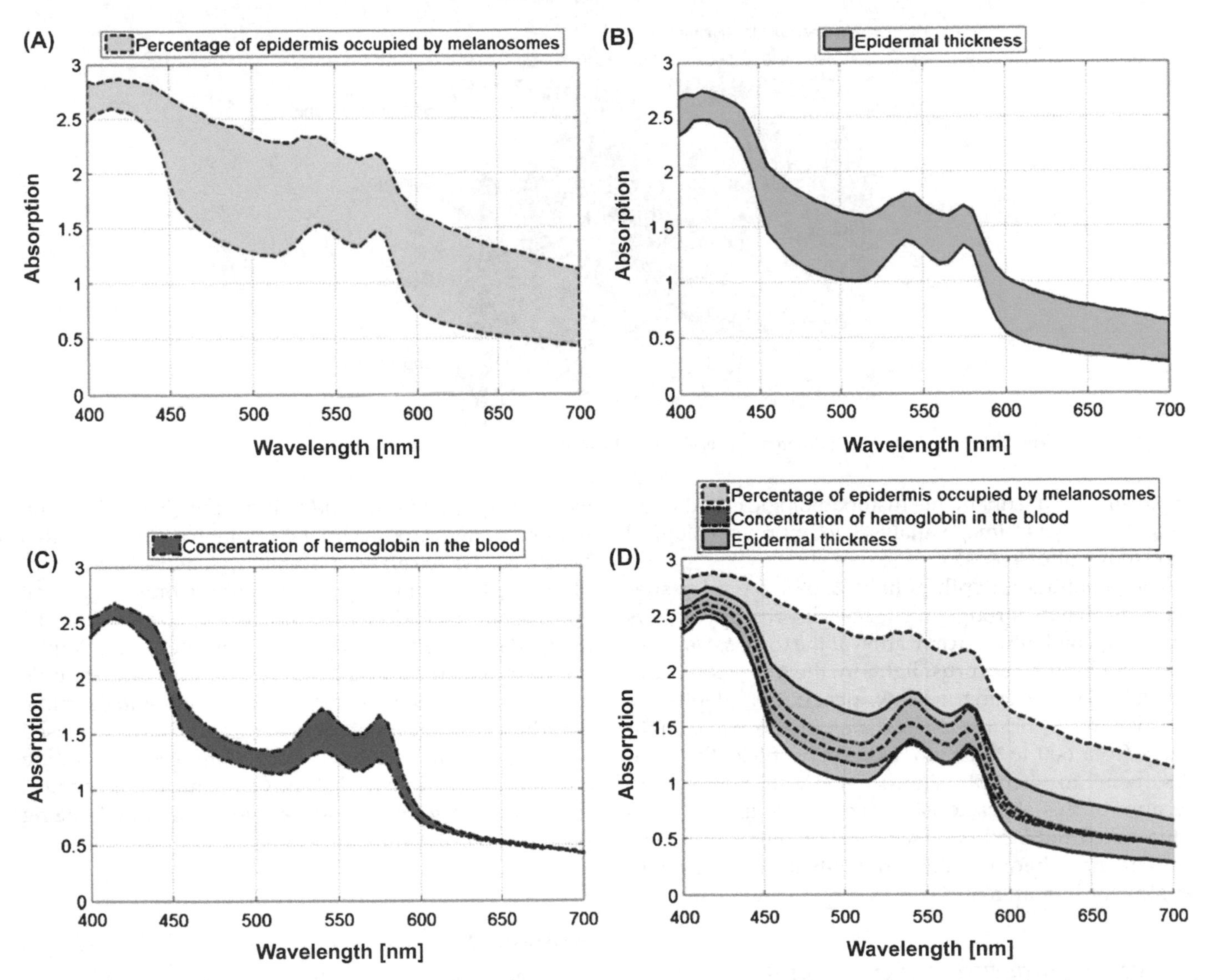

FIGURE 16.2 Skin absorption spectra at various skin geometry and chromophore concentration variations. (A) Variations in the percentage of epidermis occupied by melanosome. (B) Epidermal thickness variations. (C) Variations in concentration of hemoglobin in the blood. (D) Overlapping skin absorption spectral regions at all three skin property variations described in parts A, B, and C.

and varying the epidermal thickness from 50 to 150 μm. Fig. 16.2C displays the skin spectral absorption change resulting from changing the concentration of hemoglobin in the blood from 100 to 200 g/L. Parameters that remain constant are the percentage of epidermis occupied by melanosome, epidermal thickness, and concentration of hemoglobin in the blood at 1.3%, 100 μm, and 147 g/L, respectively. Fig. 16.2D shows the complexity of absorption spectra resulting from changing any of three skin parameters. As can be seen, different combinations of melanin, hemoglobin, or epidermal thickness values can have similar absorption spectra patterns.

Light entering tissues can also have its wavelength distribution shifted by interaction at the atomic or molecular levels, producing fluorescence or Raman signals [83]. Interpreting these changes can provide diagnostically useful information about the underlying structure of the tissue, provided that there is a plausible biological rationale for the change. Changes in the spectral characteristics in different wavelength regions produce a distinguishable spectral signature that reflects the underlying biology.

Spectral imaging technology has a unique capability for skin characterization because it can take advantage of the spatial relationships among the different tissue absorption spectra in a neighborhood. Spectral data cube analysis can incorporate complex spectral–spatial models that can provide more accurate classification of image features specific to a targeted disease. The technology unlocks new capabilities in medicine by which spatial and functional relationships among biologically active molecules can be observed, helping to noninvasively identify and quantify changes in living

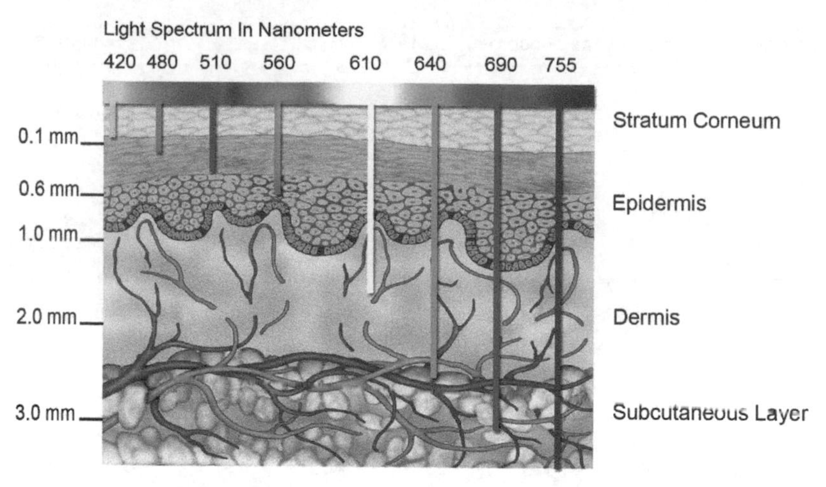

FIGURE 16.3 The penetration of light of different wavelengths in the skin.

organisms, and enhancing histopathological and fluorescent biomarker image analyses to improve biological knowledge of diseases.

The penetration depth of light into biological tissues depends on how strongly the tissue absorbs and scatters light. High melanin concentration at the topmost layer of skin (epidermis) absorbs light in the ultraviolet and visible range, leading to low penetration depth for wavelengths shorter than 600 nm. In the wavelength range from 600 to 1300 nm, skin has sufficiently weak absorbers to permit significant light penetration. Because of this characteristic, this wavelength range is often called the *therapeutic window* [79]. Fig. 16.3 shows a schematic of light penetration depth at different wavelengths for human skin.

HYPERSPECTRAL/MULTISPECTRAL IMAGING SYSTEMS

The acquisition of spectrally resolved image sets can be implemented by different techniques. The spectral discrimination devices can be inserted either in the illumination path or in the imaging path. The illumination arm of a spectral imaging system can be spectrally scanned, providing information, in reflection or absorbance, similar to what can be achieved by spectrally filtering the detection arm. In a simple approach, by employing a tunable light source, the illuminating light is varied either continuously or discontinuously through the spectrum. The wavelength selection can be done by using diffraction gratings, filter wheels, or tunable filters such as acoustooptic tunable filters (AOTFs) or liquid crystal tunable filters (LCTFs). Tunable filters can offer rapid and/or random selection of narrow spectral bands of light in a "band-sequential" operation. In order to detect simultaneously all the desired wavelengths at higher spectral resolution, one can use a dispersive element equipped with one or more slits in combination with scanning across spatial coordinates. Such a spatial scanning system can be designed to scan point by point or line by line. Another approach offers limited spectral and spatial resolution with the benefit of acquiring the image data cube in a snapshot. Computed tomography imaging spectrometer (CTIS) systems are an example of this nonscanning technique. In the following section, we will review the most interesting and commonly used imaging systems to record spectral images, elaborating on their advantages and disadvantages for biomedical applications. Fig. 16.4 illustrates typical spectral imaging systems used in skin analysis.

Whiskbroom

In whiskbroom (also called *flying spot*) imaging technology [50] (see Fig. 16.4A), scanning mirrors are often used to raster the field of view in both x and y directions, and the reflected light is dispersed by a prism or diffracted by a grating and recorded by a linear array detector. The spectral image data cubes (x, y, and λ) can be acquired by scanning over the two-dimensional scene (x, y). In this point-scanning method, two-axis motorized positioning mirrors are usually needed to finish the image acquisition, which makes the hardware configuration more complex and acquisition more time consuming. High acquisition speed and high sensitivity can be achieved by the spectral confocal laser-scanning method, in which samples are illuminated and images acquired through confocal or conjugate pinholes and multiple lasers or wavelength dispersive spectrophotometers are used to get spectral information [15]. Confocal methods provide multiple advantages, including the ability to control depth of field, elimination of the x-y scanning system, reducing signal contribution from outside the focal plane, and collecting

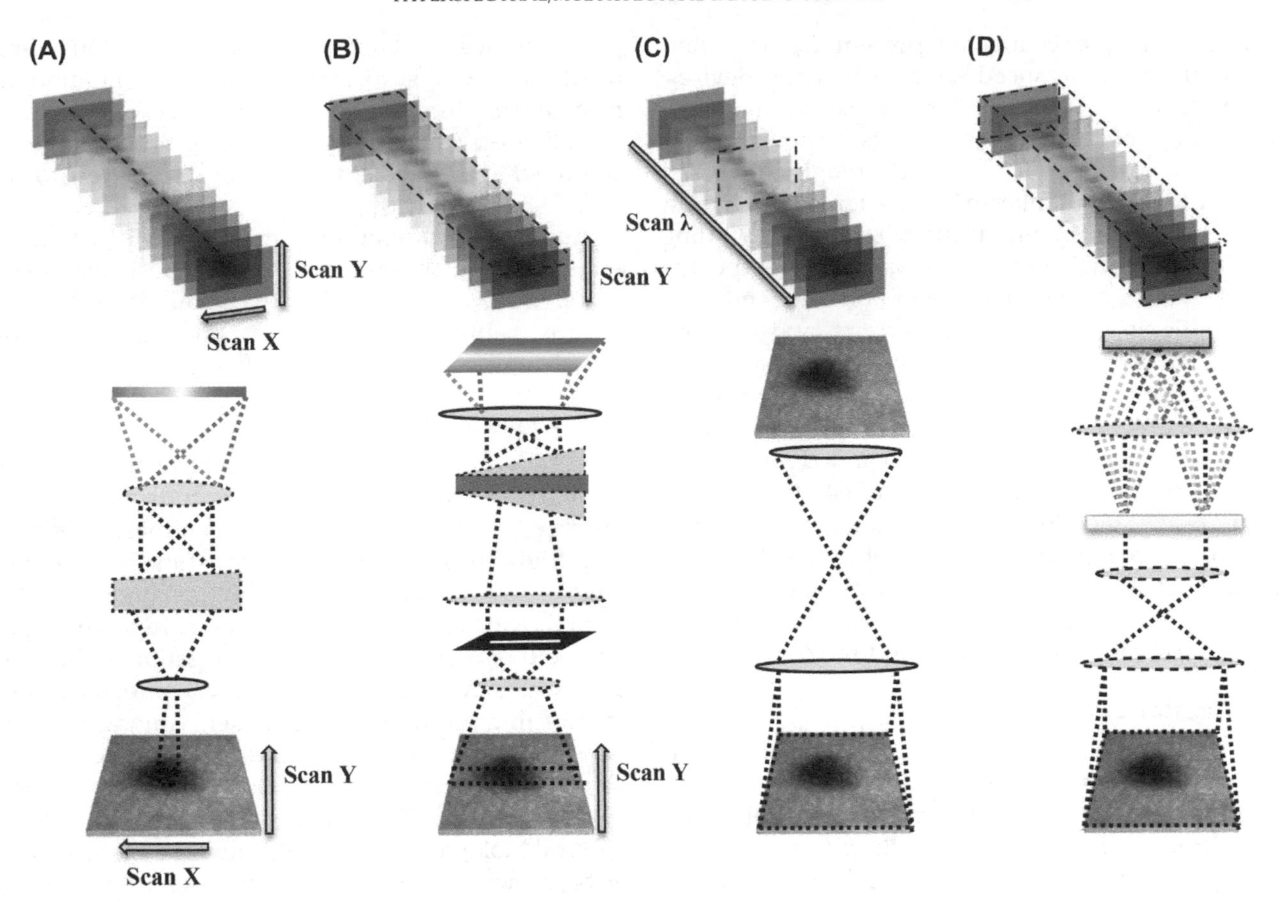

FIGURE 16.4 Typical spectral imaging techniques. (A) Whiskbroom. (B) Pushbroom. (C) Staring array. (D) Snapshot.

serial optical sections from thick specimens to form four-dimensional data (three-dimensional spatial and fourth spectral dimension).

Pushbroom

In recent years more sensitive two-dimensional sensors with fast acquisition have been introduced, providing technology platforms that extend the whiskbroom and single-slit spectrometer to line-scanning systems. In the pushbroom approach (see Fig. 16.4B), the object is wide-field illuminated, and scanning of either the object (pushbroom imaging spectrometer) or the slit (moving slit spectrometer) is performed to obtain the three-dimensional hyperspectral data cube [58]. The spatial coordinates of the object along with slit direction and spectral information are recorded in the respective row or column of the two-dimensional sensors. Pushbroom HSI systems are highly rugged, stable, and provide high spectral resolution with rapid acquisition. The pushbroom technique also allows reduced illumination (line illumination). This reduces the heat load and background scattering in the sample compared with other imaging techniques.

Staring Array

Staring array technology employs band-sequential spectral imagers that record grayscale images by scanning the illumination or remitted light, either continuously or discontinuously, through a number of wavelengths (see Fig. 16.4C). The result is always an image stack that constitutes a spectral cube and can be further analyzed for relevant information. The most common methods for wavelength scanning are described in the following sections.

Filter Wheels

Whereas slit spectrometers measure the whole spectrum at once and scan the spatial domain, band-sequential spectrometers record the whole image at once and scan the spectral domain. The most basic approach to wavelength scanning MSI is the use of a set of narrow band-pass filters mounted on a spinning disk (filter wheel). As the disk rotates in synchrony with the frame rate of the charge-coupled device (CCD) detector, different filters perpendicularly cross the optical path, leading to wavelength selection. Although they do not provide spectral imaging in the strict sense of more or less continuous high-resolution spectral content, filter changers represent the main alternative to all more advanced spectral imaging devices. They are often used for fluorescence imaging. An advantage of this approach is its basic simplicity and relatively low cost. Of all the technologies discussed here, fixed wavelength filter-based systems are the most likely to find their way into turnkey systems (including clinical applications). Potential disadvantages include their inflexibility; new dyes or fluorescent markers may be difficult to add to the repertoire, and there remains the possibility of image shift associated with each change of filter, although with careful engineering this can be minimized or, if necessary, corrected ex post facto. Such systems are not well suited for spectral analysis of complex samples with signals that are not made up of combinations of well-characterized fluorophores or chromophores. The spectral resolution is determined by the bandwidth and the number of the filters used. For example, a spectral resolution of 5 nm over a spectral range of 200 nm would require 40 filters.

Liquid Crystal Tunable Filters

LCTFs select the wavelengths they transmit based on the same principles as fixed wavelength interference filters, but with faster (~ tens of milliseconds) electronically controlled tuning. They have the advantage of no moving parts and random wavelength access within their spectral range. LCTFs consist of a number of liquid crystal layers, each of which pass a number of different frequencies; stacking them results in a single dominant transmission band, along with much smaller side bands. Typically they comprise $m+1$ polarizers, separated by m layers of liquid crystals (m is typically 3–10) and sandwiched between birefringent crystals. The optical path difference in birefringent crystals is dependent on crystal thickness and the refractive index difference between the ordinary and extraordinary light rays produced at the wavelength (λ) of incident illumination. Transmission of light through the crystal is dependent on the phase delay created by the difference in propagation speed between the extraordinary and ordinary rays. Crystals are often selected for a binary sequence of retardation so that transmission is maximal at the wavelength determined by the thickest crystal retarder. Other stages in the filter serve to block the transmission of unwanted wavelengths. All these yield versatile and relatively fast wavelength selection, without much image distortion or shift. Each filter assembly can span approximately one octave of wavelength (eg, 400–800 nm) and their useful range can extend into the near-infrared range. They can be introduced either in the illumination (or excitation) pathway, the emission pathway, or both. Throughput is a problem, in that half the light corresponding to one polarization state is lost automatically and peak transmission of the other half probably does not exceed 40% at best. Out-of-band rejection is not sufficient to prevent excitation light from leaking into the emission channel without the use of a dichroic mirror or cross-polarization [48]. A small controller box in addition to a PC is needed to drive an LCTF assembly.

Another consideration is that bandwidth is not the same at all wavelengths. The bandpass increases as wavelength increases, with bandwidth at longer wavelengths being as much as 3 times that at shorter wavelengths.

Acoustooptic Tunable Filters

AOTFs provide electronically controllable, solid-state wavelength tunability of light from ultraviolet to near-infrared, with random-access, bandpass variability, and high throughput [15]. They function based on light–sound (photon–phonon) interactions in a tellurium dioxide crystal, and are therefore tunable at speeds limited by the speed of sound propagation in the crystal. This yields typical wavelength switching times in the tens of microseconds, making them among the fastest wavelength switching devices available. These features have led to the use of AOTFs in a wide variety of spectroscopic applications. Interest in AOTFs for multispectral biological and biomedical imaging has, however, been more recent. After remote sensing-type imaging applications for ground-based and planetary targets were demonstrated in the 1970s and 1980s, several groups have reported breadboard AOTF imaging demonstrations [7,61], as well as their use in biologically relevant experiments such as fluorescence [47] and Raman [65] microscopy of biological samples.

In an AOTF, filtered light of narrow spectral bandwidth is angularly deflected away from the incident beam at the output of the crystal. The central wavelength of this filtered beam is determined by the acoustic frequency of the AOTF; this wavelength can be changed within approximately 25 μs to any other wavelength within the system range. High-end AOTFs involve additional optics compared with LCTFs, and require higher power and more involved electronics. The payoff is much faster switching and the ability to vary not only the wavelength, but also the bandwidth and the intensity of the transmitted light. Experiments involving luminescence lifetimes or very rapid acquisition of multiple wavelengths are possible using this technology. However, imaging applications have been impeded for many years by the fact that usually only one polarization state is available, that there is some image shift when wavelengths are changed, intrinsic image blur is present, and out-of-band rejection is no greater than 10^{-2}–10^{-3}. Approaches yielding solutions to some of these problems have been described previously [76]. Improvements include transducer apodization (in the emission path) and the use of two AOTFs in tandem (in the excitation path) to further improve out-of-band rejection [12]. *Apodization* refers to a technique in which the transducer is sectioned into a number of discrete slices with the relative amplitudes of the electronic signal used to drive each arranged to generate an acoustic spatial profile within the crystal, which optimizes the shape of the filter spectral bandpass. With proper apodization (11 electronically-isolated, individually-driven slices with a slice-to-slice separation of ~50 μm), Wachman and Farkas obtained a greater than 10 dB decrease in the out-of-band light between the primary side lobes, with even greater decreases obtainable in specific regions. In addition, they developed an AOTF illumination system that can be fiber-coupled to any microscope (or other optical system) to provide speed and spectral versatility for excitation as well as detection. This source consists of a 500-W short-arc xenon lamp beam followed by a pair of identical AOTFs placed in series, with the output of the second AOTF coupled into a fiber. This crystal configuration enables both polarizations to be filtered twice and recombined with a minimum of extra optics. The double filtering greatly reduces both the out-of-band light levels (lower than 10^{-4}) relative to the peak. This system has been tested for its suitability for both bright-field and fluorescence applications. More recent developments include the use of a supercontinuum laser light source, which can provide, because of the short duration of the laser pulses (several ps), not only wavelength selection but also multiphoton excitation of the sample and time-resolved measurements (like fluorescence-lifetime imaging microscopy, for example).

Digital Micromirror Devices

Fig. 16.5 shows an example of an HSI system (SkinSpect, Spectral Molecular Imaging Inc., Beverly Hills, CA, USA) for skin analysis using the OneLight system. The OneLight Spectra (OneLight Corporation, Vancouver, Canada) is a digitally-controlled spectrally programmable light engine. The SkinSpect system consists of a console and a hand-piece probe. The SkinSpect console employs a spectrally programmable digital light source, the OneLight Spectra digital micromirror device-based system incorporating a xenon arc lamp, and two LED sources for fluorescence excitation. The hand piece contains two cameras, a central chassis, a beam splitter, and fiber guides that directed the light from the console illumination source to a fixture that positions this assembly at the correct depth to illuminate the tissue surface. Linear polarizers are placed in front of the fiber optics to allow only linearly polarized light to illuminate the tissue surface. The two cameras each have a polarization filter installed and oriented orthogonally to one another. This configuration captures linear polarization images of the skin (including light reflected or remitted from both surface and deeper layers of tissue), and cross-polarization images where remitted light reaching the sensor has had its polarization changed by about 90° because of tissue scattering and therefore approximately 95% of the light comes from the deeper layers of the tissue. A synchronized image acquisition by the two cameras generates two images of an area of skin about 11 mm × 16 mm in size in both parallel and cross-polarizations.

Snapshot

Spectral imaging can be accomplished using a number of techniques that disperse the spectral information onto detectors either in sequential mode or, more recently, simultaneously (see Fig. 16.4D). The methods presented in the previous sections are all based on spatial/wavelength scan across the respective dimension. Even though fast scanners exist, limitations imposed by the low number of photons detected in many applications makes spectral imaging performed in scanning mode time costly. Many fast biological

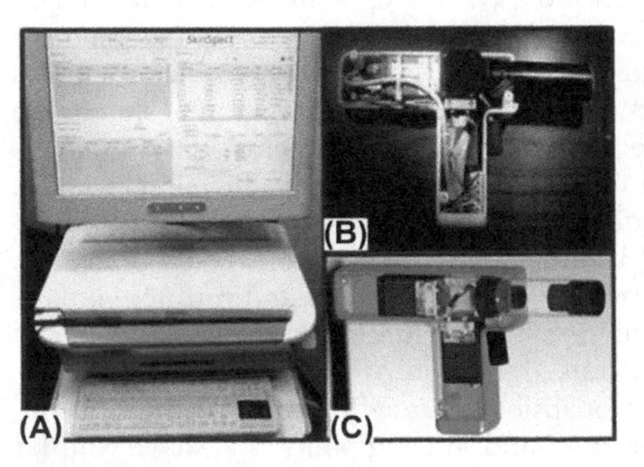

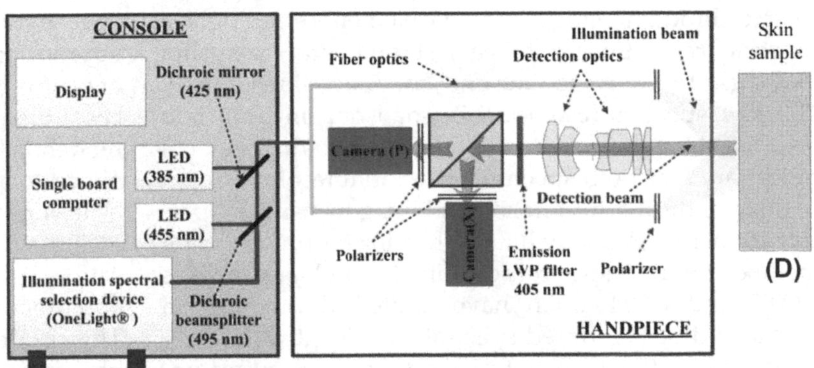

FIGURE 16.5 (A) SkinSpect (Spectral Molecular Imaging Inc., Beverly Hills, CA, USA) research prototype and the (B) handpiece module. (C) The CAD design of handpiece module. (D) Block diagram of SkinSpect research prototype system. *LWP*, Long wave pass.

processes cannot, therefore, be observed with scanning devices. In order to simultaneously record the spectral and spatial information on a CCD chip, CTIS has been developed [14,73,74]. In its most common design, a computer-generated hologram disperser is used to distribute various diffraction orders of the primary scene over a large two-dimensional CCD array (Fig. 16.4D). The position and intensity of the resulting multiple image mosaic reflect the spectral content of the original scene; the spectral content of the image is reconstructed using an iterative multiplicative algebraic reconstruction algorithm. The advantage of this approach is that spatial and spectral information can be acquired with a single image exposure, without any necessity for scanning in either the spatial or wavelength domain.

Therefore acquisition times as short as 50 ms (for bright samples) and 2 s (for dim samples) can be used. The requirement is, however, a much larger CCD array size, because the CCD must capture not only the primary image but all of its higher order diffraction images and the reconstruction of the spectral data cube from the dispersed images is computationally intense.

SYSTEM PERFORMANCE

Spatial and Spectral Properties

All whiskbroom, pushbroom, and snapshot systems offer wide spectral rage because they use dispersive elements such as gratings or prisms. Staring systems usually offer less spectral range, but on the other hand, offer spectral selectability and the option of using different wavelengths with different spectral bandwidths compared with the other technologies. This can be useful to achieve consistent signal-to-noise ratio for different spectral measurements because the intensity and exposure time can be varied at each individual spectral band.

The spectral imagers using whiskbroom and pushbroom techniques using dispersive elements offer highest spectral resolution, whereas staring and snapshot methods offer intermediate and low spectral resolution. The low spectral resolution in snapshot imaging is a result of the compromise of using the two-dimensional sensor area for both spectral and spatial resolution.

Optical throughput in spectral imagers using dispersive elements is usually higher than in spectral scanning methods because spectral scanning technologies such as AOTF and LCTF often have limited throughput in visible and near-infrared spectral wavelengths.

Except for the snapshot method, which acquires image cubes instantly, image cube collection in spectral imaging systems using electronics-based scanning (eg, staring) are faster than mechanical scanning procedures (pushbroom).

Image Data Cube Processing for Skin Analysis

There are a variety of algorithms that have been used to quantify skin chromophores that employ tissue light-transport models. Various forward models can be employed ranging from Beer–Lambert [44,68] and Kubelka–Munk [75], to the approximation of the radiative transfer equation [81]. The governing equations for light transfer through tissue can be solved using Monte Carlo [77], finite element [29], or discrete ordinate [20] methods. These approaches vary in terms of computational speed. Real-time algorithms usually are associated with relatively simple models such as ratiometric analysis [9,28]. Real-time computation (30–1000 ms) is better for extracting high-resolution skin chromophore two-dimensional maps from three-dimensional spectral image stacks with millions of voxels. These rapid quantification algorithms range from ratiometric calculations of skin reflectance maps at various wavelengths to Beer–Lambert [4] or two-flux Kubelka–Munk models (up to a few minutes) for homogenous turbid media [2,41]. Alternatively, models of light propagation can accommodate heterogeneity by incorporating two or more layers. This typically increases complexity by enabling prediction of layer thicknesses as well as chromophore concentrations for each specific tissue layer [54]. The complex geometry of skin requires computationally intensive nonlinear regression (eg, Levenberg–Marquardt [84]) techniques to fit the measured spectral signature with the estimated spectral signature derived from the related forward model.

Effect of Number of Wavelengths

To accurately estimate skin anatomical and physiological composition, we need to make sufficient detailed measurements and we have to incorporate them into an appropriate tissue model. There are two major sources of confusion preventing precise estimation of skin tissue constituents: limited measurements and "over-modeling" of tissue structure, and with SkinSpect we aimed to address both.

It has been shown that more measurements, including more wavelength bands and multiple polarizations, will reduce measurement cross-talk that can obscure accurate definition of tissue composition [63]. This is demonstrably true with major skin chromophores (melanin and hemoglobin) that have similar rise and fall absorption trends in visible and near-infrared wavelengths. An example of this (Fig. 16.6)

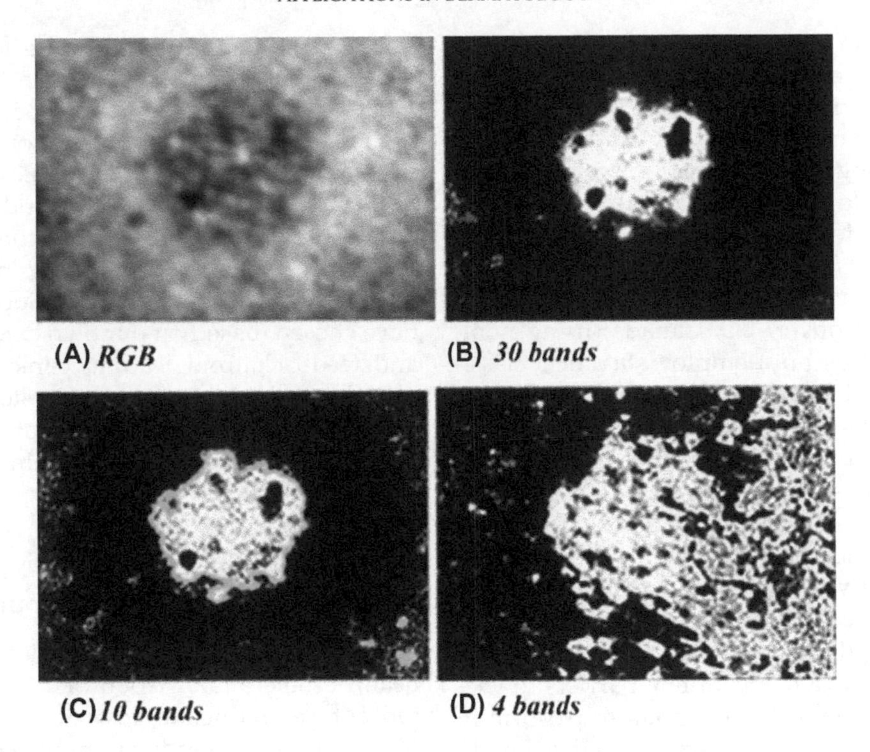

FIGURE 16.6 Skin pigmented lesion margin detection using spectral correlation analysis; dependence on the number of wavebands used. *RGB*, Red, green, blue.

illustrates the relative power of spectral segmentation of a nonuniform pigmented lesion boundary using simple correlation analysis but with different numbers of wavebands. It shows that fewer and wider bands, as used with the SIAscope (MedX Health Inc., Mississauga, ON, Canada) (four bands, see Fig. 16.3D) provide poorer segmentation when they are applied to the same lesion using the same discrimination method. The differences between MelaFind's 10 bands (see Fig. 16.3C) and our 30-band approach (see Fig. 16.3B) are more subtle but nonetheless significant, and may result in different outcomes in lesion boundary specifications.

APPLICATIONS IN DERMATOLOGY

Melanoma

Skin cancer is by far the most common of all cancers. Melanoma accounts for less than 5% of skin cancer cases but causes the large majority of skin cancer deaths, because it is an increasingly lethal form of skin cancer [30], especially when detected in later stages. Survival rates strongly favor early diagnosis, ranging from 98.2% for early, primary site detection to at best 15.1% for late or metastasized detection, during a recent 5-year study [56]. To decrease melanoma-related deaths, early detection is vital but current diagnostic practice suffers from inadequate specificity, leading to unnecessary biopsies.

The present common standard in melanoma patient care is a dermatologist's visual examination using either the Glasgow seven-point checklist or the American Cancer Society's ABCDs (asymmetry, irregular borders, more than one or uneven distribution of color, or a large (greater than 6 mm) diameter) of pigmented lesions [64], in which the practitioner looks for abnormalities in mole shape, size, and color. Around 2 million biopsies are performed annually to detect melanoma, and the vast majority of them (well over 80%) show lesions that are benign [26]. An approach to enhance evaluation can include a dermoscope with (low-power) magnification and/or specific illumination [5]. Many dermatologists are also adopting new dermoscopes that provide different wavelengths of light, higher magnification, or more advanced computer algorithms to provide enhanced diagnostic information.

Recently, more complex imaging/sensing systems that quantify anatomical and physiological information about skin constituents, such as SIAscope IV [46], have been developed. Instrumentation employing intracutaneous analysis with SIAscopy is designed to provide mapping of skin chromophores (melanin, hemoglobin, and collagen). It was reported to have high sensitivity of 96.2% but the specificity was low at 56.8% [46]. A recent independent study published in 2013 [63] concluded that: (1) SIAscopy-based results have low diagnostic accuracy for melanoma, (2) single SIAscopic features do not provide reliable diagnostic information relating to the lesions' internal structure on histopathological examination, and (3) SIAscopy cannot be used as a guide for detecting the position of maximum lesion depth (for histopathological examination).

Systems such as MelaFind (STRATA Skin Sciences, Horsham, PA , United States) (using more wavelengths compared with SIAscopy) and Verisante Aura (Verisante Technology, Inc., Vancouver, BC, Canada)(using nonimaging Raman spectroscopy) employ statistical classifiers, sometimes called "blackbox" methods, to analyze optical measurements [21,39]. Whereas these systems provide high sensitivity, they fail to achieve a desirable level of specificity. As studies with these devices progress from smaller to larger populations, they have shown significant reductions in specificity. Telling examples are the specificity reductions for MelaFind from 85% [10] to 9.9% [6] and for Verisante Aura from 78% [83] to 15% [39]. As reflected in these studies, the statistical classification approach encounters barriers to success when promising clinical devices fail to perform in larger studies. The system our group is developing works with tissue model-based algorithms and does not require larger population trials to generate learning and testing sets like those employing statistical classifiers.

Tissue quantification results from single-mode optical systems have shown that it is very difficult to characterize tissue properties with adequate sensitivity and specificity when using only one optical mode [13]. Instead, by using multiple optical modes and software that enables both correction and cross-validation, we can improve tissue characterization accuracy and achieve higher sensitivity and specificity. We believe the concerted use of multiple imaging modalities not only makes analysis less complicated, but also strengthens and simplifies it by allowing use of measurement correlations that both constrain and verify decomposition algorithms.

Based on this approach, we have integrated into a single instrument and procedure a combination of hyperspectral diffuse reflectance and autofluorescence imaging with polarization control of both illumination and image capture. These capabilities are realized in SkinSpect; the multimode dermoscopy device we have developed.

We use a model-based approach to quantify biologically and physiologically plausible features from our measurements. Because we use multiple optical modes of measurement, we can cross-check (eg, hemoglobin quantification in visible wavelengths correlated to quantification in near-infrared wavelengths) and cross-validate (eg, validating melanin concentration measurements from absorption against their expected effect on skin autofluorescence in normal and pigmented areas). These proposed multimode measurements can overcome misestimations (eg, melanin–hemoglobin misestimation [41]), and/or cross-talk specific to any individual mode. The key to this innovation is that we simultaneously measure skin in multiple imaging modes and at many wavelengths (>30) and apply fast model-based algorithms that are more accurate [67] at distinguishing and quantifying basic biological and anatomical features. The multimode measurements we have chosen (based on our own extensive investigations and feedback from leading clinicians) and our tissue model-based algorithms also allow measurement of lesion depth. This is important for melanoma risk assessment and can help determine whether melanoma is progressing through the basal layer and into the dermis.

Chronic Skin Ulcers and Wound Healing

Chronic wounds are a significant and costly global health problem [57]. Wound healing issues present significant yet reducible costs to the healthcare system. Many chronic wounds are now classified as a medical error, the cost of which must be born by hospitals, as well as costs associated with malpractice suits. Wound management and documentation is a significant part of electronic medical records. Hyperspectral wound imaging can provide an effective way to document and assess healing progress. The most common chronic wounds are diabetic foot ulcers, pressure ulcers, and venous stasis ulcers.

Diabetic foot ulcers (DFUs): More than 25.8 million people suffer from diabetes in the United States. There is a 15–25% lifetime risk for foot ulcer development in diabetic patients. The estimated annual provider cost for diabetic foot ulcers is $11 billion. Mean healing time for DFUs is 140 days. Mortality was estimated to be 11.6% in 2008, and in 2007 more than 100,000 diabetic patients had a foot amputation [43].

Pressure ulcers: These are caused by loss of circulation from prolonged skin pressure over a bony prominence. There were 503,300 pressure ulcer hospital stays in the United States, with an average stay duration of 14 days in 2006. Most patients are discharged to long-term care or home care, where treatment continues while the pressure ulcer heals. Time to heal is dependent on the stage of the pressure ulcer. Stage 1 requires 1–7 days; stage 2, 5–90 days; stage 3, 30–180 days; and stage 4, 180–360 days [66].

Venous stasis ulcer: Venous ulcers develop in areas with poor circulation, especially in the lower leg. Venous return (the return of venous blood toward the heart) is poor, and these ulcers account for 40–70% of lower extremity wounds. About 600,000 people seek treatment yearly. The mean healing time for this type of ulcer is 168 days and the prevalence 0.18–1.3% in adults.

Assessment and classification of chronic wounds is required for Medicare reimbursement in the United States, and assessment protocol tools such as the Waterlow, Norton, and Braden tools are commonly used worldwide. Assessment is usually performed by nursing staff and involves rating patient pain; assessing wound appearance, odor, and fluid exudate; and measuring size. Measurements are usually done with a ruler and the caregiver must measure in multiple directions, because size is a key part of determining the billing code. These measurements are now being incorporated in electronic medical record systems using smart phones and tablets, but many could be automated with HSI systems [8].

Advanced Treatments for Chronic Wounds

There is an array of competing technologies for wound treatment, primarily in the form of moist wound dressings. These include hydrogels, hydrocolloids, alginates, foams, and transparent films. They are now often combined with antimicrobial agents such as silver, iodine, or honey to combat infection. The use of antimicrobial dressings can reduce the need for systemic antibiotics.

There is good potential for HSI in wound treatment management because it provides spatial and functional information that can be used for measuring and assessing healing.

Clinically relevant chromophores include oxyhemoglobin, deoxyhemoglobin, melanin, bilirubin, lipid, and water. Clinically relevant fluorophores include collagen, flavin adenine dinucleotide, nicotinamide adenine dinucleotide, and porphyrin.

The healing process starts with vascular response followed by inflammation, and then moves into the proliferation phase and finally the remodeling phase. The vascular response involves formation of a fibrin mesh, trapping blood cells; in the inflammation phase, blood vessel and capillaries dilate to bring cells and chemicals necessary for healing [59]. This creates local heat and pressure, and pain and swelling occurs. In the proliferative phase, collagen is produced by fibroblasts to rebuild the damaged tissue. Granulation, contraction, and epithelization occur as scar tissue is formed. In the remodeling phase, vascularization decreases, causing the scar tissue to appear whiter, and collagen cross-links increase as the collagen matures and gains strength. The dynamic changes described involve a number of chromophores, fluorophores, and light scattering structures that can be interrogated using HSI. Pilon et al. have used HSI between 450 and 700 nm (using OxyVu, HyperMed, Inc., Burlington, MA) and a spectral separator (LCTF-10-20, CRI Inc., Woburn, MA). This HSI system assesses tissue oxygenation and epidermal thickness to help predict and understand ulceration mechanisms [82].

When wounds are chronic, it is often because the body is unable to provide the necessary rebuilding materials because of degraded circulatory and lymphatic systems. This can be further compromised by infection of the wound site. HSI can also be used to assess infection by measuring the porphyrin fluorescence produced by many bacteria.

Bruise

Skin trauma, usually from a blunt object impact resulting in bruising, is a common injury producing visually apparent color changes in the tissue over time. Because the color changes in the tissue progress from the time of impact until complete healing, there has been much interest in using color measurement and spectroscopy of tissue to determine when an injury occurred. There are many reasons why this information can be important, criminal investigations being one example.

These color changes are the result of well-understood biological processes and can also be measured using HSI [52,53,72]. Bruises can be quite colorful, and the mix of colors are due to the chromophores oxyhemoglobin (red), deoxyhemoglobin (blue), methemoglobin (brown), carboxyhemoglobin (bright red), melanin (brown), biliverdin (green), and bilirubin (yellow) [32]. The layered nature of skin adds other issues that affect color as a result of scattering properties that affect shorter wavelengths more than longer wavelengths and can exaggerate or attenuate bruise color perception depending on the depth of injury.

Immediately after trauma, the capillaries in the dermis and subdermis rupture and extravasated blood begins to pool in the nearby tissue. The oxygenated hemoglobin in the extravasated blood initially appears red but begins to turn blue as deoxygenation of the hemoglobin occurs. Macrophages begin to break down the hemoglobin into globulin and the green chromophore biliverdin. Biliverdin is rapidly converted into bilirubin, which is responsible for the yellow color in bruises. Production of carbon monoxide during catabolism of hemoglobin can result in some oxyhemoglobin being converted to bright red carboxyhemoglobin. Alternatively, the iron molecule in hemoglobin may become oxidized, producing brown methemoglobin [31].

Some color changes, apparent by direct viewing or in color photography, can appear similar but may be actually caused by different processes. An example of this is the green color in bruises, which may be the result of biliverdin absorption, but can also be a combination of bilirubin and deoxyhemoglobin absorption. HSI can be used to better differentiate the state of the tissue by allowing more accurate determination of the contribution of individual chromophores to bruise color and the distribution of tissue damage. In general, the trend in the literature is that color measurement alone is insufficient to accurately determine time of injury. Progression of color changes over time combined with accurate determination of chromophore concentration by spectroscopy and HSI are expected to provide more accurate assessment of these injuries.

Burn Injury

Thermal damage to skin can be caused in a variety of ways. According to the American Burn Association, based on the ambulatory care statistics from the Centers for Disease Control and Prevention, 486,000 people seek treatment for burn injuries every year. Of these, about 40,000 have injuries requiring medical treatment and about 3300 cases result in death [51]. It is generally agreed that for deciding optimal treatment and management, monitoring of early-stage burns is necessary. Biopsy and histology are both invasive and have limitations because of burn heterogeneity. Spectroscopy and HSI can noninvasively provide information about burned tissue health by measuring clinically relevant chromophores such as oxyhemoglobin, deoxyhemoglobin, lipid, water, and derived values such as tissue oxygen saturation [1,60]. When combined with polarization, fluorescence, or Raman HSI, additional information, including depth distribution of damage to circulatory systems, collagen structures, and extracellular matrix can be determined [27]. This allows highly localized real-time assessment of overall tissue status and provides valuable information in managing burns in the acute phase, as well as in the recovery and healing phases, when the problems are similar in many ways to those of chronic wound management as previously described. Skin burns are usually characterized by the depth of the injury. Superficial burns involve injury to the top epidermis layer. They are characterized by a reddish nonblistering appearance (eg, sunburns) cause by increased blood flow to the dermis. They are often sensitive to touch and heal in less than 14 days.

Full thickness burns extend beyond the epidermal and dermal layers of the skin into the subcutaneous layer. They typically have three zones, including areas of coagulation, stasis, and hyperemia. Areas of coagulation are necrotic and must be removed. Areas of hyperemia will recover. Areas of stasis can progress to either necrosis or to hyperemia depending on development or prevention of vascular occlusion. This is influenced by presence of prostaglandins, histamines, and bradykinin and the development of edema resulting from increased vascular permeability. If ischemia persists, a zone of stasis will become necrotic [23].

Superficial partial thickness and deep partial thickness burns are burns in which damage extends to a fraction of the dermal layer. Superficial partial thickness burns extend to only the upper layers of the papillary dermis, and depending on the extent of damage and remaining vasculature, these injuries may naturally heal in 2–3 weeks with minimal to no scarring. They often have a mottled pink and white appearance that can blanch with pressure and are less sensitive to pinpricks than normal.

Deep-partial thickness burns extend in depth to the reticular dermis and can often be found mixed with portions of noncharred full-thickness burns. These burns often require excision and grafting for optimal treatment. They also have a mottled pink and white appearance that can blanch with pressure and are less sensitive to pinpricks [11].

Partial thickness burns are therefore challenging to identify based on clinical impression. Depending on the clinician's experience, visual assessment has been shown to have a field accuracy of about 50–76%. Overestimation of burn severity results in unnecessarily invasive surgical treatment and prolonged hospitalization, whereas underestimation results in treatment delays, extended hospital stays, and increased chances of contracture and hypertrophic scar formation [25,49].

As can be seen from the foregoing descriptions of burn injuries and the effect on tissues, HSI systems based on techniques used in research can potentially provide a fast and efficient way to monitor the complex and dynamic metabolic and structural effects of burns on skin in order to guide treatment at the point of care.

Eczema

Eczema affects a substantial proportion of the population. The prevalence of eczema/atopic dermatitis in the United States is 10.7% in children and about 10.2% in adults. About a one-third have moderate to severe eczema/dermatitis requiring systemic therapy [22]. Common triggers are irritants like soaps and detergents, shampoos, dish-washing liquids, bubble bath; disinfectants like chlorine; contact with juices from fresh fruits, meats, and vegetables; allergens such as house dust mites, pets (cats > dogs), pollens (seasonal), and molds; dandruff; microbes such as certain bacteria like *Staphylococcus aureus*, viruses, and certain fungi; temperature such as hot weather, high or low humidity, perspiration from exercise; and food allergens such as dairy products, eggs, nuts and seeds, soy products, and wheat.

It manifests as an itchy red rash and can appear all over the body. Symptoms are dry, sensitive skin; intense itching; red, inflamed skin; recurring rash; scaly areas; rough, leathery patches; oozing or crusting; areas of swelling; and dark-colored patches of skin. These features can be imaged using HSI and MSI [69]. Changes to the skin from eczema primarily affect the barrier function of the skin, disrupting the protection provided by the epidermis and the basement membrane. Lesions begin as erythematous papules, which coalesce to form plaques that may display weeping, crusting, or scales. The change in color from erythema is one feature that can be measured with HSI. The capillary dilation characteristic of erythema changes the skin diffuse reflectance because of absorption in the blue and red wavelengths by hemoglobin. Development of edema changes both the absorption and scattering properties of tissue [17]. Analysis of spatial features such as texture provides other measures to characterize the skin.

Other Skin Imaging Applications

We highlighted here some of the high-volume applications of spectral imaging in dermoscopy, but the survey is definitely not exhaustive. Spectral methods have additionally been used for such diverse applications as skin chromophore mapping [15], histopathological analysis, and other promising areas that we hope to review as they become more mainstream.

SUMMARY

The capacity of optical imaging methods to characterize biological tissue in general, and skin in particular, with high resolution and discrimination ability provides a very valuable tool in the reliable, noninvasive assessment of differences between normal and abnormal states, thus allowing meaningful clinical applications, including early diagnosis of disease. For such approaches to fully realize their potential, one must start from a good understanding of tissue properties and an equally advanced understanding of the tools deployed. Certain difficult problems may require the combined use of several imaging methods and molecular-level understanding, and HSI is emerging as one of the more powerful technologies in this regard, and was reviewed here.

References

[1] Afromowitz MA, Callis JB, Heimbach DM, DeSoto LA, Norton MK. Multispectral imaging of burn wounds: a new clinical instrument for evaluating burn depth. Biomed Eng IEEE Trans 1988;35(10):842–50.
[2] Anderson RR, Parrish JA. The optics of human skin. J Invest Dermatol 1981;77(1):13–9.
[3] Angelopoulou E. Understanding the color of human skin. In: Photonics West 2001-Electronic Imaging. International Society for Optics and Photonics; 2001. p. 243–51.
[4] Attas M, et al. Visualization of cutaneous hemoglobin oxygenation and skin hydration using near-infrared spectroscopic imaging. Skin Res Technol 2001;7:238–45.
[5] Bafounta ML, Beauchet A, Aegerter P, Saiag P. Is dermoscopy (epiluminescence microscopy) useful for the diagnosis of melanoma? Results of a meta-analysis using techniques adapted to the evaluation of diagnostic test. Arch Dermatol 2001;137(10):1343–50.
[6] Bergstrom KG. MelaFind was approved by FDA: where does it fit in dermatology? J Drugs Dermatol 2012;11(3):420–2.
[7] Cui Y, Cui D, Tang JH. Study on the characteristics of an imaging spectrum system by means of an acousto-optic tunable filter. Opt Eng 1993;32(11):2899–903.
[8] Denstedt M, Pukstad BS, Paluchowski LA, Hernandez-Palacios JE, Randeberg LL. Hyperspectral imaging as a diagnostic tool for chronic skin ulcers. In: SPIE BiOS. International Society for Optics and Photonics; 2013. 85650N.
[9] Diebele I, et al. Clinical evaluation of melanomas and common nevi by spectral imaging. Biomed Opt Express 2012;3:467–72.
[10] Elbaum M, Kopf AW, Rabinovitz HS, Langley RG, Kamino H, Mihm Jr MC, et al. Automatic differentiation of melanoma from melanocytic nevi with multispectral digital dermoscopy: a feasibility study. J Am Acad Dermatol 2001;44(2):207–18.
[11] Enoch S, Roshan A, Shah M. Emergency and early management of burns and scalds. BMJ 2009;338:b1037.
[12] Farkas DL, Du C, Fisher GW, Lau C, Niu W, Wachman ES, Levenson RM. Non-invasive image acquisition and advanced processing in optical bioimaging. Comput Med Imag Graphics April 30, 1998;22(2):89–102.
[13] Farkas DL, Becker D. Applications of spectral imaging: detection and analysis of human melanoma and its precursors. Pigm Cell Res 2001;14(1):2–8.
[14] Ford BK, Volin CE, Murphy SM, Lynch RM, Descour MR. Computed tomography-based spectral imaging for fluorescence microscopy. Biophys J 2001;80(2):986–93.
[15] Fujimoto JG, Farkas D. Biomedical optical imaging. Oxford University Press; 2009.
[16] Gao L, Kester RT, Tkaczyk TS. Compact image slicing spectrometer (ISS) for hyperspectral fluorescence microscopy. Opt Express 2009;17(15):12293–308.
[17] Stamatas GN, Southall M, Kollias N. In vivo monitoring of cutaneous edema using spectral imaging in the visible and near infrared. J Invest Dermatol 2006;126:1753–60.
[18] Goetz AFH, Vane G, Solomon JE, Rock BN. Imaging spectrometry for earth remote sensing. Science 1985;228(4704):1147–53.
[19] Gowen AA, O'Donnell CP, Cullen PJ, Downey G, Frias JM. Hyperspectral imaging: an emerging process analytical tool for food quality and safety control. Trends Food Sci Technol 2007;18(12): 590–8.
[20] Guo Z, Kim K. Ultrafast-laser-radiation transfer in heterogeneous tissues with the discrete-ordinates method. Appl Opt 2003;42: 2897–905.
[21] Gutkowicz-Krusin D, Elbaum M, Jacobs A, Keem S, Kopf AW, Kamino H, et al. Precision of automatic measurements of pigmented skin lesion parameters with a MelaFind multispectral digital dermoscope. Melanoma Res 2000;10(6):563–70.
[22] Hanifin JM, Reed ML. A population-based survey of eczema prevalence in the United States. Dermatitis June 2007;18(2):82–91.
[23] Hettiaratchy S, Dziewulski P. Pathophysiology and types of burns. BMJ 2004;328(7453):1427–9.
[24] Jacques SL. Origins of tissue optical properties in the UVA, visible, and NIR regions. OSA TOPS Adv Opt Imag Photon Migr 1996;2: 364–9.
[25] Jeng JC, Bridgeman A, Shivnan L, Thornton PM, Alam H, Clarke TJ, et al. Laser Doppler imaging determines need for excision and grafting in advance of clinical judgment: a prospective blinded trial. Burns 2003;29(7):665–70.

[26] Jones TP, Boiko PE, Piepkorn MW. Skin biopsy indications in primary care practice: a population-based study. J Am Board Farm Pract 1996;9(6):397–404.

[27] Kaiser M, et al. Noninvasive assessment of burn wound severity using optical technology: a review of current and future modalities. Burns J Int Soc Burn Inj 2011;37(3):377–86. 1083–3668.

[28] Kapsokalyvas D, et al. Spectral morphological analysis of skin lesions with a polarization multispectral dermoscope. Opt Express 2013;21:4826–40.

[29] Katika KM, Pilon L. Steady-state directional diffuse reflectance and fluorescence of human skin. Appl Opt 2006;45:4174–83.

[30] Kopf AW, Rigel DS, Friedman RJ. The rising incidence and mortality rate of malignant melanoma. J Dermatol Surg Oncol 1982;8: 760–1.

[31] Langois NEI. The science behind the quest to determine the age of bruises: a review of the English language literature. Forensic Sci Med Pathol 2007;3:241–51.

[32] Langois NEI, Gresham GA. The ageing of bruises: a review and a study of the colour changes with time. Forensic Sci Int 1991;50: 227–38.

[33] Leachtenauer JC, Driggers RG. Surveillance and reconnaissance imaging systems: modeling and performance prediction. Artech House; 2001.

[34] Lee HS, Younan NH, King RL. Hyperspectral image cube compression combining JPEG-2000 and spectral decorrelation. In: Geoscience and remote sensing symposium, 2002. IGARSS'02. 2002 IEEE International, vol. 6. IEEE; 2002. p. 3317–9.

[35] Li Q, He X, Wang Y, Liu H, Xu D, Guo F. Review of spectral imaging technology in biomedical engineering: achievements and challenges. J Biomed Opt 2013;18(10):100901.

[36] Lorente D, Aleixos N, Gómez-Sanchis J, Cubero S, García-Navarrete Or L, Blasco J. Recent advances and applications of hyperspectral imaging for fruit and vegetable quality assessment. Food Bioprocess Technol 2012;5(4):1121–42.

[37] Lu G, Fei B. Medical hyperspectral imaging: a review. J Biomed Opt 2014;19(1):010901.

[38] Lu R, Chen Y-R. Hyperspectral imaging for safety inspection of food and agricultural products. In: Photonics East (ISAM, VVDC, IEMB). International Society for Optics and Photonics; 1999. p. 121–33.

[39] Lui H, Zhao J, McLean D, Zeng H. Real-time Raman spectroscopy for in vivo skin cancer diagnosis. Cancer Res 2012;72(10): 2491–500.

[40] Lyon RC, Lester DS, Neil Lewis E, Lee E, Yu Lawrence X, Jefferson EH. Near-infrared spectral imaging for quality assurance of pharmaceutical products: analysis of tablets to assess powder blend homogeneity. AAPS PharmSciTech 2002;3(3):1–15.

[41] MacKinnon NB, et al. In vivo skin chromophore mapping using a multimode imaging dermoscope (SkinSpect). Proc SPIE 2013; 8587:85870U.

[42] Deleted in review.

[43] Margolis DJ, Malay DS, Hoffstad OJ, et al. Economic burden of diabetic foot ulcers and amputations: data points #3 [Internet]. In: Rockville MD, editor. Agency for Healthcare Research and Quality; 2008. Available from: http://www.effectivehealthcare.ahrq.gov.proxy1.lib.uwo.ca [accessed 20. 05. 16].

[44] Martinez L. A non-invasive spectral reflectance method for mapping blood oxygen saturation in wounds. In: Proc. of the 31st Applied Imagery Pattern Recognition Workshop; 2002. p. 112–6.

[45] Massone C, Di Stefani A, Soyer HP. Dermoscopy for skin cancer detection. Curr Opin Oncol 2005;17(2):147–53.

[46] Moncrieff M, Cotton S, Claridge E, Hall P. Spectrophotometric intracutaneous analysis: a new technique for imaging pigmented skin lesions. Br J Dermatol 2002;146(3):448–57.

[47] Morris HR, Hoyt CC, Treado PJ. Imaging spectrometers for fluorescence and Raman microscopy: acousto-optic and liquid crystal tunable filters. Appl Spectrosc 1994;48(7):857–66.

[48] Morris HR, Hoyt CC, Miller P, Treado PJ. Liquid crystal tunable filter Raman chemical imaging. Appl Spectrosc 1996; 50(6):805–11.

[49] Niazi ZB, Essex TJ, Papini R, Scott D, McLean NR, Black MJ. New laser Doppler scanner, a valuable adjunct in burn depth assessment. Burns 1993;19(6):485–9.

[50] Winter EM. The development of a hyperspectral sensor: a data processing viewpoint. In: Aerospace Conference, IEEE Proc 4; 2001. p. 4–1979.

[51] Palmer QB. National Burn Repository: report of data from 2002 to 2011. Am Burn Assoc 2012.

[52] Randeberg LL, Baarstad I, Løke T, Kaspersen P, Svaasand LO. Hyperspectral imaging of bruised skin. In: Biomedical Optics 2006. International Society for Optics and Photonics; 2006. 60780O.

[53] Randeberg LL, Larsen ELa P, Svaasand LO. Characterization of vascular structures and skin bruises using hyperspectral imaging, image analysis and diffusion theory. J Biophoton 2010;3(1–2): 53–65.

[54] Saager RB, Truong A, Cuccia DJ, Durkin AJ. Method for depth-resolved quantitation of optical properties in layered media using spatially modulated quantitative spectroscopy. J Biomed Opt 2011;16:077002.

[55] Schmitt JM, Kumar G. Turbulent nature of refractive-index variations in biological tissue. Opt Lett 1996;21(16):1310–2.

[56] SEER Stat Fact Sheets: Melanoma of the skin, http://seer.cancer.gov/.

[57] Sen CK, Gordillo GM, Roy S, Kirsner R, Lambert L, Hunt TK, et al. Human skin wounds: a major and snowballing threat to public health and the economy. Wound Rep Regen November 1, 2009; 17(6):763–71.

[58] Shaw GA, Burke HH. Spectral imaging for remote sensing. Linc Lab J November 1, 2003;14(1):3–28.

[59] Stamatas GN, Kollias N. In vivo documentation of cutaneous inflammation using spectral imaging. J Biomed Opt 2007;12(5): 051603.

[60] Stamatas GN, Costas JB, Kollias N. Hyperspectral image acquisition and analysis of skin. In: Biomedical Optics 2003. International Society for Optics and Photonics; 2003. p. 77–82.

[61] Suhre DR, Gottlieb MS, Taylor LH, Melamed NT. Spatial resolution of imaging noncolinear acousto-optic tunable filters. Opt Eng 1992;31(10):2118–21.

[62] Sun D-W, editor. Hyperspectral imaging for food quality analysis and control. Elsevier; 2010.

[63] Terstappen K, Suurkula M, Hallberg H, Ericson MB, Wennberg AM. Poor correlation between spectrophotometric intracutaneous analysis and histopathology in melanoma and nonmelanoma lesions. J Biomed Opt 2013;18(6):061223.

[64] Thomas L, Tranchand P, Berard F, Secchi T, Colin C, Moulin G. Semiological value of ABCDE criteria in the diagnosis of cutaneous pigmented tumors. Dermatology 1998;197(1):11–7.

[65] Treado PJ, Ira WL, Neil Lewis E. High-fidelity Raman imaging spectrometry: a rapid method using an acousto-optic tunable filter. Appl Spectrosc 1992;46(8):1211–6.

[66] VanGilder C, MacFarlane GD, Meyer S. Results of nine international pressure ulcer prevalence surveys: 1989 to 2005. Ostomy Wound Manag February 2008;54(2):40–54.

[67] Vasefi F, MacKinnon NB, Farkas DL. Toward in-vivo diagnosis of skin cancer using multimode imaging dermoscopy. II. Molecular mapping of highly pigmented lesions. Proc SPIE 2014; 8947–9018 (in press).

[68] Vasefi F, MacKinnon N, Farkas DL. Toward in vivo diagnosis of skin cancer using multimode imaging dermoscopy. II. Molecular mapping of highly pigmented lesions. In: SPIE BiOS 2014;4: 89470J.
[69] Veien NK, Hattel T, Laurberg G. Hand eczema: causes, course, and prognosis I. Contact Dermat 2008;58:330–4. http://dx.doi.org/10.1111/j.1600-0536.2008.01345.x.
[70] Vo-Dinh T. A hyperspectral imaging system for in vivo optical diagnostics. Eng Med Biol Mag IEEE 2004;23(5):40–9.
[71] Vogel A, Chernomordik VV, Riley JD, Hassan M, Amyot F, Dasgeb B, et al. Using noninvasive multispectral imaging to quantitatively assess tissue vasculature. J Biomed Opt 2007;12(5): 051604.
[72] Vogeley E, Pierce M, Bertocci G. Experience with wood lamp illumination and digital photography in the documentation of bruises on human skin. Arch Pediatr Adolesc Med 2002;156(3): 265–8. http://dx.doi.org/10.1001/archpedi.156.3.265.
[73] Volin CE, Ford BK, Descour MR, Garcia JP, Wilson DW, Maker PD. High-speed spectral imager for imaging transient fluorescence phenomena. Appl Opt 1998;37(34):8112–9.
[74] Volin CE, Garcia JP, Dereniak EL, Descour MR, Hamilton T, McMillan R. Midwave-infrared snapshot imaging spectrometer. Appl Opt 2001;40(25):4501–6.
[75] Vyas S, Banerjee A, Burlina P. Estimating physiological skin parameters from hyperspectral signatures. J Biomed Opt 2013;18: 057008.
[76] Wachman ES, Niu WH, Farkas DL. AOTF microscope for imaging with increased speed and spectral versatility. Biophys J September 1997;73(3):1215.
[77] Wang L-H, Jacques SL, Zheng L-Q. MCML - Monte Carlo modeling of photon transport in multi-layered tissues. Comput Meth Prog Bio 1995;47:131–46.
[78] Weber JR, Cuccia DJ, Tromberg BJ. Modulated imaging in layered media. In: Conf Proc IEEE Eng Med Biol Soc (No. Suppl.); August 30, 2006. p. 6674–6.
[79] Parrish JA. New concepts in therapeutic photomedicine; photochemistry, optical targeting and the therapeutic window. J Invest Dermatol 1981;77(1):45–50.
[80] Yaroslavsky AN, Priezzhev AV, Rodriguez J, Yaroslavsky IV, Battarbee H. Optics of blood. In: Handbook of optical biomedical diagnostics; 2002. p. 169–216.
[81] Yudovsky D, Pilon L. Retrieving skin properties from in vivo spectral reflectance measurements. J Biophoton 2011;4:305–14.
[82] Yudovsky D, Nouvong A, Pilon L. Hyperspectral imaging in diabetic foot wound care. J Diabet Sci Technol 2010;4(5):1099–113.
[83] Zhao J, Lui H, McLean DI, Zeng H. Real-time Raman spectroscopy for non-invasive skin cancer detection: preliminary results. Conf Proc IEEE Eng Med Biol Soc 2008;2008:3107–9.
[84] Zonios G, Bykowski J, Kollias N. Skin melanin, hemoglobin, and light scattering properties can be quantitatively assessed in vivo using diffuse reflectance spectroscopy. J Invest Dermatol 2001; 117(6):1452–7.

Further Reading

[1] Alexandrescu D. Melanoma costs: a dynamic model comparing estimated overall costs of various clinical stages. Dermatol Online J 2009;15(11).
[2] American Cancer Society. Melanoma skin cancer overview. 2011. http://www.cancer.org/.
[3] Bailey EC, Sober AJ, Tsao H, Mihm Jr MC, Johnson Jr TM. Fitzpatrick's dermatology in general medicine, 8e [chapter 124]. cutaneous melanoma; 2012.
[4] Benelli C, Roscetti E, Dal Pozzo V, Gasparini G, Cavicchini S. The dermoscopic versus the clinical diagnosis of melanoma. Eur J Dermatol 1999;9(6):470–6.
[5] Bohnert M, Baumgartner R, Pollak S. Spectrophotometric evaluation of the colour of intra- and subcutaneous bruises. Int J Legal Med 2000;113:343–8.
[6] Centers for Disease Control/National Highway Traffic Safety Administration. Setting the national agenda for injury control in the 1990s. U.S. Department of Health and Human Services, Public Health Service, CDC; 1992.
[7] Claridge E, Cotton S, Hall P, Moncrieff M. From colour to tissue histology: physics-based interpretation of images of pigmented skin lesions. Med Image Anal 2003;7(4):489–502.
[8] Costs of Cancer Care report from NCI, Cancer Trends Progress Report – 2011/2012 Update. National Cancer Institute. http://progressreport.cancer.gov/.
[9] Dicker DT, Lerner J, Van Belle P, Guerry 4th D, Herlyn M, Elder DE, El-Deiry WS. Differentiation of normal skin and melanoma using high resolution hyperspectral imaging. Cancer Biol Ther 2006; 5(8):1033–8.
[10] Du Vivier AW, Williams HC, Brett JV, Higgins EM. How do malignant melanomas present and does this correlate with the seven-point checklist? Clin Exp Dermatol 1991;16(5):344–7.
[11] Dwyer PJ, DiMarzio CA. Hyperspectral imaging for dermal hemoglobin spectroscopy. In: SPIE's International symposium on optical science, engineering, and instrumentation. International Society for Optics and Photonics; 1999. p. 72–82.
[12] Gat N. Imaging spectroscopy using tunable filters: a review. In: AeroSense 2000. International Society for Optics and Photonics; 2000. p. 50–64.
[13] Higgins EM, Hall P, Todd P, Murthi R, Du Vivier AW. The application of the seven-point check-list in the assessment of benign pigmented lesions. Clin Exp Dermatol 1992;17(5):313–5.
[14] Ilias MA, Häggblad E, Anderson C, Göran Salerud E. Visible hyperspectral imaging evaluating the cutaneous response to ultraviolet radiation. In: Biomedical Optics (BiOS) 2007. International Society for Optics and Photonics; 2007. p. 644103.
[15] MacKinnon N, Vasefi F, Farkas DL. Toward in-vivo diagnosis of skin cancer using multimode imaging dermoscopy. I. clinical system development and validation. In: SPIE BiOS. International Society for Optics and Photonics; 2014. 89470I.
[16] McGovern TW, Litaker MS. Clinical predictors of malignant pigmented lesions: a comparison of the Glasgow seven-point checklist and the American Cancer Society's ABCDs of pigmented lesions. J Dermatol Surg Oncol 1992;18(1):22–6.
[17] Tsumura N, Kawabuchi M, Haneishi H, Miyake Y. Mapping pigmentation in human skin from a multi-channel visible spectrum image by inverse optical scattering technique. J Imag Sci Technol 2001;45:444–50.
[18] Vasefi F, MacKinnon N, Farkas DL. Toward in vivo diagnosis of skin cancer using multimode imaging dermoscopy. II. Molecular mapping of highly pigmented lesions. In: SPIE BiOS. International Society for Optics and Photonics; 2014. 89470J.
[19] Yudovsky D, Durkin AJ. Spatial frequency domain spectroscopy of two layer media. J Biomed Opt 2011;16:107005.
[20] Zeng H, MacAulay CE, Palcic B, McLean DI. Monte Carlo modeling of tissue autofluorescence measurement and imaging. In: SPIE OE/LASE'94; 1994. p. 94–104.
[21] Zuzak KJ, Schaeberle MD, Gladwin MT, Cannon RO, Levin IW. Noninvasive determination of spatially resolved and time-resolved tissue perfusion in humans during nitric oxide inhibition and inhalation by use of a visible-reflectance hyperspectral imaging technique. Circulation 2001;104(24):2905–10.

CHAPTER

17

Diffuse Reflectance Spectroscopy and Imaging

A.J. Moy, J.W. Tunnell

University of Texas at Austin, Austin, TX, United States

OUTLINE

INTRODUCTION

Because of its location on the outside of the human body, conditions of the skin are easily observed and are typically diagnosed by a dermatologist upon visual observation. Skin, because of its ease of accessibility, is very amenable to interrogation with optical methods, which provide a promising pathway to gain insight into skin physiology and pathology. This chapter will introduce the optical method of diffuse reflectance and its application in dermatology. We begin with the basic principles underlying light–tissue interaction and diffuse reflectance, followed by a discussion of the techniques of diffuse reflectance spectroscopy (DRS) and diffuse reflectance imaging (DRI) and the instrumentation required for these techniques. Finally, we review some examples of current applications of diffuse reflectance and future directions of diffuse reflectance in dermatology.

PRINCIPLES OF DIFFUSE REFLECTANCE

Light incident on skin can interact with the tissue by the processes of light scattering and light absorption [1–3]. Light scattering in tissue is the random trajectory, or diffusion, of incident photons as they undergo multiple elastic scattering off of the cells, extracellular matrix, blood vessels, and other tissue components that comprise skin. After a random number of scattering events, the photons are then either remitted or absorbed. Light absorption in tissue occurs when the energy of incident photons is absorbed and, by conservation of energy, converted into heat. The interaction between light and skin [4,5] is typically described and quantified by the following parameters: scattering coefficient (μ_s), absorption coefficient (μ_a), and anisotropy (g). The anisotropy is a measure of the amount of forward scattering that occurs in the tissue and is used to describe the scattering of light in tissue

Imaging in Dermatology
http://dx.doi.org/10.1016/B978-0-12-802838-4.00017-0

by the reduced scattering coefficient (μ_s'), which is a function of μ_s and g. Together, μ_s' and μ_a are the intrinsic optical properties of the tissue.

Conceptually, μ_s' and μ_a describe the average number of scattering and absorption events, respectively, a photon undergoes in a given distance in the tissue. Both optical properties are wavelength dependent; that is, the scattering and absorption properties change with wavelength. The units of μ_s' and μ_a are in inverse length, typically mm^{-1}. In most biological tissue, $\mu_s' > \mu_a$ and, as a result, light transport is said to be in the scattering-dominated diffusion regime. Skin, however, is unique in that it is a relatively thin tissue. This presents issues with assumptions made in the models of light propagation in tissue as well as stipulations on hardware, specifically the distance between the light source and detector. Both of these issues differentiate diffuse reflectance measurements of skin with other applications in the breast and brain. These aspects are discussed in more detail later in the chapter. Practically, the effects of optical properties in skin are easily visualized; skin that appears lighter in appearance typically has lower absorption, whereas skin that is darker in appearance has higher absorption.

Light scattering and absorption in tissue is determined by measurement of the diffuse reflectance (R_d), which is the amount of light remitted after subsurface interaction with the skin. Because skin, as well as many biological tissues, is heavily forward scattering, the incident light forward scatters into the tissue and continues scattering before being remitted or absorbed. The light that remits after interacting with the tissue is measured in the diffuse reflectance, which can be quantified in terms of light scattering and absorption, and ultimately used to determine the concentration of typical components in skin or chromophores.

Diffuse reflectance functions as a form of quantitative spectroscopy and is analogous to absorption spectroscopy in the ultraviolet (UV) and visible spectra. Absorption spectroscopy is used to determine the optical absorption of a solute in solution by passing light through a cuvette filled with the solution. The optical absorption can then be used to determine the solute concentration. The distance that the light travels through the cuvette, or path length (z_0), is used in an equation known as Beer's law, which relates the incident light intensity on the cuvette to the transmitted light intensity after passing through the solution-filled cuvette. The path length of absorption spectroscopy is analogous to the distance between the light source and detector (ρ_o) in diffuse reflectance. Where diffuse reflectance differs is that: (1) measurements are taken in a reflectance geometry instead of a transmission geometry and thus requires a probe, and (2) optical scattering is accounted for, whereas optical scattering is typically neglected in absorption spectroscopy. This analogy between absorption spectroscopy and diffuse reflectance is illustrated in Fig. 17.1.

From a clinical perspective, many skin lesions have altered optical properties that occur in the presence of abnormal tissue. For example, some lesions may appear darker in appearance than the healthy surrounding skin,

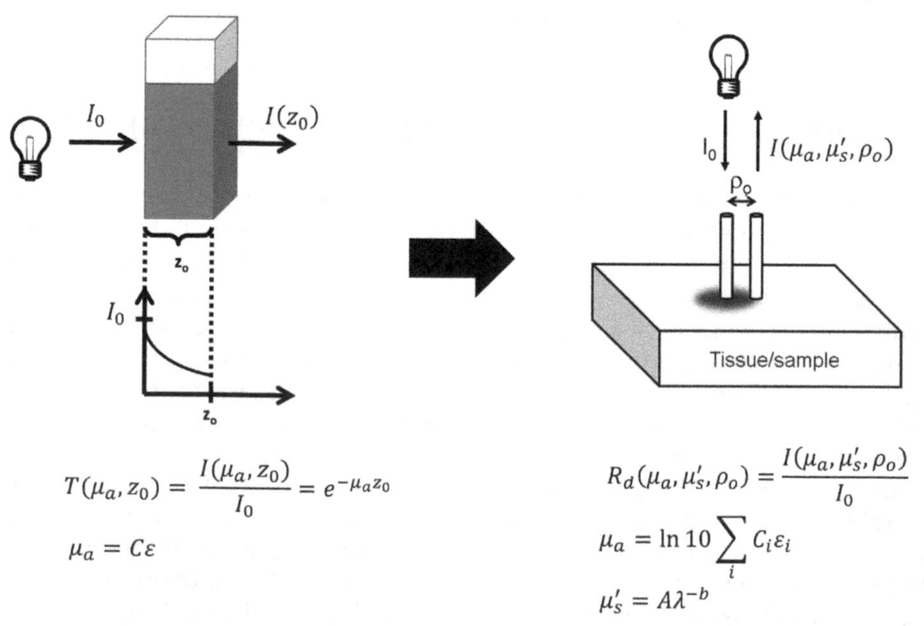

FIGURE 17.1 Comparison of absorption spectroscopy and diffuse reflectance spectroscopy. Absorption spectroscopy utilizes a transmission geometry, whereas diffuse reflectance spectroscopy uses fiber optics to collect light in a reflectance geometry. The path length of the cuvette (z_0) is analogous to the source–detector separation (ρ_0). Measurement of transmitted light in absorption spectroscopy allows quantification of optical absorption, and measurement of diffuse reflected light in diffuse reflectance spectroscopy allows quantification of optical absorption and optical scattering, μ_s' = reduced scattering coefficient, μ_a = absorption coefficient, λ = wavelength.

which can be attributed to abnormal tissue that has higher light absorption. Lesions that present a darker appearance include several types of benign nevi, or moles, as well as malignant melanomas. Other lesions may be a particular color in appearance, which suggests a low optical absorption of that particular color or a change in the optical scattering. For example, basal cell carcinomas alter the organization of cells and collagen in the dermis that results in a decrease in optical scattering as the disease progresses [6]. These changes in the optical properties of the lesion can be quantified by measurements of the diffuse reflectance and compared with the optical properties of normal skin to assess disease state.

The strength of measuring diffuse reflectance to calculate optical properties of skin is the ability to quantify the changes in optical properties and then correlate these changes to the biological changes in the skin [7]. Optical scattering in skin is caused by the index of refraction inhomogeneity between the different cell types and structures [8]. Changes in optical scattering are quantified in terms of the mean scatterer size and scatterer density and are correlated to cell and tissue morphology changes consistent with disease progression. The major biological components of skin that contribute to optical absorption are called chromophores, and these are melanin, oxyhemoglobin, deoxyhemoglobin, and water, as well as carotene and bilirubin [7,9–11]. Each of these components has a specific amount, or concentration, in normal skin. Measurement of diffuse reflectance in skin can determine the normal, or baseline, concentration of each of these components and facilitate comparison with the measurements from a suspicious lesion. Fig. 17.2 is a plot of the absorption spectra of typical skin chromophores and the typical scattering spectrum of skin.

Measurement of Diffuse Reflectance

Measurement of diffuse reflectance is straightforward and requires a measurement of reflected light intensity spectra from the skin (I_{skin}), a background intensity spectra measurement ($I_{\text{background}}$), and a measurement of the reflected light intensity spectra from a reflectance standard (I_{standard}). The wavelength range typically starts in the visible (350–650 nm) and extends to the near-infrared (650–1000 nm). The intensity spectra from skin is calibrated against the spectra from a known standard, and the standard measurement must also be properly scaled to account for the reflectance standard used. In this way, the reflectance represents the inherent characteristic of the tissue independent of the light source spectrum.

Diffuse reflectance from measurements of intensity is given as

$$R_{\text{d}}(\lambda) = \frac{I_{\text{skin}}(\lambda) - I_{\text{background}}(\lambda)}{I_{\text{standard}}(\lambda) - I_{\text{background}}(\lambda)} \qquad (17.1)$$

Determination of Optical Properties

The optical properties of the skin sample can be determined from the diffuse reflectance, because R_{d} is a function of μ_{s}' and μ_{a} as stated previously. To accomplish this, several approaches have been used and extensively studied [12], including analytical models that describe the propagation of light in tissue, computational simulations of the light propagation in tissue, and the use of precalculated tables, or lookup tables, used to interpolate the optical properties. These approaches are referred to as *inverse model calculations*, in which the independent variables (μ_{s}' and μ_{a}) are

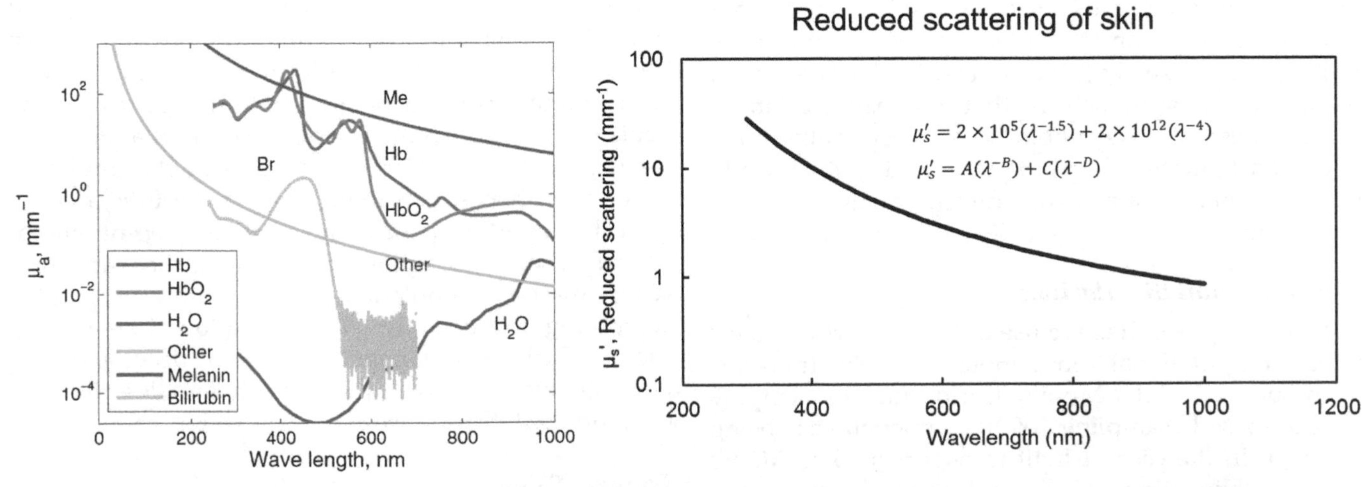

FIGURE 17.2 Plots of absorption spectra of typical chromophores in skin (*L*) and reduced scattering coefficient in skin (*R*). *Absorption spectra figure reprinted from Kim et al. Physiol Meas 2012;33:159-75 and reduced scattering expression courtesy of* Oregon Medical Laser Center News. *Jan 1998, Steven L. Jacques.*

TABLE 17.1 Comparison of Methods to Determine Optical Properties from Diffuse Reflectance

Method	Advantages	Disadvantages
Analytical model	Computationally simple	Not valid for skin
		Difficult to account for probe geometry
Computational simulation	Valid for skin	Computationally intensive and time consuming
	Very accurate results	Not all system variations can be accounted for
lookup table	Valid for skin	Somewhat labor intensive to generate
	Accounts for probe geometry self-calibrating and accounts for system variation	

calculated from the known result (R_d). Each of these methods has strengths and weaknesses that are described in Table 17.1, and a more detailed discussion of each is presented in the following sections.

Analytical Model

Analytical models of light propagation in tissue are based on the radiative transport equation (RTE), which describes the transfer of energy from electromagnetic radiation. In the case of light propagation in tissue, the RTE can be simplified to the diffusion approximation form of the RTE based on the assumption that scattering is much higher than absorption ($\mu_s' > \mu_a$), which is typically the case in biological tissue [13,14]. This results in the diffusion equation, which describes light propagation in tissue. Applying appropriate boundary conditions to the diffusion equation [15] yields an analytical solution for the diffuse reflectance as a function of μ_s' and μ_a, which results in computation of the optical properties based on the measured diffuse reflectance. Analytical model approaches to skin optical property calculations have been used extensively [7]. One drawback to the use of the analytical approach is that the assumption of $\mu_s' > \mu_a$ does not always hold in skin, especially in the case of dark, highly absorbing lesions [16]. As a result, this has necessitated the development of other approaches.

Computational Simulations

Another approach is the use of Monte Carlo model–based computational simulations of light transport in tissue [17–21]. Monte Carlo models employ statistics-based sampling of the phenomena being modeled. In the case of light transport in skin, Monte Carlo modeling simulates the trajectory of each photon as it propagates in the tissue. Based on the optical properties of skin, each photon has a certain probability, or weight, that it will scatter or be absorbed and this weight changes as the photon propagates further into the skin. This process iterates until the photon is either absorbed or remitted and is repeated for several million photons to ensure statistical accuracy of the simulated light transport in skin. Monte Carlo simulations are used to calculate the theoretical diffuse reflectance measured by the system, taking into account the system geometry and skin optical properties. The calculation of the diffuse reflectance is a forward model calculation, because the inputs (optical properties) are specified to determine the output (diffuse reflectance). The theoretical diffuse reflectance is then compared, or fit, with the measured sample diffuse reflectance to determine the similarity between them. The Monte Carlo simulation is then iteratively run with different combinations of optical properties until the measured sample diffuse reflectance and theoretical diffuse reflectance converge, yielding the optical properties of the measured sample. Whereas Monte Carlo–based computational models result in very accurate calculations of the sample optical properties, long computation times are often necessary because of the need to perform several simulations with different combinations of optical properties. One strategy to speed up the Monte Carlo simulation is the use of scaled Monte Carlo, in which a single simulation is run and then scaled to encompass the full range of optical properties [22,23]. In addition, recent efforts have focused on employing graphics processing unit (GPU)–based simulations that utilize parallel computing to speed up the simulation time [24,25]. Monte Carlo simulations have also been used to generate lookup tables [26], which are described in detail in the following section.

Lookup Tables

To address the aforementioned limitation of the diffusion approximation, Rajaram et al. proposed the concept of using a lookup table [16]. The lookup table method of determining optical properties involves the generation of a large table of predetermined diffuse reflectance values mapped to known optical properties. The measured sample diffuse reflectance can be searched for or interpolated from the lookup table, along with the corresponding sample optical properties. Generation of the lookup table can be performed by Monte Carlo–based computational simulations [26,27] or empirical measurements of laboratory-created objects, or "phantoms" with skin-like optical properties [16,28]. Monte Carlo simulation–based lookup tables are similar in principle to using Monte Carlo simulations to determine optical properties, except that several forward-model Monte Carlo simulations with different optical properties are initially run to generate the lookup table.

Measurement-based lookup tables are based on direct measurement of the diffuse reflectance from tissue phantoms, which are objects with known optical properties. Because phantoms are created in the laboratory, the optical properties can be carefully controlled to span a large range of physiological optical properties. Serial measurements are made on the created phantoms and the lookup table generated by mapping the measured diffuse reflectance of the phantoms to the optical properties.

The primary advantage of using a lookup table is that the optical property determination is very fast, because the interpolation is not computationally intensive. Each method to generate the lookup table, however, has specific aspects that must be carefully considered. A Monte Carlo simulation–based lookup table can be very accurate because the specific parameters of the optical system can be incorporated into the simulation; however, the simulations may be computationally intensive and time consuming. Measurement-based lookup tables inherently account for the probe geometry, which is an important consideration because complex probe geometries may be difficult to simulate. In addition, measurement-based lookup tables are calibrated against the system and account for any other system variations because the measurements are performed using the system itself, but can also be labor intensive in phantom creation and time consuming in data acquisition.

DIFFUSE REFLECTANCE METHODS

Diffuse reflectance can be measured using spectroscopic or imaging methods. The difference between the two methods is in the data output. DRS yields a spectral plot or graph of the diffusely reflected light, whereas diffuse reflectance imaging yields individual images at a specified wavelength interval within the wavelength range of interest. Each method is discussed in the following sections.

Diffuse Reflectance Spectroscopy

The basic tenet of DRS is the collection of diffusely reflected light and subsequent spectral analysis with a spectrometer. The instrumentation required for in vivo DRS consists of a light source and spectrometer interfaced to a computer, both of which are coupled by optical fibers to an optical probe (Fig. 17.3). Data acquisition involves placement of the probe directly on the sample, illumination of the sample through the probe, and

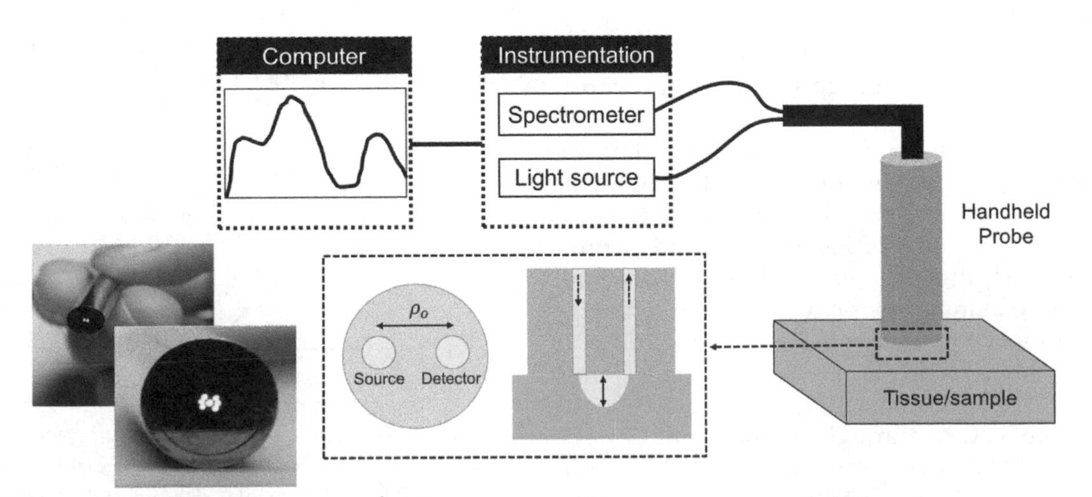

FIGURE 17.3 Schematic of typical diffuse reflectance spectroscopy (DRS) system, showing the hand-held probe, separation between source and detector fibers (ρ_0), light source, spectrometer, and spectral output, and images of a typical diffuse reflectance probe. *Probe images are reprinted and adapted from Rajaram et al. Laser Surg Med 2010;42:876–87.*

subsequent spectra collection by the probe to the spectrometer. Several types of broadband light sources are sufficient for DRS, including halogen lamps, xenon lamps, and white LEDs. Of note is that each of these light sources has different emitted light spectra, so the light source should be chosen carefully. The choice of spectrometer is also straightforward, ranging from high sensitivity spectrometers to cheaper nonimaging spectrometers. The primary factor in the high cost of imaging spectrometers is the charge-coupled device (CCD) camera, which is typically equipped with a cooled CCD for measurements requiring high sensitivity and high signal-to-noise ratio. The advent of inexpensive, fiber-coupled small-form–factor nonimaging spectrometers have made them a popular choice in research applications.

The most important instrumentation concern is the fiber-optic probe. Many researchers have developed probes of many different geometries and complexity to accommodate their measurements [29]. While the simplest probe consists of two optical fibers, one for light delivery and the other for light collection, other designs are much more elaborate, consisting, for example, of a single delivery fiber and several collection fibers surrounding the delivery fiber. In addition to the geometry, the type, size, and material of the fiber optic cables used are important considerations. Because a primary concern of most DRS applications in dermatology is collecting sufficient light from the skin, the optical fibers are predominantly multimode fibers ranging from 50 to 600 μm in diameter. In addition, the fiber material should also be considered, based on the wavelength range of the collected light in order to maximize collection. Finally, the distance between the delivery fiber (source) and collection fiber (detector) in the probe, or the source-detector separation, is important because this dictates the sampling depth of the light into the tissue. Because most DRS applications in skin are concerned with interrogating changes in optical properties of the epidermis and dermis, short source–detector separations (1 mm or less) are necessary. A recent report using Monte Carlo modeling to predict sampling depth of probes with different source–detector separations demonstrated that a source–detector separation of 500 μm resulted in a predicted sampling depth of 240 μm in a highly absorbing tissue [30]. Lastly, many probes have incorporated optical elements to focus the delivered and collected light for optimized measurements. Generally, most forms of DRS rely on the fundamental technique of light delivery and collection through a fiber-optic–based probe, with variations in probe geometry [31] and wavelength range of interest (ie, visible or near-infrared).

Diffuse Reflectance Imaging

Whereas DRS yields spectral plots of single point measurements of diffuse reflectance, DRI further extends DRS by yielding wide-field spectral images of diffuse reflectance. Similar to DRS, DRI is a quantitative technique that allows both optical absorption and optical scattering to be determined from images of diffuse reflectance. This differs from other wide-field imaging methods, such as multispectral imaging, which are largely qualitative and do not enable quantitation of optical properties. The instrumentation for DRI consists of a light source, a CCD camera coupled to a lens, and optical filters. Multispectral cameras, which utilize liquid crystal tunable filters, are well suited for DRI but are expensive. Alternatively, cost-effective approaches include using a filter wheel to acquire each spectrally separated image, or using individual light sources centered at a particular wavelength, such as LEDs, to sequentially illuminate the tissue, resulting in spectral images spanning the entire wavelength spectrum of interest. In addition, the use of CCD cameras yields spectral images with a large field of view, high spatial resolution, noncontact image acquisition, and a long working distance, as dictated by the camera lens.

Two noteworthy implementations of DRI are spatial frequency domain imaging (SFDI) [32–35], and a handheld, point-scanning imaging device [36,37]. SFDI utilizes the concept of structured illumination, where a digital micromirror device is used to project a spatially varying, sinusoidal pattern of bright and dark light bands on the tissue to construct a map of tissue optical properties (Fig. 17.4). Several structured illumination patterns of increasing sinusoidal frequency, or spatial frequency, are projected onto the tissue. At each spatial frequency, the projected pattern is linearly translated, or phase shifted, and a spectral image is acquired at each of these phase shifts. These phase-shifted images are then reconstructed into the spectrally resolved, diffuse reflectance images at each spatial frequency. The use of different

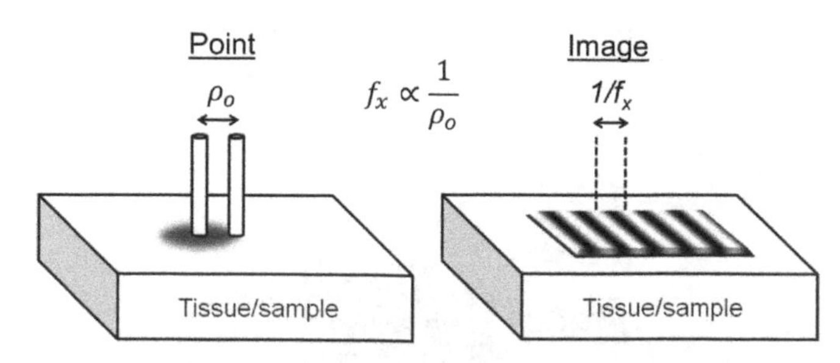

FIGURE 17.4 Analogy between point-based (DRS) and image-based [spatial frequency domain imaging (SFDI)] methods to acquire diffuse reflectance. The source–detector separation (ρ_0) in DRS is expressed as the frequency of the spatial pattern, or spatial frequency (f_x), in SFDI. ρ_0 is analogous to the reciprocal of f_x.

spatial frequencies results in the ability to interrogate different depths of tissue and to separate absorption from scattering. Of note is that, although the primary advantage of SFDI is a large field of view (dictated by the camera configuration), optical properties are computed for each image pixel using either the diffusion approximation approach or Monte Carlo simulations [38], which can lead to long computation times. Efforts to mitigate this have included Monte Carlo model calculations on the GPU [24], as well as lookup table–based interpolation [39] of optical properties.

The handheld imaging device is a point-scanning fiber-optic–based device that has optical elements to enable imaging and galvanometer scanning mirrors for two-dimensional scanning, effectively automating a probe-based point measurement. Light is delivered by an optical fiber to the scanning mirrors and scanned onto an imaging lens before illuminating the tissue. The reflected light is then collected and focused by the imaging lens onto the scanning mirrors, where it is descanned and directed to the collection fiber by a beam splitter. Because the diffusely reflected light is collected by fiber optic technology, this, in effect, results in confocal imaging ability as well as high spectral resolution. Optical properties are then computed using a lookup table approach [16]. A key difference between the hand-held, point-scanning device and SFDI is that SFDI has higher spatial resolution through the use of high-resolution CCD cameras, but lower spectral resolution because of limitations in spectral filters used with cameras. By contrast, the hand-held point-scanning device has high spectral resolution because of the use of a spectrometer, but longer acquisition times owing to limitation in scanning speed.

APPLICATIONS OF DIFFUSE REFLECTANCE IN DERMATOLOGY

The accessibility of skin makes it an ideal organ for optical interrogation with DRS and DRI. DRS data acquisition is performed with fiber-optic–based probes typically placed in contact with the skin, whereas for DRI, images of the skin are acquired. In addition, because most skin conditions involve discolored lesions that are identifiable by eye, measurements of diffuse reflectance to quantify optical properties are simple to set up and require little to no patient preparation. Whereas diffuse reflectance can be employed to investigate virtually all abnormal skin conditions, dermatology application areas of particular interest include skin cancer diagnosis, optical property changes in cutaneous vascular lesions [port wine stains (PWSs)] after laser surgery, assessment of erythema, UV protection efficacy of sunscreens, and hemodynamic changes in burn wounds.

Skin Cancer

The identification and diagnosis of skin cancers is the most extensively studied and most significant clinical application of diffuse reflectance in dermatology. One of the major challenges in skin cancer diagnosis is the ability to determine which suspicious lesions are malignant and which are benign. Suspicious lesions typically derive from the cell types that originate in the epidermis and can develop into or already indicate one of the major types of skin cancer: melanoma or nonmelanocytic skin cancers (NMSCs), which comprise basal cell carcinoma (BCC) and squamous cell carcinoma (SCC).

Diagnosis of suspicious skin lesions first involves visual recognition by a dermatologist, followed by surgical excision of the suspicious tissue, or biopsy, for microscopic analysis and eventual diagnosis by a dermatopathologist. A drawback of the biopsy is that it is an invasive procedure that, depending on the extent of the biopsy, can require wound management and affect quality of life. In addition, because biopsy of the lesion is necessary for diagnosis, biopsies are routinely performed in cases where the lesion is eventually determined to be benign, and therefore unnecessary. Finally, biopsies are expensive, costing roughly $350 each, including laboratory pathological study [40]. Because skin cancer diagnosis relies on the qualitative observations and experience of the dermatologist and dermatopathologist, as well as the invasive and expensive biopsy, there is a need to develop a noninvasive, cost-effective method that allows the extraction of specific, intrinsic, and quantitative information about the lesion. This type of information may lead to more efficient diagnosis, elimination of unnecessary biopsies, better treatment outcomes, and an overall better quality of life for the patient. Measurements of diffuse reflectance can elucidate and quantify subtle changes in optical properties between malignant and benign lesions that can guide the decision to biopsy and the ultimate diagnosis. DRS has been used in clinical settings to quantify changes in optical properties in both melanoma and NMSC.

Marchesini et al. was the first to report using fiber-optic probe–based DRS between 420 and 780 nm on 31 melanomas and 31 benign pigmented lesions. They used a discriminant analysis on the spectra and classified melanoma and benign lesions with 90.3% sensitivity and 77.4% specificity [41]. Wallace et al. built upon this study by using a probe with 18 delivery fibers and 12 collection fibers to collect to measure diffuse reflectance between 320 and 1100 nm on 121 lesions (15 melanomas, 32 compound nevi). They used a similar discriminant analysis as Marchesini et al. to classify lesions as melanoma or benign, resulting in 100% sensitivity and 84%

specificity [42]. Murphy et al. acquired diffuse reflectance spectra using a 6-around-1 probe (six collection fibers surrounding one delivery fiber) geometry in the wavelength range of 550–1000 nm on 120 pigmented lesions, of which 64 were diagnosed as melanoma [43]. Based in large part on these previous studies, Zonios et al. proposed using DRS to quantify melanin absorption as a diagnostic marker for melanoma, the first reference in the literature to looking at optical properties from diffuse reflectance spectra [44]. This was further investigated by Marchesini et al. in a study of 1671 pigmented lesions, of which 288 were melanomas [45]. Lim et al. collected spectra from 12 melanomas and 17 pigmented lesions using a combination of DRS, fluorescence spectroscopy, and Raman spectroscopy and classified melanoma and pigmented lesions with a sensitivity of 100% and specificity of 100% [46].

McIntosh et al. obtained diffuse reflectance spectra from 130 skin lesions in the extended wavelength range of 400–2500 nm. Spectra were acquired from several different types of lesions, including BCC, actinic lentigines, dysplastic melanocytic nevi, benign nevi, actinic keratosis (AK), and seborrheic keratosis. Difference spectra from normal skin demonstrated spectral changes between each lesion and normal skin [47]. Rajaram et al. collected spectra from 48 lesions comprising BCC, SCC, and AK using a system combining DRS with fluorescence spectroscopy. Diffuse reflectance spectra were analyzed using a lookup table to extract optical properties, and physiological information gained from both modalities was used to construct discriminant analysis-based disease classifiers. Rajaram et al. reported 94% sensitivity and 89% specificity in distinguishing BCC from normal skin and 100% sensitivity and 50% specificity in distinguishing SCC from AK [6]. Thompson et al. reported using DRS and fluorescence spectroscopy to measure spectra from eight BCCs with a sensitivity of 100% and specificity of 71% [48]. Lim et al. reported using a combination of DRS, fluorescence spectroscopy, and Raman spectroscopy to collect spectra from 57 SCC and BCC and 14 AK with a sensitivity of 95% and specificity of 71% in distinguishing SCC and BCC from AK [46].

Another fiber-optic–based spectroscopy method used to examine melanomas and NMSCs is oblique incidence DRS [49,50]. This method involves an alteration to the fiber geometry in the probe, where the source fibers are oriented at an angle to the skin surface. Garcia-Uribe et al. used oblique incidence DRS to calculate optical properties from 144 skin lesions, including 16 melanomas, and reported increased optical scattering and absorption in malignant lesions [31]. A follow-up study [51] in which optical properties were measured in 678 pigmented and nonpigmented lesions yielded 90% sensitivity and 90% specificity in diagnosis of melanoma, and 92% sensitivity and 92% specificity in distinguishing BCC and SCC from precancerous lesions.

Port Wine Stain

PWSs are cutaneous hypervascularized malformations [52] that are typically reddish-purple in appearance. These lesions can present anywhere on the body and often present on the face, which can lead to psychosocial issues. PWSs are typically treated with pulsed lasers to photocoagulate the abnormal blood vessels and efficacy of treatment is normally determined by visual observation of the lesion lightening in color. Measurement of diffuse reflectance has been applied to quantify changes in lesion color and a few examples are presented.

Sheehan-Dare and Cotterill evaluated PWS lightening with different laser treatments using a spectrophotometer-based system to measure diffuse reflectance [53]. Lister et al. discussed quantification of PWS optical properties from diffuse reflectance measurements by using light propagation–model approaches [54]. Jung et al. used cross-polarized DRI to quantify erythema and melanin content of PWSs after treatment [55,56]. The use of cross-polarizers eliminated glare, which resulted in accurate color assessment. Mazhar et al. used SFDI, a wide-field DRI method discussed previously, in a pilot study to determine physiological changes in PWS after treatment [56,57]. Using SFDI, the diffuse reflectance was measured and optical properties were calculated using a lookup table approach. By fitting the optical absorption to a linear combination of known optical absorbers in the skin, Mazhar et al. found an increase in deoxyhemoglobin and total hemoglobin concentration and a reduction in oxygen saturation after PWS treatment [57].

Erythema

Erythema is redness of the skin caused by injury or another inflammation-causing condition. Often presenting as a rash, erythema can be caused by environmental factors, infection, or overexposure to the sun (ie, sunburn). Measurement of diffuse reflectance has a direct application in assessing erythema and a few examples are outlined. Kollias et al. used DRS to determine oxyhemoglobin and deoxyhemoglobin concentrations in irritant-induced erythema and found that oxyhemoglobin concentration increased linearly with increasing irritant concentration, whereas deoxyhemoglobin concentration did not change [58]. Stamata and Kollias investigated the contributions of oxyhemoglobin, deoxyhemoglobin, and melanin to perceived skin color after UV irradiation. Using DRS, they found that hemoglobin concentration correlated well with clinical observation of erythema, whereas melanin concentration was correlated with skin pigmentation, or tanning. Deoxyhemoglobin concentration, as a metric of blood pooling, was also found to contribute to perceived skin color appearance [59]. In a later study, Stamatas et al. used a commercial DRS instrument to determine oxyhemoglobin, deoxyhemoglobin, and melanin concentrations in order to assess erythema and skin pigmentation induced by UV irradiation [60]. Papazoglou et al. also used DRS and fluorescence spectroscopy to assess UV irradiation–induced erythema. They determined oxyhemoglobin and deoxyhemoglobin concentrations from diffuse reflectance measurements and compared this with expression of biological markers of skin damage [61].

Evaluation of Sunscreen

Sunscreens are chemically formulated oil-in-water emulsions, or lotions, applied topically to the skin that protect against harmful UV radiation from the sun [62]. The two common bands of the UV spectrum that are involved in sunburn or other sun-induced damage to the skin are ultraviolet A (UVA) (315–400 nm) and ultraviolet B (UVB) (280–325 nm), although the majority of damage is caused by UVA radiation. Sunscreen efficacy is typically reported in terms of sun protection factor, which is a measure of the UV radiation that is blocked from the skin. Measurement of diffusely reflected UV light from skin with applied sunscreen is a straightforward and noninvasive method to assess the UV protection efficacy of sunscreens. Smith et al. measured in vivo diffuse reflectance to compare the UVA protection performance of several different sunscreen formulations [63]. Moyal et al. also used DRS to compare the in vivo UVA protection performance of different sunscreen formulations with in vitro experiments and found that the in vivo diffuse reflectance measurements confirmed the in vitro results, demonstrating the potential of DRS as an effective method for assessing UVA sunscreen protection in vivo [64]. Gillies et al. conducted a study using DRS, measuring diffuse reflectance from the skin of 20 subjects before and after application of sunscreen. The study showed a linear response in UVA protection with increased concentration of sunscreen formulation [65]. Ruvolo et al. conducted a follow-up study to compare in vivo results of UVA protection obtained from DRS measurements with UV protection results from in vitro measurements [66]. In a recent study, Ruvolo et al. reported on an improved approach using in vivo DRS to study sunscreen protection from both the UVA and UVB portions of the UV spectrum [67].

Burns

Burn wounds of the skin and underlying tissue are the result of thermal overexposure from heat, or even cold, and can also be caused by electricity, chemical contact, or ionizing radiation exposure. Whereas the presentation of the burn is generally easily observed, less clear is the degree of the burn, which is dependent on the depth of tissue injury and subsequent treatment and management. Assessment of burn degree is typically done by clinical examination, which is subject to the clinician's experience, and can sometimes involve an invasive tissue biopsy. There is considerable interest in noninvasive, optical methods of assessing burn wounds [68] that can provide more detailed information. Interrogation of tissue biochemical components, such as hemoglobin concentration and oxygen saturation, may provide a quantitative way to assess burn degree. Applications of probe-based near-infrared DRS (NIRS) and SFDI in interrogating burn wounds have been reported. Sowa et al. conducted a pilot study using NIRS to assess changes in burn wound hemodynamics during the early postburn period [69] by quantifying oxygen saturation, blood volume, and water content in a porcine model of burns. In a follow-up study, Sowa et al. used a different method of analysis to classify burn depth based on measured diffuse reflectance spectra acquired from burns of known depths [70]. Cross et al. measured diffuse reflectance spectra using NIRS and near-infrared imaging spectroscopy in 22 burn patients and quantified total hemoglobin concentration and oxygen saturation in both superficial and full-thickness burns [71]. In a different study, Cross et al. used NIRS to assess edema, or swelling, in burn wounds by quantifying tissue water content in superficial and partial thickness burns [72]. SFDI applications in assessing burn wounds have also been reported. Nguyen et al. conducted a pilot study of SFDI on a rodent model of graded burn wounds and monitored changes in water concentration, deoxyhemoglobin concentration, and optical scattering [73]. They found that the change in each of these parameters immediately postburn differed depending on the type of burn (superficial vs. partial thickness). Mazhar et al. followed up this work by using a commercial SFDI to monitor hemodynamic changes in a porcine model of graded burn wounds and correlated the optical property measurements with histological results [74]. Ponticorvo et al. expanded this work by combining SFDI [75] with laser speckle imaging [76], another noninvasive wide-field imaging technique discussed in greater detail in Chapter 8, to acquire tissue hemodynamic information from a porcine model of graded burn wounds.

CONCLUSION

Applications of diffuse reflectance in dermatology highlighted in this chapter have foreshadowed future research efforts. The most promising future direction is to combine diffuse reflectance measurements with other methods, such as fluorescence spectroscopy [6,48,61,77] in order to construct a more complete biological picture of the pathology of various skin conditions and disease states. Other spectroscopic methods [78,79] have previously been discussed in the context of skin cancer diagnosis and these methods can be applied to the application areas discussed in this chapter. One particular area of promise in this concept of multimodal spectroscopy is the combination of DRS, fluorescence spectroscopy, and Raman spectroscopy, a method described in further detail in Chapter 12. Lim et al. conducted a clinical study using these three spectroscopy methods to obtain spectra from 137 skin lesions [46]. Using a combination of data from all three spectroscopy methods, the authors reported 100% sensitivity and 100% specificity in classifying melanoma from benign pigmented lesions and 95% sensitivity and 71% specificity in classifying NMSC from AK. Sharma et al. reported on an improved combined instrument that facilitates simultaneous acquisition of DRS, fluorescence, and Raman spectroscopy from a single instrument and single probe [80]. This combined spectroscopy approach has the potential to achieve the ultimate goal of early skin cancer detection and diagnosis. A large clinical trial of the approach, initiated by the authors of this chapter, is set to commence in the near future. In addition, improvements in the light propagation model approaches that better incorporate skin physiology [30,81–83] may result in more accurate determinations of skin optical properties and more detailed insight into skin pathology.

Although the application of diffuse reflectance in dermatology is largely in the research stage, as of this writing, a number of commercial devices that utilize measurements of diffuse reflectance from skin have appeared on the market, including the Philips BiliChek bilirubinometer (Philips North America Corp., Andover, MA, United States), a noninvasive light-based instrument to detect skin bilirubin concentration, MedX SIMSYS-MoleMate (MedX Health Corp., Mississauga, ON, Canada), a skin imaging device to detect potentially malignant lesions, and MELA Sciences MelaFind (STRATA Skin Sciences, Horsham, PA, United States), another skin imaging device for melanoma detection. Each device has U.S. Food and Drug Administration approval and is commercially available in both the US and European markets. These initial commercial platforms utilizing diffuse reflectance demonstrate that there is significant interest in bringing diffuse reflectance–based optical technologies to the clinic and that there is much potential to bring devices and technologies in the research and development stage to market.

In this chapter, the concept of diffuse reflectance and its underlying physical principles, different methods to measure diffuse reflectance including probe-based spectroscopy and camera-based imaging, clinical applications of diffuse reflectance, and future directions to be explored have been presented. The application of diffuse reflectance in dermatology has yielded quantitative information and biological insight into the physiology of skin and pathology of skin disease. The ease of accessibility of skin and the simplistic instrumentation necessary for measurement of diffuse reflectance facilitates widespread adoption and application of the technique. The fiber-optic–based optical probes used in the majority of diffuse reflectance measurements enables flexibility in acquiring data from multiple locations and its ease of use facilitates clinical utility. In addition, the development of noncontact, wide-field DRI methods allows for efficient data collection over large areas of interest. The future of diffuse reflectance in dermatology lies in the combination of diffuse reflectance with other optical spectroscopic methods to obtain complete, quantitative tissue information for diagnostic purposes and even in monitoring response to treatment. Ultimately, diffuse reflectance has the potential to provide a quick, easy, noninvasive method of interrogating skin physiology that can impact modern healthcare.

List of Acronyms and Abbreviations

RTE Radiative transport equation
GPU Graphics processing unit
DRS Diffuse reflectance spectroscopy
LED Light-emitting diode
CCD Charge-coupled device
DRI Diffuse reflectance imaging
SFDI Spatial frequency domain imaging
NMSC Nonmelanocytic skin cancer
BCC Basal cell carcinoma
SCC Squamous cell carcinoma
AK Actinic keratosis
SK Seborrheic keratosis
PWS Port wine stain
UV Ultraviolet

UVA Ultraviolet A
UVB Ultraviolet B
NIRS Near-infrared diffuse reflectance spectroscopy
FDA U.S. Food and Drug Administration

Glossary

Chromophore A biomolecule constituent of tissue that has a distinct absorption spectrum.
Multispectral imaging An imaging technique to acquire image data at specific wavelengths of light.
Beer's law A physical principle that describes the attenuation of light by some material or sample through which the light is transmitted.
Monte Carlo modeling A computational algorithm that utilizes repeated random sampling to simulate a physical or mathematical process.
Digital micromirror device A chip-based device invented by Texas Instruments (Dallas, TX, United States) that contains several hundred thousand mirrors that can be individually controlled to project a specified pattern.
Graphics processing unit A computer chip dedicated solely to image and graphics processing tasks.
Sensitivity A statistical term that measures the true positive rate, or the proportion of true positive outcomes that are correctly identified in a study.
Specificity A statistical term that measures the true negative rate, or the proportion of true negative outcomes that are correctly identified in a study.

References

[1] Anderson RR, Parrish JA. The optics of human skin. J Invest Dermatol July 1981;77(1):13–9.
[2] Van Gemert MJ, Jacques SL, Sterenborg HJ, Star WM. Skin optics. IEEE Trans Biomed Eng December 1989;36(12):1146–54.
[3] Richards-Kortum R, Sevick-Muraca E. Quantitative optical spectroscopy for tissue diagnosis. Annu Rev Phys Chem 1996; 47:555–606.
[4] Jacques SL. Optical properties of biological tissues: a review. Phys Med Biol June 2013;58(11):R37–61.
[5] Wilson BC, Jacques SL. Optical reflectance and transmittance of tissues: principles and applications. IEEE J Quan Electron December 1990;26(12):2186–99.
[6] Rajaram N, Reichenberg JS, Migden MR, Nguyen TH, Tunnell JW. Pilot clinical study for quantitative spectral diagnosis of non-melanoma skin cancer. Lasers Surg Med December 2010;42(10): 716–27.
[7] Zonios G, Bykowski J, Kollias N. Skin melanin, hemoglobin, and light scattering properties can be quantitatively assessed in vivo using diffuse reflectance spectroscopy. J Invest Dermatol December 2001;117(6):1452–7.
[8] Schmitt JM, Kumar G. Turbulent nature of refractive-index variations in biological tissue. Opt Lett August 15, 1996;21(16):1310–2.
[9] Kollias N. The physical basis of skin color and its evaluation. Clin Dermatol August 1995;13(4):361–7.
[10] Young AR. Chromophores in human skin. Phys Med Biol May 1997;42(5):789–802.
[11] Bashkatov AN, Genina EA, Kochubey VI, Tuchin VV. Optical properties of human skin, subcutaneous and mucous tissues in the wavelength range from 400 to 2000 nm. J Phys Appl Phys August 7, 2005;38(15):2543.
[12] Wilson RH, Mycek M-A. Models of light propagation in human tissue applied to cancer diagnostics. Technol Cancer Res Treat April 2011;10(2):121–34.
[13] Farrell TJ, Patterson MS, Wilson B. A diffusion theory model of spatially resolved, steady-state diffuse reflectance for the noninvasive determination of tissue optical properties in vivo. Med Phys August 1992;19(4):879–88.
[14] Kienle A, Lilge L, Patterson MS, Hibst R, Steiner R, Wilson BC. Spatially resolved absolute diffuse reflectance measurements for noninvasive determination of the optical scattering and absorption coefficients of biological tissue. Appl Opt May 1, 1996; 35(13):2304–14.
[15] Haskell RC, Svaasand LO, Tsay T-T, Feng T-C, Tromberg BJ, McAdams MS. Boundary conditions for the diffusion equation in radiative transfer. J Opt Soc Am A October 1, 1994;11(10):2727–41.
[16] Rajaram N, Nguyen TH, Tunnell JW. Lookup table-based inverse model for determining optical properties of turbid media. J Biomed Opt October 2008;13(5):050501.
[17] Wilson BC, Adam G. A Monte Carlo model for the absorption and flux distributions of light in tissue. Med Phys December 1983; 10(6):824–30.
[18] Palmer GM, Ramanujam N. Monte Carlo-based inverse model for calculating tissue optical properties. I. Theory and validation on synthetic phantoms. Appl Opt February 10, 2006;45(5):1062–71.
[19] Wang L, Jacques SL, Zheng L. MCML–Monte Carlo modeling of light transport in multi-layered tissues. Comput Methods Programs Biomed July 1995;47(2):131–46.
[20] Graaff R, Koelink MH, de Mul FF, Zijistra WG, Dassel AC, Aarnoudse JG. Condensed Monte Carlo simulations for the description of light transport. Appl Opt February 1, 1993;32(4): 426–34.
[21] Kienle A, Patterson MS. Determination of the optical properties of turbid media from a single Monte Carlo simulation. Phys Med Biol October 1996;41(10):2221–7.
[22] Liu Q, Ramanujam N. Scaling method for fast Monte Carlo simulation of diffuse reflectance spectra from multilayered turbid media. J Opt Soc Am A Opt Image Sci Vis April 2007;24(4): 1011–25.
[23] Martinelli M, Gardner A, Cuccia D, Hayakawa C, Spanier J, Venugopalan V. Analysis of single Monte Carlo methods for prediction of reflectance from turbid media. Opt Express September 26, 2011;19(20):19627–42.
[24] Yang O, Choi B. Accelerated rescaling of single Monte Carlo simulation runs with the graphics processing unit (GPU). Biomed Opt Express 2013;4(11):2667–72.
[25] Jacques SL. Coupling 3D Monte Carlo light transport in optically heterogeneous tissues to photoacoustic signal generation. Photoacoustics September 10, 2014;2(4):137–42.
[26] Hennessy R, Lim SL, Markey MK, Tunnell JW. Monte Carlo lookup table-based inverse model for extracting optical properties from tissue-simulating phantoms using diffuse reflectance spectroscopy. J Biomed Opt March 2013;18(3):037003.
[27] Wen X, Zhong X, Yu T, Zhu D. A Monte Carlo based lookup table for spectrum analysis of turbid media in the reflectance probe regime. Quan Electron July 31, 2014;44(7):641.
[28] Nichols BS, Rajaram N, Tunnell JW. Performance of a lookup table-based approach for measuring tissue optical properties with diffuse optical spectroscopy. J Biomed Opt May 2012;17(5): 057001.
[29] Utzinger U, Richards-Kortum RR. Fiber optic probes for biomedical optical spectroscopy. J Biomed Opt January 2003; 8(1):121–47.
[30] Hennessy R, Goth W, Sharma M, Markey MK, Tunnell JW. Effect of probe geometry and optical properties on the sampling depth

for diffuse reflectance spectroscopy. J Biomed Opt 2014;19(10): 107002.
[31] Garcia-Uribe A, Smith EB, Zou J, Duvic M, Prieto V, Wang LV. In-vivo characterization of optical properties of pigmented skin lesions including melanoma using oblique incidence diffuse reflectance spectrometry. J Biomed Opt February 2011;16(2): 020501.
[32] Cuccia DJ, Bevilacqua F, Durkin AJ, Tromberg BJ. Modulated imaging: quantitative analysis and tomography of turbid media in the spatial-frequency domain. Opt Lett June 1, 2005;30(11):1354–6.
[33] Cuccia DJ, Bevilacqua F, Durkin AJ, Ayers FR, Tromberg BJ. Quantitation and mapping of tissue optical properties using modulated imaging. J Biomed Opt April 2009;14(2):024012.
[34] Weber JR, Cuccia DJ, Durkin AJ, Tromberg BJ. Noncontact imaging of absorption and scattering in layered tissue using spatially modulated structured light. J Appl Phys May 15, 2009;105(10):102028.
[35] Yang B, Sharma M, Tunnell JW. Attenuation-corrected fluorescence extraction for image-guided surgery in spatial frequency domain. J Biomed Opt August 2013;18(8):80503.
[36] Bish SF, Rajaram N, Nichols B, Tunnell JW. Development of a noncontact diffuse optical spectroscopy probe for measuring tissue optical properties. J Biomed Opt December 2011;16(12): 120505.
[37] Bish SF, Sharma M, Wang Y, Triesault NJ, Reichenberg JS, Zhang JXJ, et al. Handheld diffuse reflectance spectral imaging (DRSI) for in-vivo characterization of skin. Biomed Opt Express February 1, 2014;5(2):573–86.
[38] Gardner AR, Venugopalan V. Accurate and efficient Monte Carlo solutions to the radiative transport equation in the spatial frequency domain. Opt Lett June 15, 2011;36(12):2269–71.
[39] Erickson TA, Mazhar A, Cuccia D, Durkin AJ, Tunnell JW. Lookup-table method for imaging optical properties with structured illumination beyond the diffusion theory regime. J Biomed Opt June 2010;15(3):036013.
[40] Susman E. Non-melanoma skin cancer on the rise. Oncol Times March 2011;33(5):42–3.
[41] Marchesini R, Cascinelli N, Brambilla M, Clemente C, Mascheroni L, Pignoli E, et al. In vivo spectrophotometric evaluation of neoplastic and non-neoplastic skin pigmented lesions. II. Discriminant analysis between nevus and melanoma. Photochem Photobiol April 1992;55(4):515–22.
[42] Wallace VP, Crawford DC, Mortimer PS, Ott RJ, Bamber JC. Spectrophotometric assessment of pigmented skin lesions: methods and feature selection for evaluation of diagnostic performance. Phys Med Biol March 2000;45(3):735–51.
[43] Murphy BW, Webster RJ, Turlach BA, Quirk CJ, Clay CD, Heenan PJ, et al. Toward the discrimination of early melanoma from common and dysplastic nevus using fiber optic diffuse reflectance spectroscopy. J Biomed Opt December 2005;10(6): 064020.
[44] Zonios G, Dimou A, Bassukas I, Galaris D, Tsolakidis A, Kaxiras E. Melanin absorption spectroscopy: new method for noninvasive skin investigation and melanoma detection. J Biomed Opt February 2008;13(1):014017.
[45] Marchesini R, Bono A, Carrara M. In vivo characterization of melanin in melanocytic lesions: spectroscopic study on 1671 pigmented skin lesions. J Biomed Opt February 2009;14(1):014027.
[46] Lim L, Nichols B, Migden MR, Rajaram N, Reichenberg JS, Markey MK, et al. Clinical study of noninvasive in vivo melanoma and nonmelanoma skin cancers using multimodal spectral diagnosis. J Biomed Opt 2014;19(11):117003.
[47] McIntosh LM, Summers R, Jackson M, Mantsch HH, Mansfield JR, Howlett M, et al. Towards non-invasive screening of skin lesions by near-infrared spectroscopy. J Invest Dermatol January 2001;116(1):175–81.
[48] Thompson AJ, Coda S, Sørensen MB, Kennedy G, Patalay R, Waitong-Brämming U, et al. In vivo measurements of diffuse reflectance and time-resolved autofluorescence emission spectra of basal cell carcinomas. J Biophotonics March 2012;5(3):240–54.
[49] Mehrübeoğlu M, Kehtarnavaz N, Marquez G, Duvic M, Wang LV. Skin lesion classification using oblique-incidence diffuse reflectance spectroscopic imaging. Appl Opt January 1, 2002;41(1): 182–92.
[50] Garcia-Uribe A, Kehtarnavaz N, Marquez G, Prieto V, Duvic M, Wang LV. Skin cancer detection by spectroscopic oblique-incidence reflectometry: classification and physiological origins. Appl Opt May 1, 2004;43(13):2643–50.
[51] Garcia-Uribe A, Zou J, Duvic M, Cho-Vega JH, Prieto VG, Wang LV. In vivo diagnosis of melanoma and nonmelanoma skin cancer using oblique incidence diffuse reflectance spectrometry. Cancer Res June 1, 2012;72(11):2738–45.
[52] Ortiz AE, Nelson JS. Port-wine stain laser treatments and novel approaches. Facial Plast Surg FPS December 2012;28(6):611–20.
[53] Sheehan-Dare RA, Cotterill JA. Copper vapour laser treatment of port wine stains: clinical evaluation and comparison with conventional argon laser therapy. Br J Dermatol May 1993;128(5):546–9.
[54] Lister T, Wright P, Chappell P. Spectrophotometers for the clinical assessment of port-wine stain skin lesions: a review. Lasers Med Sci May 2010;25(3):449–57.
[55] Jung B, Choi B, Durkin AJ, Kelly KM, Nelson JS. Characterization of port wine stain skin erythema and melanin content using cross-polarized diffuse reflectance imaging. Lasers Surg Med 2004; 34(2):174–81.
[56] Sharif SA, Taydas E, Mazhar A, Rahimian R, Kelly KM, Choi B, et al. Noninvasive clinical assessment of port-wine stain birthmarks using current and future optical imaging technology: a review. Br J Dermatol December 2012;167(6):1215–23.
[57] Mazhar A, Sharif SA, Cuccia JD, Nelson JS, Kelly KM, Durkin AJ. Spatial frequency domain imaging of port wine stain biochemical composition in response to laser therapy: a pilot study. Lasers Surg Med October 2012;44(8):611–21.
[58] Kollias N, Gillies R, Muccini JA, Uyeyama RK, Phillips SB, Drake LA. A single parameter, oxygenated hemoglobin, can be used to quantify experimental irritant-induced inflammation. J Invest Dermatol March 1995;104(3):421–4.
[59] Stamatas GN, Kollias N. Blood stasis contributions to the perception of skin pigmentation. J Biomed Opt April 2004;9(2): 315–22.
[60] Stamatas GN, Zmudzka BZ, Kollias N, Beer JZ. In vivo measurement of skin erythema and pigmentation: new means of implementation of diffuse reflectance spectroscopy with a commercial instrument. Br J Dermatol September 2008;159(3):683–90.
[61] Papazoglou E, Sunkari C, Neidrauer M, Klement JF, Uitto J. Noninvasive assessment of UV-induced skin damage: comparison of optical measurements to histology and MMP expression. Photochem Photobiol February 2010;86(1):138–45.
[62] Gasparro FP. Sunscreens, skin photobiology, and skin cancer: the need for UVA protection and evaluation of efficacy. Environ Health Perspect March 2000;108(Suppl. 1):71–8.
[63] Smith GJ, Miller IJ, Clare JF, Diffey BL. The effect of UV absorbing sunscreens on the reflectance and the consequent protection of skin. Photochem Photobiol February 2002;75(2):122–5.
[64] Moyal D, Refrégier J-L, Chardon A. In vivo measurement of the photostability of sunscreen products using diffuse reflectance spectroscopy. Photodermatol Photoimmunol Photomed February 2002;18(1):14–22.
[65] Gillies R, Moyal D, Forestier S, Kollias N. Non-invasive in vivo determination of UVA efficacy of sunscreens using diffuse reflectance spectroscopy. Photodermatol Photoimmunol Photomed August 2003;19(4):190–4.

[66] Ruvolo E, Chu M, Grossman F, Cole C, Kollias N. Diffuse reflectance spectroscopy for ultraviolet A protection factor measurement: correlation studies between in vitro and in vivo measurements. Photodermatol Photoimmunol Photomed December 2009;25(6):298–304.

[67] Ruvolo Junior E, Kollias N, Cole C. New noninvasive approach assessing in vivo sun protection factor (SPF) using diffuse reflectance spectroscopy (DRS) and in vitro transmission. Photodermatol Photoimmunol Photomed August 2014;30(4):202–11.

[68] Kaiser M, Yafi A, Cinat M, Choi B, Durkin AJ. Noninvasive assessment of burn wound severity using optical technology: a review of current and future modalities. Burns J Int Soc Burn Inj May 2011;37(3):377–86.

[69] Sowa MG, Leonardi L, Payette JR, Fish JS, Mantsch HH. Near infrared spectroscopic assessment of hemodynamic changes in the early post-burn period. Burns J Int Soc Burn Inj May 2001; 27(3):241–9.

[70] Sowa MG, Leonardi L, Payette JR, Cross KM, Gomez M, Fish JS. Classification of burn injuries using near-infrared spectroscopy. J Biomed Opt October 2006;11(5):054002.

[71] Cross KM, Leonardi L, Payette JR, Gomez M, Levasseur MA, Schattka BJ, et al. Clinical utilization of near-infrared spectroscopy devices for burn depth assessment. Wound Repair Regen Off Publ Wound Heal Soc Eur Tissue Repair Soc June 2007; 15(3):332–40.

[72] Cross KM, Leonardi L, Gomez M, Freisen JR, Levasseur MA, Schattka BJ, et al. Noninvasive measurement of edema in partial thickness burn wounds. J Burn Care Res Off Publ Am Burn Assoc October 2009;30(5):807–17.

[73] Nguyen JQ, Crouzet C, Mai T, Riola K, Uchitel D, Liaw L-H, et al. Spatial frequency domain imaging of burn wounds in a preclinical model of graded burn severity. J Biomed Opt June 2013;18(6): 66010.

[74] Mazhar A, Saggese S, Pollins AC, Cardwell NL, Nanney L, Cuccia DJ. Noncontact imaging of burn depth and extent in a porcine model using spatial frequency domain imaging. J Biomed Opt August 2014;19(8):086019.

[75] Ponticorvo A, Burmeister DM, Yang B, Choi B, Christy RJ, Durkin AJ. Quantitative assessment of graded burn wounds in a porcine model using spatial frequency domain imaging (SFDI) and laser speckle imaging (LSI). Biomed Opt Express October 1, 2014;5(10):3467–81.

[76] Huang Y-C, Tran N, Shumaker PR, Kelly K, Ross EV, Nelson JS, et al. Blood flow dynamics after laser therapy of port wine stain birthmarks. Lasers Surg Med October 2009;41(8):563–71.

[77] Borisova E, Troyanova P, Pavlova P, Avramov L. Diagnostics of pigmented skin tumors based on laser-induced autofluorescence and diffuse reflectance spectroscopy. Quan Electron June 30, 2008;38(6):597.

[78] Calin MA, Parasca SV, Savastru R, Calin MR, Dontu S. Optical techniques for the noninvasive diagnosis of skin cancer. J Cancer Res Clin Oncol July 2013;139(7):1083–104.

[79] Drakaki E, Vergou T, Dessinioti C, Stratigos AJ, Salavastru C, Antoniou C. Spectroscopic methods for the photodiagnosis of nonmelanoma skin cancer. J Biomed Opt June 2013;18(6): 061221.

[80] Sharma M, Marple E, Reichenberg J, Tunnell JW. Design and characterization of a novel multimodal fiber-optic probe and spectroscopy system for skin cancer applications. Rev Sci Instrum August 2014;85(8):083101.

[81] Yudovsky D, Nguyen JQM, Durkin AJ. In vivo spatial frequency domain spectroscopy of two layer media. J Biomed Opt October 2012;17(10):107006.

[82] Sharma M, Hennessy R, Markey MK, Tunnell JW. Verification of a two-layer inverse Monte Carlo absorption model using multiple source-detector separation diffuse reflectance spectroscopy. Biomed Opt Express December 2, 2013;5(1):40–53.

[83] Hennessy R, Markey MK, Tunnell JW. Impact of one-layer assumption on diffuse reflectance spectroscopy of skin. J Biomed Opt February 1, 2015;20(2):27001.

CHAPTER

18

Spectral Imaging in Dermatology

D. Ho[1,2], *E. Kraeva*[1], *J. Jagdeo*[1,2,3], *R.M. Levenson*[4]

[1]Sacramento VA Medical Center, Mather, CA, United States; [2]University of California at Davis, Sacramento, CA, United States; [3]State University of New York Downstate Medical Center, Brooklyn, NY, United States; [4]UC Davis Medical Center, Sacramento, CA, United States

OUTLINE

OUTLINE

Scope: Spectral imaging is taken to include conventional wide-field microscopy, confocal microscopy, fluorescence lifetime microscopy, reflectance spectroscopy, Fourier-transform infrared (FTIR) imaging, multiphoton microscopy and its technological (albeit not spectral) companion, second harmonic generation (SHG), and pump-probe microscopy, along with different variations of Raman spectroscopy. Whereas there has been significant effort spent seeking to exploit intrinsic (agent-free) contrast mechanisms, the increased specificity and brightness of exogenous contrast agents have encouraged their exploration. An interesting theme is the

Imaging in Dermatology
http://dx.doi.org/10.1016/B978-0-12-802838-4.00018-2

continued need for morphology-based analysis in addition to non–imaging-based detection and quantitation of particular analytes.

Purpose: Basic science research, improved histological diagnosis, and preliminary and conformational studies in vitro that will motivate and validate related in vivo diagnostic or screening applications for the detection and treatment of skin conditions.

Applications: Spectral imaging may be used for skin cancer, skin diseases, processes, such as aging and sun damage, and skin interaction with cosmetics, cosmeceuticals, pharmaceuticals, and other applied agents; assistance in diagnosis of skin diseases; and monitoring of response to therapy real-time in a noninvasive, repeatable, and time-efficient manner.

INTRODUCTION

This chapter will discuss the application of different optical approaches to the characterization of skin that has been taken from a patient, volunteer, or experimental animal, and studied ex vivo. Despite the fact that such samples can be prepared histologically, with fixation, thin-sectioning, and morphological or molecular staining, and subsequently studied with standard or confocal microscopes at high spatial resolution, the (patho)physiology of skin is sufficiently complex that the aid of additional optical techniques has been sought. The challenges being addressed extend from molecular, structural, and functional characterization to automated diagnosis, resolution of ambiguous or contentious classification dilemmas (typically associated with melanotic lesions), and postmortem investigations. Techniques used to help in these efforts range from conceptually simple reflectance and transmissive or fluorescence spectroscopy in the visible + range to imaging spectroscopy in the mid-IR and multiphoton, lifetime, Raman, pump-probe approaches, and combinations thereof, along with the requisite analytical mathematical and image processing. In many cases, these ex vivo studies were used to generate preliminary results that would motivate applying a given technique to the task of in vivo screening, diagnosis, or surgical guidance. In addition to taking advantage of endogenous contrast, it is also possible to perform multiplexed immunostaining and utilize spectral imaging to help resolve the location and intensity of spectrally and/or spatially overlapping reporter chromogens or fluorophores. As the latter approach is not unique to skin samples, it will be discussed only briefly.

SKIN SPECIMENS EX VIVO

Skin to be studied ex vivo can be derived from biopsies, disposable specimen from surgical procedures, autopsy sources, or animal models. The two major distinct regions in skin are a relatively thin (100–150 μm) epidermal layer containing a number of sublayers and multiple cell types, and a thicker (150–400 μm) dermal layer composed largely of extracellular material, including collagen and elastin. These two regions combined make skin a challenging optical milieu, with layers differing in refractive index [1] and containing strong absorbers, multiple scatterers, and a variety of sources of autofluorescent signals. Nevertheless, the observation that skin lesions, including cancer and precancer, evince optically detectable changes in molecular species and structures have motivated much of the work described in this chapter.

Once the skin samples are collected, variations in how they are processed before microscopy can have a dramatic effect on their optical properties. The skin can be:

1. Interrogated immediately, or held in physiological buffers in an attempt to preserve metabolic status, and examined en face (epidermis oriented closest to the optical path) as a surrogate for in vivo studies, or alternatively, oriented "horizontally" so that all layers are simultaneously accessible. Optical properties relevant to in vivo skin studies are covered in much greater detail.
2. Fixed in formalin or other preservatives that maintain the architectural properties of the tissue but potentially alter small (often autofluorescent) molecule distribution, as well as skin mechanical properties.
3. Processed and sectioned into micron-thick slices for conventional histology and single and multiplexed immunostaining, which can be facilitated by using spectral approaches to resolve spectrally and spatially overlapping molecular distribution.

SPECTRAL IMAGING OF HISTOLOGICAL SECTIONS WITH CONVENTIONAL MICROSCOPES

The methodology of this most straightforward spectral technique to be discussed springs directly from foundational work in astronomy and remote earth sensing and depends on analysis of distinct spectral properties that results from interaction of incident electromagnetic waves with an object's intrinsic physical qualities [2,3]. In some implementations, spectra are obtained with high resolution across a continuous spectral region; in others, more information-efficient approaches and individual, ideally informative, bands are acquired discontinuously [3,4]. During the 1990s, these remote sensing approaches were adapted for use in microscopy, with commercially available tunable filter, interferometric, and wavelength dispersive instrumentation, accompanied by reasonably tractable acquisition, analysis, and display software.

Applications of these approaches include extraction of additional information from conventional histology stains, such as hematoxylin and eosin (H&E), along with molecular multiplexing. In later sections we will give examples of the use of techniques for biochemical characterization of unstained specimens.

Spectral Imaging in Stained Skin Sections

Recently, Li et al. utilized a custom-designed molecular spectral imaging system based on an acoustooptic tunable filter to discriminate tissue structures in rat skin ex vivo [4]. Absorbance spectral data from 80 single-band images of H&E-stained skin sections, spanning 550–1000 nm, were compiled into pseudocolor images for precise visualization of skin structures (Fig. 18.1). These corresponded well with manually delineated regions from the original H&E images. A similar strategy was applied using a simultaneous (not band-sequential) acquisition approach with a prism and reflector spectroscopy imaging system, which generates near-continuous spectra over a 400–800-nm wavelength range as a transilluminated stained slide is physically translated on a microscope stage. This instrumentation, assisted by spectral waveform cross-correlation analysis, was used to discriminate normal skin from benign and melanoma lesions [3]. These promising results were eventually translated into in vivo approaches to lesion characterization [5].

There is a lack of consensus on how to achieve definitive classification criteria and outcome predictions in characterizing and discriminating Spitz nevi and melanoma [6–8]. A study by Gaudi et al. reported unique spectral signatures that may be useful for separating varieties of Spitz nevi [2]. A total of 102 human H&E-stained specimens were used to build a reference library and generated clustering data required for discrimination of tissue samples. All sample groups shared common spectral signals at 496, 536, and 838 nm, whereas each group contained at least one unique spectral signal, with spitzoid melanomas exhibiting the largest set of unique spectral signals. These discriminating parameters may have clinical significance because ambiguous and/or borderline cases of Spitz nevi may lead to inconsistent diagnosis among dermatopathologists, although the challenge of arriving at definitive ground-truth for method development remains.

Several factors were identified that may improve discrimination of tissue samples. Higher magnification (×63 vs. ×10 objective) can yield spectral signatures from small features that are less likely to be affected by contributions from mixed pixels. Second, a large spectral reference library may be critical, because tissue samples may only contain small differences that may not be adequately identified with limited discriminating parameters. Third, consistent sample preparation, which involves such variables as staining time and sample thickness, is critical, because large spectral variations can occur with samples that deviate from the typical 4–6-μm thickness or were processed with 10% or more variability in staining times. These observations emphasize problems that can affect techniques that rely on exogenous staining rather than endogenous contrast, and that seem to have interfered with significant progress with staining-based spectral classification (not otherwise informed by morphology).

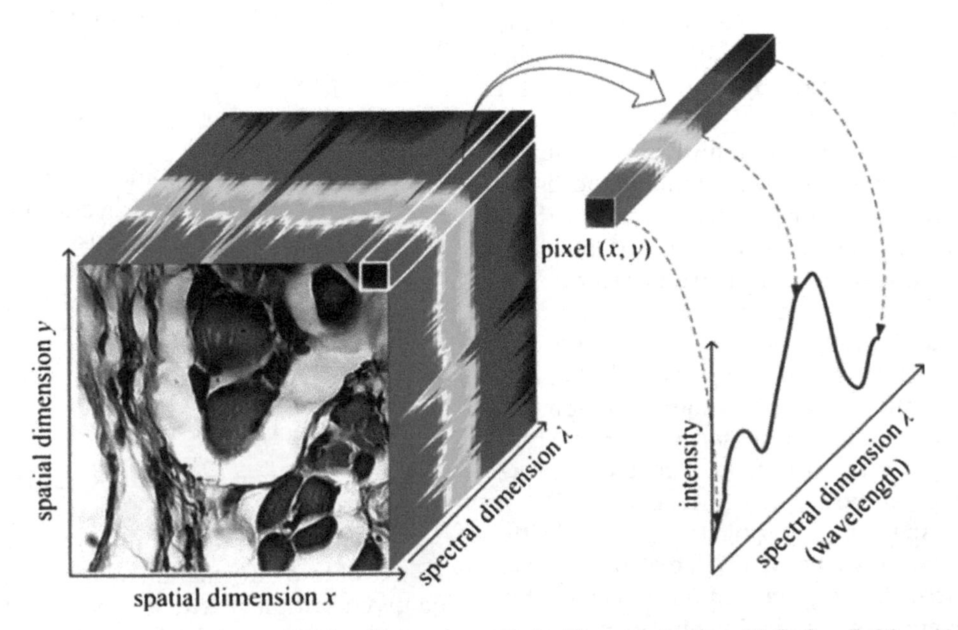

FIGURE 18.1 Molecular hyperspectral data cube of a skin section. *Used with permission from Li Q, Sun Z, Wang Y, Liu H, Guo F, Zhu J. Histological skin morphology enhancement base on molecular hyperspectral imaging technology. Skin Res Technol [Official Journal of International Society for Bioengineering and the Skin (ISBS) [and] International Society for Digital Imaging of Skin (ISDIS) [and] International Society for Skin Imaging (ISSI)] 2014; 20(3):332–40.*

It is a bit of a stretch to include the following report in this section, because the primary imaging technique was a fairly straightforward combination of reflectance and confocal imaging of fresh skin resection specimens. The spectral angle comes from the use of spectral analysis of H&E-stained slides to generate color-matching algorithms that were used to convert the confocal images into H&E look-alikes [9]. Fluorescence images highlighting acridine orange–stained nuclei were recolored purple, and reflectance images generated by remitted photons from cytoplasm and dermis were colored pink. The resulting images were of sufficient quality for identification of normal skin structures as well as the presence of malignant lesions. Recoloration was not perfect and the images appeared as red and blue, whereas true H&E images are generally purple and pink.

It is possible to extract additional information from H&E-stained slides using multispectral imaging to perform "pseudohistochemistry." For example, fibrosis is an extremely important process to detect and quantify, and although it is straightforward to perform one of a number of connective tissue histochemical stains, such as Masson trichrome, it is possible to highlight collagen deposition using its distinctive tinctorial properties with multispectral imaging, classification, and pseudocoloring, as demonstrated by Bautista et al. on liver. However, the method should work similarly with skin specimens [10]. Alternatively, spectral imaging can be used to enhance the information content and quantitative accuracy of collagen detection when true collagen histochemical staining is performed. Turner et al. demonstrated discrimination of types I and III collagen in rat tail and human scar tissue stained with the seldom used Herovici polychrome method via spectral imaging and analysis, with validation through comparison with true immunostaining for these components [11].

Finally, for completion, the use of multispectral imaging to enable multiplexed molecular staining and tissue characterization should be mentioned. This is a burgeoning field with recent developments of matched kits and spectral slide-scanning instrumentation and software good for up to six or seven probes simultaneously [12,13]. Additional methods for multiplexing that include serial staining, imaging, and destaining, for example, are commercialized or under development, resulting in a number of (not necessarily multispectral) tools available that can provide from 6 to 100 or more labels localized even at the single-cell level [14–16].

Unfortunately, spectral imaging is not able to perform a skin-specific task that would be of great potential value to the dermatology community, namely the reliable discrimination of melanin absorbance from 3,3′-diaminobenzidine absorbance in immunohistochemistry-stained skin specimens. These are both brown pigments that acquire their color from a combination of scattering and absorbance. Consequently, they do not necessarily obey the Beer–Lambert law, because their spectra can and do overlap and thus, in many cases, cannot be reliably distinguished or unmixed [17].

REFLECTANCE AND ABSORBANCE SPECTROSCOPY OF INTRINSIC SKIN CONTRAST

The optical process of reflectance is related to light scattering and absorption properties and is dependent on the sample tissue's morphology and biochemical composition, respectively. Reflectance and absorbance studies in vitro can provide information on a sampled tissue's physiological state and structural composition, and can serve to help motivate optical biopsy approaches as adjuncts to conventional biopsy for extraction of important skin diagnostic parameters [18,19].

Reflectance Spectroscopy in Forensic Medicine

One somewhat unusual application of reflectance spectroscopy is for characterization of the hemoglobin reoxygenation process in postmortem human skin, as reported by Belenki et al. [20]. Because livor mortis is important for coroners and forensic investigators, an accurate and precise model is necessary to quantify the dynamics of hemoglobin reoxygenation as an indirect measure for time of death in legal and forensic medicine. Under cold ambient temperature, the appearance of livor mortis changes to a bright red or pink color and is associated with changes in the reflectance curve at wavelengths in the 500–600 -nm range. The relationship between hemoglobin-bound oxygen [oxyhemoglobin (HbO_2)] and skin temperature immediately postmortem requires several hours after death to achieve a steady state, because low levels of HbO_2 in the body gradually equilibrate via molecular diffusion with relatively high levels of O_2 in the atmosphere [21,22].

Infrared Spectroscopy

The discovery of infrared (IR) radiation in early 1800s provided the groundwork for development of IR spectroscopy, a distinct technique that utilizes the IR optical range to obtain information on the chemical composition of a given sample, because spectral features correspond to unique vibrational frequencies that correlate with distinct functional groups [23–25]. Quantitative information (ie, concentration) can be associated to the intensity of specific IR absorption peaks with appropriate calibration [26]. The relatively recent development of FTIR spectroscopy has often replaced filter- or gratings-based techniques, owing to advantages in flexibility, shorter data acquisition time, lower detection thresholds, and improved resolution [25], although subsequent developments in tunable lasers in this spectral range also appear promising.

FTIR spectroscopic imaging may be used to identify specific molecular components in a tissue sample in a nondestructive manner and without chemical labeling. Cellular content and conformational information with proteins, nucleic acids, carbohydrates, and lipids may be extracted [27]. FTIR can be applied to thin-sectioned histology preparations of skin tissue in attempts to provide more accurate, more reproducible, and more quantitative assessments than can be achieved through the expert but subjective interpretation of conventionally stained slides by pathologists and other skilled observers.

FTIR spectroscopy has been used to characterize engineered human skin substitutes and has successfully demonstrated similar lipid organization, protein conformation, and composition between human skin substitutes and healthy human skin [28]. In agreement with histology, the stratum corneum (SC) and dermis appear to be homogenous layers, whereas the living epidermis (LE) appears heterogeneous as keratinocytes differentiate from basal stem cells in the stratum basalis to corneocytes in the SC. Spectral analysis reported proteins with high α-helix conformation in the SC that is likely indicative for keratin, and type I collagen in the dermis [28]. FTIR spectroscopy demonstrated the potential for analysis of cellular content in human skin substitute and healthy human skin without histology preparation, and may be utilized in future ex vivo and in vitro research studies.

Researchers have also reported use of FTIR spectroscopy to study hair follicle cells [24]. Based on the presence and abundance of different functional groups, different layers of the hair follicle, such as the dermal papilla, connective tissue sheath, and subcutis may be visualized [24]. FTIR spectroscopy may be useful in identifying single stem cells in the hair bulb in order to elucidate relevant pathophysiology associated with alopecia and other hair-related diseases, as well as assisting in treatment monitoring.

Fourier-Transform Infrared Spectroscopy and Skin Neoplasia

The gold standard for diagnosis of nonmelanoma skin cancers (NMSCs) and melanoma is histopathological assessment of a biopsy with H&E staining, with interpretation of cellular morphology dependent on individual dermatopathologists' criteria and skill sets. FTIR spectroscopy may be an excellent adjunct to this methodology. In a preliminary study, a prediction model was created using 15 nondewaxed formalin-fixed, paraffin-embedded (FFPE) NMSC sections and two healthy tissue samples; when applied to four additional NMSC samples, the resulting pseudocolor maps correlated well with histology images [29]. FTIR spectroscopy was also tested on FFPE basal cell carcinoma (BCC) samples and successfully detected presence of tumor cells [30]. Paraffin presents intense vibrations in the IR region and a preprocessing method, extended multiplicative signal correction, was employed to correct for the spectral contributions from paraffin. The skin layers, SC, LE, and dermis, were successfully discriminated on pseudocolor maps and all BCCs were identified in all 10 samples [30]. These results suggest the potential of a fast, noninvasive, and nondestructive optical modality for detection of BCC on unstained sections (Fig. 18.2).

FTIR spectroscopy may also be helpful in the discrimination of melanoma from benign nevi. Tfayli et al. examined thin-sectioned preparations of nondewaxed FFPE melanomas and benign nevi. Four spectral windows sensitive to absorption by melanin, DNA, and other tissue components were shown to be sufficient to generate information that could discriminate melanoma from benign nevi in a small set of samples [31]. More recent work from this group and others suggests the potential utility of FTIR approaches to stain-free histology, although the continuing requirement for advanced statistical multivariate data processing suggests that the signals are not as robust as could be desired [32].

Kong et al. investigated an engineered tissue model of melanoma using FTIR contrast to track the growth of cultured melanoma cells and the interactions between tumor cells and stroma [33] (Fig. 18.3). Whereas FTIR spectroscopy detected no significant spectral differences in malignant melanocytes that correlated with a propensity for tumor progression, a significant spectral difference in proteins of the connective tissue and extracellular matrix appeared that may be related to changes in stromal cells [33]. This spectral change occurred between day 12 and day 16, which correlated with the histological observation of tumor invasion that occurred around day 12. This is potentially of considerable biological significance, because the tumor microenvironment has been demonstrated to play a key part in tumor progression [34].

However, FTIR does present some technological hurdles that impede its adoption into the research and

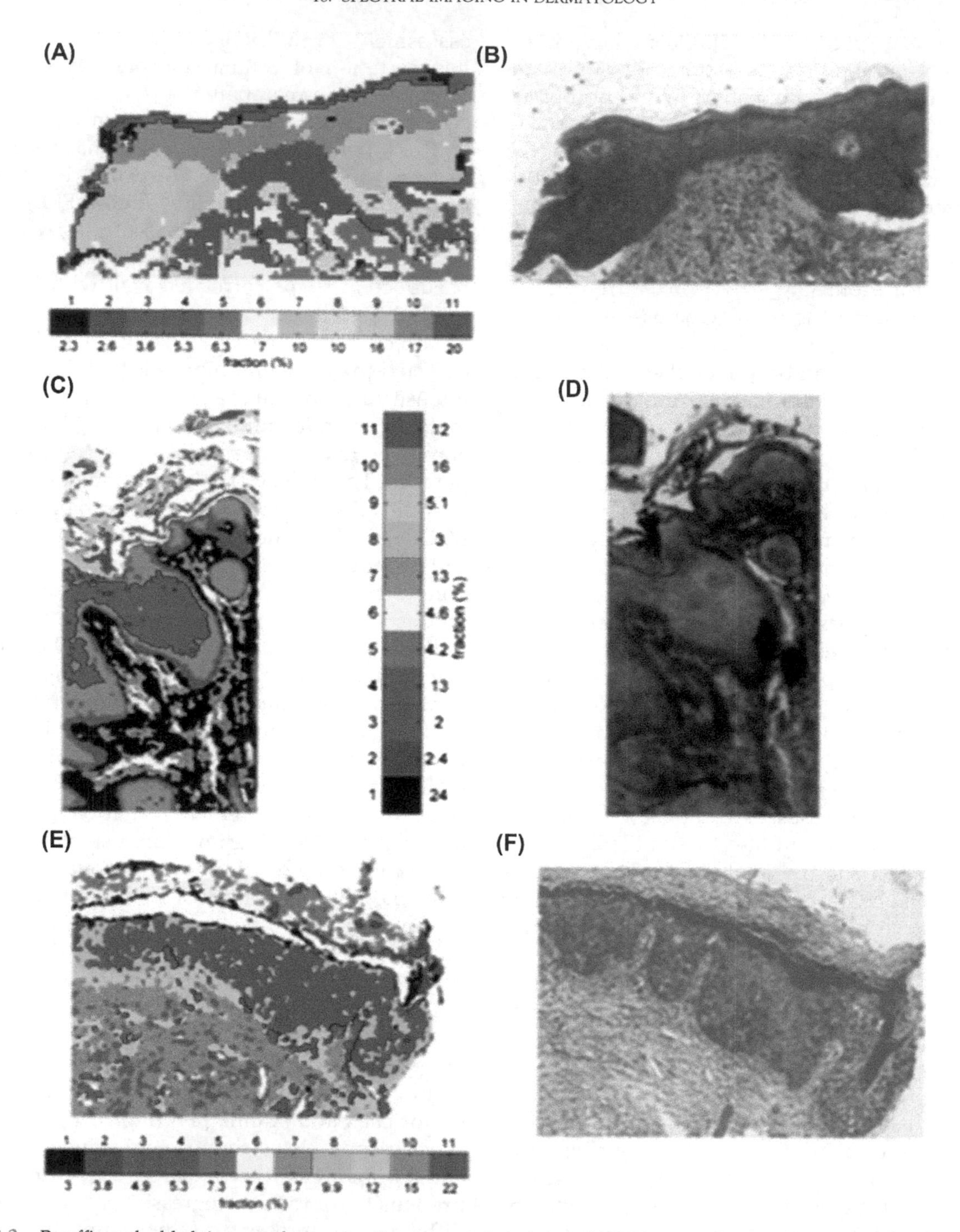

FIGURE 18.2 Paraffin-embedded tissue analysis using Fourier-transform infrared (FTIR) spectroscopy and K-means clustering. (A) Pseudocolor image from basal cell carcinoma (BCC). (C) Pseudocolor image from squamous cell carcinoma (SCC). (E) Pseudocolor image from Bowen disease. (B, D, and F) Corresponding hematoxylin and eosin (H&E)-stained sections of (A), (C), and (E), respectively. *Used with permission from Ly E, Piot O, Durlach A, Bernard P, Manfait M. Differential diagnosis of cutaneous carcinomas by infrared spectral micro-imaging combined with pattern recognition. The Analyst 2009;134(6):1208–14.*

clinical arenas, in part because full-range spectra must be acquired even if only scattered bands are relevant, and the technologies for broadband illumination and near-infrared (NIR) detection have inherent limitations. Alternative approaches that use quantum cascade tunable laser sources to achieve spectral resolution have been developed. These provide a considerable sensitivity advantage over FTIR [35], and can provide convincing real-time image navigation and photon-efficient detection [36,37].

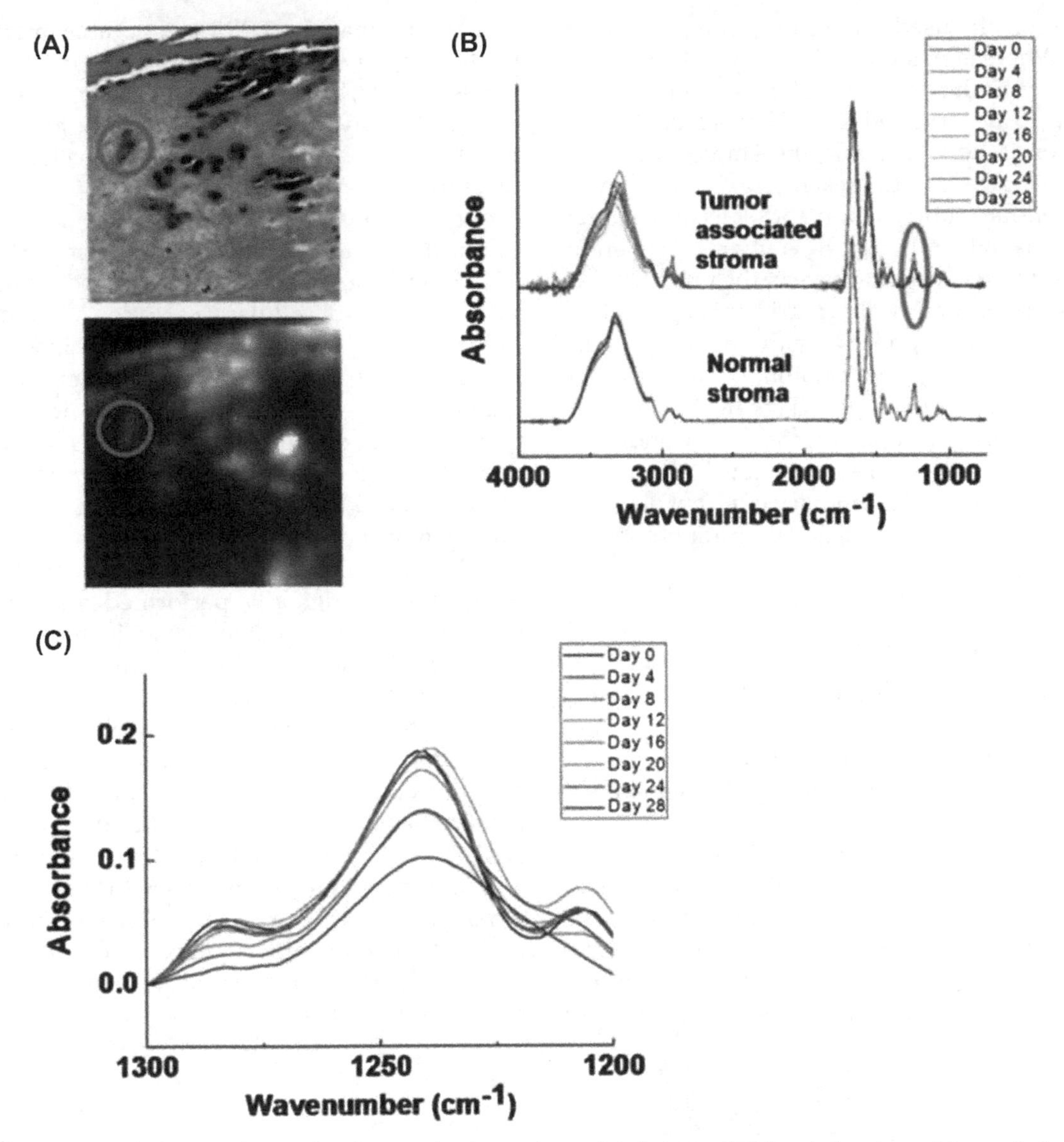

FIGURE 18.3 Temporal behavior of normal and tumor (melanoma)-associated stroma. (A) Spectral regions of tumor (*red circle*) and tumor-associated stroma are extracted for analysis of temporal changes. (B) Spectral analysis of normal and tumor-associated stroma demonstrates likely differences between the two, especially including significant changes in the 1200–1300 -cm^{-1} (*blue circle*) region. (C) Spectral detail of the 1200–1300 -cm^{-1} region indicates that the biochemical changes likely occurred around day 16. *Used with permission from Kong R, Reddy RK, Bhargava R. Characterization of tumor progression in engineered tissue using infrared spectroscopic imaging. The Analyst 2010;135(7):1569–78.*

MULTIPHOTON MICROSCOPY AND AUTOFLUORESCENCE

Fluorescence, a process of absorption and nearly immediate emission of light by various chemical moieties, was a phenomenon of continuous interest to the scientific community from the 19th century onwards. The technique of fluorescence microscopy first emerged in the early 1910s and underwent significant improvements over succeeding decades. Tissue autofluorescence is the natural emission of light by endogenous fluorophores, present in a variety of cells and cellular structures, and detectable at microscopic resolution. Autofluorescence signals can deliver important biochemical and tissue structural information directly from freshly obtained material, thereby foregoing time-consuming fixation and staining procedures used in conventional histological preparations. Spectroscopic analysis of endogenous autofluorescence signals, as outlined in pioneering work by Britton Chance beginning in the 1970s [38], can provide insight into physiological and pathological states of tissue not readily available from conventional H&E staining [39,40].

Single-photon, conventional fluorescence techniques are typically limited by the modest penetration depth of light in the visible and near-ultraviolet (UV) wavelengths typically used for exciting endogenous signals in vitro. Whereas this limitation may not be a problem when examining excised skin specimens, the sample needs to be oriented such that all depths are accessible to excitation light and the emitted light can be collected. However, to reach subsurface layers in vivo using en-face illumination and detection geometries, excitation light in the IR range is generally deployed in order to take advantage of the much smaller effects of absorption and scattering in this spectral region. To excite fluorophores with light of skin-penetrating wavelength, two-photon or multiphoton techniques have been developed (Fig. 18.4). A helpful guide to advances in multiphoton microscopy and other techniques can be found in a recent review [41]; an additional valuable resource is a summary of recent developments (and challenges) in the field of multiphoton and SHG optical imaging [42].

Two-Photon Excited Fluorescence

Two-photon excited fluorescence (TPEF) imaging, invented in the early 1990s by Denk, Strickler, and Webb, uses two photons in a high-flux beam that can arrive at the target molecule essentially simultaneously. The two photons combine to excite an electron into a higher state as if a single, more energetic (shorter wavelength) photon had in fact arrived [43]. TPEF has tendency to cause less photobleaching and photodamage than single-photon techniques, such as confocal microscopy that use visible and UV light as excitation sources [44–46]. Near-IR light is able to penetrate up to 1 mm into tissue, making TPEF a favorable method if relatively deep tissue imaging with depth discrimination is desired [43,47,48]. However, photons emitted when electrons decay to their ground state tend to be in the visible range, and in order to be collected, the photons must traverse structures that may scatter or absorb their energy, thus making it challenging to acquire the signal (output) compared to the excitation (input).

"Monochrome" TPEF has been effective at generating images with subcellular resolution that can provide much of the architectural information assessed on routine histological examination, generating diagnostic quality images ex vivo [49]. In the next sections, we will focus on exploring whether particular spectral components of the TPEF signal can provide additional or more nuanced cellular information in addition to high-quality morphological histology.

TPEF combined with excitation-emission spectral analysis was originally performed with point probes [50], and has been extended to include imaging that can be used to describe skin structures both qualitatively and quantitatively. Specific constituents in human skin ex vivo that were shown to contribute the bulk of the fluorescence signals include tryptophan within keratinocytes (emission peak, 425 nm), elastin (475 nm) [51], nicotinamide adenine dinucleotide (phosphate) NAD(P)H (475 nm) [52], keratin in the SC [53], and possibly melanin (550 nm) [54,55] (Figs. 18.5 and 18.6). An attempt to categorize all species of autofluorescent molecules in nude mouse skin added potential contributions from sebum and flavin proteins [56], but the authors cautioned that it was difficult to arrive at

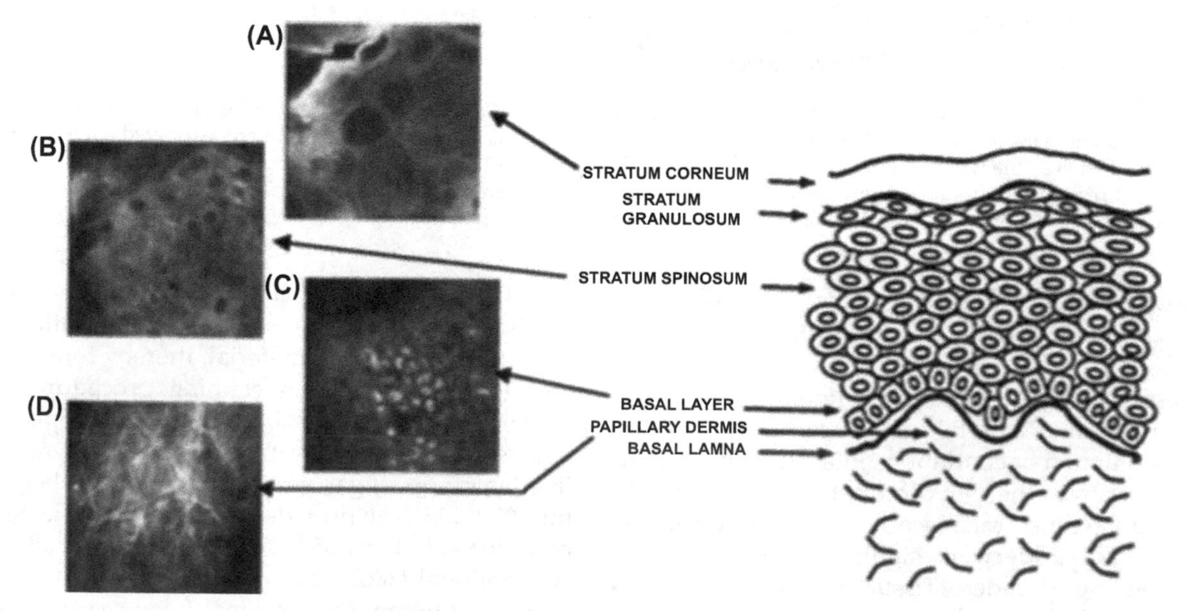

FIGURE 18.4 Basic physiology of human skin and corresponding two-photon images of several layers. (A) Stratum corneum at 5 μm, (B) Stratum spinosum at 20 μm, (C) Basal layer at 35 μm, and (D) Dermis at 75 μm. Images taken with 2 mW external to specimen. *Used with permission from Laiho LH, Pelet S, Hancewicz TM, Kaplan PD, So PT. Two-photon 3-D mapping of ex vivo human skin endogenous fluorescence species based on fluorescence emission spectra. J Biomed Opt 2005;10(2):024016.*

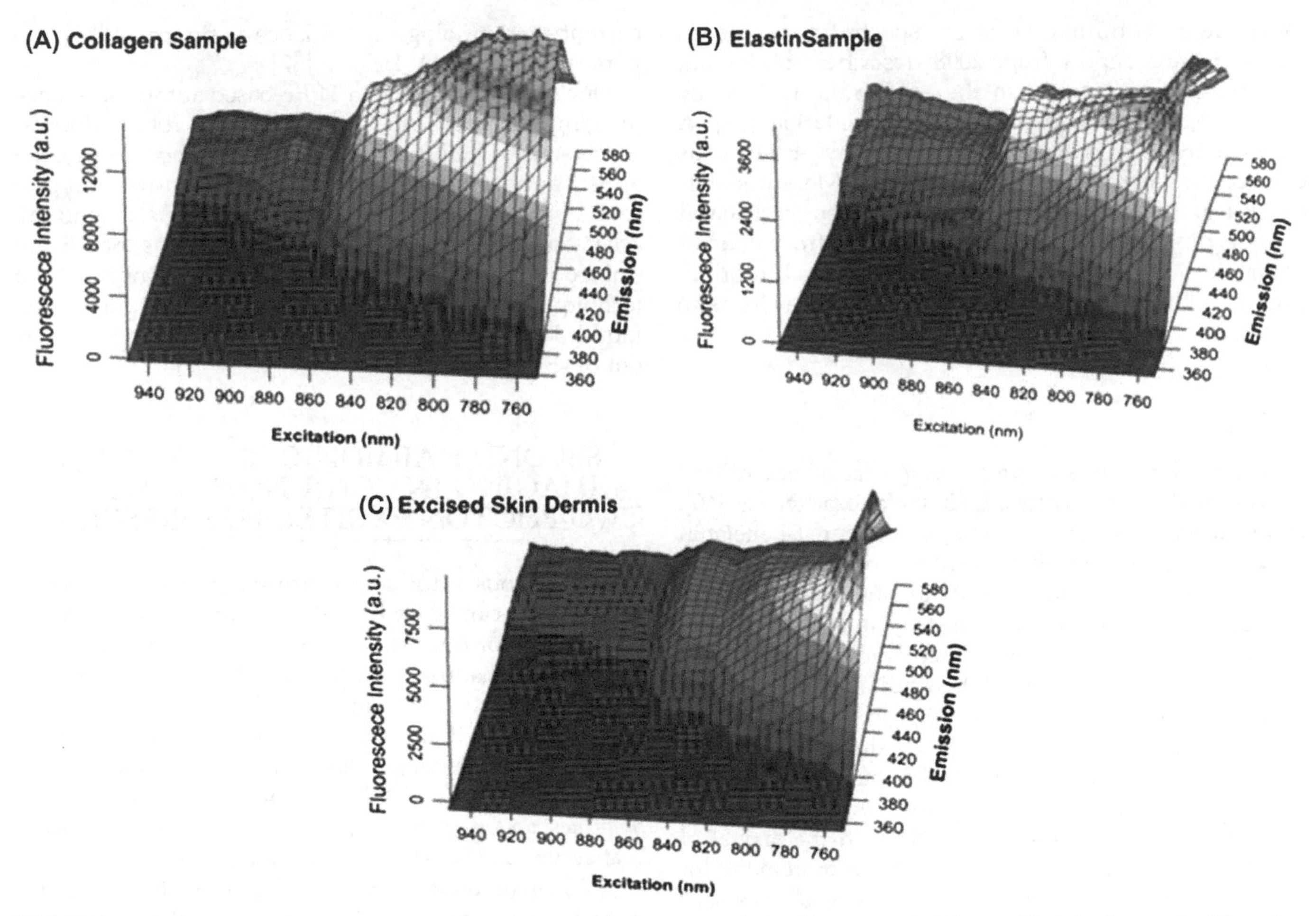

FIGURE 18.5 Three-dimensional excitation-emission matrices plot of a purified collagen sample (A) a purified elastin sample (B), and an excised human skin dermis (C). *a.u.*, Arbitrary units. *Used with permission from Chen J, Lee A, Zhao J, et al. Spectroscopic characterization and microscopic imaging of extracted and in situ cutaneous collagen and elastic tissue components under two-photon excitation. Skin Res Technol [Official Journal of International Society for Bioengineering and the Skin (ISBS) [and] International Society for Digital Imaging of Skin (ISDIS) [and] International Society for Skin Imaging (ISSI)] 2009;15(4):418–26.*

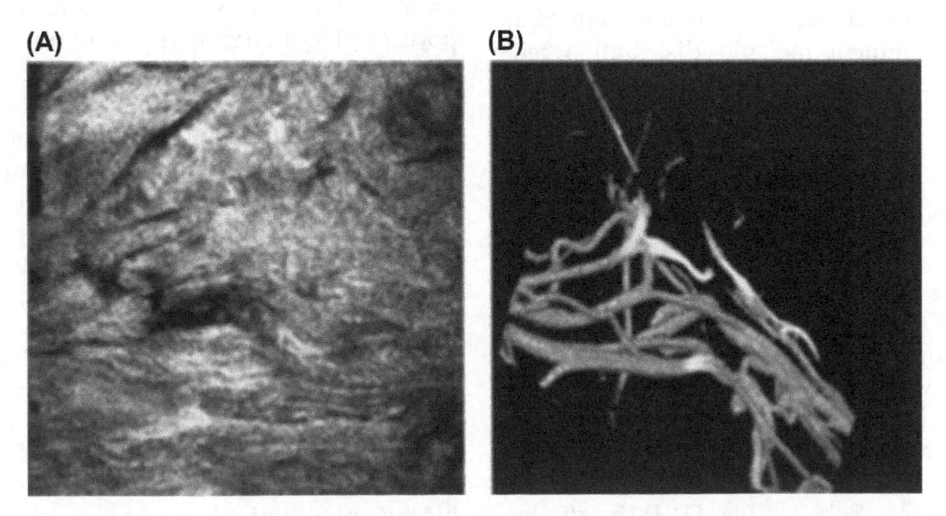

FIGURE 18.6 Dermis three-dimensional reconstruction at emission wavelengths of 400 nm, signal attributed to collagen (A), and 475 nm, signal attributed to elastin (B). *Used with permission from Laiho LH, Pelet S, Hancewicz TM, Kaplan PD, So PT. Two-photon 3-D mapping of ex vivo human skin endogenous fluorescence species based on fluorescence emission spectra. J Biomed Opt 2005;10(2):024016.*

independent confirmation of the spectral assignments. An intriguing report from 2008 describes similar but not overlapping spectra in the epidermis induced by either single-photon or two-photon excitation [57]. A noteworthy point is that mice on chlorophyll-based diets can accumulate a red skin fluorophore. Measurements based upon fluorescence intensity alone suggest that it may be possible to distinguish malignant from nonmalignant lesions, but additional morphological information on cell size, shape, and clustering still appears to be essential [58,59].

Melanin

Melanin is obviously a prominent skin constituent, and is associated (perhaps causally) in melanomagenesis [60]. Unfortunately, at least for imaging scientists, melanin proves to be not autofluorescent (or only very weakly autofluorescent) when excited in the visible range, although it is apparently possible to induce bright yellow autofluorescence of melanin by combining exposure to peroxide compounds with UV irradiation [61]. Native melanin autofluorescence, however, can be generated using femtosecond-pulse excitation or single-photon NIR illumination [62,63]. It is thought that melanin autofluorescence may be induced by stepwise two-photon excitation, which allows for a brief interval in the arrival of the two photons, as opposed to the requirement for near-simultaneous cooccurrence that seems necessary for exciting other cellular fluorophores. Under conditions of nanosecond irradiation, with a relatively lower total photon flux, melanin autofluorescence becomes more readily detectable and, intriguingly, the peak melanin emission from malignant melanomas differs from that of benign nevi, possibly reflecting alterations in the pheomelanin and eumelanin contributions [64]. Similar findings using pump-probe imaging for enhancement of the spectral signal to segment melanin distribution have been reported (see Pump-Probe Microscopy section, below).

Two-Photon Excited Fluorescence and Reactive Oxygen Species

TPEF can be used to spectroscopically detect reactive oxygen species (ROS) that are generated by UVB irradiation. The technique employs an indicator, dihydrorhodamine 123, which becomes fluorescent upon interaction with a variety of ROS [65]. One significant and worrisome finding that emerged from work in this area is that commonly used UV blockers may actually generate considerable ROS under UVB irradiation, above the level in untreated skin, and may thus paradoxically increase risk of skin cancer [66], although current epidemiological evidence supporting this hypothesis is generally lacking [67].

One major deficiency in TPEF-based autofluorescence imaging is that cell nuclei do not return robust fluorescence signals [68]. Consequently, unstained cell nuclei appear as nonfluorescent voids in most studies and present a challenge for viewers (ie, pathologists) who are accustomed to appreciating cell nuclei as positively stained objects in H&E images. (This is in contrast to a technique based on photoacoustic signal generation induced by DNA absorbance in the UV range that is out of scope of this topic.) [69].

SECOND HARMONIC GENERATION IMAGING IN CONJUNCTION WITH TWO-PHOTON EXCITED FLUORESCENCE

Endogenous autofluorescence spectra lack distinctive features and can be both weak and broad, which may be a challenge for dissection of fluorescence mixtures into individual constituents even in combination with multispectral analysis. Fortunately, SHG imaging can provide high-resolution detail of structural components, and is routinely used in conjunction with multiphoton fluorescence microscopy [47,48]. SHG, often also referred to as *frequency doubling*, is a nonlinear optical phenomenon that serves as the basis for this imaging modality. SHG signal is generated when two photons of the same frequency simultaneously pass through polarizable noncentrosymmetric material, such as collagen, combining together to create a single new photon. This new photon has twice the energy and half the wavelength compared to its original components [48,70,71].

SHG imaging was utilized in microscopy initially by Hellwarth and Christensen in 1974 [72], with additional improvements by Sheppard et al. [73] The earliest reports of SHG imaging application in biology date back to 1986, when Freund et al. investigated collagen fiber organization and structure in rat tail tendon [74].

The required lasers and optical train are common to both SHG and TPEF, with the optimal illumination wavelength for SHG collagen excitation around 810 nm and for TPEF excitation from 750 to 850 nm [47,48,70,75–77]. Consequently, high-resolution, three-dimensional images of important skin components can be generated without use of extrinsic stains. Furthermore, advanced polarization methods can provide additional detailed molecular information on collagen fibril structure [78]. Parenthetically, it is worth noting that birefringence imaging, which can be implemented with inexpensive liquid crystal polarization devices, incoherent illumination, and simple wide-field optics, can generate images very similar to SHG from thin-sectioned tissues [79].

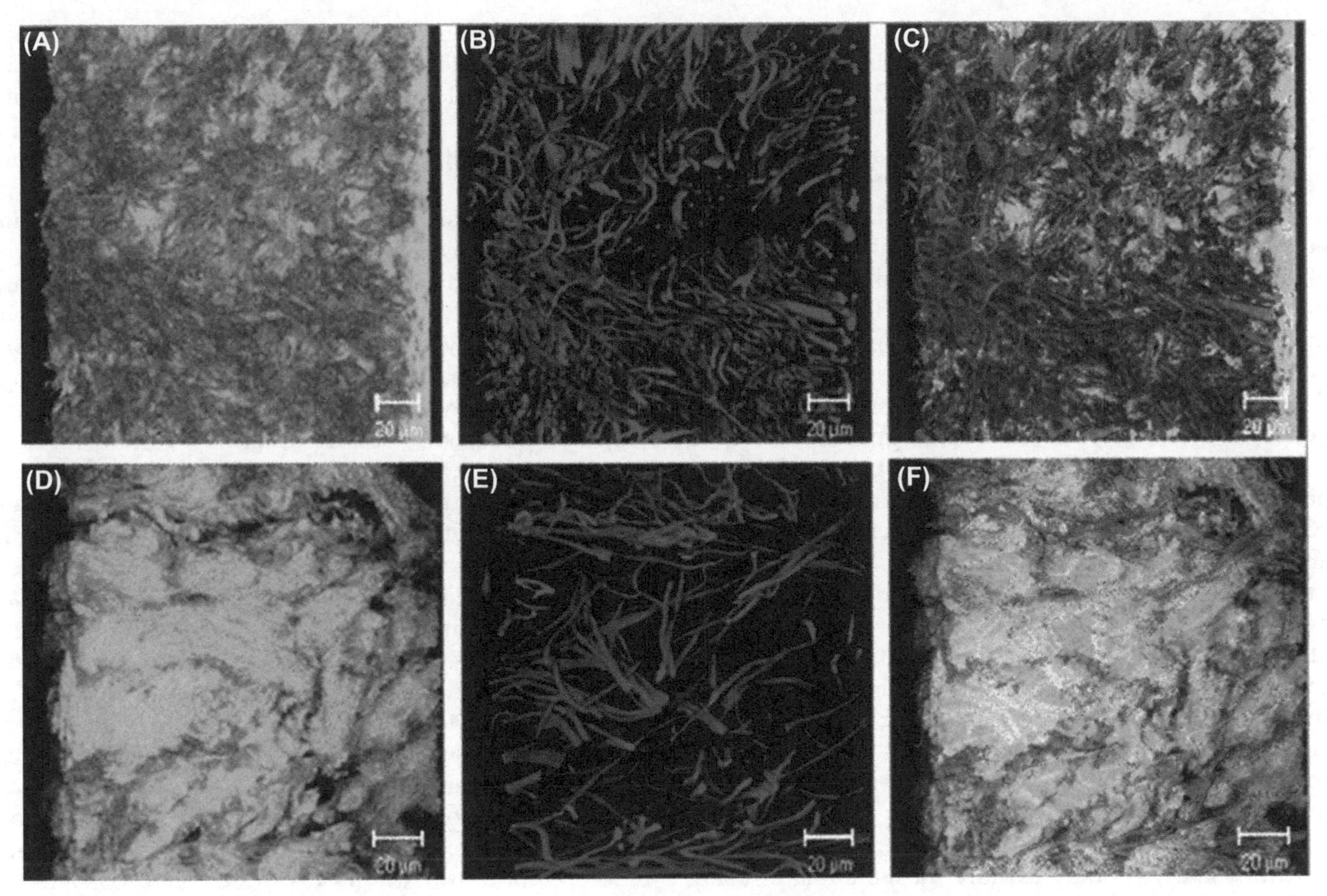

FIGURE 18.7 Three-dimensional (3D) images of collagen and elastin fibers in hypertrophic scar tissue and in normal skin dermis. (A) 3D collagen SHG image *(green)* in hypertrophic scar tissue. (B) 3D elastin two-photon excited fluorescence (TPEF) image *(red)* in hypertrophic scar tissue. (C) High-contrast TPEF/SHG image obtained by overlaying (A) and (B). (D) 3D collagen SHG image *(green)* in normal skin dermis. (E) 3D elastin TPEF image *(red)* in normal skin tissue. (F) High-contrast TPEF/SHG image obtained by overlaying (D) and (E). *Used with permission from Chen G, Chen J, Zhuo S, et al. Nonlinear spectral imaging of human hypertrophic scar based on two-photon excited fluorescence and second-harmonic generation. Br J Dermatol 2009;161(1):48–55.*

Non–cancer-related studies have combined SHG and TPEF to yield structural assessment of collagen and elastin fibers in human healthy skin and hypertrophic scar tissue ex vivo [48] (Fig. 18.7). Collagen arrangement in normal skin was observed to be well regulated with preferred orientation, whereas in hypertrophic scar tissue it was relatively disorganized with prominent swirling. Elastin fibers in hypertrophic scars were also abnormally abundant and fragmented. Further details on collagen fiber spacing, elastic fiber diameters, and ratio of collagen to elastin were also accessible. Lin et al. described discernible differences in collagen-to-elastin ratios between healthy skin and hypertrophic scars that could serve as a metric for characterizing lesion severity and/or tracking response to therapy [80]. Another potential application of TPEF and SHG may be in dermatological and/or plastic surgery to ensure complete excision of fibrotic tissue and minimize removal of healthy tissue [48].

A combination of TPEF imaging, SHG, and spectral analysis generates biochemical and morphological information on collagen, elastin, Nicotinamide adenine dinucleotide (NADH) and flavin adenine dinucleotide (FAD) that may be useful for assessment of dermally focused diseases. SHG with TPEF can likely serve as a clinically relevant biomarker to track response to therapy, particularly of skin fibrosis disorders including hypertrophic scars, keloids, and chronic graft-versus-host disease. In addition, researchers who investigate mitochondrial changes are interested in accurate detection and quantification of NADH or FAD based on signals generated using TPEF and SHG. For example, if NADH and FAD signals (corresponding to peaks at 470 and 530 nm, respectively) are measured, the ratio of NADH to FAD appears to be a good indicator of fibroblast metabolic state [81]. However, there are methodological issues in standardizing techniques looking at metabolite fluorescence in ex vivo skin specimens. The samples may have been devascularized for varying periods of time and/or maintained in saline or buffer media at temperatures ranging from 4 to 37°C. As reported by Thomas et al., the relevance of ex vivo measurements to what may obtain in vivo remains to be firmly established [42].

Collagen density and fiber orientation properties have been implicated in various malignant diseases. Studies have reported striking differences in the major direction of collagen fibers surrounding normal and malignant breast tissue [82,83]; collagen surrounding various skin cancers displays characteristic fragmentation and disordering [84]. Different metrics of skin health and/or the presence of malignancy have been constructed by combining TPEF signals and SHG signals, deriving significance from examination of various computed intensity ratios. When these two methods are employed on the same specimen, it can be shown that the fluorescence signal generated by TPEF in the dermis predominantly originates from elastin, whereas SHG images can be solely attributed to signals from collagen fibers. The collagen-to-elastin ratio has been shown to decrease in the setting of esophageal neoplasia, attributed to SHG signal loss caused by collagen fragmentation [85]. However, this field is complicated by the wide variety of collagen phenotypes that can accompany different epithelial cancers. Future researchers may explore the potential of using TPEF and SHG for investigation of various skin fibrosis disorders, and as a tool for monitoring response to treatment, because there is no preferred method or imaging modality to obtain collagen information other than biopsy. There is promise that collagen abnormalities, possibly in conjunction with simultaneous elastin alterations, could be biologically significant and optically detectable in advance of the frank cellular morphology changes that accompany later stages in cancer development.

FLUORESCENCE LIFETIME IMAGING

Fluorescence lifetime imaging (FLI) is a relatively new technique that characterizes fluorescent emissions by onset and decay time parameters as opposed to fluorescence intensity. Although the concept of fluorescence lifetime has been recognized and studied for decades, it was not until 1988 that it was used in microscopic imaging for the first time by König et al., when various technical advances of that time, including data computation and analysis needed for the rather complex process of FLI, became available [86,87]. An image obtained using FLI is essentially a plot of fluorescent lifetime (the amount of time it takes a population of fluorophores to undergo exponential decay) measured pixel by pixel after pulsed laser-induced excitation. The rate of decay, and therefore fluorescence lifetime, is an intrinsic property of the sample and is largely independent of excitation conditions, such as excitation light wavelength, duration of excitation, and fluorophore concentration [86,88,89].

As with emission spectra, fluorescence lifetimes can be affected by local environmental factors including solvent composition, pH, and energy transfer to other molecules, but may convey additional information if attention is paid to potential perturbants. These unique characteristics make FLI a valuable tool for evaluating both composition and function of the tissue of interest [88]. When FLI was used to explore the composition of the dermis ex vivo by deconvolving decay curves into their major constituents, it was possible to confirm previous results obtained from multiphoton microscopy and SHG studies with respect to the abundance and distribution of collagen and elastin [90–92]. Additional studies are required for full characterization of in vivo human skin because differences may be present compared with ex vivo tissue sample, as a result of the potential for protein degradation, metabolic alterations in small molecules, and the effects of excision, storage, and fixation techniques.

Instrumentation in FLI can be complex, because it involves pulsed excitation coupled with time-resolved detection, and as with all advanced microscopy techniques, requires care in set-up and interpretation [93]. To acquire information about fluorescence lifetime, either single photon or multiphoton excitation microscopy can be utilized; multiphoton microscopy, however, is generally preferred because of its enhanced special resolution [86]. In principle, multiphoton fluorescence lifetime (MFLT) imaging is similar to TPEF, but may involve two (or more) photon excitation events to generate time-resolved autofluorescence emission [50,53,94–96].

Dimitrov et al. used "conventional" multiphoton fluorescence spectral imaging to analyze a variety of skin lesions, and reported detection of spectral differences between benign and malignant melanocytic lesions [49]. Spectral fluorescence analysis detected a peak around 470 nm in all melanocytic skin lesions, whereas malignant melanomas exhibited increased emission around 550 nm [49,97]. Selective fluorescence excitability at 800 nm revealed atypical, pleomorphic, and highly luminescent melanocytes in malignant melanoma. The distribution of keratinocytes and melanocytes correlated well with those seen on conventional histopathological sections. However, additional information proved to be available from fluorescence lifetime decay measurements; these appeared to be dependent on intracellular quantities of melanin. These findings demonstrate that MFLT may provide important additional parameters for distinguishing melanoma from healthy skin and nonmalignant mimics, potentially improving the accuracy of biopsy results, although morphological parameters remain an essential part of the diagnostic process [49] (Fig. 18.8).

Patalay et al. have explored the use of MFLT for discriminating signals arising from normal keratinocytes and those from BCCs based on longer mean fluorescence lifetimes in the latter, with a sensitivity and specificity of 79% and 93%, respectively. This finding suggested that MFLT may prove to be a useful noninvasive imaging adjunct to histopathology [92] (Fig. 18.9).

Extending MFLT approaches for in vivo use has some challenges, including limited imaging depth in the skin, effect of patient age, and the long image acquisition time required [92]. Areas with surface irregularities, such as ears, may also pose as a challenge to MFLT imaging [97]. However, enhanced spectral analytic algorithms may allow fully automated image acquisition and may reduce processing time in the future.

PUMP-PROBE MICROSCOPY

Pump-probe microscopy is another emerging nonlinear technique that offers additional opportunities for high-resolution, depth-resolved molecular imaging. The main strength of this technique is that it gives access to a number of light–matter interactions in addition to those available with conventional nonlinear methods. Pump-probe microscopy is not restricted to processes that generate new colors of light, as in TPEF. Instead, the molecular signals are collected by monitoring changes in the amplitude of the probe beam caused by the effect of the pump beam on optical properties of the specimen, such as ground-state bleaching, excited state absorption, and stimulated emission. In addition, the ultrafast (femtosecond to picosecond) lifetimes of these excited and ground states can be assessed by gating techniques. This method has the same advantages of other nonlinear methods, such as robustness to scattering and inherent depth sectioning. In practice,

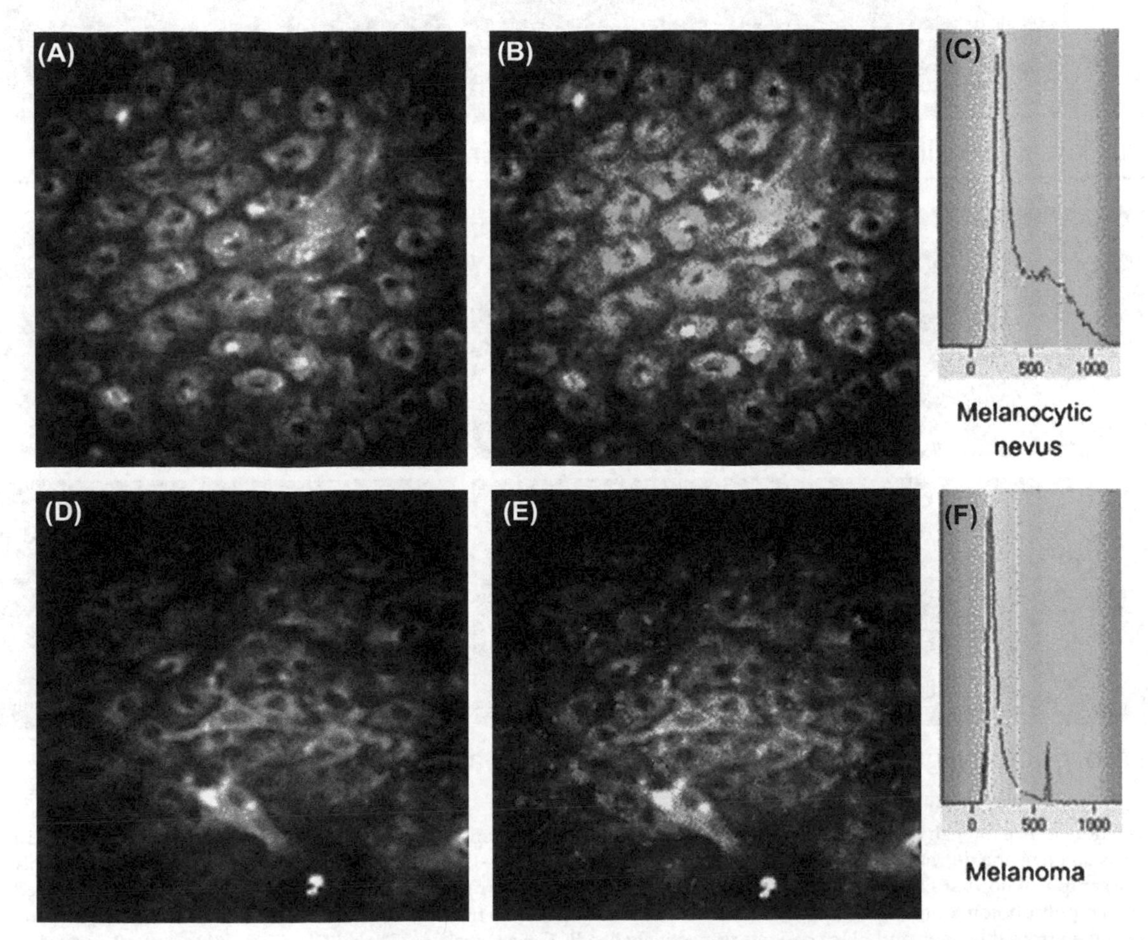

FIGURE 18.8 False color-coded fluorescence lifetime images (B and E) of different lesions with corresponding intensity images (A and D). (B) Nevus. (E) Melanoma. (C and F) Lifetime distribution histograms. Color range: −250 ps (*red*) to 1250 ps (*blue*). Excitation wavelength: 760 nm. *Used with permission from Dimitrow E, Riemann I, Ehlers A, et al. Spectral fluorescence lifetime detection and selective melanin imaging by multiphoton laser tomography for melanoma diagnosis. Exp Dermatol 2009;18(6):509–15.*

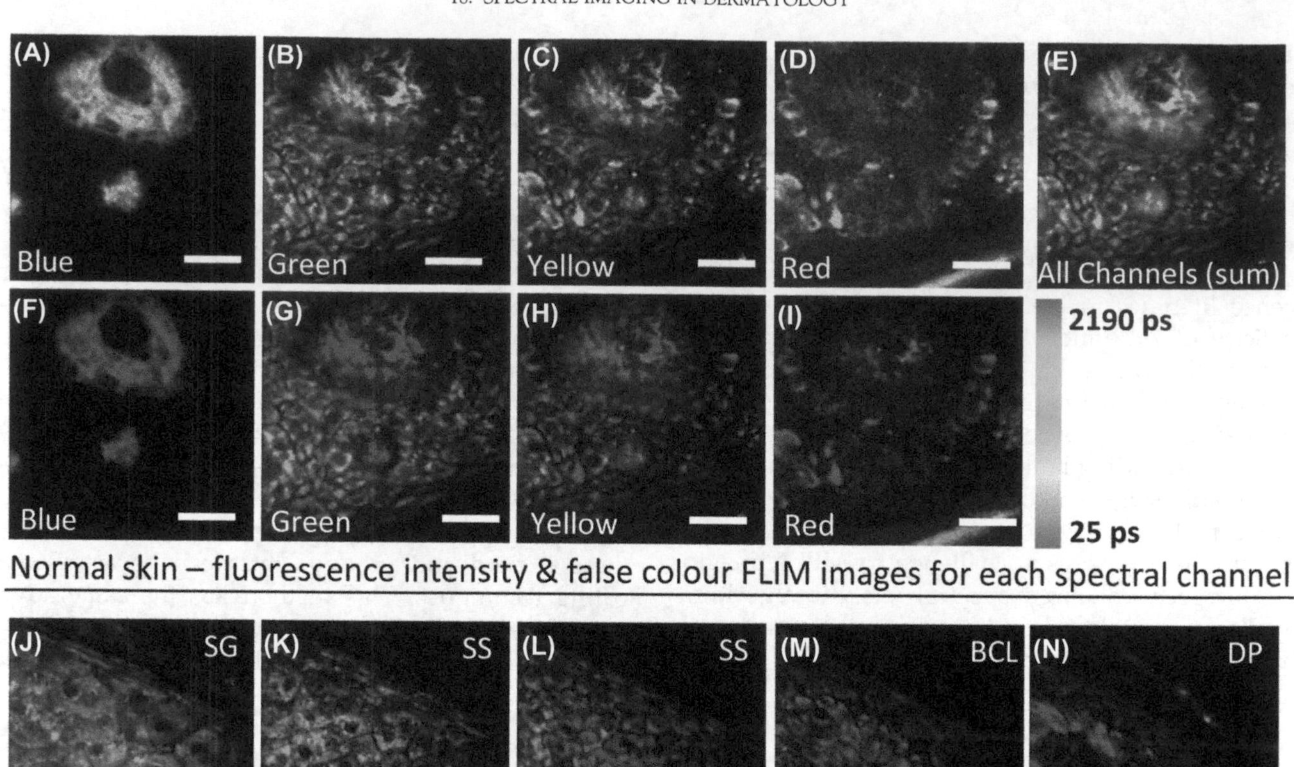

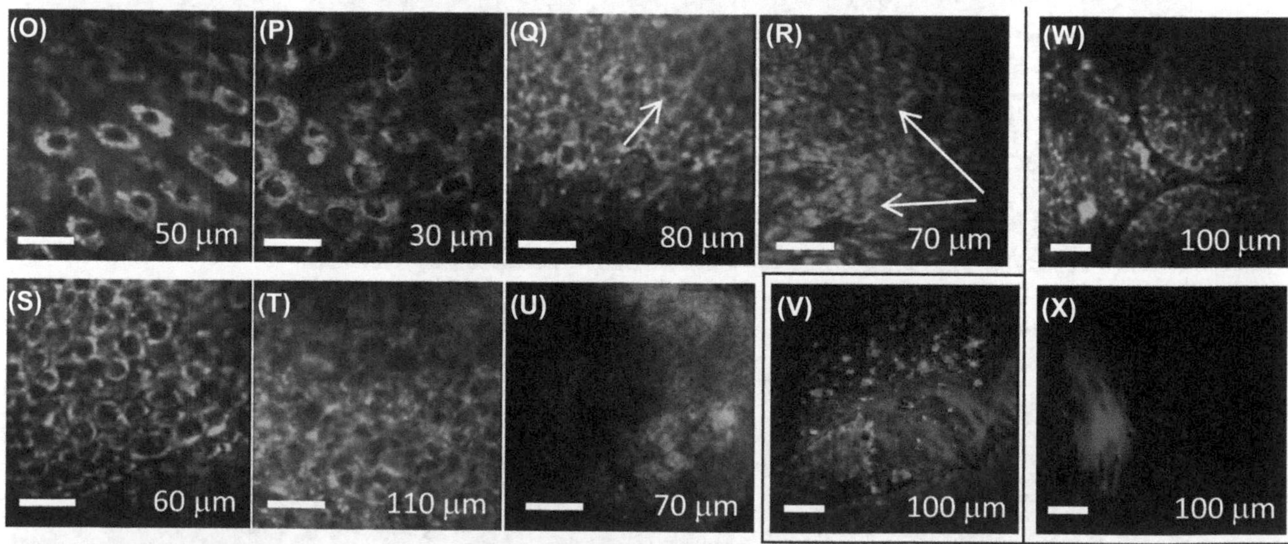

FIGURE 18.9 Multispectral fluorescence intensity and fluorescence lifetime images acquired from normal skin and basal cell carcinomas (BCCs). (A–I) Fluorescence intensity and false color fluorescence lifetime images from a single field of view acquired at a depth of 110 μm with all spectral channels taken near a dermal papilla from normal skin. (J–N) Fluorescence lifetime images taken from the green channel only of different depths within a sample of normal skin. (O–U) Fluorescence lifetime images taken from the green channel illustrating visual architectural features seen in BCC using multiphoton tomography. (V) Fluorescence lifetime image taken from the blue channel of a BCC. (W and X) Paired fluorescence lifetime images taken from the green and blue channels respectively of a BCC nest. Scale = 25 μm. *BCL*, Basal cell layer; *DP*, dermal papilla; *FLIM*, fluorescence lifetime imaging microscopy; *SG*, stratum granulosum; *SS*, stratum spinosum. *Used with permission from Patalay R, Talbot C, Alexandrov Y, et al. Multiphoton multispectral fluorescence lifetime tomography for the evaluation of basal cell carcinomas. PLoS One 2012;7(9):e43460.*

implementation of pump-probe microscopy is similar to standard laser scanning microscopy; the main changes in the system include the presence of two colors, modulation of the pump (eg, 2 MHz), and lock-in detection [98,99]. With further delineation of the precise wavelength requirements, considerably simpler instrumentation can be implemented.

Pump-probe microscopy can provide insight into the biochemical composition of melanins and their response to disease. For example, the ultrafast lifetimes are sensitive not only to the type of melanin (eumelanin and pheomelanin), but also to the oxidation state, metal content, and aggregate size [100]. This method has been used to image thin, unstained biopsy slides to help differentiate melanomas from cutaneous melanocytic nevi [101], and conjunctival melanoma from primary acquired melanosis [101], as well as providing novel insight into the metastatic potential of invasive cutaneous melanomas [102]. Extending the examination to the subcellular scale, Simpson et al. were able to distinguish malignant melanocytes from nonmalignant macrophage-derived melanophages [101]. Although this approach is promising, the clinical significance of alterations in eumelanin-pheomelanin ratios remains to be firmly established.

RAMAN SPECTROSCOPY

The discovery of inelastic light scattering properties coined the *Raman effect* because of the significant contributions from Indian physicist C.V. Raman. In 1928, a Nobel Prize was awarded to Dr. Raman for this revelation and laid the foundations for development of yet another important method for molecular characterization based on measurements of vibrational and rotational chemical characteristics [103]. For biological samples, Raman spectroscopy often utilizes a laser that outputs in the NIR wavelength region to interact with the sample, causing an exchange of energy and scattering of the laser beam light. Variations in change of energy are dependent on the biochemical makeup of the sample, and information, such as chemical integrity [104], hydration state [105], and oxidative stress [106] of the sample may be recorded.

One promising application of Raman spectroscopy is for monitoring and tracking of transdermal drug delivery. Currently a technique known as *tape stripping* is the gold standard. Layers of SC are progressively removed with tape and the amount of penetrated drug is quantified using various analytical methods. However, tape stripping is invasive, limits repeatability, and does not allow real-time tracking of drug penetration [107]. Raman spectroscopy may replace tape stripping for in vitro and in vivo clinical experiments, because it allows for characterization of particle (drug) penetration behavior in real time in a noninvasive and label-free manner [108,109].

Investigators have implemented Raman spectroscopy for tracking diffusion of water and β-carotene in ex vivo human skin [110]. Raman signal was visualized up to 40 μm into the skin, encompassing the SC and top layers of the LE. An H&E-stained human skin sample was used as a reference image to correlate with Raman spectroscopic images. Raman signals from the SC were mainly fluorescence-free, whereas deeper regions of the visualized LE had greater amount of interfering fluorescence, which may be a limitation. Results from water diffusion experiments indicated homogenous distribution of water throughout a tissue sample. On the other hand, β-carotene diffusion experiments demonstrated a diffusion gradient, with rapid loss of diffusion after the initial 10 μm of the tissue sample because of its high lipophilicity and poor skin permeability [111]. Raman spectroscopy has also been used to track and visualize small particle diffusion in the SC and top LE, a practically significant capability because the SC is one of the major barriers to transdermal drug delivery [112].

Sodium dodecyl sulfate (SDS) is a widely used component of many skin-care products, but its distribution is of also of interest because studies have reported that SDS may increase epidermal water loss and cause irritation and inflammation [113–115]. Mao et al. reported successful tracking of the penetration of SDS into human and porcine skin using Raman and IR spectroscopy (Fig. 18.10). The authors suggest that the lipid packing in the SC of porcine skin is less ordered than in human skin and is more permissive for diffusion [116].

Characterization and diffusion of different molecular-weight formulations of hyaluronic acid (HA) have also been studied in human skin samples using Raman spectroscopy, with the (not surprising) finding that diffusion rates were inversely proportional to molecular weight [117]. An increased understanding of HA diffusion properties in the skin is useful, because many skin care products are HA-based [118,119].

Raman Spectroscopy and Skin Neoplasia

In addition to applications in drug delivery studies, Raman spectroscopy may provide high-quality information useful in discriminating healthy from neoplastic tissues. In one study, a classification model for discrimination of BCC from healthy tissue using Raman spectroscopy was created using 329 Raman spectra measured in skin samples from 20 patients. Key criteria for segmentation of various tissue regions, such as BCC,

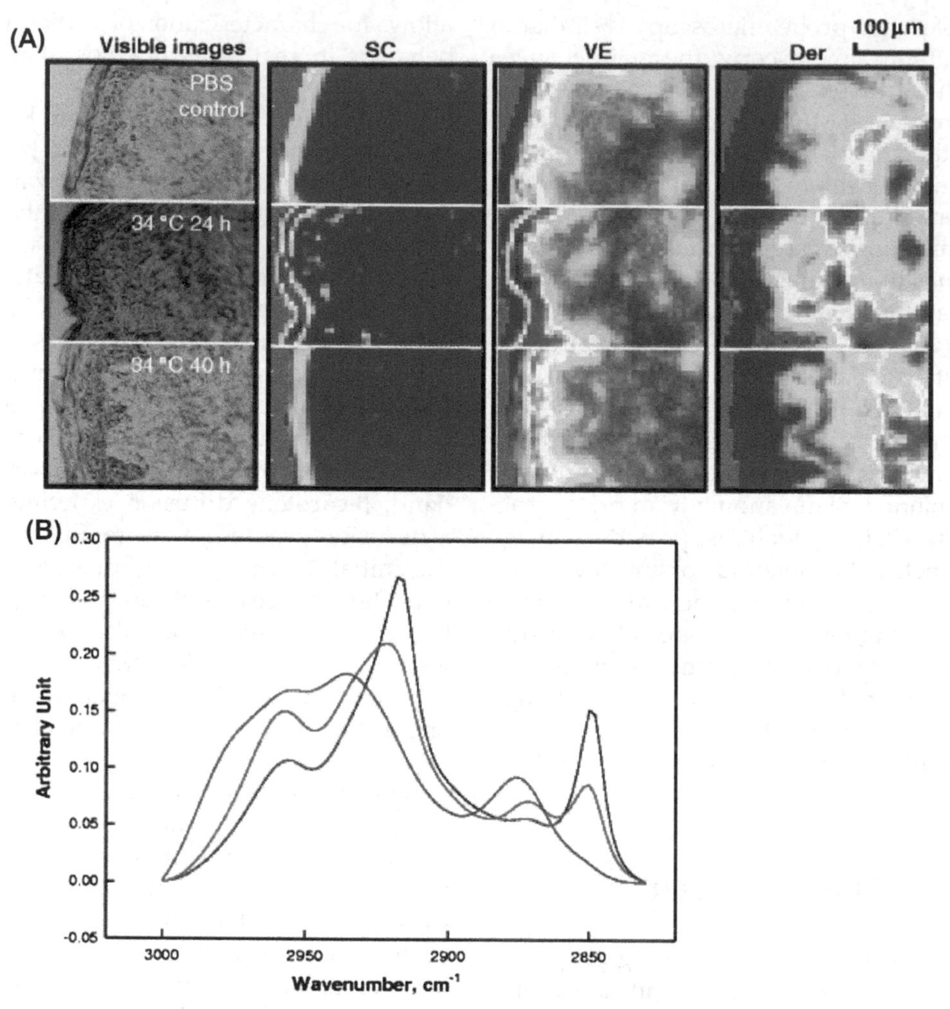

FIGURE 18.10 (A) Microscopic images and factor analysis score images (color coding of scores: *red* > *yellow* > *blue*) depicting different regions of human skin from a phosphate-buffered saline *(PBS)* control experiment (24-h treatment at 34°C) and after acyl chain perdeuterated sodium dodecyl sulfate (SDS-d25) treatment for 24 and 40 h at 34°C. The skin surface is located at the left of each image plane. Factor analysis was conducted over the C–H stretching region (2830 to 3000 cm^{-1}). (B) Factor loadings map to high scores in the following regions of human skin: *Der*, dermis *(blue)*; SC, stratum corneum *(black)*; *VE*, viable epidermis *(red)*. *Used with permission from Mao G, Flach CR, Mendelsohn R, Walters RM. Imaging the distribution of sodium dodecyl sulfate in skin by confocal Raman and infrared microspectroscopy. Pharm Res 2012;29(8):2189–201.*

epidermis, or dermis were identified from the classification model, which was then applied to blind testing on six additional skin specimens. This classification model demonstrated sensitivity and specificity scores of 90% and 85%, respectively [120], and generated pseudocolor classification images that correlate fairly closely with histopathological findings. This study demonstrated the potential utility of Raman spectroscopy in detection or diagnosis of BCC from healthy tissue ex vivo, pointing to possible value for using this modality in the diagnosis and monitoring of BCC in vivo.

Raman Spectroscopy Developments

Long image acquisition time is a key limitation to conventional Raman spectroscopy for large tissue specimens. However, Rowlands et al. described a time-efficient sampling method that may reduce processing time up to 30-fold [121]. Selective points on the tissue are sampled and the algorithm determines whether additional points are required based on sample point differences. This technique is particularly useful for specimens where there is a large uniform area with tumor or specific cells of interest in a well-defined area. Pseudocolor images demonstrated excellent correlation with histopathological findings and retained sufficient spectral signal-to-noise ratio to identify individual tissue structures.

Coherent anti-Stokes Raman spectroscopy (CARS) is a technique that utilizes paired laser beams to generate a coherent signal, resulting in a signal that is much stronger than conventional Raman signals [122]. Marks et al. utilized CARS with nonlinear interferometric vibrational imaging [123] (Fig. 18.11) to evaluate

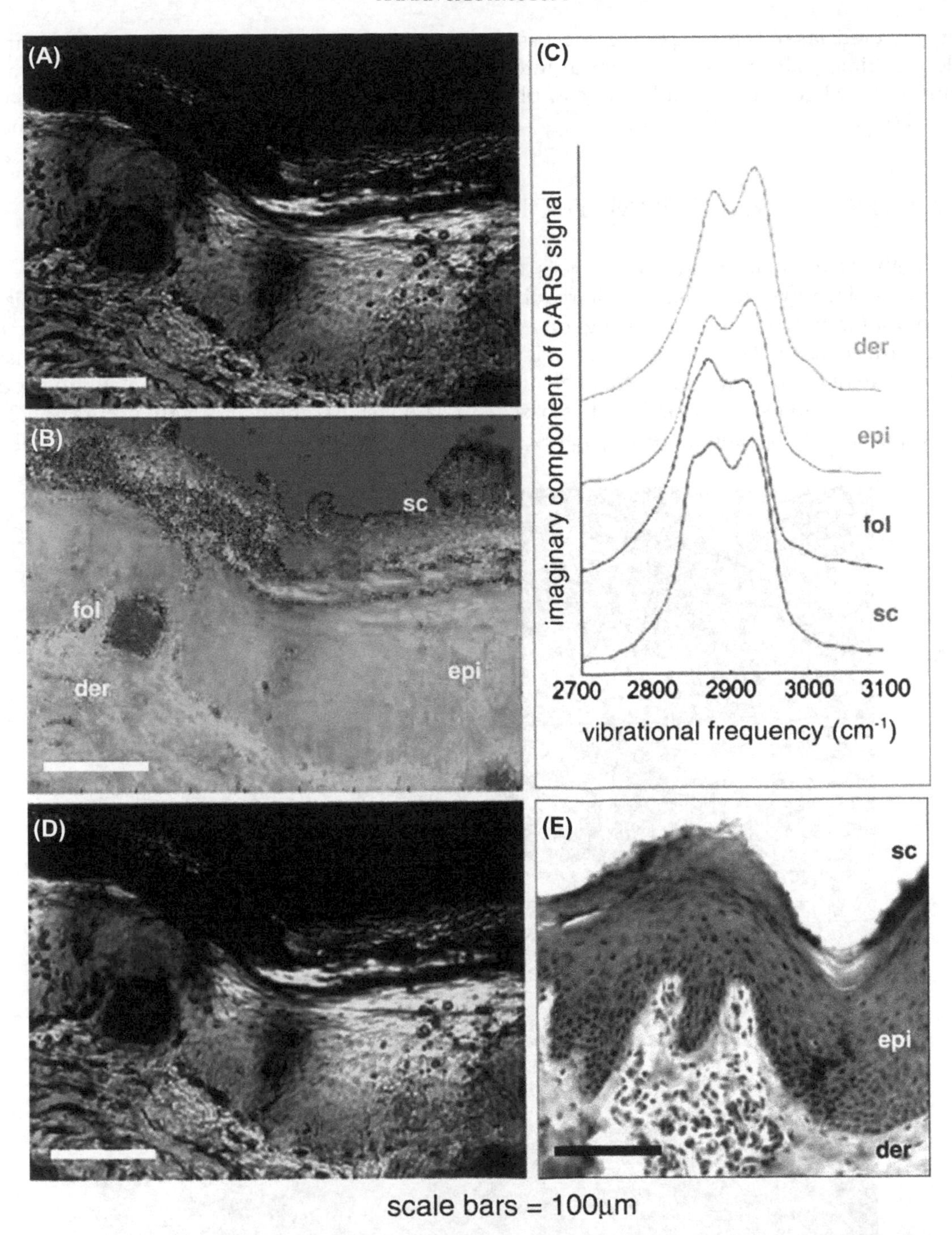

FIGURE 18.11 Molecular imaging of cutaneous tissue. (A) Intensity image (total integrated spectral power). (B) Nonlinear interferometric vibrational imaging (NIVI) composite showing discrimination of stratum corneum *(sc)*, epidermis *(epi)*, dermis *(der)*, and hair follicle *(fol)*. (C) NIVI spectra for each domain in *B*, as obtained by cluster analysis. Each spectrum is the result of averaging the spectra of the members of most prevalent cluster in regions of 20×20 pixels^2 within each domain. (D) NIVI image showing both structural and molecular compositions. (E) hematoxylin and eosin (H&E) histology of a section from the same region. Scale = 100 μm. *CARS*, Coherent anti-Stokes Raman spectroscopy. *Used with permission from Benalcazar WA, Boppart SA. Nonlinear interferometric vibrational imaging for fast label-free visualization of molecular domains in skin. Anal Bioanal Chem 2011;400(9):2817–25.*

the molecular composition of porcine skin ex vivo, and reported high discriminating power amongst dermis, epidermis, and stroma, in addition to discriminating lipids located in adipocytes versus sebaceous glands [124]. A number of ex vivo as well as in vivo applications are possible [125], although development of much higher speed Raman microscopy methods may expand applications even further [126]. One limitation of Raman spectroscopy is relatively shallow imaging depth achieved, which may be addressed by using higher-powered lasers and exposing the tissue sample for longer times, although the chemical integrity of the sample may become altered because of photobleaching and/or associated water loss.

COMPUTATIONAL ANALYTICAL TOOLS

Tools and methods for computational analysis are continuously under improvement because it requires tremendous computer power and processing time to output spectral information. For instance, analysis of TPEF data is dependent on intensity differences of autofluorescence and SHG signals, whereas MFLT is based on intrinsic properties of the substrate of interest. It may be challenging to identify and segment similar fluorescence signals caused by limitations of analytic algorithms, resulting in loss of important spectral information. This section highlights some recent advances of different analytical tools to enhance various spectral modalities.

Linear Unmixing

The most common approach to analysis of multiple and overlapping spectral features is linear unmixing as described by Zimmerman et al., who emphasized the influence of noise, spectral channel width (resolution), and relative signal intensities with accuracy [127]. For these methods, accurate estimates of the spectral shape of the

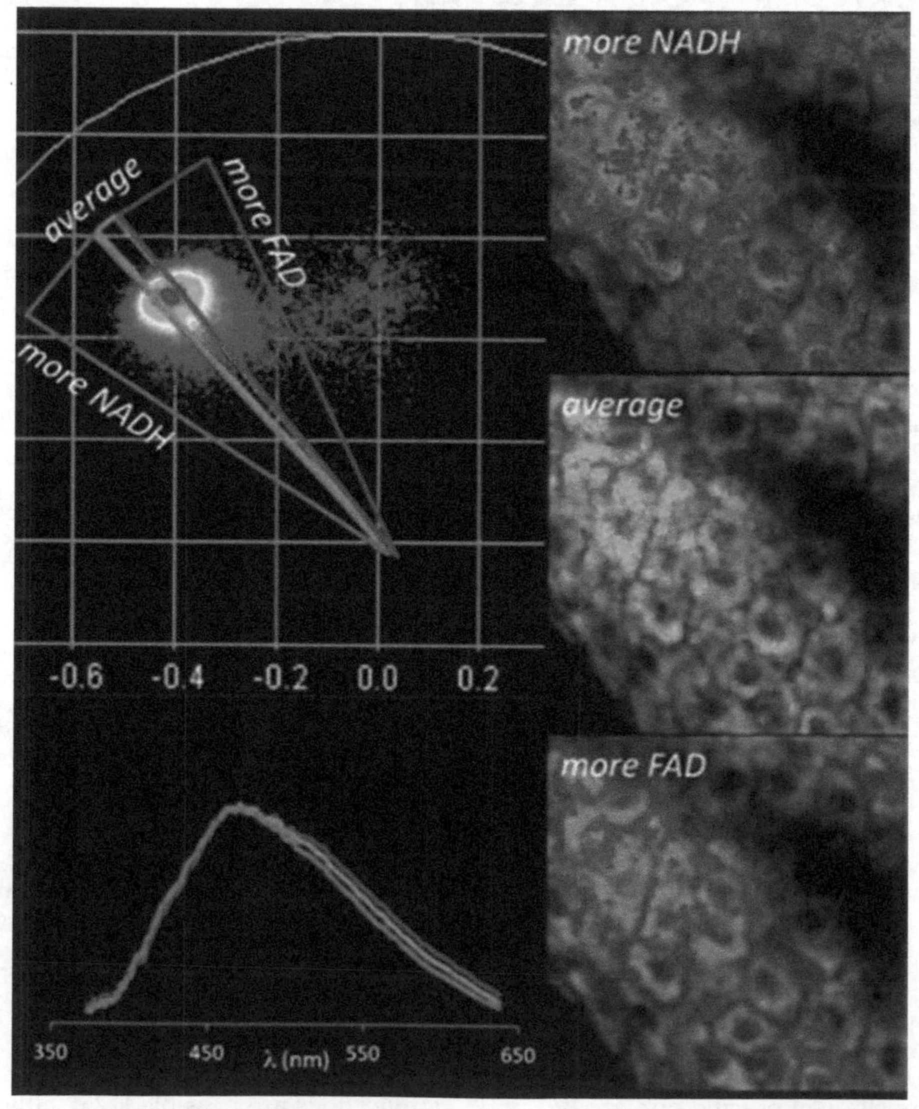

FIGURE 18.12 Spectral phasor plot of image of epidermal cells in in vivo human skin (minimal tanning). For three regions of interest (ROIs) (*blue*, mainly NADH; *green*, average; and *red*, mainly FAD), the average spectra are included. The pixels in these ROIs are highlighted in their corresponding colors in the grayscale intensity images. Experimental condition: excitation power = 20 mW, acquisition time = 128 μs per pixel (6.5 seconds per image), and excitation wavelength is 760 nm. The detection window ranges from 700 nm (channel 1) to 350 nm (channel 128); the wavelength is not linear with the channel number in the prism-based spectrograph. *FAD*, Flavin adenine dinucleotide; *NADH*, nicotinamide adenine dinucleotide. *Used with permission from Fereidouni F, Bader AN, Colonna A, Gerritsen HC. Phasor analysis of multiphoton spectral images distinguishes autofluorescence components of in vivo human skin. J Biophotonics 2014;7(8):589–96.*

end members into the respective signals being partitioned are required. Unfortunately, it is not always possible to know, a priori, the shape of all spectral components. Various methods, automated or semiautomated, have been applied and the results demonstrated varying effects. These methods include blind source separation that can be improved if various constraints are applied [128], as well as methods for manual or semiautomated "subtracting" contaminating spectra from each other to estimate the pure components present [129]. Additional methods are referenced in the review by Thomas et al. [42].

Phasor Analysis

A recently developed method, spectral phasor analysis, may simplify and improve analysis of complex spectral images [130]. Spectral phasor analysis, in which a Fourier transform is used to a "phasor plot," a graphical representation or "map" of spectral properties, with benefit that the subsequent computations require only simple arithmetic operations to unmix spectral channels (Fig. 18.12).

Phasor analysis was utilized to examine in vivo human skin and demonstrated the potential to segment structures, such as SC, LE, melanized epidermal cells, and SHG signal of dermal collagen. Epidermal cells with no melanin were correctly discriminated against epidermal cells with melanin, and autofluorescent species including NAD(P)H, keratin, FAD, melanin, collagen, and elastin were correctly discriminated [131]. A different research group reported discrimination of eumelanin in the epidermis from eumelanin in macrophages with phasor-based methods [101] (Fig. 18.13).

Denoising

Because many spectral imaging techniques are saddled with the need to collect many channels as quickly as possible, achieving a good signal-to-noise ratio can be a challenge. Fortunately, there have been recent developments in denoising mathematics [126] that seem well suited for use with spectral data sets. It appears that automated detection of spectral features can be greatly enhanced by prior (accurate) denoising.

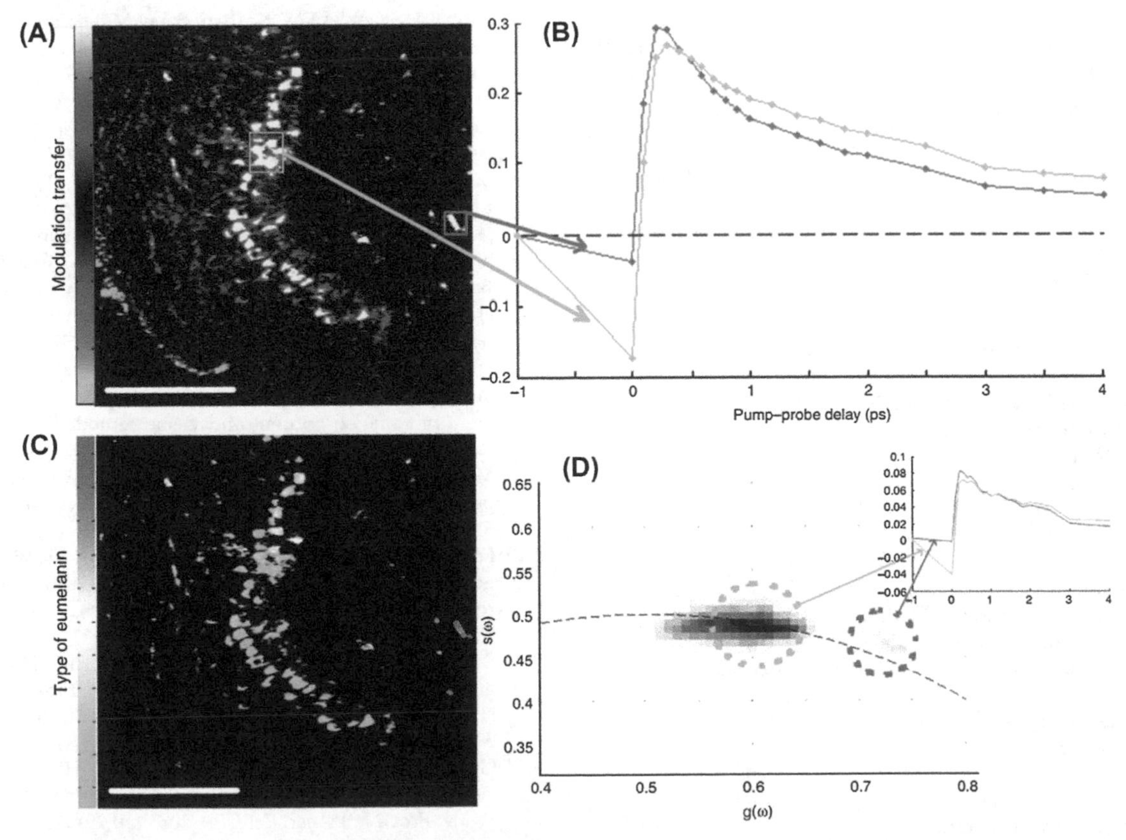

FIGURE 18.13 Melanin in macrophages differentiated from epidermal melanin. (A) Malignant melanoma imaged at approximately 300 fs pump-probe delay. Skin surface is on the left with the surgical ink appearing *blue*. False color scheme is the sign of the signal: *yellow* for positive and *cyan* for negative. (B) Delay traces averaged over two regions of interest in *A* of corresponding colors. (C) Image of (A) recolored using phasor analysis at 0.1 THz. (D) Phasor plot showing all pixels from the six images that contain both epidermal and macrophage melanin. Pixels in (C) are colored *magenta* or *cyan* according to which population they fall into (or a combination of the colors if the pixel is in both populations). The inset graph shows the average spectrum of pixels in each population from the six images. Scale = 100 μm. *Used with permission from Simpson MJ, Wilson JW, Phipps MA, Robles FE, Selim MA, Warren WS. Nonlinear microscopy of eumelanin and pheomelanin with subcellular resolution. J Invest Dermatol 2013; 133(7):1822–6.*

CONCLUSION

Spectral imaging has demonstrated promising preliminary results for evaluation of ex vivo healthy skin and skin pathological conditions derived from both human sources and animal models. Compared with conventional histologically processed specimens, viewed by eye or conventional light microscopes, images acquired by the almost bewildering variety of microscopy techniques described can contain important, novel, and interpretable data that can aid clinicians and scientists in the understanding, diagnosis, and treatment of dermatological disorders. By combining novel imaging methods with appropriate computer algorithms, the required time and labor can be minimized and interobserver agreement enhanced. That said, some practical notes of caution are in order. The path from laboratory demonstration to clinical use can be fraught with difficulties, as evidenced by the few imaging alternatives that have made it to commercialization and utilization. It is not enough to demonstrate high specificity and sensitivity, for example. Actual multiinstitutional clinical trials that provide evidence of improved clinical outcomes, compared with the standard of care, are necessary, but even these are not sufficient. In addition, it is likely that affordability, ease of use, and the availability of a nonmysterious explanation of how such systems might actually work would be required before skeptical, conservative, cost-sensitive, and busy clinicians (and the US Food and Drug Administration) move toward adoption for clinical use.

List of Abbreviations

BCC Basal cell carcinoma
CARS Coherent anti-Stokes Raman spectroscopy
FAD Flavin adenine dinucleotide
FFPE Formalin-fixed, paraffin-embedded
FLI Fluorescence lifetime imaging
FTIR Fourier-transform infrared
HA Hyaluronic acid
H&E Hematoxylin and eosin
HbO_2 Oxyhemoglobin
IR Infrared
LE Living epidermis
MFLT Multiphoton fluorescence lifetime
NADH Nicotinamide adenine dinucleotide
NAD(P)H Nicotinamide adenine dinucleotide (phosphate)
NIR Near-infrared
NMSC Nonmelanoma skin cancer
ROS Reactive oxygen species
SC Stratum corneum
SDS Sodium dodecylsulfate
SHG Second harmonic generation
TPEF Two-photon excited fluorescence
UV Ultraviolet

Acknowledgments

We would like to thank the following for generous help in preparation of this manuscript: Francisco Robles, Warren S. Warren, Martin C. Fischer, and James Mansfield.

References

[1] Tearney GJ, Brezinski ME, Southern JF, Bouma BE, Hee MR, Fujimoto JG. Determination of the refractive index of highly scattering human tissue by optical coherence tomography. Opt Lett 1995;20(21):2258.
[2] Gaudi S, Meyer R, Ranka J, et al. Hyperspectral imaging of melanocytic lesions. The Am J Dermatopathol 2014;36(2):131–6.
[3] Dicker DT, Lerner J, Van Belle P, et al. Differentiation of normal skin and melanoma using high resolution hyperspectral imaging. Cancer Biol Ther 2006;5(8):1033–8.
[4] Li Q, Sun Z, Wang Y, Liu H, Guo F, Zhu J. Histological skin morphology enhancement base on molecular hyperspectral imaging technology. Skin Res Technol 2014;20(3):332–40. Official Journal of International Society for Bioengineering and the Skin (ISBS) [and] International Society for Digital Imaging of Skin (ISDIS) [and] International Society for Skin Imaging (ISSI).
[5] Dicker DT, Kahn N, Flaherty KT, Lerner J, El-Deiry WS. Hyperspectral imaging: a non-invasive method of imaging melanoma lesions in a patient with stage IV melanoma, being treated with a RAF inhibitor. Cancer Biol Ther 2011;12(4):326–34.
[6] Barnhill RL, Argenyi ZB, From L, et al. Atypical spitz nevi/tumors: lack of consensus for diagnosis, discrimination from melanoma, and prediction of outcome. Hum Pathol 1999;30(5): 513–20.
[7] Shoo BA, Sagebiel RW, Kashani-Sabet M. Discordance in the histopathologic diagnosis of melanoma at a melanoma referral center. J Am Acad Dermatol 2010;62(5):751–6.
[8] Lodha S, Saggar S, Celebi JT, Silvers DN. Discordance in the histopathologic diagnosis of difficult melanocytic neoplasms in the clinical setting. J Cutan Pathol 2008;35(4):349–52.
[9] Bini J, Spain J, Nehal K, Hazelwood V, DiMarzio C, Rajadhyaksha M. Confocal mosaicing microscopy of human skin ex vivo: spectral analysis for digital staining to simulate histology-like appearance. J Biomed Opt 2011;16(7):076008.
[10] Bautista PA, Yagi Y. Multispectral enhancement method to increase the visual differences of tissue structures in stained histopathology images. Anal Cell Pathol 2012;35(5–6):407–20.
[11] Turner NJ, Pezzone MA, Brown BN, Badylak SF. Quantitative multispectral imaging of Herovici's polychrome for the assessment of collagen content and tissue remodelling. J Tissue Eng Regen Med 2013;7(2):139–48.
[12] Mansfield JR. Multispectral imaging: a review of its technical aspects and applications in anatomic pathology. Vet Pathol 2014; 51(1):185–210.
[13] Feng Z, Moudgil T, Cheng A, et al. Utilizing quantitative immunohistochemistry for relationship analysis of tumor microenvironment of head and neck cancer patients. J ImmunoTherapy Cancer 2014;2(Suppl. 3):P258.
[14] Schubert W, Gieseler A, Krusche A, Serocka P, Hillert R. Next-generation biomarkers based on 100-parameter functional super-resolution microscopy TIS. N Biotechnol 2012;29(5): 599–610.
[15] Gerdes MJ, Sevinsky CJ, Sood A, et al. Highly multiplexed single-cell analysis of formalin-fixed, paraffin-embedded cancer tissue. Proc Natl Acad Sci USA 2013;110(29):11982–7.
[16] Zrazhevskiy P, True LD, Gao X. Multicolor multicycle molecular profiling with quantum dots for single-cell analysis. Nat Protoc 2013;8(10):1852–69.

[17] Taylor CR, Levenson RM. Quantification of immunohistochemistry—issues concerning methods, utility and semiquantitative assessment II. Histopathology 2006;49(4):411—24.
[18] Wallace MB, Wax A, Roberts DN, Graf RN. Reflectance spectroscopy. Gastrointest Endosc Clin N Am 2009;19(2):233—42.
[19] Garcia-Uribe A, Zou J, Duvic M, Cho-Vega JH, Prieto VG, Wang LV. In vivo diagnosis of melanoma and nonmelanoma skin cancer using oblique incidence diffuse reflectance spectrometry. Cancer Res 2012;72(11):2738—45.
[20] Belenki L, Sterzik V, Schulz K, Bohnert M. Analyzing reflectance spectra of human skin in legal medicine. J Biomed Opt 2013; 18(1):17004.
[21] Bohnert M, Schulz K, Belenkaia L, Liehr AW. Re-oxygenation of haemoglobin in livores after post-mortem exposure to a cold environment. Int J Leg Med 2008;122(2):91—6.
[22] Watchman H, Walker GS, Randeberg LL, Langlois NE. Re-oxygenation of post-mortem lividity by passive diffusion through the skin at low temperature. Forensic Sci Med Pathol 2011;7(4):333—5.
[23] Herschel W. Experiments on the refrangibility of the invisible rays of the sun. Phil Trans R Soc Lond 1800;90:284—92.
[24] Lau K, Hedegaard MA, Kloepper JE, Paus R, Wood BR, Deckert V. Visualization and characterisation of defined hair follicle compartments by Fourier transform infrared (FTIR) imaging without labelling. J Dermatol Sci 2011;63(3):191—8.
[25] Derrick MR, Stulik D, Landry JM. Infrared spectroscopy in conservation science. Los Angeles: Getty Conservation Institute; 1999.
[26] Stuart BH. Infrared spectroscopy: fundamentals and applications. Wiley & Sons; 2004.
[27] Vishwasrao HD, Heikal AA, Kasischke KA, Webb WW. Conformational dependence of intracellular NADH on metabolic state revealed by associated fluorescence anisotropy. J Biol Chem 2005;280(26):25119—26.
[28] Leroy M, Lafleur M, Auger M, Laroche G, Pouliot R. Characterization of the structure of human skin substitutes by infrared microspectroscopy. Anal Bioanal Chem 2013;405(27):8709—18.
[29] Ly E, Piot O, Durlach A, Bernard P, Manfait M. Differential diagnosis of cutaneous carcinomas by infrared spectral micro-imaging combined with pattern recognition. The Analyst 2009; 134(6):1208—14.
[30] Ly E, Piot O, Wolthuis R, Durlach A, Bernard P, Manfait M. Combination of FTIR spectral imaging and chemometrics for tumour detection from paraffin-embedded biopsies. The Analyst 2008; 133(2):197—205.
[31] Tfayli A, Piot O, Durlach A, Bernard P, Manfait M. Discriminating nevus and melanoma on paraffin-embedded skin biopsies using FTIR microspectroscopy. Biochim Biophys Acta 2005; 1724(3):262—9.
[32] Sebiskveradze D, Gobinet C, Cardot-Leccia N, et al. Infrared spectral microimaging: a new tool to characterise the tissue features in skin cancers of melanoma type. In: 1st International Conference on Bioimaging, BIOIMAGING 2014-Part of 7th International Joint Conference on Biomedical Engineering Systems and Technologies, BIOSTEC 2014. Angers, Loire Valley: SciTePress; 2014. p. 59—65.
[33] Kong R, Reddy RK, Bhargava R. Characterization of tumor progression in engineered tissue using infrared spectroscopic imaging. The Analyst 2010;135(7):1569—78.
[34] van Kempen LC, Ruiter DJ, van Muijen GN, Coussens LM. The tumor microenvironment: a critical determinant of neoplastic evolution. Eur J Cell Biol 2003;82(11):539—48.
[35] Childs DTD, Hogg RA, Revin DG, Rehman I, Cockburn JW, Matcher SJ. Sensitivity advantage of QCL tunable-laser mid-infrared spectroscopy over FTIR spectroscopy. Appl Spectrosc Rev 2015;50(10).
[36] Bassan P, Weida MJ, Rowlette J, Gardner P. Large scale infrared imaging of tissue micro arrays (TMAs) using a tunable Quantum Cascade Laser (QCL) based microscope. The Analyst 2014; 139(16):3856—9.
[37] Yeh K, Kenkel S, Liu JN, Bhargava R. Fast infrared chemical imaging with a quantum cascade laser. Anal Chem 2015;87(1): 485—93.
[38] Chance B. Pyridine nucleotide as an indicator of the oxygen requirements for energy-linked functions of mitochondria. Circ Res 1976;38(5 Suppl. 1):I31—8.
[39] Levitt JM, McLaughlin-Drubin ME, Munger K, Georgakoudi I. Automated biochemical, morphological, and organizational assessment of precancerous changes from endogenous two-photon fluorescence images. PLoS One 2011;6(9):e24765.
[40] Zipfel WR, Williams RM, Christie R, Nikitin AY, Hyman BT, Webb WW. Live tissue intrinsic emission microscopy using multiphoton-excited native fluorescence and second harmonic generation. Proc Natl Acad Sci USA 2003;100(12): 7075—80.
[41] Yew E, Rowlands C, So PT. Application of multiphoton microscopy in dermatological studies: a mini-review. J Innov Opt Health Sci 2014;7(5):1330010.
[42] Thomas G, van Voskuilen J, Gerritsen HC, Sterenborg HJ. Advances and challenges in label-free nonlinear optical imaging using two-photon excitation fluorescence and second harmonic generation for cancer research. J Photochem Photobiol B Biol 2014;141:128—38.
[43] Denk W, Strickler JH, Webb WW. Two-photon laser scanning fluorescence microscopy. Science 1990;248(4951):73—6.
[44] Lerner JM, Zucker RM. Calibration and validation of confocal spectral imaging systems. Cytometry A 2004;62(1):8—34.
[45] Buehler C, Kim KH, Dong CY, Masters BR, So PT. Innovations in two-photon deep tissue microscopy. IEEE Eng Med Biol Mag 1999;18(5):23—30.
[46] Xu C, Zipfel W, Shear JB, Williams RM, Webb WW. Multiphoton fluorescence excitation: new spectral windows for biological nonlinear microscopy. Proc Natl Acad Sci USA 1996;93(20): 10763—8.
[47] Chen J, Lee A, Zhao J, et al. Spectroscopic characterization and microscopic imaging of extracted and in situ cutaneous collagen and elastic tissue components under two-photon excitation. Skin Res Technol 2009;15(4):418—26.
[48] Chen G, Chen J, Zhuo S, et al. Nonlinear spectral imaging of human hypertrophic scar based on two-photon excited fluorescence and second-harmonic generation. Br J Dermatol 2009; 161(1):48—55.
[49] Dimitrow E, Ziemer M, Koehler MJ, et al. Sensitivity and specificity of multiphoton laser tomography for in vivo and ex vivo diagnosis of malignant melanoma. J Invest Dermatol 2009; 129(7):1752—8.
[50] Masters BR, So PT, Gratton E. Multiphoton excitation fluorescence microscopy and spectroscopy of in vivo human skin. Biophys J 1997;72(6):2405—12.
[51] Konig K, Schenke-Layland K, Riemann I, Stock UA. Multiphoton autofluorescence imaging of intratissue elastic fibers. Biomaterials 2005;26(5):495—500.
[52] Huang S, Heikal AA, Webb WW. Two-photon fluorescence spectroscopy and microscopy of NAD(P)H and flavoprotein. Biophys J 2002;82(5):2811—25.
[53] Pena A, Strupler M, Boulesteix T, Schanne-Klein M. Spectroscopic analysis of keratin endogenous signal for skin multiphoton microscopy. Opt Express 2005;13(16):6268—74.
[54] Laiho LH, Pelet S, Hancewicz TM, Kaplan PD, So PT. Two-photon 3-D mapping of ex vivo human skin endogenous fluorescence species based on fluorescence emission spectra. J Biomed Opt 2005;10(2):024016.

[55] Hoffmann K, Stucker M, Altmeyer P, Teuchner K, Leupold D. Selective femtosecond pulse-excitation of melanin fluorescence in tissue. J Invest Dermatol 2001;116(4):629–30.
[56] Radosevich AJ, Bouchard MB, Burgess SA, Chen BR, Hillman EM. Hyperspectral in vivo two-photon microscopy of intrinsic contrast. Opt Lett 2008;33(18):2164–6.
[57] Zheng W, Wu Y, Li D, Qu JY. Autofluorescence of epithelial tissue: single-photon versus two-photon excitation. J Biomed Opt 2008;13(5):054010.
[58] Paoli J, Smedh M, Wennberg AM, Ericson MB. Multiphoton laser scanning microscopy on non-melanoma skin cancer: morphologic features for future non-invasive diagnostics. J Invest Dermatol 2008;128(5):1248–55.
[59] Paoli J, Smedh M, Ericson MB. Multiphoton laser scanning microscopy: a novel diagnostic method for superficial skin cancers. Semin Cutan Med Surg 2009;28(3):190–5.
[60] Morgan AM, Lo J, Fisher DE. How does pheomelanin synthesis contribute to melanomagenesis? Two distinct mechanisms could explain the carcinogenicity of pheomelanin synthesis. Bioessays 2013;35(8):672–6.
[61] Elleder M, Borovansky J. Autofluorescence of melanins induced by ultraviolet radiation and near ultraviolet light. A histochemical and biochemical study. Histochem J 2001;33(5):273–81.
[62] Kerimo J, Rajadhyaksha M, DiMarzio CA. Enhanced melanin fluorescence by stepwise three-photon excitation. Photochem Photobiol 2011;87(5):1042–9.
[63] Chen R, Huang Z, Lui H, et al. Monte Carlo simulation of cutaneous reflectance and fluorescence measurements: the effect of melanin contents and localization. J Photochem Photobiol B, Biol 2007;86(3):219–26.
[64] Leupold D, Scholz M, Stankovic G, et al. The stepwise two-photon excited melanin fluorescence is a unique diagnostic tool for the detection of malignant transformation in melanocytes. Pigment Cell Melanoma Res 2011;24(3):438–45.
[65] Hanson KM, Clegg RM. Observation and quantification of ultraviolet-induced reactive oxygen species in ex vivo human skin. Photochem Photobiol 2002;76(1):57–63.
[66] Hanson KM, Gratton E, Bardeen CJ. Sunscreen enhancement of UV-induced reactive oxygen species in the skin. Free Radic Biol Med 2006;41(8):1205–12.
[67] Green A, Williams G, Neale R, et al. Daily sunscreen application and betacarotene supplementation in prevention of basal-cell and squamous-cell carcinomas of the skin: a randomised controlled trial. Lancet 1999;354(9180):723–9.
[68] Chen J, Wong S, Nathanson MH, Jain D. Evaluation of Barrett esophagus by multiphoton microscopy. Arch Pathol Lab Med 2014;138(2):204–12.
[69] Yao DK, Maslov K, Shung KK, Zhou Q, Wang LV. In vivo label-free photoacoustic microscopy of cell nuclei by excitation of DNA and RNA. Opt Lett 2010;35(24):4139–41.
[70] Campagnola PJ, Loew LM. Second-harmonic imaging microscopy for visualizing biomolecular arrays in cells, tissues and organisms. Nat Biotechnol 2003;21(11):1356–60.
[71] Pantazis P, Maloney J, Wu D, Fraser SE. Second harmonic generating (SHG) nanoprobes for in vivo imaging. Proc Natl Acad Sci USA 2010;107(33):14535–40.
[72] Hellwarth R, Christensen P. Nonlinear optical microscope using second harmonic generation. Appl Opt 1975;14(2):247–8.
[73] Sheppard C, Gannaway J, Kompfner R, Walsh D. The scanning harmonic optical microscope. Quan Electron IEEE J 1977;13(9):912.
[74] Freund I, Deutsch M. Macroscopic polarity of connective tissue is due to discrete polar structures. Biopolymers 1986;25(4):601–6.
[75] Stoller P, Reiser KM, Celliers PM, Rubenchik AM. Polarization-modulated second harmonic generation in collagen. Biophys J 2002;82(6):3330–42.
[76] Cox G, Kable E, Jones A, Fraser I, Manconi F, Gorrell MD. 3-Dimensional imaging of collagen using second harmonic generation. J Struct Biol 2003;141(1):53–62.
[77] Williams RM, Zipfel WR, Webb WW. Interpreting second-harmonic generation images of collagen I fibrils. Biophys J 2005;88(2):1377–86.
[78] Keikhosravi A, Bredfeldt JS, Sagar AK, Eliceiri KW. Second-harmonic generation imaging of cancer. Methods Cell Biol 2014;123:531–46.
[79] Mansfield JR, Hoyt C, Levenson RM. Visualization of microscopy-based spectral imaging data from multi-label tissue sections. Curr Protoc Mol Biol 2001 [chapter 14] John Wiley & Sons, Inc.
[80] Lin SJ, Jee SH, Kuo CJ, et al. Discrimination of basal cell carcinoma from normal dermal stroma by quantitative multiphoton imaging. Opt Lett 2006;31(18):2756–8.
[81] Zhuo S, Chen J, Jiang X, Cheng X, Xie S. Visualizing extracellular matrix and sensing fibroblasts metabolism in human dermis by nonlinear spectral imaging. Skin Res Technol 2007; 13(4):406–11.
[82] Provenzano PP, Eliceiri KW, Campbell JM, Inman DR, White JG, Keely PJ. Collagen reorganization at the tumor-stromal interface facilitates local invasion. BMC Med 2006;4(1):38.
[83] Conklin MW, Eickhoff JC, Riching KM, et al. Aligned collagen is a prognostic signature for survival in human breast carcinoma. Am J Pathol 2011;178(3):1221–32.
[84] Xiong SY, Yang JG, Zhuang J. Nonlinear spectral imaging of human normal skin, basal cell carcinoma and squamous cell carcinoma based on two-photon excited fluorescence and second-harmonic generation. Laser Phys 2011;21(10):1844–9.
[85] Zhuo S, Chen J, Xie S, Hong Z, Jiang X. Extracting diagnostic stromal organization features based on intrinsic two-photon excited fluorescence and second-harmonic generation signals. J Biomed Opt 2009;14(2):020503.
[86] Berezin MY, Achilefu S. Fluorescence lifetime measurements and biological imaging. Chem Rev 2010;110(5):2641–84.
[87] Konig K, So PT, Mantulin WW, Tromberg BJ, Gratton E. Two-photon excited lifetime imaging of autofluorescence in cells during UVA and NIR photostress. J Microsc 1996;183(Pt 3):197–204.
[88] Galletly NP, McGinty J, Dunsby C, et al. Fluorescence lifetime imaging distinguishes basal cell carcinoma from surrounding uninvolved skin. Br J Dermatol 2008;159(1):152–61.
[89] Chen Y, Periasamy A. Characterization of two-photon excitation fluorescence lifetime imaging microscopy for protein localization. Microsc Res Tech 2004;63(1):72–80.
[90] Weinigel M, Breunig HG, Uchugonova A, Konig K. Multipurpose nonlinear optical imaging system for in vivo and ex vivo multimodal histology. J Med Imaging (Bellingham) 2015;2(1):016003.
[91] Pelet S, Previte MJ, Laiho LH, So PT. A fast global fitting algorithm for fluorescence lifetime imaging microscopy based on image segmentation. Biophys J 2004;87(4):2807–17.
[92] Patalay R, Talbot C, Alexandrov Y, et al. Multiphoton multispectral fluorescence lifetime tomography for the evaluation of basal cell carcinomas. PLoS One 2012;7(9):e43460.
[93] Jonkman J, Brown CM, Cole RW. Chapter 7, Quantitative confocal microscopy: beyond a pretty picture. In: Jennifer CW, Torsten W, editors. Methods in cell biology. Academic Press; 2014. p. 113–34.
[94] Zoumi A, Yeh A, Tromberg BJ. Imaging cells and extracellular matrix in vivo by using second-harmonic generation and two-photon excited fluorescence. Proc Natl Acad Sci USA 2002; 99(17):11014–9.
[95] Zipfel WR, Williams RM, Webb WW. Nonlinear magic: multiphoton microscopy in the biosciences. Nat Biotechnol 2003; 21(11):1369–77.

[96] Konig K, Riemann I. High-resolution multiphoton tomography of human skin with subcellular spatial resolution and picosecond time resolution. J Biomed Opt 2003;8(3):432–9.

[97] Dimitrow E, Riemann I, Ehlers A, et al. Spectral fluorescence lifetime detection and selective melanin imaging by multiphoton laser tomography for melanoma diagnosis. Exp Dermatol 2009; 18(6):509–15.

[98] Matthews TE, Piletic IR, Selim MA, Simpson MJ, Warren WS. Pump-probe imaging differentiates melanoma from melanocytic nevi. Sci Translational Med 2011;3(71):71–115.

[99] Robles FE, Wilson JW, Fischer MC, Warren WS. Phasor analysis for nonlinear pump-probe microscopy. Opt Express 2012; 20(15):17082–92.

[100] Simpson MJ, Wilson JW, Robles FE, et al. Near-infrared excited state dynamics of melanins: the effects of iron content, photodamage, chemical oxidation, and aggregate size. J Phys Chem A 2014;118(6):993–1003.

[101] Simpson MJ, Wilson JW, Phipps MA, Robles FE, Selim MA, Warren WS. Nonlinear microscopy of eumelanin and pheomelanin with subcellular resolution. J Invest Dermatol 2013;133(7): 1822–6.

[102] Robles FE, Deb S, Wilson JW, et al. Pump-probe imaging of pigmented cutaneous melanoma primary lesions gives insight into metastatic potential. Biomed Opt Express 2015;6(9): 3631–45.

[103] Ellis DI, Cowcher DP, Ashton L, O'Hagan S, Goodacre R. Illuminating disease and enlightening biomedicine: Raman spectroscopy as a diagnostic tool. The Analyst 2013;138(14):3871–84.

[104] Zhang G, Moore DJ, Sloan KB, Flach CR, Mendelsohn R. Imaging the prodrug-to-drug transformation of a 5-fluorouracil derivative in skin by confocal Raman microscopy. J Invest Dermatol 2007;127(5):1205–9.

[105] Chrit L, Bastien P, Biatry B, et al. In vitro and in vivo confocal Raman study of human skin hydration: assessment of a new moisturizing agent, pMPC. Biopolymers 2007;85(4):359–69.

[106] Ermakov IV, Ermakova MR, Gellermann W, Lademann J. Noninvasive selective detection of lycopene and beta-carotene in human skin using Raman spectroscopy. J Biomed Opt 2004;9(2): 332–8.

[107] Jacobi U, Weigmann HJ, Ulrich J, Sterry W, Lademann J. Estimation of the relative stratum corneum amount removed by tape stripping. Skin Res Technol 2005;11(2):91–6. Official Journal of International Society for Bioengineering and the Skin (ISBS) [and] International Society for Digital Imaging of Skin (ISDIS) [and] International Society for Skin Imaging (ISSI).

[108] Xiao C, Moore DJ, Rerek ME, Flach CR, Mendelsohn R. Feasibility of tracking phospholipid permeation into skin using infrared and Raman microscopic imaging. J Invest Dermatol 2005;124(3):622–32.

[109] Caspers PJ, Lucassen GW, Carter EA, Bruining HA, Puppels GJ. In vivo confocal Raman microspectroscopy of the skin: noninvasive determination of molecular concentration profiles. J Invest Dermatol 2001;116(3):434–42.

[110] Ashtikar M, Matthaus C, Schmitt M, Krafft C, Fahr A, Popp J. Non-invasive depth profile imaging of the stratum corneum using confocal Raman microscopy: first insights into the method. Eur J Pharm Sci 2013;50(5):601–8. Official Journal of the European Federation for Pharmaceutical Sciences.

[111] Antille C, Tran C, Sorg O, Saurat JH. Penetration and metabolism of topical retinoids in ex vivo organ-cultured full-thickness human skin explants. Skin Pharmacol Physiol 2004; 17(3):124–8.

[112] Vogt A, Rancan F, Ahlberg S, et al. Interaction of dermatologically relevant nanoparticles with skin cells and skin. Beilstein J Nanotechnol 2014;5:2363–73.

[113] de Jongh CM, Jakasa I, Verberk MM, Kezic S. Variation in barrier impairment and inflammation of human skin as determined by sodium lauryl sulphate penetration rate. Br J Dermatol 2006; 154(4):651–7.

[114] van der Valk PG, Nater JP, Bleumink E. Skin irritancy of surfactants as assessed by water vapor loss measurements. J Invest Dermatol 1984;82(3):291–3.

[115] Torma H, Lindberg M, Berne B. Skin barrier disruption by sodium lauryl sulfate-exposure alters the expressions of involucrin, transglutaminase 1, profilaggrin, and kallikreins during the repair phase in human skin in vivo. J Invest Dermatol 2008; 128(5):1212–9.

[116] Mao G, Flach CR, Mendelsohn R, Walters RM. Imaging the distribution of sodium dodecyl sulfate in skin by confocal Raman and infrared microspectroscopy. Pharm Res 2012;29(8): 2189–201.

[117] Essendoubi M, Gobinet C, Reynaud R, Angiboust JF, Manfait M, Piot O. Human skin penetration of hyaluronic acid of different molecular weights as probed by Raman spectroscopy. Skin Res Technol 2015;22(1):55–62. Official Journal of International Society for Bioengineering and the Skin (ISBS) [and] International Society for Digital Imaging of Skin (ISDIS) [and] International Society for Skin Imaging (ISSI).

[118] Pavicic T, Gauglitz GG, Lersch P, et al. Efficacy of cream-based novel formulations of hyaluronic acid of different molecular weights in anti-wrinkle treatment. J Drugs Dermatol: JDD 2011; 10(9):990–1000.

[119] Duranti F, Salti G, Bovani B, Calandra M, Rosati ML. Injectable hyaluronic acid gel for soft tissue augmentation: a clinical and histological study. Dermatol Surg 1998;24(12):1317–25.

[120] Larraona-Puy M, Ghita A, Zoladek A, et al. Development of Raman microspectroscopy for automated detection and imaging of basal cell carcinoma. J Biomed Opt 2009;14(5):054031.

[121] Rowlands CJ, Varma S, Perkins W, Leach I, Williams H, Notingher I. Rapid acquisition of Raman spectral maps through minimal sampling: applications in tissue imaging. J Biophotonics 2012;5(3):220–9.

[122] Petrov GI, Arora R, Yakovlev VV, Wang X, Sokolov AV, Scully MO. Comparison of coherent and spontaneous Raman microspectroscopies for noninvasive detection of single bacterial endospores. Proc Natl Acad Sci USA 2007;104(19): 7776–9.

[123] Marks DL, Boppart SA. Nonlinear interferometric vibrational imaging. Phys Rev Lett 2004;92(12):123905.

[124] Benalcazar WA, Boppart SA. Nonlinear interferometric vibrational imaging for fast label-free visualization of molecular domains in skin. Anal Bioanal Chem 2011;400(9): 2817–25.

[125] Breunig HG, Buckle R, Kellner-Hofer M, et al. Combined in vivo multiphoton and CARS imaging of healthy and disease-affected human skin. Microsc Res Tech 2012;75(4):492–8.

[126] Liao C-S, Choi JH, Zhang D, Chan SH, Cheng J-X. Denoising stimulated Raman spectroscopic images by total variation minimization. J Phys Chem C 2015;119(33):19397–403.

[127] Zimmermann T, Rietdorf J, Pepperkok R. Spectral imaging and its applications in live cell microscopy. FEBS Lett 2003;546(1): 87–92.

[128] Neher RA, Mitkovski M, Kirchhoff F, Neher E, Theis FJ, Zeug A. Blind source separation techniques for the decomposition of multiply labeled fluorescence images. Biophys J 2009;96(9): 3791–800.

[129] Mansfield JR, Gossage KW, Hoyt CC, Levenson RM. Autofluorescence removal, multiplexing, and automated analysis methods for in-vivo fluorescence imaging. J Biomed Opt 2005; 10(4):41207.

[130] Fereidouni F, Bader AN, Gerritsen HC. Spectral phasor analysis allows rapid and reliable unmixing of fluorescence microscopy spectral images. Opt Express 2012;20(12):12729–41.

[131] Fereidouni F, Bader AN, Colonna A, Gerritsen HC. Phasor analysis of multiphoton spectral images distinguishes autofluorescence components of in vivo human skin. J Biophotonics 2014; 7(8):589–96.

C H A P T E R

19

Applications of Multiphoton Microscopy in Dermatology

P.T.C. So[1,2,3], E. Yew[3], C. Rowlands[1]

[1]Massachusetts Institute of Technology; [2]Laser Biomedical Research Center; [3]Singapore-MIT Alliance for Science and Technology (SMART) Center

O U T L I N E

INTRODUCTION

Skin of human and many common animal models is comparatively thin with dermal–epidermal junction (DEJ) positioning within about 100 μm from the surface. The shallow tissue depth allows relatively easy imaging of skin structures based on ballistic photons [5]. Among different optical imaging modalities, the use of multiphoton microscopy to study skin physiology was accomplished in 1997 [6]. The use of multiphoton microscopy in dermatological studies has steadily increased since then (Fig. 19.1) with applications ranging from skin cancer diagnosis to the study of drug and nanoparticle delivery.

The application of multiphoton microscopy in dermatology has been previously reviewed. Schenke-Layland et al. provide an extensive summary up to 2006 [7] and there is a shorter review by Lin, Jee, and Dong with a more clinical perspective [5]. More recently, Wang, König, and Halbhuber published a review focused primarily on ophthalmology but with a substantial section on skin [8]. In another work, König

Imaging in Dermatology
http://dx.doi.org/10.1016/B978-0-12-802838-4.00019-4

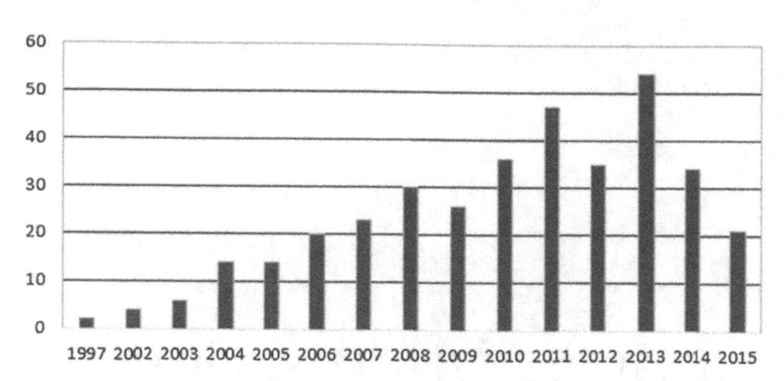

FIGURE 19.1 Number of journal publications of multiphoton microscopy imaging in the dermatology area.

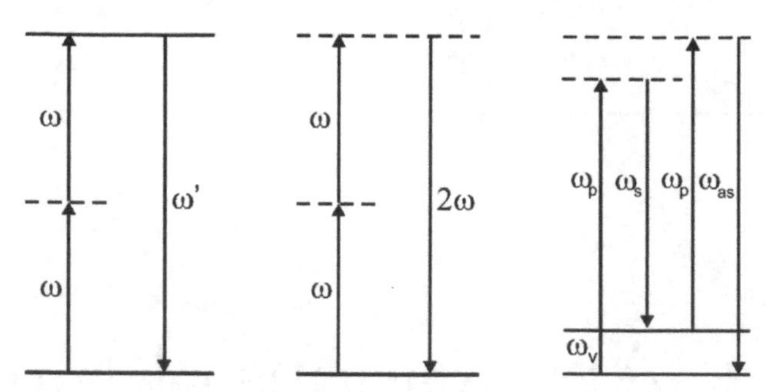

FIGURE 19.2 Many nonlinear optical processes provide different contrast mechanisms for multiphoton microscopy in dermatological imaging. Left: two-photon excitation (TPE), where two excitation photons couple ground state with an electronic excited state *(solid line)*. Center: second harmonic generation (SHG), where two excitation photons couple ground state with a virtual excited state *(dotted line)*. Right: coherent anti-Stokes Raman scattering (CARS), where ground state and a vibrational excited state are coupled via several photons and virtual excited states. ω; ω'; ω_{as}; ω_p; ω_s; ω_v.

et al. discussed the use of multiphoton imaging in drug delivery research, with emphasis on the treatment of photoaging [9]. The use of multiphoton fluorescence lifetime imaging microscopy for skin imaging has attracted recent interest, with reviews by Cicchi and Pavone [10], Seidenari et al. [11], and König [12]. Campagnola and Dong have summarized the use of second harmonic generation (SHG) in disease diagnosis, including its use in dermatology [13]. Perry, Burke, and Brown have published a review on the use of multiphoton imaging in cancer research, which contains a lengthy discussion on skin cancers [14]. We published a brief review of this field in 2013 and this chapter is an expanded version of that work [15].

This chapter will cover basic theories underlying multiphoton spectroscopy and microscopic imaging. We will discuss how skin physiology impacts multiphoton imaging with emphasis on endogenous contrast agents and photodamage mechanisms. We will subsequently present some of the most important variants of multiphoton microscopy designs. Finally, we will review a selection of popular applications of multiphoton imaging in dermatology: skin cancers, other skin diseases, skin immunological processes, aging, and regeneration, and the transport of molecules and nanoparticles through the skin.

MULTIPHOTON MICROSCOPY SPECTROSCOPY AND IMAGING THEORY FUNDAMENTALS

Multiphoton Spectroscopy

A variety of nonlinear optical processes may be utilized as contrast mechanisms in multiphoton microscopy. Excellent textbooks on nonlinear spectroscopy have been published and are indispensable references for researchers who seek to advance multiphoton microscopy technology [16,17]. Some of the most commonly used nonlinear optical processes in microscopy imaging including two-photon fluorescence, SHG, and coherent anti-Stokes Raman scattering (CARS) (Fig. 19.2). It is impossible to exhaustively discuss the spectroscopic properties of all nonlinear optical processes in this review; we will focus on the most commonly used modality, two-photon excitation (TPE) fluorescence, and other modalities will be introduced as needed in subsequent sections. This review of TPE spectroscopy will highlight many relevant spectroscopic properties that are often common among many nonlinear optical processes.

One-photon linear fluorescence excitation involves a fluorescent molecule absorbing a single excitation photon that promotes the molecule from the lowest vibronic level of the electronic ground state to a vibronic level in the first electronic excited state (Fig. 19.3). After thermal relaxation on a picosecond time scale, the excited molecule loses energy and relaxes to the lowest vibronic level of the electronic excited state. The excited state lifetime of most fluorophores is on the order of several nanoseconds. Afterwards, the excited molecule may return to a vibronic level in the ground electronic state with the emission of a fluorescence photon through a radiative process. The molecule again rapidly thermalizes and reaches the lowest vibronic level of the ground electronic state again. Because thermal relaxation processes are involved, the fluorescence photon always has lower energy than the excitation photon. This observation is called the *Stokes shift* and is important for imaging as even weak fluorescence signal can be easily observed in the presence of strong excitation light by the virtual of the color difference. The TPE process is similar, except the molecule is excited by the almost

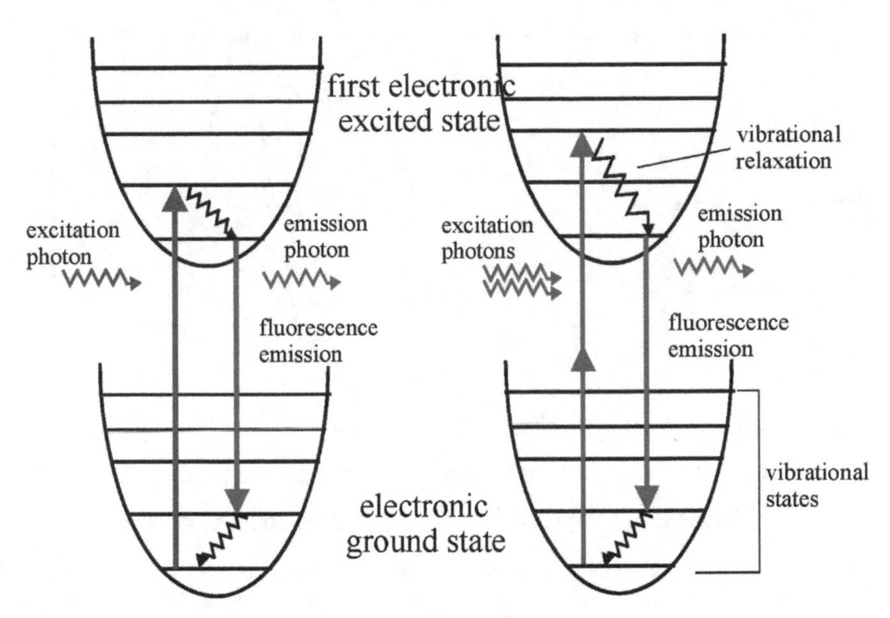

FIGURE 19.3 Detailed Jablonski diagrams depicting one-photon fluorescence excitation (left) and two-photon fluorescence excitation (right) processes. *Wiggled black arrows* indicate thermal vibrational relaxation processes.

simultaneous absorption of two lower-energy infrared photons via a virtual intermediate state. Because the virtual intermediate state has a short lifetime, on the order of femtoseconds, the two infrared photon must be almost simultaneously present to interact with the molecule requiring the use of ultrafast lasers. The ultrafast light sources for TPE are typically in the near-infrared spectral range that is approximately twice as long as the wavelength range of one-photon excitation light sources.

Two-Photon Excitation Fluorescence Microscopy

An important feature of TPE is that it is three-dimensional (3D), localized at the focus of a high numerical aperture (NA) objective. This localization is a result of TPE requiring simultaneous absorption of two excitation photons and the probability of TPE per molecule, $P_{2\gamma}(t)$, is proportional to the square of instantaneous excitation photon flux, $I_{ex-1\gamma}(t)$. For comparison, the probability of one-photon excitation, $P_{1\gamma}(t)$ is proportional to the flux of the excitation light:

$$P_{1\gamma}(t) \propto I_{ex-1\gamma}(t),$$

$$P_{2\gamma}(t) \propto \left[I_{ex-1\gamma}(t)\right]^2.$$

Using the diffraction theory of light, one can theoretically calculate the optical intensity distribution in the vicinity of the focal point of a microscope objective. Given the quadratic TPE probability, the fluorescence emission distribution goes as the square of the excitation light intensity distribution. The volumetric shape of TPE fluorescence, the two-photon point spread function (PSF), along the radial and axial directions are

$$F_{2\gamma}(v, u=0) = \left[\frac{J_1(v)}{v}\right]^4 \quad \text{where } v = \frac{2\pi \text{NA}}{\lambda} r,$$

$$F_{2\gamma}(v=0, u) = \left[\frac{\sin(u/4)}{(u/4)}\right]^4 \quad \text{where } u = \frac{2\pi \text{NA}^2}{\lambda} z,$$

where v and u are the generalized coordinates for r and z respectively, λ is the wavelength of light, and NA is the numerical aperture. The full width at half maximum (FWHM) along the radial direction is inversely proportional to the NA and the FWHM along the axial direction is inversely proportional to the square of the NA.

For 800-nm excitation light focused with a 1.0-NA objective, the resultant fluorescence emission PSF has FWHM values of 0.3 and 1.0 μm in radial and optical axes, respectively (Fig. 19.4). The volume of TPE fluorescence is typically a fraction of a femtoliter. TPE has depth discrimination along the axial direction because the quadratic relationship suppresses TPE from out-of-focus regions. For a uniform specimen, over 80% of the fluorescence is emitted within a 1-μm thick slab at the focal plane. For comparison, in one-photon fluorescence, the total fluorescence emitted is constant for each axial section because one-photon excitation probability is proportional to the number of excitation photons crossing each axial section, and that is a constant in the absence of absorption.

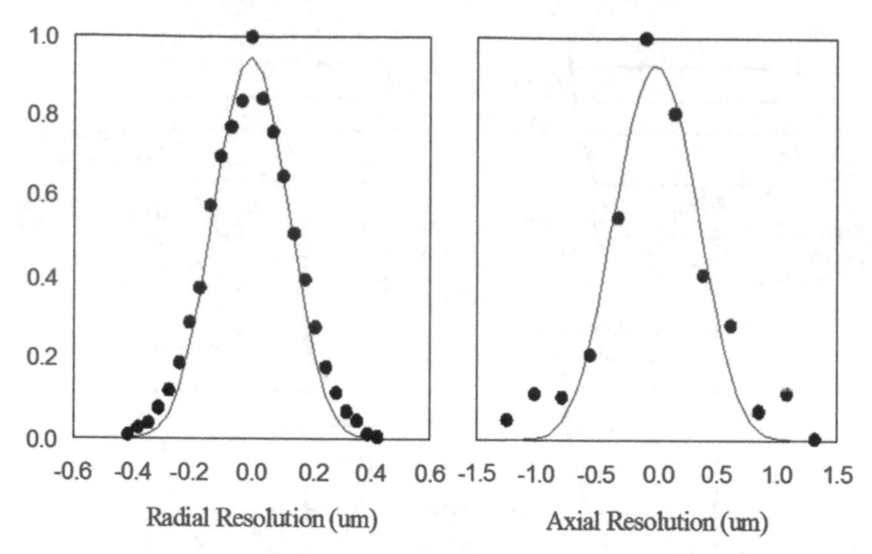

FIGURE 19.4 Point spread function (PSF) of two-photon excitation (TPE) fluorescence for 800-nm excitation and 1.0 numerical aperture (NA) objective. *Y*-axis corresponds to intensity in arbitrary units *(a.u.)* normalized to unity. Left and right figures correspond to radial and axial distributions. The *black data points* are experimental measurement, whereas the *blue line* corresponds to theoretical prediction.

TPE requires high excitation flux on the order of GW/μm^2 within the excitation volume because it is a third-order nonlinear process. This high photon flux is typically achieved based on temporally localizing light intensity using an ultrafast laser with a pulse train of femtosecond pulses and spatially localizing light intensity using a high NA objective.

The rate of TPE per molecule, $fl_{2\gamma}(t)$ [photon/s/molecule] is proportional to the square of excitation photon flux, $I(t)$ [photon/cm^2/s] and its coefficient is the two-photon absorption cross-section of the fluorophore, δ_a [photon/cm^4/s]:

$$fl_{2\gamma}(t) = \delta_a I(t)^2 = \delta_a \left[\frac{\gamma(t)}{A}\right]^2.$$

The excitation photon flux, $I(t)$, is equal to the photon arrival rate, $\gamma(t)$ [photon/s] divided by the excitation volume cross-section area, A [cm^2] that the photons pass through. The two-photon absorption cross-section, δ_a, is a characteristic of fluorophores with values for typical organic fluorophores on the order of 1–30 Göppert-Mayer (GM) units, where 1 GM corresponds to 10^{-50} photon/cm^4/s. At the focal plane of the excitation focus, the excitation cross-section area is inversely proportional to the square of the NA of light that is typically near unity for efficient multiphoton excitation:

$$A \propto \frac{1}{NA^2}.$$

Therefore the two-photon fluorescence rate of a molecule is

$$fl_{2\gamma}(t) \propto \delta_a NA^4 \gamma(t)^2.$$

The photon arrival rate, $\gamma(t)$, can be greatly increased by the usage of the femtosecond pulsed lasers generating a continuous train of pulses, in contrast to the more common continuous wave (CW) laser that has a constant intensity.

In the case of CW laser, photon rate $\gamma_{cw}(t)$ [photon/s] is proportional to its average power, P_0 [W]:

$$\gamma_{cw}(t) = P_0.$$

In case of pulsed laser with a pulse width, τ [s] and a pulse repetition rate, c [1/s], the photon arrival rate for each pulse period, $\gamma_p(t)$, can be approximately related to the average power as

$$\gamma_p(t) = \frac{P_0}{\tau f} \quad \text{for } 0 < t < \tau,$$

$$\gamma_p(t) = 0 \quad \text{for } \tau < t < \frac{1}{f}.$$

For a pulse laser, the photon rate during the pulse is boosted by a factor of $\frac{1}{\tau f}$ during the pulse. One may compare the fluorescence signal per unit time induced by CW versus pulse lasers:

$$fl_{2\gamma,cw} \propto \delta_a NA^4 P_0^2,$$

$$fl_{2\gamma,p} \propto \delta_a NA^4 f \int_0^{\frac{1}{f}} \left[\gamma_{pulse}(t)\right]^2 dt = \delta_a NA^4 f\tau \left[\frac{P_0}{\tau f}\right]^2 = \frac{\delta_a NA^4 P_0^2}{\tau f}.$$

The TPE rate per molecule is increased by factor of $\frac{1}{\tau f}$ with the pulsed laser. For the commonly used titanium-sapphire laser, the pulse width, τ, is 100 fs and the pulse repetition rate, τf, is approximately 100 MHz. The enhancement is approximately 10^5 times the pulsed laser compared with the CW laser.

Because TPE occurs only during the pulse, also important to calculate the probability of TPE for each pulse, P_p, photon/(molecule pulse):

$$P_{\mathrm{p}} \propto \mathrm{fl}_{2\gamma,\mathrm{p}} \frac{1}{f} = \frac{\delta_{\mathrm{a}} \mathrm{NA}^4 P_0^2}{\tau f^2}.$$

In typical configuration, the input laser has average power, P_0, of 10 mW, its wavelength; a pulse width, τ, of 100 fs; and a pulse repetition rate, f, of 100 MHz. The NA of the objective is 1.0 for a typical fluorophore with a two-photon cross-section, δ_a, of 10 GM. The probability of a fluorophore molecule being excited with a single pulse can be calculated to be about 0.16. Because P_p is a probability, it has an upper bound of unity. In fact, it has been shown that excitation power levels, which induce TPE probability much greater than 0.1, will cause broadening of the excitation two-photon PSF because of the depletion of molecules in the ground state, resulting in less than quadratic excitation efficiency [18,19]. Therefore the multiphoton microscope imaging speed is optimized when sufficient power is provided to achieve TPE probability at the focal plane of about 0.1.

From this discussion, we also see that the use of higher NA objective increases the probability of TPE per molecule and improves image resolution. For a uniform specimen, a higher NA objective reduces excitation volume, $V = \frac{1}{\mathrm{NA}^4}$, because the two lateral dimensions of the volume are inversely dependent on NA, whereas the axial dimension depends on inverse square of NA. We see that excitation probability has a dependence on NA as

$$P_{\mathrm{p}} \propto \mathrm{NA}^4.$$

Because the total fluorescence generated from the excitation volume is proportional to $V \cdot P_p$, it is independent of the NA of the objective. However, the fraction of the generated fluorescence signal collected by the objective still depends on the aperture of the objective as NA^2. Therefore using high a NA objective in multiphoton microscopy is essential.

Summary of Advantages of Multiphoton Excitation Microscopy

Given the photophysical properties and imaging properties of multiphoton excitation, multiphoton fluorescence microscopy has a number of advantages:

1. Because multiphoton microscopy uses excitation wavelength in the near-infrared spectral range, the excitation light falls within the optical window of tissue where tissue scattering and absorption are both minimized; therefore multiphoton imaging has better image penetration depth as compared with one-photon excitation.
2. Because optical interaction in a multiphoton microscope is localized to a subfemtoliter volume, it has inherent 3D resolution.
3. It is important that the 3D localization of multiphoton excitation be dependent on the excitation process and not on the detection process. The detection of the fluorescence signal can be accomplished by a large single-pixel detector that is relatively immune to the scattering of the emission photons.
4. The localization of the excitation volume is further important in that photodamage of biological specimens and photobleaching of the fluorophores are restricted to the focal point and the out-of-plane region is not affected.
5. Multiphoton excitation wavelengths are typically red-shifted to about twice the one-photon excitation wavelengths. The wide separation between the excitation and emission spectra ensures that the excitation light can be rejected without filtering out any of the fluorescence photons.

SKIN PHOTOPHYSICAL AND PHOTOCHEMICAL PROPERTIES AND MULTIPHOTON DERMAL IMAGING

The physiology and optical properties of the skin have been studied extensively. Knowledge of the optical properties of the skin allows better design of multiphoton microscopy technologies for dermal imaging.

Skin Physiology and Endogenous Contrast Agents

Skin is a complex organ consisting of two physiologically distinct regions [20]. The epidermis is the region near the skin surface and consists of stratified cell layers. The epidermis is fairly thin, on the order of 50–150 μm for humans, and it can be much thinner for some animals such as mice. The stratum corneum is the outermost layer of the epidermis, composed of five or six layers of cornified dead cells that are in the process of being sloughed off. The stratum corneum forms an important barrier, protecting the organism from the environment. The next two layers are made up of living keratinocytes and are called the *stratum granulosum* and the *stratum spinosum*. The keratinocytes in the stratum spinosum tend to more be polyhedral in shape. The bottom layer of the epidermal region consists of basal cells. Basal cells are cuboid in morphology and they are the germinative cells in the skin, and are constantly in the process of proliferating and differentiating into new cells that migrate toward the surface. The epidermis contains four main cell types. The majority are keratinocytes, with the few remaining percentages being dendritic cells: Langerhans cells, melanocytes, and Merkel cells. Melanocytes produce the pigment melanin, which is formed in vesicles called *melanosomes*. Melanin is not a homogenous chemical species but can be roughly classified as either eumelanin or pheomelanin, each with distinct spectroscopic properties. Some melanosomes are transferred from the melanocytes into the basal epithelial cells. The epidermis also contains other physiological structures, including free nerve endings, hair, sweat, and apocrine glands.

Below the epidermis is the dermal–epidermal junction (DEJ). The DEJ is typically not flat but has an undulating, rete-ridge morphology. The DEJ is lined with the basement membrane and is distinguished by having a high concentration of collagen IV and VII. The basement membrane in DEJ serves to anchor the basal cell stratum in the epidermis to the extracellular matrix structures in the dermis.

The dermal region is located below the DEJ. The dermis consists mostly of extracellular matrix tissue, including collagen and elastin fibers, and a sparse population of cells including fibroblasts, macrophages, and adipocytes. There are also many important functional organs in the dermis, including hair follicles, sweat, sebaceous and apocrine glands, nerve fibers, and their receptors. Studies have shown that the skin stem cells are located in niches close to the base of the hair follicle. These stem cells are responsible for the repair of the skin upon injury. The vasculature and the lymphatic vessels can also be found within the dermis. The capillary loops of the vasculature extending into the tips the rete ridge supply oxygen to the living cells both in the dermis and the epidermis. The thickness of the dermis ranges from hundreds of micrometers to millimeters, with the actual thickness depending on the location on the body and the species of the organism.

Many skin morphological structures can be imaged based on endogenous contrast agents. Keratinocytes and other cells in the epidermis are autofluorescent. Cellular autofluorescence in general is mostly the result of two classes of endogenous fluorophores: reduced pyridine nucleotides {nicotinamide adenine dinucleotide (phosphate) [NAD(P)H]}, and oxidized flavin proteins [flavin adenine dinucleotide (FAD)]. Both FAD and NAD(P)H are localized in the cellular mitochondria, but NAD(P)H is also distributed through the cytosol. These endogenous fluorophores not only allow delineation of epidermal morphology but are also important for monitoring skin metabolism. Over half a century ago, Chance and coworkers showed that cellular metabolism can be noninvasively monitored by redox fluorometry based on NAD(P)H and FAD fluorescence intensity ratios [21,22]. Endogenous fluorescence imaging of the structural protein keratin that is abundant in the stratum corneum has also been reported [23]. Because of the importance of melanoma as a disease, melanin contained in melanocytes and other epidermal cells has also been an important target for multiphoton imaging. A recent study [24] based on pump-probe transient absorption (TA) imaging has demonstrated the distinction between pheomelanin and eumelanin; their relative abundances has been postulated to be an oncogenic factor [25]. Other reports suggest that melanin can be imaged based on a stepwise-multiphoton excitation process [26].

In the dermis, collagen and elastin are also observable based on their fluorescence. However, several isoforms of collagen, including type I, often produce a higher SHG signal. As discussed earlier, SHG is a nonlinear process similar to the two-photon fluorescence process, except that the two excitation photons are coupled via two virtual intermediate states. The emitted SHG photons have energy equal exactly to the sum of excitation photon energies. Because of parity considerations, SHG signal can only be generated by molecules that lack centrosymmetry. Because SHG scattering from multiple molecules is coherent, SHG signal strength is strongly dependent on whether these noncentrosymmetric molecules are organized in quasiregular crystalline structure. In a biological specimen, the number of structures that can generate SHG efficiency are few, but collagen is one of the most important examples. Because collagen I molecules in the dermis lack centrosymmetry and exhibit a crystalline organization, collagen fibers in the dermis can be readily visualized. The combination of the SHG signal with polarization-resolved imaging can further provide information on the elements of the χ^2 susceptibility matrix, which in turn can provide more detailed information regarding the molecular level organization of the collagen fibrils, such as their chirality. It should be noted that SHG is not generated from all isoforms of collagen. It should also be noted that SHG signal is the quadratic function of the number of molecules and their packing. Therefore whereas SHG is a powerful method to qualitative monitor collagen morphology, the magnitude of SHG signal is not a quantitative indicator of collagen molecular concentration in tissue.

Besides endogenous contrast agents, most of the conventional organic or genetic fluorophores can also be used with a multiphoton microscope. The wavelength for TPE is approximately twice the one for single-photon excitation. However, this rule of thumb is not always true for all probes because the quantum mechanical process of TPE is different from that of single-photon excitation. The two-photon cross-sections (δ_a) of good probes, such as fluorescein, rhodamine, and fluorescent proteins, are in the range of 1–50 GM. The only exceptions with much higher two-photon cross-sections, on the order of 10,000 GM, are quantum dots and conjugated polymer nanoparticles [27,28].

The stratified structure of the epidermis has negative impact on optical imaging of skin morphology because different layers have very distinct refractive indices. Tearney and coworkers, [29] using optical coherence tomography, found that the refractive indices in the stratum corneum, epidermis, and dermis have values of 1.51, 1.34, and 1.41, respectively. The stratum corneum has an effective reflective index close to oil, whereas the epidermis has an index close to water, and the dermis' index lies in between. Skin, as a layered structure with different refractive indices, induces in spherical aberration, causing a loss in TPE efficiency and image resolution. One should also note that spherical aberration shifts the actual focal depth away from the location expected in free space, resulting in a distortion of the 3D image cube. In other more homogenous tissues, spherical aberration can be mostly eliminated by choosing a microscope objective that matches the refractive index of the tissue. However, in heterogenous structures such as the skin, index matching is not effective and significant aberration remains, even with the use of objectives equipped with correction collars [30].

Key Multiphoton Photodamage Mechanisms in the Skin

Possible photodamage is another limitation imposed on multiphoton skin imaging by tissue physiology. As discussed previously, the 3D confinement of the multiphoton excitation volume results in reduced photodamage in thick tissue compared with most one-photon modalities such as confocal microscopy. However, at the focal volume where photochemical interactions still occur, multiphoton processes can still cause considerable photodamage. Today, three multiphoton photodamage mechanisms are well recognized:

1. Oxidative photodamage can be caused by two-photon or higher excitation of endogenous and exogenous fluorophores, with a photodamage pathway similar to that of ultraviolet irradiation. These fluorophores act as photosensitizers in photooxidative processes [31,32]. Photoactivation of these fluorophores results in the formation of reactive oxygen species, which trigger the subsequent biochemical damage cascade in cells. Some studies found that the degree of photodamage follows a quadratic dependence on excitation power, indicating that two-photon processes are the primary damage mechanism [33–37]. Flavin-containing oxidases have been identified as one of the primary endogenous targets for photodamage [33]. However, there are also studies that show that oxidative photodamage may also result from higher photon processes [38,39].
2. Photodamage may also be caused by mechanisms resulting from the high peak power of the femtosecond laser pulses. Dielectric breakdown can readily occur at high peak power [35]. Although multiphoton induced dielectric breakdown is a negative for imaging applications, this process has found use in many biotechnology applications [40,41].
3. One- and two-photon absorption of high-power infrared radiation may also produce thermal damage. The temperature change resulting from two-photon absorption has been estimated to be on the order of 1 mK for typical excitation power, and has been shown to be insignificant [42,43]. However, in the presence of a strong infrared absorber such as melanin [44,45], there can be appreciable heating caused by one-photon absorption. Thermal damage has been observed in the basal layer of human skin when irradiated by lasers with high average excitation powers [46]. Masters et al. subsequently performed an in-depth study, firmly establishing one-photon absorption by melanin as the primary photodamage mechanism that limits the maximum power that can be used for skin imaging [47].

Even in the absence of strong absorbers such as melanin, one-photon absorption by water may still present an issue for deep imaging, especially with the recent advent of a deep-imaging approach based on three-photon excitation in the 1600-nm spectral range, where aqueous absorbance is much higher [48].

DESIGN CONSIDERATIONS OF MULTIPHOTON MICROSCOPY

The root of nonlinear optical microscopy lies in the distant past with Maria Göppert-Mayer in the 1930s, [49] when she predicted the theoretical possibility of nonlinear optical excitation. Because of a lack of appropriate light sources, these nonlinear processes were not experimentally verified until 1960s, when high intensity laser light sources were introduced [50–52]. The application of nonlinear optical microscopy and microanalysis in biology and medicine remained essentially dormant until the paper by Denk, Webb, and coworkers in 1990 unequivocally demonstrated the potentials of this exciting new tool

[53]. Although the theoretical and practical foundations of nonlinear microscopy are previously recognized by microscopists such as Sheppard and coworkers [54], a number of factors made 1990s the "right time" for the emergence of this field. The most decisive factor is clearly the invention of robust titanium–sapphire solid state femtosecond lasers at that time. The innovations in laser sources continues to drive this field today. Another less considered technical factor that makes nonlinear optical imaging and spectroscopy feasible for biology and medicine is clearly the availability of high-sensitivity and low-noise photodetectors. Nonlinear processes are inherently "weak" and many biomedical applications of nonlinear optics do not become feasible until measurements with approximately single-photon sensitivity have become routine.

The demonstration of noninvasive imaging of living embryos by Denk and coworkers firmly established multiphoton microscopy as the method of choice for the high-resolution study of optically thick specimens [55]. Piston and coworkers performed the first multiphoton ex vivo tissue study by imaging the structures of rabbit cornea [56]. Given the optical accessibility of the skin, it is not surprising that the second application of multiphoton microscopy in tissue focused on the skin [6]. In this study, Masters and coworkers demonstrated that multiphoton microscopy can image the skin down to a penetration depth of 150 μm, resolving all the stratified cellular layers in the epidermis and partly into the dermis. In the stratum spinosum, granulosum, and the basal layer, living keratinocytes were observed; the cytosol and the mitochondrial structures were imaged using NAD(P)H fluorescence. Cellular nuclei could be seen as circular voids where fluorescent species are mostly absent. Morphological changes of keratinocytes from a cuboidal geometry (in the basal layer) to a flattened geometry (in the stratum spinosum) were observed, consistent with previous histological studies and confocal microscopy. A bright fluorescent signal was also observed in the stratum corneum that may be assigned to keratin today but was not identified in the work of Masters and coworkers. The rete-ridge morphology of the DEJ was also reconfirmed in this study. In the dermis, a signal from extracellular matrix was also clearly observed, showing the characteristic fibrous structure. Finally, it is important to note that this study demonstrated the first in vivo human application of multiphoton microscopy.

Multiphoton microscopy not only allows the 3D imaging of skin morphology with submicron resolution, but the incorporation of spectroscopic measurement further allows quantification of the tissue biochemical state. Excitation/emission spectroscopy and lifetime resolved spectroscopy are two of the most widely used spectroscopic modalities for skin characterization. The utility of both spectroscopic techniques for skin studies were both first demonstrated by Masters and coworkers in the initial multiphoton study of skin [6]. In this study, emission spectra and lifetime decay kinetics were measured at selected points in the skin, but no spectrally-resolved imaging was performed. In the epidermis, the emission spectral measurements and the fluorescence-lifetime resolved measurements both established that NAD(P)H is primarily responsible for the fluorescent signals from living keratinocytes, when excited at approximately 800 nm. In the dermis, the emission spectra show characteristic sharp emission peaks of SHG superimposed on a broad background fluorescence that today are well recognized as the SHG from collagen and the endogenous fluorescence from both collagen and elastin; however, Masters and coworkers did not make the correct spectral assignments in this early study. Subsequently, Laiho and coworkers extended excitation/emission spectroscopy studies in the skin from point measurements to 3D-resolved imaging at several excitation and emission wavelengths [57]. They further applied chemometric analysis in order to identify the principal components responsible for the skin endogenous fluorescent signal, resolving contributions from NAD(P)H, collagen, and elastin that are well accepted today. They also assigned components corresponding to tryptophan and melanin that are less well supported today. With further instrument improvement, Radosevich and coworkers extended excitation/emission spectroscopic measurement to demonstrate that up to eight different luminescence components can be resolved in the skin [58]. The eight components correspond to fluorescence signals from NAD(P)H, collagen, elastin, keratin, sebum, and flavin protein, combined with second harmonic signals originating from collagen and keratin. Subsequently, excitation/emission spectrally resolved multiphoton imaging has been applied in a broad range of skin physiology studies [59–63]. After Masters et al. demonstrated lifetime-resolved spectroscopy at selected points in the skin, multiphoton microscopes with the capability for lifetime-resolved measurements also saw rapid development. Koenig and Riemann demonstrated the first lifetime-resolved imaging of the skin based on a time-correlated single-photon counting (TCSPC) approach [64]. In conjunction with the development of low-cost, easy-to-use TCSPC modules that can be readily integrated into multiphoton microscopes [65], the work of Konig et al. established TCSPC as the method of choice for lifetime-resolved skin imaging. Today, the acquisition of 3D multiphoton image stacks of the skin with lifetime-resolved spectroscopic information at every voxel has become feasible, allowing very sensitive discrimination of tissue physiological and pathological states [66–69]. Other nonlinear optical modalities have also found unique applications in dermal studies. CARS has been shown to be a powerful method to monitor the dermal transport of small drug molecules [70], and imaging based on time-resolved transient pump-probe absorption microscopy allows the detection and discrimination of different melanin species [24,71]. The combination of multiphoton microscopy with other imaging modalities, such as confocal microscopy [72], optical coherence tomography [73,74], and ultrasound imaging [69], has also been shown to be very useful in the study of skin.

The majority of the aforementioned work in multiphoton dermal imaging is based on raster scanning of a single excitation focus to induce two-photon fluorescence emission. Given the prevalence and importance of point-scanning systems, we will review their design issues in detail in the first subsection. Because SHG and third harmonic generation (THG) are also often readily detected simultaneously with fluorescence in the same point-scanning systems, we will combine the discussions of these modalities in the same section while pointing out important differences. Some of the most exciting advances in multiphoton microscopy is the development of systems based on other nonlinear processes, allowing access of different molecular contrast mechanisms including CARS, stimulated Raman scattering (SRS), and TA. All three methods share common pump-probe imaging geometry and we will provide a brief discussion of these important new variants in the next subsection. Finally, improving imaging speed is important for eventual clinical use of multiphoton microscopy methodology. We will close with a final subsection on a review of a high-throughput, temporal-focusing, wide-field design.

Raster Point Scanning Multiphoton Fluorescence, Second Harmonic Generation, and Third Harmonic Generation Microscopes

The typical schematic of a multiphoton fluorescence microscope is shown in Fig. 19.5. The excitation light source is often a mode-locked titanium–sapphire pulse laser with about 100 fs pulse width and 80 MHz pulse repetition rate. The excitation wavelength is tunable from 700 nm to slightly over 1000 nm. Both the polarization and the power of excitation light at the specimen are controlled by a half-wave plate and a Glan-Thomson polarizer. Raster scanning is accomplished via a computer-controlled galvanometric x-y scanner and a piezo-driven objective translator. The excitation beam is coupled into a microscope via a modified epiluminescence light path. A beam expander ensures that the beam overfills the objective's back aperture to achieve diffraction-limited focusing. The scanned excitation beam is reflected by a dichroic filter into the back aperture of the high NA objective and focused into the specimen. The emission light is collected by the same objective. Unlike a typical confocal microscope, the emission light is not descanned to minimize light loss in the detection path. The emitted light is coupled to the detector via a few lenses as relays. The detector used is typically a high-sensitivity photomultiplier tube (PMT). The detector is placed in the conjugate plane of the specimen to make sure that the emission beam location is stationary on the detector. Because scanning only causes the incidence angle of the emission beam to vary at the conjugate plane, this minimizes any image artifacts caused by beam movement on the detector cathode surface that often has nonuniform sensitivity. The PMT can be a readout in either analog or digital mode. In general, we prefer using a digital single-photon counting detector because it minimizes detector electronic noise.

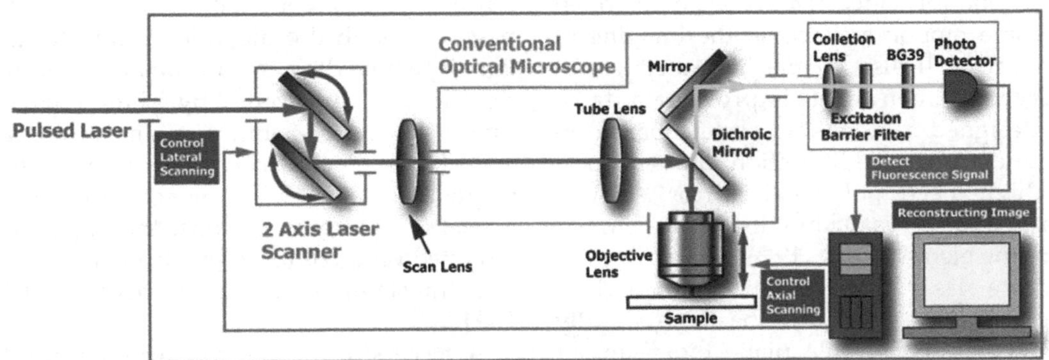

FIGURE 19.5 Schematic diagram of a typical point scanning multiphoton microscope. *BG39*

SHG and THG signal can often be detected in a multiphoton microscope designed for fluorescence. As discussed previously, SHG is produced only in biological crystals of molecules with no centrosymmetry [75]. For skin, SHG allows imaging of important collagen morphology in vivo. THG signal is not molecular-specific but is produced in any interface where there is an index of refraction discontinuity [76,77]. SHG and THG are coherent emission processes; therefore the emission is directional, depending on molecular orientation and distribution, as well as light focusing geometry. Measuring SHG and THG emissions along forward and backward directions provides important information of molecular conformation and organization. For ex vivo imaging, multiphoton microscopes have been constructed with detection paths along both epi- and trans-directions to measure both forward and backward harmonic emission. This geometry is clearly impossible for skin imaging in vivo. Although SHG emission is mostly forwardly directed, epi-detected SHG signal is nonetheless often quite strong because of the emitted photons being multiply scattered in the thick specimen and a substantial fraction of the photons are emitted backward. SHG and THG always occur exactly at one-half or one-third of the excitation wavelength. Because very short emission signal can be heavily absorbed in biological specimens, it is often necessary to shift the excitation wavelength farther into the infrared region, especially in the case of THG. Finally, it should be noted that SHG and THG are instantaneous processes that involve only virtual states. Therefore in the absence of fluorescence absorption, SHG and THG imaging do not induce oxidative damage in the specimen. Finally, because these processes are instantaneous, stronger SHG and THG signals can be induced by increasing excitation intensity up to the limit of specimen damage; in contrast, fluorescence signal strength is always limited by the nanosecond scale lifetime of a typical fluorophore that determines how often the fluorophore can be recycled to produce fluorescent light.

The incorporation of video-rate capacity is often useful to overcome motion artifacts in the imaging of patients and animals. It also enable high-throughput imaging of a large specimen area to improve data statistics. We have designed a video-rate system based on raster scanning of a single-diffraction limited focal spot using a high-speed polygonal mirror. A two-photon microscope can be easily modified to accommodate rapid scanning elements (Fig. 19.6). A femtosecond titanium–sapphire laser beam is used to induce two-photon fluorescence. The laser beam is rapidly raster-scanned across the sample plane by means of two different scanners. A fast-rotating polygonal mirror accomplishes high-speed line-scanning (x-axis) and a slower galvanometer-driven scanner with about 1-kHz bandwidth correspondingly deflects the line-scanning beam along the sample's y-axis. The spinning disc of the polygonal mirror is comprised of metal mirror facets arranged contiguously around the perimeter of the disc. The facets repetitively deflect the laser beam over a specific angular range, correspondingly scanning multiple lines every revolution; the number of lines scanned is equal to the number of mirror facets. The use of polygonal or resonant scanners can improve imaging speed to a little over video rate. However, because the use of a faster scanner reduces pixel residence time, the gain in imaging speed is exactly traded off with a reduction in image signal-to-noise ratio (SNR). In this system, two lenses between the scanners function jointly as a relay element that projects the excitation beam deflected by the polygonal mirror onto a stationary point at the center of the y-axis scan mirror. The microscope is placed so that its telecentric plane intersects with the stationary point at the y-axis scan mirror. The laser beam enters the epiluminescence light path of a microscope. The beam is reflected by the dichroic mirror towards the objective, and is focused into the specimen. The induced fluorescence signal is collected by the same objective and passes through the dichroic mirror. Residual scattered light is removed by an additional barrier filter. The fluorescence can be recorded by a high-sensitivity imaging camera or a PMT running in analog mode. In general, a PMT is preferred because it is insensitive to image degradation caused by the scattering of emitted photons. For 3D volume scans, the objective is typically mounted on a computer-controlled piezoelectric objective translator with an approximate bandwidth of 1 kHz.

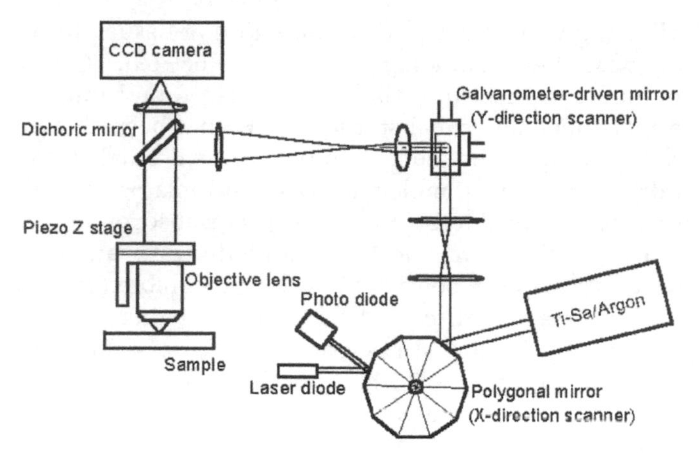

FIGURE 19.6 Design of a two-photon video rate microscope. *CCD*, Charge-coupled device; *Ti:Sa*, titanium:sapphire.

A successful method to overcome speed and SNR is the use of multifocal excitation. By parallelizing the excitation and detection process, fully utilizing the

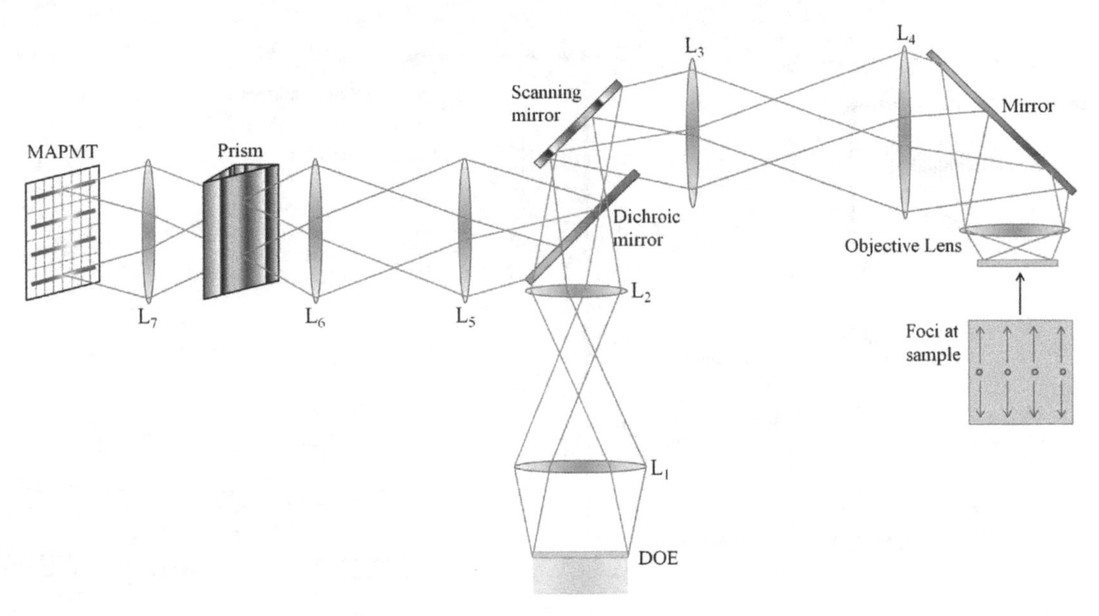

FIGURE 19.7 Schematic diagram of a typical spectral-resolved multifocal multiphoton microscope (MMM). *DOE; L; MAPMT.*

available power of the excitation laser, speed improvement over point scanning by almost 100 times can be achieved with negligible reduction in SNR. Many generations of multifocal multiphoton microscopes (MMMs) have been designed [78–81]. Fig. 19.7 shows the optical configuration of a spectral-resolved MMM. A femtosecond light source was used. The excitation laser beam was split into 8 × 1 beamlets with a diffractive optical element. The beamlets were sent to the scanning mirrors via relay lenses in a 4F configuration and scanned together. The beamlets were expanded to slightly overfill the back aperture of the objective lens. The 8 × 1 excitation foci were generated in a sample and raster scanned to cover the field of view. The emission photons were collected by the same objective lens, delivered along the same optical path as the excitation beam path, and descanned by the scanning mirrors. The stationary emission light after descanning was separated from the excitation laser light by a dichroic mirror into the detection light path. For this spectral resolved design, the emitted light is spectrally separated by a prism pair with antireflection coating. For the spectral decomposition, a diffraction grating is an alternative choice. However, the grating generally suffers from more light loss and significant variation of diffraction efficiency across the spectral range. In contrast, a prism has greater efficiency and more uniform sensitivity over a broader spectral range. However, prisms are less dispersive than gratings. Therefore in order to achieve the designed spectral dispersion power in the instrument we have utilized a pair of prisms. The spatially separated and spectrally dispersed emission light was focused onto a multianode PMT array. A barrier filter was installed in front of the multianode PMT array to completely eliminate the excitation light and pass only the emission photons in the designed spectral range. The multianode PMT has 8 × 8 channels, therefore providing eight simultaneous spectral channels and allowing eight foci to be scanned in parallel for 8 times speed improvement over single focus scanning. For nonspectral resolved imaging, 8 × 8 foci can be generated to achieve almost 64 times speed improvement.

Nonlinear Raman Microscopy for Skin Imaging

Multiphoton CARS microscopy, SRS microscopy, and TA microscopy all share a common design theme based on pump-probe excitation processes. Spontaneous Raman scattering microscopy is a powerful label-free bioimaging technique that can measure the specific vibronic signatures of molecules. However, spontaneous Raman signal is often weak, resulting in very slow imaging speed. CARS microscopy and SRS microscopy were developed to provide higher sensitivity and higher speed imaging. Both methods are based on multiphoton-coherent driven transition and hence have significantly higher emission rates than spontaneous process [82–86]. CARS emission is relatively easy to detect because the signal photons have different color from the pump and probe spectra. However, the CARS spectrum is different from its corresponding spontaneous Raman spectrum because of the presence of a nonresonant background that complicates spectral assignment, causes difficulties in image interpretation, and limits detection sensitivity [87]. SRS also provides a much stronger signal than spontaneous Raman scattering and it does not have a

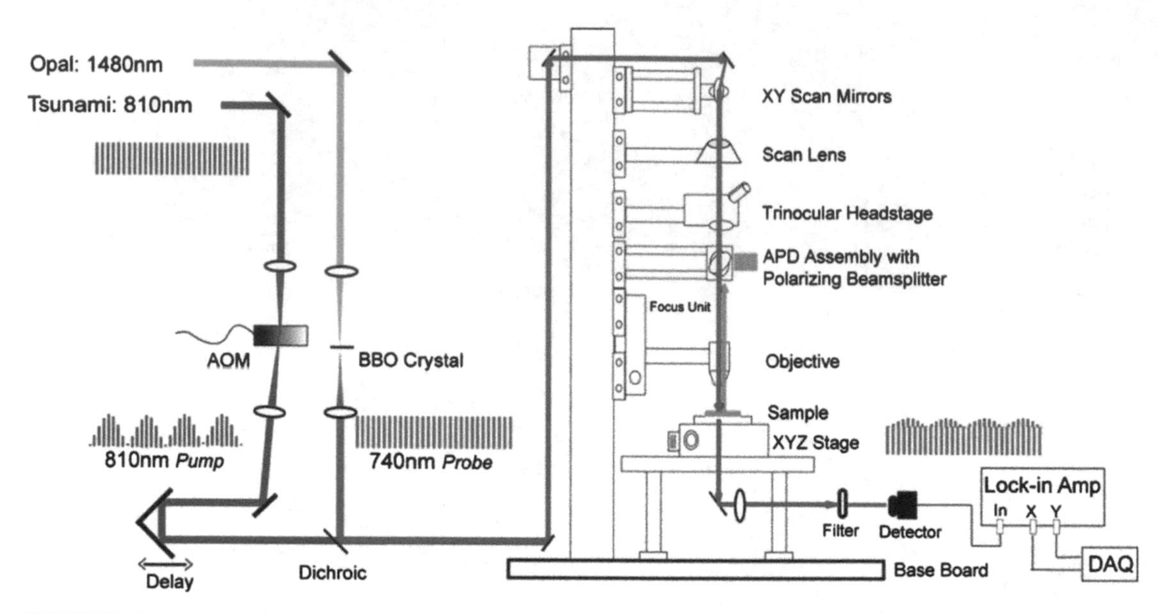

FIGURE 19.8 Schematic diagram of a pump-probe transient absorption (TA) microscope. *AOM; APD; BBO; DAQ.*

nonresonant background, providing an accurate vibronic spectrum. The difficulty in SRS measurement lies in the presence of large background that comes from the strong excitation-beam background that must be rejected by signal processing. A TA microscope is also a pump-probe technique and is useful for quantifying electronic state relaxation of molecules that has very low radiative decay probability.

A representation pump-probe TA microscope is shown in Fig. 19.8 [24,88]. For nonfluorescent absorbers such as many melanins, they can be imaged based on their absorbance. Absorbance is measured by using two excitation lasers. Ignoring the possible presence of other excited-state processes, one laser provides the pump excitation, transiently depleting the ground state of the molecule and resulting in a very small increase in the transient transmission of the probe beam. For TA measurements, the pump and probe beams may either have the same color or have different colors. In this case, the pump beam is provided by a titanium-sapphire oscillator, whereas the probe beam at a slightly different wavelength is provided by a frequency-doubled optical parametric oscillator (OPO). Given the typical abundance of absorbers in the excitation volume, the TA signal corresponds to an increase that is typically less than 10^{-5} of the average probe beam intensity. In order to separate this miniscule TA signal from the background probe beam, the pump beam is intensity modulated with an acoustooptical modulator at a high frequency. The modulation frequency is important and is typically chosen to be well above $1/f$ noise of the lasers, on the order of 2 MHz in this case. Bandwidth-narrowing detection allows the TA signal to be isolated from the probe-beam background using a fast lock-in amplifier. The rest of the pump-probe TA microscope design is similar to that of a point-scanning multiphoton fluorescence microscope.

The design of a CARS microscope is substantially similar to that of a TA microscope in some ways (Fig. 19.9) [89]. The pump and probe beams are again provided by a titanium–sapphire oscillator and a frequency-doubled OPO. Unlike the TA system, these lasers are operated in picosecond mode because of the high spectral resolution requirement of Raman imaging, requiring narrower bandwidth picosecond lasers to be used. Because CARS signal has a different color than pump and probe beams, the detection of CARS signal requires no high sensitivity lock-in scheme but can be easily isolated with spectral selection using a high-quality, sharp-edge interference filter. The rest of the system is the fast polygonal-mirror–based video-rate scanning microscope that was described extensively in the last subsection. The design of the SRS microscope essentially combines the picosecond excitation lasers of the CARS microscope with the high-sensitivity lock-in detection system of the TA microscope and will not be described in detail in this chapter (Fig. 19.10) [87].

Temporal Focusing Wide-Field Multiphoton Microscopy

A relatively new approach for high-throughput imaging that has not yet been tested for dermatological applications is 3D-resolved wide-field imaging based on temporally focused multiphoton excitation. Fig. 19.11 demonstrates the underlying concepts behind spatial

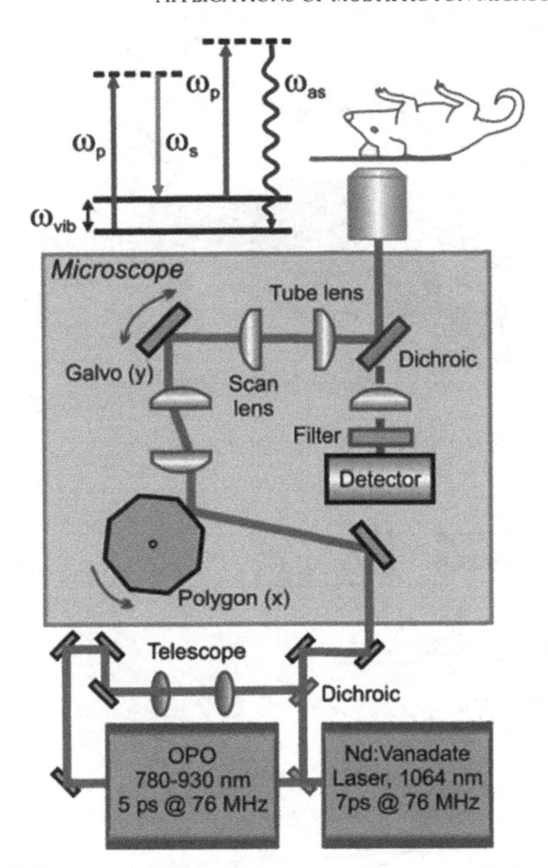

FIGURE 19.9 Schematic diagram of a coherent anti-Stokes Raman scattering (CARS) microscope. *Nd*, Neodymium-doped; *OPO*, optical parametric oscillator; ω_{as}; ω_p; ω_s; ω_{vib}.

focusing and temporal focusing. In Fig. 19.11(A), the optical pulse is focused laterally in the spatial dimension, traveling along in the axial direction. Note that its temporal pulse width is kept constant. The intensity at the focal spot reaches a maximum. A nonlinear optical process such as two-photon absorption is proportional to the power of the intensity, resulting in

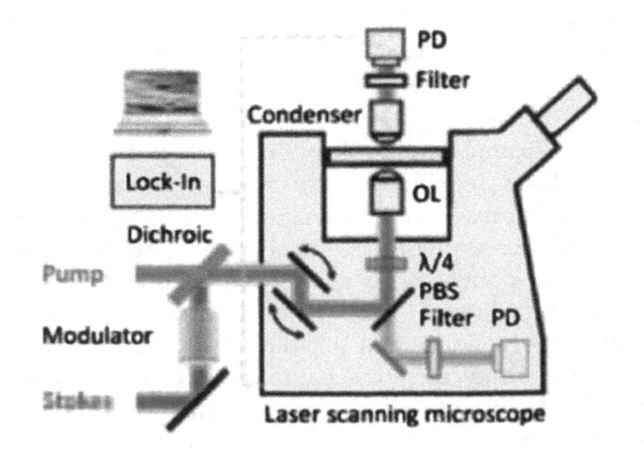

FIGURE 19.10 Schematic diagram of a stimulated Raman scattering (SRS) microscope. *OL*; *PBS*; *PD*; λ, wavelength of light.

optical sectioning. For a high NA objective, submicron lateral and axial resolution can be achieved. With temporal focusing, the optical pulse travels in the axial direction without changing the beam diameter, unlike spatial focusing [see Fig. 19.11(B)] [90,91]. However, the instantaneous intensity will be maximized at the focal plane if the temporal width of the optical pulse can be manipulated spatially such that it is minimized at the focal plane. This approach allows wide-field imaging with depth discrimination.

The trick to generating temporal focusing is to control the temporal width of ultrafast pulses. The temporal width (τ) of ultrafast pulse is related to its spectral bandwidth ($\Delta\lambda$); the time-bandwidth product ($\tau\Delta\lambda$) is a constant for transform-limited pulses. Therefore the generation of broadened pulses can be accomplished by spectrally limiting their bandwidth. A very simple geometry to accomplish this task has been proposed by Silberberg and coworkers [91] (Fig. 19.12). In case of the aberration-free lens, optical path lengths from the back focal plane to the front focal plane are same as in a 4F system, independent of the incidence angle of the ray relative to the optical axis. The temporal focusing principle can be realized in a 4F system by introducing a dispersive element, such as a diffraction grating, in the back focal plane of tube lens by sending different spectral components of an optical pulse to different directions (different incident angles relative to the optical axis). According to the definition of being aberration-free (including chromatic aberration), these different spectral components have the same path lengths to the front focal plane where they recombine (see Fig. 19.12). When all the wavelength components recombine, the ultrafast pulses retain their narrow femtosecond pulse width. For the 4F system, this path-length matching property is true only between the back focal plane of tube lens and the front focal plane of the objective, but not anywhere else. Therefore outside the focal plane, the path lengths for the different spectral components are different, the pulses broaden, and the multiphoton excitation efficiency decreases.

APPLICATIONS OF MULTIPHOTON MICROSCOPY IN DIFFERENT AREAS OF DERMATOLOGICAL STUDIES

While many researchers have contributed to this field, one may recognize the works of König and his coworkers as being singularly substantial. König and coworkers have contributed to almost every aspect of dermatological research based on multiphoton microscopy. More importantly, this group was the first to successfully commercialize multiphoton dermatological

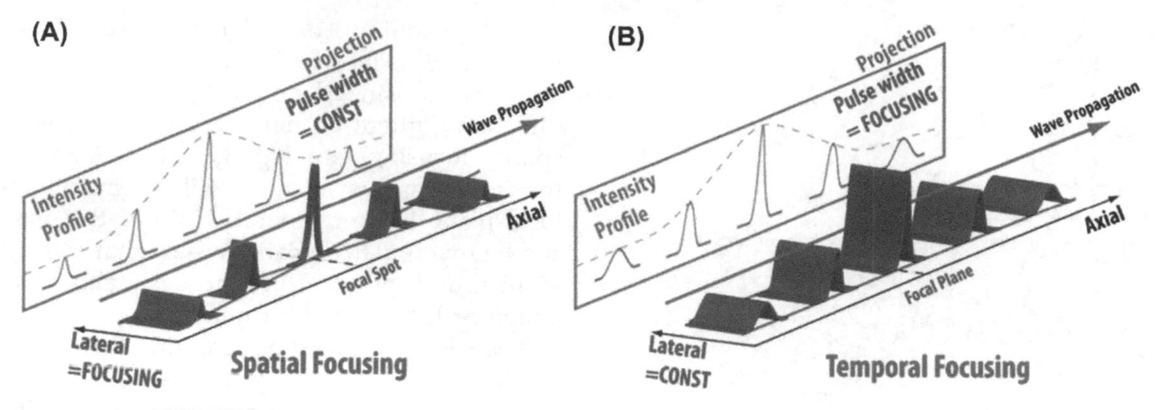

FIGURE 19.11 Pictorial description of (A) spatial focusing and (B) temporal focusing.

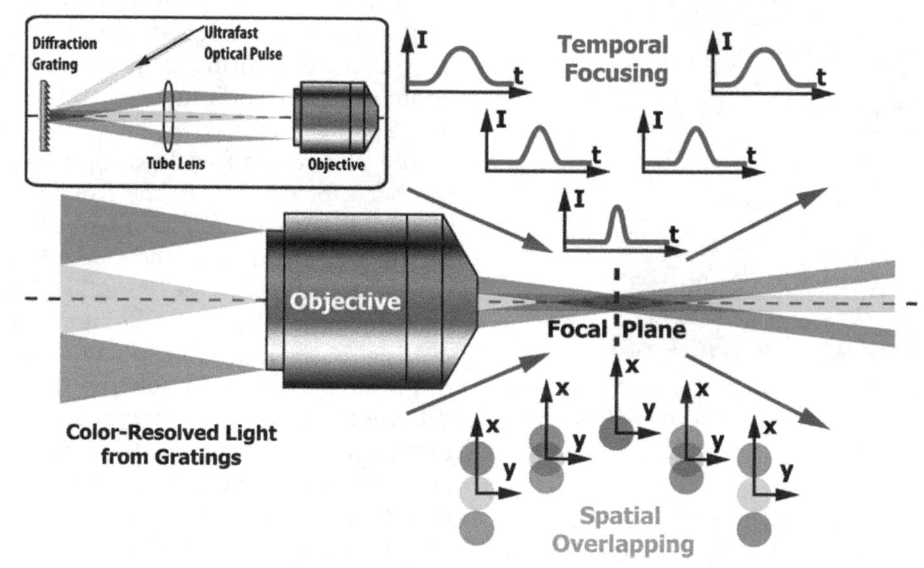

FIGURE 19.12 A pictorial description of temporal focusing multiphoton microscopy. Inset shows how the color-resolved light is generated before the objective. *I*; *t*.

instrumentation that is compatible with clinical use. Their work has established the safety and efficacy of multiphoton imaging in the skin [68]. They have further obtained regulatory approval for the clinical use of these instruments in the European Union and several other countries. The availability of commercial instruments has significantly increased the routine use of multiphoton microscopic imaging by clinicians; the resultant progress in this area has been summarized in several articles [92–96].

Skin Cancers

Minimally invasive diagnosis of skin cancer is a major reason for developing multiphoton skin imaging. The major types of skin, ranked by decreasing levels of incidence, are: basal cell carcinoma (originating from cells that make up the stratum basale of the epidermis), squamous cell carcinoma (originating from squamous cells, which make up the major part of the epidermis), and malignant melanoma (originating from melanocytes) [97]. Despite the comparatively low proportion of malignant melanoma cases, it has by far the highest mortality rate at approximately 20% in the United States [98].

Early examples of multiphoton imaging for skin cancer include Teuchner et al., who measured the two-photon fluorescence of melanin in excised skin tissue [99,100]. Subsequently, Skala et al. [101] studied a hamster cheek pouch cancer model where tumors were biopsied and imaged in 3D. From these images, five features were identified for distinguishing normal, precancerous, and cancerous (squamous cell carcinoma) tissue. Skala and coworkers extended this work to include fluorescence lifetime metabolic imaging based on NAD(P)H and FAD emission, [102] noting metabolism was different in tumors when compared with normal

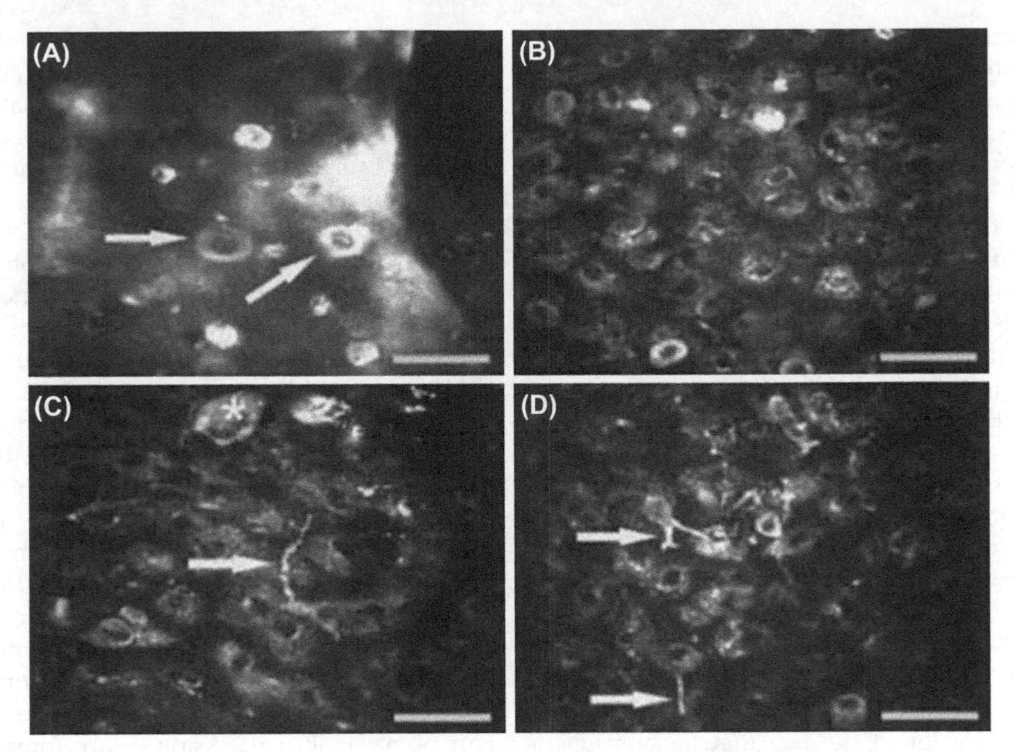

FIGURE 19.13 Optical sections of malignant melanoma *(arrows)* [4].

tissue [103]. Further development of mouse models of carcinoma remains relevant today, with a recent new xenograft mouse model that may be used to reduce the need for clinical samples of melanoma tissue [104]. For studying these skin cancer animal models, dedicated imaging stations have been built to enable long-term observation. For example, time-lapse imaging of tumor-specific CD8 T cells tagged with green fluorescent protein was performed in vivo over several days using microtattoos for image registration [105].

Although the use of animal models is helpful, the study of human tissues is an important step towards eventual clinical applications. Most of these studies are based on two-photon fluorescence of endogenous fluorophores. In a proof-of-concept study on basal cell carcinoma by Cicchi et al. [106], an increase in fluorescence intensity was observed in cancerous tissue taken from a single patient. In another early study, malignant melanoma was imaged based on fluorescence emissions from NAD(P)H and elastin, and SHG signal from collagen. Melanoma cells were observed to fluoresce much brighter than the surrounding cells [107]. A much larger trial, with 250 patients, was also performed for imaging melanoma [108]. The increased fluorescence from cancerous melanocytes was confirmed, and morphological differences could also be seen. A study by Zhang et al. further confirmed that the morphology of cancerous melanocytes was different; the cells were more elongated than normal, and the melanocytes appeared to be migrating together. Even though only one patient was in the study, this was the first time that physiological alterations in melanoma had been observed in vivo [109]. By providing images in en face geometry on nonmelanoma skin cancer, Paoli et al. [110] and Ericson et al. [111] reported that morphological features noted in histopathology could also be observed using multiphoton imaging. In another large study, 115 patients were recruited to study the sensitivity and specificity of two-photon fluorescence for imaging melanoma [4]; values of up to 95% sensitivity and up to 97% specificity were reported (Fig. 19.13). The overarching message of these reports seems to be that multiphoton fluorescence imaging may be a promising noninvasive approach providing morphological and molecular information that appears to be diagnostic and provides similar information as excisional biopsy.

Many studies further showed that the addition of spectroscopic quantification, especially metabolic assays, may improve diagnostic power. Fluorescence lifetime imaging [71,112–114] and spectrally-resolved en face two-photon fluorescence [115] have shown promises of better classification of tissue samples. Multiphoton tomography and lifetime imaging were also used to image nevi in an attempt to diagnose malignant melanoma [116]. A 37-patient study on the use of spectrally-resolved fluorescence lifetime imaging to diagnose melanoma concluded that whereas different melanized cell types (such as keratinocytes or melanocytes) could be readily distinguished using a single spectral measurement, morphological data was still necessary to distinguish benign from malignant melanocytic skin lesions [117]. Morphological differences in basal cell carcinoma and squamous cell carcinoma cells were also observed by Xiong et al. using SHG and two-photon fluorescence [118]. Because SHG is mostly generated by ordered collagen type I, it can be used to distinguish ordered collagen in healthy skin vs. disordered collagen in tumor regions. These two techniques were also used by Chernyavskiy et al., [119] who combined them with confocal reflectance and fluorescence imaging to image the effects of microwave-induced hyperthermia in mice as a treatment for melanoma.

Several other new contrast mechanisms for skin cancer diagnosis are in the early stages of development. Excited-state absorption has been shown to be effective for imaging melanin and melanoma diagnosis by Teuchner et al. [120] Pump-probe optical coherence microscopy was used to image melanoma by Wan and Applegate [121,122], and melanoma was also studied using two-photon photoacoustic microscopy by Oh et al. [123], who exploited the fact that melanin has a high two-photon absorption cross-section. CARS has been combined with SHG and two-photon fluorescence by Vogler et al. in order to image basal cell carcinoma [124]. Chen et al. used SHG and THG excited at 1230 nm by a Cr:Forsterite laser studying many different skin features and diseases; their use of longer excitation wavelength has also resulted in greater tissue penetration depth [125]. An intriguing study has shown that a higher proportion of pheomelanin in skin is positively correlated with skin cancer [126]. This result offers the possibility to distinguish cancer cells based on spectroscopy alone, rather than having to interpret the morphological data. Fu et al. developed novel NA microscope [127], demonstrating that the two types of melanin can be distinguished in hair, skin, and phantoms consisting of capillary tubes filled with red hair, black hair, or Rhodamine 6G [88]. It was discovered that the excited-state lifetime for eumelanin was much longer, permitting the two species to be imaged separately. This was exploited by Matthews et al. to image pigmented lesions that were excised during biopsy [24]. It was found that there was significantly increased eumelanin in the melanoma samples, and that this fact (combined with other morphological observations) could be used to identify all the melanomas while excluding 75% of the dysplastic nevi, albeit with a sample size of only 21 patients. Further studies have been performed using mouse models [128], as well as a study in humans that attempted to image melanogenesis on excised tissue samples [129]. Fluorescence lifetime imaging has also been used to separate melanoma from nevi [130], and attempts have also been made to use ordinary two-photon fluorescence [131,132] along with a demonstration of the use of two−spectral-channel lifetime imaging to classify nevi and basal cell carcinoma [67]. Although the identification of melanoma based on the spectral differences between pheomelanin and eumelanin is far from proven, this approach does show significant early promise.

Whereas the majority of multiphoton microscopy skin cancer research is focused on melanoma, other types of cancer are investigated as well. Cicchi et al. used multiphoton imaging to image a number of conditions; besides melanoma, they studied basal cell carcinoma, scarring, and keloid formation [133,134]. More recently, a number of studies on the diagnosis of basal cell carcinoma in excised tissue were performed by groups affiliated with König, employing multiphoton tomography [135], a combination of multiphoton tomography and fluorescence lifetime [136], just fluorescence lifetime [66], and a combination of multiphoton and optical coherence tomography [137], as well as comparing multiphoton tomography and confocal reflectance for the diagnosis of basal cell carcinoma [138,139]. Aside from more conventional skin cancers, Hoeller et al. investigated imaging of cutaneous T-cell lymphoma [140]. The investigation was performed by imaging fluorescently-labeled malignant T cells in a mouse model, and several conclusions were drawn as to the biological methods by which malignant T cells adhere to E-selectin in the skin.

The papers cited in the previous paragraph use mostly endogenous contrast agents in order to perform cancer imaging. The use of exogenous contrast agents may provide added spectral contrast to better highlight disease features that cannot be observed using endogenous signals alone, improving sensitivity, ease of use, and reducing cost. 5-Aminolevulinic acid (ALA) was used by Cicchi et al. during a larger study on the imaging of basal cell carcinoma [141,142], as well as by Riemann et al. [143]; ALA is a precursor to protoporphyrin IX, which is highly fluorescent and accumulates in tumor cells. Gold nanorods, mostly based on immune-targeting, are also being pursued on account of their strong luminescence and biocompatibility; Durr et al. have spearheaded this research [144−146], which has also been pursued by the group of Tunnell [147,148]. Although the use of exogenous contrast agents may enhance diagnostic sensitivity, regulatory issues are major hurdles for clinical translation.

Whereas conventional histopathology remains the gold standard for any clinical diagnosis, including skin cancer, the accuracy of histopathological results depends greatly on the skill of the pathologist. A number of research groups are working towards making pathological analysis more quantitative. In the diagnosis of skin lesions based on multiphoton imaging, several research groups advocate the creation of simple, sensitive, and robust diagnosis measures. The Multi-Fluorescence to SHG Index has been proposed as a means of distinguishing basal cell carcinoma from normal tissue [149,150], and was used to locate precancerous melanocytes in mice [151]. The Autofluorescence to SHG Index has also been proposed, and was tested on dorsal skin-fold chambers in nude mice [152]. Levitt et al. developed a much more complex series of image-processing metrics for two-photon fluorescence that were used on a tissue model [153]. Unfortunately, as with many of these studies, very few of them perform well enough to warrant a large clinical trial in order to evaluate whether they can provide a sensitivity and specificity comparable to, or above, that of a trained pathologist.

Two studies highlight concerns related to using multiphoton imaging for skin cancer diagnosis. Kantere et al. have noted that multiphoton protoporphyrin IX fluorescence does not increase tumor contrast; they recommend one-photon anti-Stokes fluorescence instead [154]. Nadiarnykh et al. have an even more striking conclusion: that with a diffraction-limited focal spot and a peak power of around 1 kW, significant DNA damage can occur by multiphoton absorption, as measured by the presence of cyclobutane-pyrimidine dimers [155]. The effect was strongly dependent on wavelength, with 695-nm excitation being particularly damaging, and wavelengths longer than 780 nm being less so. This is broadly consistent with the results of Le Harzic et al., who noted that 1064-nm fs laser pulses were much less damaging than wavelengths of 532 nm and shorter [156], and has clear implications for the clinical use of multiphoton imaging in cancer diagnosis.

Overall, the use of multiphoton imaging methods for cancer diagnosis and monitoring shows considerable promise. Many studies have demonstrated that tumors can be identified through a variety of different contrast mechanisms. However, the limited field of view and imaging speed mostly preclude the use of multiphoton imaging for large area surveillance applications. Instead, it may possibly find use in providing noninvasive diagnosis of suspicious nevi or other neoplasms (found by less specific but faster modalities), especially in locations such as the head and neck, where the consequences of prophylactic surgical excision are more pronounced. In addition, multiphoton imaging may also find use in margin determination, where surgeons need effective imaging tools to determine whether they have completely resected the whole lesion. The availability of clinically compatible instruments from the Koenig group goes a long way in proving that these techniques can be applied not just in a laboratory environment but work acceptably well in the clinic, and the increasing number of clinical trials that involve a substantial number of patients should provide a statistically significant evaluation of the utility of multiphoton imaging as a clinically useful tool for skin cancer diagnosis.

Other Dermatological Diseases

Besides cancer, multiphoton imaging has been used in the diagnosis and monitoring of many different skin pathologies. In the clinic it has been used to image diseases such as Jadassohn-Pellizzari anetoderma [157], scleroderma [158], lymphedema [159], and actinic keratosis, [160] and imaging of atopic dermatitis was investigated in a mouse model [77]. Some of these diseases result in a change in skin collagen and elastin and these two major dermal components can be readily distinguished by SHG and two-photon autofluorescence, respectively. Jadassohn-Pellizzari anetoderma is characterized by a loss of dermal elastin, whereas scleroderma is characterized by the abnormal accumulation of collagen. Similarly, one of the most important markers for lymphedema progression can be the extent of collagen restructuring. Atopic dermatitis can result in hyperkeratosis (a thickening of the stratum corneum) and fibrosis of the upper dermis, both of which can be successfully imaged. Huck et al. have further investigated imaging methods of atopic dermatitis via incorporating fluorescence lifetime imaging to measure the proportion of free vs. protein-bound NAD(P)H as a measure of cellular activity in 20 patients and 20 control subjects [9,161]. Actinic keratosis has been imaged in vivo, with the increased average nuclear diameter being observable both in histopathology and two-photon fluorescence [160]. Larger studies have been performed by König and coworkers, who imaged a variety of different diseases such as seborrheic keratoses, angioma, actinic keratoses, psoriasis, pemphigus vulgaris, scarring, and autoimmune bullous skin diseases [162,163]. Skin disease caused by infectious agents has also been monitored by multiphoton imaging. Lin et al. noted that fungal infections can be monitored by two-photon fluorescence; *Microsporum canis*, for example, is highly autofluorescent and could be readily distinguished from the stratum corneum in a mouse model [164].

An important function of the skin is forming a barrier against infectious agents in the environment. Multiphoton microscopy techniques have been developed to study immunological cells in the skin [165,166]. In transgenic mice, neutrophils and perivascular macrophages expressing fluorescent proteins, Abtin and coworkers demonstrated that during

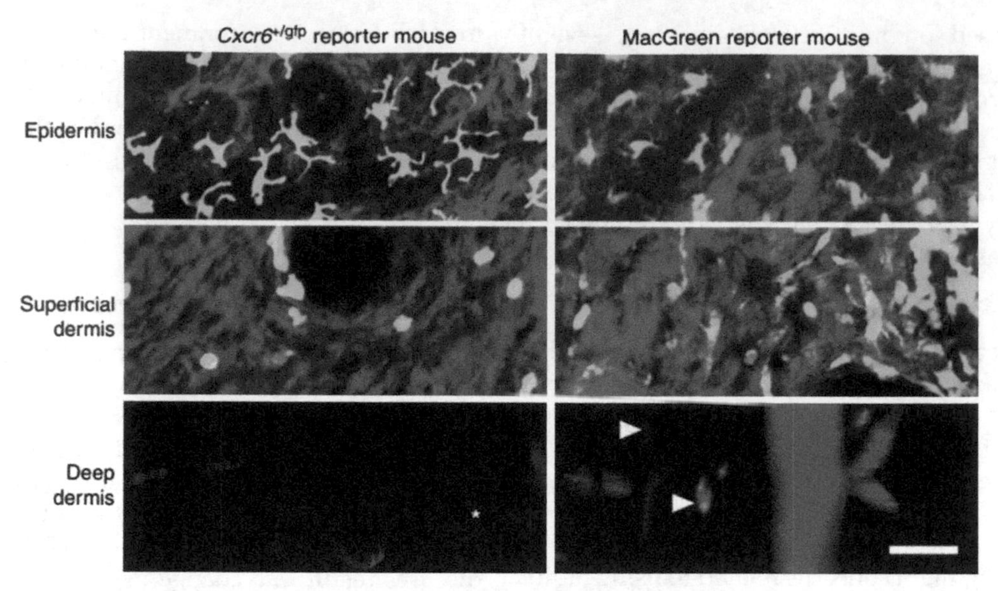

FIGURE 19.14 Immune cell distribution in skin [1].

bacterial infection, neutrophil transmigration into the skin are mediated by the presence of perivascular macrophages. Roediger and coworkers showed that group 2 innate lymphoid cells are important to regulate inflammation response in the skin and cytokine stimulation can lead to population expansion of this class of innate lymphoid cells, resulting in spontaneous dermatitis [167]. It has been recognized that skin responses to pathogens and inflammation depends on skin location; Tong and coworkers investigated this site specificity by creating an atlas of densities of macrophages, T cells, and mast cells at the ear, back, footpad, and tail of transgenic mice by combining information gained from flow cytometry and multiphoton microscopy [1] (Fig. 19.14). They also found that group 2 innate lymphoid cell density is animal age–dependent. In a passive cutaneous anaphylaxis model, the investigators further demonstrated that mast cell activation is site specific upon antigen-specific immunoglobulin E transdermal injection.

Certain systemic disorders can also be studied by how they alter dermal structures. Dong and coworkers have studied diabetes mellitus where protein glycation is postulated to be a major cause of complications caused by this disease and aging. Glycation can be observed by the increased autofluorescence resulting from the advanced glycation endproducts and slight reduction in second harmonic generation. They have quantified glycation processes in the skin, as well as in the cornea and aorta [168,169]. Marfan syndrome is a genetic disorder of extracellular matrix protein; Cui and coworkers showed that Marfan syndrome can be detected in the skin through imaging of morphological changes in collagen and elastin within the dermal extracellular matrix [171].

Skin Aging Studies

Aside from disease diagnosis and treatment, cosmetic and plastic surgery industries have considerable interest in multiphoton imaging for the investigation of chronological aging and extrinsic (often photoinduced) aging of skin. Lin et al. first attempted to create a measure that correlated with the chronological age of a subject; the SHG to Autofluorescence Aging Index of Dermis (SAAID) offered a simple means to estimate the age of skin by taking the ratio of SHG to autofluorescence [172,173]. This measure has been supported by a study that noted that there was a difference in SAAID scores between men and women of the same age [174], and another study that confirmed the correlation with age by measuring sites on the face [175]. The discrepancy between men and women was overcome by a more subjective score, the multiphoton laser-scanning tomography–based Dermis Morphology Score, which also had a better correlation with age than SAAID [176]. Depth-resolved measurements of SAAID were taken by Kaatz et al. in order to quantify to what extent the measured depth varied with imaging depth [177]. The difficulty in using SAAID to distinguish between photoaged and nonphotoaged skin was noted by Sanchez et al. and an improvement, via incorporating lifetime imaging of NAD(P)H, was made [3]. Lifetime imaging was also used by Koehler et al. to diagnose dermal elastosis, which is often a sign of extrinsic aging [178] (Fig. 19.15).

Other researchers have proposed similar measures; Puschmann et al. proposed the elastin-to-collagen ratio, which explicitly includes autofluorescence contributions from just elastin by manual masking of the image, and also takes the ratio of the area of each skin component

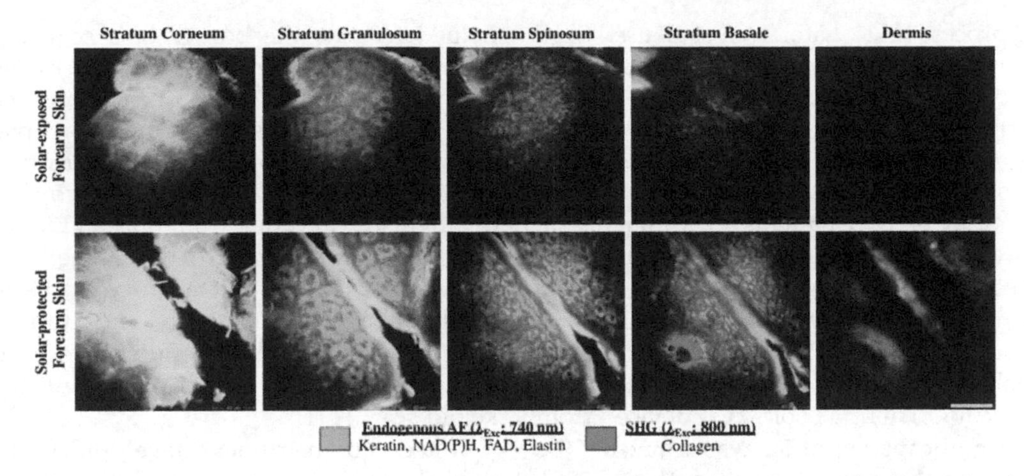

FIGURE 19.15 Solar effect on skin aging [3]. *AF*; *FAD*, flavin adenine dinucleotide; *NAD(P)H*, nicotinamide adenine dinucleotide (phosphate); *SHG*, second harmonic generation; λ_{Exc}.

as opposed to the fluorescent intensity [179]. Wu et al. proposed a measurement based on the fast Fourier transform of the SHG image [180] or the gray-level cooccurrence matrix [181]. Cicchi et al. have also proposed a similar method based on the fast Fourier transform [182].

Other studies were more qualitative, seeking to investigate the changes that occur during chronological aging or photoaging. Koehler et al. have investigated the acceleration of aging induced by sunbath; although the sample was too small to quantify the damage from the sunbath, differences were observed between the dorsal and volar forearm, demonstrating the effect that sun exposure has on skin [183]. Similar observations were made by Benati et al. [184] and Baldeweck et al., who performed 3D-resolved measurements to further investigate the differences between the sun-exposed and unexposed skin [185]. Decenciere et al. extended this 3D methodology even further, developing a segmentation algorithm that could be used to quantify the size and shape of various skin components, and in turn be used to estimate the effect of aging [186]. Takeshi et al. used polarization-resolved SHG to show that wrinkles in skin were aligned with the underlying collagen fibers [187,188], and Lutz et al. argued that collagen cross-linking could be quantified in vitro by noting the increase in SHG and a decrease in the fluorescence lifetime [189,190].

The therapeutic effects of certain skin treatments have also been investigated using multiphoton imaging. Pena et al. investigated a potential treatment for wrinkles, by noting that fibroblasts can cause contraction in the skin. The effect of Y-27,632, a RhoA-kinase inhibitor, was investigated and found to have potential as a means to inhibit this contraction [191,192]. Bazin et al. also investigated a potential antiwrinkle treatment consisting of soy and jasmine extracts showing a statistically significant increase in dermal collagen content based on two-photon fluorescence and SHG imaging [242].

Aside from biochemical agents, the aftereffects of laser treatment for wrinkles have been investigated using two-photon fluorescence and SHG; laser fractional microablative rejuvenation involves using a laser to induce a thermal shock to fibroblasts in the skin, which then produce more collagen. This increased collagen production can be imaged using multiphoton microscopy [193–195]. A very similar study was previously performed by Tsai et al., investigating the effect of erbium:yttrium aluminum garnet laser irradiation on skin for the treatment of skin hyperplasia and tumors [196].

Skin Regeneration Studies

Wound healing and dermal regeneration processes have also been studied with multiphoton imaging, because type I collagen can be imaged particularly well using SHG imaging. As the structure and morphology of the collagen that forms around the wound plays a large part in determining whether a scar forms, or whether the scar is normal, atrophic, hypertrophic or keloid, this application is particularly suited to SHG imaging. Initial studies focused on observing the wound healing process in animal models. Navarro et al. imaged the different stages of skin wound closure in guinea pig models using two-photon fluorescence at several time points after full-thickness wounds were induced surgically. Growth of blood vessels and collagen fibers was observed [197]. This was pursued further, incorporating SHG imaging and imaging the wound with greater time resolution and over a longer period in order to yield more insight into the wound-healing process [131]. Later, Luo et al. used SHG and image analysis to investigate wound healing in KunMing mice over 14 days [198].

Human scarring has been studied with ex vivo specimens. Brewer et al. imaged samples excised from normal and keloid scars. Although the study was very limited (with a total of two patients), a difference in collagen density was observed but the trend was contrary to expectations, with a greater collagen density observed in the normal scar as opposed to the keloid one [199]. Meshkinpour et al. used SHG to image biopsies taken from keloid and hypertrophic scars undergoing treatment with the ThermaCool TC device from Thermage Inc. Significant variation was found in the collagen structure of the four biopsied patients [200]. This finding was echoed by Da Costa et al., who found a swirling collagen structure in keloids, as opposed to a more wavy structure in normal skin [201].

In vivo study of human scarring was first studied by Riemann et al., imaging a biopsy scar of a single patient over a period of 60 days with images taken every 1–3 days after surgery. Two-photon fluorescence and SHG were both employed, and the organization of the new collagen fibers was noted [202]. Zhu et al. imaged much older scars by sampling from women who had previously undergone caesarian section. Their data showed a slight decrease in elastin fluorescence and SHG from collagen as the wound aged [203].

Several researchers have noted that the ratio of two-photon fluorescence to SHG can be used to classify scars [204,205]; however, because collagen structure is deemed to be important in wound healing, several researchers have attempted to quantify the degree of order within a collagen structure. Chen et al. contrasted three different approaches: an edge-detecting filter to determine the gradient at each pixel, a simple threshold to measure the collagen density for the image, and a complex semiarbitrary geometrical morphology approach. These were then weighted to form a final measurement [206]. Some collagen classification metrics that are discussed in the skin aging section also find applications in wound healing. Cicchi et al. also assessed three different approaches, including employing a gray-level cooccurrence matrix (which was studied further by Ferro et al. [207]), a fast Fourier transform, and the SAAID [182]. Rather than determining one superior technique, it was found that each was effective at a particular length scale. Jiang et al. used a fast Fourier transform as well, in order to define a Collagen Orientation Index and a Bundle Distance, which were then used to characterize collagen in the deep, middle, and superficial dermis of keloid tissue excised from patients undergoing reconstructive surgery [208]. Taking these approaches further, it is possible that image processing combined with multiphoton imaging can help determine the border of a scar and thereby aid the surgeon in determining where to excise or intervene. Chen et al. used two-photon fluorescence and SHG to image skin samples taken from six patients, five of whom had hypertrophic scars. After processing the images, a number of features were proposed in order to distinguish scar tissue from normal tissue [209]. Shortly afterwards, another set of researchers discovered that the volume density of elastin could be used to distinguish keloid, hypertrophic, and normal scars [210], and that the two-photon fluorescence and SHG from collagen could be used to distinguish atrophic and keloid scars [211].

In terms of instrument development, Ping-Jung et al. published two papers demonstrating polarization-resolved SHG and showed that the second-order susceptibility ratio d_{33}/d_{31} could be used to distinguish normal tissue from keloid, morphea, and dermal elastolysis [212,213]. The most significant developments in applying multiphoton imaging to patients in vivo have been made primarily by König and coworkers; in particular, the use of a gradient-index (GRIN) lens to allow imaging within atrophic scars and other recessed skin features, and to image ulcers in vivo [68,214–216]. Later a GRIN lens with a higher NA of 0.8 was introduced, providing increased spatial resolution [217].

The skin is known to possess adult stem cells, specifically within the hair follicles. These stem cells may be found in the bulge area as well as in the dermal papilla. Using two-photon microscopy, Rompola et al. studied the growth regulation of these stem cells in vivo in mice [218]. Through the observation of hair follicle regeneration, they found that there exists a spatial organization to these stem cell progeny divisions. Likewise, cell-to-cell signaling allows for coordinated and rapid movement of the follicle. Through targeted laser ablation, Rompola et al. also showed that the mesenchyme plays an important role in hair regeneration. Similarly, Liu et al. studied the pluripotency of the nestin-expressing stem cells found within the bulge area and the dermal papilla [219] (Fig. 19.16). These cells migrate from the bulge area to the dermal papilla, suggesting that the bulge area is the source of skin stem cells [220]. By seeding these cells on Gelfoam and subsequently transplanting them into mice with spinal cord injury, it was observed that the transplanted cells migrated towards adjacent spinal cord segments. Mice that were transplanted with these stem cells experienced plantar placing of the affected paw within 3 days of transplantation, whereas the negative control group transplanted with only Gelfoam took 7 days. Full recovery took at least 28 days for the mice transplanted with stem cells, whereas only a few mice in the untransplanted group achieved locomotor recovery.

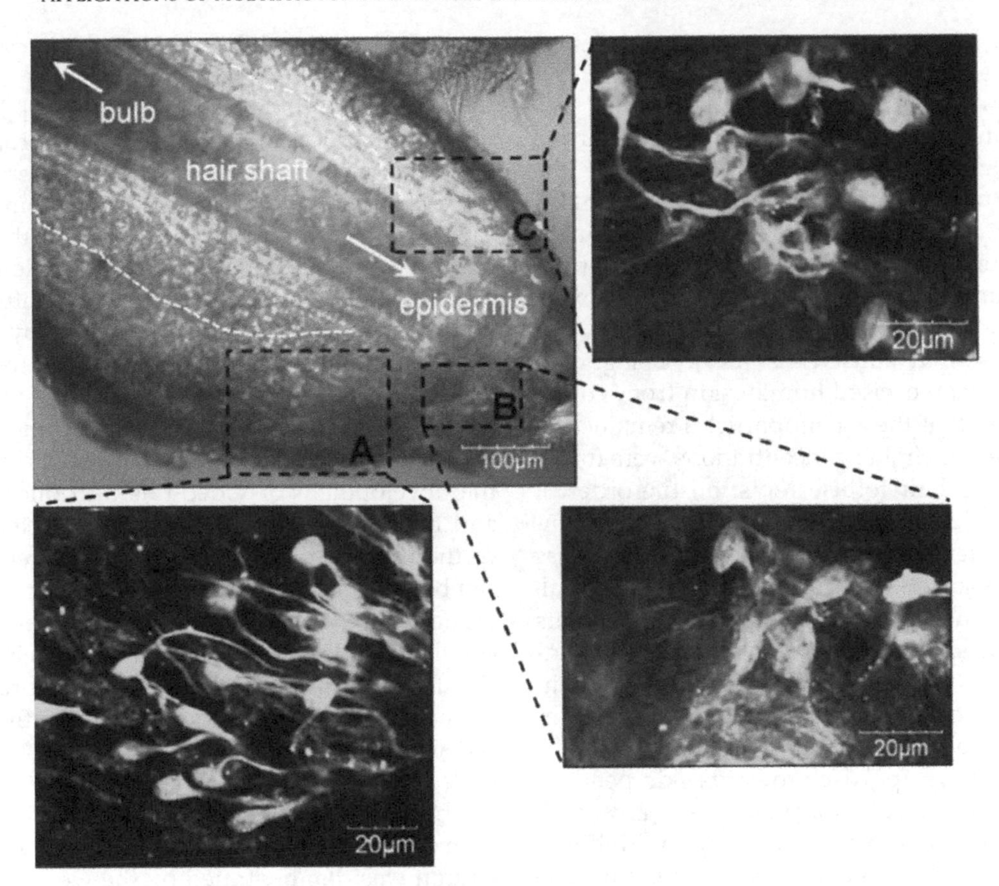

FIGURE 19.16 Distribution of stem cells near hair bulge in response to skin wounding.

Skin injury may also damage sensory nerve structures in the skin. The skin contains thermoreceptors, nociceptors, and touch receptors. Using transgenetic animals expressing fluorescent proteins, sensory nerves in mouse skin has been imaged to a depth of about 300 μm [221]. Yuryev and Khiroug took nerve regeneration study in skin one step further by using multiphoton optical dissection to cut nerve fibers in the skin of transgenic mice expressing fluorescent proteins [222]. They are able to observe the degradation and the subsequent regeneration of the damaged nerve, demonstrating intravital monitoring of the reinnervation process in vivo for almost 2 weeks.

Transdermal Transport of Drugs, Cosmetics, Sunscreens, and Nanoparticles

The skin forms a natural barrier that protects the body by keeping potentially toxic substances out. This barrier comprises of a physical layer (the stratum corneum), as well as immunological and enzymatic defenses [223]. Many cosmetic and pharmacological products are designed and sold for topical application, and their efficacy sometimes depends on their penetration into the interior of the skin. The dermal distribution of these products has been studied, and much work has gone into designing formulations that can effectively overcome the skin barrier [224].

Grewal and coworkers were the first to demonstrate that distribution of penetrated particles in skin can be noninvasively visualized using multiphoton microscopy [225]. They further demonstrated that the fluorescently-labeled dextran distribution can be modulated by topical application of different enhancers. Subsequently Yu et al. studied the distribution of fluorescent hydrophobic and hydrophilic particles in the skin, both before and after treatment with oleic acid, a common enhancer. Coupled with biochemical diffusion rate measurement data, they first quantitatively extracted transport parameters such as concentration gradient enhancement factor and the probe vehicle-to-skin partition–coefficient enhancement factor. They further proved that hydrophobic and hydrophilic agents penetrate through the skin stratum corneum and the epidermis through different routes [226]. Yu and coworkers also showed that because of skin heterogeneity, high-throughput large-area multiphoton imaging is critical to minimize errors in determining penetrant transport properties [227].

Finally, they have also quantitatively evaluated the effect of ultrasound in skin transport enhancement [228].

Besides transport, the safety of topically applied cosmetics, lotions, and creams is of key importance for the cosmetic and drug industry. Furthermore with the development of nanoparticles for a variety of applications, the biosafety of these particles resulting from inadvertent absorption through skin is also an important concern. For example, nanoparticles such as zinc oxide or titanium oxide are between 20 and 30 nm in size and are often used in sunscreens [229]. Using two-photon microscopy on excised human skin from volunteers, it was found that these nanoparticles remained in the stratum corneum. Higher concentrations were found in the skin folds or hair follicle roots: on the order of 800 particles/μm^3 [230,231]. Pigment particles remaining in the skin after tattooing represent another class of common nanoparticles in human skin. König et al. investigated how nanoparticles from tattoo pigments could be imaged, showing that they could be distinguished from other autofluorescent species in the skin by their fluorescence lifetime and emission spectrum [232]. Efforts have also been made to devise a multidimensional quantitative approach towards skin penetration of pharmacological formulations [233–235]. Recently, a multimodal approach utilizing multiphoton imaging has seen early clinical trials as a way of performing optical biopsies, as well as testing the efficacy of cosmetics [234,235]. In addition, Saar and coworkers used SRS to investigate, noninvasively and label-free, the penetration of drugs into the skin [2] (Fig. 19.17). Using ibuprofen and ketoprofen in propylene glycol applied to an excised mouse ear, they were able to image the distribution of the applied drug from the surface to the subcutaneous fat. In their study they found that both drugs penetrated through the intercellular lipids of the stratum corneum and the hair shafts. Saar and coworkers were able to use SRS microscopy to track chemical uptake and transport kinetics. They found proof that transport through the stratum corneum was slower, taking over 2 h, compared with penetration through the hair shaft, which reached steady-state in 26 min. With the development of video-rate SRS microscopy [2,236] and the label-free capability of SRS, real-time tracking of the efficacy of cosmetic and sunscreen compounds can be imaged with both high spatial and temporal resolution. For monitoring the transdermal transport of small molecules, the use of a molecule's intrinsic Raman signature is a superior method over fluorescent labeling because the chemical properties of the fluorescent label can significantly alter the transport mechanism of small molecules (see Fig. 19.16).

The importance of being able to study the efficacy of such compounds noninvasively as well as with high resolution was demonstrated by the work by Hanson and Clegg [237]. In their study, they observed and quantified

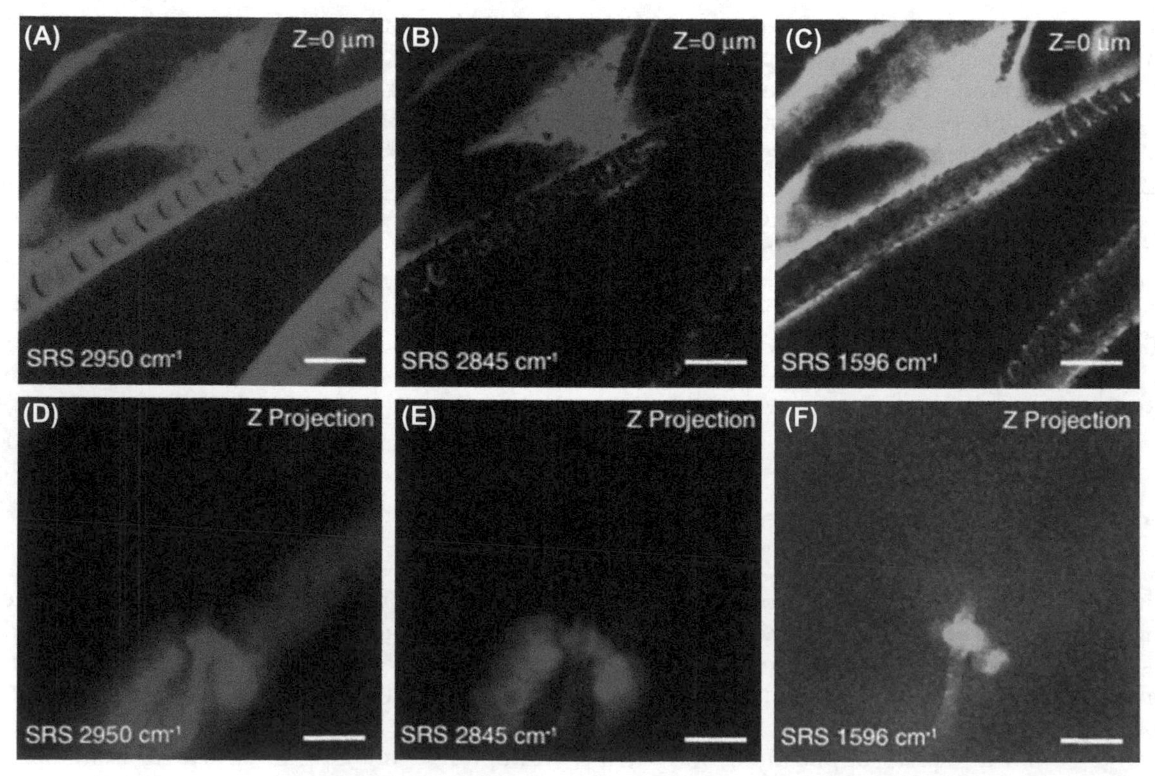

FIGURE 19.17 Stimulated Raman scattering (SRS) monitoring of trans-retinol skin penetration [2].

the generation of reactive oxygen species in ex vivo skin irradiated with ultraviolet (UV) light. In this study, ex vivo human skin samples were incubated with dihydrorhodamine-123, which only converts to a fluorescent form (rhodamine-123) after reacting with reactive oxygen species (ROS). The samples were then irradiated with varying doses of UVB and the amount of rhodamine-123 generated imaged with a two-photon microscope and quantified. It was found that for an average adult-sized face exposed for 2 h to UVB generated 14.7×10^{-3} mol of ROS (as measured by their reaction with dihydrorhodamine-123 to form fluorescent rhodamine-123) in the stratum corneum. Another 10^{-4} mol of ROS were generated in the lower epidermal strata. A subsequent study further revealed that some UV blockers actually increased the amount of ROS generated, and therefore increased the chances of skin cancer [238]. Because pH gradient may affect the transport of polar chemical species, a high-resolution study of the stratum corneum pH gradient was also performed, and it was found that the acidity decreases with increasing depth [239].

The effect of common chemical warfare agents on the skin are clearly of importance from the standpoint of protection and treatment. Werrlein, Braue, and Dillman have investigated the effects of sulfur mustard, a potent vesicant, on human epidermal keratinocyte cultures [240]. They noted a disruption to actin filaments, large punctuate inclusions, and a lack of stress fibers in exposed cells as compared with controls.

CONCLUSION

Whereas skin is one of the most accessible organs in the body, the physiology and pathology of the skin is complex and far from fully understood. The development of powerful imaging tools based on multiphoton techniques enables in vivo, minimally invasive imaging of skin physiology throughout the epidermis and into a substantial fraction of the dermis. Studies such as the in vivo imaging of stem cells and their physiological functions in the native skin environment by the König group will provide significant insights for stem cell technology and regenerative medicine [241]. Similarly, skin pathologies are medically important. Although chronic diseases, such as dermatitis, are not life threatening, they can significantly compromise the quality of life for patients. Melanoma, the most dangerous form of skin cancer, can be very effectively cured if the lesion is diagnosed early, but the 10-year mortality rate can be 70% higher after metastasis has occurred. Interestingly, a recent report has shown in a mouse model that melanoma may develop from "invisible" nevi that contain difficult-to-visualize lightly colored eumelanin instead of the darker pheomelanin [25]. The possibility of the presence of melanoma-causing invisible nevi in the population with light skin color is an important medical hypothesis that should be investigated. Recent multiphoton imaging technologies [24] that can effectively distinguish eumelanin from pheomelanin can play an important role in these studies, and may play an important future diagnostic role if this hypothesis is validated. Lastly, skin products, used for cosmetic, sun-protection, antiaging, or regeneration purposes, are commercial products that we use daily. Despite their financial significance, the efficacy of many of these products is mostly judged subjectively. Importantly, the toxicology evaluation of many of these products is often phenomenological and sometimes lacking the vigor of modern physiological investigation. Multiphoton microscopy enables the in vivo study of many of these products on animal models and human volunteers and is promising to change the research of this large and commercially important field. Common to all these many skin-related biomedical applications, the availability and further development of cutting edge multiphoton microscopy technology will have impact on the field of dermatology.

Acknowledgments

The authors acknowledge support from NIH -5-O41-EB015871-27, DP3-DK101024 01, 1-U01-NS090438-01, 1-R01-EY017656 -0,6A1, 1-R01-HL121386-01A1, the Singapore-MIT Alliance 2, the Biosym IRG of Singapore-MIT Alliance Research and Technology Center, Hamamatsu Inc., Samsung GRO Program, the MIT Skoltech Initiative, and the Koch Institute for Integrative Cancer Research Bridge Initiative. CJR is further grateful for a Wellcome Trust MIT Postdoctoral Research Fellowship to carry out this research.

References

[1] Tong PL, et al. The skin immune atlas: three-dimensional analysis of cutaneous leukocyte subsets by multiphoton microscopy. J Invest Dermatol 2015;135(1):84–93.
[2] Saar BG, et al. Video-rate molecular imaging in vivo with stimulated Raman scattering. Science 2010;330(6009):1368–70.
[3] Sanchez WY, et al. Changes in the redox state and endogenous fluorescence of in vivo human skin due to intrinsic and photo-aging, measured by multiphoton tomography with fluorescence lifetime imaging. J Biomed Opt 2013;18(6):061217.
[4] Dimitrow E, et al. Sensitivity and specificity of multiphoton laser tomography for in vivo and ex vivo diagnosis of malignant melanoma. J Invest Dermatol 2009;129(7):1752–8.
[5] Lin S-J, Jee S-H, Dong C-Y. Multiphoton microscopy: a new paradigm in dermatological imaging. Eur J Dermatol 2007;17(5).
[6] Masters BR, So PT, Gratton E. Multiphoton excitation fluorescence microscopy and spectroscopy of in vivo human skin. Biophys J 1997;72(6):2405–12.
[7] Schenke-Layland K, et al. Two-photon microscopes and in vivo multiphoton tomographs: powerful diagnostic tools for tissue engineering and drug delivery. Adv Drug Deliv Rev 2006;58(7).

[8] Wang BG, Koenig K, Halbhuber KJ. Two-photon microscopy of deep intravital tissues and its merits in clinical research. J Microscopy-Oxford 2010;238(1).

[9] Koenig K, et al. 5D-intravital tomography as a novel tool for non-invasive in-vivo analysis of human skin. Proc SPIE 2010; 7555(Journal Article):75551I.

[10] Cicchi R, Pavone FS. Non-linear fluorescence lifetime imaging of biological tissues. Anal Bioanal Chem 2011;400(9).

[11] Seidenari S, et al. Multiphoton laser microscopy and fluorescence lifetime imaging for the evaluation of the skin. Dermatol Res Pract 2012;2012(Journal Article).

[12] Koenig K. Clinical multiphoton FLIM tomography. Proc SPIE 2012;8226(Journal Article):82260H.

[13] Campagnola PJ, Dong C-Y. Second harmonic generation microscopy: principles and applications to disease diagnosis. Laser Photon Rev 2011;5(1):13–26.

[14] Perry SW, Burke RM, Brown EB. Two-photon and second harmonic microscopy in clinical and translational cancer research. Ann Biomed Eng 2012;40(2):277–91.

[15] Yew E, Rowlands C, So PT. Application of multiphoton microscopy in dermatological studies: a mini-review. J Innov Opt Health Sci 2014;7(5):1330010.

[16] Shen YR. The principles of nonlinear optics. New York: Wiley Interscience; 2002.

[17] Boyd R. Nonlinear optics. New York: Academic Press; 2008.

[18] Nagy A, Wu J, Berland KM. Characterizing observation volumes and the role of excitation saturation in one-photon fluorescence fluctuation spectroscopy. J Biomed Opt 2005;10(4):44015.

[19] Cianci GC, Wu J, Berland KM. Saturation modified point spread functions in two-photon microscopy. Microsc Res Tech 2004; 64(2):135–41.

[20] Anderson RR, Parrish JA. The optics of human skin. J Invest Dermatol 1981;77:13–9.

[21] Masters B, Chance B. Redox confocal imaging: intrinsic fluorescent probes of cellular metabolism. Fluorescent Luminescent Probes Biol Activity 1993:44–56.

[22] Chance B. Pyridine nucleotide as an indicator of the oxygen requirements for energy-linked functions of mitochondria. Circ Res 1976;38(5 Suppl. 1):I31–8.

[23] Pena A, et al. Spectroscopic analysis of keratin endogenous signal for skin multiphoton microscopy: erratum. Opt Express 2005;13(17):6667.

[24] Matthews TE, et al. Pump-probe imaging differentiates melanoma from melanocytic nevi. Sci Transl Med 2011;3(71): 71ra15.

[25] Mitra D, et al. An ultraviolet-radiation-independent pathway to melanoma carcinogenesis in the red hair/fair skin background. Nature 2012;491(7424):449–53.

[26] Lai Z, et al. Stepwise multiphoton activation fluorescence reveals a new method of melanin detection. J Biomed Opt 2013;18(6): 061225.

[27] Larson DR, et al. Water-soluble quantum dots for multiphoton fluorescence imaging in vivo. Science 2003;300(5624):1434–6.

[28] Rahim NAA, et al. Conjugated polymer nanoparticles for two-photon imaging of endothelial cells in a tissue model. Adv Mater 2009;21(34):3492–6.

[29] Tearney GJ, et al. Determination of the refractive index of highly scattering human tissue by optical coherence tomography. Opt Lett 1995;20(21):2258.

[30] Lo W, et al. Spherical aberration correction in multiphoton fluorescence imaging using objective correction collar. J Biomed Opt 2005;10(3):034006.

[31] Tyrrell RM, Keyse SM. New trends in photobiology the interaction of UVA radiation with cultured cells. J Photochem Photobiol B: Biol 1990;4(4):349–61.

[32] Keyse SM, Tyrrell RM. Induction of the heme oxygenase gene in human skin fibroblasts by hydrogen peroxide and UVA (365 nm) radiation: evidence for the involvement of the hydroxyl radical. Carcinogenesis 1990;11(5):787–91.

[33] Hockberger PE, et al. Activation of flavin-containing oxidases underlies light-induced production of H_2O_2 in mammalian cells. Proc Natl Acad Sci USA 1999;96(11):6255–60.

[34] Koester HJ, et al. Ca^{2+} fluorescence imaging with pico- and femtosecond two-photon excitation: signal and photodamage. Biophys J 1999;77(4):2226–36.

[35] Konig K, et al. Two-photon excited lifetime imaging of autofluorescence in cells during UVA and NIR photostress. J Microsc 1996; 183(Pt 3):197–204.

[36] Konig K, et al. Pulse-length dependence of cellular response to intense near-infrared laser pulses in multiphoton microscopes. Opt Lett 1999;24(2):113–5.

[37] Sako Y, et al. Comparison of two-photon excitation laser scanning microscopy with UV-confocal laser scanning microscopy in three-dimensional calcium imaging using the fluorescence indicator Indo-1. J Microsc 1997;185(Pt 1):9–20.

[38] Patterson GH, Piston DW. Photobleaching in two-photon excitation microscopy. Biophys J 2000;78(4):2159–62.

[39] Hopt A, Neher E. Highly nonlinear photodamage in two-photon fluorescence microscopy. Biophys J 2001;80(4):2029–36.

[40] Tirlapur UK, Konig K. Femtosecond near-infrared laser pulses as a versatile non-invasive tool for intra-tissue nanoprocessing in plants without compromising viability. Plant J 2002;31(3):365–74.

[41] Uchugonova A, et al. Targeted transfection of stem cells with sub-20 femtosecond laser pulses. Optics Express 2008;16(13): 9357–64.

[42] Schonle A, Hell SW. Heating by absorption in the focus of an objective lens. Opt Lett 1998;23(5):325–7.

[43] Denk W, Sugimori M, Llinas R. Two types of calcium response limited to single spines in cerebellar Purkinje cells. Proc Natl Acad Sci USA 1995;92(18):8279–82.

[44] Jacques SL, et al. Controlled removal of human stratum corneum by pulsed laser. J Invest Dermatol 1987;88(1):88–93.

[45] Pustovalov V. Initiation of explosive boiling and optical breakdown as a result of the action of laser pulses on melanosome in pigmented biotissues. Quan Electronics 1995;25(11): 1055–9.

[46] Buehler C, et al. Innovations in two-photon deep tissue microscopy. IEEE Eng Med Biol Mag 1999;18(5):23–30.

[47] Masters BR, et al. Mitigating thermal mechanical damage potential during two-photon dermal imaging. J Biomed Opt 2004;9(6): 1265–70.

[48] Horton NG, et al. Three-photon microscopy of subcortical structures within an intact mouse brain. Nat Photon 2013;7(3).

[49] Göppert-Mayer M. Über Elementarakte mit zwei Quantensprüngen. Ann Phys (Leipzig) 1931;5:273–94.

[50] Franken PA, et al. Generation of optical harmonics. Phys Rev Lett 1961;7:118.

[51] Kaiser W, Garrett CGB. Two-photon excitation in $CaF_2:Eu^{2+}$. Phys Rev Lett 1961;7:229–31.

[52] Singh S, Bradley LT. Three-photon absorption in naphthalene crystals by laser excitation. Phys Rev Lett 1964;12:162–4.

[53] Denk W, Strickler JH, Webb WW. 2-Photon laser scanning fluorescence microscopy. Science 1990;248(4951):73–6.

[54] Sheppard CJR, et al. Scanning harmonic optical microscope. IEEE J Quan Electronics 1977;13(9):D100.

[55] Denk W, Strickler JH, Webb WW. Two-photon laser scanning fluorescence microscopy. Science 1990;248:73–6.

[56] Piston DW, Masters BR, Webb WW. Three-dimensionally resolved NAD(P)H cellular metabolic redox imaging of the in situ cornea with two-photon excitation laser scanning microscopy. J Microsc 1995;178(Pt 1):20–7.

[57] Laiho LH, et al. Two-photon 3-D mapping of ex vivo human skin endogenous fluorescence species based on fluorescence emission spectra. J Biomed Opt 2005;10(2):024016.

[58] Radosevich AJ, et al. Hyperspectral in vivo two-photon microscopy of intrinsic contrast. Opt Lett 2008;33(18):2164–6.

[59] Chen J, et al. Spectroscopic characterization and microscopic imaging of extracted and in situ cutaneous collagen and elastic tissue components under two-photon excitation. Skin Res Technology 2009;15(4).

[60] Breunig HG, Studier H, Koenig K. Excitation-wavelength dependence of multiphoton excitation of fluorophores of human skin in vivo. Proc SPIE 2010;7548(Journal Article):754806.

[61] Yu Y, et al. Imaging-guided two-photon excitation-emission-matrix measurements of human skin tissues. J Biomed Opt 2012;17(7).

[62] Palero JA, et al. Three-dimensional multiphoton autofluorescence spectral imaging of live tissues. Proc SPIE 2006;6191(Journal Article):61910J.

[63] Palero JA, et al. Spectrally-resolved multiphoton imaging of post-mortem biopsy and in-vivo mouse skin tissues. Proc SPIE 2007;6442(Journal Article):64421B.

[64] Konig K, Riemann I. High-resolution multiphoton tomography of human skin with subcellular spatial resolution and picosecond time resolution. J Biomed Opt 2003;8(3):432–9.

[65] Becker W, et al. Fluorescence lifetime imaging by time-correlated single-photon counting. Microsc Res Tech 2004;63(1).

[66] Patalay R, et al. Multiphoton multispectral fluorescence lifetime tomography for the evaluation of basal cell carcinomas. PLoS One 2012;7(9):e43460.

[67] Patalay R, et al. Quantification of cellular autofluorescence of human skin using multiphoton tomography and fluorescence lifetime imaging in two spectral detection channels. Biomed Opt Express 2011;2(12):3295–308.

[68] Konig K, et al. Clinical two-photon microendoscopy. Microsc Res Tech 2007;70(5):398–402.

[69] Konig K, et al. Clinical application of multiphoton tomography in combination with high-frequency ultrasound for evaluation of skin diseases. J Biophotonics 2010;3(12):759–73.

[70] Breunig G, et al. Clinical multiphoton and CARS microscopy. Proc SPIE 2012;8226(Journal Article):822623.

[71] Fu D, et al. Two-color, two-photon, and excited-state absorption microscopy. J Biomed Opt 2007;12(5):054004.

[72] Chen W-L, et al. Single-wavelength reflected confocal and multiphoton microscopy for tissue imaging. J Biomed Opt 2009;14(5): 054026.

[73] Yazdanfar S, et al. Multifunctional imaging of endogenous contrast by simultaneous nonlinear and optical coherence microscopy of thick tissues. Microsc Res Tech 2007;70(7).

[74] Koenig K, et al. Current developments in clinical multiphoton tomography. Proc SPIE 2010;7569(Journal Article):756915.

[75] Campagnola PJ, Loew LM. Second-harmonic imaging microscopy for visualizing biomolecular arrays in cells, tissues and organisms. Nat Biotechnol 2003;21(11):1356–60.

[76] Chu SW, et al. High-resolution simultaneous three-photon fluorescence and third-harmonic-generation microscopy. Microsc Res Tech 2005;66(4):193–7.

[77] Lee JH, et al. Noninvasive in vitro and in vivo assessment of epidermal hyperkeratosis and dermal fibrosis in atopic dermatitis. J Biomed Opt 2009;14(1):014008.

[78] Bewersdorf J, Pick R, Hell SW. Multifocal multiphoton microscopy. Opt Lett 1998;23:655–7.

[79] Kim K, et al. Multifocal multiphoton microscopy based on multianode photomultiplier tubes. Optics Express 2007;15(18):11658–78.

[80] Cha JW, et al. Non-descanned multifocal multiphoton microscopy with a multianode photomultiplier tube. AIP Adv 2015; 5(8):084802.

[81] Cha JW, et al. Spectral-resolved multifocal multiphoton microscopy with multianode photomultiplier tubes. Opt Express 2014;22(18):21368–81.

[82] Evans CL, Xie XS. Coherent anti-Stokes Raman scattering microscopy: chemical imaging for biology and medicine. Annu Rev Anal Chem 2008;1(1):883–909.

[83] Begley RF, Harvey AB, Byer RL. Coherent anti-Stokes Raman spectroscopy. Appl Phys Lett 1974;25(7):387–90.

[84] Zheltikov AM. Coherent anti-Stokes Raman scattering: from proof-of-the-principle experiments to femtosecond CARS and higher order wave-mixing generalizations. J Raman Spectrosc 2000;31(8–9):653–67.

[85] Petibois C. Imaging methods for elemental, chemical, molecular, and morphological analyses of single cells. Anal Bioanal Chem 2010;397(6):2051–65.

[86] Tolles WM, et al. A review of the theory and application of coherent anti-Stokes Raman spectroscopy (CARS). Appl Spectrosc 1977;31(4):253–71.

[87] Freudiger CW, et al. Label-free biomedical imaging with high sensitivity by stimulated Raman scattering microscopy. Science 2008;322(5909):1857–61.

[88] Fu D, et al. Probing skin pigmentation changes with transient absorption imaging of eumelanin and pheomelanin. J Biomed Opt 2008;13(5):054036.

[89] Evans CL, et al. Chemical imaging of tissue in vivo with video-rate coherent anti-Stokes Raman scattering microscopy. Proc Natl Acad Sci USA 2005;102(46):16807–12.

[90] Durst ME, Zhu G, Xu C. Simultaneous spatial and temporal focusing for axial scanning. Optics Express 2006;14(25): 12243–54.

[91] Oron D, Tal E, Silberberg Y. Scanningless depth-resolved microscopy. Optics Express 2005;13(5):1468–76.

[92] Koenig K. Multiphoton tomography for tissue engineering. Proc SPIE 2008;6858(Journal Article):68580C.

[93] Koenig K, et al. Invited review: two-photon scanning systems for clinical high-resolution in vivo tissue imaging. Proc SPIE 2008; 6860(Journal Article):686014.

[94] Koenig K, et al. Clinical multiphoton tomography and clinical two-photon microendoscopy. Proc SPIE 2009;7183(Journal Article):718319.

[95] Koenig K, et al. In vivo multiphoton tomography in skin aging studies. Proc SPIE 2009;7161(Journal Article):71610H.

[96] Koenig K. New developments in multimodal clinical multiphoton tomography. Proc SPIE 2011;7903(Journal Article): 790305.

[97] Leiter U, Garbe C. Epidemiology of melanoma and nonmelanoma skin cancer: the role of sunlight. Adv Exp Med Biol 2008; 624(Journal Article).

[98] Marks R. An overview of skin cancers: incidence and causation. Cancer 1995;75(2).

[99] Teuchner K, et al. Femtosecond two-photon excited fluorescence of melanin. Photochem Photobiol 1999;70(2):146–51.

[100] Teuchner K, et al. Fluorescence studies of melanin by stepwise two-photon femtosecond laser excitation. J Fluorescence 2000; 10(3):275–81.

[101] Skala MC, et al. Multiphoton microscopy of endogenous fluorescence differentiates normal, precancerous, and cancerous squamous epithelial tissues. Cancer Res 2005;65(4).

[102] Skala MC, et al. In vivo multiphoton fluorescence lifetime imaging of protein-bound and free nicotinamide adenine dinucleotide in normal and precancerous epithelia. J Biomed Opt 2007; 12(2):024014.

[103] Skala MC, et al. In vivo multiphoton microscopy of NADH and FAD redox states, fluorescence lifetimes, and cellular morphology in precancerous epithelia. Proc Natl Acad Sci USA 2007;104(49).

[104] Wilson JW, et al. In vivo pump-probe microscopy of melanoma and pigmented lesions. Proc SPIE 2012;8226(Journal Article): 822602.

[105] Entenberg D, et al. Multimodal microscopy of immune cells and melanoma for longitudinal studies. Proc SPIE 2006;6081(Journal Article):60810A.

[106] Cicchi R, et al. Multiphoton imaging of basal cell carcinoma (BCC). Proc SPIE 2006;6090(Journal Article):60900O.

[107] Riemann I, et al. In vivo multiphoton tomography of inflammatory tissue and melanoma. Proc SPIE 2005;5686(Journal Article).

[108] Koenig K, et al. In vivo multiphoton tomography of skin cancer. Proc SPIE 2006;6089(Journal Article):60890R.

[109] Zhang K, et al. Bipolar cellular morphology of malignant melanoma in unstained human melanoma skin tissue. J Biomed Opt 2009;14(2):024042.

[110] Paoli J, et al. Multiphoton laser scanning microscopy on non-melanoma skin cancer: morphologic features for future non-invasive diagnostics. J Invest Dermatol 2008;128(5).

[111] Ericson MB, et al. Two-photon microscopy of non-melanoma skin cancer: initial experience and diagnostic criteria ex vivo. Proc SPIE 2007;6630(Journal Article):66300U.

[112] Patalay R, et al. Fluorescence lifetime imaging of skin cancer. Proc SPIE 2011;7883(Journal Article):78830A.

[113] Patalay R, et al. Non-invasive imaging of skin cancer with fluorescence lifetime imaging using two photon tomography. Proc SPIE 2011;8087(Journal Article):808718.

[114] Riemann I, et al. Non-invasive analysis/diagnosis of human normal and melanoma skin tissues with two-photon FLIM in vivo. Proc SPIE 2008;6842(Journal Article):684205.

[115] De Giorgi V, et al. Combined non-linear laser imaging (two-photon excitation fluorescence microscopy, fluorescence lifetime imaging microscopy, multispectral multiphoton microscopy) in cutaneous tumours: first experiences. J Eur Acad Dermatol Venereol 2009;23(3).

[116] Arginelli F, et al. High resolution diagnosis of common nevi by multiphoton laser tomography and fluorescence lifetime imaging. Skin Res Technology 2013;19(2):194–204.

[117] Dimitrow E, et al. Spectral fluorescence lifetime detection and selective melanin imaging by multiphoton laser tomography for melanoma diagnosis. Exp Dermatol 2009;18(6).

[118] Xiong SY, Yang JG, Zhuang J. Nonlinear spectral imaging of human normal skin, basal cell carcinoma and squamous cell carcinoma based on two-photon excited fluorescence and second-harmonic generation. Laser Phys 2011;21(10).

[119] Chernyavskiy O, et al. Imaging of mouse experimental melanoma in vivo and ex vivo by combination of confocal and nonlinear microscopy. Microsc Res Tech 2009;72(6).

[120] Teuchner K, Mory S, Leupold D. A mobile, intensified femtosecond fiber laser based TPF spectrometer for early diagnosis of malignant melanoma. Proc SPIE 2004;5516(Journal Article): 63–71.

[121] Wan Q, Applegate BE. Multiphoton coherence domain molecular imaging with pump-probe optical coherence microscopy. Opt Lett 2010;35(4):532–4.

[122] Wan Q, Applegate BE. Multiphoton coherence domain molecular imaging. 2010.

[123] Oh J-T, et al. Three-dimensional imaging of skin melanoma in vivo by dual-wavelength photoacoustic microscopy. J Biomed Opt 2006;11(3):034032.

[124] Vogler N, et al. Discrimination of skin diseases using the multimodal imaging approach. Proc SPIE 2012;8427(Journal Article): 842710.

[125] Chen S-Y, et al. In vivo virtual biopsy of human skin by using noninvasive higher harmonic generation microscopy. IEEE J Selected Top Quan Electronics 2010;16(3).

[126] Takeuchi S, et al. Melanin acts as a cause an atypical potent UVB photosensitizer to mode of cell death in murine skin. Proc Natl Acad Sci USA 2004;101(42).

[127] Fu D, et al. Two-color excited-state absorption imaging of melanins. Proc SPIE 2007;6424(Journal Article):642402.

[128] Matthews TE, et al. In vivo and ex vivo epi-mode pump-probe imaging of melanin and microvasculature. Biomed Opt Express 2011;2(6):1576–83.

[129] Leupold D, et al. The stepwise two-photon excited melanin fluorescence is a unique diagnostic tool for the detection of malignant transformation in melanocytes. Pigment Cell Melanoma Res 2011;24(3):438–45.

[130] Cicchi R, et al. Multidimensional custom-made non-linear microscope: from ex-vivo to in-vivo imaging. Appl Phys B-Lasers Opt 2008;92(3):359–65.

[131] Eichhorn R, et al. Early diagnosis of melanotic melanoma based on laser-induced melanin fluorescence. J Biomed Opt 2009;14(3): 034033.

[132] Krasieva TB, et al. Two-photon excited fluorescence spectroscopy and imaging of melanin in-vitro and in-vivo. Proc SPIE 2012;8226(Journal Article):82262S.

[133] Cicchi R, et al. Multidimensional two-photon imaging of diseased skin. Proc SPIE 2008;6859(Journal Article):685903.

[134] Cicchi R, et al. Non-linear laser imaging of skin lesions. Proc SPIE 2007;6633(Journal Article):66330Z.

[135] Seidenari S, et al. Diagnosis of BCC by multiphoton laser tomography. Skin Res Technology 2013;19(1):E297–304.

[136] Seidenari S, et al. Multiphoton laser tomography and fluorescence lifetime imaging of basal cell carcinoma: morphologic features for non-invasive diagnostics. Exp Dermatol 2012;21(11): 831–6.

[137] Alex A, et al. Three-dimensional multiphoton/optical coherence tomography for diagnostic applications in dermatology. J Biophotonics 2013;6(4):352–62.

[138] Ulrich M, et al. In vivo detection of basal cell carcinoma: comparison of a reflectance confocal microscope and a multiphoton tomograph. J Biomed Opt 2013;18(6):61229.

[139] Manfredini M, et al. High-resolution imaging of basal cell carcinoma: a comparison between multiphoton microscopy with fluorescence lifetime imaging and reflectance confocal microscopy. Skin Res Technology 2013;19(1):E433–43.

[140] Hoeller C, et al. In vivo imaging of cutaneous T-cell lymphoma migration to the skin. Cancer Res 2009;69(7).

[141] Cicchi R, et al. Multidimensional non-linear laser imaging of basal cell carcinoma. Opt Express 2007;15(16).

[142] Cicchi R, et al. Time-resolved multiphoton imaging of basal cell carcinoma. Proc SPIE 2007;6442(Journal Article):64421I.

[143] Riemann I, et al. Multiphoton tomography of skin tumors after ALA application. Proc SPIE 2007;6424(Journal Article):642405.

[144] Durr NJ, et al. Gold nanorods for optimized two-photon luminescence imaging of cancerous tissue. Proc SPIE 2007;6641(Journal Article):66410O.

[145] Durr NJ, et al. Two-photon luminescence imaging of cancer cells using molecularly targeted gold nanorods. Nano Lett 2007;7(4).

[146] Durr NJ, et al. Two-photon luminescence imaging of cancerous tissue using gold nanorods as bright contrast agents. Proc SPIE 2007;6630(Journal Article):66300Q.

[147] Park J, et al. Intra-organ biodistribution of gold nanoparticles using intrinsic two-photon-induced photoluminescence. Lasers Surg Med 2010;42(7).

[148] Puvanakrishnan P, et al. Narrow band imaging of squamous cell carcinoma tumors using topically delivered anti-EGFR antibody conjugated gold nanorods. Lasers Surg Med 2012;44(4).

[149] Lin S-J, et al. Discrimination of basal cell carcinoma from normal dermal stroma by quantitative multiphoton imaging. Opt Lett 2006;31(18).

[150] Lin S-J, et al. Quantitative multiphoton imaging for guiding basal-cell carcinoma removal. Proc SPIE 2007;6424(Journal Article):642404.

[151] Wang C-C, et al. Early development of cutaneous cancer revealed by intravital nonlinear optical microscopy. Appl Phys Lett 2010;97(11):113702.

[152] Wang C-C, et al. Utilizing nonlinear optical microscopy to investigate the development of early cancer in nude mice in vivo. Proc SPIE 2007;6630(Journal Article):66300Y.
[153] Levitt JM, et al. Automated biochemical, morphological, and organizational assessment of precancerous changes from endogenous two-photon fluorescence images. PLoS One 2011;6(9): e24765.
[154] Kantere D, et al. Anti-stokes fluorescence from endogenously formed protoporphyrin IX: implications for clinical multiphoton diagnostics. J Biophotonics 2013;6(5):409–15.
[155] Nadiarnykh O, et al. Carcinogenic damage to deoxyribonucleic acid is induced by near-infrared laser pulses in multiphoton microscopy via combination of two- and three-photon absorption. J Biomed Opt 2012;17(11):116024.
[156] Le Harzic R, et al. Nonlinear optical endoscope based on a compact two axes piezo scanner and a miniature objective lens. Optics Express 2008;16(25):20588–96.
[157] Zhao J, et al. Jadassohn-Pellizzari anetoderma: study of multiphoton microscopy based on two-photon excited fluorescence and second harmonic generation. Eur J Dermatol 2009;19(6).
[158] Lu K, et al. Multiphoton laser scanning microscopy of localized scleroderma. Skin Res Technology 2009;15(4).
[159] Wu X, et al. Real-time in vivo imaging collagen in lymphedematous skin using multiphoton microscopy. Scanning 2011;33(6).
[160] Koehler MJ, et al. Keratinocyte morphology of human skin evaluated by in vivo multiphoton laser tomography. Skin Res Technology 2011;17(4).
[161] Huck V, et al. Intravital multiphoton tomography as an appropriate tool for non-invasive in vivo analysis of human skin affected with atopic dermatitis. Proc SPIE 2011;7883(Journal Article):78830R.
[162] Koehler MJ, et al. Clinical application of multiphoton tomography in combination with confocal laser scanning microscopy for in vivo evaluation of skin diseases. Exp Dermatol 2011; 20(7).
[163] Koenig K, et al. Clinical optical coherence tomography combined with multiphoton tomography for evaluation of several skin disorders. Proc SPIE 2010;7554(Journal Article):75542I.
[164] Lin SJ, et al. Multiphoton fluorescence and second harmonic generation microscopy of different skin states. Proc SPIE 2005; 5686(Journal Article).
[165] Jain R, Weninger W. Shedding light on cutaneous innate immune responses: the intravital microscopy approach. Immunol Cell Biol 2013;91(4):263–70.
[166] Kabashima K, Egawa G. Intravital multiphoton imaging of cutaneous immune responses. J Invest Dermatol 2014;134(11): 2680–4.
[167] Roediger B, et al. Cutaneous immunosurveillance and regulation of inflammation by group 2 innate lymphoid cells. Nat Immunol 2013;14(6):564–73.
[168] Ghazaryan AA, et al. Multiphoton imaging and quantification of tissue glycation. Proc SPIE 2011;7895(Journal Article):789509.
[169] Tseng J-Y, et al. Multiphoton spectral microscopy for imaging and quantification of tissue glycation. Biomed Opt Express 2011;2(2).
[170] Deleted in review.
[171] Cui JZ, et al. Quantification of aortic and cutaneous elastin and collagen morphology in Marfan syndrome by multiphoton microscopy. J Struct Biol 2014;187(3):242–53.
[172] Lin S-J, et al. Evaluating cutaneous photoaging by use of multiphoton fluorescence and second-harmonic generation microscopy. Opt Lett 2005;30(17):2275–7.
[173] Lin S-J, et al. Monitoring photoaging by use of multiphoton fluorescence and second harmonic generation microscopy. Proc SPIE 2006;6078(Journal Article):607803.
[174] Koehler MJ, et al. In vivo assessment of human skin aging by multiphoton laser scanning tomography. Opt Lett 2006;31(19).
[175] Sugata K, et al. Evaluation of photoaging in facial skin by multiphoton laser scanning microscopy. Skin Res Technology 2011; 17(1).
[176] Koehler MJ, et al. Morphological skin ageing criteria by multiphoton laser scanning tomography: non-invasive in vivo scoring of the dermal fibre network. Exp Dermatol 2008;17(6).
[177] Kaatz M, et al. Depth-resolved measurement of the dermal matrix composition by multiphoton laser tomography. Skin Res Technology 2010;16(2):131–6.
[178] Koehler MJ, et al. Non-invasive evaluation of dermal elastosis by in vivo multiphoton tomography with autofluorescence lifetime measurements. Exp Dermatol 2012;21(1):48–51.
[179] Puschmann S, et al. Approach to quantify human dermal skin aging using multiphoton laser scanning microscopy. J Biomed Opt 2012;17(3):036005.
[180] Wu S, et al. Quantitative analysis on collagen morphology in aging skin based on multiphoton microscopy. J Biomed Opt 2011; 16(4):040502.
[181] Wu S, et al. The analysis of aging skin based on multiphoton microscopy. Proc SPIE 2010;7845(Journal Article):78450S.
[182] Cicchi R, et al. Scoring of collagen organization in healthy and diseased human dermis by multiphoton microscopy. J Biophotonics 2010;3(1–2).
[183] Koehler MJ, et al. Intrinsic, solar and sunbed-induced skin aging measured in vivo by multiphoton laser tomography and biophysical methods. Skin Res Technology 2009;15(3).
[184] Benati E, et al. Quantitative evaluation of healthy epidermis by means of multiphoton microscopy and fluorescence lifetime imaging microscopy. Skin Res Technology 2011;17(3).
[185] Baldeweck T, et al. In vivo multiphoton microscopy associated to 3D image processing for human skin characterization. Proc SPIE 2012;8226(Journal Article):82263O.
[186] Decenciere E, et al. Automatic 3D segmentation of multiphoton images: a key step for the quantification of human skin. Skin Res Technology 2013;19(2):115–24.
[187] Yasui T, Takahashi Y, Araki T. Polarization-resolved second-harmonic-generation imaging of photoaged dermal collagen fiber. Proc SPIE 2009;7183(Journal Article):71831X.
[188] Yasui T, et al. Observation of dermal collagen fiber in wrinkled skin using polarization-resolved second-harmonic-generation microscopy. Opt Express 2009;17(2).
[189] Lutz V, et al. Impact of collagen crosslinking on the second harmonic generation signal and the fluorescence lifetime of collagen autofluorescence. Skin Res Technology 2012;18(2).
[190] Lutz V, et al. Collagen crosslink status analysed in vitro using second harmonic generation (SHG) and fluorescence lifetime imaging (FLIM). Proc SPIE 2012;8207(Journal Article):820703.
[191] Pena AM, et al. Multiphoton microscopy of engineered dermal substitutes: assessment of 3D collagen matrix remodeling induced by fibroblasts contraction. Proc SPIE 2010;7548(Journal Article):754802.
[192] Pena A-M, et al. Multiphoton microscopy of engineered dermal substitutes: assessment of 3-D collagen matrix remodeling induced by fibroblast contraction. J Biomed Opt 2010;15(5): 056018.
[193] Cicchi R, et al. In-vivo multiphoton imaging of collagen remodeling after microablative fractional rejuvenation. Proc SPIE 2011; 7883(Journal Article):78830V.
[194] Gong W, Xie S, Huang Y. Evaluating thermal damage induced by pulsed light with multiphoton microscopy. Proc SPIE 2009; 7161(Journal Article):71610X.
[195] Cicchi R, et al. In vivo TPEF-SHG microscopy for detecting collagen remodeling after laser micro-ablative fractional resurfacing treatment. Proc SPIE 2011;8087(Journal Article): 80871B.
[196] Tsai T-H, et al. Monitoring laser-tissue interaction by non-linear optics. Proc SPIE 2008;6842(Journal Article):684202.

[197] Navarro FA, et al. Two photon confocal microscopy in wound healing. Proc SPIE 2001;2(19).
[198] Luo T, et al. Visualization of collagen regeneration in mouse dorsal skin using second harmonic generation microscopy. Laser Phys 2009;19(3).
[199] Brewer MB, et al. Multiphoton imaging of excised normal skin and keloid scar: preliminary investigations. Proc SPIE 2004; 5312(Journal Article).
[200] Meshkinpour A, et al. Treatment of hypertrophic scars and keloids with a radiofrequency device: a study of collagen effects. Lasers Surg Med 2005;37(5).
[201] Da Costa V, et al. Nondestructive imaging of live human keloid and facial tissue using multiphoton microscopy. Arch Facial Plast Surg 2008;10(1).
[202] Riemann I, et al. In vivo multiphoton tomography of skin during wound healing and scar formation. Proc SPIE 2007;6442(Journal Article):644226.
[203] Zhu X, et al. Characteristics of scar margin dynamic with time based on multiphoton microscopy. Lasers Med Sci 2011;26(2): 239–45.
[204] Chen J, et al. Multiphoton microscopy study of the morphological and quantity changes of collagen and elastic fiber components in keloid disease. J Biomed Opt 2011;16(5):051305.
[205] Zhu X, et al. Marginal characteristics of skin scarred dermis quantitatively extracted from multiphoton microscopic imaging. Proc SPIE 2010;7845(Journal Article):784528.
[206] Chen G, et al. Texture analysis on two-photon excited microscopic images of human skin hypertrophic scar tissue. 2008.
[207] Ferro DP, et al. Nonlinear optics for the study of human scar tissue. Proc SPIE 2012;8226(Journal Article):82263J.
[208] Jiang XS, et al. Monitoring process of human keloid formation based on second harmonic generation imaging. Laser Phys 2011;21(9).
[209] Chen G, et al. Nonlinear spectral imaging of human hypertrophic scar based on two-photon excited fluorescence and second-harmonic generation. Br J Dermatol 2009;161(1).
[210] Chen S, et al. Differentiating keloids from normal and hypertrophic scar based on multiphoton microscopy. Laser Phys 2010; 20(4).
[211] Zhu X, et al. Quantification of scar margin in keloid different from atrophic scar by multiphoton microscopic imaging. Scanning 2011;33(4):195–200.
[212] Su P-J, et al. Discrimination of collagen in normal and pathological skin dermis through second-order susceptibility microscopy. Opt Express 2009;17(13).
[213] Su P-J, et al. Discrimination of collagen in normal and pathological dermis through polarization second harmonic generation. Proc SPIE 2010;7569(Journal Article):75692A.
[214] Schenkl S, et al. Rigid and high NA multiphoton fluorescence GRIN-endoscopes. Proc SPIE 2007;6631(Journal Article):66310Q.
[215] Ehlers A, et al. In vivo multiphoton endoscopy of endogenous skin fluorophores. Proc SPIE 2007;6442(Journal Article):64421Y.
[216] Koenig K, et al. Clinical in vivo two-photon microendoscopy for intradermal high-resolution imaging with GRIN optics. Proc SPIE 2007;6442(Journal Article):644215.
[217] Weinigel M, et al. Compact clinical high-NA multiphoton endoscopy. Proc SPIE 2012;8217(Journal Article):821706.
[218] Rompolas P, et al. Live imaging of stem cell and progeny behaviour in physiological hair-follicle regeneration. Nature 2012; 487(7408):496–9.
[219] Liu F, et al. The bulge area is the major hair follicle source of nestin-expressing pluripotent stem cells which can repair the spinal cord compared to the dermal papilla. Cell Cycle 2011; 10(5):830–9.
[220] Uchugonova A, et al. The bulge area is the origin of nestin-expressing pluripotent stem cells of the hair follicle. J Cell Biochem 2011;112(8):2046–50.
[221] Sevrain D, et al. Two-photon microscopy of dermal innervation in a human re-innervated model of skin. Exp Dermatol 2013; 22(4):290–1.
[222] Yuryev M, Khiroug L. Dynamic longitudinal investigation of individual nerve endings in the skin of anesthetized mice using in vivo two-photon microscopy. J Biomed Opt 2012;17(4):046007.
[223] Baroli B. Penetration of nanoparticles and nanomaterials in the skin: fiction or reality? J Pharm Sci 2010;99(1):21–50.
[224] Barry BW. Breaching the skin's barrier to drugs. Nat Biotechnol 2004;22(2):165–7.
[225] Grewal BS, et al. Transdermal macromolecular delivery: real-time visualization of iontophoretic and chemically enhanced transport using two-photon excitation microscopy. Pharm Res 2000;17(7):788–95.
[226] Yu B, et al. Visualization of oleic acid-induced transdermal diffusion pathways using two-photon fluorescence microscopy. J Invest Dermatol 2003;120(3):448–55.
[227] Yu B, et al. Topographic heterogeneity in transdermal transport revealed by high-speed two-photon microscopy: determination of representative skin sample sizes. J Invest Dermatol 2002; 118(6):1085–8.
[228] Kushner JT, et al. Dual-channel two-photon microscopy study of transdermal transport in skin treated with low-frequency ultrasound and a chemical enhancer. J Invest Dermatol 2007; 127(12):2832–46.
[229] Nohynek GJ, Dufour EK, Roberts MS. Nanotechnology, cosmetics and the skin: is there a health risk? Skin Pharmacol Physiol 2008;21(3):136–49.
[230] Song Z, et al. Characterization of optical properties of ZnO nanoparticles for quantitative imaging of transdermal transport. Biomed Opt Express 2011;2(12):3321–33.
[231] Zvyagin AV, et al. Imaging of zinc oxide nanoparticle penetration in human skin in vitro and in vivo. J Biomed Opt 2008;13(6): 064031.
[232] Koenig K. Multiphoton tomography of intratissue tattoo nanoparticles. Proc SPIE 2012;8207(Journal Article):82070S.
[233] Labouta HI, et al. Combined multiphoton imaging-pixel analysis for semiquantitation of skin penetration of gold nanoparticles. Int J Pharmaceutics 2011;413(1–2):279–82.
[234] Roberts MS, et al. Non-invasive imaging of skin physiology and percutaneous penetration using fluorescence spectral and lifetime imaging with multiphoton and confocal microscopy. Eur J Pharmaceutics Biopharmaceutics 2011;77(3):469–88.
[235] Stracke F, et al. Multiphoton microscopy for the investigation of dermal penetration of nanoparticle-borne drugs. J Invest Dermatol 2006;126(10):2224–33.
[236] Ozeki Y, et al. High-speed molecular spectral imaging of tissue with stimulated Raman scattering. Nat Photon 2012;6(12):845–51.
[237] Hanson KM, Clegg RM. Observation and quantification of ultraviolet-induced reactive oxygen species in ex vivo human skin. Photochem Photobiol 2002;76(1):57–63.
[238] Hanson KM, Gratton E, Bardeen CJ. Sunscreen enhancement of UV-induced reactive oxygen species in the skin. Free Radic Biol Med 2006;41(8):1205–12.
[239] Hanson KM, et al. Two-photon fluorescence lifetime imaging of the skin stratum corneum pH gradient. Biophys J 2002;83(3): 1682–90.
[240] Werrlein RJ, Braue CR, Dillman JF. Multiphoton imaging the disruptive nature of sulfur mustard lesions. Proc SPIE 2005; 5700(Journal Article).
[241] Uchugonova A, et al. The bulge area is the origin of nestin-expressing pluripotent stem cells of the hair follicle. J Cell Biochem 2011;112(8):2046–50.
[242] Bazin R, Flament F, Colonna A, Le Harzic R, Bückle R, Piot B, et al. Clinical study on the effects of a cosmetic product on dermal extracellular matrix components using a high-resolution multiphoton tomograph. Skin Res Tech 2010;16(3):305–10.

CHAPTER

20

Nonlinear Microscopy in Clinical Dermatology

R. Cicchi[1,2], D. Kapsokalyvas[2], F.S. Pavone[1,2]

[1]National Institute of Optics, National Research Council (INO-CNR), Sesto Fiorentino, Italy; [2]European Laboratory for Non-linear Spectroscopy (LENS), Sesto Fiorentino, Italy

OUTLINE

INTRODUCTION

Modern microscopic methodologies provide imaging tools for a noninvasive, high-resolution, label-free deep imaging of skin that offer the potential for an "optical biopsy". For example, two-photon fluorescence (TPF) microscopy [1] provides high-resolution deep optical imaging of tissues without any exogenously added probe (Fig. 20.1A). TPF intrinsically offers several advantages compared with single-photon techniques, including higher spatial resolution, intrinsic optical sectioning capability, reduced photodamage and phototoxicity, and deeper penetration depth within biological tissues [2]. Further, biological tissues contain a variety of intrinsically fluorescent molecules [reduced nicotinamide adenine dinucleotide (NADH), tryptophan, keratins, melanin, elastin, cholecalciferol, and others], which can be imaged by TPF microscopy [3–5]. In particular, the fluorescence signal provided by mitochondrial NADH can be used for a morphological characterization of epithelial layers, as demonstrated by studies performed on ex vivo tissue samples [6,7], fresh biopsies [8–12], and also in vivo on both animals [13] and humans [14–18]. Additional morphological information can be provided by second harmonic generation (SHG) microscopy [19–29]. SHG microscopy offers the direct imaging of anisotropic molecules inside cells [19,20] and tissues such as collagen fibers [22]. SHG has also been used for investigating orientation of collagen fibers and their structural changes in healthy tissues such as human dermis [10,24,28,30,31] or cornea [25,27,32] versus in the tumor microenvironment [33–35] (see Fig. 20.1B). The combination of TPF and SHG microscopy represents a powerful tool for imaging skin dermis, because elastin and collagen can be respectively imaged by TPF and SHG microscopy [4]. In particular, it has been used for monitoring collagen alterations in dermal disorders [28] or at the tumor–stroma interface

Imaging in Dermatology
http://dx.doi.org/10.1016/B978-0-12-802838-4.00020-0

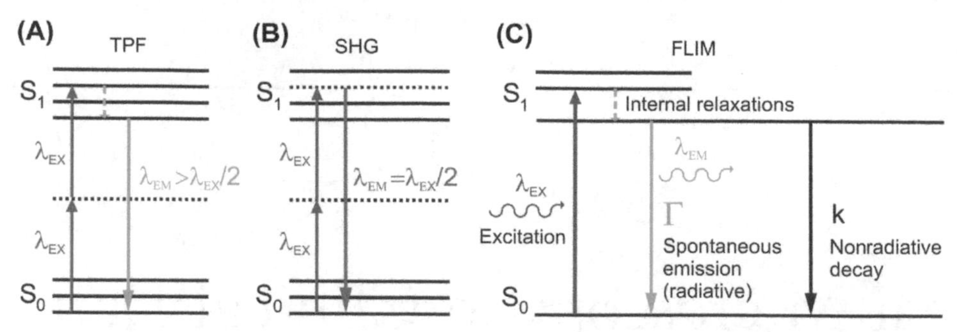

FIGURE 20.1 **Nonlinear microscopy diagrams.** (A) Energy transition diagram for two-photon absorption and fluorescence *(TPF)*. Two-photon energy level diagram shows the simultaneous absorption of two photons and the possibility to observe the transition from the ground level S_0 to the excited level S_1, through an intermediate virtual state. (B) Energy diagram for second-harmonic generation *(SHG)*. The excited state can be a virtual level or a real level. In the latter case, the SHG signal is enhanced by resonance. (C) Energy diagram for fluorescence lifetime imaging microscopy *(FLIM)*. In evidence, the two decay rate constants for radiative and nonradiative decays, respectively: Γ and k.

[33–35], as well as to monitor skin aging by measuring the collagen/elastic fiber content [36–38]. Fluorescence lifetime imaging microscopy (FLIM) is an additional microscopy technique, which can characterize a biological sample based on the measurement of the fluorescence decay rate [39,40] (see Fig. 20.1C). FLIM can be utilized to study protein localization [41] or the fluorescent molecular environment [42], and to provide functional information about tissue conditions [16,17,39,40,43–45]. FLIM can be used to detect cellular differentiation inside epithelia, as demonstrated by studies performed on cell cultures [46], fresh biopsies [8,11,12], and in vivo [18]. NADH emission decay by means of time-resolved analysis can be utilized to gain functional information on the examined tissue [46–48], as demonstrated by previous studies on both normal and diseased skin [16,17,45].

In this chapter, after having described the techniques used from a theoretical point of view, we will focus on epidermal imaging, showing how it is possible to differentiate various epidermal layers in vivo by using TPF microscopy with time-resolved detection of fluorescence. As expected, cells located in the deepest epidermal layers have the strongest metabolic activity, whereas the activity reduces when moving towards the epidermal surface. Such an approach can be used for characterizing epithelial tissues in various physiological conditions and has the potential to detect pathological conditions at a very early stage, as demonstrated by studies performed on cell cultures [46,49], fresh biopsies [8,11,12], and also in vivo [18]. Then, moving to dermal imaging, we will present two different applications of nonlinear imaging of dermis: in vivo optical characterization of psoriatic skin using nonlinear microscopy and follow-up observation of collagen remodeling within dermis after laser resurfacing. In particular, differences between healthy and psoriatic skin can be observed in vivo by nonlinear microscopy, which highly correlates with common histology. In particular, in psoriasis we observed a drastically different morphology of epithelium with respect to healthy skin that includes a thickening of corneum layer, a disorganization of corneocytes, and a more sparse arrangement of keratinocytes. Differences were observed also at the dermal level in terms of an increased density and penetration depth of dermal papillae in psoriasis with respect to healthy skin. In addition, combined TPF–SHG microscopy can be performed in vivo at the dermal level with the aim of characterizing collagen abundance and its organization in the inner forearm of subjects treated with laser resurfacing. Both qualitative and quantitative analyses demonstrated a stronger collagen synthesis and remodeling on treated subjects, with a strong dependence of effectiveness with age.

In conclusion, the imaging modalities presented here represent promising tools to be used for both diagnostic and follow-up purposes in dermatology and they could be a part of clinical dermatological settings in the near future.

THEORETICAL CONSIDERATIONS

Two-Photon Fluorescence Microscopy

The most popular nonlinear microscopy technique, able to provide morphological information on the sample under examination, is represented by TPF microscopy. TPF microscopy is based on a nonlinear optical process in which a fluorescent molecule simultaneously interacts with two photons in the same quantum event. In particular, by taking advantage of the nonlinear properties of this optical process, an electronic transition can be excited by the simultaneous absorption of two photons, each of them having one-half the transition energy, instead of through absorption of a single photon, having energy resonant with the transition. It has to be noted that such an excitation process does not consist in two separated absorption events, as in a two-step absorption, but has to be considered as a single quantum event. A theoretical description of the process can be provided using the quantum time-dependent perturbation theory. In fact, if we consider the Hamiltonian function of the system:

$$H = H_0 + \lambda V \quad (20.1)$$

where H_0 is the Hamiltonian of the unperturbed molecule and λV is the perturbation represented by the electromagnetic field. At the first order in the λ-expansion we find the single-photon transition probability; at the second order in the λ-expansion we find the two-photon transition probability. The second order is calculated as a combination of two single-photon transitions through an intermediate virtual state. The transition probability for single- and two-photon transition can be expressed as follows:

$$\text{single} - \text{photon} \quad \left|c_k^{(1)}(t)\right|^2 \propto E^2 = I \quad (20.2)$$

$$\text{two} - \text{photon} \quad \left|c_k^{(2)}(t)\right|^2 \propto E^4 = I^2 \quad (20.3)$$

where E is the lectric field and I is its intensity. At the second order of the perturbation theory, the two-photon transition probability scale is with the square of the light intensity, instead of with intensity, as demonstrated for single-photon transition.

Such a property offers several advantages for tissue imaging with respect to single-photon excitation. First, the nonlinear dependence of the signal on the excitation light intensity allows selectively exciting only molecules located in an extremely confined volume around the focal point. The direct consequence is an intrinsically high spatial resolution in comparison with other conventional microscopy techniques that employ similar excitation wavelengths. Second, the axial confinement of the excitation volume intrinsically provides optical sectioning capability, because it prevents the interaction with molecules located "out-of-focus." This feature allows reducing photodamage and phototoxic effect with respect to confocal microscopy. Third, nonlinear microscopy employs near-infrared (NIR) laser wavelengths for exciting electronic transitions with absorption bands located in the visible region of the spectrum. This feature offers deep label-free imaging in biological tissues [2,5]. In fact, every biological tissue intrinsically contains a certain amount of fluorescent molecules with excitation bands located in the visible range. This allows for imaging a biological tissue without adding any probe. In addition, the use of NIR wavelengths offers the advantage of a reduced scattering and hence a deeper penetration inside optically turbid samples as biological tissues. Further, scattered photons do not contribute to the excitation because they are unable to reach the excitation volume. This results in a spatial resolution that remains almost unchanged when imaging deep into an optically dense specimen [50]. For these reasons, nonlinear microscopy allows label-free optical sectioning with subdiffraction limited resolution in deep tissue.

Second Harmonic Generation Microscopy

In addition to TPF microscopy, morphological information on biological tissues can be obtained by taking advantage of the capability of some biological molecules to perform harmonic up-conversion of the exciting electromagnetic field. Harmonic up-conversion is an optical phenomenon involving coherent radiative scattering, whereas fluorescence generation involves incoherent radiative absorption and reemission. Because of its coherent nature, harmonic radiation is directional and depends critically on the spatial extent of the emission source. A detailed description of the SHG theory is provided in a study by Mertz and Moreaux [51]. A common relevant feature with respect to two-photon excitation, treated in the previous section, is the nonlinearity of the process that makes the total signal depending on the fourth power of the electromagnetic driving field amplitude. The most widespread microscopic technique based on harmonic up-conversion is represented by SHG microscopy [52,53].

SHG microscopy is based on a nonlinear second-order scattering process that occurs in molecules satisfying particular requirements. At the molecular level, only molecules having a large hyperpolarizability are able to generate SHG. This condition is necessary, but not sufficient for having an SHG signal. In fact, the coherence of the SHG process and the fact that in general an ensemble of molecules is excited have to be considered; hence coherent effects play a crucial role. When considering an ensemble of SHG emitting molecules, excited by the same field, SHG fields radiated by individual molecules are subjected to interference and hence the total radiated field strongly depends on the mutual phases of single emitters within the ensemble. For example, for randomly oriented molecules, each molecule scatters light with a random phase so that when performing coherent summation of the fields, different contribution destructively interferes, giving rise to a negligible SHG intensity. On the other hand, if we have an ensemble of aligned molecules, their phase relationship is well defined and the scattered fields constructively interferes, giving rise to a strong SHG signal. In this depiction, it is easy to understand that the additional requirement for SHG consists in having a particular structural organization of individual molecular emitters at the focal volume scale. In particular, it is required that the sample under investigation has a structural anisotropy at the focal volume scale. These conditions are satisfied in anisotropic biological molecules such as collagen [22,54,55], muscle [26,55–59], or microtubules [56], making SHG microscopy an ideal tool for imaging and probing the organization of these biological systems. SHG microscopy is extremely powerful for imaging connective and collagen-rich tissues such as the cornea [32,57], tendons [58,59], and arteries [60]. SHG microscopy can be used for the investigation of both structural organization and fibrillar organization of collagen in human dermis [10,24,30,61,62], keloid scars [28,63,64], fibrosis [65–67], thermally treated samples [25,31,68–70], and also in the tumor microenvironment [33–35,71–74].

Fluorescence Lifetime Imaging Microscopy

Both conventional and nonlinear fluorescence microscopy consist detect a steady-state fluorescence signal. Information is generated via observation of different levels of fluorescence intensity in different regions of the sample. However, the fluorescence level detected depends on several parameters, such as differences in concentration of endogenous fluorophores, variations in tissue morphology that affects the propagation of light, and differences in the microenvironment potentially affecting the quantum yield of the excited fluorophores. In this depiction, the contrast level achievable with this approach is in general not so high, leading to difficulties in discriminating diseased tissue, especially for a premalignant or early tumor stage. Further, the steady-state approach is only able to provide morphological information. An additional approach, also useful for obtaining functional information on the sample when using fluorescence microscopy, is based on the measurement of fluorescence dynamics. Functional information on a tissue can be obtained by temporal analysis of the fluorescence emission decay. This approach is the basis of the FLIM technique. FLIM consists of creating image contrast using fluorescence lifetime instead of fluorescence intensity as in steady-state measurements. Lifetime measurements offer several advantages with respect to steady-state measurements. It does not depend on light losses within the tissue, is very sensitive to sample microenvironment and to molecular energy exchanges, and is almost independent from excitation intensity. FLIM has been already demonstrated as a powerful technique in providing functional information about tissue conditions [16,17,39,40,43–45]. In particular, FLIM was successfully employed to characterize various tissues and to probe cellular differentiation in epithelium, as demonstrated by studies performed on cell cultures [46], fresh biopsies [8], and also in vivo [18,75].

IMAGING OF EPIDERMIS

In Vivo Imaging of Epidermis

TPF microscopy allows performing in-vivo imaging of the epidermis with subcellular spatial resolution, as shown in the images in Fig. 20.2. The resolution is high enough for resolving epithelial cells in various epidermal layers by taking advantage of the emission of mitochondrial NADH. Cells appear with a fluorescent cytoplasm and a dark nucleus. The discrimination of various epidermal layers can be performed on a morphological basis. The granular layer (see Fig. 20.2A) shows larger cells containing the characteristic granuli and emitting a lower average fluorescence with respect to other epidermal cells. Then, going deeper into skin (spinous layer), cells appear smaller in size, more round in shape and they are in average more fluorescent as well as tightly packed (see Fig. 20.2B) with respect to the upper layer. Finally, the basal layer (see Fig. 20.2C) contains the smaller epidermal cells having the largest metabolic activity, corresponding to a higher fluorescent signal, as shown in Fig. 20.2C. At this depth, images start exhibiting some extremely bright spots, probably corresponding to melanin granules. Even if the probability of generating reactive oxygen species by prolonged multiphoton excitation inside skin has already been demonstrated to be not higher than normal sun exposure [76], at this depth it is recommended to limit the exposure to a minimum because the basal layer is the most absorbing region inside the skin.

Characterization of Epidermal Layers

The characterization of epidermal layers can be performed via FLIM imaging. Typical mean fluorescence lifetime distributions, obtained from images acquired at 20, 40, and 60 μm depth from skin surface, using an excitation wavelength of 740 nm, are shown in Figs. 20.2G and H. It can be noted that the distribution of spinous and basal layers are well separated from one another, whereas the granular layer distribution is a mix of the previous two. The differentiation of various epidermal layers in terms of protein and cytokeratin content could relate to the measured differences in terms of fluorescence lifetime. In fact, even if the main contribution to the detected autofluorescence should arise from NADH and flavin adenine dinucleotide (FAD), the observed differences could also be the result of the differences in cytokeratin content in each layer. In particular, the granular layer is characterized by the presence of loricrin and profilaggrin, the spinous layer by cytokeratins 1 and 5, and the basal layer by cytokeratins 5 and 14. A spectroscopic and lifetime analysis on all these purified molecules would give new insights but would not definitely clarify the correspondence between cytokeratin composition and measured fluorescence lifetime in each epidermal layer, because the fluorescence lifetime depends on the molecular environment. Therefore, the three epidermal

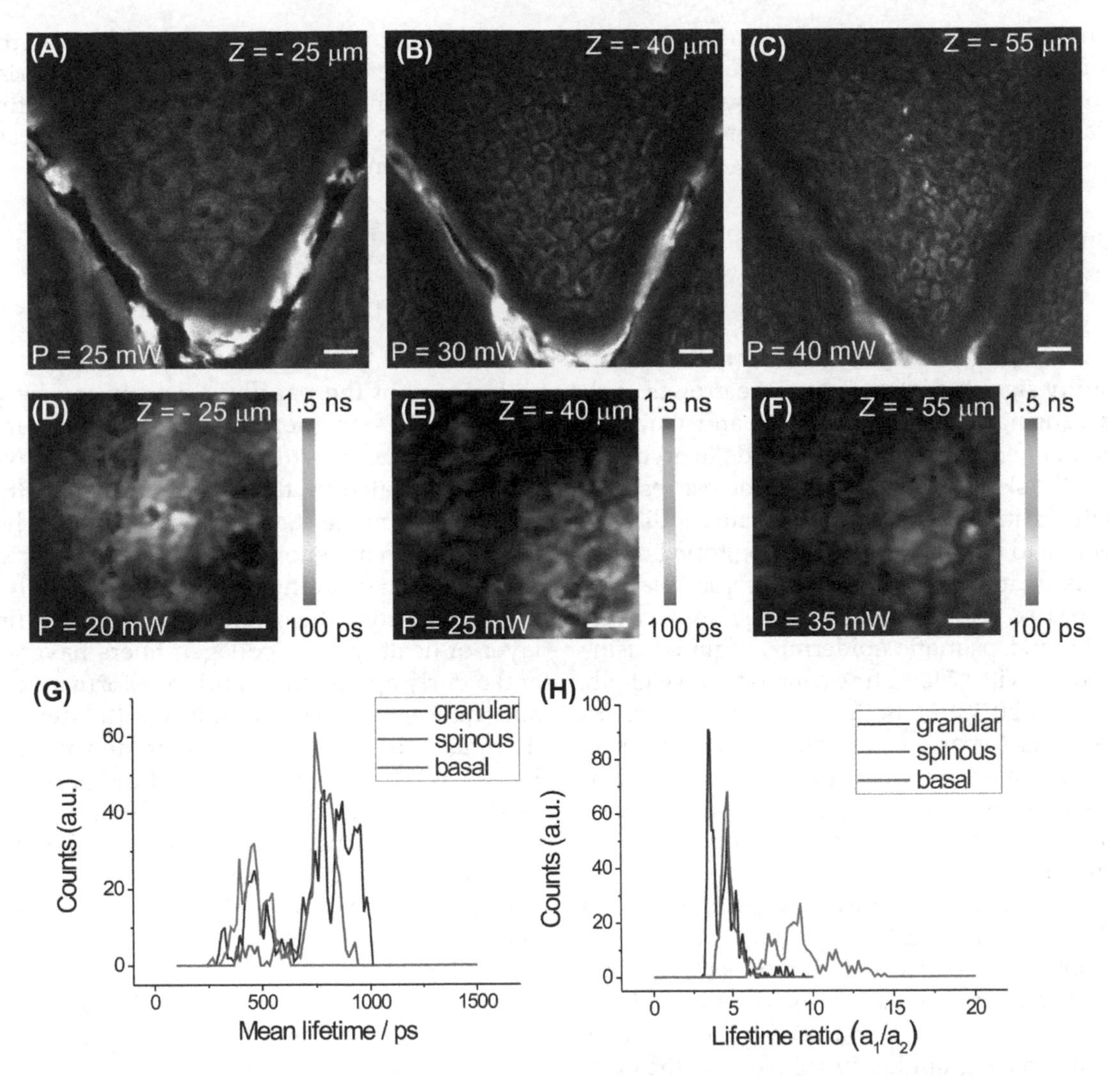

FIGURE 20.2 **TPF imaging of epidermis.** Scale = 20 μm. Two-photon fluorescence images acquired from the inner forearm of a healthy male volunteer at: 25 μm depth from skin surface (A) 40 μm depth from skin surface (B) and 55 μm depth from skin surface (C). FLIM images (D, E, and F), approximately correspond to the fluorescence steady-state images in (A–C). Scale = 6 μm. Excitation wavelength, 740 nm. An inset in each image *(lower left corner)* indicates laser power measured at the objective output. Mean cellular lifetime distribution (G) and mean cellular lifetime components ratio (H) of the three different epidermal layers were obtained after system response deconvolution and double-exponential fitting. *Figure modified from Cicchi R, Kapsokalyvas D, Pavone FS. Clinical nonlinear laser imaging of human skin: a review* BioMed Res Int 2014;2014:903589.

layers imaged can be characterized and discriminated on the basis of the ratio of lifetime components. In fact, by performing a fit of the measured fluorescence decay using a double-exponential decay function, and considering that NADH fluorescence is the main component of the endogenous TPF signal, the ratio of lifetime components can be correlated to tissue metabolism. NADH has a short lifetime in its free state and a much longer lifetime in its protein-bound state [47,48]. The fluorescence lifetime of protein-bound NADH depends on the molecule to which it is bound, and the changes in the binding site of NADH, connected with tumor development, can be potentially probed by measuring its lifetime. All these features can be used to optically monitor the metabolic state of a tissue and to potentially detect cancer at a very early stage on the basis of the fluorescence lifetime components ratio, as demonstrated by Skala and coauthors on cultivated living tissues [46,49]. The example shown in Fig. 20.2 demonstrates that the lifetime components ratio moves to higher values from skin surface to the deeper layers of epidermis. The result shown in Fig. 20.2H confirms that the ratio of lifetime components could be taken as an indicator of tissue metabolic activity. Along these lines, basal cells were found to have the highest ratio, corresponding to a higher metabolic activity. Hence a detailed characterization and differentiation of various epidermal layers, useful for diagnostics, can be obtained by analyzing the decay of NADH autofluorescence using FLIM. Considering that an altered metabolic activity of cells is very often precursor of a diseased state, these two parameters offer the potential to be used for diagnosing altered physiological conditions at a very early stage.

In Vivo Optical Characterization of Psoriatic Epidermis

Psoriasis is a skin disease with periods of remission and of exacerbation. It occurs when the immune system sends out faulty signals that speed up the growth cycle of skin cells, causing an excessive growth and abnormal differentiation of keratinocytes, which leads to erythema and scaling of the skin. As a result of the intense research focusing on the treatment and cure of this chronic disease, it is fundamental to have the means of monitoring the effect of various treatments at the microscopic level. An example with two typical image stacks, respectively from a healthy and psoriatic epidermis, acquired using TPF microscopy with 740-nm excitation wavelength, are shown in Fig. 20.3 [94]. Both image stacks were acquired in the inner forearm of the examined subjects. In healthy skin, the fluorescence originating from the corneum layer is very strong (see Fig. 20.3A) and it appears to be uniform without a characteristic morphology. On the other hand, in psoriatic skin the fluorescence intensity is lower and a characteristic punctuated pattern appears (see Fig. 20.3I). This atypical morphology of corneum layer is probably because psoriatic keratinocytes are not completely differentiated; hence it could be considered a characteristic feature of psoriasis. Typical big cells with the characteristic granular morphology in the cytoplasm are found in healthy epidermis (see Fig. 20.3B) at the level of stratum granulosum. On the other hand, this layer is very thin and in some cases is even absent in psoriatic skin, with cells (see Fig. 20.3J) having a very small cytoplasm. When moving down the focal plane, in healthy skin cells a typical spiny morphology is found (see Fig. 20.3C), corresponding to the stratum spinosum, which is a relatively extended layer with cells emitting a more uniform fluorescence compared with the outer layer. Skin surface cells at the psoriatic lesion (see Fig. 20.3K) appear quite different with respect to healthy cells. They have a very small cytoplasmic area, bigger nuclei, and are sparsely located within the lesion. At the depth of the basal layer, healthy cells (see Fig. 20.3D) have smaller size, are denser, and emit a strong fluorescence. In psoriatic lesions (see Fig. 20.3L), the basal layer is not clearly distinguishable in a single image because of the typical wavy morphology of the dermo–epidermal junction and the overexpression of dermal papillae. Nevertheless it is possible to identify basal cells that are arranged in circle around the formation of the dermal papillae. Cells show a very small cytoplasmic area and their packing is even denser, compared with the outer layer. Another characteristic feature of psoriatic skin at this depth is the presence of dermal papillae that infiltrate deep inside the epidermis (see Fig. 20.3E–P), as described in the following section.

IMAGING OF DERMIS

In Vivo Optical Characterization of Psoriatic Dermis

Imaging of the papillary dermis can be performed down to 0.2-mm depth from the skin surface using SHG microscopy with 900-nm excitation wavelength, as demonstrated by the images in Fig. 20.3E–P, where two typical image stacks, acquired within healthy and psoriatic skin are shown. The images cover on average 70 μm, which can range from approximately 100–170-μm depth from the skin surface. In the first dermal layer of healthy skin, collagen fibers have a small size and a curly appearance, and they form a very complex and dense network. At this depth, dermal papillae dominate the images and their density is high (see Fig. 20.3E). Dark regions around the papillae are occupied with epidermal cells proliferating inside the dermis. At a depth of 130 μm from the skin surface, most of the dermal papillae disappear in the dense collagen network (see Fig. 20.3F). When moving deeper, collagen fibers gradually increase in size and the collagen network appears with a better contrast. The network is less complex and the direction of the fibers is more ordered. At the depth of 170 μm, the quality of the image starts to degrade because of scattering (see Fig. 20.3G). A drastically different dermal morphology is found within psoriatic dermis. First, the density of dermal papillae is higher compared with healthy skin. The presence of papillae is still evident when moving deeper into the tissue, where the space around them starts to be filled with collagen. At a depth of 170 μm from the skin surface, dermis starts having a similar morphology with respect to healthy skin. However, the fine collagen network of interwoven curly fibers below the dermo–epidermal junction that is seen in healthy skin is not visible in psoriatic lesions. From the acquired images, it looks like the formation of papillae in psoriasis starts at depths around 170 μm below the skin surface, whereas in healthy skin the formation starts at a depth of around 115 μm. The characteristic feature of psoriatic skin (see Fig. 20.3M–O) is the elongated dermal papillae. This feature becomes better visible in a three-dimensional reconstructed image, shown in Fig. 20.3P. Examination of the three-dimensional reconstructed image reveals that papillae in psoriatic lesions have a length of around 100 μm, which is much longer than

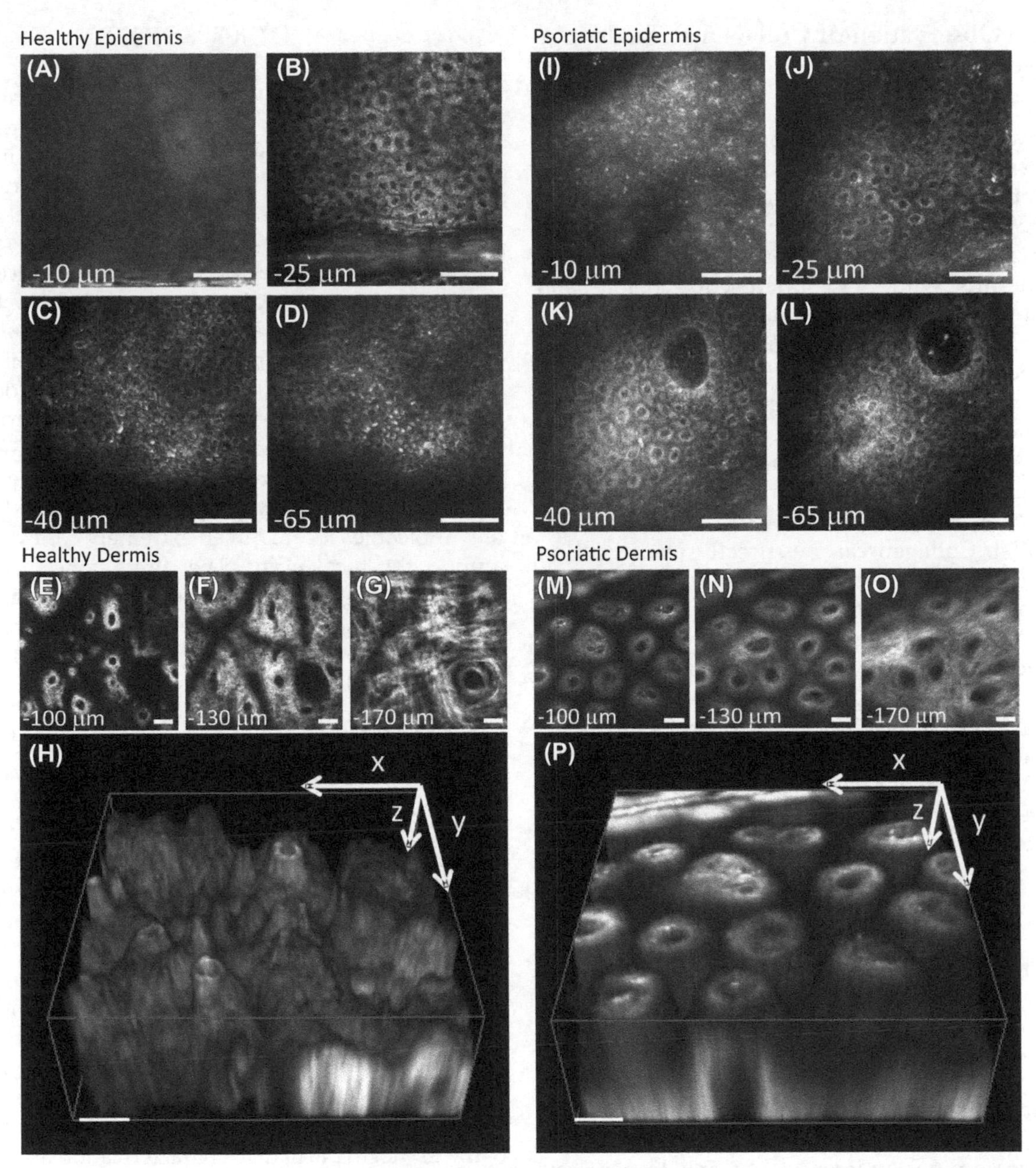

FIGURE 20.3 **TPF imaging of healthy and psoriatic skin.** Two-photon fluorescence images of the epidermal layers, acquired from the inner forearm of healthy (A–D) and psoriatic (I–L) skin. Scale = 50 μm. Excitation wavelength, 740 nm. SHG images of collagen from healthy papillary dermis (E–G) and from psoriatic papillary dermis (M–O). Scale bars: 50 μm. Excitation wavelength: 900 nm. An inset in each image *(lower left corner)* indicates depth of recording from skin surface. Note the three-dimensional reconstruction of the dermal layer of healthy (H) and psoriatic (P) skin. The volumes are rotated 30 degrees in the *y–z* plane in order to highlight dermal papillae. *Figure modified from Kapsokalyvas D, Cicchi R, Bruscino N, et al. In-vivo imaging of psoriatic lesions with polarization multispectral dermoscopy and multiphoton microscopy. Biomed Opt Express 2014;5: 2405–19.*

the length of papillae in healthy skin (around 30 μm). In addition, in psoriasis, dermal papillae are larger in diameter compared with healthy skin. A quantitative evaluation of this feature was obtained by measuring the cross-sectional surface occupied by dermal papillae at around 10 μm below their tips in both healthy and psoriatic skin. In conclusion, SHG microscopy within dermis allows noninvasive and quantitative characterization of dermal morphology. In the example shown, this imaging technique allowed measuring a length of the dermal papillae that is 60% longer in psoriasis and almost doubled in size, in comparison with healthy skin. The quantitative assessment of the dermal morphology could be useful for a follow-up study of the effectiveness of a treatment and whether could help in personalizing treatments.

Follow-Up Observation of Collagen Remodeling After Laser Resurfacing

An additional example of the potential application offered by nonlinear microscopy for noninvasive characterization of dermal morphology is presented in this section. In particular, a microscopic observation of the effects on collagen caused by microablative laser resurfacing treatment was performed by imaging dermis with SHG microscopy before and 40 days after the treatment. The acquired SHG images were visually examined for extracting information on collagen fibers and amorphous component appearances that are related to skin aging. In particular, as the age increases, an increase in collagen fiber thickness and density and a reduction of the amorphous component (mainly composed by hyaluronic acid and glycosaminoglycans) that affects tissue hydration are expected [77]. Collagen can be directly visualized on SHG images, whereas the increased scattering that gives the images a more "cloudy" appearance can indirectly give an indication about the amorphous component abundance [38]. For youngest people examined (age <35 years), the images acquired after treatment (Fig. 20.4D,E, and F) did not show any significant modification with respect to the corresponding images acquired before treatment (see Fig. 20.4A,B, and C). For middle-aged subjects (35–60 years; see Fig. 20.4G–L), a slight increase of collagen fiber density was observed in the images taken after the treatment with respect to those ones taken before. In addition, the amorphous component was increased as demonstrated by a more cloudy appearance of the images. For oldest subjects (age >60 years; see Fig. 20.4M–R), collagen fibers strongly increased in density and the amorphous component had undergone a drastic improvement. Further, both epidermal thickness and the number of dermal papillae increased. All of the observed features were in agreement with a rejuvenating effect, demonstrating that nonlinear microscopy is able to noninvasively image collagen morphology at a microscopic level and it can highlight subtle changes in collagen organization, such as those induced by fractional laser resurfacing treatment. The effectiveness of the treatment seems to increase with age, although better statistics on a larger number of volunteers would be beneficial to confirm this tendency. The relatively small effectiveness observed on young and middle-aged volunteers could be attributed to the particular anatomical site chosen for this study, the inner forearm. This site is not much photoexposed or pigmented. These features help the acquisition of quality images. On the other hand, the skin aging of this site is usually less with respect to the other anatomical sites because of reduced photoexposure.

DISCUSSION

In this chapter, we have highlighted the capability of nonlinear microscopy for clinical dermatological imaging of both the epidermis and the dermis in vivo. In particular, both cells in the epidermis and dermal collagen morphology can be noninvasively imaged and characterized by nonlinear microscopy, providing high-resolution images of the skin morphology, which can be used for both diagnostic and therapy follow-up purposes. Although skin morphology can be noninvasively imaged with high-resolution by various laser scanning imaging techniques, such as confocal reflectance microscopy [78–90], the nonlinear approach offers several advantages with respect to the confocal reflectance. First, the nonlinear dependence of the signal on the excitation light intensity allows selectively exciting only molecules located in an extremely confined volume around the focal point. The direct consequence is an intrinsically high spatial resolution in comparison with other conventional microscopy techniques that employ similar excitation wavelengths. Second, a confocal pinhole rejecting out-of-focus light is not required in nonlinear microscopy, allowing for a three-dimensional scanning of the specimen using a reduced exposure with respect to confocal microscopy. Third, every biological tissue intrinsically contains certain amount of fluorescent molecules and SHG emitters that can be excited using nonlinear microscopy. This allows imaging the morphology of a biological tissue without the addition of probes. Finally, the use of particular contrast methods based on nonlinear excitation, such as SHG and FLIM, allows a more specific identification of molecular species in skin, such as collagen in dermis and other nucleotides in epidermis that allow a more exhaustive characterization of both epidermis and dermis. In fact, a selective imaging and spectroscopy of these molecules can provide information about the physiological condition of the tissue. In particular, characterization and differentiation of various epidermal layers, useful for diagnostics, can be obtained by analyzing the decay of NADH autofluorescence using FLIM, as described in the section titled Second Harmonic Generation Microscopy. In particular, the mean fluorescence lifetime of NADH and the ratio of fast to slow fluorescence lifetime components can be taken as indicator of the metabolic state of cells. Considering that an altered metabolic activity of cells is very often precursor of a diseased state, these two parameters offer the potential to be used for diagnosing altered physiological conditions at a very early stage.

The potential clinical dermatological applications of nonlinear microscopy are not limited to skin diagnostics, but they also extend to treatment follow-up observation. These nonlinear imaging techniques can be used for

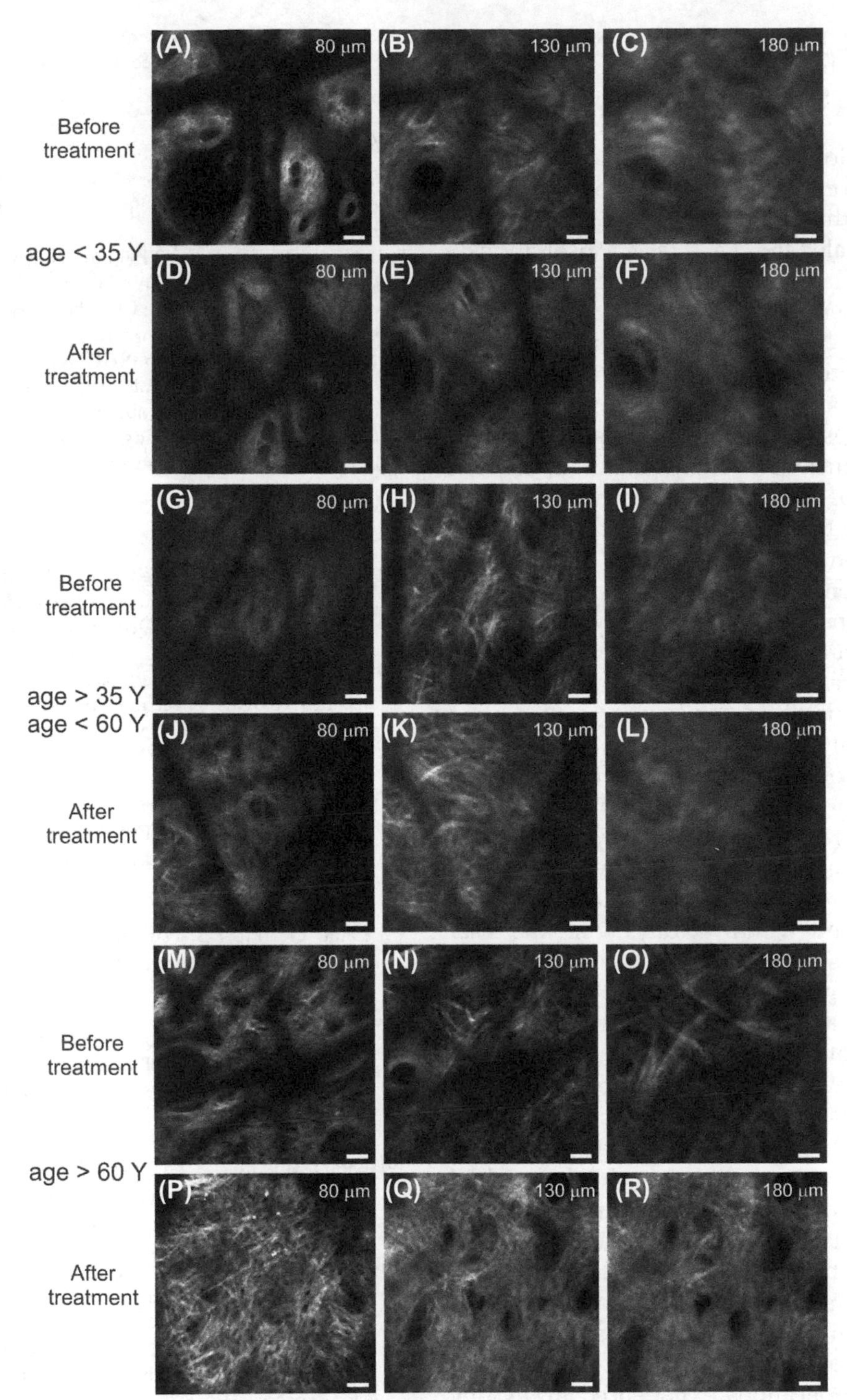

FIGURE 20.4 **SHG imaging of collagen after laser resurfacing.** SHG images of human dermis taken at 80, 130, and 180 μm from skin surface on the inner forearm of healthy volunteers before and 40 days after microablative fractional laser resurfacing treatment. *Top*: representative images of subjects with aged 35 years, taken before (A, B, and C) and 40 days after the treatment (D, E, and F). *Middle*: representative images of subjects aged 35–60 years, taken before (G, H, and I) and 40 days after treatment (J, K, and L). *Bottom*: representative images of subjects aged >60 years, taken before (M, N, and O) and 40 days after treatment (P, Q, and R). Excitation wavelength, 900 nm. Scale = 4.0 μm. *Figure modified from Cicchi R, Kapsokalyvas D, Troiano M, et al. In vivo non-invasive monitoring of collagen remodelling by two-photon microscopy after micro-ablative fractional laser resurfacing.* J Biophoton 2014;7:914–25.

in vivo determination of the characteristic micromorphology of psoriasis, highlighting at the epidermal level the sparsely located cells with small cytoplasm compared with healthy cells. Further, the more pronounced epidermal proliferation and the dilated papillae of the dermis, typical of psoriasis, were imaged at high resolution. In particular, a three-dimensional reconstruction of dermal papillae revealed the characteristic psoriatic skin morphology, where dilated and elongated dermal papillae could be observed. A quantitative measurement demonstrated that the length of dermal papillae is 60% longer in psoriasis and almost doubled in size compared with healthy skin. A potential quantitative monitoring of a treatment effect on psoriasis could be based on the measurement of the nucleus-to-cytoplasm ratio in the epidermis and on papillae size in the dermis. Apart from monitoring the effect of experimental treatments, the presented imaging techniques could also be used for the personalization of existing treatments.

Further, we demonstrated the capability of nonlinear microscopy for imaging dermis in vivo after microablative fractional laser resurfacing treatment. In particular, for the first time nonlinear microscopy was used in vivo with the final goal of monitoring a laser-based treatment [38]. The obtained results have shown that nonlinear microscopy is able to noninvasively provide a quantitative measurement of the efficacy of the resurfacing treatment. The production of new collagen, as well as an increase in the amount of dermal amorphous component, was found within 40 days from the laser treatment. The effects caused by the treatment can be evaluated qualitatively by visual examination of SHG images of collagen, and quantitatively by measuring the relative amount of SHG and TPF signals.

In conclusion, the methodologies described here could become a powerful tool to be used in a dermatological setting for both early diagnosis and therapy follow-up purposes. Emerging technologies for ultrafast pulsed laser sources, potentially cheaper than the usual solid state Ti:Sapphire oscillator, can help nonlinear laser scanning microscopy to become more and more popular among medical doctors with the final goal of being recognized as a standard clinical imaging method. On the basis of the results described here, and of several other successful dermatological applications experienced in vivo by means of nonlinear microscopy in recent years [16–18,37,38,45,91,92], we believe that in the near future nonlinear laser scanning microscopy will find a stable place in the clinical dermatological setting.

List of Acronyms and Abbreviations

FAD Flavin adenine dinucleotide
FLIM Fluorescence lifetime imaging microscopy
NADH Reduced form of nicotinamide adenine dinucleotide
NIR Near-infrared
SHG Second harmonic generation
TPF Two-photon fluorescence

Glossary

Nonlinear microscopy Laser scanning imaging techniques in which the interaction between light and the specimen is mediated by two or more photons.
Two-photon fluorescence (TPF) microscopy A laser scanning fluorescence microscopy technique in which the excitation of the fluorescent molecule is provided by the simultaneous interaction with two photons in the same quantum event.
Second harmonic generation (SHG) microscopy A laser scanning microscopy technique in which the contrast is provided by the harmonic up-conversion capability of particular molecules.
Fluorescence lifetime imaging microscopy (FLIM) A time-resolved implementation of laser scanning fluorescence microscopy.

References

[1] Denk W, Strickler JH, Webb WW. Two-photon laser scanning fluorescence microscopy. Science 1990;248:73–6.
[2] Helmchen F, Denk W. Deep tissue two-photon microscopy. Nat Methods 2005;2:932–40.
[3] Zoumi A, Yeh A, Tromberg BJ. Imaging cells and extracellular matrix in vivo by using second harmonic generation and two-photon excited fluorescence. Proc Natl Acad Sci U S A 2002;99:11014–9.
[4] Zipfel WR, Williams RM, Christie R, Nikitin AY, Hyman BT, Webb WW. Live tissue intrinsic emission microscopy using multiphoton-excited native fluorescence and second harmonic generation. Proc Natl Acad Sci U S A 2003;100(12):7075–80.
[5] Zipfel WR, Williams RM, Webb WW. Nonlinear magic: multiphoton microscopy in the biosciences. Nat Biotechnol 2003; 21(11):1369–77.
[6] Laiho LH, Pelet S, Hancewicz TM, Kaplan PD, So PTC. Two-photon 3-D mapping of ex vivo human skin endogenous fluorescence species based on fluorescence emission spectra. J Biomed Opt 2005;10:024016.
[7] Cicchi R, Pavone FS. Non-linear fluorescence lifetime imaging of biological tissues. Anal Bioanal Chem 2011;400(9):2687–97.
[8] Cicchi R, Massi D, Sestini S, et al. Multidimensional non-linear laser imaging of basal cell carcinoma. Opt Express 2007;15(16): 10135–48.
[9] Paoli J, Smedh M, Wennberg A-M, Ericson MB. Multiphoton laser scanning microscopy on non-melanoma skin cancer: morphologic features for future non-invasive diagnostics. J Invest Dermatol 2008;128:1248–55.
[10] Cicchi R, Sestini S, De Giorgi V, Massi D, Lotti T, Pavone FS. Nonlinear laser imaging of skin lesions. J Biophotonics 2008; 1(1):62–73.
[11] Cicchi R, Crisci A, Cosci A, et al. Time- and spectral-resolved two photon imaging of healthy bladder mucosa and carcinoma in situ. Opt Express 2010;18:3840–9.
[12] Cicchi R, Sturiale A, Nesi G, et al. Multiphoton morpho-functional imaging of healthy colon mucosa, adenomatous polyp and adenocarcinoma. Biomed Opt Express 2013;4(7):1204–13.
[13] Palero JA, de Bruijn HS, van der Ploeg van den Heuvel A, Sterenborg HJCM, Gerritsen HC. Spectrally resolved multiphoton imaging of in vivo and excised mouse skin tissues. Biophys J 2007; 93:992–1007.
[14] Masters BR, So PTC, Gratton E. Optical biopsy of in vivo human skin: multi-photon excitation microscopy. Lasers Med Sci 1998;13: 196–203.
[15] Masters BR, So PTC. Confocal microscopy and multi-photon excitation microscopy of human skin in vivo. Opt Express 2001;8:2–9.

[16] Konig K, Riemann I. High-resolution multiphoton tomography of human skin with subcellular spatial resolution and picosecond time resolution. J Biomed Opt 2003;8(3):432–9.

[17] Dimitrow E, Riemann I, Ehlers A, et al. Spectral fluorescence lifetime detection and selective melanin imaging by multiphoton laser tomography for melanoma diagnosis. Exp Dermatol 2009; 18(6):509–15.

[18] Konig K. Clinical multiphoton tomography. J Biophotonics 2008; 1(1):13–23.

[19] Moreaux L, Sandre O, Charpak S, Blanchard-Desce M, Mertz J. Coherent scattering in multi-harmonic light microscopy. Biophys J 2001;80:1568–74.

[20] Campagnola PJ, Loew LM. Second-harmonic imaging microscopy for visualizing biomolecular arrays in cells, tissues and organisms. Nat Biotechnol 2003;21:1356–60.

[21] Campagnola PJ, Millard AC, Terasaki M, Hoppe PE, Malone CJ, Mohler WA. Three-dimensional high-resolution second-harmonic generation imaging of endogenous structural proteins in biological tissues. Biophys J 2002;81:493–508.

[22] Roth S, Freund I. Second-harmonic generation in collagen. J Chem Phys 1979;70:1637–43.

[23] Williams RM, Zipfel WR, Webb WW. Interpreting second-harmonic generation images of collagen I fibrils. Biophys J 2005; 88:1377–86.

[24] Stoller P, Reiser KM, Celliers PM, Rubenchik AM. Polarization-modulated second harmonic generation in collagen. Biophys J 2002;82:3330–42.

[25] Matteini P, Ratto F, Rossi F, et al. Photothermally-induced disordered patterns of corneal collagen revealed by SHG imaging. Opt Express 2009;17(6):4868–78.

[26] Vanzi F, Sacconi L, Cicchi R, Pavone FS. Protein conformation and molecular order probed by second-harmonic generation microscopy. J Biomed Opt 2012;17:060901.

[27] Matteini P, Cicchi R, Ratto F, et al. Thermal transitions of fibrillar collagen unveiled by second-harmonic generation microscopy of corneal stroma. Biophysical J 2012;103(6):1179–87.

[28] Cicchi R, Kapsokalyvas D, De Giorgi V, et al. Scoring of collagen organization in healthy and diseased human dermis by multiphoton microscopy. J Biophoton 2010;3:34–43.

[29] Cicchi R, Matthaus C, Meyer T, et al. Characterization of collagen and cholesterol deposition in atherosclerotic arterial tissue using non-linear microscopy. J Biophotonics 2014;7:135–43.

[30] Yasui T, Tohno Y, Araki T. Characterization of collagen orientation in human dermis by two-dimensional second-harmonic-generation polarimetry. J Biomed Opt 2004;9:259–64.

[31] Sun Y, Chen WL, Lin SJ, et al. Investigating mechanisms of collagen thermal denaturation by high resolution second-harmonic generation imaging. Biophys J 2006;91:2620–5.

[32] Han M, Giese G, Bille JF. Second harmonic generation imaging of collagen fibrils in cornea and sclera. Opt Express 2005;13: 5791–7.

[33] Brown EB, McKee T, DiTomaso E, et al. Dynamic imaging of collagen and its modulation in tumors in vivo using second-harmonic generation. Nat Med 2003;9:796–800.

[34] Lin SJ, Jee SH, Kuo CJ, et al. Discrimination of basal cell carcinoma from normal dermal stroma by quantitative multiphoton imaging. Opt Lett 2006;31:2756–8.

[35] Provenzano PP, Eliceiri KW, Campbell JM, Inman DR, White JG, Keely PJ. Collagen reorganization at the tumor–stromal interface facilitates local invasion. BMC Med 2006;4(1):38.

[36] Lin SJ, Wu RJ, Tan HY, et al. Evaluating cutaneous photoaging by use of multiphoton fluorescence and second-harmonic generation microscopy. Opt Lett 2005;30:2275–7.

[37] Koehler MJ, König K, Elsner P, Buckle R, Kaatz M. In vivo assessment of human skin aging by multiphoton laser scanning tomography. Opt Lett 2006;31:2879–81.

[38] Cicchi R, Kapsokalyvas D, Troiano M, et al. In vivo non-invasive monitoring of collagen remodelling by two-photon microscopy after micro-ablative fractional laser resurfacing. J Biophotonics 2013.

[39] Tadrous PJ. Methods for imaging the structure and function of living tissues and cells. 2. Fluorescence lifetime imaging. J Pathol 2000;191(3):229–34.

[40] Tadrous PJ, Siegel J, French PMW, Shousha S, Lalani EN, Stamp GWH. Fluorescence lifetime imaging of unstained tissues: early results in human breast cancer. J Pathol 2003;199(3):309–17.

[41] Chen Y, Periasamy A. Characterization of two-photon excitation fluorescence lifetime imaging microscopy for protein localization. Microsc Res Tech 2004;63(1):72–80.

[42] Hanson KM, Behne MJ, Barry NP, Mauro TM, Gratton E, Clegg RM. Two-photon fluorescence lifetime imaging of the skin stratum corneum pH gradient. Biophys J 2002;83:1682–90.

[43] Bastiaens PIH, Squire A. Fluorescence lifetime imaging microscopy: spatial resolution of biochemical processes in the cell. Trends Cell Biol 1999;9(2):48–52.

[44] Cubeddu R, Pifferi A, Taroni P, et al. Fluorescence lifetime imaging: an application to the detection of skin tumors. IEEE J Sel Top Quant 1999;5(4):923–9.

[45] Dimitrow E, Ziemer M, Koehler MJ, et al. Sensitivity and specificity of multiphoton laser tomography for in vivo and ex vivo diagnosis of malignant melanoma. J Invest Dermatol 2009;129(7):1752–8.

[46] Skala MC, Riching KM, Bird DK, et al. In vivo multiphoton fluorescence lifetime imaging of protein-bound and free nicotinamide adenine dinucleotide in normal and precancerous epithelia. J Biomed Opt 2007;12(2).

[47] Lakowicz JR, Szmacinski H, Nowaczyk K, Johnson ML. Fluorescence lifetime imaging of free and protein-bound NADH. Proc Natl Acad Sci U S A 1992;89:1271–5.

[48] Li D, Zheng W, Qu JY. Time-resolved spectroscopic imaging reveals the fundamental of cellular NADH fluorescence. Opt Lett 2008;33:2365–7.

[49] Skala MC, Riching KM, Gendron-Fitzpatrick A, et al. In vivo multiphoton microscopy of NADH and FAD redox states, fluorescence lifetimes, and cellular morphology in precancerous epithelia. Proc Natl Acad Sci U S A 2007;104(49):19494–9.

[50] Centonze VE, White JG. Multiphoton excitation provides optical sections from deeper within scattering specimens than confocal imaging. Biophys J 1998;75:2015–22.

[51] Mertz J, Moreaux L. Second-harmonic generation by focused excitation of inhomogeneously distributed scatterers. Opt Commun 2001;196:325–30.

[52] Gannaway JN, Sheppard CJR. Second-harmonic imaging in the scanning optical microscope. Opt Quant Elect 1978;10:435–9.

[53] Sheppard CJR, Kompfner R, Gannaway J, Walsh D. Scanning harmonic optical microscope. IEEE J Quan Electron 1977;13E: 100D.

[54] Freund I, Deutsch M, Sprecher A. Optical second-harmonic microscopy, crossed-beam summation and small-angle scattering in rat-tail tendon. Biophys J 1986;50:693–712.

[55] Cicchi R, Vogler N, Kapsokalyvas D, Dietzek B, Popp J, Pavone FS. From molecular structure to tissue architecture: collagen organization probed by SHG microscopy. J Biophoton 2012.

[56] Dombeck DA, Kasischke KA, Vishwasrao HD, Ingelsson M, Hyman BT, Webb WW. Uniform polarity microtubule assemblies imaged in native brain tissue by second-harmonic generation microscopy. Proc Natl Acad Sci U S A 2003;100:7081–6.

[57] Yeh A, Nassif N, Zoumi A, Tromberg BJ. Selective corneal imaging using combined second harmonic generation and two-photon excited fluorescence. Opt Lett 2002;27:2082–4.

[58] Stoller P, Kim BM, Rubenchik AM, Reiser KM, Da Silva LB. Polarization-dependent optical second-harmonic imaging of a rat-tail tendon. J Biomed Opt 2002;7(2):205–14.

[59] Theodossiou T, Thrasivoulou C, Ekwobi C, Becker D. Second harmonic generation confocal microscopy of collagen type I from rat tendon cryosections. Biophys J 2006;91:4665–77.

[60] Zoumi A, Lu X, Kassab GS, Tromberg BJ. Imaging coronary artery microstructure using second-harmonic and two-photon fluorescence microscopy. Biophys J 2004;87:2778–86.

[61] Su PJ, Chen WL, Hong JB, et al. Discrimination of collagen in normal and pathological skin dermis through second-order susceptibility microscopy. Opt Express 2009;17:11161–71.

[62] Pena A, Fagot D, Olive C, et al. Multiphoton microscopy of engineered dermal substitutes: assessment of 3-D collagen matrix remodeling induced by fibroblast contraction. J Biomed Opt 2010;15:056018.

[63] Chen J, Zhuo S, Jiang X, et al. Multiphoton microscopy study of the morphological and quantity changes of collagen and elastic fiber components in keloid disease. J Biomed Opt 2011;16:051305.

[64] Medyukhina A, Vogler N, Latka I, et al. Automated classification of healthy and keloidal collagen patterns based on processing of SHG images of human skin. J Biophoton 2011;4:627–36.

[65] Strupler M, Pena AM, Hernest M, et al. Second harmonic imaging and scoring of collagen in fibrotic tissues. Opt Express 2007;15:4054–65.

[66] Guilbert T, Odin C, Le Grand Y, et al. A robust collagen scoring method for human liver fibrosis by second harmonic microscopy. Opt Express 2010;18:25794–807.

[67] Gailhouste L, Grand Y, Odin C, et al. Fibrillar collagen scoring by second harmonic microscopy: a new tool in the assessment of liver fibrosis. J Hepathol 2010;52:398–406.

[68] Lin SJ, Hsiao CY, Sun Y, et al. Monitoring the thermally induced structural transitions of collagen by use of second-harmonic generation microscopy. Opt Lett 2005;30:622–4.

[69] Theodossiou T, Rapti GS, Hovhannisyan V, Georgiou E, Politopoulos K, Yova D. Thermally induced irreversible conformational changes in collagen probed by optical second harmonic generation and laser-induced fluorescence. Lasers Med Sci 2002;17:34–41.

[70] Lo W, Chang YL, Liu JS, et al. Multimodal, multiphoton microscopy and image correlation analysis for characterizing corneal thermal damage. J Biomed Opt 2009;14:054003.

[71] Guo Y, Savage HE, Liu F, Schantz P, Ho PP, Alfano RR. Subsurface tumor progression investigated by noninvasive optical second harmonic tomography. Proc Natl Acad Sci U S A 1999;96:10854–6.

[72] Han X, Burke RM, Zettel ML, Tang P, Brown EB. Second harmonic properties of tumor collagen: determining the structural relationship between reactive stroma and healthy stroma. Opt Express 2008;16:1846–59.

[73] Raja AM, Xu S, Sun W, et al. Pulse-modulated second harmonic imaging microscope quantitatively demonstrates marked increase of collagen in tumor after chemotherapy. J Biomed Opt 2010;15:056016.

[74] Nadiarnykh O, LaComb RB, Brewer MA, Campagnola PJ. Alteration of the extracellular matrix in ovarian cancer studied by second harmonic generation imaging microscopy. BMC Cancer 2010;10:94.

[75] Palero JA, Bader AN, de Bruijn HS, van den Heuvel AV, Sterenborg HJCM, Gerritsen HC. In vivo monitoring of protein-bound and free NADH during ischemia by nonlinear spectral imaging microscopy. Biomed Opt Express 2011;2(5):1030–9.

[76] Fischer F, Volkmer B, Puschmann S, et al. Risk estimation of skin damage due to ultrashort pulsed, focused near-infrared laser irradiation at 800 nm. J Biomed Opt 2008;13(4).

[77] Ghersetich I, Lotti T, Campanile G. Hyaluronic acid in cutaneous intrinsic aging. Int J Dermatol 1994;33:119–22.

[78] Gonzalez S, Rajadhyaksha M, Rubinstein G, Anderson RR. Characterization of psoriasis in vivo by reflectance confocal microscopy. J Med 1999;30(5–6):337–56.

[79] Huzaira M, Rius F, Rajadhyaksha M, Anderson RR, Gonzalez S. Topographic variations in normal skin, as viewed by in vivo reflectance confocal microscopy. J Invest Dermatol 2001;116(6):846–52.

[80] Gonzalez S, Tannous Z. Real-time, in vivo confocal reflectance microscopy of basal cell carcinoma. J Am Acad Dermatol 2002;47(6):869–74.

[81] Wang LT, Demirs JT, Pathak MA, Gonzalez S. Real-time, in vivo quantification of melanocytes by near-infrared reflectance confocal microscopy in the Guinea pig animal model. J Invest Dermatol 2002;119(2):533–5.

[82] Gonzalez S, Gilaberte-Calzada Y, Gonzalez Rodriguez A, Torres A, Mihm Jr MC. In vivo reflectance-mode confocal scanning laser microscopy in dermatology. Adv Dermatol 2004;20:371–87.

[83] Yamashita T, Kuwahara T, Gonzalez S, Takahashi M. Non-invasive visualization of melanin and melanocytes by reflectance-mode confocal microscopy. J Invest Dermatol 2005;124(1):235–40.

[84] Agero AL, Busam KJ, Benvenuto-Andrade C, et al. Reflectance confocal microscopy of pigmented basal cell carcinoma. J Am Acad Dermatol 2006;54(4):638–43.

[85] Scope A, Benvenuto-Andrade C, Agero AL, et al. In vivo reflectance confocal microscopy imaging of melanocytic skin lesions: consensus terminology glossary and illustrative images. J Am Acad Dermatol 2007;57(4):644–58.

[86] Gonzalez S, Gilaberte-Calzada Y. In vivo reflectance-mode confocal microscopy in clinical dermatology and cosmetology. Int J Cosmet Sci 2008;30(1):1–17.

[87] Gonzalez S. Confocal reflectance microscopy in dermatology: promise and reality of non-invasive diagnosis and monitoring. Actas dermo-sifiliograficas 2009;100(Suppl. 2):59–69.

[88] Ulrich M, Lange-Asschenfeldt S, Gonzalez S. Clinical applicability of in vivo reflectance confocal microscopy in dermatology. G Ital Dermatol Venereol 2012;147(2):171–8.

[89] Ulrich M, Lange-Asschenfeldt S, Gonzalez S. The use of reflectance confocal microscopy for monitoring response to therapy of skin malignancies. Dermatol Pract Concept 2012;2(2):202a10.

[90] Venturini M, Arisi M, Zanca A, et al. In vivo reflectance confocal microscopy features of cutaneous microcirculation and epidermal and dermal changes in diffuse systemic sclerosis and correlation with histological and videocapillaroscopic findings. Eur J Dermatol 2014.

[91] Koehler MJ, Hahn S, Preller A, et al. Morphological skin ageing criteria by multiphoton laser scanning tomography: non-invasive in-vivo scoring of the dermal fibre network. Exp Dermatol 2008;17:519–23.

[92] Koehler MJ, Preller A, Kindler N, et al. Intrinsic, solar and sunbed-induced skin aging measured in vivo by multiphoton laser tomography and biophysical methods. Skin Res Tech 2009;15:357–63.

[93] Cicchi R, Kapsokalyvas D, Pavone FS. Clinical nonlinear laser imaging of human skin: a review. Biomed Res Int 2014.

[94] Kapsokalyvas D, Cicchi R, Bruscino N, et al. In-vivo imaging of psoriatic lesions with polarization multispectral dermoscopy and multiphoton microscopy. Biomed Opt Express 2014;5(7):2405–19.

CHAPTER

21

Noninvasive Topical In Vivo Imaging of Skin: Confocal Reflectance Microscopy and Polarized Light Imaging

S.L. Jacques

Oregon Health & Science University Portland, OR, United States

OUTLINE

INTRODUCTION

Imaging the skin relies on contrast to provide detailed features. Otherwise the skin looks like a homogenous material with nothing to see. Visually, there is contrast because of: (1) surface roughness (glare), (2) superficial absorption by blood and melanin that attenuates back-scattered light from the dermis (sharp shadows), and (3) deeper perturbations caused by variation in light scattering and vasculature that affect the dermal back-scatter (diffuse shadows). Hence, the dermatologist may see the normal stratum corneum or erosion of the epidermis, the reticular pigment pattern of melanin in the epidermis, superficial blood vessels, such as in basal cell carcinoma, regions of erythema, the blue veil of a deep thick melanoma, and white areas caused by epidermal thickening or dermal fibrosis.

Topical optical imaging of skin using light-scattering techniques can provide image contrast based on either tissue architecture or intrinsic optical properties. One can recognize *boundaries* between different tissue types, as well as recognize the *brightness* and *texture* of each region of homogenous tissue type. This chapter discusses two label-free methods of obtaining such contrast: (1) confocal reflectance (CR), which includes both confocal scanning laser microscopy (CSLM) and optical coherence tomography (OCT), and (2) polarized light imaging. The term *label-free* means no dyes or stains are needed for image contrast. There are other ways to obtain label-free contrast, such as autofluorescence,

Imaging in Dermatology
http://dx.doi.org/10.1016/B978-0-12-802838-4.00021-2

Raman spectra, second and third harmonic generation, and photoacoustic imaging. This discussion uses skin as an example.

CONFOCAL REFLECTANCE

In CR, light is delivered to a focus within the skin and light backscattered from the focus is collected and focused through a pinhole, or into an optical fiber serving as a pinhole, to reach a detector. The term *confocal* means that only light originating from within the focus inside the tissue reaches the "focus" of the pinhole. Light scattered from tissue regions outside the focus do not properly refocus through the pinhole, and hence this light is rejected. Consequently, the signal from the focus is not contaminated by extraneous light scattered from outside the focus. By scanning the focus over a range of depths, z, and over the coordinates x and y of the field of view at each depth z, the reflectance signal $R(x,y,z)$ is generated.

The purpose of CR is to generate an image based on the light scattering properties of the structures within a skin site. Hence, the image is a structural image rather than a molecular image that indicates the presence of certain molecules (eg, fluorescence, Raman). Structures can be planar boundaries that efficiently backscatter light, such as the air–stratum corneum surface, or local regions of a particular tissue type, such as epidermis, papillary dermis, and reticular dermis. Tissue-type regions differ in their backscatter efficiencies that depend on the spatial distribution of mass, which is related to optical refractive index. The brightness of a tissue strongly depends on the size distribution of scattering "particles," or alternatively, depends on the spatial frequency spectrum of a continuous mass density distribution. The texture depends on variations in optical backscatter on mesoscopic and larger scales.

The classic form of CR is often called *CSLM*. There is also a form of CR that is called *OCT*, which adds broadband interferometry to the classical CR method. *Broadband* means that a band of wavelengths are used, which allows improved spatial resolution during interferometric imaging. Interferometry lowers the noise floor, thereby improving the signal-to-noise ratio to allow deeper imaging.

Interferometry can also detect motion. Movement is a powerful contrast parameter. OCT can detect the moving red blood cells within blood vessels and detect the movement of cells (the osmotic swelling of cells, the undulation of cell membranes, and the movement of transport machinery on the cytoskeleton [1]).

Two Types of Confocal Reflectance

This section describes the two methods of CR (CSLM and OCT), outlines the analysis of signals, and illustrates with an image. The basic experimental setup for CR is shown in Fig. 21.1. Fig. 21.1A depicts the classic CSLM setup. Fig. 21.1B depicts the classic OCT setup.

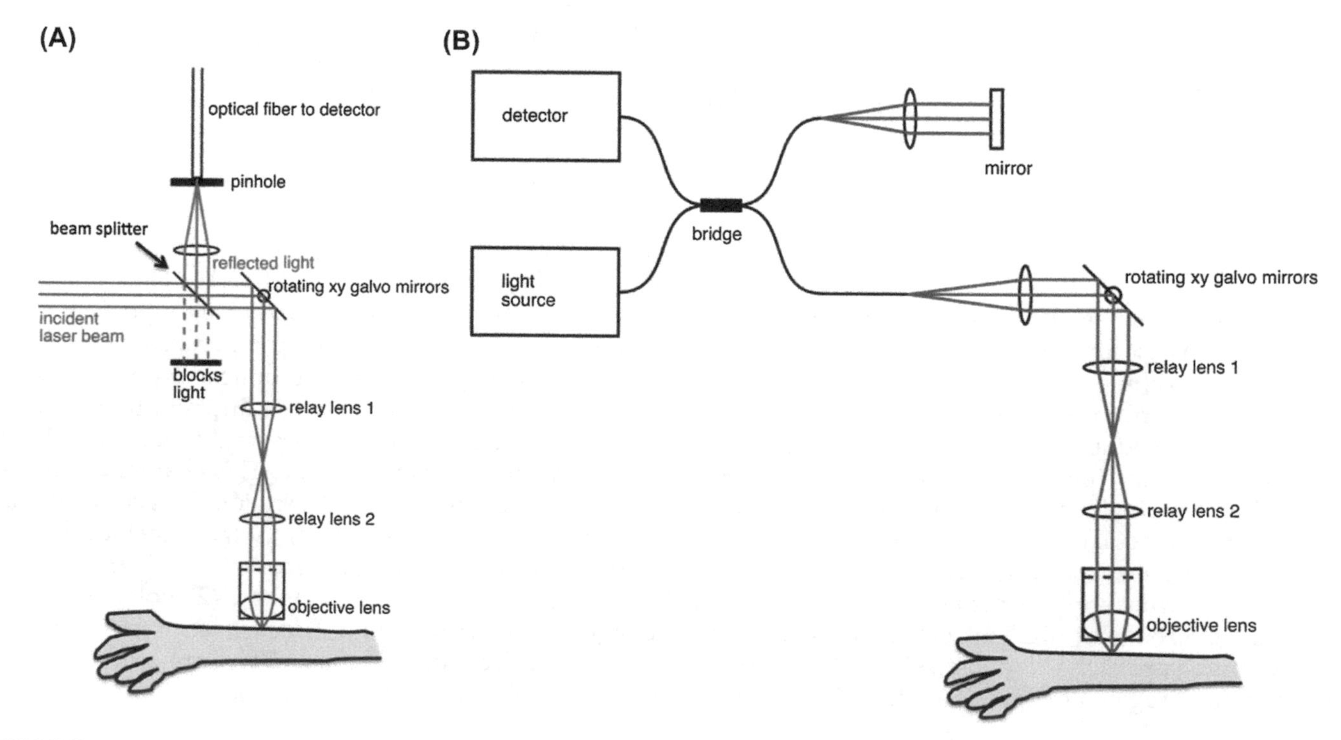

FIGURE 21.1 Confocal reflectance (CR) systems. (A) Confocal scanning laser microscopy (CSLM). (B) Optical coherence tomography (OCT).

Confocal Scanning Laser Microscopy

CSLM uses light intensity reflected from the focus within the skin for imaging. Fig. 21.1A shows the basic experimental setup.

In Fig. 21.1A, a collimated light source, usually a laser, is delivered to a pair of rotating x,y galvo mirrors, which redirect the collimated beam of light by 90 degrees, plus/minus small angles (θ,ϕ) controlled by the galvo mirrors. The beam travels from the mirrors through two relay lenses that send collimated light into the rear port of the objective lens at the angles θ and ϕ. The angles of entry into the objective lens control the lateral movement (x,y) of the focus. The objective lens is moved vertically by a piezo lens holder (not shown) that moves the focus as a function of depth (z).

The light backscattered from the focus that enters the objective lens will return through the same path to be recollimated by the x,y galvo mirrors. The reflected beam is partially reflected by a beam splitter and focused by a lens into a pinhole. Behind the pinhole is an optical fiber that carries the light to a detector.

Classically, CSLM uses a pinhole instead of an optical fiber to select the light that reaches the detector. However, a single-mode optical fiber can serve the same function. Using single-mode optical fibers allows both CSLM and OCT to be conducted with the same set-up (see Optical Coherence Tomography section).

CSLM can use any wavelength of collimated light. Shorter wavelengths in the ultraviolet (UV) to blue spectral range can be used, which improves image resolution. Axial resolution is $\Delta z = 2\lambda/\mathrm{NA}^2$ and lateral x,y resolution is $\Delta x = \Delta y = \lambda/(2\mathrm{NA})$, where NA is the numerical aperture of the objective lens and λ is wavelength. Longer wavelengths can go deeper in the skin, sacrificing resolution.

Fig. 21.2 shows an example of a CSLM image that used a blue (488 nm) laser to image mouse skin in a black (B6) mouse with melanoma [2]. The surface and thin epidermis show as a bright superficial layer, with a darker dermis spotted with bright melanoma cells. The melanosomes within the melanoma cells scatter very strongly in all directions and, in particular,

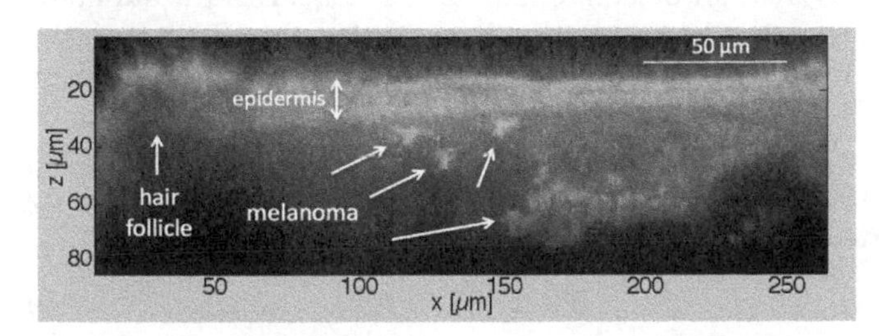

FIGURE 21.2 Confocal reflectance (CR) image of mouse skin (black B6 mouse with melanoma). *Labeled arrows* point to melanoma cells with melanosomes that scatter brightly.

backscatter toward the objective lens. Hence melanoma cells with increased melanosome density give a strong CR signal.

The CSLM detector records the intensity of light reflected by the skin, typically in volts or counts (V). A signal reflected by the skin (V_{skin}) and by a standard reflector, such as a water–glass interface (V_{wg}) are acquired. The reflectance of the water–glass interface (R_{wg}) is known:

$$R_{\mathrm{wg}} = \left(\frac{n_{\mathrm{glass}} - n_{\mathrm{water}}}{n_{\mathrm{glass}} + n_{\mathrm{water}}}\right)^2 = \left(\frac{1.52 - 1.33}{1.52 + 1.33}\right)^2 = 0.0044, \quad (21.1)$$

therefore, the reflectance of skin (R_{skin}) is calculated, as

$$R_{\mathrm{skin}} = \frac{V_{\mathrm{skin}}}{V_{\mathrm{wg}}} R_{\mathrm{wg}}. \quad (21.2)$$

Optical Coherence Tomography

OCT uses the interference of light reflected from the skin and light reflected from a mirror for imaging. The basic experimental setup for OCT is shown in Fig. 21.1B, which is identical to Fig. 21.1A once the light hits the galvo mirrors and is directed to/from the skin. OCT uses an optical fiber bridge to couple the light source to the galvo mirrors in the sample arm, and to couple the light source to a mirror in the reference arm. Light returning from the two arms is again split by the bridge, with a portion of the light reflected by both the sample and reference arms reaching the detector, where the signals interfere.

The OCT detector records a signal (V), which fluctuates if the difference in sample arm and reference arm pathlengths (ΔL) changes or the wavelength (λ) changes. The distance from the bridge to the focus within the skin is L_1, and the optical distance to the mirror is L_2. ΔL equals $L_1 - L_2$. The OCT signal is

$$V = P_0\big(\sqrt{R_{\mathrm{skin}}} + \sqrt{R_{\mathrm{reference}}} + 2\sqrt{R_{\mathrm{skin}}R_{\mathrm{reference}}}\cos(2\pi\Delta Ln/\lambda)\big), \quad (21.3)$$

where P_0 is the incident power, and R_{skin} and $R_{\mathrm{reference}}$ are the reflectances of intensity from the sample arm (skin) and reference arm (mirror), respectively. The oscillatory component of the signal, $2P_0 \times$ the square root of $R_{\mathrm{skin}} \times R_{\mathrm{reference}} \cos(2\pi\Delta Ln/\lambda)$, is isolated, either by subtraction of $P_0 \times$ (the square root of R_{skin} + the square root of $R_{\mathrm{reference}}$), or by digital filtering that isolates the frequency of the cosine function and detects its amplitude. The oscillatory signal indicates the detected electric field. Squaring this field yields the light intensity (I) reflected from both the skin and reference mirror: $I_{\mathrm{skin}} = V^2 = 4P_0^2\ R_{\mathrm{skin}}R_{\mathrm{reference}}$. To isolate R_{skin}, the oscillatory signal of a skin measurement, I_{skin}, and the oscillatory signal of a measurement on a standard reflector, such as a water–glass interface, I_{wg}, are acquired. The skin reflectance is calculated as

$$R_{skin} = \frac{V_{skin}^2}{V_{wg}^2} R_{wg} = \frac{I_{skin}}{I_{wg}} R_{wg}. \tag{21.4}$$

OCT is typically implemented in three ways:

1. Time-resolved OCT uses a broadband light source. The mirror in the reference arm is moved as the detector records intensity. An oscillatory component to the OCT signal is generated during mirror movement, A_{osc}cos(4 (n (L1 – -L2) / (), and the amplitude of oscillation (Aosc) is only significant at the tissue depth where $L_1 \approx L_2$, typically within ±5 μm. Hence as the mirror moves, the depth position being detected moves, and a $V(z)$ depth profile is generated.
2. Spectral-domain OCT has a fixed mirror and the detector uses a diffraction grating to disperse the wavelengths onto a line camera to yield a spectrum. Reflectance from a particular depth specifies a path length L_1, as mentioned previously. The path length to the reference mirror (L_2) is fixed. So ΔL becomes linearly related to a particular depth within the skin. This ΔL sets the frequency of an oscillation in reflected signal across the detected spectrum, $I(\lambda)$. Measurements of $I(\lambda)$ are mapped by interpolation to yield $I(k)$, using equally spaced steps in k, where k equals $2\pi n/\lambda$. Alternatively, hardware that triggers acquisition off of equally spaced k values can directly acquire $I(k)$. Either way produces $I(k)$, which is then Fourier transformed to yield the depth profile, $V(2nz)$, of reflected electric field amplitude, which is recorded as $V_{skin}(z)$.
3. Swept-wave OCT uses a single wavelength light source that can be swept over a range of wavelengths and detected by a simple intensity detector. A fixed mirror is used and the detector records intensity as the wavelength is swept to produce $I(\lambda)$. The system is analyzed the same as spectral-domain OCT (see previous paragraph).

If one blocks the beam in the reference arm by placing a block in front of the mirror, the detector will only see the nonoscillatory term, P_0R_{skin}, which is equivalent to a CSLM measurement. Therefore both CSLM and OCT measurements can be made with the system in Fig. 21.3.

The "coherence gate" of OCT is at the depth position where $L_1 \approx L_2$, and OCT signal is generated. To optimize the OCT signal, the objective lens can be moved axially to position its focus at the coherence gate, a technique called *focus tracking*. An objective lens with a moderately high NA can provide good x,y resolution while the coherence gate provides z resolution.

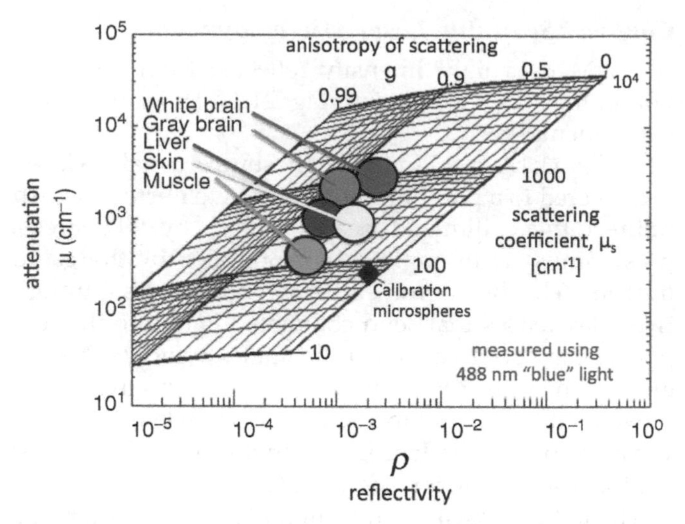

FIGURE 21.3 Analysis grid showing experimental values of μ versus ρ (attenuation versus reflectivity) for a series of optical properties, μ_s versus g (scattering coefficient versus anisotropy of scattering). Measured μ and ρ for various fresh mouse tissues are superimposed on the grid. *Figure adapted from Gareau DS. In vivo confocal microscopy in turbid media [Ph.D. thesis]. Oregon Health & Science University; 2006.*

Tissue Optical Properties

The optical scattering properties of tissues determine the strength and depth-dependent attenuation of the CR signal. Hence the signal changes as the focus translates from one tissue type (eg, epidermis) to another (eg, dermis), yielding a demarcation between epidermis and dermis that is visually apparent in images. Consequently the architecture of the tissue can be recognized.

However, it is also possible to analyze the signal within a region of one tissue type to specify the optical properties of that tissue type. For example, McLaughlin et al. [3] have reported using optical attenuation of OCT signal as a contrast parameter to improve the contrast between normal breast lymph nodes and cancer in the nodes. This section describes how to specify tissue optical scattering properties using CR.

When an objective lens moves closer to the skin surface, the focus moves deeper into the skin, and the observed reflectance signal, R, falls. Moving the focus deeper into tissues increases the difficulty for photons to propagate to the focus and back again so as to reach the pinhole. This attenuation can be summarized by the simple expression,

$$R(z_f) = \rho e^{-\mu Z_f}. \tag{21.5}$$

The depth position of the focus is z_f. The ρ (dimensionless]) denotes reflectivity of the skin. The μ (cm^{-1})

denotes the attenuation coefficient that describes attenuation as photons travel to/from the focus and the skin surface. This section describes how the factors ρ and μ govern the experimentally observed behavior $R(z_f)$. Both parameters (ρ,μ) are affected by the size distribution of scattering particles, ie, the local fluctuations in the density of tissue on the 10–10-μm scale, especially when the size of scattering particles (mitochondria, nuclei, cells) and the wavelengths of light are comparable (300–1300 nm).

The two optical scattering properties of interest are the scattering coefficient, μ_s (cm^{-1}), and the anisotropy of scattering, g (dimensionless).

Scattering Coefficient

The μ_s (cm^{-1}) indicates the number of scattering events per unit length of photon travel within a tissue. The μ_s equals the product of the cross-sectional area of scatterers (cm^2) and number density of scatterers ($\#/cm^3$). Larger particles increase μ_s, but a higher density of smaller particles can also increase μ_s. There is a relationship between the size distribution of scatterers and μ_s.

The Anisotropy

The anisotropy, g, indicates how forward directed the angle of scattering is. The value of g equals the average projection of the scattered trajectory on the photon's initial trajectory, ie, the average of $\cos(\theta)$ over all angles where there is the deflection angle of scatter [$g = \cos(\theta)$]. Very forward-directed scatter that causes only a slight average deflection angle, ie, a small (θ), yields a value of g close to 1. Isotropic scatter is equal in all directions, and forward and backward projections cancel, therefore $g \approx 0$. The parameter g is especially sensitive to the size of scatterers in a tissue, rising from 0 to approximately 0.95 as the particle's diameter increases from structures smaller than 100 nm to 1-μm structures.

When light is focused to a depth, z_f, in a tissue, the light may scatter multiple times yet still reach the focus, if the scatter is very forward directed. The ability of photons to reach the focus attenuates as $\exp(-\mu z_f)$, where $\mu = a\mu_s + \mu_a \approx a\mu_s$. The attenuation by the absorption coefficient μ_a is usually negligible compared with the effect of scattering coefficient μ_s. The ability of an incident photon to reach the focus equals the ability of an escaping photon to travel from the focus to the surface, escape the tissue, and enter the objective lens at an angle that allows reaching the pinhole at the detector. Hence a factor of two scales of attenuation accounts for the in/out round trip. The NA of the objective lens allows delivery and collection of light over a range of angles, which increases the average photon path length relative to the depth position z_f. Hence, a collection geometry factor, G, also scales the attenuation. The factor G is between 1 and 1.4 for low to high NAs. In summary, the attenuation of light reaching the focus and returning to the pinhole is

$$\mu = a\mu_s 2G. \tag{21.6}$$

McLaughlin et al. [3] used the factor μ to achieve an enhanced contrast between normal tissue (breast lymph nodes) and cancer.

Once the photon reaches the focus, the chance of scattering within the focus is $\mu_s \Delta z$ (dimensionless), where Δz is the axial extent of the focus. (In OCT, Δz is the smaller of the coherence gate or the axial resolution imposed by the NA of the objective lens.) Once a photon has scattered within the focus, only a fraction, b, of the photons have a trajectory that will reenter the objective lens. The factor b is calculated based on the angular scattering function, $p(\theta)$, where θ is the deflection angle of a scattering event. The Henyey-Greenstein function [4] serves well as an approximate scattering function for tissues, because multiple scattering washes out slight variations in $p(\theta)$. Integration of $p(\theta)$ over the backscattered angles that enter the objective lens yields a value for b. The value of b increases for very small scattering structures that scatter isotropically and hence backscatter toward the objective lens. Large structures scatter in a forward direction, away from the objective lens. The sensitivity of b to particle size greatly affects the value of ρ; hence there is a relationship between ρ and g.

For example, the nucleus of a cell in a CR image is dark, because forward scatter by the large nuclear structures does not return light into the objective lens. Small structures like the cytoskeleton scatter isotropically, which sends significant light back into the objective lens for detection. The cytoplasm of cells looks bright in CR images.

In summary, the fraction of light reaching the focus and reflected by the focus at a trajectory that would enter the objective lens and reach the pinhole detector if tissue scattering were ignored is written as

$$\rho = \mu_s \Delta z b. \tag{21.7}$$

Combining these factors together, the overall behavior of CR signal as a function of the depth of the focus [see Eq. (21.5)] is obtained.

The experimentally observed parameters μ and ρ map into the tissue optical properties μ_s and g using the given equations. Fig. 21.3 summarizes this relationship as a grid of μ vs ρ values for a set of μ_s vs. g properties. A set of tissue types freshly excised from mouse skin are superimposed on the grid [5]. Most tissues have g values that fall in the 0.6–0.9 range. Notice that the variation in g correlates with large variation in ρ. Experimentally, ρ is very sensitive to changes in g.

To further illustrate the usefulness of analyzing CR images in terms of the μ and ρ parameters and the μ_s and g optical properties, three examples are mentioned.

First, Samatham et al. [6] imaged the skin of normal mice and mice with a single gene mutation, osteogenesis imperfecta, which interferes with the ability of collagen fibrils to assemble into larger collagen fiber bundles. The g value of the mutant skin was lower (more isotropic scattering) than the g value for normal skin (more forward-directed scattering) because the collagen fibrils are much smaller than the blue wavelength of light used in the experiment, which increased the backscatter into the objective lens. Hence ρ increased.

Second, Levitz et al. [7,8] imaged collagen gels seeded with smooth muscle cells. Over 5 days, the cells remodeled the gel and constricted the gel into a small button of gel. The g value had decreased after 5 days because matrix metalloproteinases produced by the cell had degraded the collagen fiber bundles into fibrils, which scattered more isotropically. Hence, ρ increased.

Third, Samatham et al. [9] imaged mouse skin dermal samples before and after soaking for 1 h in glycerol to achieve optical clearing. After glycerol treatment, the skin was optically cleared. One could read a newspaper through the skin. Was this change because of a decrease in μ_s or an increase in g? The answer was consistently that g increased toward a very forward-directed scattering, whereas μ_s did not change significantly. Hence ρ decreased. Such forward-directed scatter clarified the skin. The simple explanation commonly cited is that glycerol index matched the collagen fibers to the matrix medium. However, this mechanism did not explain the phenomenon for skin, although it is likely the correct mechanism for soft tissues. A computer simulation of index matching simply dropped the value of μ_s but did not strongly affect g. A current working hypothesis is that glycerol desiccates the skin, causing a subtle shift in the packing of fibrils within the fiber bundles that enables constructive interference between wavelets of light scattered by the fibrils, which yields a forward-directed scatter. This behavior is analogous to how the cornea of the eye is clear as a result of active control of corneal hydration, which maintains the corneal collagen fibers in an ordered structure. Upon death, the cornea loses its control of hydration and becomes cloudy.

Motion as a Contrast Parameter

The interferometric nature of OCT allows sensitivity to subtle movements within a tissue. Movement can shift the phase of reflected light. Because the wavelength of light is typically 0.4–1.3 μm, shifts in the phase of a reflective wave can have submicrometer and even subnanometer sensitivity. Any axial shift in a scatterer's position affects ΔL in Eq. (21.3) and hence affects the phase of the oscillatory interference. A version of OCT called *Doppler OCT* can detect the movement of red blood cells (RBCs) in vessels. Whereas a simple OCT structural image may not show blood vessels that get lost in the background noise, Doppler OCT uses RBC movement as a contrast parameter to show the blood vessels in the image.

Nolte et al. in 2010 [1] used a form of Doppler OCT to monitor vibrations in cultured organoids, ie, groups of cells (0.1–1 mm in diameter) within a soft gel. A Fourier transform of the interferometric signal over time yields a power spectrum of frequencies of movement in the cells. A running display of such spectra versus time (a spectrogram) shows vibrations in the 0.01 -Hz range as a result of osmotic swelling/contraction of cells in the 0.1 Hz-range because of undulations of cell membranes, and in the 1 Hz-range because of micromotor movements on the cytoskeleton. Drugs can induce unique dynamic changes in the spectrogram pattern that serve as drug fingerprints to identify the type of drug action on the organoids.

OCT vibrometry can document the mechanical properties of skin lesions and normal skin. OCT can detect small movements in response to applied stress. How biomechanical properties change with age and actinic sun damage is an area of ongoing research. Biomechanical properties may prove to be useful contrast parameters for imaging.

POLARIZED LIGHT IMAGING

Polarized light imaging can characterize the structure of a tissue in terms of the size and concentration of scattering particles and the degree of birefringent fibers like collagen and actin–myosin. Polarized light can also gate collected photons to accept superficially scattered photons and reject deeply penetrating multiply scattered photons in order to image superficial skin structure.

What is Polarized Light?

Polarized light refers to the coordination of the orientation of oscillation of the electric fields of a population of photons. When the oscillations of all photons are aligned, the light is polarized. If the orientation of the oscillation of the photons becomes randomized, the light is unpolarized. The electric fields of individual photons still oscillate, but the population of photons is no longer synchronized.

The electric field can be broken into two components: an x-axis component and a y-axis component. The x and y components have both an amplitude and a phase of oscillation. These components vectorially add to yield the net electric field. In linear polarization, the x and y components are "in phase," whereas in elliptical polarization the x and y components are "out of phase." The special case where the x and y components are 90 degrees out of phase is called *circular polarization.*

Because experiments measure intensity (I), it is customary to specify the intensities of four types of linear polarization:

1. **I_x polarization**: If the x-component amplitude is dominant and the y-component amplitude is zero, the oscillation is along the x-axis and the population of photons is in a state of horizontal linear polarization (I_x).
2. **I_y polarization**: If the y-component dominates and the x-component is zero, then the state is vertical linear polarization (I_y).
3. **I_{+45} polarization**: If the x and y components are equal, then the state is +45-degree linear polarization (I_{+45}).
4. **I_{-45} polarization**: If the x and y components are equal in magnitude but different in sign, then the state is −45-degree linear polarization (I_{-45}).

There are also two types of circular polarization. If the phase of the x and y components of the electric field are out of phase, then the net electric field vector of the photon rotates like a corkscrew as it propagates.

1. **I_R polarization**: If the y-phase leads the x-phase by 90 degrees, then the rotation is a right-handed screw and the state is called *right circular polarization* (I_R).
2. **I_L polarization**: If the y-phase lags the x-phase by 90 degrees, then the rotation is a left-handed screw and the state is called *left circular polarization* (I_L).

The polarization state of any population of photons can be described by a vector of four values, called a *Stokes vector* (S), which depends on the balance between the measured intensities I_x and I_y, between I_{+45} and I_{-45}, and between I_R and I_L:

$$S = \begin{vmatrix} I \\ Q \\ U \\ V \end{vmatrix} = \begin{vmatrix} I_x + I_y \\ I_x - I_y \\ I_{+45} - I_{-45} \\ I_R - I_L \end{vmatrix} \tag{21.8}$$

In S, I denotes total intensity, Q is the balance between I_x and I_y, U is the balance between I_{+45} and I_{-45}, and V is the balance between I_R and I_L. For example, the light that skips off the surface of water or roads is horizontally polarized ($I_x = 1$ and $I_y = 0$). Therefore $I_x + I_y$ equals 1, $I_x - I_y$ equals 1, and U and V equal 0. The light would be described as $S = [1\,1\,0\,0]^T$, where T indicates *transpose*, such that the vector is oriented vertically. A right circularly polarized light, where $I_R - I_L = 1$ and $I_x = I_y = 1/2$ such that $I_x + I_y = 1$, would be described as $S = [1\,0\,0\,1]^T$ and left circularly polarized light would be $[1\,0\,0\,1]^T$.

To experimentally generate polarized light, linear polarization filters and quarter-wave retarders are used:

1. **I_x**: Light is passed through a linear polarization filter oriented horizontally. Only photons with electric fields oscillating horizontally will pass the filter.
2. **I_y**: Light is passed through a linear polarization filter oriented vertically.
3. **I_{+45}**: Light is passed through a linear polarization filter oriented at +45 degrees (positive x and positive y components).
4. **I_{-45}**: Light is passed through a linear polarization filter oriented at −45 degrees (positive x and negative y components).
5. **I_R**: Light is passed through a linear polarization filter oriented at +45 degrees and then passed through a quarter-wave retarder whose slow axis (highest n) is oriented horizontally, which delays the x relative to the y component of the electric field by 90 degrees, yielding right circularly polarized light.
6. **I_L**: Light is passed through a +45-degree linear polarization filter, then passed through a quarter-wave retarder with a slow axis that is oriented vertically, which delays the y relative to the x component of the electric field by 90 degrees, yielding left circularly polarized light.

In a similar way, one can experimentally measure the intensities of the six types of polarization for any light by detecting the light after it passes through an appropriately oriented linear polarizer or through a quarter-wave retarder followed by the appropriately oriented linear polarizer.

The change in polarization state as light passes through a medium (or tissue) is described by a Mueller matrix (M) that characterizes the medium. An input Stokes vector $S_{in} = [I_{in}\ Q_{in}\ U_{in}\ V_{in}]^T$ is transformed by M to yield an S_{out}:

$$S_{out} = \begin{vmatrix} I_{out} \\ Q_{out} \\ U_{out} \\ V_{out} \end{vmatrix} = \begin{vmatrix} M_{11} & M_{12} & M_{13} & M_{14} \\ M_{21} & M_{22} & M_{23} & M_{24} \\ M_{31} & M_{32} & M_{33} & M_{34} \\ M_{41} & M_{42} & M_{43} & M_{44} \end{vmatrix} \times \begin{vmatrix} I_{in} \\ Q_{in} \\ U_{in} \\ V_{in} \end{vmatrix} \tag{21.9}$$

There is ongoing work on how the Mueller matrix can characterize tissues. For example, Antonelli et al. in 2011 [10] reported on the M for light reflected from normal and cancerous colon tissue (early stage cancer growth in the mucosa but not in the underlying tissue), which is summarized as

$$\frac{M}{M_{11}} = \begin{vmatrix} 1 & 0 & 0 & 0 \\ 0 & 0.15 & 0 & 0 \\ 0 & 0 & 0.15 & 0 \\ 0 & 0 & 0 & 0.05 \end{vmatrix}_{\text{normal}},$$

$$\frac{M}{M_{11}} = \begin{vmatrix} 1 & 0 & 0 & 0 \\ 0 & 0.40 & 0 & 0 \\ 0 & 0 & 0.40 & 0 \\ 0 & 0 & 0 & 0.15 \end{vmatrix}_{\text{cancer}} \quad (21.10)$$

The nonzero diagonal elements indicate a strong depolarization of the light. Antonelli et al. reported that this pattern is true for many tissues. The depolarization by cancerous colon is less than depolarization by normal colon, indicated by the higher value of elements in M for cancerous colon relative to normal colon. The depolarization by a collagenous tissue would be greater than for a cellular tissue; therefore the result may indicate growth of a cellular mass that displaces collagenous tissue.

Both M_{22} and M_{33} are less than M_{44}, which indicates that circularly polarized light is depolarized faster than linearly polarized light, a behavior caused by scatterers much smaller than the wavelength of light. With larger scatterers that are comparable to or greater than the wavelength of light, the opposite is true. Linearly polarized light is depolarized faster than circularly polarized light. In summary, there is diagnostic potential in observing the Mueller matrix that characterizes reflectance from a tissue, and ongoing work is developing this method.

Depolarization by Various Tissues

The depolarization of light depends on both the scattering properties of a tissue and the heterogeneity of birefringent fibers in the tissue.

Depolarization by Scattering

Scattering deflects the trajectory of a photon into a new direction. The probability of a photon scattering in a particular direction depends on the orientation of its electric field and the size of the scatterers encountered in the tissue. As a population of photons scatters, the orientation of electric fields in the population randomizes and polarization is lost. The scattering coefficient μ_s is a key factor in depolarization. However, the anisotropy of scattering, g, also matters, because forward-directed scattering (eg, $g = 0.9$) does not depolarize as fast as isotropic scattering ($g \approx 0$). The combined parameter $\mu_s(1 - g)$ is called the *reduced scattering coefficient* and is a better parameter for predicting how quickly light will be depolarized as it propagates through tissue. Typically, linearly polarized light incident on a tissue becomes depolarized at a distance of about $2/[\mu_s(1 - g)]$ from the point of entry into the tissue.

Depolarization by Birefringent Fibers

Birefringent fibers, like collagen and actin myosin, can also depolarize light if the fibers are oriented in a heterogenous fashion. Birefringence occurs when the refractive index of a medium is different in one direction (eg, x) relative the orthogonal direction (eg, y). As a polarized population of synchronized photons propagates in a birefringent but nonscattering medium, the one direction of electric field oscillation (eg, x) is slowed relative to the orthogonal direction (eg, y) of oscillation because of different refractive indices in the two directions. The speed of light is slower when a photon's electric field is oriented parallel to the fiber length versus perpendicular to the fibers. The x and y components of the electric fields of the photons get out of phase, and the net electric field corkscrews as it propagates. This is called *elliptical polarization*, and in the case where the x and y components are 90 degrees out of phase, the light is circularly polarized. If a linearly polarized population of photons propagates through a purely nonscattering birefringent material, like a birefringent crystal, the photons will be cyclically converted from linear polarization to circular polarization then to linear polarization again. However, if photons propagate through a heterogenous tissue, where there are local regions of birefringence with each region oriented in a different direction, the photons will be partially converted toward circular polarization by one birefringent region, but not returned toward linear polarization by the next region with its different orientation of birefringence. If the beam of light is larger than the size scale of the local regions of birefringent fibers (typically 10–100 μm), the population of photons will encounter different orientations of fibers and grow increasingly out of synchronization. The light will be depolarized by birefringence, although no scattering occurs.

Putting the two mechanisms of depolarization together, a net rate of loss of polarization, P, versus length of photon travel, L, can be described:

$$P(L) = e^{-\mu_{LP}L} \quad (21.11a)$$

and

$$P(L) = e^{-\mu_{CP}L}, \quad (21.11b)$$

where μ_{LP} (cm^{-1}) is the coefficient for depolarization of linear polarization and μ_{CP} (cm^{-1}) is the coefficient for depolarization of circular polarization.

It is difficult to parse the coefficient of depolarization into the component caused by scattering and the component resulting from heterogeneous birefringence. Jacques [11] reported that the rate of depolarization of liver, muscle, and skin scaled 1:20:100, which suggested that birefringence is a very important factor. In a review of the brief literature on polarization properties of tissues, Jacques [12] used computer simulations to show that the two components (scattering and birefringence) are both significant contributors to depolarization. He also summarized the relationship between μ_s and μ_{LP} for several tissues as reported in the literature, as depicted in Fig. 21.4 (adapted from that review). Fig. 21.4 illustrates that although tissues may have somewhat similar μ_s values, the μ_{LP} values can vary by orders of magnitude. An efficiency of depolarization per scattering event (k) links μ_s and μ_{LP}:

$$\mu_{LP} = k\mu_s.$$

with k equal to approximately 0.0015 for liver, 0.31 for muscle, 0.091 for skin, 0.41 for myocardium, 0.8 for tendon, and 0.84 for fat. This wide range for k, whereas μ_s is relatively constant, is consistent with either: (1) the value of g ranging widely and being largely responsible for depolarization by its influence on the parameter $\mu_s(1 - g)$, or (2) the increased amount of birefringent fibers (collagen, actin–myosin) in some tissues, or (3) a combination of both.

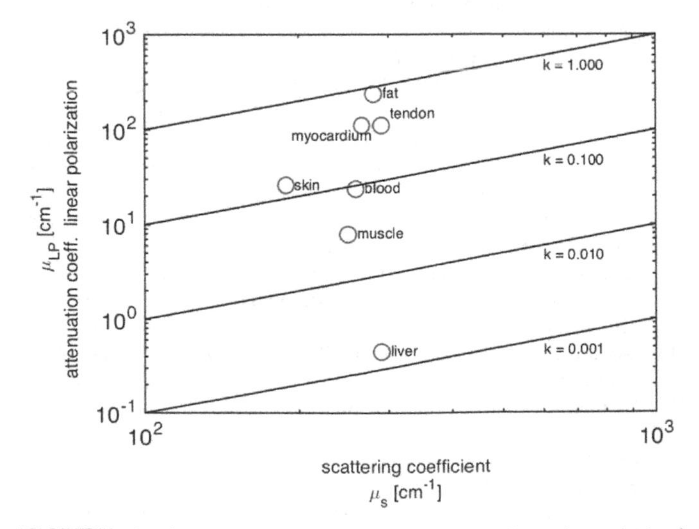

FIGURE 21.4 Depolarization coefficient for linearly polarized light, μ_{LP} (cm^{-1}), vs. the scattering coefficient, μ_s (cm^{-1}), where $\mu_{LP} = k\mu_s$, and k is an efficiency of depolarization per scattering event. Values for various tissues are superimposed. *Figure adapted from Jacques SL, Roman JR, Lee K. Imaging superficial tissues with polarized light. Lasers Surg Med 2000;26:119–29.*

Imaging Superficial Skin Layers with Polarized Light

A simple polarized light camera emphasizes another aspect of polarized light imaging, namely the ability to select superficially scattered photons and reject deeper multiply scattered photons [13,14]. When incident linearly polarized light (eg, I_x) illuminates a tissue, the superficially backscattered light has undergone just one or a few scattering events and retains some of the polarization of the incident light. Let this superficial backscatter of I_x light be called S. The deeply penetrating light has been multiply scattered and its polarization randomized, so it has equal amounts of I_x and I_y. Let this deeply penetrating backscattered light be called D. A camera images the reflected light through a linear polarizer oriented parallel to the x-axis, which yields an image I_{par}:

$$I_{par} = S + 1/2D.$$

A second image is acquired with the linear polarizer oriented parallel to the y-axis, which yields an image I_{per}:

$$I_{per} = 1/2D.$$

The difference image isolates an image based only on the superficially scattered light:

$$I_{diff} = I_{par} - I_{per} = S.$$

Such a difference image may capture only 5–10% of the total reflected light, but the image will be restricted to the superficial tissue layers. The depth sensitivity of the image depends on the amount of birefringence in the tissue. Difference images using midvisible light (yellow to red) involve the upper 100 μm in skin, 500 μm in muscle, and 1.2 mm in liver. In skin, the I_{par} and I_{par} both look featureless because of the dominant ½D component, but the difference image I_{diff} shows the "fabric pattern" of the papillary dermis, ie, the natural weave of collagen fiber bundles in normal skin. The growth of cancer penetrates and remodels the collagen structure in the papillary dermis and disrupts the fabric pattern of normal skin, transforming the skin into a more homogeneous structure.

Samatham et al. [15] used I_{diff} to find the margins of skin cancer during Mohs surgery. The cancer disrupted the fabric pattern, and one could see the cancer margin as a line between a more homogenous cancer region and the fabric pattern of normal skin. The report compared the margin assessments of the Mohs surgeon who visually observed the lesions, the assessments by Samatham using the I_{diff} images, and the dermatopathologist's report. The study compared four quadrants around the periphery of each lesion, yielding 48 determinations. A negative assessment meant the margins were expected to be cancer free. The Mohs surgeon had 14 false negatives and 34 true negatives for a negative predictive value (NPV) of $34/48 = 71\%$. The surgeon never excises less than the expected margin, so evaluation of false positives and true positives was not possible. The assessment using I_{diff} was 10 true positives, 18 false positives, 1 false negative, and 19 true negatives, for an NPV of $19/20 = 95\%$. The sensitivity was $10/(10 + 1) = 91\%$. The specificity was $19/(19 + 18) = 51\%$. This was an early result on a limited number of subjects, but encourages further work on imaging skin cancer margins using I_{diff}.

CONCLUSION

Optical CR imaging is based on light scattering and provides a noninvasive label-free imaging modality that characterizes the architecture and structural composition of a tissue, such as skin, in contrast to indicating the chemical composition of a tissue. Boundaries between tissue types are visualized, indicating the tissue architecture. The brightness and texture within local regions of a particular tissue type characterize the region. Structures on the 10 nm−10 -μm scale affect photons with wavelengths in the UV, visible, and infrared wavelength ranges. The interaction between structures and photons is strongest when the ratio of structure size to wavelength is in the 0.1−1 range, ie, structures on the 50−500 -nm scale for midvisible light. Scattering can discriminate the status of collagen fiber bundles, because small fibrils scatter isotropically and large fiber bundles scatter in a forward direction.

Polarized light imaging is sensitive to both the scattering and the birefringence of a tissue. If polarized light is incident on a tissue, scattering by tissue structures depolarizes the light as it propagates in a tissue. A heterogeneous distribution of birefringent fibers, such as collagen and actin−myosin also depolarizes light. Shifts from a collagenous tissue to a cellular tissue will change the transport of polarized light through the tissue, which can be monitored by changes in the Mueller matrix. Polarized light imaging (I_{diff}) can also select for superficially scattered photons and reject deeply penetrating photons that have been multiply scattered. Such superficial I_{diff} images show the fabric pattern of the superficial roughly 100 μm of papillary dermis. Cancer will disrupt this pattern, allowing cancer margins to be visualized. Other factors that influence dermal remodeling, such as age and actinic damage, are expected to influence polarized light images.

Taken together, CR and polarized light imaging are useful modalities for monitoring the status of and changes in skin structure.

Glossary

CR Confocal reflectance.
CSLM Confocal scanning laser microscopy.
g Anisotropy of scatter (dimensionless).
M Mueller matrix, describing transport of polarized light.
OCT Optical coherence tomography.
S Stoke's vector, describing state of polarized light.
μ_a Absorption coefficient (cm^{-1}).
μ_{LP} Attenuation coefficient (cm^{-1}) for linearly polarized light.
μ_s Absorption coefficient (cm^{-1}).

References

[1] Nolte D, An R, Turek J, Jeong K. Tissue dynamics spectroscopy for phenotypic profiling of drug effects in three-dimensional culture. Biomed Opt Express 2012;3(12):2825−41.
[2] Gareau DS, Merlino G, Corless C, Kulesz-Martin M, Jacques SL. Noninvasive imaging of melanoma with reflectance mode confocal scanning laser microscopy in a murine model. J Invest Dermatol 2007;27(9):2184−90.
[3] McLaughlin RA, Scolaro L, Robbins P, Saunders C, Jacques SL, Sampson DD. Parametric imaging of cancer with optical coherence tomography. J Biomed Opt 2010;15(4):0546029.
[4] Henyey LG, Greenstein JL. Diffuse radiation in the galaxy. Astrophys J 1941;93:70−83.
[5] Gareau DS. In vivo confocal microscopy in turbid media [Ph.D. thesis]. Oregon Health & Science University; 2006.
[6] Samatham R, Jacques SL, Campagnola P. Optical properties of mutant vs wildtype mouse skin measured by reflectance-mode confocal scanning laser microscopy (rCSLM). J Biomed Opt 2008;13:041309.
[7] Levitz D, Hinds MT, Choudhury N, Tran NT, Hanson SR, Jacques SL. Quantitative characterization of developing collagen gels using optical coherence tomography. J Biomed Opt 2010; 15(2):026019.
[8] Levitz D, Hinds MT, Ardeshiri A, Hanson SR, Jacques SL. Nondestructive label-free monitoring of local smooth muscle cell remodeling of collagen gels using optical coherence tomography. Biomaterials 2010;31(32):8210−7.
[9] Samatham R, Phillips KG, Jacques SL. Assessment of optical clearing agents using reflectance-mode confocal scanning laser microscopy. J Innovative Opt Health Sci 2010;3(3):183−8.
[10] Antonelli MR, Pierangelo A, Novikova T, Validire P, Benali A, Gayet B, et al. Impact of model parameters on Monte Carlo simulations of backscattering Mueller matrix images of colon tissue. Biomed Opt Express 2011;2(7):1836−51.
[11] Jacques SL, Roman JR, Lee K. Imaging superficial tissues with polarized light. Lasers Surg Med 2000;26:119−29.
[12] Jacques SL. Polarized light imaging of biological tissues. In: Boas D, Ramanujam N, Pitris C, editors. Handbook of biomedical optics. Boca Raton (London, New York): CRC Press; 2010.
[13] Jacques SL, Ramella-Roman JC, Lee K. Imaging skin pathology with polarized light. J Biomed Opt 2002;7:329−40.
[14] Ramella-Roman JC, Lee K, Prahl SA, Jacques SL. Design, testing, and clinical studies of a handheld polarized light camera. J Biomed Opt 2004;9(6):1305−10.
[15] Samatham R, Lee K, Jacques SL. Clinical study of imaging skin cancer margins using polarized light imaging. Proc SPIE 2012; 820700−1.

CHAPTER

22

Polarization Optical Imaging of Skin Pathology and Ageing

A.N. Yaroslavsky[1,2], X. Feng[1], V.A. Neel[2]

[1]University of Massachusetts, Lowell, MA, United States; [2]Massachusetts General Hospital, Boston, MA, United States

OUTLINE

INTRODUCTION

Polarization optical imaging is a powerful tool for the noninvasive assessment of skin pathology and ageing. Polarization characterizes electromagnetic waves, such as light, by specifying the direction of the wave's electric field oscillations. Natural light is not polarized. When the electric field vector of the light oscillates in a single, fixed plane all along the beam, the light is defined as linearly polarized; when the plane of the electric field rotates, the light is defined as elliptically polarized because the electric field vector traces out an ellipse at a fixed point in space as a function of time; and when both axes of an ellipse are equal, the light is considered as circularly polarized (Fig. 22.1).

The use of polarized light in biology was pioneered by Brewster in 1815, who conducted the first experiments on the depolarization of light exhibited by various mineral, animal, and plant structures, including the eye lens [1]. Similar investigations by others followed [2,3]. These works contributed to the accumulation of a large base of empirical knowledge, which combined with modern technological developments in photonics, have led to a deeper understanding of polarized light interactions with biological tissues. In 1976, Bickel et al. [4] reported that the analysis of the complete polarization

Imaging in Dermatology
http://dx.doi.org/10.1016/B978-0-12-802838-4.00022-4

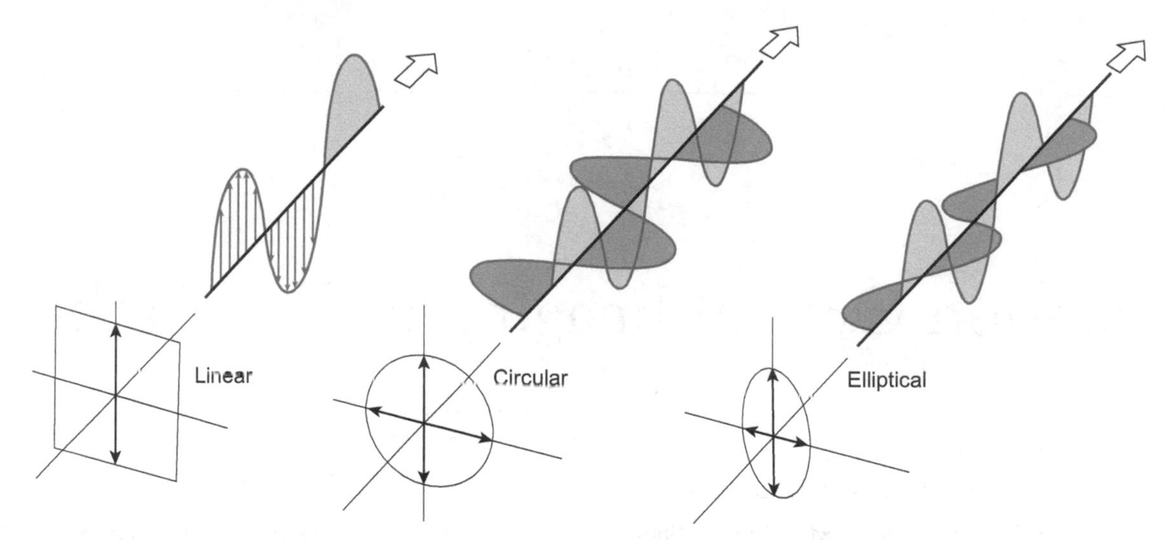

FIGURE 22.1 Polarization states of light.

state of light, elastically remitted from biological media, contains structural information about the interrogated system. This discovery opened the door to the structural analysis of many different biological tissues [5–10] and to the first efforts to diagnose different diseases [11–16]. In the field of dermatology, the use of polarized light for examining human skin was initiated by Philp et al. in 1988 [17] and by Anderson in 1991 [18], who noticed that viewing skin through a linear polarizer, under linearly polarized illumination, yields different and complementary information. When the planes of polarization are parallel, images enhance surface detail, such as wrinkles (Fig. 22.2A). When the planes are orthogonal, an enhanced view of deeper structures, such as vasculature and pigmented lesions, is obtained (see Fig. 22.2B).

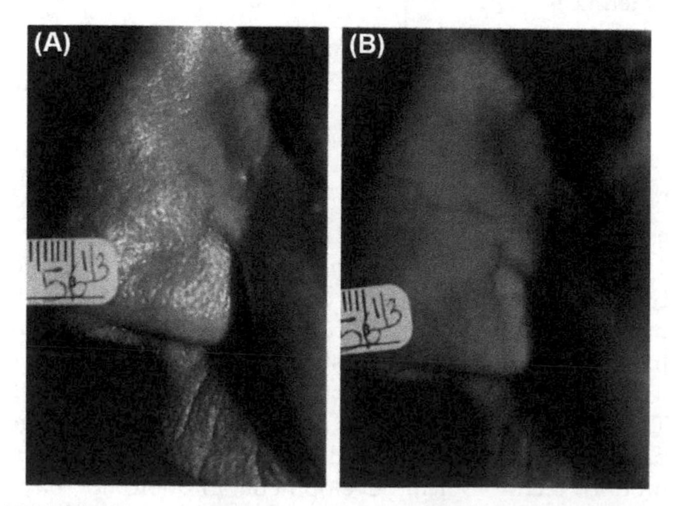

FIGURE 22.2 Polarization photography of skin. (A) Picture taken with linearly polarized illumination and co-polarized detection. (B) Picture taken with linearly polarized illumination and cross-polarized detection. *From Anderson RR. Polarized light examination and photography of the skin. Arch Dermatol 1991;127(7):1000–5.*

Based on this observation, the first polarized dermatoscope was introduced in 2001. Currently, cross-polarized white light imaging is routinely used for detecting and diagnosing skin lesions (see Fig. 22.3A). Because deep scattered light reveals the internal structure of the suspicious skin areas, dermoscopy provides the doctor with additional diagnostically relevant information. The primary goal of dermoscopy is to make a decision as to whether a suspicious lesion needs to be biopsied or removed. In this respect, conventional dermoscopy can be as useful as polarization-enhanced dermoscopy. However, the use of a conventional dermatoscope requires an immersion fluid and contact with skin to improve optical coupling. With the advent of a polarization-enhanced device, dermoscopic examination has become more convenient because physicians no longer have to apply compression to the skin (Fig. 22.3). Because there is no physical contact, application of the immersion oil, alcohol, or water to improve light coupling before examining each lesion is no longer required. In addition, because skin is not compressed during examination, there is no more risk of missing blood vessels. Nonetheless the specificity of diagnosing skin cancer even with the use of cross-polarized light remains lower than desired [19]. Therefore continuing efforts to develop more elaborate polarization-enhanced techniques for light diagnostic of skin diseases and ageing are ongoing.

The fluorescence phenomenon was first observed from quinine by G.G. Stokes in 1852 [20], who also measured polarization of the emitted signal. However, because of the long lifetime of the fluorophore (~20 ns), he found that emission was completely depolarized. Thus the discovery of fluorescence polarization (FP) was delayed by almost half a century. In 1920, F. Weigert observed, and in 1922 Vavilov and Levshin

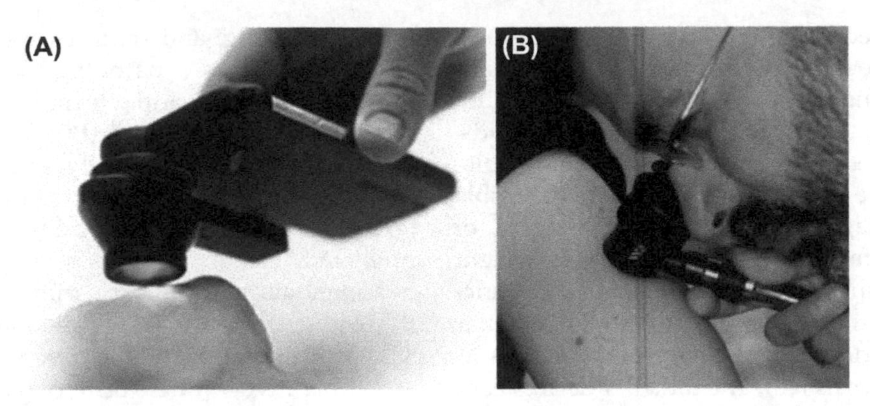

FIGURE 22.3 (A) Polarized dermatoscope. (B) Conventional contact dermatoscope.

conducted comprehensive study on FP from various dye solutions and the change of FP with temperature and viscosity [21,22]. The theory of FP was formalized by F. Perrin in 1926 [23]. However, it was not until 1951 that FP was first applied in biochemistry by G. Weber [24,25]. Since then a multitude of studies employed this phenomenon for biological investigations, such as the antigen–antibody reaction [26], macromolecule binding [27], membrane fluidity [28,29], and local viscosity [30]. Several studies attempted to use differences in FP between normal and tumor tissue for cancer detection [31,32]. Determination of fetal lung maturity by measuring the FP value of the amniotic fluid used to be the method of choice in a large number of perinatal units because of its simplicity and relatively high predictive value [33,34]. Recently, there have been several publications on the use of exogenous FP imaging for the intraoperative delineation of breast cancer [16,35]. Nevertheless utilization of FP in medicine remains very limited. In dermatology the first work reporting exogenous FP imaging of nonmelanoma skin cancer (NMSC) margins was in 2004 [36]. Since then there have been consistent efforts to translate this technology to dermatological surgery [37–39].

OPTICAL INTERACTIONS OF LIGHT WITH SKIN

The Structure of Skin

Human skin has three primary layers: epidermis, dermis, and hypodermis (Fig. 22.4), all of which contain appendages, such as hair follicle and sweat glands [40]. The stratum corneum is the outermost layer of the epidermis, and is accretion of the lipid and protein remnant of dead keratinocytes [41]. It serves as a barrier that restricts the permeability of water and biomolecules both in and out of the body. The average thickness of stratum corneum is 0.015 mm [41].

The average thickness of epidermis is approximately 0.1 mm. However, on the face it may be as thin as 0.02 mm, whereas on the soles of the feet it is as thick as 1–5 mm. The epidermis consists of up to 90% keratinocytes, which function as a barrier, keeping harmful substances out and preventing water and other essential substances from escaping the body. The other 10% of epidermal cells are melanocytes, which manufacture and distribute melanin, the protein that adds pigment to skin and protects the body from ultraviolet (UV) rays. Melanin is one of the strongest skin chromophores, with the absolute refractive index of 1.7. Absorption and scattering of epidermis in the visible spectral range is defined almost exclusively by its melanin. Epidermis contains five layers, including the stratum corneum, stratum lucidum, stratum granulosum, stratum spinosum, and stratum basale. Two thin layers, the stratum lucidum and the stratum granulosum, reside under the stratum corneum. The stratum spinosum is located under the stratum granulosum. It is the thickest layer of epidermis. The stratum basale is the innermost epidermal layer that contains melanocytes and a single layer of basal cells (basal keratinocytes).

Dermis is the connective tissue layer of human skin that is located under the epidermis. It is composed of gellike and elastic materials, water, and, primarily,

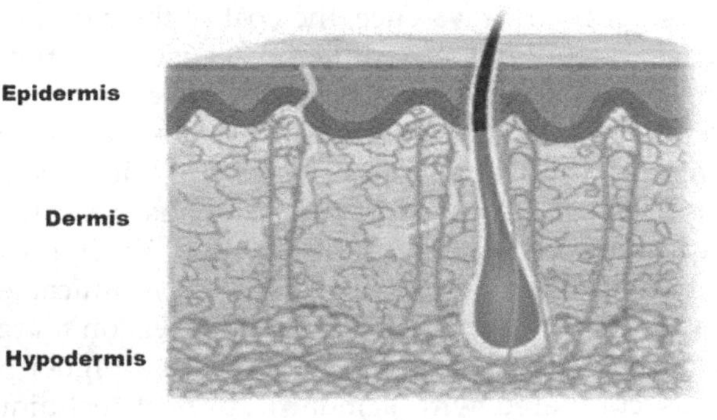

FIGURE 22.4 Human skin.

type I collagen. Embedded in this layer are systems and structures common to other organs, such as lymph channels, blood vessels, nerve fibers, and muscle cells, but unique to the dermis are hair follicles, sebaceous glands, and sweat glands. Blood or, more precisely, hemoglobin, defines the absorptive properties of dermis in the visible spectral range. Collagen is the major component of dermis, accounting for 77% of the fat-free dry weight of skin [42]. The dermis can be recognized as two functional substructures: the papillary layer and the reticular layer. The papillary dermis consists of thin collagen fibers and capillary vessels [40]. The reticular dermis consists of thick collagen fibers and blood vessels. The average thickness of the dermis layer is around 2 mm across the body.

The hypodermis is the deepest layer of skin that includes connective tissue intermixed with energy-storing adipocytes or fat cells. Fat cells are grouped together in clusters held in place by fibrous bands called *septae*. The hypodermis is generously supplied with blood vessels, ensuring a quick delivery of stored nutrients. The hypodermis serves as the energy reservoir that provides thermal insulation of the body. Thickness of hypodermis is highly variable.

Skin Cancers

Skin cancer is more common than all other cancers combined [43]. It includes melanoma and nonmelanoma cancers.

NMSCs, ie, basal cell and squamous cell carcinomas, account for approximately 97% of all skin cancers [44,45]. They are a major cause of morbidity in the fair-skinned population. Most of these cancers are curable by surgery when detected early. The lesions usually appear later in life in areas that have received the most sun exposure. Whereas only 20% of nonmelanoma cancers are squamous cell carcinomas (SCCs), they tend to be more aggressive than basal cell cancers. They are more likely to invade fatty tissues beneath the skin and, although rare, can metastasize to lymph nodes and distant organs. In most cases, nonmelanoma cancers are often disfiguring but rarely fatal. However, because of their prevalence, the cost of their treatment reaches $4.8 billion per year [43]. Because these tumors often occur on the face and rarely metastasize, it is important to spare as much healthy tissue as possible to reserve appearance and function. However, in many cases the contrast of the lesions is poor, which complicates visual tumor localization during treatment. Mohs micrographic surgery [46] is a clinical technique that allows complete control of excision margins during the operation. Mohs surgery has a higher cure rate as compared with standard surgical techniques, but is used only in 25% of cases because it is expensive, time consuming, and requires special training by surgeons. It requires a pathology laboratory adjacent to the operation room and a technician to prepare the sections. Basal cell carcinoma (BCC) has been increasing at a dramatic rate. Statistically, every fourth Caucasian will develop at least one lesion during their lifetime [47]. Thus NMSCs are becoming a major public health problem.

Melanoma is the most serious form of skin cancer. It accounts for only 3% of all skin cancers but causes 83% of skin cancer deaths. Even though melanoma is a comparatively rare form of skin cancer, it is the fourth most commonly diagnosed form of cancer for men and the fifth most commonly diagnosed cancer for women in the United States. In its advanced stages, it spreads to other parts of the body, where it becomes more difficult to treat and can be fatal. The incidence rates of melanoma in the United States have increased almost tenfold since 1975 and are rising faster than any other cancer [48]. Melanoma is quickly becoming a very serious clinical problem in the fields of dermatology and surgical oncology.

The most common treatment of both melanoma and NMSCs is the removal of the lesions, usually by surgical excision. Currently most cancers are removed without intraoperative margin control. After the surgery is completed and the resulting wound is closed, the tissue is sent for histopathological analysis. Postoperative methods of cancer delineation involve sampling and examine only 0.01% of the surgical margin [49,50]. For example, the "bread loaf" method uses vertical sectioning of the excised tissue and is prone to sampling errors, which may lead to cancer recurrence and metastases. If cancerous cells are detected in the pathology slides, the patient has to be brought back to the surgical suite, the wound has to be reopened, and more tissue has to be excised. This repetitive procedure doubles the cost of the treatment and involves psychological stress to the patient. New methods to address this problem are sorely needed.

Skin Ageing

Cutaneous ageing is a cumulative process that depends on both intrinsic and extrinsic factors. Intrinsic ageing results from natural changes in skin with the passage of time. *Extrinsic ageing* refers to changes largely caused by UV radiation from chronic sun exposure. Extrinsic ageing is superimposed on intrinsic ageing, and is believed to be the major cause of cutaneous ageing. Visually, skin ageing is characterized by wrinkle formation, tissue laxity, and increased pigmentation. However, most age-related structural changes occur in the dermis. Thinning of the dermis layer [51], degradation of the elastin–collagen network [52,53], and atrophy of the dermis [54] are observed in aged skin. Histochemical and biochemical analysis showed decreased amount of collagen in aged skin. It has been reported that UV radiation induces proteolytic degradation of mature collagen as well as inhibition of ongoing collagen synthesis [55,56]. Reduced synthesis and increased fragmentation of collagen fibrils lead to low collagen content in aged skin [9,52,57,58]. Because collagen is the major component of the dermis, changes in collagen structure and content contribute to wrinkle formation and age-related skin diseases.

The Optical Properties of Skin

Light propagation in skin is determined by its optical properties [59], namely, by the refractive index, the absorption coefficient, μ_a, the scattering coefficient, μ_s, and the scattering phase function, $f(\mu)$ (μ is the cosine of the scattering angle). Another commonly used quantity is the transport scattering coefficient, also called the reduced scattering coefficient. It is determined as $\mu_s' = \mu_s(1-g)$, where g is the average cosine of the scattering angle. Absorption and scattering coefficients are defined as the probability for the photon to be absorbed and scattered per unit length, respectively. The *scattering phase function* describes the angular distribution of the scattered light. Since the advent of light treatments, the optical properties of skin have been studied extensively [60–71].

Absorption and scattering coefficients of skin layers are shown in Figs. 22.5–22.7 [67]. The figures demonstrate that the scattering of all skin layers decreases with the increasing wavelength. The scattering of epidermis is considerably higher than the scattering of dermis and hypodermis. The optical properties of epidermis in the visible spectral range are determined by melanin content [72].Absorption of melanin monotonously decreases with the increase of wavelength. Therefore the influence of melanin on epidermal properties is more pronounced at shorter wavelengths. In the dermis, scattering is predominantly caused by collagen [67]. The bundles of collagen can be appreciated in the confocal image of the dermis (see Fig. 22.6B). Hemoglobin dominates absorption properties of dermis and fat in the visible spectral range. Hemoglobin absorption peaks around 410, 577, and 595 nm appear consistently in their spectra. Despite the higher extinction coefficient of hemoglobin, as compared with melanin, at the Soret absorption band around 410 nm the number of photons that penetrate into the dermis at this wavelength is insufficient for visualization of dermal blood because of the high attenuation of the incident light by the melanin of epidermis. At 577–595 nm, melanin absorption and scattering are reduced and tissue penetration is increased.

Reflectance Polarization Imaging

As discussed, while propagating in human skin, the light is being scattered and absorbed. Scattering dominates absorption by at least one order of magnitude in the optical spectral range. Most photons are scattered elastically, ie, without a change of the wavelength. A small fraction is scattered inelastically, ie, with the change of wavelength. Inelastic scattering includes fluorescence, phosphorescence, and Raman. In this chapter, we focus on reflectance and FP imaging, which

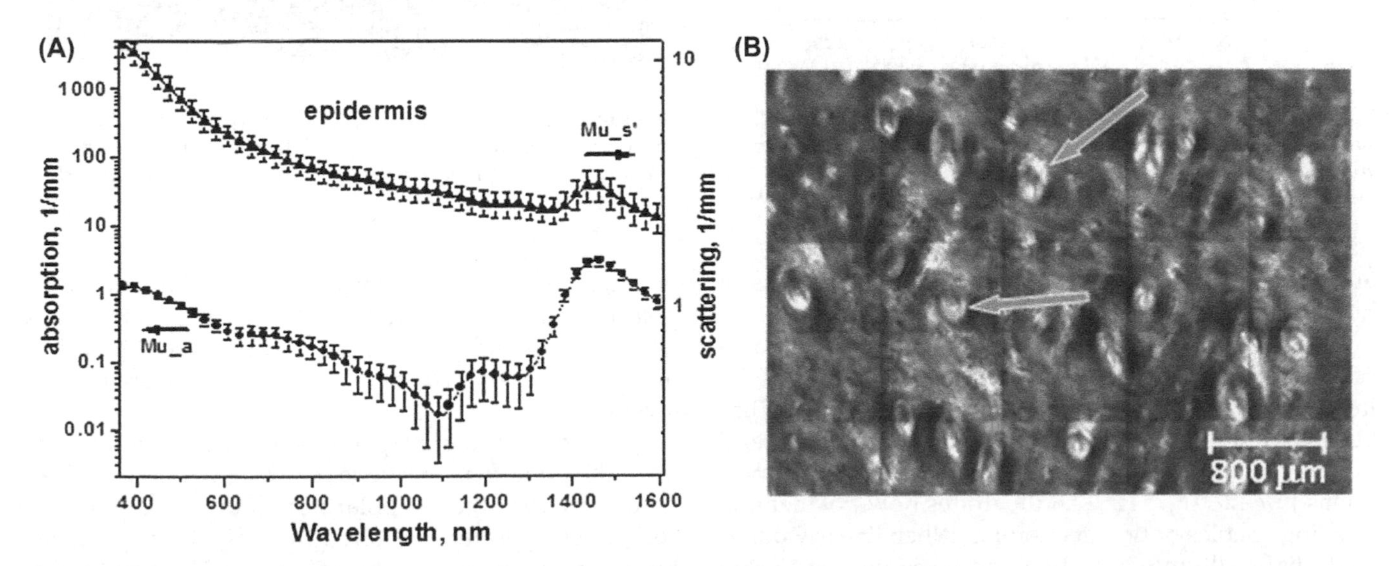

FIGURE 22.5 (A) Optical properties of epidermis. *Triangles*, Reduced scattering coefficients; *circles*, absorption coefficients; *bars*, standard errors. Averaged over seven samples. (B) Typical confocal image of epidermis. *Arrows* point to hair follicles. *From Salomatina E, Jiang B, Novak J, Yaroslavsky AN. Optical properties of normal and cancerous human skin in the visible and near-infrared spectral range. J Biomed Opt 2006;11(6):064026.*

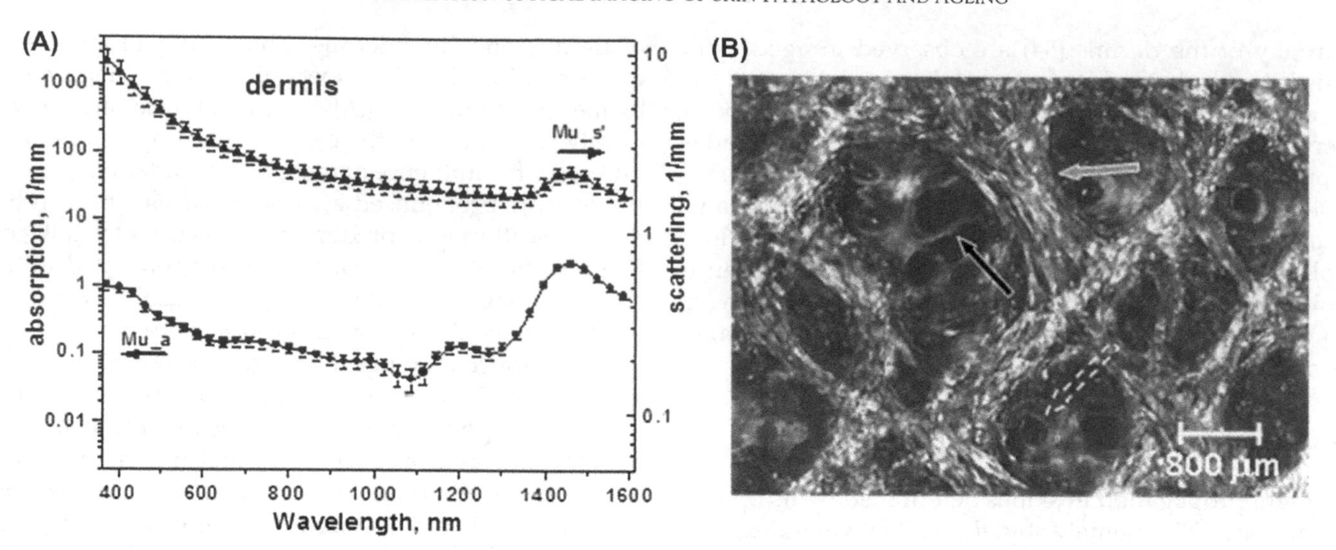

FIGURE 22.6 (A) Optical properties of dermis. *Triangles*, Reduced scattering coefficients; *circles*, absorption coefficients; *bars*, standard errors. Averaged over eight samples. (B) Typical confocal image of dermis. The *gray arrow* points to a collagen-elastin bundle; the *black arrow* points to a sebaceous gland; the *dashed arrow* points to a hair shaft. *From Salomatina E, Jiang B, Novak J, Yaroslavsky AN. Optical properties of normal and cancerous human skin in the visible and near-infrared spectral range. J Biomed Opt 2006;11(6):064026.*

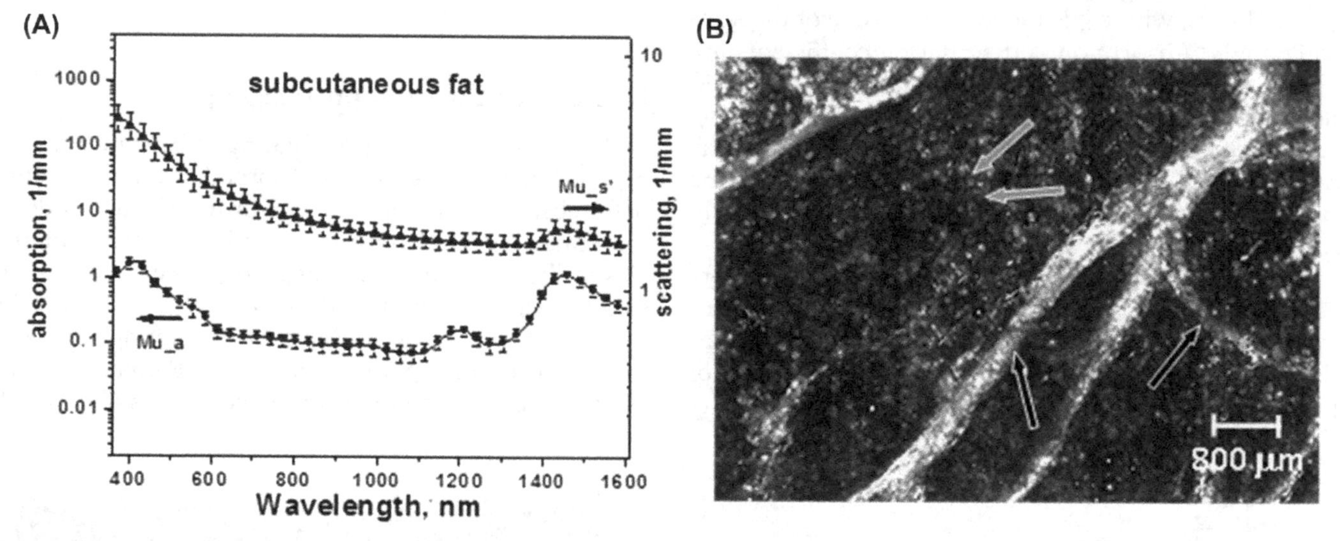

FIGURE 22.7 (A) Optical properties of subcutaneous fat. *Triangles*, Reduced scattering coefficients, *circles*, absorption coefficients, *bars*, standard errors. Averaged over 10 samples. (B) Typical confocal image of subcutaneous fat. The *gray arrows* point to fat cells/adipocytes; the *black arrow* points to connective tissue septum. *From Salomatina E, Jiang B, Novak J, Yaroslavsky AN. Optical properties of normal and cancerous human skin in the visible and near-infrared spectral range. J Biomed Opt 2006;11(6):064026.*

are utilized to assess the state and morphology of human skin.

Reflectance imaging relies on the detection of backscattered photons of the same frequency as that of the incident light. Reflectance polarization imaging has been commonly used to provide higher resolution and higher contrast imaging of patients and biological specimens [5,7,12–16]. Fig. 22.8 illustrates how polarization imaging enables optical sectioning. When linearly polarized light illuminates biological tissue, scattering randomizes the light remitted from the deeper tissue layers, whereas single backscattered photons preserve polarization of the incident beam. Thus imaging cross-polarized light remitted from tissue eliminates signal from the superficial tissues, because cross-polarized light returns predominantly from the deeper tissue layers.

The single scattered photons return predominantly from the superficial tissue layer. As a result, the difference between the co-polarized and cross-polarized components of the light remitted by tissue contains information on the superficial tissue layer only, thus enabling optical sectioning. The imaging depth of the superficial layer depends on the wavelength of the

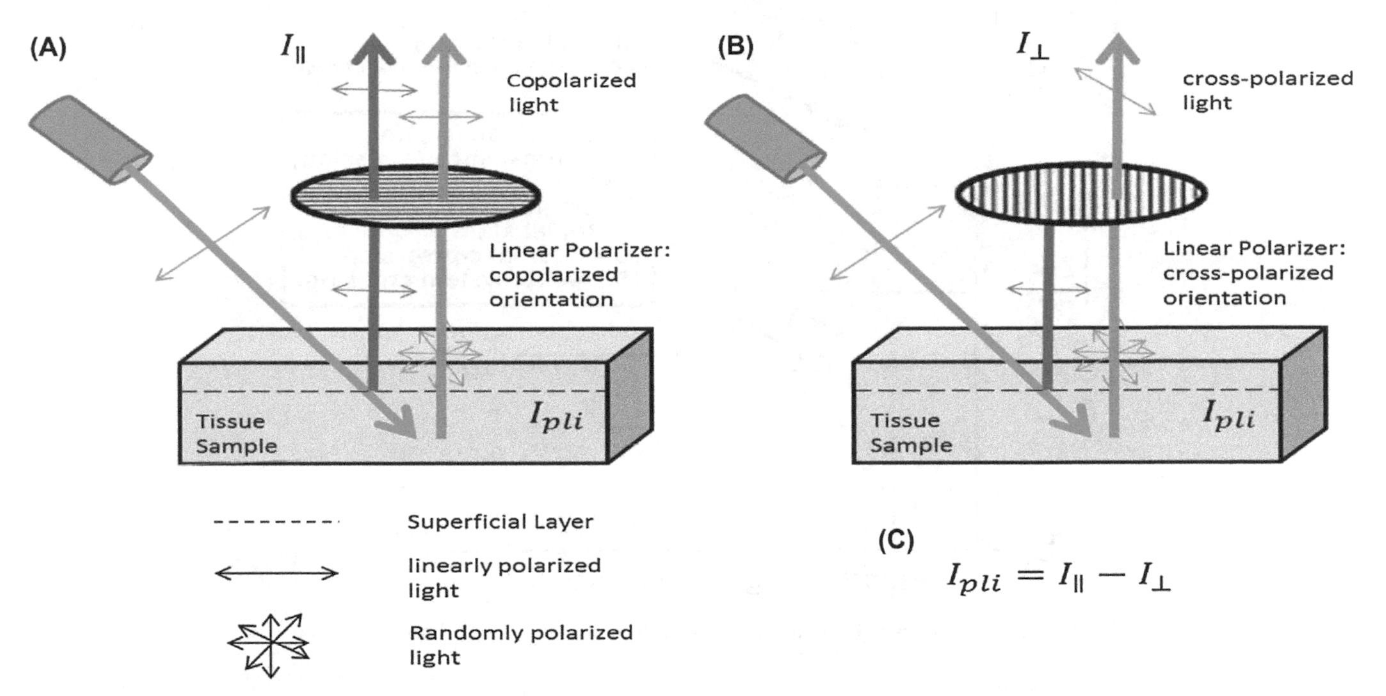

FIGURE 22.8 Reflectance polarization imaging. Green shows incident linearly polarized light. Red indicates the co-polarized light remitted from the superficial layers of the sample. Orange indicates light remitted from deeper tissues. Light is randomly polarized before the polarizer and linearly polarized after the linear polarizer. (A) Imaging co-polarized light ($I_{\parallel}$) remitted from tissue. (B) Imaging cross-polarized light ($I_{\perp}$) remitted from tissue eliminates signal from the superficial tissue layer. (C) Difference image (I_{pli}) is obtained by subtracting cross-polarized light from co-polarized light remitted from tissue. It enables visualization of the superficial tissue layer (optical sectioning).

incident light and tissue optical properties, ie, the scattering coefficient, μ_s, and the anisotropy factor, g, of the investigated medium, and can be expressed as: $D = 1/[\mu_s(1 - g)]$ [14]. Using optical properties of skin from the literature [67,73,74], it is possible to estimate the dependence of the imaging depth of the skin images on the illumination wavelength. This dependence is presented in Fig. 22.9. It shows that in skin, the imaging depth increases with the wavelength from approximately 60 to about 225 μm over the wavelength range of 350–750 nm.

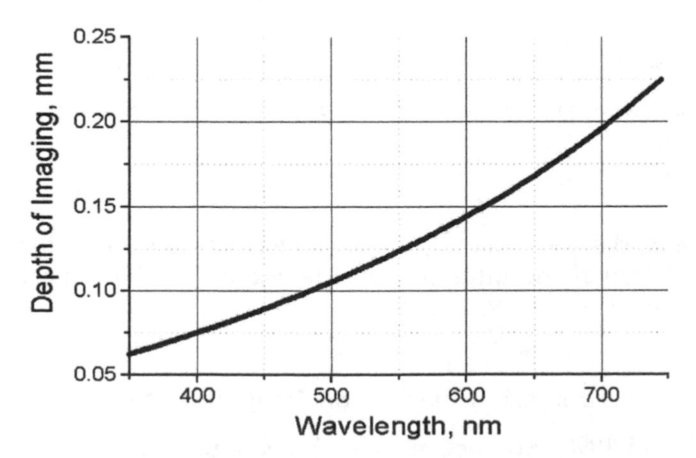

FIGURE 22.9 Dependence of the imaging depth (superficial image section thickness) on the wavelength of imaging light estimated using the known optical properties of skin. *From Yaroslavsky AN, Neel V, Anderson RR. Demarcation of nonmelanoma skin cancer margins in thick excisions using multispectral polarized light imaging. J Invest Dermatol 2003;121(2):259–66.*

Fluorescence Polarization Imaging

In fluorescence imaging, incident light photons are absorbed by a molecule, and an orbital electron is excited to a higher quantum state (Fig. 22.10). The electron can undergo nonradiative transitions before emitting a fluorescent photon as the orbital electron relaxes to the ground state. The fluorescence lifetime is defined as the average time the electron remains in an excited state before emitting a photon. The fluorescent photon has in general a longer wavelength than the incident light, owing to energy loss during the fluorescence lifetime.

FP, or anisotropy, quantifies polarization of the fluorescence emission with respect to the polarization of the incident light. It is determined by the rotational diffusion of a fluorophore during the lifetime of an excited state. Polarization of the fluorescent photons can be affected by several factors including binding, viscosity, and fluorescence lifetime (Fig. 22.11). If a molecule is free to rotate during the lifetime of the excited state, the emitted light will be depolarized, whereas if rotational motion is restricted or fluorescence lifetime of the excited state is decreased, than the emitted light

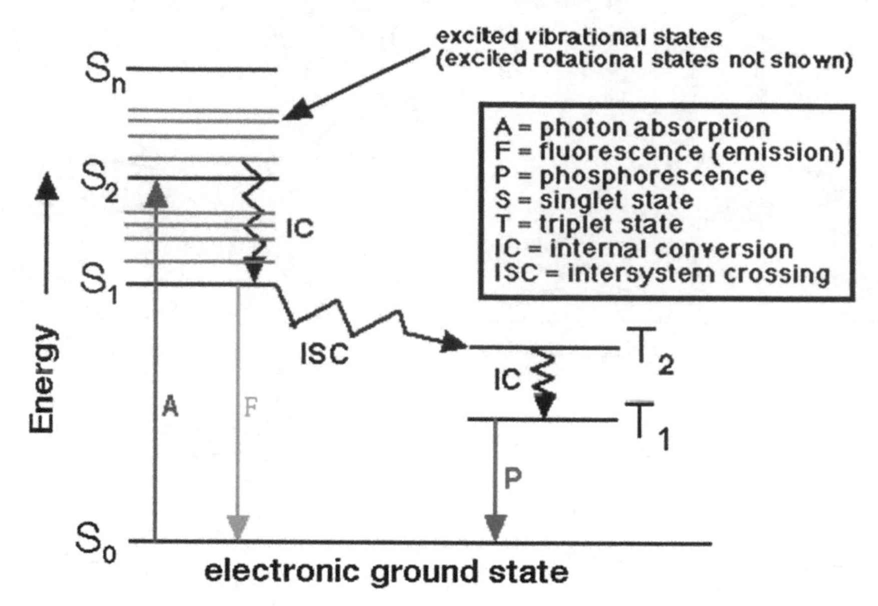

FIGURE 22.10 Jablonski diagram illustrating photon absorption and fluorescence emission.

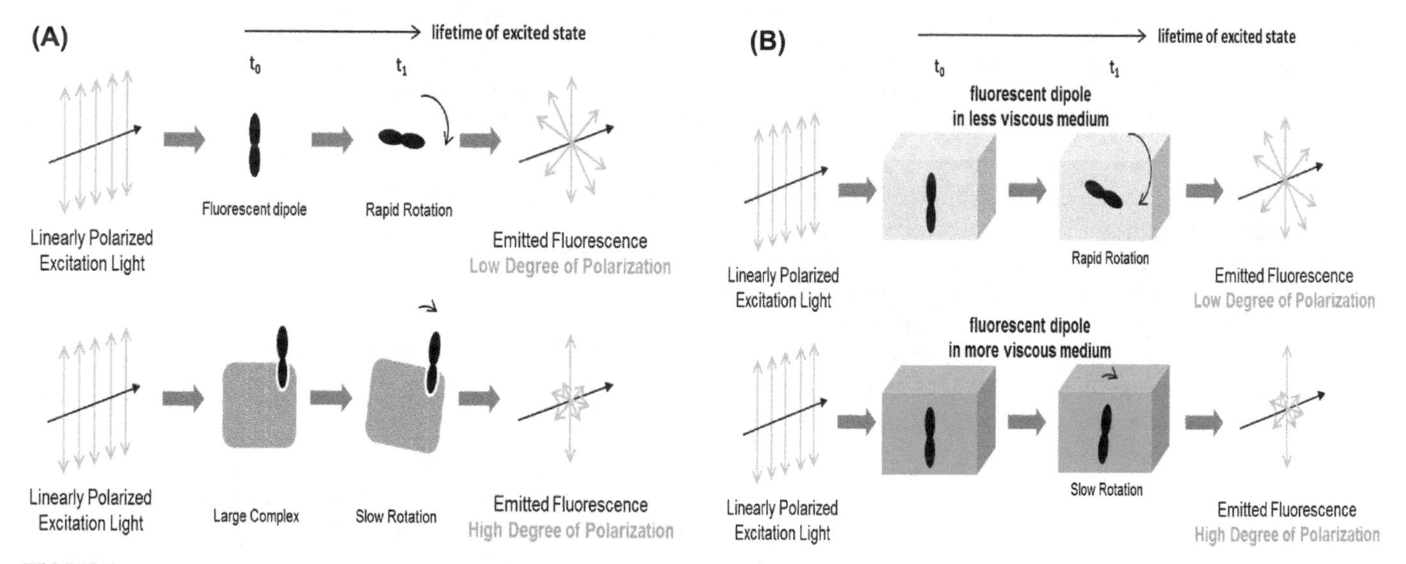

FIGURE 22.11 (A) Diagram demonstrating dependence of fluorescence polarization (FP) on fluorophore binding. (B) Diagram demonstrating dependence of FP on viscosity.

remains polarized. Binding and viscosity changes can restrict rotational motion whereas fluorescence lifetime will affect the amount of time available to alter FP.

REFLECTANCE POLARIZATION IMAGING OF HUMAN SKIN

Polarization optical imaging is perfectly suited for imaging skin structure, aging, and pathological states. Since 1995, considerable effort has been dedicated by different investigators to the development of polarization-enhanced optical technologies for the detection and intraoperative delineation of skin cancer and ageing.

Polychromatic Polarization Imaging of Skin

In 1998, S.L. Jacques and K. Lee reported a camera system based on polarized light for visualizing skin cancer margins [75]. In this paper, the authors illuminated the skin with linearly polarized incoherent white light, and recorded backscattered light with parallel (I_{par}) and perpendicular (I_{per}) polarization with respect to

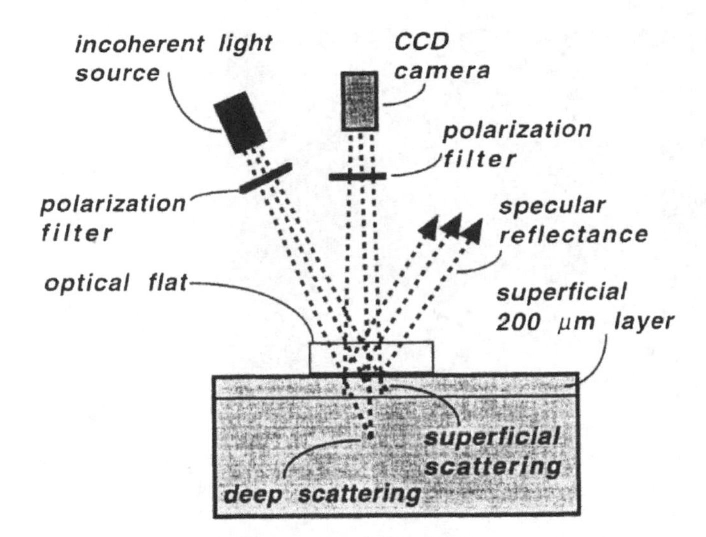

FIGURE 22.12 Proposed polarized imaging system proposed in Ref. [71] for delineating skin cancer margin. *CCD*, Charge-coupled device.

the incident light. The scheme is shown below in Fig. 22.12. In order to maximize the details of superficial layer, Jacques and Lee calculated a polarization image (Pol) by $\frac{I_{par}-I_{per}}{I_{par}+I_{per}}$, which subtracts out light scattered from deeper layers.

Fig. 22.13 demonstrates the ability of the Pol to visualize sclerosing BCC. The normal white light image (A) does not identify the lesion. However, the Pol (B) shows the texture of the superficial skin and indicates the margin of the lesion (white arrows).

In another work reported in 2002, Jacques and Lee added a 500-nm long-pass transmission filter and a collimating lens to their original setup, and imaged other types of lesions [13]. Results in this study further expanded the applications of polarized light imaging. Example images are shown in Figs. 22.14–22.17. Fig. 22.14 shows I_{per} and Pol images of a compound nevus. The lesion can be identified as a melanin pigmented region with atypical features in the I_{per} image (A). The Pol (B) reveals some structural changes in the lesion. Fig. 22.15 shows images of actinic keratosis. The I_{per} image does not clearly demonstrate the lesion. The Pol identifies the lesion as a dark region. Fig. 22.16 shows images of a malignant BCC. The lesion is clearly distinguished in the Pol as a region with higher polarization signal. The lesion also shows different structure as compare with that of surrounding healthy skin. Fig. 22.17 shows images of a SCC. The Pol detects the altered skin structure in the lesion and indicates the margin.

In all cases, the Pol was able to discriminate the lesion as an area where normal dermal structure has been disrupted. As a particular application, information provided by a Pol may be used to guide the surgical excision of skin cancer.

Narrow Band Polarization Imaging of Skin Cancer

Polarized white light imaging enables gross demarcation of skin lesions. However, the resolution of the method is not sufficient for morphological assessment of skin cancer and skin appendages, such as hair follicles or sebaceous and eccrine glands. Mohs micrographic surgery (the gold standard of skin cancer excision techniques) utilizes frozen hematoxylin and eosin (H&E) histopathology for the accurate intraoperative delineation of cancer margins. Thus it is desirable that a competing imaging technique had the resolution approaching that of light microscope–based histology.

Joseph et al. [15] used cross-polarized and superficial polarized light imaging for emphasizing morphology of

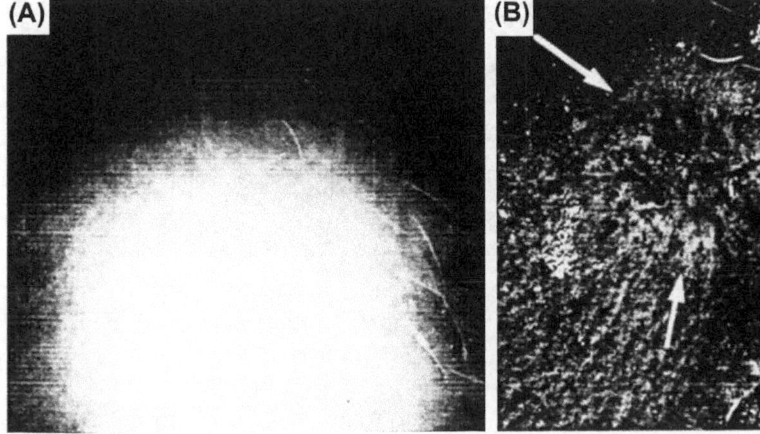

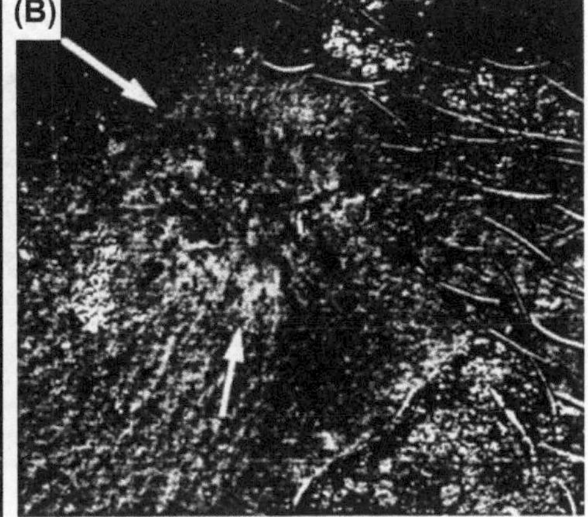

FIGURE 22.13 Normal light image (A) and polarization image (Pol) (B) of sclerosing basal cell carcinoma (BCC). *Results were obtained from J acques SL, Lee K. Polarized video imaging of skin. Proc SPIE 1998;3245:356–62.*

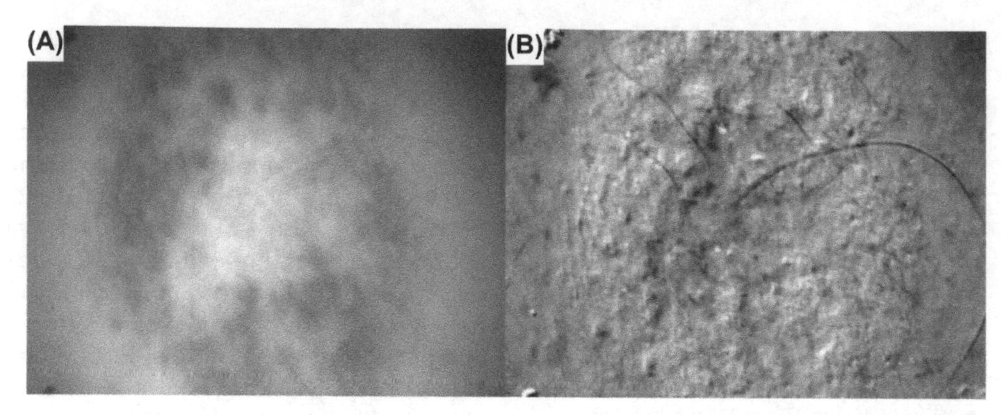

FIGURE 22.14 Backscattered light with perpendicular polarization (I_{per}) (A) and polarization image (Pol) (B) of compound nevus with atypical features. *Results were obtained from Jacques SL, Ramella-Roman JC, Lee K. Imaging skin pathology with polarized light. J Biomed Opt 2002;7(3): 329–40.*

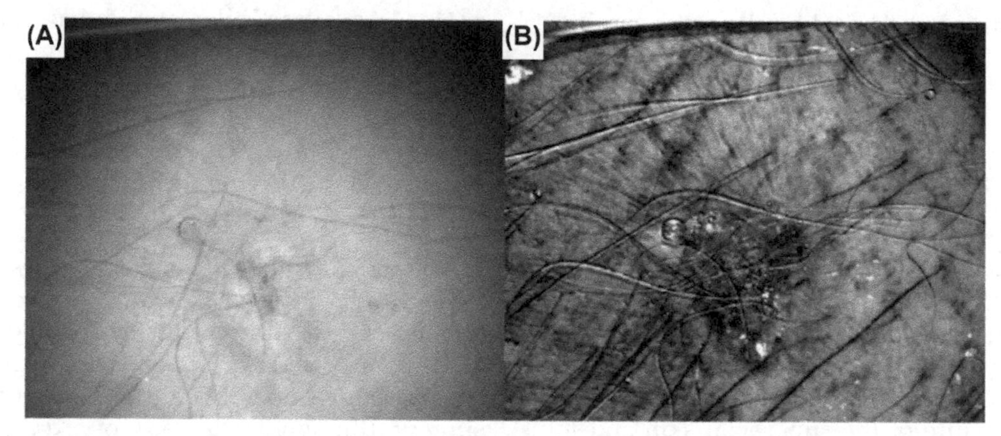

FIGURE 22.15 Backscattered light with perpendicular polarization (I_{per}) (A) and polarization image (Pol) (B) of actinic keratosis. *Results were obtained from Jacques SL, Ramella-Roman JC, Lee K. Imaging skin pathology with polarized light. J Biomed Opt 2002;7(3):329–40.*

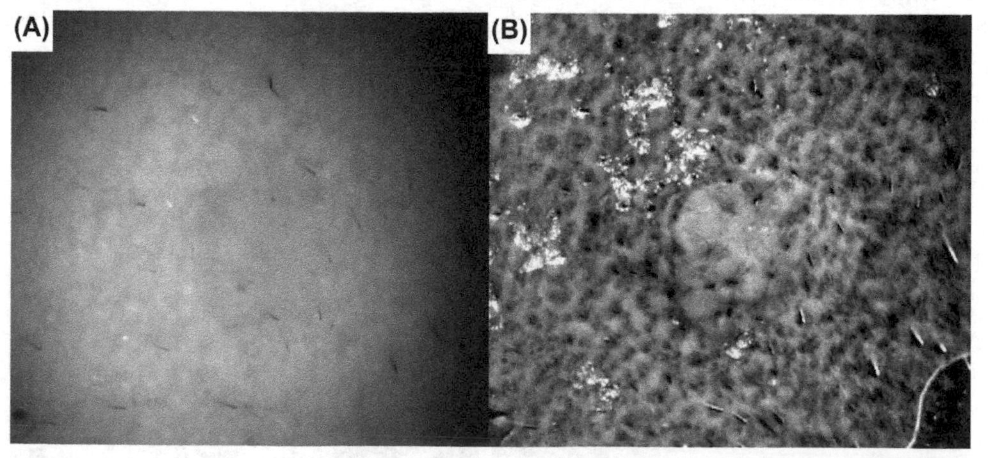

FIGURE 22.16 Backscattered light with perpendicular polarization (I_{per}) (A) and polarization image (Pol) (B) of malignant basal cell carcinoma (BCC). *Results were obtained from Jacques SL, Ramella-Roman JC, Lee K. Imaging skin pathology with polarized light. J Biomed Opt 2002;7(3): 329–40.*

skin structures in the images of excisions with residual cancers. The optical set-up employed for polarization imaging was very similar to the one used in studies from Jacques et al. [13,75], with one notable difference, ie, a narrow bandpass filter, centered at 440 nm [full width at half maximum (FWHM): 10 nm] was introduced into the path of the incident light to provide monochromatic illumination of the specimens. Pols

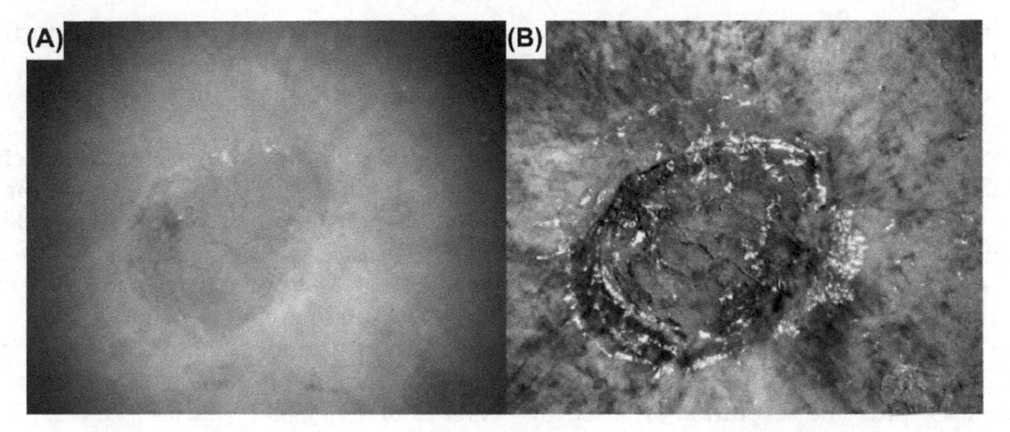

FIGURE 22.17 Backscattered light with perpendicular polarization (I_{per}) (A) and polarization image (Pol) (B) of squamous cell carcinoma (SCC). *Results were obtained from Jacques SL, Ramella-Roman JC, Lee K. Imaging skin pathology with polarized light. J Biomed Opt 2002;7(3):329–40.*

were processed using the formula: PLI $= I_{co} - I_{cross}$, where PLI is the polarization light image, I_{co} is the co-polarized image, and I_{cross} is the cross-polarized image.

Fresh thick cancer specimens were obtained from Mohs micrographic surgeries. For imaging, the specimens were covered with a 1-mm cover glass. To prevent dehydration during the experiment, the samples were placed on a gauze soaked in pH balanced (pH 7.4) saline solution. Frozen H&E sections, processed from the imaged tissue, were used for evaluation of the results yielded by polarization imaging.

Example optical reflectance cross-polarized and polarization difference images from a BCC from study by Joseph et al. [15] are presented in Fig. 22.18A and B,

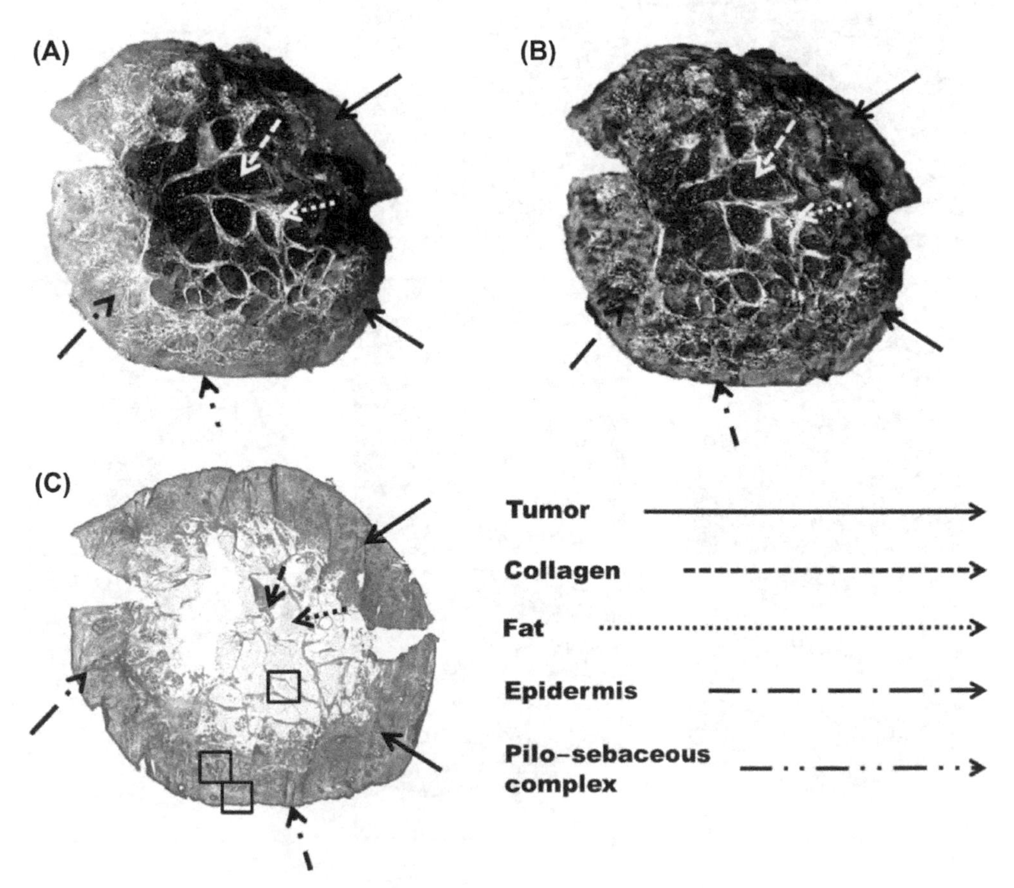

FIGURE 22.18 Specimen with infiltrative basal cell carcinoma (BCC). (A) The cross-polarized optical image. (B) The polarized light image. (C) The hematoxylin and eosin (H&E) stained histology of a 5-μm frozen section. *From Joeseph CS, Patel R, Neel VA, Giles RH, Yaroslavsky AN. Imaging of* ex vivo *nonmelanoma skin cancers in the optical and terahertz spectral regions optical and terahertz skin cancer imaging. J Biophotonics 2014;7(5):295–303.*

respectively. The comparison of optical images to histology (see Fig. 22.18C) demonstrates close resemblance of the morphological appearance of small skin structures. This improvement in the resolution of the resulting images, as compared with studies from Jacques et al. [13,75], is to be expected if we compare the transversal resolution of white and blue light images. The lateral resolution of both types of images can be as good as approximately 12 μm. In the transversal to the surface of skin direction, polarization difference imaging at 440 nm provides optical sectioning to about 50–70 μm [14], as compared with white light that yields at best 250 μm. The resolution afforded by blue light polarization optical imaging is demonstrated in the magnified sections of regions outlined in histology (see Fig. 22.18, boxes) in Fig. 22.19. The optical images clearly show morphological features, such as the epidermis (dash–dot arrow), the pilo–sebaceous complex (dash–dot–dot arrow), subcutaneous fat (dot–dot arrow), as well as highly reflective collagen strands (dash–dash arrow). The tumor region (solid arrow) is characterized by a distortion of normal structures, such as the loss of skin appendages, and collagen appears as a homogenous dark area as seen in Fig. 22.19D and E.

Fig. 22.20 shows a representative specimen with SCC. Comparison of the optical images (see Fig. 22.20A and B) with H&E histopathology presented in Fig. 22.20C demonstrates that the location, size, and shape of cancer (solid arrow), as in the case of BCC, are correctly identified. The tumor area is dark, indicating a lack of collagen and loss of normal skin structure. In Fig. 22.21, higher magnification optical images of adipose tissue (see Fig. 22.21A–C), hair follicles (see Fig. 22.21D–F), tumor lobule (see Fig. 22.21G–I), and sebaceous glands (see Fig. 22.21J–L) are compared with respective structures in the H&E histopathology image. An impressive resemblance in the appearance of optical and histological images can be well appreciated.

In total, nine specimens from nine patients were evaluated in the study [15]. In each case, polarization-enhanced blue light optical images showed good

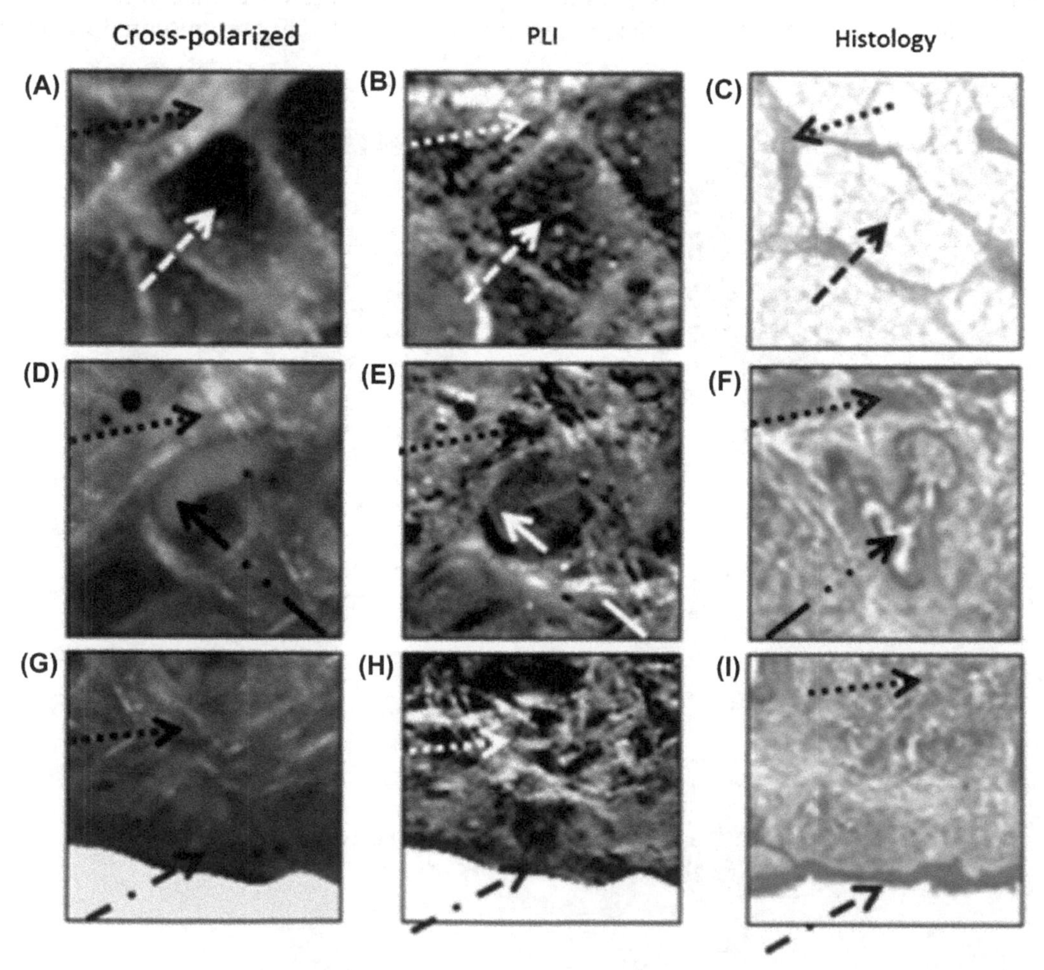

FIGURE 22.19 Comparison of morphological features outlined by boxes in the histology image of the infiltrative BCC specimen shown in Fig. 22.18C. *PLI*, Polarization light image. *From Joeseph CS, Patel R, Neel VA, Giles RH, Yaroslavsky AN. Imaging of* ex vivo *nonmelanoma skin cancers in the optical and terahertz spectral regions optical and terahertz skin cancer imaging. J Biophotonics 2014;7(5):295–303.*

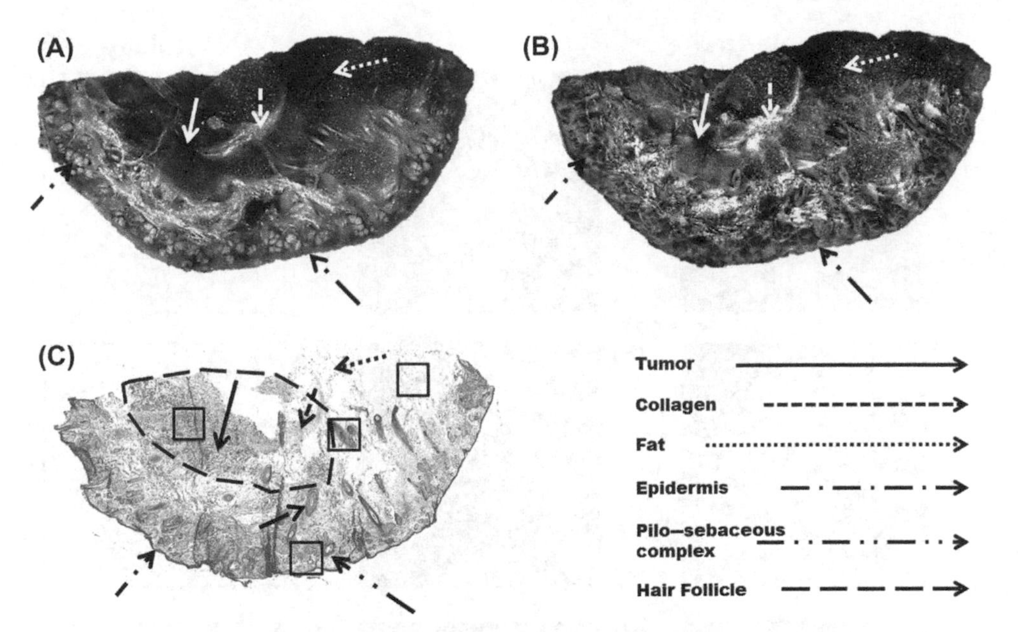

FIGURE 22.20 Specimen with squamous cell carcinoma (SCC). (A) Cross-polarized optical image. (B) Polarized light image. (C) Hematoxylin and eosin (H&E) stained histology of a 5-μm frozen section. *From Joeseph CS, Patel R, Neel VA, Giles RH, Yaroslavsky AN. Imaging of* ex vivo *nonmelanoma skin cancers in the optical and terahertz spectral regions optical and terahertz skin cancer imaging. J Biophotonics 2014;7(5):295–303.*

correlation with histopathology and remarkable resolution for such a simple technique. Nevertheless the contrast of cancer in the images required improvement.

Narrow Band Polarization Imaging of Collagen

In 2015, Feng et al. [9] used an approach similar to that described in section "Narrow Band Polarization Imaging of Skin Cancer" to evaluate dermal structures in vivo. The experimental system is presented in Fig. 22.22. A filtered xenon arc lamp emission was employed for illumination. Linear polarizers were introduced into the pathways of the incident and reflected light. Cross-polarized images were acquired with a charge-coupled device (CCD) camera coupled with a 0.5× lens.

To optimize imaging depth for visualizing dermal collagen, the authors calculated scattering coefficients and imaging depth between 400 and 700 nm in human dermis (Fig. 22.23). Because the cumulative thickness of stratum corneum and epidermis of facial skin is approximately 100 μm, the authors suggested that the cross-polarized images at 440 nm should provide appropriate depth for imaging papillary dermis. As shown in Fig. 22.23, longer and shorter wavelengths may highlight the presence of blood, because of the double-peaked oxyhemoglobin absorption band at 570 and 590 nm, and absorption of the Soret band around 410 nm. With the system operating at 440 nm (FWHM = 10 nm), in vivo collagen images of the facial skin of 17 subjects aged between 24 and 65 years were acquired noninvasively. The subjects had skin types I, II, and III, according to the Fitzpatrick classification [76].

Collagen content was quantitatively evaluated as follows. The images were thresholded to a level set at 40% brightness (Fig. 22.24). This threshold level allows for the discrimination of collagen bundles, removing signal from the noncollagen area, eg, the space between collagen bundles, sebaceous gland, and hair follicles. Collagen content was then calculated using the formula:

$$\text{Collagen content} = (1 - \text{Threshold Value}) \times 100\%.$$

Fig. 22.25 shows example images of a 35-year-old subject. Structures of collagen network and highly reflective collagen fibers can be clearly appreciated (see Fig. 22.25B). Fig. 22.25C shows the intensity histogram of the image, where mean pixel value and FWHM were determined.

For the analysis, study subjects were grouped into three age cohorts of volunteers: 24–29 years old, 35–43 years old, and 50–65 years old. The authors found decreased collagen content, decreased mean pixel value, and increased FWHM of the intensity histogram that correlated with age (Fig. 22.26). Fig. 22.26A shows the age-dependent collagen content changes observed in the study. A comparison of the imaging results with data from an immunohistochemical evaluation shows a good correlation. The decrease of the mean pixel values and increase of FWHM with age indicates partial loss of compactness and reflectivity of dermal collagen, a result reported in previous investigations [52,57,58].

The authors also compared cross-polarized 440-nm wide-field collagen images to the ex vivo reflectance confocal images acquired from the dermal side of skin. An in vivo wide-field cross-polarized reflectance image

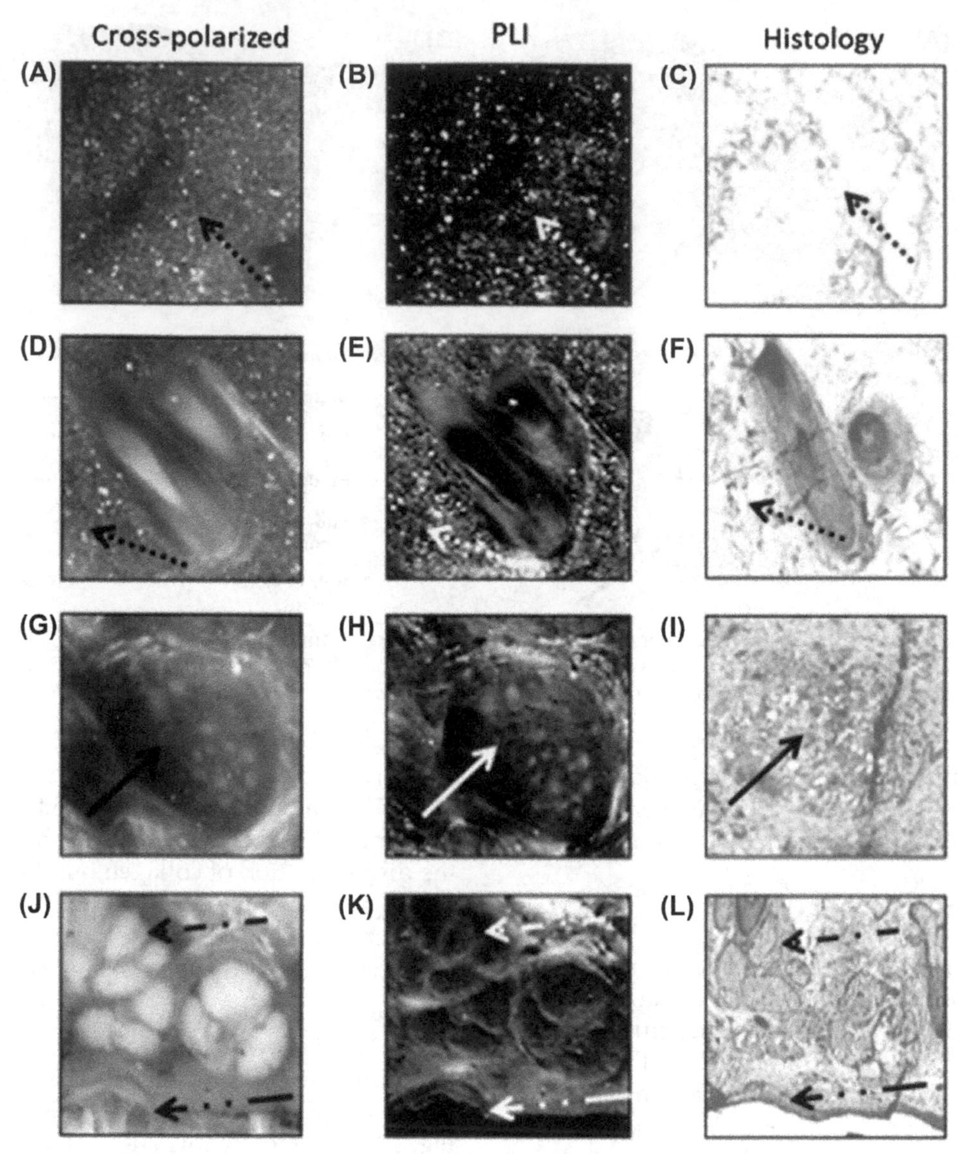

FIGURE 22.21 Comparison of magnified optical and histology images from boxes in the histology image shown in Fig. 22.20C. *PLI*, Polarization light image. *From Joeseph CS, Patel R, Neel VA, Giles RH, Yaroslavsky AN. Imaging of* ex vivo *nonmelanoma skin cancers in the optical and terahertz spectral regions optical and terahertz skin cancer imaging. J Biophotonics 2014;7(5):295–303.*

at 440 nm (A) and ex vivo confocal reflectance image (B) are shown in Fig. 22.27. The confocal image shows collagen bundles (white arrows) and hair follicles (gray arrows). The cross-polarized 440-nm wide-field image demonstrates a similar pattern with collagen network (white arrows) interweaved with hair follicles (gray arrows). Comparison of the images demonstrates the ability of cross-polarized in vivo images acquired from the epidermal side to reveal the structure of collagen bundles and network with sufficient detail, comparable with those yielded by ex vivo confocal images acquired from the dermal side of skin. It is worth noting that the resolution of the cross-polarized in vivo images is sufficient for the quantitative assessment of the average diameter of collagen bundles. The 100-μm diameter determined from the image is consistent with previous findings [77]. Thus cross-polarized 440-nm imaging enabled quantitative and noninvasive evaluation of dermal collagen. The authors have also demonstrated that an unsophisticated, in-house–built imaging device was capable of real-time visualization and quantitation of dermal structures with the resolution down to tens of microns over the wide fields of approximately 4 cm^2 in healthy volunteers with skin types I–III without biopsy or laser exposure. In addition, they were able to quantify collagen content using image analysis and establish the trends in age-related changes of collagen structure and content.

In the following study, Feng et al. [10] used the same imaging device and image analysis to evaluate collagen

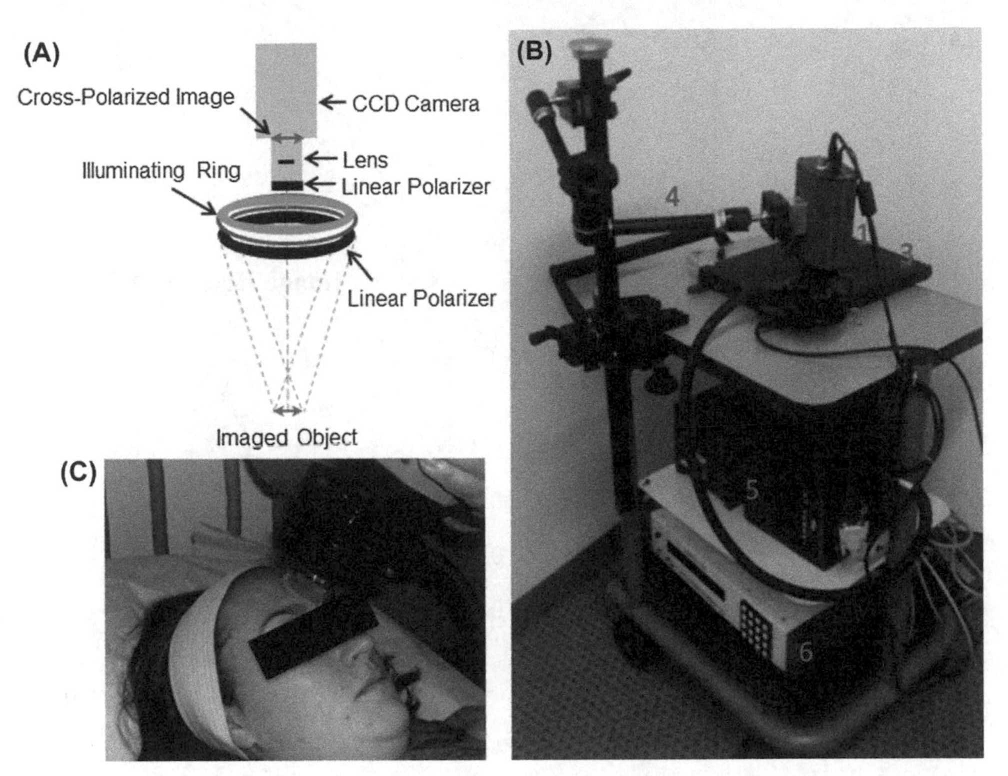

FIGURE 22.22 Proposed polarization enhanced wide-field imaging system proposed in Ref. [9]. Schematic (A) and photograph (B) of the system. 1-CCD camera, 2-illuminator, 3-computer, 4-articulating arm, 5-lamp, 6-controller. (C) Digital picture of imaging process. *CCD*, Charge-coupled device.

changes after nonablative fractional laser treatment. Previously, clinical efficacy studies were performed using clinical photography and/or ex vivo histological evaluation. Clinical photography is low resolution and subjective, whereas 440-nm cross-polarized wide-field imaging provides objective quantitative results in vivo, as well as lateral resolution down to 12 μm. In comparison to ex vivo histological evaluation, imaging is entirely noninvasive and enables real-time image acquisition and analysis. Moreover, because of the noninvasive nature of the imaging method, baseline assessment and treatment evaluation could be performed on the same volume of skin. Fig. 22.28 demonstrates imaging results and ability of the technique to quantitatively detect increased collagen content as early as 2 weeks after the treatments (see Fig. 22.28B and D). Increased collagen reflectivity and more compact collagen network can be observed from the posttreatment collagen image (see Fig. 22.28D). Quantitative analysis revealed that collagen content of this subject increased by 15% after 2 weeks of daily treatment.

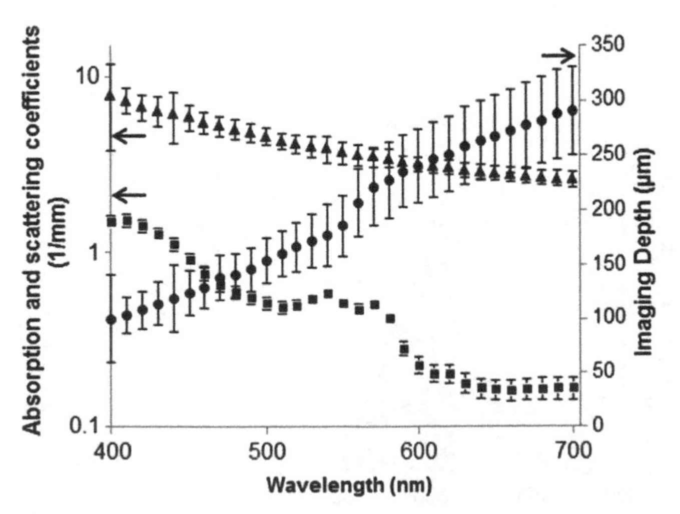

FIGURE 22.23 Optical properties of human dermis. Triangles, scattering coefficients; *squares*, absorption coefficients; *circles*, imaging depth. *From Feng X, Patel R, Yaroslavsky AN. Wavelength optimized cross-polarized wide-field imaging for noninvasive and rapid evaluation of dermal structures. J Biophotonics 2015;8(4):324–31.*

In total, seven out of eight subjects in the study by Feng et al. [10] showed varying degrees of improvement in collagen content, ranging from 1% to 26%. Efficacy of the treatment reported in the study by Feng et al. [10] correlates well with findings of other clinical trials [78–81].

This study revealed that the polarization-enhanced wide-field imaging method is capable of quantitative monitoring of changes in collagen structure. Unlike ex vivo histopathology, imaging is harmless, noninvasive, and can be performed in real time. Other

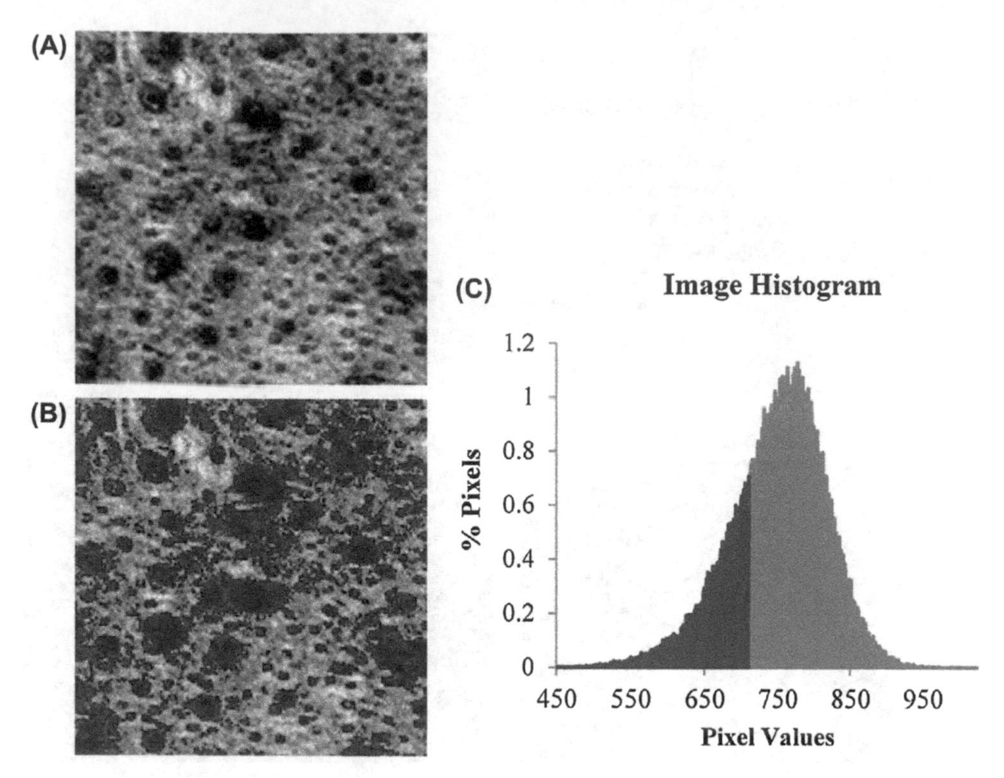

FIGURE 22.24 Quantitative images processing described in Refs. [9,10]. (A) Cross-polarized wide-field image, field of view = 4.7 × 4.7 mm. (B) Threshold image. (C) Image histogram.

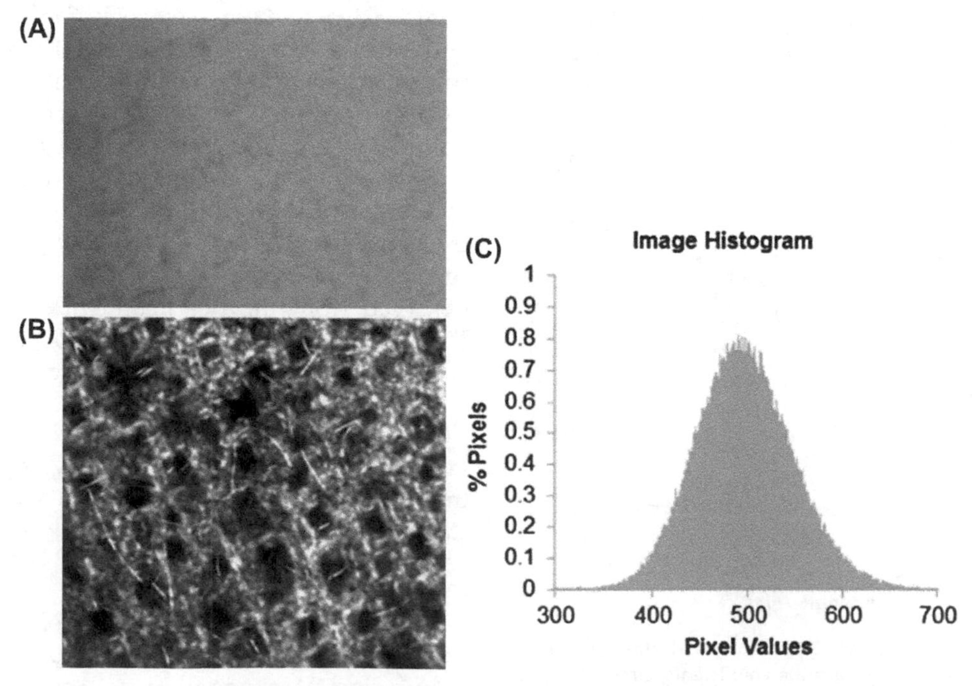

FIGURE 22.25 (A) Photograph of a 35-year-old subject. (B) Cross-polarized 440-nm wide-field image. (C) Collagen image histogram. *Images were obtained from Feng X, Patel R, Yaroslavsky AN. Wavelength optimized cross-polarized wide-field imaging for noninvasive and rapid evaluation of dermal structures. J Biophotonics 2015;8(4):324–31.*

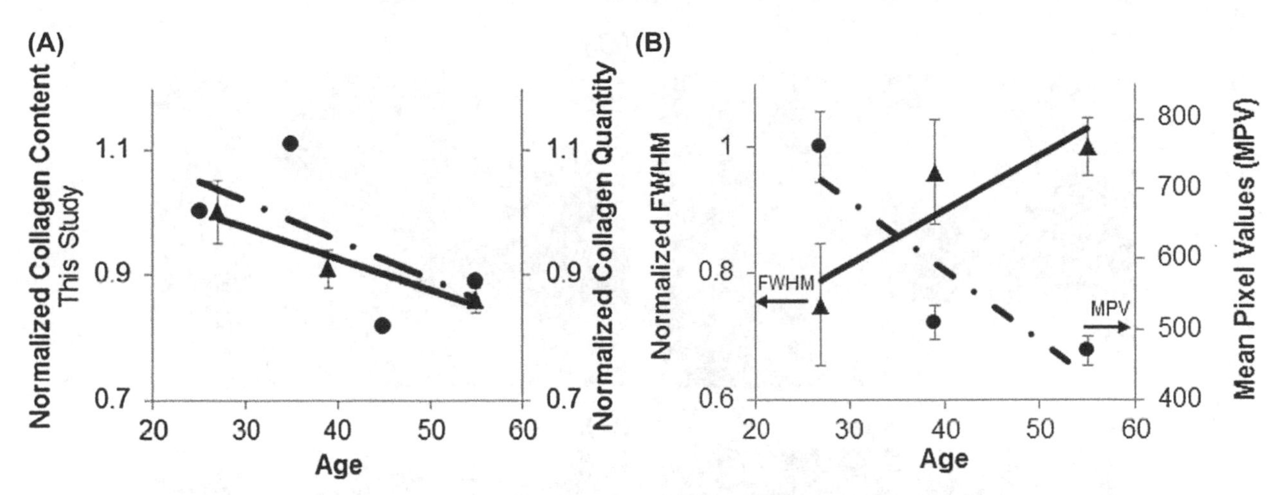

FIGURE 22.26 Quantitative collagen evaluation. (A) Collagen content versus age. *Triangles*, Results from cross-polarized 440 nm wide-field image analysis; *circles*, results from histopathology study. (B) Full width at half maximum (FWHM) (*triangles*) and mean pixel value (*circles*) of collagen image versus age. *From Feng X, Patel R, Yaroslavsky AN. Wavelength optimized cross-polarized wide-field imaging for noninvasive and rapid evaluation of dermal structures. J Biophotonics 2015;8(4):324–31.*

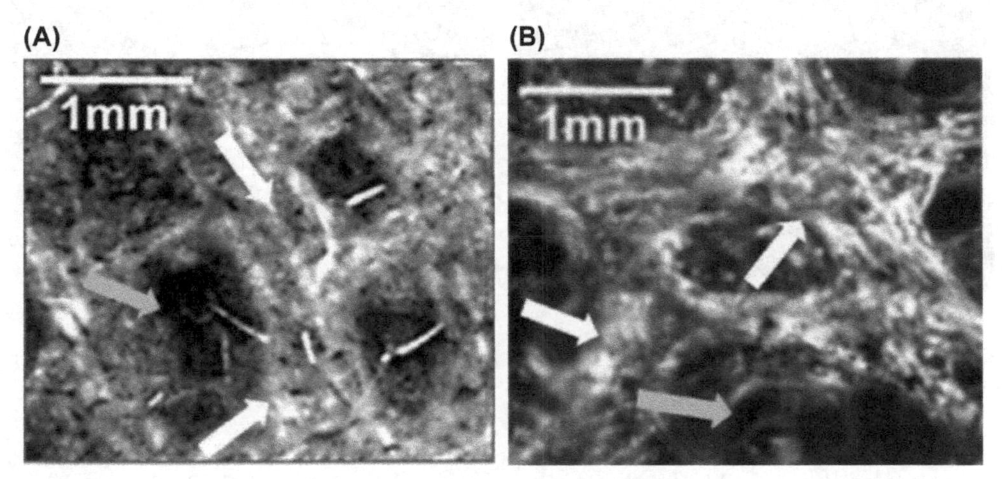

FIGURE 22.27 (A) In vivo collagen image acquired with polarization enhanced wide-field imaging system at 440 nm. (B) Ex vivo confocal image acquired from the dermal side. *White arrows*, collagen network; *gray arrows*, hair follicles. *Images were obtained from Feng X, Patel R, Yaroslavsky AN. Wavelength optimized cross-polarized wide-field imaging for noninvasive and rapid evaluation of dermal structures. J Biophotonics 2015; 8(4):324–31.*

noninvasive imaging methods, such as confocal microscopy, optical coherence tomography (OCT) [58], and second harmonic microscopy [57] have been employed for evaluation of dermal collagen. These modalities provide higher spatial resolution, but are also characterized by a very limited field of view. In addition, they are quite expensive to implement and maintain. Moreover, their use requires a considerable degree of training. Therefore the inexpensive and easy-to-use system that provides a field of view of several centimeters, 12–20 -μm lateral resolution, and real-time imaging capability may offer some advantages over the other sophisticated imaging methods in clinical use. Future investigations will show if this promising technology will make its way into dermatological practice.

Dye-Enhanced Spectrally-Resolved Polarization Imaging of Skin

In 2003, Yaroslavsky et al. [14] utilized dye-enhanced polarization imaging for the delineation of NMSCs. The authors understood that because of low scattering of basal and SCC [67] and variable melanin content in these tumors, high intrinsic contrast imaging of NMSC is not feasible. Thus it was decided to use intravital dyes, ie, methylene blue (MB) and toluidine blue (TB) to enhance the contrast of cancerous tissue. These phenothiazinium dyes have been extensively used for staining various tumors in vivo and ex vivo [82–87]. MB and TB have similar chemical structure and similar physicochemical properties. Their blue color is determined by the strong absorption band in the 550–700 nm region. In contrast,

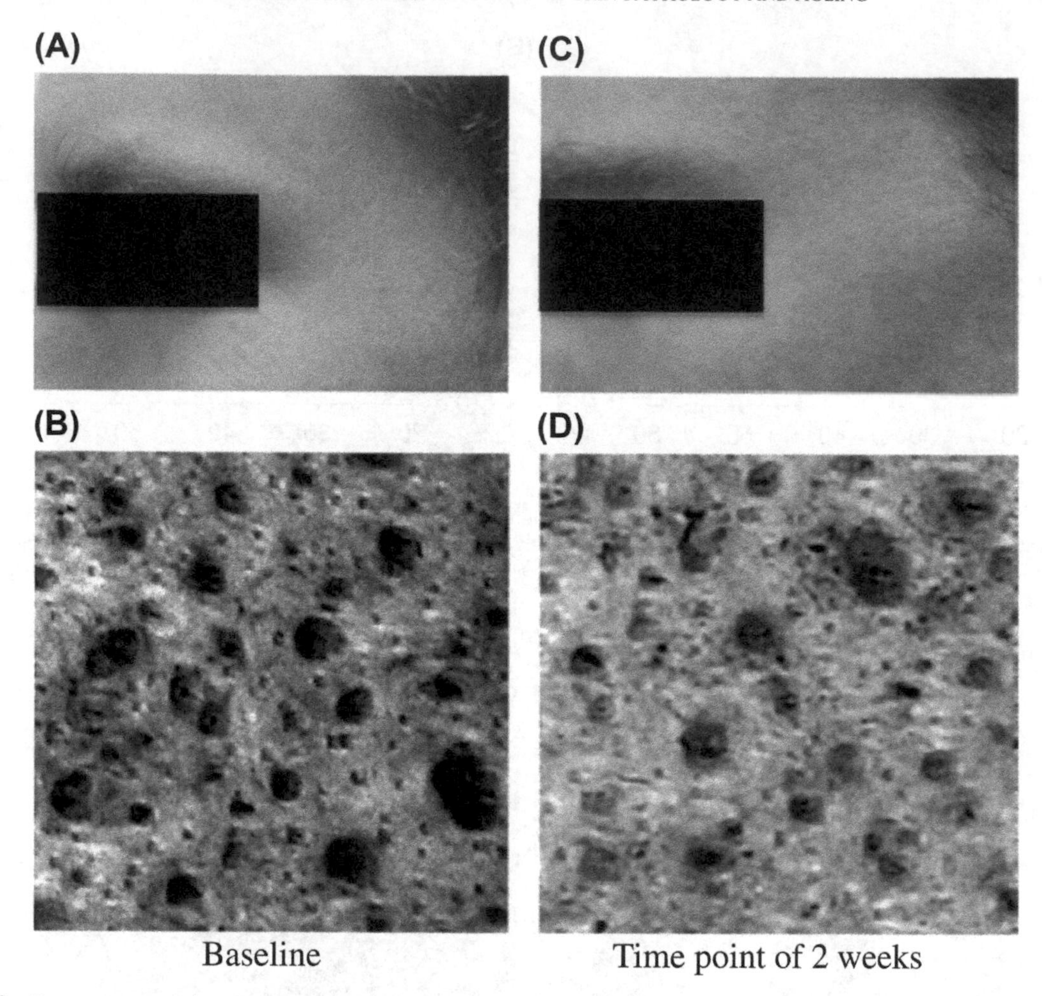

FIGURE 22.28 Pretreatment and posttreatment images of a 25-year-old subject. (A) Pretreatment photograph. (B) Pretreatment collagen image. (C) Posttreatment photograph. (D) Posttreatment collagen image.

because of the presence of melanin and hemoglobin, absorption in skin is maximal around 410 nm. Absorption spectra of MB, TB, and human skin are presented in Fig. 22.29, which demonstrates that spectrally resolved imaging in the visible spectral range is capable of demarcating the areas of blood absorption from those of dye localization. This technique may be particularly valuable for in vivo intraoperative tumor margin demarcation, because traces of blood are always present in the surgical bed, making white light inspection of the tumor margins complicated.

The authors have assembled a laboratory set-up for rapid imaging at five wavelengths, including 410, 600, 610, 620, and 710 nm. Discarded tumor material was received from Mohs micrographic surgeries under an institutional review board–approved protocol. The tissue was stained for up to 5 min in 0.01–0.05% pH balanced aqueous dye solution. Then the specimens were briefly rinsed in saline solution. For imaging, the tissue was placed in a Petri dish on gauze soaked in saline solution, and covered with a cover slip. Copolarized and cross-polarized images of thick skin excisions were acquired before and after staining with TB or MB. These images were processed and superficial images for four wavelengths, including 410, 600, 610, and 620 nm, were obtained and analyzed. To obtain the superficial images, the difference images, $I_\delta = I_{co} - I_{cross}$ were calculated for each wavelength. In order to reject the background signal, the resulting images for wavelengths, λ, of 410, 600, 610, 620 nm, were obtained using the following equation: $I_\Delta{}^\lambda = I_\delta{}^\lambda - I_\delta{}^{\lambda r}$, where $I_\Delta{}^\lambda$ is the resulting image, I_δ is the difference image, $\lambda_r = 710$ nm. Resulting images were analyzed and compared to histopathology.

The authors confirmed that the difference images at 410 nm were capable of emphasizing or rejecting the signal from blood and emphasize tissue morphology, thus helping with tumor delineation (Fig. 22.30). However, the most important finding was that exogenous contrast agents significantly increased the contrast of cancer and, in combination with image subtraction, enabled high contrast and sufficient resolution for visualization of tumors and normal skin morphology

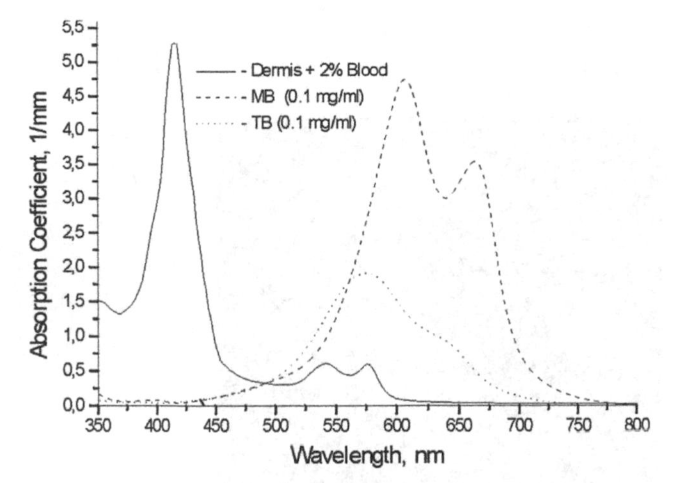

FIGURE 22.29 Absorption spectra of dermis stained with methylene blue (MB) and/or toluidine blue (TB). The dyes exhibit strong absorption in the 550–700 -nm region. In contrast, absorption in skin, determined by two main chromophores, melanin and hemoglobin, exhibits maximum around 400 nm. Therefore spectrally resolved imaging in the range from 400 to 700 nm can delineate the areas of enhanced blood absorption and the areas of enhanced dye absorption. *From Yaroslavsky AN, Neel V, Anderson RR. Demarcation of nonmelanoma skin cancer margins in thick excisions using multispectral polarized light imaging. J Invest Dermatol 2003;121(2):259–66.*

(Fig. 22.31). The investigators studied 45 skin excisions. Out of these samples, in 41 cases polarized light images correlated well with H&E frozen sections. In the remaining four cases, they reported partial correlation. The authors noted that polarized light images looked very much like standard Mohs micrographic–surgery frozen H&E histopathology. Hair follicles, collagen, sebaceous glands, and fat were visible in detail. They could be clearly discriminated from the tumors, which were very dark because of the increased uptake of the dye. This indicated that the lateral resolution of approximately 30 μm and the optical section thickness of approximately 150 μm were sufficient for intraoperative detection of tumor margins. Combination of tissue staining with polarized light limited the imaged volume to the superficial tissue layers and ensured that sufficient contrast was achieved for reliable differentiation between tumor and surrounding tissue. Finally, the investigators concluded that multispectral dye-enhanced polarized light macroimaging could delineate the margins of nonmelanoma cancers, including the morpheaform BCC, for which demarcation is especially challenging even in histopathology.

FLUORESCENCE POLARIZATION IMAGING OF HUMAN SKIN

Exogenous Fluorescence Polarization of Methylene Blue (MB) and Toluidine Blue (TB) for Delineation of Nonmelanoma Skin Cancers

Exogenous FP imaging for skin cancer detection was first introduced by Yaroslavsky et al. in 2004 [36]. In this work, the investigators proposed to utilize FP imaging for delineating NMSC margins. Similarly to earlier work, [14] they used two phenothiazine dyes, MB and TB, as exogenous fluorophores. A schematic diagram of the imaging experiment is shown in Fig. 22.32.

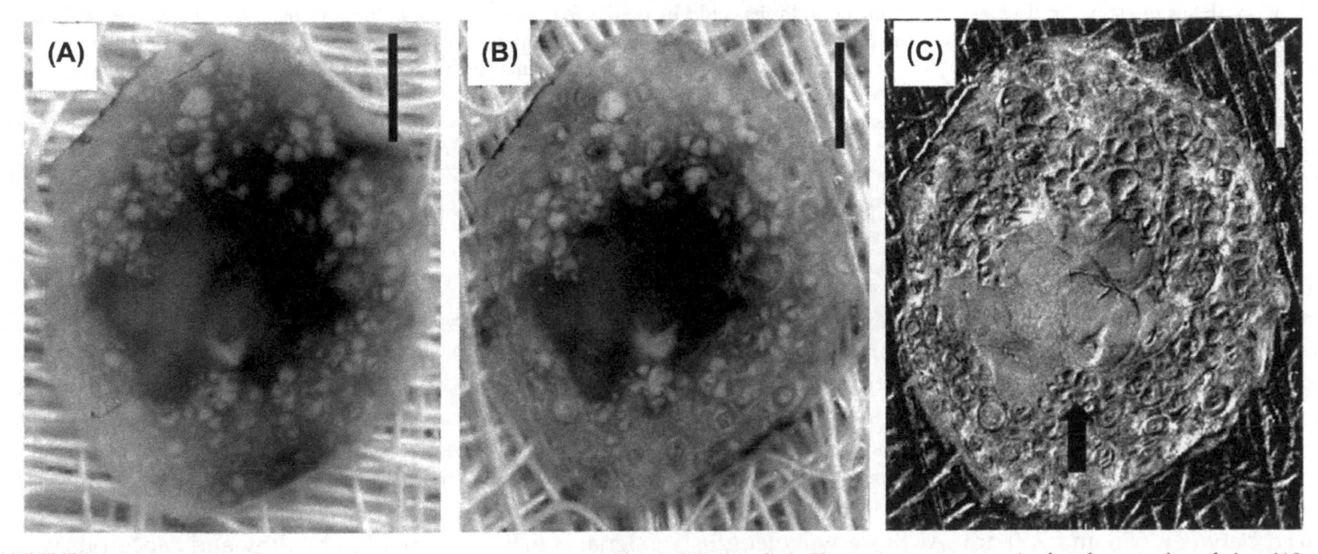

FIGURE 22.30 Images of skin with infiltrative basal cell carcinoma (BCC) (site: lip). The images were acquired at the wavelength $\lambda = 410$ nm, scale = 5 mm. In the conventional images, acquired before (A) and after toluidine blue (TB) staining (B), blood appears dark. The tumor is not apparent in the images because of high blood content and absorption (A and B) and negligible absorption of TB (B) at 410 nm. (C) In contrast, the tumor can be clearly delineated as a structureless area in the superficial image, I_{Δ}^{410}, at 410 nm (*arrow*). *From Yaroslavsky AN, Neel V, Anderson RR. Demarcation of nonmelanoma skin cancer margins in thick excisions using multispectral polarized light imaging. J Invest Dermatol 2003;121(2):259–66.*

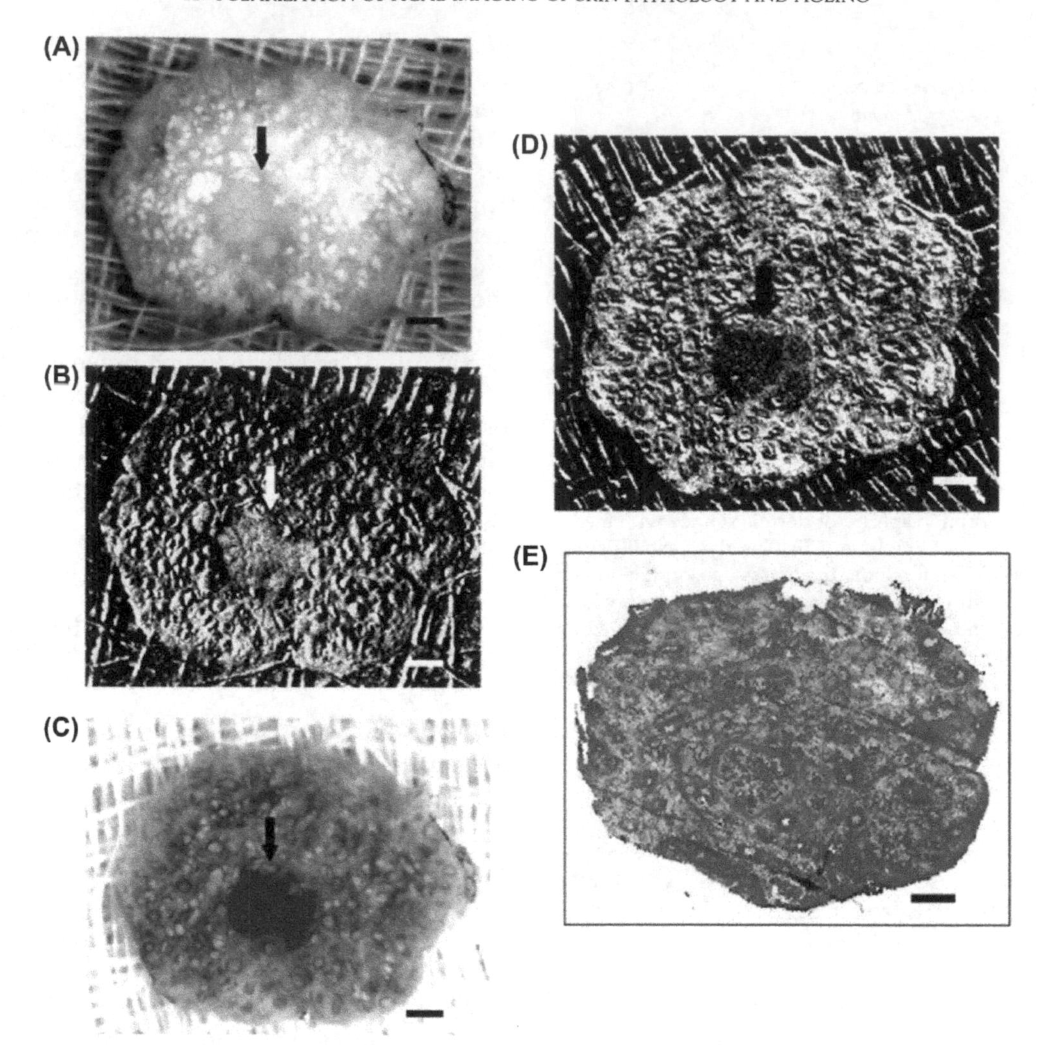

FIGURE 22.31 Images of skin with nodular and micronodular basal cell carcinoma (BCC) (site: nose). The images were acquired at the wavelength $\lambda = 620$ nm, scale = 1 mm. (A) Tumor margins are difficult to identify in the conventional image of the tissue acquired before dye application. (B) In the superficial image the tumor boundaries could be delineated even without staining (*arrow*). In the conventional (C) and superficial, I_{Δ}^{620} (D) images of the same specimen after toluidine blue (TB) staining the tumor is very dark and can be easily demarcated. The location and the shape of the tumor in B, C, and D (*arrows*) compare well with frozen hematoxylin and eosin (H&E) (E); *red line* outlines the tumor margins. The detailed examination shows that in the superficial image acquired before staining and in the conventional image of the stained specimen the tumor appears as a single nest, whereas the superficial image of stained tissue reveals three closely seated tumor lobules. Frozen H&E (E) confirms that the number and location of the tumor lobules were identified accurately in the image (D) and proves that polarization light imaging enables imaging of the superficial tissue layer only. *From Yaroslavsky AN, Neel V, Anderson RR. Demarcation of nonmelanoma skin cancer margins in thick excisions using multispectral polarized light imaging. J Invest Dermatol 2003;121(2):259–66.*

Fluorescence of the sample was excited with light from a xenon-arc lamp combined with interference filters. As shown in Fig. 22.32, two linearly polarizing filters, P1 and P2, were employed for polarization imaging. A CCD camera equipped with fluorescence filters was used to register co-polarized and cross-polarized signals. Fluorescence of TB and MB was excited at 577 and 620 nm, respectively. Fluorescent images were acquired between 650 and 750 nm. At these wavelengths, contribution of endogenous fluorophores can be considered negligible.

The authors imaged freshly excised thick BCC and SCC samples and observed higher FP in cancerous tissue. Example images from their work are presented in Figs. 22.33–22.35. Fig. 22.33 shows images of a nodular BCC. Fig. 22.33A is the reflectance image of the sample before staining, which reveals that contrast between normal and cancerous tissue is not sufficient for reliable tumor discrimination. Fig. 22.33B shows the fluorescence emission image of this sample after TB staining. The figure demonstrates that a high level of fluorescence signal is exhibited by both healthy and cancerous tissue. Similar to the reflectance image, TB fluorescence emission image does not demarcate the tumor.

Fig. 22.34 shows the pseudocolor TB FP image (FPI) (A) and a H&E histopathology section (B) of the same

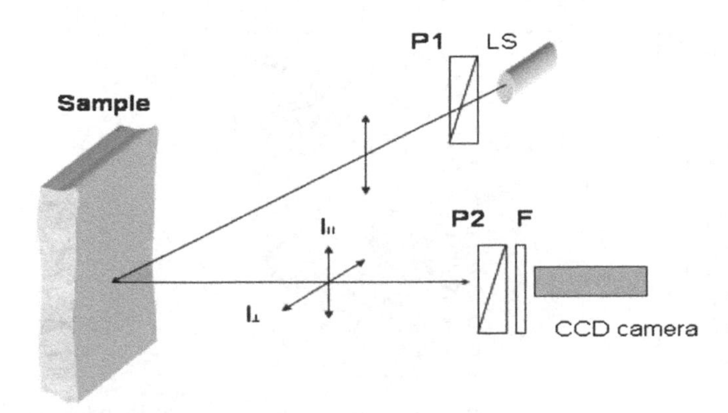

FIGURE 22.32 Schematic diagram of fluorescence polarization (FP) imaging experiment introduced in Ref. [36]. *CCD*, Charge-coupled device; *P1* and *P2*, two linearly polarizing filters.

sample. Comparison with the section shows good correlation between the bright regions in FPI and the tumor outlined in histopathology by the surgeon. The gray arrow shows the larger tumor lobule and green arrow shows the smaller tumor lobule. Average FP value of the cancer area is 2.7 times higher than that of normal area, which provides sufficient contrast for cancer delineation.

Fig. 22.35 shows a BCC sample stained with MB. Fig. 22.35A is the pseudocolor FPI and Fig. 22.35B is the histological frozen section. FPI shows high contrast between cancerous and healthy tissue. The region with high fluorescence signal correlates well with the tumor outlined by the surgeon.

The authors quantitatively evaluated FP values for all samples stained with TB and MB (Fig. 22.36). For samples stained with TB, the average FP of cancerous tissue was about 2.5 times higher than that of normal tissue. Similarly, average FP of MB-stained cancerous tissue is approximately 2.75 times higher than that of MB-stained normal tissue. A Student's *t*-test for two independent populations was performed to analyze the data. Differences in FP between cancerous and healthy tissue were significant for both MB ($p < 0.005$) and TB ($p < 0.001$).

Based on the results, Yaroslavsky et al. concluded that FP imaging may prove to be a real-time intraoperative tool for delineating NMSC during surgeries. As

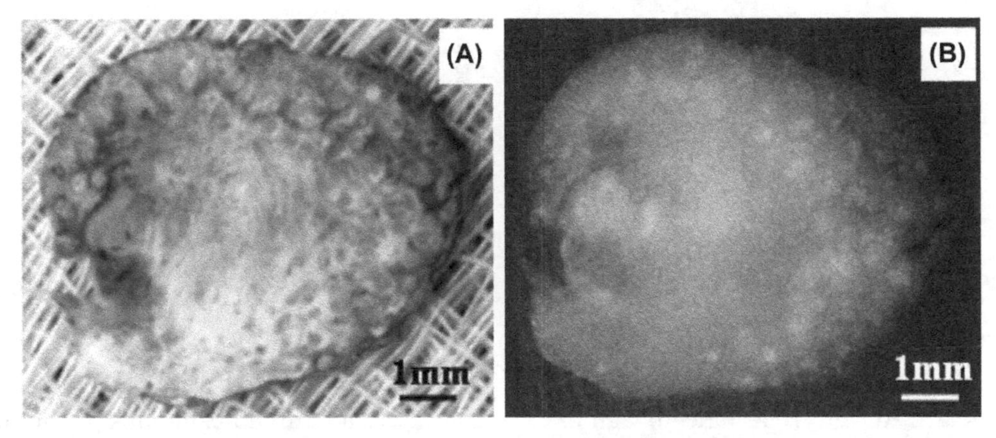

FIGURE 22.33 Images of a nodular basal cell carcinoma (BCC). (A) 577-nm reflectance image of unstained sample. (B) Fluorescence emission image of sample stained with 0.05 mg/mL toluidine blue (TB) solution. *Results were obtained from Yaroslavsky AN, Neel V, Anderson RR. Fluorescence polarization imaging for delineating nonmelanoma skin cancers. Opt Lett 2004;29(17):2010–12.*

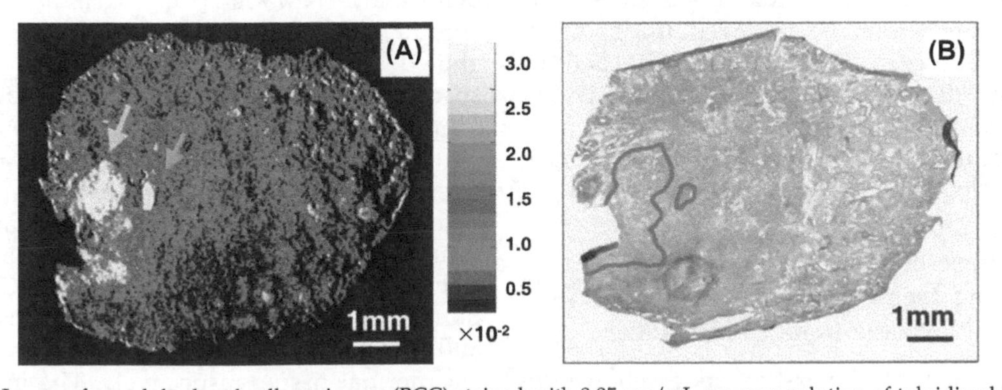

FIGURE 22.34 Image of a nodular basal cell carcinoma (BCC) stained with 0.05 mg/mL aqueous solution of toluidine blue (TB). (A) Pseudocolor fluorescence polarization (FP) image. (B) Histological frozen section. *Results were obtained from Yaroslavsky AN, Neel V, Anderson RR. Fluorescence polarization imaging for delineating nonmelanoma skin cancers. Opt Lett 2004;29(17):2010–12.*

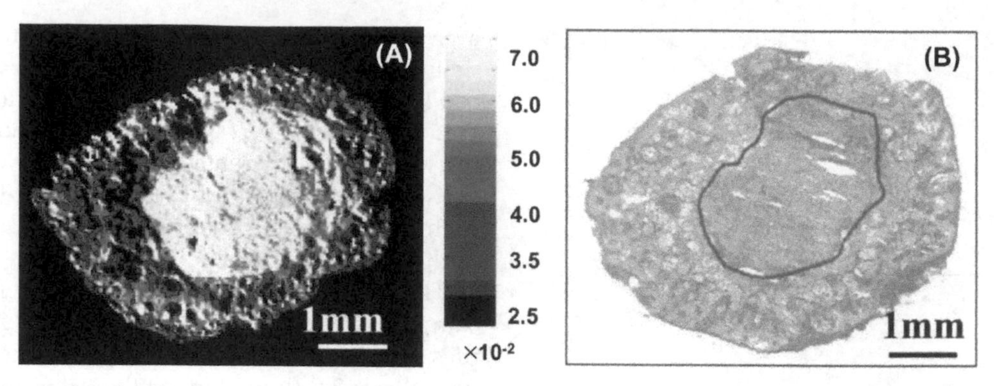

FIGURE 22.35 Images of a basal cell carcinoma (BCC) stained with 0.25 mg/mL aqueous solution of methylene blue (MB). (A) Pseudocolor fluorescence polarization (FP) image. (B) Histological frozen section. *Results were obtained from Yaroslavsky AN, Neel V, Anderson RR. Fluorescence polarization imaging for delineating nonmelanoma skin cancers. Opt Lett 2004;29(17):2010–12.*

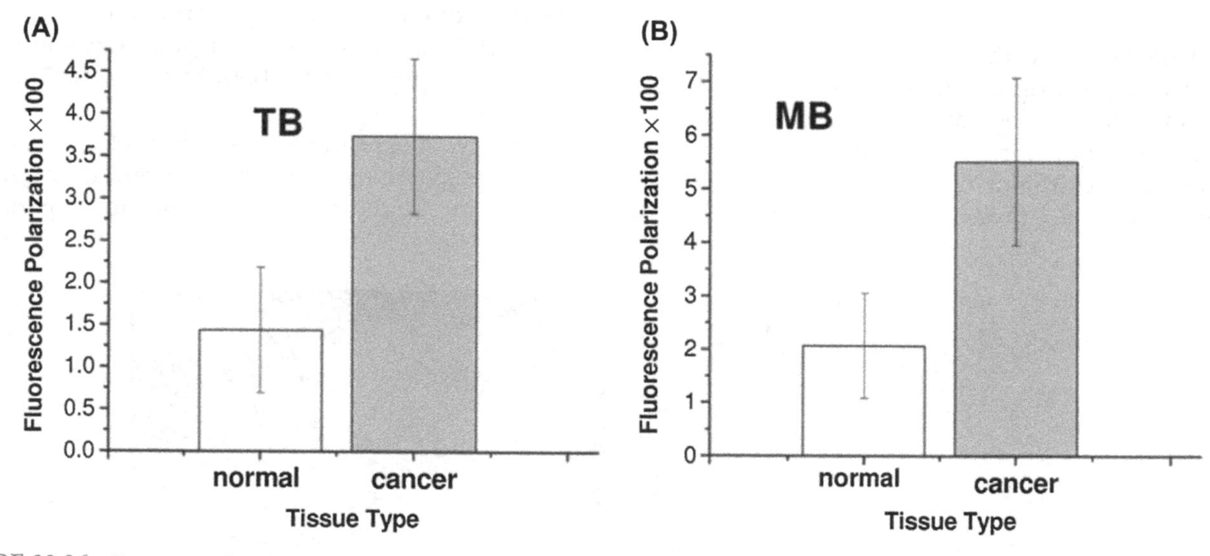

FIGURE 22.36 Exogenous fluorescence polarization (FP) of cancerous and normal tissue stained with toluidine blue (TB) (A) and methylene blue (MB) (B).

compared with polarized light reflectance imaging, FP has a significant advantage: it can provide quantitative results. Even though MB and TB are preferentially retained in cancerous tissue, some normal skin structures, such as hair follicles accumulate some dye. Cellular resolution cannot be achieved using digital macroimaging. However, if FP values of MB and TB were significantly different in normal and tumor tissues, it could resolve the ambiguity and deliver a cost-effective and accurate method for nonmelanoma cancer demarcation during surgeries.

Exogenous Fluorescence Polarization of Tetracycline and Demeclocycline for Delineation of Nonmelanoma Skin Cancers

In 1957, Rall et al. were the first to report yellow fluorescence from breast tumors following tetracycline (TCN) administration [88]. Since then, multiple clinical trials have attempted using this phenomenon for cancer diagnostics [88–92] and therapy [91,93,94].

Yaroslavsky et al. explored the use of TCN and demeclocycline (DMN) fluorescence and FP imaging for delineating NMSCs [95]. Excitation and emission spectra of TCN and DMN are shown in Fig. 22.37A and B. For the experiments, freshly excised thick specimens were imaged before and after TCN/DMN staining. Fluorescence was excited by linearly polarized light centered at 390 nm (FWHM = 10 nm), and registered in the range of 450–650 nm. Size, shape, and contrast of the tumor in optical images were compared with those in histological samples. Contrast of tumor with respect to normal skin (CNL) was determined by: $CNL = \frac{AVFP_{cancer}}{AVFP_{normal}} \times 100$, where $AVFP_{cancer}$ and $AVFP_{normal}$ are the average values of FP of cancer and normal tissue, respectively. The investigators also compared surface areas of tumor in FPI (S_{fpi}) and histology (S_h). They reasoned that in order to completely

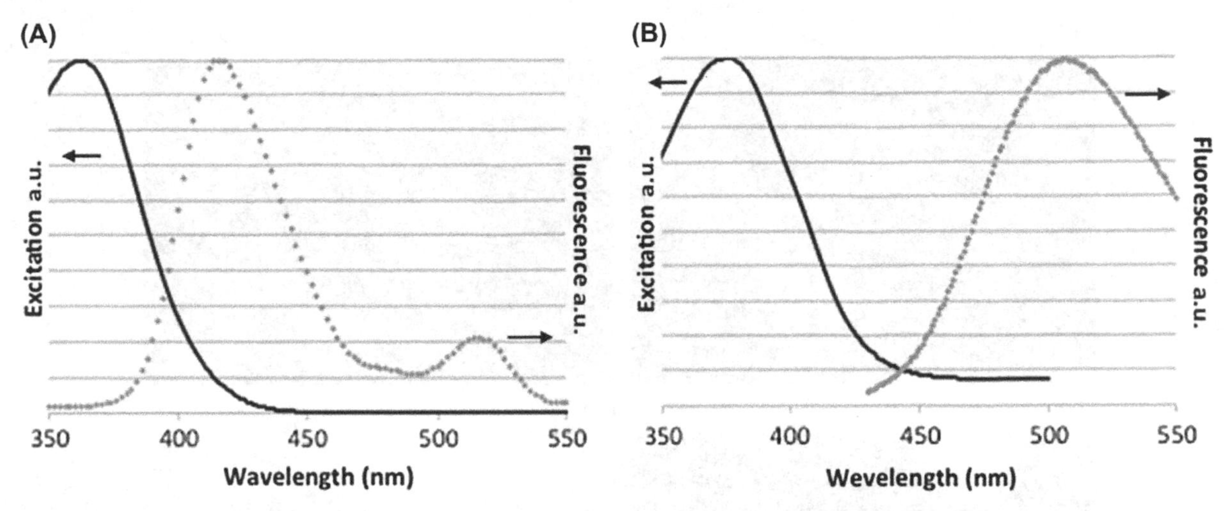

FIGURE 22.37 Excitation and emission spectra of tetracycline (TCN) (A) and demeclocycline (DMN) (B). Both at 1 mg/mL (in Dulbecco's phosphate-buffered saline (DPBS)), pH 7.4. *Black solid lines*, excitation spectra; *gray dotted lines*, emission spectra.

remove tumor using image-guided surgery, S_{fpi} should be equal to or up to 15% greater than S_h.

The investigators reported that tumor boundaries could not be reliably visualized using fluorescence emission imaging (Fig. 22.38A). In contrast, FP imaging was capable of accurate tumor margin delineation. As demonstrated in Fig. 22.38B, a tumor is much brighter as compared with normal skin in FPI. It can also be appreciated that the shape, size, and location of the tumor delineated in FPI (see Fig. 22.38B) correlate well with those in the histopathological sample (see Fig. 22.38C).

Several endogenous fluorophores in skin, eg, collagen and porphyrins, exhibit fluorescence when excited at 390 nm. To examine the possible effect of intrinsic fluorescence on imaging, the investigators acquired endogenous fluorescence images, $I_{\parallel}^{en}$ and $I_{\perp}^{en}$, from tissues. After that, the specimens were stained for 5 min with 1–2 or 0.25–2 mg/mL TCN or DMN solutions, respectively. Mixed (endogenous and exogenous) fluorescence images, $I_{\parallel}^{ee}$ and $I_{\perp}^{ee}$, were acquired after staining. Endogenous (P^{en}) and mixed (P^{ee}) FPIs were generated in real time. Exogenous fluorescence emission (I^{ex}) and FPI images were calculated using the following equations:

$$I^{ex} = I^{ee} - I^{en},\ P^{i} = \left(I_{\parallel}^{i} - I_{\perp}^{i}\right)/\left(I_{\parallel}^{i} + I_{\perp}^{i}\right),\ \text{where i} = \text{ex, en, ee}.$$

Fig. 22.39 shows example endogenous, mixed, and exogenous FPIs of a sample with micronodular BCC. Bright areas in Fig. 22.39A correspond to the collagen in the dermis. Even though Fig. 22.39B is not corrected for endogenous fluorescence, the tumor can be distinguished in the image, exhibiting a much higher FP signal, as compared with surrounding healthy skin. It is worth noting that the resolution of the image is sufficient for morphological analysis. AVFP in normal and cancerous areas in Fig. 22.39A–C were quantitatively evaluated and summarized in Fig. 22.39E. Exogenous AVFP of TCN is more than five times higher in tumor compared with in healthy skin, whereas endogenous AVFP is slightly higher in normal skin. Image analysis shows that CNLs of mixed and exogenous FPIs are

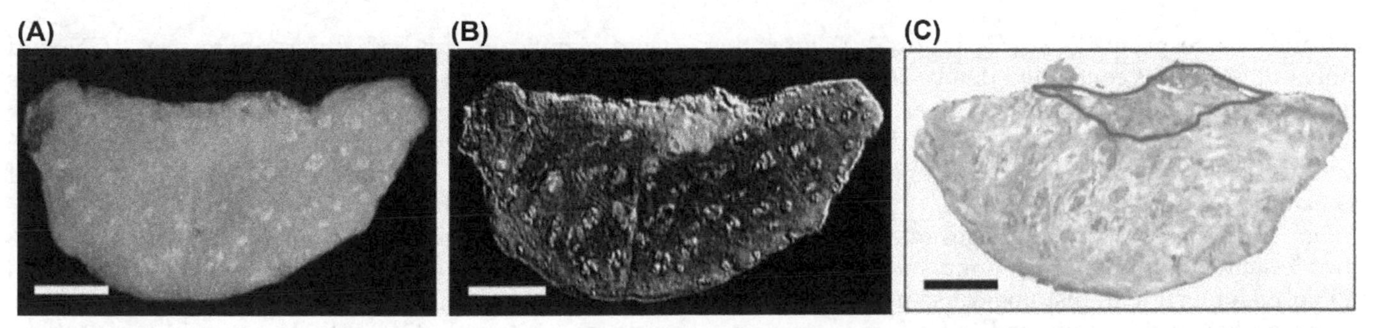

FIGURE 22.38 Example images of a nodular basal cell carcinoma (BCC) sample. (A) Fluorescence emission image after staining with 2 mg/mL tetracycline (TCN). (B) TCN fluorescence polarization (FP) image. (C) Hematoxylin and eosin (H&E) frozen histological section. *From Yaroslavsky AN, Salomatina EV, Neel V, Anderson R, Flotte T. Fluorescence polarization of tetracycline derivatives as a technique for mapping nonmelanoma skin cancers. J Biomed Opt 2007;12(1):014005*

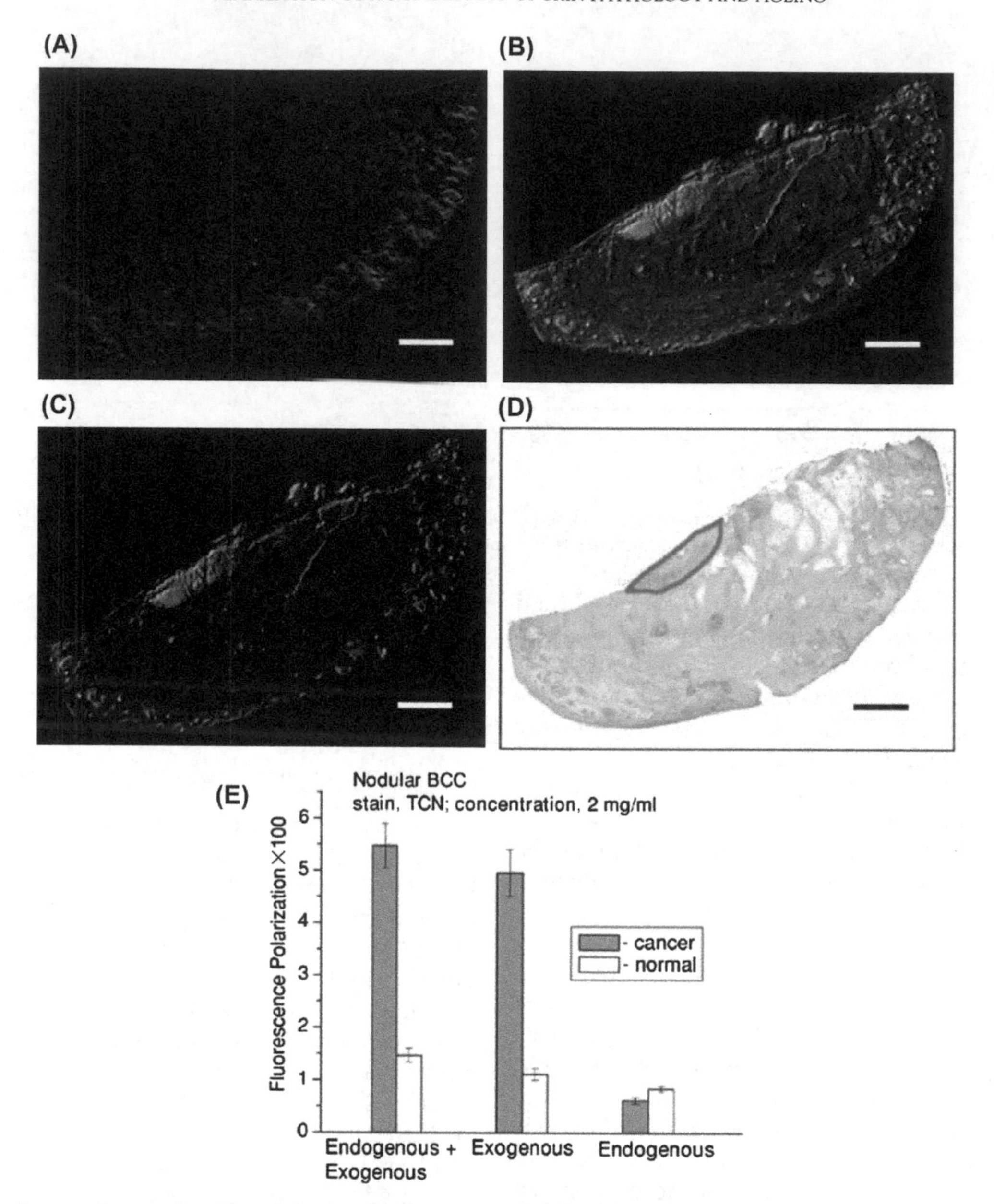

FIGURE 22.39 Images of a sample with nodular basal cell carcinoma (BCC), scale = 2 mm. (A) Endogenous fluorescence polarization (FP) image. (B) Mixed FP image, acquired after staining with 2 mg/mL tetracycline (TCN). (C) Net exogenous FP. (D) Hematoxylin and eosin (H&E) frozen histological section. (E) Average value of fluorescence polarization (AVFP) measured from A–C.

4.00 and 3.84, respectively. S_{fpi}/S_h ratio is 1.02. From this analysis the authors concluded that because there are no significant differences in contrast, size, and shape of tumor in the mixed and exogenous FPIs, deconvolution of the mixed (endogenous plus exogenous) FPI is not necessary.

Fig. 22.40 shows example images of an invasive SCC sample stained with DMN. Tumor areas identified from FPI correlate well with histological samples.

The author also investigated the effect of stain concentration on FP imaging. A student's *t*-test was performed to analyze the data. They have found that AVFP differences between cancerous and normal skin were significant for both fluorophores, ie, TCN and DMN, and for all concentrations of the stains ($p < 0.001$) (Fig. 22.41A and B).

In total, the authors examined 79 specimens with 86 tumors, including 73 cases of BCC and 13 cases of SCC. FP imaging of TCN and DMN was successful in accurate and high-contrast delineation of cancer in 88% and 94% of BCCs and SCCs, respectively. This is an impressive outcome, especially considering that some discrepancies with histopathology could result from the differences in the thickness of the optical image (~70 μm) and 5-μm thick histopathology. As compared with the phenothiazine fluorophores, MB and TB, which

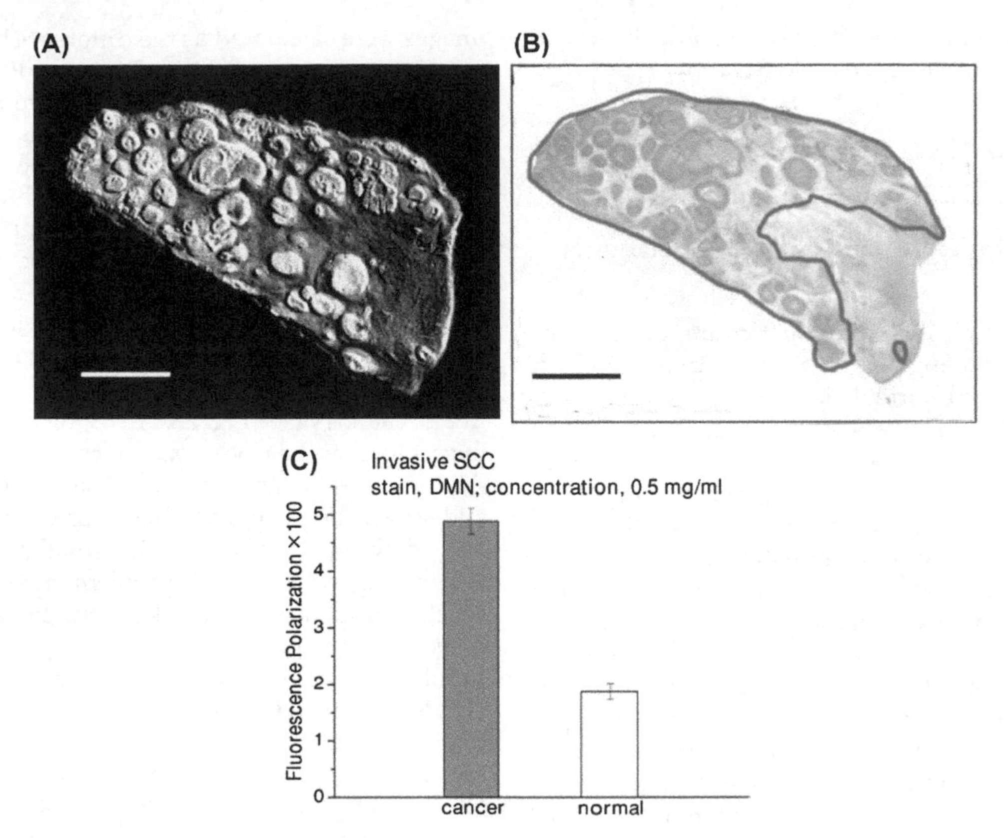

FIGURE 22.40 Example of a sample with invasive squamous cell carcinoma (SCC), scale = 1 mm. (A) Fluorescence polarization (FP) images after staining with 0.5 mg/mL demeclocycline (DMN). (B) Hematoxylin and eosin (H&E) frozen histological section. (C) Average value of fluorescence polarization (AVFP) measured from A. $CLN^{ee} = 3.02$, $S_{fpi}/S_h = 1.06$.

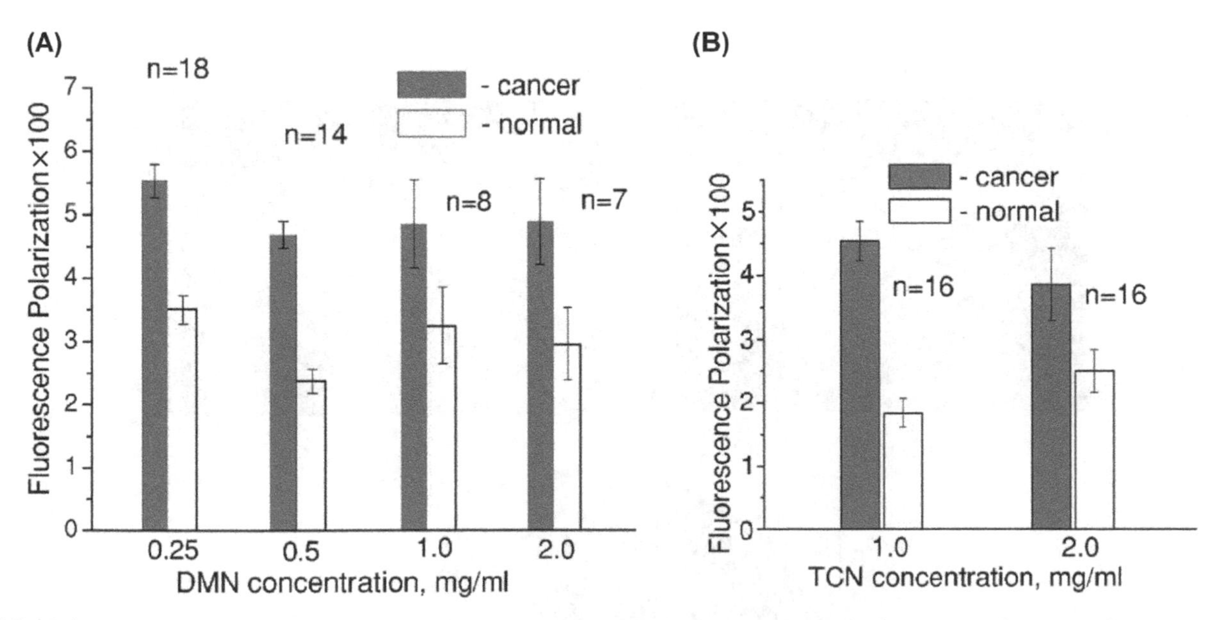

FIGURE 22.41 Effect of demeclocycline (DMN) (A) and tetracycline (TCN) (B) concentration on the contrast of cancer with respect to healthy tissue.

have to be injected into the lesion, TCNs possess a clear advantage of being accepted antibiotics, often used to treat skin infections. Thus TCNs may be introduced to the patient systemically. They are known to accumulate in the tumors; therefore measurable concentration levels are likely attainable. In summary, the investigators concluded that the use of safe and readily available contrast agents, such as TCN and DMN, as well as the simplicity of the proposed method, should help with clinical translation of the technology.

MULTIMODAL IMAGING OF SKIN CANCERS

Multimodal Optical Imaging of Nonmelanoma Skin Cancers

Salomatina et al. [38] combined multimodal reflectance and FP imaging with spectroscopic analysis to delineate BCC from normal skin at a single pixel level. Gross demarcation of NMSCs using reflectance and FP imaging has proven to be accurate in several studies [14,36,95]. However, because of the lack of morphological information using an approach similar to that of histopathology, correct discrimination of single pixels in wide-field macroimages is not feasible. To remedy this deficiency, the investigators proposed to use spectral analysis of single pixels in the image. For the experiments, fresh BCC samples were obtained from Mohs micrographic surgeries. Before imaging, samples were stained with 0.2 mg/mL aqueous MB solution for 2 min and then briefly rinsed in saline solution (pH 7.4). Co-polarized and cross-polarized reflectance images were acquired at 35 imaging wavelengths from 395 to 735 nm at 10-nm intervals. Superficial reflectance images were calculated as the difference between the co-polarized and cross-polarized images. Fluorescence images were excited at 615 nm and registered in the range between 660 and 750 nm. FPI was calculated as: $F = (F_{co} - F_{cross}) \times 100/(F_{co} + F_{cross})$, where F is FPI, and F_{co} and F_{cross} are co-polarized and cross-polarized fluorescence emission images. Example color-coded quantitative superficial reflectance and FPIs of a BCC sample are shown in Fig. 22.42A and B. Cancer can be identified in the images as regions with low absolute reflectance and high FP. Location, size, and shape of the tumor in optical images correlate well with those in the histology (see Fig. 22.42C). However, as was noted in previous studies [36], some normal structures, such as epidermis (solid arrows), hair follicles (dotted arrow), and sebaceous glands (dashed arrows) retain some dye and may be confused with malignant structures. Attempting to resolve this problem, the author analyzed spectral responses of all skin structures presented in reflectance images. Analysis was performed with a copolarized reflectance image at each wavelength. To enable absolute quantification of diffuse reflectance, a calibrated gray reference (reflectance value ~35% in the range from 395 to 735 nm) was placed in the field of view. Absolute diffuse reflectance and optical density were calculated using the following equation, $R^{\lambda} = 0.35R_s^{\lambda}\big/R_{ref}^{\lambda}$ and $OD^{\lambda} = \log(1/R^{\lambda})$, where R^{λ} and

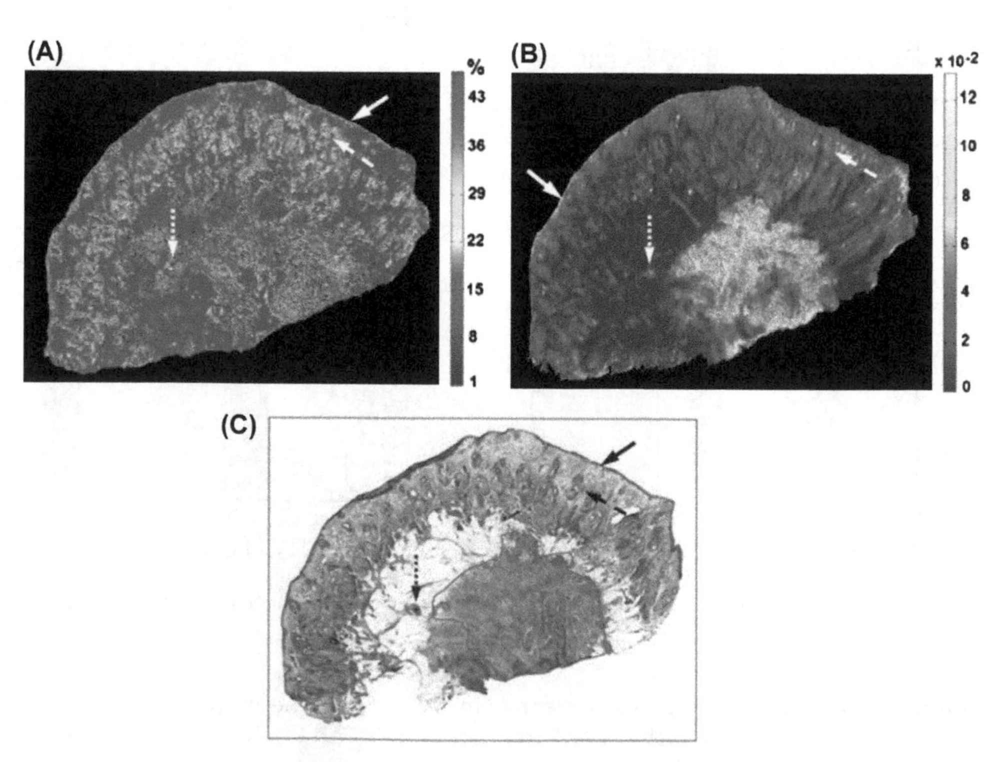

FIGURE 22.42 (A) Color-coded 665-nm superficial reflectance image. (B) Methylene blue (MB) fluorescence polarization (FP) image. (C) Hematoxylin and eosin (H&E) frozen histology of a basal cell carcinoma (BCC) sample.

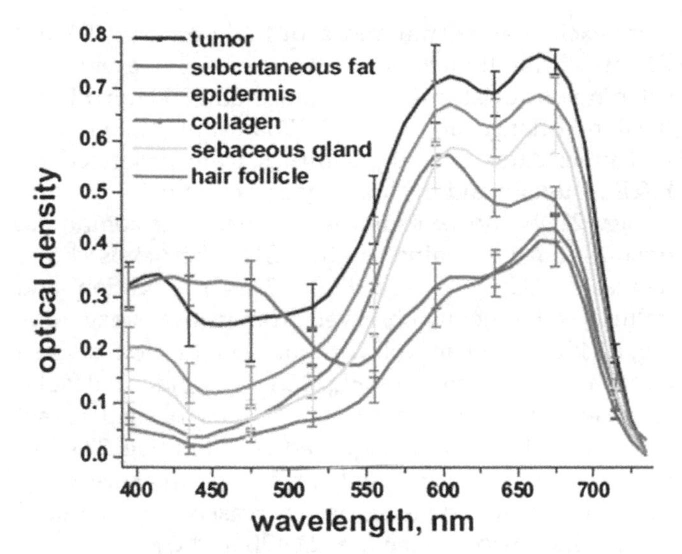

FIGURE 22.43 Average optical density spectra of cancer and normal structures. *Results were obtained from Salomatina E, Muzikansky A, Neel V, Yaroslavsky AN. Multimodal optical imaging and spectroscopy for the intraoperative mapping of nonmelanoma skin cancer. J Appl Phys 2009;105(10):102010.*

R_s^λ are the absolute and relative diffuse reflectance of the skin structure, R_{ref}^λ is the relative diffuse reflectance of the reference, and OD^λ is the optical density of the skin structure at wavelength λ. Each experimental spectrum was normalized by the value of the diffuse reflectance measured at 735 nm.

Optical density spectra of different structures in the MB-stained BCC sample are shown in Fig. 22.43. Because MB concentration is the highest in cancer, the tumor spectrum exhibits high optical density around MB absorption peaks at 615 and 665 nm. As shown in Fig. 22.43, optical density of collagen and fat is low in the MB absorption wavelength range, and can be easily distinguished from cancer. Epidermis spectrum shows a more pronounced peak around 615 nm as compared with that around 665 nm, whereas tumor exhibits the opposite distribution of the two peaks. This feature allows for discriminating epidermis from cancer. Spectral response of hair follicles and sebaceous glands are similar to that of a tumor, with lower optical density in the wavelength range between 615 and 665 nm.

In total, the authors investigated 20 samples. For each sample they analyzed optical density spectra for all skin structures. Then the responses obtained for each structure of each sample were averaged over 20 samples. The results are summarized in Table 22.1, columns 2

TABLE 22.1 Spectral Regions of Maximal Contrast Between Cancerous and Normal Skin Structures Determined by an Unpaired Two-Tailed *t*-Test (Significance Level = 0.05)

	Optical Density		Derivatives	
Tumor vs. Healthy Structure Pairs	**Wavelength Range of Statistically Significant Difference (nm)**	***p*-Values**	**Wavelength Range of Statistically Significant Difference (nm)**	***p*-Values**
Collagen	[595–725]	$0 \le p \le 0.00045$	595	$p = 0.0114$
			625	$p = 0.0095$
			635	$p = 0.0021$
			[675–725]	$0 \le p \le 0.036$
Epidermis	[615–715]	$0 \le p \le 0.019$	[595–665]	$0 \le p \le 0.0017$
			[685–725]	$0 \le p \le 0.000001$
Subcutaneous fat	[595–725]	$0 \le p \le 0.000001$	595	$p = 0.00031$
			[675–725]	$0 \le p \le 0.017$
Hair follicle	–	–	615	$p = 0.035$
			705	$p = 0.021$
			725	$p = 0.009$
Sebaceous gland	[625–695]	$0.001 \le p \le 0.032$	645	$p = 0.034$
			[685–705]	$0.0008 \le p \le 0.035$
			725	$p = 0.011$

Results were obtained from Salomatina E, Muzikansky A, Neel V, Yaroslavsky AN. Multimodal optical imaging and spectroscopy for the intraoperative mapping of nonmelanoma skin cancer. J Appl Phys 2009;105(10):102010.

and 3. An unpaired two-tailed *t*-test at a significance level of 0.05 was performed for each cancer–normal tissue pair. Statistical analysis confirmed that spectra of most normal structures are significantly different from that of tumor in the wavelength range from 615 to 700 nm, with the exception of hair follicles, which cannot be distinguished from cancer using optical density analysis.

In order to reliably discriminate hair follicles from tumor, the author further analyzed the wavelength derivatives of the optical densities. Differences between cancerous and normal structures are statistically evaluated and summarized in Table 22.1.

The investigators reported that wavelength derivatives of optical densities of hair follicles were significantly different from those of tumor at 615, 705, and 735 nm. Interestingly, derivatives of other normal skin structures have shown significant differences from those of tumor, at 665 nm.

The authors concluded that both types of images, ie, reflectance and FP, could be successfully used for gross demarcation of BCCs. They noted that reflectance Pols highlighted skin morphology, whereas FPIs displayed higher contrast of cancerous tissue as compared with reflectance imaging. Quantitative spectral analysis of the optical densities and their wavelength derivatives has shown promise for discriminating tumor from normal tissue at a single-pixel level. The investigators emphasized that as spectral responses of all the normal skin structures investigated were significantly different from those of cancer at 615 and 665 nm, quick spectroscopic analysis at only two interrogation wavelengths should be sufficient to allow diagnostic accuracy. The reduction of the number of imaging wavelengths from 35 to 2, combined with the ability of this multimodal technique to accurately diagnose single pixels, may be critical for successful clinical translation of the technology.

Multimodal Optical Imaging of Melanoma

Tannous et al. reported on the evaluation of TCN and MB for demarcation of melanoma using reflectance and FP imaging [37]. This was a small blinded pilot trial with six skin excisions, diagnosed as suspicious for melanoma. Excisions were obtained after surgery and each was bisected into two halves through the middle (Fig. 22.44). One half of the sample was stained with 0.2 mg/mL MB and another half was stained with 2 mg/mL TCN.

Bisected tumor faces were imaged with the system shown in Fig. 22.45. Coreflectance and cross-reflectance images were acquired between 390 and 750 nm with the step of 40 nm. TCN fluorescence images were excited at 390 nm and acquired between 450 and 700 nm. MB fluorescence images were excited at 630 nm and registered between 660 and 750 nm. Superficial reflectance images and FPIs were generated in real time. After imaging, samples were processed for H&E histology and evaluated by pathologist.

Fig. 22.46 shows example images of a compound dysplastic nevus stained with TCN. The nevus shows increased TCN FP (see Fig. 22.46C). Location and volume of the nevus observed from optical image (see Fig. 22.46C) correlate well with those in the H&E-stained histopathological sample (Fig. 22.46D). Epidermal (solid arrow) and the dermal part of the nevus (dashed arrow) can be visualized from magnified TCN FPI (Fig. 22.47A). Regions with high FP signal correspond to areas where the melanocyte density is increased in standard histological analysis (see Fig. 22.47B and C).

Fig. 22.48 shows the reflectance and FPI of an MB-stained malignant melanoma sample. After MB staining, the tumor can be clearly distinguished in optical images, appearing dark in the reflectance image (see Fig. 22.48B) and bright in FPI (see Fig. 22.48C). Both reflectance images and FPIs correlate well with histology (see Fig. 22.48D). Comparison between magnified optical images (Fig. 22.49A and B) and histology (see Fig. 22.49C) of the MB-stained tissue demonstrates that optical images provide sufficient resolution and contrast to delineate the melanoma lesion.

The results of the pilot trial are summarized in Table 22.2. In all the studied cases, optical images of the tissues stained using MB correlated well with histopathology. TCN FPIs correlated well with histopathology for dysplastic nevi. In one out of the two melanoma cases, the results were ambiguous. TCN also enhanced the contrast of the scar tissue within the excised specimen. It should be noted that MB has an important advantage over TCN. Unlike TCN, MB absorbs in the visible spectral range and therefore it enhances the contrast of lesions in the reflectance as well as in fluorescence images (see Figs. 22.48 and 22.49).

Using optical multimodal, reflectance, and fluorescence imaging permits tumor demarcation and rapid assessment of large surgical fields without altering tissue structure or affecting standard H&E postoperative processing. The authors hypothesized that other methods of application, such as injection of MB or oral ingestion of TCN, could further improve image quality by possibly distinguishing nonspecific contrast uptake in nontumor cells. The investigators noted that macroimaging does not afford the cellular detail seen in H&E stains, but that degree of detail is generally not needed for accurate tumor removal. The magnified images of the lesions presented in Figs. 22.47 and 22.49 reveal sufficient resolution for the gross demarcation of tumor nests during surgery.

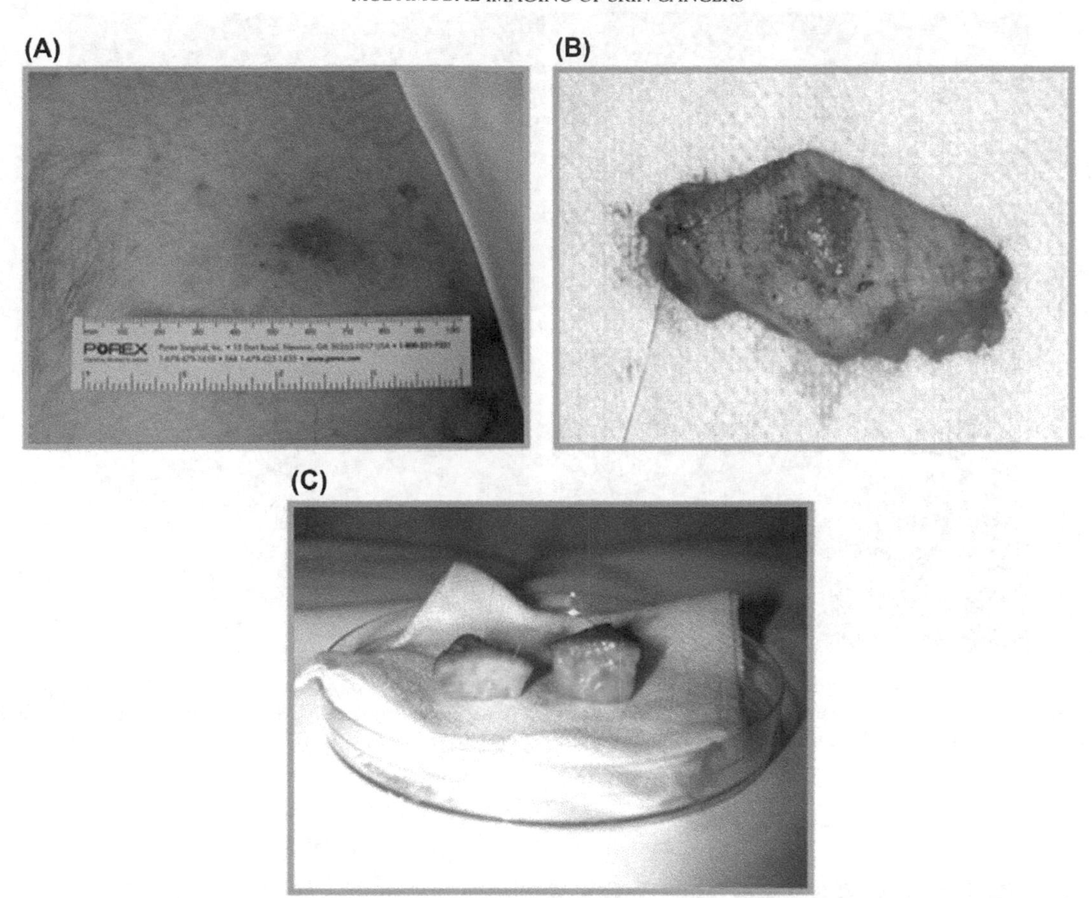

FIGURE 22.44 (A) Digital picture of the region suspicious for melanoma. (B) Elliptically excised sample. (C) Two halves of the bisected sample. *From Tannous Z, Al-Arashi M, Shah S, Yaroslavsky AN. Delineating melanoma using multimodal polarized light imaging. Lasers Surg Med 2009; 41(1):10–16.*

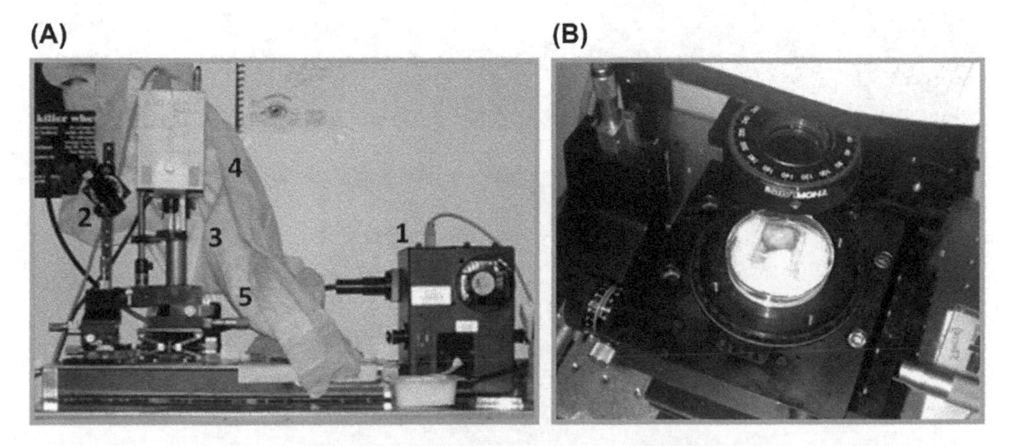

FIGURE 22.45 (A) Polarization imaging system. (1) Xenon-arc lamp with a filter wheel; (2) illuminator with diffuser and polarizer; (3) analyzer and fluorescence filter; (4) charge-coupled device (CCD) camera with imaging lens; (5) sample stage. (B) Sample prepared in Petri dish for imaging. *From Tannous Z, Al-Arashi M, Shah S, Yaroslavsky AN. Delineating melanoma using multimodal polarized light imaging. Lasers Surg Med 2009; 41(1):10–16.*

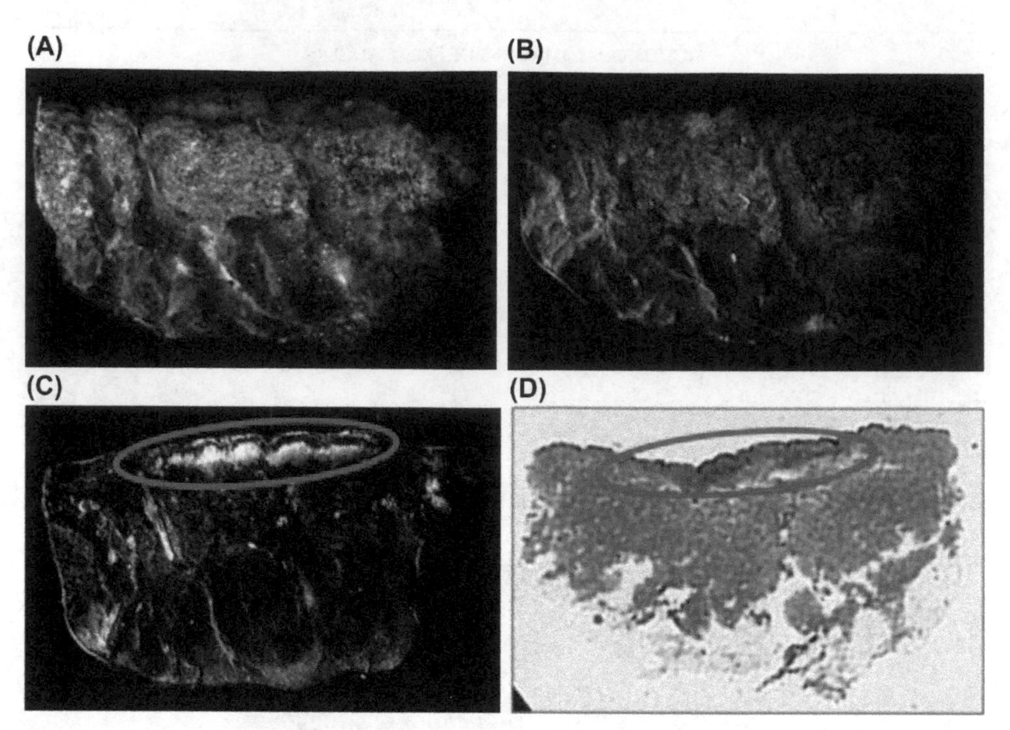

FIGURE 22.46 Images of a sample with compound dysplastic nevus stained with tetracycline (TCN). (A) 390-nm superficial reflectance image, no stain. (B) Reference fluorescence polarization (FP) image, no stain. (C) TCN FP image. (D) Hematoxylin and eosin (H&E) histopathology. *Results were obtained from Tannous Z, Al-Arashi M, Shah S, Yaroslavsky AN. Delineating melanoma using multimodal polarized light imaging. Lasers Surg Med 2009;41(1):10–16.*

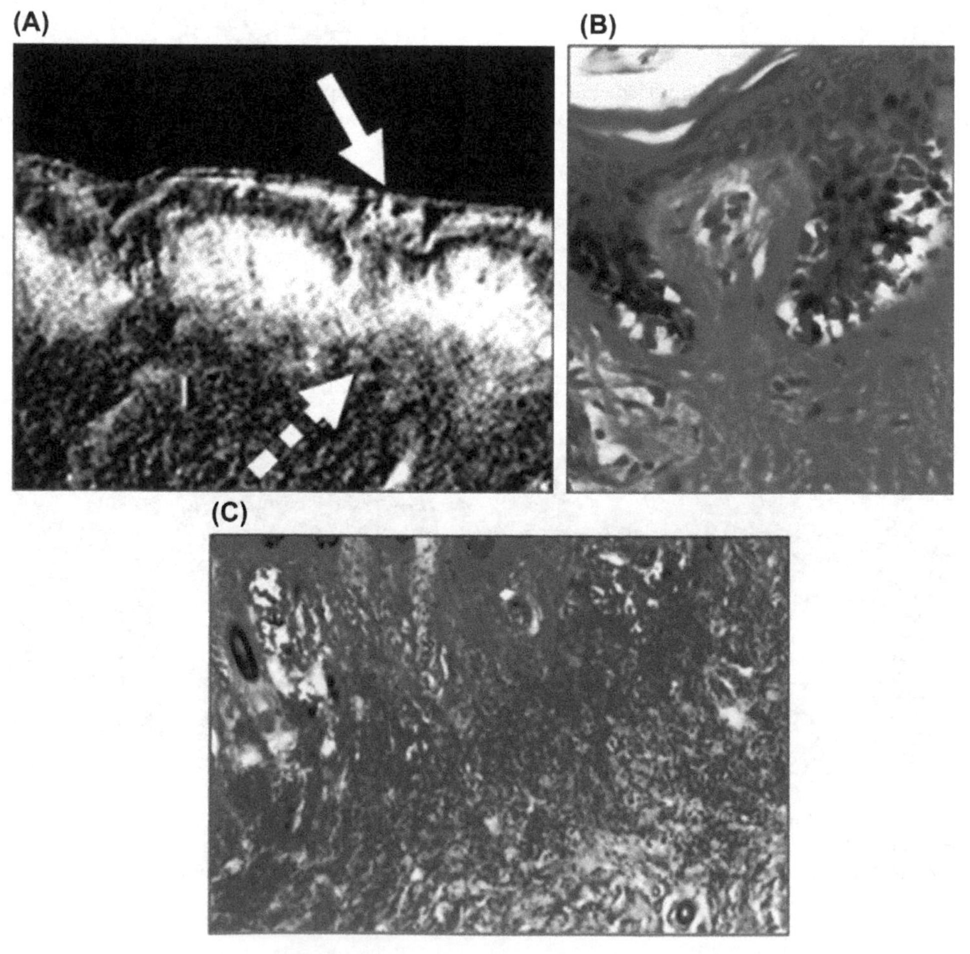

FIGURE 22.47 (A) Magnified tetracycline (TCN) fluorescence polarization (FP) image of a compound dysplastic nevus. (B) Epidermal and (C) dermal component of the nevus in hematoxylin and eosin (H&E) histology. *From Tannous Z, Al-Arashi M, Shah S, Yaroslavsky AN. Delineating melanoma using multimodal polarized light imaging. Lasers Surg Med 2009;41(1):10–16.*

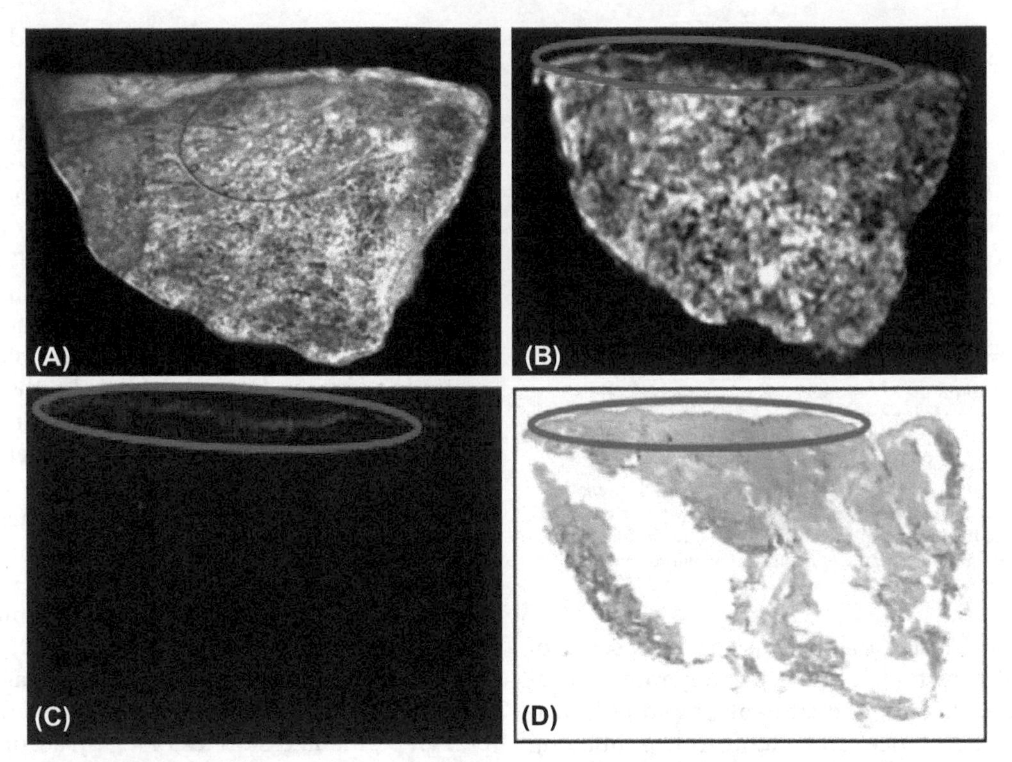

FIGURE 22.48 Images of a malignant melanoma stained with methylene blue (MB). (A) 670-nm superficial reflectance image without staining. (B) 670-nm superficial reflectance image after MB staining. (C) MB fluorescence polarization (FP) image. (D) Hematoxylin and eosin (H&E) histopathology. *Results were obtained from Tannous Z, Al-Arashi M, Shah S, Yaroslavsky AN. Delineating melanoma using multimodal polarized light imaging. Lasers Surg Med 2009;41(1):10–16.*

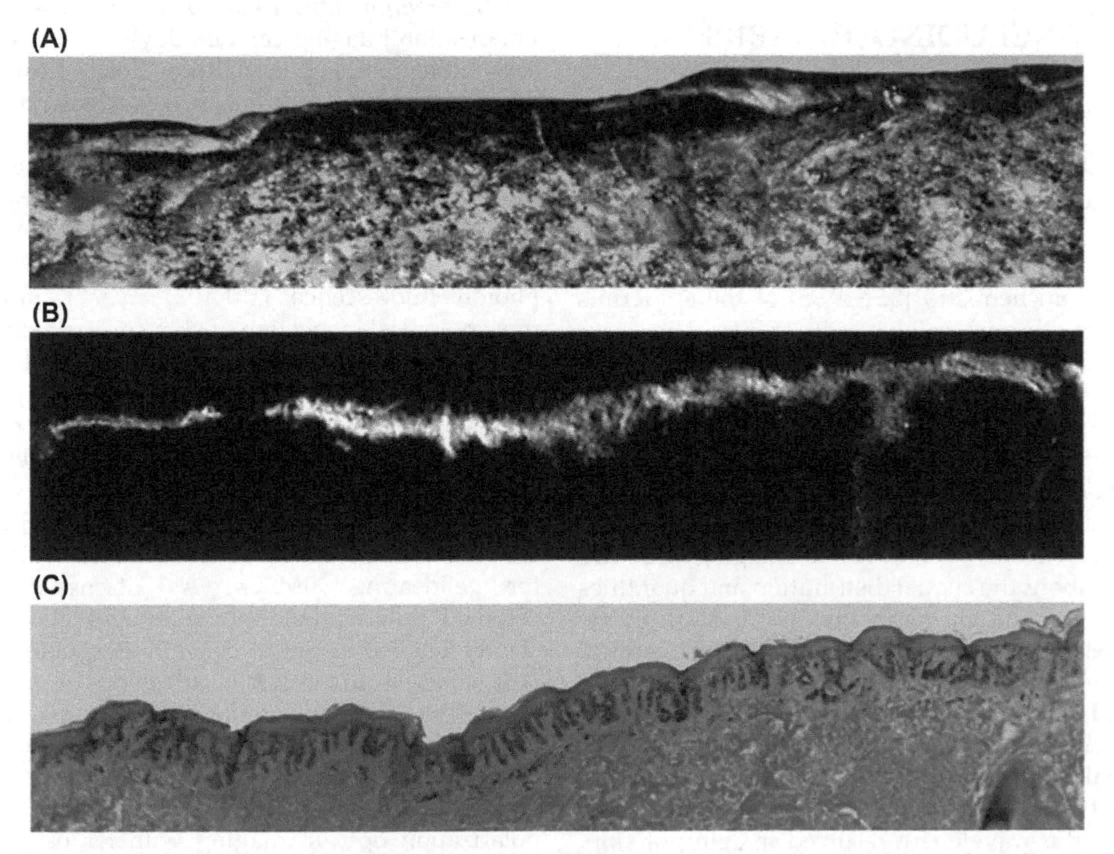

FIGURE 22.49 (A) Magnified methylene blue (MB) reflectance and (B) fluorescence polarization (FP) images of a malignant melanoma. (C) Magnified hematoxylin and eosin (H&E) histology. *From Tannous Z, Al-Arashi M, Shah S, Yaroslavsky AN. Delineating melanoma using multimodal polarized light imaging. Lasers Surg Med 2009;41(1):10–16.*

TABLE 22.2 Study Outcomes

#	Tumor Type	Results		
		TCN FPI	MB PLI	MB FPI
1	Compound dysplastic nevus	+	+	+
2	Scar	−	+	+
3	Compound dysplastic nevus	+	+	+
4	Compound dysplastic nevus	+	+	+
5	Melanoma in situ	−/+	+	+
6	Malignant melanoma	+	+	+

+, Absence/presence and localization of tumor was identified correctly; −, absence/presence and localization of tumor was identified incorrectly; −/+, presence and location of tumor was identified correctly; however, strong fluorescence polarization signal coming from healthy structures was also registered; *MB FPI*, methylene blue fluorescence polarization; *MB PLI*, methylene blue reflectance polarization light image; *TCN FPI*, tetracycline fluorescence polarization.

The authors concluded that because of the absence of any standard of care for real-time intraoperative guidance methods for melanoma treatments, rapid polarization enhanced reflectance and fluorescence imaging could be very valuable tool for assessing melanocytic lesions and guiding excision of malignant melanocytic lesions.

CONCLUDING REMARKS

The skin biopsy is the mainstay for diagnosis of dermatological disease and is often used for monitoring treatment responses in dermatology. As discussed herein, reflectance and fluorescence polarized light techniques offer the advantages of real-time, noninvasive viewing of skin morphology [5,7,9,10,13–15,17,18,75] and probing biochemistry [36–39,95] of the epidermis and papillary dermis safely and rapidly. There are currently available polarization optical devices that are used in translational and clinical research to aid in the noninvasive diagnosis of dermatological diseases, following disease progression and treatment effects over time [14,15,36–39,95]. Analysis of the propagation of polarized light and its depolarization combined with spectrally-resolved reflectance imaging provides information about the spatial distribution and quantities of the scatterers and chromophores [9,10,14,15]. FP responses yield quantitative data on the biochemical composition of the interrogated biotissue [36–39,95]. The structure, biochemical environment, and chromophore content and quantity differ depending on the age and health of human skin. Thus selective targeting and contrast enhancement of pathological tissue can be achieved by the wavelength-resolved imaging of skin. For example, considering that melanin exhibits maximum absorption in the UV and blue spectral ranges, whereas blood preferentially absorbs blue and yellow light, imaging protocols can be devised that target pigmented lesions, skin vasculature, or collagen [9,10]. Furthermore, to enhance the differences in optical signals from normal and diseased tissues exogenous intravital dyes and fluorophores, such as MB [35–39], TB [36,39], antibiotics of TCN family [37], aminolevulinic acid [96,97], fluorescein [98], and indocyanine green [99,100] can be utilized. Whereas the endogenous and exogenous chromophores determine skin absorption, scattering largely determines the depths to which light penetrates through skin because it dominates absorption in the visible and spectral range by at least one order of magnitude [67]. Spectral dependence of scattering enables optical sectioning in skin, ie, imaging comparatively thin tissue layers (~75–200 μm in the visible spectral range). Combining a large field of view and sufficient spatial resolution expedites examination of large surfaces, thus facilitating tumor margin delineation [13–15,36–38,95]. Polarized light imaging has been shown to successfully discriminate malignant and benign pigmented skin lesions [95], accurately delineate nonmelanoma and melanoma skin cancer margins [13–15,36–38], and rapidly and noninvasively monitor skin structure after laser procedures, photodynamic therapy, and pharmacological treatments [10].

The inherent limitations of the macroscopic optical polarization imaging can be addressed by combining it with other imaging modalities. For example, the lack of cellular resolution in macroscopic polarized light images makes accurate and specific diagnosis of diseases requiring analysis of intracellular morphology, such as melanoma, difficult. Combining polarization widefield imaging with high-resolution optical modalities [16,35], such as confocal microscopy [39] and two-photon fluorescence [101,102] may permit in vivo morphological inspection of skin similar to that attained in histopathology, while preserving all the benefits of macroimaging, such as large field of view and relative insensitivity to the movement of the imaged object. Similarly, polarization imaging readily lends itself to a combination with nonoptical modalities, such as terahertz imaging. For example, Joseph et al. [15] proposed to combine cross-polarized terahertz and optical imaging for delineating NMSCs. As discussed, spectrally resolved polarization optical images allow for sufficiently detailed inspection of skin morphology, but often lack contrast for reliable detection of cancer nests. Conversely, terahertz imaging yields excellent tumor contrast but lacks spatial resolution for definitive diagnosis. Thus low intrinsic contrast of NMSCs in polarization imaging can be improved by combining polarization optical imaging with terahertz interrogation of skin.

It should be mentioned that this chapter covers just a small selection of optical polarization imaging techniques. Two notable examples of the technologies that were not included in the chapter are polarization-sensitive OCT [103,104] and complete Mueller matrix analysis of human skin [105,106]. Polarization-sensitive OCT images birefringence in the skin, which is attributed to the regular arrangements of collagen fibers in the dermis. Thus it allows for detection and quantitation of changes in the structural integrity of collagen scaffolding, which is demonstrated in cancerous [15,107] and thermally damaged skin [108]. Mueller matrix analysis requires interrogation of skin using linear, circular, and elliptically polarized light. The incident and remitted Stokes vectors are measured and the complete 16-element Mueller matrix of the interrogated tissue is recovered. Obviously, this is a more thorough approach to skin diagnostics, potentially enabling complete biophysical characterization of human skin.

In summary, polarized light optical imaging, enabled by advances in photonics technology, provides important information about skin morphology, function, composition, biochemistry, and its responses to medical treatments. Novel clinical applications are being developed and improved to offer safer, more accurate, and cost-effective approaches to diagnostics and intraoperative monitoring. Advanced polarization-enhanced multimodal approaches, combining the strengths of each single technology, are being explored, implemented, and tested in clinic. Rapid development of robust and, at the same time, sophisticated methods that use polarization of light for skin monitoring and characterization make polarization optical imaging indispensable for translational dermatological research and bring it closer to clinical adoption. Polarized light imaging holds the potential to significantly augment existing standard-of-care methods employed in dermatology for evaluating skin condition, detecting, and treating diseases.

References

[1] Brewster D. Experiments on the depolarization of light as exhibited by various mineral, animal, and vegetable bodies, with a reference of the phenomena to the general principles of polarization. Phil Trans R Soc Lond 1815;(Pt. 1):21–53.

[2] von Erlach C. Mikroskopische Beobachtungen über organische Elementarteile bei polarisiertem Licht. Arch f Anat u Physiol 1847:313.

[3] Cogen DG. Some ocular phenomena produced with polarized light. Arch Opthalmol 1941;25(3):391–400.

[4] Bickel WS, Davidson JF, Huffman DR, Kilkson R. Application of polarization effects in light scattering: a new biophysical tool. Proc Natl Acad Sci USA 1976;73(2):486–90.

[5] Demos SG, Alfano RR. Optical polarization imaging. Appl Opt 1997;36(1):150–5.

[6] Gurjar RS, Backman V, Perelman LT, Georgakoudi I, Badizadegan K, Itzkan I, et al. Imaging human epithelial properties with polarized light-scattering spectroscopy. Nat Med 2001; 7(11):1245–8.

[7] Jacques SL, Roman JR, Lee K. Imaging superficial tissues with polarized light. Lasers Surg Med 2000;26(2):119–29.

[8] Schulz B, Chan D, Backstrom J, Rubhausen M. Hydration dynamics of human fingernails: an ellipsometric study. Phys Rev E 2002;65:061913.

[9] Feng X, Patel R, Yaroslavsky AN. Wavelength optimized cross-polarized wide-field imaging for noninvasive and rapid evaluation of dermal structures. J Biophotonics 2015;8(4):324–31.

[10] Feng X, Doherty S, Yaroslavsky I, Altshuler G, Yaroslavsky AN. Polarization enhanced wide-field imaging for evaluating dermal changes caused by non-ablative fractional laser treatment. Lasers Surg Med 2015. http://dx.doi.org/10.1002/lsm.22390.

[11] Qi C, Zhu W, Niu Y, Zhuang HG, Zhu GY, Meng YH, et al. Detection of hepatitis B virus markers using a biosensor based on imaging ellipsometry. J Viral Hepat 2009;16(11):822–32.

[12] Backman V, Wallace MB, Perelman LT, Arendt JT, Gurjar R, Müller MG, et al. Detection of preinvasive cancer cells. Nature 2000;406:35–6.

[13] Jacques SL, Ramella-Roman JC, Lee K. Imaging skin pathology with polarized light. J Biomed Opt 2002;7(3):329–40.

[14] Yaroslavsky AN, Neel V, Anderson RR. Demarcation of nonmelanoma skin cancer margins in thick excisions using multispectral polarized light imaging. J Invest Dermatol 2003;121(2): 259–66.

[15] Joseph CS, Patel R, Neel VA, Giles RH, Yaroslavsky AN. Imaging of *ex vivo* nonmelanoma skin cancers in the optical and terahertz spectral regions optical and terahertz skin cancer imaging. J Biophotonics 2014;7(5):295–303.

[16] Patel R, Khan A, Wirth D, Kamionek M, Kandil D, Quinlan R, et al. Multimodal optical imaging for detecting breast cancer. J Biomed Opt 2012;17(6):066008.

[17] Philp J, Carter NJ, Lenn CP. Improved optical discrimination of skin with polarized light. J Soc Cosmet Chem 1988;39:121–32.

[18] Anderson RR. Polarized light examination and photography of the skin. Arch Dermatol 1991;127(7):1000–5.

[19] Argenziano G, Fabbrocini G, Carli P, De Giorgi V, Sammarco E, Delfino M. Epiluminescence microscopy for the diagnosis of doubtful melanocytic skin lesions: comparison of the ABCD rule of dermatoscopy and a new seven point checklist based on pattern analysis. Arch Dermatol 1998;134:1563–70.

[20] Stokes GG. On the change of refrangibility of light. Phil Trans R Soc Lond 1852;142:463–562.

[21] Weigert F. Über polarisiertes fluoreszenz. Verh d D Phys Ges 1920;1:100–2.

[22] Vavilov SI, Levshin VL. Zur frage uber polarisierte fluoreszenz von farbstofflosungen. I Phys Z 1922;23:173–6.

[23] Perrin F. Polarization de la luniere de fluorescence. Vie moyenne de molecules dans l'etat excite. J Phys Radium 1926;7(12): 390–401.

[24] Weber G. Polarization of the fluorescence of macromolecules. 1 Theory and experimental method. Biochem J 1952;51:145–55.

[25] Weber G. Polarization of the fluorescence of macromolecules. 2 Fluorescent conjugates of ovalbumin and bovine serum albumin. Biochem J 1952;51:155–67.

[26] Dandliker WB, Feigen GA. Quantification of the antigen-antibody reaction by the polarization of fluorescence. Biochem Biophys Res Commun 1961;5:299–304.

[27] Waston RA, Landon J, Shaw EJ, Smith DS. Polarisation fluoroimmunoassay of gentamicin. Clin Chim Acta 1976;73(1):51–5.

[28] Fox MH, Delohery TM. Membrane fluidity measured by fluorescence polarization using an EPICS V cell sorter. Cytometry 1987; 8(1):20–5.

[29] Beccerica E, Piergiacomi G, Curatola G, Ferretti G. Influence of auranofin on lymphocyte membrane fluidity in rheumatoid arthritis. A fluorescence polarization study. Scand J Rheumatol 1989;18(6):413–8.
[30] Hashimoto Y, Shinozaki N. Measurement of cytoplasmic viscosity by fluorescence polarization in phytohemagglutinin-stimulated and unstimulated human peripheral lymphocytes. J Histochem Cytochem 1988;36(6):609–13.
[31] Tata DB, Foresti M, Cordero J, Tomashefsky O, Alfano MA, Alfano RR. Fluorescence polarization spectroscopy and time-resolved fluorescence kinetics of native cancerous and normal rat kidney tissues. Biophys J 1986;50(3):463–9.
[32] Pu Y, Wang WB, Das BB, Achilefu S, Alfano RR. Time-resolved fluorescence polarization dynamics and optical imaging of Cytate: a prostate cancer receptor-targeted contrast agent. Appl Opt 2008;47(13):2281–9.
[33] Barkai G, Mashiach S, Modan M, Serr DM, Lanir D, Lusky A, et al. Clin Chem 1983;29(2):264–7.
[34] Molcho J, Avraham H, Cohen-Luria R, Parola AH. Intrinsic fluorescence polarization of amniotic fluid: evaluation of human fetal lung maturity. Photochem Photobiol 2003;78(2):105–8.
[35] Patel R, Khan A, Quinlan R, Yaroslavsky AN. Polarization-sensitive multimodal imaging for detecting breast cancer. Cancer Res 2014;74(17):4685–93.
[36] Yaroslavsky AN, Neel V, Anderson RR. Fluorescence polarization imaging for delineating nonmelanoma skin cancers. Opt Lett 2004;29(17):2010–2.
[37] Tannous Z, Al-Arashi M, Shah S, Yaroslavsky AN. Delineating melanoma using multimodal polarized light imaging. Lasers Surg Med 2009;41(1):10–6.
[38] Salomatina E, Muzikansky A, Neel V, Yaroslavsky AN. Multimodal optical imaging and spectroscopy for the intraoperative mapping of nonmelanoma skin cancer. J Appl Phys 2009; 105(10):102010.
[39] Al-Arashi MY, Salomatina E, Yaroslavsky AN. Multimodal confocal microscopy for diagnosing nonmelanoma skin cancers. Lasers Surg Med 2007;39(9):696–705.
[40] Marks Jr JG, Miller JJ, editors. Lookingbill and Marks' principles of dermatology. Elsevier Inc.; 2013. p. 1–10.
[41] Montagna W, Parakkal PF, editors. The structure and function of skin. Academic Press Inc.; 1974. p. 18–30.
[42]] Saladin K, editor. Human anatomy. McGraw-Hill Inc.; 2007. p. 130–5.
[43] Guy Jr GP, Machlin SR, Ekwueme DU, Yabroff KR. Prevalence and cost of skin cancer treatment in the US, 2002–2006 and 2007–2011. Am J Prev Med 2015;48:183–7.
[44] Dahl E, Aberg M, Rausing EL. Basal cell carcinoma: an epidemiologic study in a defined population. Cancer 1992; 70(1):104–8.
[45] Casson P. Basal cell carcinoma. Clin Plast Surg 1980;7:301–11.
[46] Mohs FE. Chemosurgery – a microscopically controlled method of cancer excision. Arch Surg 1941;42(2):279–95.
[47] Pervan V, Cohen LH, Jaftha T, editors. Oncology for health-care professionals. Juta & Co, Ltd.; 1995. p. 491.
[48] Melanoma skin cancer. American Cancer Society, http://www.cancer.org/acs/groups/cid/documents/webcontent/003120-pdf.pdf; 2015.
[49] Snow SN, Mikhail GR. Mohs micrographic surgery. The University of Wisconsin Press; 2004. p. 56–7.
[50] Robinson JK, Hanke CW, Siegel DM, Fratila A, Bhatia AC, Rohrer TE. Surgery of the skin: procedural dermatology. Elsevier Inc.; 2015. pp. 704s.
[51] Branchet MC, Boisnic S, Frances C, Robert AM. Skin thickness changes in normal aging skin. Gerontology 1990;36(1):28–35.
[52] El-Domyati M, Attia S, Saleh F, Brown D, Birk DE, Gasparro F, et al. Intrinsic aging vs. photoaging: a comparative histopathological, immunohistochemical, and ultrastructural study of skin. Exp Dermatol 2002;11(5):398–405.
[53] Uitto J. Connective tissue biochemistry of the aging dermis. Age-associated alterations in collagen and elastin. Clin Geriatr Med 1989;5(1):127–47.
[54] Lapiere CM. The aging dermis: the main cause for the appearance of 'old' skin. Br J Dermatol 1990;122(Suppl. 35):5–11.
[55] Fisher GJ, Kang S, Varani J, Bata-Csorgo Z, Wan Y, Datta S, et al. Mechanisms of photoaging and chronological skin aging. Arch Dermatol 2002;138(11):1462–70.
[56] Fisher GJ, Talwar HS, Lin J, Lin P, McPhillips F, Wang Z, et al. Retinoic acid inhibits induction of c-Jun protein by ultraviolet radiation that occurs subsequent to activation of mitogen-activated protein kinase pathways in human skin *in vivo*. J Clin Invest 1998;101(6):1432–40.
[57] Yasui T, Yonetsu M, Tanaka R, Tanaka Y, Fukushima S, Yamashita T, et al. *In vivo* observation of age-related structural changes of dermal collagen in human facial skin using collagen-sensitive second harmonic generation microscope equipped with 1250-nm mode-locked Cr:Forsterite laser. J Biomed Opt 2013;18(3):031108/1–031108/10.
[58] Neerken S, Lucassen GW, Bisschop MA, Lenderink E, Nuijs TA. Characterization of age-related effects in human skin: a comparative study that applies confocal laser scanning microscopy and optical coherence tomography. J Biomed Opt 2004;9(2):274–81.
[59] Ishimaru A. Wave propagation and scattering in random media. Academic Press; 1978. p. 66.
[60] Bashkatov AN, Genina EA, Kochubey VI, Tuchin VV. Optical properties of human skin, subcutaneous and mucous tissues in the wavelength range from 400 to 2000 nm. J Phys D Appl Phys 2005;38(15):2543–55.
[61] Graaff R, Dassel AC, Koelink MH, de Mul FF, Aarnoudse JG, Zijistra WG. Optical properties of human dermis in vitro and in vivo. Appl Opt 1993;32(4):435–47.
[62] Jacques SL, Alter CA, Prahl SA. Angular dependence of HeNe laser light scattering by human dermis. Lasers Life Sci 1987;1: 309–33.
[63] Marchesini R, Bertoni A, Andreola S, Melloni E, Sichirollo AE. Extinction and absorption coefficients and scattering phase functions of human tissues in vitro. Appl Opt 1989;28(12):2318–24.
[64] Muller G, Roggan A, editors. Laser-induced interstitial thermotherapy. SPIE Press; 1995.
[65] Peters VG, Wyman DR, Patterson MS, Frank GL. Optical properties of normal and diseased human breast tissues in the visible and near infrared. Phys Med Biol 1990;35(9):1317–34.
[66] Prahl S. Light transport in tissue [Ph.D. dissertation]. University of Texas at Austin; 1988.
[67] Salomatina E, Jiang B, Novak J, Yaroslavsky AN. Optical properties of normal and cancerous human skin in the visible and near-infrared spectral range. J Biomed Opt 2006;11(6):064026.
[68] Simpson CR, Kohl M, Essenpreis M, Cope M. Near-infrared optical properties of *ex vivo* human skin and subcutaneous tissues measured using the Monte Carlo inversion technique. Phys Med Biol 1998;43(9):2465–78.
[69] Troy TL, Thennadil SN. Optical properties of human skin in the near infrared wavelength range of 1000 to 2200 nm. J Biomed Opt 2001;6(2):167–76.
[70] van Gemert MJ, Jacques SL, Sterenborg HJCM, Star WM. Skin optics. IEEE Trans Biomed Eng 1989;36(12):1146–54.
[71] Wan S, Anderson RR, Parrish JA. Analytical modeling for the optical properties of the skin with in vitro and in vivo applications. Photochem Photobiol 1981;34(4):493–9.

[72] Anderson RR, Parrish JA. The optics of human skin. J Invest Dermatol 1981;77(1):13–9.
[73] Svaasand LO, Norvang LT, Fiskerstrand EJ, Stopps EKS, Berns MW, Nelson JS. Tissue parameters determining visual appearance of normal skin and port-wine stains. Lasers Med Sci 1995;10(1):55–65.
[74] Douven LFA, Lucassen GW. Retrieval of optical properties of skin from measurement and modelling the diffuse reflectance. Proc SPIE 2000;3914:312–23.
[75] Jacques SL, Lee K. Polarized video imaging of skin. Proc SPIE 1998;3245:356–62.
[76] Pathak MA. J Invest Dermatol 2004;122:xx–xxi.
[77] Mason TJ, editor. Advances in sonochemistry, vol. 5. JAI Press, Inc.; 1999. p. 251–2.
[78] Ong MW, Bashir SJ. Fractional laser resurfacing for acne scars: a review. Br J Dermatol 2012;166(6):1160–9.
[79] Alexiades-Armenakas MR, Dover JS, Arndt KA. The spectrum of laser skin resurfacing: nonablative, fractional, and ablative laser resurfacing. J Am Acad Dermatol 2008;58(5):719–37.
[80] Geronemus RG. Fractional photothermolysis: current and future applications. Lasers Surg Med 2006;38(3):169–76.
[81] Leyden J, Stephens TJ, Herndon Jr JH. Multicenter clinical trial of a home-use nonablative fractional laser device for wrinkle reduction. J Am Acad Dermatol 2012;67(5):975–84.
[82] Eisen GM, Montgomery EA, Azumi N, Hartmann D-P, Bhargava P, Lippman M, et al. Qualitative mapping of Barrett's metaplasia: a prerequisite for intervention trials. Gastrointest Endosc 1999;50:814–8.
[83] Fedorak IJ, Ko TC, Gordon D, Flisak M, Prinz RA. Localization of islet cell tumors of pancreas: a review of current techniques. Surgery 1993;113:242–9.
[84] Fukui I, Yokokawa M, Mitani G, et al. *In vivo* staining test with methylene blue for bladder cancer. J Urol 1983;130:252.
[85] Gill WB, Huffman JL, Lyon ES, Bagley DH, Schoenberg HW, Straus II FH. Selective surface staining of bladder tumors by intravesical methylene blue with enhanced endoscopic identification. Cancer 1984;53:2724–7.
[86] Kaisary AV. Assessment of radiotherapy in invasive bladder carcinoma using in vivo methylene blue staining technique. Urology 1986;28:100–2.
[87] Canto MI, Setrakian S, Petras RE, Blades E, Chak A, Sivak Jr MV. Methylene blue selectively stains metaplasia in Barrett's esophagus. Gastrointest Endosc 1996;44:1–7.
[88] Rall DP, Loo TL, Lane M, Kelly MG. Appearance and persistence of fluorescent material in tumor tissue after tetracycline administration. J Natl Cancer Inst 1957;19(1):79–85.
[89] McLeary JR. The use of systemic tetracyclines and ultraviolet in cancer detection. Am J Surg 1958;96:415.
[90] Holman BL, Kaplan WD, Dewanjee MK, Fliegel CP, Davis MA, Skarin AT, et al. Tumor detection and localization with ^{99m}Tc-tetracycline. Radiology 1974;112:147–53.
[91] Davis RC, Wood P, Mendelsohn ML. Localization and therapeutic potential of tritiated tetracycline in rodent tumors. Cancer Res 1997;37:4539–45.
[92] Olmedo-Garcia N, Lopez-Prats F. Tetracycline fluorescence for the preoperative localization of osteoid osteoma of the triquetrum. Acta Orthop Belg 2002;68:306–9.
[93] Van den Bogert C, Dontje BH, Kroon AM. Doxycycline in combination chemotherapy of a rat leukemia. Cancer Res 1988;48: 6689–90.
[94] Duivenvoorden WCM, Vukmirovic P, Lhotak S, Seidlitz E, Hirte HW, Tozer RG, et al. Doxycycline decreases tumor burden in a bone metastasis model of human breast cancer. Cancer Res 2002;62:1588–91.
[95] Yaroslavsky AN, Salomatina EV, Neel V, Anderson R, Flotte T. Fluorescence polarization of tetracycline derivatives as a technique for mapping nonmelanoma skin cancers. J Biomed Opt 2007;12(1):014005.
[96] Ulrich M, Klemp M, Darvin ME, Konig K, Lademann J, Meinke MC. In vivo detection of basal cell carcinoma: comparison of a reflectance confocal microscope and a multiphoton tomograph. J Biomed Opt 2013;18:612291–7.
[97] Manfredini M, Arginelli F, Dunsby C, French P, Talbot C, Koenig K, et al. High-resolution imaging of basal cell carcinoma: a comparison between multiphoton microscopy with fluorescence lifetime imaging and reflectance confocal microscopy. Skin Res Technol 2013;19:E433–43.
[98] Robertson TA, Bunel F, Roberts MS. Fluorescein derivatives in intravital fluorescence imaging. Cells 2013;2:591–606.
[99] Alander JT, Kaartinen I, Laakso A, Patila T, Spillmann T, Tuchin VV, et al. A review of indocyanine green fluorescent imaging in surgery. Int J Biomed Imaging 2012;2012:940585.
[100] Marshall MV, Rasmussen JC, Tan I, Aldrich MB, Adams KE, Wang X, et al. Near-infrared fluorescence imaging in humans with indocyanine green: a review and update. Open Surg Oncol J 2010;2(2):12–25.
[101] Ericson MB, Simonsson C, Guldbrand S, Ljungblad C, Paoli J, Smedh M. Two-photon laser-scanning fluorescence microscopy applied for studies of human skin. J Biophotonics 2008;1(4): 320–30.
[102] Patalay R, Talbot C, Alexandrov Y, Lenz MO, Kumar S, Warren S, et al. Multiphoton multispectral fluorescence lifetime tomography for the evaluation of basal cell carcinomas. PLoS One 2012; 7(9):e43460.
[103] Pierce MC, Strasswimmer J, Park BH, Cense B, de Boer JF. Advances in optical coherence tomography imaging for dermatology. J Invest Dermatol 2004;123:458–63.
[104] Sattler E, Kastle R, Welzel J. Optical coherence tomography in dermatology. J Biomed Opt 2013;18:061224.
[105] Sun M, He H, Zeng N, Du E, Guo Y, Liu S, et al. Characterizing the microstructures of biological tissues using Mueller matrix and transformed polarization parameters. Biomed Opt Express 2014;5(12):4223–34.
[106] Yao G, Wang LV. Two-dimensional depth-resolved Mueller matrix characterization of biological tissue by optical coherence tomography. Opt Lett 1999;24:537–9.
[107] Wu P, Hsieh T, Tsai Z, Liu T. *In vivo* quantification of the structural changes of collagens in a melanoma microenvironment with second and third harmonic generation microscopy. Sci Rep 2015;5:8879.
[108] Kaiser M, Yafi A, Cinat M, Choi B, Durkin AJ. Noninvasive assessment of burn wound severity using optical technology: a review of current and future modalities. Burns 2011;37:377–86.

CHAPTER

23

Mechanical Characterization of Skin Using Surface Acoustic Waves

C. Li[1,2], G. Guan[1,2], R. Wang[2], Z. Huang[1]

[1]University of Dundee, Dundee, Scotland, United Kingdom; [2]University of Washington, Seattle, WA, United States

OUTLINE

INTRODUCTION

The alteration of mechanical properties with the change of tissue condition is commonly observed in tissue pathologies, such as in skin disorders. Thus assessing skin mechanical properties is useful in improving our understanding of skin pathophysiology, which will aid medical diagnosis and treatment of dermatological diseases, such as skin cancer and scleroderma [1–3]. Early detection of skin cancer is important because treatment at its early stages can improve 5-year survival rates and increases the chance for a cure [4,5]. In early stages of skin cancer, the lesion is often confined within the epidermis and dermis (1-mm thickness), with relatively small alterations of Young's modulus (ie, tensile modulus; elastic modulus is force per unit area that is needed to stretch or compress a material sample) compared with normal skin tissues. A diagnosis at this stage can improve the survival rate by up to 90–100% [6]. However, if diagnosed at a later stage, the tumor will have already invaded subcutaneous fat with a drastic change of stiffness, and in such cases the prognosis is poor [7–9]. Tissue geometry and stiffness are therefore important parameters for the clinical prognosis of skin diseases. Most skin diseases are generally diagnosed qualitatively based on a visual inspection and/or palpation by a dermatologist with clinical experience in the field. During manual palpation, pathological tissue regions can be identified by having a different strain response to an imposed stress compared with the surrounding healthy tissue. The magnitude of the strain response depends on the nature of the pathological condition. This method of diagnosis has high variability among dermatologists; therefore a sensitive, nondestructive and noninvasive method that is capable of assessing the skin's mechanical properties as well as its geometry is needed.

Imaging in Dermatology
http://dx.doi.org/10.1016/B978-0-12-802838-4.00023-6

A number of elastography technologies have been developed for qualitative and quantitative assessment of tissue mechanical properties [10–15]. The main idea in elastography is to use a sensitive device to quantify the image of mechanical disturbance, which is induced directly or indirectly by a mechanical stimulation, such as compression, vibration, or acoustic radiation [10]. Ultrasound elastography [11–13] and magnetic resonance elastography [14,15] are the most common methods in medical diagnosis that involve either ultrasound or magnetic resonance imaging to measure the passive tissue disturbances, from which the mechanical properties of tissue are obtained. Despite their success in cardiac applications and detection of certain types of cancers, these methods are difficult to quantify skin mechanical properties because of their low spatial resolution, which is not sufficient to detect small lesions in thin skin layers. In addition, the methods described involve tracking low-frequency shear wave that propagates within the body and thus may present limitations in the assessment of material surface.

Surface acoustic wave (SAW) technology has been used in industrial applications, such as analyzing surface structure, composition, geometry, roughness, flatness, and elastic properties of metallic specimens [16–20]. SAW methods are mainly used to evaluate the mechanical properties of materials because it has the advantage of quantitatively assessing Young's modulus. SAW has great superiority in the characterization of skin because: (1) the domain wave energy locates at near-surface region, and (2) the propagation of SAW in layered materials shows a dispersive behavior, where dispersion means that different frequency components have different phase velocities, which penetrate into different depths of the tissue. The phase velocity values are directly related to Young's modulus, and so the quantitative elasticity information of skin layers the SAW propagates into can be obtained. To detect the SAW, the most common method is to employ an ultrasound transducer [21], which requires physical contact with the sample. This requirement leads to a number of drawbacks: the sensing area is limited by the transducer, wave energy leakage occurs at the tissue-transducer boundary, and wave distortion will be caused by the weight of the transducer on the sample. To mitigate these problems, a preferred method is to use a noncontact and nondestructive approach to detect the SAW. One of such methods is optical interrogation. Optical interrogation has been widely used because it is noncontact and remote; therefore no surface loading is required. The sensitivity of optical measuring systems is inherently high, which allows detection of significantly small displacement. Optical interrogation techniques can also provide a broad detection bandwidth. Furthermore, as a remote sensing approach, it provides the ability to access the samples in a hostile field environment, and is generally not sensitive to surface orientation.

Optical coherence tomography (OCT) [10,22–29] is a promising noninvasive, noncontact imaging technique capable of providing microstructural information of the tissue with high resolution. Phase-sensitive OCT (PhS-OCT) has been developed for the detection of small displacement and vibration and offers distinct features in comparison with ultrasound and magnetic resonance as a displacement detector and sensor, including:

1. Ultrahigh imaging resolution: PhS-OCT has an imaging spatial resolution of 2–15 μm. Based on this, PhS-OCT has the potential to resolve the microscopic heterogeneity in the assessment of morphological and geometrical properties of tissues.
2. High acquisition speed: PhS-OCT provides two-dimensional (2D) image acquisition rates in the range of 10–100s of kHz, and a sample rate as high as ~92 kHz, and potentially even higher.
3. High mechanical sensitivity: PhS-OCT can measure a smaller displacement of up to picometer. This ultrahigh sensitivity has the potential to detect mechanical waves with very small amplitudes that propagate on a material.

This chapter aims to develop innovative noninvasive systems that combine the impulse-stimulated SAW method with PhS-OCT that allows the rapid functional characterization of different layers of in vivo human skin tissue. Phantoms are made using different concentrations of agar solution of different layer thickness to mimic the elastic properties typically found in skin. The behaviors of SAW in the soft materials are observed and analyzed. Dispersion phase-velocity curves are calculated to obtain the elastic properties from well-defined layers, including different kinds of agar phantoms and different sites of in vivo human skin, which exhibit different mechanical properties. The PhS-OCT system also provides depth-resolved microstructure information of the interrogated sample to assist the elasticity evaluation of the heterogeneous tissue.

BACKGROUND AND CHARACTERIZATION OF SURFACE ACOUSTIC WAVES

The application or variation of a force on a body produces a stress and strain response, and might generate mechanical waves that behave differently in the body, eg, the travel path and directions, partial trajectory, travel speed, and energy distribution. These are body waves and surface waves. Body waves, ie, longitudinal waves and shear waves, propagate deep into the interior of material. When the medium is bounded by a free surface, SAWs are generated and propagate near the surface and do not irradiate energy toward the interior. SAWs induce a particle motion in the vertical plane that contains the direction of propagation. In a homogeneous body with free surface, SAWs have a retrograde elliptic orbit. If there are variations of the elastic properties with depth, SAWs become dispersive; a SAW with a different wavelength propagates with different velocities. This section discusses the background and characterizations of SAWs including generation methods, velocity, frequency range, dispersion, and behavior when they come across a material with transverse elasticity change.

Surface Acoustic Wave Generation

Q-Switched Laser Pulse

When a material is illuminated with a short laser pulse, the energy is absorbed through various mechanisms. In lower laser energy illumination, the mechanism will not alter the tested material. This is called the *thermoelastic mechanism* [21]. The laser-generated SAW by the thermoelastic mechanism is a complex process that involves the optical, thermal, and mechanical properties of the material being tested. A pulsed laser beam impinges on a material and is partially absorbed by it, which results in a rapid increase in temperature of the irradiated area that in turn causes a rapid thermal expansion. Because there is no restraining force to the surface of the sample, the sample is free to expand outward, away from the bulk of the material. The material is not free to expand parallel to the surface of the sample, which leads to a generation of strong elastic wave pulses. The generated elastic waves act radially outward from the center of the heated zone in the plane of the surface. The consequence is the generation of ultrasonic waves that propagate within the material, such as body waves, and propagate on the material surface, such as SAWs. The amplitude of the SAW is linearly proportional to the power density of the laser pulse. In addition, for efficient SAW generation, the duration of the optical pulse should be sufficiently short so that both thermal and stress confinements take place [21].

The advantages of laser-induced SAWs include: (1) it is noncontact (coupling problems are eliminated and no surface loading), (2) it can be remote (access to the sample in hostile environments), (3) it has rapid scanning capability, (4) it allows operation on geometrically irregular specimens, and (5) it uses broadband light. The disadvantages of such methods include: (1) it has a relatively high cost, (2) the generation efficiency is a function of material optical absorption properties, and (3) there is a requirement for laser safety precautions.

Laser-induced SAW technology is not commonly used in dermatology applications, because it is still a problem to directly apply this method to the tissue, owing to the complex mechanisms and potential safety issues when applying a high energy laser pulse, eg, tissue thermal damage. Meanwhile, direct application of a laser pulse to the skin will not generate ideal SAWs with detectable amplitude, because skin is highly scattering and has low/variable absorptive materials. Thus laser-induced SAW techniques will not be discussed in this chapter. However, a recent study proposes a novel, yet practical solution for laser-induced SAWs used in tissue elasticity measurement. A thin layer of black agar membrane is applied to the tissue of interest for: (1) acting as a surface shield to protect tissue from heat generation caused by the irradiation of laser pulse, and (2) increasing the absorption of laser pulse in order to generate SAWs with better signal-to-noise ratio (SNR) [30].

Shaker

Impulse-stimulated SAW can also be produced by a mechanical shaker. Compared with the high-energy laser source, a mechanical shaker inducing SAWs is a safer, more portable and simpler approach. It directly passes the vibration to samples and generates waves. In addition, the amplitude of SAWs generated by mechanical shaker is easily controlled and enhanced. The SAW amplitude is controlled to avoid the possibility that the tissue response would be out of the linear elastic region. However, the disadvantages of using a mechanical shaker to induce SAWs include the need for contact with the sample during detection, and the necessity for surface loading (the additional load generated on the skin surface when in contact with the mechanical shaker), which is inapplicable in certain clinical settings. It thus has specific requirements for tissue conditions and geometry. In addition, it has limited frequency bandwidth (refer to section "Frequency Range" for more details).

Velocity

The most commonly used SAW is stimulated by impulse stimulation. An impulse-responded SAW is normally bipolar, strong in amplitude, broadband in frequency, and propagated on the surface of a material. On an ideal homogeneous flat elastic solid, SAWs show no dispersion and travels at a constant velocity as described in Eq. (23.1) [16,21,30–33].

$$C_R \approx \frac{0.87 + 1.12v}{1 + v} \sqrt{\frac{E}{2\rho(1+v)}} \tag{23.1}$$

where C_R is the phase velocity, E is Young's modulus, ν is Poisson's ratio, and ρ is the density of the material. From Eq. (23.1) it can be observed that Young's modulus has the most significant impact on SAW velocity in soft tissues, because the density and Poisson's ratio in soft tissue do not vary much (0.45–0.49 for Poisson's ratio and 1000–1400 kg/m^3 for density). Thus in most of SAW applications in dermatology, the density and Poisson's ratio are obtained from the literature to estimate Young's modulus of the targeted material.

Frequency Range

With impulse stimulation, a broadband SAW can be generated. The maximum frequency component is shown in Eq. (23.2) (where r_0 indicates the radius of pulse stimulation) [34]:

$$f_{\max} = \frac{2\sqrt{2}C_R}{\pi r_0} \quad (23.2)$$

SAWs have broad bandwidth of frequency (up to 6 MHz in hard solids, eg, metal and 5–50 kHz in soft biology tissues), and different frequency components indicate the corresponding depth propagated into the sample surface. The maximum frequency component that a SAW signal can reach is proportional to the SAW velocity of the material, and inversely proportional to the radius of pulse stimulation. Thus stiffer materials have broader frequency bands than softer materials. In addition, the frequency bandwidth that the stimulator can generate also decides the maximum frequency a SAW signal can reach. The shorter the impulse stimulation, the broader frequency bandwidth SAW can be generated. The balance between material stiffness, stimulator radius, and stimulator pulse time is important.

Dispersion

SAWs propagate on stress-free surfaces, for example, the air–solid interface. The particle displacement decays exponentially with depth and becomes negligible for a penetration depth of more than a few wavelengths. The particle motion can be decomposed into two orthogonal components, one in the direction of SAW propagation and one perpendicular to the free surface. In homogeneous materials, SAW is nondispersive and therefore its phase velocity is independent of the frequency. However, a layered material perturbs SAWs' propagation at the surface. The propagation of SAWs in layered materials shows a dispersive behavior, where dispersion means that in SAWs, the different frequency components have different phase velocities. For a multilayer medium, in which different layers have different elastic properties, the phase velocity of SAW is influenced by the mechanical properties of all the layers it penetrates. By analysis of the phase velocity curve, the true elasticity conditions of a material can be revealed, where a phase-velocity dispersion curve can be defined as the phase velocity associated with the maximum spectral density at each frequency; that is, at each frequency, the phase velocity at which most of the energy propagates. The elastic properties that affect the phase-velocity dispersion curve include not only Young's modulus, Poisson's ratio, and the density of each layer, but also the thickness of each layer. In this case, SAWs with shorter wavelengths (higher frequency) penetrate in shallow depth with the phase velocity depending on the superficial layers. On the other hand, SAWs with longer wavelengths (lower frequency) penetrate deeper in the material because the phase velocity will be influenced by the elastic properties of the deeper layers.

It is this characteristic that allows SAWs to be used for material characterization of thin-layered materials, because waves of different frequencies can be used for the characterization of different lengths of scales. The probing depth can be estimated by Eq. (23.3) [30,35]:

$$z \approx \lambda = C_R/f \quad (23.3)$$

where f is the signal frequency.

Fig. 23.1A shows the typical SAW signal recorded from one-layer ~3.5% agar–agar phantom [36]. The detecting point was first located at a position ~0.5 mm away from the excitation laser beam, and then moved at ~0.5 mm/step to ~3 mm away. It is clear that the SAW is moving away from the laser-excitation position. Because the sample was a one-layer homogeneous sample, no velocity dispersion was found in the detected waveforms; ie, no wave distortion occurred. On the other hand, Fig. 23.1B shows the typical SAW signals of two-layer agar–agar phantoms, with ~2% agar as the superficial layer and ~3.5% agar as the substrate layer with 1mm/step from 1 to 6 mm away from stimulation. The high-frequency signal that the green arrow points to was a high-frequency thermal expansion caused by the heating of laser source. The red arrow points to the occurrence of wave dispersion, which becomes more apparent at the position of 6 mm. Fig. 23.1C compares the phase-velocity dispersion curves of two kinds of double-layer phantom: ~2% agar on ~3.5% agar phantom (green) and ~5% agar on ~3.5% agar phantom (blue). The result from one-layer 3.5% agar–agar phantom (red) is also plotted for comparison. The phase velocity dispersion curves of double-layer phantom are no longer a straight line. ~5% agar on ~3.5% agar phantom has an initial phase velocity of ~13.08 ± 1.22 m/s, which matches very well with one-layer 3.5% agar–agar phantom. However, with the increase of the frequency, the phase velocity increases to ~20.30 m/s, which indicates the phase velocity of ~5% agar. In the case of ~2% agar on ~3.5% agar phantom, initial phase velocity agrees with that of ~3.5% agar–agar phantom and then drops to

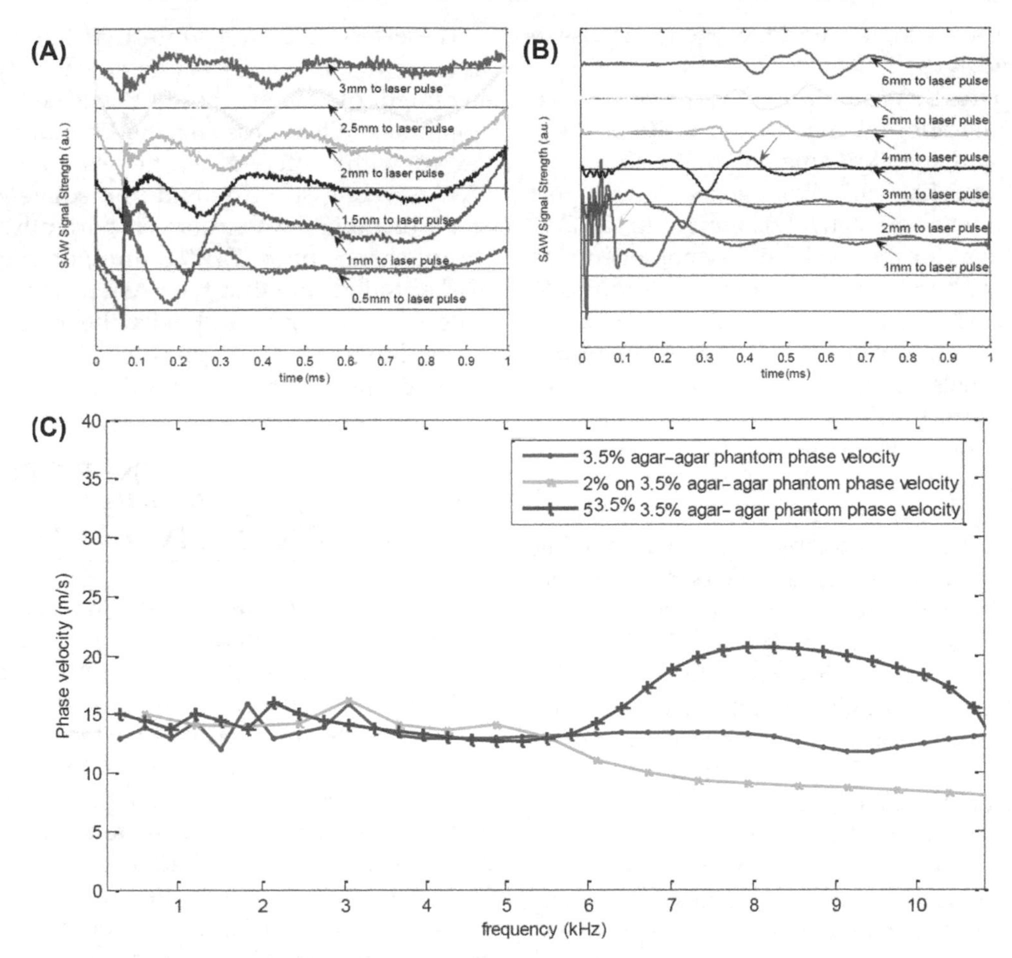

FIGURE 23.1 (A) Surface acoustic wave (SAW) signal of one-layer 3.5% agar–agar phantom with the distance of 0.5 mm (top) to 3 mm (bottom) to laser pulse, with 0.5 mm/step. (B) SAW signal of double-layer 2% agar on 3.5% agar phantom with the distance of 1–6 mm to laser pulse, with 1 mm/step. (C) Comparison of phase-velocity dispersion between one-layer 3.5% agar–agar phantom and two-layer 2% on 3.5% agar–agar phantom, and 5% on 3.5% agar–agar phantoms. Each SAW signal is purposely shifted vertically by equal distance in order to better illustrate the results captured from different positions. *a.u.*, Arbitrary units.

~7.91 m/s with the increase of the frequency, which indicates the phase velocity of ~2% agar. These results are in good agreement with theoretical expectations; ie, the phase velocity in the lower frequency region indicates the mechanical properties of the substrate layer, whereas the phase velocity in a higher frequency indicates the mechanical properties of the upper layer. The slope is not discussed in this paper, because it is mainly influenced by the thickness of each layer.

Behavior with Transverse Elasticity Alteration

In realistic in vivo situations, the tissue would often exhibit lateral heterogeneity. In this case, when the acoustic waves travel to the boundary between two materials, the acoustic energy is either reflected or transmitted depending on the material acoustic impedances. The acoustic impedance is the product of material density and acoustic wave velocity. The bigger difference in acoustic impedance results in a bigger percentage of energy that is changed at the interface or boundary between one medium and another. When a wave travels from material 1 to material 2, it is required that:

$$\frac{A_t}{A_i} = \frac{2z_1}{z_1 + z_2} = \frac{2\rho_1 c_1}{\rho_1 c_1 + \rho_2 c_2} \tag{23.4}$$

where A_t (A_i) is the transmitted (incident) wave amplitude, z, ρ, and c are the acoustics impedance, density, and the acoustic wave velocity of the medium, with the subscripts 1 and 2 representing material 1 and material 2, respectively. When the acoustic impedance of material 1 is higher than that of the material 2, the transmitted wave amplitude would be amplified. Otherwise, it would be considerably decreased. Although P-wave velocity and S-wave velocity have been extensively studied, no record of SAWs is found in literature. It is important to study the behavior of SAWs traveling through the interface (boundary) at which the alteration of elastic property occurs, so that the feasibility of the SAW method to detect the localized change of elasticity within a mechanically heterogeneous tissue could be assessed.

Fig. 23.2 compares SAW amplitude change with transverse elasticity alteration obtained from finite element (FE) simulation and the experimental results from the single-layer and mechanically homogeneous agar gels (for 3% and 1% agar, respectively). From Fig. 23.2, it can be observed that for the mechanically homogeneous phantom of either 3% or 1% agar gel, the SAW amplitude adheres to approximately an exponential attenuation when it travels away from the origin. The SAW amplitude initially parallels the amplitude from the homogeneous phantom until the SAW arrives at the interface. After the SAW crosses the interface, its amplitude is increased by 150% when traveling from the 1% agar side to the 3% agar side (see Fig. 23.2A). In contrast, the SAW amplitude is decreased by 50% when traveling in the opposite direction (see Fig. 23.2B). Both the experimental and FE simulation results matches well with the theoretical expectations predicted by Eq. (23.4), which indicates that the wave amplitude would increase by 150% when passing through the interface from 3% agar to 1% agar, and drop by 50% when crossing the interface from 1% agar to 3% agar. Note here that in the theoretical estimation using Eq. (23.1), the necessary parameters were estimated based on a previous study [37]. The SAW velocity of 1% agar was 5 m/s and that of 3% agar was 12 m/s, and the densities of 1% and 3% agar gel were 1020 and 1060 kg/m^3, respectively. Note also that when traveling from the stiff side to the soft side, the SAW amplitude did not increase to its maximum immediately (see Fig. 23.2A). This was most likely because of the localization error in which the interface was judged visually.

These results demonstrate that the SAW method is sensitive to the lateral change of elasticity in tissues, as measured by either the SAW velocity or the SAW amplitude. When crossing the boundary between two tissues with different elasticities to each other, the SAW traveling speed would immediately adapt to that of the material. And more importantly, the observed abrupt change of the SAW amplitude can be taken as a marker to indicate that the SAW has traveled from one material to another and tells the relative location of material interface to the shaker, serving as a potentially useful purpose in biomedical diagnosis.

EXPERIMENTAL GENERATION AND DETECTION OF SURFACE ACOUSTIC WAVE ON IN VIVO HUMAN SKIN

A specially designed shaker was used as the SAW stimulator, which included a signal generator and a single-element piezoelectric ceramic with a metal rod that directly connected to the piezoelectric ceramic as the shaker head (length of 20 mm and diameter of 2 mm). During the experiments, the shaker head was directly in touch with the sample to pass the vibration from the piezoelectric ceramic to stimulate the SAW on the sample surface. The stimulus was applied with a 45-degree angle to the tissue surface, which provided equal longitudinal and shear energy to the sample and improved the amplitude of the SAW. Compared with the point source, the line source (provided by the metal rod) improved the SNR, and reduced the attenuation of the SAW, which enabled the SAW to propagate longer distances [32,34–39]. The shaker created a maximum

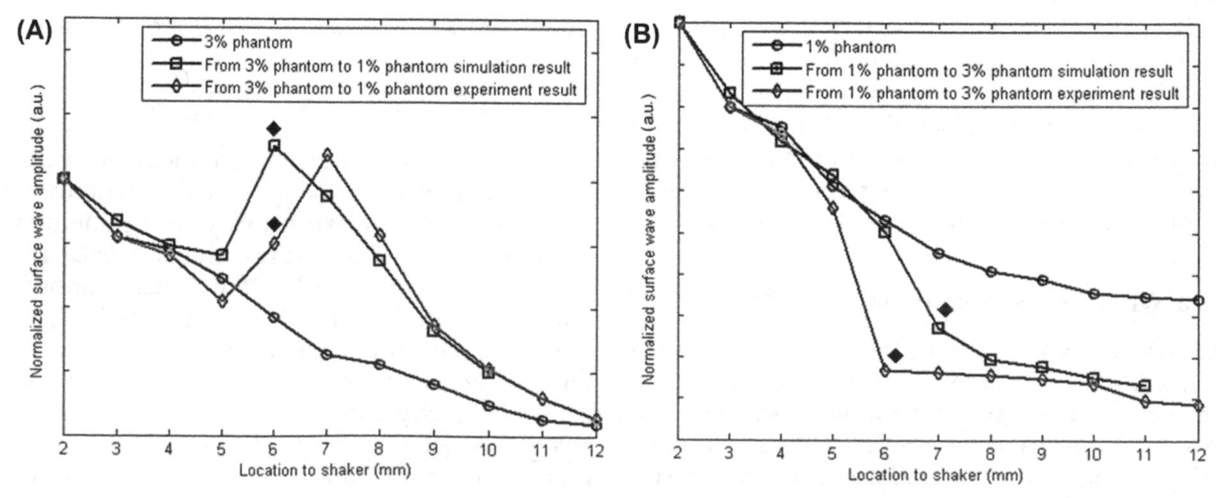

FIGURE 23.2 The surface acoustic wave (SAW) amplitude evaluated from the experiments and finite element (FE) simulations. (A) SAW travels from 3% agar to 1% agar. (B) SAW travels from 1% agar to 3% agar. The *diamonds* indicate the interface position.

tissue displacement of 100 nm in the axial direction. The tissue displacement is proportional to the SAW amplitude. The amplitude has no influence to the shape and velocity of SAW. To analyze all the frequencies from the SAW, the shaker is required to generate a pulse with a short duration. For this study an external trigger was used to control the shaker, which generated pulses of 20 Hz with a 0.2% duty cycle, producing frequencies up to 10 kHz.

Detection of SAWs was performed using a PhS-OCT system. The OCT imaging system can provide the structural image of the samples as a function of depth, ie, the thickness of each layer. This system allows imaging the tissue's inner layer structures as well as its elastic properties simultaneously.

The PhS-OCT system (Fig. 23.3A) employed a spectral-domain OCT system with a center wavelength of 1310 nm and a bandwidth of 46 nm from a

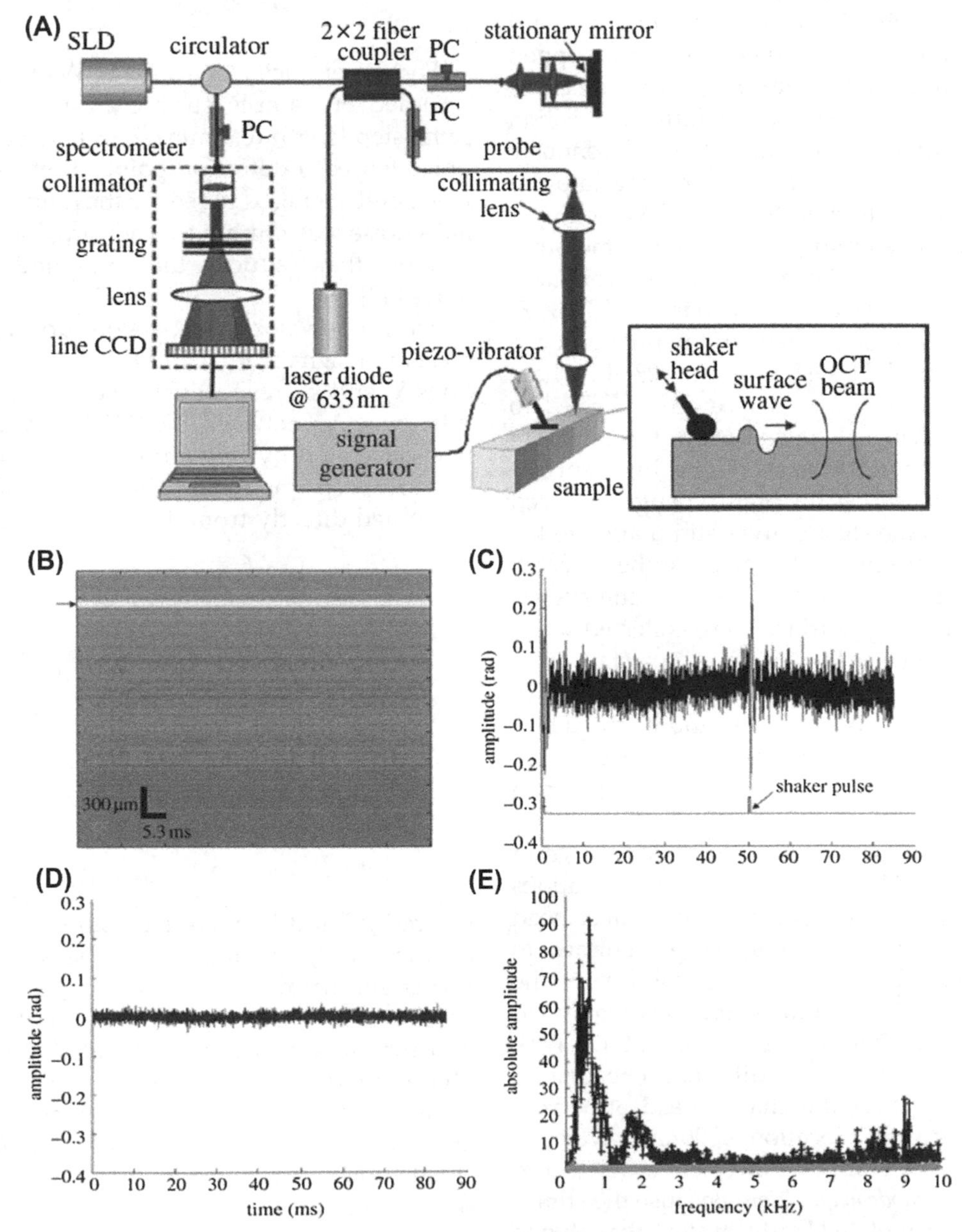

FIGURE 23.3 (A) Set-up of phase-sensitive optical coherence tomography (PhS-OCT) system. (B) Amplitude data of motion mode (M-mode) PhS-OCT image at the detection point. (C) Phase change of the SAW signal and waveform of the shaker pulses. (D) Phase change of the detected system noise. (E) Frequency contribution of the system noise and the detected surface wave signal. *CCD*, charge-coupled device; *OCT*, optical coherence tomography; *SLD*, superluminescent diode.

superluminescent diode (SLD; DenseLight, Ltd.) as the light source. It provided an axial resolution of 15 μm in air (10 μm within the skin, assuming the refractive index is 1.4). Via an optical circulator, the light from the SLD broadband light source was split into two paths in a 50/50 fiber–based Michelson interferometer. One beam was coupled onto a stationary reference mirror and the second was focused onto the phantom samples via an objective lens. The focal length of the objective lens was 50 mm to provide a transverse resolution of 18 μm. The coupler recombined the backscattered light from the sample arm and the reflected light from the reference arm into a home-built, high-speed spectrometer via the optical circulator. The interference light was then coupled into a fast spectrometer equipped with a 14-bit, 1024 pixel Indium gallium arsenide (InGoAs) line-scan camera with a maximum acquisition rate of 47 kHz. A computer was used to synchronously control the acquisition of the camera and the impulse excitation of the shaker. In this study, the SAW stimulation system and camera are triggered at the same time.

An example of how SAW can be obtained from agar phantom by a PhS-OCT system is shown in Fig. 23.3B–E. The motion mode (M-mode) image of the phantom can be obtained with the OCT system in Fig. 23.3B, where the black arrow indicates the surface data selected to analyze the phase change. The surface of the phantom has the strongest signal because of the high surface reflectivity. Fig. 23.3C shows the phase change observed at the surface. When the shaker gives stimulation to the phantom surface, a strong SAW signal can be detected with a very short time delay. To determine the system noise, data from the stage surface were collected when no sample was present (see Fig. 23.3D). The system noise detected from the sample stage was much lower than the detected SAW signal in amplitude and its frequency contribution was negligible, as observed in Fig. 23.3E. The measured SNR at a 0.5 -mm axial depth position was 100 dB.

To calculate the phase-velocity dispersion curve of the shaker-induced SAWs, several detecting locations with known separations are required. The shaker head needs to be in contact with the sample surface to generate the SAWs. To maintain the stability of the generated SAWs, the shaker and sample were mounted on a translation stage. During the experiment, the PhS-OCT sample beam was moved to different detecting locations while maintaining the shaker head at a fixed location. At each detecting location, 4000 lines were acquired over time at a sampling frequency of 47 kHz, also known as an *M-mode acquisition*. Because the stimulation had a frequency of 20 Hz, the system was able to detect two phase changes (see the bottom line in Fig. 23.3C) with a time gap of 50 ms. SAW displacements were calculated using [37]:

$$\Delta Z = \frac{\Delta\varphi\lambda}{4\pi n} \tag{23.5}$$

where $\Delta\varphi$ is the detected phase change, λ is the central wavelength of the PhS-OCT system (1310 nm), and n is the index of refraction of the sample.

SIGNAL PROCESSING OF SURFACE ACOUSTIC WAVE PHASE VELOCITY DISPERSION CURVE

For the elasticity analysis, SAWs were detected and recorded on sample surface at six locations with the same step length (0.5 mm/step) between adjacent locations. For each detection point, 10 measurements were made and averaged to reduce the random noise. The signal's noise was minimized by using the Hilbert–Huang method that reduces the low- and high-frequency noise [40].

The phase-velocity dispersion curve between any two measured signals, $y_1(t)$ and $y_2(t)$ corresponding to locations x_1 and x_2, respectively, were analyzed. The phase difference $\Delta\varphi$ between the SAW signals $y_1(t)$ and $y_2(t)$ was calculated by determining the phase of the cross-power spectrum $Y_{12}(f)$. The phase difference can be computed directly from the two complex spectra:

$$y_1 \xrightarrow{F} Y_1(f) = A_1(f)e^{i\phi_1(f)} \tag{23.6}$$

$$y_2 \xrightarrow{F} Y_2(f) = A_2(f)e^{i\phi_2(f)} \tag{23.7}$$

and it gives the phase of the cross-power spectrum, $Y_{12}(f)$:

$$Y_{12}(f) = Y_1(f)\cdot\overline{Y_2(f)} = A_1A_2e^{i(\varphi_2-\varphi_1)} \tag{23.8}$$

$$\Delta\phi = \phi_2 - \phi_1 \tag{23.9}$$

where $Y_1(f)$ and $Y_2(f)$ are the Fourier transformations of $y_1(t)$ and $y_2(t)$, A_1 and A_2 are the amplitude of cross-power spectrum and $\Delta\varphi = \varphi_1 - \varphi_2$ is the phase difference between the measured signals $y_1(t)$ and $y_2(t)$. Given two receivers at a known distance ΔX, the measured phase difference is 2π when the propagating wave has a wavelength that equals the distance ΔX. More generally, the ratio between the phase difference and 2π equals the ratio between the distance and the wavelength:

$$\Delta\varphi/2\pi = \Delta X/\lambda \tag{23.10}$$

If the distance between the receivers and the phase difference between the two corresponding signals has been measured, the wavelength can be computed as

$$\lambda = \Delta X \cdot 2\pi / \Delta\phi \qquad (23.11)$$

The phase velocity is then obtained, given the frequency f, as:

$$C_R = \lambda \cdot f \qquad (23.12)$$

Thus the phase velocity can be expressed as:

$$C_R = \Delta X \cdot 2\pi \cdot f / \Delta\varphi \qquad (23.13)$$

The phase velocity is then computed from the phase difference between the signals at two receivers. Both the autocorrelation spectrum and the phase velocity dispersion curves are the key parameters for our analyses, because the former provides the available frequency range of the signals, whereas the latter provides the elastic and structural information of the samples. The cut-off frequency for the surface waves has been defined at −20 dB from the maximum of the autocorrelation spectrum, where the uncertainty of the dispersion curves are increased [35]. The final phase-velocity dispersion curve was determined by averaging all the phase velocities between two detection points.

IN VIVO HUMAN SKIN ELASTICITY MEASUREMENT BY SURFACE ACOUSTIC WAVE METHOD

In vivo experiments were carried on 11 healthy human volunteers with an age span of 25–45 years (4 females and 7 males). These experiments were undertaken at room temperature and humidity. Measurements were obtained from two skin sites: the forearm and the palm. The subjects were asked to keep the arm (palm) stable during the six measurements (6 × 85 ms). A complete experiment lasted less than 5 min. The shaker head was gently pressed over the skin, and no discomfort was felt by the subjects.

Fig. 23.4A and B shows the typical SAW from in vivo human skin for the forearm and palm, respectively. The detection points varied from 2 to 12 mm away from the shaker head in 2-mm increments. The SAW signal on the human skin attenuated faster than in the phantoms, which may be a result of the high viscosity and influence of the microvasculature under the skin.

The phase velocity dispersion curves of the palm and forearm from one female subject are plotted in Fig. 23.4C. The value of the phase velocity in the low frequency for the palm and forearm is similar. At 1 kHz, the phase velocity represents the subcutaneous fat layer,

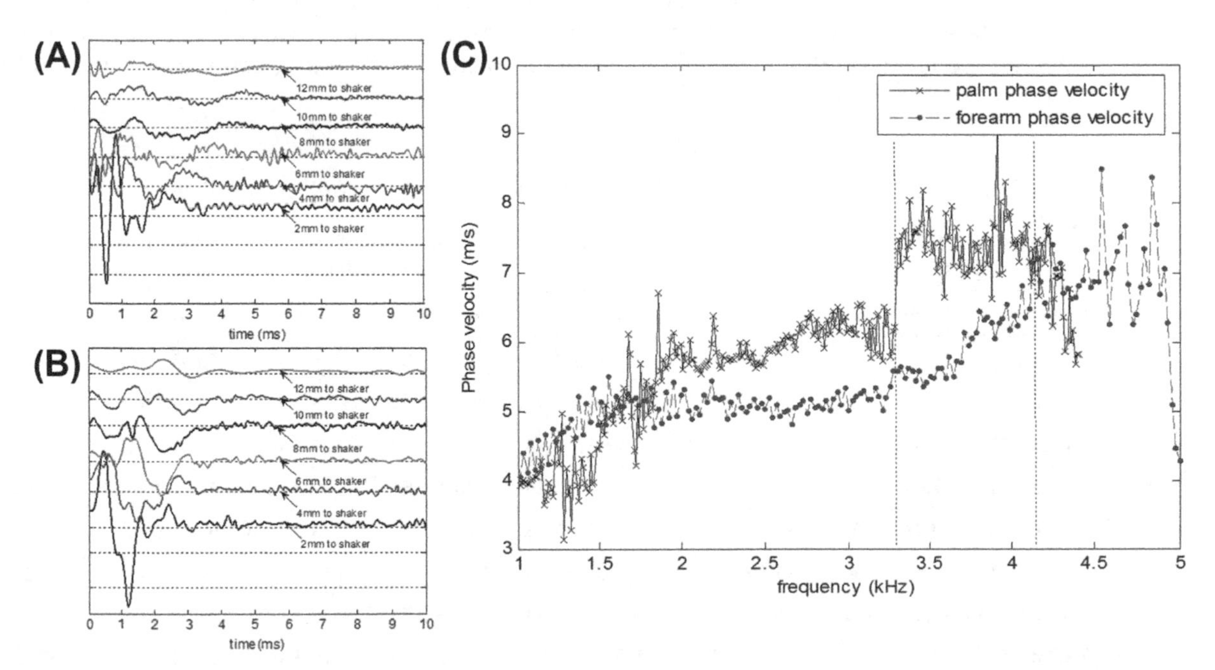

FIGURE 23.4 (A) Typical SAW signals from in vivo human forearm skin and (B) palm skin between 2 and 12 mm away from to the shaker head, in 2 mm steps. (C) Comparison of phase-velocity dispersion curves between the palm and forearm (*dotted line* shows the beginning frequency content of the dermis layer).

with a value of 4 m/s. The phase velocity increases at higher frequencies, which represents the dermis layer of the skin. The phase velocity reaches a plateau of 7 m/s at 4.1 kHz in the forearm and 7.5 m/s at 3.3 kHz in the palm (the plateau is marked with vertical dotted lines in Fig. 23.4C). These values indicate that the dermis' Young's modulus of the palm is higher than that of the forearm, and the thickness of the palm dermis layer is thicker than that in the forearm, which is indicated from the phase velocity curve reaching saturation at a lower frequency.

Young's modulus of the epidermis cannot be detected by the current system because the layer is thin and superficial, which requires collecting data at a higher frequency range. It is estimated that the shaker pulse will require a pulse of 20 kHz, and the PhS-OCT system will require a sampling rate of 100 kHz, which is higher than the current system range (47 kHz). This study has value because most skin diseases and aging of the skin cause alterations in the mechanical properties of the dermis layer.

The measured phase velocities for the palm and forearm skin of all 11 subjects from the experiments described in this chapter are summarized in Table 23.1. To calculate Young's modulus, the phase velocity value at 1 kHz is chosen for the fat layer, and the phase velocity of the plateau is used for the dermis layer. Based on previous studies, we can assume that Poisson's ratio of human skin tissue is 0.48. We used 1116 kg/m^3 as the dermis density and 971 kg/m^3 as the subcutaneous fat density. Therefore with these measured phase-velocity values, we can calculate Young's modulus.

Fig. 23.5 presents the statistics of calculated Young's modulus of dermis and subcutaneous fat layers from 11 female and male subjects' forearms and palms. As can be found in the figure, Young's modulus of dermis layer (200 kPa) is much higher than that of subcutaneous fat layer (50 kPa). Young's modulus of dermis is higher than forearm. In addition, the females have softer dermis and subcutaneous fat than males.

The results show that the SAW method combined with PhS-OCT is sensitive enough to differentiate Young's modulus between different genders at different skin sites. However, the analysis is not integrated. Factors like age, body mass index, race, and the degree of hydration will influence the skin elasticity. Because of the limited number of subjects and incomplete detailed information of subjects, only the differences between female and male palm and forearm skin have been analyzed.

Although the current system is capable of evaluating Young's modulus from different layers, there are some limitations. The skin layer boundaries are not flat surfaces and the tissue microstructures (such as blood vessels) are ignored. Therefore there is a higher uncertainty in distinguishing the frequency contents at each layer in the phase-velocity curve. The Young's moduli presented in Table 23.1 are estimates obtained from averaged phase velocity values, using values for the Poisson's ratio and density that were found in literature. However, the standard deviations of the phase velocities were small compared with the expected stiffness changes that were found in skin diseases, eg, basal cell carcinoma tumors being 3–50% stiffer compared with healthy skin.

TABLE 23.1 The Phase Velocity (Averaged) and Estimated Young's Modulus of 11 Subjects in Palm and Forearm

Subject (F, female M, male)	Age	Dermis Phase Velocity (m/s)		Dermis Young's Modulus (kPa)		Subcutaneous Fat Phase Velocity (m/s)		Subcutaneous Fat Young's Modulus (kPa)	
		Forearm	Palm	Forearm	Palm	Forearm	Palm	Forearm	Palm
1F	27	7.02 ± 1.14	7.50 ± 0.55	180.38 ± 4.76	205.89 ± 1.11	4.00 ± 0.38	4.25 ± 0.69	50.95 ± 0.46	57.52 ± 1.51
2F	26	6.45 ± 0.87	7.94 ± 0.52	152.27 ± 2.77	230.75 ± 0.99	3.92 ± 0.27	4.05 ± 0.81	48.93 ± 0.23	52.23 ± 2.09
3F	25	7.17 ± 1.25	7.58 ± 1.15	176.54 ± 5.37	280.96 ± 8.73	3.64 ± 0.95	3.71 ± 0.51	45.50 ± 3.09	49.93 ± 0.94
4F	26	7.94 ± 0.61	7.64 ± 0.96	216.49 ± 1.28	212.01 ± 3.16	4.26 ± 0.54	4.39 ± 0.69	62.32 ± 1.00	69.99 ± 1.51
5M	45	7.37 ± 0.93	8.29 ± 0.61	198.81 ± 3.17	251.55 ± 1.36	4.18 ± 0.41	4.38 ± 0.78	55.64 ± 0.54	61.09 ± 1.94
6M	36	7.24 ± 0.61	7.69 ± 0.48	191.86 ± 1.36	216.45 ± 0.84	4.25 ± 0.35	4.56 ± 0.64	57.52 ± 0.39	66.22 ± 1.30
7M	26	8.85 ± 0.79	10.02 ± 0.51	286.68 ± 2.28	368.22 ± 0.98	4.17 ± 0.55	4.03 ± 0.73	55.37 ± 0.96	51.72 ± 1.69
8M	29	7.47 ± 0.58	8.06 ± 0.39	191.62 ± 1.15	235.96 ± 0.55	3.27 ± 0.85	4.15 ± 0.39	36.72 ± 2.48	62.55 ± 0.55
9M	27	8.52 ± 0.96	8.79 ± 0.62	249.28 ± 3.16	280.63 ± 1.39	4.37 ± 1.63	4.79 ± 0.41	65.58 ± 9.12	83.33 ± 0.61
10M	28	8.76 ± 1.12	8.85 ± 0.85	263.51 ± 4.31	284.48 ± 2.62	4.27 ± 0.56	4.63 ± 0.78	62.61 ± 1.08	77.86 ± 1.94
11M	28	8.61 ± 0.95	9.23 ± 0.78	254.57 ± 3.09	315.50 ± 2.21	4.22 ± 0.84	4.52 ± 1.54	61.15 ± 2.42	74.21 ± 8.61

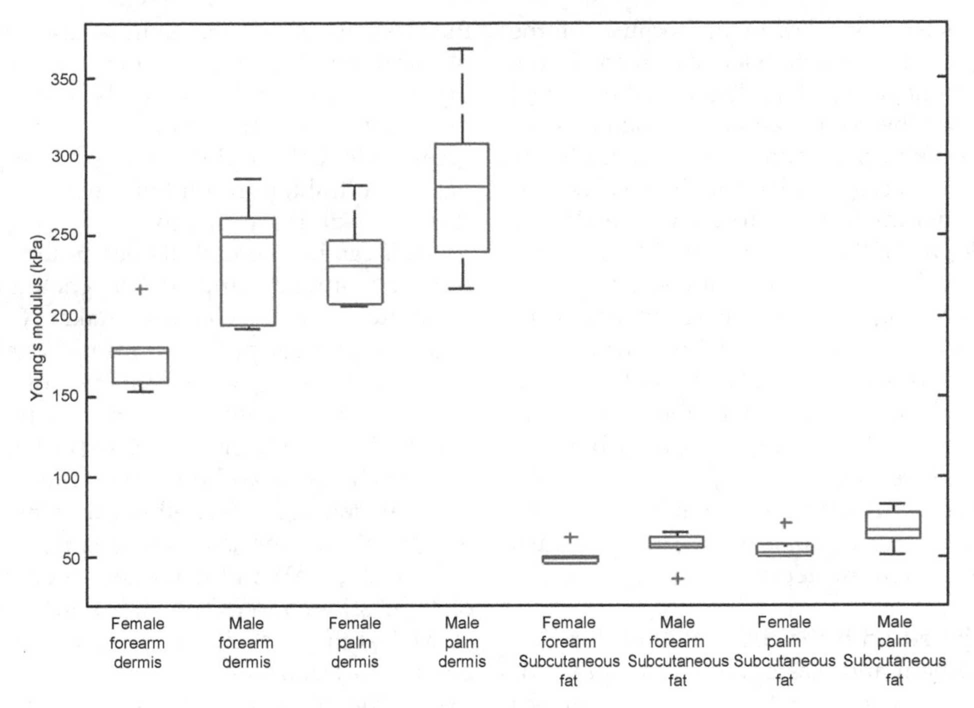

FIGURE 23.5 Statistical results of Young's modulus on forearm and palm dermis and subcutaneous fat in female and male groups.

The minimum SAW velocity that the current system can measure is 0.011 m/s, determined by the time of SAW traveling between adjacent sampling locations (0.5 mm), corresponding to Young's modulus 0.45 Pa, which represents the system sensitivity [41]. Thus the system is sensitive enough to detect very low elasticity (3% stiffer to healthy skin in basal cell carcinoma tumors).

Whereas the SAW method is applicable to detect skin elasticity change, the elastography of tissue can provide direct visualization of the tissue stiffness in aid of clinical diagnosis. The main concept is to estimate the Young's modulus curve as a function of frequency based on the phase-velocity data calculated from two adjacent locations, resulting in an *A* line of SAW elastography. After all the adjacent locations are evaluated, a *B*-scan elastographic image can be built, with its horizontal axis representing the spatial extent (ie, distance, mm) over the sample and the vertical axis representing the SAW frequency (Hz) that is related to the depth.

The minimum depth that the SAW method can detect is related to the stimulator radius. Because the shaker had 2-mm radius, the minimum depth of elastography by system was thus 1 mm. On the other hand, the maximum depth of SAW elastography can be calculated from Eq. (23.3) by the lowest frequency content and its corresponding SAW velocity. Here, the lowest frequency (typically 1 kHz) was defined by −10 dB from the maximum of the autocorrelation spectrum. The maximum depth that the current system can sense is 5 mm, because the SAW velocity was typically 5 m/s at 1 kHz in this study.

The proposed SAW elastography method [41] was tested on a human forearm skin in vivo with a hard nodule in the dermis layer, which can be observed by OCT image Fig. 23.6A. From the OCT image, it can be seen that the hard nodule is 0.5 mm beneath the surface,

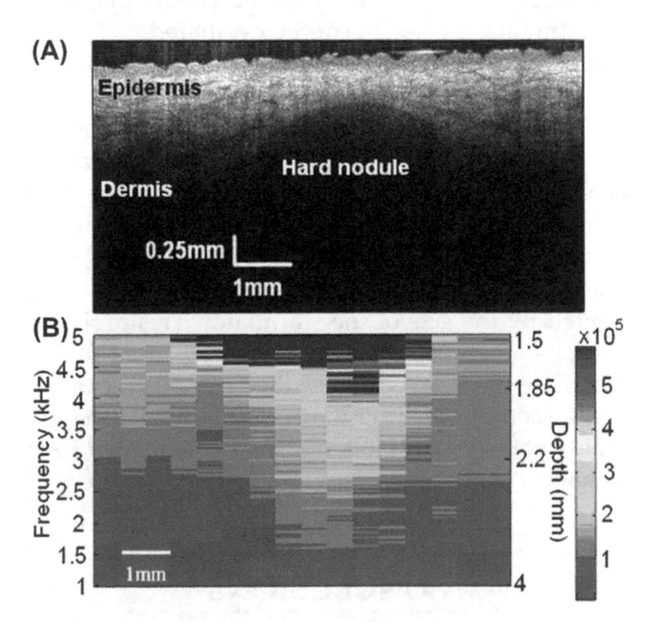

FIGURE 23.6 (A) Optical coherence tomography (OCT) image of human forearm skin in vivo with a hard nodule in the middle. (B) The resulting elastogram.

but to provide its size was difficult because of the limited imaging depth, which was less than 1 mm. The elastographic image (see Fig. 23.6B) distinguished the stiffness between the nodule and background tissue. At normal human skin, the Young's moduli gradually decrease from the surface (200 kPa from 5 to 3 kHz, corresponding to the depths from 1.5 to 2.2 mm, ie, dermis layer) into the depth (100 kPa from 3 to 1 kHz, corresponding to the depths from 2.2 to 4 mm, ie, subcutaneous fat layer). The Young's moduli measured for dermis and subcutaneous fat layers were consistent with those reported in the previous study. The hard nodule (shown in red) can be easily distinguished in the elastogram, because it has a much higher Young's modulus than surrounding tissues and reaches to a depth of 1.85 deep inside the skin. However, the current system had difficulty providing the mechanical property of the epidermis layer because the minimum depth that the system can measure was 1 mm.

PhS-OCT combined with the SAW method for skin elasticity measurement and elastography has some advantages as well as some limitations. The advantages are: (1) the elastography depth is much higher than traditional OCE because the system can image the tissue at 5-mm depth (2) it provides quantitative elastography, and (3) because of the dispersion of SAWs, full information of different layers of skin can be obtained with one data set. The limitations of the system include:

1. The spatial resolution of the SAW elastography is much lower than that of OCE. This is because the latter relies on OCT to perform the strain-rate imaging, whereas the former requires a certain distance for the SAW filter to sufficiently disperse so that the phase velocity can be evaluated.
2. Mapping from the phase velocity to the depth information in the SAW elastography is unfortunately a nonlinear process. This needs to be further studied in order to provide the elastography in a linear geometrical scale so that it can be directly presented alongside with the OCT imaging for ease of use in biomedical applications.
3. The minimum sensing depth of the current system is limited by the size of the stimulator. This problem is perhaps amendable if a laser source is used to stimulate the SAW, because the laser beam can be easily focused, leading to dramatically increased high-frequency contents within the generated SAW signals.

CONCLUSIONS

The SAW method, as one transient elastography method, has the potential to quantitatively measure the elasticity of different skin layers and different kinds of soft tissues [42–44]. However, among different imaging and elastography methods, optical imaging combined with the transient elastography method has great potential in dermatological applications. OCT has an ultrahigh resolution, which can differentiate different skin layers clearly.

SAW can be successfully generated on soft solid material by different approaches, including high energy laser pulse and mechanical shaker. Compared with a high-energy laser pulse, for safety considerations a mechanical shaker is more suitable for generating SAW propagation on in vivo human skin. It is still a problem to directly apply laser pulse to tissue because of the complex mechanisms and potential safety issues, e.g. tissue thermal damage. Meanwhile, in many cases, a high-energy stimulating laser pulse is necessary to generate a detectable SAW in the far field relative to the location of laser stimulation. The stimulation method of SAW as well as the elasticity of the sample will influence the maximum frequency range of a SAW. The balance between the elasticity of testing material, stimulation method and detection method should be balanced before any SAW research takes place.

SAW phase velocities, as well as SAW amplitudes are sensitive to elasticity change in both axial and transverse directions. The SAW phase velocity is related to the material elasticity; higher SAW phase velocity indicates stiffer material. By the signal processing method proposed in this study, a SAW phase-velocity curve can be calculated and inverted to reveal the mechanical properties of different models of phantoms (homogeneous, layered, and with elasticity change in both axial and transverse directions).

SAW methods combined with a PhS-OCT system can provide both structural images and elastography of in vivo human skin. It can differentiate skin elasticity and thickness of different subjects. Because of its high sensitivity, the SAW method combined with PhS-OCT is feasible in dermatological applications, eg, diagnosis of skin cancer, monitoring skin aging, and the effect of different treatment options for reversing signs of aging.

The work presented here shows the potential of the SAW method for the in vivo characterization of human skin elasticity measurement and elastography. There are several straightforward studies to carry out as next steps. The work carried out in this research involves the development of SAW phase-velocity dispersion curves from the generated SAW, which gives quantitative information on changes in the geometrical and mechanical properties of the layers. However, the signal processing is not real time. Higher data processing speed is also required to provide results for a real-time quantitative elastography on the detecting site of the skin tissue, which will be beneficial in clinical applications. The study aims to make simultaneous quantification of the elastic modulus and geometrical parameters; the elastography will merge with the real-time high-resolution OCT imaging of the tested samples. The inverted elasticity information should be simultaneously processed and sent to the real-time OCT imaging software. With the depth–frequency relationship of the SAW method, the elasticity information of a specific depth can be defined in OCT imaging. This will be a challenging stage of the project. In addition, 2D SAW elastography has been presented in this study successfully, and in future research efforts it would be interesting to test three-dimensional elastography for better localization of the skin lesions in a bulk volume of skin tissue.

References

[1] Gennisson J, Baldeweck TT, Catheline MS, Fink M, Sandrin L, Cornillon C, et al. Assessment of elastic parameters of human skin using dynamic elastography. IEEE Trans Ultrason Ferroelectr Freq Control 2004;51(8):908–89.

[2] Agache PG, Monneur C, Leveque JL, De Regal J. Mechanical Properties and Young's modulus of human skin in vivo. Arch Dermatol Res 1980;269:221–32.

[3] Zhang X, Kinnick RR, Pittelkow MR, Greenleaf JF. Skin viscoelasticity with surface wave method. IEEE Int Ultrason Symp Proc 2008. Paper. 0156.

[4] Melanoma skin cancer. American Cancer Society; 2011. http://www.cancer.org/acs/groups/cid/documents/webcontent/003120-pdf.

[5] Skin cancer. American Cancer Society; 2007. http://www.cancer.org/acs/groups/content/@nho/documents/document/skincancerpdf.pdf.

[6] Ciarletta P, Foret L, Ben Amar M. The radial growth phase of malignant melanoma: multi-phase modelling, numerical simulations and linear stability analysis. J R Soc Interface 2010;8(56):345–68.

[7] Tilleman TR, Tilleman MM, Neumann MH. The elastic properties of cancerous skin: Poisson's ratio and Young's modulus. Isr Med Assoc J 2004;6(12):753–5.

[8] Williams M, Ouhtit A. Towards a better understanding of the molecular mechanisms involved in sunlight-induced melanoma. J Biomed Biotechnol 2005;2005(1):57–61.

[9] Allen AC, Spitz S. Malignant melanoma; a clinicopathological analysis of the criteria for diagnosis and prognosis. Cancer 1953; 6(1):1–45.

[10] Sun C, Standish B, Yang VX. Optical coherence elastography: current status and future applications. J Biomed Opt 2011;16(4): 043001.

[11] Evans A, Whelehan P, Thomson K, McLean D, Brauer K, Purdie C, et al. Quantitative shear wave ultrasound elastography: initial experience in solid breast masses. Breast Cancer Res 2010;12(6): R104.

[12] Rago T, Santini F, Scutari M, Pinchera A, Vitti P. Elastography: new developments in ultrasound for predicting malignancy in thyroid nodules. J Clin Endocrinol Metab 2007;92(8):2917–22.

[13] Rivaz H, Boctor EM, Choti MA, Hager GD. Real-time regularized ultrasound elastography. IEEE Trans Med Imaging 2011;30(4): 928–45.

[14] Venkatesh SK, Yin M, Glockner JF, Takahashi N, Araoz PA, Talwalkar JA, et al. MR elastography of liver tumors: preliminary results. AJR Am J Roentgenol 2008;190(6):1534–40.

[15] McKnight AL, Kugel JL, Rossman PJ, Manduca A, Hartmann LC, Ehman RL. MR elastography of breast cancer: preliminary results. AJR Am J Roentgenol 2002;178(6):1411–7.

[16] Schneider D, Schwarz TA. Photoacoustic method for characterising thin films. Surf Coat Tech 1997;91:136–46.

[17] Wang HS, Fleming S, Law S, Huang T. Selection of appropriate laser parameters for launching surface acoustic waves on tooth enamel for non-destructive hardness measurement. In: IEEE Australian Conference of Optical Fibre Technology/Australian Optical Society (ACOFT/AOS); 2006.

[18] Huang QJ, Cheng Y, Liu XJ, Xu XD, Zhang SY. Study of the elastic constants in a $La_{0.6}Sr_{0.4}MnO_3$ film by means of laser-generated ultrasonic wave method. Ultrasonics 2006;44(Supp. 1):e1223–7.

[19] Reverdy F, Audoin B. Ultrasonic measurement of elastic constant of anisotropic materials with laser source and laser receiver focused on the same interface. J Appl Phys 2001;90(9):4829–35.

[20] Ridgway P, Russo R, Lafond E, Jackson T, Zhang X. A sensor for laser ultrasonic measurement of elastic properties during manufacture. In: Proceedings of 16th WCNDT 2004-World Conference on NDT; 2004. Paper 466.

[21] Scruby CS, Drain LE. Laser ultrasonics: techniques and applications. London: Taylor & Francis; 1990. p. 325–35. ISBN-10: 0750300507.

[22] Fercher AF. Optical coherence tomography: development, principles, applications. Z Med Phys 2010;20(4):251–76.

[23] Tomlins PH, Wang RK. Theory, developments and applications of optical coherence tomography. J Phys D Appl Phys 2005;38(15): 2519–35.

[24] Welzel J. Optical coherence tomography in dermatology: a review. Skin Res Technol 2001;7(1):1–9.

[25] Chan RC, Chau AH, Karl WC, Nadkarni S, Khalil AS, Iftimia N, et al. OCT-based arterial elastography: robust estimation exploiting tissue biomechanics. Opt Express 2004;12(19):4558–72.

[26] Wang RK, Ma Z, Kirkpatrick SJ. Tissue Doppler optical coherence elastography for real time strain rate and strain mapping of soft tissue. Appl Phys Lett 2006;89(14):144103.

[27] Liang X, Oldenburg AL, Crecea V, Chaney EJ, Boppart SA. Optical micro-scale mapping of dynamic biomechanical tissue properties. Opt Express 2008;16(15):11052–65.

[28] Kennedy BF, Hillman TR, McLaughlin RA, Quirk BC, Sampson DD. In vivo dynamic optical coherence elastography using a ring actuator. Opt Express 2009;17(24):21762–72.

[29] Kennedy BF, Liang X, Adie SG, Gerstmann DK, Quirk BC, Boppart SA, et al. In vivo three-dimensional optical coherence elastography. Opt Express 2011;19(7):6623–34.

[30] Li C, Guan G, Zhang F, Nabi G, Wang RK, Huang Z. Laser induced surface acoustic wave combined with phase sensitive optical coherence tomography for superficial tissue characterization: a solution for practical application. Biomed Opt Express 2014;5(5): 1403–18.

[31] Lee YC, Kim JO, Achenbach JD. Measurement of elastic constants and mass density by acoustic microscopy. IEEE Ultrason Symp 1993;1:607–12.

[32] Hurley DH, Spicer JB. Line source representation for laser-generated ultrasound in an elastic transversely isotropic half-space. J Acoust Soc Am 2004;116:2914–22.

[33] Neubrand A, Hess P. Laser generation and detection of surface acoustic waves: elastic properties of surface layers. J Appl Phys 1992;71:227–38.

[34] Sohn Y, Kirshnaswamy S. Mass spring lattice modelling of the scanning laser source technique. Ultrasonics 2002;39:543–51.

[35] Wang HC, Fleming S, Lee YC, Law S, Swain M, Xue J. Laser ultrasonic surface wave dispersion technique for non-destructive evaluation of human dental enamel. Opt Express 2009;17:15592–607.

[36] Li CH, Huang ZH, Wang RKK. Elastic properties of soft tissue-mimicking phantoms assessed by combined use of laser

ultrasonics and low coherence interferometry. Opt Express 2011; 19(11):10153–63.

[37] Li C, Guan G, Reif R, Huang Z, Wang RK. Determining elastic properties of skin by measuring surface waves from an impulse mechanical stimulus using phase-sensitive optical coherence tomography. J R Soc Interface 2012;9:831–41.

[38] Doyle PA, Scala CM. Near-field ultrasonic Rayleigh waves from a laser line source. Ultrasonics 1996;34:1–8.

[39] Kenderian S, Djordjevic BB, Green Jr RE. Point and line source laser generation of ultrasound for inspection of internal and surface flaws in rail and structural materials. Res Nondestr Eval 2001;13:189–200.

[40] Sun W, Peng Y, Xu J. A de-noising method for laser ultrasonic signal based on EMD. J Sandong Univ 2008;38:1–6.

[41] Li C, Guan G, Cheng X, Huang Z, Wang RK. Quantitative elastography provided by surface acoustic waves measured by phase-sensitive optical coherence tomography. Opt Lett 2012;37(4):722–4.

[42] Li C, Guan G, Zhang F, Song S, Wang RK, Huang Z, et al. Quantitative elasticity measurement of urinary bladder wall using laser-induced surface acoustic waves. Biomed Opt Express 2014; 5(12):4313–28.

[43] Li C, Guan G, Huang Z, Johnstone M, Wang RK. Non-contact all-optical measurement of corneal elasticity. Opt Lett 2012;37: 1625–7.

[44] Li C, Guan G, Huang Z, Wang RK. Evaluating elastic properties of heterogeneous soft tissue by surface acoustic waves detected by phase-sensitive optical coherence tomography. J Biomed Optics 2012;17(5).

CHAPTER

24

Photoacoustic Tomography in the Diagnosis of Melanoma

Y. Zhou, L.V. Wang

Washington University in St. Louis, St. Louis, Missouri, United States

OUTLINE

INTRODUCTION

Melanoma is the most deadly form of skin cancer. In the United States alone, in 2014 about 76,100 cases were diagnosed and 9,710 deaths were caused by melanoma [1]. Although melanoma accounts for less than 2% of all skin cancers, it causes more than 75% of skin-cancer–related deaths [2]. In addition, the incidence of melanoma has been increasing dramatically in the last few years. In men, melanoma is increasing more rapidly than any other malignancy and in women it is increasing more rapidly than any other malignancy except lung cancer [3]. It is estimated that the lifetime risk of developing melanoma for persons born in the United States in 2014 will be 1 in 42 [2]. Early diagnosis is the key, because localized melanoma is curable and metastatic melanoma becomes fatal. Recent data show that for localized melanoma, the 5-year survival rate is 98%, but the survival rate declines to 16% for metastatic melanoma [2].

The current standard of care for melanoma diagnosis involves visual examination of the skin and recognition of gross lesion characteristics, biopsy of suspicious lesions, and subsequent histological identification. For the initial visual diagnosis, the *ABCDE* rule is a guideline to distinguish between benign and malignant lesions. A malignant lesion usually has an asymmetric (*A*) shape, an irregular border (*B*), multiple colors (*C*), a diameter (*D*) greater than 6 mm, and an elevation (*E*) above the skin. However, not all melanomas present with all the criteria. In addition, some unrelated dermatological disorders, such as seborrheic keratosis, can meet all the criteria. Thus a follow-up biopsy and histological examination of a suspicious lesion are needed to confirm the diagnosis and also to determine the severity of the melanoma. Nevertheless, visual inspection of the

Imaging in Dermatology
http://dx.doi.org/10.1016/B978-0-12-802838-4.00024-8

suspicious lesions is important, especially when there are multiple nevi.

Once the suspicion of melanoma is confirmed, further diagnosis may be needed, depending on the histological features of the melanoma, including its thickness, ulceration, and mitotic rate. Among these features, thickness is the most critical. If the tumor thickness is less than 0.75 mm, regardless of other characteristics, no sentinel lymph node (SLN) biopsy is generally recommended [3]. Otherwise, SLN biopsy is usually performed to determine whether the tumor is already in transit. If the SLN biopsy result is positive, which means the melanoma may have already spread locally or globally, a whole-body examination is needed to locate the metastatic tumors.

During the diagnostic procedure, biopsy plays a key role in primary tumor identification and thickness measurement, and also in identifying metastatic tumors in SLNs. Although patients with a suspicious lesion should undergo an excisional biopsy with negative margins, there are many cases where excisional biopsy might be undesirable. These cases include but are not limited to lesions on the face, a distal digit, and the sole of the foot, and very large lesions [3]. In addition, excisional primary tumor biopsy may alter the local lymphatic system and subsequently affect possible later SLN dissection results. Thus partial biopsies (incisional or punch biopsy) instead of excisional biopsies are widely used to determine tumor characteristics. In fact, in 2013, 57% of the melanoma patients in the United States were diagnosed based on partial biopsies [4]. However, partial biopsy samples only part of the tumor, which may lead to an inaccurate measurement of tumor features (particularly the thickness) and thus to an inaccurate diagnosis. Therefore, in vivo imaging of the melanoma to measure tumor features would greatly improve the diagnostic accuracy.

Combining optical excitation and acoustic detection, photoacoustic tomography (PAT) is suitable for in vivo melanoma imaging [5–7]. In PAT, the target is illuminated either by a pulsed or intensity-modulated continuous-wave light source [7–11]. After the absorption of the light, an initial temperature rise leads to a pressure rise, which propagates as a photoacoustic (PA) wave and finally is detected by an ultrasonic transducer [7,12]. Based on optical absorption contrast, PAT has imaged a variety of exogenous absorbers such as dyes [13–15] and nanoparticles [16–18]. PAT has also imaged numerous endogenous absorbers, including hemoglobin [8], melanin [19], water [20], DNA and RNA [21], lipid [22], carboxyhemoglobin [23], bilirubin [24], myoglobin [25], cytochrome C [26], and methemoglobin [27]. Because melanoma mainly consists of melanin, PAT can detect it with high sensitivity. Based on acoustic detection, PAT can image at depths with high spatial resolution. So far, in vivo melanoma thicknesses of more than 7 mm have been successfully measured by PAT [28].

In this chapter, we discuss the state-of-the-art roles of PAT in the diagnosis of primary and metastatic melanoma. We first introduce the major embodiments of PAT for melanoma-related work. Then we discuss current applications of PAT in melanoma diagnosis, including measuring melanoma thicknesses, monitoring its rates of growth, locating SLNs, label-free screening for SLNs, and detecting melanoma circulating tumor cells (CTCs). Next, we point out potential applications such as metastatic tumor detection, label-free histology, and noninvasive primary tumor examination. Finally we summarize the application of PAT in the diagnosis of primary and metastatic melanoma and suggest future research directions.

MAJOR EMBODIMENTS OF PHOTOACOUSTIC TOMOGRAPHY FOR THE DIAGNOSIS OF MELANOMA

According to the image formation mechanisms, PAT can be classified into three major embodiments [7]: raster-scanning photoacoustic microscopy (PAM) [9,11,29], reconstruction-based photoacoustic computed tomography (PACT) [8,30–32], and circumferential section–scanning PA endoscopy [33–36]. Among the three embodiments, PAM and PACT have been widely studied for the diagnosis of primary and metastatic melanoma. Thus in this section, only these two embodiments are discussed in detail.

Fig. 24.1 is a schematic of a typical PAM system [11]. The pump laser, consisting of a neodymium-doped yttrium aluminium garnet laser and a second harmonic generator, generates light at 532 nm. For deep PAM imaging, a dye cell generates light at longer wavelengths, such as 650 nm, where light suffers less attenuation in tissue than at 532 nm [37]. For quantitative studies, a photodiode measures the light-intensity fluctuation of the laser for calibration. In most cases, fiber delivery of light rather than free-space delivery is preferred, making it much easier to assemble the scanning head. A conical lens and mirrors form a dark field illumination with an incident angle of about 45 degrees, which can reduce the strong interfering signals generated at the skin surface. To detect PA signals with high sensitivity, a spherically focused ultrasonic transducer with water or ultrasound (US) gel coupling is used. One laser pulse generates a one-dimensional (1D) image (A-line), and two-dimensional (2D) transverse scanning forms the final three-dimensional (3D) volumetric image. In this type of PAM system, the acoustic focus is tighter than the optical focus. Thus the lateral

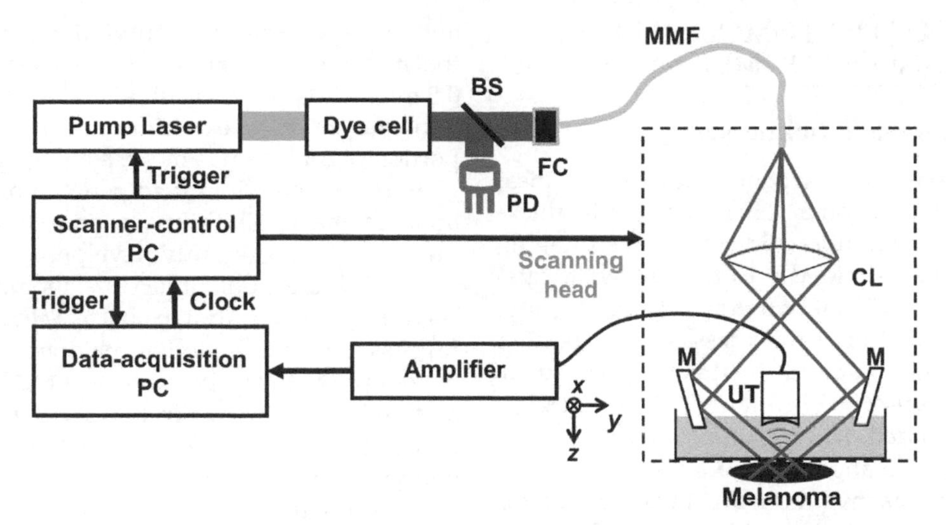

FIGURE 24.1 Typical photoacoustic microscopy (PAM) system. *BS*, Beam splitter; *CL*, conical lens; *FC*, fiber coupler; *M*, mirror; *MMF*, multimode fiber; *PD*, photodiode; *UT*, ultrasonic transducer.

resolution is determined by the spherically focused ultrasonic transducer. Typically, the lateral resolution is given by $0.72\lambda/\mathrm{NA}$, where λ is the central wavelength of the ultrasonic transducer, and *NA* is the numerical aperture of the transducer [38]. The ultrasonic transducer also determines the axial resolution of this PAM system, which can be expressed as $0.88c/\Delta f$, where c is the speed of sound in soft tissue, and Δf is the frequency bandwidth of the transducer [39,40]. To date, by using a 50-MHz central frequency transducer with a 70% bandwidth and an NA of 0.44, a lateral resolution of 45 μm and an axial resolution of 15 μm have been achieved, with an imaging depth of 3 mm [29]. To achieve greater penetration, ultrasonic transducers with lower central frequencies (eg, 10 or 5 MHz) can be used. For instance, using a 5-MHz central frequency transducer, the imaging depth was extended to 4 cm, but the lateral resolution was relaxed to 560 μm [41]. In this case, because the resolution is within the resolving capability of human eyes, the imaging technology is called *photoacoustic macroscopy* (*PAMac*).

Different from PAM, PACT typically employs a linear or ring array transducer for signal detection [31,42,43]. As seen in Fig. 24.2, most of the parts in a typical linear-array–based PACT system are similar to a PAM system, except for two major differences: First, fiber bundles are used instead of a single fiber because light with higher pulse energy is used in a PACT system; second, a linear-array transducer is used. Thus each laser pulse generates a 2D PA image, and 1D scanning is enough to provide a 3D volumetric image. Similar to the PAM system, the spatial resolutions of this typical PACT system are determined by the ultrasonic transducer.

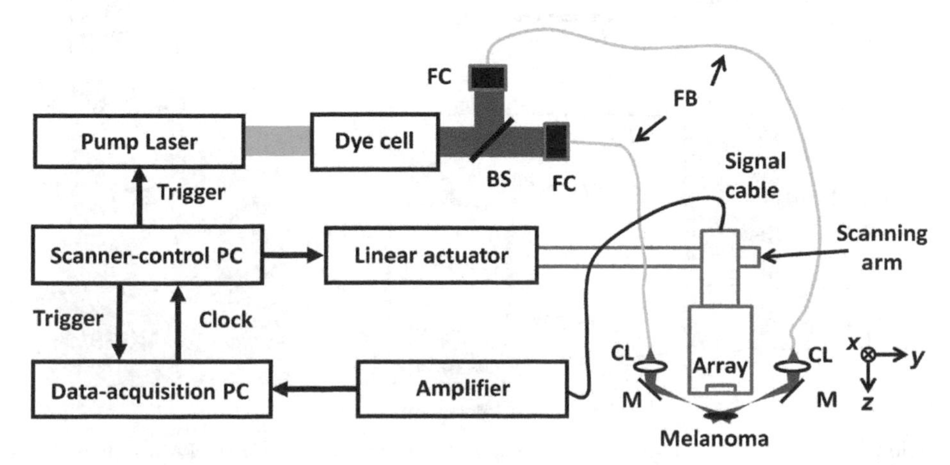

FIGURE 24.2 Typical linear-array–based photoacoustic computed tomography (PACT) system. *BS*, Beam splitter; *CL*, cylindrical lens; *FB*, fiber bundle; *FC*, fiber coupler; *M*, mirror.

PHOTOACOUSTIC TOMOGRAPHY IN THE DIAGNOSIS OF MELANOMA

Measuring Melanoma Thicknesses

As mentioned in the introduction, thickness is critical in the diagnosis of melanoma. Depending on its thickness, a tumor, T, in the tumor–node–metastasis staging system can be further divided into the following categories: T1, ≤ 1.0 mm; T2, 1.0–2.0 mm; T3, 2.0–4.0 mm; and T4, ≥ 4.0 mm [3]. Tumor thicknesses are associated with different prognoses and treatment procedures. For example, to excise a T1 tumor, a 1-cm surgical margin is recommended, whereas a 2-cm margin is recommended for both T3 and T4 tumors [3].

Immediately after its invention, PAM was applied for melanoma thickness measurement. In 2006, Oh et al. imaged skin melanomas in vivo by using a table-top PAM system [19]. In an initial proof of feasibility, black ink–filled tubes with different diameters were used to mimic melanomas of different sizes. The surrounding medium was made of 1% intralipid and 10% gelatin to mimic normal biological tissue. The results showed that both the top and bottom boundaries of the melanoma phantoms could be detected. In addition, the measured thicknesses of the melanoma phantoms agreed well with the known values. However, because light suffered strong attenuation in the melanoma phantom, the signals from the top boundary were very strong, but those from the bottom boundary were very weak. In the end, the maximum thickness of the detected melanoma phantoms was limited to 1.27 mm. The authors also performed in vivo experiments in which the maximum thickness of the melanomas was only 0.5 mm. Thus the measured thicknesses in both phantom and in vivo experiments covered only a small portion of melanomas in the staging system.

To overcome the limited penetration problem and to broaden the applicability to various anatomical sites, the same group recently developed a hand-held PAM system [44] that could image melanomas in all the tumor classifications. As shown in Fig. 24.3A, the hand-held PAM system shares similar major components with the table-top PAM system except for the light illumination mechanism. As shown in Fig. 24.3B, instead of conventional oblique illumination (45-degree incidence angle), normal illumination was employed to avoid the strong optical attenuation in melanoma by passing the light around it. By using Monte Carlo simulations, the authors showed that with the new illumination mechanism, the average signal-to-noise ratio of the melanoma at the bottom boundary was improved by more than 1600 times. A photograph of the handheld probe is shown in Fig. 24.3C.

Phantom experiments showed the hand-held PAM system's deep melanoma detection capability. As shown in Fig. 24.4, melanoma phantoms with three different diameters (7.0, 9.5, and 14.0 mm) were prepared with varied thicknesses. Strong PA signals can be observed from both the top and bottom boundaries. As shown in Fig. 24.4D, all the measured thicknesses are in accordance with the actual values, and the minimum and maximum measured thicknesses are 0.7 and 4.1 mm,

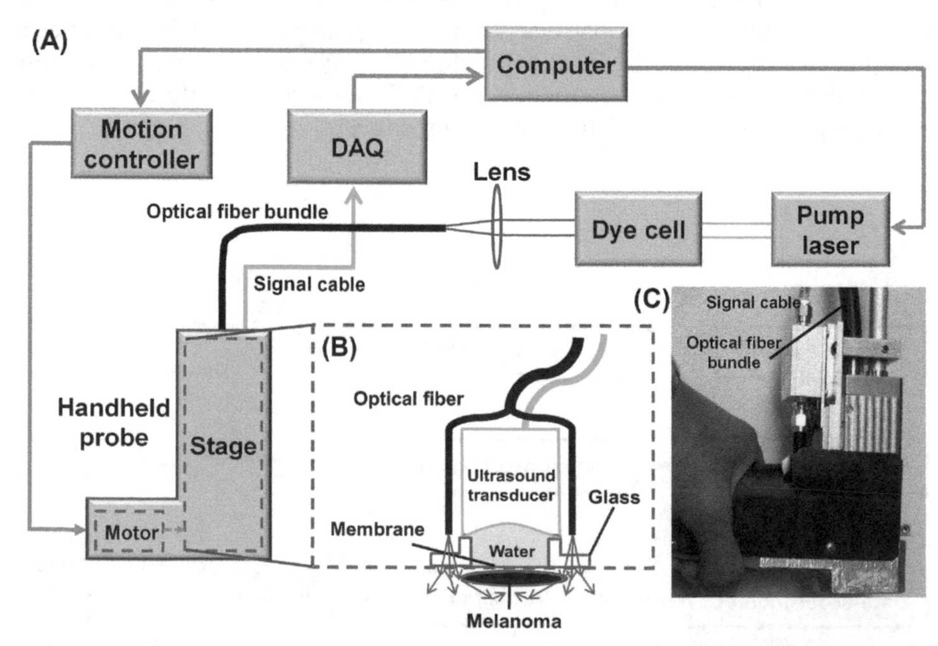

FIGURE 24.3 Hand-held photoacoustic microscopy (PAM) system. (A) Schematic of the handheld PAM system. (B) Components held by the translation stage in the hand-held probe. (C) Photograph of the hand-held probe. *DAQ*, Data-acquisition system. *Figure reprinted with permission from Zhou Y, Xing W, Maslov KI, Cornelius LA, Wang LV. Handheld photoacoustic microscopy to detect melanoma depth in vivo. Opt Lett 2014;39(16): 4731–4.*

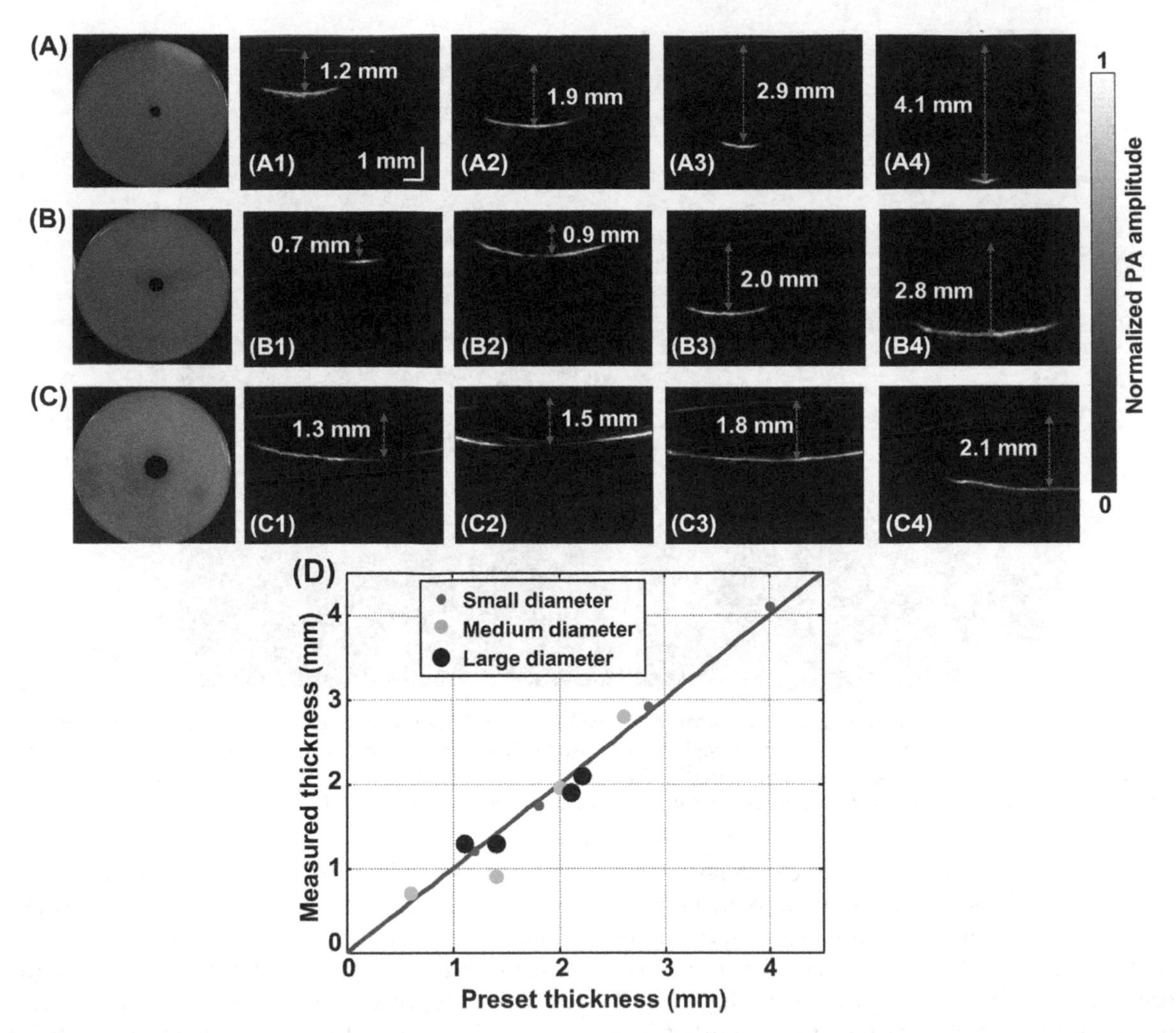

FIGURE 24.4 Handheld photoacoustic (PA) microscopy of melanoma phantoms. Photos of the melanoma phantoms with diameters of 7 mm (A), 9.5 mm (B), and 14 mm (C). The numbered pieces (A1–A4, B1–B4, and C1–C4) of each part are PA images of melanoma phantoms as shown in (A), (B), and (C) with varied thicknesses, respectively. (D) Measured thicknesses versus actual values. *Figure reprinted with permission from Zhou Y, Xing W, Maslov KI, Cornelius LA, Wang LV. Handheld photoacoustic microscopy to detect melanoma depth in vivo. Opt Lett 2014;39(16):4731–4.*

respectively. Thus melanomas in all the tumor classifications in the staging system were covered.

In the in vivo experiment, melanoma was introduced by subcutaneously injecting B16 melanoma cells into the dorsal side of a nude mouse. As shown in Fig. 24.5, both the top and bottom boundaries of the melanoma can be clearly imaged. The measured thickness was 3.66 mm, which was close to the invasively confirmed thickness (3.75 mm). In addition, as shown in Fig. 24.5C and D, the contours of the melanoma are similar in its photograph and the PA image, further proving the feasibility of PAM for melanoma detection in vivo. With its convenient hand-held design, this PAM system is highly promising for clinical studies.

Detecting Rates of Growth

Rate of growth (ROG) has been proposed as an important feature of malignant tumors, including melanoma [45–47]. Various biological markers such as melanoma thickness, mitotic rate, and ulceration are used to quantify the ROG of melanomas. However, all these measurements are indirect, and there is still little direct information on the ROG of the melanomas. One roadblock is the lack of a method for noninvasive melanoma volume quantification. Because melanoma may have irregular shapes, to accurately quantify its volume the whole boundary needs to be visualized.

Recently, a linear transducer array–based PACT system was applied to measure melanoma volume and ROG [28]. The transducer array, with a central frequency of 21 MHz and a 55% bandwidth, contained 256 elements and measured 23 × 3 mm. Each element of the transducer array was cylindrically focused, and the focal distance was 15 mm. Because the acceptance angle of the transducer array was large (~85 degrees), nearly the entire melanoma boundary could be detected in 3D.

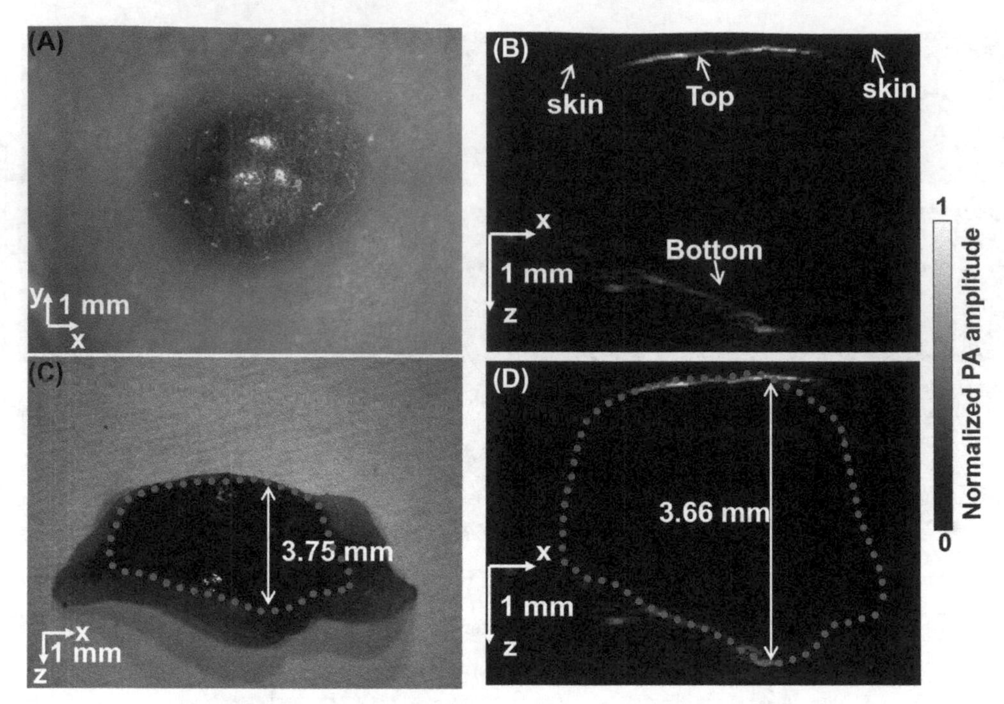

FIGURE 24.5 Hand-held photoacoustic microscopy (PAM) of melanoma in a nude mouse in vivo. (A) Photograph of the melanoma in vivo. (B) Photoacoustic (PA) image of the melanoma. (C) Photograph of the excised melanoma. The *red dots* outline the boundary of the melanoma. (D) The same PA image as in (B). The *red dots* are the same as in (C). *Figure reproduced with permission from Zhou Y, Xing W, Maslov KI, Cornelius LA, Wang LV. Handheld photoacoustic microscopy to detect melanoma depth in vivo. Opt Lett 2014;39(16):4731–4.*

Ex vivo experiments were performed to show the capability of the PACT system for melanoma volume measurement. B16 melanoma cells were subcutaneously injected into the dorsal surface of nude mice, and the tumors were allowed to grow for 15 days. Fresh melanomas excised from the mice were embedded in a tissue-mimicking environment that contained an agar and intralipid mixture. Right after the PA measurements, the melanomas were carefully removed from the mixture and corresponding gross images were taken with a table-top microscope for comparison. As shown in Fig. 24.6, there is an excellent correlation between the PA and gross images, demonstrating the ability of this system to accurately measure melanomas. For quantitative comparison of the melanoma volume, a graduated cylinder was used to provide the gold standard. As shown in Fig. 24.6D, the volume calculated by PACT is in good accord with the standard measurements.

In other experiments, the same B16 cells were injected into nude mice. In vivo measurements were made at days 3 and 6 after the injection. The same melanoma was imaged and its volume was calculated to determine the ROG. As shown in Fig. 24.7, the detected melanoma can be clearly differentiated from the surrounding tissues. The volumes of the melanoma at days 3 and 6 were measured to be 22.3 and 71.9 mm^3, respectively. The calculated ROG was 16.5 mm^3/day over the 3-day interval.

With the ability to image entire melanomas in vivo, PAT has successfully measured both the thickness and volume of melanomas. Thus the ROG, a significant feature of melanoma, can be directly calculated. However, the reported work measured an average ROG over only a short time and not long enough to study the full growth behavior of melanoma. In addition, the relationship between ROG and other tumor features, such as thickness, is still unclear. With the ability to image whole melanomas in vivo, PAT can answer these questions in the future, and ROG may prove to be superior to tumor thickness as a diagnostic feature.

Locating Sentinel Lymph Nodes

To perform SLN biopsy, the SLN first needs to be located. The most commonly used method for SLN mapping is planar lymphoscintigraphy, in which a radioactive tracer substance is injected, followed by scintigraphic imaging with a gamma probe [48]. For areas where the gamma probe has difficulty in correctly localizing the SLN because of the vicinity of the primary tumor and the SLN, single photon emission computed tomography (SPECT)/computed tomography (CT) is used [48]. However, both scintigraphic imaging and SPECT/CT are based on ionizing radiation and raise concerns about safety. Thus a safer alternative with similar detection ability is preferred, and PAT is a good

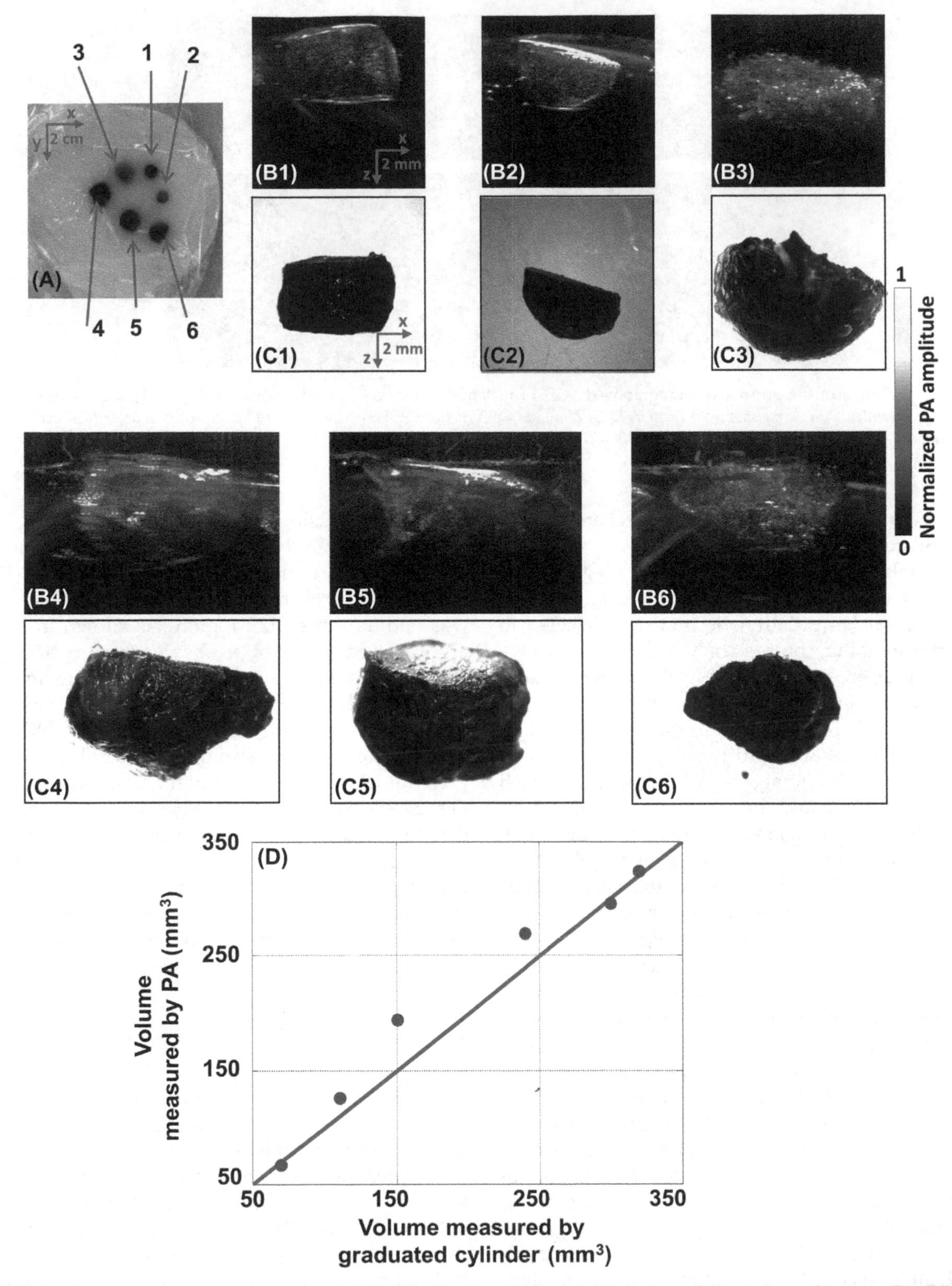

FIGURE 24.6 Photoacoustic computed tomography (PACT) of ex vivo melanomas. (A) A photo of the ex vivo melanomas. (B1–B6) Melanoma images captured by PACT and, (C1–C6) a standard microscope that correspond to melanomas 1 to 6 in (A). (D) Photoacoustic (PA)–measured volumes versus standard measurements. *Blue dots*, PA measurements; *red lines*, ideal fit if the PA measurements are identical to the standard measurements. *Figure reproduced with permission from Zhou Y, Li G, Zhu L, Li C, Cornelius LA, Wang LV. Handheld photoacoustic probe to detect both melanoma depth and volume at high speed in vivo. J Biophotonics 2015;1(7).*

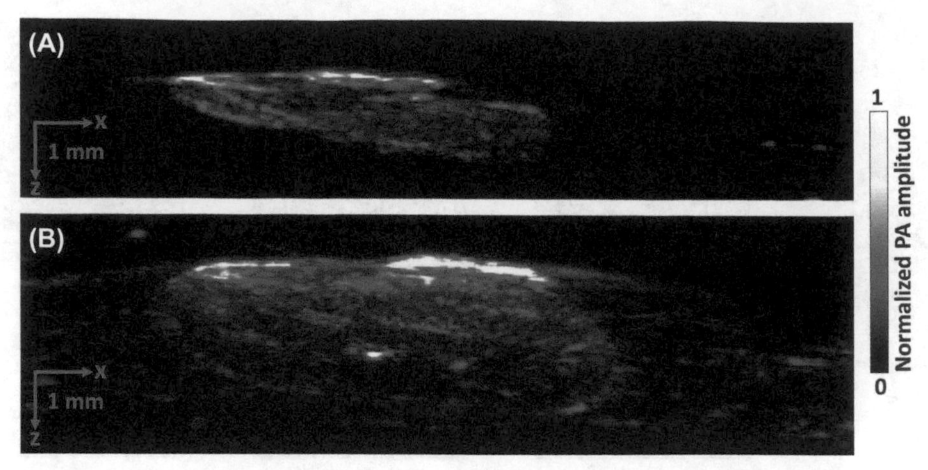

FIGURE 24.7 Photoacoustic computed tomography (PACT) of melanoma in vivo in a nude mouse at day 3 (A) and day 6 (B). *PA*, Photoacoustic. *Figure reprinted with permission from Zhou Y, Li G, Zhu L, Li C, Cornelius LA, Wang LV. Handheld photoacoustic probe to detect both melanoma depth and volume at high speed in vivo. J Biophotonics 2015;1(7).*

candidate because of its nonionizing mechanism and ability to image deeply.

In 2008, a table-top PAMac system with a 5-MHz central frequency ultrasonic transducer was used for SLN detection in a rat [14]. Methylene blue was injected to improve the imaging contrast of the SLN. Fig. 24.8A and B are photographs of the SLN region taken before and after PA imaging. Note that after the imaging, the skin was removed to identify the location of the SLN, as shown in Fig. 24.8B. Before methylene blue was injected, a control PA image was obtained, in which only blood vessels could be detected, as shown in Fig. 24.8C. Because methylene blue molecules are small (only 0.6 nm in diameter and 1.6 nm in length), their diffusion coefficient is large, based on the Einstein relation. Thus the SLN appeared in the image with high contrast right after injection (see Fig. 24.8D). The contrast-to-noise ratios of the vasculature and the SLN were calculated to be around 61 and 146, respectively. At 50 min after dye injection, another PA image was acquired with a relocated field of view to center the SLN. The SLN was still clearly observed, as shown in Fig. 24.8E. To check the feasibility of detecting the SLN at depths, chicken breast tissue was placed on top of the rat, and a PA image was acquired. As shown in Fig. 24.8F, although the vessels are faded, the SLN is still indicated by strong signals. Fig. 24.8G shows that the depth of the SLN is around 18 mm, which is greater than the mean depth of SLNs in humans (~12 mm). Thus PAT can detect SLNs in humans in vivo, enabling more accurate SLN biopsies and more accurate metastatic diagnoses of melanomas.

To further show the potential of PA imaging–guided SLN biopsy, a hand-held probe modified from a clinical US array system was applied for SLN detection [49]. Image-guided needle biopsies were successfully performed by using this hand-held PA system. Indocyanine green (ICG) dye was injected into a rat to increase the PA image contrast. To test the clinical feasibility of this technique on humans, a 2-cm thick chicken tissue was laid atop the SLN region. As shown in Fig. 24.9A and B, the SLN cannot be observed before ICG injection but becomes dominant in the image 10 min after ICG injection. On the other hand, based only on the US image, it is hard to see the SLN, as shown in Fig. 24.9C. The needle used for biopsy is also imaged with high contrast by PAT, but with low contrast in the US image, as shown in Fig. 24.9D and E, respectively. These results indicated that PAT could be used for image-guided SLN biopsy with high accuracy. It is worth mentioning that other PA systems (eg, a 2D transducer array) [43,50] and other contrast agents (eg, nanoparticles) [51–53] can also be used for SLN detection, but are not discussed here because the basic principles are the same.

Screening for Sentinel Lymph Nodes

Biopsy followed by hematoxylin and eosin staining is the standard of care in assessing metastatic melanomas in SLNs. However, as discussed, incomplete examination of the SLN may miss melanoma cells and thus lead to misdiagnosis of metastasis. To address this problem, high-frequency US imaging has been considered, but the false negative rate, as high as 10%, was considered unacceptable [54]. For complete SLN assessment, PAT, with its high sensitivity for melanoma imaging, is a promising modality [55–57].

Recently, Jose et al. identified melanoma cells hidden in a pig lymph node with a typical PACT system [56]. As shown in Fig. 24.10 A and C, two agar gel beads embedded with different number densities of melanoma cells were prepared to mimic different metastatic clumps

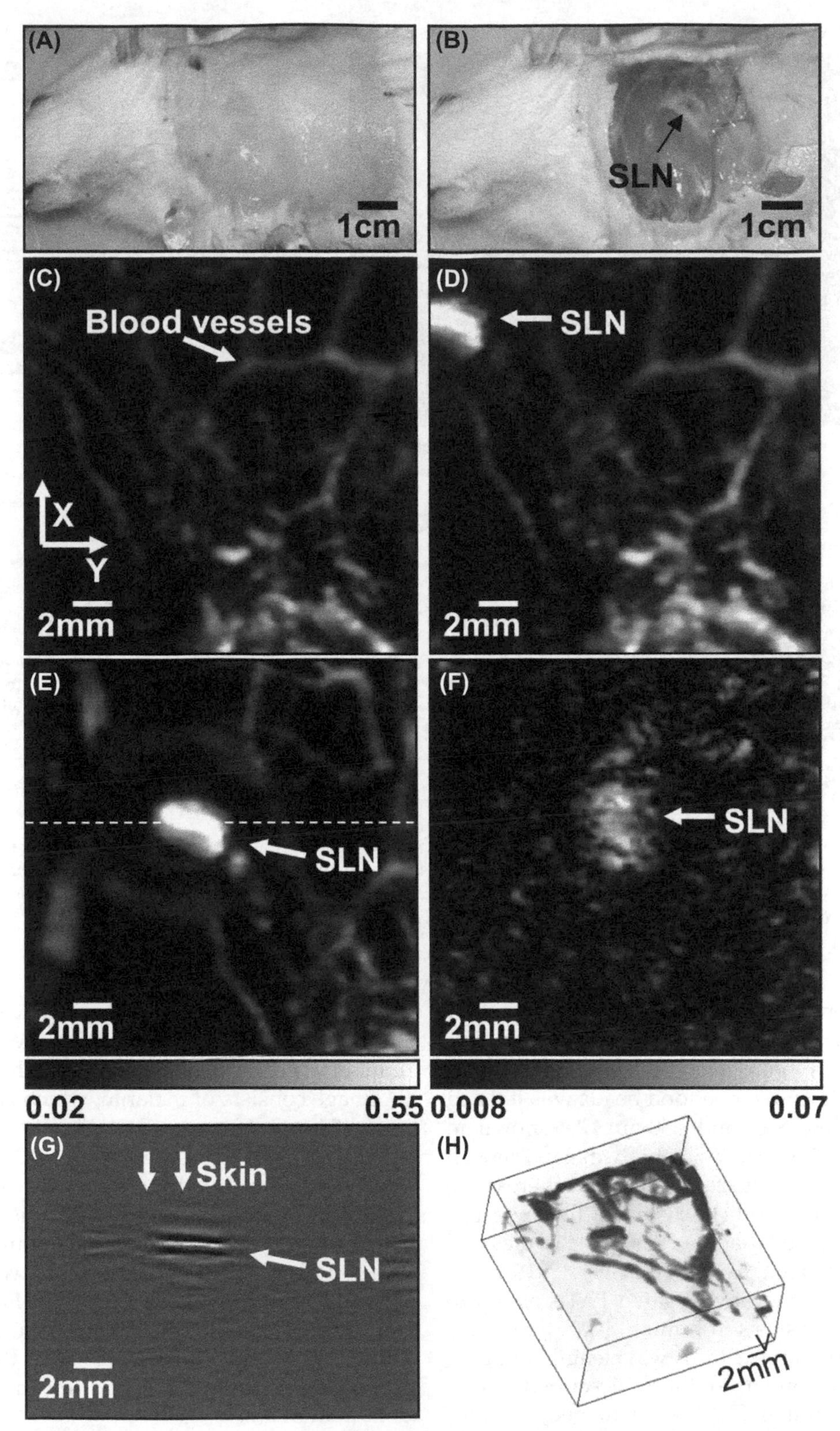

FIGURE 24.8 Photoacoustic (PA) macroscopy of a sentinel lymph node (SLN) in a rat in vivo. Photos of the SLN region taken before (A) and after (B) PA imaging. PA images before (C) and after (D) methylene blue injection. (E) PA image about 50 min after methylene blue injection, with scanning head repositioned. (F) PA image of SLN beneath overlaid chicken breast tissue. (G) B-scan image along the *dashed line* in (E). (H) Volumetric image of SLN. The *color bars* represent PA amplitude. (C and D) Share the same color bar range as (E). *Figure reprinted with permission from Song KH, Stein EW, Margenthaler JA,Wang LV. Noninvasive photoacoustic identification of sentinel lymph nodes containing methylene blue in vivo in a rat model. J Biomed Opt 2008;13(5):054033.*

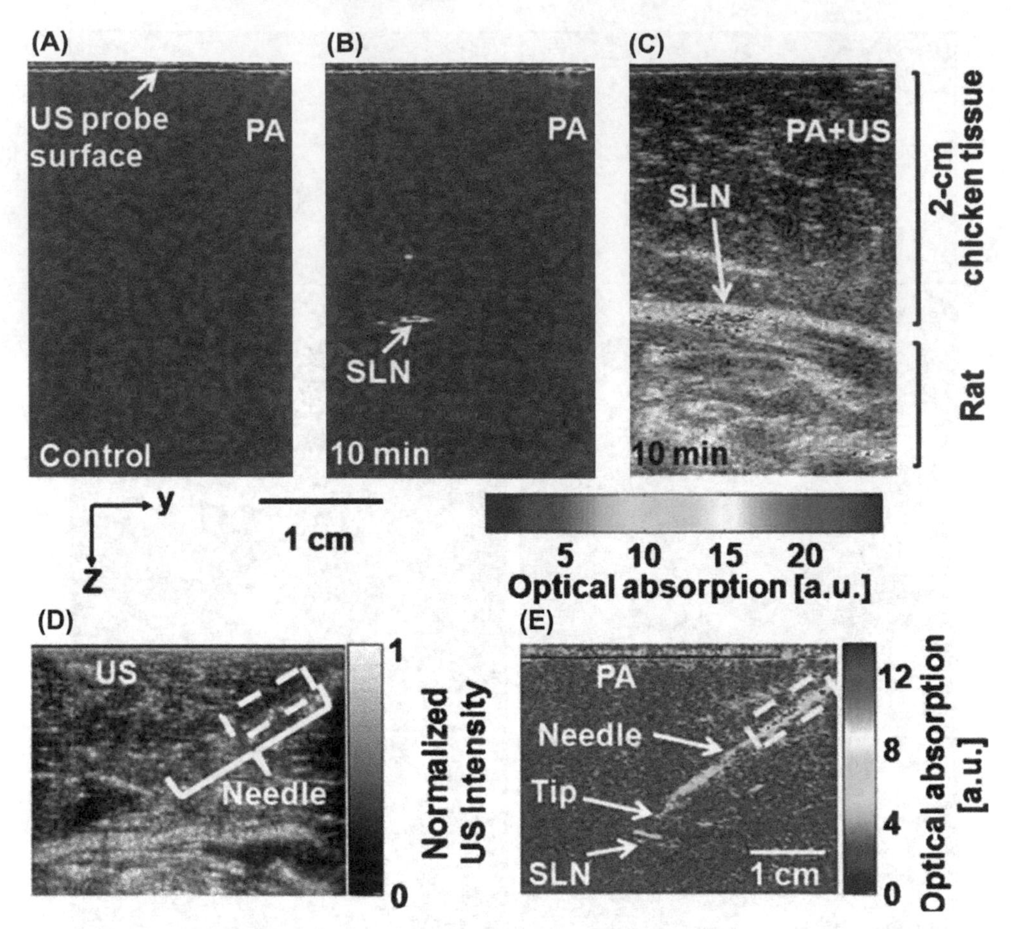

FIGURE 24.9 In vivo photoacoustic (PA) imaging for sentinel lymph node (SLN) needle biopsy guidance. PA image before (A) and 10 min after (B) indocyanine green (ICG) dye injection. (C) Coregistered PA (*pseudocolor*) and ultrasound (US) (*gray scale*) images. Needle biopsy of an SLN guided by US (D) and PA imaging (E). *a.u.*, Arbitrary unit. *Figure reprinted with permission from Kim C, Erpelding TN, Maslov KI, Jankovic L, Akers WJ, Song L, et al. Handheld array-based photoacoustic probe for guiding needle biopsy of sentinel lymph nodes. J Biomed Opt 2010;15(4):046010.*

in an SLN. One density was $5 \times 10^4/\mu L$, and the other was $5 \times 10^3/\mu L$. A control image was acquired before inserting the beads into the SLN, as shown in Fig. 24.10B. The SLN with embedded beads was imaged by PACT at 720, 760, 800, and 850 nm. As shown in Fig. 24.10D and E, the bead with high melanoma cell density could be clearly detected by PACT, whereas the low-density bead could barely be distinguished. Spectroscopic analysis confirmed that the PA signals in the bead region were from melanin, as shown in Fig. 24.10F.

The same group further investigated PAT's ability to detect melanoma metastasis in human SLNs [58]. They used a PACT system and a multiwavelength imaging processing method similar to those described previously. Two resected human SLNs were imaged; one contained metastatic melanomas, but the other did not. Although multiwavelength PACT images were acquired for component analysis, high-resolution US images were simultaneously acquired to provide anatomy information. As shown in Fig. 24.11A, the US image of the tumor SLN shows a round node. In the corresponding PA image, only the surface of the node can be detected, as shown in Fig. 24.11B. Based on the multiwavelength unmixing procedure, a portion of the structure in the PA image consists of melanin, which is marked in green in Fig. 24.11C. The corresponding pathological image of the malignant node is shown in Fig. 24.11G, where the melanin is found in the upper part of the node. Thus the PA measurement was in accordance with the standard pathological result. The same unmixing procedure was conducted on the benign SLN, which mainly provided blood contrast, as shown in Fig. 24.11D–F and confirmed by the pathological image (see Fig. 24.11H). Thus based on multiwavelength PA measurement, differentiating between blood and melanin is possible in human SLNs.

Further research is required to improve the sensitivity of PAT for melanoma detection, as well as its specificity in the presence of other absorbers, such as blue dye for lymphatic mapping. Blind studies could further test the accuracy of PAT-based SLN screening. With the ability to image deeply, PAT could potentially

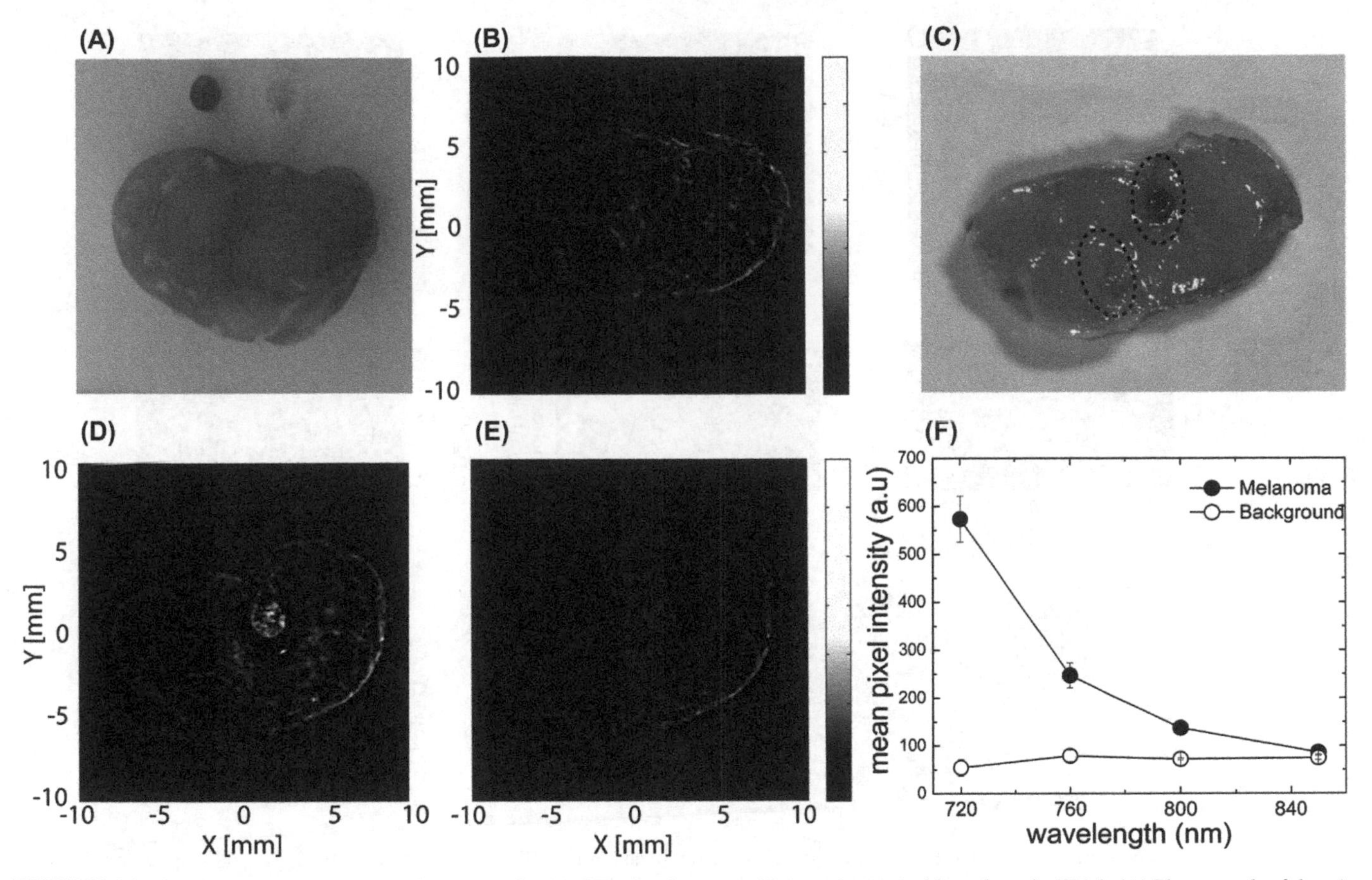

FIGURE 24.10 Photoacoustic computed tomography (PACT) of melanoma cells in a pig sentinel lymph node (SLN). (A) Photograph of the pig SLN and beads with embedded melanoma cells. (B) PACT of the SLN without inserted beads. (C) Photograph of the SLN after beads were inserted (*dotted ovals*). PACT images of the SLN with inserted beads, at 720 (D) and 800 nm (E). (F) Spectroscopic analysis of the PACT images of the SLN. *a.u.*, Arbitrary unit; *f*, frequency. *Figure reprinted with permission from Jose J, Grootendorst DJ, Vijn TW, Wouters MW, van Boven H, van Leeuwen TG, et al. Initial results of imaging melanoma metastasis in resected human lymph nodes using photoacoustic computed tomography. J Biomed Opt 2011;16(9):119801.*

be applied for detecting melanoma metastasis in SLNs in vivo.

Monitoring Circulating Tumor Cells

The high mortality rate of melanoma is attributed mainly to its high propensity for metastasis. The presence of CTCs has been suggested as an early predictor for metastatic development, motivating intensive studies for detection and isolation of CTCs in recent years. The current standard for CTC detection is based on ex vivo blood tests. However, the limited blood sample volume dramatically decreases the effective ability to detect CTCs in the whole body. An in vivo CTC detection technique that could perform whole body blood sampling would be of great interest.

Based on PA sensing, photoacoustic flow cytometry (PAFC) has been developed for CTC monitoring [59–62]. As shown in Fig. 24.12A, in vivo PAFC excites selected blood vessels with short laser pulses, followed by time-resolved PA measurements by an ultrasonic transducer placed on the skin. Based on the difference in the corresponding PA signals, pure blood and blood containing melanoma cells can be distinguished, as shown in Fig. 24.12B. In an in vivo experiment, 10^4 B16 melanoma cells were injected into a tail vein of a mouse. As shown in Fig. 24.12C, PA signal spikes were observed shortly after the injection of melanoma cells, indicating that PAFC successfully detected CTCs in vivo. To detect CTCs originating from a primary tumor, an abdominal wall blood vessel in a mouse was monitored for CTCs on the fourth day after inoculation with B16 melanoma cells. As shown in Fig. 24.12D, frequent PA signal spikes were detected, very likely from the CTCs. Compared with the results from direct CTC injection, the PA spikes were less frequent, which may be because of the reduced CTC density in the blood, and their magnitudes relative to the background were smaller, which may be the result of the larger detection voxel. Because blood vessels are more deeply embedded in humans than mice, in human studies the detection voxel would be larger and the fractional PA signal changes caused by CTCs would be smaller. One solution is to tag CTCs with a magnetosensitive contrast agent and then magnetically concentrate them in the detection voxel, thus increasing the PA signal changes [61]. The magnetic contrast agent is conjugated

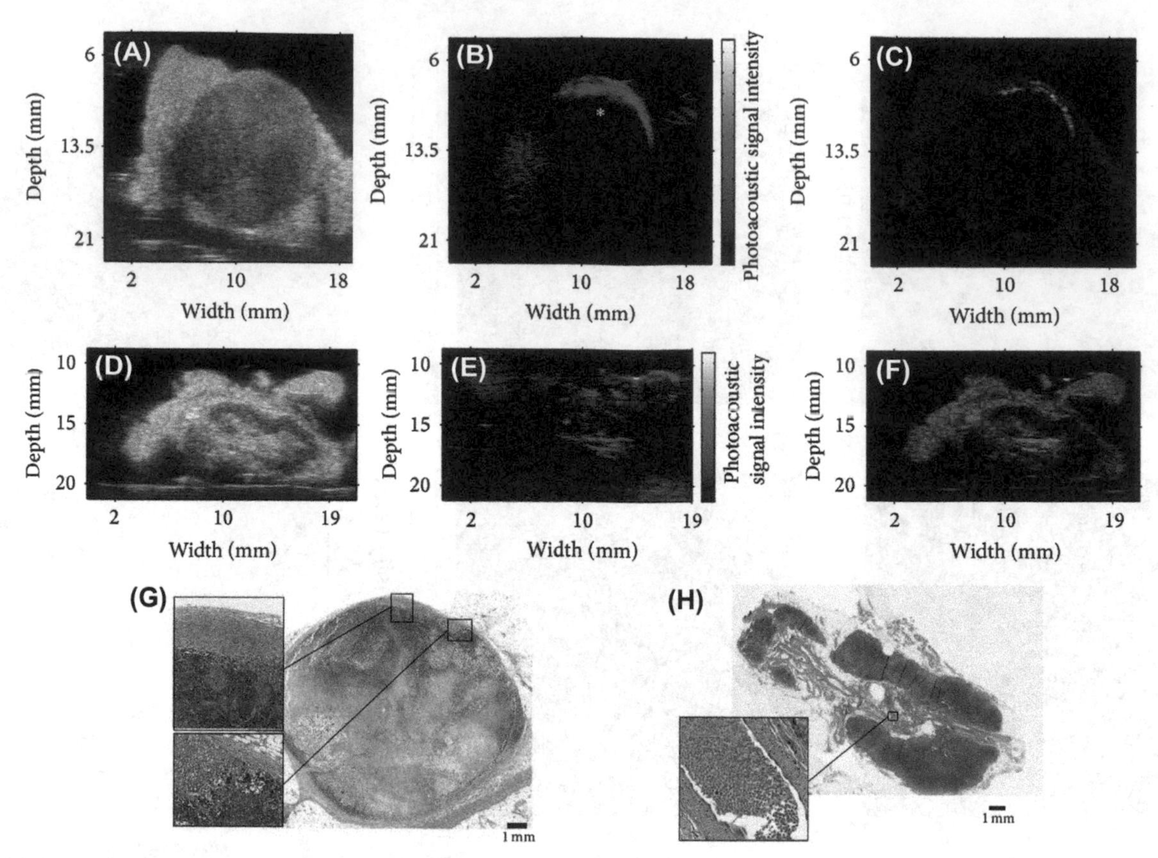

FIGURE 24.11 Photoacoustic computed tomography (PACT) of two resected human SLNs. One is malignant and the other is benign. The first and the second rows are images of the malignant and benign nodes, respectively. (A and D) high-frequency ultrasound (US) images. (B and E) PACT images. (C and F) The calculated melanin and blood components from multiwavelength PACT images overlaid with the ultrasound images. Histological images of the malignant (G) and benign (H) SLNs. *Figure reprinted with permission from Langhout GC, Grootendorst DJ, Nieweg OE, Wouters M, Van der Hage J, Jose J, et al. Detection of melanoma metastases in resected human lymph nodes by noninvasive multispectral photoacoustic imaging. Int J Biomed Imaging 2014;2014:163652.*

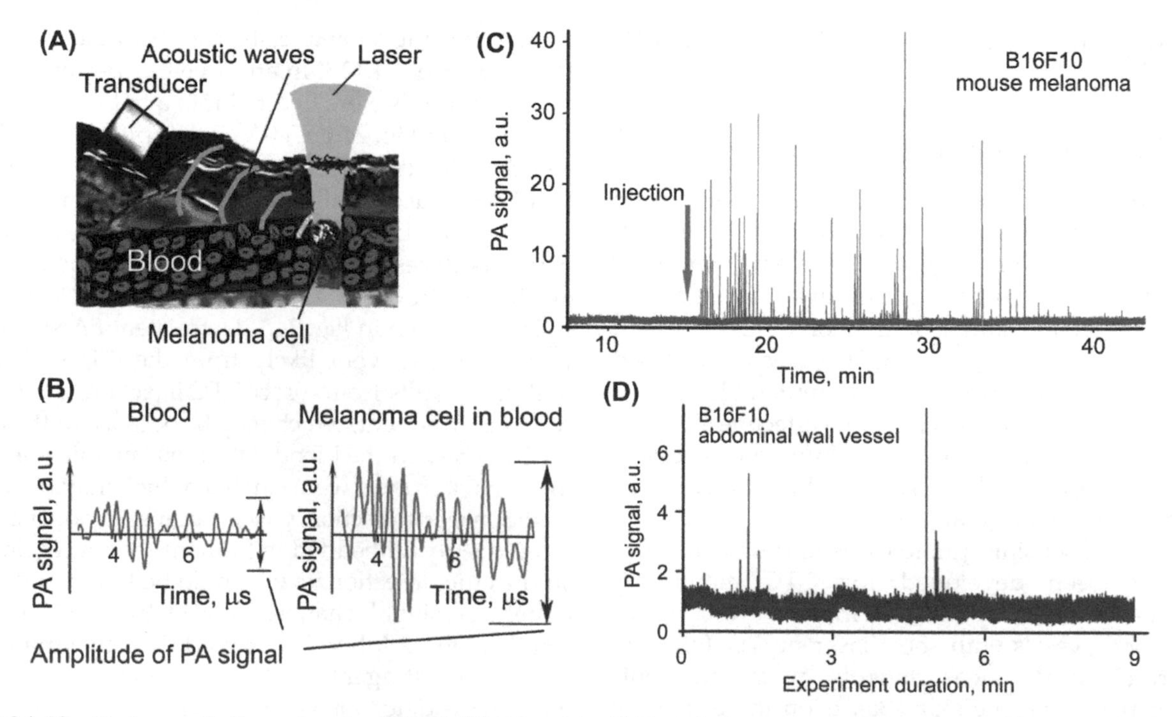

FIGURE 24.12 Photoacoustic flow cytometry (PAFC). (A) Schematic. (B) Typical photoacoustic (PA) response of pure blood and blood with melanoma cells. (C) Typical PAFC results for circulating tumor cells (CTCs) in vivo in a mouse ear. The CTCs were injected into a tail vein. (D) Representative PAFC results of CTCs in vivo in a tumor-bearing mouse. *a.u.*, Arbitrary unit. *Figure reproduced with permission from Nedosekin DA, Sarimollaoglu M, Ye JH, Galanzha EI, Zharov VP. In vivo ultra-fast photoacoustic flow cytometry of circulating human melanoma cells using near-infrared high-pulse rate lasers. Cytom Part A 2011;79A(10):825–33.*

to the urokinase plasminogen activator, which binds specifically to the urokinase plasminogen activator receptors that are expressed on many kinds of cancer cells. However, the potential toxicity of such an exogenous contrast agent hinders its applications in humans. Thus further improvement of PAFC is most likely to focus on increasing the sensitivity of label-free PAT to CTCs.

FURTHER POTENTIAL APPLICATIONS

Noninvasive Primary Tumor Examination

Invasive excision of a suspicious lesion followed by histological examination, has been the standard of care in tumor screening for a long time. Although high diagnostic accuracy can be achieved, problems do arise. For example, for patients with multiple suspicious lesions, multiple invasive excisions need to be performed, which is painful and may cause cosmetic concerns. In addition, for deeply seated lesions, precise excision is difficult without causing significant damage to the surrounding healthy tissue. Thus a noninvasive tumor screening method with high accuracy is preferred.

Because of the high correlation between tumor behavior and hemodynamics, such as angiogenesis and hyperoxia, measuring vasculature changes (both structural and functional) may provide noninvasive tumor screening. PAT, with high sensitivity to hemoglobin, can measure multiple vessel properties, including their diameter, flow rate, oxygen saturation, and metabolic rate of oxygen (MRO_2) [7,63–67]. Notably, MRO_2 in tumor tissue is much higher than in normal healthy tissue, because the cancer cells are multiplying rapidly. Although PAT has measured MRO_2 in mice, such measurement still has not been achieved in humans. One major challenge is measuring the flow rate in human blood vessels. Because human blood vessels are deeper than mouse vessels, current PA methods for in vivo flow measurement in mice cannot be translated to humans. In addition, recent methods proposed for deep flow measurements in PAT are still limited to phantoms [68,69]. Nevertheless, these problems are being tackled and a new cross-correlation–based method may achieve in vivo human blood flow measurement by increasing the detection time and thus provide a potential solution for noninvasive in vivo MRO_2 measurement and melanoma screening [70,71].

Label-Free Histology

The current standard for confirming melanoma is histological examination, which requires tissue fixing, dehydrating, clearing, infiltrating, embedding, sectioning, staining, and fluorescence or absorption imaging [72]. In addition to the long and complicated process, histological imaging examines only a portion of the target, which may lead to false diagnoses. Because the nuclei of cancer and normal cells have highly different morphologies, pathologists regard these characteristics as hallmarks of tumor grade. A label-free imaging modality that can detect cell nuclei may play a significant role in pathology. Because the main components in cell nuclei, DNA and RNA, have strong absorption in the ultraviolet (UV) band, PAT can image cell nuclei without labeling, via UV excitation [21,26]. Fig. 24.13A and C are PA images of cell nuclei taken at 266 nm in the small intestine of a mouse, and Fig. 24.13B and D are the corresponding histological images. The PA images show high correlation with standard histological measurements, with correlation coefficients of 0.88 between Fig. 24.13A and B, and 0.83 between Fig. 24.13C and D. With the ability to measure cell nuclei without staining, PAT may provide an intraoperative and real-time examination of the whole suspicious lesion and SLN, which would greatly decrease processing time and increase diagnostic accuracy.

Metastatic Tumor Detection

A positive SLN biopsy suggests that the melanoma has started metastasis locally or at a distance. To locate metastatic melanomas, routine imaging investigations, including CT, positron emission tomography (PET)/CT, and magnetic resonance imaging, are recommended [3]. Among these, PET/CT has the highest sensitivity in detecting metastases. However, its potentially carcinogenic ionizing radiation remains a concern.

Recently, nonionizing small-animal whole-body imaging has been achieved in vivo by using a ring-shape PACT system [73,74]. A 512-element full-ring transducer array with 5-MHz central frequency (80% bandwidth) and 50-mm ring diameter was used for PA signal detection. As shown in Fig. 24.14, detailed vasculature can be clearly seen within multiple organs, including the brain (see Fig. 24.14A), the liver (see Fig. 24.14B), and the kidneys (see Fig. 24.14C). In addition, the spinal cord, stomach, and gastrointestinal tract can also be detected by their surrounding microvasculature. Major blood vessels such as the vena cava are also clearly visible, as shown in Fig. 24.14B. In this study, light at 532 nm (see Fig. 24.14A) or 760 nm (see Fig. 24.14B–D) was used, and the main contrast was endogenous hemoglobin. For bladder imaging, an exogenous contrast, IRDye800, was injected into the mouse. Light at the dye's absorption peak, 776 nm, was used for illumination. As shown in Fig. 24.14D, the bladder can be clearly detected with strong contrast because it is filled with the dye. The whole body results show that PACT can image all the major organs, with

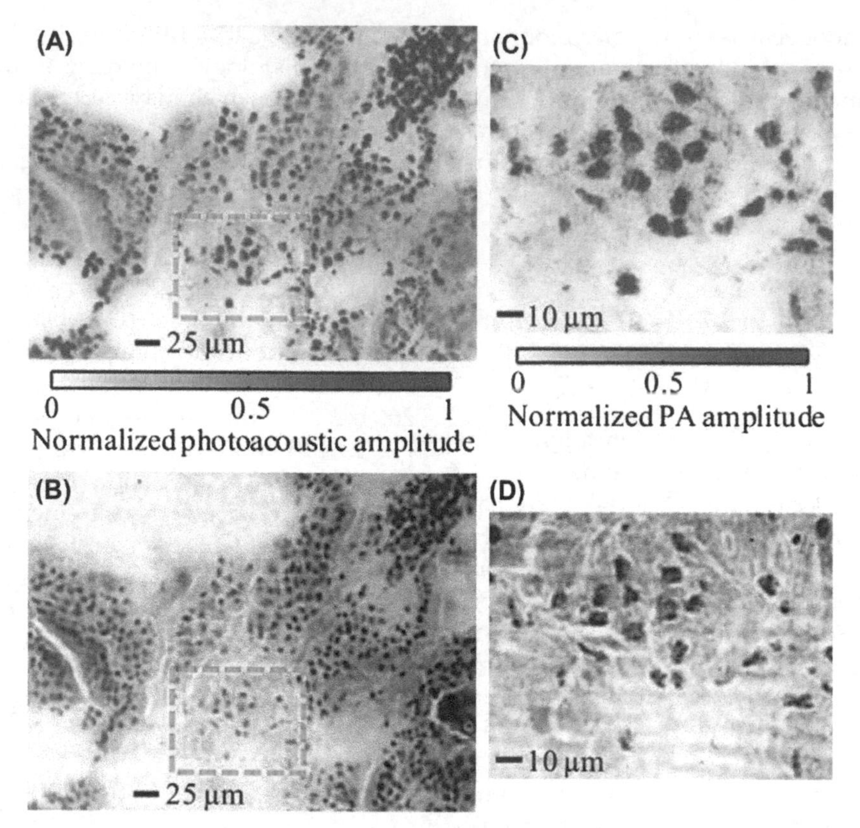

FIGURE 24.13 Label-free photoacoustic tomography (PAT) of cell nuclei in a mouse small intestine. Cell nuclei images acquired by PAT (A) and histology (B). (C and D) Close-ups of the area enclosed by *dashed lines* in (A and B) respectively. *Figure reprinted with permission from* Yao DK, Maslov K, Shung KK, Zhou QF, Wang LV. In vivo label-free photoacoustic microscopy of cell nuclei by excitation of DNA and RNA. Opt Lett 2010;35(24):4139–41.

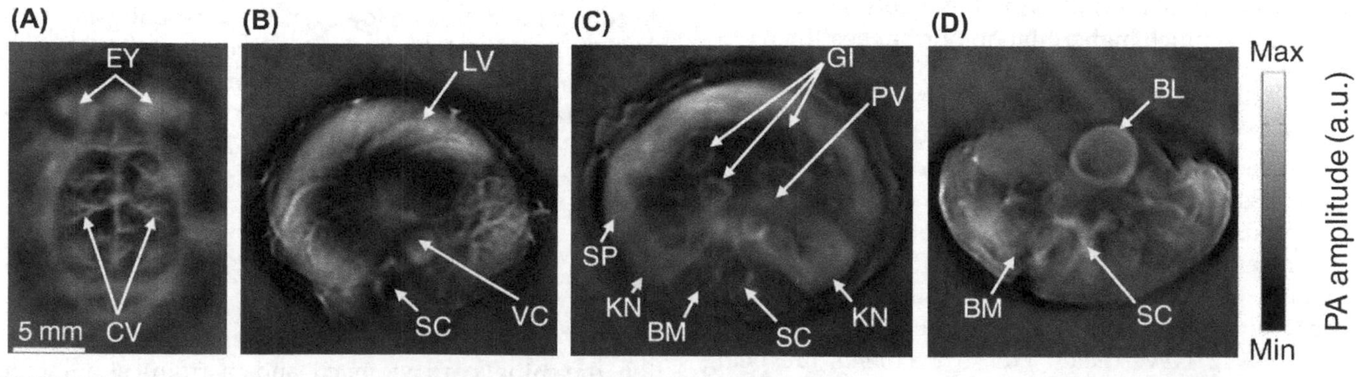

FIGURE 24.14 In vivo photoacoustic computed tomography (PACT) of a nude mouse at various anatomical locations. (A) The brain, (B) The liver, (C) The kidneys, and (D) The bladder. *a.u.*, Arbitrary unit; *BL*, bladder; *BM*, backbone muscle; *CV*, cortical vessels; *EY*, eyes; *GI*, gastrointestinal tract; *KN*, kidney; *LV*, liver; *PA*, photoacoustic; *PV*, portal vein; *SC*, spinal cord; *SP*, spleen; *VC*, vena cava. *Figure reprinted with permission from Xia J, Wang LV. Small-animal whole-body photoacoustic tomography: a review. IEEE Trans Biomed Eng 2014;61(5):1380–9.*

either endogenous hemoglobin contrast or exogenous dye contrast. Because melanin has a broad and strong absorption spectrum, PACT can image it with high contrast. Thus we can expect that whole-body imaging of both primary melanoma and its metastatic tumors will be achieved in the future.

CONCLUSIONS

In summary, PAT has been widely applied in the diagnosis of both primary and metastatic melanoma. Current applications include the measurement of melanoma thicknesses and rates of growth, detection of SLN location, label-free SLN examination, and melanoma CTC monitoring. Promising results have been shown, yet additional work is still needed to further mature and translate these techniques towards clinical demonstrations. Potential applications, including metastatic tumor detection, label-free histology, and primary tumor screening, would significantly improve the diagnosis of primary and metastatic melanoma. Although the potential applications have not been fully demonstrated, there are no fundamental obstacles. Moreover, because many tumors have similar behavior, all the applications discussed are potentially useful for diagnosing other tumors and their metastases.

Acknowledgments

The authors thank Professor James Ballard for close manuscript editing. This work was supported in part by National Institutes of Health grants DP1 EB016986 (NIH Director's Pioneer Award) and R01 CA186567 (NIH Director's Transformative Research Award). L.W. has a financial interest in Microphotoacoustics, Inc. and Endra, Inc., which, however, did not support this work.

References

[1] Siegel R, Ma JM, Zou ZH, Jemal A. Cancer statistics, 2014. Ca-Cancer J Clin 2014;64(1):9–29.
[2] Cancer Facts & Figures 2015. 2015 http://www.cancer.org.
[3] NCCN Guidelines Version 2. 2015 Melanoma. National Comprehensive Cancer Network; 2015.
[4] Mills JK, White I, Diggs B, Fortino J, Vetto JT. Effect of biopsy type on outcomes in the treatment of primary cutaneous melanoma. Am J Surg 2013;205(5):585–90.
[5] Zhang C, Maslov KI, Wang LV. Subwavelength-resolution label-free photoacoustic microscopy of optical absorption in vivo. Opt Lett 2010;35(19):3195–7.
[6] Favazza CP, Jassim O, Cornelius LA, Wang LV. In vivo photoacoustic microscopy of human cutaneous microvasculature and a nevus. J Biomed Opt 2011;16(1):016015.
[7] Wang LV, Hu S. Photoacoustic tomography: in vivo imaging from organelles to organs. Science 2012;335(6075):1458–62.
[8] Wang X, Pang Y, Ku G, Xie X, Stoica G, Wang LV. Noninvasive laser-induced photoacoustic tomography for structural and functional in vivo imaging of the brain. Nat Biotechnol 2003;21(7):803–6.
[9] Maslov KI, Zhang HF, Hu S, Wang LV. Optical-resolution photoacoustic microscopy for in vivo imaging of single capillaries. Opt Lett 2008;33(9):929–31.
[10] Maslov KI, Wang LV. Photoacoustic imaging of biological tissue with intensity-modulated continuous-wave laser. J Biomed Opt 2008;13(2):024006.
[11] Maslov KI, Stoica G, Wang LV. In vivo dark-field reflection-mode photoacoustic microscopy. Opt Lett 2005;30(6):625–7.
[12] Wang LV. Multiscale photoacoustic microscopy and computed tomography. Nat Photon 2009;3(9):503–9.
[13] Ku G, Wang LV. Deeply penetrating photoacoustic tomography in biological tissues enhanced with an optical contrast agent. Opt Lett 2005;30(5):507–9.
[14] Song KH, Stein EW, Margenthaler JA, Wang LV. Noninvasive photoacoustic identification of sentinel lymph nodes containing methylene blue in vivo in a rat model. J Biomed Opt 2008;13(5):054033.
[15] Yao J, Maslov KI, Hu S, Wang LV. Evans blue dye-enhanced capillary-resolution photoacoustic microscopy in vivo. J Biomed Opt 2009;14(5):054049.
[16] Kim C, Favazza CP, Wang LV. In vivo photoacoustic tomography of chemicals: high-resolution functional and molecular optical imaging at new depths. Chem Rev 2010;110(5):2756–82.
[17] Kim C, Cho EC, Chen JY, Song KH, An L, Favazza CP, et al. In vivo molecular photoacoustic tomography of melanomas targeted by bioconjugated gold nanocages. Acs Nano 2010;4(8):4559–64.
[18] Cai X, Li W, Kim CH, Yuan YC, Wang LV, Xia Y. In vivo quantitative evaluation of the transport kinetics of gold nanocages in a lymphatic system by noninvasive photoacoustic tomography. Acs Nano 2011;5(12):9658–67.
[19] Oh JT, Li ML, Zhang HF, Maslov KI, Stoica G, Wang LV. Three-dimensional imaging of skin melanoma in vivo by dual-wavelength photoacoustic microscopy. J Biomed Opt 2006;11(3):034032.
[20] Xu Z, Li C, Wang LV. Photoacoustic tomography of water in phantoms and tissue. J Biomed Opt 2010;15(3):036019.
[21] Yao DK, Maslov KI, Shung KK, Zhou QF, Wang LV. In vivo label-free photoacoustic microscopy of cell nuclei by excitation of DNA and RNA. Opt Lett 2010;35(24):4139–41.
[22] Wang HW, Chai N, Wang P, Hu S, Dou W, Umulis D, et al. Label-free bond-selective imaging by listening to vibrationally excited molecules. Phys Rev Lett 2011;106(23):238106.
[23] Chen Z, Yang S, Xing D. In vivo detection of hemoglobin oxygen saturation and carboxyhemoglobin saturation with multiwavelength photoacoustic microscopy. Opt Lett 2012;37(16):3414–6.
[24] Zhou Y, Zhang C, Yao DK, Wang LV. Photoacoustic microscopy of bilirubin in tissue phantoms. J Biomed Opt 2012;17(12):126019.
[25] Zhang C, Cheng YJ, Chen J, Wickline SA, Wang LV. Label-free photoacoustic microscopy of myocardial sheet architecture. J Biomed Opt 2012;17(6):060506.
[26] Zhang C, Zhang YS, Yao DK, Xia Y, Wang LV. Label-free photoacoustic microscopy of cytochromes. J Biomed Opt 2013;18(2):020504.
[27] Tang M, Zhou Y, Zhang R, Wang LV. Noninvasive photoacoustic microscopy of methemoglobin in vivo. J Biomed Opt 2015;20(3):036007.
[28] Zhou Y, Li G, Zhu L, Li C, Cornelius LA, Wang LV. Handheld photoacoustic probe to detect both melanoma depth and volume at high speed in vivo. J Biophotonics 2015;1(7).
[29] Zhang HF, Maslov KI, Stoica G, Wang LV. Functional photoacoustic microscopy for high-resolution and noninvasive in vivo imaging. Nat Biotechnol 2006;24(7):848–51.
[30] Zemp RJ, Bitton R, Li M, Shung KK, Stoica G, Wang LV. Photoacoustic imaging of the microvasculature with a high-frequency ultrasound array transducer. J Biomed Opt 2007;12(1):010501.
[31] Song L, Maslov K, Bitton R, Shung KK, Wang LV. Fast 3-D dark-field reflection-mode photoacoustic microscopy in vivo with a 30-MHz ultrasound linear array. J Biomed Opt 2008;13(5):054028.
[32] Kruger RA, Lam RB, Reinecke DR, Del Rio SP, Doyle RP. Photoacoustic angiography of the breast. Med Phys 2010;37(11):6096–100.
[33] Li C, Yang JM, Chen RM, Yeh C, Zhu L, Maslov KI, et al. Urogenital photoacoustic endoscope. Opt Lett 2014;39(6):1473–6.
[34] Yang JM, Chen RM, Favazza C, Yao J, Li C, Hu Z, et al. A 2.5-mm diameter probe for photoacoustic and ultrasonic endoscopy. Opt Express 2012;20(21):23944–53.
[35] Yang JM, Favazza C, Chen RM, Yao J, Cai X, Maslov KI, et al. Simultaneous functional photoacoustic and ultrasonic endoscopy of internal organs in vivo. Nat Med 2012;18(8):1297–302.
[36] Yang JM, Li C, Chen RM, Zhou QF, Shung KK, Wang LV. Catheter-based photoacoustic endoscope. J Biomed Opt 2014;19(6):066001.
[37] Wang LV, Wu H. Biomedical optics: principles and imaging. WILEY; 2007.

[38] Stein EW, Maslov KI, Wang LV. Noninvasive, in vivo imaging of the mouse brain using photoacoustic microscopy. J Appl Phys 2009;105(10):10202701–5.
[39] Zhang C, Zhou Y, Li C, Wang LV. Slow-sound photoacoustic microscopy. Appl Phys Lett 2013;102(16):163702.
[40] Zhang C, Maslov KI, Yao J, Wang LV. In vivo photoacoustic microscopy with 7.6-μm axial resolution using a commercial 125-MHz ultrasonic transducer. J Biomed Opt 2012;17(11):116016.
[41] Song KH, Wang LV. Deep reflection-mode photoacoustic imaging of biological tissue. J Biomed Opt 2007;12(6):060503.
[42] Song LA, Maslov KI, Shung KK, Wang LV. Ultrasound-array–based real-time photoacoustic microscopy of human pulsatile dynamics in vivo. J Biomed Opt 2010;15(2):021303.
[43] Song L, Kim C, Maslov KI, Shung KK, Wang LV. High-speed dynamic 3D photoacoustic imaging of sentinel lymph node in a murine model using an ultrasound array. Med Phys 2009;36(8): 3724–9.
[44] Zhou Y, Xing W, Maslov KI, Cornelius LA, Wang LV. Handheld photoacoustic microscopy to detect melanoma depth in vivo. Opt Lett 2014;39(16):4731–4.
[45] Staley J, Grogan P, Samadi AK, Cui HZ, Cohen MS, Yang XM. Growth of melanoma brain tumors monitored by photoacoustic microscopy. J Biomed Opt 2010;15(4):040510.
[46] Liu W, Dowling JP, Murray WK, McArther GA, Thompson JF, Wolfe R, et al. Rate of growth in melanomas: characteristics and associations of rapidly growing melanomas. Arch Dermatol 2006;142(12):1551–8.
[47] Lin M, Mar V, McLean C, Kelly J. Melanoma rate of growth measured using sequential biopsies. J Dtsch Dermatol Ges 2013; 11:98.
[48] Schadendorf D, Kochs C, Livingstone E. Handbook of cutaneous melanoma. Springer; 2013.
[49] Kim C, Erpelding TN, Maslov KI, Jankovic L, Akers WJ, Song L, et al. Handheld array-based photoacoustic probe for guiding needle biopsy of sentinel lymph nodes. J Biomed Opt 2010;15(4): 046010.
[50] Wang Y, Erpelding TN, Jankovic L, Guo Z, Robert JL, David G, et al. In vivo three-dimensional photoacoustic imaging based on a clinical matrix array ultrasound probe. J Biomed Opt 2012; 17(6):061208.
[51] Pan DPJ, Cai X, Yalaz C, Senpan A, Omanakuttan K, Wickline SA, et al. Photoacoustic sentinel lymph node imaging with self-assembled copper neodecanoate nanoparticles. Acs Nano 2012; 6(2):1260–7.
[52] Pramanik M, Song KH, Swierczewska M, Green D, Sitharaman B, Wang LV. In vivo carbon nanotube-enhanced non-invasive photoacoustic mapping of the sentinel lymph node. Phys Med Biol 2009; 54(11):3291–301.
[53] Song KH, Kim CH, Cobley CM, Xia Y, Wang LV. Near-infrared gold nanocages as a new class of tracers for photoacoustic sentinel lymph node mapping on a rat model. Nano Lett 2009; 9(1):183–8.
[54] Sanki A, Uren RF, Moncrieff M, Tran KL, Scolyer RA, Lin HY, et al. Targeted high-resolution ultrasound is not an effective substitute for sentinel lymph node biopsy in patients with primary cutaneous melanoma. J Clin Oncol 2009;27(33):5614–9.
[55] McCormack D, Al-Shaer M, Goldschmidt BS, Dale PS, Henry C, Papageorgio C, et al. Photoacoustic detection of melanoma micrometastasis in sentinel lymph nodes. J Biomech Eng 2009;131(7): 074519.
[56] Jose J, Grootendorst DJ, Vijn TW, Wouters MW, van Boven H, van Leeuwen TG, et al. Initial results of imaging melanoma metastasis in resected human lymph nodes using photoacoustic computed tomography. J Biomed Opt 2011;16(9):119801.
[57] Grootendorst DJ, Jose J, Wouters MW, van Boven H, Van der Hage J, Van Leeuwen TG, et al. First experiences of photoacoustic imaging for detection of melanoma metastases in resected human lymph nodes. Laser Surg Med 2012;44(7):541–9.
[58] Langhout GC, Grootendorst DJ, Nieweg OE, Wouters M, Van der Hage J, Jose J, et al. Detection of melanoma metastases in resected human lymph nodes by noninvasive multispectral photoacoustic imaging. Int J Biomed Imaging 2014;2014:163652.
[59] Nedosekin DA, Sarimollaoglu M, Ye JH, Galanzha EI, Zharov VP. In vivo ultra-fast photoacoustic flow cytometry of circulating human melanoma cells using near-infrared high-pulse rate lasers. Cytom Part A 2011;79A(10):825–33.
[60] Galanzha EI, Shashkov EV, Spring PM, Suen JY, Zharov VP. In vivo, noninvasive, label-free detection and eradication of circulating metastatic melanoma cells using two-color photoacoustic flow cytometry with a diode laser. Cancer Res 2009;69(20):7926–34.
[61] Galanzha EI, Shashkov EV, Kelly T, Kim JW, Yang LL, Zharov VP. In vivo magnetic enrichment and multiplex photoacoustic detection of circulating tumour cells. Nat Nanotechnol 2009;4(12):855–60.
[62] Galanzha EI, Shashkov EV, Kokoska MS, Myhill JA, Zharov VP. In vivo non-invasive detection of metastatic melanoma in vasculature and sentinel lymph nodes by photoacoustic cytometry. Laser Surg Med 2008:81.
[63] Zhou Y, Yi X, Xing W, Hu S, Maslov KI, Wang LV. Microcirculatory changes identified by photoacoustic microscopy in patients with complex regional pain syndrome type I after stellate ganglion blocks. J Biomed Opt 2014;19(8):086017.
[64] Zhou Y, Yao J, Maslov KI, Wang LV. Calibration-free absolute quantification of particle concentration by statistical analyses of photoacoustic signals in vivo. J Biomed Opt 2014;19(3):037001.
[65] Liang J, Zhou Y, Winkler AW, Wang L, Maslov KI, Li C, et al. Random-access optical-resolution photoacoustic microscopy using a digital micromirror device. Opt Lett 2013;38(15):2683–6.
[66] Yao J, Maslov KI, Zhang Y, Xia Y, Wang LV. Label-free oxygen-metabolic photoacoustic microscopy in vivo. J Biomed Opt 2011; 16(7):076003.
[67] Yao J, Wang LV. Transverse flow imaging based on photoacoustic Doppler bandwidth broadening. J Biomed Opt 2010;15(2):021304.
[68] Wang L, Xia J, Yao J, Maslov KI, Wang LV. Ultrasonically encoded photoacoustic flowgraphy in biological tissue. Phys Rev Lett 2013; 111(20):204301.
[69] Tay J, Liang J, Wang LV. Amplitude-masked photoacoustic wavefront shaping and application in flowmetry. Opt Lett 2014;39(19): 5499–502.
[70] Zhou Y, Liang J, Maslov KI, Wang LV. Calibration-free in vivo transverse blood flowmetry based on cross correlation of slow time profiles from photoacoustic microscopy. Opt Lett 2013; 38(19):3882–5.
[71] Liang J, Zhou Y, Maslov KI, Wang LV. Cross-correlation–based transverse flow measurements using optical resolution photoacoustic microscopy with a digital micromirror device. J Biomed Opt 2013;18(9):096004.
[72] Ghaznavi F, Evans A, Madabhushi A, Feldman M. Digital imaging in pathology: whole-slide imaging and beyond. Annu Rev Pathol-Mech 2013;8:331–59.
[73] Xia J, Wang LV. Small-animal whole-body photoacoustic tomography: a review. IEEE Trans Biomed Eng 2014;61(5):1380–9.
[74] Xia J, Chatni MR, Maslov KI, Guo Z, Wang K, Anastasio M, et al. Whole-body ring-shaped confocal photoacoustic computed tomography of small animals in vivo. J Biomed Opt 2012;17(5): 050506.

CHAPTER

25

Ultrasound Imaging in Dermatology

X. Wortsman

University of Chile, Santiago, Chile

OUTLINE

INTRODUCTION

According to PubMed indexed literature, the first use of ultrasound in dermatology was reported by Meyer et al. in the year 1951 [1]. In 1979, Miller et al. published the use of pulsed ultrasound in dermatology [2]. During the 80s and 90s, the development of high-frequency probes allowed several groups of researchers from different countries to explore more clinical applications of ultrasound such as tumors and inflammatory conditions, as well as to assess some sonographic variations according to age [3–8]. In 2004, the first experience on the use of variable high-frequency ultrasound probes in dermatological conditions was published [9] and since then, there has been an explosive growth of applications that include common dermatological conditions such as benign and malignant tumors, inflammatory diseases, and nail and scalp entities, as well as cosmetic applications. The higher the frequency of the probes, the lower the penetration. Therefore in contrast with fixed high-frequency probes, variable frequency probes present the advantage of defining both the skin and deeper layers with high definition and additionally show the blood flow patterns of the tissue through their color Doppler application.

The aim of the sonographic examination is to provide relevant anatomical data, different from and complementary to that already deduced by the naked eye of a

Imaging in Dermatology
http://dx.doi.org/10.1016/B978-0-12-802838-4.00025-X

well-trained physician [10]. This ultrasound information should ideally be capable of supporting diagnosis and guiding management. Moreover, the imaging data may be used for achieving an early diagnosis and improving the cosmetic prognosis by potentially decreasing serial biopsies and increasing diagnostic precision.

The advantages of ultrasound are its real-time capacity, the good balance between resolution and penetration (resolution:100 μm/pixel axial and 90 μm/pixel lateral; penetration: 0.1–60 -mm depth), the provision of qualitative and quantitative data on the tissue and its blood flow, the multiaxial capability of observation that includes depth, the lack of effects from radiation, and the no confinement of the patient to a reduced space [10,11]. This potential for deep penetration without losing definition is an advantage when compared with other imaging modalities used in dermatology such as confocal microscopy (penetration ≤0.3 mm) or optical coherence tomography (penetration ≤2 mm) [10]. The latter point may be critical when studying malignant skin tumors because imaging devices that present shallow penetration may underestimate the depth of the lesion. Moreover, the commercially available computed tomography (CT), positron emission tomography-CT (PET-CT) and magnetic resonance imaging (MRI) equipment do not show good sensitivity for discriminating the skin layers, especially in superficial lesions measuring ≤ 5 mm [10]. Nevertheless, CT, PET-CT and MRI may support the study of advanced skin tumors and their staging. On the other hand, ultrasound can show common primary dermatological conditions with high definition as well as support a locoregional staging.

Currently, the use of ultrasound in dermatology is limited by the observation of epidermal-only lesions measuring ≥0.1 mm and its inability to detect pigments such as melanin [12]. Also it should be kept in mind that ultrasound does not see structures obscured by air or bone. Thus sonography may be limited, for example, for the study of deep retropharyngeal lymph nodes or brain tissue [13]. However, sonography may characterize superficial lesions, detect the mass effect in pigmented tumors, and study soft-tissue lymph-node chains. Other potential limitation is that this imaging method requires both an operator trained in dermatological conditions and an adequate device, which may be not available in all institutions. This could be also limited in places where the acquisition of the data is performed by one operator (for example, a technician) and the report is written by another (for example, a medical doctor). In such applications, it is strongly recommended that the individual in charge of the report should be aware of the clinical appearance of the lesion and the history of the patient [10–12,14].

The sonographic report may provide the information on the nature of the lesion (solid or cystic), the actual extent in all axes (cm or mm), the anatomical layers involved, the lesional and perilesional vascularity (hypovascular or hypervascular, thick afferent or efferent vessels), the blood flow characteristics (arterial, venous, or capillary vessels), the thickness and velocity of the vessels (cm/s), the relevant neighboring structures (ie, muscles, cartilage, glands, vessels), the stage of the condition (eg, proliferative, partial regression, total regression; active, inactive or atrophy phase), the severity [through the assessment of the degree of involvement and/or sonographic scorings (SOS)], whether the lesion is suggestive of benignancy or malignancy, and, lastly, some differential diagnoses through the assessment of echostructural patterns [10–16]. Because of its ability to identify various aforementioned pathophysiological processes, ultrasound has become a widely used imaging technique in the clinical setting.

TECHNICAL CONSIDERATIONS

The ideal requirements for performing a dermatological ultrasound examination are multichanneled equipment with a ≥15-MHz compact linear or linear high-frequency probe(s) and an operator trained in dermatological pathologies and the ultrasound technique. In children ≤4 years old, for sedation purposes, commonly chloral hydrate (50 mg/kg) may be orally administered 30 min before the examination [10–12]. This would prevent the development of artifacts in the screen derived from movement or crying of the child. Because ultrasound is usually performed in a low-light room, management of the lights in the examination room should be carried out when dealing with multiple lesions so that the lights are turned on to locate the probe properly in all lesions. All examinations are performed under a protocol that includes grayscale and color Doppler ultrasound (for studying vascularity) with spectral curve analysis of the vessels (for detecting the type of flow, arterial or venous, and their velocity). The studies are performed in at least two perpendicular axes, and provide the echostructure of the tissue [anechoic (ie, echoes absent: black); hypoechoic (ie, low intensity echoes: gray), hyperechoic (ie, high intensity echoes: white) or isoechoic (ie, similar echoes)] in comparison with the neighboring structures.

The lesion is measured in all axes, as is the thickness and velocity (cm/s) of the vessels. Any relevant structure in the vicinity and its spatial relation with the lesion is reported, which includes thick vessels or neural bundles close to the lesion. The involvement of deep structures such as muscle, tendons, or glands is also mentioned. Therefore besides the description of the type and nature of the lesion (ie, solid or cystic, hypovascular or hypervascular), an anatomical "GPS-like map" is provided to the clinician, who can use this information for planning a surgery or monitor a treatment. In addition, this objective noninvasive information can be used in research and clinical trials [10–16].

Thus sonography has been proven useful for discriminating lesional from nonlesional tissue, dermatological from pseudodermatological conditions, and endogenous from exogenous components (eg, fillers or foreign bodies) [12].

Like any other imaging modality, ultrasound requires formal training of the operator for distinguishing the common patterns of presentation of dermatological entities. In addition, close work with the clinical and dermatopathological team is recommended. In the presence of these ideal conditions, and with the knowledge of its limitations, sonography can show high accuracy for diagnosing and monitoring dermatological entities and may become the primary imaging technique in dermatological practice [12–16].

In the following sections, the anatomy and most common dermatological applications of ultrasound are described.

NORMAL ANATOMY

Skin

The sonographic appearance of the skin layers varies according to the corporal site. Thus the nonglabrous skin (ie, not that of the palms and soles) shows a monolaminar bright hyperechoic epidermis whose echogenicity is mainly provided by the keratin content in the stratum corneum. The dermis appears as a hyperechoic band, less bright than the epidermis. The dermal echogenicity is mainly provided by the collagen. With age, a subepidermal hypoechoic band appears in the upper dermis of sun-exposed regions, called a *subepidermal low-echogenicity band,* which corresponds to the deposit of glycosaminoglycans in the skin, a sonographic sign of photoaging. The hypodermis, also called *subcutaneous tissue,* shows as a hypoechoic layer with hyperechoic septa, and this echogenicity is provided by fatty lobules with fibrous septa in between. The glabrous skin (ie, palms and soles) presents a thicker epidermis that shows as a bilaminar bright hyperechoic layer because of its more prominent keratin content. The thickness of the dermis varies according to the region, being thinner in the face and ventral forearm, and thicker in the dorsum. This may be relevant if we are dealing with a facial skin tumor, because it can be easier to infiltrate deeper layers.

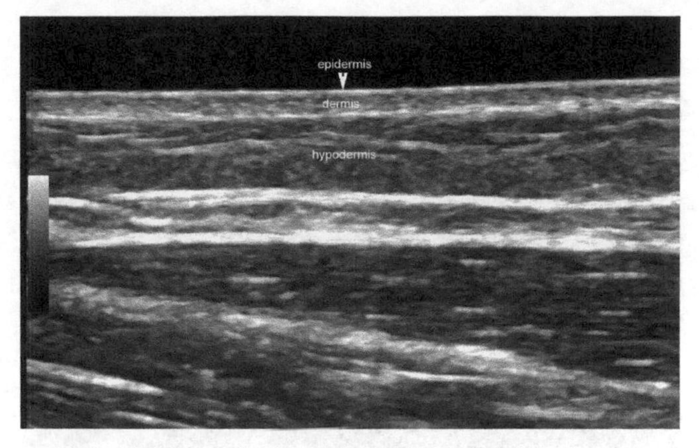

FIGURE 25.1 Normal sonographic anatomy of the skin.

Usually low-velocity arterial and venous vessels run through the hypodermis [10–12,16,17] (Fig. 25.1).

Nail

The nail unit is composed of the ungual and periungual regions and comprises an enthesis closely related to the distal insertion of the extensor tendon and the distal interphalangeal joint. Thus the nail or ungual region shows a bilaminar hyperechoic layer in the surface that corresponds to the ungual plate. This laminar structure shows an outer layer called the *dorsal plate* and an inner layer named the *ventral plate.* In between these plates there is an anechoic or hypoechoic virtual interplate space. Underlying the nail plate there is a hypoechoic region called the *nail bed.* The matrix region is located in the proximal area of the nail bed, which tends to show hyperechogenicity in comparison with the rest of the nail bed. At the bottom there is a hyperechoic line that corresponds to the bony margin of the distal phalanx. The periungual region presents the appearance of nonglabrous skin in the dorsum and of glabrous skin in the pulp of the finger. On color Doppler examination, it is possible to detect the low flow arterial and venous vessels located in the nail bed that run on top of the bony margin of the distal phalanx. The distal insertion of the extensor and flexor tendons present a hyperechoic fibrillar pattern that may turn to slightly hypoechoic as a result of the anisotropic artifact caused by the oblique axis of the tendons [10,11,18,19] (Fig. 25.2).

Hair

The hair presents two main parts: the hair follicle located in the dermis and the hair tract, or shaft, on the surface. The hair follicles appear as hypoechoic oblique dermal bands (Fig. 25.3), and the hair tracts show a mostly trilaminar hyperechoic appearance in the frontal

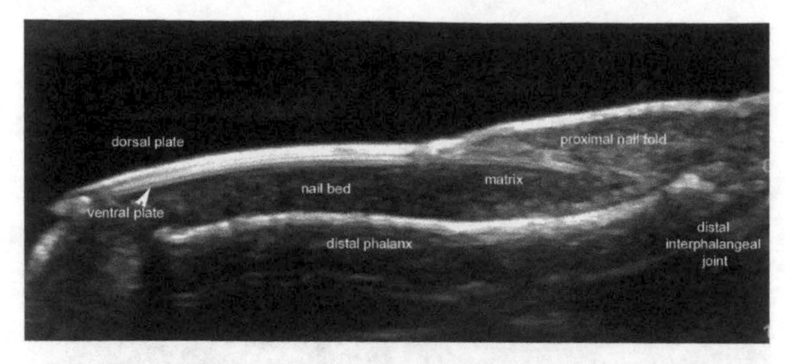

FIGURE 25.2 Normal sonographic anatomy of the nail.

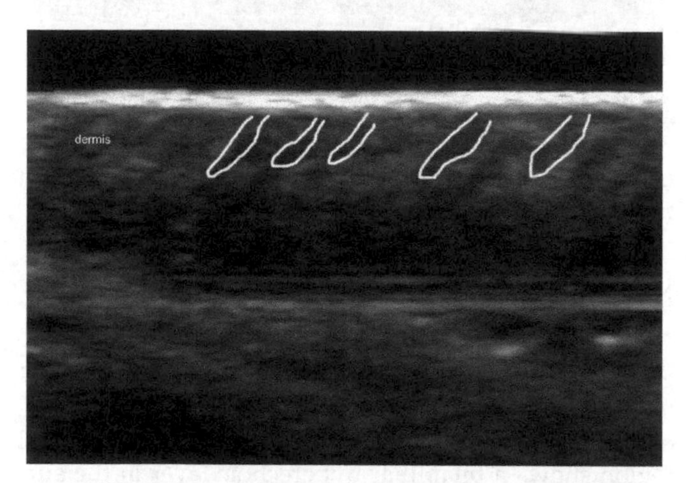

FIGURE 25.3 Normal hair follicles (*outlined*) in dermis.

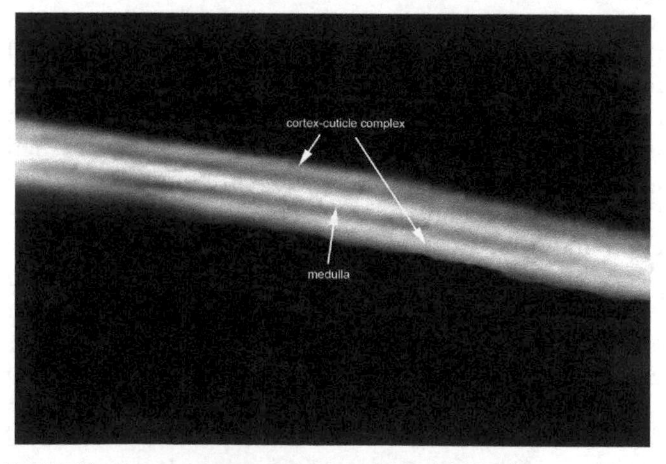

FIGURE 25.4 Normal sonographic anatomy of a scalp hair tract (frontal region). Notice the trilaminar hyperechoic appearance.

region (Fig. 25.4). However, in the occipital region a mixed pattern with 80% trilaminar and 20% bilaminar hair tracts has been reported. These bilaminar hair tracts might correspond with the villus type of hair, also present in the eyelashes and eyebrows and in the rest of the body. Nevertheless, in these latter corporal segments, the pattern of the hair shows as a monolaminar hyperechoic layer. It is also possible to detect low-flow arterial and venous vascularity in the hypodermis underlying the hair follicles [11,20–24].

COMMON APPLICATIONS

There are several applications of ultrasound in common dermatological conditions. For academic purposes, the topics have been separated into the following sections.

Tumors

Benign Tumors

Cystic Tumors

Epidermal Cyst Epidermal cysts are common causes of referral for an ultrasound examination because they tend to become inflamed and rupture. Histologically they are composed of the implantation of epidermal elements in the dermis and/or hypodermis. They show stratified squamous epithelium with a granular layer and contain keratin. Upon rupture this keratinous material is spread in the periphery, causing a foreign body type of inflammatory reaction. On ultrasound the appearance of epidermal cysts differs according to the phase. If the cyst is intact it shows as a well-defined, round or oval-shaped, anechoic structure located in the dermis and/or hypodermis with posterior acoustic enhancement artifact (ie, hyperechoic white reverberation deep to the structure that is typically seen in fluid-filled conditions). The cyst may present an anechoic tract connecting to the subepidermal region called *punctum*. Giant epidermal cysts may also turn hypoechoic with anechoic filiform bands because of compacted keratin deposits and a higher presence of cholesterol crystals, which has been called a "pseudotestis" appearance because of the resemblance to the sonographic pattern of the testes. During the inflammatory process the cyst may increase its inner echogenicity and peripheral blood flow. On rupture, the borders of the cyst become irregular and this hypoechoic keratinous material can be detected in the periphery, usually producing a "foreign body-like" reaction. A key sonographic sign for diagnosing epidermal cysts is to look for the posterior acoustic enhancement artifact typically seen in cystic structures [10,11,25–27] (Fig. 25.5–25.7).

Pilonidal Cyst This cyst is typically seen in individuals exposed to chronic pressure on the sacral region and thought to be caused by ingrown hair. The classic location of these cysts is the intergluteal region. It was named "Jeep disease" because of its presence in the soldiers of World War II. It is composed of a nest of hyperechoic linear tracts that correspond to fragments of hair tracts trapped in the dermis and hypodermis surrounded by hypoechoic inflammatory tissue and mature squamous epithelium (detected on histology). Usually, these cysts connect to the base of widened regional

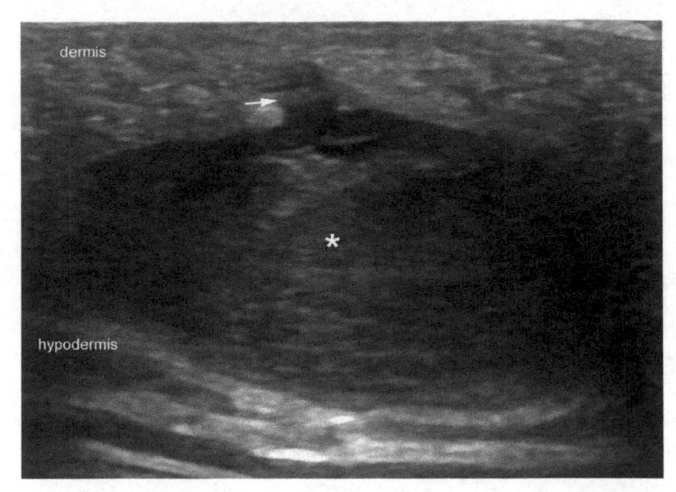

FIGURE 25.5 **Intact epidermal cyst.** Ultrasound (grayscale, transverse view) shows dermal and hypodermal hypoechoic oval-shaped structure (*). Notice the communicating hypoechoic tract (*arrow*) from the cyst to the upper dermis.

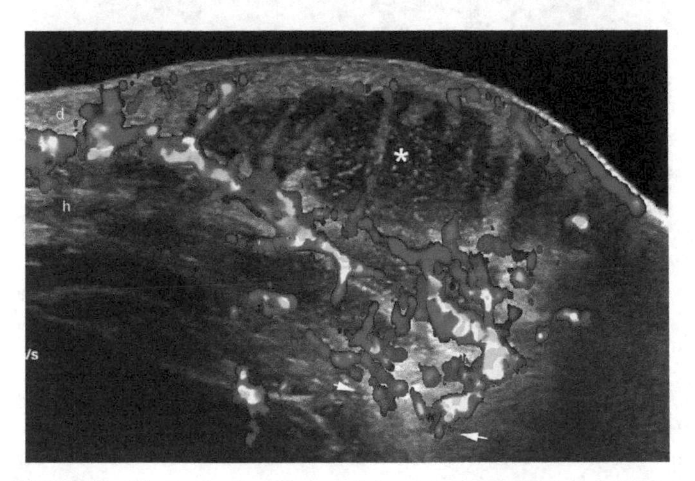

FIGURE 25.6 **Ruptured and inflamed epidermal cyst.** Color Doppler ultrasound (longitudinal view) demonstrates ill-defined, hypoechoic structure (*) in the dermis and hypodermis. Prominent vascularity (*colors*) is detected in the periphery of the cyst. *d*, Dermis; *h*, hypodermis.

dermal hair follicles. On color Doppler, it is possible to detect the hypervascularity in the periphery caused by frequent inflammation and/or infection, which can turn the cyst into an abscess. Sonography can demonstrate the nature, actual axis, and extent of the cyst, which may support surgical planning [10,11,28,29] (Fig. 25.8).

Solid Tumors

Pilomatrixoma This benign tumor is also called *calcifying epitheliomas of Malherbe*, or *pilomatricoma*. It is most often seen on the face and extremities of children and young adults. The rate of error reported in the clinical diagnosis of pilomatrixomas has been as high as 56% [10,11,30]; therefore ultrasound has an active role in the

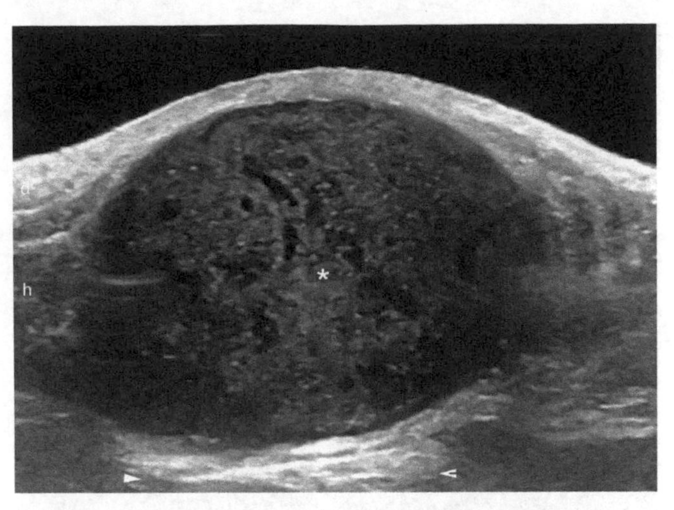

FIGURE 25.7 **Epidermal cyst with "pseudotestis" appearance.** Ultrasound (grayscale, transverse view, color filter) shows a well-defined, oval-shaped hypoechoic structure with anechoic bands. The cyst is located in the hypodermis and slightly emerges into the dermis. Notice the posterior acoustic reinforcement artifact (*arrowheads*) typically seen in the fluid-filled structures. *d*, Dermis; *h*, hypodermis.

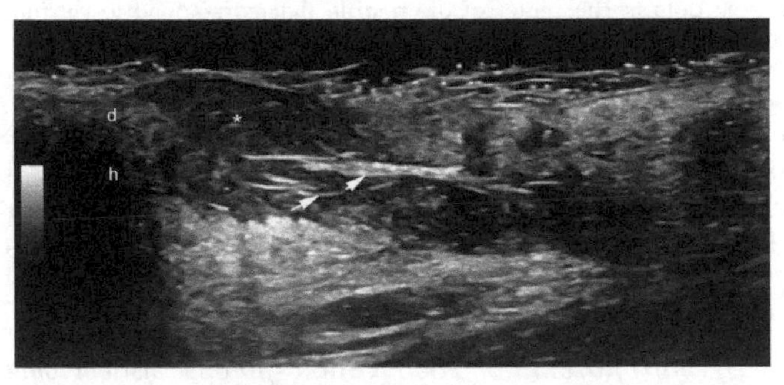

FIGURE 25.8 **Pilonidal cyst.** Ultrasound (grayscale, longitudinal view, color filter) demonstrates dermal and hypodermal hypoechoic structure (*) with hyperechoic lines that correspond to hair tract fragments (*arrows*). *d*, Dermis; *h*, hypodermis.

diagnosis. Histologically, pilomatrixomas contain basaloid and ghost cells, eosinophilic keratinous debris, and calcifications. On sonography the classical appearance is the "target type" that presents a hypoechoic rim and hyperechoic center with multiple hyperechoic dots that correspond to the calcium deposits. According to the size and degree of calcification, these tumors may show posterior acoustic shadowing (ie, anechoic black band at the bottom of the calcified structure owing to the blocking of the passage of the sound waves by the calcium deposits). Some pilomatrixomas contain low levels of calcium and others are completely calcified. Usually, pilomatrixomas are located in the dermis and hypodermis and commonly show vascularity in the center and periphery with slow-flow arterial and venous vessels. Occasionally these tumors may present hypervascularity and clinically mimic a vascular tumor. Also they may show internal bleeding and show as an

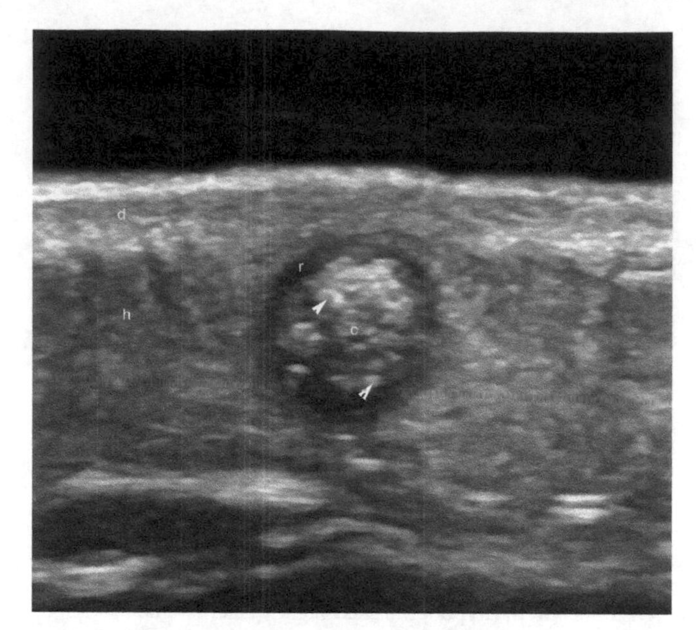

FIGURE 25.9 **Pilomatrixoma.** Ultrasound (grayscale, transverse view) shows target type of nodule with a hypoechoic rim (*r*) and hyperechoic center (c) located in the hypodermis. Notice the hyperechoic dots in the center of the nodule that correspond to calcium deposits. *d*, Dermis; *h*, hypodermis.

anechoic cystic tumor with an eccentric hypoechoic nodule. Thus ultrasound can be critical in these unusual clinical presentations, and the key sonographic sign for diagnosing this entity is the presence of calcifications [10,11,29–33] (Fig. 25.9).

Dermatofibroma Also called *fibrous histiocytoma* and *histiocytoma cutis*, this fibrous tumor is most commonly seen in the trunk and extremities of women. It is not clear if their origin is a reaction to trauma such an insect bite or is an actual neoplastic disorder. However, they tend not to regress over time. On sonography, dermatofibromas appear as ill-defined hypoechoic and heterogeneous dermal structures that usually cause distortion and enlargement of the regional hair follicles. Vascularity may be variable and they tend to show thin, slow-flow arterial and venous vessels; however, some of them can be hypovascular with no detectable blood flow on color Doppler. Deposits of calcium have not been reported in dermatofibromas [11,28] (Fig. 25.10).

Lipoma These are the most common soft-tissue tumors and are composed of mature fatty cells. The name of the tumor varies according to the tissue that accompanies these fatty cells. Thus if fibrous tissue is present, the tumor is named *fibrolipoma*, and if capillary vessels are attached to the fatty cells, the tumor is named *angiolipoma*. In most cases, these tumors are single; however, they may be multiple. On sonography, the presentation also changes according to the accompanying

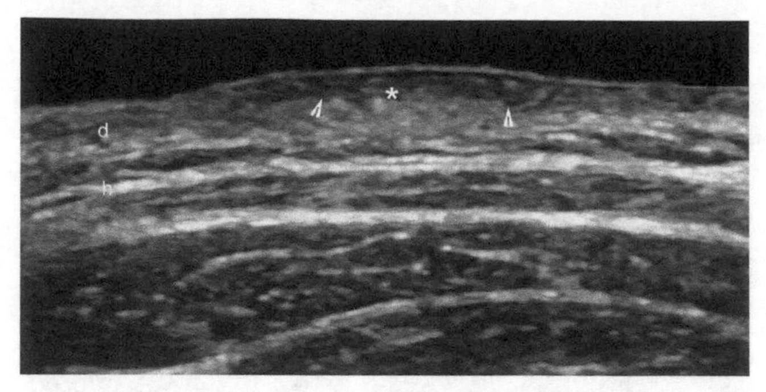

FIGURE 25.10 **Dermatofibroma.** Ultrasound (grayscale, transverse view, color filter) demonstrates ill-defined hypoechoic thickening of the dermis (*) with distortion and widening of the regional hair follicles (*arrowheads*). *d*, Dermis; *h*, hypodermis.

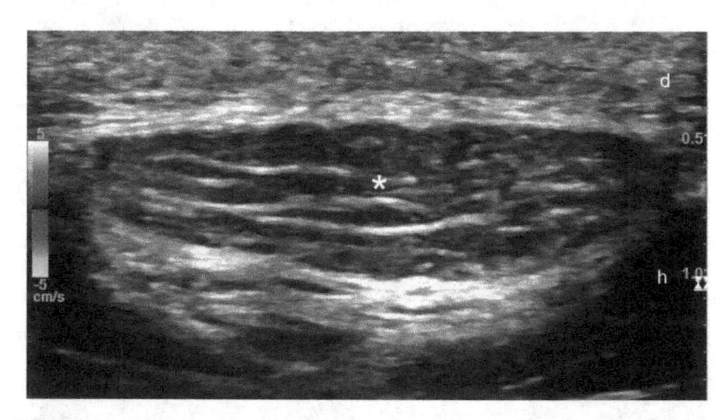

FIGURE 25.11 **Lipoma.** Ultrasound (color Doppler, transverse view) shows well-defined, oval shaped, hypoechoic and hypovascular hypodermal structure (*) with hyperechoic septa. *d*, Dermis; *h*, hypodermis.

tissue; thus fibrolipomas tend to show hypoechogenicity and angiolipomas present hyperechogenicity. However, in general, lipomatous tumors appear as well-defined, round or oval-shaped structures located in the dermis and/or hypodermis that commonly present hyperechoic septa and follow the axis of the skin layers (Fig. 25.11). Although lipomas are benign entities, they can be in risky locations such as the neck or temple regions, where they can be close to thick vessels or nerves. Sometimes these tumors can be found under the epicranius muscle in the frontal region; they are called *subgaleal lipomas* and mimic a cyst or an exostosis. Commonly they are hypovascular, and the presence of heterogenicity and hypervascularity within the tumor would raise the suspicion of a malignant transformation [28,34–36].

Malignant Tumors

Nonmelanocytic Skin Cancer

This is the most common cancer in humans and often affects areas of the body highly exposed to the sun such as the face. Nonmelanocytic skin cancer (NMSC) is composed of basal cell carcinoma (BCC) and squamous cell carcinoma (SCC), with BCC being the most common type. Although NMSC is usually not lethal, it can be highly disfiguring. The correlation of tumor depth between ultrasound and histology has been reported as excellent [37].

Both NMSC types show as hypoechoic and/or heterogeneous oval-shaped lesions with irregular or lobulated borders, commonly located in the dermis. However, BCC classically presents hyperechoic spots within the tumor and the density of these spots seems to be related to the high or low risk of recurrent histological subtypes [10,11,37–41]. BCCs with higher numbers of hyperechoic spots (≥7) have been reported to correlate well with more aggressive histological subtypes such as micronodular, morpheiform, sclerosing, infiltrating, and metatypical subtypes [42]. SCCs do not show hyperechoic spots, tend to be more infiltrating than BCCs, and extend to the underlying muscle or cartilage. Ultrasound can detect the NMSC lesion, define its extent, including depth, and support the prediction of type (BCC or SCC) and also the subtype (high or low risk of recurrence of BCC), as well as demonstrate the involvement of deeper layers. On color Doppler, NMSC shows the vascularity within the lesions with slow-flow arterial and venous vessels, with SCC usually being more hypervascular than BCC [40,41] (Figs. 25.12 and 25.13).

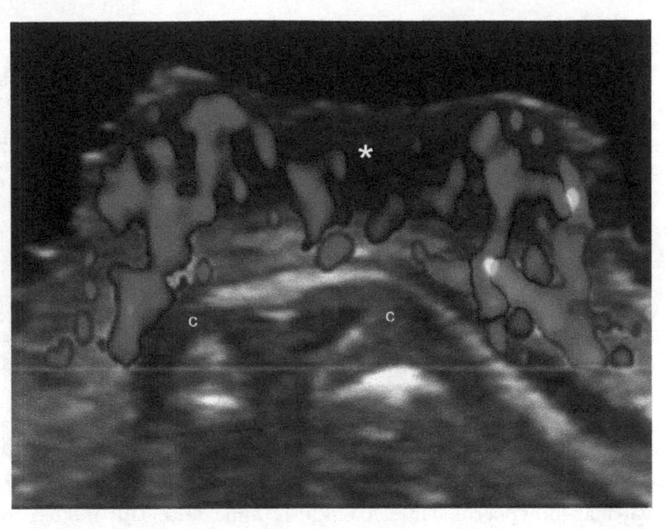

FIGURE 25.13 **Squamous cell carcinoma (SCC).** Color Doppler ultrasound (transverse view, tip of the nose) shows hypervascular (*colors*) hypoechoic mass (*) that involves the dermis. The nasal cartilages (c) are unremarkable.

Melanoma

Despite the fact that ultrasound cannot detect pigments such as melanin, sonography may show the solid component and depth of the lesion.

Primary Tumor Melanoma is the most aggressive type of skin cancer; however, it is not the most common type. Its recurrence has been reported to be as high as 46.1% in melanomas of the head and neck region [43,44]. The prognosis is given by the depth of tumoral invasion (Breslow index). Therefore the support of sonography may provide relevant anatomical data and provide a sonographic version of this index. It has also been reported that ultrasound may support the differentiation between melanomas that measure less or more than 1 mm [45,46]. This may support critical decisions such as the noninvasive assessment of the free margins in surgery, the size of the excision, and the performance of a sentinel node procedure, which is usually indicated in melanomas measuring greater than 1 mm. On ultrasound, melanomas show as well-defined, hypoechoic and/or heterogeneous, oval-shaped, and commonly fusiform structures located in the dermis and/or hypodermis. Often, they present high vascularity because of their high angiogenic tendency and also may affect deeper layers [10,11,44] (Fig. 25.14). It has been reported that the degree of malignancy may correlate well with the degree of vascularity, which has been mostly assessed through color Doppler and contrast-enhanced ultrasound studies. In addition, ultrasound may assist in the diagnosis of amelanotic melanomas [47–50].

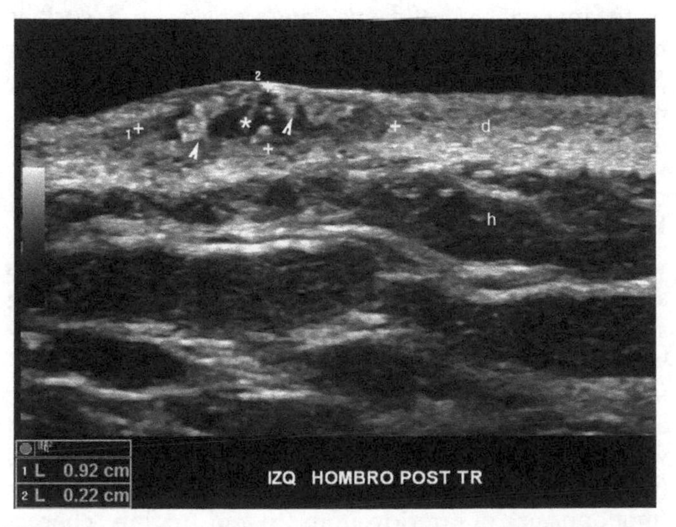

FIGURE 25.12 **Basal cell carcinoma (BCC).** Ultrasound (grayscale, transverse view, left dorsum) demonstrates 9.2-mm transverse × 2.2-mm depth oval-shaped, hypoechoic dermal lesion (*) with hyperechoic spots (*arrowheads*). *d*, Dermis; *h*, hypodermis.

Locoregional Staging It has been reported that ultrasound is much more sensitive than clinical examination alone for diagnosis of melanoma metastasis [51]. Sonography has been reported as sensitive for diagnosing satellite (<2 cm from the primary tumor), in-transit (≥2 cm from the primary tumor), and nodal metastasis [10,11,51–53]. Satellite or in-transit metastases show on ultrasound as hypoechoic nodules sometimes with irregular borders and hyperechogenicity of

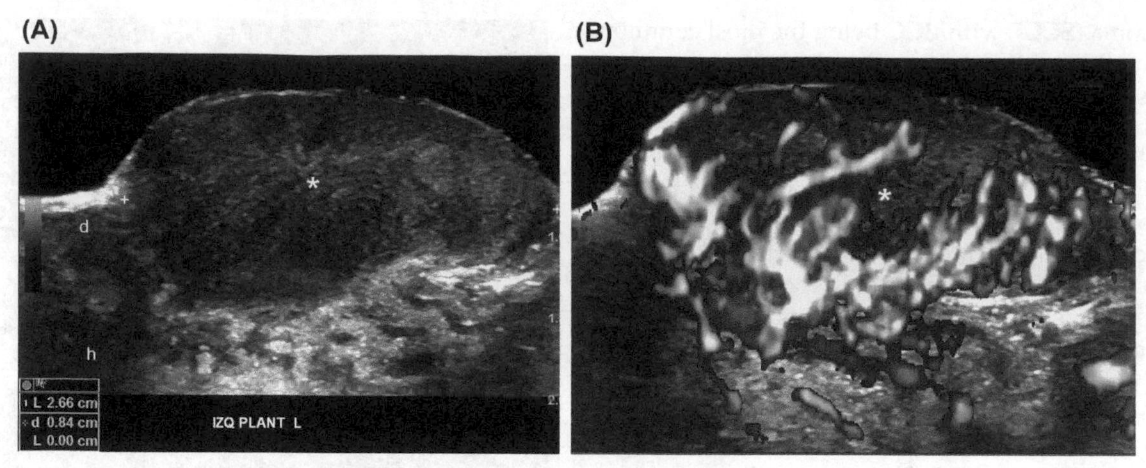

FIGURE 25.14 **Melanoma (longitudinal views, plantar region).** (A) Top (grayscale), and (B) bottom (power Doppler) demonstrate fusiform oval-shaped hypoechoic mass (*) located in the dermis (d) and hypodermis (h). Notice the strong vascularity (B, *color*) within the mass.

the surrounding hypodermis caused by perilesional edema [52,54–57]. Anechoic areas may also be detected within these lesions, most probably because of increased transmission of the sound in hypercellular regions [58]. These metastatic sites commonly show hypervascularity with thick, tortuous vessels [52]. There are certain sonographic signs in the lymph nodes that may suggest benign or malignant nature (Table 25.1) [52]. The locoregional staging should follow the lymphatic drainage chains according to the anatomical location of the primary tumor. Ultrasound may guide percutaneous fine-needle aspiration, cytology, or biopsy of the lymph nodes [59–62].

Vascular Anomalies

These are common causes of referral in the pediatric population. Hemangiomas and vascular malformations present different clinical origins and evolution, and they also differ in their treatment and prognosis. Therefore differentiation of these two entities can be of paramount importance for proper management.

TABLE 25.1 Sonographic Differences Between Benign and Malignant Lymph Nodes

Morphology	Benign	Malignant
Shape	Oval	Round
Center	Hyperechoic	Hypoechoic/anechoic
Cortex thickening	Diffuse	Nodular, eccentric
Ratio longitudinal/transverse	>2	<2
Vascularity	Regular, central	Irregular, peripheral
Cortex vessels	Absent/few	Present/prominent

Hemangiomas

These vascular benign tumors are the most common soft-tissue lesions in the pediatric population and tend to grow rapidly during the first year and then start a slow involution period. On sonography, it is possible to detect and measure the hemangioma, as well as define its phase of progression. Usually, hemangiomas in the proliferative phase show as an ill-defined hypoechoic mass with prominent blood flow that presents arterial and venous vessels and some arteriovenous shunts (Fig. 25.15A and B). In the partial regression phase, the echogenicity becomes mixed and the hemangioma presents hypoechoic and hyperechoic regions. The vascularity also decreases in this phase. In the total regression phase, the hemangioma turns hyperechoic and hypovascular [63–65]. Ultrasound may support the follow-up course of treatment and provide anatomical information on the involvement of deeper layers or structures such as muscle, cartilage, glands, or the ocular globe. The latter data may be critical in hemangiomas located in the face region [63]. Finally, sonography allows noninvasive monitoring of the propranolol treatment [66].

Vascular Malformations

Vascular malformations are errors of morphogenesis and they do not compose an actual tumor. They can be subgrouped according to the type of vessel (arterial, venous, capillary, or lymphatic) involved and as high- (arterial or arteriovenous) or low-flow (venous, capillary, or lymphatic). Color Doppler ultrasound with spectral curve analysis can show the arterial (with systolic and diastolic component), venous (monophasic), or arteriovenous (to-and-fro flow) characteristics (Fig. 25.16). Most vascular malformations show as a nest of anechoic tubular structures or lacunar spaces. Capillary vascular malformations may show

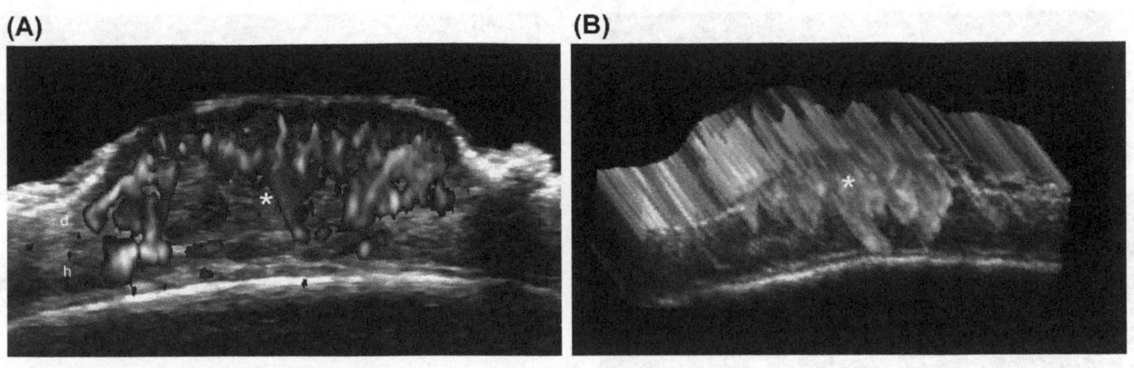

FIGURE 25.15 **Hemangioma in proliferative phase (transverse view, scalp).** (A) Power Doppler ultrasound. (B) Three-dimensional power Doppler reconstruction. Both show an ill-defined, dermal and hypodermal, hypoechoic, and hypervascular mass (*). *d*, Dermis; *h*, hypodermis.

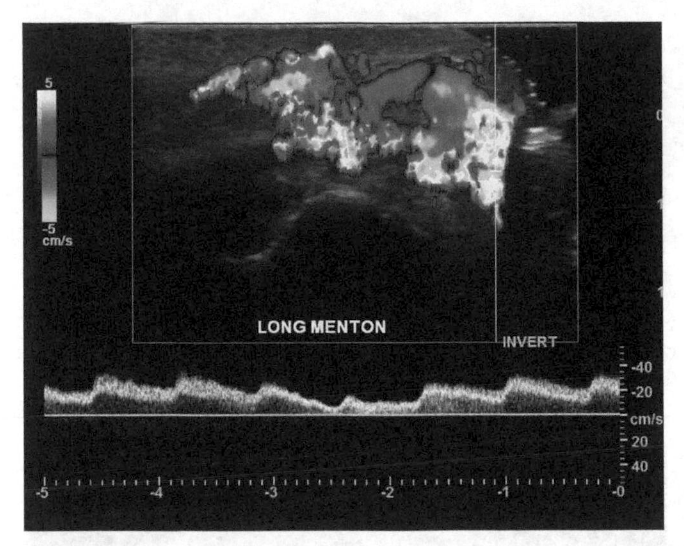

FIGURE 25.16 **High-flow arteriovenous vascular malformation (longitudinal view chin region).** Color Doppler with spectral curve analysis demonstrates hypervascular nest of vessels that present to-and-fro arterialized venous flow.

only epidermal thickening, decreased echogenicity of the upper dermis, or increased echogenicity of the hypodermis. Occasionally in very superficial lesions there are no abnormalities detected under ultrasound. This may happen with flat malformations that involve only the epidermis and/or measure less than 0.1 mm. Sonography may support the diagnosis, detect the extent of the malformation, and also potentially guide a percutaneous sclerosing therapy [63,64].

Inflammatory Diseases

These are several inflammatory diseases in dermatology that can show abnormalities on ultrasound. However, we will review the most common causes of referral.

Psoriasis

This is an autoimmune inflammatory chronic disorder that affects the skin, nail, tendons, joints, and bone. The use of sonography is usually focused in the assessment of the activity and severity of the disease [67,68]. Psoriatic cutaneous plaques appear on ultrasound as thickening of the epidermis and decreased echogenicity of the upper dermis. Occasionally, undulation of the epidermis may also be seen. On color Doppler study, dermal hypervascularity with slow-flow arterial and venous vessels may be detected in the plaques (Fig. 25.17). In the nail, the most common psoriatic changes on sonography are thickening and decreased echogenicity of the nail bed, focalized hyperechoic deposits in the ventral plate, loss of definition of the ventral plate, thickening of the dorsal and ventral plates, and wavy plates (Fig. 25.18). In addition, increased vascularity may be detected in the nail bed, most often in the proximal part affecting the matrix region [68]. The enthesis (ie, insertion sites of the tendons) may present thickening and hypoechogenicity or heterogeneous echogenicity. The joints can show increased anechoic fluid because of synovitis. The bony margins may show erosions commonly at the interphalangeal joints of the fingers and toes. Inflammatory signs in the median nerve have also been described and appear on color Doppler as increased intraneural vascularity, which may

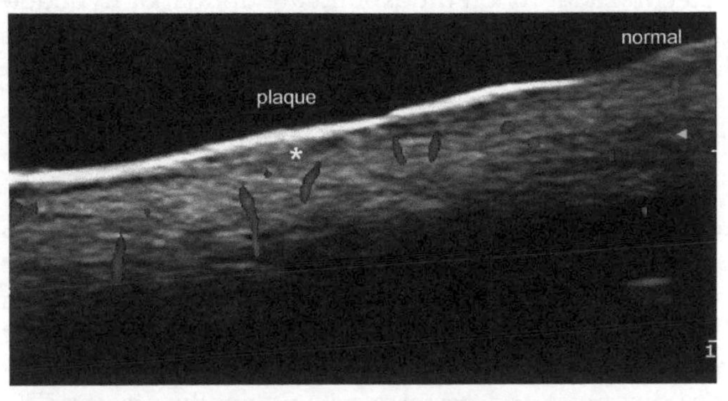

FIGURE 25.17 **Psoriasis skin plaque.** Power Doppler shows thickening of the epidermis and dermis with hypoechoic upper dermal band (*) and dermal hypervascularity in the plaque region. Notice the difference with the normal skin in the right aspect of the figure.

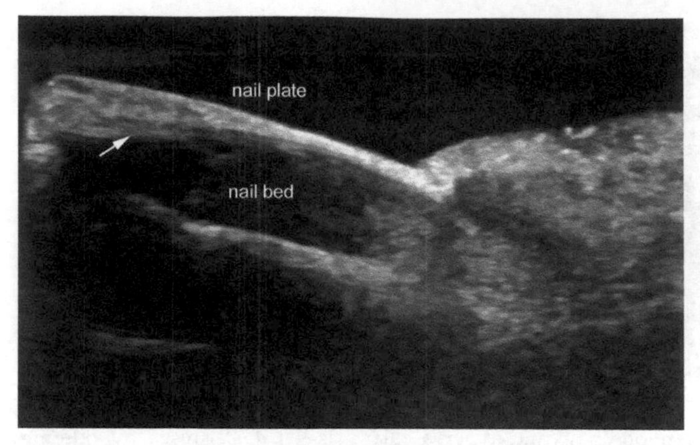

FIGURE 25.18 **Psoriasis nail.** Ultrasound (longitudinal view, right thumb) demonstrates thickening and decreased echogenicity of the nail bed and hyperechoic deposit (*arrow*) in the distal nail plate.

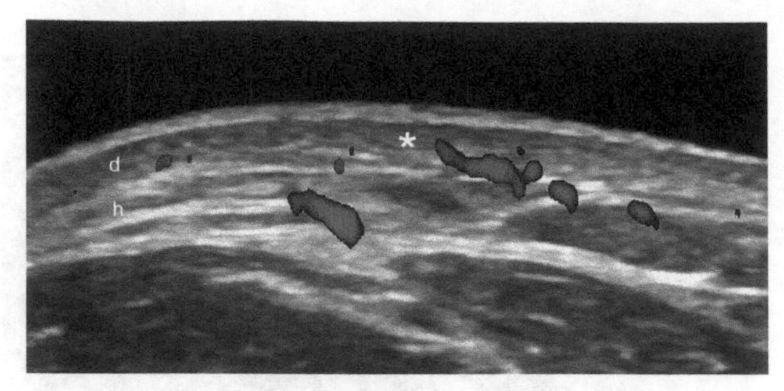

FIGURE 25.19 **Active morphea.** Color Doppler ultrasound (transverse view, right dorsal forearm) shows thickening and decreased echogenicity of the dermis (*) and dermal hypervascularity (*colors*). *d*, Dermis; *h*, hypodermis.

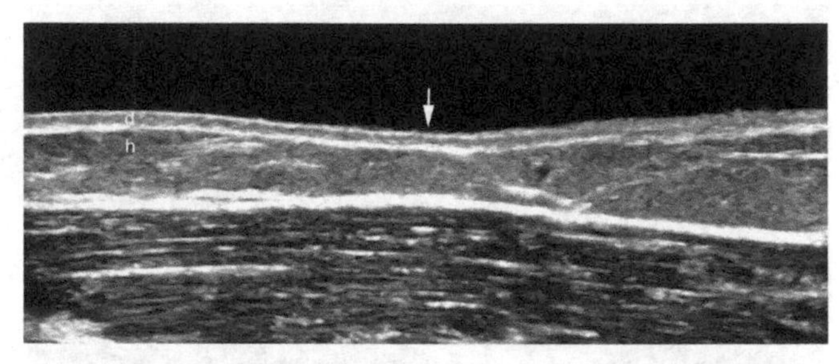

FIGURE 25.20 **Atrophic morphea.** Grayscale ultrasound (longitudinal view, left arm, color filter) demonstrates thinning (*arrow*) of the dermis and hypodermis. *d*, Dermis; *h*, hypodermis.

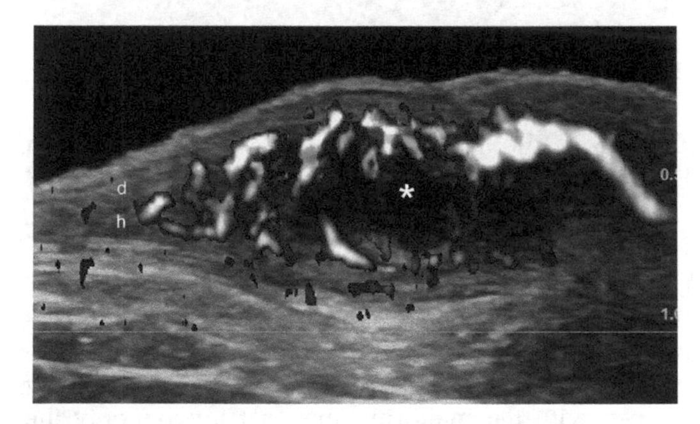

FIGURE 25.21 **Hidradenitis suppurativa (HS) fluid collection.** Power Doppler (transverse view, right axillary region) shows dermal and hypodermal hypoechoic structure (*) with hypervascularity in the periphery (*color*). *d*, Dermis; *h*, hypodermis.

also be detected on MRI. Importantly, all these anatomical changes are susceptible to monitoring by sonography [68–76].

Morphea

This connective tissue disease affects the skin by provoking extensive fibrosis with thick bundles of collagen. It is also named *localized* or *cutaneous scleroderma*. On sonography, it is possible to distinguish the different phases of activity. During the active phase, there is thickening and decreased echogenicity of the dermis. Increased echogenicity of the hypodermis can also be detected. On color Doppler, dermal and hypodermal hypervascularity are usually seen in the active phase. In the atrophy phase, there is increased echogenicity and thinning of the dermis, as well as thinning or lack of the fatty tissue of the dermis, sometimes with almost direct contact between the dermis and the muscle layer. Hypovascularity of the dermis and hypodermis can be detected in this phase. The inactive phase is considered to exist by default when the activity and atrophy criteria are not accomplished. The most sensitive signs for detecting activity on ultrasound are dermal or hypodermal hypervascularity and increased echogenicity of the hypodermis [68,77–80] (Figs. 25.19 and 25.20).

Hidradenitis Suppurativa

Also called *acne inversa*, this chronic inflammatory disease is characterized by recurrent nodules, collections of pus, and fistulous tracts in the skin. The most common locations are the axillary and groin regions. However, inframammary, buttock, vulvar, and other corporal segments may be affected. Ultrasound can support diagnosis and perform a staging of the disease. The most common sonographic signs of hidradenitis suppurativa (HS) are widening of the hair follicles, anechoic or hypoechoic pseudocysts, anechoic or hypoechoic fluid collections, and fistulae (Figs. 25.21 and 25.22). The sonographic diagnostic criterion for HS is the presence of 3 or more of these signs. Commonly, these fluid collections and fistula are connected to the base of widened hair follicles and present hyperechoic hair tracts. In spite of the extensive inflammatory process, the regional lymph nodes are rarely enlarged in HS, unless the patient is at an advanced stage and/or with accompanying infection of the fluid collections (abscess) or fistulae. The sonographic staging of HS is named *sonographic scoring*

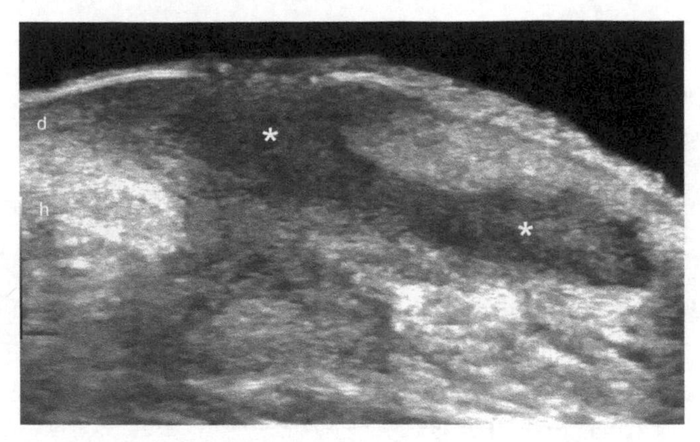

FIGURE 25.22 **Hidradenitis suppurativa (HS) fistulous tract.** Ultrasound (grayscale, transverse view, intergluteal region) shows hypoechoic tract (*) running through the dermis (d) and hypodermis (h).

TABLE 25.2 Sonographic Scoring of Hidradenitis Suppurativa[a]

Stage I: Single fluid collection and/or dermal changes[b] affecting a single body segment[c] (either one side or bilateral) without fistulous tracts
Stage II: Two to four fluid collections and/or a single fistulous tract with dermal changes affecting up to two body segments (either one side or bilateral)
Stage III: Five or more fluid collections and/or two or more fistulous tracts with dermal changes and/or involvement of three or more body segments (either one side or bilateral)

[a]Extracted from *Wortsman X, Moreno C, Soto R, Arellano J, Pezo C, Wortsman J. Ultrasound in-depth characterization and staging of hidradenitis suppurativa. Dermatol Surg 2013;39:1835e42.*
[b]*Dermal changes include hypoechoic or anechoic pseudocystic nodules, widening of the hair follicles, and/or alterations in the dermal thickness or echogenicity.*
[c]*Body segments are defined as the anatomic area affected, eg, axillary, groin, breast, buttock, etc.; involvement can be on just one side, or bilateral.*

of hidradenitis suppurativa (*SOS-HS*). There are three sonographic stages of severity according to the presence of fluid collections and fistula that are described in Table 25.2. The sonographic examination may be used for monitoring the treatment of patients noninvasively and objectively. It can provide information on the nature and extent of the lesions and also assess the degree of vascularity, which may be a potent indicator of activity [68,81–85].

Hematomas–Seromas–Abscesses

Anechoic or hypoechoic fluid collections are commonly seen after trauma and/or surgery. When the collections are complicated with infection, a pus-filled cavity called an *abscess* may appear. On ultrasound, abscesses tend to have internal echoes caused by debris and some hyperechoic septa (Fig. 25.23). Increased blood flow is commonly seen in the periphery of abscesses. Sonography allows the confirmation of the diagnosis, provides the measurements of these collections, may guide a percutaneous drainage, and can provide noninvasive monitoring of anatomical changes over time [68,86].

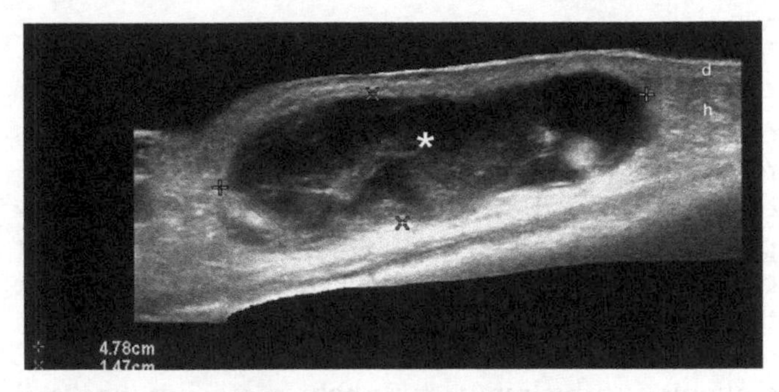

FIGURE 25.23 **Hematoma.** Ultrasound (grayscale, longitudinal view, left dorsal forearm) demonstrate 4.78-cm long dermal and hypodermal, oval-shaped anechoic fluid collection (*) with floating echoes (debris). A posterior acoustic reinforcement artifact is detected at the bottom of the collection.

Plantar Warts

These are infectious lesions originated by the human papilloma virus. Because of their painful symptoms, they can be clinically confused with a foreign body or a Morton neuroma. On sonography they show as hypoechoic, fusiform epidermal and dermal structures. Often hypervascularity is detected in the lesion or sublesional site and the degree of vascularity is commonly related to the degree of pain felt by the patient. Usually the higher the lesional vascularity, the higher the degree of pain mentioned by the patient. In addition, distention with fluid of the regional plantar bursae can be found underlying the wart site. Therefore the inflammatory process may require a multispecialist approach for treating all these anatomical changes [67,68,87,88] (Fig. 25.24).

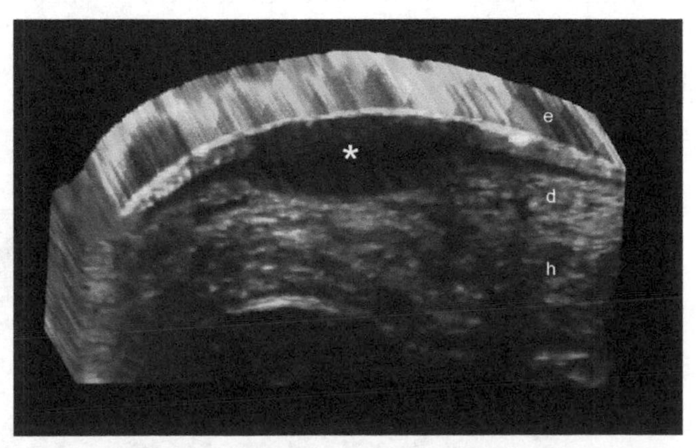

FIGURE 25.24 **Plantar wart.** Three dimensional reconstruction (grayscale with color filter, transverse view) presents epidermal and dermal fusiform oval shaped hypoechoic lesion. *d*, Dermis; *e*, epidermis; *h*, hypodermis.

Nail Conditions

Glomus Tumor

These are derived from the neuromyoarterial plexus of the nail. Usually glomus tumors are very painful, especially when the nail is exposed to cold. On sonography, they show as a hypoechoic solid nodule, most commonly in the proximal nail bed. On color Doppler these tumors tend to show prominent vascularity with slow-flow arterial and venous vessels. Commonly, scalloping of the hyperechoic bony margin of the distal phalanx is detected. According to the degree of affection of the matrix region, it can be possible to detect thickening and irregularities of the nail plate in the same axis as the tumor [89–92] (Fig. 25.25).

Subungual Granuloma

This is composed of chronic inflammatory tissue usually mixed with prominent scarring that causes a subungual pseudotumor. On sonography there is thickening or mixed thickening and thinning of the nail bed, usually affecting the matrix region. Diffuse hypovascularity or hypervascularity may be detected in the nail bed. Irregularities such as wavy shape and thickening of the nail plate are common findings. The bony margin of the distal phalanx is commonly unremarkable [18,19] (Fig. 25.26).

Subungual Exostosis

These are benign outgrowths of bony and/or cartilaginous tissue that extend from the bony margin of the distal phalanx and protrude into the nail bed. On ultrasound, exostoses appear as hyperechoic bands connected to the hyperechoic bony margin that displace the nail plate upward. These bands can produce a posterior acoustic shadowing artifact because of their calcium component. Commonly, thickening and hypoechogenicity of the nail bed is found as a result of the secondary inflammatory process and scarring. Hypovascularity or hypervascularity may be detected in the vicinity of the exostosis [18,19,92] (Fig. 25.27).

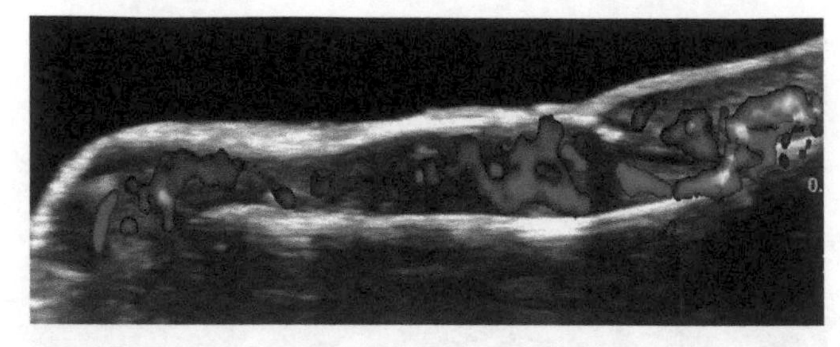

FIGURE 25.26 **Subungual granuloma.** Color Doppler ultrasound (longitudinal view, right thumb) shows thickening of the proximal nail bed that involves the matrix region and displaced upward the proximal nail plate. Notice the thickening of the nail plate. The hyperechoic bony margin of the distal phalanx is unremarkable.

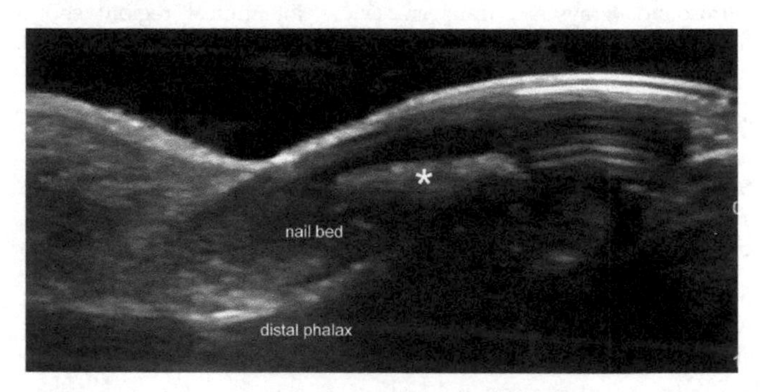

FIGURE 25.27 **Subungual exostosis.** Ultrasound (grayscale, longitudinal view, right big toenail) presents hyperechoic band (*) that emerges from the bony margin of the distal phalanx and protrude into the nail bed.

Myxoid or Synovial Cysts

These can be generated by synovial fluid leaking or synovial proliferation into the periungual region. On sonography, these cysts appear as well-defined, oval-shaped, anechoic structures (Fig. 25.28). Usually, there is a thin and tortuous anechoic tract that connects the cyst with the distal interphalangeal joint. In contrast with mucous cysts, myxoid or synovial cysts tend not to involve the nail bed. The differential diagnosis between these types of cysts is commonly performed by histology [18,19].

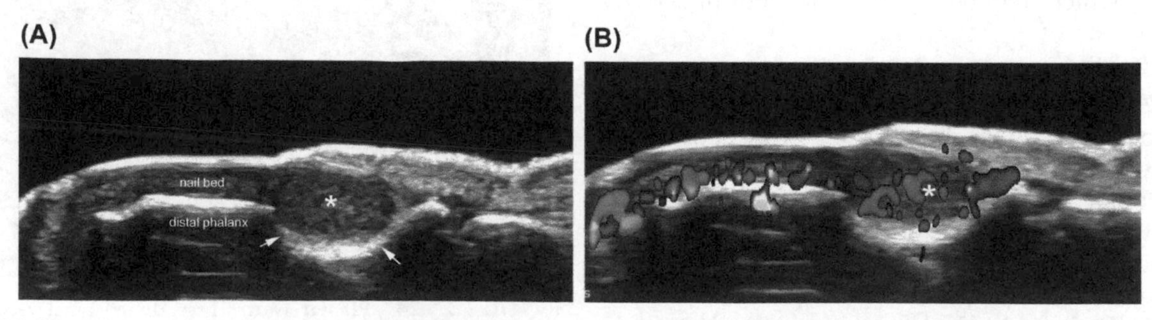

FIGURE 25.25 **Glomus tumor.** (A) Grayscale ultrasound and (B) color Doppler ultrasound (longitudinal view, left little finger) show oval shaped, hypoechoic nodule in the proximal part of the nail bed (*). Notice the scalloping of the bony margin (*arrows* in A) and the hypervascularity within the nodule (B). The nail plate appears as a thick and monolaminar layer (dystrophic) because of involvement of the matrix region.

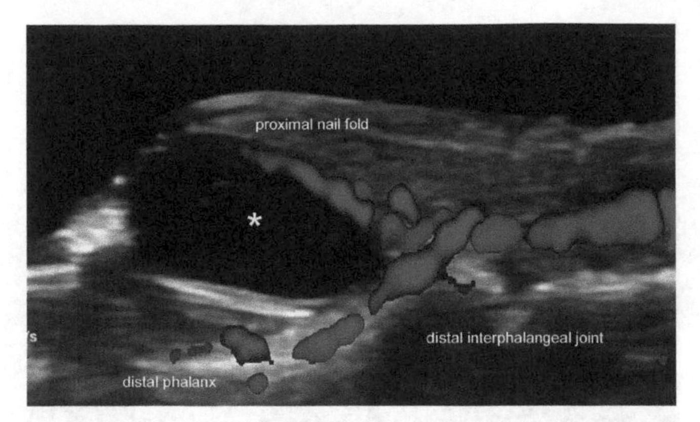

FIGURE 25.28 **Synovial cyst.** Color Doppler ultrasound (longitudinal view, left index finger) shows well-defined, oval-shaped, anechoic structure (*) in the proximal nail fold that compress the proximal nail bed. No vascularity is detected within the cyst. In the periphery there is vascularity.

Scalp and Hair Conditions

Androgenetic Alopecia

The early diagnosis of androgenetic alopecia (AGA) may be difficult and the tests for assessing the involvement of AGA, such as trichogram, can be painful. On sonography, it is possible to detect the decrease in the density of the hair follicles, with the remaining follicles showing significantly lower depth. Additionally, in AGA there is an increase in mixed pattern of hair tracts with trilaminar and bilaminar (nonmedular villus) morphology, compared with normal individuals whose hair tracts show mostly trilaminar appearance (Fig. 25.29). This imaging test may support early diagnosis and allow the monitoring of patients [20,21,24].

Tumors

Trichilemmal Cysts

These cysts are lined with cuboidal epidermal cells and do not present a granular layer. Their most common location is the scalp, where they often produce focal sites of alopecia and lumps. Trichilemmal cysts contain keratin and sometimes hair fragments. On sonography, they show as well defined, round or oval-shaped anechoic dermal and/or hypodermal structures, commonly with echoes or debris and sometimes with hyperechoic fragments of hair tracts (Fig. 25.30). On inflammation, it is possible to detect increased vascularity in the periphery of the cysts [20,24].

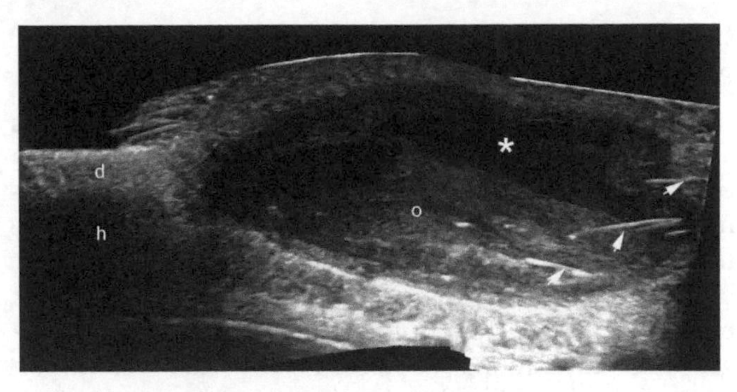

FIGURE 25.30 **Trichilemmal cyst.** Grayscale ultrasound (color filter) shows dermal and hypodermal, oval-shaped, well-defined structure (*) with hypoechoic echoes (o) and some hyperechoic linear fragments of hair tracts (*arrows*). *d*, Dermis; *h*, hypodermis.

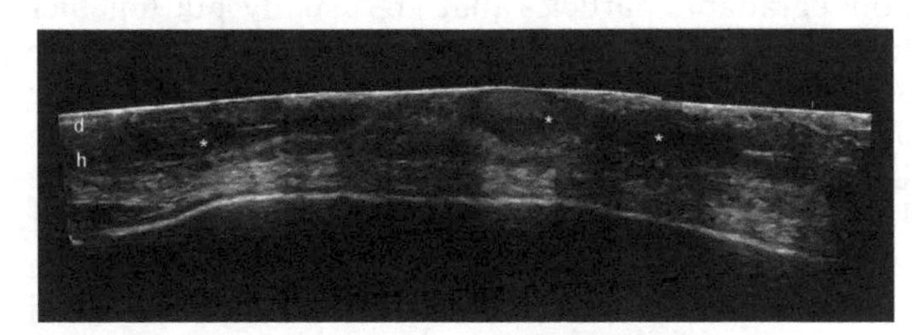

FIGURE 25.31 **Perifolliculitis capitis abscedens et suffodiens.** Grayscale ultrasound (longitudinal view, scalp region) demonstrate fistulous tracts (*) running through the dermis (d) and hypodermis (h).

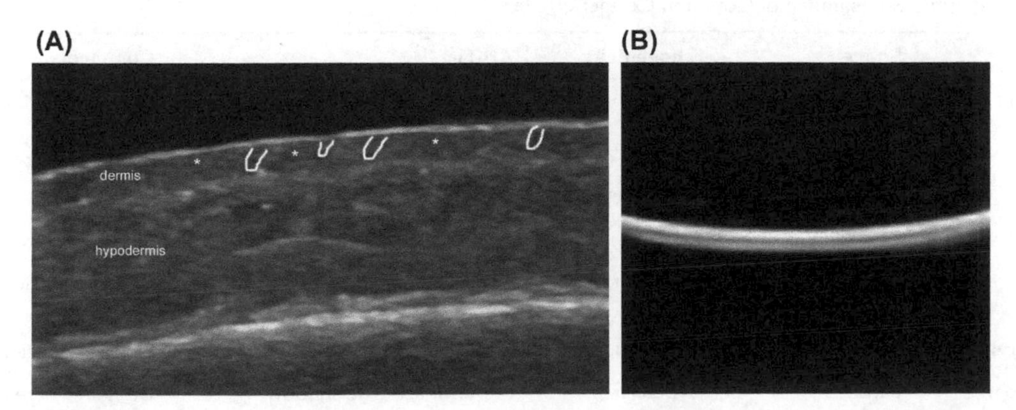

FIGURE 25.29 **Androgenetic alopecia (AGA) (frontal region of the scalp).** (A) Grayscale ultrasound demonstrates decreased density and depth of the hair follicles (*outlined*) with interfollicular wide spaces (*) in the dermis. (B) Grayscale ultrasound shows bilaminar hyperechoic hair tracts.

Inflammation

Perifolliculitis Capitis Abscedens et Suffodiens

This is a dissecting inflammatory entity that causes a patched type of alopecia. On sonography it is possible to detect the multiple dermal and hypodermal fluid collections. It is also possible to detect the fistulae underlying these sites of alopecia, which may be connected between each other and also to the base of widened hair follicles (Fig. 25.31). The sonographic alterations detected in this condition commonly resemble the changes previously described for HS [20,24].

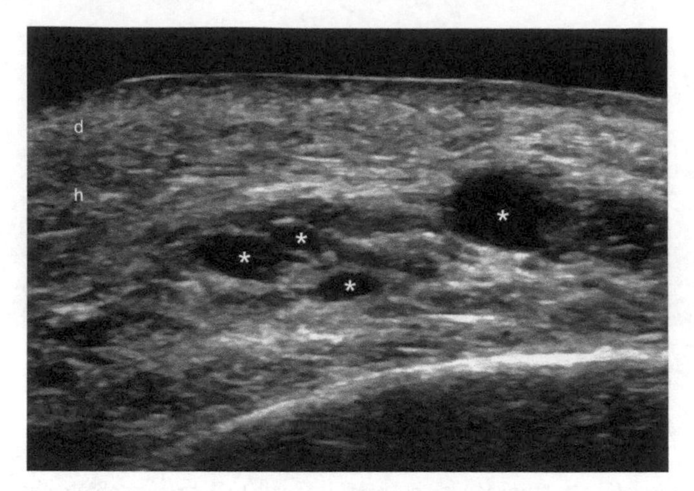

FIGURE 25.32 **Hyaluronic acid.** Grayscale ultrasound (transverse view, right cheek, color filter) shows oval shaped, anechoic hypodermal pseudocystic structures (*). *d*, Dermis; *h*, hypodermis.

Cosmetic Applications

Fillers

Cosmetic fillers are materials used for treating wrinkled and sagging skin. Sonography can identify the most common types of fillers and detect the extent of the deposits. These fillers can be divided into biodegradable products such as pure hyaluronic acid and synthetic or semisynthetic materials. The most common representatives of synthetic fillers are silicone (pure or oil forms), polymethylmethacrylate, calcium hydroxyapatite, and polyacrylamide. Semisynthetic fillers are usually a mix of degradable and nondegradable particles that are usually put together to generate long-lasting results. An example of these semisynthetic materials is high-density hyaluronic acid. There are fillers that are not approved by the US Food and Drug Administration, such as silicone. However, this is used in other countries on a legal or illegal basis. The sonographic appearance of the most common fillers is shown in Table 25.3 [93–98] (Figs. 25.32–25.35).

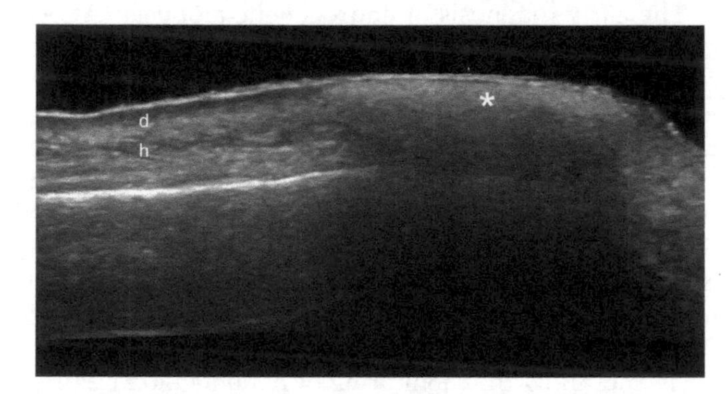

FIGURE 25.33 **Silicone oil.** Grayscale ultrasound (longitudinal view, glabella region) demonstrates dermal and hypodermal hyperechoic deposit with posterior acoustic reverberance, also called "snow storm" appearance. *d*, Dermis; *h*, hypodermis.

TABLE 25.3 Sonographic Appearance of Common Cosmetic Fillers

Filler	Shape	Echogenicity	Artifact	Comments
Pure hyaluronic acid	Round or oval shaped	Anechoic	Posterior acoustic enhancement	Decrease in size in 3–6 months
High-density pure hyaluronic acid	Round or oval shaped	Anechoic or hypoechoic	Posterior acoustic enhancement	Do not significantly change in 18 months
Pure silicone	Oval shaped	Anechoic	Posterior acoustic enhancement	Do not change over time
Silicone oil	Diffuse	Hyperechoic	Reverberance, "snow storm"	Do not significantly change over time
Polymethylmethacrylate	Diffuse	Hyperechoic	Mini–comet tail	Do not change over time
Calcium hydroxyapatite	Bandlike	Hyperechoic	Posterior acoustic shadowing	Do not change over time
Polyacrylamide	Round or oval shaped	Anechoic	Posterior acoustic enhancement	Do not significantly change in 18 months

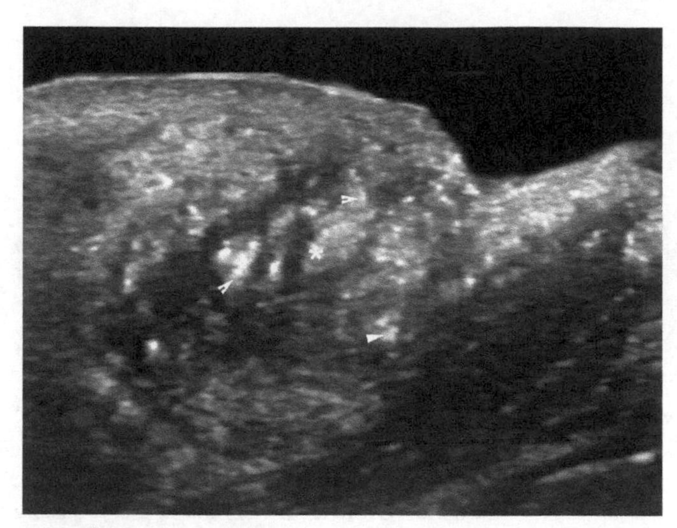

FIGURE 25.34 **Polymethylmethacrylate.** Grayscale ultrasound (transverse view; right nasofold line) demonstrates dermal and hypodermal hyperechoic deposits with posterior mini–comet tail artifact (*arrowhead*).

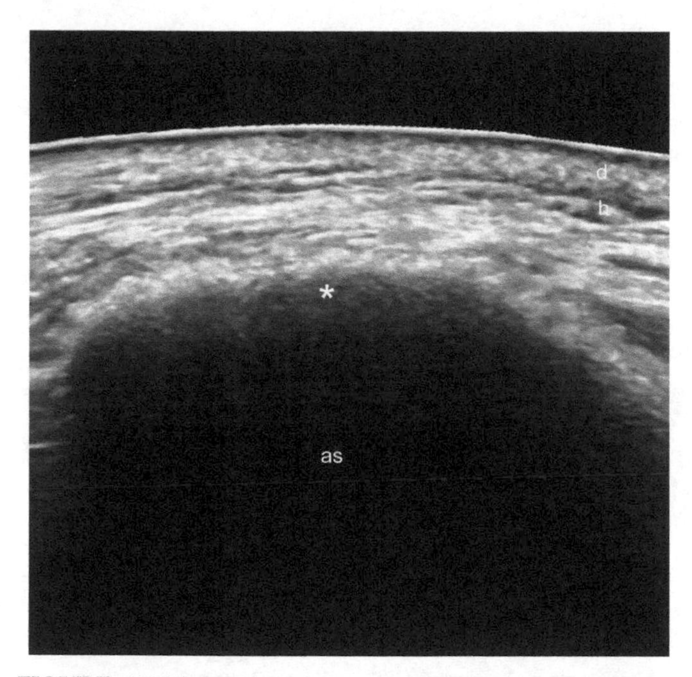

FIGURE 25.35 **Calcium hydroxyapatite.** Grayscale ultrasound (longitudinal view, left cheek, color filter) shows hypodermal hyperechoic band (*) with posterior acoustic shadowing artifact (as). *d*, Dermis; *h*, hypodermis.

CONCLUSION

Ultrasound in dermatology can be a potent diagnostic tool when certain conditions are met, such as a trained operator and an adequate machine. It allows live and safe interaction with the patient, and provides a wide range of anatomical information on common dermatological diseases without penetration issues. This high-definition imaging technique may enhance clinical diagnostic power by supporting an earlier and more precise diagnosis, and this technology is currently available worldwide for daily practice.

References

[1] Meyer J, Sans G, Rodallec C. Ultrasonics in dermatology. Bull Soc Fr Dermatol Syphiligr 1951;58:266–7.

[2] Alexander H, Miller DL. Determining skin thickness with pulsed ultra sound. J Invest Dermatol 1979;72:17–9.

[3] Serup J, Staberg B. Ultrasound for assessment of allergic and irritant patch test reactions. Contact Dermat 1987;17:80–4.

[4] Gropper CA, Stiller MJ, Shupack JL, Driller J, Rorke M, Lizzi F. Diagnostic high-resolution ultrasound in dermatology. Int J Dermatol 1993;32:243–50.

[5] Harland CC, Bamber JC, Gusterson BA, Mortimer PS. High frequency, high resolution B-scan ultrasound in the assessment of skin tumours. Br J Dermatol 1993;128:525–32.

[6] Seidenari S, Pagnoni A, Di Nardo A, Giannetti A. Echographic evaluation with image analysis of normal skin: variations according to age and sex. Skin Pharmacol 1994;7:201–9.

[7] Stiller MJ, Gropper CA, Shupack JL, Lizzi F, Driller J, Rorke M. Diagnostic ultrasound in dermatology: current uses and future potential. Cutis 1994;53:44–8.

[8] El Gammal S, El Gammal C, Kaspar K, Pieck C, Altmeyer P, Vogt M, et al. Sonography of the skin at 100 MHz enables in vivo visualization of stratum corneum and viable epidermis in palmar skin and psoriatic plaques. J Invest Dermatol 1999; 113:821–9.

[9] Wortsman XC, Holm EA, Wulf HC, Jemec GB. Real-time spatial compound ultrasound imaging of skin. Skin Res Technol 2004;10: 23–31.

[10] Wortsman X. Common applications of dermatologic sonography. J Ultrasound Med 2012;31:97–111.

[11] Wortsman X. Ultrasound in dermatology: why, how, and when? Semin Ultrasound CT MR 2013;34:177–95.

[12] Wortsman X, Wortsman J. Clinical usefulness of variable-frequency ultrasound in localized lesions of the skin. J Am Acad Dermatol 2010;62:247–56.

[13] Ying M, Bhatia KS, Lee YP, Yuen HY, Ahuja AT. Review of ultrasonography of malignant neck nodes: greyscale, Doppler, contrast enhancement and elastography. Cancer Imaging 2014; 13:658–69.

[14] Wortsman X. How to start on skin, nail and hair ultrasound: guidance and protocols. In: Wortsman X, Jemec GBE, editors. Dermatologic ultrasound with clinical and histologic correlations. 1st ed. NY: Springer; 2013. p. 597–607.

[[15] Echeverría-García B, Borbujo J, Alfageme F. The use of ultrasound imaging in dermatology. Actas Dermosifiliogr 2014;105: 887–90.

[16] Wortsman X. Sonography of cutaneous and ungual lumps and bumps. Ultrasound Clin 2012;7:505–23.

[17] Wortsman X, Wortsman J, Carreño L, Morales C, Sazunic I, Jemec GBE. Sonographic anatomy of the skin, appendages and adjacent structures. In: Wortsman X, Jemec GBE, editors. Dermatologic ultrasound with clinical and histologic correlations. 1st ed. NY: Springer; 2013. p. 15–35.

[18] Thomas L, Vaudaine M, Wortsman X, Jemec GBE, Drape JL. Imaging the nail unit. In: Baran R, de Berker D, Holzberg M, Thomas L, editors. Baran & Dawber's diseases of the nails and their management. 4th ed. Wiley; 2012. p. 132–53.

[19] Wortsman X. Sonography of the nail. In: Wortsman X, Jemec GBE, editors. Dermatologic ultrasound with clinical and histologic correlations. 1st ed. NY: Springer; 2013. p. 419–76.

[20] Wortsman X, Wortsman J, Matsuoka L, Saavedra T, Mardones F, Saavedra D, et al. Sonography in pathologies of scalp and hair. Br J Radiol 2012;85:647–55.

[21] Wortsman X, Guerrero R, Wortsman J. Hair morphology in androgenetic alopecia: sonographic and electron microscopic studies. J Ultrasound Med July 2014;33:1265–72.

[22] Garrido-Colmenero C, Arias-Santiago S, Aneiros Fernández J, García-Lora E. Trichoscopy and ultrasonography features of aseptic and alopecic nodules of the scalp. J Eur Acad Dermatol Venereol December 10, 2014. http://dx.doi.org/10.1111/jdv.12903.

[23] Imafuku K, Hata H, Kitamura S, Iwata H, Shimizu H. Ultrasound B-mode and elastographic findings of mixed tumour of the skin on the scalp. J Eur Acad Dermatol Venereol July 30, 2014. http://dx.doi.org/10.1111/jdv.12644

[24] Wortsman J. Sonography of the scalp and hair. In: Wortsman X, Jemec GBE, editors. Dermatologic ultrasound with clinical and histologic correlations. 1st ed. NY: Springer; 2013. p. 477–503.

[25] Yuan WH, Hsu HC, Lai YC, Chou YH, Li AF. Differences in sonographic features of ruptured and unruptured epidermal cysts. J Ultrasound Med 2012;31:265–72.

[26] Huang CC, Ko SF, Huang HY, Ng SH, Lee TY, Lee YW, et al. Epidermal cysts in the superficial soft tissue: sonographic features with an emphasis on the pseudotestis pattern. J Ultrasound Med 2011;30:11–7.

[27] Kim HK, Kim SM, Lee SH, Racadio JM, Shin MJ. Subcutaneous epidermal inclusion cysts: ultrasound (US) and MR imaging findings. Skeletal Radiol 2011;40:1415–9.

[28] Wortsman X, Bouer M. Common benign non-vascular skin tumors. In: Wortsman X, Jemec GBE, editors. Dermatologic ultrasound with clinical and histologic correlations. 1st ed. NY: Springer; 2013. p. 119–75.

[29] Mentes O, Oysul A, Harlak A, Zeybek N, Kozak O, Tufan T. Ultrasonography accurately evaluates the dimension and shape of the pilonidal sinus. Clinics (Sao Paulo) 2009;64:189–92.

[30] Roche NA, Monstrey SJ, Matton GE. Pilomatricoma in children: common but often misdiagnosed. Acta Chir Belg 2010;110:250–4.

[31] Choo HJ, Lee SJ, Lee YH, Lee JH, Oh M, Kim MH, et al. Pilomatricomas: the diagnostic value of ultrasound. Skeletal Radiol 2010;39:243–50.

[32] Solivetti FM, Elia F, Drusco A, Panetta C, Amantea A, Di Carlo A. Epithelioma of Malherbe: new ultrasound patterns. J Exp Clin Cancer Res 2010;29:42.

[33] Wortsman X, Wortsman J, Arellano J, Oroz J, Giugliano C, Benavides MI, et al. Pilomatrixomas presenting as vascular tumors on color Doppler ultrasound. J Pediatr Surg 2010;45:2094–8.

[34] Fornage BD, Tassin GB. Sonographic appearances of superficial soft tissue lipomas. J Clin Ultrasound 1991;19:215–20.

[35] Wagner JM, Lee KS, Rosas H, Kliewer MA. Accuracy of sonographic diagnosis of superficial masses. J Ultrasound Med 2013;32:1443–50.

[36] Hung EH, Griffith JF, Ng AW, Lee RK, Lau DT, Leung JC. Ultrasound of musculoskeletal soft-tissue tumors superficial to the investing fascia. AJR Am J Roentgenol 2014;202:W532–40.

[37] Bobadilla F, Wortsman X, Muñoz C, Segovia L, Espinoza M, Jemec GBE. Pre-surgical high resolution ultrasound of facial basal cell carcinoma: correlation with histology. Cancer Imaging 2008;22:163–72.

[38] Barcaui Ede O, Carvalho AC, Valiante PM, Barcaui CB. High-frequency ultrasound associated with dermoscopy in pre-operative evaluation of basal cell carcinoma. An Bras Dermatol 2014;89:828–31.

[39] Hernández-Ibáñez C, Aguilar-Bernier M, Fúnez-Liébana R, Del Boz J, Blázquez N, de Troya M. The usefulness of high-resolution ultrasound in detecting invasive disease in recurrent basal cell carcinoma after nonsurgical treatment. Actas Dermosifiliogr 2014;105:935–9.

[40] Wortsman X, Carreño L, Morales C. Skin cancer: the primary tumors. In: Wortsman X, Jemec GBE, editors. Dermatologic ultrasound with clinical and histologic correlations. 1st ed. NY: Springer; 2013. p. 249–82.

[41] Wortsman X. Sonography of facial cutaneous basal cell carcinoma: a first-line imaging technique. J Ultrasound Med 2013;32:567–72.

[42] Wortsman X, Vergara P, Castro A, Saavedra D, Bobadilla F, Sazunic I, et al. Ultrasound as predictor of histologic subtypes linked to recurrence in basal cell carcinoma of the skin. J Eur Acad Dermatol Venereol 2015;29:702–7.

[43] Nazarian LN, Alexander AA, Rawool NM, Kurtz AB, Maguire HC, Mastrangelo MJ. Malignant melanoma: impact of superficial US on management. Radiology 1996;199:273–7.

[44] Wortsman X. Sonography of the primary cutaneous melanoma: a review. Radiol Res Pract 2012;2012:814396. http://dx.doi.org/10.1155/2012/814396.

[45] Music MM, Hertl K, Kadivec M, Pavlović MD, Hocevar M. Pre-operative ultrasound with a 12-15 MHz linear probe reliably differentiates between melanoma thicker and thinner than 1 mm. J Eur Acad Dermatol Venereol 2010;24:1105–8.

[46] Crisan M, Crisan D, Sannino G, Lupsor M, Badea R, Amzica F. Ultrasonographic staging of cutaneous malignant tumors: an ultrasonographic depth index. Arch Dermatol Res 2013;305:305–11.

[47] Lassau N, Koscielny S, Avril MF, Margulis A, Duvillard P, De Baere T, et al. Prognostic value of angiogenesis evaluated with high-frequency and color Doppler sonography for preoperative assessment of melanomas. AJR Am J Roentgenol 2002;178:1547–51.

[48] Lassau N, Lamuraglia M, Koscielny S, Spatz A, Roche A, Leclere J, et al. Prognostic value of angiogenesis evaluated with high-frequency and colour Doppler sonography for preoperative assessment of primary cutaneous melanomas: correlation with recurrence after a 5 year follow-up period. Cancer Imaging 2006;6:24–9.

[49] Srivastava A, Woodcock JP, Mansel RE, Webster DJ, Laidler P, Hughes LE, et al. Doppler ultrasound flowmetry predicts 15 year outcome in patients with skin melanoma. Indian J Surg 2012;74:278–83.

[50] Kato M, Mabuchi T, Yamaoka H, Ikoma N, Tamiya S, Ozawa A, et al. Diagnostic usefulness of findings in Doppler sonography for amelanotic melanoma. J Dermatol 2013;40:700–5.

[51] Krüger U, Kretschmer L, Thoms KM, Padeken M, Peter Bertsch H, Schön MP, et al. Lymph node ultrasound during melanoma follow-up significantly improves metastasis detection compared with clinical examination alone: a study on 433 patients. Melanoma Res 2011;21:457–63.

[52] Catalano O, Voit C. Locoregional staging of melanoma. In: Wortsman X, Jemec GBE, editors. Dermatologic ultrasound with clinical and histologic correlations. 1st ed. NY: Springer; 2013. p. 293–343.

[53] Catalano O, Siani A. Cutaneous melanoma: role of ultrasound in the assessment of locoregional spread. Curr Probl Diagn Radiol 2010;39:30–6.

[54] Kunte C, Schuh T, Eberle JY, Baumert J, Konz B, Volkenandt M, et al. The use of high-resolution ultrasonography for preoperative detection of metastases in sentinel lymph nodes of patients with cutaneous melanoma. Dermatol Surg 2009;35:1757–65.

[55] Catalano O, Setola SV, Vallone P, Raso MM, D'Errico AG. Sonography for locoregional staging and follow-up of cutaneous melanoma: how we do it. J Ultrasound Med 2010;29:791–802.

[56] Nazarian LN, Alexander AA, Kurtz AB, Capuzzi Jr DM, Rawool NM, Gilbert KR, et al. Superficial melanoma metastases: appearances on gray-scale and color Doppler sonography. AJR Am J Roentgenol 1998;170:459–63.

[57] Alexander AA, Nazarian LN, Capuzzi Jr DM, Rawool NM, Kurtz AB, Mastrangelo MJ. Color Doppler sonographic detection of tumor flow in superficial melanoma metastases: histologic correlation. J Ultrasound Med 1998;17:123–6.
[58] Catalano O, Voit C, Sandomenico F, Mandato Y, Petrillo M, Franco R, et al. Previously reported sonographic appearances of regional melanoma metastases are not likely due to necrosis. J Ultrasound Med 2011;30:1041–9.
[59] Ulrich J, van Akkooi AJ, Eggermont AM, Voit C. New developments in melanoma: utility of ultrasound imaging (initial staging, follow-up and pre-SLNB). Expert Rev Anticancer Ther 2011;11:1693–701.
[60] Voit C, Van Akkooi AC, Schäfer-Hesterberg G, Schoengen A, Kowalczyk K, Roewert JC, et al. Ultrasound morphology criteria predict metastatic disease of the sentinel nodes in patients with melanoma. J Clin Oncol 2010;28:847–52.
[61] Voit CA, Gooskens SL, Siegel P, Schaefer G, Schoengen A, Röwert J, et al. Ultrasound-guided fine needle aspiration cytology as an addendum to sentinel lymph node biopsy can perfect the staging strategy in melanoma patients. Eur J Cancer 2014;50:2280–8.
[62] Voit CA, van Akkooi AC, Eggermont AM, Schäfer-Hesterberg G, Kron M, Ulrich J, et al. Fine needle aspiration cytology of palpable and nonpalpable lymph nodes to detect metastatic melanoma. J Natl Cancer Inst 2011;103:1771–7.
[63] Peer S, Wortsman X. Hemangiomas and vascular malformations. In: Wortsman X, Jemec GBE, editors. Dermatologic ultrasound with clinical and histologic correlations. 1st ed. NY: Springer; 2013. p. 183–248.
[64] Paltiel HJ, Burrows PE, Kozakewich HP, Zurakowski D, Mulliken JB. Soft-tissue vascular anomalies: utility of US for diagnosis. Radiology March 2000;214(3):747–54.
[65] Dubois J, Patriquin HB, Garel L, Powell J, Filiatrault D, David M, et al. Soft-tissue hemangiomas in infants and children: diagnosis using Doppler sonography. AJR Am J Roentgenol 1998;171: 247–52.
[66] Kuntz AM, Aranibar L, Lobos N, Wortsman X. Color Doppler ultrasound follow-up of infantile hemangiomas and peripheral vascularity in patients treated with propranolol. Pediatr Dermatol 2015;32:468–75.
[67] Wortsman X, Gutierrez M, Saavedra T, Honeyman J. The role of ultrasound in rheumatic skin and nail lesions: a multi-specialist approach. Clin Rheumatol 2011;30:739–48.
[68] Wortsman X, Carreño L, Morales C. Inflammatory diseases of the skin. In: Wortsman X, Jemec GBE, editors. Dermatologic ultrasound with clinical and histologic correlations. 1st ed. NY: Springer; 2013. p. 73–117.
[69] Grassi W, Gutierrez M. Psoriatic arthritis: need for ultrasound in everyday clinical practice. J Rheumatol Suppl 2012;89:39–43.
[70] Gutierrez M, Filippucci E, De Angelis R, Salaffi F, Filosa G, Ruta S, et al. Subclinical entheseal involvement in patients with psoriasis: an ultrasound study. Semin Arthritis Rheum 2011;40: 407–12.
[71] Gutierrez M, De Angelis R, Bertolazzi C, Filippucci E, Grassi W, Filosa G. Clinical images: multi-modality imaging monitoring of anti-tumor necrosis factor α treatment at the joint and skin level in psoriatic arthritis. Arthritis Rheum 2010;62:3829.
[72] Gutierrez M, De Angelis R, Bernardini ML, Filippucci E, Goteri G, Brandozzi G, et al. Clinical, power Doppler sonography and histological assessment of the psoriatic plaque: short-term monitoring in patients treated with etanercept. Br J Dermatol 2011;164:33–7.
[73] Gutierrez M, Filippucci E, Salaffi F, Di Geso L, Grassi W. Differential diagnosis between rheumatoid arthritis and psoriatic arthritis: the value of ultrasound findings at metacarpophalangeal joints level. Ann Rheum Dis 2011;70:1111–4.
[74] Tehranzadeh J, Ashikyan O, Anavim A, Shin J. Detailed analysis of contrast-enhanced MRI of hands and wrists in patients with psoriatic arthritis. Skeletal Radiol 2008;37:433–42.
[75] De Agustín JJ, Moragues C, De Miguel E, Möller I, Acebes C, Naredo E, et al. A multicentre study on high-frequency ultrasound evaluation of the skin and joints in patients with psoriatic arthritis treated with infliximab. Clin Exp Rheumatol 2012;30: 879–85.
[76] Naredo E, Möller I, de Miguel E, Batlle-Gualda E, Acebes C, Brito E, et al. Ultrasound School of the Spanish Society of Rheumatology and Spanish ECO-APs Group. High prevalence of ultrasonographic synovitis and enthesopathy in patients with psoriasis without psoriatic arthritis: a prospective case-control study. Rheumatology (Oxford) 2011;50:1838–48.
[77] Wortsman X, Wortsman J, Sazunic I, Carreño L. Activity assessment in morphea using color Doppler ultrasound. J Am Acad Dermatol 2011;65:942–8.
[78] Li SC, Liebling MS, Haines KA, Weiss JE, Prann A. Initial evaluation of an ultrasound measure for assessing the activity of skin lesions in juvenile localized scleroderma. Arthritis Care Res (Hoboken) 2011;63:735–42.
[79] Porta F, Kaloudi O, Garzitto A, Prignano F, Nacci F, Falcini F, et al. High frequency ultrasound can detect improvement of lesions in juvenile localized scleroderma. Mod Rheumatol 2014; 24:869–73.
[80] Pérez-López I, Garrido-Colmenero C, Ruiz-Villaverde R, Tercedor-Sánchez J. Ultrasound monitoring of childhood linear morphea. Actas Dermosifiliogr 2015;106:340–2.
[81] Wortsman X, Jemec GBE. High frequency ultrasound for the assessment of hidradenitis suppurativa. Dermatol Surg 2007;33:1–3.
[82] Wortsman X, Revuz J, Jemec GBE. Lymph nodes in hidradenitis suppurativa. Dermatology 2009;219:32–41.
[83] Kelekis NL, Efstathopoulos E, Balanika A, Spyridopoulos TN, Pelekanou A, Kanni T, et al. Ultrasound aids in diagnosis and severity assessment of hidradenitis suppurativa. Br J Dermatol 2010;162:1400–2.
[84] Wortsman X, Moreno C, Soto R, Arellano J, Pezo C, Wortsman J. Ultrasound in-depth characterization and staging of hidradenitis suppurativa. Dermatol Surg 2013;39:1835–42.
[85] Zarchi K, Yazdanyar N, Yazdanyar S, Wortsman X, Jemec GB. Pain and inflammation in hidradenitis suppurativa correspond to morphological changes identified by high-frequency ultrasound. J Eur Acad Dermatol Venereol 2015;29:527–32.
[86] Wortsman X, Holm EA, Gniadecka M, Wulf HC, Jemec GBE. Real time spatial compound imaging of skin lesions. Skin Res Technol 2004;10:23–31.
[87] Wortsman X, Sazunic I, Jemec GBE. Sonography of plantar warts. J Ultrasound Med 2009;28:787–93.
[88] Wortsman X, Jemec GBE, Sazunic I. Anatomical detection of inflammatory changes associated to plantar warts. Dermatology 2010;220:213–7.
[89] Matsunaga A, Ochiai T, Abe I, Kawamura A, Muto R, Tomita Y, et al. Subungual glomus tumour: evaluation of ultrasound imaging in preoperative assessment. Eur J Dermatol 2007;17: 67–9.
[90] Chen SH, Chen YL, Cheng MH, Yeow KM, Chen HC, Wei FC. The use of ultrasonography in preoperative localization of digital glomus tumors. Plast Reconstr Surg 2003;112:115–9.
[91] Wortsman X, Jemec GBE. Role of high variable frequency ultrasound in preoperative diagnosis of glomus tumors: a pilot study. Am J Clin Dermatol 2009;10:23–7.
[92] Wortsman X, Wortsman J, Soto R, Saavedra T, Honeyman J, Sazunic I, et al. Benign tumors and pseudotumors of the nail: a novel application of sonography. J Ultrasound Med 2010;29: 803–16.

[93] Young SR, Bolton PA, Downie J. Use of high-frequency ultrasound in the assessment of injectable dermal fillers. Skin Res Technol 2008;14:320–3.
[94] Schelke LW, Van Den Elzen HJ, Erkamp PP, Neumann HA. Use of ultrasound to provide overall information on facial fillers and surrounding tissue. Dermatol Surg 2010;36(S3):1843–51.
[95] Grippaudo FR, Mattei M. The utility of high-frequency ultrasound in dermal filler evaluation. Ann Plast Surg 2011;67:469–73.
[96] Wortsman X, Wortsman J, Orlandi C, Cardenas G, Sazunic I, Jemec GB. Ultrasound detection and identification of cosmetic fillers in the skin. J Eur Acad Dermatol Venereol 2012;26:292–301.
[97] Wortsman X, Wortsman J. Polyacrylamide fillers on skin ultrasound. J Eur Acad Dermatol Venereol 2012;26:660–1.
[98] Wortsman X, Wortsman J. Sonographic outcomes of cosmetic procedures. AJR Am J Roentgenol 2011;197:W910–8. http://dx.doi.org/10.2214/AJR.11.6719.

CHAPTER

26

Optoacoustic Imaging of Skin

M. Schwarz[a], J. Aguirre[a], M. Omar, V. Ntziachristos

Institute of Biological and Medical Imaging, Helmholtz Zentrum München, Neuherberg, Germany

OUTLINE

INTRODUCTION TO THE OPTOACOUSTIC (PHOTOACOUSTIC) TECHNIQUE

In this section, we present and describe the basis of the optoacoustic imaging techniques, focusing on the unique advantages that they offer for dermatology.

General Overview of the Technique

Optical imaging techniques play a predominant role in the imaging portfolio available for biomedical research and clinical diagnostics. For example, every day hundreds of biopsied samples are analyzed under the light of optical microscopes in hospitals worldwide to accurately diagnose inflammatory, cancerous, or immunological diseases, among many others. Furthermore several types of macroscopic optical techniques are used in everyday clinical practice to calculate arterial oxygen saturation, characterizing a wide variety of metabolic and physiological processes. As another example, molecular imaging optical probes based on fluorophores have innumerable basic research, preclinical, and potential clinical applications such as gene expression profiling, determination of protein function, characterization of metabolic and physiological processes, drug discovery, therapeutic evaluation in animal models, or clinical diagnosis of breast cancer and inflammatory diseases.

In summary, optical imaging tools provide anatomical, functional, and molecular images that drive large areas of biomedical research and modern clinical practice. However, because of the strong photon-scattering nature of biological tissues, the performance

[a] These authors contributed equally to this work.

Imaging in Dermatology
http://dx.doi.org/10.1016/B978-0-12-802838-4.00026-1

of optical imaging tools drastically degrades at depths greater than a few hundred micrometers [1], even though light in the near-infrared (NIR) region can penetrate through several centimeters of tissue. Therefore tissues must be excised and sliced to be imaged under the microscope; otherwise the resolution of macroscopic optical imaging techniques is restricted to a few millimeters in deep tissues [1].

In this context, optoacoustic imaging emerged around 2005, offering advancement in the field of optical imaging as a result of its unprecedented resolution with regard to depth capabilities [2–4]. Optoacoustic imaging is a noninvasive optical modality that is able to retrieve the three-dimensional (3D) distribution of light energy deposition noninvasively and in vivo. It is based on the photoacoustic effect, which is produced when biological tissue is irradiated with a short light pulse, leading to a thermoelastic expansion that induces the emission of an acoustic wave. Using ultrasonic transducers and an appropriate image reconstruction scheme, the distribution of the absorbed light energy deposition can be obtained with ultrasonic-scale spatial resolution in the case of deep tissues (>1 mm), and optical resolution in the case of superficial tissues (<1 mm). As an example, structures situated a few centimeters deep in tissue can be imaged with 200 μm resolution, whereas structures situated at depths below 1 mm can be imaged with a resolution on the order of micrometers [4].

Motivation for Optoacoustic Imaging in Dermatology: Optoacoustic Mesoscopy

The human skin is a complex, multilayered structure. Its thickness ranges from 1 to 4 mm, and hosts a large variety of chemical environments and vasculature regions [5]. Skin diseases heavily impact society; as an example the most common form of cancer in the United States is skin cancer, whereby approximately 20% of all Americans will develop skin cancer in the course of their lifetime [6]. Optical imaging plays a central role in the diagnostic portfolio of dermatology in the form of dermatoscopy and optical microscopy techniques. Dermatoscopy provides two-dimensional (2D) "photographic views" of skin lesions. Optical microscopy instead attempts to resolve lesions three-dimensionally and gain more information on the extent and infiltration of the disease. Techniques such as nonlinear microscopy methods and confocal microscopy are used in dermatology clinics [7–9]. However, they are fundamentally limited by light scattering in tissue. Therefore the penetration depth only reaches a few hundred micrometers.

Optical coherence tomography (OCT) is another modality of optical microscopy that offers higher penetration depths by using light in the NIR–infrared region, which is less sensitive to light scattering. OCT can reach up to 2 -mm depth with a resolution around 1–10 μm [10,11]. However, the contrast mechanism of OCT is based on light reflection. Generally, in optical imaging, if the contrast mechanism for image generation is light absorption, several wavelengths can be used for data acquisition (multispectral imaging). After image acquisition, the relative concentration of different chromophores and fluorophores can be quantified, thanks to their unique spectral signature, enabling functional and molecular imaging.

OCT can use several acquisition and postprocessing strategies to obtain spectral information; however, it faces difficulties in the quantification of relative concentrations between optical absorbers [12]. The modern imaging portfolio lacks a technique capable of high-resolution imaging across the entire depth of skin that can also provide direct absorption optical contrast of tissue optical absorbers. On the other hand, histopathological analysis of biopsied skin is necessary to accurately diagnose a high number of skin diseases [13]. However, histological analysis is an invasive, slow, and expensive medical process.

Optoacoustic mesoscopy may provide an alternative to the current skin imaging methods because of its unique combination of direct optical absorption contrast and high penetration depth. *Mesoscopic techniques* in this context refer to systems capable of high-resolution imaging at depths typically ranging from 0.5 to 5 mm, ie, going beyond historical limits of the optical microscopy techniques [1], which in the case of organs like the skin, implies imaging its whole depth.

OPTOACOUSTIC MESOSCOPY SYSTEMS FOR SKIN IMAGING

In this section we review the state of the art of optoacoustic methods available for skin imaging, putting into perspective the latest system developed in our group, which is described in detail.

Overview of Early Developments

Optoacoustic tomography has found applications in dermatology since the late 1990s. Several groups have developed optoacoustic imaging devices that allow for the visualization of human skin to a depth of several millimeters. Most of the developed imaging platforms share a common design, which consists of a light delivery system and a single piezoelectric transducer that is raster-scanned over the skin surface. Alternative reported approaches are based on a transparent Fabry Perot detector as well as a linear transducer array for parallel acquisition of many channels simultaneously.

The detection frequency band of the transducer is crucial, because it determines the sizes of the objects that will be visible in the reconstructed images. Essentially, small objects emit optoacoustic signals with stronger high-frequency components than bigger objects.

In 1999, the first optoacoustic imaging device applied to human skin achieved in-depth resolution of 10–15 μm and light penetration of up to 4 mm, but suffered from a low lateral resolution of 200 μm [14]. To improve upon the lateral resolution of the first optoacoustic systems, a focused ultrasound transducer with a central frequency of 50 MHz and darkfield illumination was introduced in 2006, providing lateral resolution of approximately 45 μm in the focal region [15]. With this set-up, volumetric images of human cutaneous microvasculature have been imaged [16].

The first optoacoustic imaging systems in human skin showed the strong potential of optoacoustics in dermatology, providing direct contrast of the human vascular network. However, the development status of laser sources and ultrasound sensors has so far prevented optoacoustics to find application in dermatology. The main limiting components were first, low pulse repetition frequency (PRF) of single-wavelength lasers; second, the bulkiness of the optoacoustic systems; and, third, the moderate lateral resolution. All of these setups employed laser sources with a PRF of 10 Hz. Thus the acquisition time of a region of interest (ROI) measuring $8 \times 8\ mm^2$ took approximately 50 min [17]. Long acquisition times lead involuntarily to motion artifacts when imaging live samples and made imaging in the clinical context cumbersome. The bulky and fixed systems rendered certain skin areas on a human body inaccessible for imaging. The lateral resolution of close to 45 μm was sufficient for imaging the large vessels of the lower dermis and hypodermis, but insufficient for imaging the microvasculature of the epidermal–dermal junction. Nevertheless, within the last decade, technological advancements have pushed the limits of optoacoustic imaging in dermatology.

More advanced laser sources, operating at PRFs of up to 3 kHz, have reduced acquisition times dramatically [18]. With faster excitation sources, the maximal PRF and thus acquisition times are mainly limited by the maximum permissible laser power on the skin surface, as specified for example by the American National Standards Institute (ANSI) laser safety standards [19].

Miniaturization of optoacoustic systems into hand-held devices has also progressed in the last years, although no hand-held device has reached a resolution better than 90 μm. Two transportable skin imaging systems have been recently developed/described [20,21]. The first system was based on a transducer with central frequency of 25 MHz, designed for detection of melanoma depth that was used to visualize the boundaries of B16 melanoma tumors in mice [20]. The second system was based on a lead zirconate titanate (PZT) transducer with a central frequency of 35 MHz and was capable of imaging the large vessels of the lower dermis and hypodermis in humans [21].

In terms of resolution, the latest optoacoustic mesoscopy device, developed by our group, has shown a significant improvement in the visualization of human microvasculature of the papillary dermis. The system is based on a 100-MHz wideband transducer and achieved an isotropic in-plane resolution of 18 μm and in-depth resolution of 4 μm [22]. With this system, key anatomical vascular features at different layers of the skin, including the epidermal–dermal junction, have been revealed [23].

For the sake of completeness, two optoacoustic systems deviating from the single piezo transducer design shall be mentioned as well, both providing in-depth resolution of 20 μm and in-plane resolution of less than 100 μm. In the first deviation, a transparent interferometric Fabry Perot detector was employed to visualize the large vessels of the reticular dermis and hypodermis [24]. The transparency of the Fabry Perot detector enabled an integration into a multimodal optoacoustic and OCT scanner [25]. In the second deviation, a linear array of transducers was used to image several millimeters of tissue burn and of superficial skin lesions [26,27].

A Comprehensive Approach: Broadband Raster-Scan Optoacoustic Mesoscopy

Imaging human skin with optoacoustics is based on photoabsorbing species in the visible and NIR imaging window, comprising melanin, oxyhemoglobin, deoxyhemoglobin, water, fat, and protein. The size of human blood vessels in the skin ranges from thin capillaries with an internal endothelial tube diameter of 4–6 μm [28], located in the dermal papillae, to larger vessels with a diameter of up to 100 μm in the mid and deep dermis [5]. Thus the most comprehensive approach to imaging human skin should: (1) provide a resolution of several micrometers, imaging anatomical structures in the range of 5–100 μm; and, (2) penetrate several millimeters into tissue. Given these premises, broadband raster-scan optoacoustic mesoscopy (RSOM) is to date the most promising optoacoustic imaging platform in dermatology.

The in vivo skin images shown in this chapter were acquired with the most advanced RSOM system

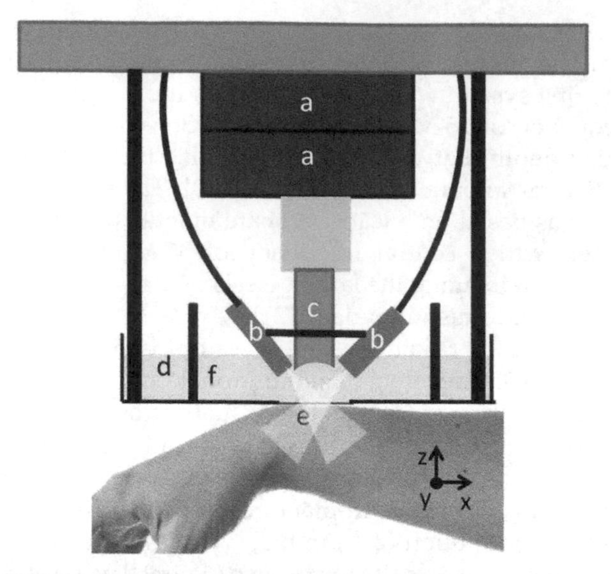

FIGURE 26.1 **System set-up.** Scheme of the system: motorized stages (a), ends of the fiber bundles (b), transducer (c), interface unit (d), plastic membrane (e), and water (f).

developed in our group, [22,29] which is sketched in Fig. 26.1. The main component of the system was a lithium nitrate spherically focused ultrasound transducer with a central frequency of 100 MHz, a relative bandwidth of 105%, and an f-number of approximately 1. The incoming signal from the sensor was passed through a low noise amplifier (63 dB, AU-1291, Mited, Hauppauge, New York, United States) and collected afterwards by high-speed digitizer (CS122G1, Gage, Lockport, Illinois, United States; 12 bit resolution; max sampling rate, 2 GS/s) for further processing on the computer. A single wavelength laser was used, operating at 532 nm, pulsing with a frequency of up to 2 kHz, and a laser pulse width of less than 1 ns (Wedge HB.532, Bright-solutions). The laser light was delivered to the skin using three optical fiber bundles, achieving a circularly illuminated area of approximately 40 mm^2 at the skin surface. The PRF and per-pulse energy were chosen in order to conform to the ANSI laser safety limits for human skin [19].

The transducer–illumination unit was raster-scanned in tandem over the ROI using two piezoelectric stages (Physik Instrumente, KG, Karlsruhe, Germany). A custom-made interface unit (IU) held the human tissue in position and coupled the transducer–illumination component to human skin. The IU consisted of a water reservoir for ultrasonic coupling of the detector to human tissue. The IU had a rectangular hole, which held the sample in the center of the imaging field and was sealed by a thin plastic membrane. Acoustic gel was used as coupling medium between the plastic membrane and the skin surface.

SKIN IMAGING WITH BROADBAND RASTER-SCAN OPTOACOUSTIC MESOSCOPY

In this section we showcase the imaging potential of the latest RSOM system developed in our group.

Anatomical Human Skin Imaging

The anatomical imaging capabilities of the RSOM system are showcased in Fig. 26.2, showing an image of the palm of the hand of a healthy volunteer. One acquisition was made acquiring data over an area of 8 × 8 mm. In Fig. 26.2, maximum intensity projections (MIPs) at different skin depths, as well as an axial view, are shown. Different structures can be distinguished with high resolution and the epidermal thickness can be estimated from the axial view. Epidermal thickness was estimated to be approximately 200 μm and was found to be in accordance with values found in the literature [30]. The displayed features in the top MIP (see Fig. 26.2B) corresponds to the epidermal–dermal junction. The stripes correspond to the shape of dermal papillae, which in the palm of the hand follow the shape of the outer epidermal ridges. The bright dots along the stripes are in agreement with the superficial loops of the capillary beds, appearing similar to what have also been recently identified by other techniques like OCT [31]. The image of the middle MIP (see Fig. 26.2C) shows many smaller vessels of the horizontal plexus situated closer to the epidermal-dermal junction and the connections between capillary loops. The bigger vessels located below in the horizontal plexus appear clearly in the lower MIP (see Fig. 26.2D).

As outlined in Section "A Comprehensive Approach: Broadband Raster-Scan Optoacoustic Mesoscopy," the skin vasculature has a layered structure. Most of the vessels are located within a horizontal plexus, situated up to 2 mm below the skin. From this plexus, arterial capillaries protrude towards the epidermal–dermal junction, forming capillary loops, which provide oxygen and nutrients to the dermal papillae. Capillary loops are thinner (7–10 -μm outer diameter and an endothelial tube diameter of 4–6 μm). Within the horizontal plexus, most of the vessel diameters are in the range of 17–26 μm. Within the mid and deep dermis, vessels are larger, reaching a maximum of 100 μm in diameter [5]. The broadband characteristics of the transducer described in Section "A Comprehensive Approach: Broadband Raster-Scan Optoacoustic Mesoscopy" are responsible for the unique scalability properties of the imaging system. *Scalability* herein refers to the capability of the system to measure the incoming high-frequency

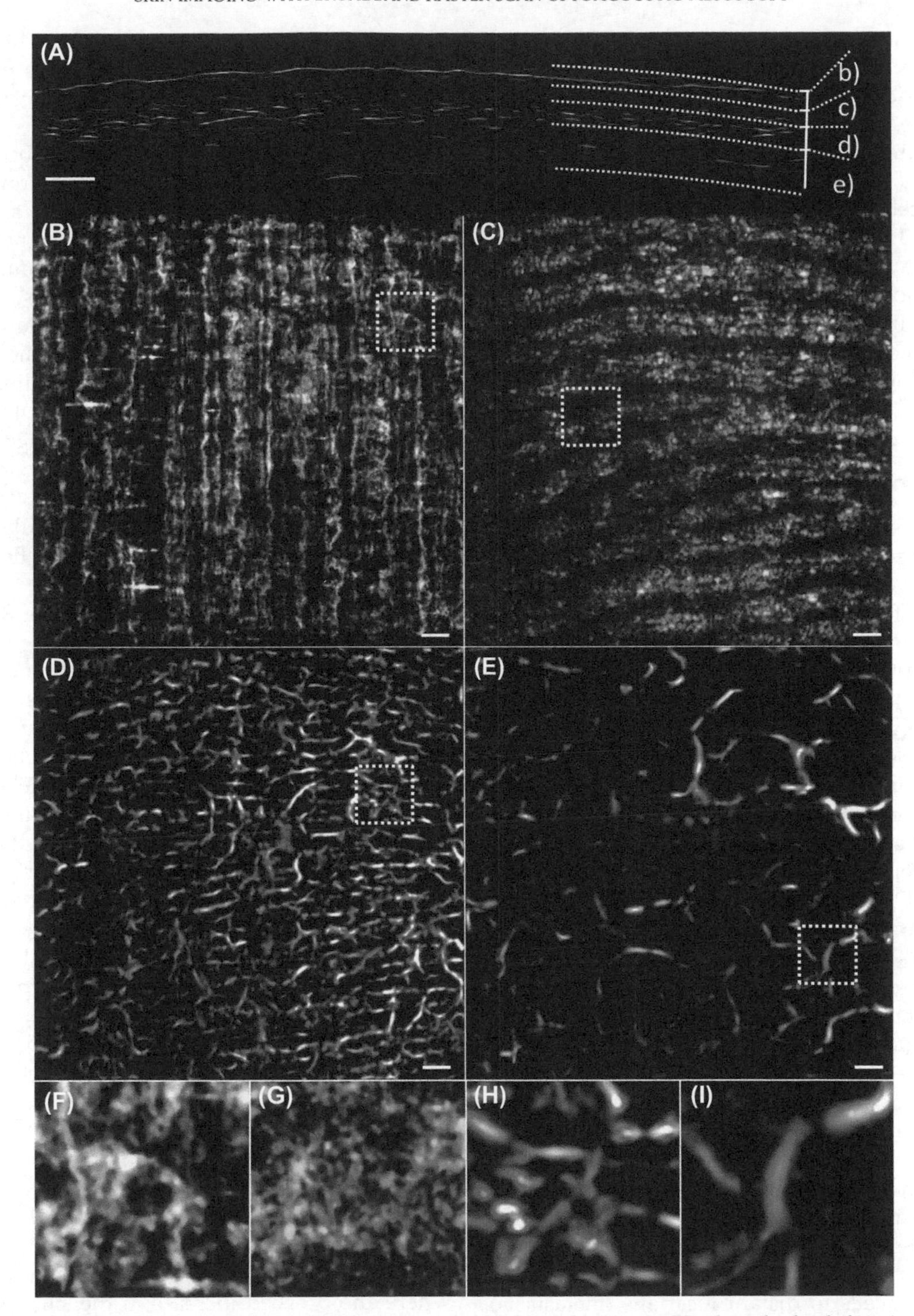

FIGURE 26.2 **Different layers of human skin.** (A) Cross section through human skin. (b–e) MIPs along the depth direction within the regions marked in (A). (B) Stratum corneum of the epidermis. (C) Vascular fingerprint of the dermal papillae. (D) Microvasculature of the subpapillary plexus. (E) Deep vessels of the papillary dermis. (F–I) Zoomed-in images of the region marked by the *white dashed boxes* in (B–E) are shown. (D and E) The original image (*neon red*) is overlaid with an image filtered for vessels in *white*. Scale = 500 μm.

signal from the approximately 10-μm vessels situated close to the epidermal–dermal junction with a high-frequency component, and the ultrasound signals generated by the deeper and larger vessels with lower-frequency content. As already outlined in Section "A Comprehensive Approach: Broadband Raster-Scan Optoacoustic Mesoscopy," previous optoacoustic imaging systems do not achieve the same scalability as

RSOM. Such systems are capable to resolve the vessels of the horizontal plexus, but are not able to resolve the subepidermal ridges and the vessels structure in the dermal papillae because of the high-frequency nature of these structures.

Furthermore, cross-sectional views reconstructed with RSOM nicely reveal features as a function of depth (see Fig. 26.2). The illumination herein was at a wavelength of 532 nm. Illumination in the visible maximizes label-free imaging of the skin but limits the penetration depth achieved because of the strong light attenuation in tissues. The selection of a wavelength closer to the NIR region would enhance the penetration depth.

Angiogenesis of Skin Tumors in Mice

Detailed statistics on cancer, and, of particular relevance skin cancer, are well documented for the US population. The most common form of skin cancer is basal cell carcinoma, followed by squamous cell carcinoma. Malignant melanoma accounts for less than 5% of all skin cancers but is responsible for the vast majority (more than 75%) of skin cancer deaths [32,33]. The lifetime risk of developing melanoma is 1 in 50 in the United States [34] and 1 in 55 for men in the United Kingdom [35]. In Great Britain the incidence rates of melanoma has increased by a factor of seven in men and by a factor of four in women within 33 years. Early detection of melanomas is crucial because the overall 5-year survival rate drops from approximately 98% (early detection) to 62% (disease has reached the lymph nodes) and 15% (disease has metastasized to distant organs) [32]. In clinical practice the diagnosis of melanoma skin cancer is based on superficial features: asymmetry, boundary, coloring, diameter, and evolution. However, essential information on the melanoma is encoded in depth, such as the vascular network surrounding the melanoma as well as the infiltration depth; the latter is a significant factor in the prognosis of recurrence and metastasis [36–38].

The vascular network feeding tumors is very important, because it supplies the tumor with nutrients and oxygen [39] and late in disease progression carries the migrating tumor cells to distant locations (metastasis). As no living cell can exist at a distance farther than about100 μm away from the vascular network, tumors rapidly induce new vessels to feed tumor cell growth in what is commonly known as *angiogenesis* [39]. Studies of angiogenesis, vascular function, and microenvironment are being pursued using multiple approaches. For example, histological and molecular methods readily provide quantitative analyses at tissue, cellular, and molecular levels, both in preclinical and in clinical studies. However, these techniques are not suitable for dynamic or functional studies, and are highly invasive. On the other hand, imaging techniques provide noninvasive or minimally invasive dynamic measurements of physiological functions in real time.

Several imaging techniques have been developed over the past three decades and can be roughly divided into two groups: (1) techniques such as fluorescence molecular tomography [40] and magnetic resonance imaging [41] that image the bulk tumor properties, with low resolution, and (2) techniques such as intravital microscopy (IVM), which image only small and superficial fields of view (ie, not the whole tumor), but with high cellular level resolution. Techniques such as optical frequency domain imaging (OFDI) [42] have also been developed, which can image deeper than IVM with high spatial resolution. These techniques can image marginally beyond 1 mm, and are capable of imaging only the anatomy of the vasculature. In addition, OFDI is relatively slow, where a field of view of $5 \times 4\ mm^2$ takes over 10 min to image.

In order to demonstrate the ability of RSOM to characterize angiogenesis in skin tumors, our group has studied angiogenesis of B16F10 melanoma cells (0.5×10^6 cells in 25 μL phosphate-buffered saline solution) in an 8-week old female Hsd:Athymic Nude-Foxn1^{nu} mouse. The melanoma cells were injected into the mammary fat pad, and melanoma tumor growth and vascular changes were monitored by scanning the mouse four times over nine days. A mouse bed was designed for the in vivo experiments in order to keep the mouse's head above water while covering the tumor area by several millimeters of water. The water was heated to maintain the animal's body temperature at 36°C. All procedures were approved by the district government of Upper Bavaria.

To take full advantage of the broadband capabilities of our detector, we divided the detection bandwidth into low- and high-frequency ranges [22,29,43]. The low-frequency range from 10 to 30 MHz resolves large vascular structures (represented in red; Fig. 26.3), whereas the high-frequency range from 30 to 90 MHz resolves smaller vessels (represented in green; see Fig. 26.3). Fig. 26.3 shows tumor growth over time. The lower row of Fig. 26.3 shows angiogenesis after day 4, where an observed increase in the small vasculature is observed, represented by the green color. This observation fits the previous findings that angiogenesis develops at day 5 or 6 after injection. The tumor area was scanned on days 2, 4, 7, and 9 after melanoma cell injection. Fig. 26.3 shows partial MIPs along the depth direction starting at a depth of about 900 μm below the surface of the skin. In this way the tumor boundary as well as the vascular network surrounding the tumor is visualized on the same image. The subfigures of Fig. 26.3 show a strong increase in the size of the

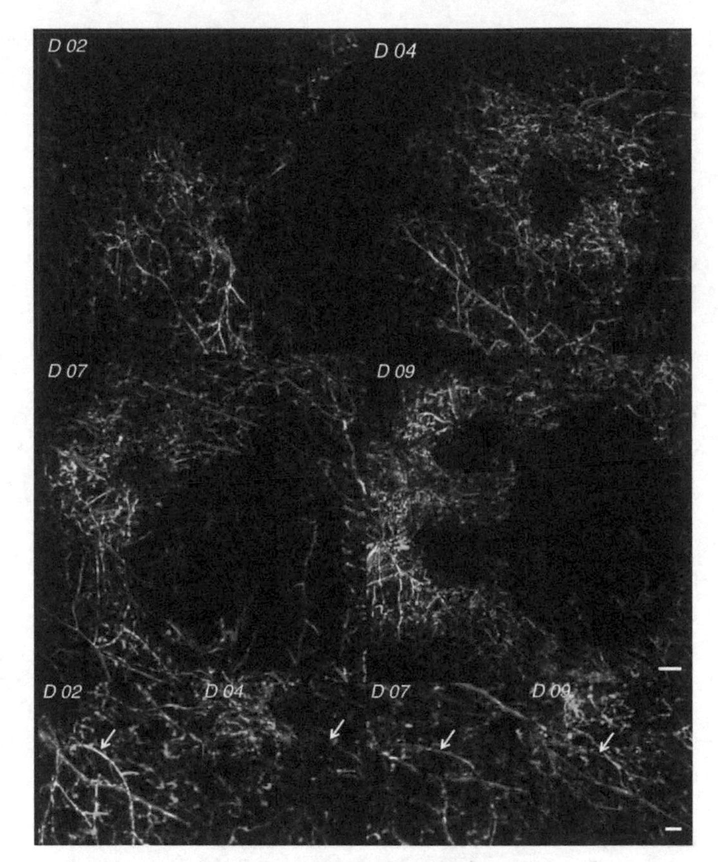

FIGURE 26.3 **Angiogenesis in B16F10 tumors.** Follow-up observation of tumor growth over the course of 9 days; the bottom row shows a zoom-in of the same region. On day 7, an increase in the microvasculature (*green*) is observed. The *arrow* points to the same big vessel observed, which is used as a location indicator.

nonvascularized spot, indicated by the thick white arrows, representing the location of the tumor. The inset on every subfigure shows a zoom-in of the same vascular structures in proximity to the tumor. Over time the two thick vessels, marked by the white arrows, shift while small vessels start to grow.

Based on these data, RSOM shows strong potential for longitudinal studies of tumor formation resolving growth-related changes of the tumor area and the surrounding area at high resolution. Because of the generally higher signal-to-noise ratio of low-frequency structures and the stronger acoustic attenuation of high frequencies, the high frequencies will be masked by the low frequencies. By dividing the detection bandwidth into low- and high-frequency bands, the visualization of the small structures was improved.

Thus RSOM has potential for a range of preclinical studies, for example, those aimed at understanding tumor angiogenesis, progression of the disease, and potential treatments. In the clinical context, RSOM imaging could provide important depth information that would help in the diagnosis of skin cancer without the need for invasive approaches.

DISCUSSION AND CONCLUSION: CHALLENGES AND PERSPECTIVES OF RASTER-SCAN OPTOACOUSTIC MESOSCOPY

Multispectral Raster-Scan Optoacoustic Mesoscopy

When using illumination sources operating at single wavelength, optoacoustic techniques reveal anatomical images that contain strong photoabsorbers [3]. Depending on the selected wavelength, the image will show mainly the distribution of oxyhemoglobin and deoxyhemoglobin, as well as other strong tissue photoabsorbers such as melanin, lipids, or water. However, the reconstructed images from a single wavelength illumination represent the sum of absorption by all optical absorbers, but do not resolve each of them separately. Multispectral optoacoustic tomography (MSOT) was conceived with the aim of spectral identification of photoabsorbing molecules distributed in tissue [44,45]. To this end, the object to be imaged is illuminated sequentially with light sources operating at different wavelengths, selected to provide the spectral signature of each photoabsorber. Applying spectral unmixing algorithms, the spatial distribution and relative amounts of each absorber can be determined [46,47]. MSOT has greatly expanded the imaging capabilities of optoacoustic tomography, enabling the imaging of extrinsic and intrinsic chromophores and fluorophores, as well as nanoparticles [48–51].

In summary, MSOT provides anatomical, functional images of vasculature as well as intrinsic and extrinsic chromophores. From these images, hemodynamics or oxygen saturation can be derived, thereby suggesting countless applications in preclinical research and clinical diagnosis.

The quantification of tissue chromophore properties has great potential in the diagnosis and therapy of skin diseases, particularly in the context of the chromophores melanin and hemoglobin. The amount of eumelanin compared with pheomelanin has been shown to vary between melanomas and nonmalignant nevi [52]. Furthermore, dysplastic melanocytic nevi contain significantly higher amount of pheomelanin, and are, according to the National Cancer Institute, more likely to develop into melanoma [53]. The impact of changes in hemoglobin-related oxygen saturation in the microenvironment of the tumor is of major interest, with new opportunities for cancer therapy arising therefrom and entire books written on the subject [54,55]. It was shown that two hypoxia-inducible factors play a critical role in the progression of melanoma [56]. Hypoxia is not only found in skin cancer, but also in systemic sclerosis [57], and oxygen saturation levels in ischemic wounds play an important role in the healing process [58].

Several efforts are being devoted to develop multispectral raster-scan optoacoustic mesoscopy systems, eg, towards broadband RSOM systems that have multispectral capabilities, enabling molecular and functional imaging of the skin with the high spatial resolution and depth of imaging advantages of RSOM.

Instrumentation Challenges

From an instrumentation point of view, RSOM of the skin depends on the integration of several key technologies at the peak of their technical abilities, namely the laser and transducers. Regarding the laser source, it must fulfill the so-called "stress confinement condition" [59]. This condition implies that the pulse length should be shorter than the time it takes for the generated photoacoustic wave to travel through the volume elements in which the sound is generated. When this condition is not fulfilled, the frequency content of the optoacoustic signal is modified. The modification can be modelled by the convolution of the signal that would be generated by a Dirac delta pulse and the actual pulse length [60]. For skin imaging purposes, pulses of length around 1–2 ns were useful to image the structures of the smaller skin vessels [61]. The theoretical resolution limit for a 1-ns and a 2-ns pulse is of 1.5 and 3 μm, respectively, which can be compared with the size of the smaller vessels in the skin, which are 7 μm in diameter.

Another technical consideration is that the energy per pulse must be sufficient to generate an optoacoustic wave with enough amplitude to be detected. For the system discussed here, the required pulse energy is 1 mJ to ensure enough energy after coupling losses with the optical fiber bundle. Furthermore the laser should be tunable in the future to allow spectral measurements that contain molecular and functional information. Last but not least, the PRF should be of the order of several 100 Hz to achieve fast acquisitions while fulfilling the safety protocols.

Single-wavelength Q-switched diode-pumped solid-state lasers that emit at 532 nm that fulfill both the energy and pulse length requirements are provided by several companies. However, tunable lasers with such characteristics cannot so far be found, and single-wavelength lasers with wavelengths different from 532 nm are only manufactured in a customized way, carrying a big cost.

The raster-scan configuration comes with a clear drawback because acquisition times are always on the order of minutes. Fast imaging configurations are desirable in clinics for the sake of the comfort of patients and the imaging system user. Furthermore the faster the acquisition, the fewer motion-related artifacts are expected on the image. Optoacoustic systems that use planar or linear arrays of fixed PZT detection elements are able to achieve higher imaging speed. In this case the laser repetition rate is the limiting factor for the acquisition time. The resolution of such imaging systems lies within the range of the size of each detection element. Current ultrasonic PZT detectors can be manufactured at a size of several tens of micrometers; however, their small size is compromised by weak sensitivity, which limits the quality of the images. Capacitive micromachined ultrasonic transducers (CMUTs) are transducers with a detection principle based on changes in capacitance, in contrast to traditional transducers, which rely on the piezoelectric effect. CMUTs are built of silicon, allowing construction of very dense 2D and 3D arrays, which may have right sensitivity and broad bandwidth. Several research efforts are being pursued to incorporate CMUT technology in optoacoustic skin imaging.

The directivity of the ultrasound detector is another major limitation when imaging with focused transducers in a raster-scan configuration. When a large object composed of individual sub–resolution-sized absorbers is illuminated with a short laser pulse, the individual absorbers emit optoacoustic waves with a strong phase correlation, leading to a series of interference effects that determines the final shape of the optoacoustic wave [62]. When the absorbers are located in elongated structures, the interference effects lead to waves with a high directivity perpendicular to the structure. For this reason, the performance of RSOM systems depends on the maximum acceptance angle of the spherically focused transducers [62]. Currently available transducers provide a maximum acceptance angle of 60 degrees. Increasing the numerical aperture and thus the acceptance angle, will have a direct impact on the quality of the reconstructed images. Therefore industrial technological efforts are also devoted to improve upon the numerical aperture of the transducers.

Computational Challenges

RSOM is computationally intensive because images are reconstructed tomographically, ie, the focused ultrasound transducer detects signals that lie within a cone-shaped sensitivity field. The opening angle of the detection cone is determined by the numerical aperture of the transducer. For a 5 × 5-mm^2 area with an acquisition grid spacing of 10 μm, 251,001 positions are taken into account, each containing more than 1000 time samples.

Images are generated from the collective datasets using one of many different reconstruction approaches.

The most popular reconstruction algorithm is the backprojection formula owing to its easy implementation, robust reconstructed images, and relatively fast reconstruction times [63]. The backprojection formula is based on analytical inversions of the optoacoustic signals into absorption maps. The formula is accurate for an infinite number of perfect point detectors of infinite bandwidth, covering a solid angle of 2π. Limitations in the number of projections (ie, angular coverage), the bandwidth of the detector, and the acceptance angle of the detector lead to a decrease in image quality. Because the backprojection formula is a linear problem, the reconstruction algorithm can be easily parallelized on a graphics processing unit (GPU), providing $5 \times 5 \times 2$ mm^3 reconstructed volumes within few minutes.

The so-called "model-based" reconstruction algorithms represent an alternative to backprojection algorithms [64,65]. They are performed by minimizing the least square error between the measured signals and the signals predicted by an optoacoustic propagation model. The optoacoustic propagation model can include several physical parameters that otherwise cannot be included in the backprojection formulation [66]. Generally, backprojection algorithms limit the quantification capacity of the spectral unmixing algorithms, because negative values appear in the reconstructed data set, which have no physical meaning. However, implementation of model-based reconstruction algorithms to RSOM represents a great computational challenge. Storage of the matrix associated to the physical model requires several thousands of gigabytes, and the solution of the least squares problem requires computational times of the order of weeks. Whereas strategies based on symmetries have resolved the storage-size challenge [67], several efforts are being directed towards reducing the reconstruction times, namely via parallelization strategies on GPUs.

Translation to the Clinic

It has been demonstrated in the laboratory that RSOM can provide high-resolution anatomical images of the skin. The images show different skin layers, including the epidermis, papillary dermis, and lower dermis. The optoacoustic technique overcomes the long-standing light-diffusion resolution limit, achieving resolutions of few tens of micrometers to several millimeters deep in tissue, going beyond the limitations of optical imaging, which acts as the current portfolio for skin imaging.

Therefore, it is expected that RSOM will have a broad impact in the dermatology clinical practice for the diagnosis of a wide range of diseases that are characterized by vascular changes, including systemic sclerosis [68,69], psoriasis [70,71], and collagenosis [72], as well as for assessing the infiltration depth of skin lesions such as melanoma skin cancer. To this end, efforts in technological development are being pursued with aims to miniaturize the hardware to allow handheld operation and easy scanning access to desired skin areas. The data acquisition times must be reduced in order to make the system useable in a clinical context. Because RSOM is a tomographical imaging method, it is also vulnerable to motion artifacts in live tissue samples. Thus hardware and software methods must be advanced in order to provide robust imaging results despite motion. Adding multispectral feature will boost the clinical potential of the technique.

List of Acronyms and Abbreviations

CMUT Capacitive micromachined ultrasonic transducer
GPU Graphics processing unit
IU Interface unit
MIP Maximum intensity projection
PRF Pulse repetition frequency
PORH Postocclusive reactive hyperemia
PZT Lead zirconate titanate
ROI Region of interest
RSOM Raster-scan optoacoustic mesoscopy

References

[1] Ntziachristos V. Going deeper than microscopy: the optical imaging frontier in biology. Nat Methods 2010;7(8):603–14.
[2] Ntziachristos V, Ripoll J, Wang LV, Weissleder R. Looking and listening to light: the evolution of whole-body photonic imaging. Nat Biotechnol 2005;23(3):313–20.
[3] Ntziachristos V, Razansky D. Molecular imaging by means of multispectral optoacoustic tomography (MSOT). Chem Rev 2010;110(5):2783–94.
[4] Wang LV, Hu S. Photoacoustic tomography: in vivo imaging from organelles to organs. Science 2012;335(6075):1458–62.
[5] Braverman IM. The cutaneous microcirculation: ultrastructure and microanatomical organization. Microcirculation 1997;4(3): 329–40.
[6] Robinson JK. Sun exposure, sun protection, and vitamin D. JAMA 2005;294(12):1541–3.
[7] Koenig K, Dimitrow E, Kaatz M. Clinical multiphoton tomography of malignant melanoma. J Am Acad Dermatol 2012;66(4): Ab142.
[8] Nehal KS, Gareau D, Rajadhyaksha M. Skin imaging with reflectance confocal microscopy. Semin Cutan Med Surg 2008;27(1): 37–43.
[9] Nehal KS, Gareau D, Rajadhyaksha M, editors. Skin imaging with reflectance confocal microscopy: seminars in cutaneous medicine and surgery. Elsevier; 2008.
[10] Mogensen M, Thrane L, Jorgensen TM, Andersen PE, Jemec GB. OCT imaging of skin cancer and other dermatological diseases. J Biophotonics 2009;2(6–7):442–51.
[11] Enfield J, Jonathan E, Leahy M. In vivo imaging of the microcirculation of the volar forearm using correlation mapping optical coherence tomography (cmOCT). Biomed Opt Express 2011;2(5): 1184–93.

[12] Gambichler T, Jaedicke V, Terras S. Optical coherence tomography in dermatology: technical and clinical aspects. Arch Dermatol Res 2011;303(7):457–73.
[13] The Lewen Group I, Association TSfIDaTAAoD. The burden of skin diseases 2005. 2005.
[14] Karabutov AA, Savateeva EV, Oraevsky AA, editors. Imaging of layered structures in biological tissues with optoacoustic front surface transducer. BiOS'99 International Biomedical Optics Symposium. International Society for Optics and Photonics; 1999.
[15] Zhang HF, Maslov K, Stoica G, Wang LV. Functional photoacoustic microscopy for high-resolution and noninvasive in vivo imaging. Nat Biotechnol 2006;24(7):848–51.
[16] Zhang HF, Maslov K, Li ML, Stoica G, Wang LV. In vivo volumetric imaging of subcutaneous microvasculature by photoacoustic microscopy. Opt Express 2006;14(20):9317–23.
[17] Zhang HF, Maslov K, Wang LV. In vivo imaging of subcutaneous structures using functional photoacoustic microscopy. Nat Protoc 2007;2(4):797–804.
[18] Favazza CP, Jassim O, Cornelius LA, Wang LV. In vivo photoacoustic microscopy of human cutaneous microvasculature and a nevus. J Biomed Opt 2011;16(1):016015.
[19] American National Standards Institute, The Laser Institute of America. American National Standard for safe use of lasers: approved March 16, 2007. Orlando, FL: The Laser Institute of America; 2007. XIII, 249 S. p.
[20] Zhou Y, Xing WX, Maslov KI, Cornelius LA, Wang LHV. Handheld photoacoustic microscopy to detect melanoma depth in vivo. Opt Lett 2014;39(16):4731–4.
[21] Bost W, Lemor R, Fournelle M. Optoacoustic imaging of subcutaneous microvasculature with a class one laser. IEEE Trans Med Imaging 2014;33(9):1900–4.
[22] Omar M, Soliman D, Gateau J, Ntziachristos V. Ultrawideband reflection-mode optoacoustic mesoscopy. Opt Lett 2014;39(13): 3911–4.
[23] Schwarz M, Omar M, Buehler A, Aguirre J, Ntziachristos V. Implications of ultrasound frequency in optoacoustic mesoscopy of the skin. IEEE Trans Med Imaging 2014.
[24] Zhang EZ, Laufer JG, Pedley RB, Beard PC. In vivo high-resolution 3D photoacoustic imaging of superficial vascular anatomy. Phys Med Biol 2009;54(4):1035–46.
[25] Zhang EZ, Povazay B, Laufer J, Alex A, Hofer B, Pedley B, et al. Multimodal photoacoustic and optical coherence tomography scanner using an all optical detection scheme for 3D morphological skin imaging. Biomed Opt Express 2011;2(8):2202–15.
[26] Schwarz M, Buehler A, Ntziachristos V. Isotropic high resolution optoacoustic imaging with linear detector arrays in bi-directional scanning. J Biophotonics 2014;9999(9999).
[27] Vionnet L, Gateau J, Schwarz M, Buehler A, Ermolayev V, Ntziachristos V. 24 MHz scanner for optoacoustic imaging of skin and burn. IEEE Trans Med Imaging 2013.
[28] Yen A, Braverman IM. Ultrastructure of the human dermal microcirculation: the horizontal plexus of the papillary dermis. J Invest Dermatol 1976;66(3):131–42.
[29] Omar M, Gateau J, Ntziachristos V. Raster-scan optoacoustic mesoscopy in the 25–125 MHz range. Opt Lett 2013;38(14): 2472–4.
[30] Whitton JT, Everall JD. The thickness of the epidermis. Br J Dermatol 1973;89(5):467–76.
[31] Blatter C, Weingast J, Alex A, Grajciar B, Wieser W, Drexler W, et al. In situ structural and microangiographic assessment of human skin lesions with high-speed OCT. Biomed Opt express 2012;3(10):2636–46.
[32] Society AC. Cancer facts and figures 2013. Atlanta: American Cancer Society; 2013.
[33] Ekwueme DU, Guy Jr GP, Li C, Rim SH, Parelkar P, Chen SC. The health burden and economic costs of cutaneous melanoma mortality by race/ethnicity-United States, 2000 to 2006. J Am Acad Dermatol 2011;65(5 Suppl. 1):S133–43.
[34] Howlader N, Noone A, Krapcho M, Neyman N, Aminou R, Waldron W, et al. SEER cancer statistics review, 1975–2008. Bethesda, MD: National Cancer Institute; 2011.
[35] Uk CRUK, Cancer R. Skin cancer incidence statistics [Web page]. UK: Cancer Res; 2013 [updated 04/09/2013 16:13]. Available from: http://www.cancerresearchuk.org/cancer-info/cancerstats/types/skin/incidence/.
[36] Clark Jr WH, From L, Bernardino EA, Mihm MC. The histogenesis and biologic behavior of primary human malignant melanomas of the skin. Cancer Res 1969;29(3):705–27.
[37] Breslow A. Thickness, cross-sectional areas and depth of invasion in the prognosis of cutaneous melanoma. Ann Surg 1970;172(5): 902–8.
[38] Garbe C, Buttner P, Bertz J, Burg G, d'Hoedt B, Drepper H, et al. Primary cutaneous melanoma: prognostic classification of anatomic location. Cancer 1995;75(10):2492–8.
[39] Hanahan D, Weinberg RA. The hallmarks of cancer. Cell 2000; 100(1):57–70.
[40] Ale A, Ermolayev V, Herzog E, Cohrs C, de Angelis MH, Ntziachristos V. FMT-XCT: in vivo animal studies with hybrid fluorescence molecular tomography–X-ray computed tomography. Nat Methods 2012;9(6):615–20.
[41] Barrett T, Brechbiel M, Bernardo M, Choyke PL. MRI of tumor angiogenesis. J Magn Reson Imaging 2007;26(2):235–49.
[42] Vakoc BJ, Lanning RM, Tyrrell JA, Padera TP, Bartlett LA, Stylianopoulos T, et al. Three-dimensional microscopy of the tumor microenvironment in vivo using optical frequency domain imaging. Nat Med 2009;15(10):1219–23.
[43] Omar M, Schwarz M, Soliman D, Symvoulidis P, Ntziachristos V. Pushing the optical imaging limits of cancer with multi-frequency-band raster-scan optoacoustic mesoscopy (RSOM). Neoplasia 2015;17(2):208–14.
[44] Razansky D, Buehler A, Ntziachristos V. Volumetric real-time multispectral optoacoustic tomography of biomarkers. Nat Protoc 2011;6(8):1121–9.
[45] Razansky D, Vinegoni C, Ntziachristos V. Multispectral photoacoustic imaging of fluorochromes in small animals. Opt Lett 2007;32(19):2891–3.
[46] Tzoumas S, Deliolanis NC, Morscher S, Ntziachristos V. Unmixing molecular agents from absorbing tissue in multispectral optoacoustic tomography. IEEE Trans Med Imaging 2014;33(1):48–60.
[47] Glatz J, Deliolanis NC, Buehler A, Razansky D, Ntziachristos V. Blind source unmixing in multi-spectral optoacoustic tomography. Opt Express 2011;19(4):3175–84.
[48] Buehler A, Kacprowicz M, Taruttis A, Ntziachristos V. Real-time handheld multispectral optoacoustic imaging. Opt Lett 2013; 38(9):1404–6.
[49] Buehler A, Dean-Ben XL, Claussen J, Ntziachristos V, Razansky D. Three-dimensional optoacoustic tomography at video rate. Opt Express 2012;20(20):22712–9.
[50] Buehler A, Herzog E, Razansky D, Ntziachristos V. Video rate optoacoustic tomography of mouse kidney perfusion. Opt Lett 2010;35(14):2475–7.
[51] Ntziachristos V. Emerging optoacoustic methods in bio-optics. J Biophotonics 2013;6(6–7):473–4.
[52] Matthews TE, Piletic IR, Selim MA, Simpson MJ, Warren WS. Pump-probe imaging differentiates melanoma from melanocytic nevi. Sci Transl Med 2011;3(71).
[53] Salopek TG, Yamada K, Ito S, Jimbow K. Dysplastic melanocytic nevi contain high levels of pheomelanin: quantitative comparison of pheomelanin/eumelanin levels between normal skin, common nevi, and dysplastic nevi. Pigment Cel Res 1991;4(4):172–9.
[54] Melillo G. Hypoxia and cancer: biological implications and therapeutic opportunities. Springer; 2013.

[55] Brown JM. Tumor hypoxia in cancer therapy. Methods Enzymol 2007;435:297–321.
[56] Steunou AL, Ducoux-Petit M, Lazar I, Monsarrat B, Erard M, Muller C, et al. Identification of the hypoxia-inducible factor 2alpha nuclear interactome in melanoma cells reveals master proteins involved in melanoma development. Mol Cell Proteomics 2013;12(3):736–48.
[57] Silverstein JL, Steen VD, Medsger TA, Falanga V. Cutaneous hypoxia in patients with systemic sclerosis (scleroderma). Arch Dermatol 1988;124(9):1379–82.
[58] Modarressi A, Pietramaggiori G, Godbout C, Vigato E, Pittet B, Hinz B. Hypoxia impairs skin myofibroblast differentiation and function. J Invest Dermatol 2010;130(12):2818–27.
[59] Diebold GJ, Sun T, Khan MI. Photoacoustic monopole radiation in one, two, and three dimensions. Phys Rev Lett 1991;67(24): 3384–7.
[60] Wang LV, Wu H-I. Biomedical optics: principles and imaging. Hoboken, NJ: Wiley-Interscience; 2007. XIV, 362 S. p.
[61] Aguirre J, Schwarz M, Soliman D, Buehler A, Omar M, Ntziachristos V. Broadband mesoscopic optoacoustic tomography reveals skin layers. Opt Lett 2014;39(21):6297.
[62] Gateau J, Chaigne T, Katz O, Gigan S, Bossy E. Improving visibility in photoacoustic imaging using dynamic speckle illumination. arXiv preprint arXiv:13080243. 2013.
[63] Xu M, Wang LV. Universal back-projection algorithm for photoacoustic computed tomography. Phys Rev E, Stat Nonlinear, Soft Matter Phys 2005;71(1 Pt 2):016706.
[64] Rosenthal A, Razansky D, Ntziachristos V. Fast semi-analytical model-based acoustic inversion for quantitative optoacoustic tomography. IEEE Trans Med Imaging 2010;29(6):1275–85.
[65] Dean-Ben XL, Buehler A, Ntziachristos V, Razansky D. Accurate model-based reconstruction algorithm for three-dimensional optoacoustic tomography. IEEE Trans Med Imaging 2012;31(10): 1922–8.
[66] Rosenthal A, Ntziachristos V, Razansky D. Model-based optoacoustic inversion with arbitrary-shape detectors. Med Phys 2011; 38(7):4285–95.
[67] Aguirre J, Giannoula A, Minagawa T, Funk L, Turon P, Durduran T. A low memory cost model based reconstruction algorithm exploiting translational symmetry for photoacoustic microscopy. Biomed Opt Express 2013;4(12):2813–27.
[68] Wigley FM. Vascular disease in scleroderma. Clin Rev Allergy Immunol 2009;36(2–3):150–75.
[69] Herrick AL. Vascular function in systemic sclerosis. Curr Opin Rheumatol 2000;12(6):527–33.
[70] Nestle FO, Kaplan DH, Barker J. Psoriasis. New Engl J Med 2009; 361(5):496–509.
[71] Archid R, Patzelt A, Lange-Asschenfeldt B, Ahmad SS, Ulrich M, Stockfleth E, et al. Confocal laser-scanning microscopy of capillaries in normal and psoriatic skin. J Biomed Opt 2012;17(10): 101511.
[72] Claussen M, Riemekasten G, Hoeper M. Pulmonary arterial hypertension in collagenoses. Z für Rheumatologie 2009;68(8): 630–8.

CHAPTER

27

Use of Total Body Photography and Serial Digital Dermoscopy in Dermatology

B.W. Petersen, H.W. Higgins II

Brown University, Providence, RI, United States

OUTLINE

INTRODUCTION

Melanoma is estimated to make up 4.6% of all new cancer cases (76,100) and is responsible for 1.7% of all cancer deaths annually (9710) [1].

The 5-year survival rate for melanoma is currently 91.3%. However, survival varies widely based on stage at diagnosis with a 5-year survival rate of 98.1%, 62.6%, and 16.1% for individuals with localized disease (stages 0–II), regional spread to lymph nodes (stage III), and distant cancer metastasis (stage IV), respectively [1]. Despite efforts to develop new therapies, the prognosis for individuals with advance disease remains poor [2].

Even with the improvement in the 5-year survival rate (81.1% in 1975 to 92.8% in 2006), mortality has increased with rising incidence. Over the past 35 years, the number of new cases of melanoma has increased from 7.9 per 100,000 to 22.7 cases per 100,000, whereas the mortality has increased from 2.1 to 2.7 per 100,000. Melanoma now ranks fifth behind colon and rectal cancer in new cases per year [1].

At the time of diagnosis, 84% of patients are determined to have localized disease confined to the primary site, 9% have regional disease with spread to lymph nodes, 4% have distant metastatic disease, and 3% of cases are unknown [1].

Improving the detection of melanoma and thereby diagnosing patients at an early stage can greatly reduce patient mortality [3–5]. Breitbart et al. concluded that early detection is the most effective intervention in improving melanoma prognosis [6]. Although no consensus has been reached on how to improve early detection, a number of methods have been utilized to try and improve screening and diagnosis of melanoma [7].

Advancements in imaging technologies including digital photography and digital dermoscopy have led to interest in the effectiveness and practicality of incorporating new technologies to improve skin screening and melanoma detection. For the purposes of this chapter the technology can be divided into two main components: total body photography (TBP) and serial digital dermoscopy (SDD).

Imaging in Dermatology
http://dx.doi.org/10.1016/B978-0-12-802838-4.00027-3

TABLE 27.1 Advantages of Total Body Photography

Aids in detection of de novo melanomas

- Total body photography (TBP) and digital dermoscopy use in a high-risk population resulted in early detection of melanomas with a low rate of excisions [14].
- TBP in high-risk subjects was associated with lower biopsy rates and nevus-to-melanoma biopsy ratios, and facilitated detection of new or changing lesions [7].
- Baseline TBP and follow-up examinations enabled early diagnosis of melanomas [13].

Aids in the detection of melanoma arising in nevi that were not previously concerning

- TBP in high-risk patients helped in detection of changes in size, shape, and color that led to the diagnosis of malignant melanoma [17].
- TBP and digital dermoscopy use in a high-risk population resulted in early detection of melanomas arising in previously benign-appearing nevi [14].
- Serial photography every 3–6 months in high-risk patients resulted in increased detection of invasive melanomas [18].
- TBP was helpful in detecting change in previously benign lesions [22].

Aids in detection of slowly changing melanomas detected only on TBD

- TBP allowed for detection of increased numbers of melanoma, of which the majority arose de novo [13].
- TBP in high-risk patients helped in detection of changes in size, shape, and color that led to the diagnosis of malignant melanoma [17].
- TBP was helpful in detecting change in previously benign lesions [22].
- TBP and screening resulted in earlier detection of melanoma [62].
- TBP-assisted follow-up examination helped detect new and subtly changing melanomas [20].

Aids in increased detection of malignant melanoma in situ with lower incidence of malignant melanoma, and decreased average malignant melanoma tumor thickness

- TBP allowed for earlier detection of melanoma and lower biopsy rates [21].
- Baseline TBP and follow-up examination enabled early diagnosis of melanomas [13].

Lower biopsy rates or high melanoma-to-nonmelanoma biopsy ratios

- TBP in high-risk subjects was associated with lower biopsy rates and nevus-to-melanoma biopsy ratios, and facilitated detection of new or changing lesions [7].
- Using TBP at baseline and dermoscopy was associated with low biopsy rates and early detection of melanomas [21].
- Baseline TBP and follow-up examination is more cost-effective in detecting and preventing melanoma than prophylactic excision of dysplastic nevi [13].
- TBP resulted in reduced excision of benign lesions [23].
- TBP in high-risk subjects was associated with lower biopsy rates and nevus-to-melanoma biopsy ratios, and facilitated detection of new or changing lesions [63].

Increases sensitivity and specificity of self-skin examinations

- TBP at baseline improved the diagnostic accuracy of self-skin examinations on the back and chest or abdomen and improved detection of changing and new moles [24].

Decreases patient worry/anxiety

- TBP helps decrease worry about developing melanoma in at-risk patients monitored by a pigmented-lesion clinic [25].

This chapter will explore TBP and SDD and their use by dermatology providers.

TOTAL BODY PHOTOGRAPHY

TBPuses photography of the body in standard poses with the goal of imaging the entirety of a patient's skin [8]. These images can be on film or, increasingly, digital. They can be printed or stored electronically for future reference. There are large variations in protocols for TBP, including the number of photographs and poses used to image the patient's skin (range 6–85 images) [8]. As a result, the images can vary from full-scale to reduced size [8]; however, this technology does not include stored dermoscopic images.

This technique was first reported by Atkinson et al. and Slue et al. in 1987 and 1988, respectively, with the intended use of monitoring dysplastic nevi for change as well as monitoring for new lesions [9,10]. The underlying premise is that the majority of melanomas arise either de novo or as benign-appearing lesions that were not of clinical concern at the time of imaging. Comparing these lesions with baseline photographs helps facilitate the identification of new lesions, lesions with significant changes, and lesions growing at a notably faster rate than surrounding lesions [7,11–13]. Some of these changes are subtle and therefore could be missed by patients and examiners [11].

The usefulness of this technology has strong anecdotal evidence. Some clinicians report that detecting new or changing nevi is extremely difficult when relying solely on memory [14–16]. Multiple studies have shown that providers think this technology is helpful in detecting new and changing melanocytic nevi, thereby aiding in the diagnosis of melanoma [10,11,17,18], with some providers suggesting that the technology may increase sensitivity and specificity of melanoma diagnosis [11]. A study of academic dermatologists by Nehal et al. found that 87% of respondents found TBP to be useful in managing patients with dysplastic nevi, 80% thought TBP helped with detecting melanoma, 81% felt TBP decreased patient anxiety, and 73% thought TBP prevented unnecessary biopsies [19].

Whereas some of these anecdotal findings are well accepted, supporting nonanecdotal evidence is less robust (Table 27.1).

A number of studies have documented that TBP increases sensitivity of melanoma detection [3,13,17,20–22]. This is achieved in a number of ways. Several pigmented lesion studies have shown that approximately 30% of melanomas are recognized solely based on TBP [13,17,22], whereas other studies show improvements in sensitivity by documenting that patients undergoing TBP have a higher incidence of melanoma in situ, lower incidences of invasive melanoma, and decreased average tumor thickness [13,20,21]. It is extrapolated that because this technology can assist with early detection, it may increase a dermatologist's sensitivity in detecting melanoma.

The use of TBP also increases the ratio of malignant to benign biopsies [13,20,21,23], presumably leading to an improvement in specificity. To date, there are no studies available to provide specific sensitivity and specificity data on TBP.

In addition to these benefits, TBP decreases biopsy rates [7,13,21], improves the sensitivity and specificity of patient skin self-examinations (60–72.4% and 96.2–98.4%, respectively) [11,24], decreases patient anxiety [25], and is particularly helpful in older populations who are less likely to have benign changing or new nevi [21].

TBP has numerous limitations. As mentioned, the evidence supporting the efficacy of TBP is limited [26,27]. It is time consuming, adding 20–30 min to initial visits and 10–20 min to follow-up visits. In order to reposition the patient in the same positions as the original photographs, extra time is required to pose patients in the exact positions [7,8,11,14,28,29]. This increased time requirement can take away from consideration of other important clinical factors [30]. The usefulness of the images is dependent upon their quality. Therefore pictures require trained office staff with the proper equipment and lighting or a professional photographer. The costs associated with these requirements are typically not reimbursed by insurers [11,26,27,29,31,32].

In addition, there is a wide range in numbers of photographs taken (6–85 images) and a wide range in the resolution of these images. They do not carry the same detail as SDD. Each photograph taken for TBD may have his or her own technique and poses for taking the images. Thus images are not standardized and may not be comparable between patients.

There are no consensus guidelines on who should be monitored with TBP and it is therefore not possible to calculate the number of patients who need to be monitored with TBP to diagnose one melanoma. Categories of individuals who could be considered for TBP include those with a personal history of melanoma, strong family history, a large number of common nevi, those with atypical nevi, those with an increased risk of melanoma, high-risk patients, those with atypical nevi and a history or melanoma, and those with "complex" skin examinations [8,28,33,34]. In addition, certain groups such as those younger than 30 years old and pregnant women are more likely to have new or changing lesions [11,21,30]. It is unclear how helpful TBP is in these subgroups.

There are no formal criteria for determining which new or changing lesions need to be biopsied at follow-up examination [7,26,27,33]. There is no standard follow-up period for these patients (ranges 3–12 months) [7,18,35] and Goodson et al. noted that only 43% of the patients in their study returned for follow-up examination [7].

TABLE 27.2 Limitations of Total Body Photography

Limited evidence [26,27]
Time consuming [7,8,14,28–30]
Nonreimbursed costs
Requires professional photographer
Variability in resolution
Lack of guidelines on who should utilize total body photography (TBP)
Lack of agreement on criteria for which new or changing lesions should be biopsied at follow-up examination
Subjectivity of follow-up period
Noncompliance with follow-up visit

TBP has distinct advantages over visual screening alone, in particular aiding in detection of melanomas on previously nonconcerning skin, increased detection of thinner lesions, lower biopsy rates, and the improvement of self–skin examination. However, these benefits are restricted by increased time and cost associated with each visit, limited resolution, lack of consensus on screening and future biopsy criteria, and the very real risk that a patient may never return for a follow-up visit (Tables 27.1 and 27.2).

SERIAL (SEQUENTIAL) DIGITAL DERMOSCOPY

SDD is the capture and sequential monitoring of dermoscopic images of single or multiple lesions. For the purpose of this chapter, we will focus on the use of SDD with reference to pigmented lesions, but the same technique could be used to monitor other clinically relevant lesions as well. These images can be taken on film or, now predominately, digitally. They can be printed or stored electronically for future reference.

The use of SSD for monitoring melanocytic lesions was first reported by Stolz et al. and Braun et al. in the 1990s [36,37]. These initial studies have been followed by additional studies, which showed that the utilization of SSD has both advantages and limitations (Tables 27.3 and 27.4).

These advantages are distinct from those of live dermoscopy. The benefits can be categorized as related to improvements in provider monitoring of lesions and indirect benefits. Haenssle et al. reported that the use of SSD was particularly helpful in the evaluation of high-

TABLE 27.3 Advantages of Serial Digital Dermoscopy

Improvements in provider monitoring
Helpful in evaluating high-risk patients [38]
Improves sensitivity over dermoscopy alone [38–41]
Aids in the identification of featureless melanomas [15,38,41–48]
Aids in identification of dermoscopic changes in nevi, especially dysplastic nevi [32,41,46,47]
Decreases unnecessary biopsies [32,41,45–47,49]
Indirect benefits
Improves self-examination training [41,51]
Improves patient–doctor relationship [41]
Increases compliance with short-term follow-up visits [45,46]

risk patients and increased the sensitivity and positive predictive value of melanoma screening at follow-up visits [38]. SDD is extremely sensitive to the detection of changes over time [32], with multiple studies documenting an increase in sensitivity over dermoscopy alone [38–41]. Given that these studies were in vivo and nonconcerning lesions were not biopsied for histopathological diagnosis, sensitivities cannot be calculated. Specificity varies but multiple studies have documented values between 83% and 84% [42,43].

By evaluating stored dermoscopy images of melanomas that were excised at serial follow-up examination without the benefit of the initial stored images, Kittler et al. and Haenssle et al. were able to show that these melanomas would likely not have been biopsied without the information regarding change. Therefore, SDD helped detect melanoma during an inconspicuous period when the melanomas had no dermoscopy-specific features [40]. Additional studies support these findings; specifically, incipient melanomas may lack dermoscopic features associated with melanoma [15,44], and SSD can aid in the identification of changes in these featureless melanomas or increase early stage detection [15,38,41–43,45–48].

TABLE 27.4 Limitations of Serial Digital Dermoscopy

Time consuming: 30–60 min per visit [38,39,41,43,47]
Require additional staff time [48]
Increased costs [39,52]
Requires expert examiners [39,48,46,51]
Image problems [42]
Cannot assist in the diagnosis of de novo melanomas [32]
Cannot assist in the diagnosis of nevi not previously imaged [32]
No consensus regarding which lesions should be imaged [50,54–16]
No consensus regarding changes requiring biopsy at follow-up visit [32,37,38,41,48,51]
Still need to perform a full skin examination [42,51]
Danger of monitoring melanoma instead of biopsy at first visit [40]
Lack of consensus on follow-up periods [41–43,45,48,56]
Low patient compliance with follow up [48]

The utilization of SSD also aids in the identification of dermoscopic changes at follow-up visits, especially in dysplastic nevi [32,41,46,47].

In addition to assisting with which lesions need to be biopsied, numerous studies report that the use of SDD also decreased cases of unnecessary excisions [32,41,45–47,49]. Some authors argue that given the low rate of change appreciated in atypical appearing nevi, the reassurance of patients and decreased biopsy rate is the most prevalent application of the SDD to atypical nevi [50]. The benefits also extend into improvements in self-examination training [41,51], patient–doctor relationships [41], and compliance with short-term follow-up visits [45,46].

However, the benefits associated with the utilization of SDD are not without limitations. Similar to TBP, the use of SDD is time consuming. Office visits can take an additional 30–60 min per patient [38,39,41,43,47] and require additional staff for taking and managing images [48]. All of these requirements increase the cost–benefit ratio for this technology [39,52]. The use of this technology requires examiners who are highly experienced in dermoscopy [39,46,48,51]. Problems exist surrounding functional confounders, including changes occurring from distortion of the skin or other errors in photography [42]. The application of SDD is also limited by melanomas that arise in nevi that were not initially imaged and those that arise de novo in normal appearing skin [32].

It is unclear how many lesions need to be monitored to detect a single melanoma, because the risk of melanoma developing in an existing nevus is fairly low, with recent estimates ranging from one melanoma arising in thousands of regular melanocytic nevi to hundreds of atypical nevi [52,53]. Providers still need to perform a full skin examination of all lesions, including nonimaged lesions [42,51]. There is a possibility that equivocal or suspicious lesions may be monitored digitally instead of biopsied at the patient's initial visit [40]. Furthermore there is an absence of specific guidelines stratifying whether these lesions should be biopsied, monitored with SSD, or monitored without SSD [50,54–56].

Some studies show that many lesions eventually biopsied because of changes noted on SDD may have been biopsied even if SDD was not utilized, because patients also identified them as changing or concerning [7]. In addition, there is controversy regarding appropriate follow-up periods (3–12 months) after initial screening [41–43,45,48,56]. This variation is based on different assumptions regarding whether melanomas are fast growing, allowing recognition of change over a series of months, or whether they are slow growing, requiring prolonged surveillance [45].

Kittler et al. have examined this issue and suggested that follow-up times should vary [40]. The short-term, 3-month follow-up visit should be utilized to monitor individual concerning lesions that lack specific dermoscopic features of melanoma [40], with biopsy of any changing lesions [40,43,51]. Long-term monitoring at 6–12 months is recommended for patients with multiple atypical nevi or other high-risk individuals [40]. However, there is no consensus regarding what changes require biopsy, with some providers excising lesions with any change and other providers only excising lesions that meet specific criteria at long-term follow-up visits [32,37,38,41,48,51].

Patient compliance with return visits also makes it difficult to study the ideal time until the follow-up examination. Rates for follow-up clinical visits vary by length of SDD follow-up checks. Compliance with 3-, 6-, and 12-months follow-up visits vary from 84% to 63% and 30%, respectively [45]. This casts additional doubts on effective verses ideal follow-up times.

Even when patients keep scheduled appointments there is a lack of consensus on what long-term monitoring is needed after the initial and first return visit before malignancy can be excluded [38,41,42].

Similarly to TBP, SDD has distinct advantages over visual screening alone. It increases the sensitivity and specificity of screening, can capture melanomas before dermoscopic features of melanoma become apparent, decreases unnecessary excisions, improves doctor–patient relationships, and increases compliance with follow-up visits. Unfortunately, this technology is also has a great deal of limitations. It is very time consuming, requires expert operators, relies on patient compliance with follow-up examinations, does not assist with melanomas arising de novo or in previously nonconcerning nevi, and has a lack of consensus regarding criteria for use, follow-up care, and biopsy.

TOTAL BODY PHOTOGRAPHY WITH SERIAL DIGITAL DERMOSCOPY

Given the previously mentioned limitations of TBP and SDD, a number of authors have advocated a combined two-step approach with either the selection of screening technique based on the unique characteristics of the patient or the incorporation of both imaging modalities at each visit [14,47,50,57–59]. The combined technique improved the diagnosis of de novo and featureless melanomas over dermoscopy alone [14], located lesions that were thinner than melanomas detected by visual screening alone [59], and may be more sensitive and specific than either method alone [47].

Whereas such a two-step approach reduces some limitations unique to each modality, such as the lack of evaluation of de novo or previously nonconcerning lesions on SDD and the lack of resolution on TBP, it has the potential to worsen others. The combination of both technologies is more time consuming than the use of either alone and would compound costs. In addition, combination use does not solve problems of criteria for monitoring and biopsy and continues to rely heavily on patient compliance.

CONCLUSION

Melanoma is increasing in incidence and now ranks as the fifth most common cancer. Mortality of melanoma increases with advanced disease and early disease detection improves prognosis. As such, there is an increased focus on melanoma screening.

TBP and SDD have distinct advantages over melanoma screening without the aid of previous images. Most importantly, these screening techniques improve sensitivity and specificity of melanoma detection and lead to the diagnosis of early staged disease. Given the central role early detection plays in survival these benefits cannot be under estimated. However, the current limitations of these technologies undermine their practicality.

Given the incidence of melanoma among Caucasians, one would have to image 10,000 patients yearly to capture one melanoma [30]; therefore guidelines are needed to determine which patients may benefit from these procedures and which would not. This is extremely difficult because of the lack of guidelines for screening of melanoma in general [60] and the lack of general acceptance of what an appropriate dollars per quality-adjusted life year ($/QALY) gained would be for cancer screening. The most recent study related to visual melanoma screening states that in the US population, one-time visual screening for patients over 50 years of age and screening every 2 years for siblings of patients with melanoma have $/QALY ratios comparable to colorectal, breast, and cervical cancer screening [61]. At the time of publication, the authors were not aware of any studies estimating $/QALY for TBP- or SDD-based screening.

The increased time, resources, and lack of reimbursement are major barriers to widespread utilization of TBP and SDD. Although there are a number of imaging and evaluation products specifically aimed at decreasing the time needed for each visit, these improvements themselves have significant costs. There are currently products on the marked designed to streamline initial imaging and follow-up visits. Technological advances that allow for further increases in accuracy and faster comparison combined with policy changes to allow for reimbursement could dramatically change the prevalence of these screening techniques. However, at this phase it is unclear whether the use of either of any of these techniques will become widely available.

References

[1] Howlader NNA, Krapcho M, Garshell J, Miller D, Altekruse SF, Kosary CL, et al., editors. SEER Cancer Statistics Review. Bethesda, MD: National Cancer Institute; 1975-2011. http://seer.cancer.gov/csr/1975_2011/.

[2] Tsao H. Management of cutaneous melanoma. N Engl J Med 2004; 351(10):998.

[3] Tucker MA, Fraser MC, Goldstein AM, Elder DE, Guerry D, Organic SM. Risk of melanoma and other cancers in melanoma-prone families. J Invest Dermatol March 1993;100(3):S350–5.

[4] Kopf AW, Welkovich B, Frankel RE, et al. Thickness of malignant-melanoma: global analysis of related factors. J Dermatol Surg Oncol April 1987;13(4):345.

[5] Blum A, Brand CU, Ellwanger U, et al. Awareness and early detection of cutaneous melanoma: an analysis of factors related to delay in treatment. Br J Dermatol November 1999;141(5):783–7.

[6] Breitbart EW, Waldmann A, Nolte S, et al. Systematic skin cancer screening in Northern Germany. J Am Acad Dermatol February 2012;66(2):201–11.

[7] Goodson AG, Florell SR, Hyde M, Bowen GM, Grossman D. Comparative analysis of total body and dermatoscopic photographic monitoring of nevi in similar patient populations at risk for cutaneous melanoma. Dermatol Surg July 2010;36(7):1087–98.

[8] Shriner DL, Wagner RF, Glowczwski JR. Photography for the early diagnosis of malignant-melanoma in patients with atypical moles. Cutis November 1992;50(5):358–62.

[9] Atkinson JM, From L, Boyer R. A new method of photo-documentation for the follow-up of dysplastic naevi. J Audiov Media Med 1987 January 1987;10(1):12–4.

[10] Slue W, Kopf AW, Rivers JK. Total-body photographs of dysplastic nevi. Arch Dermatol August 1988;124(8):1239–43.

[11] Halpern AC. Total body skin imaging as an aid to melanoma detection. Semin Cutan Med Surg March 2003;22(1):2–8.

[12] Salerni G, Carrera C, Lovatto L, et al. Characterization of 1152 lesions excised over 10 years using total-body photography and digital dermatoscopy in the surveillance of patients at high risk for melanoma. J Am Acad Dermatol November 2012;67(5): 836–45.

[13] Kelly JW, Yeatman JM, Regalia G, Mason G, Henham AP. A high incidence of melanoma found in patients with multiple dysplastic naevi by photographic surveillance. Med J Aust August 1997; 167(4):191–4.

[14] Salerni G, Carrera C, Lovatto L, et al. Benefits of total body photography and digital dermatoscopy ("two-step method of digital follow-up") in the early diagnosis of melanoma in patients at high risk for melanoma. J Am Acad Dermatol July 2012;67(1): E17–27.

[15] Skvara H, Teban L, Fiebiger M, Binder M, Kittler H. Limitations of dermoscopy in the recognition of melanoma. Arch Dermatol February 2005;141(2):155–60.

[16] Rhodes AR. Intervention strategy to prevent lethal cutaneous melanoma: use of dermatologic photography to aid surveillance of high-risk persons. J Am Acad Dermatol August 1998;39(2):262–7.

[17] Tiersten AD, Grin CM, Kopf AW, et al. Prospective follow-up for malignant melanoma in patients with atypical-mole (dysplastic-nevus) syndrome. J Dermatol Surg Oncol January 1991;17(1): 44–8.

[18] Mackie RM, McHenry P, Hole D. Accelerated detection with prospective surveillance for cutaneous malignant-melanoma in high-risk groups. Lancet June 1993;341(8861):1618–20.

[19] Nehal KS, Oliveria SA, Marghoob AA, et al. Use of and beliefs about baseline photography in the management of patients with pigmented lesions: a survey of dermatology residency programmes in the United States. Melanoma Res April 2002;12(2): 161–7.

[20] Feit NE, Dusza SW, Marghoob AA. Melanomas detected with the aid of total cutaneous photography. Br J Dermatol April 2004; 150(4):706–14.

[21] Banky JP, Kelly JW, English DR, Yeatman JM, Dowling JP. Incidence of new and changed nevi and melanomas detected using baseline images and dermoscopy in patients at high risk for melanoma. Arch Dermatol August 2005;141(8):998–1006.

[22] Rivers JK, Kopf AW, Vinokur AF, et al. Clinical characteristics of malignant melanomas developing in persons with dysplastic nevi. Cancer March 1990;65(5):1232–6.

[23] Hanrahan PF, D'Este CA, Menzies SW, Plummer T, Hersey P. A randomised trial of skin photography as an aid to screening skin lesions in older males. J Med Screen 2002;9(3):128–32.

[24] Oliveria SA, Chau D, Christos PJ, Charles CA, Mushlin AI, Halpern AC. Diagnostic accuracy of patients in performing skin self-examination and the impact of photography. Arch Dermatol January 2004;140(1):57–62.

[25] Moye MS, King SMC, Rice ZP, et al. Effects of total-body digital photography on cancer worry in patients with atypical mole syndrome. JAMA Dermatol February 2015;151(2):137–43.

[26] Mayer JE, Swetter SM, Fu T, Geller AC. Screening, early detection, education, and trends for melanoma: current status (2007–2013) and future directions. I. Epidemiology, high-risk groups, clinical strategies, and diagnostic technology. J Am Acad Dermatol October 2014;71(4):12.

[27] Rice ZP, Weiss FJ, DeLong LK, Curiel-Lewandrowski C, Chen SC. Utilization and rationale for the implementation of total body (digital) photography as an adjunct screening measure for melanoma. Melanoma Res October 2010;20(5):417–21.

[28] Guitera P, Menzies SW. State of the art of diagnostic technology for early-stage melanoma. Expert Rev Anticancer Ther May 2011;11(5):715–23.

[29] Coverman MH. Total-body photographs of dysplastic nevi: who pays. Arch Dermatol April 1989;125(4):565.

[30] De Giorgi V, Grazzini M, Rossari S, Gori A, Scarfi F, Lotti T. Total body photography versus digital dermoscopic follow-up in the diagnosis of pigmented lesions. Dermatol Surg March 2011; 37(3):406–7.

[31] Slue WE. Total-body photographs of dysplastic nevi: reply. Arch Dermatol April 1989;125(4):566–7.

[32] Fuller SR, Bowen GM, Tanner B, Florell SR, Grossman D. Digital dermoscopic monitoring of atypical nevi in patients at risk for melanoma. Dermatol Surg October 2007;33(10):1198–205.

[33] Halpern AC. The use of whole body photography in a pigmented lesion clinic. Dermatol Surg December 2000;26(12):1175–80.

[34] Esmaeili A, Scope A, Halpern AC, Marghoob AA. Imaging techniques for the in vivo diagnosis of melanoma. Semin Cutaneous Med Surg March 2008;27(1):2–10.

[35] Rigel DS, Rivers JK, Kopf AW, et al. Dysplastic nevi: markers for increased risk for melanoma. Cancer January 15, 1989;63(2):386–9.

[36] Stolz W, Schiffner R, Pillet L, et al. Improvement of monitoring of melanocytic skin lesions with the use of a computerized acquisition and surveillance unit with a skin surface microscopic television camera. J Am Acad Dermatol August 1996;35(2):202–7.

[37] Braun RP, Lemonnier E, Guillod J, Skaria A, Salomon D, Saurat JH. Two types of pattern modification detected on the follow-up of benign melanocytic skin lesions by digitized epiluminescence microscopy. Melanoma Res October 1998;8(5):431–7.
[38] Haenssle HA, Krueger U, Vente C, et al. Results from an observational trial: digital epiluminescence microscopy follow-up of atypical nevi increases the sensitivity and the chance of success of conventional dermoscopy in detecting melanoma. J Invest Dermatol May 2006;126(5):980–5.
[39] Haenssle HA, Vente C, Bertsch HP, et al. Results of a surveillance programme for patients at high risk of malignant melanoma using digital and conventional dermoscopy. Eur J Cancer Prev April 2004;13(2):133–8.
[40] Kittler H, Guitera P, Riedl E, et al. Identification of clinically featureless incipient melanoma using sequential dermoscopy imaging. Arch Dermatol September 2006;142(9):1113–9.
[41] Robinson JK, Nickoloff BJ. Digital epiluminescence microscopy monitoring of high-risk patients. Arch Dermatol January 2004; 140(1):49–56.
[42] Altamura D, Avramidis M, Menzies SW. Assessment of the optimal interval for and sensitivity of short-term sequential digital dermoscopy monitoring for the diagnosis of melanoma. Arch Dermatol April 2008;144(4):502–6.
[43] Menzies SW, Gutenev A, Avramidis M, Batrac A, McCarthy WH. Short-term digital surface microscopic monitoring of atypical or changing melanocytic lesions. Arch Dermatol December 2001; 137(12):1583–9.
[44] Menzies SW, Ingvar C, Crotty KA, McCarthy WH. Frequency and morphologic characteristics of invasive melanomas lacking specific surface microscopic features. Arch Dermatol October 1996; 132(10):1178–82.
[45] Argenziano G, Mordente I, Ferrara G, Sgambato A, Annese P, Zalaudek I. Dermoscopic monitoring of melanocytic skin lesions: clinical outcome and patient compliance vary according to follow-up protocols. Br J Dermatol August 2008;159(2):331–6.
[46] Bauer J, Blum A, Strohhacker U, Garbe C. Surveillance of patients at high risk for cutaneous malignant melanoma using digital dermoscopy. Br J Dermatol January 2005;152(1):87–92.
[47] Malvehy J, Puig S. Follow-up of melanocytic skin lesions with digital total-body photography and digital dermoscopy: a two-step method. Clin Dermatol May–June 2002;20(3):297–304.
[48] Kittler H, Pehamberger H, Wolff K, Binder M. Follow-up of melanocytic skin lesions with digital epiluminescence microscopy: patterns of modifications observed in early melanoma, atypical nevi, and common nevi. J Am Acad Dermatol September 2000; 43(3):467–76.
[49] Tromme I, Sacre L, Hammouch F, et al. Availability of digital dermoscopy in daily practice dramatically reduces the number of excised melanocytic lesions: results from an observational study. Br J Dermatol October 2012;167(4):778–86.
[50] Kittler H. Digital dermoscopic monitoring of atypical nevi in patients at risk for melanoma: commentary. Dermatol Surg October 2007;33(10):1205–6.
[51] Schiffner R, Schiffner-Rohe J, Landthaler M, Stolz W. Long-term dermoscopic follow-up of melanocytic naevi: clinical outcome and patient compliance. Br J Dermatol July 2003;149(1):79–86.
[52] Bauer J, Garbe C. Risk estimation for malignant transformation of melanocytic nevi. Arch Dermatol January 2004;140(1):127.
[53] Tsao H, Bevona C, Goggins W, Quinn T. The transformation rate of moles (melanocytic nevi) into cutaneous melanoma: a population-based estimate. Arch Dermatol March 2003;139(3):282–8.
[54] Bowling J, Argenziano G, Azenha A, et al. Dermoscopy key points: recommendations from the International Dermoscopy Society. Dermatology 2007;214(1):3–5.
[55] Stanganelli I, Ascierto P, Bono R, et al. Impact of mole mapping in the Italian health system. Dermatology May 2013;226:13–7.
[56] Carli P, Ghigliotti G, Gnone M, et al. Baseline factors influencing decisions on digital follow-up of melanocytic lesions in daily practice: an Italian multicenter survey. J Am Acad Dermatol August 2006;55(2):256–62.
[57] Moloney FJ, Guitera P, Coates E, et al. Detection of primary melanoma in individuals at extreme high risk a prospective 5-year follow-up study. JAMA Dermatol August 2014;150(8):819–27.
[58] Lucas CR, Sanders LL, Murray JC, Myers SA, Hall RP, Grichnik JM. Early melanoma detection: nonuniform dermoscopic features and growth. J Am Acad Dermatol May 2003; 48(5):663–71.
[59] Rademaker M, Oakley A. Digital monitoring by whole body photography and sequential digital dermoscopy detects thinner melanomas. J Prim Health Care December 2010;2(4):268–72.
[60] Heymann WR. Screening for melanoma. J Am Acad Dermatol January 2007;56(1):144–5.
[61] Losina E, Walensky RP, Geller A, et al. Visual screening for malignant melanoma: a cost-effectiveness analysis. Arch Dermatol January 2007;143(1):21–8.
[62] Masri GD, Clark Jr WH, Guerry Dt, Halpern A, Thompson CJ, Elder DE. Screening and surveillance of patients at high risk for malignant melanoma result in detection of earlier disease. J Am Acad Dermatol 1990;22:1042–8.
[63] Yao XY, Fernandes BJ. The relation of deoxyribonucleic acid contents and nuclear estrogen receptors in breast cancers. Zhonghua bing li xue za zhi Chin J Pathol 1991;20:28–31.63.

CHAPTER

28

Functional MRI Advances to Reveal the Hidden Networks Behind the Cerebral Processing of Itch

Alexandru D.P. Papoiu

Therapeutics Inc., San Diego, CA, United States

OUTLINE

INTRODUCTION

Itching and scratching are so closely intertwined that the currently accepted definition of itch formulated by Hafenreffer, which dates from 1660 [1], actually includes both. Itch is an undesired, particularly intrusive, irritative sensation, which rapidly and irresistibly triggers a scratching response.

In the past two decades, state-of-the-art neuroimaging techniques have been developed and have becomemore widely available to probe the mysteries of brain function. The cerebral processes underlying itch perception have not escaped this veritable methodological revolution in neuroscience. Advances in functional magnetic resonance imaging (fMRI) have enabled the visualization and mapping of brain responses evoked by itch stimulation. Two major fMRI methods are available: blood oxygen level dependent (BOLD) fMRI and perfusion fMRI, also known as arterial spin labeling (ASL). Of the two, ASL is well-suited to capture the long-lasting effect of itch on cerebral activity. ASL can be performed either as a two-dimensional echo planar

Imaging in Dermatology
http://dx.doi.org/10.1016/B978-0-12-802838-4.00028-5

imaging (EPI) technique or a three-dimensional (3D) spiral method, such as the 3D gradient echo and spin echo (GRASE) combined with periodically rotated overlapping parallel lines with enhanced reconstruction (PROPELLER) method [2]. Both modalities have been used successfully in the study of itch. As an example, we have employed ASL to analyze and compare the patterns of cerebral activation evoked by histamine and cowhage itches [3], or the cerebral mechanism of itch relief provided by active scratching, in contrast with passive scratching [4]. Thanks to the inherent methodological advantage that arises from the ability to quantify cerebral perfusion in absolute terms, ASL enables the comparative analysis of itch responses in healthy individuals and chronic itch sufferers such as patients with atopic dermatitis (AD) [5] or pruritus of end-stage renal disease (ESRD) [6]. We have recently employed ASL in a pharmacological fMRI study aiming to identify the essential brain structures that mediate itch inhibition exerted by the opioid drug butorphanol [7].

Electrophysiological studies using antidromic stimulation in nonhuman primates (macaques) showed that itch information is transmitted from the spinal cord to ventrobasal and posterior thalamic nuclei via the anterolateral funiculus [8,9]. The suprathalamic, and possibly parathalamic, projection of itch information to cortical areas and other brain structures has been largely inferred from neuroimaging studies. Irrespective of the technical principle used to generate contrast [positron emission tomography (PET), or BOLD or ASL fMRI], imaging studies consistently showed that itch stimulation induces a complex, multidimensional cerebral response, represented in multiple cortical and subcortical regions that process sensory-discriminative, cognitive, affective, and memory-related dimensions of this sensory experience. Itch information registers in the primary and secondary somatosensory areas (S1 and S2), but it is also distributed widely at cortical and subcortical levels. Almost invariably, itch engages associative parietal regions of the supramarginal gyrus, the angular gyrus, and the precuneus (the medial posterior parietal cortex). Itch stimulation also activates insula, a salience and interoceptive center, and the neighboring claustrum, a thin sheet of gray matter (GM) implicated in the detection of fast stimuli and multisensory integration [10]. The multiplicity of reciprocal claustrocortical connections and the extensive functional relationships of the claustrum with many other structures implicated in itch processing, such as the anterior cingulate cortex (ACC), the hippocampus, entorhinal cortex, amygdala, and septum, suggest that claustrum is a pivotal region engaged in the formation and modulation of itch perception. Other higher-order associative cortical areas that appear to play a relevant function in encoding cognitive aspects of itch are located in the frontal lobe: medial and dorsolateral prefrontal cortex, superior frontal gyrus, Brodmann areas 8, 44–45, temporal gyri, including superior temporal, middle temporal, the superior parietal lobule and temporooccipital areas of the fusiform gyrus, as well as the visual associative cortex. The highly emotional nature of itching translates into the activation of deep-seated areas of the cingulate cortex, amygdala, hippocampus, and mammillary bodies situated along the Papez circuit [3,11]. The therapeutic management of clinically persistent forms of itch, commonly referred to as *chronic itch*, is still challenging, indicating that itch is a primary sensation not easily suppressible. From a fundamental point of view, the essential question functional neuroimaging is tasked to answer is which brain areas are critically involved in the formation of itch sensation, and among them, which ones could become suitable targets for medical or pharmacological intervention. Several psychophysical and neuroimaging studies have sought to uncover clues at the higher levels of the CNS to decipher a central mechanism for itch inhibition. Various experimental interventions to relieve itch were used: passive and active scratching, acupuncture or thermal modulation, and even mirror scratching [4,12–17]. The results of our recent fMRI study of active self-scratching performed in healthy individuals suggested that the reward-processing areas of the midbrain, ventral striatum, and associated structures may encode not only the pleasurable aspect of scratching, but may be critically involved in the execution of itch relief [4]. Notably, areas of the brain related to reward processing express high levels of opioid receptors. Therefore an alternative experimental approach we have used in a pharmacological fMRI study of itch inhibition was to stimulate intrinsic brain opioid receptors by exogenous administration of pharmacological agents, such as butorphanol, a drug that is, in fact, effective clinically in the treatment of refractory pruritus [18]. Butorphanol is a mixed-action opioid drug (μ antagonist/κ agonist) with a more pronounced affinity towards κ opioid receptors. κ-Agonists have been well-studied and demonstrated numerous times previously to exert a powerful antipruritic effect [19–22]. The κ-agonist nalfurafine is approved for the treatment of uremic pruritus in Japan, and it is being investigated in a multicenter trial in the United States. μ-Receptor agonists such as morphine have a pruritogenic effect that accompany their analgesic action, whereas μ-receptor antagonists and κ-agonists produce the opposite result (relieve pruritus). Administered epidurally, butorphanol blocks the itch induced by the epidural injection of morphine [23]. Currently, the interaction of opioid signaling pathways with itch transduction mechanisms is a hot area of research [24,25a].

To visualize by fMRI the cortical representation of itch, various itch inducers have been used. Histamine, spicules of the velvet bean (cowhage, *Mucuna pruriens*), or allergens in susceptible individuals (atopics) are among the commonly agents used in humans.

NEUROPHYSIOLOGICAL ASPECTS OF ITCH TRANSMISSION

The elucidation of itch neurophysiology was complicated by the baffling multiplicity of substances that can induce it. In humans, itch can be triggered by small molecules such as histamine, substance P, serotonin, acetylcholine, β-alanine, endothelin-1, neuropeptides such as BAM8–22, and the antimalarial drug chloroquine, but also larger ligands such as cytokines or enzymes (proteases). The receptors for these various ligands are expressed not only in the epidermal/dermal nerve fiber endings, but also in a myriad of cell types such as keratinocytes and immune system cells present in the skin. The currently established consensus is that itch is transmitted via distinct histamine-dependent and histamine-independent pathways, through unmyelinated C and thinly-myelinated Aδ fibers that synapse in lamina I and II of the dorsal horn of the spinal cord. One particular form of histamine-independent itch that has been more intensively studied is the cowhage-induced itch. This form of itch is triggered when skin comes in contact with the spicules covering the pods of this tropical plant. As originally postulated by Shelley and Arthur in 1955 [25b,c], these spicules act as microneedles that upon skin contact release a cysteine protease called *mucunain*. Mucunain was isolated as a 36 kDa protein and its targets were identified [26] as protease activated receptors PAR2 and PAR4. PAR2 is expressed in the nerve terminals in the skin and on keratinocytes. (PAR4 has not yet been identified in human skin).

Histamine is the prototypical experimental pruritogen, traditionally associated with urticaria and allergic responses. However, PAR2-mediated itch is thought to more closely resemble certain forms of chronic pruritus of pathological origin. Several studies have implicated the PAR2 itch pathway in atopic eczema [27] or pruritus of ESRD. Increased tryptase serum levels have been described in chronic kidney disease patients on hemodialysis, correlating with itch severity, as well as in patients with AD [28,29]. Common features, but also significant differences in the cortical processing of histamine and cowhage/PAR2 itch modalities have been described [3]. A high expression level of PAR2 in the skin of ESRD patients suffering with pruritus has been reported [30]. Supporting these observations, we have discovered a significantly different pattern of cerebral processing for cowhage (PAR2-mediated) itch, but not of histamine itch in ESRD patients receiving hemodialysis and suffering with chronic itch, when analyzed in comparison with healthy individuals [6]. Indeed, the lack of therapeutic efficacy of the currently available antihistamines in chronic pruritus has been well-known and has constituted a major argument in favor of the idea that many forms of pruritus encountered in clinical practice may have a fundamentally different, histamine-independent mechanism.

Brain imaging studies have long attempted to decipher mechanisms of itch processing at the highest levels of the CNS, with the secondary aim to understand neural mechanisms that can be used for itch therapy.

Types of Information fMRI Can Provide

- Anatomical mapping identifies the key brain areas serving a certain central nervous system function or activity. It is important to be able to identify the brain structure where a certain response occurs. Functional MRI offers a noninvasive unparalleled opportunity and a transparent window into brain function, surpassing anatomical barriers.
- Functional MRI can explore dynamic brain responses triggered by various sensory experiences (such as itch or pain) in experimentally controlled conditions, and reveal cognitive operations performed during task performance (eg, in relation to itch, active scratching), which represents the core principle of functional imaging.
- Functional MRI enables testing potential correlations between cerebral responses and experimentally measurable variables defining the phenomenon under study. Brain imaging is assisted by sophisticated statistical approaches to obtain additional information through regression analysis. The intricate mechanism of itch processing could be better understood through an in-depth analysis of the activation signal in relation to the itch intensity ratings reported inside the scanner. The FMRIB Software Library (FSL; University of Oxford; Oxford, England, UK) is equipped with complex analytical capabilities that allow testing the influence of covariates of interest. The results are ultimately expressed as geometric loci on maps of brain areas linked to the factor studied.
- With resting state fMRI, it is now feasible to quantify cerebral blood flow (CBF) in baseline conditions in healthy individuals as well as in patients with various pathological conditions, which further allows a statistically competent comparison [6].
- All MRI imaging techniques using functional paradigms also record high-resolution structural images of the

brain. These structural images are used for anatomical registration (localization) of the functional signal, but they can be used separately to investigate changes in GM or white matter structure (density) in different states or groups (patients or healthy individuals). For example, we have investigated changes in GM density in ESRD patients with chronic pruritus in comparison with an age- and gender-matched group of healthy volunteers. These types of analyses were also performed extensively in patients with chronic pain [31].

- Pharmacological fMRI represents the combination of functional neuroimaging with the administration of centrally acting drugs, and it is used to identify drug targets in the brain and to explore the underlying mechanism of action. These studies typically demand a rigorous, controlled, randomized design, including the use of proper controls: a placebo and a perfusion control scan for the drug's effect on cerebral circulation, in addition to a baseline perfusion control.

Neuroimaging Techniques That Have Been Used to Investigate the Brain Responses Elicited by Itch

Positron emission tomography (PET)
Functional MRI
Magnetoencephalography (MEG)
The combination of MEG with fMRI
Near-infrared (NIR) spectroscopy

A BRIEF CHRONOLOGY OF BRAIN IMAGING STUDIES DEDICATED TO THE EXPLORATION OF ITCH PROCESSING

Historically, the first neuroimaging study dedicated to itch processing in the brain used PET with ^{15}O-butanol as a radiotracer infused intravenously in systemic circulation [32]. Generally, PET follows the distribution, tissue uptake, and, depending on the source, tissue utilization of radioactive (β^+) tracers emitting positrons, which are unstable, short-lived radioactive isotopes, eg, ^{15}O, ^{18}F, or ^{11}C, synthetically/chemically inserted into various carrier molecules (such as H_2O or glucose). The principle employed in functional PET neuroimaging is that the signal is proportionally correlated with the cerebral metabolic activity, owing to cerebral activity–hemodynamic coupling [33]. With a few subtle differences, the same principle applies to fMRI. Among the significant limitations of PET is that it is an invasive technique dependent on injectable radiotracers with a very short half-life ($t_{½}$; for ^{15}O, $t_{½} = 122$ s); hence experiments have to be carried out in a limited temporal window of opportunity because of the fast decay of the positron-emitting sources. PET is prohibitively expensive and involves some degree of radiation exposure to subjects. PET studies have evaluated changes in cerebral perfusion that occurred while subjects experienced itch, in contrast to a baseline (resting) state. The first PET study of itch described activations in the motor areas, including the supplementary motor area and premotor areas, which were interpreted as reflecting the intention or the urge to scratch, along with responses in the ACC (BA 24), inferior parietal lobule, and cerebellum. Other PET studies of itch that followed confirmed and expanded the list of brain structures involved in itch processing to include the contralateral somatosensory cortex and the prefrontal cortex [34–36].

Blood Oxygen Level Dependent Functional Magnetic Resonance Imaging

BOLD (fMRI) emerged as a novel technique of brain imaging study in the mid-1990s, and it was swiftly employed for the investigation of pain. It took about a decade before the first peer-reviewed BOLD article on itch was published [37], and it was soon followed by others [38–40]. The main technical challenge encountered in BOLD studies was the need to adapt the short temporal duration of the BOLD signal (or contrast) to the prolonged nature of the itch sensation, which evokes a long-lasting response in the brain, usually in the order of minutes. Several approaches were used to create a somewhat artificial "on and off" itch switch to induce itch responses and then shut them down in order to obtain a sufficient contrast that could be used in data analysis. These approaches used either counter-stimulation by pain stimuli, anesthetics (eg, lidocaine) delivered by skin microdialysis, thermal modulation (cold stimuli), or passive scratching performed by investigators. These combined interventions, however, were likely to complicate matters because the convoluted experimental designs may have impacted the brain responses under observation. The precise relationship between the BOLD signal and cerebral activity was not completely understood for many years and has been the subject of intense investigation and debate. The BOLD signal emerges from changes in the concentration of paramagnetic deoxyhemoglobin in cerebral venous circulation, indirectly reflecting a feedback hemodynamic adaptation to cerebral metabolic activity. Parallel studies employing electrophysiological recordings and BOLD suggested that the BOLD signal correlates with local field potentials and mostly reflects incoming information (input) and local, intracortical processing rather than the output of the brain area in question, or spiking activity [41]. Despite being performed in small groups of volunteers, the picture emerging from initial BOLD studies of itch was relatively consistent with the findings from PET studies. However, PET and BOLD fMRI studies did not always observe some of the most likely expected brain regions to be activated, such as S1 or S2, or the thalamus.

Arterial Spin Labeling

The introduction of ASL for the study of itch solved most if not all of the technical problems that have limited the applicability of BOLD fMRI. ASL has been developed through technical improvements in signal acquisition design and analytical breakthroughs in programming and image reconstruction, such as those developed by Roberts et al. [42] and Buxton et al. [43]. Before being used for the study of itch, ASL was successfully employed to visualize the brain processing of pain [44].

ASL fMRI studies rely on the general principle and physiological observation that CBF adapts proportionally to changes in neuronal activity [33]. ASL is a noninvasive technique that can quantitatively measure CBF and its dynamic changes that parallel transitions through various experimental states. ASL uses water molecules in the blood as an endogenous tracer to measure CBF. ASL contrast can also detect brain lesions, tumors, or cerebrovascular diseases. Originally ASL employed two-dimensional EPI trajectories and later, spiral acquisition trajectories. Newer techniques such as 3D GRASE PROPELLER have gained traction because of a higher signal-to-noise ratio (sensitivity) and better spatial coverage, enabling the imaging of the whole brain. A previous limitation of 3D GRASE, the through-plane blurring caused by T2 decay, was overcome by the implementation of the PROPELLER trajectory, without sacrificing perfusion sensitivity or increasing scan time [2]. 3D GRASE PROPELLER technology demonstrated reduced through-plane blurring, improved anatomical details, high repeatability, and robustness against motion, suitable for the brain imaging of itch and active scratching in human subjects. Compared with traditional contrast-based methods, ASL is noninvasive, repeatable, and quantifies CBF without using ionizing radiation as PET does. MRI contrast agents or radioactive tracers are not needed for ASL, an advantage in patients with renal failure or pediatric patients. The first ASL study of itch allowed a comparison of brain responses evoked by itch in healthy individuals and patients with AD [5].

It has been pointed out previously that ASL has certain advantages over BOLD fMRI [45a,b]: (1) The absolute measurement of CBF is useful in the study of itch where healthy volunteers and patients are compared, enabling the study of brain mechanisms in chronic conditions; (2) brain activation measured by ASL is less susceptible to intersubject variability; (3) ASL has a stable (predictable) "noise" over a larger frequency spectrum, being suitable for the study of long-term changes in brain function over periods longer than a few minutes [46], fitting for studying itch, a typically prolonged sensory experience; and (4) signal changes detected by ASL are considered to have a superior spatial and temporal resolution in comparison with BOLD [47,48]. Other brain imaging modalities such as MEG have been used [49]. MEG traces the changes in the electromagnetic field produced by the activity in the cerebral tissue. Caveats include the fact that despite its unsurpassed temporal resolution of the order of milliseconds, MEG has a limited ability to detect signal transversally oriented to the probe or to precisely identify its anatomical substrate. MEG is also less suited to study deep-seated structures, which are now known to be implicated in itch processing, such as the septal area, basal forebrain, hippocampus, amygdala, or midbrain. The brain processes underlining itch and pain have been investigated by NIR spectroscopy, but an association with the anatomical regions of the brain was difficult to achieve [50].

CEREBRAL AREAS SIGNIFICANTLY INVOLVED IN ITCH PROCESSING

Irrespective of the technique used, neuroimaging studies showed that brain responses evoked by experimental itch are complex and multidimensional. This was not entirely surprising, because the cortical representation of pain ("the pain matrix") had a similar wide distribution [31]. As a bothersome, intrusive sensation, itch prompts several arms of the cerebral response to deploy almost simultaneously to serve the refocusing of attention, planning motor action, and actively seeking itch relief. Irrespective of the investigative modality, imaging studies showed that motor, cognitive, emotional, mnemonic, and motivational processes are initiated and operate together in a cohesive response.

The Differential Brain Processing of Histamine and Cowhage Itches

Common features and notable differences were observed in the cerebral processing of these distinct itch modalities. Using ASL as an imaging technique, we were able to identify significant differences in the patterns of brain activity evoked by histamine and cowhage itch in healthy volunteers [3]. To date, no other study has succeeded in identifying significant differences in the cerebral processing of different itch modalities. In a BOLD fMRI study at a 3-T field, Leknes and colleagues induced itch either via histamine or allergens, but cerebral representation maps have not been compared [39]. In our ASL study, the brain activation evoked by these two itch pathways overlapped to a substantial extent, consolidating the idea that a core network is involved in itch processing, but also presented distinguishable particularities (Fig. 28.1). The pattern of brain activation captured by ASL for histamine itch was in agreement with previous PET and BOLD fMRI results reported in the literature [12,32,34–36,39,49]; in addition, we showed activations in multiple thalamic nuclei, and S1 and S2, which were not reported consistently before our study. Cowhage itch evoked a more extensive and higher activation of the insular cortex and claustrum, basal ganglia, thalamus, pulvinar, and ACC (BA 24) than histamine itch.

Somatosensory Areas

S1 and S2 have been invariably found activated by ASL imaging studies of itch, whereas the initial PET and BOLD fMRI studies did not reach a clear consensus on the involvement of S1 and S2. Activation of S1, contralateral to the stimulation side, and of S2, bilaterally, has been invariably detected by ASL, as expected for a sensory modality projected via the thalamus. Current experimental data indicate that projection of itch information to the postcentral gyrus (S1) is consistent with the classic distribution of the sensory (Penfield) homunculus, although a comprehensive analysis of the entire body surface has not been yet performed.

The Role of the Parietal and Frontal Associative Areas in Itch Processing

In most brain imaging studies, irrespective of the method used, the superior parietal lobule, along with the supramarginal and angular gyri of the parietal cortex were invariably involved in itch processing. These areas are specialized in spatial recognition and body image representation, and enjoy multiple connections with other associative areas of the brain; it is likely, therefore, that they assist to localize the stimulus within the body scheme. Activations of the inferior, middle, and superior frontal gyri, which serve higher-degree executive functions, cognitive processes, planning, decision making, and self-reflection, were also activated.

The Anterior Cingulate Cortex

The ACC has been constantly found to be engaged by itch processing. The ACC occupies a unique position in the brain, connecting the emotional limbic system with the cognitive prefrontal cortex. ACC includes BAs 24, 25, 32, and 33, but is also subdivided according to morphological and connectivity criteria into the rostral and dorsal ACC. The rostral ACC is further divided into the pregenual and subgenual ACC in rapport with genu corporis callosum. The posterior part of the dorsal ACC is called the *midcingulate cortex* (*MCC*). The pregenual ACC has extensive connections with areas processing emotion (amygdala), autonomic functions (lateral hypothalamus and brainstem),

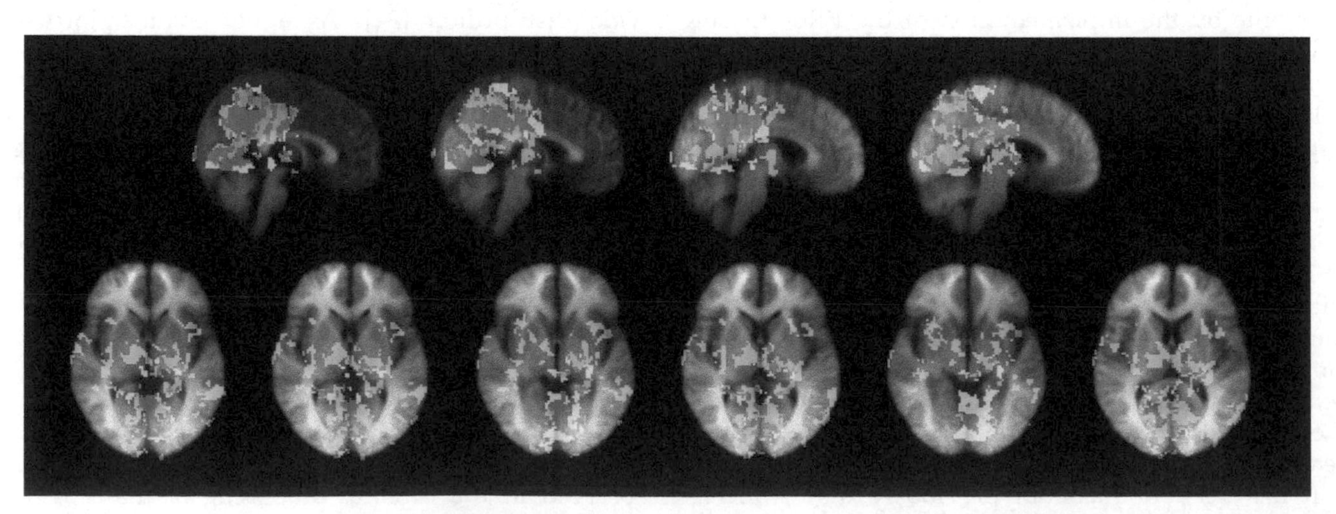

FIGURE 28.1 Processing of histamine and cowhage itches. Histamine itch, green; cowhage itch, purple; overlap of cerebral areas involved in the processing of both itch modalities, red (coactivation).

memory (hippocampal region), and reward (orbitofrontal cortex, ventral striatum). The MCC has extensive connections with cognitive areas (dorsolateral prefrontal cortex), motor areas, and thalamic nuclei. Interestingly, the midcingulate part of the ACC displayed an increased GM density in ESRD patients experiencing chronic itch.

Neuroimaging studies have consistently identified various segments of the ACC being engaged in itch processing, depending on the context. BA 25 (subgenual ACC) is unique in its expression of high-density of serotonin receptors. The subgenual ACC is one the areas proposed to mediate histamine itch relief induced by butorphanol, along with the septal nuclei and nucleus accumbens (NAc). The subgenual ACC has strong connections with the amygdala and projects to the ventral striatum and autonomic centers in the hypothalamus and brainstem.

Brodmann areas 24, 25, and 32 of the ACC are stimulated by experimental itch induction, and are deactivated by self-scratching an itch. The activation of ACC reflects processing of emotional and motivational aspects, motor-planning, and cognitive evaluation of itch. Another role attributed to the ACC is the evaluation of reward. Itch relief is perceived as a form of reward when produced by self-scratching and is prominently represented in the ACC. Whereas different subdivisions of the ACC activated by itch were differentially impacted by active scratching because some were activated and others were deactivated, virtually all ACC subdivisions were correlated significantly with the pleasurability of scratching and with itch relief (Table 28.1). In pathological itch conditions, a high level of perfusion in the pregenual ACC (BA 24) was found at rest in ESRD patients with persistent itch. ACC is connected with the insula, claustrum, the prefrontal cortex and the thalamus, areas that appear essential in the formation of itch perception. Subgenual ACC (BA 25), a formation mediating the butorphanol inhibition of itch [7], is connected with NAc and other reward processing structures such as the ventral tegmental area (VTA). Its involvement may explain the intense craving for relief as the driving force of the itch—scratch cycle, which is observed in chronic itch patients, and suggests a common neurophysiological basis for the persistent itch and scratching, on a par with other highly addictive experiences. ACC activations were correlated with itch intensity and the degree of disease severity in atopic eczema patients alongside with insula [5]. A dysfunction of central circuits processing the emotional aspects related to itch may create an imbalance in the reward controlled by the ACC in regulating craving, motivation, and addiction, and could lead to excessive (unrewarding/unbalanced) deleterious scratching, as seen in many, if not all, clinical forms of chronic pruritus. ACC is also connected with the hippocampus, which has an essential function in the processing of nociceptive and pruritoceptive information. The posterior cingulate cortex has also been implicated in itch processing. Its main roles are considered to relate to the evaluation of sensory experiences, memory, and cognition [3].

Insula

The insula is a cortical region linked with salience detection, self-awareness, interoception, pain processing, and addiction [51]. The insula is a major hub for visceroceptive or interoceptive inputs, and an essential component of the pain matrix, being involved in the assessment of nociceptive stimulus intensity. Experimentally induced histamine and cowhage itch extensively activated the anterior and posterior divisions of the insula. Our analyses of baseline cerebral perfusion in resting conditions in ESRD patients (undergoing hemodialysis) experiencing chronic pruritus, found insula activated bilaterally (highly perfused) at rest (Fig. 28.2).

Claustrum

Among the areas discovered to be involved by itch processing, the claustrum commands special attention. It is useful to discuss the claustrum in the context of recent discoveries related to its role in sensory processing, detection of salient stimuli, and sensory-motor integration. The claustrum is a thin GM structure, intercalated between insula and putamen and surrounded by external and extreme capsules in the human brain [52]. Several studies have showed that claustrum is associated with itch processing [3,4,7,38]. The claustrum's extensive cortical connectivity and specialization in detecting fast sensory stimuli is fitting for a region involved in itch sensing. The ability to compare and integrate several sensory modalities has been attributed to the claustrum, because it is connected to almost all areas of the cortex, including the somatosensory and motor cortices, thalamus, and limbic structures such as the cingulate cortex, hippocampus, septum, and amygdala. The claustrum is activated more extensively by a transient, fluctuating itch stimulus, such as cowhage, than by a constant stimulus like histamine [3] . In neuroimaging studies, the activation of the claustrum and insula was correlated with perceived itch intensity; in addition, discrete claustral zones were invariably activated during itch stimulation irrespective of itch stimulus intensity. The claustrum is among the very few brain areas (that also include the insula, S2, and the angular gyrus) that showed this dual pattern of activation. The insula and claustrum activate when

TABLE 28.1 Regions of Interest Involved in the Brain Processing of Itch and the Impact of Active or Passive Scratching on Their Activity. The Correlations Between Changes in Cerebral Perfusion Evoked by Scratching, the Associated Pleasurability, and Itch Relief Were Evaluated by Regression Analyses

Brain area	Itch	Actively scratching an itch	Passively scratching an itch	Correlation with pleasurability of active scratching[a]	Correlation with itch relief induced by active scratching
Primary somatosensory area	Activated	Activated	Activated	Inverse Brodmann area (BA) 3; Z = 4.8	Inverse BA 1,2,3; Z = 4.4
Secondary somatosensory area	Activated	Activated	Activated	Inverse BA 40; Z = 2.3	Inverse BA 40; Z = 3.2
Primary motor area	Activated	Activated	Deactivated	Inverse BA 4; Z = 5.7	Inverse BA 4; Z = 5.4
Supplementary motor area	Activated	Activated		Inverse BA 5,6,7 Z = 4.3	Inverse BA 5,6,7; Z = 5.5
Prefrontal cortex					
Dorsolateral	Activated	Activated		Positive BA 9,46; Z = 10 Inverse BA 8; Z = 6.9	Inverse BA 8; Z = 4.8
Frontal pole	Activated	Deactivated	Deactivated	Positive Extensive BA 10; Z = 10.0	
Orbitofrontal	Activated	Deactivated	Deactivated	Positive BA 11; Z = 8.1	
Ventrolateral	Activated	Deactivated	Deactivated	Positive BA 45; Z = 3.2	
Anterior cingulate cortex	Activated	Activated or deactivated in different regions	Deactivated	Positive BA 32,33; Z = 5.7	Positive BA24; Z = 3.5
Posterior cingulate cortex	Activated	Activated	Activated	Inverse (Extensive) BA 23, 31; Z = 5.4	Inverse (Extensive) BA 23, 31; Z = 9.7
Precuneus	Activated	Activated	Activated	Positive BA 31; Z = 8.3 Inverse BA 7; Z = 9.3	Positive BA 7; Z = 4.2 Inverse BA 31; Z = 8.8
Parahippocampus	Activated	Deactivated	Activated	Positive Z = 10.8	Positive Z = 9.0
Hippocampus	Activated	Deactivated	Activated	Positive Z = 4.7	Inverse Z = 5.2
Amygdala	Activated	Deactivated		Inverse Z = 5.4	Inverse Z = 4.6

Insula	Activated	Deactivated	Deactivated		
Thalamus	Activated	Activated	Deactivated		
Ventroposterolateral nucleus				Positive Z = 2.8	
Ventroposteromedial nucleus				Positive Z = 2.7	Positive Z = 2.6
Lateral posterior nucleus				Positive Z = 7.2	Positive Z = 3.6
Lateral dorsal nucleus				Positive Z = 7.9	Positive Z = 3.5
Anterior nucleus				Positive Z = 6.8	Positive Z = 6.3
Mediodorsal nucleus				Positive Z = 7.4	Positive Z = 7.7
Pulvinar				Positive Z = 4.6	Positive Z = 2.7
Striatum					
Caudate		Activated		Positive (Caudate head) Z = 7.9	Positive (Caudate body) Z = 4.2
Putamen	Activated		Deactivated	Inverse Z = 4.3	Positive Z = 4.9
Nucleus accumbens		Deactivated[b]			
Globus pallidus	Activated			Positive Z = 6.0	Positive Z = 4.5
Substantia nigra				Positive Z = 4.4	Positive Z = 4.8
Subthalamic nucleus			Activated		
Midbrain					
Ventral tegmental area		Deactivated[b]		Positive Z = 4.7	Positive Z = 6.0
Red nucleus		Deactivated[b]			
Dorsal raphe nucleus		Deactivated[b]		Positive Z = 4.1	Positive Z = 6.5
Periaqueductal gray		Deactivated[b]		Positive Z = 4.1	

[a]*Brain areas correlated with itch relief and scratching pleasurability are shown. The highest Z scores >2.3 are presented for most significantly correlated clusters.*
[b]*Deactivated in a higher-level contrast analysis in comparison with the itch condition.*

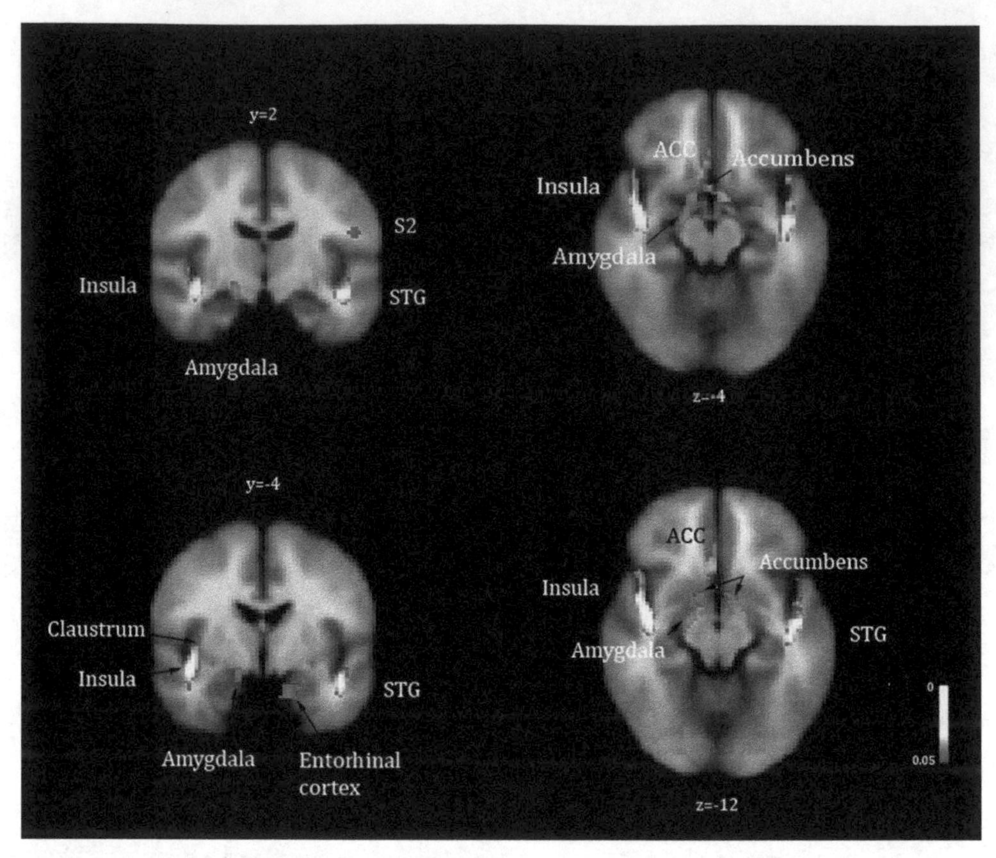

FIGURE 28.2 Increased levels of cerebral perfusion were found at baseline in end-stage renal disease (ESRD) patients with chronic pruritus in comparison with levels in healthy individuals in the insula, claustrum, anterior cingulate cortex (ACC), amygdala, entorhinal cortex (EC), secondary somatosensory area (S2), and nucleus accumbens (NAc). STG, Superior temporal gyrus. Arterial spin labeling (ASL) fMRI: $p < .05$. Montreal Neurological Institute (MNI) standard space coordinates (x, y, z). *© Papoiu AD, Emerson NM, Patel TS, Kraft RA, Valdes-Rodriguez R, Nattkemper LA, et al. Voxel-based morphometry and arterial spin labeling fMRI reveal neuropathic and neuroplastic features of brain processing of itch in end-stage renal disease. Reproduced from J Neurophysiol October 1, 2014;112(7):1729–38.*

itch intensity varies and are fully activated bilaterally when histamine and cowhage stimuli are combined. These features suggest a principal role in the processing of itch for the insula and claustrum. It is noteworthy that the claustrum registers unimodal sensory information from various sources: auditory, visual, or somatosensory [10], in a manner suggesting a main switchboard. It was hypothesized that the claustrum not only detects but also coordinates salient inputs [53]. Through its multiple, complex, and extensive connections with many areas of the cortex, it was suggested that claustrum could support the formation of conscious awareness [54].

We have recently discovered that claustrum was extensively deactivated (its perfusion was significantly decreased) by intranasal administration of the κ-opioid agonist butorphanol, a drug with a powerful antipruritic action [7]. After administration of butorphanol to healthy individuals, the perception of histamine itch was virtually suppressed, whereas cowhage itch intensity was reduced by 35%. An independent validation of these results is provided by the remarkably high density of κ-opioid receptors in the claustrum, possibly having the highest concentration (density) in the human brain when expressed per number of neuronal cell bodies [55].

By virtue of its extensive connections and intermediate position between ascending sensory pathways and the cortex, the claustrum could serve as a processing hub for itch, linking sensory information with motor, motivational, mnemonic, and emotional functions. The claustrum is connected with the somatosensory cortex, with the motor areas, prefrontal areas, the thalamus, the ACC, the septal area, the hippocampus and the entorhinal cortex, as well as with nucleus accumbens (NAc). These associations could serve multiple functional relationships relevant to itch processing. The claustrum maintains connections with the core and the shell subdivisions of the NAc [56]. In pathological settings, the claustrum was found to be less perfused in patients with Creutzfeldt-Jakob disease suffering with severe itch, which suggests a regulatory role for the claustrum in itch perception [57]. Recent findings suggested that the claustrum may play a major role in coordinating or integrating sensory perception, possibly binding perceptual constructs into a cohesive field of conscious awareness. The role of the claustrum has been compared with that of an orchestra conductor [54]. One way the claustrum may coordinate multiple sensory inputs could be organized by frequency codes or oscillations [58]. This putative function may be coordinated through synchronized slow oscillations known as the *theta rhythm*, a prominent feature of the septohippocampal structures to which claustrum is also connected [59]. Buzsaki observed that "coherent oscillations of cell assemblies during theta rhythm provide an ideal mechanism for temporal coding and decoding" [60,61].

An interesting addition to the various experimental approaches tested to relieve itch used the optical illusion of scratching in a mirror. Drawing its inspiration from the techniques introduced by Ramachandran in relieving phantom limb pain (which frequently includes phantom itch), the experiment was designed in such a way that the image (in effect, the optical illusion) of scratching, performed in reality on the contralateral arm was projected by mirror to be perceived as performed on the other arm that was itching [16]. This study reported that itch intensity was attenuated by about 20% by mirror scratching. This approach provides a glimpse of higher order integrative processes that may contribute to the perceptual integration of itch sensation.

Nucleus Accumbens and Itch Modulation

The intricate mechanism of itch modulation was dissected in a pharmacological fMRI study where inhibition of itch was achieved with butorphanol, a centrally acting opioid [7]. This study confirmed that κ-agonist opioids do suppress histamine itch, in accordance with recent experimental findings [25]. The analysis of fMRI data indicated that histamine itch inhibition was mediated by the NAc and septal nuclei and was paralleled by deactivation of the claustrum, putamen, and insula [7] (Fig. 28.3). These observations are in agreement with pertinent findings in the literature: the ability of κ-opioid receptors to exert itch inhibition, the high level of expression of κ-opioid receptors in the claustrum, and the rich content of opioid receptors in the septal area and NAc. The latter were previously described to play a role in general anesthesia and analgesia/antinociception, respectively. In this study we identified the medial septal nuclei, nuclei of the diagonal band of Broca, and the basal nucleus of Meynert, together with the NAc (also a septal nucleus, its name literally meaning "leaning against the septum") as the structures activated during the inhibition of itch, therefore possibly mediating itch inhibition.

Besides the novelty of this finding, an examination of the roles of the septal nuclei and their interconnections with hippocampus and amygdala could offer further clues for devising a strategy to treat itch, especially if a selective targeting of these structures could be achieved. This finding confirms and expands the role of NAc and reward circuits in antinociception to antipruritoception. The fact that the NAc has an antinociceptive (analgesic) function has been established previously [62]. It was not known previously that the NAc could exert a suppressive effect against itch when mediating the action of centrally acting (pharmacological) opioids or otherwise.

Multiple arguments support the NAc involvement in itch processing. The NAc was first mentioned in relation to itch in a BOLD fMRI study of histamine and allergen-induced itch [39] when it was found as one of the areas where activation was correlated with itch intensity. The interpretation offered at that time was that activation of the NAc reflected a high degree of craving for itch relief (similarly to the way activation of motor-related cortices was described as reflecting the urge to scratch). The implication in chronic pruritus is that when baseline cerebral perfusion was compared with healthy volunteers, NAc was found to be highly perfused (activated) at rest in ESRD patients receiving hemodialysis and suffering

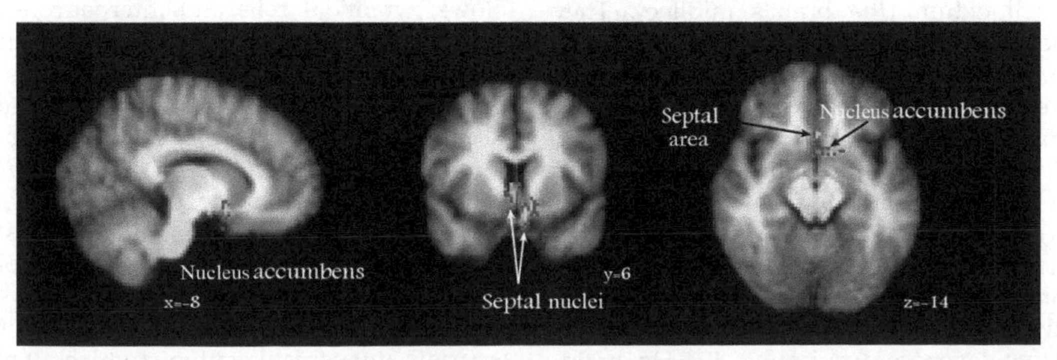

FIGURE 28.3 The cerebral mechanism underlying butorphanol's suppression of histamine itch was analyzed by a general linear model paired *t*-test (contrast) of perfusion weighted images between the following states: [(histamine + drug) versus (drug)] versus [(histamine + placebo) versus (placebo)]. This analysis showed that the inhibition of histamine itch by butorphanol was paralleled by the activation of nucleus accumbens (NAc) and septal nuclei (blue). Montreal Neurological Institute (MNI) standard space coordinates (x, y, z).

with chronic itch, alongside the insula, ACC, claustrum, hippocampus, and amygdala [6] (Fig.28.2). The NAc was activated during the inhibition of histamine itch by butorphanol; a dose of 1 mg administered intranasally, leading to the perception of histamine itch being virtually (99%) suppressed.

The NAc expresses three major types of opioid receptors: μ, κ, and δ. It was demonstrated previously that κ-opioid receptor stimulation leads to a decrease of dopamine release in the NAc, and that stimulation of μ receptors in the NAc induces analgesia, which is mediated by dopamine release [62]. Therefore it is tempting to infer that κ-opioid–decreased release of dopamine in the NAc may represent the molecular mechanism for itch inhibition at this level. However, things are never that simple in the brain; κ-opioid receptor activation at the level of the NAc also decrease γ-aminobutyric acid (GABA) and glutamate release [63]. What is certain is that several reports have confirmed that κ-opioid agonists are capable of exerting itch inhibition via a central mechanism (within the CNS), as does the natural κ-agonist dynorphin at the spinal level [19,20,64].

Septal Nuclei

The fMRI study of centrally mediated butorphanol inhibition of histamine itch also brought into forefront the intriguing area of the septal nuclei, which were known for their involvement in reward, pleasurable experiences, and addiction, as demonstrated in the classic neurophysiological experiments of Olds and Milner [64b]. On one hand, these observations strengthen our proposition that the brain's reward circuits could effectively mediate itch relief [4]. It could also suggest even more tantalizing possibilities; considering that one of the most extensively described functions of the septal nuclei is to initiate and drive the hippocampal theta rhythm, this raises the possibility that the theta rhythm could be relevant for itch perception. Septal nuclei are heterogeneous discrete GM structures situated next to the septum pellucidum (the brain's midline). They include the medial septal nucleus, nuclei of the diagonal band of Broca, and the basal nucleus of Meynert. One of the most interesting questions emerging from the implication of septal nuclei in itch inhibition is whether there is a connection between the synchronous oscillatory neuronal activity observed in the septal area and hippocampus known as the *theta rhythm* and itch processing. In humans and primates, theta oscillations are observed not only in cortical areas but in several other subcortical regions, mostly in the hippocampus, ACC, and entorhinal and prefrontal cortices (pulsating 4–8 Hz brainwaves in humans; this band is 4–12 Hz in rodents). Interestingly, the theta rhythm has been intensely studied for its relationship with pain processing [65–70]. Theta rhythm has been associated with learning and working memory, spatial orientation tasks, performance of intense mental tasks, and also with rapid eye movement (REM) sleep. Interestingly, considering the strong connection of the claustrum with itch perception, a recent study showed that the claustrum is essential for the generation of REM sleep, which features dominant theta waves [71]. In turn, ACC has been also found to display sustained, synchronized theta rhythmic activity in humans. The frontal-medial theta rhythm originates in the ACC [70], and is correlated with the theta rhythm of the prefrontal cortex. This recently published study indicated that theta phase locking of ACC neurons (and their synchronization with medial thalamus) was modulated by visceral noxious stimulation. An insight into the significance of this oscillation in relation to sensory processing was provided by studies of Alonso and Garcia-Austt, Bragin et al., Chrobak and Buzsáki, and Deshmukh et al. [72–76]. Hinman condensed these findings, stating that "during the awake state, (hippocampal) CA1 theta reflects the processing of sensory inputs filtered by neocortical associative networks and provided directly by the theta-related discharge of entorhinal cortical (EC) neurons" [77]. The hippocampal CA1 signal also reflects the theta-related output of the CA3 hippocampal network [78], which is presumed to store and output patterns of activity in relation to previous experiences [79]. CA3 and EC inputs provide theta frequency synaptic potentials to the dendritic field of CA1 neurons, and the precise timing of both inputs plays a fundamental role in the amplitude modulation of the signal [80,81]. The theta signal fluctuates in amplitude and frequency depending on sensory, associative, and motor inputs. The association between theta rhythm and sensory information processing, attention to novelty, or meaningful stimuli during the awake state and REM sleep has been proposed [77]. Theta phase precession of action potentials supports the idea that *phase coding* is used by the hippocampus and associated structures. Buszáki hypothesized that this mechanism allows pyramidal cells and interneurons to beat at a slightly different frequency and phase than the "master clock" given by the theta rhythm. As a result, the same author proposed that phase differences could segregate assemblies of neurons assigned to different representations [59]. We envision that itch can be one of these representations encoded in reference to the theta rhythm ie, that the sensation of itch could be constructed as a distinct perception through a unique phase coding signature.

Opioids, endocannabinoids, and the N-methyl-D-aspartate antagonist ketamine (which all produce analgesia) influence the theta rhythm in the hippocampus. Morphine, a μ-opioid receptor (MOR) agonist, was reported to influence dorsal hippocampal CA1 pyramidal cell suppression partly via an effect on septohippocampal processing [65]. Ketamine disrupts theta synchrony across the septotemporal axis of the CA1 region of hippocampus [77]. Another study suggested that δ-opioid inputs from the pedunculopontine tegmental nucleus may regulate the theta rhythm [82,83]. On the other hand, the hippocampal theta is activated by serotoninergic inputs exerted via 5HT1A receptors [83]. Interestingly, a MEG study reported that there was an increase in theta band power after receiving a reward [84].

To conclude, the role of septal nuclei in itch perception modulation brings into focus the potential connection of the septohippocampal theta rhythm with itch processing. Moreover, it was previously observed that sensory stimulation resets the pace of theta oscillations in the septal nuclei [85]. Neurons oscillating in phase lock with septal nuclei were found in the VTA and dorsal raphé nucleus (DRN) [86]. In our fMRI studies of itch modulation, these structures were deactivated by active self-scratching. More precisely, their perfusion in the itch condition was greater when contrasted with scratching an itch condition; this effect was observed with active scratching of an itch, but not with passive scratching. The significance of the engagement of these structures in itch modulation is difficult to dismiss or ignore, because both are major sources of dopamine and serotonin, respectively, occupying crucial positions within the dopaminergic and serotoninergic circuits of the brain.

THE ROLE OF BRAIN'S REWARD CIRCUITS IN ITCH RELIEF

We have recently proposed that the pleasurability of scratching an itch has found a neurobiological basis in the engagement of the reward circuits (Fig. 28.4). The VTA of the midbrain and the substantia nigra, areas which produce a significant amount of dopamine, were implicated in the relief of itch produced by active scratching.

Dopamine is the major neurotransmitter transported from the VTA to the end-stations of the mesolimbic and mesocortical circuits (the accumbens and the prefrontal cortex, respectively). Therefore it was obvious that reward-processing areas were engaged during the inhibition of itch by active scratching. Moreover, regression analyses of CBF data suggested that structures of dopaminergic mesolimbic circuits controlling reward were also significantly involved in the modulation of itch perception [4,7]. Dopaminergic circuits are under endogenous opioidergic and serotoninergic influences

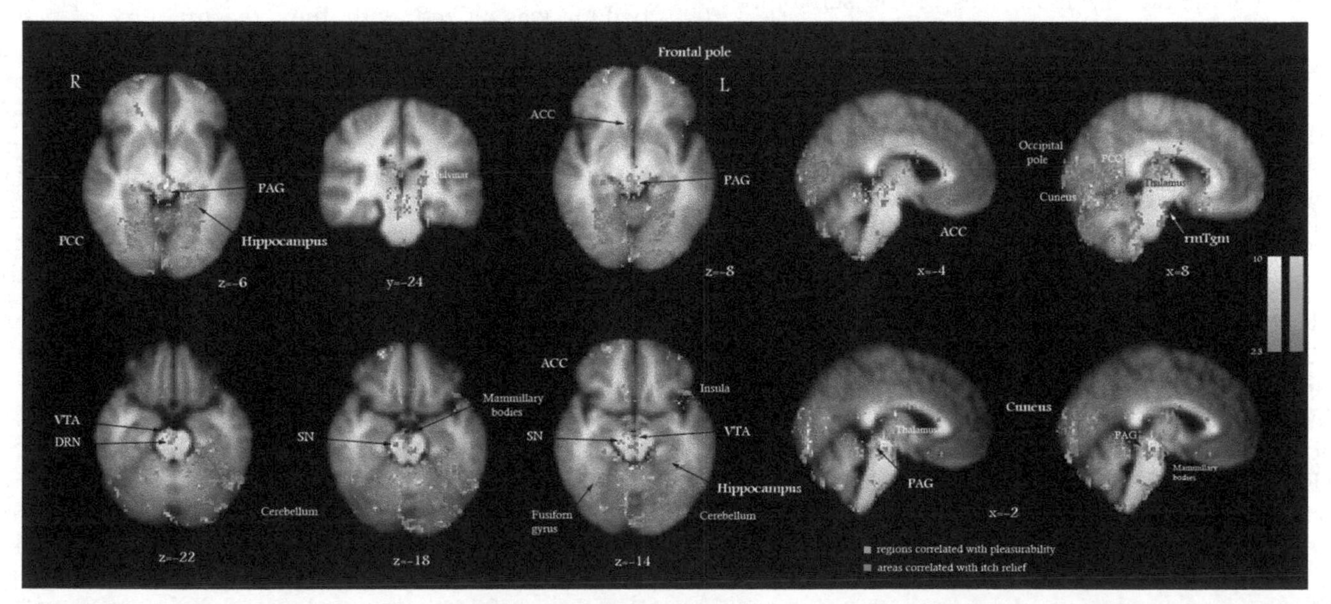

FIGURE 28.4 Brain areas significantly correlated with the pleasurability of scratching (pink) and with itch relief (blue) induced by actively scratching an itch, as identified by multiple regression analysis. Visual analog scale ratings of pleasurability and itch relief (calculated as the difference in itch ratings before and after scratching) were used as covariates of interest. Areas significantly correlated with a Z score >2.3, $p < .05$ are displayed. The color tones displayed correspond to Z score values as depicted in the color bars (at the same scale). Montreal Neurological Institute (MNI) standard space coordinates (x, y, z). *ACC*, anterior cingulate cortex; *DRN*, dorsal raphe nucleus; *L*, left; *PAG*, periaqueductal gray; *PCC*, posterior cingulate cortex; *R*, right; *rmTgm*, rostromedial tegmentum; *SN*, substantia nigra; *VTA*, ventral tegmental area. *© Papoiu AD, Nattkemper LA, Sanders KM, Kraft RA, Chan YH, Coghill RC, et al. Brain's reward circuits mediate itch relief. A functional MRI study of active scratching. Reproduced from PLoS One December 6, 2013;8(12):e82389. http://dx.doi.org/10.1371/journal.pone.0082389.g005.*

and are regulated through reciprocal connections with the medial frontal cortex. Septal nuclei, NAc and ACC are interlinked, and further connected with the insula, as well as with the thalamus (anterior and mediodorsal nuclei). The septal area may serve as the origin of an inhibitory, descending pathway that ultimately projects to spinal neurons in the dorsal horn. Recent findings demonstrate that stimulation of the septal area can modulate the activity of neurons in the dorsal horn [87]. Powerful central antinociceptive actions initiated at the level of the NAc descend to influence spinal modulatory processes [88]. It is possible that these descending pathways are mediated by, or connected with the VTA and/or the DRN. Noradrenergic descending pathways originating in the locus coeruleus are known to exert pain inhibition. A study in rodents demonstrated that this circuit also plays a role in itch modulation [89]. Neurotoxic destruction of catecholaminergic neurons in the spinal cord enhanced itch-related behaviors, implying that descending noradrenergic neurons inhibit spinal itch signaling [90]. DRN and periaqueductal gray (PAG) were significantly correlated with the pleasurability of active scratching (with high Z scores; see Table 28.1). This encouraged us to advance the hypothesis that the top-down inhibitory circuit deployed through PAG may operate in reverse to the mechanism used in pain modulation [4].

Dopaminergic circuits that are involved in itch modulation and in the rewarding outcome of scratching are under the regulatory influences of the serotoninergic system. The septal area of the ACC (subgenual area BA 25), as well as the claustrum, ACC, and VTA express numerous serotonin receptor types and receive serotoninergic influences from the DRN. In a neuroimaging study of active scratching, we showed significant changes in the perfusion of DRN of the midbrain during itch modulation [4]. DRN is a major source of serotonin and occupies a central position in the regulation of brain's serotoninergic circuits. Recent studies in the field consolidate the idea that central serotoninergic pathways are involved in itch processing [91,92]. The claustrum receives a diffuse, evenly distributed serotoninergic input from DRNs [93,94a] and expresses several subtypes of serotonin receptors, including 5HT1A, 5HT1F, 5HT2A, and 5HT2C receptors [94b–e]. 5HT2C receptors in particular were discovered to exert control over the theta rhythm of the septohippocampal axis [95]. It was recently discovered that the claustrum along with supramammillary nuclei activate the cortex during REM sleep, which electroencephalographically features prominent theta rhythms [71].

Besides the activation of the cingulate cortex, the engagement of the amygdala, hippocampus, and mammillary bodies in itch processing suggests the full participation of the limbic system network described as the *Papez circuit* [4,11]. The hippocampus and amygdala have been described extensively to play significant roles in nociceptive processing [96]. An area that perhaps has not received sufficient attention previously for its highly likely important role in itch processing is the hippocampus.

OTHER SUBCORTICAL CENTERS INVOLVED IN ITCH PROCESSING

The basal ganglia, particularly the putamen and the lentiform nucleus, were observed often in numerous fMRI studies of itch. The putamen was almost invariably found to be involved in the brain responses induced by itch [3,14,17]. Studies of itch attenuation by acupuncture showed that the putamen was deactivated concomitantly with itch reduction, and a similar finding was obtained in a neuroimaging study of itch relief induced by passive scratching. Basal ganglia exert various roles from coordinating motor activity to processing motivation, reward, and pleasure. Our regression analyses indicated that the putamen was among the areas correlated with the itch relief exerted by self-scratching.

The cerebellum is featured prominently in many neuroimaging studies of itch, suggesting that its role likely surpasses the classical motor coordination function, ie, the coordination of the scratching response. The cerebellum was linked with the pleasurability of active scratching, and its deactivation has also been found to be correlated with the inhibition of cowhage itch exerted by butorphanol [7]. The cerebellum expresses a high level of opioid receptors, which raises the possibility that this structure is an important neural relay conveying the rewarding aspect of scratching. Regression analyses showed extensive areas correlated with pleasurability of scratching, scattered widely over the entire cerebellum [4].

BRAIN PROCESSING OF ITCH IN PATIENTS WITH CHRONIC PRURITUS

When experimental itch is induced in patients with preexisting chronic pruritus, the correlations between brain activation and itch intensity vary with the underlying context of the disease and also differ in their relationship with disease severity [97]. In AD patients, activation of the ACC and dorsolateral prefrontal cortex was directly correlated with disease severity as evaluated by clinical measures such as Eczema Area and Severity Index, whereas histamine itch was correlated with activations in the ACC and insula [5]. In a PET study, cerebral activations evoked by itch appeared slightly different in healthy individuals (where itch activated somatosensory and motor cortices, midcingulate gyrus, and prefrontal cortex) vs. AD (thalamus, somatosensory, motor and prefrontal cortex, and cerebellum). Itch-induced activations were higher in AD in the thalamus and caudate and globus pallidus [36]. The pattern of associations between activations and perceived itch intensity was found somewhat to differ from healthy volunteers [5]. For reasons that remain obscure, differences in the brain responses evoked by histamine and cowhage itch as described in healthy individuals were not replicated in chronic itch sufferers, although the methodology was identical. Stimulus-specific activations appeared differentially nuanced in a different pathological context [6].

In a recent brain imaging study, we investigated both structural and functional factors that could influence itch processing in ESRD patients with chronic pruritus, in contrast with healthy volunteers. GM density changes in patients with chronic kidney disease were described in several reports [98a–c]. Using a voxel-based morphometric-FSL analysis, we have discovered an extensive decrease in GM density in frontal, parietal, temporal cortices in ESRD, accompanied by significant increases in the NAc, ACC, brainstem, amygdala, and hippocampus [6]. Notably, all areas with increased GM density were previously described to be involved in itch processing. The increase in GM density in the NAc, ACC, brainstem, amygdala, and hippocampus in ESRD patients with chronic itch could have been a result of continuous pruritoceptive stimulation rather than a neuropathic cause of chronic pruritus. These findings could also suggest that ascending brainstem pathways that project to the ventral striatum and septal area may be important channels for itch transmission as alternative or parallel routes, beneath the spinothalamocortical projection to S1 and S2.

We have also measured cerebral perfusion patterns at rest using ASL fMRI in ESRD patients with chronic pruritus and contrasted them with perfusion in healthy volunteers (using a permutation-based method called *randomize*). We have found significant (persistent) perfusion increases at baseline in the insular cortex, ACC, claustrum, amygdala, hippocampus, and NAc in ESRD (Fig. 28.2). When itch was induced experimentally in these patients, using either histamine or cowhage, the activation of these areas was positively correlated with perceived itch intensity. Interestingly, the cortical representation of cowhage itch appeared altered in ESRD, being less extensive in S1, S2, the precuneus, and the insula (Fig. 28.5), although no significant differences were found for histamine itch (when compared with healthy volunteers). In ESRD patients with chronic pruritus, certain brain activations were directly or inversely correlated with itch intensity, suggesting a dual, complex modulation of itch perception. These findings could signify that certain cortical networks may suppress or diminish itch perception, whereas others may amplify it.

These complex features of itch processing were unique to patients with chronic kidney disease who had chronic pruritus and were not observed previously either in healthy individuals or in other groups of patients with chronic itch [97]. We proposed that a neocortical functional reorganization occurs in ESRD, induced by GM density changes, creating an "altered" environment that limits the brain activation induced by cowhage itch. Because these significant differences implicated selectively the cowhage itch pathway, it was suggested that the nonhistaminergic PAR2-mediated itch pathway could have been overstimulated in ESRD patients (as a preexisting condition), leading to a tonic inhibition at the cortical level. This possibility is in agreement with observations of increased serum levels of tryptase (a PAR2 ligand) reported in ESRD [28] and the (over)expression of PAR2 in the skin of chronic kidney disease patients [30].

We inferred that a neocortical functional reorganization pressured by GM changes, in analogy with adaptive cortical changes observed in chronic pain states, occurred in ESRD and led to a different pattern of modulation of itch sensation, observable when itch was elicited exogenously in a chronic itch environment. In other words, the neocortex in patients used to coping with the daily experience of pruritus responded differently to the exogenous induction of cowhage itch. Regression analyses suggested that modulatory neurocircuits limited the extent of activation of S1, S2, the precuneus, and the insula; the magnitude of their activation was inversely correlated with perceived itch intensity. It is unclear at this time whether these changes are specific to the ESRD condition where they were found, or if they could be meaningful for the brain modulation of itch in more general terms.

CONTAGIOUS ITCH

"Contagious itch" is an intriguing phenomenon reported often in daily life and in medical settings that has been recently confirmed in several controlled studies in humans and nonhuman primates [99a–g]. The elucidation of central mechanisms underlying this phenomenon remains an outstanding challenge. If this mystery could be solved, it may provide clues for the treatment of itch by targeting the CNS centers or relays involved. Contagious itch manifests as a surreptitious feeling of itch induced in an observer while looking at other people scratching, or in people being

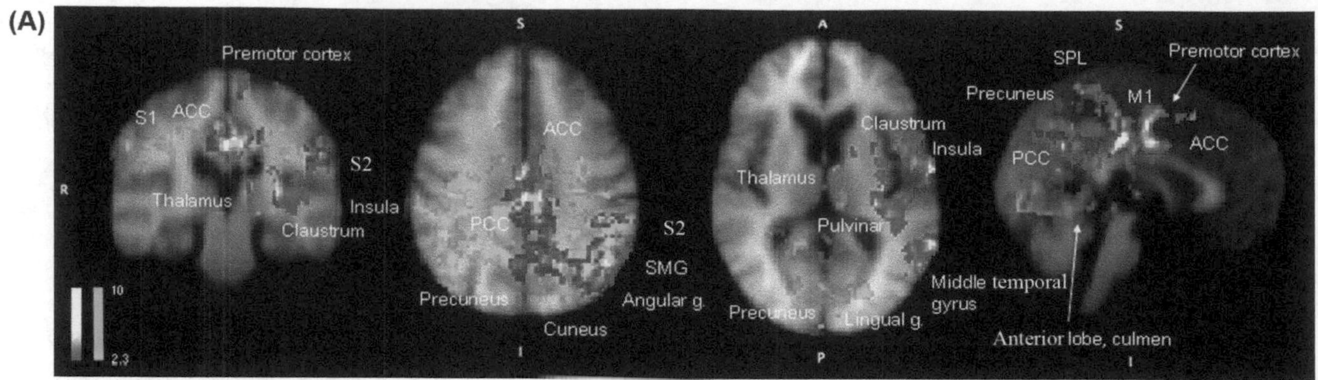

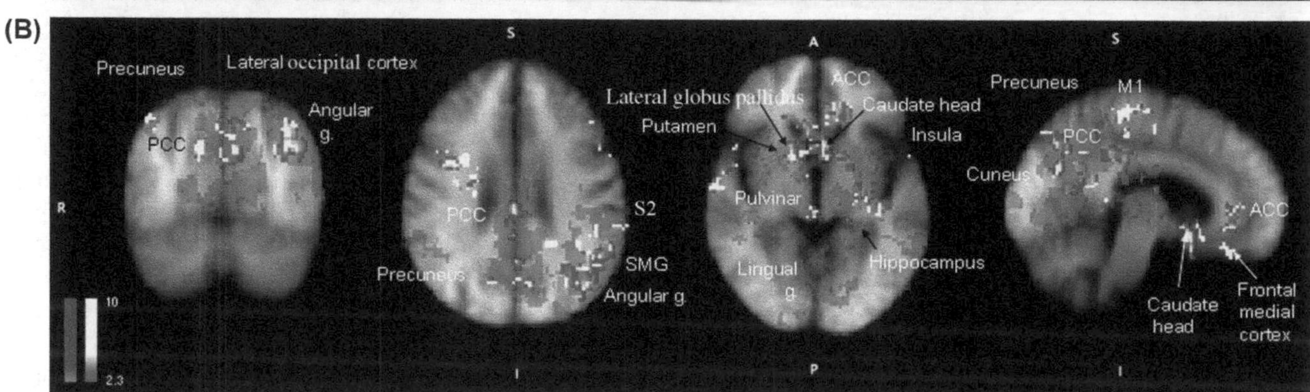

FIGURE 28.5 (A) Histamine itch in ESRD patients (red) induces an extensive, contralateral activation of the insula, claustrum, anterior cingulate cortex (ACC), hippocampal formation, and secondary somatosensory cortex (S2) in comparison with healthy subjects (green). (B) An overlay of the significant responses evoked by cowhage itch in ESRD patients and healthy subjects. In ESRD patients (red-yellow) cowhage induced a significant activation of the ACC (Brodmann areas 24 and 25), lateral globus pallidus, and putamen, and a limited involvement of the insular cortex in contrast with healthy subjects (blue). The activation of precuneus and somatosensory areas (S1 and S2) was also blunted/reduced in ESRD patients. *M1*, primary motor cortex; *SPL*, superior parietal lobule; *SMG*, supramarginal gyrus; *PCC*, posterior cingulate cortex. $Z > 2.3$; $p < .05$. (*© Papoiu AD, Emerson NM, Patel TS, Kraft RA, Valdes-Rodriguez R, Nattkemper LA, et al. Voxel-based morphometry and arterial spin labeling fMRI reveal neuropathic and neuroplastic features of brain processing of itch in end-stage renal disease. Figure part B adapted from J Neurophysiol 1, 2014; 112(7):1729–38*).

presented with visual or mental cues merely suggestive of itch. For lack of a better explanation, it was proposed that contagious itch could be analogous to other socially contagious behaviors (yawning), which might be correlated with empathy or proneness to neurosis and possibly mediated by "mirror neurons." Contagious itch is significantly easier to induce in AD sufferers than in healthy volunteers. [99b,d] Interestingly, the itch induced by visual cues has a scattered, wide body distribution. The cerebral mechanism of itch "triggered by sight" or by mental suggestion is poorly understood. A brain imaging study aimed to identify the neural brain networks involved in the generation of contagious itch and suggested that BA 44 and the premotor cortex (BA 6) acted as mediators [99c]. From a fundamental neuroscientific point of view, it would be useful to identify the brain centers that can support the generation of a somatic sensation in the absence of peripheral stimulation. In analogy with the notions of placebo and nocebo, we proposed a new term *pruricebo [99b]*, to bring attention to the unique features of this intriguing phenomenon consisting of itch induced by mere visual or mental cues. Contagious itch/pruricebo appears as a rather mental or psychological form of itch (centrally produced) not caused by peripheral stimulation of the skin. A unique name should help consolidate the notion this is a distinct entity from nocebo. The induction of itch purely by mental suggestion or visual cues must have a central (CNS) mechanism, because it occurs in the absence of specific pruritogenic stimulation. *Nocebo* refers to pain or unpleasantness. We have good reasons to distinguish between pain and itch; thus a term that conflates the two would only confuse matters (if not becoming simply misleading). More importantly, the phenomenology of contagious itch is sufficiently unique and intriguing in its own right to deserve a name of its own. It is not clearly established whether the experience of pain can be induced by mental suggestion or visual cues; therefore it appears unique to itch.

CONCLUSIONS

Our current understanding of itch processing in the brain is evolving and may continue to progress rapidly. About 2 years ago we wrote that the existence of a dedicated itch center in the brain remained elusive [100]. New data emerging from a pharmacological fMRI study suggested that suppression of histamine itch by butorphanol could be controlled, in fact, by a set of relatively discrete, well-defined structures, such as the NAc and septal nuclei. Because it is clear that in the complex architecture of the brain, no particular structure functions in isolation, any finding of this nature is likely pointing just to the tip of the iceberg.

In summary, brain imaging performed by ASL fMRI provided interesting clues related to the intimate neural networks modulating itch perception in the central nervous system.

(1) Studies of active scratching an itch as well as of pharmacological itch inhibition in humans suggest a role for dopaminergic structures, which are also reward-processing structures, in itch relief, including the VTA, the NAc, and the subgenual area of the ACC (BA 25). (2) The serotoninergic structure of the dorsal raphe nucleus could also be involved in itch inhibition, effectively modulating itch perception during active scratching. The notion that serotoninergic circuits play a role in itch processing is getting strong support [91,92]. (3) Deeply hidden areas residing close to the septum in the basal forebrain, such as septal nuclei and the nucleus accumbens appear to mediate the inhibition of itch by mixed-action opioid butorphanol. (4) A confirmation of the role played by the claustrum in itch perception came forth. The claustrum, a structure expressing a remarkably high level of κ-opioid receptors is extensively deactivated by butorphanol during itch inhibition.

The overall picture offered by brain imaging of itch is a large, tridimensional puzzle waiting to be fully deciphered. One of the present challenges is to understand how different cerebral areas work together, and to identify the molecular and cellular nature of their connections. The interplay between excitatory and inhibitory inputs in the areas identified by fMRI as engaged during itch processing has not been elucidated. Some of the outstanding questions are: What is the exact mechanism underlying the NAc and septal nuclei modulation of hippocampal processing of itch information? What is the relationship between theta rhythm and itch processing? Is there a role for claustrum in regulating the theta rhythm? Are there alternative or parallel pathways of itch transmission besides the spinothalamocortical route? Do ascending brainstem pathways play a role in itch processing? It is likely that a deeper understanding of itch processing could be achieved by integrating neuroimaging findings into their larger neurobiological context, related to the function of the claustrum, septal area, and hippocampus in processing sensory information.

References

[1] Hafenreffer S. In: Kuhnen B, editor. De pruritu, in Nosodochium, in quo cutis, eique adaerentium partium, affectus omnes, singulari methodo, et cognoscendi et curandi fidelissime traduntur; 1660. p. 98–102. Ulm, Germany.

[2] Tan H, Hoge WS, Hamilton CA, Günther M, Kraft RA. 3D GRASE PROPELLER: improved image acquisition technique for arterial spin labeling perfusion imaging. Magn Reson Med 2011;66(1):168–73.

[3] Papoiu AD, Coghill RC, Kraft RA, Wang H, Yosipovitch G. A tale of two itches: common features and notable differences in brain activation evoked by cowhage and histamine induced itch. Neuroimage February 15, 2012;59(4):3611–23.

[4] Papoiu AD, Nattkemper LA, Sanders KM, Kraft RA, Chan YH, Coghill RC, et al. Brain's reward circuits mediate itch relief. A functional MRI study of active scratching. PLoS One December 6, 2013;8(12):e82389. http://dx.doi.org/10.1371/journal.pone.0082389.g005.

[5] Ishiuji Y, Coghill RC, Patel TS, Oshiro Y, Kraft RA, Yosipovitch G. Distinct patterns of brain activity evoked by histamine-induced itch reveal an association with itch intensity and disease severity in atopic dermatitis. Br J Dermatol 2009; 161:1072–80.

[6] Papoiu AD, Emerson NM, Patel TS, Kraft RA, Valdes-Rodriguez R, Nattkemper LA, et al. Voxel-based morphometry and arterial spin labeling fMRI reveal neuropathic and neuroplastic features of brain processing of itch in end-stage renal disease. J Neurophysiol October 1, 2014; 112(7):1729–38.

[7] Papoiu AD, Kraft RA, Coghill RC, Yosipovitch G. Butorphanol suppression of histamine itch is mediated by nucleus accumbens and septal nuclei: a pharmacological fMRI study. J Invest Dermatol February 2015;135(2):560–8.

[8] Davidson S, Zhang X, Yoon CH, Khasabov SG, Simone DA, Giesler Jr GJ. The itch-producing agents histamine and cowhage activate separate populations of primate spinothalamic tract neurons. J Neurosci 2007;27(37):10007–14.

[9] Davidson S, Zhang X, Yoon CH, Khasabov SG, Simone DA, Giesler Jr GJ. Pruriceptive spinothalamic tract neurons: physiological properties and projection targets in the primate. J Neurophysiol September 2012;108(6):1711–23.

[10] Remedios R, Logothetis NK, Kayser C. Unimodal responses prevail within the multisensory claustrum. J Neurosci 2010;30(39): 12902–7.

[11] Papez JW. A proposed mechanism of emotion (1937). J Neuropsychiatry Clin Neurosci Winter 1995;7(1):103–12.

[12] Mochizuki H, Tashiro M, Kano M, Sakurada Y, Itoh M, Yanai K. Imaging of central itch modulation in the human brain using positron emission tomography. Pain 2003;1:339–46.

[13] Yosipovitch G, Ishiuji Y, Patel TS, Hicks MI, Oshiro Y, Kraft RA, et al. The brain processing of scratching. J Invest Dermatol 2008; 1:1806–11.

[14] Vierow V, Fukuoka M, Ikoma A, Dörfler A, Handwerker HO, Forster C. Cerebral representation of the relief of itch by scratching. J Neurophysiol 2009;102(6):3216–24.

[15] Pfab F, Valet M, Sprenger T, Huss-Marp J, Athanasiadis GI, Baurecht HJ, et al. Temperature modulated histamine-itch in lesional and nonlesional skin in atopic eczema: a combined psychophysical and neuroimaging study. Allergy January 2010; 65(1):84–94.

[16] Helmchen C, Palzer C, Münte TF, Anders S, Sprenger A. Itch relief by mirror scratching: a psychophysical study. PLoS One 2013; 8(12):e82756.

[17] Napadow V, Li A, Loggia ML, Kim J, Schalock PC, Lerner E, et al. The brain circuitry mediating antipruritic effects of acupuncture. Cereb Cortex 2014;24(4):873–82.

[18] Dawn AG, Yosipovitch G. Butorphanol for treatment of intractable pruritus. J Am Acad Dermatol 2006;54(3):527–31.

[19] Ko MC, Lee H, Song MS, Sobczyk-Kojiro K, Mosberg HI, Kishioka S, et al. Activation of kappa-opioid receptors inhibits pruritus evoked by subcutaneous or intrathecal administration of morphine in monkeys. J Pharmacol Exp Ther 2003;305(1): 173–9.

[20] Ko MC, Husbands SM. Effects of atypical kappa-opioid receptor agonists on intrathecal morphine-induced itch and analgesia in primates. J Pharmacol Exp Ther 2009;328(1):193–200.

[21] Inan S, Cowan A. Kappa opioid agonists suppress chloroquine-induced scratching in mice. Eur J Pharmacol 2004;502(3):233–7.

[22] Cowan A, Kehner GB, Inan S. Targeting itch with ligands selective for κ opioid receptors. Handb Exp Pharmacol 2015;226: 291–314.

[23] Yokoyama Y, Yokoyama T, Nagao Y, Nakagawa T, Magaribuchi T. Treatment of epidural morphine induced pruritus with butorphanol. Masui 2009;58(2):178–82.

[24] Liu XY, Liu ZC, Sun YG, Ross M, Kim S, Tsai FF, et al. Unidirectional cross-activation of *GRPR* by *MOR1D* uncouples itch and analgesia induced by opioids. Cell 2011;147(2):447–58.

[25] [a] Kardon AP, Polgár E, Hachisuka J, Snyder LM, Cameron D, Savage S, et al. Dynorphin acts as a neuromodulator to inhibit itch in the dorsal horn of the spinal cord. Neuron 2014;82(3): 573–86.
[b] Shelley WB, Arthur RP. Studies on cowhage (*Mucuna pruriens*) and its pruritogenic proteinase, mucunain. AMA Arch Derm 1955;72(5):399–406.
[c] Shelley WB, Arthur RP. Mucunain, the active pruritogenic proteinase of cowhage. Science 1955;122(3167):469–70.

[26] Reddy VB, Iuga AO, Shimada SG, LaMotte RH, Lerner EA. Cowhage-evoked itch is mediated by a novel cysteine protease: a ligand of protease-activated receptors. J Neurosci April 23, 2008;28(17):4331–5.

[27] Steinhoff M, Neisius U, Ikoma A, Fartasch M, Heyer G, Skov PS, et al. Proteinase-activated receptor-2 mediates itch: a novel pathway for pruritus in human skin. J Neurosci 2003;23(15): 6176–80.

[28] Dugas-Breit S, Schöpf P, Dugas M, et al. Baseline serum levels of mast cell tryptase are raised in hemodialysis patients and associated with severity of pruritus. J Dtsch Dermatol Ges 2005;3: 343–7.

[29] Kawakami T, Kaminishi K, Soma Y, Kushimoto T, Mizoguchi M. Oral antihistamine therapy influences plasma tryptase levels in adult atopic dermatitis. J Dermatol Sci 2006;43:127–34.

[30] Kim H, Jeong S, Jeong M, Ahn J, Moon S, Lee S. The relationship of PAR2 and pruritus in end stage renal disease patients and the clinical effectiveness of soybean extracts containing moisturizer on epidermal permeability barrier in end stage renal disease patients. J Invest Dermatol 2010;130:S56 [abstract].

[31] Apkarian AV, Bushnell MC, Treede RD, Zubieta JK. Human brain mechanisms of pain perception and regulation in health and disease. Eur J Pain 2005;1:463–84.

[32] Hsieh JC, Hagermark O, Stahle-Backdahl M, Ericson K, Eriksson L, Stone-Elander S, et al. Urge to scratch represented in the human cerebral cortex during itch. J Neurophysiol 1994; 1:3004–8.

[33] Roy CS, Sherrington CS. On the regulation of blood flow in the brain. J Phys 1896;1:85–108.

[34] Darsow U, Drzezga A, Frisch M, Munz F, Weilke F, Bartenstein P, et al. Processing of histamine-induced itch in the human cerebral cortex: a correlation analysis with dermal reactions. J Invest Dermatol Symp Proc 2000;1:1029–33.

[35] Drzezga A, Darsow U, Treede RD, Siebner H, Frisch M, Munz F, et al. Central activation by histamine-induced itch: analogies to pain processing: a correlational analysis of O^{15} H_2O positron emission tomography studies. Pain 2001;1:295–305.

[36] Schneider G, Stander S, Burgmer M, Driesch G, Heuft G, Weckesser M. Significant differences in central imaging of histamine-induced itch between atopic dermatitis and healthy subjects. Eur J Pain 2008;12(7):834–41.

[37] Walter B, Sadlo MN, Kupfer J, Niemeier V, Brosig B, Stark R, et al. Brain activation by histamine prick test-induced itch. J Invest Dermatol 2005;125(2):380–2.

[38] Herde L, Forster C, Strupf M, Handwerker HO. Itch induced by a novel method leads to limbic deactivations a functional MRI study. J Neurophysiol 2007;1:2347–56.

[39] Leknes SG, Bantick S, Willis CM, Wilkinson JD, Wise RG, Tracey I. Itch and motivation to scratch: an investigation of the central and peripheral correlates of allergen- and histamine-induced itch in humans. J Neurophysiol 2007;1:415–22.

[40] Valet M, Pfab F, Sprenger T, Woller A, Zimmer C, Behrendt H, et al. Cerebral processing of histamine-induced itch using short-term alternating temperature modulation: an fMRI study. J Invest Dermatol 2008;1:426–33.

[41] Logothetis NK. The neural basis of the blood-oxygen-level–dependent functional magnetic resonance imaging signal. Philos Trans R Soc Lond B Biol Sci 2002;357(1424):1003–37.

[42] Roberts DA, Detre JA, Bolinger L, Insko EK, Leigh Jr JS. Quantitative magnetic resonance imaging of human brain perfusion at 1.5 T using steady-state inversion of arterial water. Proc Natl Acad Sci USA 1994;91(1):33–7.

[43] Buxton RB, Frank LR, Wong EC, Siewert B, Warach S, Edelman RR. A general kinetic model for quantitative perfusion imaging with arterial spin labeling. Magn Reson Med 1998;40(3):383–96.

[44] Owen DG, Bureau Y, Thomas AW, Prato FS, St Lawrence KS. Quantification of pain-induced changes in cerebral blood flow by perfusion MRI. Pain 2008;136(1–2):85–96.

[45] [a] Detre JA, Rao H, Wang DJ, Chen YF, Wang Z. Applications of arterial spin labeled MRI in the brain. J Magn Reson Imaging 2012;35(5):1026–37.
[b] Detre JA, Zhang W, Roberts DA, Silva AC, Williams DS, Grandis DJ, et al. Tissue specific perfusion imaging using arterial spin labeling. NMR Biomed 1994;7(1–2):75–82.

[46] Aguirre GK, Detre JA, Zarahn E, Alsop DC. Experimental design and the relative sensitivity of BOLD and perfusion fMRI. Neuroimage 2002;15(3):488–500.

[47] Silva AC, Lee SP, Yang G, Iadecola C, Kim SG. Simultaneous blood oxygenation level-dependent and cerebral blood flow functional magnetic resonance imaging during forepaw stimulation in the rat. J Cereb Blood Flow Metab 1999;19(8):871–9.

[48] Duong TQ, Kim DS, Uğurbil K, Kim SG. Localized cerebral blood flow response at submillimeter columnar resolution. Proc Natl Acad Sci USA 2001;98(19):10904–9.

[49] Mochizuki H, Inui K, Tanabe HC, Akiyama LF, Otsuru N, Yamashiro K, et al. Time course of activity in itch-related brain regions: a combined MEG-fMRI study. J Neurophysiol 2009; 102(5):2657–66.

[50] Lee CH, Sugiyama T, Kataoka A, Kudo A, Fujino F, Chen YW, et al. Analysis for distinctive activation patterns of pain and itchy in the human brain cortex measured using near infrared spectroscopy (NIRS). PLoS One October 3, 2013;8(10):e75360.

[51] Menon V, Uddin LQ. Saliency, switching, attention and control: a network model of insula function. Brain Struct Funct 2010; 214(5–6):655–67.

[52] Mathur BN. The claustrum in review. Front Syst Neurosci 2014; 8:48.

[53] Smith JB, Radhakrishnan H, Alloway KD. Rat claustrum coordinates but does not integrate somatosensory and motor cortical information. J Neurosci 2012;32(25):8583–8.

[54] Crick FC, Koch C. What is the function of the claustrum? Philos Trans R Soc Lond B Biol Sci June 29, 2005;360(1458):1271–9.

[55] Peckys D, Landwehrmeyer GB. Expression of mu, kappa, and delta opioid receptor messenger RNA in the human CNS: a 33P in situ hybridization study. Neuroscience 1999;88(4): 1093–135.

[56] Usunoff KG, Schmitt O, Lazarov NE, Itzev DE, Rolfs A, Wree A. Efferent projections of the claustrum to the dorsal, and ventral striatum, substantia nigra, ventral tegmental area and parabrachial nuclei: retrograde tracing studies in the rat. C R Acad Bulg Sci 2008;61(6):817–30.

[57] Cohen OS, Chapman J, Lee H, Nitsan Z, Appel S, Hoffman C, et al. Pruritus in familial Creutzfeldt-Jakob disease: a common symptom associated with central nervous system pathology. J Neurol 2011;258(1):89–95.

[58] Smythies J, Edelstein L, Ramachandran V. Hypotheses relating to the function of the claustrum. II. Does the claustrum use frequency codes? Front Integr Neurosci 2014;8:7.

[59] Buzsáki G. Theta oscillations in the hippocampus. Neuron January 31, 2002;33(3):325–40.

[60] Lisman JE, Idiart MA. Storage of 7 ± 2 short-term memories in oscillatory subcycles. Science 1995;267:1512–5.

[61] Wallenstein GV, Hasselmo ME. GABAergic modulation of hippocampal population activity: sequence learning, place field development, and the phase precession effect. J Neurophysiol 1997;78:393–408.

[62] Altier N, Stewart J. Dopamine receptor antagonists in the nucleus accumbens attenuate analgesia induced by ventral tegmental area substance P or morphine and by nucleus accumbens amphetamine. J Pharmacol Exp Ther 1998;285(1):208–15.

[63] Hjelmstad GO, Fields HL. Kappa opioid receptor activation in the nucleus accumbens inhibits glutamate and GABA release through different mechanisms. J Neurophysiol May 2003;89(5): 2389–95.

[64] [a] Lee H, Naughton NN, Woods JH, Ko MC. Effects of butorphanol on morphine-induced itch and analgesia in primates. Anesthesiology 2007;107(3):478–85.
[b] Olds J, Milner P. Positive reinforcement produced by electrical stimulation of septal area and other regions of rat brain. J Comp Physiol Psychol 1954;47(6):419–27.

[65] Khanna S, Zheng F. Morphine reversed formalin-induced CA1 pyramidal cell suppression via an effect on septo-hippocampal neural processing. Neuroscience 1999;89(1):61–71.

[66] Sarnthein J, Jeanmonod D. High thalamo-cortical theta coherence in patients with neurogenic pain. Neuroimage 2008;39(4):1910–7.

[67] Wang J, Li D, Li X, Liu FY, Xing GG, Cai J, et al. Phase-amplitude coupling between θ and γ oscillations during nociception in rat electroencephalography. Neurosci Lett July 20, 2011;499(2): 84–7.

[68] Liu CC, Chien JH, Kim JH, Chuang YF, Cheng DT, Anderson WS, et al. Cross-frequency coupling in deep brain structures upon processing the painful sensory inputs. Cereb Cortex 2013; 23(10):2437–47.

[69] Leblanc BW, Lii TR, Silverman AE, Alleyne RT, Saab CY. Cortical theta is increased while thalamo-cortical coherence is decreased in rat models of acute and chronic pain. Pain 2014;155(4):773–82.

[70] Wang J, Cao B, Yu TR, Jelfs B, Yan J, Chan RH, et al. Theta-frequency phase-locking of single anterior cingulate cortex neurons and synchronization with the medial thalamus are modulated by visceral noxious stimulation in rats. Neuroscience 2015;298:200–10.

[71] Renouard L, Billwiller F, Ogawa K, Clément O, Camargo N, et al. The supramammillary nucleus and the claustrum activate the cortex during REM sleep. Sci Adv 2015;1(3):e1400177.

[72] Alonso A, Garcia-Austt E. Neuronal sources of theta rhythm in the entorhinal cortex of the rat. II. Phase relations between unit discharges and theta field potentials. Exp Brain Res 1987;67: 502–9.

[73] Bragin A, Jando G, Nadasdy Z, Hetke J, Wise K, Buzsaki G. Gamma (40–100 Hz) oscillation in the hippocampus of the behaving rat. J Neurosci 1995;15:47–60.

[74] Chrobak JJ, Buzsáki G. Selective activation of deep layer (V-VI) retrohippocampal neurons during hippocampal sharp waves in the behaving rat. J Neurosci 1994;14:6160–70.

[75] Chrobak JJ, Buzsáki G. Gamma oscillations in the entorhinal cortex of the freely-behaving rat. J Neurosci 1998;18:388–98.

[76] Deshmukh SS, Yoganarasimha D, Voicu H, Knierim JJ. Theta modulation in the medial and lateral entorhinal cortices. J Neurophysiol 2010;104:994–1006.

[77] Hinman JR, Penley SC, Escabí MA, Chrobak JJ. Ketamine disrupts theta synchrony across the septotemporal axis of the CA1 region of hippocampus. J Neurophysiol 2013;109(2):570–9.

[78] Kocsis B, Bragin A, Buzsáki G. Interdependence of multiple theta generators in the hippocampus: a partial coherence analysis. J Neurosci 1999;19:6200–12.

[79] O'Reilly RC, McClelland JL. Hippocampal conjunctive encoding, storage, and recall: avoiding a trade-off. Hippocampus 1994;4: 661–82.

[80] Ang CW, Carlson GC, Coulter DA. Hippocampal CA1 circuitry dynamically gates direct cortical inputs preferentially at theta frequencies. J Neurosci 2005;25:9567–80.

[81] Sabolek HR, Penley SC, Hinman JR, Bunce JG, Markus EJ, Escabí M, et al. Theta and gamma coherence along the septotemporal axis of the hippocampus. J Neurophysiol 2009;101:1192–200.

[82] Leszkowicz E, Kuśmierczak M, Matulewicz P, Trojniar W. Modulation of hippocampal theta rhythm by the opioid system of the pedunculopontine tegmental nucleus. Acta Neurobiol Exp (Wars) 2007;67(4):447–60.

[83] Marrosu F, Cozzolino A, Puligheddu M, Giagheddu M, Di Chiara G. Hippocampal theta activity after systemic administration of a non-peptide delta-opioid agonist in freely-moving rats: relationship to D1 dopamine receptors. Brain Res 1997; 776:24–9.

[84] Doñamayor N, Schoenfeld MA, Münte TF. Magneto- and electroencephalographic manifestations of reward anticipation and delivery. Neuroimage 2012;62(1):17–29.

[85] Buzsáki G, Grastyán E, Tveritskaya IN, Czopf J. Hippocampal evoked potentials and EEG changes during classical conditioning in the rat. Electroencephalogr Clin Neurophysiol 1979; 47(1):64–74.

[86] Bland BH. The physiology and pharmacology of hippocampal formation theta rhythms. Prog Neurobiol 1986;26(1):1–54.

[87] Hagains CE, He JW, Chiao JC, Peng YB. Septal stimulation inhibits spinal cord dorsal horn neuronal activity. Brain Res 2011; 1382:189–97.

[88] Tambeli CH, Levine JD, Gear RW. Centralization of noxious stimulus-induced analgesia (NSIA) is related to activity at inhibitory synapses in the spinal cord. Pain 2009;143(3):228–32.

[89] Kuraishi Y. Noradrenergic modulation of itch transmission in the spinal cord. Handb Exp Pharmacol 2015;226:207–17. Springer Verlag.

[90] [a] Gotoh Y, Andoh T, Kuraishi Y. Noradrenergic regulation of itch transmission in the spinal cord mediated by α-adrenoceptors. Neuropharmacology 2011;61(4):825–31.
[b] Gotoh Y, Omori Y, Andoh T, Kuraishi Y. Tonic inhibition of allergic itch signaling by the descending noradrenergic system in mice. J Pharmacol Sci 2011;115(3):417–20.

[91] Zhao ZQ, Liu XY, Jeffry J, Karunarathne WK, Li JL, Munanairi A, et al. Descending control of itch transmission by the serotonergic system via 5-HT1A-facilitated GRP-GRPR signaling. Neuron 2014;84(4):821–34.

[92] Morita T, McClain SP, Batia LM, Pellegrino M, Wilson SR, Kienzler MA, et al. HTR7 mediates serotonergic acute and chronic itch. Neuron 2015;87(1):124–38.

[93] Baizer JS. Serotonergic innervation of the primate claustrum. Brain Res Bull 2001;55:431–4.

[94] [a] Rahman FE, Baizer JS. Neurochemically defined cell types in the claustrum of the cat. Brain Res 2007;1159:94–111.
[b] Pompeiano M, Palacios JM, Mengod G. Distribution of the serotonin 5-HT2 receptor family mRNAs: comparison between 5-HT2A and 5-HT2C receptors. Brain Res Mol Brain Res 1994;23: 163–78.
[c] Wright DE, Seroogy KB, Lundgren KH, Davis BM, Jennes L. Comparative localization of serotonin1A, 1C and 2 receptor subtype mRNAs in rat brain. J Comp Neurol 1995;351:357–73.
[d] Mengod G, Vilaró MT, Raurich A, López-Giménez JF, Cortés R, Palacios JM. 5-HT receptors in mammalian brain: receptor autoradiography and in situ hybridization studies of new ligands and newly identified receptors. Histochem J 1996; 28:747–58.
[e] Pasqualetti M, Ori M, Castagna M, Marazziti D, Cassano GB, Nardi I. Distribution and cellular localization of the serotonin type 2C receptor messenger RNA in human brain. Neuroscience 1999;92:601–11.

[95] Sörman E, Wang D, Hajos M, Kocsis B. Control of hippocampal theta rhythm by serotonin: role of 5–HT2c receptors. Neuropharmacology 2011;61(3):489–94.

[96] Khanna S. Nociceptive processing in the hippocampus and entorhinal cortex: neurophysiology and pharmacology. In: Gebhart GF, Schmidt RF, editors. Encyclopedia of pain. Berlin Heidelberg: Springer; 2013. p. 2198–201.

[97] Papoiu AD, et al. Differential processing of cowhage and histamine itch in health and disease: insights into brain processing of chronic pruritus revealed by arterial spin labeling fMRI. Acta Derm Venereol 2011;2011(91):612 [abstract].

[98] [a] Zhang LJ, Wen J, Ni L, Zhong J, Liang X, Zheng G, et al. Predominant gray matter volume loss in patients with end-stage renal disease: a voxel-based morphometry study. Metab Brain Dis 2013;28(4):647–54.
[b] Qiu Y, Lv X, Su H, Jiang G, Li C, Tian J. Structural and functional brain alterations in end stage renal disease patients on routine hemodialysis: a voxel-based morphometry and resting state functional connectivity study. PLoS One 2014;9(5):e98346.
[c] Prohovnik I1, Post J, Uribarri J, Lee H, Sandu O, Langhoff E. Cerebrovascular effects of hemodialysis in chronic kidney disease. J Cereb Blood Flow Metab 2007;27(11):1861–9.

[99] [a] Niemeier V, Kupfer J, Gieler U. Observations during itch-inducing lecture. Dermatol Psychosomatics 2000;1:15–8.
[b] Papoiu AD, Wang H, Coghill RC, Chan YH, Yosipovitch G. Contagious itch in humans: a study of visual 'transmission' of itch in atopic dermatitis and healthy subjects. Br J Dermatol June 2011;164(6):1299–303.
[C] Holle H, Warne K, Seth AK, Critchley HD, Ward J. Neural basis of contagious itch and why some people are more prone to it. Proc Natl Acad Sci USA 2012;109(48):19816–21.
[d] Feneran AN, O'Donnell R, Press A, Yosipovitch G, Cline M, Dugan G, et al. Monkey see, monkey do: contagious itch in nonhuman primates. Acta Derm Venereol 2013;93(1):27–9.
[e] Lloyd DM, Hall E, Hall S, McGlone FP. Can itch-related visual stimuli alone provoke a scratch response in healthy individuals? Br J Dermatol 2013;168(1):106–11.
[f] Schut C, Bosbach S, Gieler U, Kupfer J. Personality traits, depression and itch in patients with atopic dermatitis in an experimental setting: a regression analysis. Acta Derm Venereol 2014;94(1):20–5.
[g] Ogden J, Zoukas S. Generating physical symptoms from visual cues: an experimental study. Psychol Health Med 2009; 14(6):695–704.

[100] Mochizuki H, Papoiu ADP, Yosipovitch G. The brain processing of itch and scratching. In: Itch: mechanism and treatment. CRC Press, Taylor & Francis; 2014. p. 391–407 [chapter 23].

CHAPTER

29

Magnetic Resonance Microscopy of Skin

K.E. Göbel[1,2]

[1]University Medical Center Freiburg, Freiburg im Breisgau, Germany; [2]University of Freiburg, Freiburg im Breisgau, Germany

OUTLINE

OUTLINE

There is a growing interest to extend the resolution of conventional magnetic resonance (MR) systems from millimeter to micrometer range in order to resolve the structure of small planar samples, as for instance the human skin, with its cellular and matrix network. If spatial resolution, contrast, and sensitivity are sufficient, MR microscopy can be used as a powerful instrument for in vivo investigation of the human skin as an alternative/additive to invasive histopathological sampling, which remains the current method of choice for highly resolved morphological characterization.

This chapter gives an introduction into the field of high-resolution magnetic resonance imaging (MRI) of skin, its current research topics, and the latest results. It is structured into four subtopics starting with an introduction section on the basics of MR microscopy and followed by a section on the status quo in skin MRI. A detailed review of the literature categorized into low- and high-field MRI introduces the expert community and its research interests. Selected images represent highlights in the field.

The chapter concludes with promising future perspectives concerning fundamental research and clinical applications.

INTRODUCTION

A new imaging technique, namely MRI, was born by Lauterbur's breakthrough in 1973 [1], which is today one of the most important methods for tissue imaging for clinical as well as preclinical purposes. In contrast to other imaging methods such as radiography or computed tomography, this noninvasive technique shows exceedingly high contrast for the differentiation of soft tissues, as well as information about diffusion and perfusion, physiological and functional processes, movement, and, furthermore, metabolic activities.

The fast progress in the domain of genetic and biotechnology demands the development of high-resolution techniques downscaling to molecular and cellular levels. The basis of MR microscopy was set with the first publication of a single-cell image by Aguayo et al. in 1986 [2]. It was obtained at 9.5 T

Imaging in Dermatology
http://dx.doi.org/10.1016/B978-0-12-802838-4.00029-7

(corresponding to 400 MHz proton resonance frequency) with strong 20 G/cm gradients, a 5-mm diameter solenoidal coil, and the sample, an ovum from *Xenopus laevis*, contained within a glass capillary. The in-plane resolution was $10 \times 13\ \mu m^2$ and 250 μm in the slice direction. Since this historic publication, much smaller and dedicated radio frequency (RF) coils in combination with stronger gradients were developed considerably improving spatial resolution [3]. The progress so far in electrical and computer engineering allows isotropic resolution down to 3 μm under long scanning time conditions of several days [4,5]. Diffusion-weighted measurements determine the properties of tissue structure because its sensitivity to translational motions of water protons are even possible in single neurons [6]. Furthermore, comparative studies between MR microscopy and traditional histology have demonstrated the ability of this three-dimensional (3D) technique to serve as a noninvasive alternative to invasive histology [7–11].

There is no clear definition of the term *MR microscopy*. A quick look at the dimensions makes this clear. MRI covers a huge range in patient/sample size; with higher resolution the typical linear voxel dimensions may decrease from 1 mm in the human brain to about 16 μm in a frog embryo or even smaller for a fixed specimen. If the voxel size decreases, the volume and thus the amount of water molecules responsible for the signal in proton imaging, and correspondingly the number of detectable spins, which are the basis for MRI or nuclear magnetic resonance (NMR) methods, become several orders of magnitude smaller. There is no clear definition in the literature but generally it is said that if the spatial resolution is of the order of 100 μm and smaller for fixed tissue, then we are in the domain of MR microscopy [12–14].

The key obstacle to overcome when imaging small volumes is the decrease in signal-to-noise ratio (SNR), because signal intensity decreases with the third power of the linear resolution [15]. As a consequence, owing to the low sensitivity of MR microscopy, extremely long acquisition times are required, because SNR is also proportional to the square root of the measurement time. For example, an isotropic doubling of the resolution resulting in an 8-fold reduction in SNR would then require a 64-fold increase in acquisition time to recover SNR losses. Dedicated highly sensitive RF coils also become necessary to minimize SNR losses [16]. In earlier studies, only solenoidal coils were used to obtain higher resolution in the μm range [4,17]. Novel modern microfabrication techniques, for example photolithography, consecutive etching, or automatic wire-bonding methods, [18] allow the fabrication of individual microcoils. New MR detectors with various geometries (eg, RF surface microcoils [19–24], array microcoils [25], on-chip systems [26], and cryogenic and superconducting systems) have been developed with working volumes designed to alleviate such limitations and maximize signal detection obtaining ideal image quality [5,6,17,27]. Another method to increase SNR is to increase the main magnetic field strength B_0; however, achievable fields are still quite limited, particularly for clinical applications [28]. A strong and rapidly switching gradient system is needed as well to attain high resolution.

MR imaging of the skin is very challenging because dedicated approaches are required to overcome low sensitivity and contrast of standard MR investigations applied at microscale. A reason for that is the special geometry of the skin, a kind of bicurved 2D sheet embedded in 3D space, with layers of large lateral dimensions (millimeter to centimeter) and micrometer thickness demanding an extremely high resolution over a large field of view (FOV). The advantage of MR skin microscopy is the possibility to investigate a large variety of parameters with high sensitivity and contrast mechanisms characterizing the skin, especially its water properties, for instance structural composition, metabolism, and diffusion, which is not accessible to other microimaging modalities.

In terms of the range of applications there is a clear distinction between clinical applications on patients and preclinical skin MR on small animals or tissue samples. Clinical applications have to be conducted on clinical scanners at comparative low fields with field strengths up to 3 T. Gradient systems are designed for the much lower spatial resolution necessary to cover typical applications for macroscopic imaging of organs and organ systems. Commercially available RF coils typically cover several centimeters and are much too large for dedicated applications to the skin.

Examinations on small animals and tissue samples allow the use of smaller magnets with much higher field strength (up to 20 T). Gradient systems are smaller and much more powerful, and small RF coils suitable to achieve microscopic resolution are commercially available.

STATUS QUO FOR IN VIVO SKIN MAGNETIC RESONANCE IMAGING ON HUMANS

MR investigation of the skin started to develop in the 1990s and since then has been pursued by a small research community, but it has not yet been widely used in dermatology. The FOV in MRI is wider, exceeding the imaging range of ultrasound and of optical microscopy techniques. Compared with other noninvasive imaging techniques [29], MRI offers a variety of tissue contrasts, producing volumetric images of many anatomical regions in vivo with a better soft-tissue contrast and penetration depth than ultrasound [30]. This gives MRI the ability to measure the size and extent of skin tumors before surgery and to monitor cutaneous lesions without resorting to traditional incisional biopsy and alteration of the skin itself. The need for side-effect–prone surgical procedures is reduced and this allows for improved noninvasive disease follow-up observation. This technique provides detailed information about the anatomical and biochemical parameters of the tissues and distinguishes diseased tissue from the surrounding healthy tissue.

With the advent of significant developments in physics, electronics, and computer engineering, several imaging modalities were recently integrated in the skin analysis research framework to improve the diagnostic accuracy of skin diseases and allow the morphological characterization of in vivo skin tumors, cutaneous lesions, or inflammatory infiltrates in the skin. Thus MRI has gained importance as a clinical diagnostic method because it is safe, noninvasive, and precise.

The review article of Cal et al. [31] gives an overview of advanced methods for in vivo skin investigations, further introducing techniques such as MRI, electron paramagnetic resonance, Doppler flowmetry, and time domain reflectometry. In Stefanowska et al. [32], the multiple possibilities rising from the use of MRI to explore skin structure noninvasively are briefly reviewed.

LOW-FIELD MAGNETIC RESONANCE IMAGING: FIRST STEPS TOWARDS CLINICAL ROUTINE

In the early 1990s, Bittoun and Richard and their team started MRI of the human skin using field strengths between 0.1 and 1.5 T and have thus pioneered the field [33–42]. Bittoun et al. designed a surface gradient coil for 1.5 T connected to a whole-body system permitting an increase in resolution up to 30 × 80 μm^2 in the direction perpendicular to the skin surface. MRI confirmed for the first time the layered structure of skin, namely the epidermis as a thin, bright outer layer and the dermis as a hypointense inner layer. Hair follicles penetrating the epidermis and other cutaneous structures could be precisely visualized. MRI of the skin has been performed by several investigators [43–45]. Several groups started to use skin MRI for the diagnosis of tumors [46–53] or unilateral lymphedema, [54] applying adapted sequences and dedicated coils.

In 1997, Song et al. obtained images of calf skin after about 10 min' acquisition time with a voxel volume of 19 × 78 × 800 μm^3 and using a 1 × 1 cm^2 surface coil on a 1.5-T machine [55]. Weis et al. [56,57] demonstrated

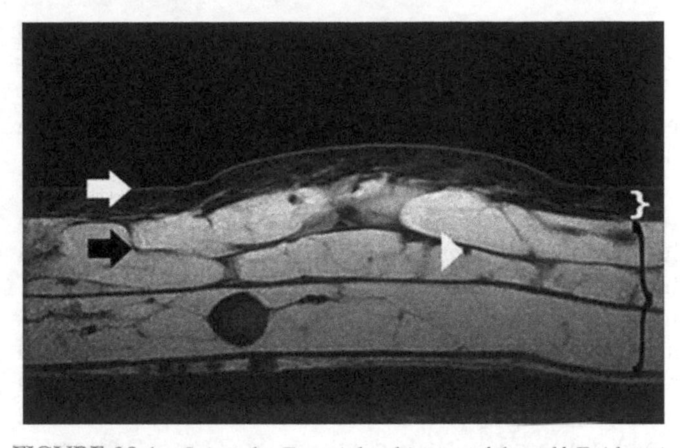

FIGURE 29.1 Spin-echo T_1-weighted image of the calf. Epidermis (*white arrow*), dermis (*white curly bracket*), hypodermis (*black curly bracket*), an interlobular septum (*black arrow*), and a septal vessel (*white arrowhead*) are visible [62]. *Reprinted with kind permission from Springer Science and Business Media.*

how MRI is used for the general morphological characterization of the skin, showing its three main layers, and for studying various skin-related pathological states, such as the early diagnosis of melanoma.

Dias et al. reported the first use of the MR profiling with Gradient At Right Angles to Field technology for in vivo and in vitro skin hydration studies. In clinical application, this practice would allow clinicians to perform follow-up examinations of cutaneous lesions and monitor them during the therapy process [58–60]. Barral et al. [61] exploited an increase in SNR in the hypodermis and dermis by increasing the field strength from 1.5 to 3 T. Besides that, effects of motion are analyzed and a method for navigator-based motion correction was applied to focus blurred images. Aubry and his team developed a receiver coil for a 3-T system, allowing the detailed examination of skin (Fig. 29.1) [62]. The epidermis appears as a thin superficial layer with strong positive contrast. The dermis underneath is microstructurally defined as a dense network of connective tissue made of elastic and collagen fibers. The deepest layer of the skin, the hypodermis, delivers very high signal. It is mainly composed of fat grouped in lobules. Because of the increased magnetic field strength, the acquisition time could be shortened and the spatial resolution and SNR increased. There were artifacts occurring, which is not surprising at higher magnetic fields; however, this did not influence the image quality and thus the evaluation of the image. The higher field strength enables imaging of 87–180 μm^2 in-plane resolution during scanning times of 15–20 min. Thirty-six volunteers were included into this study examining calf and face sections in routine practice. This study showed that although cutaneous 3-T MRI is feasible, higher resolution is, however, required to obtain a reliable analysis.

Besides morphological examinations of the skin, MRI can be used for investigations of human skin microvasculature. Small vessels down to 100 μm could be extracted using a normal conducting surface coil in a 3-T system [63].

The use of superconducting or cooled probes is routinely applied in high-resolution NMR and MRI [64–66]. Laistler et al. [67] implemented a 12.4-mm high-temperature superconducting (HTS) coil at 1.5 T for in vivo skin imaging of the calf. The aim of this study was to improve spatial resolution in clinical MRI systems by using the high RF sensitivity of a small HTS coil. The improvement of SNR by a factor of 32 was compared with a commercial 23-mm transmit/receive copper microcoil. 3D gradient-echo (GE) images identifying fine anatomical skin structures were acquired with isotropic resolutions of 80 μm in about 30 min, including three averages. Fig. 29.2 shows different images acquired in vivo during this study. In Fig. 29.2A–D, four different slices at different depths highlight fine structures such as areolae cutaneae on the outermost layer of hairy skin, hair follicles appearing with a high contrast at a depth of about 700 μm, thin vessels at about 1.1 mm depth, and larger vessels in the hypodermis located at about 3.4 mm, respectively. Fig. 29.2E and F have an isotropic lower resolution of 100 μm. The remaining images, Fig. 29.2G and H, were acquired with the copper coil as reference measurement. Tremendous differences in image quality are clearly visible.

Specifically, the higher the spatial resolution, the lower the volume coverage; therefore the FOV and the penetration depth are very limited in all microscopy cases. Stratum corneum and dermis are the most interesting layers for clinical issue, but because of their short T_2 signal, most sequences are limited. Consequently, many of the microscopy techniques remain adapted only for in vitro and ex vivo investigations.

HIGH-FIELD MAGNETIC RESONANCE IMAGING: AN EX VIVO APPROACH

In addition to human subjects, diseased skin was investigated in vivo in small animal models [68,69] and in vitro in small skin patches and cell culture [11,70]. Generally, these kinds of microimaging experiments are performed on high-resolution NMR spectrometers, which are upgraded with additional imaging software and with a special imaging probehead. Microscopic surface coils with dimensions in the micrometer range and thus with consequently high sensitivity are already available on the market. Even on conventional animal MRI systems, MR microscopy can be performed because of the optimization of highly sensitive microcoils [71].

Kinsey et al. [70] acquired high-resolution images of in vitro preparations of hydrated hairless rat skin at 14.1 T. Major anatomical features such as stratum corneum, epidermis, sebaceous glands, cell layers, and hair follicles, as well as fatty deposits, were comparable to electron microscopy results. Water mobility data could be acquired by calculated diffusion maps, allowing the examination of transdermal transport processes for in vitro skin samples. Gadolinium contrast-enhanced MRI became a common diagnostic tool for the evaluation of skin cancer [72,73]. Furthermore this delayed contrast MRI method enables the visualization of abnormalities in the epidermis because the thickening of epidermis is an index for mice skin viability [74].

First comparisons with the current gold standard, the histomorphological investigation of skin biopsies, were performed by Aubry et al. [75]. Thirty keratinocytic skin tumor samples were imaged on a 7-T MRI system and subsequently compared with histopathology results. Lesions could clearly be delineated with good correlation to the hematoxylin and eosin (H&E) stainings.

Several physical modifications can be employed to further increase the SNR in MR microscopy of the skin, namely reducing the size of the receiver coil, increasing the gradient strength, and combining receiving microcoils in a phased-array configuration that is adapted for various skin areas of the body. Combining the high sensitivity of smaller surface coils with the volume coverage of a large volume resonator, Roemer et al. introduced the so-called "phased-array" coil design at the macroscale in 1990 [76]. Overall, this kind of array coil found application in experiments with an interest in high spatial resolution obtained during reasonably short scanning time, as demonstrated in Schmitt et al., [77] where a phased array of 128 coils (Ø = 7.5 cm of each coil) was used for human cartilage imaging. In Gruschke et al., [27] the idea of Roemer was applied in a drastically reduced size. With this design, high-resolution MRI over a large area, meaning an enhanced FOV and parallel MR spectroscopy acquisition and thus enabling applications such as thin-film characterization, skin morphology, or cellular microbiology, were targeted. Göbel et al. [11,78,79] performed first feasibility studies with this dedicated array, presenting detailed MR images of healthy and affected skin biopsies that were subsequently compared with corresponding H&E stains. The axial image in Fig. 29.3 was performed on a healthy skin sample taken from the tail bone region using a multislice GE sequence. High-resolution images ($30 \times 30 \times 100\ \mu m^3$) were obtained in 26 min and 26 s, enabling delineation of the epidermis and dermis through an axial cut. The epidermis appears as a positive contrast horizontal

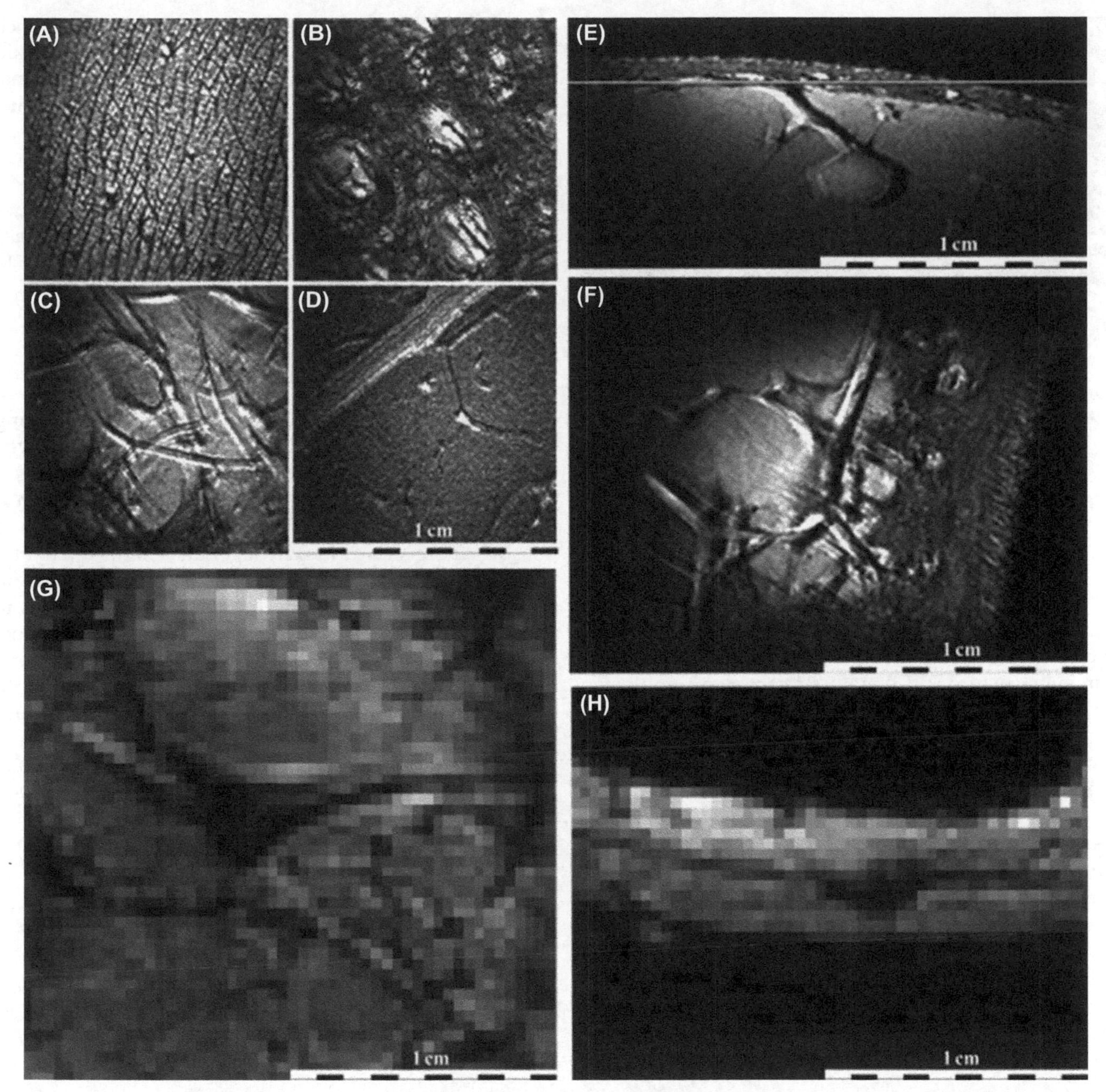

FIGURE 29.2 In vivo skin images obtained with a superconducting coil (A–F) and a commercially available copper microscopy coil (G and H). Images were acquired with isotropic resolutions of 80 (A–D), 100 (E and F), and 400 μm (G and H). (A) The skin surface featuring areolae cutaneae (*rhombic structure*). The dark spots correspond to positions where hair penetrates through the surface. (B) Pilosebaceous units (hair follicles) appearing as bright round shapes at a depth of approximately 700 μm (C) The deep dermal plexus at approximately 1.1 mm from the surface; thin vessels (hyperintense lines) can be seen. (D) At a depth of approximately 3.4 mm, a larger vessel in the hypodermis is displayed. The lumen (depth ≈ 1 mm) and vessel walls approximately 250 μm thick can be distinctly observed. (E) Axial slice perpendicular to the skin surface featuring a cut through a larger vessel (depth ≈ 2 mm). The vessel wall can be differentiated from the lumen. A smaller vessel ($d \approx 400$ μm) connects the large vessel to the deep vascular plexus. (F) Sagittal slice containing vessels of the deep vascular plexus at the border of dermis and hypodermis. (G) Sagittal slice containing vessels in hypodermis. (H) Axial slice perpendicular to the skin surface only showing hypodermis because of insufficient spatial resolution [67]. *Reprinted with permission.*

line on the outer surface of the skin. The dermis underneath is microstructurally defined as a dense network of connective tissue made of elastic and collagen fibers. This justifies the hypointense MR signal of the dermis obtained that is mainly the result of the short T_2 values of its fiber network.

Histological investigations performed on the same skin sample confirm the MRI results, proving the

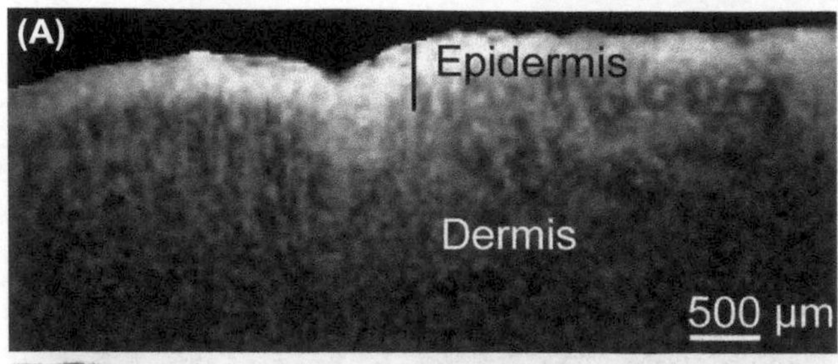

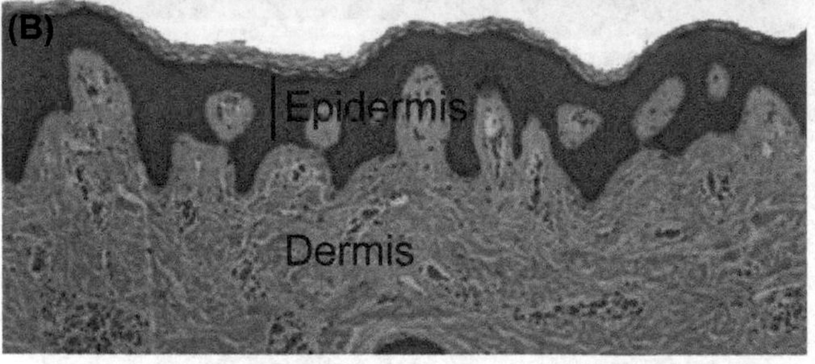

FIGURE 29.3 (A) Representative axial MR image revealing the epidermis and dermis (resolution $30 \times 30 \times 100\ \mu m^3$, $T_{scan} = 26$ min, 26 s). (B) Corresponding hematoxylin and eosin (H&E) staining of the same skin sample validating the MRI results. The epidermis is indicated by a *line* [11]. *Reprinted with permission.*

potential of this approach to noninvasively delineate the skin substructure and therefore to act as a complementary technique to histology.

A GE sequence was employed to image the skin in a plane parallel to the coil surface with a resolution of $45 \times 45 \times 200\ \mu m^3$ in 4 min and 19 s, and $40 \times 40 \times 140\ \mu m^3$ in 13 min and 7 s, respectively. In this case, lower spatial resolutions were chosen because of the large in-plane dimension of the skin that required larger FOV in both read and phase directions. To be able to obtain the same resolution as for the axial case, the number of phase-encoding steps needed to be increased significantly to cover the entire FOV, leading to a very long scanning time. Therefore lower resolutions were preferred, and they proved to be sufficient to clearly distinguish the main features of the skin structure such as hair follicles and areolar-cutaneous regions at the skin surface. Fig. 29.4 shows the detailed view of a coronal MR image. In addition to the surrounding epithelial sheath, the medulla and cortex are clearly highlighted.

This work represents the first study that uses a phased-array of microcoils tailored for MR-based investigation of the human skin. These in vitro techniques lend themselves particularly well to analysis of disease mechanisms at the cellular and molecular level and to the testing of therapeutic applications.

CONCLUSION

High-resolution MRI has become a ground-breaking technique in diverse fields in clinical medicine. However, SNR is the most significant limiting factor in MR microimaging. To date, in vivo resolutions are still lower than resolutions acquired in ex vivo samples. Dedicated devices such as those shown in Schwaiger and Blümich [80] and adapted MR sequences and methods may help MRI to become one of the most useful noninvasive methods in the field of skin imaging, and with that a powerful tool for investigations of the human skin anatomy, metabolism, and diffusion to study diverse skin-related pathological changes such as diagnosis of melanoma [34] and hidradenitis suppurativa [81–83], examination of skin vascularization [84,85], skin tumor angiogenesis, [86] and inherited skin fragility disorders like epidermolysis bullosa [68].

Glossary

Axial plane One of the three basic orthogonal slice orientations. Generally, this is the plane perpendicular to the long axis of the human body (head-to-foot).

Coronal plane One of the three basic orthogonal slice orientations defined by the head-to-foot and left-to-right directions in the human body.

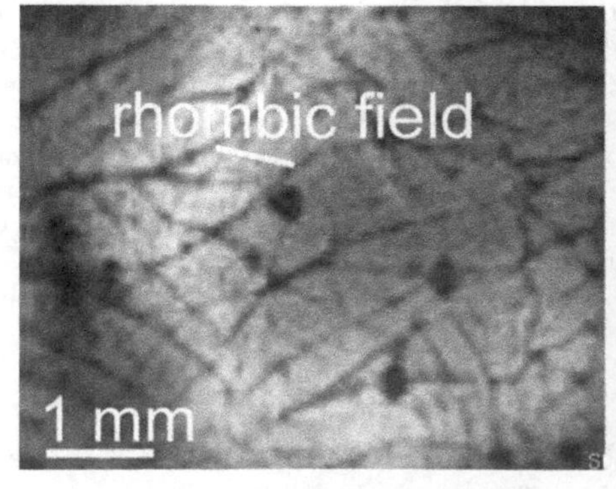

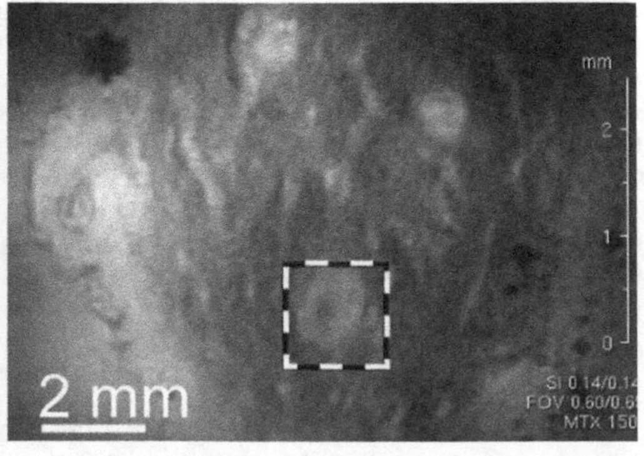

FIGURE 29.4 Coronal MR images depicting rhombic fields (areolae cutaneae) on the skin surface; black, round-shaped spots corresponding to the hair shaft locations illustrate the follicular orifices. The white round areas (*dashed box*) represent the outer epithelial sheath of hair follicles [11]. *Reprinted with permission.*

Echo time Time between middle of 90° pulse and echo production.
Field of view Rectangular region where magnetic resonance imaging (MRI) data is acquired.
Gradient echo (GE) pulse sequence Relatively fast pulse sequence that relies on gradient reversal to rephase the transverse magnetization.
Magnetic resonance imaging (MRI) Use of the principle of magnetic resonance to acquire images of the body noninvasively.
Radio frequency (RF) coil RF coils are used for transmitting and/or receiving magnetic resonance signals. Common coil configurations are birdcage coils, solenoid coils, and saddle coils.
Repetition time Time period between the successive applications of a pulse sequence.
Sagittal plane This is one of the three basic orthogonal slice orientations and defined by the head-to-foot and anterior-to-posterior directions in the human body.
Signal-to-noise-ratio (SNR) Relative contributions of a detected signal and random superimposed noise; an important factor for resolution and image quality.
Spin echo pulse sequence Radio frequency (RF) pulse sequence with a 90° excitation pulse followed by a 180° refocusing pulse; not very sensitive to field inhomogeneities and chemical shift effects.
T_1, spin−lattice, or transversal relaxation time The characteristic time constant for spins to interact with the external magnetic field. The z magnetization will grow to 63% of its final maximum value in a certain time, T_1.
T_2, spin−spin, or transverse relaxation time The characteristic time constant for loss of phase coherence among interacting spins at a certain angle to the static magnetic field. The x−y magnetization will lose 69% of its initial value in a certain time, T_2.

List of Acronyms and Abbreviations

3D Three-dimensional
EB Epidermolysis bullosa
FOV Field of view
GARField Gradient at Right Angles to Field
GE Gradient echo sequence
H&E Hematoxylin and eosin stain
HTS High-temperature superconducting
MRI Magnetic resonance imaging
RF Radio frequency
SE Spin echo sequence
SNR Signal-to-noise ratio
T Tesla

References

[1] Lauterbur P. Image formation by induced local interactions: examples employing nuclear magnetic resonance. Nature 1973;242: 190−1.
[2] Aguayo JB, Blackband SJ, Schoeniger J, Mattingly MA, Hintermann M. Nuclear magnetic resonance imaging of a single cell. Nature July 10, 1986;322(6075):190−1.
[3] Neuberger T, Webb A. Radiofrequency coils for magnetic resonance microscopy. NMR Biomed November 1, 2009;22(9):975−81.
[4] Ciobanu L, Seeber DA, Pennington CH. 3D MR microscopy with resolution 3.7 μm by 3.3 μm by 3.3 μm. J Magn Reson September 2002;158(1−2):178−82.
[5] Ciobanu L, Pennington CH. 3D micron-scale MRI of single biological cells. Solid State Nucl Magn Reson January 2004;25(1−3): 138−41.
[6] Grant SC, Buckley DL, Gibbs S, Webb AG, Blackband SJ. MR microscopy of multicomponent diffusion in single neurons. Magn Reson Med 2001;46(6):1107−12.
[7] Meadowcroft MD, Connor JR, Smith MB, Yang QX. Magnetic resonance imaging and histological analysis of beta-amyloid plaques in both human Alzheimer's disease and APP/PS1 transgenic mice. J Magn Reson Imaging JMRI May 2009;29(5):997−1007.
[8] Flint JJ, Lee CH, Hansen B, Fey M, Schmidig D, Bui JD, et al. Magnetic resonance microscopy of mammalian neurons. NeuroImage July 15, 2009;46(4):1037−40.
[9] Blackwell ML, Farrar CT, Fischl B, Rosen BR. Target-specific contrast agents for magnetic resonance microscopy. NeuroImage June 2009;46(2):382−93.
[10] Göbel K, Leupold J, Dhital B, LeVan P, Reisert M, Gerlach J, et al. MR microscopy and DTI of organotypic hippocampal slice cultures. In: Proceedings of the 22nd Annual Meeting of ISMRM. Milan, Italy; 2014.
[11] Göbel K, Gruschke OG, Leupold J, Kern JS, Has C, Bruckner−Tuderman L, et al. Phased-array of microcoils allows MR microscopy of ex vivo human skin samples at 9.4 T. Skin Res Technol 2015;21(1):61−8.
[12] Eccles CD, Callaghan PT. High-resolution imaging: the NMR microscope. J Magn Reson 1969 June 15, 1986;68(2):393−8.
[13] Benveniste H, Blackband S. MR microscopy and high resolution small animal MRI: applications in neuroscience research. Prog Neurobiol August 2002;67(5):393−420.
[14] Gimi B. Magnetic resonance microscopy: concepts, challenges, and state-of-the-art. Methods Mol Med 2006;124:59−84.
[15] Hoult DI, Richards RE. The signal-to-noise ratio of the nuclear magnetic resonance experiment, 1976. J Magn Reson San Diego Calif 1997 1976;24:71−85.
[16] Hoult DI, Lauterbur P. The sensitivity of the zeugmatographic experiment involving human samples. J Magn Reson 1979;34: 425−33.
[17] Lee SC, Kim K, Kim J, Lee S, Han Yi J, Kim SW, et al. One micrometer resolution NMR microscopy. J Magn Reson San Diego Calif 1997 June 2001;150(2):207−13.
[18] Kratt K, Badilita V, Burger T, Korvink JG, Wallrabe U. A fully MEMS-compatible process for 3D high aspect ratio micro coils obtained with an automatic wire bonder. J Micromech Microeng January 1, 2010;20(1):015021.
[19] Hyde JS, Jesmanowicz A, Kneeland JB. Surface coil for MR imaging of the skin. Magn Reson Med 1987;5(5):456−61.
[20] Dechow J, Lanz T, Stumber M, Forchel A, Haase A. Preamplified planar microcoil on GaAs substrates for microspectroscopy. Rev Sci Instr November 2003;74(11):4855.
[21] Eroglu S, Gimi B, Roman B, Friedman G, Magin RL. NMR spiral surface microcoils: design, fabrication, and imaging. Concepts Magn Reson Part B Magn Reson Eng January 1, 2003;17B(1):1−10.
[22] Massin C, Vincent F, Homsy A, Ehrmann K, Boero G, Besse P-A, et al. Planar microcoil-based microfluidic NMR probes. J Magn Reson October 2003;164(2):242−55.
[23] Massin C, Eroglu S, Vincent F, Gimi BS, Besse P-A, Magin RL, et al. Planar microcoil-based magnetic resonance imaging of cell: transducers, solid-state sensors, actuators and microsystems. In: 12th International Conference on, 2003, vol. 2; 2003. p. 967−70.
[24] Baxan N, Rabeson H, Pasquet G, Châteaux J-F, Briguet A, Morin P, et al. Limit of detection of cerebral metabolites by localized NMR spectroscopy using microcoils. Comptes Rendus Chim April 2008; 11(4−5):448−56.
[25] Anders J, Chiaramonte G, SanGiorgio P, Boero G. A single-chip array of NMR receivers. J Magn Reson San Diego Calif 1997 December 2009;201(2):239−49.
[26] Badilita V, Kratt K, Baxan N, Mohmmadzadeh M, Burger T, Weber H, et al. On-chip three dimensional microcoils for MRI at the microscale. Lab Chip June 7, 2010;10(11):1387−90.

[27] Gruschke OG, Baxan N, Clad L, Kratt K, von Elverfeldt D, Peter A, et al. Lab on a chip phased-array MR multi-platform analysis system. Lab Chip 2012;12(3):495.

[28] Beuf O, Jaillon F, Saint-Jalmes H, Small-animal MRI. signal-to-noise ratio comparison at 7 and 1.5 T with multiple-animal acquisition strategies. Magn Reson Mater Phys Biol Med September 7, 2006;19(4):202–8.

[29] Smith L, Macneil S. State of the art in non-invasive imaging of cutaneous melanoma. Skin Res Technol Off J Int Soc Bioeng Skin ISBS Int Soc Digit Imaging Skin ISDIS Int Soc Skin Imaging ISSI August 2011;17(3):257–69.

[30] Liffers A, Vogt M, Ermert H. In vivo biomicroscopy of the skin with high-resolution magnetic resonance imaging and high frequency ultrasound. Biomed Tech (Berl) May 2003;48(5):130–4.

[31] Cal K, Zakowiecki D, Stefanowska J. Advanced tools for in vivo skin analysis. Int J Dermatol May 2010;49(5):492–9.

[32] Stefanowska J, Zakowiecki D, Cal K. Magnetic resonance imaging of the skin. J Eur Acad Dermatol Venereol 2010;24(8):875–80.

[33] Querleux B, Yassine MM, Darrasse L, Saint-Jalmes H, Sauzade M, Leveque JL. Magnetic resonance imaging of the skin: a comparison with the ultrasonic technique. Bioeng Skin 1988;4(1):1–14.

[34] Bittoun J, Saint-Jalmes H, Querleux BG, Darrasse L, Jolivet O, Idy-Peretti I, et al. In vivo high-resolution MR imaging of the skin in a whole-body system at 1.5 T. Radiology August 1, 1990;176(2): 457–60.

[35] Richard S, Querleux B, Bittoun J, Idy-Peretti I, Jolivet O, Cermakova E, et al. In vivo proton relaxation times analysis of the skin layers by magnetic resonance imaging. J Invest Dermatol July 1991;97(1):120–5.

[36] Richard S, Querleux B, Bittoun J, Jolivet O, Idy-Peretti I, de Lacharriere O, et al. Characterization of the skin in vivo by high resolution magnetic resonance imaging: water behavior and age-related effects. J Invest Dermatol 1993 Mai;100(5):705–9.

[37] Querleux B, Richard S, Bittoun J, Jolivet O, Idy-Peretti I, Bazin R, et al. In vivo hydration profile in skin layers by high-resolution magnetic resonance imaging. Skin Pharmacol Off J Skin Pharmacol Soc 1994;7(4):210–6.

[38] Bittoun J, Querleux B, Jolivet O, Richard SB. Microscopy imaging of the skin in-vivo by using a high gradient intensity and a narrow bandwidth. In: Proceedings of the 3rd Annual Meeting of ESMRMB. Nice, France; 1995.

[39] Hawnaur JM, Dobson MJ, Zhu XP, Watson Y. Skin: MR imaging findings at middle field strength. Radiology December 1996; 201(3):868–72.

[40] Querleux B, Cornillon C, Jolivet O, Bittoun J. Anatomy and physiology of subcutaneous adipose tissue by in vivo magnetic resonance imaging and spectroscopy: relationships with sex and presence of cellulite. Skin Res Technol 2002;8(2):118–24.

[41] Querleux B. Magnetic resonance imaging and spectroscopy of skin and subcutis. J Cosmet Dermatol 2004;3(3):156–61.

[42] Bittoun J, Querleux B, Darrasse L. Advances in MR imaging of the skin. NMR Biomed 2006;19(7):723–30.

[43] Mirrashed F, Sharp JC. In vivo morphological characterisation of skin by MRI micro-imaging methods. Skin Res Technol 2004; 10(3):149–60.

[44] Denis A, Loustau O, Chiavassa-Gandois H, Vial J, Lalande Champetier de Ribes C, Railhac JJ, et al. High resolution MR imaging of the skin: normal imaging features. J Radiol August 2008;89(7–8 Pt 1):873–9.

[45] Sans N, Faruch M, Chiavassa-Gandois H, de Ribes CLC, Paul C, Railhac J-J. High-resolution magnetic resonance imaging in study of the skin: normal patterns. Eur J Radiol November 2011;80(2): e176–181.

[46] Zemtsov A, Reed J, Dixon L. Magnetic resonance imaging evaluation helps to delineate a recurrent skin cancer present under the skin flap. J Dermatol Surg Oncol June 1992;18(6):508–11.

[47] Mäurer J, Knollmann FD, Schlums D, Garbe C, Vogl TJ, Bier J, et al. Role of high-resolution magnetic resonance imaging for differentiating melanin-containing skin tumors. Invest Radiol November 1995;30(11):638–43.

[48] Ono I, Kaneko F. Magnetic resonance imaging for diagnosing skin tumors. Clin Dermatol August 1995;13(4):393–9.

[49] Drapé JL, Idy-Peretti I, Goettmann S, Guérin-Surville H, Bittoun J. Standard and high resolution magnetic resonance imaging of glomus tumors of toes and fingertips. J Am Acad Dermatol October 1996;35(4):550–5.

[50] El Gammal S, Hartwig R, Aygen S, Bauermann T, el Gammal C, Altmeyer P. Improved resolution of magnetic resonance microscopy in examination of skin tumors. J Invest Dermatol June 1996;106(6):1287–92.

[51] Mäurer J, Hoffmann KT, Lissau G, Schlums D, Felix R. High-resolution magnetic resonance imaging for determination of thickness and depth of invasion of skin tumours. Acta Derm Venereol November 1999;79(6):478–9.

[52] Rajeswari MR, Jain A, Sharma A, Singh D, Jagannathan NR, Sharma U, et al. Evaluation of skin tumors by magnetic resonance imaging. Lab Invest September 1, 2003;83(9):1279–83.

[53] Hong H, Sun J, Cai W. Anatomical and molecular imaging of skin cancer. Clin Cosmet Investig Dermatol October 7, 2008;1:1–17.

[54] Idy-Peretti I, Bittoun J, Alliot FA, Richard SB, Querleux BG, Cluzan RV. Lymphedematous skin and subcutis: in vivo high resolution magnetic resonance imaging evaluation. J Invest Dermatol May 1998;110(5):782–7.

[55] Song HK, Wehrli FW, Ma J. In vivo MR microscopy of the human skin. Magn Reson Med Off J Soc Magn Reson Med Soc Magn Reson Med February 1997;37(2):185–91.

[56] Weis J, Ericsson A, Hemmingsson A. Chemical shift artifact-free microscopy: spectroscopic microimaging of the human skin. Magn Reson Med 1999;41(5):904–8.

[57] Weis J, Ericsson A, Åström G, Szomolanyi P, Hemmingsson A. High-resolution spectroscopic imaging of the human skin. Magn Reson Imaging February 2001;19(2):275–8.

[58] Dias M, Hadgraft J, Glover PM, McDonald PJ. Stray field magnetic resonance imaging: a preliminary study of skin hydration. J Phys Appl Phys February 21, 2003;36(4):364.

[59] Backhouse L, Dias M, Gorce JP, Hadgraft J, McDonald PJ, Wiechers JW. GARField magnetic resonance profiling of the ingress of model skin-care product ingredients into human skin in vitro. J Pharm Sci 2004;93(9):2274–83.

[60] Ciampi E, van Ginkel M, McDonald PJ, Pitts S, Bonnist EYM, Singleton S, et al. Dynamic in vivo mapping of model moisturiser ingress into human skin by GARfield MRI. NMR Biomed 2011; 24(2):135–44.

[61] Barral JK, Bangerter NK, Hu BS, Nishimura DG. In vivo high-resolution magnetic resonance skin imaging at 1.5 T and 3 T. Magn Reson Med 2010;63(3):790–6.

[62] Aubry S, Casile C, Humbert P, Jehl J, Vidal C, Kastler B. Feasibility study of 3-T MR imaging of the skin. Eur Radiol March 11, 2009; 19(7):1595–603.

[63] Laistler E, Loewe R, Moser E. Magnetic resonance microimaging of human skin vasculature in vivo at 3 Tesla. Magn Reson Med 2011;65(6):1718–23.

[64] Wright A, Song HK, Wehrli FW. In vivo MR micro imaging with conventional radiofrequency coils cooled to 77 degrees K. Magn Reson Med 2000;43(2):163–9.

[65] Ginefri J-C, Darrasse L, Crozat P. High-temperature superconducting surface coil for in vivo microimaging of the human skin. Magn Reson Med 2001;45(3):376–82.

[66] Darrasse L, Ginefri J-C. Perspectives with cryogenic RF probes in biomedical MRI. Biochimie September 2003;85(9):915–37.

[67] Laistler E, Poirier-Quinot M, Lambert SA, Dubuisson R-M, Girard OM, Moser E, et al. In vivo MR imaging of the human

skin at subnanoliter resolution using a superconducting surface coil at 1.5 Tesla. J Magn Reson Imaging JMRI February 2015; 41(2):496–504.

[68] Fritsch A, Loeckermann S, Kern JS, Braun A, Bösl MR, Bley TA, et al. A hypomorphic mouse model of dystrophic epidermolysis bullosa reveals mechanisms of disease and response to fibroblast therapy. J Clin Invest May 1, 2008;118(5):1669–79.

[69] Canuto HC, Fishbein KW, Huang A, Doty SB, Herbert RA, Peckham J, et al. Characterization of skin abnormalities in a mouse model of osteogenesis imperfecta using high resolution magnetic resonance imaging and Fourier transform infrared imaging spectroscopy. NMR Biomed 2012;25(1):169–76.

[70] Kinsey ST, Moerland TS, McFadden L, Locke BR. Spatial resolution of transdermal water mobility using NMR microscopy. Magn Reson Imaging 1997;15(8):939–47.

[71] Weber H, Baxan N, Paul D, Maclaren J, Schmidig D, Mohammadzadeh M, et al. Microcoil-based MRI: feasibility study and cell culture applications using a conventional animal system. Magn Reson Mater Phys Biol Med February 18, 2011;24(3):137–45.

[72] Bond JB, Haik BG, Mihara F, Gupta KL. Magnetic resonance imaging of choroidal melanoma with and without gadolinium contrast enhancement. Ophthalmology April 1991;98(4):459–66.

[73] Mäurer J, Strauss A, Ebert W, Bauer H, Felix R. Contrast-enhanced high resolution magnetic resonance imaging. Melanoma Res [Internet]. LWW. 2000. Available from:, http://journals.lww.com/melanomaresearch/Fulltext/2000/02000/Contrast_enhanced_high_resolution_magnetic.6.aspx.

[74] Sharma R. Gadolinium toxicity: epidermis thickness measurement by magnetic resonance imaging at 500 MHz. Skin Res Technol 2010;16(3):339–53.

[75] Aubry S, Leclerc O, Tremblay L, Rizcallah E, Croteau F, Orfali C, et al. 7-Tesla MR imaging of non-melanoma skin cancer samples: correlation with histopathology. Skin Res Technol 2012;18(4): 413–20.

[76] Roemer PB, Edelstein WA, Hayes CE, Souza SP, Mueller OM. The NMR phased array. Magn Reson Med Off J Soc Magn Reson Med Soc Magn Reson Med November 1990;16(2):192–225.

[77] Schmitt M, Potthast A, Sosnovik DE, Polimeni JR, Wiggins GC, Triantafyllou C, et al. A 128-channel receive-only cardiac coil for highly accelerated cardiac MRI at 3 Tesla. Magn Reson Med Off J Soc Magn Reson Med Soc Magn Reson Med June 2008;59(6): 1431–9.

[78] Göbel K. Development of dedicated methods for mr microscopy of the human skin using phased-array microcoils [Diploma thesis]. University of Konstanz; 2012.

[79] Göbel K, Gruschke OG, Leupold J, Kern JS, Has C, Korvink JG, et al. MR microscopy of diseased human skin using phased-array of microcoils at 9.4 T: first results. 2013. Salt Lake City, USA.

[80] Schwaiger A, Blümich B. Biomedizinische Anwendungen der NMR-MOUSE 25.11.2002 [Internet]. Publikationsserver der RWTH Aachen University; 2003. Available from: http://publications.rwth-aachen.de/record/52070/files/Schwaiger_Andrea.pdf.

[81] Alharbi Z, Kauczok J, Pallua N. A review of wide surgical excision of hidradenitis suppurativa. BMC Dermatol June 26, 2012; 12(1):9.

[82] Griffin N, Williams AB, Anderson S, Irving PM, Sanderson J, Desai N, et al. Hidradenitis suppurativa: MRI features in anogenital disease. Dis Colon Rectum June 2014;57(6):762–71.

[83] Kelly AM, Cronin P. MRI features of hidradenitis suppurativa and review of the literature. Am J Roentgenol November 1, 2005; 185(5):1201–4.

[84] Csernok E, Gross WL. Primary vasculitides and vasculitis confined to skin: clinical features and new pathogenic aspects. Arch Dermatol Res September 2000;292(9):427–36.

[85] Bley TA, Uhl M, Venhoff N, Thoden J, Langer M, Markl M. 3-T MRI reveals cranial and thoracic inflammatory changes in giant cell arteritis. Clin Rheumatol March 2007;26(3):448–50.

[86] Döme B, Paku S, Somlai B, Tímár J. Vascularization of cutaneous melanoma involves vessel co-option and has clinical significance. J Pathol July 2002;197(3):355–62.

CHAPTER

30

The Role of Magnetic Resonance Imaging in the Management of Anogenital Hidradenitis Suppurativa

O. Westerland[1], *G. Charles-Edwards*[1,2], *V. Goh*[1,2], *N. Griffin*[1]

[1]Guy's and St Thomas' NHS Foundation Trust, London, United Kingdom; [2]King's College London, London, United Kingdom

OUTLINE

INTRODUCTION

Hidradenitis suppurativa (HS), also known as acne inversa or Verneuil's disease [1], is a chronic inflammatory dermatological condition that affects apocrine-gland–bearing skin. It is characterized by inflammatory nodules, abscesses, sinus tracts, and fistulas, with progression to tissue scarring in chronic disease [2]. HS is a relatively common condition, with a reported prevalence of approximately 1% [3], is more common in females (female:male ratio = 3:1), and predominantly affects young to middle-aged adults with postpubertal onset. Disease severity is usually greatest in males. Despite its prevalence and distinct clinical features, the diagnosis of HS is often delayed, particularly outside of the specialist dermatology setting [2].

Until recently, the assessment of disease activity in HS was mostly clinical. However, it is becoming increasingly recognized that magnetic resonance imaging (MRI) can help in the delineation of tracts in patients with anogenital disease and hence direct both medical therapy and the mapping of anogenital disease before surgery. This chapter discusses the epidemiology of HS and the role of MRI in the management of patients with the anogenital form of disease.

ETIOLOGY

The etiology of HS is incompletely understood and is most likely multifactorial. Approximately 40% of patients report having an affected family member, and

Imaging in Dermatology
http://dx.doi.org/10.1016/B978-0-12-802838-4.00030-3

mutations in the gamma-secretase gene have been identified in a subset of HS patients [4]. At present an autosomal-dominant mutation with incomplete penetrance is a favored theory [5]; however, further research is needed in this area.

HS is also thought to have an autoimmune basis, supported by the fact that it is associated with conditions such as Crohn's disease, ulcerative colitis, spondyloarthropathies, and pyoderma gangrenosum [6]. There have also been several studies that have shown that severe HS may respond to immunomodulatory treatments such as anti-tumor necrosis factor (TNF)-α inhibitors licensed for treatment of psoriasis (infliximab, adalimumab, and etanercept) and more recently ustekinumab, a human anti-p40 monoclonal antibody [7].

HS is also thought to have an endocrine basis, given its onset is usually postpubertal (third decade of life), with active disease during reproductive years. Some patients also report premenstrual flare-ups and reduction in disease severity during pregnancy. However, it should be noted that the potential hormonal mechanisms in HS are incompletely understood. Serum androgen levels in HS patients are not elevated in relation to control subjects, and it is hypothesized that there is probable increased end-organ sensitivity to circulating serum androgen levels in HS patients [6].

Smoking and obesity are recognized risk factors for the development of HS and correlate positively with clinical severity. Elevated serum nicotine levels are thought to promote epidermal hyperplasia and follicular plugging. Nicotine also has several proinflammatory effects; for example, it promotes mast cell degranulation and potentiates the effects of neutrophils [3]. However, smoking cessation has not been shown to ameliorate disease severity once there are established clinical features of HS [8]. The contribution of obesity to HS pathogenesis is thought to be as a result of increased tissue stress, promoting formation of sinus tracts. The warm, humid environment created by skin folds is also thought to contribute to secondary bacterial infections. Finally, obese patients have increased circulating levels and local tissue concentrations of proinflammatory cytokines [eg, TNF-α, interleukin (IL)-1, and IL-6], and this likely contributes to the excessive immune response to follicular rupture [9].

PATHOGENESIS

HS affects the apocrine gland–bearing areas, primarily the axillary and inguinal regions, but the mammary, inframammary, perineal, and perianal regions and trunk may also be affected. In females the axillary and inguinal areas are most often involved. Perineal and perianal involvement is more common in men, as well as more unusual sites of involvement (eg, retroauricular region and chest). The primary pathogenic insult in HS affects the folliculopilosebaceous unit with secondary involvement of the apocrine glands [6]. In HS, keratin plugging of follicles results in follicular distension followed by follicular rupture. Follicular rupture leads to spillage of keratin/exudative products into the surrounding tissues and triggers an inflammatory response, with secondary subcutaneous inflammation, edema, and abscess formation. Moderate and severe forms of HS are additionally characterized by sinus tract and fistula formation, with scarring and tissue contractures observed in the chronic phase. HS, acne conglobata, dissecting cellulitis of the scalp, and pilonidal abscess together form the acne tetrad because they share a common underlying pathogenesis of follicular occlusion [10].

CLINICAL MANIFESTATIONS

Patients with HS present with severe pain as a result of inflammatory lesions/abscesses. Scarring in chronic HS may result in contractures and impaired mobility. HS patients may also suffer isolation, impaired interpersonal relationships, and depression due to the malodorous discharge associated with this disease. As a result, HS patients generally have higher scores on Dermatology Life Quality Index (DLQI) testing in comparison with patients with other chronic skin conditions (eg, acne vulgaris and psoriasis), indicating greater quality of life impairment [11]. HS is also associated with other conditions (eg, spondyloarthropathies and Crohn's disease). Patients with HS may sometimes be misdiagnosed with perianal Crohn's disease. The fact that HS and Crohn's disease may coexist (in one study, 17% of patients with Crohn's disease had symptoms of HS) [12] leads to further diagnostic difficulty.

DIAGNOSIS

HS is a clinical diagnosis, and three of the following features need be present: (1) typical lesions are present (ie, painful nodules, abscesses, sinus tracts, scars, and tombstone comedos), (2) lesions are distributed in typically affected areas (eg, axillae, groin, perineal, perianal, buttocks, and inter/inframammary folds), and (3) disease shows chronicity and recurrence [13]. Infundibular plugging, cyst formation, and epidermal psoriasiform hyperplasia comprise the major histological findings [3].

ASSESSMENT OF DISEASE SEVERITY

HS severity is measured using several assessment tools, the most well known being the Hurley assessment scale [14]. Patients with Hurley stage I disease (mild disease) have single or multiple abscesses, without sinus tracts or cicatrization/scarring. Hurley stage II (moderate) disease is associated with recurrent abscesses, sinus tracts, and scarring; however, lesions are widely separated. In Hurley stage III (severe) disease there is diffuse skin involvement with multiple complex, interconnecting sinus tracts and abscesses across the entire affected area. Most patients have Hurley stage I or II disease. The modified Sartorius scale is another measure of disease severity that is often used in the clinical trial setting [8,15]. This involves counting the number of active lesions in each region (and spacing between lesions); therefore it is more time-consuming and perhaps less suitable for routine use in the clinical setting. HS severity may also be assessed using DLQI testing.

ROLE OF IMAGING

Until recently there has been a limited role for imaging in the investigation and management of patients with HS. However, in the past two decades MRI has become established as the noninvasive method of choice for evaluating perianal fistula disease in patients with either Crohn's disease or de novo perianal sepsis [16]. Advantages include the lack of ionizing radiation, its multiplanar capability, and its excellent contrast and resolution, thereby facilitating the delineation of tracts and their relationship to the anal sphincter complex. Our institution is a tertiary referral center for HS, and over the past 5 years we have used pelvic MRI in the investigation of patients with anogenital HS. Our experience of MRI in anogenital HS has been recently published [17]. MRI helps in identifying the distribution and number of subcutaneous sinus tracts and the presence of anal sphincter involvement, abscesses, and supralevator disease. It can help determine if surgical intervention is required to drain abscesses or debride or place setons for the control of perianal fistulas. The use of MRI may distinguish HS from Crohn's disease (in which additional features such as proctitis or small bowel inflammation may be seen). It can also be used in the reassessment of anogenital HS after either surgical or medical treatment.

Magnetic Resonance Imaging Protocol

The protocol at our institution is as in Table 30.1. Patients are imaged in the supine position using a combination of phased-array surface and spine coils. No

TABLE 30.1 Parameters for Magnetic Resonance Imaging Protocol

Parameter	Sagittal STIR	Axial oblique STIR	Coronal oblique STIR	Fat-suppressed echoplanar diffusion-weighted imaging	Axial T2 small FOV
TR (ms)	4000	4000	4000	4400	6360
TE (ms)	26	26	26	70	103
TI (ms)	150	150	150	–	–
Flip angle (degrees)	153	153	153	–	180
TA (min)	2.04	2.52	2.20	4.12	3.57
Averages	2	2	2	4	3
Slices	23 × 3 mm (1-mm gap)	30 × 3 mm (1-mm gap)	30 × 4 mm (1-mm gap)	35 × 5 mm (1.5-mm gap)	35 × 3 mm (1-mm gap)
FOV (mm)	300 × 291	300 × 281	300 × 281	300 × 300	220 × 220
Acquired spatial resolution (mm)	0.9 × 0.9 × 3	0.9 × 0.9 × 3	0.9 × 0.9 × 3	2.1 × 2.6 × 5	0.4 × 0.4 × 3
Parallel imaging factor	2	2	2	2	2
Bandwidth (Hz/Px)	220	220	220	1318	199
Turbo factor	10	10	10	–	16
b-values	–	–	–	0, 100, 400, 700, 1000	–

FOV, field of view; *STIR*, short tau inversion recovery; *TE*, echo time; *TI*, inversion time; *TR*, repetition time.

patient preparation is required. Imaging is performed on a 1.5-T magnetic resonance (MR) scanner, which is the magnet strength most frequently used in institutions for perianal fistula imaging. Although an improved signal-to-noise ratio from higher strength clinical magnets (3 T) can be traded off for increased spatial resolution, disadvantages include increased susceptibility artifacts, field inhomogeneity, and specific absorption rates [18].

Some institutions use an endoanal coil to increase spatial resolution [19]. However, it is poorly tolerated by patients with severe anal disease and gives a more limited field of view than the phased array coil; hence it is not routinely used.

Our protocol uses a T2-weighted short tau inversion recovery (STIR) sequence that provides fat-suppressed T2-weighted images. There is good contrast delineation between the high STIR signal tracts and the surrounding low-signal fat within the subcutaneous tissues, ischioanal fossa, and supralevator space. STIR sequences are performed in three planes with the axial and coronal oblique images obtained perpendicular and along the long axis of the anal canal. It is important to ensure that the axial and coronal imaging planes are correctly planned from the sagittal STIR sequence so that the exact relationship of tracts to the anal sphincter complex can be seen, especially if fistulating disease is present. It is sometimes difficult to determine if a tract is active or fibrotic on T2-weighted fat-suppressed images, although a typical fibrotic tract lacks the hyperintense T2 signal of fluid that is seen within an active tract [20]. To help clarify, T1-weighted fat-suppressed gadolinium-enhanced images can also be performed. Active tracts are said to show wall enhancement with the center of the tracts remaining hypointense because of fluid, whereas fibrotic tracts filled with granulation tissue uniformly enhance [21]. Abscesses are also well demonstrated with the use of intravenous gadolinium, appearing as rim-enhancing fluid collections. Some studies have looked at the use of dynamic contrast-enhanced imaging in patients with perianal fistula disease unrelated to HS. Intensity curves have been shown to correlate with the degree of inflammation [22]. However, this approach is time consuming. At our institution, we do not routinely use gadolinium-enhanced sequences because it is felt that the STIR images are sufficient in most cases.

MR fistulography has been reported as a useful adjunct to standard MR fistula protocols and may improve delineation of the internal opening [23]. The technique essentially involves cannulation of the external opening using an infant feeding tube and instillation of gadolinium (1 mL in 20 mL of normal saline). The hyperintense fistulous tract is then clearly delineated on T1-weighted fat-suppressed sequences [23]. Although MR fistulography is a recognized technique for patients with simple perianal fistula disease unrelated to HS, it is likely to have a limited role in HS, in which anal sphincter involvement is less common. The presence of multiple sinus tracts in patients with severe anogenital HS would also make cannulation of all of the sinus openings impractical.

Our inclusion of a high-resolution small field-of-view T2-weighted sequence in the MR protocol is optional. We perform it in the axial plane through the anal canal because it gives better delineation of the normal anal sphincter anatomy compared with T2-weighted fat-suppressed sequences in which signal (other than from fluid-filled tracts) is largely homogenized. It can also help clarify if small high-signal structures seen on the STIR sequences are due to sinus tracts or vessels because the latter appear as high signal on STIR images but low signal on turbo-spin–echo sequences.

It is increasingly being recognized that diffusion-weighted imaging (DWI) can help evaluate perianal fistula activity with lower apparent diffusion coefficients (ADCs) (ie, greater restriction of water diffusion) in patients with active disease [24]. In our experience, the sinus tracts in HS typically appear as high signal on the diffusion-weighted images with correspondinglow signal on the ADC map. Fused axial T2/diffusion-weighted images can facilitate interpretation of the STIR sequences (Fig. 30.1). Diffusion tensor imaging (DTI) has been mostly used in imaging the central nervous system to characterize normal and diseased cerebral and spinal white matter tracts and to a lesser extent the fiber orientation of striated skeletal muscle [25]. It describes the directionality of water diffusion resulting from the anisotropic internal microstructure within tissues. In DTI this directional dependence of the water diffusion in each voxel is modeled by a diffusion tensor, from which so-called fiber tracts can be reconstructed. It has been shown that DTI with fiber tractography can be used to visualize the three-dimensional (3D) orientation of muscle fibers in the pelvic floor [26], and it may facilitate the 3D modeling of the anal sphincter. In HS patients it may also delineate active sinus tracts (Fig. 30.2).

The use of whole-body DWI is an attractive noninvasive, nonionizing technique to demonstrate all sites of HS involvement. It has been used to demonstrate sites of involvement in patients with various malignancies (eg, lymphoma and myeloma) [27,28]. However, at present it is unlikely to be routinely used in HS because this disorder is primarily a cutaneous disease with little visceral or skeletal involvement. It also lacks

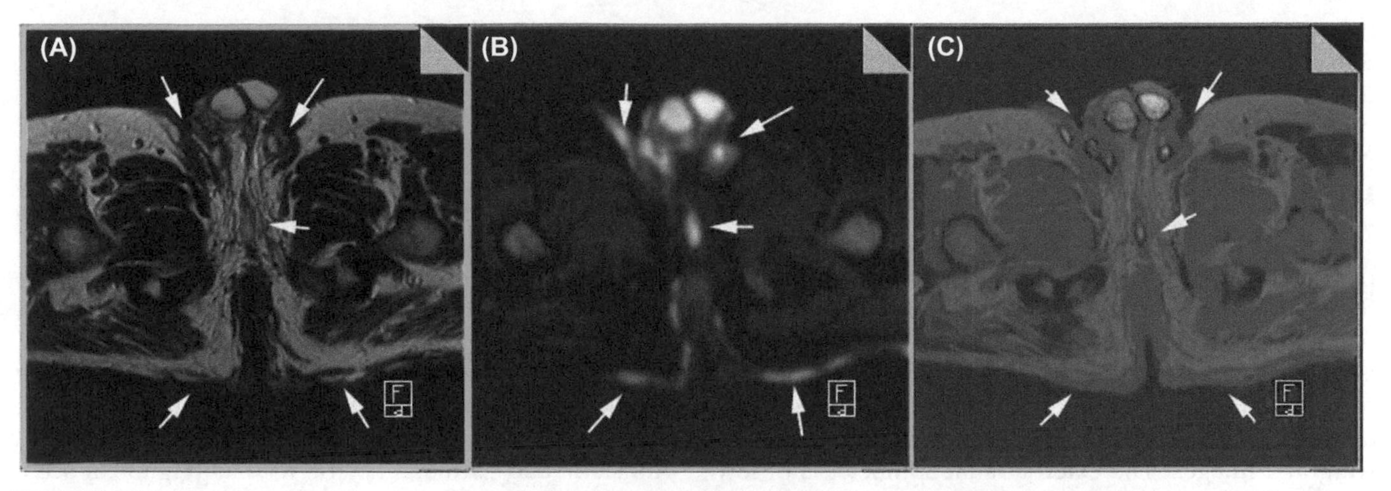

FIGURE 30.1 Example of fused axial T2/diffusion-tensor imaging (DWI). Sinus tracts are seen in the buttocks, perineum, and groins (*arrows*) on the (A) high-resolution axial T2-weighted image, (B) DWI (B1000), and corresponding (C) fused image, where tracts are highlighted in *red*.

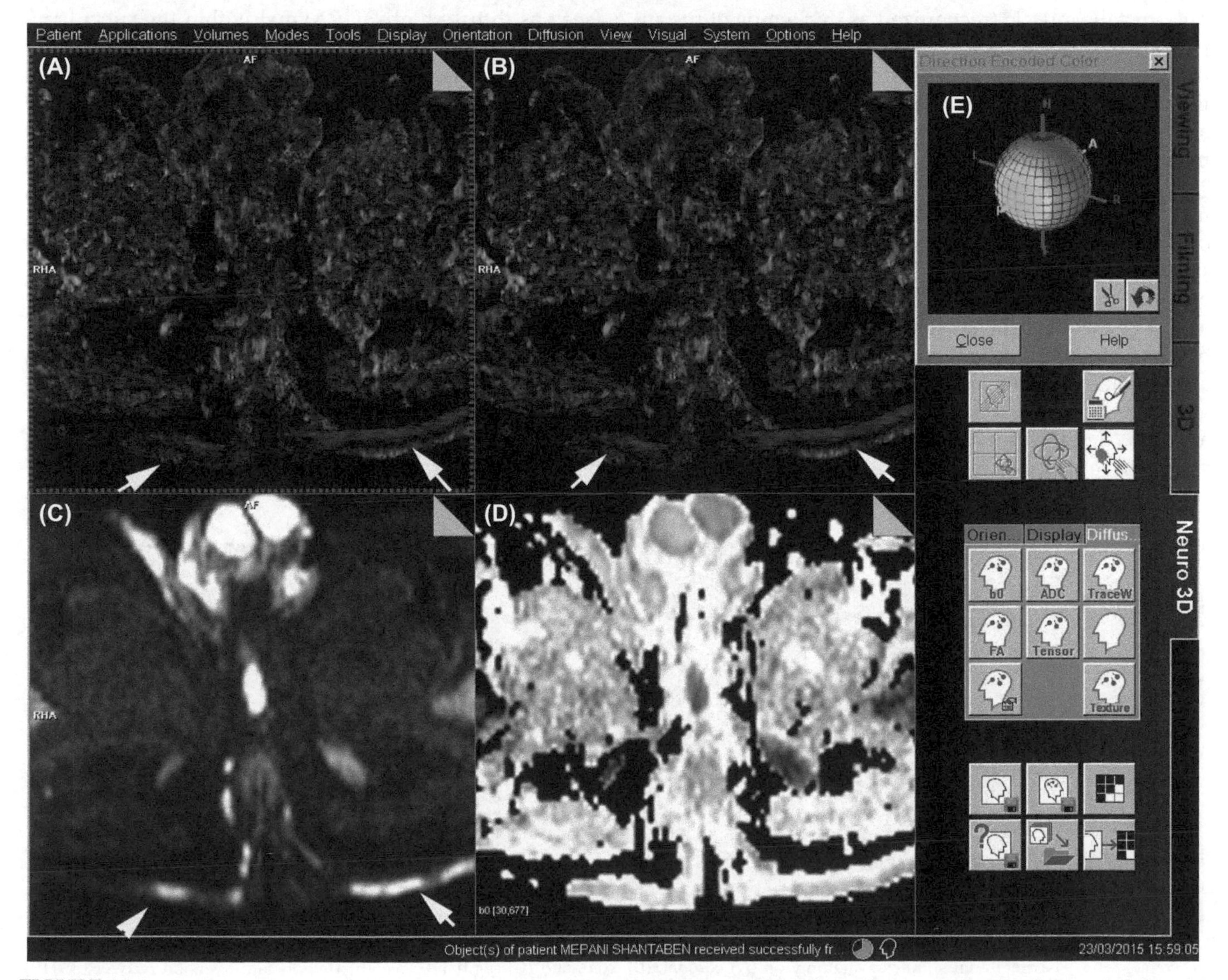

FIGURE 30.2 Example of the use of diffusion tensor imaging of the anal sphincter complex in hidradenitis suppurativa. Color-coded (A) axial plane fractional anisotropy and (B) relative anisotropy maps with corresponding (C) diffusion-weighted imaging and (D) ADC map. The sinus tracts (*arrows*) in the buttocks show a left-right directionality according to the color map (E).

the spatial resolution that dedicated pelvic MRI has in the delineation of tracts in patients with anogenital disease.

Normal Anal Anatomy on Magnetic Resonance Imaging

The anal canal extends from the anal verge to the puborectalis muscle and forms the most distal part of the gastrointestinal tract. It is made up of the internal and external anal sphincter. The internal anal sphincter is a continuation of the circular smooth muscle of the rectum and appears of relatively high signal on STIR sequences (Fig. 30.3) but low signal on T1- and T2-weighted images. The external anal sphincter is of low T1/T2/STIR signal (Fig. 30.3) and is made up of voluntary striated muscle. It lies on the outside of the internal anal sphincter and is continuous with the fibers of the puborectalis muscle (recognized as thickening of the superior fibers of the external anal sphincter), which then merges with the pelvic floor (levator ani muscle). The intersphincteric plane lies between these two sphincters and contains a continuation of the outer longitudinal muscle fibers of the rectum; this is not visible at 1.5 T. The levator plate separates the infralevator compartment (containing the anal canal) and the supralevator compartment (containing the rectum and other pelvic viscera) and appears of the same signal intensity as the external anal sphincter on all sequences, being of striated muscle. The fat on either side of the anal canal, bordered medially by the external anal sphincter, laterally by the obturator internus, and superiorly by the levator ani muscle, is known as the ischioanal fossa. This appears of high signal on T2-weighted images but low signal on fat-suppressed T2-weighted images.

Appearances of Hidradenitis Suppurativa on Pelvic Magnetic Resonance Imaging

Most patients with anogenital HS have superficial subcutaneous sinus tracts. The number and size of tracts, as well as the distribution, vary depending on the disease activity and severity. The most common sites of involvement (Fig. 30.4) on imaging [17] are on either side of the natal cleft and buttocks, with involvement of the perineum (scrotum/vulva), groins, lower abdominal wall, proximal thighs, and tissue overlying the sacrococcygeal region seen to a lesser extent. In a minority of cases there is involvement of the anal sphincter complex, with a combination of intersphincteric (Fig. 30.5) and transphincteric and extrasphincteric fistulas (Fig. 30.6) seen. Rectovaginal fistulas have also been observed, although these are rare (Fig. 30.7). Associated subcutaneous edema (Fig. 30.4B) is often appreciated in active disease (seen as diffuse high-STIR signal within the subcutaneous fat), with high-STIR signal also sometimes demonstrated in the underlying pelvic musculature (eg, gluteus maximus muscle) because of inflammatory change (Fig. 30.6). Very occasionally, if there is superadded sepsis, osteomyelitis may be seen if tracts lie in close proximity to the underlying bone [eg, tracts overlying the sacrum can lead to underlying sacrococcygeal osteomyelitis (Fig. 30.8)]. High-STIR signal is then present in the underlying bone, with corresponding low signal on T1-weighted images, if performed.

Abscesses are occasionally seen in more severe cases, recognized on STIR images as focal dilatation of sinus tracts or as separate high-signal fluid collections in the adjacent tissues (Fig. 30.6). Reactive inguinal adenopathy is a common finding in many patients with anogenital HS (Fig. 30.9).

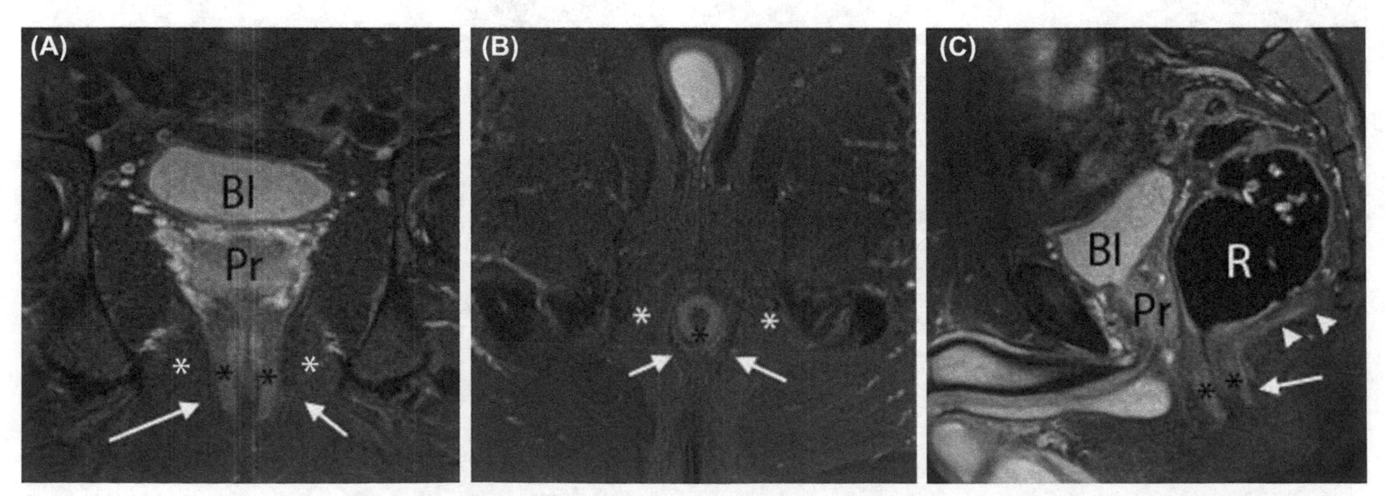

FIGURE 30.3 Normal anal sphincter anatomy on (A) coronal, (B) axial, and (C) sagittal T2-weighted short tau inversion recovery (STIR) images through the anal canal. The internal anal sphincter (IAS) (*black asterisk*) appears as higher STIR signal compared with the external anal sphincter (EAS; *arrows*). The ischioanal fossa is the triangular compartment containing fat, seen outside of the sphincter complex (*white asterisk*). The levator ani muscle (*arrowheads*) is continuous with the EAS and forms the pelvic floor. *Bl*, bladder; *Pr*, prostate; *R*, rectum.

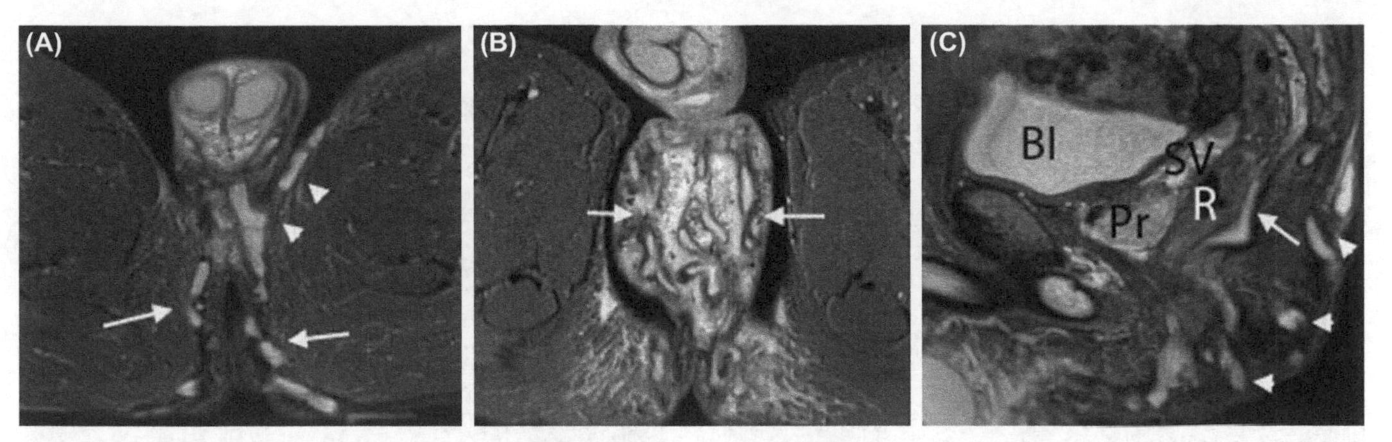

FIGURE 30.4 Typical distribution of subcutaneous sinus tracts in a patient with anogenital hidradenitis suppurativa. (A) Axial short tau inversion recovery (STIR) showing tracts on either side of the natal cleft extending into the left buttock (*arrows*), perineum, and left groin (*arrowhead*). (B) Axial STIR image in a patient with severe sinus disease involving the scrotum (*arrows*); there is subcutaneous edema seen as high-STIR signal in both buttocks. (C) Sagittal STIR image in a patient with multiple subcutaneous sinus tracts involving the perineum, natal cleft, and tissue overlying the sacrum (*arrowheads*). Further tracts are noted extending above the levator plate (*arrow*). *Bl*, bladder; *Pr*, prostate; *R*, rectum; *SV*, seminal vesicles.

It is important to try and distinguish anogenital HS from perianal Crohn's disease on MRI because management may differ and there are implications for more proximal bowel involvement in the latter. In a few cases this may be difficult on imaging criteria alone because of the overlap in imaging features. In general, sinus involvement of the perineum and inguinal and gluteal regions is unusual in perianal Crohn's disease, with most, if not all Crohn's patients with perianal disease demonstrating the presence of a fistula.

MRI can be used in the reassessment of anogenital HS after surgical or medical treatment. If there has been medical therapy, there may be a downstaging in activity of disease on follow-up imaging, with the tracts becoming lower signal on STIR images, with low-signal fibrotic walls noted. Tracts may also

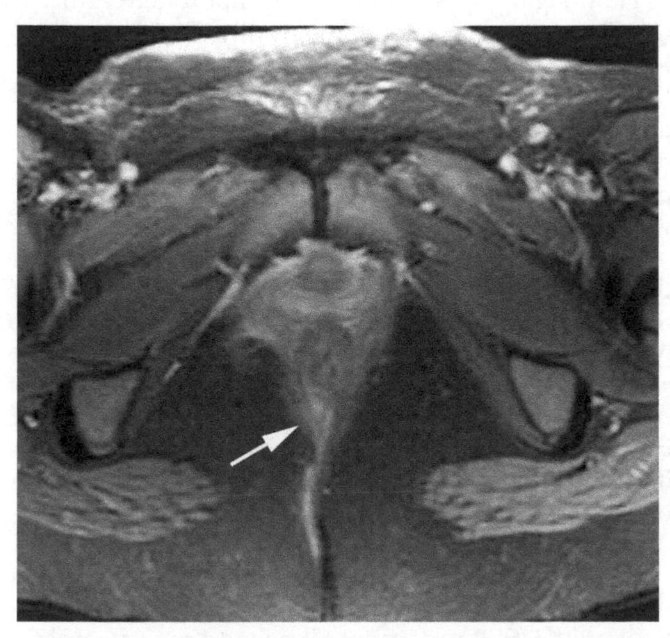

FIGURE 30.5 Axial short tau inversion recovery image in a patient with anogenital hidradenitis suppurativa and intersphincteric fistula formation (*arrow*).

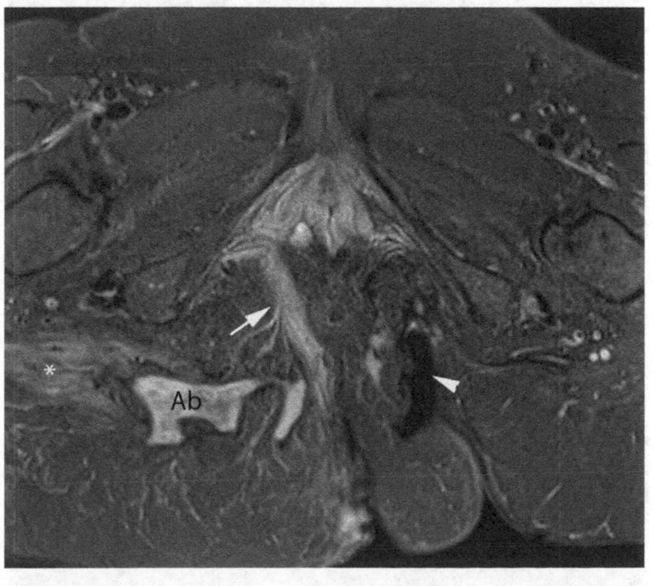

FIGURE 30.6 Axial short tau inversion recovery image in a patient with anogenital hidradenitis suppurativa and extrasphincteric fistula formation (*arrow*). There is a focal abscess collection in the right buttock (*Ab*) with myositis (*asterisk*). There is evidence of previous surgical debridement (*arrowhead*).

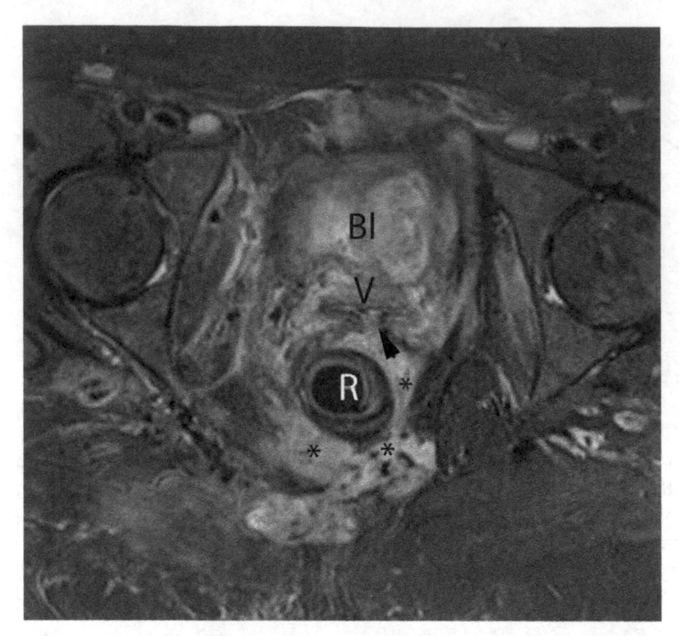

FIGURE 30.7 Axial short tau inversion recovery image in a patient with anogenital hidradenitis suppurativa with a horseshoe supralevator collection (*asterisk*) extending around the rectum and with fistulation into the posterior wall of the vagina (*arrow*). *Bl*, bladder; *R*, rectum; *V*, vagina.

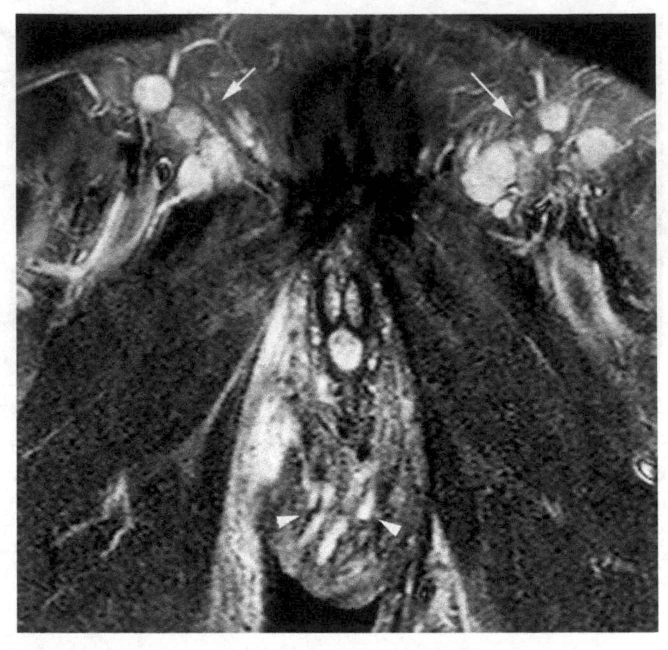

FIGURE 30.9 Coronal short tau inversion recovery image in a patient with multiple subcutaneous sinus tracts in the scrotum (*arrowheads*) and reactive bilateral inguinal adenopathy (*arrows*).

become smaller and less numerous. In limited disease, complete resolution may rarely be seen (Fig. 30.10). Edema within the surrounding subcutaneous tissues may also resolve. In some cases the disease is resistant to medical therapy and sometimes worsens, with greater fluid distension of tracts and development of abscesses.

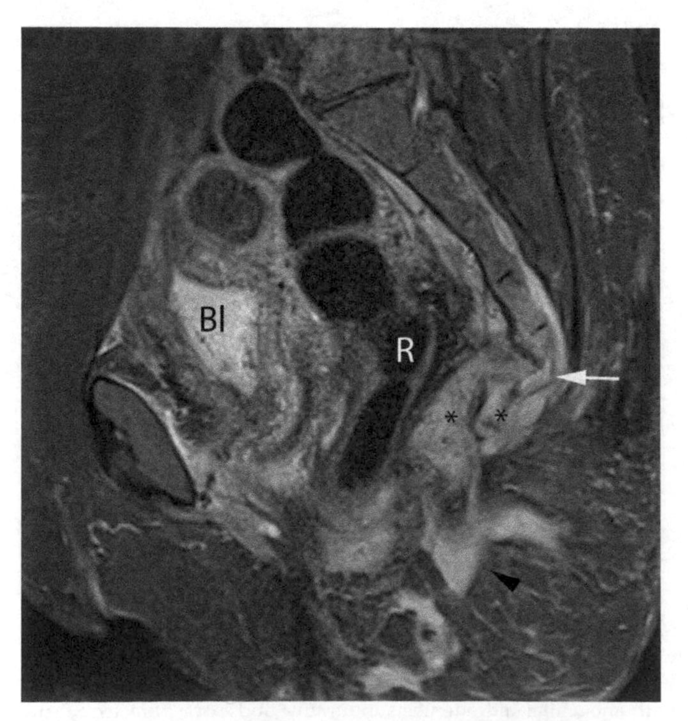

FIGURE 30.8 Sagittal short tau inversion recovery image in same patient as Fig. 30.9 with high signal seen in the coccyx (*arrow*) in keeping with osteomyelitis. *Bl*, bladder; *R*, rectum.

COMPARISON WITH OTHER IMAGING MODALITIES

HS imaging appearances on 18-fluorodeoxyglucose (FDG) positron emission tomography (PET) computed tomography (CT) have been anecdotally described. PET-CT combines the functional information obtained on a PET scanner and the anatomical information obtained on a CT scanner. FDG is the most common radiopharmaceutical used to assess glucose metabolism, with the uptake of this tracer increased at sites of inflammation or tumor. One case report has shown that a patient who had undergone PET-CT as part of the investigative process for a coexisting disease (eg, lymphoma [29]) demonstrated focal FDG avidity at sites corresponding to inflamed subcutaneous nodules. Although this is a potential tool to demonstrate sites of involvement, it is unlikely to become part of the standard imaging assessment of patients with HS because of the use of ionizing radiation and poorer contrast resolution than MRI.

There are several studies that highlight the potential utility of 3D endoanal ultrasound in evaluation of patients with HS. 3D endoanal ultrasound can detect early HS changes (eg, enlargement of the hair follicle). It can also be used to visualize collections and abscesses and

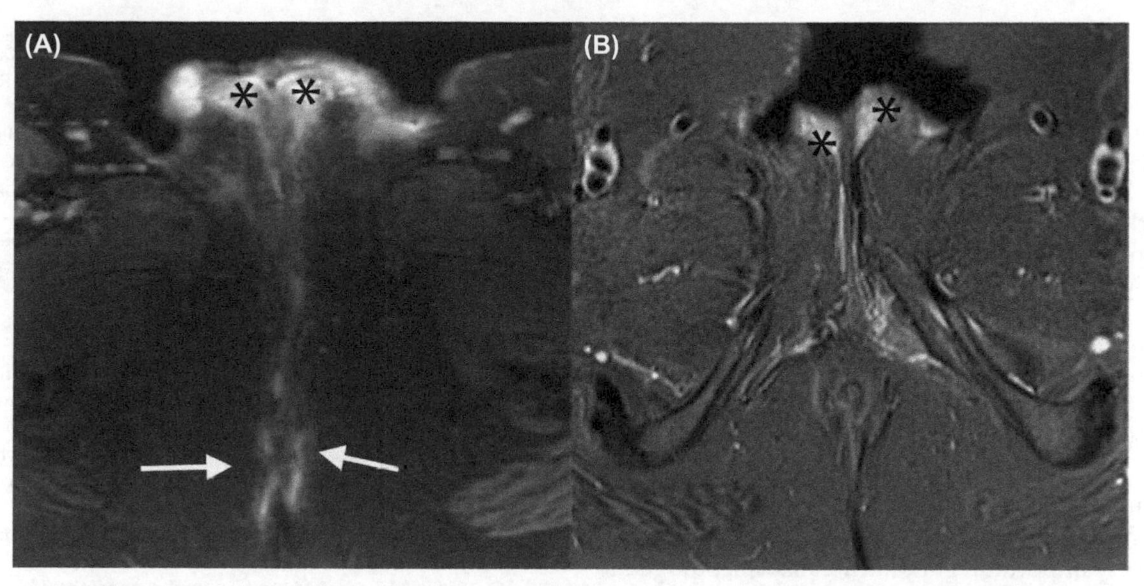

FIGURE 30.10 Example of treatment response to infliximab—axial short tau inversion recovery images (A and B) in a patient with sinus tracts related to the natal cleft (*arrows*) and vulval edema (*asterisk*). There is resolution of the tracts seen after a 24-month period with only residual vulval edema demonstrated.

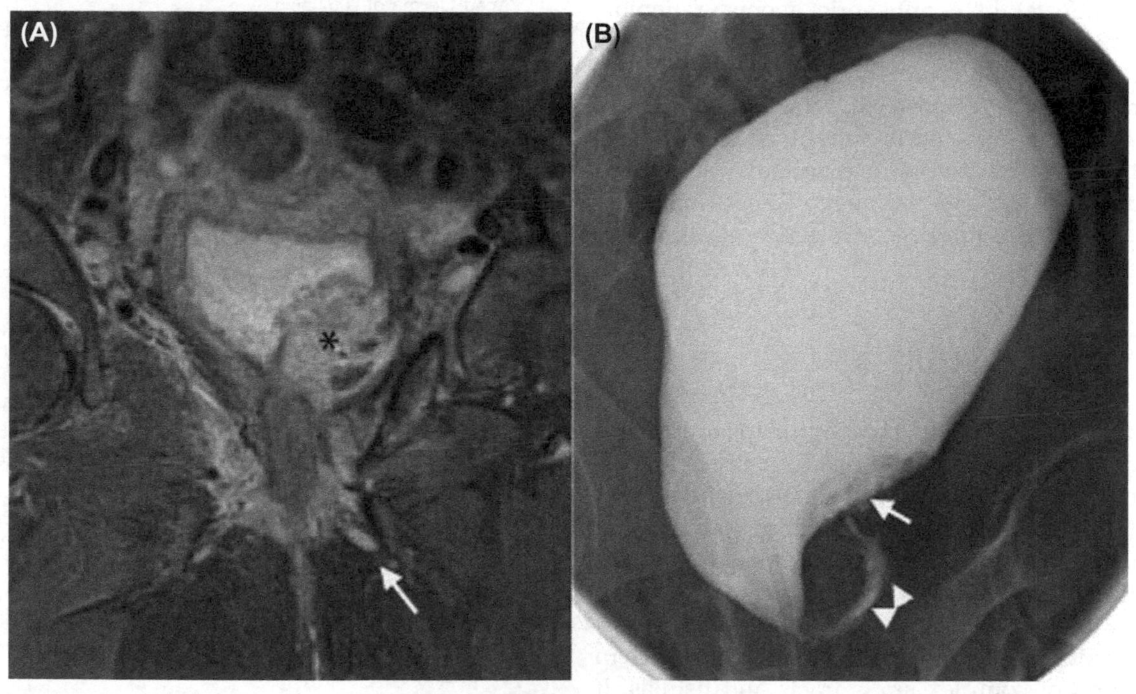

FIGURE 30.11 Example of a patient with anogenital hidradenitis suppurativa and involvement of the bladder. (A) Coronal short tau inversion recovery image shows a partially imaged tract (*arrow*) extending to the left levator plate, terminating in an abscess collection in the left bladder base (*asterisk*). *Bl*, bladder. (B) Cystogram shows corresponding extrinsic indentation of the left bladder base (*arrow*) due to the previously seen abscess collection and the delineation of a tract extending from the bladder base to the pelvic floor (*arrowheads*).

to delineate sinus tracts [30]. Advantages of ultrasound include the fact that it is easily obtainable, relatively cheap, quick to perform, and relatively well tolerated. However, it is operator dependent, less reproducible when compared with cross-sectional techniques, and may be of limited utility in patients with large body habitus and in the delineation of deep sinus tracts/fistulas. It may also be poorly tolerated in patients with painful anogenital disease.

Fluoroscopic studies (eg, barium enema) were once popular and utilized in demonstrating sinuses and fistulous tracts communicating with the rectum

[31]. However, barium enemas have a limited role in contemporary imaging of patients with anogenital HS given that most patients tend not to have rectal involvement, and the relationship of anogenital disease to surrounding structures is better demonstrated on MRI. MRI also has the advantage of no ionizing radiation. A cystogram may rarely be of use in severe anogenital HS, in which fistulating disease with the bladder is suspected (Fig. 30.11).

MANAGEMENT

There is no cure for HS at present, and management strategies are multifactorial [6]. Lifestyle measures such as smoking cessation, weight reduction, and dietary modifications (eg, reduced intake of dairy products) are extremely important. Patients are also encouraged to wear loose clothing to reduce traumatic insults to the skin. Medical treatment usually consists of antibiotics (the combination of rifampicin and clindamycin is particularly efficacious, in part as a result of the antiinflammatory effects) [32], antiseptic agents, retinoids, antiandrogens, and steroids. Immunomodulatory agents such as anti-TNF-α inhibitors have shown promise in Hurley stage III disease [7]. Patients with severe disease may require wide excision with healing via secondary intention (after antibiotic treatment of secondary infection and incision and drainage of abscesses), and some patients may require reconstructive surgical flap formation [33].

CONCLUSION

HS is a relatively common, chronic, relapsing, inflammatory skin condition characterized by painful inflammatory nodules, abscesses, sinus tracts, fistulas, and scarring, and it has a multifactorial etiology. Anogenital HS has characteristic imaging findings on MR, and the use of fat-suppressed T2-weighted sequences have been shown to be of value in confirming the diagnosis, in characterizing disease extent and severity, and in the delineation of complex sinus tracts and fistulas. It provides a noninvasive, nonionizing imaging tool to help direct management and assess treatment response in patients with this debilitating condition.

References

[1] Velpeau A. Dictionnaire de Medicine, un Repertoire General des Sciences Medicales sons le Rapport Theorique et Pratique [in French]. 2nd ed. Paris, France: Z Bechet Jeune; 1839.

[2] Dufour DN, Emtestam L, Jemec GB. Hidradenitis suppurativa: a common and burdensome, yet under-recognised inflammatory skin disease. Postgrad Med J 2014;90:216–21.

[3] Deckers IE, van der Zee HH, Prens EP. Epidemiology of hidradenitis suppurativa: prevalence, pathogenesis, and factors associated with the development of HS. Curr Derm Rep 2014;3: 54–60.

[4] Pink AE, Simpson MA, Desai N, Trembath RC, Barker JNW. Gamma-secretase mutations in hidradenitis suppurativa: new insights into disease pathogenesis. J Invest Dermatol 2012;133: 601–7.

[5] Von der Werth JM, Williams HC, Raeburn JA. The clinical genetics of hidradenitis suppurativa revisited. Br J Dermatol 2000;142: 947–53.

[6] Margesson LJ, Danby FW. Hidradenitis suppurativa. Best Pract Res Clin Obstet Gynecol 2014;28:1013–27.

[7] Kelly G, Sweeney CM, Tobin A, Kirby B. Hidradenitis suppurativa; the role of immune dysregulation. Int J Dermatol 2014;53: 1186–96.

[8] Sartorius K, Emtestam L, Jemec GBE, Lapins J. Objective scoring of hidradenitis suppurativa reflecting the role of tobacco smoking and obesity. Br J Dermatol 2009;161:831–9.

[9] Coppack SW. Pro-inflammatory cytokines and adipose tissue. Proc Nutr Soc Lond 2001;60:349–56.

[10] Plewig G, Kligman AM. Acne: morphogenesis and treatment. Berlin, Germany: Springer; 1975.

[11] Wolkenstein P, Loundou A, Barrau K, Auquier P, Revuz J, Quality of Life Group of the French Society of Dermatology. Quality of life impairment in hidradenitis suppurativa: a study of 61 cases. J Am Acad Dermatol 2007;56:621–3.

[12] Van der Zee HH, Van Der Woude CJ, Florencia EF, Prens EP. Hidradenitis suppurativa and inflammatory bowel disease: are they associated? Results of a pilot study. Br J Dermatol 2010;162: 195–7.

[13] Revuz J. Hidradenitis suppurativa. J Eur Acad Dermatol Venereol 2009;23:985–98.

[14] Hurley HJ. Axillary hyperhidrosis, apocrine bromhidrosis, hidradenitis suppurativa and familial benign pemphigus: surgical approach. In: Roenigk RK, Roenigk Jr HH, editors. Dermatologic surgery: principles and practice. 2nd ed. New York: Marcel Dekker; 1996. p. 623–45.

[15] Sartorius K, Lapins J, Emtestam L, Jemec GBE. Suggestions for uniform outcome variables when reporting treatment effects in hidradenitis suppurativa. Br J Dermatol 2003;149:211–3.

[16] Halligan S. Imaging fistula-in-ano. Clin Radiol 1998;53:85–95.

[17] Griffin N, Williams AB, Anderson S, et al. Hidradenitis suppurativa: MRI features in anogenital disease. Dis Colon Rectum 2014; 57:762–71.

[18] Lee VS, Hecht EM, Taouli B, Chen Q, Prince K, Oesingmann N. Body and cardiovascular imaging at 3.0 T. Radiology 2007;244: 692–705.

[19] deSouza NM, Gilderdale DJ, Coutts GA, Puni R, Steiner RE. MRI of fistula-in ano: a comparison of endoanal coil with external phased array coil techniques. J Comput Assist Tomogr May–June 1998;22(3):357–63.

[20] Beets Tan R, Beets G, Van D, Hoop A. Preoperative MR imaging for anal fistulas: does it really help the surgeon? Radiology 2001;218:75–84.

[21] Buchanan GN, Bartram CI, Phillips RKS, et al. Efficacy of fibrin sealant in the management of complex anal fistula. A prospective trial. Dis Colon Rectum 2003;46(9):1167–74.

[22] Torkzad MR, Karlbom U. MRI for assessment of anal fistula. Insights Imaging 2010;1:62–71.

[23] Waniczek D, Adamczyk T, Arendt J, Kluczewska E, Kozinska-Marek E. Usefulness assessment of preoperative MRI fistulography in patients with perianal fistulas. Pol J Radiol 2011;76(4):40–4.

[24] Yoshizako T, Wada A, Takahara T, et al. Diffusion-weighted MRI for evaluating perianal fistula activity: feasibility study. Eur J Radiol September 2012;81(9):2049–53.

[25] Qiu A, Mori S, Miller MI. Diffusion tensor imaging for understanding brain development in early life. Annu Rev Psychol January 3, 2015;66:853–76.
[26] Zijta FM, Froeling M, van der Paardt MP, et al. Feasibility of diffusion tensor imaging (DTI) with fibre tractography of the normal female pelvic floor. Eur Radiol June 2011;21(6):1243–9.
[27] Lin C, Luciani A, Itti E, et al. Whole-body diffusion magnetic resonance imaging in the assessment of lymphoma. Cancer Imaging September 28, 2012;12:403–8.
[28] Derlin T, Bannas P. Imaging of multiple myeloma: current concepts. World J Orthop July 18, 2014;5(3):272–82.
[29] Simpson RC, Dyer MJS, Entwisle J, Harman KE. Positron emission tomography features of hidradenitis suppurativa. Brit J Radiol 2011;84:164–5.
[30] Wortsman XC, Holm EA, Wulf HC, Jemec GBE. Real-time spatial compound ultrasound imaging of skin. Skin Res Technol 2004;10: 23–31.
[31] Nadgir R, Rubesin SE, Levine MS. Perirectal sinus tracks and fistulas caused by hidradenitis suppurativa. AJR Am J Roentgenol 2001;177:476–7.
[32] Guet-Revillet H, Coignard-Biehler H, Jais JP, et al. Bacterial pathogens associated with hidradenitis suppurativa, France. Emerg Infect Dis 2014;20(12):1990–8.
[33] Alharbi Z, Kauczok J, Pallua N. A review of wide surgical excision of hidradenitis suppurativa. BMC Dermatol 2012;12(9):1–8.

CHAPTER

31

Thermal Imaging in Dermatology

M. Bonmarin[1], F.A. Le Gal[2]

[1]Zurich University of Applied Sciences, Winterthur, Switzerland; [2]Geneva University Hospital, Geneva, Switzerland

OUTLINE

INTRODUCTION

Thermal imaging, or thermography, consists of measuring and imaging the thermal radiation emitted by every object above the absolute zero temperature [1]. Because this radiation is temperature-dependent, the infrared (IR) images recorded can be converted into temperature maps, or thermograms, allowing retrieving valuable information about the object under investigation. Thermal imaging has been known since the middle of the 20th century; recent technological achievements concerning IR imaging devices, together with the development of new procedures based on transient thermal emission measurements have revolutionized the field. Nowadays thermography is a method with advantages that are undisputed in engineering. It is routinely used for the nondestructive testing of materials [2], to investigate electronic components [3], or in the photovoltaic industry to detect defects in solar cells [4].

Despite an early interest for potential medical applications [5–7], thermal imaging is rarely used in the clinic. The main reason is probably the initial disappointing results obtained solely with static measurement procedures, where the sample is investigated in its steady state and using unhandy and performance-limited first-generation IR cameras. Fortunately those early studies have been put into perspective and medical thermal imaging is experiencing a renaissance since the late 1990s [8,9].

Among the numerous potential medical applications of thermal imaging (such as in neurology [10], oncology

Imaging in Dermatology
http://dx.doi.org/10.1016/B978-0-12-802838-4.00031-5

[11], ophthalmology [12], surgery [13], or dentistry [14]), dermatology is one of the most promising application fields. Despite its low specificity, static thermal imaging is a powerful tool to detect and characterize problems affecting the skin physiology. Indeed, abnormalities such as malignancies, inflammation, and infection usually cause localized increases in temperature that can be identified as hot spots or as asymmetrical patterns in a skin thermogram. In combination with skin thermal models, new active procedures drastically extend the capabilities of thermal imaging by allowing the retrieval of quantitative physiological information.

The goal of this chapter is to demonstrate the potential of thermal imaging for dermatological applications and to review the main investigations accomplished so far. Skin thermal properties will be briefly reviewed together with the main heat transfer models used to extract physiological parameters from experimental data. The basics of thermal radiation and thermal-imaging device technology will be presented with the different experimental procedures that have been reported in the literature. Rather than giving an exhaustive description of thermography, we aim to familiarize the reader with key concepts that will allow designing a proper experimental setup with an optimal IR camera depending on the specific application. As an illustration, two examples of the utilization of thermal imaging in skin cancer detection and burn depth evaluation will be presented. Other promising applications will be briefly outlined in the last paragraph.

SKIN THERMAL PROPERTIES

In the following paragraphs, we will briefly review the skin structure, its thermoregulation mechanisms, and the different models proposed in the literature to simulate its thermal behavior.

Skin Structure

The skin is the biggest organ of the body, with a mean surface of 2 square meters and a weight of 4–10 kg in adults (ie, around 8% of the body mass). Skin has four main functions: protection, sensation–connection, thermoregulation, and metabolism. The skin preserves the hydration of the body thanks to the stratum cornea that limits the loss of water. Moreover, it protects the body from mechanical injuries, chemical injuries, temperatures variations, ultraviolet (UV) radiations, and microorganisms [15]. Its thickness is highly variable, conferring an adapted flexibility or mechanical protection according to the needs of the different parts of the body. The skin is stratified in three main layers, ie, the epidermis, the dermis, and the hypodermis (Fig. 31.1A).

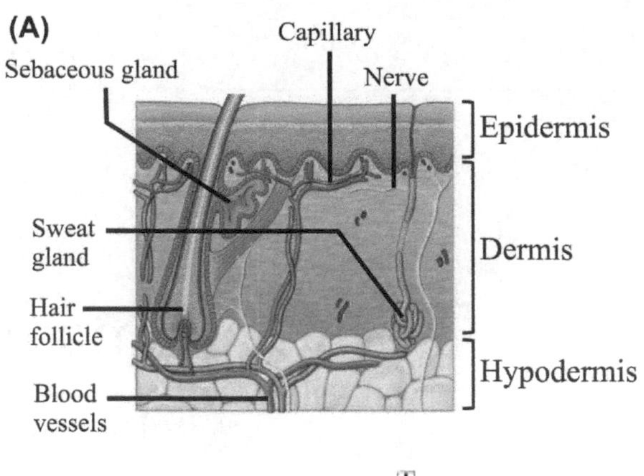

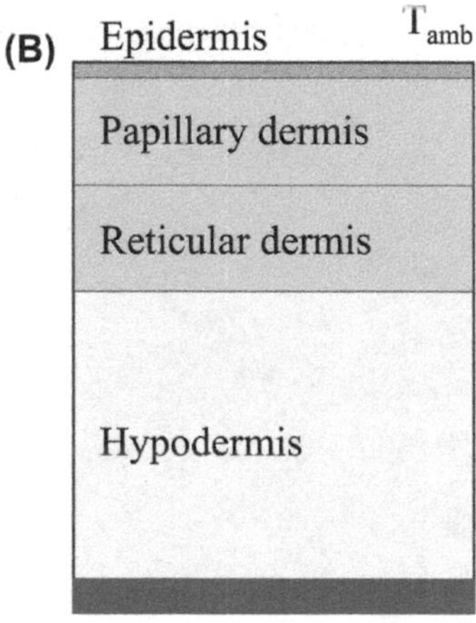

FIGURE 31.1 (A) Pictorial description of the skin structure *(Adapted from Robinson JK. Anatomy for procedural dermatology. In: Robinson JK, Hanke WC, Sengelmann R, Siegel D, editors. Surgery of the skin. 3rd ed. Philadelphia: Elsevier; 2015. p. 1–27 with permission from Elsevier.)*. The epidermis, a layer of epithelial cells, rests on the dermis, a dense connective tissue. The human skin is a highly inhomogeneous and anisotropic medium. (B) Idealized skin model according to the structure depicted in (A). The dermis is split into two layers: the papillary dermis and the reticular dermis. Some models include an additional muscle layer. T_{amb}, ambient room temperature; T_c, core temperature.

Epidermis

The epidermis is the outer cellular layer of the skin. The thickness of epidermis varies with the location; for example, the epidermis of the eyelid measures around 30–40 μm, compared with 140 μm for the buttocks, or more than 600 μm for the palms or soles [16]. Keratinocytes are the most important contingent of epidermal cells. As these cells differentiate, they move upward to the surface and their shape and their content change and form the successive layers of the epidermis. The last layer is the stratum corneum, or horny layer, made of dead keratinocytes filled with keratin, a sulfur-rich protein, and coated with lipids. Its very cohesive structure plays a major role in limiting the water loss and in the global protective barrier function of the skin. Melanocytes are located in the basal layer of the epidermis and connected to several layers of keratinocytes, thanks to their dendritic morphology. These cells secrete a pigment, the melanin, and transfer the melanin granules to keratinocytes by their dendritic processes. Langerhans cells are professional antigen-presenting cells. They play a pivotal role in the immune defenses of the body. Merkel cells (initially called *touch cells* by Merkel) are connected to sensory nerve endings and involved in fine sensitive discrimination.

Dermis

The dermis is a very dense layer that joins the epidermis at a thin basement membrane zone. Its thickness can vary from 0.4 mm (eyelids, prepuce) to 1 cm on the back. Dermis is a fibrous connective tissue that contains the blood supply for the skin. Through the very developed vascularization (around 1 m of capillaries per square centimeter of skin) the dermis brings nutrients to the epidermis, because it has no direct blood supply. The adnexae, ie, sebaceous glands that secrete the sebum, apocrine, and eccrine glands that secrete the sweat, and the hair follicles, are located in this skin layer (see Fig. 31.1A). The basis of the dermis is a supporting matrix made of mucopolysaccharides, ie, macromolecules that retain the water like sponges. Within this matrix, two kinds of fiber confer the strength properties of the skin: a great tensile strength from the collagen fibers and the elasticity from the elastic fibers. Dermal cells are principally fibroblasts that produce the collagen and the elastic fibers. Other cells like histiocytes and mast cells are involved in the immune system.

Hypodermis

Hypodermis is also called *subcutaneous fat*, though it is part of the skin. It is its innermost and thickest layer, attached to the dermis by collagen and elastic fibers. It is constituted of adipocytes organized in lobules and connective tissue containing vessels and nerves (see Fig. 31.1A). This deep layer of the skin can be extremely thin in some parts of the body (<1 mm on eyelids) or extremely thick in others (several centimeters in abdomen or buttocks), with huge variations from person to person according to the body mass index.

Skin Thermal Modeling

As discussed later in this chapter thermal imaging achieves its full potential only in combination with a proper thermal modeling of the sample under investigation. The building of such thermal models is a requirement for the extraction of quantitative parameters from the experimental data. Heat transfer phenomena taking place in the different skin layers are a mix of heat conduction processes coupled with physiological mechanisms that include blood perfusion, metabolic heat generation, and sweating. The skin is an active medium that is regulating the bodily temperature. Absorption and emission of visible, UV, or IR radiation depends on the thickness and the pigmentation of the skin. Hypodermis and hairs isolate the body from the cold. Thermal regulation can also be achieved by the variations of the diameter of the skin vessels. Indeed, the skin is highly vascularized, containing about 10% of the vessels of the all body. Thermoreceptors located in the skin detect levels and variations of temperature. If skin temperature drops, neuronal signals are sent to trigger vasoconstriction of dermal and hypodermal arterioles, limiting thermal loss by reducing the blood flow exposed to peripheral low temperatures. If the internal temperature rises, vasodilation allows heat transfers to external environment by the means of radiation, conduction, and convection. When ambient temperature is high, evaporation of sweat produced by eccrine and apocrine glands induces a lowering of the body temperature. In case of extreme temperature the metabolism will eventually slow down. Skin tissue is a complex, active, nonhomogeneous and anisotropic medium; as a result, the building of a realistic thermal skin model remains challenging.

Numerous heat-transfer skin models have been proposed in the literature (see, for example, Ref. [17] for a review on the subject). They can be classified into four categories: continuum models, vascular models, hybrid models, and models based on porous media theory. Continuum models are widely used because of their simplicity and because they can be either solved analytically or using finite elements or finite difference solvers. They are based on the Penne's bioheat equation describing the influence of blood perfusion on the skin temperature distribution in terms of volumetrically distributed heat sources [18].

$$\rho C \frac{\partial T}{\partial t} + \omega \rho_b C_b (T - T_b) - Q = k \nabla^2 T \tag{31.1}$$

where C, k, ρ, ω, and Q are respectively the specific heat, the heat conductivity, the density, the blood perfusion rate, and the metabolic heat generation. C_b, ρ_b, and T_b denote the blood specific heat, the blood density, and the blood temperature usually set to C. T represents the local tissue temperature, t denotes the time variable, and ∇^2 is the Laplace operator. Eq. 31.1 states that the rate of change of thermal energy contained in a unit volume is equal to the sum of the rates at which the thermal energy enters or leaves the volume by conduction, perfusion, and metabolic heat generation. The term $\omega \rho_b C_b (T - T_b)$ describes the volumetrically distributed heat sources or heat sinks, depending on whether the local tissue temperature is above or under blood temperature.

For a simple one-layer model where the skin is considered as a semiinfinite, homogenous medium, closed-form analytical solutions of Eq. (31.1) can be obtained for different boundary conditions [19,20]. Nonetheless, to achieve a more realistic description, skin tissue should be considered as composed of different layers, each layer exhibiting specific thermophysical properties. Besides, vasoconstriction and vasodilation mechanisms should be added to the model by making tissue perfusion a function of the local temperature. The complexity of such a system requires the use of numerical methods.

Fig. 31.1B represents the idealized model usually adopted to describe the skin structure [21]. Three different layers are considered: epidermis, dermis, and fat tissue. The size of each layer has to be adapted depending on the bodily location. Some models split the dermis into reticular and papillary dermis or take into account an additional muscle layer [22,23]. Table 31.1 summarizes the thermophysical properties of each of these layers taken from the literature.

Eq. (31.1) is numerically solved imposing appropriate boundary conditions at the domain borders and continuity conditions for the temperature and heat flux at each interface between the different tissue layers. The bottom surface of the skin is usually fixed at the core temperature $T_c = T_b = 37°C$ (see Fig. 31.1B), whereas the boundary condition for the heat transfer occurring at the skin surface is generally composed of three parts, ie, convection, radiation, and evaporation [21]:

$$-k\nabla T_s = h(T_s - T_{amb}) + \varepsilon\sigma\left(T_s^4 - T_{amb}^4\right) + Q_e \quad (31.2)$$

where T_s is the skin surface temperature, T_{amb} the ambient room temperature, h the convective heat transfer coefficient (natural or forced), ε the skin emissivity, σ the Stefan–Boltzmann constant, and Q_e the evaporative heat losses caused by sweating.

In skin thermal imaging, an IR camera measures the radiative component of Eq. (31.2). Even neglecting the atmospheric absorption happening between the skin and the camera detector, the ambient temperature, as well as the skin surface emissivity, should be known to calculate the absolute skin surface temperature T_s. This well-known difficulty in thermal imaging is discussed more in details in the "Thermal Radiation Characteristics" section.

It follows from the skin thermal model that only pathological states affecting one or several of the thermophysical parameters presented in Table 31.1, or affecting the thickness of the different skin layers, can induce a potentially measurable variation of the skin surface temperature T_s. More restricting is the nonspecificity of thermal imaging. From static skin thermograms it is difficult to differentiate the origin of thermal signals. A way to overcome this limitation is to perform active or dynamic thermography measurements where the skin surface is monitored in a transient state. For example, the convection term of Eq. (31.2) can be periodically modulated by varying the ambient temperature or the heat transfer coefficient h. We suggested that monitoring the skin surface temperature response obtained for different modulation frequencies allows differentiating thermal signals originating from perfusion variations [24].

By specifically designing an active thermography experiment, together with the development of a heat transfer model, thermal imaging has the ability to retrieve quantitative information about distinct skin thermophysical properties. Different active thermography methods are presented in the "Measurement Procedures" section.

TABLE 31.1 Thermophysical Properties of the Different Skin Layers

Tissue	C(J/kg K)	k(W/m K)	ρ(kg/m³)	ω(mL/s/mL)	Q(W/m³)
Epidermis	3600	0.25	1200	0	0
Papillary dermis	3300	0.45	1200	0.0002	370
Reticular dermis	3300	0.45	1200	0.001	370
Fat	2700	0.2	1000	0.0001	370
Muscle	3900	0.5	1100	0.003	700
Blood	3770	0.5	1100	None	None

C, specific heat; *k*, heat conductivity; *Q*, metabolic heat generation; *ρ*, density; *ω*, blood perfusion rate.

Reproduced from Bonmarin M, Le Gal FA. Lock-in thermal imaging for the early-stage detection of cutaneous melanoma: a feasibility study. Comput Biol Med 2014;47:36–43 with permission from Elsevier.

FUNDAMENTALS OF THERMAL IMAGING

A minimum understanding of thermal imaging and its limitations is a basic requirement to avoid improper utilization. We shortly discuss the basics characteristics of the thermal radiation, compare the imaging devices available on the market, and investigate the different measurement procedures that can be implemented.

Thermal Radiation Characteristics

Every object above the absolute zero temperature (−273.15°C) emits an electromagnetic radiation. The characteristics of this radiation are given by Planck's law stating that the spectral radiance M_λ of a perfect emitter, so-called "blackbody," is given by [1]

$$M_\lambda(\lambda, T) = \frac{2hc^2}{\lambda^5}\frac{1}{e^{\frac{hc}{\lambda kT}} - 1} \quad (31.3)$$

where λ is the wavelength, T is the absolute temperature, h is the Planck constant, k is the Boltzmann constant, and c is the speed of light in a vacuum.

Fig. 31.2 shows the spectral radiance of such a blackbody calculated at different equilibrium temperatures. The emitted radiation is mainly in the far-IR region of the electromagnetic spectrum, between 3 and 20 μm. Thermal imaging should not be confused with near-IR imaging methods investigating radiation with shorter wavelength, between 0.8 and 2.5 μm.

Also demonstrated in Fig. 31.2 is the shift of the spectral radiance toward the visible range of the electromagnetic spectrum at higher temperatures. This effect is referred to as the *Wien displacement law*, where the maximal emission wavelength is inversely proportional to the object's absolute temperature:

$$\lambda_{\max} = \frac{b}{T} \quad (31.4)$$

with b being the Wien displacement constant.

Integration of Eq. (31.3) over all wavelengths leads to the Stefan–Boltzmann law that states that the total emissive power of a blackbody is directly proportional to the fourth power of its temperature T:

$$M(T) = \int_0^\infty M_\lambda(\lambda, T)d\lambda = \sigma T^4 \quad (31.5)$$

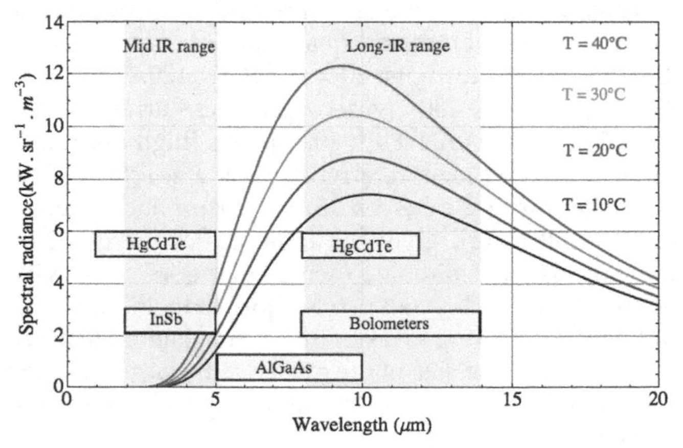

FIGURE 31.2 Spectral radiance of a blackbody at different equilibrium temperatures. Almost all the emitted radiation is in the infrared (IR) range of the electromagnetic spectrum. The spectral range of different IR detectors is indicated in yellow. Because of water vapor and carbon dioxide absorption, cameras are usually designed to work in either the mid-IR or long-IR range windows where atmospheric transmission is maximum. *AlGaAs*, aluminum gallium arsenide; *HgCdTe*, mercury cadmium telluride; *InSb*, indium antimonide.

where σ denotes the Stefan–Boltzmann constant. Most objects are not perfect emitters and their ability to emit IR radiation is described by their surface emissivity, ε, defined as the ratio of the surface object emissive power, E, to the emissive power of a blackbody:

$$E(T) = \varepsilon(\lambda)\sigma T^4 \quad (31.6)$$

The emissivity is usually wavelength-dependent. According to Kirchhoff's law, $\varepsilon(\lambda)$ equals the absorption coefficient, $\alpha(\lambda)$, for a uniform medium at temperature equilibrium. The fact that the emissive power of an object depends on its surface emissivity and not only on its temperature is one of the major limitations of thermal imaging. Fortunately human skin, independently of its pigmentation and even burned, [25] is a very good emitter with a constant emissivity close to unity in the 3–20 μm region [25,26], allowing (after proper calibration) absolute temperature calculation. To achieve the conversion from analog-to-digital units (ADUs) obtained from the camera into temperatures, polynomial expressions are commonly used, such as [2]:

$$T(^\circ\mathrm{C}) = -13.4 + 0.05 \times \mathrm{ADU} - 1.6 \times 10^{-5}\mathrm{ADU}^2 + 2.2 \times 10^{-9}\mathrm{ADU}^3 \quad (31.7)$$

Eq. (31.7) is only valid for high-emissivity materials and when atmospheric absorption is negligible. In dermatological applications, the distance between the patient skin and the IR detector is relatively small (less than 1 m), so that the latter approximation is valid. Second, it means that the skin is a very good absorber in the far-IR region and that no spurious IR reflections on its surface are expected, as is often the case with low-emissivity materials like metals.

Surface emissivity, ε, is theoretically angle-dependent, and particular care should be taken when investigating skin surfaces with large curvatures like the bridge of the nose. Nevertheless it has been demonstrated that measurement errors are likely to be small for an angle of view up to 45 degrees [27]. Although it greatly depends on the location, equilibrium temperature of nonpathological skin is around 30°C. In these conditions, maximal emission wavelength is just below 10 μm, a region well covered by current IR imaging devices.

Thermal Radiation Imaging Devices

The core component of all thermography experiments is an IR imaging system that converts the skin surface thermal emission into a measurable electrical signal. A detailed and exhaustive description of current IR detector technology goes far beyond the scope of this chapter and can be found in numerous reviews on the subject [28–30]. The aim of this paragraph is more to provide the reader with key concepts allowing an optimal IR camera selection for the desired application. The main criteria to be examined are the camera's spectral range, the detector's format, the objective, the camera's frame rate, and the camera's sensitivity. Other important factors such as price or usability are also briefly discussed.

Spectral Range

The spectral range refers to the part of the emitted electromagnetic radiation that will be integrated by the IR detector:

$$I = \int_{\lambda_1}^{\lambda_2} \varepsilon(\lambda) M_\lambda(\lambda, T) d\lambda \quad (31.8)$$

There are two general classes of IR detectors: quantum and thermal detectors, and both are used in commercial IR systems.

Thermal detectors convert the far-IR radiation into heat, causing electrically measurable changes in the resistance of a specific element called a *bolometer*. For this reason, such systems are often referred to as *microbolometer IR cameras* and are usually based on amorphous silicon or vanadium oxide.

Quantum devices transform absorbed photons energy directly into released electrons; the material band gap describes the energy necessary for the transition of a charge carrier from the valence to the conduction band. Usual materials include indium antimonide (InSb), mercury cadmium telluride (HgCdTe), and aluminum gallium arsenide (AlGaAs), also referred as *quantum well infrared photodetectors (QWIPs)*. Platinum silicide has been abandoned because of low quantum efficiency. Whereas new promising types of quantum IR detectors, called *type II super lattice detectors*, are in development, we would like to restrict this paragraph to commercially available devices. Quantum detectors are either operating in the photoconductive (the photon flux increases the number of conductive electrons, allowing more current to flow when the detective element is used in a bias circuit) or photovoltaic mode (photo excited carriers are collected by a diode junction), the second being more commonly used by commercial systems. Fig. 31.2 shows the spectral range of the different detector types. Because the thermal emission is strongly absorbed by water vapor and carbon dioxide, the IR detectors are usually designed to work within the medium wavelength IR (mid-IR) or long wavelength IR (long-IR) atmospheric transmission window (see Fig. 31.2) where absorption is minimal. The assembly in a stack arrangement of different QWIPs allows the fabrication of dual-band detectors that are sensitive in both the mid-IR and long-IR range.

Fig. 31.2 gives the impression that bolometers or HgCdTe detectors are the best options because both exhibit their spectral range where the skin thermal emission is maximal. Nevertheless microbolometers are less sensitive than quantum cameras and between HgCdTe and InSb detectors, the latter exhibit higher quantum efficiency and are thus more sensitive (they will be able to detect smaller temperature differences at the skin surface).

The choice of an optimal detector's spectral band depends on the foreseen application and measurement procedure. For microscopy experiments, lower wavelengths are highly desirable to achieve superior performances. Indeed, the diffraction-limited resolution is proportional to the wavelength. InSb cameras will achieve an approximately two times better resolution in comparison with microbolometers. For other applications, the spectral band has a limited impact when performing standard passive thermography measurements. In this case, emphasis should be given to other parameters in particular to the desired sensitivity.

Detector Format

First-generation IR imaging devices were based on a single IR detector in combination with scanner mirrors. Fig. 31.3A and B show examples of the first-generation IR camera models. All modern cameras are using so-called "staring" or focal plane array (FPA) detectors, allowing higher speed and increased reliability (see Fig. 31.3B and C). The detector format is another important parameter that will determine, in combination with the camera objective, the lateral resolution of IR images. Both microbolometers and quantum cameras are available in different formats: 80 × 60, 160 × 120, 320 × 256, 382 × 288, 640 × 480 [video graphics array(VGA)], 640 × 512, and even 1280 × 1024 pixels [high definition (HD)]. The pixel size, referred to as the *detector's pitch*, is usually varying between 15 and 40 μm and together with the fill factor, gives the total size of the detector. Large pixels are desirable to achieve superior sensitivity. On the other hand, small pixels allow reducing the integration time, leading to faster image acquisition. In addition, the larger the size of the detector, the larger the size of the required optics and the more expensive the camera objective will be.

Camera Objectives

Different IR objectives ranging from telephoto lenses to wide angles or macro objectives can be either ordered

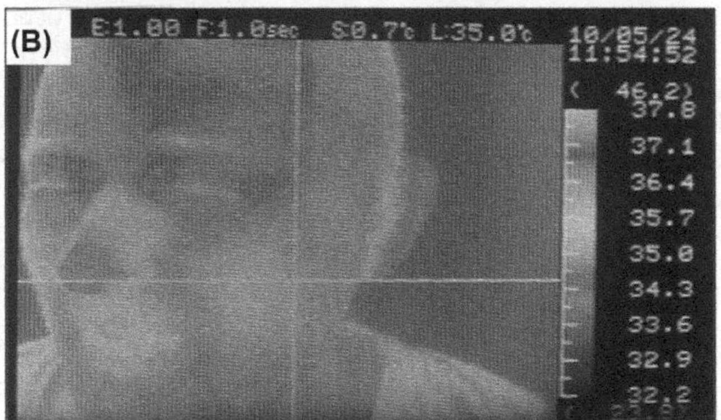

FIGURE 31.3 (A) First generation of thermal cameras (1990, Model 6T62 ThermoTracer, NEC San-ei Instruments, Japan). The camera is based on a single pixel mercury cadmium telluride (HgCdTe) detector and two galvanometric scanner gold mirrors to achieve image formation. Maximal frame rate is about 1 Hz. The camera requires liquid nitrogen cooling of the detector. (B) Typical thermogram of a face obtained with the 6T62 Thermo Tracer. (C) Microbolometer-based infrared (IR) camera (Gobi-640-GigE, Xenics nv, Belgium [81]). The 640 × 480 pixel chip does not require extra cooling. Maximal frame rate is 50 Hz, and the noise equivalent temperature difference (NETD) is about 50 mK. The camera can be easily connected to a notebook via a gigabit Ethernet (GigE) protocol. (D) Thermal image of the same face in (B), obtained with the microbolometer camera pictured in (C).

by IR camera manufacturers or acquired separately and mounted via standard bayonet or threaded mount. The field of view (FOV) is the angular extension of the observable object field. The FOV depends on the camera objective and detector size. For a given objective with focal length, f and detector size, d, the FOV is given by

$$\mathrm{FOV} = 2\tan^{-1}\left(\frac{b}{2f}\right) \tag{31.9}$$

The rectangular shape of the detector leads to different horizontal and vertical FOVs. For a given FOV, the sample area with length, l, seen by the camera at a working distance, d, is calculated by

$$l = 2d\tan\left(\frac{\mathrm{FOV}}{2}\right) \tag{31.10}$$

As an example, we mount a standard 18-mm focal lens objective on a VGA microbolometer array with a 17-μm pitch and investigate a patient's skin at a working distance of 20 cm. With such a configuration, the camera measures a skin area of 12 × 9 cm with an image resolution of 0.18 mm.

Using an appropriate objective, it is even possible to carry out thermal imaging microscopy experiments. Nevertheless the diffraction-limited maximal resolution achievable in the mid-IR range is around 5 μm and far below the performances of a visible microscope. Besides, the necessity of high-resolution systems for skin thermal imaging, at least for passive in vivo experiments, is questionable because of the lateral heat spreading into the tissue.

Camera Frame Rate

Another important parameter is the camera speed, or frame rate. The frame rate in hertz gives the number of IR images that can be acquired per second by the camera. Most microbolometers are working in the 50–60 Hz range, whereas quantum devices can go up to several 100 Hz. Microbolometers are operating in the so-called "rolling frame" readout mode and in comparison with quantum cameras working in "snapshot mode" are more difficult to trigger with an external source. Such considerations are particularly important when performing dynamic thermal imaging (DTI) where transient signals are recorded and when synchronization between camera and stimulation source is required. Nevertheless, heat conduction through human skin is a relatively slow process (in comparison with metals, for example) so that few hertz are generally sufficient for most dermatological applications.

Camera Sensitivity

The camera sensitivity describes the minimum temperature difference at the object surface that can be measured. It is usually stated by the noise equivalent temperature difference (NETD) given in degrees Kelvin. Table 31.2 shows the NETDs of the different camera types mentioned earlier. Quantum cameras can achieve an NETD as low as 17 mK (0.017°C), whereas the best microbolometers are around 30 mK.

It is important to notice that the sensitivity is not given by the NETD alone. Because of the square root behavior of the signal-to-noise-ratio (SNR) with the number of measurements, an AlGaAs camera working at 200 Hz with the same NETD of 30 mK as a microbolometer working at 50 Hz will be capable to measure 2 times smaller signals in 1 s. Besides, the detector format influences the sensitivity as well. A 640 × 512 FPA has 4 times more pixels than a 320 × 256 format. When both detectors have the same NETD and the same frame rate, reducing the resolution of the 640 × 512 array to 320 × 256 by spatial averaging will allow a gain of $\sqrt{4} = 2$ in the sensitivity.

If best performances are desirable, it is worth mentioning that camera sensitivity is often not the limiting factor in thermal imaging. In skin cancer detection, for example, differentiation between relevant signals and disturbing thermal artifacts originating from subcutaneous tissues is much more of a concern than the SNR [24].

Other Parameters

Finally, the impact of other factors like the handling of the camera, its price, or its connectivity should not be minimized. Quantum cameras require an extra cooling of their sensor (80 K for InSb and HgCdTe and up to 60 K for AlGaAs). First generation cameras were cooled using liquid nitrogen (see Fig. 31.3A). Nowadays almost all commercial models use a more practical cryogenic cooler engine, liquid nitrogen cooling being reserved for applications where vibrations are an issue as in microscopy. Because most microbolometers do not require any cooling, their size and price are drastically reduced. Microbolometer arrays (80 × 60 pixels) smaller than a dime are now available for integration into smartphones (Lepton Core, FLIR Systems, USA). Table 31.2 summarizes the characteristics of commercially available IR cameras.

Measurement Procedures

Basically, two measurement strategies can be implemented in thermal imaging: passive or active measurements. Passive thermography investigates the sample in its steady state, whereas active thermography (sometimes referred to as *DTI* in medical applications) measures transient temperatures resulting from an external thermal stimulation. The thermal modulation can be achieved by conductive or convective heat transfer or by electromagnetic absorption. Active thermography can be subdivided into pulsed, stepped, or lock-in methods, depending on the shape and duration of the stimulation signal.

Passive Thermography

Passive thermography probes samples that are naturally at a different (often higher) temperature than the surrounding environment [1,2]. Thermograms are

TABLE 31.2 Summary of the Main Characteristics of Different Thermal Imaging Detectors Available on the Market

Parameters	InSb	HgCdTe	AlGaAs	Bolometer
Spectral range	3–5 μm	1–5 and 8–12 μm	5–10 μm	8–14 μm
Largest format available (pixels)	1280 × 1024	1280 × 1024	1280 × 720	1024 × 768
Frame rate at 640 × 480 pixels	100 Hz	125 Hz	200 Hz	50 Hz
NETD	17 mK	25 mK	30 mK	30 mK
Quantum efficiency	80%	80%	10%	–
Cooling	80 K	80 K	60 K	No cooling
Readout mode	Snapshot	Snapshot	Snapshot	Rolling frame

AlGaAs, aluminum gallium arsenide; *HgCdTe*, mercury cadmium telluride; *InSb*, indium antimonide; *NETD*, noise equivalent temperature difference.

analyzed looking for abnormal temperature differences, indicating a potential problem. Passive thermal imaging is qualitative and the retrievable information is rather limited. Nevertheless it is by far the most commonly used thermal imaging procedure. Limitations arise because the skin surface temperature is largely influenced by subcutaneous factors like the presence of large blood vessels or bones. As an example, Fig. 31.3D demonstrates a typical thermogram of a face. A small hot or cold "spot" would be buried into spurious thermal signal and is not detectable. In addition, other external factors may influence passive thermography measurements such as the patient position, the recent absorption of hot or cold beverage, the time of the day, or the menstrual cycle in a woman [31]. To limit the influence of such factors, some authors attempted to elaborate strict measurement procedures [32] that unfortunately drastically reduced the feasibility of passive thermography in clinical routine activity. If passive thermography is mainly qualitative and limited by the mentioned factors, it can still be useful in dermatology as a complementary tool to identify, for example, abnormal metabolism activity or perfusion variations (see "Other Potential Applications" section for potential applications of passive thermography).

Active Thermography

In comparison with passive thermal imaging, active thermography is a much more powerful investigation tool that allows retrieving quantitative information about the sample thermal properties. In active thermography, the sample is thermally stimulated while an IR imaging device records its transient temperature for further processing (Fig. 31.4).

The thermal stimulation can be achieved with different modalities. The first active thermography setups for dermatology were based on conductive heat transfer stimulation of the skin using cold gel packs or thermalized balloons filled with alcohol [33]. This technique has the advantage to rapidly create large temperature gradients homogenously distributed, but demonstrates some major drawbacks. First, it is almost impossible to monitor the tissue surface temperature during the stimulation, limiting the different modalities of investigation. Second, an accurate synchronization between the thermal detection and stimulation is hard to achieve, although it is a requirement for the extraction of quantitative information. Alternatively, the tissue could be warmed by absorption of electromagnetic radiation. Visible light is to avoid, as the skin or lesion pigmentation would lead to selective warming of the lesion compared with the surrounding healthy tissue and would create disrupting artifacts. The use of a mid-IR radiation source, emitting, for example, at around 2–3 μm, where the absorption of the skin is less dependent on the pigmentation, is an interesting option. Because the skin possess a very high emissivity in this spectral range, no disrupting reflections of the IR radiation onto the camera objective have to be feared, as is often the case when working with low-emissivity materials and, technically, many low-cost IR sources are available. Nevertheless with such methods, it is only possible to warm the skin and therefore a smaller thermal gradient is to be expected. Although it is technically challenging, convective heat transfer is probably the optimal stimulation method for dermatological applications. A temperature-adjustable airflow can be advantageously used. Relatively large temperature gradients can be created, and the skin is always accessible for monitoring its surface temperature with an IR camera. In addition the technique stays hygienic, because no contact with the patient's skin is required.

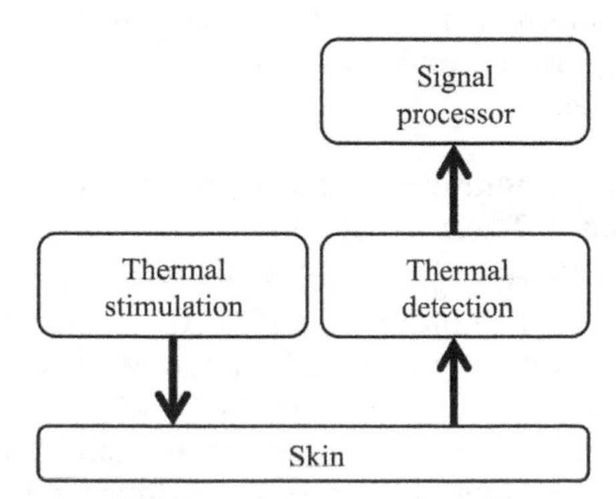

FIGURE 31.4 Schematics of active thermal imaging. The skin is thermally stimulated by either convective or conductive heat transfer, or by absorption of an electromagnetic radiation. An infrared (IR) imaging device records the skin surface transient temperature, which is further digitally processed to extract quantitative information about the tissue.

Pulsed Thermography and Stepped Thermography

One of the most popular active thermography methods is pulsed thermography (PT). PT consists of briefly heating (or cooling) the sample and recording the resulting transient surface temperature response (Fig. 31.5A). Qualitatively, PT works as follows: the sample temperature changes rapidly after the initial thermal pulse because of the propagation of the thermal front by diffusion into the sample and also because of radiation and convection losses. The way the sample retrieves its equilibrium temperature is dependent on its thermal properties. For a homogenous sample, those are the thermal effusivity, e, and in the case of biological samples, the perfusion, ω. The thermal effusivity is defined as

$$e = \sqrt{k\rho C_t} \tag{31.11}$$

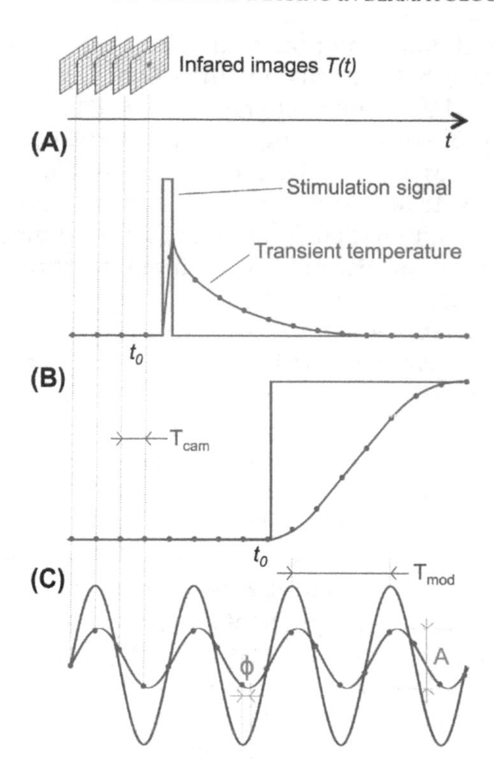

FIGURE 31.5 Schematic description of the different active thermography measurement procedures. (A) Pulsed thermography (PT). (B) Stepped thermography (ST). (C) Lock-in thermography (LIT). The *curve* in blue represents the stimulation signal, whereas red represents the skin surface transient temperature, captured by one single pixel of the infrared (IR) camera array. Active thermography requires synchronization between the stimulation source and the IR camera. *A*, amplitude image; t_0, temperature before stimulation; T_{cam}, acquisition period of the infrared imaging device; T_{mod}, modulation period; $T(t)$, time-dependent temperature of one pixel; ϕ, phase image.

where k is the thermal conductivity, ρ is the density, and C_t is the heat capacity. The time-dependent thermal contrast, $C(t)$, is calculated as

$$C(t) = \frac{T_l(t) - T_l(t_0)}{T_s(t) - T_s(t_0)} \quad (31.12)$$

where T_l and T_s describe the temperature above a potential lesion and above the surrounding skin, respectively. C is computed before the thermal stimulation at time t_0 to suppress static contributions of the environment and normalized by the behavior of healthy tissue. Computing C for each camera pixel gives a time-dependent contrast image $C(t)(x,y)$. Quantitatively, the knowledge of the evolution of the thermal contrast in conjunction with solutions derived from inverse heat transfer modeling of the skin allows extracting valuable parameters (see "Burns" section about burn depth evaluation). When performing active or passive thermography, a pure surface emission is often assumed. Nevertheless many materials, and in particular the human skin, exhibit a small wavelength-dependent IR transparency. Previous work on transient thermal emission spectroscopy suggested that useful information could be retrieved from different spectral bands [34,35]. Although an exhaustive description of IR imaging spectroscopy goes beyond the scope of this chapter devoted to thermal imaging, it is worth mentioning that dual-band IR imaging devices allow retrieval of some wavelength-dependent information. Using a dual-band QWIP IR camera, Abuhamad and Netzelmann demonstrated that the skin surface transient temperature signal resulting from an optical pulse exhibits differences in the 5 and 8 μm regions [36]. Those first experiments showed that the principle of transient emission spectroscopy can be extended to an imaging technique and add new information to active thermography measurements.

Contrary to PT, where transient temperature signals are measured after the thermal stimulation, stepped thermography (ST) records the sample surface temperature during the application of a step heating (or cooling). Fig. 31.5B pictorially describes a standard ST experiment. Again, the extraction of quantitative information relies on a heat-transfer model of the experiment. Although different from the experimental point of view, both ST and PT procedures mathematically contain exactly the same information [2] and the optimal choice depends on the application. PT is usually preferred for the investigation of fast phenomena. In the case of the skin, it is chosen to measure variations in the effusivity. ST will be optimal for the measurement of slower processes coming from skin perfusion variations.

Contrary to passive thermography, synchronization of the IR camera and the thermal stimulation source is a necessity in both PT and ST. Image acquisition of quantum cameras is easily externally triggered. For this reason, they usually work in slave mode, with the thermal stimulation system giving the master clock. On the contrary, microbolometers work in the rolling frame mode and set the master clock for the start of the stimulation. For PT and ST experiments, a camera with a high frame rate is desirable to capture fast transient signals.

Lock-In Thermography, Pulsed-phase Thermography, and Dynamic Thermal Assessment

Lock-in thermal imaging, or lock-in thermography (LIT), is a technique developed by Busse and coworkers in the early 1990s [37]. LIT works as follows: the sample is thermally stimulated at a determined modulation frequency and the simultaneously recorded IR images of the sample surface are processed digitally according to the lock-in principle (see Fig. 31.5C). The result of this demodulation is a phase image, ϕ, and an amplitude image, A. The phase image (ϕ) is a map of the phase angles (in radian) between the periodic thermal stimulation and the harmonic temperature response of the skin surface, whereas A gives the skin oscillations amplitude (degrees Kelvin). Different demodulation algorithms have been proposed in the literature [38,39]. As stated by Breitenstein, if the camera is synchronized with the stimulation source and the modulation frequency is low in comparison with the camera frame rate, synchronous narrow two-channel correlation is the optimal digital demodulation algorithm [3]. This method presents the advantage that the processing is done in real time and does not require the stocking of the IR images. Two sets of weighting factors are used, one approximating the sine function and the other one, the cosine function. The correlation takes the form:

$$S_0 = \frac{2}{nT_{\text{mod}}} \int_0^{nT_{\text{mod}}} I(t) \sin\left(\frac{2\pi}{T_{\text{mod}}} t\right) dt \tag{31.13}$$

$$S_{90} = \frac{2}{nT_{\text{mod}}} \int_0^{nT_{\text{mod}}} I(t) \cos\left(\frac{2\pi}{T_{\text{mod}}} t\right) dt \tag{31.14}$$

where $I(t)$ is the time-dependent IR image, T_{mod} is the modulation period, and n is the number of acquired modulation cycles. S_0 and S_{90} are the so-called "in-phase" and "in-quadrature" images that allow calculating the amplitude and phase image.

$$A = \sqrt{S_0^2 + S_{90}^2} \tag{31.15}$$

$$\phi = \tan^{-1}\left(\frac{S_{90}}{S_0}\right) \tag{31.16}$$

The phase image is experimentally preferred over the amplitude. First, it is not influenced by the sample surface emissivity. Second, the phase is less sensitive to inhomogeneous thermal stimulation or bad camera calibration. The direct advantage of LIT lies in the averaging nature of the technique. LIT allows detection of very small temperature gradients, even when they are concealed in a noisy background. LIT sensitivity is given by [3]

$$S = \frac{\text{NETD}}{\sqrt{f_{\text{cam}} \times t_{\text{acq}}}} \tag{31.17}$$

where f_{cam} is the camera frame rate and t_{acq} is the acquisition time. As an example, for a microbolometer camera with an NETD of 30 mK and a frame rate of 50 Hz, temperature differences of 50 μK are measurable after only 2 min. The high sensitivity of LIT can be advantageously used to reduce the amplitude of the thermal modulation applied to the skin. Large temperature gradients may influence the skin's thermophysical parameters such as blood perfusion. In addition, if the thermal modulation frequency is chosen high enough, LIT has the ability to prevent lateral heat spreading from the IR images, giving sharp thermograms. This is particularly important for locating lesion margins with high accuracy [23].

Performing lock-in experiments requires proper synchronization of the IR camera with the thermal modulation. If four images per modulation cycle are theoretically sufficient, a minimum of 10 frames per period is highly desirable for a proper digital demodulation. As a result, if not using time-consuming undersampling methods [3], current camera frame rate restricts the maximal modulation frequency to a few hertz. This is fortunately unproblematic for dermatological investigations. When the skin surface is subject to a periodic thermal modulation, highly damped thermal waves will propagate into the tissue on a distance referred to as the *thermal diffusion length* [3]. This thermal diffusion length is dependent on the thermophysical properties of the tissue, but also on the modulation frequency. The lower the frequency, the deeper the waves propagate into the skin. Using high frequencies (above a few Hz) would limit the investigation to the very superficial skin layers (stratum corneum) [23]. Although almost all theoretical investigations assume a harmonic thermal modulation, the shape of the stimulation is irrelevant, because the synchronous narrow two-channel correlation filters out the nonharmonic components of the transient skin surface if the modulation frequency is low compared with the camera frame rate. The lock-in demodulation formalism works under the assumption that the measurement is performed under quasi—steady-state conditions. In other words, the sample surface temperature oscillates with time around a steady mean temperature. In reality, at the beginning of the thermal modulation, the skin surface temperature varies during an initial nonsteady phase. This initial period or thermal relaxation time, which depends mainly on the heat transfer resistance between the skin surface and the surrounding area, will induce a phase and amplitude shift in the demodulated signal [3]. Theoretically one should wait until the quasi—steady state is reached before demodulating the thermal signal, but practically, depending on the modulation parameters, this can take several minutes. Different techniques have been proposed in the literature to solve this problem [40]. Finally, as for PT, additional information concerning the outer layers of the skin could be retrieved from LIT by using dual band or multiband IR detectors.

Pulsed-phase thermography (PPT) is a novel active thermography technique, which somehow combines the advantages of both PT and LIT. In PPT deployment, the sample is pulse-heated (or cooled) in the same manner as in a classical PT experiment, but the transient temperature signal of each pixel is analyzed in the frequency domain. Stimulating the sample with a very short pulse in the time domain is equivalent to a broad-frequency excitation. The amplitude and the phase spectrum can be calculated by a simple Fourier transformation of the transient temperature signal. In one single experiment, PPT allows retrieving the same amplitude and phase images that would be obtained sequentially by LIT with multiple modulation frequencies. However, PPT does not exhibit similar capabilities as LIT in terms of SNR. In PPT, a single transient measurement is achieved to compute the phase and amplitude spectrum, whereas in LIT, the temperature signal is averaged over many cycles.

We close this section with dynamic thermal assessment (DTA) also misguidedly referred in the literature as *DTI*, although no external thermal stimulation of the sample is required. DTA takes advantage of the natural thermal modulation of the skin, originating from pulsatile cardiogenic and neuronal control changes of the microvasculature blood flow [41]. In brief, the skin surface temperature is monitored over a time and the thermograms obtained are Fourier transformed. The Fourier transformation calculates the amplitude of frequency components ranging from millihertz to a few hertz. Frequencies of 8–815 mHz are attributed to neuronal controlling effects, whereas frequencies of 815 mHz and higher are attributed to hemodynamic effects [42]. The technique was developed by Anbar in the early 1990s and demonstrated potential for pathologies that affect any anatomical or physiological parameters of the blood supply.

APPLICATIONS IN DERMATOLOGY

Since the beginning of medical thermography, skin cancer detection and burn depth evaluation have benefitted from most investigations. Although still in the evaluation phase, both fields have started to demonstrate promising clinical records, in particular when using active measurement procedures. In the next paragraphs, we will describe in more detail the state of the art in the two domains and will conclude by briefly reviewing other potential dermatological applications of thermal imaging.

Burns

A burn is an injury caused by heat, electricity, chemicals, friction, or radiation. Being the protective envelope of the body, skin is the first organ touched by this aggression.

Burns by heat are the most common. The more energy absorbed by the skin, the greater the degree of cellular disruption that will occur and the greater the depth to which the injury will extend [43]. Heat denatures extracellular and intracellular proteins. Over 40°C, cells begin to malfunction, and over 45°C, cellular repair mechanisms fail and cells die. When temperatures reach 60°C, thrombosis occurs in vessels and tissue becomes necrotic. Burn wounds are traditionally sorted according to the following classification: I, superficial; IIa, superficial dermal; IIb, deep dermal; and III, full thickness of the skin. The healing of a burn wound depends on the amount of remaining viable epithelial cells. In superficial burns (I) where basal keratinocytes can still divide, the cicatrization will be easy. In partial thickness burns (IIa and IIb), hair follicles and glands deeply located in the dermis will be precious reservoirs of invaginated epithelial cells from which the wound will heal. In total thickness burns (III), the complete destruction of epithelial cells precludes the spontaneous wound healing and the need for early grafting. Establishing the difference between superficial dermal burns that will spontaneously heal within few weeks and deeper burns that will require surgery is of particular importance to guide the treatment decision (Fig. 31.6). The clinical evaluation of burns is mainly based on visual examination of the lesions. Histological assessment remains the gold standard but is routinely rarely used because it is invasive and time consuming [44]. Laser Doppler imaging (LDI) provides reliable results [45] but remains an expensive technology.

An interesting alternative is thermal imaging. The first attempt to use thermal imaging for burns evaluation was by Lawson and coworkers [46] in 1961. Lawson demonstrated that passive thermography could be advantageously used to predict the depth of burns in dogs. These results were reproduced with patients several years later by Mladick [47], Hackett [48], and Newman [49]. Later work by Cole showed that thermography could identify subgroups of patients with deep thermal burns that might benefit from early surgery [50]. Despite those very promising results, thermal imaging did not manage to become routinely used in the clinic. Technological challenges were probably the main reason. Cameras used at that time were expensive and not easy to manipulate. Since 1990, IR imaging devices have experienced dramatic improvements, and small microbolometer-based cameras with reasonable sensitivity are affordable. These technological improvements triggered new investigations [51]. Hardwicke et al. investigated 11 patients with burns affecting upper and lower limbs, such as the posterior and anterior trunk. IR images were taken 42 h and 5 days postburn [52]. Full thickness burns were significantly cooler than healthy skin (−2.3°C) in comparison with deep dermal burns (−1.2°C). Superficial burns exhibited small temperature differences with healthy tissue (−0.1°C) [52]. Whereas such temperature differences are relatively easy to measure with IR imaging devices, passive thermography still suffers from several

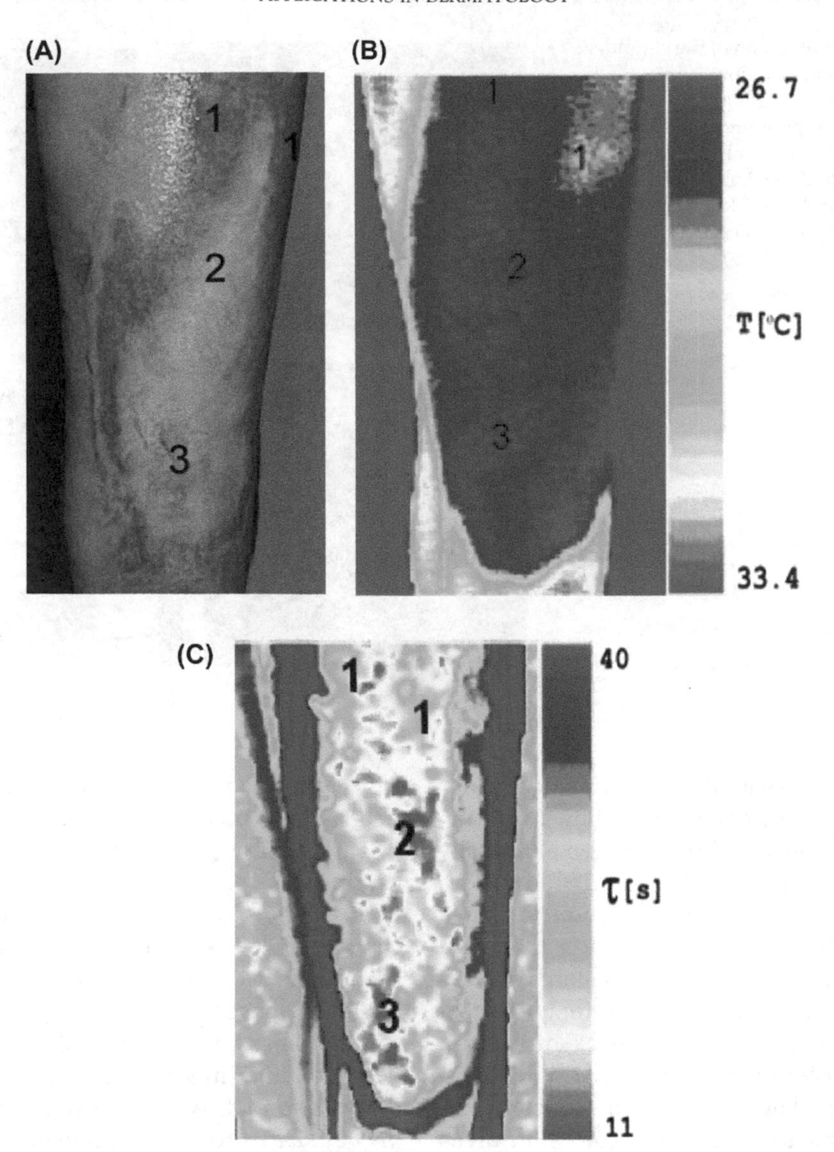

FIGURE 31.6 Visible, thermal, and thermal time constant image of a burn wound to the posterior surface of the right thigh of a 49-year-old patient: (A) Clinical burn depth evaluation of the areas marked: 1, IIa superficial dermal; 2, IIa superficial dermal and, more centrally, IIb, deep dermal; and 3, IIb deep dermal and, in the lower part, III, full thickness. (B) Static thermogram (the extremity is slightly medially rotated in comparison with the photograph). Areas 2 and 3 are the coldest and without marked internal differentiation. (C) Thermal time constant image (same rotation as in (B)). The areas with short time constants that qualified for surgery are well defined. These areas are markedly smaller when compared with areas 2 and 3 in the visible or static thermograms. *T*, temperature; τ, thermal time constant. *Reproduced from Renkielska A, Nowakowski A, Kaczmarek M, Ruminski J. Burn depths evaluation based on active dynamic IR thermal imaging: a preliminary study. Burns 2006;32(7): 867–75 with permission from Elsevier.*

limitations for accurate burn depth evaluation. First, a meticulous care about measurement conditions should be taken, eg, fixed room temperature, patient acclimatization time, and other factors. Second, there is a lack of a commonly accepted temperature-difference range for a particular burn wound depth classification [51].

These problems can be solved if active thermography procedures are used. Ruminsky is one of the pioneers of dynamic thermography for burn depth evaluation [53–55]. He demonstrated a setup based on PT: the skin surface is thermally stimulated (heated in this particular case) by a short optical pulse generated by two halogen lamps [53]. The transient skin surface temperature after the thermal stimulation is monitored by an IR camera and modeled using a simple exponential function:

$$T(t) = T_0 + \Delta T\left[\exp\left(\frac{-t}{\tau}\right)\right] \tag{18}$$

where $T(t)$ is the time-dependent temperature of one pixel, T_0 the temperature before the stimulation, ΔT the temperature rise resulting from the thermal

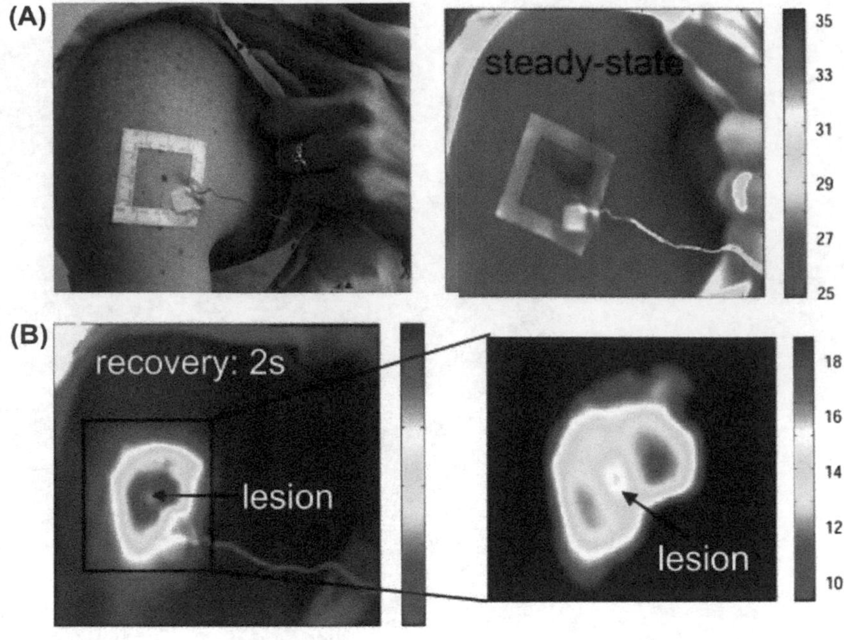

FIGURE 31.7 (A) Left: photograph of the shoulder area with a cluster of pigmented lesions and the adhesive window. Right: reference infrared (IR) image of the same region at ambient temperature. (B) Left: same area after cooling (2 s after the removal of the thermal stimulation). Right: magnified section of the melanoma lesion and its surroundings. *Reproduced from Cetinguel MP, Herman C. Quantification of the thermal signature of a melanoma lesion. Int J Therm Sci 2011;50(4): 421–31 with permission from Elsevier.*

stimulation, and τ the tissue thermal time constant. Repeating this operation for each pixel allows production of a thermal time constant image $\tau(x,y)$. Ruminsky and coworkers demonstrated that the thermal time constant is strongly correlated with the depth of the burn wound. τ had a higher value for wounds that self-healed within 3 weeks than for the healthy skin, whereas τ had a lower value for deeper burns that failed to heal within this period. Fig. 31.7 shows the example of a burn wound of the posterior surface of the right thigh of a 49-year-old patient. The thermal time constant image allows the clinician to precisely locate skin areas that will require surgery. Nowakowski enhanced this approach by considering a multilayer skin model, leading to multiple thermal time constants [56]. In parallel, alternative experimental setups based on skin convective thermal stimulation have been proposed and are under investigation [56].

Skin Cancer

Skin cancer is the most common form of all cancers. Basal cell carcinoma (BCC) and squamous cell carcinoma (SCC), two types of lesions associated with aging and sun exposure, represent the vast majority of nonmelanoma skin cancers (NMSC). BCC is by far the most common nonmelanoma cancer type. Although this cancer is rarely fatal, it can lead to dramatic disfiguring. SCC is less common but still has the potential to be fatal. Melanoma is one of the most serious forms of skin cancers because it has a potential for metastasis, which explains why melanoma is responsible for 90% of all skin cancer–related deaths. However, if melanoma is detected at an early stage, when the tumor thickness is less than 1 mm, the survival rate is excellent [57]. Skin carcinomas generate great health expenses; in the United States, skin cancer is among the five most costly cancers [58]. Therefore early detection of skin cancer is mandatory to preserve a good prognosis and to reduce health expenses.

The potential of thermal imaging for skin cancer detection and, in particular, as an early-stage melanoma diagnostic tool, was already noticed in the early 1960s [59,60]. Unfortunately, pioneer thermographic studies devoted to the subject showed disappointing results with a high percentage of false-negative outcomes [61]. Those poor results drastically reduced the enthusiasm of the medical community. Nevertheless they can be explained by several factors. First, potential thermal signals involved in early-stage lesions are small, and highly sensitive IR imaging devices were not available at the time of the first studies. Second, such small temperature differences are usually buried in larger thermal signals originating from the subcutaneous tissue. In such circumstances, small hot or cold spots would be hardly detectable even using current highly sensitive IR cameras [23].

In 1995, Di Carlo proposed to overcome those difficulties by performing active thermography measurements. He implemented a PT set-up where the skin is stimulated by the contact with a balloon filled with a thermostatic alcohol solution [33]. A few seconds after the removal of the stimulation, melanoma and hyperpigmented BCCs exhibited drastic temperature differences compared with the surrounding healthy skin. Several years later, Buzug and coworkers reproduced Di Carlo's experiment using cold gel packs and a more recent IR camera [62]. Very recently, Çetingül and Herman at John Hopkins University investigated transient thermal signals of melanoma lesions. Using a multilayer heat transfer skin model, they could extract quantitative information from the transient thermal signals [22,63]. As an example, Fig. 31.7 shows a cluster of pigmented lesions located on a shoulder and investigated with PT. Whereas the steady state thermogram (see Fig. 31.7A) does not allow any diagnostics, the melanoma lesion is clearly distinguishable on the IR image taken few seconds after the removal of the cold stimulation (see Fig. 31.7B). Çetingül and Hermann claimed that the method is able to detect Clark level I (melanoma in situ) lesions, but unfortunately the study was performed only on a limited number of patients [63].

We recently developed an LIT-based set-up specifically dedicated to the investigation of skin cancer [24,65]. In our apparatus, the skin surface temperature is periodically modulated using an airflow, while an IR camera records the emitted thermal radiation. The advantages of such a periodic modulation of the skin surface have been investigated by numerous theoretical studies [19,20,23,66,67]. Compared with the previous set-ups based on pulsed stimulation, LIT has the ability to suppress lateral heat spreading, leading to sharper thermograms [24]. Fig. 31.8 demonstrates the capabilities of LIT for the accurate detection of skin cancer lesion margins. The device is undergoing clinical testing at the Geneva University Hospital.

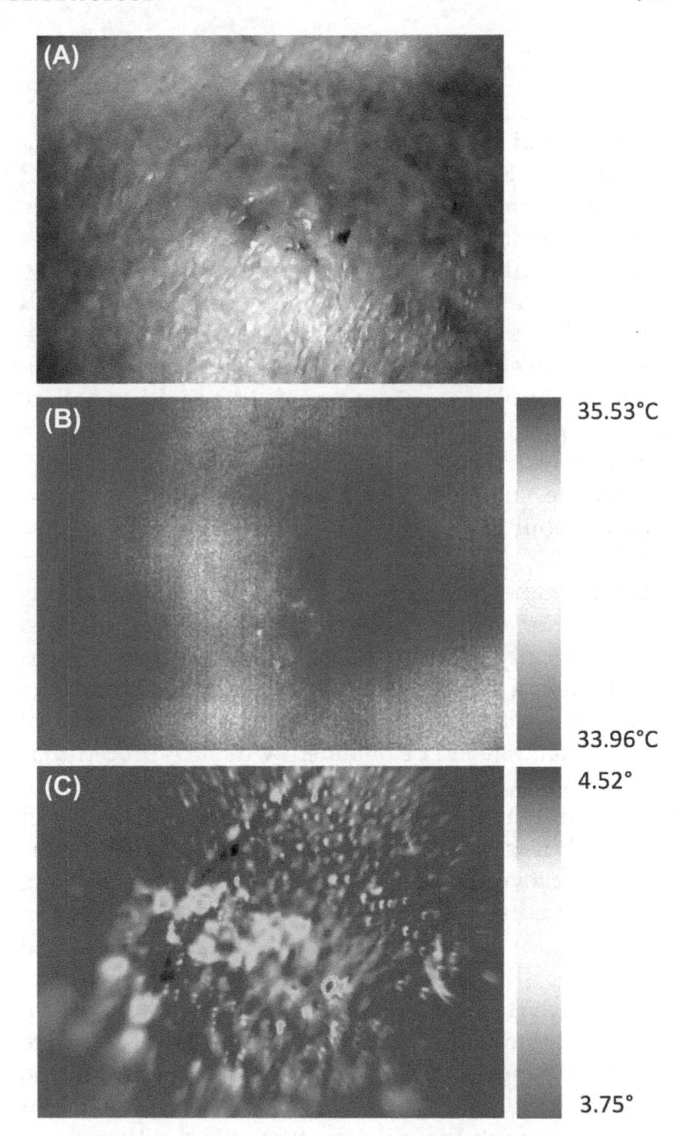

FIGURE 31.8 Visible, thermal, and phase image of a basal cell carcinoma (BCC) located on the forehead of a 79-year-old male patient. (A) Clinical image of the lesion. (B) Static thermogram of the same lesion. The margins are difficult to identify on the steady-state infrared (IR) image. (C) Pseudocolored phase image resulting from the digital lock-in demodulation. The lesion exhibit higher phase values in comparison with healthy skin and the margins of the BCC can be detected with accuracy.

Other Potential Applications

Although less investigated, other potential dermatological applications of thermal imaging have been reported in the literature. Passive thermal imaging has been, for example, employed to differentiate melanoma metastasis from benign cutaneous lesions, exhibiting excellent sensitivity and specificity for lesions larger than 15 mm [68]. Hassan and coworkers compared passive thermography with LDI for the follow-up observation of Kaposi sarcoma (KS) patients [69]. They demonstrated that thermography can be used to detect functional vascular abnormalities in KS lesions and to quantitatively assess anti-KS therapy. Santa Cruz and coworkers used PT for the follow-up monitoring of melanoma patients treated with boron neutron capture therapy (BNCT), attesting that erythema reactions to radiation and melanoma nodules have faster temperature recovery in comparison with healthy skin [70]. Di Carlo already noticed the potential of thermal imaging for the assessment of sclerodermic disorders in his initial publication [33]. Thermograms can demonstrate the dysfunctional cutaneous blood flow in response to cold stimuli. A few studies have investigated this application more in depth [71–74]. Bharara and Cobb explored active thermography procedures for the assessment of the diabetic neuropathic foot [75,76]. Laino and Di Carlo demonstrated that thermography can be considered a useful complementary method for the evaluation of patch tests [77]. Finally, in cosmetic applications, passive thermography has been used for the follow-up observation of port wine stains treated with argon laser [78,79] or to determine the severity of cellulite [31].

CONCLUSION

Thermal imaging is a noncontact and noninvasive imaging method particularly suited to dermatological investigations. Indeed, the human skin is a very good thermal emitter from which radiation can be easily captured by an IR imaging device. Passive procedures based on steady-state temperature measurements allow retrieving limited information. However, the price of IR cameras has been drastically sinking over the last decades because of technical improvements in microbolometer technology, opening new potential utilization. For example, IR camera cores could be integrated for a negligible extra cost into a standard epiluminescence microscope, giving the dermatologist additional information about the lesion perfusion and/or metabolic activity.

On the other hand, active thermal imaging permits the extraction of quantitative physiological parameters. The complexity associated with the computational thermal modeling of the human skin, together with large biological variability, however, restrains the amount of information that can be obtained with such procedures. In addition, thermal imaging is a sensitive method with the capability to detect temperature differences less than 0.001°C, but is not very specific. For those reasons, the authors believe that passive or active thermal imaging will achieve its full potential only in combination with other (optical) diagnostic methods. Among the potential candidates, IR imaging spectroscopy is an interesting option. Because the skin surface is partially transparent to IR radiation, active thermography together with the use of IR bandpass filters, Fabry-Pérot interferometer, or multiband QWIP cameras will allow retrieval of information about the skin surface composition for a more specific diagnosis.

List of Acronyms and Abbreviations

ADU Analog-to-digital unit
BCC Basal cell carcinoma
BNCT Boron neutron capture therapy
DTA Dynamic thermal assessment
DTI Dynamic thermal imaging
FOV Field of view
FPA Focal plan array
GigE Gigabit Ethernet
HD High definition
HFOV Horizontal field of view
IR Infrared
KS Kaposi sarcoma
LDI Laser Doppler imaging
LIT Lock-in thermography
Long-IR Long wavelength infrared
Mid-IR Medium wavelength infrared
NETD Noise equivalent temperature difference
NMSC Nonmelanoma skin cancer
PPT Pulsed phase thermography
PT Pulsed thermography
QWIP Quantum well infrared photodetector
SCC Squamous cell carcinoma
ST Stepped thermography
UV Ultraviolet
VFOV Vertical field of view
VGA Video graphics array

Acknowledgments

This work was supported by the Gebert Rüf Foundation (Grant Nr. GRS-072/13), the Geneva Cancer League, the Geneva University Hospital (HUG), and the Zurich University of Applied Sciences (ZHAW). The authors would like to the thank the company Xenics nv for the possibility of testing different infrared cameras and cores presented in this study, and for fruitful discussions.

References

[1] Vollmer M, Möllman K-P. Infrared thermal imaging: fundamentals, research and applications. Wiley-VCH; 2010.
[2] Maldague X. Theory and practice of infrared technology for nondestructive testing. Wiley-Interscience; 2001.
[3] Breitenstein O, Warta W, Langenkamp M. Lock-in thermography: basics and use for evaluating electronic devices and materials. Springer; 2010.
[4] Breitenstein O, Rakotoniaina JP, Al Rifai MH. Quantitative evaluation of shunts in solar cells by lock-in thermography. Prog Photovoltaics 2003;11(8):515–26.
[5] Williams KL. Infrared thermometry as a tool in medical research. Ann NY Acad Sci 1964;121:99–112.
[6] Gershon-Cohen J, Haberman-Brueschke JA, Brueschke EE. Medical thermography: a summary of current status. Radiol Clin North Am 1965;3(3):403–31.
[7] Davy JR. Medical applications of thermography. Phys Technol 1977;8(2):54–61.
[8] Jones B. A reappraisal of the use of infrared thermal image analysis in medicine. IEEE Trans Med Imaging 1998;17(6):1019–27.
[9] Diakides N, Bronzino J, Peterson D. Medical infrared imaging. CRC Press; 2007.
[10] Kateb B, Yamamoto V, Yu C, Grundfest W, Gruen JP. Infrared thermal imaging: a review of the literature and case report. Neuroimage 2009;47:154–62.
[11] Amalu W, Hobbins W, Head J, Elliot R. Infrared imaging of the breast: a review. In: Diakides M, Bronzino J, Peterson D, editors. Medical infrared imaging. CRC Press; 2007. p. 1–22.
[12] Tan JH, Ng EYK, Acharya UR, Chee C. Infrared thermography on ocular surface temperature: a review. Infrared Phys Technol 2009; 52(4):97–108.
[13] Campbell P, Roderick T. Thermal imaging in surgery. In: Diakides M, Bronzino J, Peterson D, editors. Medical infrared imaging. CRC Press; 2007. p. 1–18.
[14] Gratt B. Infrared imaging applied to dentistry. In: Diakides M, Bronzino J, Peterson D, editors. Medical infrared imaging. CRC Press; 2007. p. 1–8.
[15] van der Graaff KM, Strete D, Creek CH. Human anatomy. McGraw Hill; 2001.
[16] Lee Y, Hwang K. Skin thickness of Korean adults. Surg Radiol Anat 2002;24(3–4):183–9.
[17] Xu F, Lu TJ. Skin biothermomechanics: modeling and experimental characterization. Advances in Applied Mechanics. 2009. p. 147–248.
[18] Pennes H. Analysis of the tissue and arterial blood temperature in the resting human forearm. J Appl Physiol 1948;1:93–122.

[19] Shih T, Yuan P, Lin W, Kou H. Analytical analysis of the Pennes bioheat transfer equation with sinusoidal heat flux condition on the skin surface. Med Eng Phys 2007;29:946–53.
[20] Liu J, Xu L. Estimation of blood perfusion using phase shift in temperature response to sinusoidal heating at the skin surface. IEEE Trans Biomed Eng 1999;46(9):1037–43.
[21] Deng Z, Liu J. Mathematical modelling of temperature mapping over skin surface and its implementation in thermal disease diagnostics. Comput Biol Med 2004;34:495–521.
[22] Cetinguel MP, Herman C. A heat transfer model of skin tissue for the detection of lesions: sensitivity analysis. Phys Med Biol 2010; 55(19):5933–51.
[23] Bonmarin M, Le Gal FA. Lock-in thermal imaging for the early-stage detection of cutaneous melanoma: a feasibility study. Comput Biol Med 2014;47:36–43.
[24] Bonmarin M, Le Gal FA. A lock-in thermal imaging setup for dermatological applications. Skin Res Technology 2015;21(3): 284–90.
[25] Boylan A, Martin CJ, Gardner GG. Infrared emissivity of burn wounds. Clin Phys Physiol Meas 1992;13(2):125–7.
[26] Sanchez-Marin F, Calixto-Carrera S, Villasenor-Mora C. Novel approach to assess the emissivity of the human skin. J Biomed Opt 2009;14(2):024006.
[27] Watmough DJ, Fowler PW, Oliver R. The thermal scanning of a curved isothermal surface: implications for clinical thermography. Phys Med Biol 1970;15(1):1–8.
[28] Rogalski A. Infrared detectors. CRC Press; 2010.
[29] Jagadish C, Gunapala S, Rhiger D. Advances in infrared photodetectors. Elsevier Science; 2011.
[30] Henini M. Handbook of infrared detection technologies. Elsevier; 2002.
[31] Nkengne A, Papillon A, Bertin C. Evaluation of the cellulite using a thermal infra-red camera. Skin Res Technol 2013;19(1):231–7.
[32] Ammer K, Ring FJ. Standard procedures for infrared imaging in medicine. In: Diakides M, Bronzino JD, Peterson DR, editors. Medical infrared imaging. CRC Press; 2007. p. 1–15.
[33] Di Carlo A. Thermography and the possibilities for its applications in clinical and experimental dermatology. Clin Dermatol 1995;13(4):329–36.
[34] Notingher I, Imhof R, Xiao P, Pascut F. Spectral depth profiling of arbitrary surfaces by thermal emission decay-Fourier transform infrared spectroscopy. Appl Spectrosc 2003;57(12):1494–501.
[35] Notingher I, Imhof R. Mid-infrared in vivo depth-profiling of topical chemicals on skin. Skin Res Technol 2004;10(2):113–21.
[36] Abuhamad M, Netzelmann U. Dual-band active thermography on infrared transparent materials. Quantitative InfraRed Thermography J 2010;7(2):189–200.
[37] Busse G, Wu D, Karpen W. Thermal wave imaging with phase sensitive modulated thermography. J Appl Phys 1992;71:3962–5.
[38] Krapez JC. Compared performances of four algorithms used for digital lock-in thermography. In: Balageas D, Busse G, Carlomagno C, editors. Quantitative infrared thermography QIRT; 1998. p. 148–53. Lodz, Poland.
[39] Junyang L, Yang W, Jingmin D. Research on thermal wave processing of lock-in thermography based on analysing image sequences for NDT. Infrared Phys Technol 2010;53:348–57.
[40] Gupta R, Breitenstein O. Unsteady-state lock-in thermography: application to shunts in solar cells. Quantitative InfraRed Thermography J 2007;4(1):85–105.
[41] Anbar M. Quantitative dynamic telethermometry in medical diagnosis and management. CRC-Press; 1994.
[42] Anbar M, Milescu L, Grenn MW, Zamani K, Marino MT. Study of skin hemodynamics with fast dynamic area telethermometry (DAT). 19th Annual International Conference of the IEEE Engineering in Medicine and Biology Society. 1997. p. 644–8.
[43] Judson RT. Burns. In: Tjandra J, Clunie G, Kaye H, Smith JA, editors. Textbook of surgery. 3rd ed. John Wiley and Sons; 2005.
[44] Watts AMI, Tyler MPH, Perry ME, Roberts AHN, McGrouther DA. Burn depth and its histological measurement. Burns 2001;27(2):154–60.
[45] Hoeksema H, Van de Sijpe K, Tondu T, et al. Accuracy of early burn depth assessment by laser Doppler imaging on different days post burn. Burns 2009;35(1):36–45.
[46] Lawson RN, Wlodek GD, Webster DR. Thermographic assessment of burns and frostbite. Can Med Assoc J 1961;84: 1129–31.
[47] Mladick R, Georgiade N, Thorne F. A clinical evaluation of the use of thermography in determining degree of burn injury. Plast Reconstr Surg 1966;38(6):512–8.
[48] Hackett ME. The use of thermography in the assessment of depth of burn and blood supply of flaps, with preliminary reports on its use in Dupuytren's contracture and treatment of varicose ulcers. Br J Plast Surg 1974;27(4):311–7.
[49] Newman P, Pollock M, Reid WH, James WB. A practical technique for the thermographic estimation of burn depth: a preliminary report. Burns 1981;8(1):59–63.
[50] Cole RP, Jones SG, Shakespeare PG. Thermographic assessment of hand burns. Burns 1990;16(1):60–3.
[51] Renkielska A, Nowakowski A, Kaczmarek M, et al. Static thermography revisited: an adjunct method for determining the depth of the burn injury. Burns 2005;31(6):768–75.
[52] Hardwicke J, Thomson R, Bamford A, Moiemen N. A pilot evaluation study of high resolution digital thermal imaging in the assessment of burn depth. Burns 2013;39(1):76–81.
[53] Renkielska A, Nowakowski A, Kaczmarek M, Ruminski J. Burn depths evaluation based on active dynamic IR thermal imaging: a preliminary study. Burns 2006;32(7):867–75.
[54] Ruminski J, Kaczmarek M, Renkielska A, Nowakowski A. Thermal parametric imaging in the evaluation of skin burn depth. IEEE Trans Biomed Eng 2007;54(2):303–12.
[55] Renkielska A, Kaczmarek M, Nowakowski A, et al. Active dynamic infrared thermal imaging in burn depth evaluation. J Burn Care Res 2014;35(5):294–303.
[56] Nowakowski A. Quantitative active dynamic thermal IR-imaging and thermal tomography in medical diagnostic. In: Diakides M, Bronzino JD, Peterson DR, editors. Medical infrared imaging. CRC Press; 2007. p. 1–30.
[57] Balch C, Soong S, Gershenwald J, et al. Prognostic factors analysis of 17,600 melanoma patients: validation of the American Joint Committee on cancer melanoma staging system. J Clin Oncol 2001;19(16):3622–34.
[58] Housman TS, Feldman SR, Williford PM, et al. Skin cancer is among the most costly of all cancers to treat for the medicare population. J Am Acad Dermatol 2003;48(3):425–9.
[59] Brasfield R, Sherman R, Laughlin J. Thermography in management of cancer: a preliminary report. Ann NY Acad Sci 1964; 121(A1):235–47.
[60] Hartmann M, Kunze J, Friedel S. Telethermography in the diagnostics and management of malignant melanomas. J Dermatol Surg Oncol 1981;7(3):213–8.
[61] Cristofolini M, Perani B, Pisciolini F, Rechchia G, Zumiani G. Uselessness of thermography for diagnosis and follow-up of cutaneous malignant melanoma. Tumori 1981;67(2):141–3.
[62] Buzug TM, Schumann S, Pfaffmann L, Reinhold U, Ruhlmann J. Functional infrared imaging for skin-cancer screening. Conf IEEE Eng Med Biol Soc 2006:2766–9.
[63] Cetinguel MP, Herman C. Quantification of the thermal signature of a melanoma lesion. Int J Therm Sci 2011;50(4):421–31.
[64] Deleted in review.
[65] http://www.dermolockin.com.

[66] Deng Z, Liu J. Analytical Study on bioheat transfer problems with spatial or transient heating skin surface or inside biological bodies. J Biomech Eng 2002;124:638–50.
[67] Yuan P, Liu H, Chen C, Kou H. Temperature response in biological tissue by alternating heating and cooling modalities with sinusoidal temperature oscillation on the skin. Int Commun Heat Mass Transfer 2008;35:1091–6.
[68] Shada AL, Dengel LT, Petroni GR, Smolkin ME, Acton S, Slingluff Jr CL. Infrared thermography of cutaneous melanoma metastases. J Surg Res 2013;182(1):9–14.
[69] Hassan M, Little RF, Vogel A, Aleman K, Wyvill K, Yarchoan R, et al. Quantitative assessment of tumor vasculature and response to therapy in Kaposi's sarcoma using functional noninvasive imaging. Technol Cancer Res Treat 2004;3(5):451–7.
[70] Santa Cruz G, Bertotti J, Marin J, Gonzalez SJ, Gossio S, Alvarez D, et al. Dynamic infrared imaging of cutaneous melanoma and normal skin in patients treated with BNCT. Appl Radiat Isot 2009;67:54–8.
[71] Cutolo M, Sulli A, Smith V. Assessing microvascular changes in systemic sclerosis diagnosis and management. Nat Rev Rheumatol 2010;6(10):578–87.
[72] Murray AK, Moore TL, Manning JB, Taylor C, Griffiths CEM, Herrick AL. Noninvasive imaging techniques in the assessment of scleroderma spectrum disorders. Arthritis Rheum 2009;61(8): 1103–11.
[73] Pauling JD, Shipley JA, Harris ND, McHugh NJ. Use of infrared thermography as an endpoint in therapeutic trials of Raynaud's phenomenon and systemic sclerosis. Clin Exp Rheumatol 2012; 30(2):103–15.
[74] Merla A, Di Donato L, Di Luzio S, Farina G, Pisarri S, Proietti M, et al. Infrared functional imaging applied to Raynaud's phenomenon. IEEE Eng Med Biol Mag 2002;21(6):73–9.
[75] Bharara M, Cobb JE, Claremont DJ. Thermography and thermometry in the assessment of diabetic neuropathic foot: a case for furthering the role of thermal techniques. Int J Low Extrem Wounds 2006;5(4):250–60.
[76] Bharara M, Viswanathan V, Cobb JE. Cold immersion recovery responses in the diabetic foot with neuropathy. Int Wound J 2008; 5(4):562–9.
[77] Laino L, Di Carlo A. Telethermography: an objective method for evaluating patch test reactions. Eur J Dermatol 2010;20(2): 175–80.
[78] Patrice T, Dreno B, Weber J, Le Bodic L, Barriere H. Thermography as a predictive tool for laser treatment of port-wine stains. Plast Reconstr Surg 1985;76(4):554–7.
[79] Troilius A, Wardell K, Bornmyr S, Nilsson GE, Ljunggren B. Evaluation of port wine stain perfusion by laser Doppler imaging and thermography before and after argon-laser treatment. Acta Derm Venereol 1992;72(1):6–10.
[80] Robinson JK. Anatomy for procedural dermatology. In: Robinson JK, Hanke WC, Sengelmann R, Siegel D, editors. Surgery of the skin. 3rd ed. Philadelphia: Elsevier; 2015. p. 1–27.
[81] http://www.xenics.com/en.

CHAPTER

32

The Role of Positron Emission Tomography/Computed Tomography in Cutaneous Melanoma

A.C. Bourgeois[1], A.S. Pasciak[1,2], Y.C. Bradley[1]

[1]The University of Tennessee Graduate School of Medicine, Knoxville, TN, United States; [2]The University of Tennessee Medical Center, Knoxville, TN, United States

OUTLINE

BACKGROUND

KEY POINTS:

- Positron emission tomography (PET) typically utilizes 2-deoxy-2-(^{18}F)fluoro-D-glucose (FDG), a radiolabeled glucose molecule that reflects cellular metabolic activity.
- Computed tomography (CT) images allow for localization of PET abnormalities and also increase the validity of PET images through the process of attenuation correction.
- The average PET/CT exam requires approximately 40–60 min to acquire images and imparts more radiation exposure than CT alone.
- The degree of radiotracer uptake in tissue is referred to as FDG avidity and is quantitatively denoted with the standard uptake value (SUV).

Imaging in Dermatology
http://dx.doi.org/10.1016/B978-0-12-802838-4.00032-7

PET/CT plays a prominent role in the staging, management, and surveillance of high-risk cutaneous melanoma (CM) patients. PET/CT represents a combination of two imaging modalities and a fusion of functional and structural imaging.

The PET imaging component of a PET/CT exam is performed after intravenous (IV) administration of a radiopharmaceutical. By convention, most PET imaging in the United States is performed with FDG, a glucose analog with the positron-emitting radioactive isotope fluorine-18. In the body, ^{18}FDG enters cells in a similar fashion to glucose; however, once ^{18}FDG is phosphorylated it does not continue down the glycolytic pathway as glucose would. Instead, ^{18}FDG-6-phosphate remains permanently trapped within the cell where ^{18}F decays by positron emission [1]. A positron is the antimatter equivalent of an electron and carries a positive electrostatic charge, equal to and opposite to that of an electron [1]. After being released by the decaying ^{18}F nucleus, the positron will be drawn to a nearby electron by electrostatic attraction, and both particles will annihilate, emitting two high-energy photons in 180 degrees opposition to one another [1]. PET scanners use several static rings of detectors around the patient to record these photons in coincidence (Fig. 32.1). Because both photons are created at the same time by a single annihilation event, they arrive at detectors on opposite sides of the patient in coincidence or, within a very small time window. These photon interactions are recorded and used to identify the origin of the ^{18}F atom that decayed in the patient through a complex set of mathematical algorithms [2].

PET imaging with ^{18}FDG can provide whole-body quantification of glucose uptake in tissues. Because glucose uptake often correlates with metabolic activity in normal tissues and malignancies, PET imaging can provide a powerful tool for distinguishing normal from abnormal tissues. However, detection of malignancy with PET requires identification of these tissues based on elevated glucose metabolism relative to normal tissue background. This can be problematic in particularly hypermetabolic organs such as the brain and heart, or in organs where physiological excretion of ^{18}FDG occurs such as the gastrointestinal and genitourinary tract. PET sensitivity is also reduced in tumor subtypes that demonstrate inherently less ^{18}FDG uptake. Fortunately, CM is generally highly FDG-avid, providing high diagnostic accuracy with staging sensitivity and specificity exceeding 90% [3].

The CT component used in concert with modern PET imaging significantly improves the specificity of the PET/CT modality. PET/CT scanners feature mechanically connected PET and CT gantries, allowing for multimodality imaging while the patient rests in the same

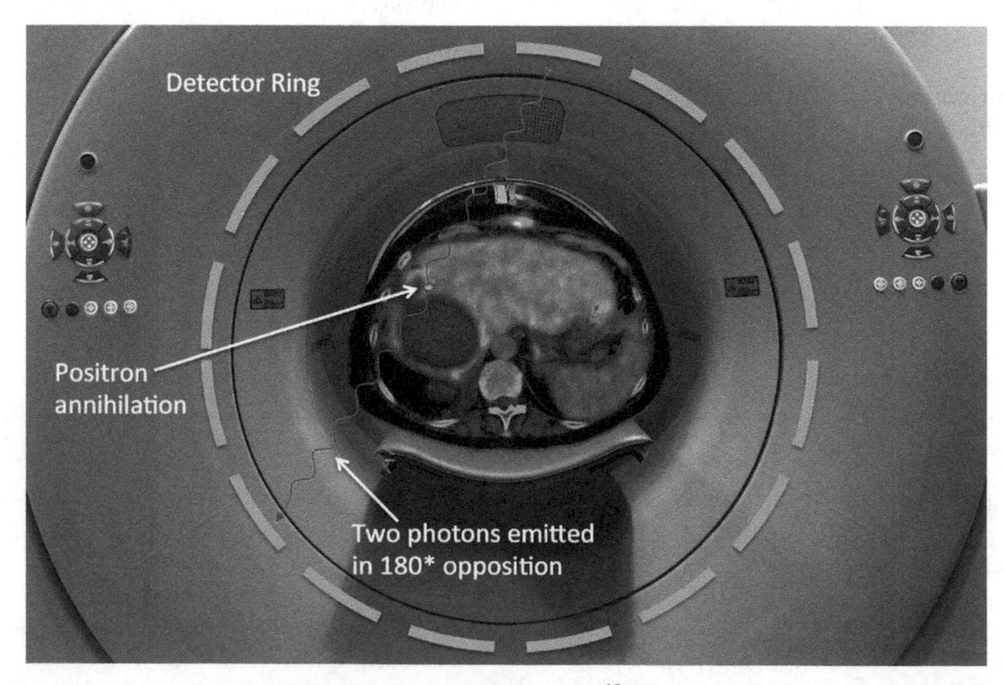

FIGURE 32.1 Illustration of a positron annihilation event after decay of an ^{18}F nucleus. Positron annihilation creates two photons that are emitted in 180 degrees opposition to each other and are detected by a ring of detectors in the positron emission tomography scanner gantry in coincidence. This patient is post-right–hepatectomy. Minimally increased areas of metabolic uptake are present from postoperative inflammation surrounding a large seroma along the right lobe remnant. Note the expected absence of any 2-deoxy-2-(^{18}F)fluoro-D-glucose uptake in the seroma.

position on the scanner bed (Fig. 32.2). This configuration provides registration of PET and CT images with the capacity to provide anatomical localization of a lesion detected on PET, improving interpretation accuracy. CT also provides data related to the photon attenuation properties of patient tissues, which combined with sophisticated image reconstruction algorithms are used to make PET/CT a quantitative imaging modality [2]. The quantitative aspect of PET/CT allows its images to be used to directly evaluate the rate of glucose metabolism in a tumor to estimate malignant potential [2]. Moreover, PET/CT can be used to quantify the changes in glucose metabolism of a patient's tumor over multiple imaging sessions to determine progression of disease or response to therapy, significantly aiding in the medical management of oncology patients.

In the setting of CM, PET/CT imaging serves as a crucial adjunct to clinical and surgical staging in many high-risk patients. This chapter provides an overview of the clinical scenarios and tumor characteristics in which PET/CT is likely to most impact clinical management while providing an explanation of its limitations and interpretive pitfalls.

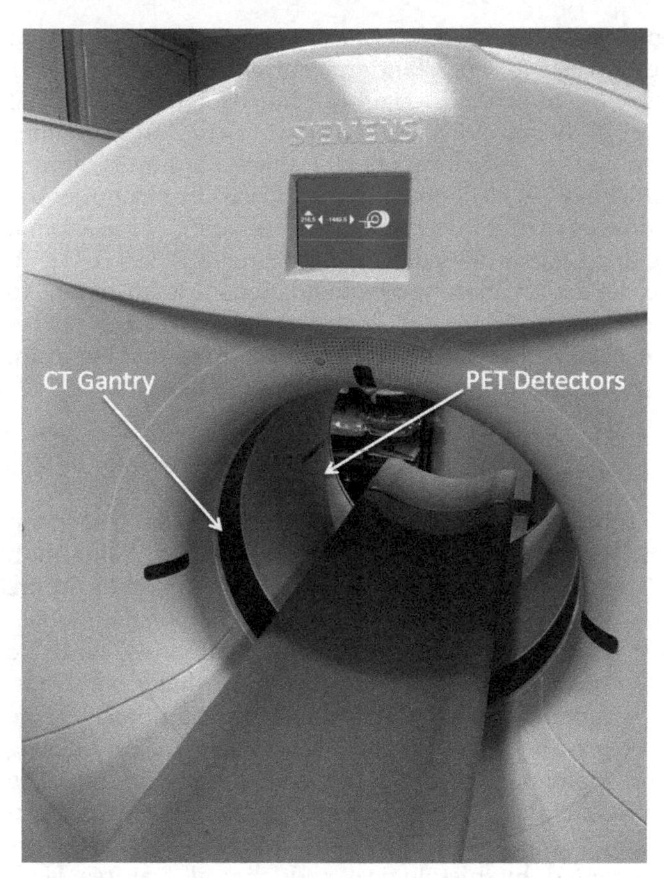

FIGURE 32.2 Modern positron emission tomography (PET)/computed tomography (CT) systems, such as this Siemens mCT Flow, utilize mechanically coupled PET and CT gantries, both serviced by the same patient bed. This allows accurate image registration and anatomical overlay of both imaging modalities.

STAGING MELANOMA AND THE ROLE OF POSITRON EMISSION TOMOGRAPHY/COMPUTED TOMOGRAPHY

KEY POINTS:

- FDG-PET/CT has relatively poor sensitivity in detecting micrometastatic disease. Thus sentinel lymph node (SLN) biopsy is the mainstay of routine locoregional nodal evaluation.
- PET/CT is most sensitive in detecting macrometastatic disease or distant metastasis and of highest clinical utility in differentiating stage III and IV disease.
- PET/CT is of little utility in early-stage (I or IIA) disease.
- Patients with high-risk histological features and/or location could potentially benefit from PET/CT evaluation.
- Co-registered PET/CT is of improved diagnostic accuracy relative to PET or CT alone in melanoma staging.

CM staging algorithms incorporate several complex histological features and clinical data points [4]. Recent research has focused on the clinical and prognostic significance of many of these factors, including SLN status, presence of tumor ulceration and satellite metastases, serum lactate dehydrogenase level, and mitotic rate [4–7]. A more simplified approach to staging is provided by the American Joint Committee of Cancer (AJCC), in which patients are categorized in three staging groups: localized (stage I and II), regional (stage III), and distant disease (stage IV) [4]. In the context of melanoma staging, PET/CT has shown limited utility in localized disease and is generally reserved for high-risk patients with either regional or distant disease [4]. In clinical practice, the settings in which PET/CT is used to evaluate CM are widely variable depending on institutional protocols, clinical preference, and access to resources. No rigid standards currently govern PET/CT use in melanoma staging; however, the National Comprehensive Cancer Network (NCCN) guidelines provide an evidence-based framework for its appropriate use. In 2013, the NCCN updated its recommendations regarding the workup and restaging of melanoma, with a section dedicated to the scope of PET/CT [4]. To contextualize these recommendations it is important to understand the general approach to clinical and pathological CM staging as well as the foundational literature upon which the NCCN guidelines are based.

Localized Disease

Clinical workup typically begins with inspection and biopsy of a suspicious cutaneous lesion. A thorough history and physical exam and complete skin exam are then performed with focus on the locoregional nodal basin, in-transit lesions, synchronous primaries, evidence of distant metastasis, and predisposing conditions. These clinical data points make up the clinical stage and greatly influence the method of surgical staging and potential need for advanced imaging [4]. The mainstay of surgical management is wide local excision of the primary lesion and surgical nodal sampling, as warranted by histological and clinical features [4,8]. Integrated clinical and surgical staging data are used to stratify risk, plan therapy, and determine need for advanced imaging.

Localized disease, including stages I and II, incorporates any histological features of the primary tumor in the absence of locoregional nodal involvement, in-transit metastasis, or distant metastases. Breslow thickness and tumor ulceration are the two most important prognostic factors in early-stage melanoma and comprise the basis of the pathological staging criteria [4]. Stage I disease includes primary lesions 1 mm or less in thickness with (Ib) or without (Ia) ulceration, as well as lesions 2 mm or less in thickness without ulceration (Ib). Stage IIa disease includes primary lesions 1–2 mm in thickness with ulceration and lesions 2–4 mm in thickness without ulceration. Stage IIb includes lesions 2–4 mm with ulceration and 4 mm or greater in thickness without ulceration. Stage IIC disease is any lesion 4 mm or greater in thickness with ulceration. Several other factors have shown prognostic significance but are not included in AJCC staging criteria [9].

The presence of nodal metastasis and the degree of nodal metastatic burden are of high clinical and prognostic significance. A negative SLN biopsy generally denotes localized (stage I or II) disease and imparts a 5-year survival rate greater than 90% [4]. According to a meta-analysis of more than 25,000 patients, the rate of disease recurrence after negative SLN biopsy is only 5% [4]. However, the presence of even one positive SLN significantly reduces 5-year survival to 62% [10]. SLN biopsy is considered the standard of care for evaluating the regional nodal basin, although no high-level evidence shows overall survival benefit from its use [11,12]. SLN biopsy offers high accuracy with false-negative rates less than 4% [12] and reduced morbidity compared with complete lymph node dissection (CLND). As a generalization, PET/CT is rarely used for metastatic surveillance in a patient with a negative SLN biopsy and is included in the NCCN staging algorithm to only evaluate specific symptoms in stages I and II [4] (Fig. 32.3).

PET and PET/CT have shown little utility as a primary screening tool for nodal metastasis, particularly in the absence of high-risk tumor histology [11–14]. A 2007 study by Kell in which 83 patients with 1.9-mm mean tumor thickness underwent SLN biopsy and preoperative PET/CT found little benefit with addition of PET/CT [15]. Of the 15 patients in this series with pathologically proven nodal metastasis, PET/CT was positive in only 2 (13.3%). A similar study by Singh corroborated these findings, demonstrating a 14.3% sensitivity of PET/CT versus SLN biopsy in evaluating the nodal status of 52 patients [16]. Finally, Bikhchandani retrospectively examined 47 patients with predominately low-risk primary head-and-neck tumors using PET/CT, finding no cases in which PET/CT detected occult regional or distant disease [17]. Although the sensitivity of the older scanners used in these articles are subpar compared with modern PET/CT, these data still clearly show inferiority of PET/CT to SLN biopsy as a primary means of nodal evaluation.

The poor sensitivity of nodal evaluation with PET imaging is largely due to low sensitivity in detecting micrometastatic disease. The ability of detecting metastatic lesions is based on the metabolic activity and size of the lesion. A high metabolic lesion, indirectly measured by mitotic rate, allows detection of a smaller lesion. In addition, the quantity of tumor has a direct correlation to detectability. A study by Crippa examined the relationship of nodal size with PET detection accuracy in 38 patients with stage III disease who underwent CLND of 56 lymph node basins [18]. Although PET yielded an aggregate 95% sensitivity and 84% specificity for nodal detection, it detected every nodal metastasis with nodal size greater than 10 mm [18]. PET also

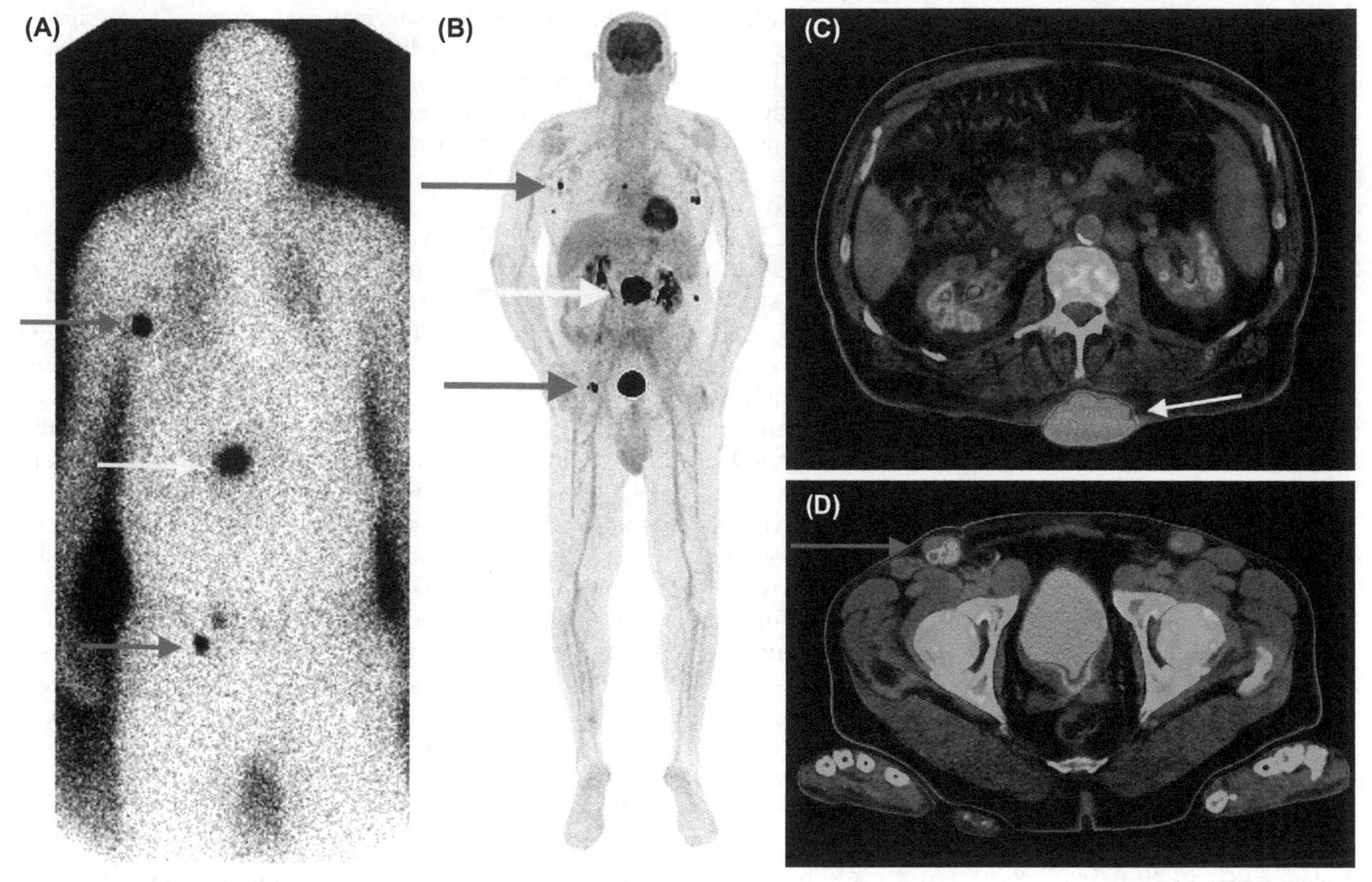

FIGURE 32.3 The benefit of lymphoscintigraphy before surgery is clearly demonstrated in this 56-year-old male patient with midline lower back melanoma. (A) The lymphoscintigraphy demonstrates the lymphatic drainage to the right axillary and right inguinal nodes. (B) Subsequent positron emission tomography (PET)/computed tomography (CT) reveals metastases localized to the exact locations found on the lymphoscintigraphy. Multiple other metastases are also visualized on the PET/CT. *Yellow arrows* localize to the primary melanoma and the *red arrows* identify metastases on the PET, maximum intensity projection (B), and fused axial PET/CT (C and D).

detected 83% of lymph nodes 5–10 mm in size and only 23% of those 5 mm or greater in size [18]. The relationship of PET detection efficiency to tumor size reflects the intrinsic limitations of current PET technology. It is possible that detection rates of smaller nodal metastases may improve as PET/CT technology and more specific radiopharmaceutical agents evolve. However, for now, SLN biopsy remains the mainstay of nodal evaluation and should not be supplanted by PET/CT in patients without clinical evidence of locoregional or distant metastatic disease [4].

High-Risk Primary Lesions and Advanced Disease

The role of advanced imaging is clearer in patients with either high-risk primary tumors or in stages III and IV disease. Patients with high-risk tumor histology may derive benefit from advanced imaging, in part because of the higher positive predictive value of imaging in higher risk primary lesions. NCCN guidelines currently suggest a role for PET/CT in the setting of thick melanoma (>1 mm thickness) and thin melanoma (0.76–1.0 mm) with aggressive features such as tumor ulceration and mitotic rate greater than 1 per millimeter [24]. However, it should be noted that these guidelines suggest that advanced imaging should only be used in these settings to evaluate specific symptoms. The potential benefit of PET/CT in high-risk lesions derives from increased detection accuracy in these patients. Iagura examined this phenomenon in 106 patients with tumors with an average Breslow thickness of 3.56 mm, noting 89.3% sensitivity and 88% specificity for melanoma detection [12]. This far exceeds the aforementioned detection accuracy when PET/CT is used as a screening tool for all patients. It is interesting to note that in this series PET/CT showed the highest sensitivity for detection of metastatic lesions in intermediate-risk primary tumors with a Breslow thickness of 1–4 mm (92.7%). It should be noted that routine PET/CT is controversial in intermediate-thickness melanoma, in part because of research suggesting high false-positive rates. In one such study, in which Barsky reviewed PET/CT use in 149 intermediate-risk patients, an 85% false positive rate was identified [19]. This led the authors to conclude that routine PET/CT in intermediate-thickness tumors could lead to unnecessary invasive and noninvasive tests.

Another possible clinical scenario in which PET/CT could provide benefit is in patients with positive SLN biopsy. Although this may be common practice at many institutions for high-risk primary lesions, current NCCN guidelines do not recommend routine use of PET/CT in this patient subset. Because patients with positive SLN biopsy typically undergo subsequent node dissection, the need for PET/CT is generally assessed based on CLND results. Because of this approach of management, limited data explore the role of PET/CT in patients with only positive SLN biopsy. A 2008 study of 30 patients who underwent PET imaging and CLDN with positive SLN biopsy found that the addition of imaging did not change management outcomes in these patients [20]. Thus the NCCN currently recommends PET/CT to evaluate specific clinical symptoms in patients with positive SLN biopsy.

The potential clinical impact of advanced imaging is clearer in patients with palpable adenopathy (clinical stage III) or with pathological stages III or IV disease. It is thought that the palpable nature of clinical stage III disease implies macroscopic metastatic burden, which is more likely to be detected by PET imaging. Several studies have noted improved positive predicted values in clinically advanced disease than when compared with PET/CT used as a screening exam for low-risk patients [16,21,22]. In clinical stage III disease, PET/CT provides relatively high sensitivity in detecting metastases, altering surgical management in a significant proportion of patients. In patients with clinically suspected stage III or IV melanoma, the 2013 NCCN guidelines now recommend fine needle aspiration of the palpable lymph nodes when feasible and PET/CT rather than standard management with SLN biopsy and CLND [4]. This is in large part due to the clinical impact of PET/CT on operative planning and outcomes. A study by Aukema was conducted among 70 clinical stage III patients who underwent preoperative imaging with PET/CT and brain MRI. PET/CT changed the intended regional nodal dissection in 26 patients (37%), with a sensitivity of 87% and specificity of 98% [23]. PET/CT results also showed prognostic significance with 84% 2-year survival in patients without additional lesions on PET/CT, compared with 56% in patients with additional metastases ($p < .001$) (Fig. 32.4).

The clinical impact of FDG imaging in clinically advanced melanoma (stage III or IV) has been corroborated in several additional studies, altering management in as many as 49% [24]. On average, PET/CT changes management in approximately 10–20% of patients [21,24] by identifying additional resectable disease or confirming nonresectable distant metastasis (stage IV).

PET/CT may also play a role in the detection of potentially resectable distant metastases and providing prognostic information [25]. Although stage IV melanoma carries poor prognosis, skin metastases are associated with a more favorable prognosis than visceral metastases [27]. In addition, resection of all radiologically and clinically apparent disease may further improve survival [27]. It is interesting to note that a small patient series has reported surgical cure with resection of limited distant metastasis [26]. Therefore the added sensitivity of PET/CT compared with CT alone in metastatic detection can be clinically relevant, even in stage IV disease.

THE ROLE OF POSITRON EMISSION TOMOGRAPHY/COMPUTED TOMOGRAPHY IN RESTAGING

KEY POINTS:

- Recent literature shows a potential role for PET/CT screening of treated patients with high metastatic burden for recurrent disease.
- PET/CT is currently recommended for evaluating symptomatic disease recurrence.
- PET/CT information yields prognostic significance in restaging and can aid surgical planning.

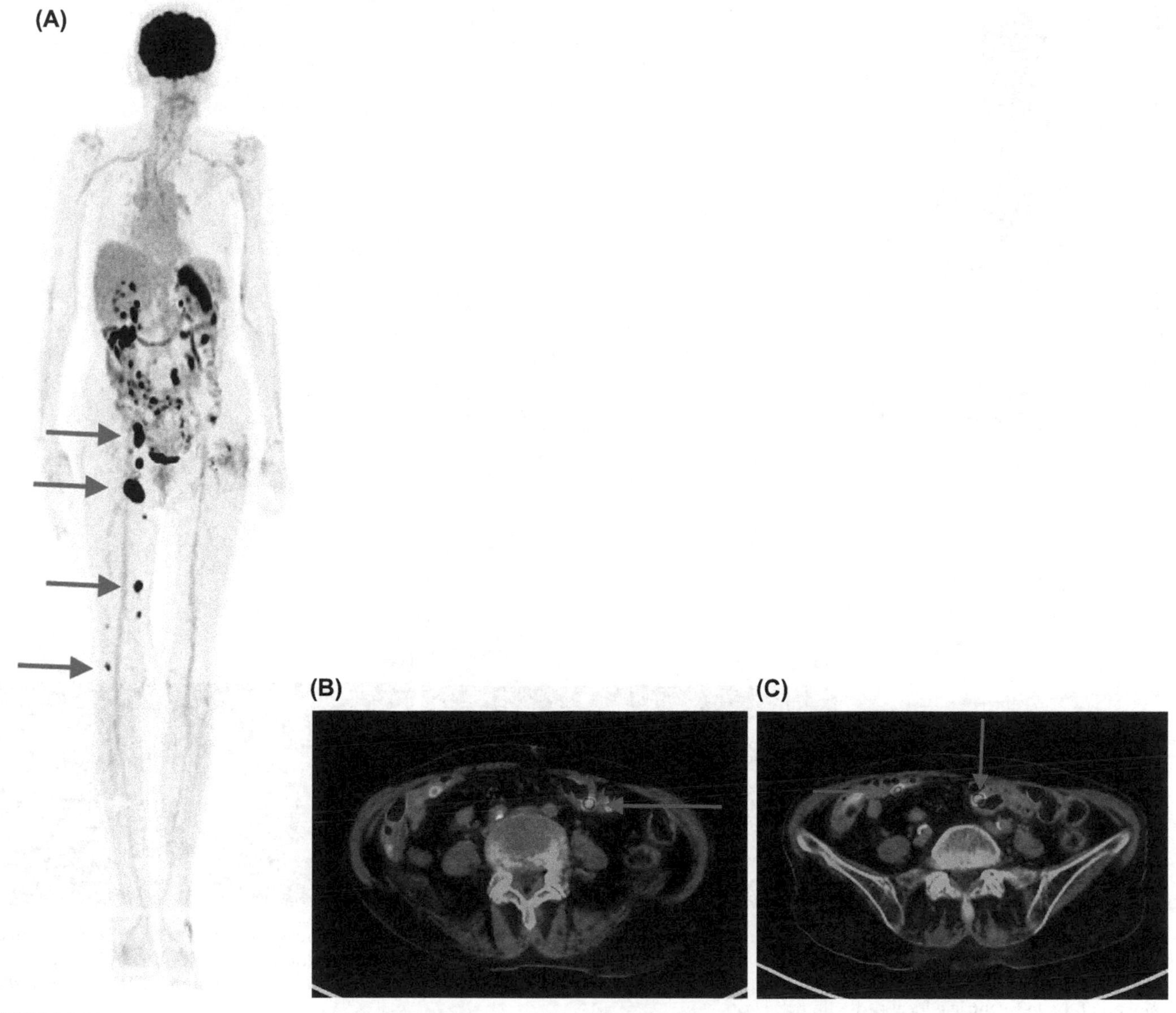

FIGURE 32.4 This 82-year-old female patient was initially thought to be stage III. On the basis of positron emission tomography (PET)/computed tomography (CT), her staging was upgraded to stage IV. (A) A faint postsurgical site (*yellow arrow*) of the primary lesion at the medial aspect of the midleg is revealed. Metastases are noted up the nodal chains and soft tissues along the affected limb. It is interesting to note that numerous metastases were found along the small bowel walls as identified by *red arrows* in panels (B and C).

Longitudinal follow-up of melanoma patients involves a multidisciplinary approach based on clinical and imaging findings. The current NCCN guidelines recommend close clinical observation for melanoma patients and do not suggest routine PET/CT surveillance for asymptomatic patients. However, a study from the Yale Melanoma Unit of 373 patients showed a survival benefit of patients who were asymptomatic at the time of recurrence, suggesting that disease outcomes can be influenced by early detection [28]. Although limited data explore the role of PET/CT in melanoma restaging of the asymptomatic patient, emerging evidence suggests that PET/CT could aid early metastatic detection. In a study of 34 patients treated for stage III melanoma, annual PET/CT follow-up detected recurrence in six of seven asymptomatic patients [9]. The only undetected recurrence in this study was a clinically evident local recurrence, perhaps obscured by regional postsurgical changes. However, the utility of PET/CT in routine postsurgical follow-up requires further confirmation, and this application currently remains institution specific (Fig. 32.5).

In the setting of clinically evident disease recurrence, advanced imaging provides the benefits of identifying the site(s) and extent of disease. For this reason, PET/CT is currently recommended by NCCN to evaluate

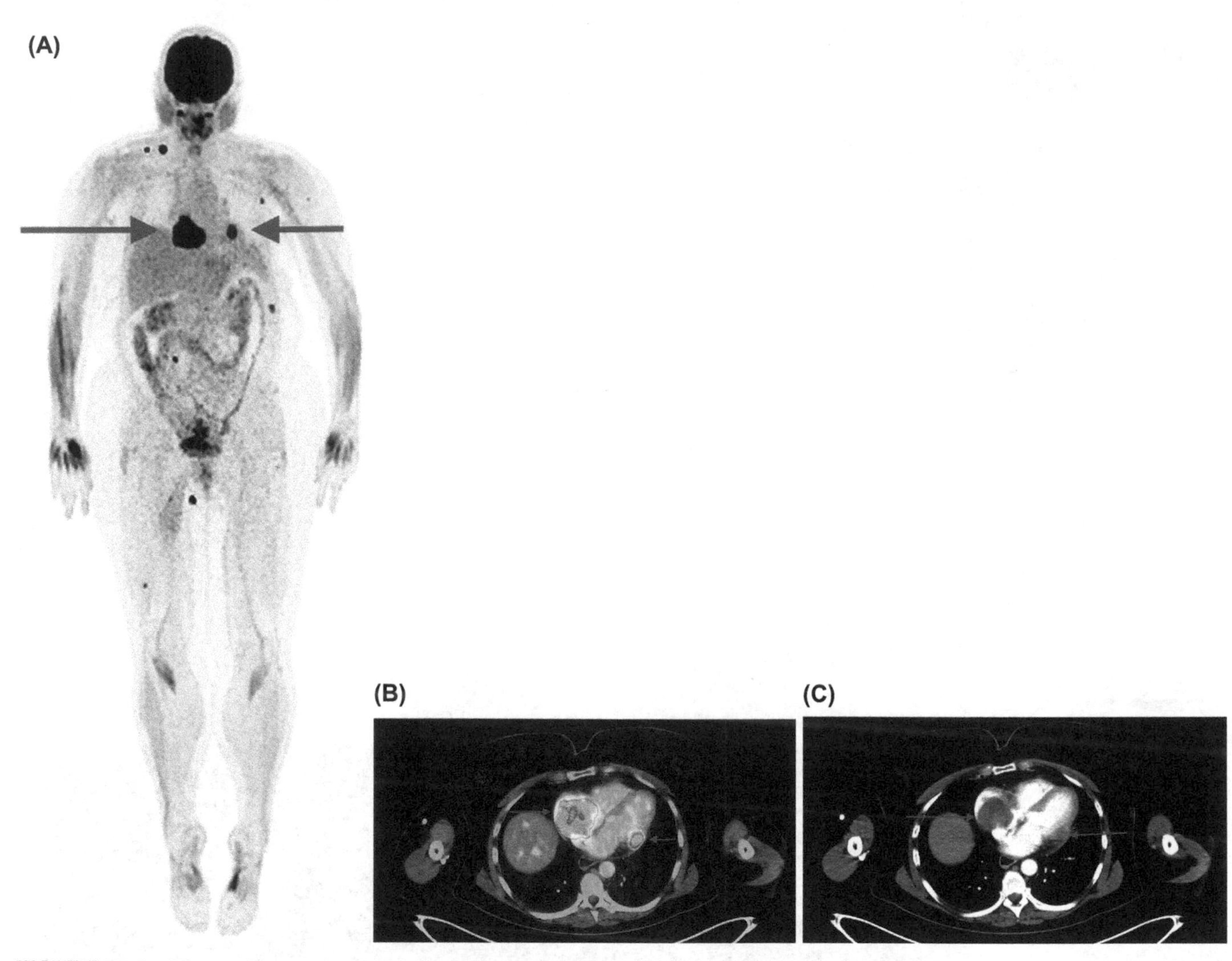

FIGURE 32.5 This is a 74-year-old male with stage IV melanoma who was restaged with positron emission tomography (PET)/computed tomography (CT) after therapy. (A) Multiple metastases are revealed throughout the body, with large central metastases. PET/CT with intravenous contrast for the CT portion (B and C) confirms a large soft tissue metastasis in the right atrium and a small soft tissue metastasis in the lateral wall of the left ventricle localized with *red arrows*.

symptomatic recurrent disease [4], largely because of its impact on clinical management. Camarago examined the clinical impact of PET/CT in melanoma recurrence on 78 patients, noting that imaging results changed the planned clinical management in 27% of [29]. In this study, imaging showed the highest utility in upstaging patients (22%) from stage III to stage IV disease, demonstrating 5% sensitivity for metastatic detection. Even in patients with distant recurrence, imaging results led to a change in the overall management of approximately 30% regarding planned surgery, radiotherapy, or chemotherapy. The major concern in the routine use of PET/CT in restaging is the possibility of false-positive imaging findings that could upstage an otherwise curable patient. False positives have proven most prevalent in patients with clinically suspected local recurrence, which suggests that a higher interpretative threshold may be warranted in this subset [30] (Fig. 32.6).

In addition to initial staging and management of recurrence, the information provided by PET/CT plays an important role in several other aspects of treatment. It has been reported that PET/CT data lead to alteration of the radiation field for up to 60% of patients undergoing radiation [30]. The metabolic activity information provided by PET also allows for noninvasive treatment response, yielding important prognostic information. This has been used in the process of tailoring treatment strategies and aids in the development of potential melanoma therapies [30]. It should be noted that the use of PET/CT in evaluating treatment response and prognostication has not proven cost-effective [30]. Conversely, it is likely cost-effective

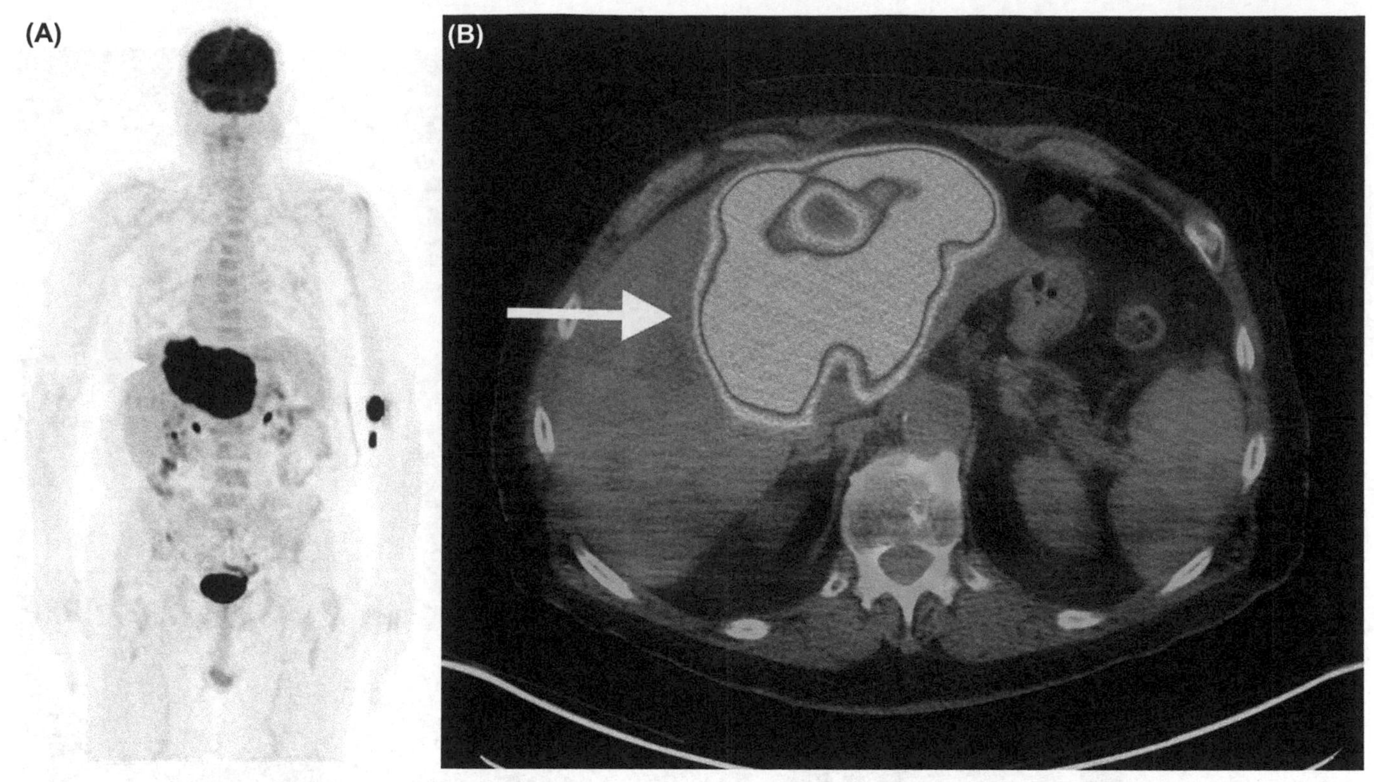

FIGURE 32.6 (A) Patient is 64-year-old male with distant history of melanoma (>10 years) and recent 1-year history of colon cancer. Positron emission tomography (PET)/computed tomography (CT) scan reveals a single large mass within the left lobe of the liver thought to be an aggressive colonic metastasis with central necrosis as demonstrated by the central absence of 2-deoxy-2-(^{18}F)fluoro-D-glucose activity (B). However, the biopsy revealed metastasis from the patient's prior melanoma.

when used for surgical planning, in which it has a proven impact on clinical management.

POSITRON EMISSION TOMOGRAPHY/COMPUTED TOMOGRAPHY TECHNIQUE

With few exceptions a PET/CT exam images a field of view that extends from the skull base to proximal femurs. Because melanoma can occur in sites outside of this field of view, such as the scalp and distal appendicular skeleton, many institutions routinely image the entire body of melanoma patients from head to toe. This ensures a standard protocol for all melanoma patients and may detect metastatic lesions outside of the field of view on a standard PET/CT exam. However, the addition of the skull and distal extremities nearly doubles the PET scanning time, limiting scanner throughput. This also increases whole-body radiation exposure. For these reasons the utility of whole-body PET imaging in melanoma has been the source of several studies. A 2007 study by Niederkohrn examined 296 patients who received whole-body PET imaging and found that no isolated metastases were detected in the skull or peripheral extremities that would affect staging and/or management [31]. The authors concluded that imaging the skull and lower extremities provides little clinical benefit. Querellou corroborated these findings with a study of 122 patients without known involvement of the lower extremities [32]. In their study only 28 patients were found to have metastatic foci, and of these 28 only 5 had abnormal findings in the lower extremities found to be equivocal or suggestive of metastatic involvement [32]. Each of these findings was subsequently found to not represent melanoma upon further clinical follow-up. This considerable false-positive rate and lack of proven benefit argue against routine whole-body imaging; however, many institutions prefer this protocol because it prevents the need for tailored protocols for primary facial/scalp and distal extremity melanoma. We have found in our institution that unless the primary lesion is located within an extremity, the appearance of metastatic lesions to extremities occurs with widely metastatic disease.

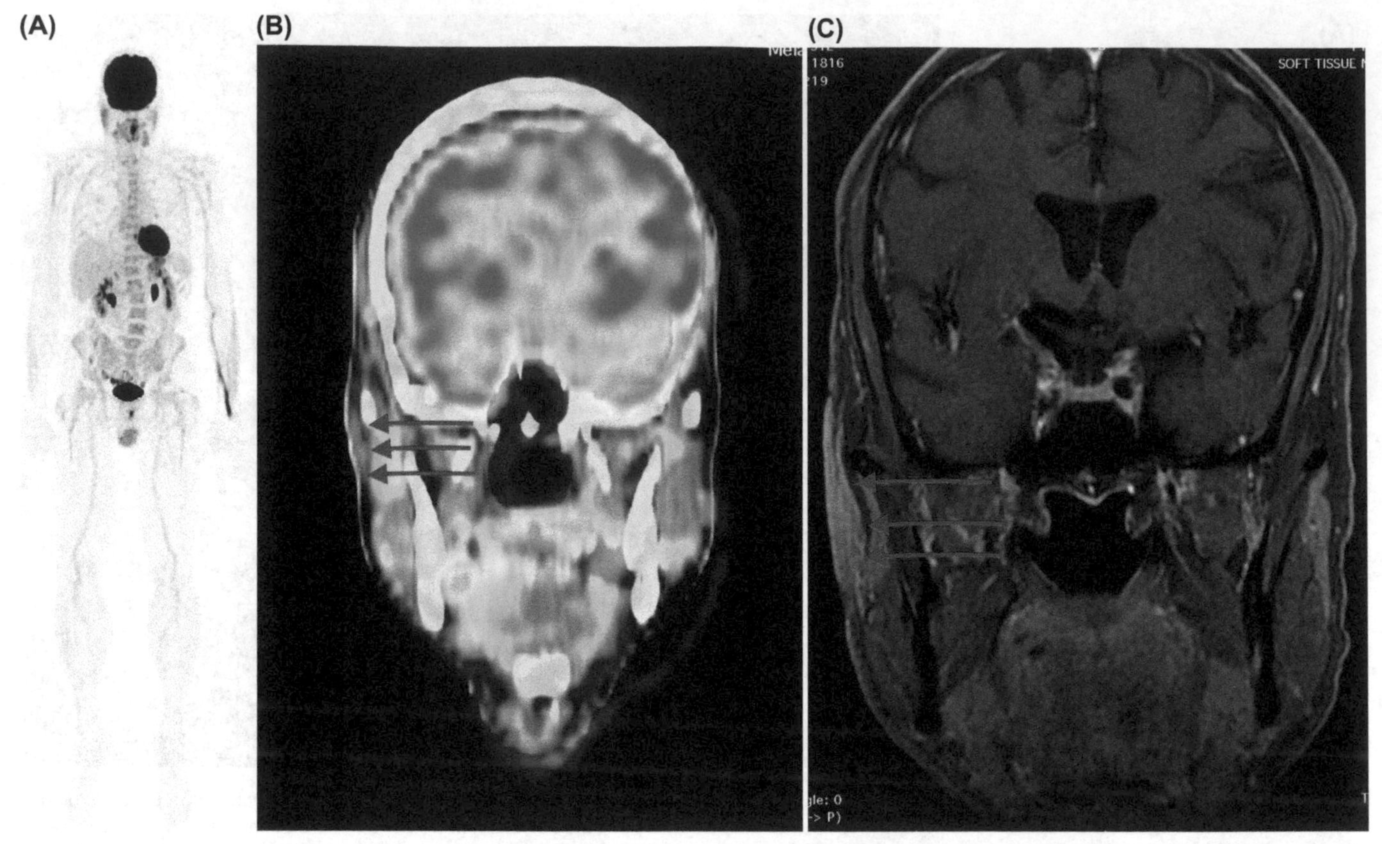

FIGURE 32.7 (A) Patient is an 80-year-old male with a large desmoplastic melanoma of the right side of face with a low mitotic rate. Positron emission tomography (PET)/computed tomography (CT) did not reveal any metastasis and did not demonstrate significant 2-deoxy-2-(^{18}F)fluoro-D-glucose uptake at the site of the primary tumor. (B) Dedicated coronal PET/CT confirmed no significant radiopharmaceutical activity. (C) However, magnetic resonance imaging with contrast revealed the extent of the melanoma, demonstrating contrast enhancement with clearly identifiable margins.

The decision to use IV and oral contrast as a part of the routine PET/CT protocol for melanoma is significant in its effect on the safety profile of the study and exam throughput, although its clinical impact remains unclear. Oral and IV contrast aim to increase the accuracy of the CT component of a PET/CT, but they remain controversial in routine use. In a study of 50 patients, Pfluger showed improved metastatic detection sensitivity of 97% with IV contrast-enhanced CT (CECT) compared with 90% in non-CECT [33]. However, this additional sensitivity of CECT did not change staging or affect clinical management in any cases [33]. The use of oral contrast in PET imaging is perhaps even more controversial. Oral contrast is commonly utilized in PET/CT, but it is reported to cause SUV overestimation up to 125% [33] in some studies. This is likely to be ameliorated by the generally high metabolic activity in macroscopic metastatic disease, and subsequent research suggests that oral contrast introduces a clinically insignificant effect [34]. Because no high-level evidence clarifies appropriate use of oral and IV contrast, it is most prudent to follow institution-specific protocols regarding its use.

Clinicians should also be aware that elevated serum blood glucose and insulin resistance can impair the accuracy of PET/CT. Circulating blood glucose competes with FDG, altering cellular uptake and potentially skewing the apparent metabolic activity of a tumor. Insulin, whether administered exogenously or intrinsically produced by the patient, promotes preferential FDG deposition in insulin-sensitive tissues. Both of these factors may alter PET/CT sensitivity and provide inconsistencies among sequential comparison exams [33]. In addition, patients should not be involved in strenuous activities within 48 h before imaging because the increased muscle uptake will decrease the tumor uptake, potentially obscuring small or lower metabolic lesions. Cold weather can also have an effect by increased metabolic activity of "brown fat," confounding the study. Brown fat activity can be mostly managed by putting patients in a warm environment or with pharmacological agents, such as benzodiazepine.

COMPARISON OF POSITRON EMISSION TOMOGRAPHY/COMPUTED TOMOGRAPHY WITH OTHER IMAGING MODALITIES

Standalone CT has been show to provide decreased diagnostic accuracy relative to combined PET/CT, and it is used to evaluate specific symptoms. In particular, CT is useful for following indeterminate small pulmonary nodules below the 6-mm size threshold for accurate PET characterization. CT also imparts a high false-positive rate. One study examining 347 patients with clinical stage III disease showed that CT identified twice as many false positives as true/false-negative melanoma lesions [29]. However, the CT component of a PET/CT is important in localizing pathology for staging and operative planning. A study by Strobel demonstrated that dedicated imaging review of the co-registered CT images significantly improves accuracy of PET/CT by detecting nonmetabolically active metastasis [35]. This is of particular importance in the evaluation of pulmonary lesions where respiration artifact decreases PET accuracy [35]. CT also allows for improved attenuation correction, helping to better differentiate true pathology from artifact, as discussed in the *Background* section.

In addition to PET/CT, magnetic resonance imaging (MRI) has proven diagnostic utility in certain clinical scenarios. MRI is often used to address specific clinical questions and helps to delineate the anatomical relationship between tumor and surrounding tissue. MRI is also the modality of choice and has documented superiority to PET/CT in evaluating lesions in the liver and brain, where background metabolic activity is a limiting factor [28]. It should be noted that brain MRI is controversial in asymptomatic patients and is not generally used as a screening tool (Fig. 32.7).

CONCLUSION

PET/CT imaging plays an important role in the initial evaluation of high-risk melanoma patients, showing the ability to change management in a significant proportion. Current guidelines stop short of recommending its use in all pathologically node-positive disease. However, PET/CT is recommended for staging patients with clinically apparent nodal or distant metastases, showing high accuracy in this population. It also plays an important role in evaluating symptomatic disease recurrence, response to therapy, and prognostication.

References

[1] Cherry SR, Sorenson JA, Phelps ME. Physics in nuclear medicine. Elsevier Health Sciences; 2012.

[2] Zaidi H. Quantitative analysis in nuclear medicine imaging. Springer Science & Business Media; 2006.

[3] Prichard RS, Hill ADK, Skehan SJ, O'Higgins NJ. Positron emission tomography for staging and management of malignant melanoma. Br J Surg 2002;89:389–96.

[4] Coit DG, Andtbacka R, Anker CJ, Bichakjian CK, Carson 3rd WE, Daud A, et al. Melanoma, version 2.2013 featured updates to the NCCN guidelines. J Natl Compr Canc Netw 2013;11:395–407.

[5] Vazquez V, Silva TB, Vieira Mde A, de Oliveira AT, Lisboa MV, de Andrade DA, et al. Melanoma characteristics in Brazil: demographics, treatment, and survival analysis. BMC Res Notes 2015;8:4.

[6] Kelderman S, Heemskerk B, van Tinteren H, van den Brom RR, Hospers GA, van den Eertwegh AF, et al. Lactate dehydrogenase as a selection criterion for ipilimumab treatment in metastatic melanoma. Cancer Immunol Immunother 2014;63:449–58.

[7] Bonnelykke-Behrndtz ML, Schmidt H, Christensen IJ, Damsgaard TE, Moller HJ, Bastholt L, et al. Prognostic stratification of ulcerated melanoma: not only the extent matters. Am J Clin Pathol 2014;142:845–56.

[8] Gallegos Hernandez JF, Nieweg OE. Cutaneous melanoma (CM): current diagnosis and treatment. Gac Med Mex 2014;150(Suppl. 2): 175–82.

[9] Abbott RA, Acland KM, Harries M, O'Doherty M. The role of positron emission tomography with computed tomography in the follow-up of asymptomatic cutaneous malignant melanoma patients with a high risk of disease recurrence. Melanoma Res 2011;21:446–9.

[10] Torre LA, Bray F, Siegel RL, Ferlay J, Lortet-Tieulent J, Jemal A, et al. Global cancer statistics, 2012. CA Cancer J Clin 2015. http://dx.doi.org/10.3322/caac.21262.

[11] El-Maraghi RH, Kielar AZ. PET vs sentinel lymph node biopsy for staging melanoma: a patient intervention, comparison, outcome analysis. J Am Coll Radiol 2008;5:924–31.

[12] Iagaru A, Quon A, Johnson D, Gambhir SS, McDougall IR. 2-Deoxy-2-[F-18]fluoro-D-glucose positron emission tomography/computed tomography in the management of melanoma. Mol Imaging Biol 2007;9:50–7.

[13] Sabel MS, Wong SL. Review of evidence-based support for pretreatment imaging in melanoma. J Natl Compr Canc Netw 2009; 7:281–9.

[14] Yancovitz M, Finelt N, Warycha MA, Christos PJ, Mazumdar M, Shapiro RL, et al. Role of radiologic imaging at the time of initial diagnosis of stage T1b-T3b melanoma. Cancer 2007;110: 1107–14.

[15] Kell MR, Ridge JA, Joseph N, Sigurdson ER. PET CT imaging in patients undergoing sentinel node biopsy for melanoma. Eur J Surg Oncol 2007;33:911–3.

[16] Singh B, Ezziddin S, Palmedo H, Reinhardt M, Strunk H, Tuting T, et al. Preoperative ^{18}F-FDG-PET/CT imaging and sentinel node biopsy in the detection of regional lymph node metastases in malignant melanoma. Melanoma Res 2008;18:346–52.

[17] Bikhchandani J, Wood J, Richards AT, Smith RB. No benefit in staging fluorodeoxyglucose-positron emission tomography in clinically node-negative head and neck cutaneous melanoma. Head Neck 2013. http://dx.doi.org/10.1002/hed.23456. n/a–n/a.

[18] Crippa F, Leutner M, Belli F, Gallino F, Greco M, Pilotti S, et al. Which kinds of lymph node metastases can FDG PET detect? A clinical study in melanoma. J Nucl Med 2000;41:1491–4.

[19] Barsky M, Cherkassky L, Vezeridis M, Miner TJ. The role of preoperative positron emission tomography/computed tomography (PET/CT) in patients with high-risk melanoma. J Surg Oncol 2014; 109:726–9.

[20] Constantinidou A, Hofman M, O'Doherty M, Acland KM, Healy C, Harries M. Routine positron emission tomography and positron emission tomography/computed tomography in melanoma staging with positive sentinel node biopsy is of limited benefit. Melanoma Res 2008;18:56–60.

[21] Bronstein Y, Ng CS, Rohren E, Ross MI, Lee JE, Cormier J, et al. PET/CT in the management of patients with stage IIIC and IV metastatic melanoma considered candidates for surgery: evaluation of the additive value after conventional imaging. AJR Am J Roentgenol 2012;198:902–8.

[22] Akcali C, Zincirkeser S, Ergajcy Z, Akcalı A, Halac M, Durak G, et al. Detection of metastases in patients with cutaneous melanoma using FDG-PET/CT. J Int Med Res 2007;35:547–53.

[23] Aukema TS, Valdes Olmos RA, Wouters MW, Klop WM, Kroon BB, Vogel WV, et al. Utility of preoperative ^{18}F-FDG PET/CT and brain MRI in melanoma patients with palpable lymph node metastases. Ann Surg Oncol 2010;17:2773–8.

[24] Gulec SA, Faries MD, Lee CC, Kirgan D, Glass C, Morton DL, et al. The role of fluorine-18 deoxyglucose positron emission tomography in the management of patients with metastatic melanoma: impact on surgical decision making. Clin Nucl Med 2003;28: 961–5.

[25]] Bradley Y. PET/CT, an Issue of radiologic clinics of North America. Elsevier Health Sciences; 2013.

[26] Andrews S, Robinson L, Cantor A, DeConti RC. Survival after surgical resection of isolated pulmonary metastases from malignant melanoma. Skin (Depth Unkn) 2006;2:7.

[27] Ho Shon IA, Chung DKV, Saw RPM, Thompson JF. Imaging in cutaneous melanoma. Nucl Med Commun 2008;29:847–76.

[28] Choi EA, Gershenwald JE. Imaging studies in patients with melanoma. Surg Oncol Clin N Am 2007;16:403–30.

[29] Etchebehere EC, Romanato JS, Santos AO, Buzaid AC, Camargo EE. Impact of [F-18] FDG-PET/CT in the restaging and management of patients with malignant melanoma. Nucl Med Commun 2010;31:925–30.

[30] Buck AK, Herrmann K, Stargardt T, Dechow T, Krause BJ, Schreyogg J. Economic evaluation of PET and PET/CT in oncology: evidence and methodologic approaches. J Nucl Med 2010;51:401–12.

[31] Niederkohr RD, Rosenberg J, Shabo G, Quon A. Clinical value of including the head and lower extremities in ^{18}F-FDG PET/CT imaging for patients with malignant melanoma. Nucl Med Commun 2007;28:688–95.

[32] Querellou S. Clinical and therapeutic impact of ^{18}F-FDG PET/CT whole-body acquisition including lower limbs on patients with malignant melanoma. Nucl Med Commun 2011;32:873.

[33] Pfluger T, Melzer HI, Schneider V, La Fougere C, Coppenrath E, Berking C, et al. PET/CT in malignant melanoma: contrast-enhanced CT versus plain low-dose CT. Eur J Nucl Med Mol Imaging 2011;38:822–31.

[34] Gorospe L, Raman S, Echeveste J, Avril N, Herrero Y, Herna Ndez S. Whole-body PET/CT: spectrum of physiological variants, artifacts and interpretative pitfalls in cancer patients. Nucl Med Commun 2005;26:671–87.

[35] Strobel K, Dummer R, Husarik DB, Perez Lago M, Hany TF, Steinert HC. High-risk melanoma: accuracy of FDG PET/CT with added CT morphologic information for detection of metastases. Radiology 2007;244:566–74.

CHAPTER

33

Molecular Imaging of Merkel Cell Carcinoma

V. Beylergil[1], *J.A. Perez*[2], *J.R. Osborne*[2,3]

[1]Metrohealth Campus of Case Western University, Cleveland, OH, United States; [2]Memorial Sloan Kettering Cancer Center, New York, NY, United States; [3]Weill-Cornell Medical College, New York, NY, United States

OUTLINE

INTRODUCTION

Merkel cell carcinoma (MCC) was first described by Cyril Toker in the 1970s [1], and viral carcinogenesis has been suggested in several publications. MCC is an aggressive skin cancer that originates from the neuroendocrine cells of the basal layer of the epidermis and it affects mostly sun-exposed areas, although it can also develop in nonexposed areas [2]. The risk factors and possible viral etiology are beyond the scope of this chapter. The authors will review molecular imaging methods currently available, taking into consideration the underlying neuroendocrine mechanism. Anatomical imaging modalities, such as computed tomography (CT) will be discussed separately in this comprehensive book.

MOLECULAR IMAGING METHODS

Metaiodobenzylguanidine Scintigraphy

Metaiodobenzylguanidine (MIBG) is a radiolabeled analogue of guanethidine that enters the cells via the norepinephrine transporter and is either stored in the cytoplasm or in secretory granules [3]. There are limited data on MIBG scintigraphy in MCC patients; however, a few studies showed MIBG uptake in MCC patients. In an earlier study, Van Moll reported ^{131}I-MIBG uptake in a patient with MCC [4]. ^{131}I-MIBG is not favorable from a dosimetry perspective and is largely replaced by ^{123}I-MIBG; however, the data and experience are scarce, although a few scattered reports suggested successful visualization with this radiopharmaceutical. For example, an eyelid MCC was visualized by Watanabe et al. [5] using ^{123}I-MIBG. The motivating force behind MIBG studies is the possibility of high-dose MIBG treatment if there is unequivocal uptake in MCC lesions; however, given the limited literature, a role for MIBG imaging cannot be advocated at this point. ^{18}F-labeled new investigational agents, such as meta-[(18) F]-fluorobenzylguanidine are being investigated, which would exploit superior resolution of positron emission tomography (PET)/CT. The authors believe that this approach may be worth revisiting if ^{18}F-labeled MIBG derivatives prove successful.

Somatostatin Receptor Scintigraphy

There are a limited number of studies in the literature investigating the role of somatostatin receptor (SSTR)

Imaging in Dermatology
http://dx.doi.org/10.1016/B978-0-12-802838-4.00033-9

scintigraphy in MCC. For example, SSTR expression was high in MCC in a series comprising 105 MCC tissue samples from 98 patients, in which Gardair et al. showed that at least one type of SSTR was expressed in approximately 80% of tumors [6]. ^{111m}In-Octreotide is a radiolabeled somatostatin analogue with the greatest affinity for SSTR2 receptors. Earlier studies using ^{111m}In-Octreotide were promising. For example, Kwekkeboom et al. reported equal or greater sensitivity compared with CT [7]. Guitara-Rovel reported that ^{111m}In-Octreotide was a specific modality for MCC; however, they did not think there was enough evidence to suggest routine use. A few other case reports have also been published, but unfortunately there has been no conclusive evidence that ^{111m}In-Octreotide performed better that conventional imaging. A more recent head-to-head comparison between 2-deoxy-2-(^{18}F)fluoro-D-glucose (FDG) PET/CT and ^{111m}In-Octreotide showed that FDG PET/CT outperformed octreotide scintigraphy in a group of nine patients [8].

The poor performance of ^{111m} In-Octreotide scintigraphy can partially be explained by the limited resolution of SPECT systems. Indeed, PET-based SSTR imaging may perform better. For example, in a group of 24 patients Buder et al. demonstrated that ^{68}GaDOTA-D-Phe1-Tyr3-octreotide or -octreotate performed better and resulted in a change of management in three patients (13%) and upstaged four patients (14%) [9]. Unfortunately, ^{68}Ga-based peptide receptor imaging is not widely available stateside, although significant efforts are underway.

SSTRs represent a potential diagnostic and therapeutic target. Reports of receptor imaging followed by peptide therapy have started to appear in the literature [10]. The authors believe that further studies are needed to investigate this theranostic approach to MCC.

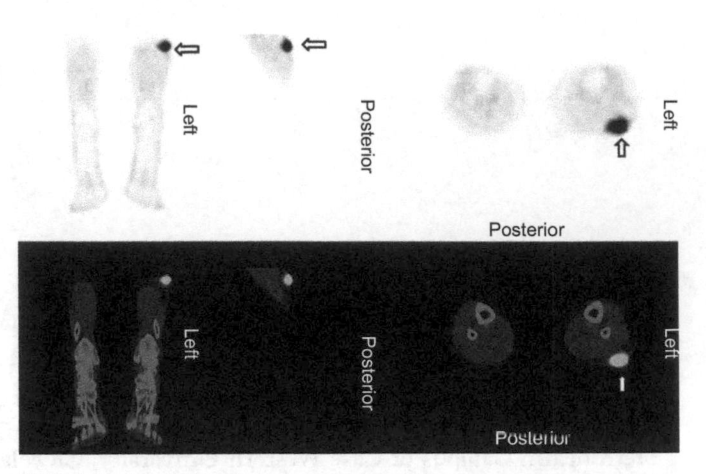

FIGURE 33.1 Transaxial, sagittal, and coronal positron emission tomography (PET) and fused PET/computed tomography images showing a biopsy-proven Merkel cell carcinoma in the left posterior thigh with high 2-deoxy-2-(^{18}F)fluoro-D-glucose avidity. White arrows demonstrate the primary lesion in scalp.

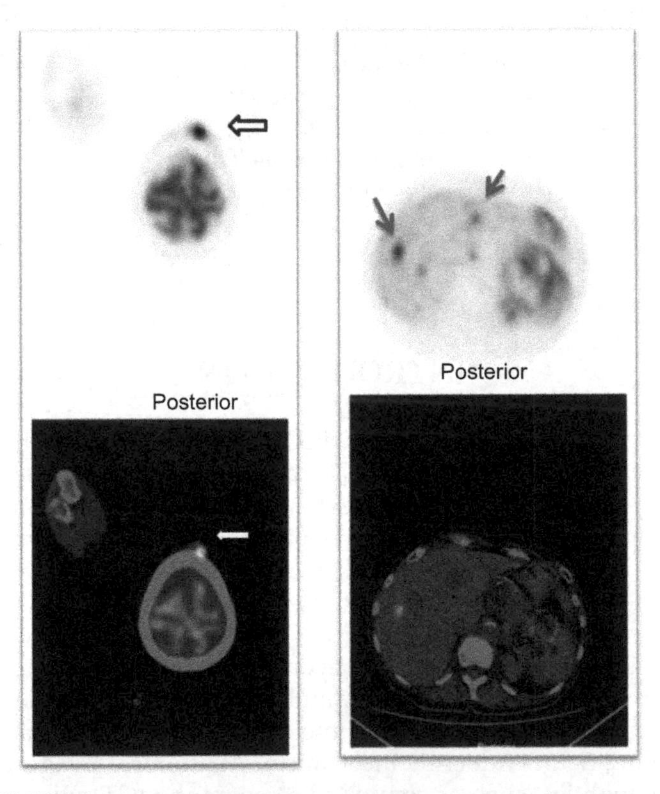

FIGURE 33.2 Transaxial positron emission tomography (PET; *upper row*) and transaxial fused PET/computed tomography images showing a 2-deoxy-2-(^{18}F)fluoro-D-glucose avid primary tumor in the scalp (*white arrows*) and bilobar hepatic metastases (*red arrows*).

2-Deoxy-2-(^{18}F)Fluoro-D-Glucose Positron Emission Tomography/Computed Tomography in Merkel Cell Carcinoma

The body of evidence is much greater for ^{18}F-FDG PET/CT, which makes this modality the current workhorse of molecular imaging in MCC. FDG PET/CT can be performed in initial staging, prognostic stratification, and restaging/evaluation of treatment response (Figs. 33.1 and 33.2).

The reported sensitivity and specificities are very high, with sensitivities ranging from 79% to 94% and specificities greater than 95% [11,12]. A recent meta-analysis including 328 patients from eligible studies showed a sensitivity of 90% and a specificity of 98% [13].

There has been evidence supporting unequivocal change in management in scenarios in which ^{18}F-FDG PET/CT was integrated into clinical decision-making. For example, in a retrospective study including 270 FDG PET/CT studies performed in 97 patients, Hawryluk et al. concluded that ^{18}F-FDG PET/CT resulted in upstaging in 16% of patients [14]. In a prospective study including 102 patients, Siva et al. observed that PET changed the management plan in 37% of patients, which was in parallel with prior studies in smaller groups [15]. They showed that PET staging results differed from conventional staging results in 22% of patients, with PET upstaging in 17% and downstaging in 3% of patients. They reiterated that ^{18}F-FDG PET/CT was important for prognostic stratification in MCC.

Very recently the same Australian group published a study that investigated the utility of PET/CT in the post-therapy setting. They showed that complete metabolic response (CMR) was associated with better overall survival. Patients without a CMR had a 15% 1-year overall survival whereas those with a CMR had an 88% 2-year overall survival. PET/CT had a high or medium impact on restaging in most patients [16].

Sentinel Node Imaging

Another area in which nuclear medicine and molecular imaging play an important role is sentinel node imaging (Fig. 33.3). The sentinel node is the first draining lymph node or group of nodes for a tumoral mass. The use of the sentinel node is not new; however, the role of sentinel node biopsy (SLNB) is not established in MCC, partly because of the low incidence of MCC and the lack of prospective studies. Although the role of SLNB is less established compared with other cancers, such as melanoma and breast cancers, and its impact on overall survival is unclear, the National Comprehensive Cancer Network guidelines still recommend SLNB in all clinically N0 patients because SLNB is an important staging tool for local control. Recently an excellent study comprising a comprehensive review of more than 4000 patients showed that patients with no nodal evaluation and positive nodes had worse survival, which emphasized the role of lymph node evaluation in MCC [17]. Although older than some of the methods in our armamentarium, SLNB is here to stay, and every effort should be made to improve the technique given the availability of new radiopharmaceuticals and possibly combined methods using fluorescence and radioactivity [18].

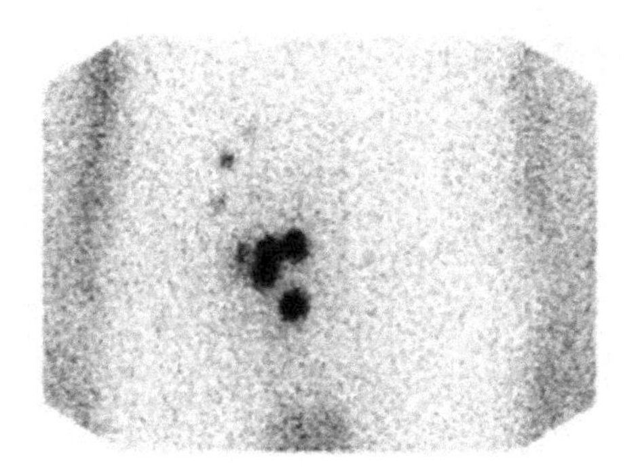

ANT PEL TRANS

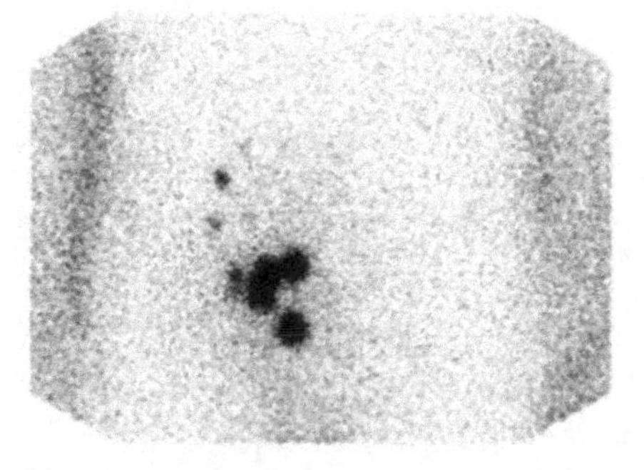

FIGURE 33.3 Sentinel node scintigraphy showing visualization of multiple right groin nodes after injection of radiolabeled colloid around the primary tumor in the right leg (not shown). *ANT PEL TRANS*; anterior pelvis spot image with Co-57 transmission source.

CONCLUSION

The authors believe that 18F-FDG PET/CT is the best available molecular imaging technique at this time [19]. Further studies are needed to elucidate the role of non-FDG PET/CT imaging, such as peptide imaging and peptide-based targeted therapies. Given the SSTR positivity, there is a potential for such theranostic strategies, as shown in several neuroendocrine tumors.

References

[1] Toker C. Trabecular carcinoma of the skin. Arch Dermatol 1972; 105(1):107–10.
[2] Marcoval J, Ferreres JR, Penin RM, Perez D, Vinals JM. Merkel cell carcinoma: differences between sun-exposed and non-sun-exposed variants–a clinical analysis of 36 cases. Dermatology (Basel, Switzerland) 2014;229(3):205–9.
[3] Carrasquillo JA, Chen CC. Molecular imaging of neuroendocrine tumors. Semin Oncol 2010;37(6):662–79.
[4] Von Moll L, McEwan AJ, Shapiro B, Sisson JC, Gross MD, Lloyd R, et al. Iodine-131 MIBG scintigraphy of neuroendocrine tumors other than pheochromocytoma and neuroblastoma. J Nucl Med Off Publ Soc Nucl Med 1987;28(6):979–88.
[5] Watanabe N, Shimizu M, Kageyama M, Kitagawa K, Hayasaka S, Seto H. 123I-MIBG SPECT of Merkel cell carcinoma. Br J Radiol 1998;71(848):886–7.
[6] Gardair C, Samimi M, Touze A, Coursaget P, Lorette G, Caille A, et al. Somatostatin receptors 2A and 5 are expressed in merkel cell carcinoma with No association with disease severity. Neuroendocrinology 2015;101(3):223–35.
[7] Kwekkeboom DJ, Hoff AM, Lamberts SW, Oei HY, Krenning EP. Somatostatin analogue scintigraphy. A simple and sensitive method for the in vivo visualization of Merkel cell tumors and their metastases. Arch Dermatol 1992;128(6):818–21.
[8] Lu Y, Fleming SE, Fields RC, Coit DG, Carrasquillo JA. Comparison of 18F-FDG PET/CT and 111In pentetreotide scan for detection of Merkel cell carcinoma. Clin Nucl Med 2012;37(8):759–62.

[9] Buder K, Lapa C, Kreissl MC, Schirbel A, Herrmann K, Schnack A, et al. Somatostatin receptor expression in Merkel cell carcinoma as target for molecular imaging. BMC cancer 2014;14:268.

[10] Schmidt MC, Uhrhan K, Markiefka B, Hasselbring L, Schlaak M, Cremer B, et al. (68)Ga-DotaTATE PET-CT followed by peptide receptor radiotherapy in combination with capecitabine in two patients with merkel cell carcinoma. Int J Clin Exp Med 2012;5(4): 363–6.

[11] Belhocine T, Pierard GE, Fruhling J, Letesson G, Bolle S, Hustinx R, et al. Clinical added-value of 18FDG PET in neuroendocrine-merkel cell carcinoma. Oncol Rep 2006;16(2): 347–52.

[12] Peloschek P, Novotny C, Mueller-Mang C, Weber M, Sailer J, Dawid M, et al. Diagnostic imaging in Merkel cell carcinoma: lessons to learn from 16 cases with correlation of sonography, CT, MRI and PET. Eur J Radiol 2010;73(2):317–23.

[13] Treglia G, Kakhki VR, Giovanella L, Sadeghi R. Diagnostic performance of fluorine-18-fluorodeoxyglucose positron emission tomography in patients with Merkel cell carcinoma: a systematic review and meta-analysis. Am J Clin Dermatol 2013;14(6):437–47.

[14] Hawryluk EB, O'Regan KN, Sheehy N, Guo Y, Dorosario A, Sakellis CG, et al. Positron emission tomography/computed tomography imaging in Merkel cell carcinoma: a study of 270 scans in 97 patients at the Dana-Farber/Brigham and Women's Cancer Center. J Am Acad Dermatol 2013;68(4):592–9.

[15] Siva S, Byrne K, Seel M, Bressel M, Jacobs D, Callahan J, et al. 18F-FDG PET provides high-impact and powerful prognostic stratification in the staging of Merkel cell carcinoma: a 15-year institutional experience. J Nucl Med Off Publ Soc Nucl Med 2013; 54(8):1223–9.

[16] Byrne K, Siva S, Chait L, Callahan J, Bressel M, Seel M, et al. 15-Year experience of 18F-FDG PET imaging in response assessment and re-staging after definitive treatment of merkel cell carcinoma. J Nucl Med Off Publ Soc Nucl Med 2015;56(9):1328–33.

[17] Tseng J, Dhungel B, Mills JK, Diggs BS, Weerasinghe R, Fortino J, et al. Merkel cell carcinoma: what makes a difference? Am J Surg 2015;209(2):342–6.

[18] Stoffels I, Leyh J, Poppel T, Schadendorf D, Klode J. Evaluation of a radioactive and fluorescent hybrid tracer for sentinel lymph node biopsy in head and neck malignancies: prospective randomized clinical trial to compare ICG-Tc-nanocolloid hybrid tracer versus Tc-nanocolloid. Eur J Nucl Med Mol Imaging 2015;42(11): 1631–8.

[19] Beylergil V, Carrasquillo JA. Molecular imaging and therapy of merkel cell carcinoma. Cancers 2014;6(2):1020–30.

CHAPTER

34

Imaging in Merkel Cell Carcinoma

L. Kadletz, S.F. Nemec, B.M. Erovic

Medical University of Vienna, Vienna, Austria

OUTLINE

INTRODUCTION

Historical Background and Basic Findings

Dr. Cyril Toker first described Merkel cell carcinoma in 1972 and characterized the tumor as a trabecular carcinoma of the skin [1–3]. In 1978, he detected dense neurosecretory granules via electron microscopy, which subsequently led to the suggestion that the trabecular carcinoma might have arisen from the neural crest [3,4].

Although there is still an ongoing debate about the cellular origin of the Merkel cell carcinoma, electron microscopy has shown that carcinoma cells that showed histological and immunohistochemical features similar to those of Merkel cells, and which are located in the basal epidermis, serve as mechanoreceptors of neuroendocrine differentiation [2]. In contrast, other studies hypothesized that Merkel cell carcinoma cells derive from pluripotent epidermal stem cells [5].

In 2008, Feng and coworkers found a polyomavirus in specimens obtained from Merkel cell carcinoma patients. This new polyomavirus was termed Merkel cell polyomavirus and has been found to play a key role in carcinogenesis and disease progression [5–8].

Clinical Appearance and Management of the Disease

With an incidence rate of 0.4 per 100,000 inhabitants in the United States, Merkel cell carcinoma is very rare compared with other skin malignancies [7–9]. At the

Imaging in Dermatology
http://dx.doi.org/10.1016/B978-0-12-802838-4.00034-0

time of diagnosis, the mean age of patients is 75 years and immunosuppression seems to have a major impact on incidence and on patient survival [2,4,10,11].

Clinically, Merkel cell carcinoma lesions appear as a reddish, shiny, and firm nodules, sometimes with teleangiectatic and seldom-ulcerous lesions in sun-exposed areas of the skin, particularly of the extremities and head and neck [7,12]. This rapidly growing and painless tumor is associated with immunosuppression (ie, human immunodeficiency virus infection or lymphoma) [8,13].

Merkel cell carcinoma has a high potential for lymphangitic metastatic spread. Recent data have shown that, at the time of initial diagnosis, 25–60% of all patients present with a palpable lymphadenopathy and 5–30% have distant metastasis to liver, bone, lung, or brain [8,14].

Yiengpruksawan and coworkers developed a staging system for Merkel cell carcinoma that represents local disease as stage I, patients with lymph node metastases as stage II, and patients with metastatic disease as stage III, respectively [15]. Moreover, staging is based on the new American Joint Committee on Cancer staging system for Merkel cell carcinoma, linked to the 5-year survival rate, and which subsequently determines the therapeutic strategy for each patient [8,16].

Surgical resection with clear margins, with or without adjuvant radiotherapy, is the treatment of choice in stage I and II disease. In particular, wide surgical margins, more than 2 cm, except on the face, are preferable [17]. Mohs surgery with close surgical margins seems to be contrary to the biology of Merkel cell carcinoma, which usually forms satellite metastases [8]. The role of sentinel lymph node biopsy is controversial, but most recent studies emphasize the value of sentinel lymph node biopsy in patients with Merkel cell carcinoma [6,18,19].

Radiotherapy, as an adjuvant treatment, is a therapeutic cornerstone of Merkel cell carcinoma treatment because of its considerably high radiosensitivity [20–22]. Recently published studies have shown that the use of adjuvant radiation therapy significantly reduces the locoregional recurrence rate, and, subsequently, has a high impact on disease-free and overall patient survival [23]. However, there is very little literature about studies of Merkel cell carcinoma treated primarily with radiation [31,32]. Primary radiotherapy is a treatment option reserved for patients who cannot tolerate wide surgical excision or for cases in which the primary tumor is not resectable [20]. Patients who are diagnosed with advanced, disseminated disease should receive chemotherapy [2,7,10].

Because Merkel cell carcinomas belong to the category of small-cell neuroendocrine tumors, chemotherapy of Merkel cell carcinoma patients is comparable to the therapeutic regimens of small-cell lung cancer. Specifically, combinations of carboplatin, cisplatin, and etoposide, or cyclophosphamide with vincristine, doxorubicin, prednisone, and bleomycin or 5-fluorouracil, are administered in advanced-stage disease [10].

Before therapy can be initiated, staging of the malignant disease is of crucial importance. In addition to clinical investigation, biopsy, and histological workup, imaging is a key step in determining the localization, size, and extent of the tumor. In general, to characterize locoregional disease, ultrasonography, computed tomography (CT), and magnetic resonance imaging (MRI) are used whereas positron emission tomography (PET) and positron emission computed tomography (PET-CT) are used to exclude distant metastatic disease. Lymphoscintigraphy and somatostatin receptor scintigraphy are used specifically for Merkel cell carcinoma patients to define locoregional lymph node metastasis [11,24,25]. However, because of the sparse incidence of Merkel cell carcinoma, there are still no standardized diagnostic or imaging algorithms as yet [24,26,27].

ULTRASONOGRAPHY

Background and Technical Aspects

Since the introduction of ultrasonography in dermatology by Harold Alexander in the late 1970s, this tool has become part of an important armamentarium in the diagnosis of dermatological diseases [28]. Ultrasonography is a ubiquitously available, noninvasive, and cost-effective imaging modality to quantify and qualify structures of skin pathologies. Recently published studies have shown that the determination of tumor extension, its invasion from the epidermis into the subdermal space, and vascularization can accurately be determined with ultrasound [29–33]. Indeed, ultrasonography is a key imaging tool for defining locoregional dissemination of the primary malignant disease, particularly lymph node metastasis of the neck, axilla, and abdomen.

To properly examine the epidermis, a high-frequency echography system is required, with linear probes and frequencies ranging between 20 and 25 MHz [31,34], whereas subdermal lesions and lymph nodes are examined with probes from 5 to 12 MHz [35]. To gain additional information about the vascularity in primary tumors and lymph nodes, Doppler ultrasonography is helpful. In particular, Doppler sonography, with frequencies between 7 and 14 MHz, determines the intensity (hyper- vs hypovascularization) and localization (peri- or intratumoral) of the vascularization of unclear lesions, which helps to stratify these lesions into malignant or benign formations.

First, a gray-scale examination to visualize the structure and internal architecture is performed, followed by a color Doppler. Color Doppler usually allows discrimination of arterial versus venous blood flow by displaying these flows as red or blue, respectively. Power Doppler sonography has a three times higher sensitivity compared with conventional color Doppler, enabling the examination of low-velocity blood flow in superficial lesions that consist of small vessels [31,36,37] Although no recent literature on Merkel cell carcinoma and power Doppler is available, nonmetastatic lymph nodes larger than 5 mm appear either avascular or with hilar vascularity [38,39]. However, most importantly, power Doppler is very useful in detecting hilar and/or peripheral vascularity in metastatic lymph nodes [38,39].

Ultrasound elastography, another noninvasive technique, is based on the fact that stiff tissue deforms less and shows less strain compared with healthy tissue. In general, metastatic lymph nodes demonstrate greater stiffness than benign lymph nodes; however, there are still no data about the use of this technique in Merkel cell carcinoma [40].

Clinical Application and Appearance

The primary goal of ultrasonography in patients with Merkel cell carcinoma is to obtain information about whether locoregional lymph node metastasis is present. Stage I is defined by primary disease without lymph node involvement, whereas stage II includes lymph node involvement but without distant metastases, and stage III includes distant metastases. Ultrasound examination might also help to determine the extension of the primary tumor into adjacent tissue, in combination with CT and MRI [31,34].

Merkel cell carcinoma initially appears as a single or multicentric, hypoechoic, irregular lesion with indistinct margins arising from the dermis [41]. The tumor is solid, without any signs of calcification, but with a significant hypervascularity of the adjacent tissue [24,42]. Although evaluation of tumor extension to the subcutis seems to be difficult because of the hypoechoic signal of the subcutis itself, it has been shown that Merkel cell carcinoma characteristically develops finger-like extensions to the subcutis and can develop subcutaneous local metastases [24,33,42]. It is interesting to note that muscular and fascial infiltration of Merkel cell carcinoma has been found only on MRI and CT imaging [26]. To avoid the pitfall of not recognizing non-Merkel cell carcinomatous lesions, such as malignant melanoma, cutaneous squamous cell carcinoma, and basal cell carcinoma, knowledge about their characteristic clinical appearance, as well as their ultrasound features, is critical. In fact, malignant melanoma appears as a hypoechoic mass with a homogeneous structure and horizontal spreading in early-stage disease, whereas Merkel cell carcinoma shows a vertical extension more like that of melanoma [33,41,43,44].

Cutaneous squamous cell carcinoma is difficult to distinguish from Merkel cell carcinoma because its sonographic appearance as a hypoechoic formation is similar to Merkel cell carcinoma. However, both lesions tend to penetrate more into the subcutaneous tissue and dermis compared with malignant melanoma [31,34,45].

Basal cell carcinoma shows multiple hyperechoic spots within the tumor [46,47], making the ultrasound feature of this lesion unique compared with Merkel cell carcinoma, malignant melanoma, and cutaneous squamous cell carcinoma. It is interesting to note that only one case report by Baek and colleagues described exactly the same ultrasonographic features in an axillary Merkel cell carcinoma mass as in a basal cell carcinoma [41]. However, that report is the only publication to describe this feature in Merkel cell carcinoma because other studies could not confirm this observation [24,42]. To complete the list of differential diagnosis of cutaneous benign lesions, epidermal cysts, dermatofibromas, or intradermal nevi appear as cystic or solid masses, encapsulated, with well-defined margins and with peritumoral vascularity on color Doppler ultrasonography [48–50].

Metastatic lymph nodes of Merkel cell carcinoma patients show the same sonomorphologic features as described for other malignancies [11,51–53]. Although nodal size, reflected by the minimum transverse diameter, strongly depends on localization, specifically, 15 mm for submental, 8 mm for retropharyngeal, 10 mm for mediastinal, and 5 mm for mesenteric lymph nodes, nodes are hypoechoic or heterogeneous in appearance with an irregular border and spherical shape, with an increased transverse diameter, increased intra- and/or peritumoral hypervascularization, and eccentric or concentric cortical thickening. The lymph nodes have no fatty hilum, and, in most cases, a central necrosis is detectable (Fig. 34.1) [38,44,54]. On power Doppler examination, 90% of benign nodes with a diameter of 5 mm and greater show hilar vascularity, or, in some cases, are even avascular. In contrast, because of neoangiogenesis and recruitment of peripheral vessels, metastatic lymph nodes appear with peripheral and mixed vascularity [38,39].

With regard to ultrasound features of distant metastasis in Merkel cell carcinoma, clinical studies are few as yet. Only case reports have been published describing multilobulated tumor masses within the heart after echocardiographic examination [51,55].

Another very important clinical aspect of ultrasonography is its use as guidance during fine needle aspiration. An ultrasound-guided biopsy may be done in

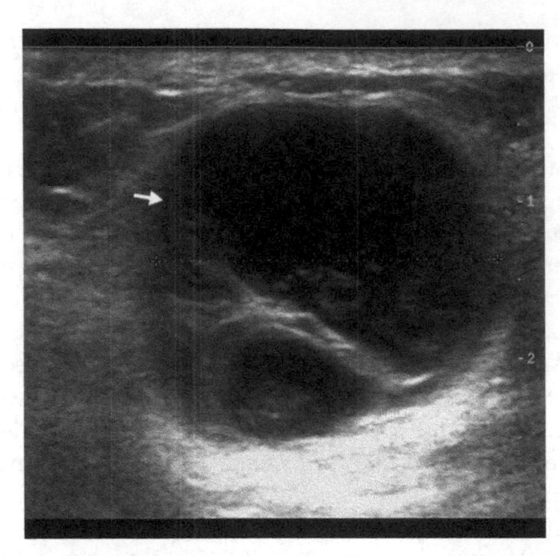

FIGURE 34.1 Ultrasonography of an inguinal lymph node in a 68-year-old male patient: a round, ill-defined (*arrow*) and significantly enlarged nodule appears as a hypoechoic lesion with central necrosis.

patients with nonpalpable lesions to evaluate abnormal findings on ultrasonography, MRI, or CT imaging. In patients with Merkel cell carcinoma, most fine-needle aspiration cytologies are obtained from cervical lymph nodes (43.5%), followed by the axilla (20.3%) and inguinal region (11.6%) [8,9,11,13,16,56]. There is only one case report of an endoscopic ultrasound-guided fine-needle aspiration of the pancreas, where a metastatic Merkel cell carcinoma lesion was diagnosed [57].

Merkel cell carcinoma aspirates show characteristically small, round tumor cells, with no or little cytoplasm, and nucleoli with a moderate anisokaryosis. However, those tumors cells are, in some cases, difficult to differentiate from melanoma or lymphoma cells [58].

Clinical Data

Because lymphoangetic spread is a frequent event in patients with this rare skin tumor, ultrasonography is a key imaging tool at the initial presentation of patients for locoregional nodal staging workup [24,59]. Because Merkel cell carcinoma tends to occur 80% of the time on the arms and the head/neck region, the first lymph node involved in metastatic disease is usually located in the axilla or the neck [8]. Clinically palpable lymphadenopathy is found in 25% of all patients, and those lymph nodes are, in general, accessible to sonographic examination [8,27,60]. With regard to the sensitivity and specificity of the ultrasound examination for the differentiation between benign and malignant lymph nodes in squamous cell carcinoma patients, recent literature shows that the sensitivity of ultrasound for submental and submandibular lymph nodes and upper jugular nodes is approximately 80%, whereas its specificity ranges between 80% and 90%. It is interesting to note that sonography of metastatic nodes located right at the jugular bulb has a significantly lower sensitivity of only 38%, with a specificity of 89% [11,24,42,53].

Considering the post-treatment follow-up of patients with Merkel cell carcinoma, no guidelines are available about the time intervals for ultrasound examination. At our institution, ultrasonography of the regional lymph nodes is obtained every 6 months within the first 2 years and then once every year.

Take-Home Message

Lymph node staging by ultrasonography in patients with Merkel cell carcinoma is of the utmost importance. Because of its wide availability, considerable cost-effectiveness, and high accuracy, local, regional dissemination of Merkel cell carcinoma, based on nodal involvement, can easily, rapidly, and accurately be diagnosed. Sonography, particularly of the neck and axilla, should be done in all patients at the first presentation. Even when no suspicious lymph nodes are detected, CT and MRI are mandatory to further exclude locoregional and distant metastatic disease. In patients with early- and advanced-stage Merkel cell carcinoma and highly suspicious lymph nodes, full body scanning with CT and MRI should also be performed.

Again, lymph nodes in patients with Merkel cell carcinoma are hypoechoic or heterogeneous in appearance, with an irregular border, a spherical shape with an increased transverse diameter, increased intra- and/or peritumoral hypervascularization, and eccentric or concentric cortical thickening. The nodes have no fatty hilum, but they usually show a central necrosis. The primary tumor appears as a single or multicentric, hypoechoic, solid, irregular lesion with irregular margins arising from the dermis. Moreover, Merkel cell carcinoma characteristically develops finger-like extensions to the subcutis and is able to develop subcutaneous local metastases.

Ultrasonography of the lymph nodes and CT and MRI each have advantages. Ultrasonography provides dynamic sequences, but the diagnosis strongly depends on the expertise of the examiner, whereas CT and magnetic resonance images are static and reflect, in addition to the lymph nodes, the whole scanned region of the body and can be presented for discussion to other colleagues and at tumor boards. Ultrasonography has no known side effects and can be repeated without limit for diagnosis and routine follow-ups, whereas CT is associated with a considerable radiation exposure for the patient. Although ultrasonography has a limited application in some patients with regard to tissue penetration and accessibility (ie, obesity), this imaging modality is always an excellent alternative for patients with known contrast agent allergy, chronic renal failure, and hyperthyroidism.

COMPUTED TOMOGRAPHY

Background and Technical Aspects

In the 1990s, the first studies to analyze CT in Merkel cell carcinoma patients were published [61,62]. Currently, contrast-enhanced CT scanning of the neck, thorax, and abdomen is key for tumor staging by (1) determining localization and extension; (2) determining whether infiltration by the primary lesion is present; and finally, and most importantly (3) detecting lymph node metastasis and/or distant dissemination [27,63].

Because Merkel cell carcinoma is a highly aggressive and rapidly metastasizing skin carcinoma, ultrasonography, followed by CT imaging of the primary tumor and the regional lymph nodes, should be performed in early- and in advanced-stage disease at the initial presentation. However, it should be emphasized that MRI is significantly superior to ultrasonography and CT scanning when determining whether there is soft tissue involvement in a Merkel cell carcinoma.

Clinical Application and Appearance

Although the diagnosis of Merkel cell carcinoma is always determined by histological workup and cytokeratin-20 positivity in immunohistochemistry, CT can provide an accurate insight into the extension of the tumor with potential infiltration of adjacent tissue and lymph node metastases.

Because of the highly aggressive biological behavior of the tumor—even in early-stage disease—CT imaging is mandatory in all patients. Rarely, when cytokeratin-20 expression is low or even absent in a specimen, thoracoabdominal CT imaging, in combination with bronchoscopy and coloscopy, has to be performed to exclude a neuroendocrine primary tumor of the lungs, colon, or pancreas [26,27,61].

The primary Merkel cell carcinoma lesion appears mostly as a round nodule extending from the dermis into the subdermal tissue specifically, on contrast-enhanced CT scans, the cutaneous mass is a delineated, isoattenuated to slightly hyperattenuated area compared with the adjacent muscles (Fig. 34.2A)

In some patients, subcutaneous linear or reticular fat-stranding, with lymphedema close to the primary tumor, is observed (Fig. 34.3A). This phenomenon can be associated with a lymphangitic invasion. It is rare that the tumor is very close or even adjacent to bone. In that case, local osteolytic destruction zones are common [64].

In up to 30% of patients, Merkel cell carcinoma develops subcutaneous satellite lesions, which, in most cases, are difficult to see or palpate. Again, CT imaging can help to determine the number of lesions, the level of extension, and whether infiltration of adjacent tissue is present. These satellite lesions are homogeneously enhanced, either with smooth edges or in the form of lobules ranging from 20 to 70 mm in size [24,27,42].

With regard to the differential diagnosis of primary skin lesions, basal cell carcinoma, which is, by far, the most common skin malignancy [66], manifests itself as an iso- or hypoattenuated tumor relative to the adjacent muscles [67]. On the other hand, Merkel cell carcinoma appears as an iso- to slightly hyperdense mass on contrast-enhanced CT images [24,27,42]. However, in the clinical routine, CT imaging will be performed only in patients with an advanced-stage basal cell carcinoma to determine the extent of the disease and possible infiltration of adjacent tissue. Malignant melanoma and cutaneous squamous cell carcinoma show imaging features similar to those of Merkel cell carcinoma. Both tend to appear as iso- to hyperdense lesions on contrast-enhanced CT scans. However, the only significant difference is the presence of satellite nodules in the adjacent subcutaneous fat in Merkel cell carcinoma patients, which are significantly less common in patients with squamous cell carcinoma or malignant melanoma [27]. In addition, cutaneous metastases from small-cell lung cancer and malignant small-cell lymphoma are very rarely, but important to differentiate from Merkel cell carcinoma into consideration for differential diagnosis. Both may show a similiar appearance on CT scans [42]. In contrast to all malignant masses, benign skin lesions, such as epidermal inclusion cysts or neurofibroma, appear as well-defined hypodense lesions with minimal or no contrast enhancement.

Because the 5-year survival rate for Merkel cell carcinoma patients with and without lymphatic dissemination is 58% and 80%, respectively, assessment of lymph node metastasis by CT imaging is of the utmost importance. Increased number and particularly nodal size may indicate lymph node involvement. The measurement is based mainly on the maximum transverse diameter or ratios of maximum longitudinal to maximum transverse diameter [68,69] (Fig. 34.3B and C). Further criteria for malignant lymphadenopathy are nodes with central necrosis (Fig. 34.3B). However, it should always be kept in mind that fatty replacement and nodal abscess may mimic metastatic lymph nodes. Therefore, the use of T1-weighted spin-echo sequences without contrast and fat saturation and fat-saturated contrast-enhanced images are critical for differentiating between malignant and nonmalignant lesions [68]. Finally, infiltration of adjacent fat planes, as an indicator of

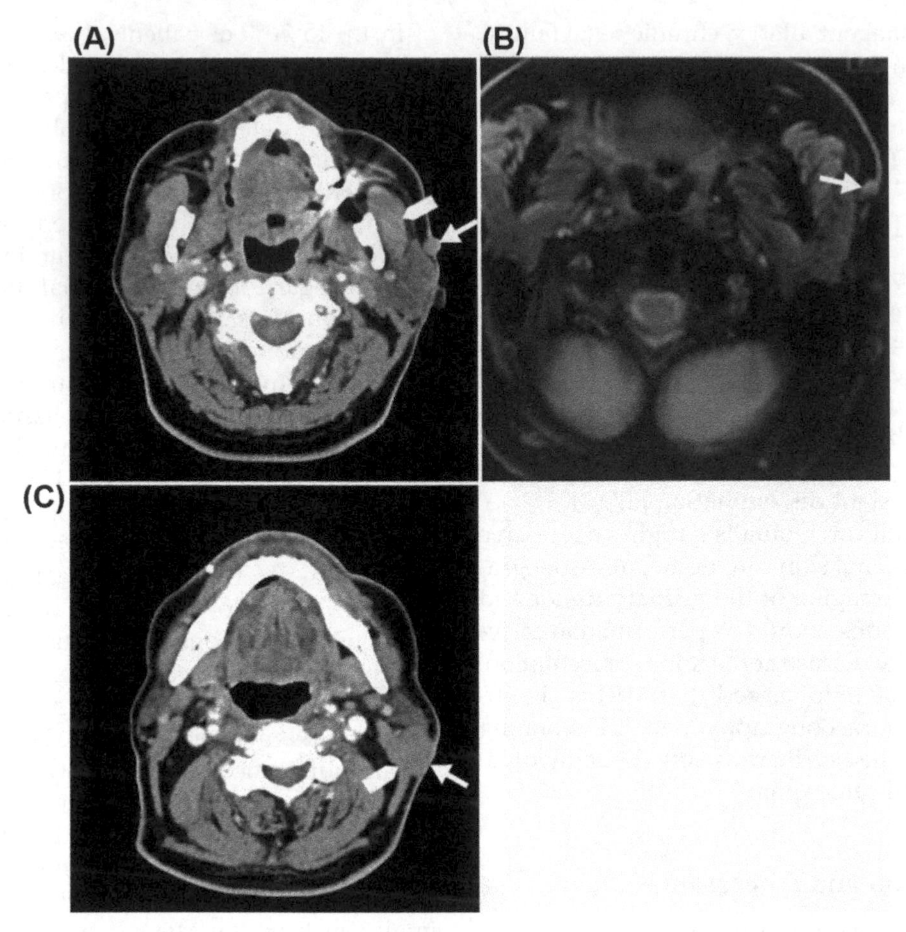

FIGURE 34.2 (A) An 84-year-old male patient with a primary Merkel cell carcinoma of the left neck. The lesion appears as a round nodule (*arrow*) extending from the dermis into the subcutaneous fat. The cutaneous mass is well demarcated and isoattenuated to slightly enhanced compared with the adjacent masseter muscle (*block arrow*) on contrast-enhanced computed tomography (CT). (B) Contrast-enhanced magnetic resonance imaging with fat saturation shows the same primary tumor (*arrow*) as in panel a. Analogous to the CT imaging, the round primary tumor is well defined and is not infiltrating the adjacent musculature. (C) The same patient demonstrates a significantly enlarged metastatic lymph node (*arrow*) on the left side, more caudal to the primary tumor. The node is ill defined and isoattenuated compared with the adjacent sternocleidomastoid and trapezius muscles. Skin invasion, as well as a deep infiltration of the subdermal tissue and both muscles (*block arrow*), reflects extracapsular spread.

extracapsular spread, is a definitive indicator of lymph node invasion [69,70] (Fig. 34.2C).

Another use of CT, and its most important application, is the evaluation of distant metastasis. In approximately 5% of all patients, metastatic spread has already occurred at the time of initial diagnosis. Cutaneous, pulmonary, hepatic, osseous, urogenital, or, less frequently, brain metastasis may occur in Merkel cell carcinoma (Fig. 34.4).

Visceral lesions of Merkel cell carcinoma tend to mimic other neuroendocrine metastatic lesions, such as small-cell lung cancer, and appear with a typical ring-like enhancement. Of note, liver lesions from Merkel cell carcinoma patients typically have a target appearance or a uniform low attenuation after contrast application [11,24,61].

Clinical Data

The most recent study performed by Colgan and colleagues included 69 patients with Merkel cell carcinoma, in which the specificity and sensitivity of CT imaging for regional lymph node involvement were analyzed. Thirty-two of 69 patients had histologically confirmed positive lymph nodes whereas 37 patients had no metastatic lymph node disease. In 15 of those 32 patients, CT also detected lymph node metastasis, resulting in a sensitivity of only 47%. However, in 36 of 37 cases CT imaging confirmed negative lymph node involvement, resulting in a specificity of 97%. The positive and negative predictive value of the CT imaging was reported as 94% and 68%, respectively. When detecting pathologic lymph nodes, F-18 fluorodeoxyglucose PET-CT scanning showed a sensitivity of 83% and specificity of 95%, resulting in a positive and negative predictive value of 91% [65]. Peloschek and coworkers used CT, MRI, and ultrasonography to assess positive lymph node basins. The sensitivity and specificity after combining all three imaging modalities was 96% and 89%, respectively [71].

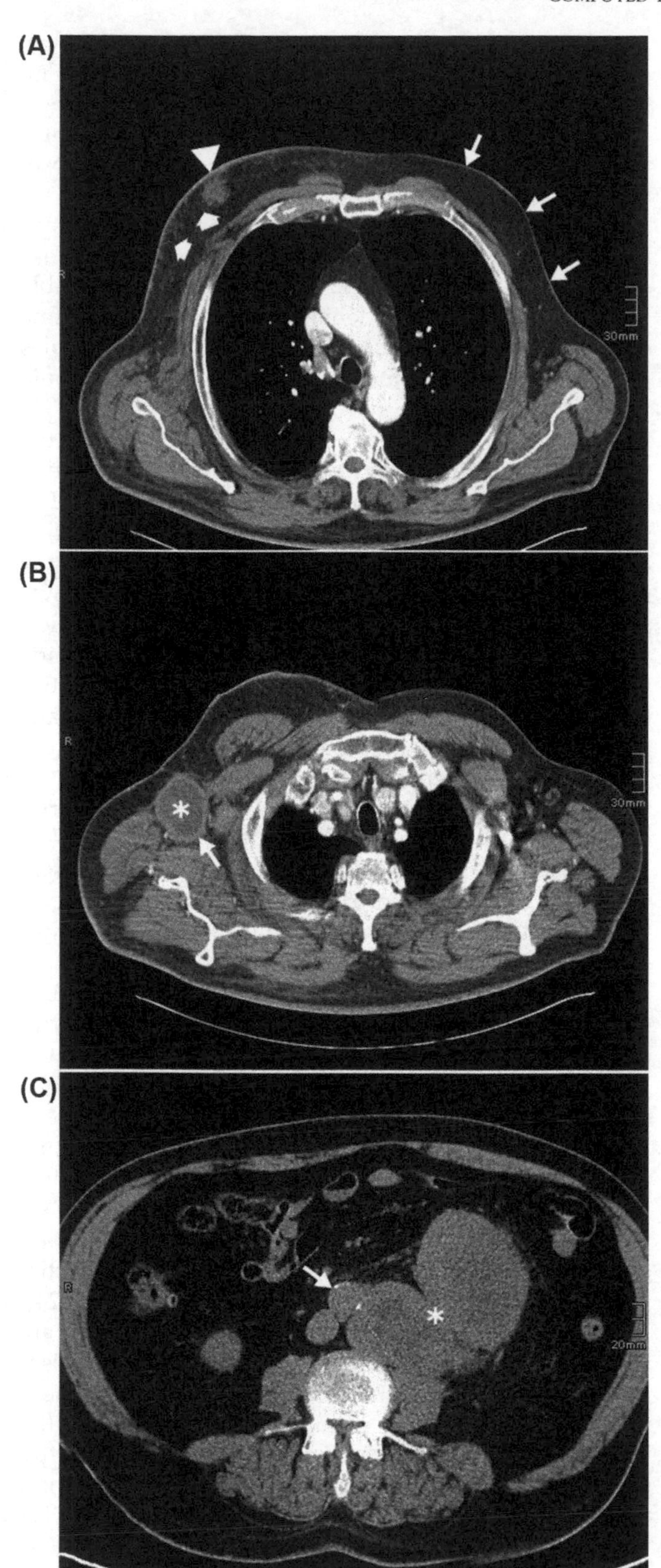

FIGURE 34.3 (A) Computed tomography (CT) imaging of a 76-year-old male patient shows a right-sided lymph node metastasis (*arrow head*) of a head-and-neck Merkel cell carcinoma. The lymph node appears as a round, speculated nodule, with irregular borders that extend from the dermis into the subdermal tissue. The cutaneous mass is again isoattenuated compared with the chest wall muscles on the contrast-enhanced CT scan. The most remarkable issue is the significant diffuse cutaneous thickening with reticular fat-stranding (*small arrows*) on the right chest wall, compared with the left side (*arrows*), where no fat-stranding can be observed. (B) The same patient has a significantly enlarged axillar lymph node (*arrow*) on the right side with central necrosis (*asterisk*). (C) Again, an enlarged para-aortal (*arrow*) lymph node conglomerate (*asterisk*) is found in the same patient as shown in panels A and B on nonenhanced CT imaging.

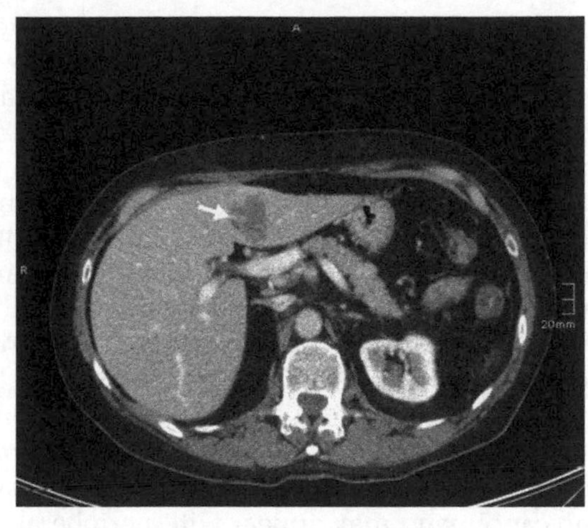

FIGURE 34.4 Computed tomography (CT) imaging was performed in a 76-year-old female patient. A single ill-defined liver metastasis (*arrow*) with central necrosis could be detected on the contrast-enhanced CT.

In addition to its importance for initial diagnostic and staging workup, CT imaging is a key imaging tool for clinical follow-up in Merkel cell carcinoma patients. Some authors propose follow-up CT scans in patients with histologically confirmed positive lymph nodes 3, 6, 12, and 18 months after initial treatment. These time intervals are based on the mean period for local or regional recurrence and the mean time for detection of distant metastasis [61]. Recent data have shown that 30–55% of all patients with Merkel cell carcinoma develop recurrent disease within the first 12 months [61,63].

Take-Home Message

CT imaging in Merkel cell carcinoma patients is a key imaging tool for tumor staging. This highly valuable imaging modality defines size, extension to the subdermal space, and localization of the primary tumor, and, specifically, locoregional and distant metastatic spread. Merkel cell carcinoma appears as an isoattenuated to slightly hyperattenuated round nodule extending from the dermis after contrast agent application. In some cases, this tumor develops subcutaneous linear or reticular fat-stranding with lymphedema close to the primary tumor.

There are no specific CT imaging features for nodal disease in Merkel cell carcinoma patients. On the basis of the maximum transverse diameter, metastatic lymph node dimensions are enlarged, the structure appears with central necrosis, and/or extracapsular spread is present.

Thoracoabdominal CT scans are strongly recommended in patients with stage II disease for detection of visceral metastasis. In particular, cutaneous, pulmonary, hepatic, osseous, urogenital, or, less frequently, brain metastasis may occur in 5% of Merkel cell carcinoma cases at initial presentation. Those lesions tend to mimic other neuroendocrine metastatic lesions, such as small-cell lung cancer, and may appear with peripheral, ring-like enhancement.

MAGNETIC RESONANCE IMAGING

Background and Technical Aspects

The first imaging study describing MRI findings in Merkel cell carcinoma patients was published in 1998 [72]. However, to date, literature on MRI in Merkel cell carcinoma patients is very sparse [26,27,73].

Compared with CT, MRI demonstrates a superior ability to reflect soft tissue compartments, providing an excellent insight into the subcutis, fascia, and musculature [26]. In particular, MRI is indicated when aiming to assess (1) extension of the primary tumor, (2) skin involvement, (3) infiltration of the fascia and musculature, and (4) lymph node metastasis [74]. Absolute contraindications for MRI include implanted pacemakers or defibrillators, specific cerebral aneurysm clips, cochlear implants, metallic foreign bodies (in particular of the eye), thin-walled intravascular catheters, and drug infusion devices.

Clinical Application and Appearance

The appearance of primary lesions from Merkel cell carcinoma on MRI depends on the size of the tumor. Small lesions, less than 15 mm in diameter, present as homogeneous with intermediate signal intensity on T1 sequences (Fig. 34.2B) and heterogeneously hyperintense on T2 sequences compared with the adjacent musculature [26,27]. Some authors have hypothesized that the signal intensity is caused by the microscopical structure of Merkel cell carcinoma. In particular, dense small blue cells with hyperchromatic nuclei and a small amount of cytoplasm are highly characteristic of Merkel cell carcinoma. For surgically challenging areas around the eye and nose, small and superficially lying structures can be scanned with microscopy-coil, high-resolution MRI to determine the deep soft tissue margins of skin tumors [75]. Because there are still no studies in Merkel cell carcinoma patients and microscopy-coil MRI, this technique may represent a new way to determine soft tissue invasion, and subsequently, the deep soft tissue margins.

On T1 sequences, lesions more than 15 mm in size show, in most cases, a heterogeneous signal and diffuse contrast enhancement [26,72]. In particular, variable heterogeneous signal intensity might be due to fine stromal septa that are observed in pathological samples of primary as well as metastatic Merkel cell carcinoma lesions.

The size and presence of necrosis are, to date, the most important radiological characteristics indicative of malignant transformation of lymph nodes. The size of the lymph nodes shows a great variability, and spherical lymph nodes larger than 10 mm in the short axis are indicative of malignant disease. An oval shape and fatty hilus are signs of benign lymph nodes, whereas extracapsular spread is specific for metastatic lymph nodes [68]. The presence of border irregularities (Fig. 34.5A and B) increases the likelihood of metastatic disease [76].MRI signal hyperintensity on T2 weighted images with corresponding signal hypointensity on T1 weighted images with or without surrounding rim enhancement indicates nodal necrosis.

Diffusion-weighted MRI studies have shown that, in general, the apparent diffusion coefficient is significantly decreased in metastatic lymph nodes compared with benign nodes. In particular, the most important advantage of diffusion-weighted MRI, compared with anatomical MRI, is its high sensitivity of 76% in detecting metastatic lymph nodes of the head and neck smaller than 1 cm. In contrast, anatomical MRI shows a sensitivity of only 7% in detecting neck lymph nodes less than 1 cm in size [77,78]. Therefore, diffusion-weighted MRI may provide additional information in the detection of nodal metastatic disease, particularly in subcentimeter-sized lymph nodes, than does anatomical MRI [77,78]. In addition, MRI is particularly suited for the detection of brain metastases [24].

Satellite nodules in subcutaneous fat tissue, which are typical of Merkel cell carcinoma, appear either as an isolated lesion or in combination with fat-stranding. These nodules are close to the primary tumor and follow the lymphatic drainage pathways. Those masses, mostly greater than 20 mm, are heterogeneous, hyperintense, and distinctly circumscribed or lobulated on T2-weighted images. In addition, they show a diffuse enhancement that is analogous to that of the primary tumor. It should be kept in mind that melanoma and myeloma, for example, also develop satellite cutaneous lesions. However, those tumors do not follow lymphatic alignment as does the Merkel cell carcinoma [79,80].

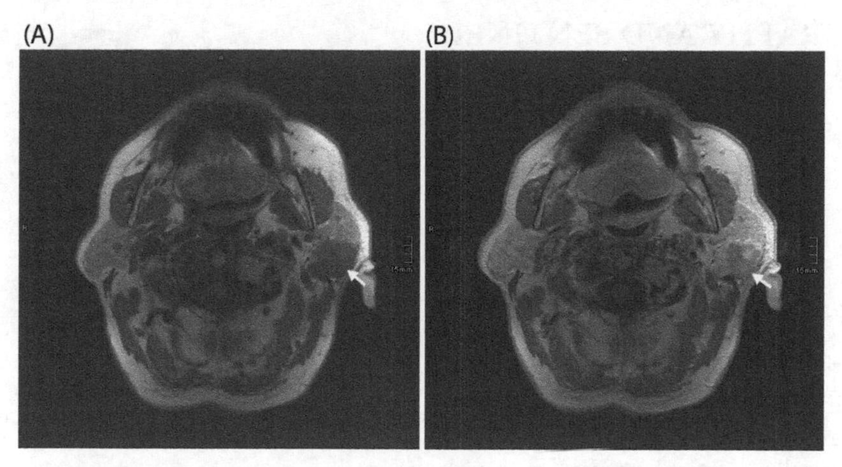

FIGURE 34.5 (A) A noncontrast-enhanced magnetic resonance (MR) image was obtained in a 75-year-old male patient with a Merkel cell carcinoma from an unknown primary. The metastatic lymph node (*arrow*) is located within the parotid gland. The T1-weighted noncontrast-enhanced MR image shows an ill-defined mass that is hypo- and isotense compared with the parotid gland and the muscles, respectively. (B) Corresponding to panel a, the T1-weighted MR image shows a lesion (*arrow*) with significant inhomogeneous contrast enhancement.

Basal cell carcinoma, malignant melanoma, and cutaneous squamous cell carcinoma have a morphologic appearance on MRI similar to that of Merkel cell carcinoma. In particular, basal cell carcinoma appears as an isointense to hyperintense lesion on T1 sequences and hyperintense on T2 scans [26,27,67,81]. Because of the paramagnetic effect of stable free radicals within the melanin, melanin induces a shortening of T1 and T2 relaxation times. Therefore, the expected signal pattern for melanoma is hyperintensity on T1-weighted images and hypointensity on T2-weighted images [27,74]. Benign skin or subdermal lesions, similar to epidermoid cysts and lipoma, are reported to be well-defined, encapsulated, T2-hyperintense masses with a very thin rim of peripheral enhancement in some cases. Myoperiocytomas are well-defined lesions, with an intense enhancement. Compared with adjacent muscles, in particular, this lesion is iso- and hyperintense on both T1-and T2-weighted images. Neurofibromas appear with a target sign that is characterized by a central low but peripheral high signal intensity on T2-weighted images [82]. Various factors, such as skin thickening, fascial edema, necrosis within the lesion, lobulation, and hemorrhage, are reported to be significant factors for malignant disease and are valuable in the differentiation between benign and malignant lesions [83].

Clinical Data

Reticular fat-stranding, also described in CT imaging, occurs in approximately 60% of all patients [11,23,24,26,42]. However, the detection rate of lymphatic involvement due to subcutaneous reticular fat-stranding is higher on MRI compared with CT [26]. Moreover, on weighted fat-suppressed sequences, reticular stranding is characterized as hyperintense linear reticulations caused by tissue congestion or edema. Reticular fat-stranding could be observed in 55–60% of patients with histologically confirmed positive lymph nodes [7,26,73]. Affected lymph nodes in Merkel cell carcinoma patients are comparatively large, with fine, compressed, retained fatty tissue. Colgan and colleagues reported the largest case series, with 10 Merkel cell carcinoma patients, the aim of which was to assess regional lymph node involvement [65]. Six of seven patients were correctly identified as lymph node-negative patients. However, MRI failed to detect histology-proven lymph node metastasis in all three lymph-node–positive Merkel cell carcinoma patients [65].

Take-Home Message

MRI in Merkel cell carcinoma patients is helpful in determining soft tissue infiltration, particularly in muscles and fascia, by the primary tumor. MRI is also useful in detecting multiple aligned subcutaneous lesions and lymph node metastasis [26,27,73]. Merkel cell carcinoma or satellite lesions, smaller than 15 mm in size, in particular, demonstrate a homogeneously intermediate signal intensity on T1, but a heterogeneously hyperintense signal intensity on T2 sequences, compared with the adjacent musculature [26,27]. However, tumor masses more than 15 mm in size show mostly a heterogeneous signal, and varying hypo- and hyperintense areas, with a diffuse contrast enhancement on T1 due to fine stromal septa.

LYMPHOSCINTIGRAPHY AND SENTINEL LYMPH NODE BIOPSY

Background and Technical Aspects

The tumor growth cascade involves local expansion followed by the spread of the tumor cells via the lymphatic or blood vessels to ultimately initiate distant metastatic disease. Merkel cell carcinoma is a highly aggressive, rapidly locally growing, and particularly lymphangitically spreading skin tumor. The sentinel lymph node is defined as the first lymph node that receives lymphatic drain from the primary tumor site [29,52], and, consequently, the first lymph node that is encountered by tumor cells. Clinical studies in melanoma, squamous cell carcinoma of the head and neck, and Merkel cell carcinoma patients have shown that (1) the presence of a positive sentinel node predicts clinical outcome and (2) when the sentinel node is negative, it is very unlikely that other lymph nodes of the same basin have been invaded [16,19,59,84].

In 1992, Donald L. Morton and colleagues introduced sentinel lymph node biopsy in a feline model [85]. Over the years this procedure has become a benchmark for staging patients with melanoma, and subsequently, for the staging of squamous cell carcinoma of the head and neck and Merkel cell carcinoma [86–88]. Thus sentinel lymph node mapping, followed by sentinel node biopsy, has been incorporated into the guidelines of the National Comprehensive Cancer Network.

Clinical Application and Appearance

Lymphoscintigraphy is performed by injecting a radiotracer into the cutis and subcutis adjacent to the primary tumor or around the excision margins. The most commonly used tracer is technetium-99m (Tc-99m) antimony sulfide colloid, which has a diameter of 10 nm. Other radiocolloids, such as Tc-99m human serum albumin, Tc-99m nanocolloid, or Tc-99m sulfur colloid, have a diameter ranging from 5 to 40 nm. All radioactively marked tracers enter the drainage system and, following the gravitation, move caudally until they reach the first collector, the sentinel lymph node or a group of lymph nodes [27,89].

Lymphoscintigraphy is associated with exposure to radioactivity. In particular, every injection results in an exposure of 30 Sv of radioactivity to the patient [89,90].

Immediately after injection, a gamma camera or scintillation camera obtains dynamic pictures every 30 s for approximately 20 min. These dynamic pictures are followed by a static image, with an acquisition time of 300 s. Single-photon emission CT provides the opportunity for three-dimensional reconstruction of the distribution of a radiotracer in the human body [90,91]. Any hot spot visualized by the gamma camera is considered as the sentinel lymph node (Fig. 34.6). If several lymph nodes of different lymphatic paths show a significant hot spot, then all are regarded as sentinel lymph nodes. In some cases the sentinel lymph node is not necessarily the closest lymph node to the neoplastic lesion.

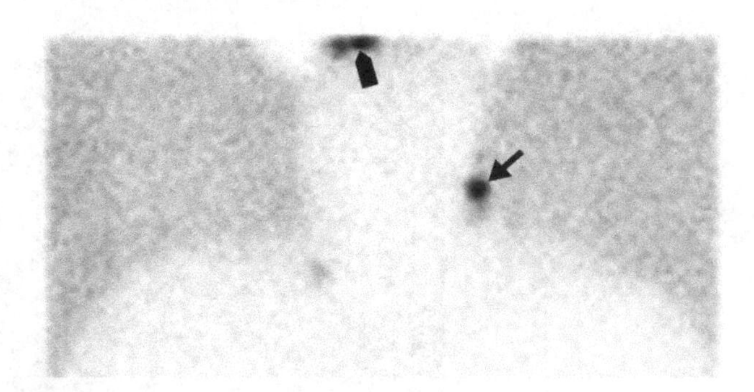

FIGURE 34.6 Lymphoscintigram of a patient with a primary tumor of the left cheek (*block arrow*), demonstrating a classical sentinel node (*arrow*) at the level of the ipsilateral omohyoid muscle.

Subsequently, the sentinel lymph node has to be marked 1 day or on the same day before a planned lymph node dissection is performed. In addition, intraoperative lymph node visualization and mapping is usually done with blue dye injection. In general, sentinel node biopsy should be performed at the time of wide local excision to minimize the risk of interrupting cutaneous lymphatic drainage.

Clinical Data

Recently published studies have shown that approximately one-third of all Merkel cell carcinoma patients have a positive sentinel lymph node at initial presentation [91,92]. Depending on the primary tumor site, the most common sites of lymphogenic metastasis are the neck, axilla, and inguinal region. In very rare cases cubital, popliteal, or otherwise located lymph nodes are revealed as the first drainage station [19,62].

Gupta and colleagues analyzed sentinel lymph node biopsies in 122 patients with Merkel cell carcinomas without palpable nodes and found lymph node metastasis in 32% of patients. When comparing melanoma with Merkel cell carcinoma patients, the incidence rate of occult nodal metastases is significantly lower in melanoma patients. Merkel cell carcinoma patients with a positive sentinel lymph node had a three-fold increased risk of developing recurrent disease [87].

In contrast, other studies have reported that 33–67% of patients with a negative sentinel lymph node biopsy developed regional nodal metastasis over the course of disease [18,19]. Moreover, one study emphasized that the only predictors of a positive sentinel lymph node biopsy are the size of the primary tumor (25% in <2-cm size tumors vs 45% in >2-cm tumors) and the presence of lymphatic invasion (4% in negative vs 55% in positive sentinel lymph node biopsies). However, in terms of positive and negative sentinel lymph node biopsies, there was no difference in disease recurrence or survival rates of Merkel cell carcinoma patients [6].

Take-Home Message

Lymphoscintigraphy, followed by sentinel lymph node biopsy, is an important diagnostic and therapeutic tool in patients with Merkel cell carcinoma. In particular, because up to one-third of all patients undergoing sentinel lymph node biopsy show positive lymph node metastasis, a thorough lymph node diagnostic workup followed by treatment is key for the prolonged disease-free and disease-specific overall survival of patients with Merkel cell carcinoma. In contrast, sentinel lymph node biopsy is a minimally invasive option in patients who present with a Merkel cell carcinoma to avoid the morbidity of elective neck dissection in patients who are sentinel node biopsy negative.

List of Acronyms and Abbreviations

ADC Apparent diffusion coefficient
AJCC American Joint Committee on Cancer
CT Computed tomography
FDG Fluorodeoxyglucose F 18
HFUS High frequency ultrasound
HIV Human immunodeficiency virus
Hz Hertz
MRI Magnetic resonance imaging
PET Positron emission tomography
PET-CT Positron emission computed tomography
SPECT Singe-photon emission computed tomography
Sv Sievert
Tc-99m Technetium-99m

Acknowledgment

The authors thank Mary McAllister (Johns Hopkins University, Baltimore, MD) for her outstanding help in editing the manuscript.

References

[1] Toker C. Trabecular carcinoma of the skin. Arch Dermatol 1972;105(1):107–10.
[2] Erovic I, Erovic BM. Merkel cell carcinoma: the past, the present, and the future. J Skin Cancer 2013;2013:929364.
[3] Tang CK, Toker C. Trabecular carcinoma of the skin: an ultrastructural study. Cancer 1978;42(5):2311–21.
[4] Brunner M, et al. Expression of hedgehog signaling molecules in Merkel cell carcinoma. Head Neck 2010;32(3):333–40.
[5] Feng H, et al. Clonal integration of a polyomavirus in human Merkel cell carcinoma. Science 2008;319(5866):1096–100.
[6] Fields RC, et al. Five hundred patients with Merkel cell carcinoma evaluated at a single institution. Ann Surg 2011;254(3):465–75.
[7] Hughes MP, et al. Merkel cell carcinoma: epidemiology, target, and therapy. Curr Dermatol Rep 2014;3:46–53.
[8] Schrama D, et al. Merkel cell carcinoma: recent insights and new treatment options. Curr Opin Oncol 2012;24(2):141–9.
[9] Hodgson NC. Merkel cell carcinoma: changing incidence trends. J Surg Oncol 2005;89(1):1–4.
[10] Desch L, Kunstfeld R. Merkel cell carcinoma: chemotherapy and emerging new therapeutic options. J Skin Cancer 2013;2013:327150.
[11] Enzenhofer E, et al. Imaging in patients with Merkel cell carcinoma. J Skin Cancer 2013;2013:973123.
[12] Rockville Merkel Cell Carcinoma Group. Merkel cell carcinoma: recent progress and current priorities on etiology, pathogenesis, and clinical management. J Clin Oncol 2009;27(24):4021–6.
[13] Hitchcock CL, et al. Neuroendocrine (Merkel cell) carcinoma of the skin. Its natural history, diagnosis, and treatment. Ann Surg 1988;207(2):201–7.
[14] Brissett AE, et al. Merkel cell carcinoma of the head and neck: a retrospective case series. Head Neck 2002;24(11):982–8.
[15] Yiengpruksawan A, et al. Merkel cell carcinoma. Prognosis and management. Arch Surg 1991;126(12):1514–9.
[16] Edge SB, Compton CC. The American Joint Committee on Cancer: the 7th edition of the AJCC cancer staging manual and the future of TNM. Ann Surg Oncol 2010;17(6):1471–4.
[17] Haymerle G, et al. Merkel cell carcinoma: overall survival after open biopsy versus wide local excision. Head Neck 2015;2015:24148.
[18] Maza S, et al. Impact of sentinel lymph node biopsy in patients with Merkel cell carcinoma: results of a prospective study and review of the literature. Eur J Nucl Med Mol Imaging 2006;33(4):433–40.
[19] Warner RE, et al. Management of merkel cell carcinoma: the roles of lymphoscintigraphy, sentinel lymph node biopsy and adjuvant radiotherapy. Ann Surg Oncol 2008;15(9):2509–18.
[20] Harrington C, Kwan W. Outcomes of Merkel cell carcinoma treated with radiotherapy without radical surgical excision. Ann Surg Oncol 2014;21(11):3401–5.
[21] Veness M, Howle J. Radiotherapy alone in patients with Merkel cell carcinoma: the Westmead Hospital experience of 41 patients. Australas J Dermatol 2015;56(1):19–24.
[22] Hruby G, Scolyer RA, Thompson JF. The important role of radiation treatment in the management of Merkel cell carcinoma. Br J Dermatol 2013;169(5):975–82.
[23] Bishop AJ, et al. Merkel cell carcinoma of the head and neck: favorable outcomes with radiotherapy. Head Neck 2015;2015:24678.
[24] Nguyen BD, McCullough AE. Imaging of Merkel cell carcinoma. Radiographics 2002;22(2):367–76.
[25] Buder K, et al. Somatostatin receptor expression in Merkel cell carcinoma as target for molecular imaging. BMC Cancer 2014;2014:26814.
[26] Anderson SE, et al. MRI of merkel cell carcinoma: histologic correlation and review of the literature. Am J Roentgenol 2005;185(6):441–8.
[27] Grandhaye M, et al. Focus on Merkel cell carcinoma: diagnosis and staging. Skeletal Radiol 2015;44(6):777–86.
[28] Alexander H, Miller DL. Determining skin thickness with pulsed ultra sound. J Invest Dermatol 1979;72(1):17–9.
[29] Gershenwald JE, et al. Multi-institutional melanoma lymphatic mapping experience: the prognostic value of sentinel lymph node status in 612 stage I or II melanoma patients. J Clin Oncol 1999;17(3):976–83.
[30] Schmid-Wendtner M-H, Burgdorf W. Ultrasound scanning in dermatology. Arch Dermatol 2005;141(2):217–24.
[31] Wortsman X. Common applications of dermatologic sonography. J Ultrasound Med 2012;31(1):97–111.

[32] Wortsman X, et al. Sonography in pathologies of scalp and hair. Br J Radiol 2012;85(1013):647–55.
[33] Wortsman X, Wortsman J. Ultrasound accuracy in the diagnosis of skin and soft-tissue lesions. Am J Roentgenol 2015;204(2):220–2.
[34] Kleinerman R, et al. Ultrasound in dermatology: principles and applications. J Am Acad Dermatol 2012;67(3):478–87.
[35] Jemec GB, et al. Ultrasound in dermatology. Part I. High frequency ultrasound. Eur J Dermatol 2000;10(6):492–7.
[36] Babcock DS, et al. Power doppler sonography: basic principles and clinical applications in children. Pediatr Radiol 1996;26(2): 109–15.
[37] Martinoli C, Derchi LE. Gain setting in power Doppler US. Radiology 1997;202(1):284–5.
[38] Ahuja AT, et al. Ultrasound of malignant cervical lymph nodes. Cancer Imaging 2008;8(1):48–56.
[39] Ying M, et al. Power Doppler sonography of normal cervical lymph nodes. J Ultrasound Med 2000;19(8):511–7.
[40] Choi YJ, et al. Ultrasound elastography for evaluation of cervical lymph nodes. Ultrasonography 2015;34(3):157–64.
[41] Baek SH, et al. Merkel cell carcinoma of the axilla and adrenal gland: a case report with imaging and pathologic findings. Case Rep Med 2015;2015:931238.
[42] Eftekhari F, et al. Merkel cell carcinoma of the skin: imaging and clinical features in 93 cases. Br J Radiol 1996;69(819):226–33.
[43] Machet L, et al. Preoperative measurement of thickness of cutaneous melanoma using high-resolution 20 MHz ultrasound imaging: a monocenter prospective study and systematic review of the literature. Ultrasound Med Biol 2009;35(9):1411–20.
[44] Righi A, et al. An ultrasonography-cytology protocol for the diagnostic management of regional nodes in a subset of patients with Merkel cell carcinoma of the skin. Br J Dermatol 2013;168(3): 563–70.
[45] Dasgeb B, et al. An introduction to primary skin imaging. Int J Dermatol 2013;52(11):1319–30.
[46] Uhara H, et al. Multiple hypersonographic spots in basal cell carcinoma. Dermatol Surg 2007;33(10):1215–9.
[47] Harland CC, et al. High frequency, high resolution B-scan ultrasound in the assessment of skin tumours. Br J Dermatol 1993; 128(5):525–32.
[48] Fornage BD, et al. Imaging of the skin with 20-MHz US. Radiology 1993;189(1):69–76.
[49] Park J, Chae IS, Kwon DR. Utility of sonoelastography in differentiating ruptured from unruptured epidermal cysts and implications for patient care. J Ultrasound Med 2015;34(7):1175–81.
[50] Oebisu N, et al. Contrast-enhanced color Doppler ultrasonography increases diagnostic accuracy for soft tissue tumors. Oncol Rep 2014;32(4):1654–60.
[51] Fong LS, et al. Right atrial Merkel cell tumour metastasis characterization using a multimodality approach. Eur Heart J 2012; 33(17):2205.
[52] Wong JH, et al. Lymphatic drainage of skin to a sentinel lymph node in a feline model. Ann Surg 1991;214(5):637–41.
[53] Zager JS, et al. Imaging of Merkel cell carcinoma. Curr Probl Cancer 2010;34(1):65–76.
[54] Norling R, et al. Staging of cervical lymph nodes in oral squamous cell carcinoma: adding ultrasound in clinically lymph node negative patients may improve diagnostic work-up. PLoS One 2014; 2014:90360.
[55] Wang LW, et al. Metastatic Merkel cell carcinoma of the heart. Eur Heart J 2014;35(41):2927.
[56] Collins BT, et al. Fine-needle aspiration of Merkel cell carcinoma of the skin with cytomorphology and immunocytochemical correlation. Diagn Cytopathol 1998;18(4):251–7.
[57] Bernstein J, et al. Endoscopic ultrasound-guided fine-needle aspiration diagnosis of Merkel cell carcinoma metastatic to the pancreas. Diagn Cytopathol 2014;42(3):247–52.
[58] Shield PW, Crous H. Fine-needle aspiration cytology of Merkel cell carcinoma-a review of 69 cases. Diagn Cytopathol 2014; 42(11):924–8.
[59] Ricard A-S, et al. Sentinel lymph node biopsy for head and neck Merkel cell carcinoma: a preliminary study. Eur Ann Otorhinolaryngol Head Neck Dis 2015;132(2):77–80.
[60] Schwartz JL, et al. Features predicting sentinel lymph node positivity in Merkel cell carcinoma. J Clin Oncol 2011;29(8): 1036–41.
[61] Gollub MJ, et al. Merkel cell carcinoma: CT findings in 12 patients. Am J Roentgenol 1996;167(3):617–20.
[62] Orr LA, et al. Computed tomography evaluation of recurrent Merkel cell carcinoma. Clin Imaging 1992;16(1):52–4.
[63] Tirumani SH, et al. Merkel cell carcinoma: a primer for the radiologist. Am J Roentgenol 2013;200(6):1186–96.
[64] Azizi L, et al. Merkel cell carcinoma: a rare cause of hypervascular nasal tumor. Am J Neuroradiol 2001;22(7):1389–93.
[65] Colgan MB, et al. The predictive value of imaging studies in evaluating regional lymph node involvement in Merkel cell carcinoma. J Am Acad Dermatol 2012;67(6):1250–6.
[66] Nakayama M, et al. Basal cell carcinoma of the head and neck. J Skin Cancer 2011;2011:496910.
[67] Baheti AD, et al. Basal cell carcinoma: a comprehensive review for the radiologist. Am J Roentgenol 2015;204(2):132–40.
[68] Mack MG, et al. Cervical lymph nodes. Eur J Radiol 2008;66(3): 493–500.
[69] Ganeshalingam S, Koh D-M. Nodal staging. Cancer Imaging 2009; 9:104–11.
[70] Chai RL, et al. Accuracy of computed tomography in the prediction of extracapsular spread of lymph node metastases in squamous cell carcinoma of the head and neck. JAMA Otolaryngol Head Neck Surg 2013;139(11):1187–94.
[71] Peloschek P, et al. Diagnostic imaging in Merkel cell carcinoma: lessons to learn from 16 cases with correlation of sonography, CT, MRI and PET. Eur J Radiol 2010;73(2):317–23.
[72] Dunlop P, et al. Merkel cell carcinoma of the abdominal wall. Skeletal Radiol 1998;27(7):396–9.
[73] Caldarone F, et al. Merkel cell carcinoma of the calf: MRI imaging of a rare tumor. Rofo 2002;174(9):1175–6.
[74] Budak MJ, et al. High-resolution microscopy-coil MR imaging of skin tumors: techniques and novel clinical applications. Radiographics 2015;35(4):1077–90.
[75] de Bondt RBJ, et al. Morphological MRI criteria improve the detection of lymph node metastases in head and neck squamous cell carcinoma: multivariate logistic regression analysis of MRI features of cervical lymph nodes. Eur Radiol 2009;19(3): 626–33.
[76] Thoeny HC, et al. Metastases in normal-sized pelvic lymph nodes: detection with diffusion-weighted MR imaging. Radiology 2014; 273(1):125–35.
[77] Thoeny HC, et al. Diffusion-weighted MR imaging in the head and neck. Radiology 2012;263(1):19–32.
[78] Patten RM, et al. Subcutaneous metastases from malignant melanoma: prevalence and findings on CT. Am J Roentgenol 1989; 152(5):1009–12.
[79] Moulopoulos LA, et al. Extraosseous multiple myeloma: imaging features. Am J Roentgenol 1993;161(5):1083–7.
[80] Hawnaur JM, et al. Skin: MR imaging findings at middle field strength. Radiology 1996;201(3):868–72.
[81] Kim J-H, et al. MR imaging manifestations of skin tumors. Eur Radiol 2008;18(11):2652–61.
[82] Zhuang KD, et al. MRI features of soft-tissue lumps and bumps. Clin Radiol 2014;69(12):568–83.
[83] Calleja M, et al. MRI of superficial soft tissue masses: analysis of features useful in distinguishing between benign and malignant lesions. Skeletal Radiol 2012;41(12):1517–24.

[84] Shaw H, Thompson J. Predicting sentinel and residual lymph node basin disease after sentinel lymph node biopsy for melanoma. Cancer 2002;94(1):283–4.
[85] Morton DL, et al. Technical details of intraoperative lymphatic mapping for early stage melanoma. Arch Surg 1992;127(4):392–9.
[86] Bredell MG. Sentinel lymph node mapping by indocyanin green fluorescence imaging in oropharyngeal cancer - preliminary experience. Head Neck Oncol 2010;2:31.
[87] Gupta SG, et al. Sentinel lymph node biopsy for evaluation and treatment of patients with Merkel cell carcinoma: the Dana-Farber experience and meta-analysis of the literature. Arch Dermatol 2006;142(6):685–90.
[88] Naehrig D, et al. Sentinel lymph node mapping for defining site and extent of elective radiotherapy management of regional nodes in Merkel cell carcinoma: a pilot case series. J Med Imaging Radiat Oncol 2014;58(3):353–9.
[89] Stoffels I, et al. Evaluation of a radioactive and fluorescent hybrid tracer for sentinel lymph node biopsy in head and neck malignancies: prospective randomized clinical trial to compare ICG-(99m)Tc-nanocolloid hybrid tracer versus (99m)Tc-nanocolloid. Eur J Nucl Med Mol Imaging 2015;2015:203466.
[90] Kretschmer L, et al. Dynamic lymphoscintigraphy and image fusion of SPECT and pelvic CT-scans allow mapping of aberrant pelvic sentinel lymph nodes in malignant melanoma. Eur J Cancer 2003;39(2):175–83.
[91] Wasserberg N, et al. Sentinel-node guided lymph-node dissection for merkel cell carcinoma. Eur J Surg Oncol 1999;25(4):444–6.
[92] Messina JL, et al. Selective lymphadenectomy in patients with Merkel cell (cutaneous neuroendocrine) carcinoma. Ann Surg Oncol 1997;4(5):389–95.

CHAPTER

35

Imaging Evaluation of Cutaneous Lymphoma Using Functional and Structural Imaging

S. Fardin, S. Gholami, T.J. Werner, A.H. Rook, A. Alavi

Hospital of the University of Pennsylvania, Philadelphia, PA, United States

OUTLINE

INTRODUCTION

After gastrointestinal tract lymphoma, which is the predominant type of extranodal non-Hodgkin lymphoma, primary cutaneous lymphoma (PCL) is the second most common group, accounting for 10% of extranodal non-Hodgkin lymphoma. Two common classifications are introduced for PCL by the European Organization for Research and Treatment of Cancer and the World Health Organization classifications [1,2]. On the basis of these classifications, there are two general types of cutaneous lymphoma, including cutaneous T cell lymphomas (CTCLs) and cutaneous B cell lymphomas, which are defined when the tumor involves skin without any extracutaneous manifestations at the time of diagnosis. CTCL is the most common subtype and accounts for approximately 65% of PCLs. In general, CTCLs are rare conditions but potentially devastating, and several reports have illustrated a three-fold increase in its incidence during the last 25–30 years in the United States and other parts of the world [3,4].

The two most common types of CTCL are mycosis fungoides (MF) and sezary syndrome (SS). MF is the most common form of CTCL and has several clinical presentations with skin symptoms appearing as patches, plaques, tumors, or erythroderma; as disease progresses, lymph node enlargement, peripheral blood involvement, and internal organ symptoms may also be seen. SS is an advanced variant of MF that is a combination of generalized erythroderma, lymphadenopathy, and atypical T cells (sezary cells) in the peripheral blood [5,6].

Similar to many other malignancies, the prognosis of PCL is based on the extent of the disease at presentation. Therefore determination of the extent of the disease and accurate staging is paramount in patients with CTCL and can help with choosing the proper treatment and predicting the outcome [7]. Given the diversity of clinical manifestations, the diagnosis and evaluation of CTCL is very challenging and may require a complex workup. The recommended evaluation and staging of CTCL currently includes complete physical examination, skin biopsy, and blood tests, including complete

Imaging in Dermatology
http://dx.doi.org/10.1016/B978-0-12-802838-4.00035-2

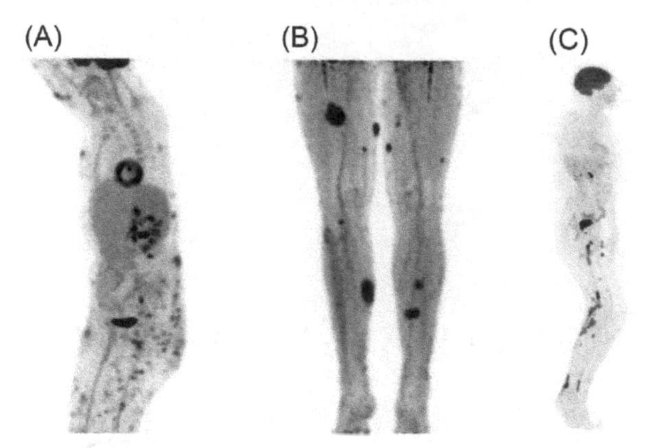

FIGURE 35.1 Malignant cells in the lymph nodes and cutaneous tissue take up ^{18}F-fluorodeoxyglucose, which is detectable by positron emission tomography scan. (A) 48-year-old female with extensive involvement of skin and lymph nodes, (B) 43-year-old female with involvement of lower extremities, and (C) 83-year-old man with involvement of pelvic girdle and lower extremities.

blood cell count, lactate dehydrogenase, liver function test, T cell receptor (TCR) gene rearrangement on skin and blood, flow cytometry, radiologic imaging studies, and possibly lymph node biopsy.

Radiologic studies that are commonly done in CTCL patients are directed to help in thoracic needle biopsy staging. In otherwise healthy patients with T1N0B0 stage who do not have any complaints about a specific organ and in selected patients with T2N0B0 disease with limited skin disease, the radiologic imaging is usually limited to chest X-ray or ultrasound of peripheral nodal groups to confirm the absence of adenopathy. However, in all patients with any disease stage worse than presumed stage 1A disease or abnormal laboratory tests, including abnormal peripheral blood flow cytometry or elevated lactate dehydrogenase, computed tomography (CT) scans of chest, abdomen, and pelvis with or without ^{18}F-fluorodeoxyglucose (FDG) positron emission tomography (PET) scan are suggested to identify lymphadenopathy and visceral involvement (Fig. 35.1) [8].

Imaging studies are very commonly used in lymphoma staging. Although CT scan has played a crucial role in staging, it has a limited value in detecting cutaneous lesions, and given its capability in detecting structural abnormalities, it cannot detect extracutaneous involvement in normal-sized lymph nodes and minimally enlarged reactive lymph nodes [6]. FDG-PET/CT has demonstrated a very high sensitivity in detecting nodal and extranodal spread of lymphomas. Moreover, it is even more accurate for detecting sites of involvements, particularly extranodal sites, than CT alone [9–11]. Therefore, as previously published, FDG-PET can be used as a valid tool for primary tumor evaluation and providing follow-up and prognostic information in lymphomas [12].

At present, PET/CT is used as the standard of care for staging, assessment of response to therapy, and detection of recurrence in most Hodgkin's and aggressive non-Hodgkin's lymphomas. However, there had been limited literature available regarding the role of PET/CT in PCL until recently [13]. The idea of use of FDG-PET in PCL was first introduced in 2002 when increased FDG uptake was noticed in CTCL cutaneous lesions [14]. It was the first time that a potential role for PET imaging in CTCL diagnosis was discussed. After this finding, some additional studies focused on the role of FDG-PET in diagnosing PCL [15]. In more recent studies FDG-PET has been shown to be of great value in staging, detecting disease recurrence, assessing response to treatment, and guiding interventions in PCL. However, it has not yet been routinely used as the standard of care in these patients [6,14–16].

MYCOSIS FUNGOID AND SEZARY SYNDROME

MF is the most common variant of CTCL and accounts for approximately 50% of all PCLs. It is clinically characterized by the evolution of patches, plaques, and tumors of the skin. Histopathology typically demonstrates epidermotropic atypical lymphocytes with cerebriform nuclei. The usual age of onset is in the 50s with a male to female ratio of 1.6:2.0. SS is defined typically by the triad of erythroderma and the presence of neoplastic T cells (sezary cells $> 1000/\mu L$) in the peripheral blood [17]. It is a rare disease and often clinically manifests with marked edema, exfoliation and lichenification, lymphadenopathy, alopecia, and onychodystrophy [18].

Evaluating and screening these patients at primary and follow-up visits are typically performed by physical examination with palpation of superficial lymph nodes for staging. However, physical examination is subjective and often underestimates the size of tumors in the primary site and lymph nodes.

Contrast-enhanced CT scan has routinely been used as part of lymphoma tumor staging to further detect lymph node and visceral involvement. FDG-PET alone can detect hypermetabolism in primary tumor and lymph node involvement even with minimum nodal involvement. Moreover, FDG-PET is a modality that is capable of demonstrating metabolic and molecular activities of the lymph nodes as well as the other organs. However, it gives very poor anatomical information regarding the specific location and size of the lesions.

Integrated PET/CT provides highly specific information regarding a combination of structural and functional features of cutaneous lymphoma and lymph node involvement. Compared with PET alone, hybrid

PET/CT provides precise localization of hypermetabolic lesions within a specific anatomical structure and has increased specificity and sensitivity. The higher sensitivity of PET/CT than that of CT scan alone is more prominent in the diagnosis of early stages of PCL. A CT scan may miss primary skin lesions that are not manifested in nodule- or mass-type histology, such as the gamma delta type or extranodal natural killer/T cell lymphoma (known as the nasal type). Moreover, some studies have shown the usefulness of FDG-PET/CT in detecting plaque cutaneous lesions. Because diagnosis by CT is only based on anatomy, normal-sized lymph nodes are commonly missed by CT and enlarged reactive lymph nodes are easily misinterpreted as being lymphomatous [19–24].

The PET component of PET/CT is very sensitive in detecting involved lymph nodes and internal tissues. Patients with no palpable lymphadenopathy whose nodes did not meet CT size criteria may have increased metabolism in a PET scan. Moreover, PET/CT may increase the specificity of detection of lymph node involvement in MF/SS. PET/CT scans have altered the treatment regimens for many patients whose previous staging was underestimated by physical exam and CT alone. The intensity of FDG uptake may correlate with histologic features of LN involvement. However, it is important to consider other clinical conditions, such as inflammation and infection because FDG uptake is nonspecific and may not be able to solely show the tumor involvement. A PET scan also provides a beneficial biopsy guide for surgeons by delineating hypermetabolic lymph nodes. However, it may not be well recognized intraoperatively which node had the highest level of FDG uptake, and surgeons must rely on their own judgment to select the best node for biopsy. Nowadays the use of an intraoperative PET probe in some centers has remarkably increased the ability to detect hypermetabolic lymph nodes and has decreased the probability of sampling error.

It has also been demonstrated that FDG/PET might be useful in the evaluation of skin involvement in MF and it clearly improves the detection of cutaneous foci of disease in comparison to CT alone because a combined PET/CT scan is capable of differentiating between the FDG uptake in skin and artifacts of a CT scan or between the uptake in nodes and subcutaneous tissues. Many studies have supported the use of PET/CT in staging MF/SS because of its ability to detect a higher number of involved lymph nodes, which can contribute to more accurate staging and prognostic information. Combining PET/CT with an intraoperative PET probe and fine-needle aspiration may increase the accuracy of surgical sampling [5].

FDG-PET/CT has shown benefits in identification of two types of MF: plaque or tumor stage type. It also has been observed that detectable FDG uptake is seen with PET/CT in SS with high tumor stage regardless of clinical erythrodermic state. Therefore, with the higher tumor stage seen in SS, PET/CT might be able to detect skin tumors very well [20,25].

NONMYCOSIS FUNGOIDES/SEZARY SYNDROME CUTANEOUS MALIGNANT LYMPHOMA

Among the non-MF/SS PCLs, there is literature on subcutaneous panniculitis-like T cell lymphoma (SPTCL). SPTCL is an uncommon and aggressive form of CTCL that comprises less than 1% of all non-Hodgkin lymphomas. It is clinically characterized by solitary or multiple subcutaneous plaques or nodules that can mimic an inflammatory panniculitis. SPTCL occurs in adults and in young children, and both sexes are equally affected. Histopathology demonstrates infiltration of subcutaneous fat by neoplastic cytotoxic T cells and many macrophages, affecting areas of adipose tissue, and may be complicated by hemophagocytic syndrome. Systemic symptoms, such as fever, fatigue, and weight loss may be present.

Only cases with α/β TCR phenotypes are currently classified as SPTCL whereas cases with γ/δ TCR phenotypes now belong to the cutaneous γ/δ T cell lymphomas, a subgroup of "primary cutaneous peripheral T-cell lymphomas, unspecified." α/β^+ T cell phenotypes are often limited to subcutaneous tissue without any dermal or epidermal involvement, and most cases have an indolent clinical course. However, γ/δ T cell phenotypes are usually not restricted to subcutaneous tissue, they also involve the dermis, and they have a particularly poor prognosis [18].

Only a few cases of SPTCL have been reportedly scanned by FDG-PET or PET/CT [26]. In some patients, similar to other types of lymphoma, it was found that physical examination, in comparison with PET/CT, greatly underestimates disease burden. On contrast-enhanced CT examination of these patients, multiple enhancing nodules may be seen in the subcutaneous layer of the diseased body site. However, these findings are not specific and are noted in inflammatory panniculitis due to systemic lupus erythematous or rheumatoid arthritis, subcutaneous metastasis from malignant melanoma or breast cancer, and nodules originating from a bacterial or fungal infection or from parasitic infestation. It has also been illustrated that CT may fail to identify these subcutaneous lesions. CT is able to estimate the thickness of lesions, whereas PET demonstrates the metabolic activity of the lesions. These subcutaneous lesions show FDG avidity and are detected with greater sensitivity by PET/CT than physical exam or CT alone

(Fig. 35.1) [19,26]. In other types of non-MF/SS cutaneous lymphoma disorders, FDG/PET may be a valuable modality in monitoring, disease staging, detecting extracutaneous involvement, and assessment of response to therapy.

EVALUATING TUMOR BIOLOGY AND BEHAVIOR AND ASSESSING RESPONSE TO TREATMENT AND OUTCOME

Evaluation of the ability of FDG-PET in characterizing the biology of a tumor, predicting the tumor behavior, and determining prognostic PET metrics has always been of high importance. FDG uptake is correlated with the grade of lymphomas. For example, in CTCL it is proposed that FDG uptake is significantly higher in MF lesions with, than without, large cell transformation. Likewise, it has been demonstrated that lymph node involvement of MF/SS with large cell transformation has higher uptake than those without this transformation [5,20].

QUANTIFICATION OF PRIMARY CUTANEOUS LYMPHOMA BY ^{18}F-FLUORODEOXYGLUCOSE POSITRON EMISSION TOMOGRAPHY/COMPUTED TOMOGRAPHY IMAGING THROUGH A GLOBAL APPROACH

There are few published studies on assessment of lymphomas using FDG-PET/CT. Moreover, few studies have compared the FDG uptake measurement indices of PET/CT in assessment of CTCL and prognosis determination. PET metrics include qualitative assessment; regional semiquantitative indices, such as the mean standardized uptake value (SUVmean), which is the average of FDG uptake activity in an area, and the maximum standardized uptake value (SUVmax), which represents the pixel with the highest FDG uptake activity; and global metrics, such as global metabolic volume and global lesion glycolysis, which represent the total lesion metabolic activity (Fig. 35.2).

There is some debate about the most reliable quantitative approach by FDG-PET/CT for monitoring CTCL and other lymphomas at different stages of the disease. The superiority of each index over the others remains controversial. In patients with lymphoma, some investigators support conventional measurements (SUVmax and SUVmean) as the best index for the response assessment whereas a few studies have demonstrated greater value of global tumor measurements in patients [27–30]. The applicability of these indices has been controversial regarding their capability in assessing the behavior of the tumor and predicting its activity, response to treatment, and future recurrence. Therefore additional studies are required to identify the most useful PET metrics [3,17].

Traditional metrics such as SUVmean and SUVmax are mostly useful in regional assessment that may show tumor activity and prognosis in a single location, but they may have less value in describing a tumor's extension and global activity. Moreover, regional indices are more prone to be affected by the nonhomogeneity seen in most solid tumors at a single time or from time to time [29]. Because cutaneous lymphomas foci can individually be unpredictable in their progression, global evaluation may be a better methodology to study global tumor activity regardless of focal tumor behaviors.

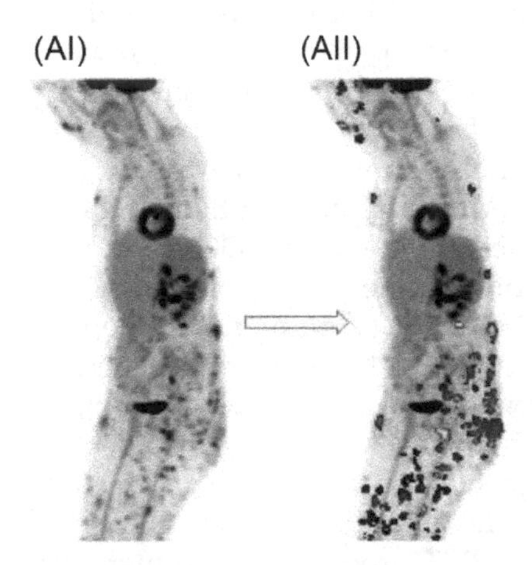

Parameter	Value
MTV	342.7
SUVmax	3.7
SUVmean	1.3
TLG	429.8

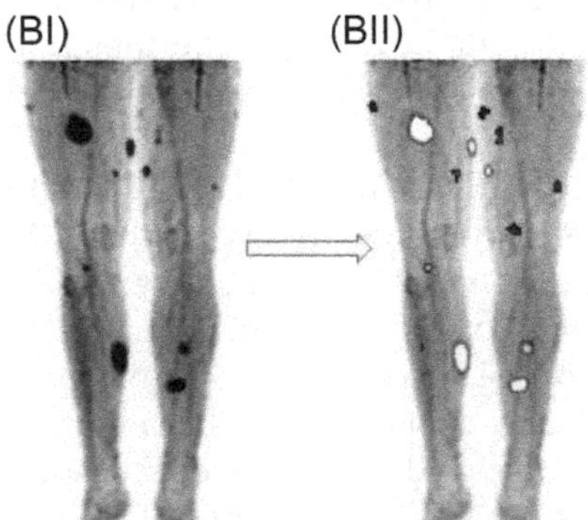

Parameter	Value
MTV	71.6
SUVmax	3.7
SUVmean	1.3
TLG	90.9

FIGURE 35.2 Image analysis in a patient with cutaneous T cell lymphoma. (AI) Tumor involvement of upper and lower extremities and abdomen and pelvis. Lesions have been segmented by quantitative analysis software (Rover) on the right side (AII). (BI) Lesions in the lower extremities. Lesions have been delineated by a quantitative analysis software (Rover; BII). The lesion metabolic activity and volume are reflected by quantitative segmentation and volumetric numbers in the tables. *MTV*, Metabolic tumor volume; *TLG* (total lesional glycolysis = MTV*mean standardized uptake value).

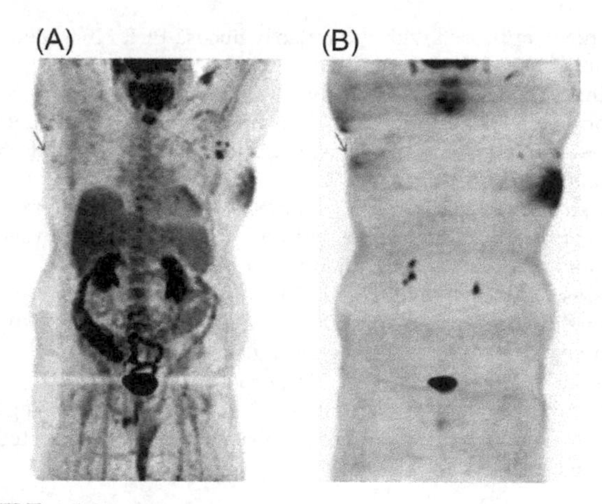

FIGURE 35.3 This patient has skin lymphoma in the right subaxillary area (*red arrows*) that did not light up in the attenuation-corrected ^{18}F-fluorodeoxyglucose–positron emission tomography (FDG-PET) scan (A) but is visualized in the nonattenuation-corrected FDG-PET scan (B).

CONCLUSION

We may wonder why we need FDG-PET/CT in assessment of cutaneous lesions when they are easily visualized. It has been shown that physical examination, in comparison with FDG-PET/CT, greatly underestimated disease burden. Although the CT component can detect the thickness of the lesion and the FDG-PET component can estimate the metabolic activity of the lesion, FDG-PET/CT might be able to provide more information about disease burden. There are some debates about the applicability of PET for detection of patchy and thin plaques because of the partial volume effect, which causes the underestimation of the FDG uptake. FDG-PET/CT nonattenuation-corrected images might overcome this problem, and it has been suggested as a more preferable tool than attenuation-corrected (AC) images (Fig. 35.3). However, no study has yet evaluated the applicability of AC images in PCL. FDG-PET/CT should be considered as a very powerful screening tool that can effectively complement the patient's history and physical examination, which are always indispensable for exact estimation of the disease burden and activity, and for ruling out infection or inflammation, which also result in high FDG uptake [25,31].

References

[1] Willemze R, et al. EORTC classification for primary cutaneous lymphomas: a proposal from the cutaneous lymphoma study group of the European Organization for Research and Treatment of Cancer. Blood 1997;90(1):354–71.

[2] Jaffe ES, et al. World Health Organization classification of neoplastic diseases of the hematopoietic and lymphoid tissues. A progress report. Am J Clin Pathol 1999;111(1 Suppl. 1):S8–12.

[3] Riou-Gotta MO, et al. Primary cutaneous lymphomas: a population-based descriptive study of 71 consecutive cases diagnosed between 1980 and 2003. Leuk Lymphoma 2008;49(8): 1537–44.

[4] Litvinov IV, et al. Demographic patterns of cutaneous T-cell lymphoma incidence in Texas based on two different cancer registries. Cancer Med 2015.

[5] Tsai EY, et al. Staging accuracy in mycosis fungoides and sezary syndrome using integrated positron emission tomography and computed tomography. Arch Dermatol 2006;142(5):577–84.

[6] Kumar R, et al. ^{18}F-fluorodeoxyglucose-positron emission tomography in evaluation of primary cutaneous lymphoma. Br J Dermatol 2006;155(2):357–63.

[7] Agar NS, et al. Survival outcomes and prognostic factors in mycosis fungoides/Sezary syndrome: validation of the revised International Society for Cutaneous Lymphomas/European Organisation for Research and Treatment of Cancer staging proposal. J Clin Oncol 2010;28(31):4730–9.

[8] Burg G, et al. Cutaneous malignant lymphomas: update 2006. J Dtsch Dermatol Ges 2006;4(11):914–33.

[9] Jerusalem G, et al. Whole-body positron emission tomography using ^{18}F-fluorodeoxyglucose for posttreatment evaluation in Hodgkin's disease and non-Hodgkin's lymphoma has higher diagnostic and prognostic value than classical computed tomography scan imaging. Blood 1999;94(2):429–33.

[10] Moog F, et al. Extranodal malignant lymphoma: detection with FDG PET versus CT. Radiology 1998;206(2):475–81.

[11] Moog F, et al. Lymphoma: role of whole-body 2-deoxy-2-[F-18]fluoro-D-glucose (FDG) PET in nodal staging. Radiology 1997;203(3): 795–800.

[12] Zinzani PL, et al. Role of [^{18}F]fluorodeoxyglucose positron emission tomography scan in the follow-up of lymphoma. J Clin Oncol 2009;27(11):1781–7.

[13] Jerusalem G, et al. Evaluation of therapy for lymphoma. Semin Nucl Med 2005;35(3):186–96.

[14] Shapiro M, et al. Assessment of tumor burden and treatment response by ^{18}F-fluorodeoxyglucose injection and positron emission tomography in patients with cutaneous T- and B-cell lymphomas. J Am Acad Dermatol 2002;47(4):623–8.

[15] Kuo PH, et al. FDG-PET/CT for the evaluation of response to therapy of cutaneous T-cell lymphoma to vorinostat (suberoylanilide hydroxamic acid, SAHA) in a phase II trial. Mol Imaging Biol 2008;10(6):306–14.

[16] Spaccarelli N, et al. Role of (18)F-fluorodeoxyglucose positron emission tomography imaging in the management of primary cutaneous lymphomas. Hell J Nucl Med 2014;17(2):78–84.

[17] Olsen E, et al. Revisions to the staging and classification of mycosis fungoides and Sezary syndrome: a proposal of the International Society for Cutaneous Lymphomas (ISCL) and the cutaneous lymphoma task force of the European Organization of Research and Treatment of Cancer (EORTC). Blood 2007;110(6): 1713–22.

[18] Willemze R, et al. WHO-EORTC classification for cutaneous lymphomas. Blood 2005;105(10):3768–85.

[19] Dan S, et al. Preliminary discussion on the value of (18)F-FDG PET/CT in the diagnosis and early staging of non-mycosis fungoides/Sezary's syndrome cutaneous malignant lymphomas. Eur J Radiol 2015;84(7):1293–8.

[20] Feeney J, et al. Characterization of T-cell lymphomas by FDG PET/CT. AJR Am J Roentgenol 2010;195(2):333–40.

[21] Kim JS, et al. Usefulness of F-18 FDG PET/CT in subcutaneous panniculitis-like T cell lymphoma: disease extent and treatment response evaluation. Radiol Oncol 2012;46(4):279–83.

[22] Mitsuhashi K, et al. Positron emission tomography revealed diffuse involvement of the lower legs and occult extracutaneous lesions in subcutaneous panniculitis-like T-cell lymphoma. Clin Nucl Med 2013;38(3):209–11.
[23] Schramm N, et al. Subcutaneous panniculitis-like T-cell lymphoma with breast involvement: functional and morphological imaging findings. Br J Radiol 2010;83(989):e90–4.
[24] Babb A, et al. Subcutaneous panniculitis-like T-cell lymphoma with extracutaneous dissemination demonstrated on FDG PET/CT. Am J Hematol 2011;86(4):375–6.
[25] Kuo PH, et al. FDG-PET/CT in the evaluation of cutaneous T-cell lymphoma. Mol Imaging Biol 2008;10(2):74–81.
[26] Kang BS, et al. Subcutaneous panniculitis-like T-cell lymphoma: US and CT findings in three patients. Skeletal Radiol 2007;36(Suppl. 1):S67–71.
[27] Berkowitz A, et al. Determination of whole-body metabolic burden as a quantitative measure of disease activity in lymphoma: a novel approach with fluorodeoxyglucose-PET. Nucl Med Commun 2008;29(6):521–6.
[28] Kim TM, et al. Total lesion glycolysis in positron emission tomography is a better predictor of outcome than the International Prognostic Index for patients with diffuse large B cell lymphoma. Cancer 2013;119(6):1195–202.
[29] Basu S, et al. FDG PET/CT methodology for evaluation of treatment response in lymphoma: from "graded visual analysis" and "semiquantitative SUVmax" to global disease burden assessment. Eur J Nucl Med Mol Imaging 2014;41(11):2158–60.
[30] Rossi C, et al. Interim ^{18}F-FDG PET SUVmax reduction is superior to visual analysis in predicting outcome early in Hodgkin lymphoma patients. J Nucl Med 2014;55(4):569–73.
[31] Bakshi A, et al. Assessing cutaneous Psoriasis activity using FDG-PET: nonattenuation corrected versus attenuation corrected PET images. Clin Nucl Med 2015.

CHAPTER

36

Imaging Cutaneous Squamous Cell Carcinoma of the Head and Neck

G.C. *Casazza*, M.M. *Monroe*

University of Utah School of Medicine, Salt Lake City, Utah, United States

OUTLINE

INTRODUCTION

Nonmelanoma skin cancers are the most common malignancies in the United States, with an estimated 3.5 million cases occurring annually [1]. Cutaneous squamous carcinoma makes up roughly 20% of all nonmelanoma skin cancers, with an estimated 700,000 cases occurring annually in the United States and an estimated incidence of up to 316 cases per 100,000 person-years for both the northern and southern United States combined [2]. Squamous cell carcinoma is the second most common carcinoma, only second to basal cell carcinoma in incidence. Basal cell carcinoma, however, is a relatively benign neoplasm with a low propensity for invasive or metastatic spread. Unlike basal cell carcinoma, cutaneous squamous cell carcinomas are associated with a substantial risk of metastasis.

At least 75% of all cases of cutaneous squamous cell carcinoma occur within the head and neck [3,4], this is likely the result of increased exposure to ultraviolet B radiation. Those living in areas with high sun exposure, such as Australia and the southern United States have a demonstrated increased risk for cutaneous squamous cell carcinoma. For each 8–10 degrees of latitude closure to the equator, there is a doubling of that risk [5]. A lifetime of sun exposure is the most important risk factor in the development of head and neck cutaneous squamous cell carcinoma. Incidence will increase with age, with a mean age at diagnosis of 72 years. Furthermore, an older age at diagnosis is typically associated with more aggressive and larger lesions at the time of presentation [6]. Males will have an increased risk for developing both invasive and in situ lesions, with overall incidences of 2154 per 100,000 persons and 736 per 100,000 persons,

Imaging in Dermatology
http://dx.doi.org/10.1016/B978-0-12-802838-4.00036-4

respectively [7]. This is in contrast to incidence in females of invasive and in situ disease with 1265 per 100,000 persons and 681 per 100,000 persons diagnosed, respectively [7]. Head and neck lesions account for 60% of invasive lesions in men compared with only 43% in women [7]. In addition, invasive lesions larger then 2.0 cm are significantly associated with male gender, and there is unfortunately a longer delay in seeking treatment [6].

Additional risk factors include chemical or radiation exposure, human papilloma virus, chronically injured skin (ie, ulcers, burns, chronic skin disease), and patients with premalignant lesions and/or history of a previously diagnosed cutaneous squamous cell carcinoma. 12–25% of patients with premalignant actinic keratoses will eventually develop squamous cell carcinoma [4]. In addition, immunosuppression has been shown to dramatically increase the risk of cutaneous squamous cell carcinoma [8,9]. Recipients of solid organ transplants have a 65–250 times increased risk of developing cutaneous squamous cell carcinoma when compared with the general population, and a subsequently increased risk of more aggressive disease, recurrence, and distant metastasis [9]. This risk correlates directly with the extent of immunosuppression.

CLINICAL PRESENTATION

Aggressive forms of cutaneous squamous cell carcinoma commonly occur on the head and neck. These lesions are typically located in high sun-exposure areas, with more aggressive lesions more commonly found on the cheeks, auricle, forehead, and temple, and less likely on the lips, nose, and neck [3]. These lesions often present as papules or plaques with varying textures and colors. Primary lesions are characteristically firm, skin-colored or pink, and smooth or hyperkeratotic in texture. Ulceration and easy bleeding of the lesion may be present, particularly with disease progression. Often patients will report persistent itching or irritation of the lesion or excessive bleeding with minor trauma to the lesion [10]. At presentation, these lesions may be fixed to the underlying bone or cartilage, likely indicating a worse prognosis and more aggressive treatment. The initial clinical examination of these patients should focus on the lesion with a critical appraisal of involved anatomical regions, lesion mobility and involvement of the underlying cartilage and bone, neurological deficits suggestive of perineural extension of the tumor, and regional lymphadenopathy. This knowledge is critical for accurate risk-stratification and determination of the need for further evaluation with anatomical imaging.

Additional information from the patient's biopsy can help with the decision as to when imaging is necessary. Evidence of lymphovascular invasion, perineural tumor spread, poorly differentiated histology, and increasing depth of invasion are all poor prognostic indicators [11]. Along with the clinical examination, these findings will help inform the decision on whether or not imaging is required.

Head and neck cutaneous squamous cell carcinomas are typically more aggressive then lesions arising on other parts of the body. In addition, there are extensive neurological, vascular, and lymphatic networks throughout the head and neck that provide numerous pathways for spread. Features of primary lesions that suggest a more aggressive course include horizontal size, depth of invasion, tumor location, and histological differentiation:

Horizontal Size

In general, lesions greater then 20 mm in horizontal size are more aggressive. Some have even identified lesions as small as 18 mm as being more aggressive [11,12]. Lesions this size will have an increased risk of invasion into the underlying adjacent structures and distant spread.

Depth of Invasion

Depth of invasion, as noted on excisional or punch biopsy, is a strong prognostic indicator suggesting aggressive disease. As in cutaneous melanoma, it has been demonstrated to be an independent risk factor for identifying aggressive disease. Lesions less then 2.0 mm in depth have a no or low risk for invasion, lesions 2.0–5.9 mm have an intermediate risk, and lesions equal or greater then 6.0 mm have a high risk for metastasis and local invasion [12,13].

Histological Differentiation

A high degree of desmoplasia is suggestive of a more aggressive lesion. These poorly differentiated lesions are typically larger in horizontal size and invade deeply into the underlying tissue, both poor prognostic indicators suggesting aggressive disease [11].

Tumor Location

Squamous cell carcinomas located on the head and neck are typically more aggressive than lesions elsewhere on the body. Lesions located on the auricle and in the periocular region tend to be more aggressive then lesions elsewhere on the head and neck. In both of these regions, the thin skin overlying bone and an extensive array of vascular, lymphatic, and neurological structures provides easy access for spread. In addition, as these lesions invade into the underlying tissue, vital structures, such as the facial nerve, orbit, and cavernous sinus are at an increased risk of damage by the tumor.

IMAGING OF THE PRIMARY LESION

A thorough clinical examination is an important step in the evaluation of all patients with cutaneous squamous cell carcinoma, because even those who are relatively asymptomatic may harbor extensive disease. Perineural invasion and cervical lymphadenopathy may not become symptomatic until a late stage. For this purpose, imaging is used to provide vital staging and prognostic information, and will help guide the most appropriate treatment.

Evaluation of Bone and Cartilage Invasion

Invasion into the underlying bone or cartilage suggests an aggressive lesion. These lesions may be fixed to the underlying bone or cartilage and are typically hypomobile on examination. Bone and cartilage invasion will characteristically occur in areas with little subcutaneous fat or muscle, such as the scalp and ear, both the pinna and external auditory canal. Fixation to the scalp with invasion into the skull is a common presentation for primary lesions of the scalp. On examination, these lesions will be fixed with decreased or absent mobility compared with the surrounding scalp. Bone invasion may be relatively asymptomatic initially; however, invasion may progress into the cranial vault and eventual spread to the meninges is possible.

Bone invasion is often the one symptom of cutaneous squamous cell carcinoma that is easily identified on physical examination; however, early bone invasion or evaluation of the extent of the disease is difficult when relying on physical examination alone. Utilization of radiographical imaging can be helpful in determining early invasion and the extent of the disease. Computed tomography (CT) is the modality of choice when evaluating a fixed or hypomobile lesion. Characteristically CT evidence of bony invasion includes destruction or increased lucency of the bone underlying the lesion, giving it moth-eaten characteristics. One of the most common regions of bony invasion for cutaneous squamous cell carcinoma is with involvement of the underlying skull in cancers located on the scalp. Early erosion of the calvarium is most commonly characterized by a loss of the smooth line of the outer cortex (Fig. 36.1). In extreme cases, the tumor may be visualized through the entire extent of the bone (Fig. 36.2). Magnetic resonance imaging (MRI) may also demonstrate features characteristic of bone invasion, such as abnormal enhancement of the bone and tissue surrounding the bone or loss of the normal marrow signal on T1 imaging.

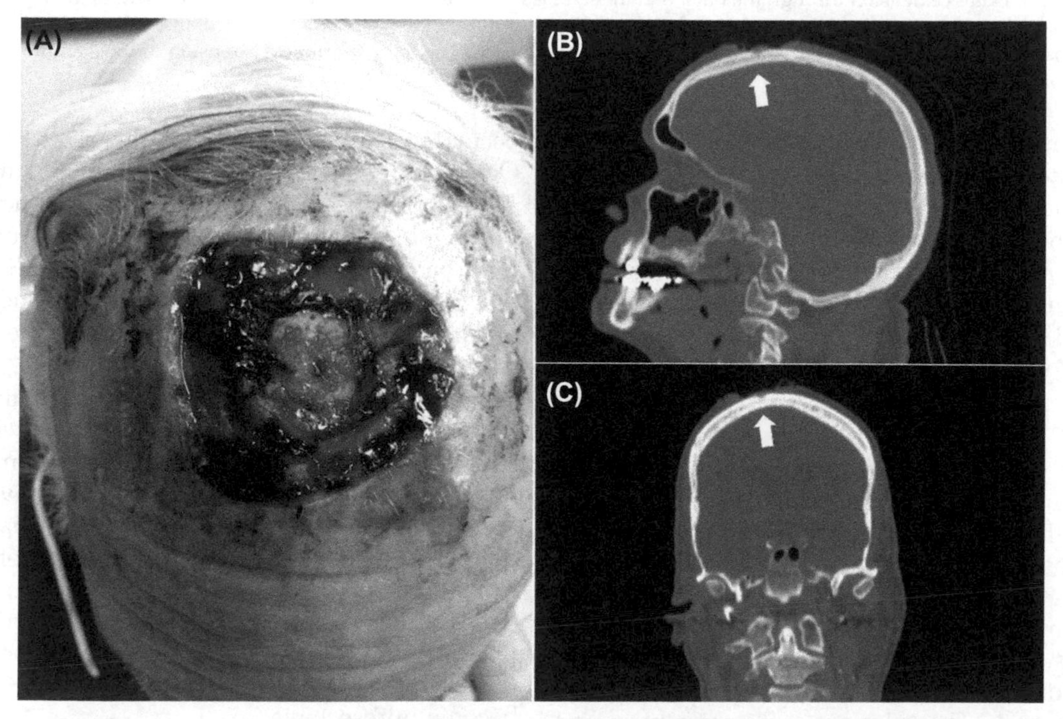

FIGURE 36.1 Recurrent squamous cell carcinoma of the scalp with invasion of the calvarium. (A) Patient had previously undergone MOHS excision, revealing obvious involvement of the underlying calvarium. (B and C) Computed tomography (CT) scan of the head demonstrating extension through the outer layer of the calvarium. Note the loss of the smooth line of the along the outer cortex of the skull (*arrows*).

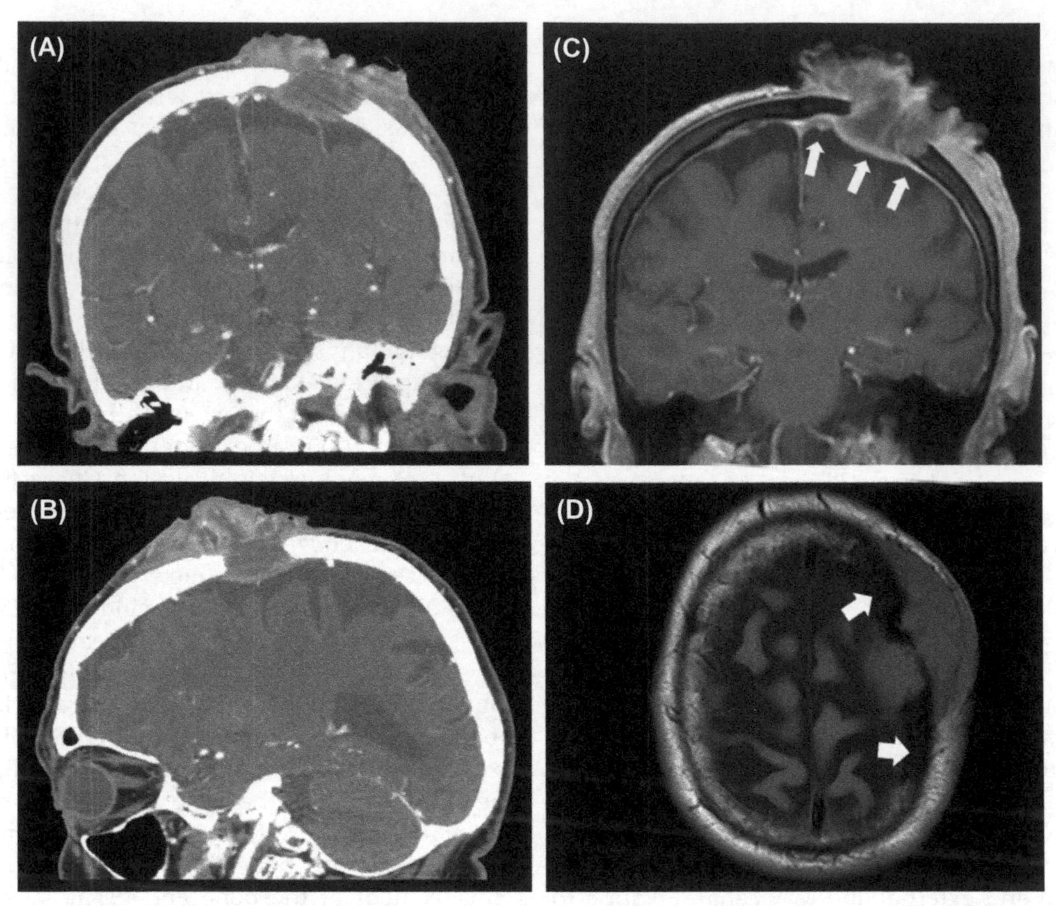

FIGURE 36.2 Squamous cell carcinoma of the scalp with full-thickness extension through the calvarium to involve the underlying dura. Computed tomography (CT) and magnetic resonance imaging (MRI) have complementary roles in the evaluation of this patient. (A and B) CT scan demonstrating full-thickness extension through the outer and inner tables of the skull. (C) T1 contrast-enhanced MRI demonstrating intracranial extension of the scalp squamous cell carcinoma to involve the underlying dura, which appears thickened and enhanced (*arrows*). (D) T1 MRI demonstrating loss of the normal marrow signal within the skull (*arrows*), suggestive of subclinical extension of the squamous cell carcinoma through the diploic layer of the skull.

If there is any concern for extension into the cranial vault, dedicated imaging of the brain should be performed. CT may demonstrate the primary lesion extending through the skull with mass effect on the brain. Early involvement may be more readily detected on MRI, where abnormal enhancement of the dural meninges may be noted (see Fig. 36.2C).

Imaging Evaluation of Perineural Tumor Spread

Perineural invasion can be a challenging diagnosis. Whereas it can be recognized clinically with obvious cranial neuropathological conditions, the difficulty comes when diagnosing clinically silent perineural invasion before surgical resection, because it has the potential to alter treatment and patient prognosis. Perineural invasion is the result of tumor spread along the epineurium, allowing for dissemination of the neoplasm throughout the head and neck. These lesions will typically spread anterograde towards the skull base and central nervous system, although retrograde spread is possible, typically occurring at junction points (ie, the trigeminal ganglion). Occasionally, spread along the epineurium will break out beyond the nerve sheath, causing macroscopic deposits of disease. The extent of perineural invasion may be difficult to assess clinically, radiographically, and surgically because of skip lesions and small rests of tumor cells propagating along small nerves that are easily missed.

Perineural invasion is a relatively common occurrence in cutaneous squamous cell carcinoma with 2.5–14% of lesions demonstrating some amount of perineural invasion [21,22]. Clinically it can have a dramatic effect on morbidity and mortality for patients when diagnosed with it. There is nearly a threefold increase in the rate of local–regional recurrence, and a 30% decrease in the overall disease-specific survival [23–26]. In addition, as these lesions extend up to and through the skull base, they have the potential to become unresectable.

Perineural invasion can be grossly separated into two categories: definite clinical and/or radiographical evidence of perineural invasion, and microscopic disease only detected after surgical excision and histopathological analysis. The more clinical or radiographical evidence of disease, the worse the prognosis is for the patient when compared with patients without demonstrated perineural invasion [27]. Primary lesions at risk for perineural invasion have similar features to what has previously been characterized as high-risk cutaneous squamous cell carcinoma. Increasing horizontal size of the primary lesion, depth of invasion, and poor histological differentiation all increase the risk for perineural invasion at the initial diagnosis [11,12]. Furthermore, recurrent lesions have the highest risk for perineural invasion, likely due to rests of tumor cells along nerves, suggesting missed perineural invasion of the initial primary lesion.

Symptoms associated with perineural invasion can be insidious, and initially can be easily missed by clinicians because patients may have vague complaints of facial weakness, numbness, or hoarseness. The most common nerves initially affected are the trigeminal nerve [cranial nerve (CN) V] and the facial nerve (CN VII) [28]. Patients with trigeminal nerve involvement will commonly have incomplete numbness and tingling of a single trigeminal nerve distribution. They may initially describe symptoms similar to formications (sensation of ants crawling on or under the skin), with eventual progression of their symptoms towards pain and complete paresthesias [28].

Involvement of the facial nerve can present with subtle weakness in a single branch of the facial nerve to complete paralysis of one side of the face. In some, spread to the facial nerve and facial weakness may develop as tumor spreads along normal anatomical connections between the trigeminal and facial nerves. In most cases, however, involvement of the facial nerve is the result of direct extension from adjacent auricular squamous cell carcinoma or regionally metastatic disease in the parotid region. It would not be surprising that some of these patients will have been mistakenly diagnosed with Bell palsy or trigeminal neuralgia in the interim. In rare and advanced cases of perineural invasion, there may be involvement of the lower cranial nerves, such as the vestibulocochlear nerve (CN VIII) with associated changes in hearing and balance; the glossopharyngeal nerve (CN IX), leading to alteration in the patient's taste; the vagus nerve (CN X), causing a hoarse voice; and the hypoglossal nerve (CN XII) leading to aberrant tongue protrusion and movement [29–31].

If there is clinical evidence of perineural invasion, there is likely also radiographical evidence of disease. It is rare to have radiographically evident disease in an asymptomatic patient. The radiographical diagnosis of perineural invasion can be challenging. The diagnosis may be missed in 50–85% of patients on pretreatment imaging [23]. This difficulty in diagnosis can be because of the small size of the affected nerves, the presence of skip lesions, or microscopic disease along the nerve that cannot be readily visualized with conventional imaging techniques. Even if there is histologically proven perineural invasion, retrospective review of pretreatment imaging may not demonstrate any clear evidence of disease [23].

MRI is the most sensitive imaging modality to evaluate for perineural invasion. Abnormal enhancement or enlargement of nerves and obliteration of the fat plane surrounding the nerve at the skull base foramen, such as the inferior alveolar canal, infraorbital foramen, and foramen rotundum, are characteristic findings on MRI of perineural invasion (Fig. 36.3). Care should be taken to also evaluate imaging for downstream effects of nerve invasion, including muscle atrophy and fatty replacement, both which can be readily identified on conventional MRI. In addition, use of high-powered scanners, such as 3.0-T scanners have been demonstrated to improve spatial resolution of head and neck structures, increasing the identification of disease [32]. Although MRI is the modality of choice when imaging specifically for perineural invasion, it cannot as easily identify the extent of perineural invasion with as much sensitivity because of skip lesions and microscopic tumor spread.

Alternative imaging options to assess a patient with suspected perineural invasion include CT and positron emission tomography (PET)–CT (PET-CT). Ultrasound lacks the resolution and ability to evaluate nerves and is not useful in the evaluation of perineural invasion. With CT imaging, although it is lacking definition to delineate small changes in soft-tissue structures, perineural invasion may still be suggested by enlargement of neural foramina or bony erosion at the periphery of the foramina. In some instances of microscopic invasion, enlargement and erosion of the neural foramen may be the only evidence of disease.

The utility of PET-CT in the diagnosis of perineural invasion is questionable [33]. PET-CT has the disadvantage of low image quality and inability to distinguish lesions within close proximity to one another. Whereas affected nerves can be as small as 0.1 mm, PET-CT is not reliable in evaluating lesions smaller then 4–8 mm in size. If the diagnosis of perineural invasion is called into question and cannot be demonstrated with conventional imaging, PET-CT may be useful in identifying an area of increased metabolism, especially at the skull base, that may suggest potential perineural invasion.

When imaging the primary lesion, evaluating the immediate surrounding tissue for any evidence of perineural invasion is crucial because it can provide important prognostic and treatment planning information with inadequate resection, leading to increased risk of

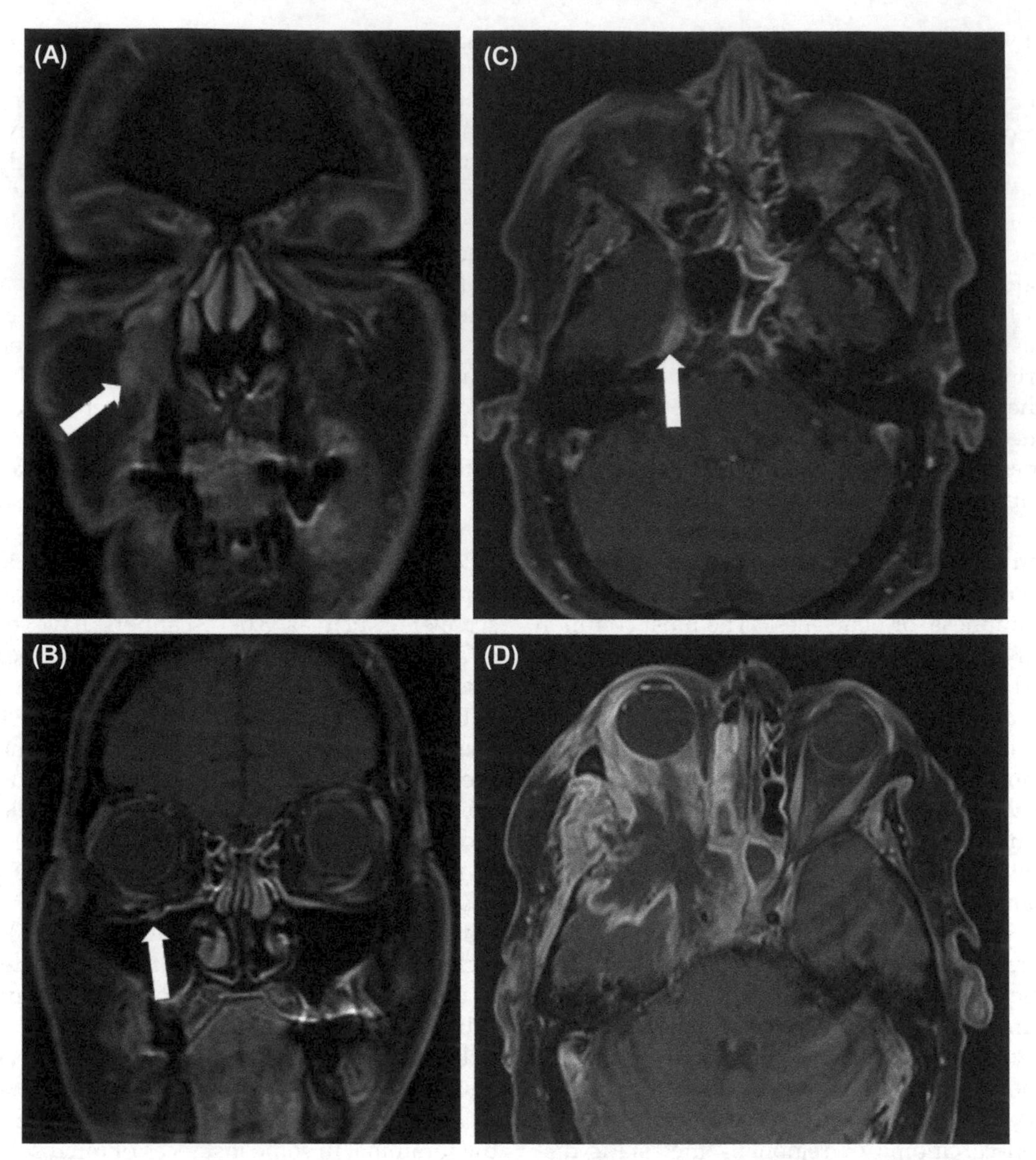

FIGURE 36.3 Squamous cell carcinoma of the right cheek with extensive perineural tumor extension. (A) Contrast-enhanced MRI of a right cheek squamous cell carcinoma demonstrating an enhancing subcutaneous mass in the medial right cheek (*arrow*). (B) Enhancement and enlargement of the right infraorbital nerve (*arrow*) consistent with perineural tumor extension. (C) Extension along the infraorbital nerve to the Meckel cave (*arrow*). (D) Despite treatment, the patient developed progressive cranial neuropathies and progression along the base of the skull and orbit.

recurrence of disease. Given the challenge of identifying direct evidence of perineural invasion, evaluating the muscles of the face, tongue, and palate for atrophy and fatty replacement may be the most reliable method for identifying evidence of disease.

Imaging Evaluation of Squamous Cell Carcinomas of the Ear

The anatomical position of the auricle makes it particularly susceptible to an increased exposure to ultraviolet radiation and more prone to actinic damage. Roughly 5–10% of cutaneous malignancies involve the external and/or the internal ear, including both the auricle and the external auditory canal [14]. The distinction between the involvement of the internal and external ear can have a dramatic impact on associated morbidity and prognosis for the patient. Tumors involving the external auditory canal have a much higher risk for spread into the parotid gland and periparotid lymph node basins, as well as the upper cervical lymph nodes because of the bony anatomy of the temporal bone [15]. Spread through the fissures of Santorini and Huschke and spread through the tympanomastoid suture can lead to dissemination into the adjacent parotid space, and the stylomastoid and jugular foramen, causing lower cranial nerve involvement (Fig. 36.4). Further spread along the skull base can lead to involvement of the petrous apex, with spreading to involve higher cranial nerves [16].

It is important to differentiate between auricular and external auditory canal involvement. The distinguishing line is at the bony cartilaginous junction of the external auditory canal; however, cartilage invasion in auricular primary lesions can make distinguishing these lesions difficult. Differentiating these two subtypes is important because the spread of disease will differ dramatically, as does the degree of surgical intervention (Fig. 36.5). The auricle drains into the preauricular and postauricular nodal basins, whereas the external auditory canal typically drains into the both the preauricular and postauricular nodal basins, the superficial parotid nodal basins, and the jugulodigastric nodes [14]. Spread of disease can vary between the two subtypes, with disease of the external auditory canal spreading directly through the fissures of Santorini and Huschke into the preparotid space with direct involvement of the parotid region (see Fig. 36.4A). The auricle does not typically have such extensions because the auricular skin is more tightly adherent in the anterior direction then the posterior direction; extension of auricular primary lesions typically spreads in an anterior to posterior direction [14,17].

The imaging modality chosen will vary depending on the location of the lesion, features of the lesion, and associated symptoms. Lesions involving the bony portion of the external auditory canal have a high propensity for invasion into and through the temporal bone. A dedicated temporal bone CT scan is the imaging modality of choice to evaluate for bone invasion. These scans typically result in high-resolution 1-mm cuts through the temporal bone, giving detailed views of the anatomy. Determining the extent of bony invasion is important when planning for surgical resection, because it can distinguish the need for a total or subtotal temporal bone resection versus a simpler sleeve resection.

As disease progresses, there may be associated cranial neuropathies caused by perineural spread of disease. Signs and symptoms associated with cranial neuropathies can be facial weakness, dysphagia, hoarseness, and facial numbness. The most common cranial nerve that can be affected is the facial nerve (CN VII). In one study, 33% of patients with primary lesions involving the external auditory canal had involvement of the facial nerve at presentation [16]. Involvement of the facial nerve can be secondary to direct extension into the middle ear to involve the tympanic portion of the facial nerve, or by invasion through the tympanomastoid sutures to involve the mastoid segment of the facial nerve as it exits the stylomastoid foramen.

As these lesions extend along the skull base, they can involve the jugular foramen, leading to symptoms of dysphagia (affecting the glossopharyngeal nerve, CN IX, and the vagus nerve, CN X), shoulder weakness (spinal accessory nerve, CN XI), and tongue deviation with extension to the hypoglossal canal (hypoglossal nerve, CN XII). Further extension through the skull base will lead to involvement of the petrous apex and possible involvement of the trigeminal nerve (CN V) and abducens nerve (CN VI), causing facial paresthesias and decreased extraocular movements, respectively. If a patient has any symptoms of cranial neuropathies, an MRI should be obtained to evaluate for the extent of perineural invasion. An MRI should be considered even in the absence of symptoms for advanced disease,

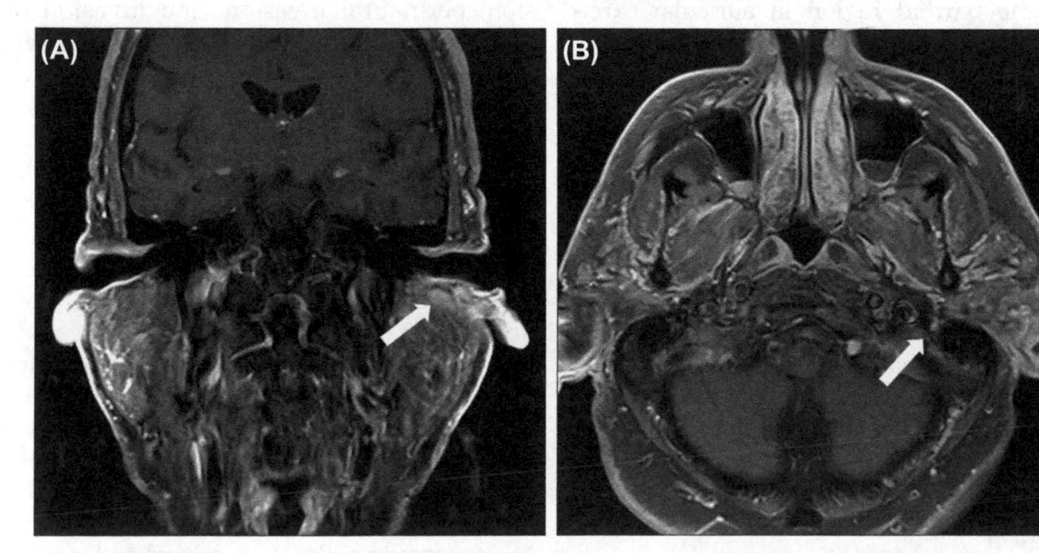

FIGURE 36.4 Contrast-enhanced T1 MRI of a left conchal bowl squamous cell carcinoma. (A) Extension through the cartilaginous external auditory canal to involve the adjacent parotid gland (*arrow*). (B) Enhancement of the facial nerve at the stylomastoid foramen (*arrow*) consistent with perineural tumor spread.

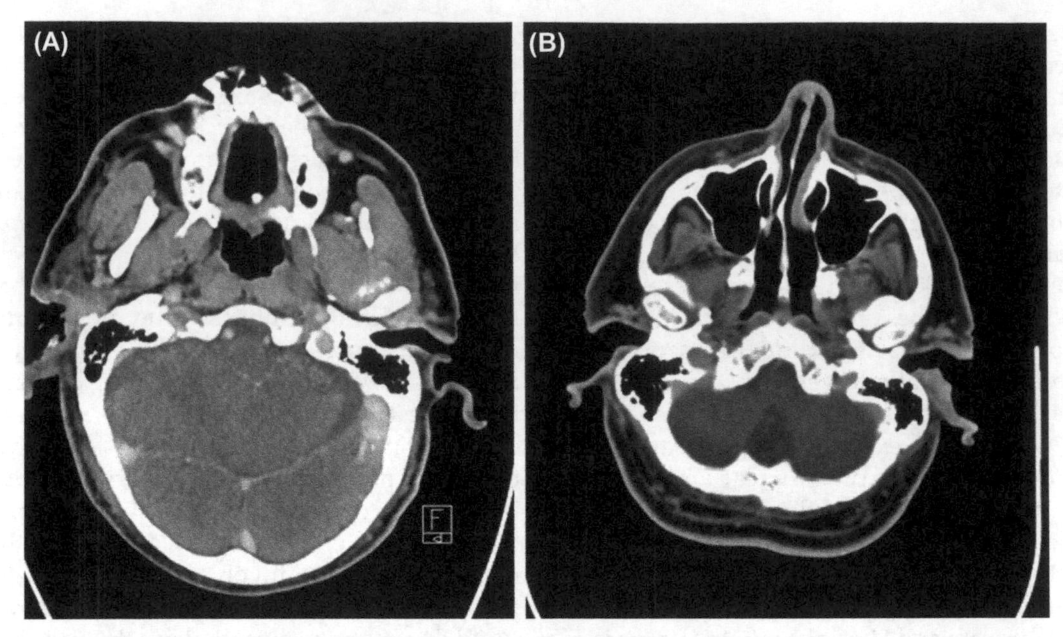

FIGURE 36.5 Two examples of CT scans of squamous cell carcinomas of the external ear. These examples demonstrate the importance of preoperative imaging to assist in determining depth of extension into the external auditory canal. In both (A) squamous cell carcinoma of the right conchal bowl, and (B) squamous cell carcinoma of the left conchal bowl, the soft-tissue mass of the tumor extends to the bony cartilaginous junction of the external auditory canal. In each of these cases a lateral temporal bone resection was required to ensure adequate surgical margins.

because early perineural invasion is typically asymptomatic.

Whereas the evaluation of the external auditory canal is relatively straightforward, when considering auricular lesions that do not extend past the bony cartilaginous junction of the external auditory canal, the evaluation can become complicated. The risk of regional spread is lower than when the primary lesion is located on the auricle. Several studies have proposed up to an 18% initial incidence of parotid involvement in auricular carcinoma [18,19]. One study demonstrated a 37.5% involvement of the parotid region in auricular carcinomas [14]. Size appeared to be the primary determinate in risk of spread to the parotid with lesions greater then 4.0 cm in horizontal size having the highest propensity to spread to the parotid space; however, all lesions greater then 2.0 cm had an increased risk of spread to the parotid region [14]. In addition, lesions with any demonstrated parotid or periparotid lymph node involvement have a significantly increased risk for both vascular involvement and perineural invasion [14,15,20]. Therefore even in isolated auricular lesions, an appropriate work-up evaluating for bony, vascular, and perineural spread is appropriate.

Imaging for Squamous Cell Carcinomas of the Periorbital Region

Cutaneous squamous cell carcinoma is the second most common cutaneous neoplasm of the periorbital region, second only to basal cell carcinoma. Cutaneous squamous cell carcinoma within the immediate periorbital region, including the upper and lower eyelid, forehead, lateral and medial canthus, and infraorbital region, pose diagnostic and treatment challenges because of the unique anatomical considerations within the periocular region and the functional and aesthetic impact of resection and reconstruction. The concern with lesions within the periorbital region is extension into the orbit. This extension can occur through a variety of pathways including direct extension, perineural invasion, and invasion of lymphatic and vascular structures. As disease spreads into the orbit, there can be further spread into the cavernous sinus, meninges, or the middle and/or posterior cranial fossa.

The rate of all forms of orbital extension is reported at up to 5.9% [34]. Direct extension into the orbit should be expected with large (ie, greater then 20 mm) ulcerative lesions in the immediate proximity to the orbit. On clinical examination, there may be ocular mobility restrictions, epiphora (uncontrolled tearing), and dysesthesia (abnormal sensation, often pain) [34]. Imaging may demonstrate mass effect within the orbit or invasion into the extraocular muscles. Whereas direct extension is a common pathway of invasion into the orbit, more commonly invasion results from perineural invasion with traveling towards the central nervous system. Most commonly, the supraorbital and infraorbital nerves are involved; however, perineural invasion of the trigeminal nerve (CN V), facial nerve (CN VII), and branches of the oculomotor, trochlear, and abducens nerves (CN III, CN IV, and CN VI, respectively) with retrograde travel into the orbit can be seen. Involvement of CN V and CN VII allows for even distant lesions throughout the head and neck to spread into and to involve the orbit.

Symptoms caused by perineural invasion of the orbit include diplopia (double vision), ophthalmoplegia (visual impairments), ptosis of the upper eyelid, and sensory deficits in the periorbital region. In addition, strabismus (misalignment of the eyes), hypoglobus or hyperglobus, and proptosis may also be noted [35]. As with symptomatic perineural invasion outside of the orbit, any evidence of these orbital symptoms signifies advanced disease. In very severe disease, there may even be evidence of orbital apex syndrome (oculomotor and ophthalmic nerve dysfunction leading to ophthalmoplegia, visual disturbances, and hypoesthesia of the upper face) or orbital fissure syndrome (involvement of CN III, CN IV, V1, and VI, and the ophthalmic vein, which leads to a combination of diplopia, paralysis of extraocular muscles, exophthalmos, and ptosis), both indicating spread into the cavernous sinus [36].

To define the detailed anatomy of the orbit, fine-cut CT or MRI is needed to evaluate for gross or microscopic orbital disease. Obvious lesions causing mass effect of the orbital structures are readily visible on both CT and MRI imaging. When there is concern for perineural invasion, fine-cut MRI, evaluating for abnormal enhancement or enlargement of nerves or fatty replacement within the extraocular muscles can be seen. CT can demonstrate bony erosion and widening of the supraorbital and infraorbital foramina, bony erosion around the superior orbital fissure, and invasion into the paranasal sinuses. If there is concern for extension into the cavernous sinus, thin-slice MRI ($\leq$3 mm) with contrast enhancement is the imaging modality of choice, because the cavernous sinus is largely composed of two layers of interposed dura. Imaging should include the orbital apex to the prepontine cistern, and thin slices should be obtained to allow visualization of individual cranial nerves. CT is best performed using intravenous contrast and sections that are less then 1 -mm thick.

Lymphatic invasion will occur in up to 24% of lesions with dissemination to both the periparotid and cervical lymph nodes [35]. Lymphatic spread is more commonly associated with lesions located on or near the medial canthus. Anatomical studies have demonstrated that the region surrounding the medial canthus will more readily drain to the submandibular nodes via the facial nodes spread along the facial vein, whereas the lateral portion of the eye will drain into the periparotid lymph node basin [37].

IMAGING FOR REGIONAL AND DISTANTLY METASTATIC DISEASE

Whereas most head and neck cutaneous squamous cell carcinomas are not associated with metastatic disease, spread to both regional lymph nodes and distant sites can occur. Lesions greater than or equal to 20 mm in the horizontal direction have been reported as a critical size for differentiating high-risk lesions from those considered to be low risk for nodal metastasis [11,12]. The first echelon for nodal metastasis differs based upon the anatomical position of the primary lesion, but commonly includes the periparotid and intraparotid lymph node basins and upper cervical lymphatics, mostly because of their proximity to high-risk locations for cutaneous squamous cell carcinoma [38].

Metastatic cutaneous squamous cell carcinoma is one of the most common parotid malignancies encountered. In patients with high-risk primary lesions, risk of parotid metastasis can run as high as 55–68% [38,39]. Clinical diagnosis of metastatic spread to the parotid region can be difficult, with occult rates of parotid metastases reported as high as 36% [38]. Therefore radiographical diagnosis confirming involvement of the parotid and cervical lymph nodes in the absence of a clinically negative neck is crucial, because these are typically the first level of metastatic spread.

If there is clinical suspicion of metastatic spread to the parotid region or cervical lymph node basin, imaging can be useful in understanding the extent of disease, or in evaluating for occult disease. Ultrasound is a useful technique for the initial assessment of parotid and cervical lymphadenopathy. Ultrasound is cheap, readily available, and avoids exposure to ionizing radiation. Ultrasound can be used to differentiate benign versus malignant lesions in up to 90% of cases, and can distinguish glandular from extraglandular masses with 98% accuracy [40]. In addition, ultrasound allows for fine-needle aspiration of lesions and guided biopsies, which can increase the diagnostic yield of the imaging. Despite the simplicity, safety, and ease of use, ultrasound is not without drawbacks. It cannot readily identify the facial nerve; it lacks the ability to evaluate deep parotid masses, pharyngeal extension, or lymphadenopathy extending into the retropharyngeal and deep neck spaces; and it cannot demonstrate evidence of any intracranial or skull base extension.

MRI is useful in distinguishing soft-tissue planes and the spatial resolution within the parotid gland. T1-weighted images can readily identify parotid lesions against the hyperintense background of parotid tissue and can easily distinguish the margins of the lesion. MRI is the imaging modality of choice when evaluating for perineural invasion, and is excellent for imaging the parapharyngeal space. As with CT, MRI can be used to evaluate lesions within the deep lobe or those that are not clinically palpable. CT is also useful to evaluate for any bony erosion or adenopathy that extends into the pharynx or deep neck. Both MRI and CT are superior to PET-CT, because PET-CT does not readily distinguish between benign and malignant lesions or provide the distinguishing detail required in the small spaces of the head and neck, especially when there are multiple lesions in proximity to one another.

Spread of disease to involve the cervical region will significantly worsen the prognosis [38,41], making accurate assessment of the neck either clinically or radiographically crucial in determining any management decisions. Traditionally the neck is divided into six levels. Whereas these levels were designed to aid in the surgical management of the neck, understanding the basic levels can guide both the clinical and radiographical interpretation and communication [42]:

Level I is divided into two groups: level IA (submental group) and level IB (submandibular group). Level IA is defined as the lymph nodes within a triangular boundary formed by the anterior belly of the digastric muscle and the hyoid bone, and level IB is defined as the lymph nodes within the boundaries of the anterior and posterior bellies of the digastric muscle, the stylohyoid muscle, and the body of the mandible.

Level II is comprised of nodes surrounding the upper one-third of the internal jugular vein extending from the skull base to the level of the inferior border of the hyoid bone. The medial border of level II is the lateral border of the sternohyoid and stylohyoid muscle, and the lateral border is the posterior border of the sternocleidomastoid muscle.

Level III, or the *midjugular group*, is the group of lymph nodes extending from the inferior border of the hyoid bone to the inferior border of the cricoid cartilage. The medial boundary is the lateral border of the sternohyoid muscle and the lateral border is the posterior border of the sternocleidomastoid muscle.

Level IV, or the *lower jugular group*, is the group of lymph nodes encompassing the lower one-third of the internal jugular vein from the inferior aspect of the cricoid cartilage to the clavicle. The medial border is the lateral border of the sternohyoid muscle, whereas the lateral border is the posterior border of the sternocleidomastoid muscle.

Level V makes up the posterior triangle of the neck. It is bordered superiorly by the convergence of the sternocleidomastoid and the trapezius muscles, inferiorly by the clavicle, medially by the posterior border of the sternocleidomastoid muscle, and laterally by the trapezius muscle.

Level VI is the anterior or central compartmental group of lymph nodes, including the pretracheal and paratracheal lymph nodes, the precricoid lymph nodes, and the perithyroid nodes. The superior border is the hyoid bone, the inferior border is the sternal notch, and the lateral boundaries are the common carotid arteries.

Whereas metastatic spread to the neck is more readily apparent on clinical examination then metastatic spread to the parotid region, evaluation of the neck for occult disease is still an important step in pretreatment planning. The patterns of cutaneous squamous cell carcinoma metastasis are predominately related to the location of the primary lesions on the head and neck [43]. Traditionally the anterior portion of the head and neck is composed of the anterior scalp and face, and the posterior portion of the head and neck is made up of the posterior scalp and neck. Levels II and III are the most commonly involved cervical levels and if there is any neck disease, these levels will almost always be involved. Isolated level I disease typically results from lesions on the anterior portion of the head and neck, whereas level I disease associated with a posterior lesion will always be associated with coinvolvement of levels II and III. Involvement of levels IV and V is more common in posterior lesions. Any involvement of levels IV and V will always be associated with levels II and III [15,38,43].

Patients with no evident lymph node metastasis to the neck can be treated with an elective neck dissection, sentinel lymph node biopsy, elective radiation, or conservative management with serial examinations. Therefore pretreatment assessment and identification of cervical lymph node metastases is crucial. Affected nodes may lie deep or may be too small to be palpated on clinical examination, making radiography necessary for the diagnosis. Although clinical examination remains the most convenient method for evaluating the neck, it does require significant skill and has a diagnostic accuracy ranging between 59% and 84% [44] because of difficulty palpating nodes less than 1 cm in diameter. Imaging can reveal these clinically silent nodes, and if properly performed has the potential to define their anatomical landmarks, giving information to classify the nodes in the neck before surgery.

Most patients with head and neck cancer will receive a CT scan or MRI evaluating the neck. Only in patients with a small primary lesion (<10 mm), where clinical suspicion for metastatic disease is low, can the decision not to image the patient be safely made. However, because of the significant prognostic implications of cervical metastases, imaging all patients, regardless of the size of the primary lesion, can be considered appropriate. CT and MRI can provide excellent anatomical detail of the head and neck. Because of its ease of use, CT with intravenous contrast enhancement is typically the chosen imaging modality to evaluate the neck. The reported sensitivity and specificity of CT in identifying cervical metastases can range from 54% to 95%, and from 39% to 100% [45,46].

Given the three-dimensional structure of the neck, classifying lymph nodes located at the junction between levels based on imaging may be challenging. Optimizing the patient before the procedure can assist in this evaluation and classification [47]. The patient's head should be in a neutral position with the hard palate perpendicular to the table and the shoulders down as far as possible. If there is no contraindication, intravenous contrast should be given to help delineate vessels from lymph nodes. Scans should encompass the entirety of the neck from the skull base to the manubrium. CT scans should be performed with no more than 2–3-mm slices, whereas MRI scans should use no thicker than 5-mm slices with a 1-mm interslice gap [47].

Ultrasound and ultrasound-guided fine-needle aspiration cytology (FNAC) have been demonstrated to be viable options when evaluating the neck for metastatic nodes because of the relative ease of use and availability. Both ultrasound and ultrasound with FNAC have been shown to have an equal ability to identify cervical metastatic disease when compared with other conventional imaging methods, such as CT, MRI, and PET-CT [48]. Furthermore, ultrasound exceeds CT imaging in the

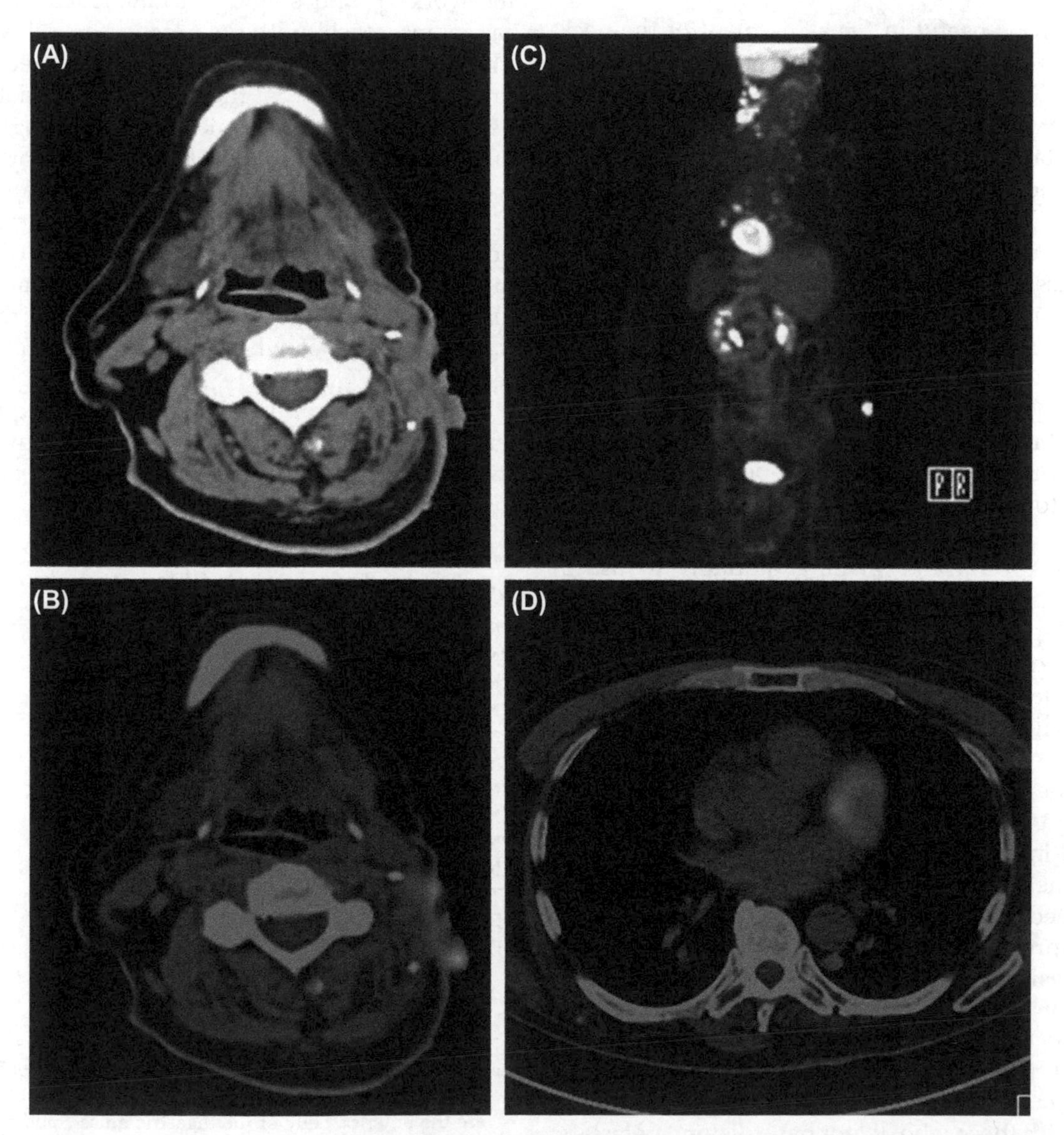

FIGURE 36.6 (A) Squamous cell carcinoma of right cheek with extensive perineural tumor extension. (B) Squamous cell carcinoma of the right cheek with extension along the infraorbital nerve to the Meckel cave (C). (D) Despite treatment, the patient developed progressive cranial neuropathies and progression along the base of the skull.

ability to delineate the internal architecture of involved nodes [49]. Ultrasonographic features, such as size (in both the short and long axis), internal matting, and changes in hilar vascular patterns, in addition to patient age over 40 years have been demonstrated to be ultrasonographic findings and clinical features associated with cervical metastases [44]. In addition, fine-needle aspiration of questionable lymph nodes for cytology has been shown to be superior to even ultrasound alone at identifying and staging the cervical lymph nodes [48]. The reported sensitivity and specificity of ultrasound-guided FNAC is 63–97%, and 74–100%, respectively [50]. Drawbacks to ultrasound and ultrasound-guided FNAC result from the dependence on the experience of the sonographer and the cytologist, and the inability to assess the retropharyngeal and mediastinal nodal basins.

PET-CT can be useful in the assessment of the neck. Although it is expensive and requires a longer scanning time and patient cooperation with fasting, it can offer important details in the assessment of cervical neck disease and distant metastasis (Fig. 36.6). With a reported sensitivity and specificity ranging from 67% to 96%, and from 82% to 100%, respectively [51–54], PET-CT can provide a unique ability to assess the primary tumor, cervical metastatic disease, potential concurrent second primary lesions, and distant metastases with a single scan. The major disadvantage of PET-CT is its limitations in resolution, which can make small tumor deposits (less then 4–5 mm) impossible to find. In addition, false-negative findings can occur because of lymphatic necrosis from tumor spread, whereas false-positives can occur secondarily to inflammatory changes or reactive lymph nodes [33].

The most appropriate imaging modality to use in assessment of regional lymph node disease will ultimately depend upon the clinical scenario and what imaging is required for the primary lesion. A combination of clinical examination and different imaging modalities will offer the most comprehensive evaluation of the neck for both staging and clinical decision making. The prognostic changes associated with cervical metastases make thorough assessment of the neck crucial in treatment planning.

Distant metastatic disease outside of the head and neck is exceedingly rare. Whereas the reported rate of metastatic spread outside of the head and neck is only 4–6%, all lesions with aggressive features should be evaluated for distant metastatic spread because of the dramatic effect on prognosis for the patient [55]. Patients with distant metastases have eight times the risk of death from cutaneous squamous cell carcinoma when compared with those who do not have distant metastatic disease [55]. The most common places for metastatic spread from cutaneous squamous cell carcinoma includes the lungs, mediastinal lymph nodes, and liver. Distant imaging should be considered in high-risk patients, particularly those with known regionally metastatic disease. Based upon the individual clinical scenario, imaging with CT, MRI, PET-CT, or chest radiography may be appropriate.

CONCLUSION

High-risk cutaneous squamous carcinomas are at substantial risk for subclinical local and regional extension of disease, knowledge of which is critical to understanding the extent of disease, determining patient prognosis, and choosing the most appropriate treatment. Multiple imaging modalities can be used in the work-up and staging of cutaneous squamous cell carcinoma, each with its strengths and weaknesses. The choice of when imaging is required and what imaging modality is the most appropriate depends heavily upon the anatomical and pathological features of the cancer, as well as the presence of absence of worrisome features during the clinical evaluation.

It is likely that future improvements in risk stratification, perhaps through the integration of molecular markers with currently used clinical, anatomical, and histopathological features, will aid in the decision making process of when and what type of imaging is required.

Future developments in imaging techniques, particularly molecular imaging, may lead to improved ability to recognize disease spread in its earlier stages when intervention is more likely to be curative.

List of Acronyms and Abbreviations

CN III Cranial nerve III
CN IV Cranial nerve IV
CN V Cranial nerve V
CN VI Cranial nerve VI
CN VII Cranial nerve VII
CN VIII Cranial nerve VIII
CN IX Cranial nerve IX
CN X Cranial nerve X
CN XI Cranial nerve XI
CN XII Cranial nerve XII
CT Computed tomography
MRI Magnetic resonance imaging
PET-CT Positron emission tomography–computed tomography

Glossary

Cervical nodal level I Divided into two groups: level IA, (submental group) and level IB (submandibular group). Level IA boundaries are the anterior belly of the digastric muscle and the hyoid bone; level IB boundaries are the anterior and posterior bellies of the digastric muscle, the stylohyoid muscle, and the body of the mandible.

Cervical nodal level II Upper one-third of the internal jugular vein from the skull base to the inferior border of the hyoid bone. The medial border is the lateral border of the sternohyoid and stylohyoid muscle, the lateral border is the posterior border of the sternocleidomastoid muscle.
Cervical nodal level III Midjugular group, extending from the inferior border of the hyoid bone to the inferior border of the cricoid cartilage. The medial boundary is the lateral border of the sternohyoid muscle and the lateral border is the posterior border of the sternocleidomastoid muscle.
Cervical nodal level IV Lower jugular group, extending from the inferior aspect of the cricoid cartilage to the clavicle. The medial border is the lateral border of the sternohyoid muscle, and the lateral border is the posterior border of the sternocleidomastoid muscle.
Cervical nodal level V Posterior triangle of the neck. Bordered superiorly by the convergence of the sternocleidomastoid and the trapezius muscles, inferiorly by the clavicle, medially by the posterior border of the sternocleidomastoid muscle, and laterally by the trapezius muscle.
Cervical nodal level VI Anterior or central compartment, includes the pretracheal and paratracheal lymph nodes, precricoid lymph nodes, and the perithyroid nodes. Superior border is the hyoid bone, inferior border is the sternal notch, lateral boundaries are the common carotid arteries.
Diplopia Double vision.
Dysesthesia Abnormal sensation, often pain.
Epiphora Uncontrolled tearing.
Formications Sensation of ants crawling on or under the skin.
Hyperglobus Superior displacement of the globe.
Hypoglobus Inferior displacement of the globe.
Ophthalmoplegia Visual impairments.
Orbital apex syndrome Oculomotor and ophthalmic nerve dysfunction leading to ophthalmoplegia, visual disturbances, and hypoesthesia of the upper face.
Orbital fissure syndrome Involvement of cranial nerves III, IV, V1, and VI, and the ophthalmic vein, leading to diplopia, paralysis of extraocular muscles, exophthalmos, and ptosis.
Proptosis Forward protrusion of the globe.
Strabismus Misalignment of the eyes.

References

[1] Rogers HW, Wienstock MA, Harris AR, Hinckley MR, Feldman SR, Fleischer AB, et al. Incidence estimate of nonmelanoma skin cancer in the United States, 2006. Arch Dermatol 2010;146(3):283–7.
[2] Karia P, Han J, Schmults CD. Cutaneous squamous cell carcinoma: estimated incidence of disease, nodal metastasis, and deaths from disease in the United States, 2012. J Am Acad Dermatol 2013;68(6): 957–66.
[3] Kyrgidis A, Tzellos TG, Kechagias N, Patriidou A, Xirou P, Kitikidou K, et al. Cutaneous squamous cell carcinoma (SCC) of the head and neck: risk factors of overall and recurrence-free survival. Eur J Cancer 2010;46(9):1563–72.
[4] Brantsch KD, Meisner C, Schönfisch B, Trilling B, Wehner-Caroli J, Röcken M, et al. Analysis of risk factors determining prognosis of cutaneous squamous-cell carcinoma: a prospective study. Lancet Oncol 2008;9(8):713–20.
[5] Scotto J, Kopf AW, Urbach F. Non-melanoma skin cancer among Caucasians in four areas of the United States. Cancer 1974;34(4): 1333–8.
[6] Renzi C, Mastroeni S, Passarelli F, Mannooranparampil TJ, Caggiati A, Potenza C, et al. Factors associated with large cutaneous squamous cell carcinomas. J Am Acad Dermatol 2010; 63(3):404–11.
[7] Nguyen KD, Han J, Li T, Qureshi AA. Invasive cutaneous squamous cell carcinoma in US health care workers. Arch Dermatol Res 2014;306(6):555–60.
[8] Lindelof B, Sigurgeirsson B, Gäbel H, Stern RS. Incidence of skin cancer in 5356 patients following organ transplantation. Br J Dermatol 2000;143(3):513–9.
[9] Cooper JZ, Brown MD. Special concern about squamous cell carcinoma of the scalp in organ transplant recipients. Arch Dermatol 2006;142(6):755–8.
[10] Alam M, Ratner D. Cutaneous squamous-cell carcinoma. N Engl J Med 2001;344(13):975–83.
[11] Jensen V, Prasad AR, Smith A, Raju M, Wendel CS, Schmelz M, et al. Prognostic criteria for squamous cell cancer of the skin. J Surg Res 2008;159(1):509–15.
[12] Rowe DE, Carroll RJ, Day CL. Prognostic factors for local recurrence, metastasis, and survival rates in squamous cell carcinoma of the skin, ear, and lip. J Am Acad Dermatol 1992;26(6): 977–90.
[13] Mourouzis C, Boynton A, Grant J, Umar T, Wilson A, MacPheson D, et al. Cutaneous head and neck SCCs and risk of nodal metastasis: UK experience. J Cranio-Maxillofac Surg 2009; 37(8):443–7.
[14] Kadakia S, Saman M, Gordin E, Marra D, Ducic Y. The role of parotidectomy in the treatment of auricular squamous cell carcinoma. Otolaryngol Head Neck Surg 2015;152(6):1048–52.
[15] Peiffer N, Kutz JW, Myers LL, Isaacson B, Sumer BD, Truelson JM, et al. Patterns of regional metastasis in advanced stage cutaneous squamous cell carcinoma of the auricle. Otolaryngol Head Neck Surg 2011;144(1):36–42.
[16] Gal TJ, Futran ND, Bartels LJ, Klotch DW. Auricular carcinoma with temporal bone invasion: outcome analysis. Otolaryngol Head Neck Surg 1999;121(1):62–5.
[17] Clark RR, Soutar DS. Lymph node metastases from auricular squamous cell carcinoma: a systemic review and meta-analysis. J Plast Reconstr Aesthet Surg 2008;61(10):1140–7.
[18] Turner SJ, Morgan GJ, Palme CE, Veness MJ. Metastatic cutaneous squamous cell carcinoma of the external ear: a high-risk cutaneous subsite. J Laryngol Otol 2010;124(1):26–31.
[19] Wermker K, Kluwig J, Schipmann S, Klein M, Schulz HJ, Hallermann C. Prediction score for lymph node metastasis from cutaneous squamous cell carcinoma of the external ear. Eur J Surg Oncol 2015;41(1):128–35.
[20] Khurana VG, Mentis DH, O'Brien CJ, Hurst TL, Stevens GN, Packham NA. Parotid and neck metastases from cutaneous squamous cell carcinoma of the head and neck. Am J Surg 1995;170(5): 446–50.
[21] Lin C, Tripcony L, Keller J, Poulsen M, Martin J, Jackson J, et al. Perineural infiltration of cutaneous squamous cell carcinoma and basal cell carcinoma without clinical features. Int J Radiat Oncol Biol Phys 2012;82(1):334–40.
[22] Han A, Ratner D. What is the role of adjuvant radiotherapy in the treatment of cutaneous squamous cell carcinoma with perineural invasion? Cancer 2007;109(6):1053–9.
[23] Lee KJ, Abemayor E, Sayre J, Bhuta S, Kirsch C. Determination of perineural invasion preoperatively on radiographic images. Otolaryngol Head Neck Surg 2008;139(2):275–80.
[24] Rapidis AD, Givalos N, Gakiopoulou H, Faratzis G, Stravrianos SD, Vilos GA, et al. Adenoid cystic carcinoma of the head and neck: clinicopathological analysis of 23 patients and review of the literature. Oral Oncol 2005;41(3):328–35.
[25] Fagan JJ, Collins B, Barnes L, D'Amico F, Myers EN, Johnson JT. Perineural invasion in squamous cell carcinoma of the head and neck. Arch Otolaryngol Head Neck Surg 1998;124(6):637–41.
[26] Ampil FL, Hardin JC, Peskind SP, Stucker FJ. Perineural invasion in skin cancer of the head and neck: a review of nine cases. J Oral Maxillofac Surg 1995;53(1):34–8.

[27] Garcia-Serra A, Hinerman RW, Mendenhall WM, Amdur RJ, Morris CG, Williams LS, et al. Carcinoma of the skin with perineural invasion. Head Neck 2003;25(12):1027–33.
[28] Mendenhall WM, Ferlito A, Takes RP, Bradford CR, Corry J, Fagan JJ, et al. Cutaneous head and neck basal and squamous cell carcinomas with perineural invasion. Oral Oncol 2012; 48(10):918–22.
[29] Murofushi T, Pohl DV, Halmagyi GM. Perineural spread of facial squamous cell carcinoma to the vestibulocochlear nerve. Otolaryngol Head Neck Surg 1997;116(3):392–4.
[30] Clouston PD, Sharpe DM, Corbett AJ, Kos S, Kennedy PJ. Perineural spread of cutaneous head and neck cancer: its orbital and central neurologic complications. Arch Neurol 1990;47(1):73–7.
[31] Begemann M, Rosenblum MK, Loh J, Kraus D, Raizer JJ. Leptomeningeal metastases from recurrent squamous cell cancer of the skin. J Neurooncol 2003;63(3):295–8.
[32] Penn R, Abemayor E, Nabili V, Bhuta S, Kirsch C. Perineural invasion detected by high-field 3.0-T magnetic resonance imaging. Am J Otolaryngol Head Neck Med Surg 2010;31(6):482–4.
[33] Purohit BS, Ailianou A, Dulguerov N, Becker CD, Ratib O, Becker M. FDG-PET/CT pitfalls in oncologic head and neck imaging. Insights Imaging 2014;5:585–602.
[34] Limawararut V, Leibovitch I, Sullivan T, Selva D. Periocular squamous cell carcinoma. Clin Exp Ophthalmol 2007;35(2):174–85.
[35] Yin VT, Merritt HA, Sniegowski M, Esmaeli B. Eyelid and ocular surface carcinoma: diagnosis and management. Clin Dermatol 2015;33(2):159–69.
[36] Veness MJ, Biankin S. Perineural spread leading to orbital invasion from skin cancer. Australas Radiol 2000;44(3):296–302.
[37] Jeong HS, Son YI, Baek CH. The pattern of lymphatic metastasis of malignant tumors in the periorbital area. Am J Otolaryngol Head Neck Med Surg 2006;27(1):5–8.
[38] Barzilai G, Greenburg E, Cohen-Kerem R, Doweck I. Pattern of regional metastases from cutaneous squamous cell carcinoma of the head and neck. Otolaryngol Head Neck Surg 2005;132(6): 852–6.
[39] Ch'ng S, Maitra A, Lea R, Brasch H, Tan ST. Parotid metastasis: an independent prognostic factor for head and neck cutaneous squamous cell carcinoma. J Plast Reconstr Aesthet Surg 2006;59(12): 1288–93.
[40] Sriskandan N, Hannah A, Howlett DC. A study to evaluate the accuracy of ultrasound in the diagnosis of parotid lumps and to review the sonographic features of parotid lesions: results in 220 patients. Clin Radiol 2010;65(5):366–72.
[41] Cerezo L, Millan I, Torre A, Aragon G, Otero J. Prognostic factors for survival and tumor control in cervical lymph node metastases from head and neck cancer: a multivariate study of 492 cases. Cancer 1992;69(5):1224–34.
[42] Robbins KT, Clayman G, Levine PA, Medina J, Sessions R, Shaha A, et al. The Committee for Head and Neck Surgery and Oncology, American Academy of Otolaryngology–Head and Neck Surgery. Neck dissection classification update: revisions proposed by the American Academy of Otolaryngology–Head and Neck Surgery. Arch Otolaryngol Head Neck Surg 2002; 128(7):751–8.
[43] Ebrahimi A, Moncrieff MD, Clark JR, Shannon KF, Gao K, Milross G, et al. Predicting the pattern of regional metastases from cutaneous squamous cell carcinoma of the head and neck based on location of the primary. Head Neck 2010;32(10):1288–94.
[44] Lai YS, Kuo CY, Chen MK, Chen HC. Three-dimensional Doppler ultrasonography in assessing nodal metastases and staging head and neck cancer. Laryngoscope 2013;123(12):3037–42.
[45] de Bondt RB, Hoeberigs MC, Nelemans PJ, Deserno WM, Peutz-Kootstra C, Kremer B, et al. Diagnostic accuracy and additional value of diffusion-weighted imaging for discrimination of malignant cervical lymph nodes in head and neck squamous cell carcinoma. Neuroradiology 2009;51(3):183–92.
[46] Yoon DY, Hwang HS, Chang SK, Rho YS, Ahn HY, Kim JH, et al. CT, MR, US,18F-FDG PET/CT, and their combined use for the assessment of cervical lymph node metastases in squamous cell carcinoma of the head and neck. Eur Radiol 2009;19(3):634–42.
[47] Som PM, Curtin HD, Mancuso AA. Imaging-based nodal classification for evaluation of neck metastatic adenopathy. AJR Am J Roentgenol 2000;174(3):837–45.
[48] Stoeckli SJ, Haerle SK, Haile SR, Hany TF, Schuknecht B. Initial staging of the neck in head and neck squamous cell carcinoma: a comparison of CT, PET/CT, and ultrasound-guided fine-needle aspiration cytology. Head Neck 2012;34(4):469–76.
[49] Sumi M, Ohki M, Nakamura T. Comparison of sonography and CT for differentiating benign from malignant cervical lymph nodes in patients with squamous cell carcinoma of the head and neck. AJR Am J Roentgenol 2001;176(4):1019–24.
[50] de Bondt RB, Nelemans PJ, Hofman PA, Casselman JW, Kremer B, van Engelshoven JM, et al. Detection of lymph node metastases in head and neck cancer: a meta-analysis comparing US, USgFNAC, CT and MR imaging. Eur J Radiol 2007;64(2):266–72.
[51] Murakami R, Uozumi H, Hirai T, Nishimura R, Shiraishi S, Ota K, et al. Impact of FDG PET/CT imaging on nodal staging for head-and-neck squamous cell carcinoma. Int J Radiat Oncol Biol Phys 2007;68(2):377–82.
[52] Jeong HS, Baek CH, Son YI, Ki Chung M, Kyung Lee D, Young Choi J, et al. Use of integrated 18F-FDG PET/CT to improve the accuracy of initial cervical nodal evaluation in patients with head and neck squamous cell carcinoma. Head Neck 2007;29(3): 203–10.
[53] Gordin A, Golz A, Keidar Z, Daitzchman M, Bar-Shalom R, Israel O. The role of FDG-PET/CT imaging in head and neck malignant conditions: impact on diagnostic accuracy and patient care. Otolaryngol Head Neck Surg 2007;137(1):130–7.
[54] Ng SH, Ko SF, Toh CH, Chen YL. Imaging of neck metastases. Chang Gung Med J 2006;29(2):119–28.
[55] Brunner M, Veness MJ, Ch'ng S, Elliott M, Clark JR. Distant metastases from cutaneous squamous cell carcinoma: analysis of AJCC stage IV. Head Neck 2013;35(1):72–5.

CHAPTER

37

Imaging Patterns of Metastatic Melanoma

B. Peters[1,2], F.M. Vanhoenacker[1,2,3]

[1]University Hospital Antwerp, Edegem, Belgium; [2]AZ Sint-Maarten, Mechelen-Duffel, Belgium; [3]University Hospital Ghent, Ghent, Belgium

OUTLINE

INTRODUCTION

Malignant melanoma is an aggressive tumor with the potential of spread to all organ systems. Dissemination can occur through local invasion, the lymphatic system, or hematogenous spread [1,2]. Almost all patients dying from melanoma have multifocal organ involvement. Although the role of medical imaging in the diagnosis of primary melanoma is rather limited, imaging may play an important role in evaluation and therapy monitoring of distant metastatic disease and assessment of local recurrence [3].

In this chapter we will discuss the merit of the different imaging techniques for assessment of metastatic melanoma and illustrate involvement of most affected organ systems.

IMAGING TECHNIQUES

Ultrasound

Ultrasound is a relatively cheap technique and has a widespread availability. Nonetheless the use of ultrasound in metastatic melanoma is limited. Ultrasound is mainly used in detecting local recurrence and in-transit metastasis (metastasis located on the lymphatic pathway between the primary lesion and its draining lymph nodes). Detection of abdominal metastasis in the liver, spleen, and lymph nodes with ultrasound is less sensitive than cross-sectional imaging techniques such as computed tomography (CT), positron emission tomography (PET) or magnetic resonance imaging (MRI) [4].

Imaging in Dermatology
http://dx.doi.org/10.1016/B978-0-12-802838-4.00037-6

A major advantage of ultrasound is the possibility to perform a fine-needle aspiration in the same session as the diagnostic examination with a reported sensitivity of 95% and specificity of 100% to diagnose in-transit or satellite metastases, and a sensitivity and specificity of 79% and 100%, respectively, to diagnose nodal basins metastasis [5,6].

Radiography

Radiography, or X-ray (XR), is a cheap technique that uses relatively low radiation doses. Nonetheless, the role of XR is limited in detecting metastatic melanoma. Studies of asymptomatic patients have shown low detection rates for occult lung metastasis and high false-positive rates [7,8]. Mooney et al. reported a cost-effective study of the life-long, systematic follow-up observation of asymptomatic patients with chest XR. The study concluded that the high cost for systematic screening outweighed any benefit for early disease detection [9].

Despite this knowledge, routine chest XR are used in follow-up examinations of asymptomatic patients in daily practice.

Evaluation of metastasis to the bone with XR is limited. Most of the lesions remain invisible until cortical destruction has occurred or at least 30% of the trabecular bone is destroyed (see section "Spine") [10].

Computed Tomography

CT has no role in the evaluation of asymptomatic patients with in situ, stage I, and stage II melanoma. In patients with stage III melanoma and palpable disease, CT may detect additional metastasis and alter therapy. In patients with advanced disease, ie, stages IIIB, IIIC, and IV, CT may be useful for surgical planning, might detect additional metastasis, and thus alter therapy [3,11].

The advantages of CT compared with other imaging techniques in evaluation of distant staging are the quick examination of large parts of the body (chest, abdomen) and different windowing settings, allowing evaluation of lymph nodes, different parenchymal organs, and bone.

In contrast to many other metastasis, metastatic melanomas are hypervascular, with specific contrast behavior in liver lesions. Hepatic metastatic melanoma are hyperdense on contrast-enhanced (CE) CT compared with the surrounding tissue on peak hepatic arterial enhancement (early arterial contrast phase). Multiphasic CE CT with at least two different phases (nonenhancement and portal venous phase or arterial and portal venous phase) can detect up to 14% more lesions compared with single CE CT in the portal venous phase [12].

CT is the preferred technique for the evaluation of lung disease [13].

A study by Gray et al. showed that the combination of serum l-lactate dehydrogenase and CT examination after bevacizumab therapy has the highest accuracy for predicting survival in patients with metastatic melanoma [14].

Magnetic Resonance Imaging

MRI is not used as a routine technique for evaluation of metastatic melanoma, but as a problem-solving technique to answer specific clinical questions such as evaluation of detailed tumor size and preoperative anatomical correlation of resectable distant metastasis, precise characterization of questionable lesions, and patients suspected for brain metastasis [3].

MRI is a cross-sectional technique that uses a magnetic field and radio waves to create detailed images of the body. Because of its superior contrast resolution, MRI is more sensitive for detecting metastasis in most systems except for lung metastasis, where CT remains the preferred modality [13]. The disadvantages of MRI are patient-related contraindications (Further discussion is beyond the scope of this chapter), its restricted availability, its limited field of view, and the relatively long duration of each examination. Detailed information about absolute and relative contraindications of MRI can be found at www.MRIsafety.com.

Whole-body (WB) MRI is a promising new technique in the management of malignant melanoma. According to a recent study comparing WB MRI with PET-CT in staging of malignant melanoma, WB MRI is superior in detecting liver, bone, and brain metastasis [15].

Melanin has a paramagnetic effect on different pulse sequences with shortening of the T1 and T2 relaxation time, resulting in hyperintensity on T1-weighted imaging (WI) and hypointensity on T2-WI [16]. Metastatic melanomas containing little or no melanin (amelanotic), on the contrary, will show nonspecific signal characteristics [16].

Positron Emission Tomography

PET is a functional imaging technique using a radioactive transducer visualizing metabolic activity of a lesion. It will differentiate cells with high metabolic activity from normal cells. The disadvantage is the poor anatomical resolution and low specificity. Tyler et al. reported a false-positive rate of 56% for staging patients with clinically evident regional lymph node or in-transit metastasis [17]. Therefore PET is mainly used in combination with CT and/or MRI.

PET-CT has the great advantage of combining the study of metabolic activity of a lesion and the high anatomical resolution of CT. In this regard, PET-CT has a higher detection rate of metastasis than CT or PET alone [3]. Although more research on the rational use of PET-CT should be performed, PET-CT may have an important additional value for characterization of indeterminate lesions on conventional imaging, for precise evaluation of the resectability of metastases, and for monitoring follow-up care after treatment [3].

PET-MRI has a similar advantage of combining detection of increased metabolic activity and high-contrast resolution, but is technically more demanding than PET-CT. Its potential future role in the evaluation of metastatic melanoma has yet to be defined. For more detailed information regarding PET imaging of melanoma see Chapter 32.

SPECT

Single photon emission CT (SPECT) uses a radioligand, which accumulates in the organ of interest. SPECT is used in the management of malignant melanoma as sentinel node imaging. It is valuable in assessment of depth of nodal uptake and thus preoperative planning.

SPECT is particularly useful in evaluating lymphatic drainage of the head and neck region, which has high variability among patients [3].

METASTATIC INVOLVEMENT OF THE DIFFERENT ORGAN SYSTEMS

Skin, Subcutaneous Tissue, and Regional Lymph Nodes

Metastasis at the skin, subcutaneous tissue, and regional lymph nodes is the most common site for recurrent disease and is seen in 59% of metastatic melanomas [18].

Local recurrence is often suspected during physical examination. Ultrasound is useful for confirmation of indeterminate subcutaneous lesions or lymph node involvement. Ultrasound-guided fine-needle aspiration may be used for subsequent histopathological confirmation of suspected local recurrence or nodal involvement. Lymphoscintigraphy can be useful for preoperative or intraoperative evaluation of the lymphatic drainage system, especially at the head and neck region.

Although lymph node and in-transit metastases are often apparent on CT, CT is not preferred for the evaluation of local involvement. The lesions are isodense to muscle on CT, often embedded in hypodense subcutaneous fat (Fig. 37.1).

(A)

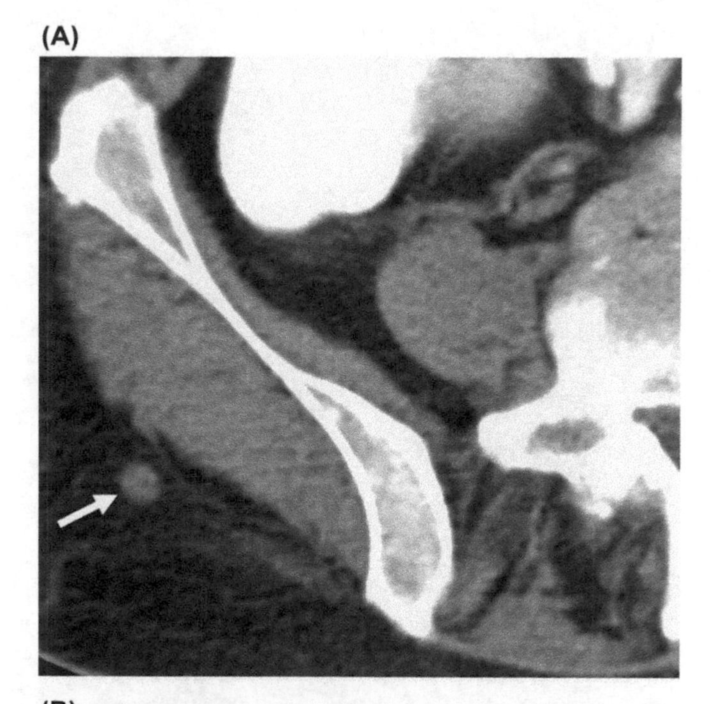

(B)

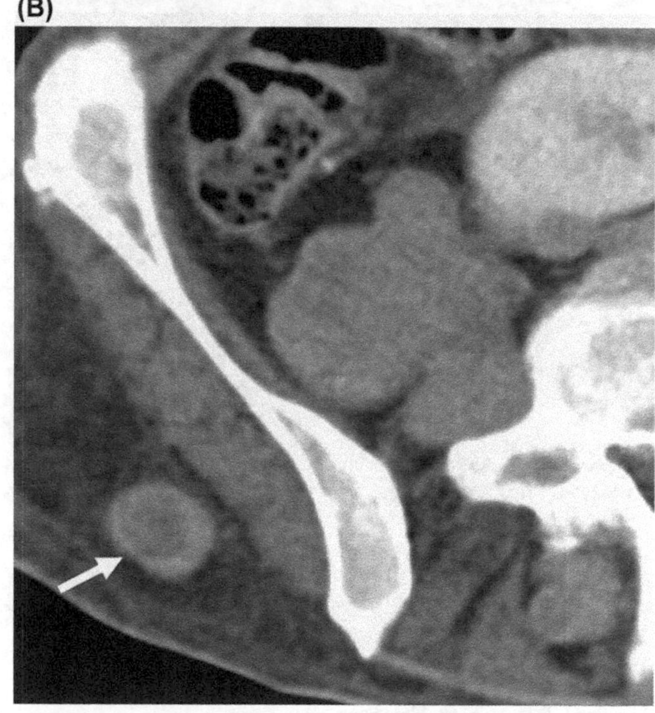

FIGURE 37.1 (A) Axial contrast-enhanced (CE) CT of the pelvis shows a subcutaneous metastasis (*white arrow*) in the gluteal region. (B) A follow-up CT after 5 months shows an enlargement of the metastasis.

Chest

The lung is the most affected organ, seen in 36% of metastatic melanoma, which can originate from the lung parenchyma (Fig. 37.2) or the pleura (Fig. 37.3) [18]. Lesions can be variable in size, ranging from less than 0.6 mm up to 5 cm. Most lesions are 1–2 cm in size. A feeding vessel is often seen, in keeping with hematogenous spread. Mediastinal or hilar lymph node enlargement is a common associated finding (Fig. 37.4) [19].

The sensitivity of XR for detection of pulmonary nodules is only 54%, with a specificity of 99% [20]. CT will detect more lesions with a sensitivity of nearly 100%, but has a specificity of only 59% [21]. For questionable pulmonary lesions, 18F-fluorodeoxyglucose PET-CT is

(A)

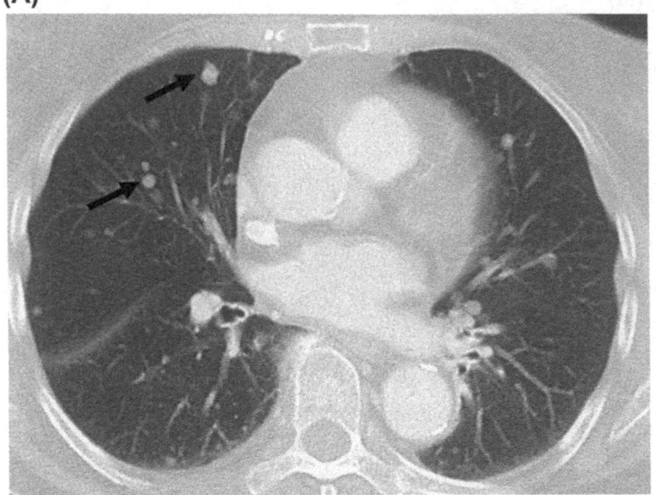

(B)

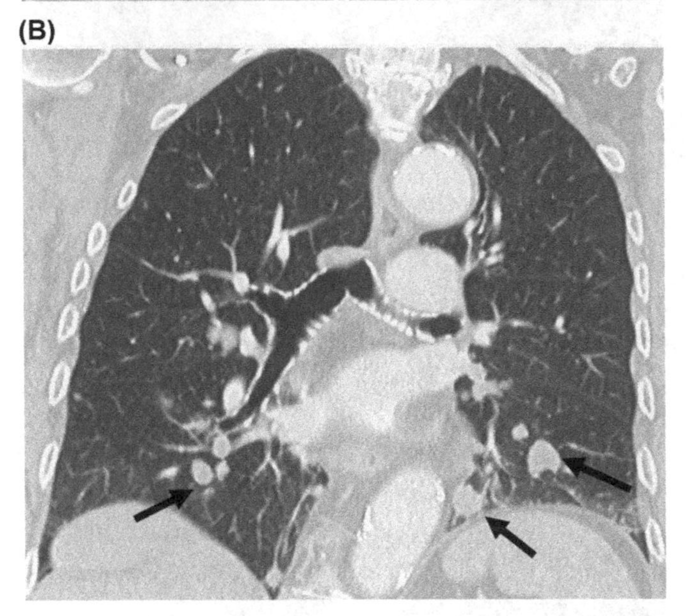

FIGURE 37.2 Axial (A) and coronal (B) reformatted CT of the chest shows bilateral pulmonary metastases of malignant melanoma (*black arrows*).

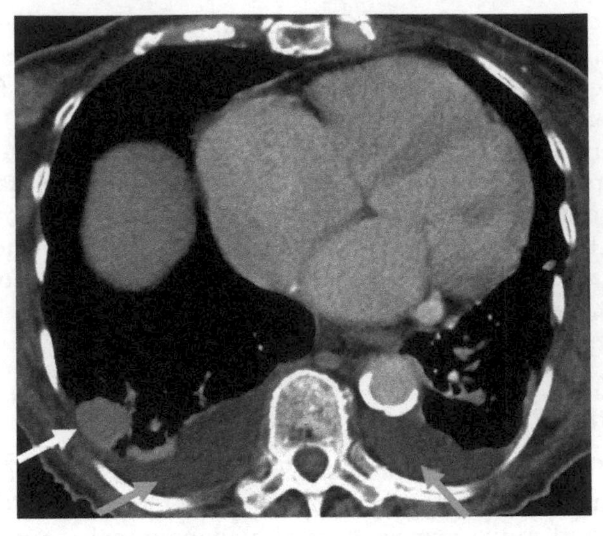

FIGURE 37.3 Axial CT of the chest shows a pleural metastasis (*white arrow*) posterior in the right lower lobe. Additional pleural effusion (*blue arrows*) is seen.

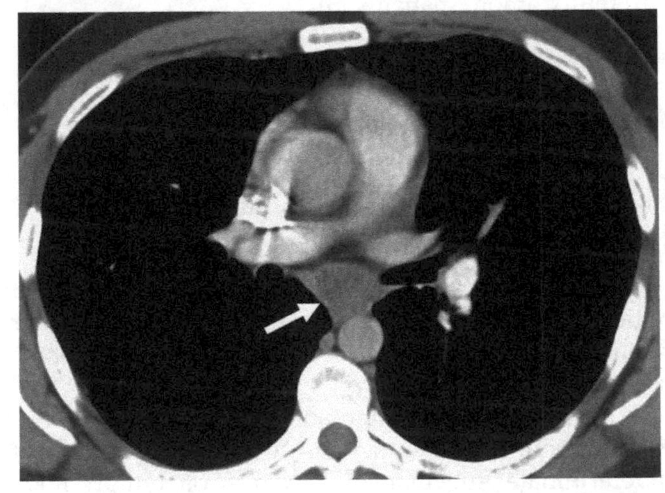

FIGURE 37.4 Axial contrast enhanced CT of the chest shows an enlarged infracarinal lymph node within station 7 (*white arrow*).

recommended, with reported sensitivity and specificity in the detection of pulmonary metastasis of 85% and 96%, respectively [22].

Central Nervous System

Brain

Metastasis to the brain is the second most common site of organ involvement, occurring in up to 20% of patients [18]. It is the most common cause of death in metastatic melanoma [23]. The lesions may be located either intraaxially (Fig. 37.5) or within the leptomeninges (Fig. 37.6). The majority of the lesions measure between 1 and 4 cm and are surrounded by marked perilesional edema. In contrast to most other brain metastases, intratumoral hemorrhage is a common and often prominent feature in metastatic melanoma (19%) (Fig. 37.7) [24]. Because of intralesional hemorrhage, patients often experience acute onset of cerebral symptoms such as headache, mental status alternation, nausea, and vomiting.

Twelve percent of patients with brain metastasis remain asymptomatic [25]. For correct detection and characterization of intralesional hemorrhage, unenhanced CT is mandatory and must be supplemented by CE CT. MRI has a higher sensitivity and specificity in detecting metastasis than CT, and should always be used if available [26]. Melanin has a specific signal behavior (see section "Magnetic Resonance Imaging") on MRI, which can differentiate melanotic metastasis

(A)

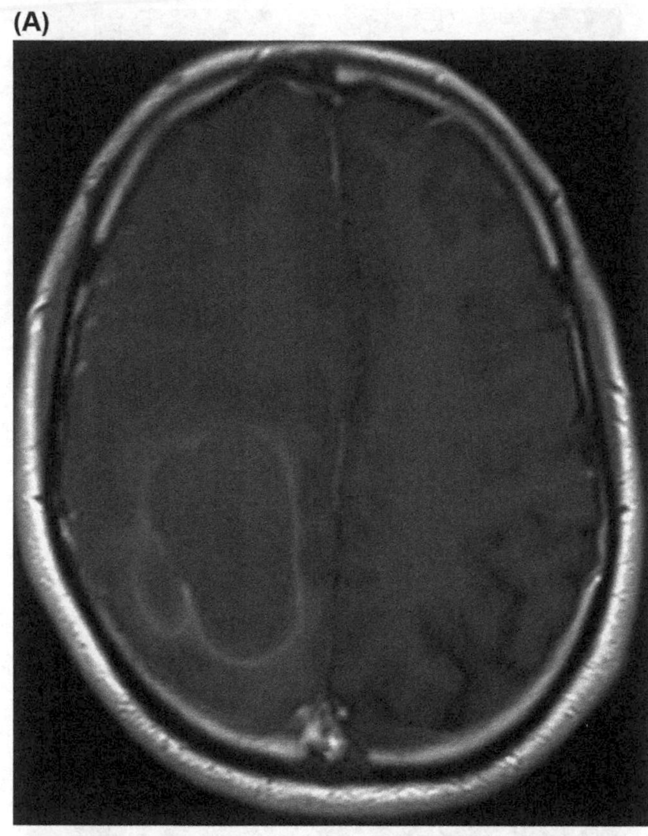

(B)

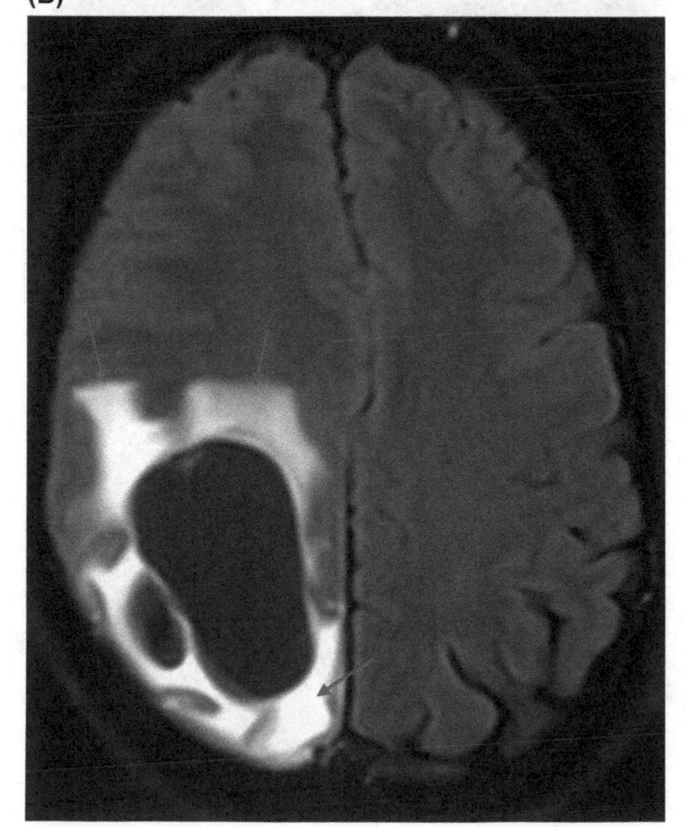

FIGURE 37.5 Intraaxial metastasis in the brain. (A) Axial T1-weighted imaging (WI) after intravenous gadolinium contrast administration shows a large ring-enhanced metastasis located in the right parietal lobe. (B) Axial fluid attenuated inversion recovery (FLAIR) (*dark fluid sequence*) shows extensive perilesional edema (*red arrows*).

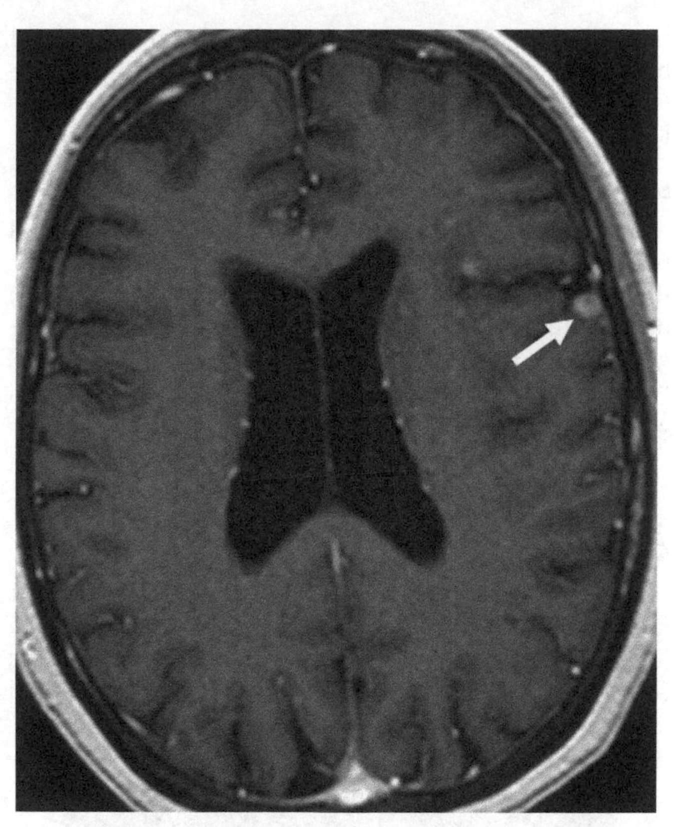

FIGURE 37.6 Leptomeningeal metastasis. Axial T1-weighted imaging (WI) after intravenous gadolinium contrast administration shows nodular enhancement of the leptomeninges at the left sylvian fissure (*white arrow*).

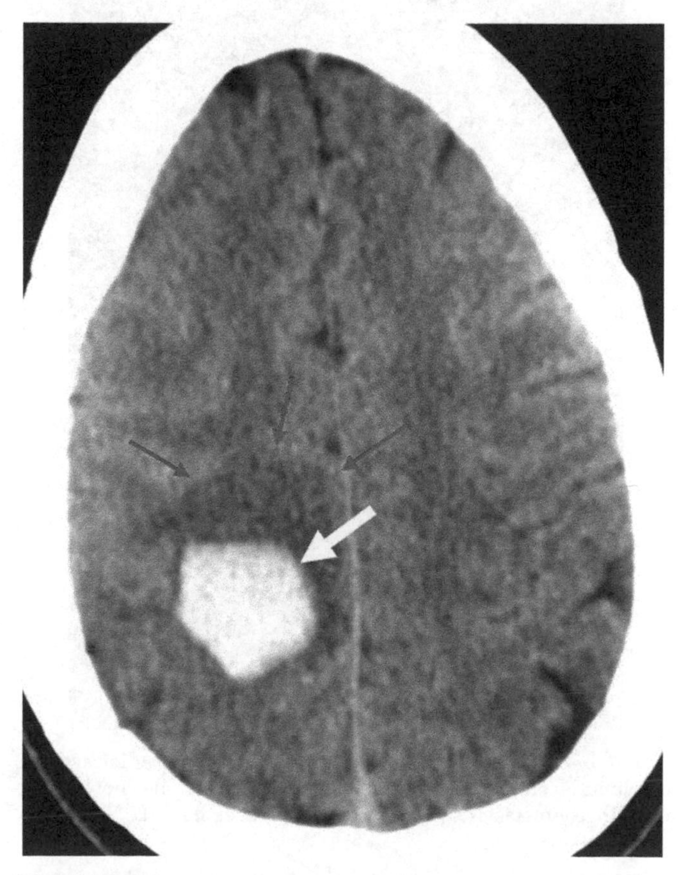

FIGURE 37.7 Axial unenhanced CT of the brain shows a metastatic melanoma with intralesional hemorrhage (*white arrow*) and perilesional edema (*red arrows*).

from other lesions. They are hyperintense on T1-WI and hypointense on T2-WI [16]. Unfortunately 12% of metastases are amelanotic with nonspecific MRI appearances [27].

Spine

Seventeen percent of metastases located in the central nervous system are spinal lesions. Based on the location, they can be classified as osseous (see section "Bone"), intramedullary, extramedullary intradural (leptomeningeal), and extradural (epidural) lesions. Most of the spinal lesions are located in the leptomeninges (Fig. 37.8). Medullary lesions are rare, and epidural lesions (Fig. 37.9) are extremely rare [28,29].

Abdomen

The liver is the third most affected organ in metastatic melanoma, accounting for 20% of metastatic melanoma [18]. Metastasis of the liver can be solitary or multiple (Fig. 37.10). Lesions may vary in size ranging from small

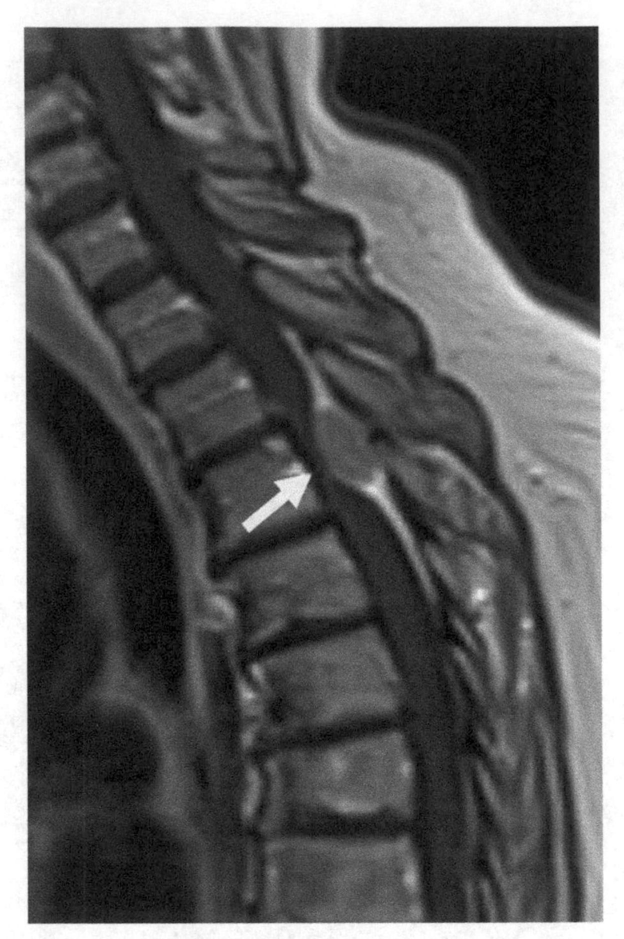

FIGURE 37.9 Sagittal T1-weighted imaging (WI) after intravenous gadolinium contrast administration of the thoracic spine shows a metastatic melanoma located in the epidural space (*white arrow*) located at the level of Th3 [28].

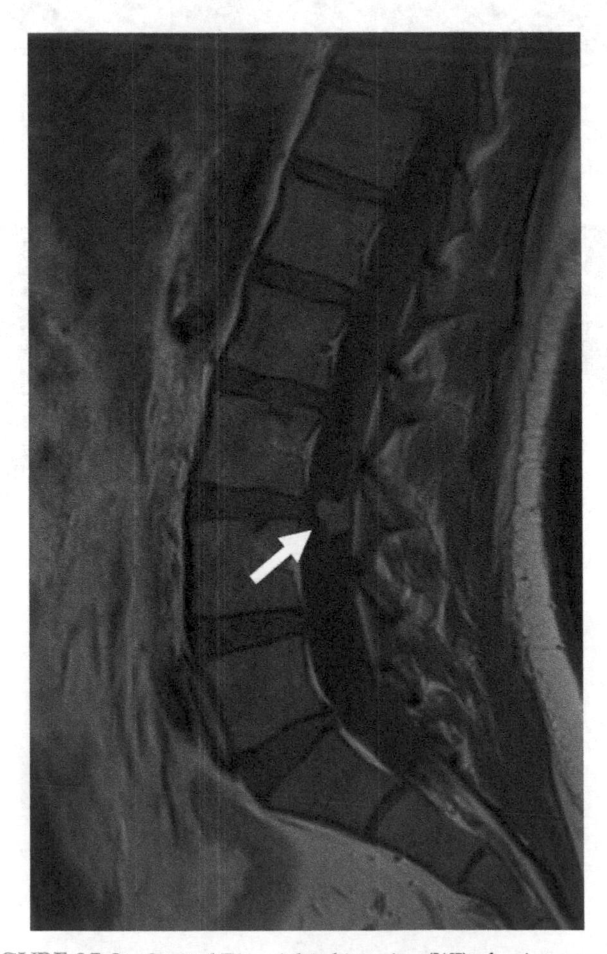

FIGURE 37.8 Sagittal T1-weighted imaging (WI) after intravenous gadolinium contrast administration of the lumbar spine shows a leptomeningeal metastasis (*white arrow*) at the level of L2–L3.

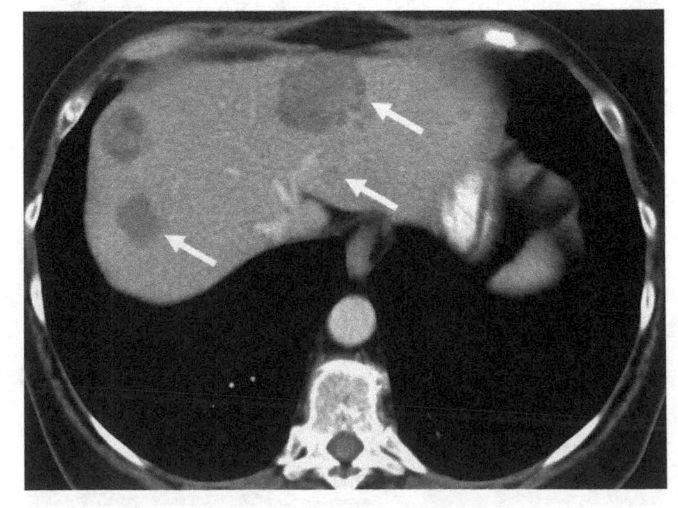

FIGURE 37.10 Axial contrast-enhanced (CE) CT of the abdomen shows multiple metastasis variable in size spread throughout the liver (*white arrows*).

lesions (<5 mm) to large lesions (>15 cm). Intralesional hemorrhage, calcifications, and necrotic areas causing lesion heterogeneity are not uncommon [19]. Administration of intravenous contrast is mandatory for both CT and MRI. Metastatic melanomas are hypervascular, and therefore appear hyperdense on arterial hepatic CE CT phase [12]. It is recommended to use a multiphasic scan for the evaluation of liver metastasis (see section "Computed Tomography") [12]. Large lesions can contain necrotic areas, appearing hypodense on CE CT (Fig. 37.11).

Whereas autopsy studies show a metastatic rate of 50% to the adrenal glands, often bilaterally, in vivo detected adrenal metastasis is rare and usually unilateral (Fig. 37.12) [30]. Metastasis to the spleen is uncommon, ranging from 1.4% to 4.5% on in vivo studies to 30% in autopsy studies (Fig. 37.13) [30–32].

Involvement of other organs such as the kidneys, pancreas, gallbladder, and bladder are rare [33,34].

Metastasis to the gastrointestinal (GI) tract and mesenterial lymph nodes (Fig. 37.14) has a high incidence in autopsy studies, but is rarely seen on imaging. The small bowel (Fig. 37.15) has the highest rate of involvement, followed by the colon (Fig. 37.16) and stomach. Involvement of the duodenum, rectum, esophagus, and anus are less common. Despite the high incidence, GI tract involvement is often asymptomatic. Ulcerated bowel lesions may cause life-threatening complications, such as GI perforation presenting as an acute abdomen [35].

A combined administration of intravenous, oral, and rectal contrast on CT yields the best results for evaluation of GI involvement.

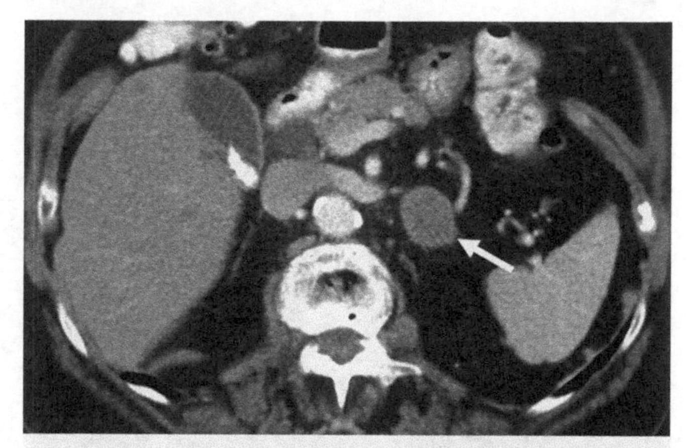

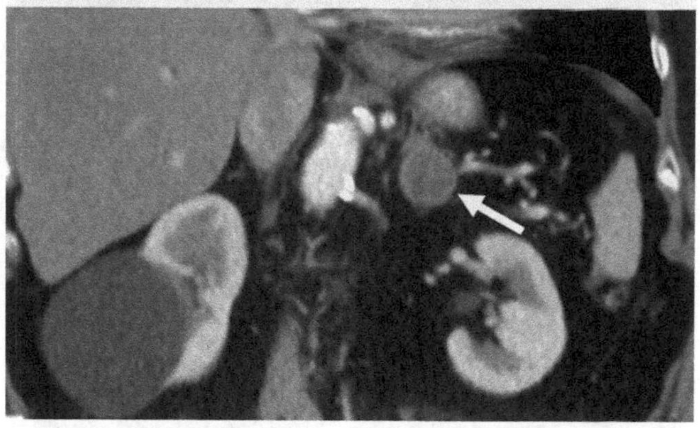

FIGURE 37.12 Axial (A) and coronal (B) contrast-enhanced (CE) CT of the abdomen shows a metastasis in the left adrenal gland (*white arrow*).

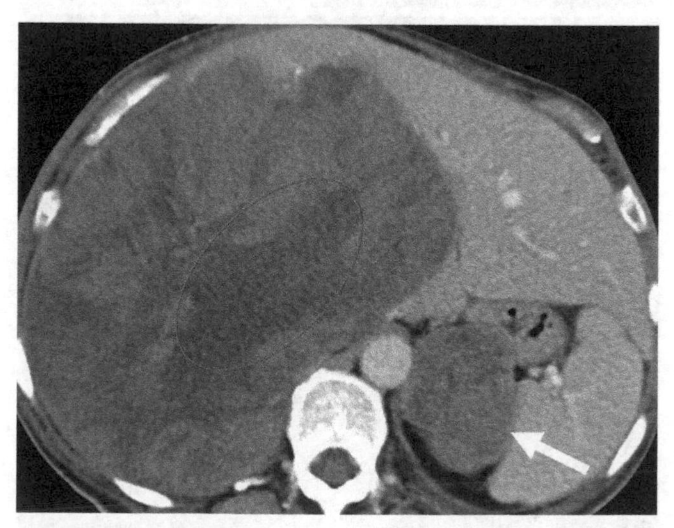

FIGURE 37.11 Axial contrast-enhanced (CE) CT shows a large heterogeneous metastasis in the right liver lobe containing large necrotic regions (*red ellipse*). An additional metastasis is seen in the left adrenal gland (*white arrow*).

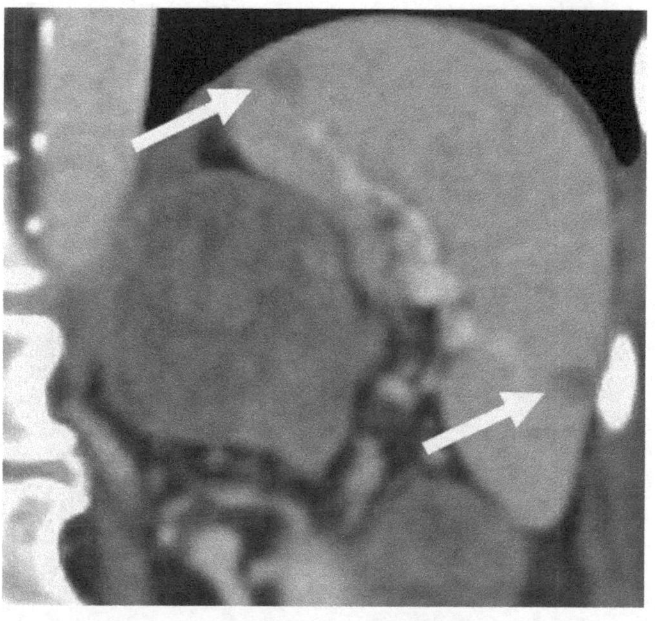

FIGURE 37.13 Axial contrast-enhanced (CE) CT of the abdomen shows two metastasis in the spleen (*white arrows*). Note the additional metastasis in the left adrenal gland and mesentery.

(A)

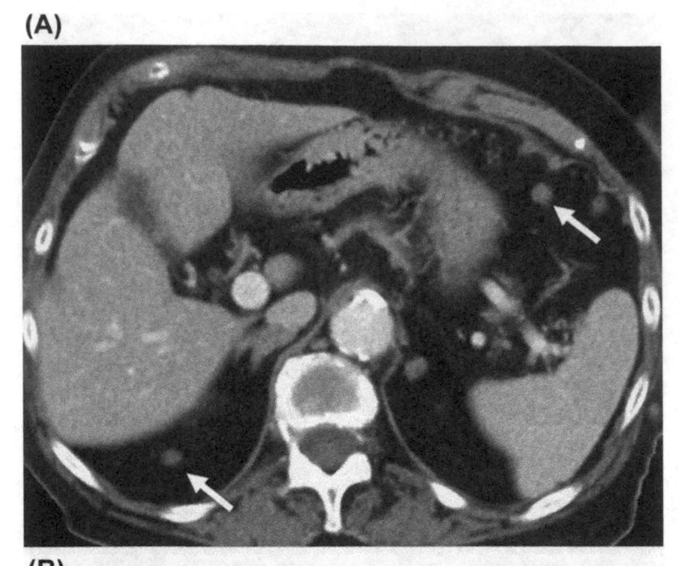

(B)

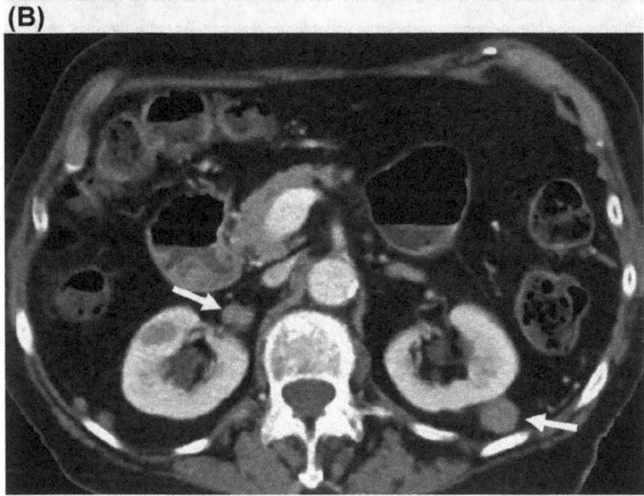

FIGURE 37.14 Axial contrast-enhanced (CE) CT of the abdomen shows multiple enlarged mesenterial (A) and retroperitoneal (B) lymph nodes (*white arrows*).

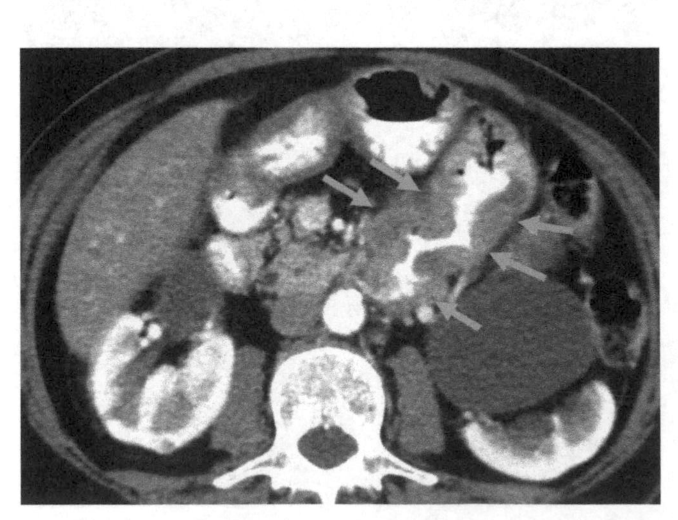

FIGURE 37.15 Axial contrast-enhanced (CE) CT of the abdomen after oral contrast administration shows diffuse enlargement of the jejunum (*blue arrows*) caused by metastatic melanoma.

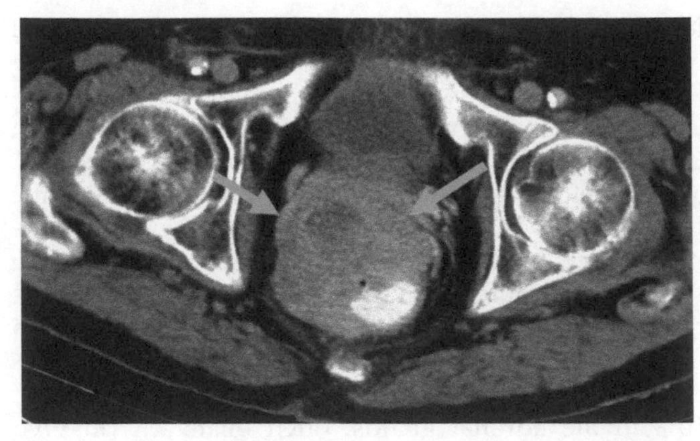

FIGURE 37.16 Axial contrast-enhanced (CE) CT of the abdomen after rectal contrast administration shows diffuse enlargement of the colon sigmoideum (*blue arrows*) as a result of metastatic melanoma.

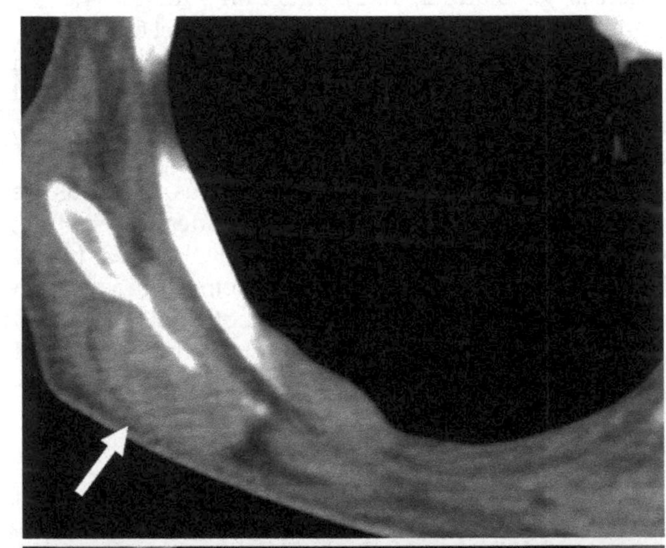

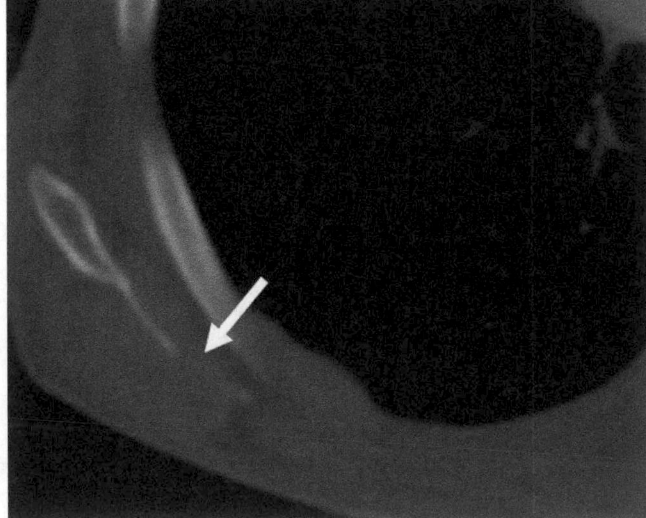

FIGURE 37.17 Axial contrast-enhanced (CE) CT in mediastinal window (A) and bone window and (B) of the chest shows a osteolytic metastasis in the right scapula with an extraosseous soft-tissue mass.

Musculoskeletal System

Bone

Metastasis to the bone is the fourth most common site and is seen in 17% of patients [18]. The most common location is by far the axial skeleton and ribs, but virtually any bone can be affected [36]. The majority of the lesions are osteolytic and slightly expansile (Fig. 37.17). Osteoblastic metastasis of melanoma is rare [37]. Although bone metastasis is relatively common, patients are often asymptomatic. The most common symptoms are caused by pathological fractures and spinal compression [37]. Lesions usually remain invisible on XR until sufficient bone destruction has occurred (see section "Radiography") (Fig. 37.18) [10]. CT (Fig. 37.19A and B) or MRI (see Fig. 37.19C) is recommended in symptomatic patients in case of suspicion of osseous metastasis involving the spine, particularly in case of neurological symptoms [38].

Muscle

The incidence of metastasis to skeletal muscle from malignant neoplasms in autopsy studies ranges from 0.8% to 16% [39]. Although seemingly common, skeletal muscle metastasis in the clinical setting is rare. This implies that muscle involvement only appears in advanced disease and is rarely isolated. The most common sites of metastasis are the muscles of the lower extremities

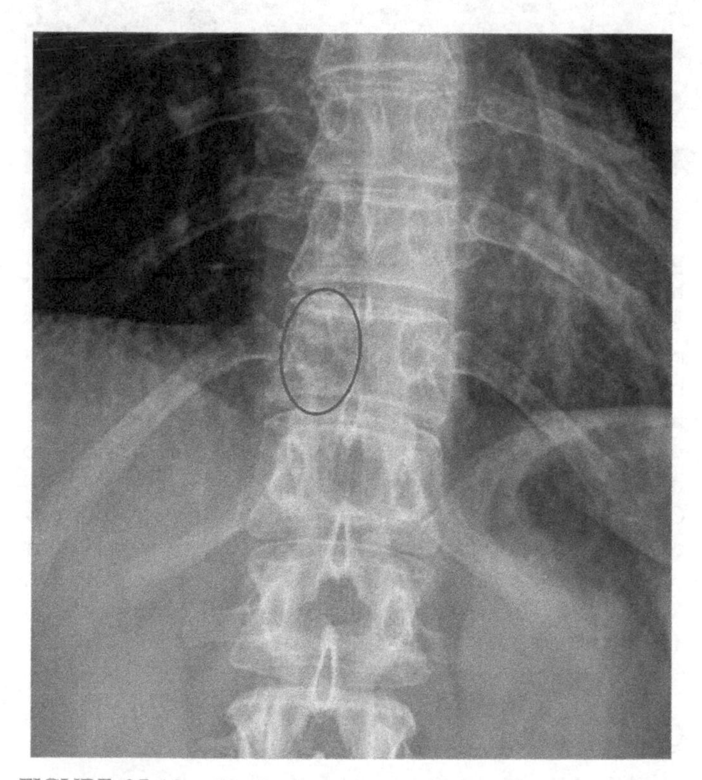

FIGURE 37.18 Osteolytic metastasis. Plain radiograph showing destruction of the cortical outline of the pedicle of Th11 (*red circle*).

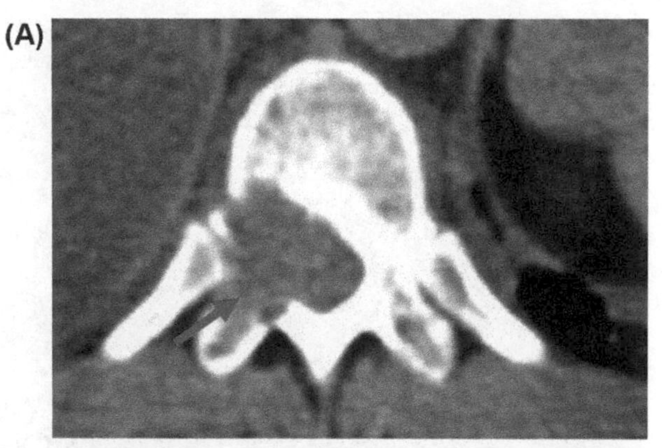

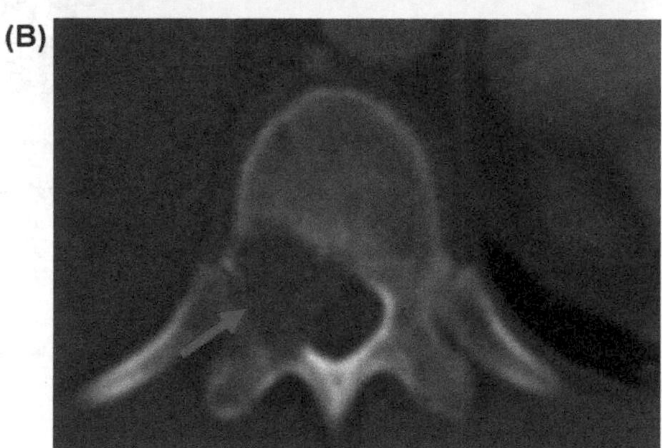

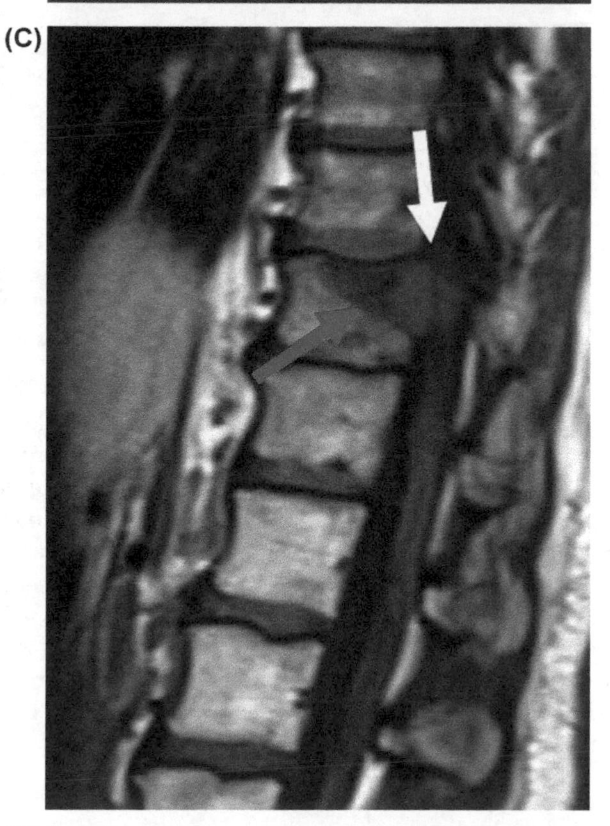

FIGURE 37.19 Osteolytic metastasis. (A and B) CT better shows the lesion's extent into the spinal canal (*red arrows*). (C) T1-weighted imaging (WI) MRI bone marrow replacement in the body of Th11 (*red arrow*) as well as lesion extension into the spinal canal (*white arrow*).

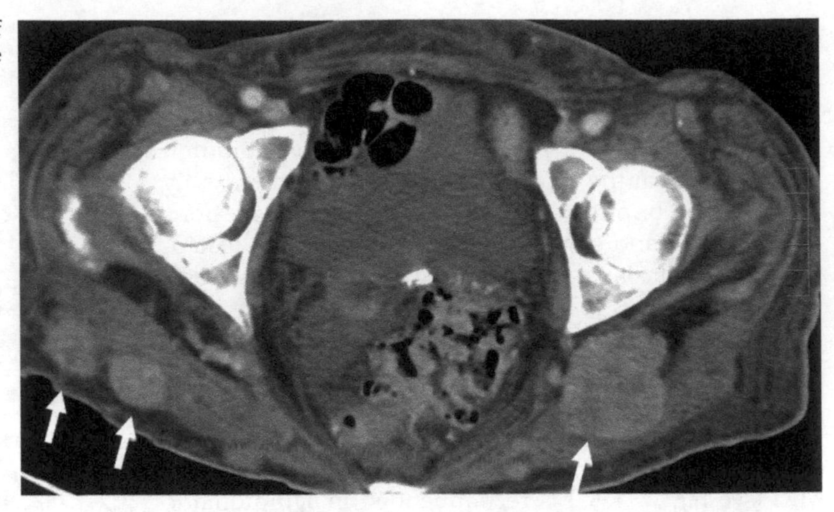

FIGURE 37.20 Axial contrast-enhanced (CE) CT of the pelvis shows multiple intramuscular metastases in the gluteus maximus muscles on both sides (*white arrows*).

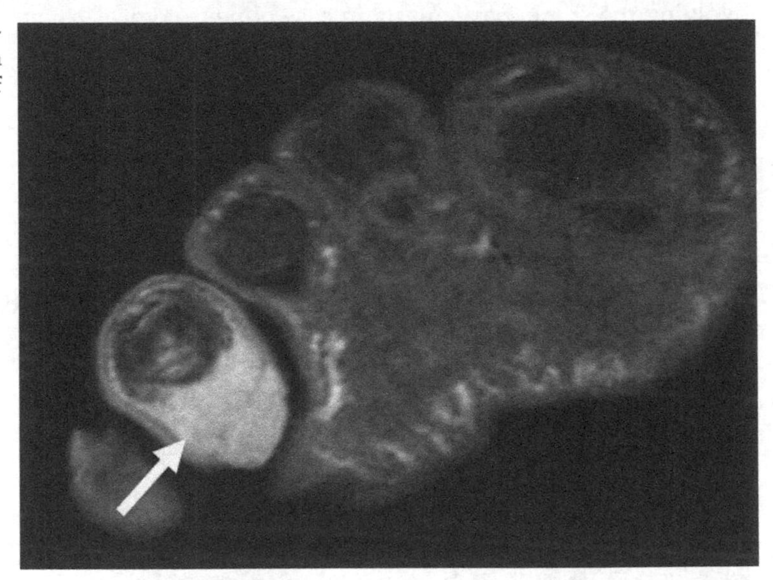

FIGURE 37.21 Coronal fat–suppressed T1-weighted imaging (WI) after intravenous gadolinium contrast administration shows a metastatic melanoma adjacent to the plantar aspect of the fourth phalanx of the right foot (*white arrow*).

(40%), the trunk (30%), and the upper extremities (26%) [40]. Unlike primary sarcomas, metastases to skeletal muscle are often painful [41]. On CT, lesions are isodense to muscle and show contrast enhancement (Fig. 37.20). MRI is the gold standard for the evaluation of muscle disease (Fig. 37.21). Melanotic metastasis will show the typical intensity of melanin (see section "Magnetic Resonance Imaging"), but biopsy is often required for the final diagnosis. Although MRI is the preferred modality for evaluation of muscular metastasis, subsequent CT or PET-CT is advisable to detect other distant metastases [13].

Eye

Primary melanoma often involves the uvea, but intraocular metastasis of malignant melanoma is rare and estimated to be 0.5–1% [42]. The differentiation between primary lesions and metastasis may be difficult. Medical history of a previous melanoma and the presence of widespread metastatic disease suggest secondary disease, although a metachronous tumor should always be considered [43]. Ultrasound is the modality of choice in the evaluation of the ocular bulb. It can detect lesions with a thickness of less than 3 mm, which are difficult to detect on other cross-sectional imaging techniques [44]. CT and preferably MRI are used to evaluate the tumor extension. Lesions are hyperdense to the vitreous humor on CT with a typical mushroomlike appearance. On MRI, they are usually moderately hyperintense on T1-WI and moderately hypointense on T2-WI [45,46].

Other

Malignant melanoma is the third most common tumor to metastasize to the breast (Fig. 37.22) [47,48].

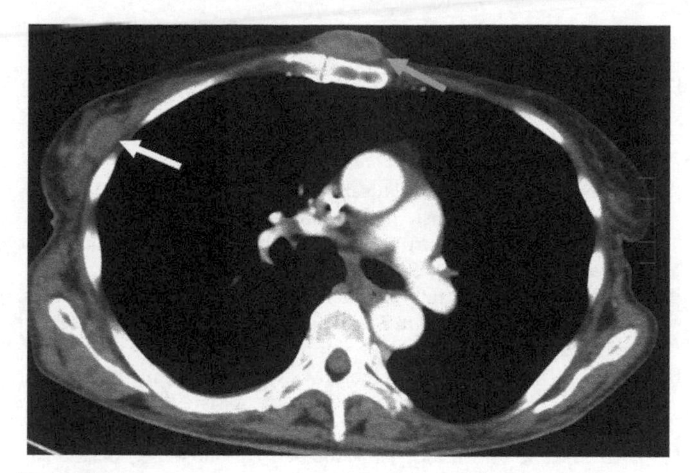

FIGURE 37.22 Axial contrast-enhanced (CE) CT of the chest shows a metastasis in the right breast (*white arrow*). Note an additional subcutaneous lesion anteriorly to the sternum (*blue arrow*).

Although the breast is relatively commonly involved, it is rarely isolated. In up to 50% of cases, metastasis occurs in the upper outer quadrant [47].

Although cardiac metastasis is rare, the prevalence of metastasis involving the heart is higher in melanoma compared with other tumors (Fig. 37.23) [49]. Melanoma of the salivary glands (Fig. 37.24) is often secondary to metastasis of a primary melanoma, although primary melanomas of the salivary glands have been reported [50,51].

To our knowledge, there are no reports in the literature of metastasis to the nasosinal cavity, although primary melanoma may be rarely located at the paranasal sinuses [52].

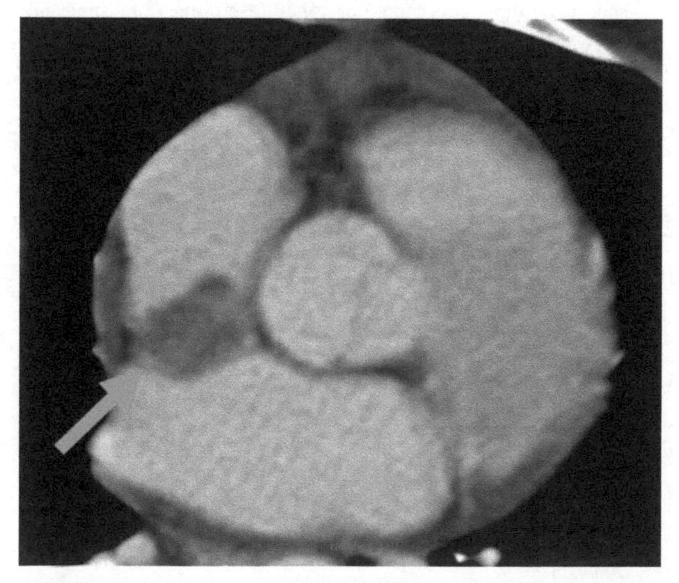

FIGURE 37.23 Axial contrast-enhanced (CE) CT of the chest shows a metastasis in the right atrium of the heart (*blue arrow*).

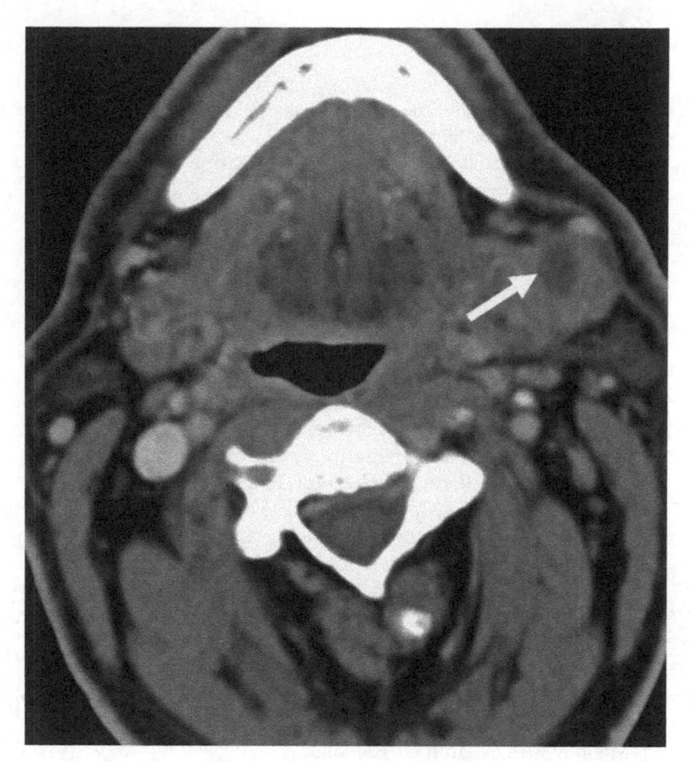

FIGURE 37.24 Axial contrast-enhanced (CE) CT of the neck shows a metastasis in the left submandibular gland (*white arrow*).

CONCLUSION

Malignant melanoma is a potentially aggressive tumor that can disseminate through local invasion, the lymphatic system, or hematogenous spread. The majority of metastasis occurs in the skin, subcutaneous tissue, and regional lymph nodes. The lung, brain, and liver are the most affected organs, but virtually any organ can be affected. Imaging plays a pivotal role in evaluation of the tumor extent, managing therapy, and monitoring the therapeutic response. Particularly, imaging has an established role in assessing symptomatic patients or patients with advanced disease. There is no evidence for screening early staged, asymptomatic patients with imaging.

List of Acronyms and Abbreviations

CE Contrast-enhanced
CT Computed tomography
GI Gastrointestinal
MRI Magnetic resonance imaging
PET Positron emission tomography
SPECT Single photon emission computed tomography
WB Whole body
WI Weighted image
XR X-ray

References

[1] Haass NK, Smalley KSM, Li L, Herlyn M. Adhesion, migration and communication in melanocytes and melanoma. Pigment Cell Res 2005;18:150–9.

[2] Zbytek B, Carlson JA, Granese J, Ross J, Mihm MC, Slominski A. Current concepts of metastasis in melanoma. Expert Rev Dermatol 2008;3:569–85.

[3] Choi EA, Gershenwald JE. Imaging studies in patients with melanoma. Surg Oncol Clin N Am 2007;16:403–30.

[4] Kinkel K, Lu Y, Both M, Warren RS, Thoeni RF. Detection of hepatic metastases from cancers of the gastrointestinal tract by using noninvasive imaging methods (US, CT, MR imaging, PET): a meta-analysis. Radiology 2002;224:748–56.

[5] Solivetti FM, Di Luca Sidozzi A, Pirozzi G, Coscarella G, Brigida R, Eibenshutz L. Sonographic evaluation of clinically occult in-transit and satellite metastases from cutaneous malignant melanoma. Radiol Med 2006;111:702–8.

[6] Voit C, Kron M, Schäfer G, et al. Ultrasound-guided fine needle aspiration cytology prior to sentinel lymph node biopsy in melanoma patients. Ann Surg Oncol 2006;13:1682–9.

[7] Terhune MH, Swanson N, Johnson TM. Use of chest radiography in the initial evaluation of patients with localized melanoma. Arch Dermatol 1998;134:569–72.

[8] Ardizzoni A, Grimaldi A, Repetto L, Bruzzone M, Sertoli MR, Rosso R. Stage I–II melanoma: the value of metastatic work-up. Oncology 1987;44:87–9.

[9] Mooney MM, Mettlin C, Michalek AM, Petrelli NJ, Kraybill WG. Life-long screening of patients with intermediate-thickness cutaneous melanoma for asymptomatic pulmonary recurrences: a cost-effectiveness analysis. Cancer 1997;80:1052–64.

[10] Dimopoulos M, Terpos E, Comenzo RL, et al. International myeloma working group consensus statement and guidelines regarding the current role of imaging techniques in the diagnosis and monitoring of multiple myeloma. Leukemia 2009;23:1545–56.

[11] Balch CM, Gershenwald JE, Soong S-J, et al. Final version of 2009 AJCC melanoma staging and classification. J Clin Oncol 2009;27: 6199–206.

[12] Blake SP, Weisinger K, Atkins MB, Raptopoulos V. Liver metastases from melanoma: detection with multiphasic contrast-enhanced CT. Radiology 1999;213:92–6.

[13] Müller-Horvat C, Radny P, Eigentler TK, et al. Prospective comparison of the impact on treatment decisions of whole-body magnetic resonance imaging and computed tomography in patients with metastatic malignant melanoma. Eur J Cancer 2006;42: 342–50.

[14] Gray MR, del CSM, Zhang X, et al. Metastatic melanoma: lactate dehydrogenase levels and CT imaging findings of tumor devascularization allow accurate prediction of survival in patients treated with bevacizumab. Radiology 2014. published online Feb 1, http://pubs.rsna.org/doi/full/10.1148/radiol.13130776.

[15] Pfannenberg C, Schwenzer N. Whole-body staging of malignant melanoma: advantages, limitations and current importance of PET-CT, whole-body MRI and PET-MRI. Radiologe 2015;55: 120–6.

[16] Isiklar I, Leeds NE, Fuller GN, Kumar AJ. Intracranial metastatic melanoma: correlation between MR imaging characteristics and melanin content. AJR Am J Roentgenol 1995;165:1503–12.

[17] Tyler DS, Onaitis M, Kherani A, et al. Positron emission tomography scanning in malignant melanoma. Cancer 2000;89: 1019–25.

[18] Balch CM, Soong SJ, Murad TM, Smith JW, Maddox WA, Durant JR. A multifactorial analysis of melanoma. IV. Prognostic factors in 200 melanoma patients with distant metastases (stage III). J Clin Oncol 1983;1:126–34.

[19] Fishman EK, Kuhlman JE, Schuchter LM, Miller JA, Magid D. CT of malignant melanoma in the chest, abdomen, and musculoskeletal system. Radiographics 1990;10:603–20.

[20] Gavelli G, Giampalma E. Sensitivity and specificity of chest X-ray screening for lung cancer. Cancer 2000;89:2453–6.

[21] Veronesi G, Bellomi M, Mulshine JL, et al. Lung cancer screening with low-dose computed tomography: a non-invasive diagnostic protocol for baseline lung nodules. Lung Cancer 2008; 61:340–9.

[22] Strobel K, Dummer R, Husarik DB, Pérez Lago M, Hany TF, Steinert HC. High-risk melanoma: accuracy of FDG PET/CT with added CT morphologic information for detection of metastases. Radiology 2007;244:566–74.

[23] Sampson JH, Carter JH, Friedman AH, Seigler HF. Demographics, prognosis, and therapy in 702 patients with brain metastases from malignant melanoma. J Neurosurg 1998;88:11–20.

[24] Atlas SW, Grossman RI, Gomori JM, et al. MR imaging of intracranial metastatic melanoma. J Comput Assist Tomogr 1987;11: 577–82.

[25] Zukauskaite R, Schmidt H, Asmussen JT, Hansen O, Bastholt L. Asymptomatic brain metastases in patients with cutaneous metastatic malignant melanoma. Melanoma Res 2013;23:21–6.

[26] Schellinger PD, Meinck HM, Thron A. Diagnostic accuracy of MRI compared to CCT in patients with brain metastases. J Neurooncol 1999;44:275–81.

[27] Velez A, Walsh D, Karakousis CP. Treatment of unknown primary melanoma. Cancer 1991;68:2579–81.

[28] Peters B, Peters R, De Praeter G, Vanhoenacker F. Epidural metastatic melanoma. Eurorad 2015:12851.

[29] De la Monte SM, Moore GW, Hutchins GM. Patterned distribution of metastases from malignant melanoma in humans. Cancer Res 1983;43:3427–33.

[30] Patel JK, Didolkar MS, Pickren JW, Moore RH. Metastatic pattern of malignant melanoma. Am J Surg 1978;135:807–10.

[31] Silverman PM, Heaston DK, Korobkin M, Seigler HF. Computed tomography in the detection of abdominal metastases from malignant melanoma. Invest Radiol 1984;19:309–12.

[32] Shirkhoda A, Albin J. Malignant melanoma: correlating abdominal and pelvic CT with clinical staging. Radiology 1987;165:75–8.

[33] Trout AT, Rabinowitz RS, Platt JF, Elsayes KM. Melanoma metastases in the abdomen and pelvis: frequency and patterns of spread. World J Radiol 2013;5:25–32.

[34] Patnana M, Bronstein Y, Szklaruk J, et al. Multimethod imaging, staging, and spectrum of manifestations of metastatic melanoma. Clin Radiol 2011;66:224–36.

[35] Schuchter LM, Green R, Fraker D. Primary and metastatic diseases in malignant melanoma of the gastrointestinal tract. Curr Opin Oncol 2000;12:181–5.

[36] Fon GT, Wong WS, Gold RH, Kaiser LR. Skeletal metastases of melanoma: radiographic, scintigraphic, and clinical review. AJR Am J Roentgenol 1981;137:103–8.

[37] Potepan P, Spagnoli I, Danesini GM, et al. The radiodiagnosis of bone metastases from melanoma. Radiol Med 1994;87:741–6.

[38] Fayad LM, Kawamoto S, Kamel IR, et al. Distinction of long bone stress fractures from pathologic fractures on cross-sectional imaging: how successful are we? AJR Am J Roentgenol 2005;185: 915–24.

[39] Viswanathan N, Khanna A. Skeletal muscle metastasis from malignant melanoma. Br J Plast Surg 2005;58:855–8.

[40] Herring CL, Harrelson JM, Scully SP. Metastatic carcinoma to skeletal muscle: a report of 15 patients. Clin Orthop Relat Res 1998:272–81.

[41] Damron TA, Heiner J. Distant soft tissue metastases: a series of 30 new patients and 91 cases from the literature. Ann Surg Oncol 2000;7:526–34.

[42] Drummond SR, Fenton S, Pantilidis EP, Harnett AN, Kemp EG. A case of cutaneous melanoma metastatic to the right eye and left orbit. Eye (Lond) 2003;17:420–2.

[43] De Bustros S, Augsburger JJ, Shields JA, Shakin EP, Pryor CC. Intraocular metastases from cutaneous malignant melanoma. Arch Ophthalmol (Chicago, Ill 1960) 1985;103:937–40.

[44] Singh P, Singh A. Choroidal melanoma. Oman J Ophthalmol 2012; 5:3–9.
[45] Heller M, Guthoff R, Hagemann J, Jend H-H. CT of malignant choroidal melanoma: morphology and perfusion characteristics. Neuroradiology 1982;23:23–30.
[46] Russo A, Mariotti C, Longo A, et al. Diffusion-weighted magnetic resonance imaging and ultrasound evaluation of choroidal melanomas after proton-beam therapy. Radiol Med 2015;120:634–40.
[47] Ravdel L, Robinson WA, Lewis K, Gonzalez R. Metastatic melanoma in the breast: a report of 27 cases. J Surg Oncol 2006;94: 101–4.
[48] Bassi F, Gatti G, Mauri E, Ballardini B, De Pas T, Luini A. Breast metastases from cutaneous malignant melanoma. Breast 2004;13: 533–5.
[49] Bussani R, De-Giorgio F, Abbate A, Silvestri F. Cardiac metastases. J Clin Pathol 2007;60:27–34.
[50] Batsakis JG, Bautina E. Metastases to major salivary glands. Ann Otol Rhinol Laryngol 1990;99:501–3.
[51] Bussi M, Cardarelli L, Riontino E, Valente G. Primary malignant melanoma arising in the parotid gland: case report and literature review. Tumori 1999;85:523–5.
[52] Yousem DM, Li C, Montone KT, et al. Primary malignant melanoma of the sinonasal cavity: MR imaging evaluation. Radiographics 1996;16:1101–10.

CHAPTER

38

From Image to Information: Image Processing in Dermatology and Cutaneous Biology

M.O. Visscher, S.A. Burkes, R. Randall Wickett, K.P. Eaton

University of Cincinnati, Cincinnati, OH, United States

OUTLINE

INTRODUCTION

Simply stated, the objective of image processing is to extract information that has been acquired with an imaging system. The term *dermatology* is used broadly, encompassing skin structure, function, diseases, cancer, melanoma, photodamage, wounds, scars, skin restoration, cosmetics, and environmental effects. This chapter examines image processing for skin conditions relative to normal, healthy skin. The skin provides a barrier to water loss and irritants, sensation, thermal regulation, resilience to trauma, and immunosurveillance [1]. It

Imaging in Dermatology
http://dx.doi.org/10.1016/B978-0-12-802838-4.00038-8

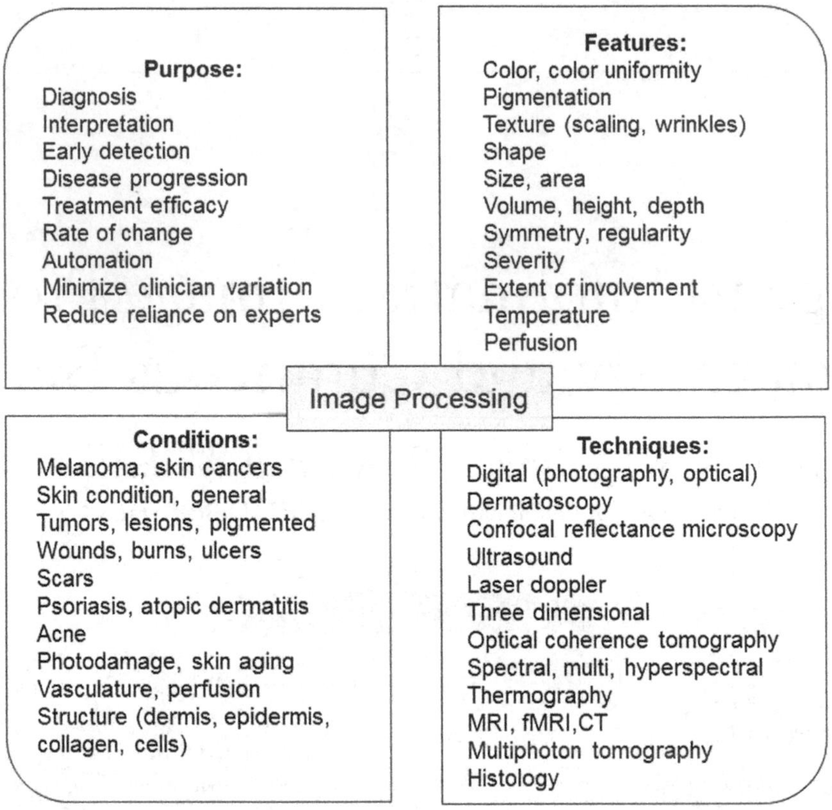

Purpose:
Diagnosis
Interpretation
Early detection
Disease progression
Treatment efficacy
Rate of change
Automation
Minimize clinician variation
Reduce reliance on experts

Features:
Color, color uniformity
Pigmentation
Texture (scaling, wrinkles)
Shape
Size, area
Volume, height, depth
Symmetry, regularity
Severity
Extent of involvement
Temperature
Perfusion

Image Processing

Conditions:
Melanoma, skin cancers
Skin condition, general
Tumors, lesions, pigmented
Wounds, burns, ulcers
Scars
Psoriasis, atopic dermatitis
Acne
Photodamage, skin aging
Vasculature, perfusion
Structure (dermis, epidermis, collagen, cells)

Techniques:
Digital (photography, optical)
Dermatoscopy
Confocal reflectance microscopy
Ultrasound
Laser doppler
Three dimensional
Optical coherence tomography
Spectral, multi, hyperspectral
Thermography
MRI, fMRI,CT
Multiphoton tomography
Histology

FIGURE 38.1 **Factors impacting image processing strategy.** Selection of image processing methods depends upon four interrelated factors: purpose, eg, clinical evaluation, research study; specific skin condition, eg, disease, wound; characteristics/features of interest; and attributes of the specific imaging technique and/or protocol. *CT,* Computed tomography; *fMRI,* functional magnetic resonance imaging; *MRI,* magnetic resonance imaging.

confers innate immunity through proinflammatory and antiinflammatory cytokines, structural proteins, lipids, antigen presenting cells, and the physical barrier of the stratum corneum (SC). Therefore skin damage is any change that disrupts normal structure and homeostasis.

Contributions from academic medical, bioengineering, and skin-care industry researchers provide an overview of important concepts, research, and development. The selection of an image processing technique depends upon the specific clinical and/or research purpose and the imaging modality or modalities being applied. Here the topic is presented in general, using specific examples from established and emerging techniques in the context of current clinical/research questions.

PURPOSE OF SKIN IMAGE PROCESSING

The objectives of skin imaging are to diagnose; differentiate; determine severity, disease progression, and treatment efficacy; assess clinical judgment; identify biological changes defining the disease; and assess the mechanism of action of a disease process [2]. Imaging can relate clinical judgment to patient response and patient expectations to clinician-perceived treatment efficacy. Selection of image processing methods depends upon interrelated factors: (1) purpose; (2) skin condition, eg, disease, wound; (3) condition characteristics; and (4) imaging technique (Fig. 38.1).

MOTIVATION FOR IMAGING IN DERMATOLOGY

Imaging "creates visual representations of objects for data extraction or medical diagnosis using a computerized technique" [3]. Visual inspection is the gold standard for skin assessment [4]. Clinical judgments depend upon perception of the skin surface [5]. Clinicians, patients, and consumers use skin color uniformity, distribution, and texture to surmise physiological health [6]. Low color variability, lighter skin, less yellow coloration, and fewer wrinkles connote younger age [7–9]. Numerous visual methods have been implemented. They typically incorporate severity of damage, eg, inflammation, and area of involvement versus normal skin and may be condition dependent [10–18]. For example, the asymmetry, border sharpness, color variegation, and diameter (ABCD) rule, Menzies score and seven-point checklist are used for melanoma [19]. Clinicians use a visual analog scale with blinded color images to decide improvement as stable/worse (0), slight (<25%), moderate (25–50%), good (50–75%), and excellent (>75%) for vascular tumors [20,21]. Clinical judgment develops from formal training, practice, and experience, with atypical cases building expertise.

Low reproducibility, variation among experts, and low reliability limit visual methods [22,23]. Accuracy is significantly impacted by skin pigmentation, varying from very light (Caucasian) to very dark (African). Evaluation of compromise in dark skin is confounded because greater pigmentation interferes with detection of erythema [24,25]. Visual assessment of neonatal jaundice underestimated serum bilirubin in 17–40% and overestimated in 5–36% of cases [26]. The agreement among four clinicians who evaluated wounds from color images was 55–76% [27]. Interobserver variability in diagnosing sebaceous neoplasms was fair to moderate [28]. Melanoma diagnostic accuracy was 56–62% for 3–5 years of experience, increasing to 80% after 10 years [29]. Visual methods are inaccurate, expert-intensive, and time consuming. [30] and may lead to delayed or ineffective treatment, infection, pain, and increased costs. The need for objective, reliable approaches for measuring skin features has motivated the development of imaging modalities.

TRENDS IN SKIN IMAGING AND IMAGE PROCESSING

Literature searches on "imaging and skin evaluation" and "image processing in skin" from 1977 to 2015 produced 1698 citations for review. The number increased substantially, with twice as many from 2006 to 2015 as from 1996 to 2005 (Fig. 38.2A). Skin cancer was the most common topic, ie, about twice the number as other conditions, followed by wounds, vasculature or perfusion, tissue structure, lesions, and general skin conditions. Digital imaging was the most common method, followed by confocal microscopy, dermoscopy, ultrasound, laser Doppler perfusion, three-dimensional (3D) imaging, optical coherence, and multispectral/spectral and thermal imaging. Most (88%) were in vivo reports, particularly since 2001. Between 2011 and 2015, hyperspectral, multiphoton tomography, optical coherence tomography (OCT), and confocal, thermal and photoacoustic imaging emerged (see Fig. 38.2B).

Publications from optical, biophysical, biomedical, engineering, and computing journals increased for 2006–2015 versus 1996–2005 with inaugural issues for *Skin Research and Technology* (1995), *Journal of Biomedical Optics* (1996), *Optics Express* (1997), *Computerized Medical Imaging and Graphics* (1999), and *Journal of Biophotonics* (2008). About 30% focused on image processing, including preprocessing, segmentation algorithms, feature extraction, automatic classification, and statistical techniques.

PROGRESS IN SKIN IMAGING TECHNIQUES

Established and newer skin imaging techniques are presented, setting the stage for image processing.

Digital Imaging

Human vision uses light via retinal cells, the optic nerve, and neural networks to examine objects [31]. Visible light interacts with melanins (brown), oxygenated hemoglobin (red), deoxyhemoglobin (blue-purple), bilirubin (yellow), and carotene (yellow) to generate perceived skin color [32]. About 5% is reflected back and approximately 95% transmitted, absorbed, or scattered [7]. The SC transmits light, subcutaneous tissue scatters light, and the epidermis and dermis absorb and scatter light [33].

Digital imaging techniques have improved the objectivity and quantitation of skin anomalies including erythema, dyspigmentation, photoaging, atrophy, wounds, lesions, and disease [23,33–39]. Greater repeatability occurred with objective assessment of infantile hemangioma (IH) lightness (3% versus 26% for subjective) [40]. Computer analysis of red channel image means and standard deviations differentiated erythrotelangiectatic rosacea, papulopustular rosacea, and seborrheic dermatitis. Clinicians successfully determined disease patterns but were not as effective as the computer for quantifying erythema [41].

Collection of digital images facilitates evaluation over time and review by multiple individuals. Standardized conditions, including distance, lighting, positioning, and color correction are necessary. Inherent skin coloration varies considerably because of differences in chromophore levels and ultraviolet exposure [42]. This variability is addressed by measuring an uninvolved site as a control. Digital imaging is used extensively, but standards have not yet been established [43]. It is also applied to analysis of tissue samples. Densities of psoriatic tissues stained for specific biomarkers were highly correlated with manual counts [44].

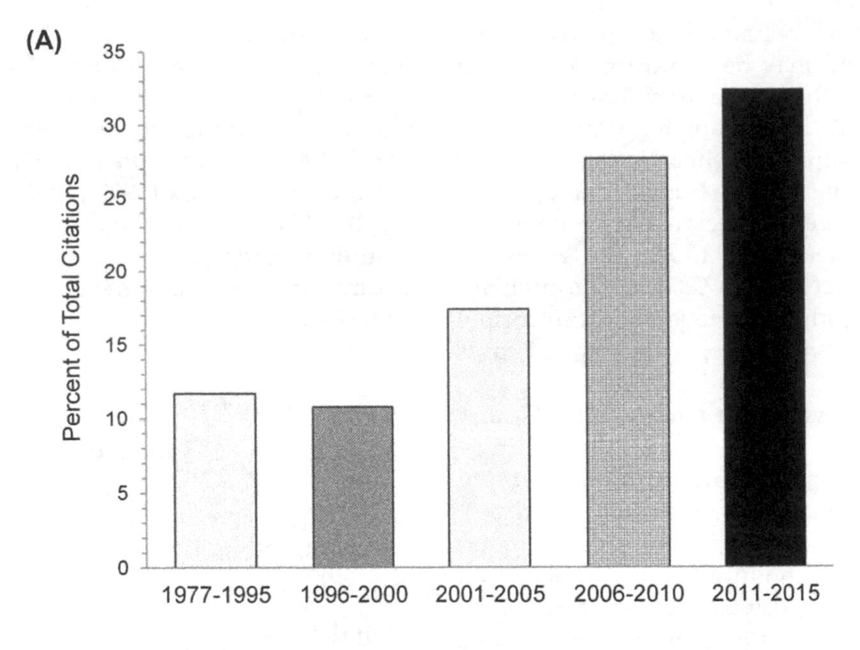

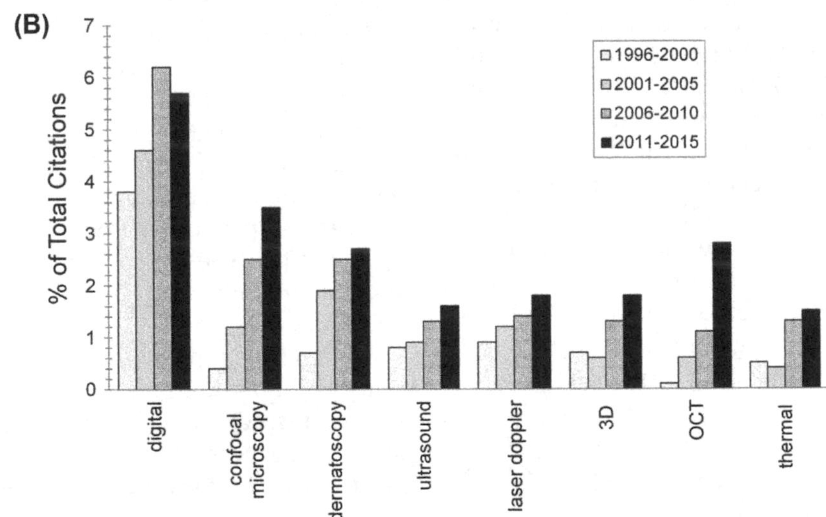

FIGURE 38.2 **Trends in skin imaging and image processing.** Literature searches on "imaging and skin evaluation" and "image processing in skin" identified 1698 topic matter citations for review. (A) Substantial increase over the past 20 years with twice the number from 2006 to 2015 versus 1996–2005. Digital imaging was the most common topic, followed by confocal microscopy, dermoscopy, ultrasound, laser Doppler perfusion imaging, three-dimensional (3D) imaging, optical coherence tomography (OCT), and multispectral/spectral and thermal imaging. (B) Increases from 2011 to 2015 versus 2006–2010 reflected the emergence of hyperspectral imaging, multiphoton tomography, OCT, and confocal, thermal, and photoacoustic methods.

Dermoscopy

Dermoscopy overcomes some limitations of digital imaging by using a liquid on the skin surface to increase light transmission into the epidermis and upper dermis. This technique can differentiate malignant versus nonmalignant lesions [45]. Image collection with polarized light removes interference from reflected light and enhances epidermal and dermal features [46,47]. Melanoma detection sensitivity increased with dermoscopy versus visual examination and accuracy improved when they were used together [48]. Improvements in melanoma diagnostic accuracy were limited to experienced clinicians [49].

Reflectance Confocal Microscopy/Confocal Scanning Laser Imaging

The number of publications on reflectance confocal microscopy (RCM) has increased markedly in the past 10 years (see Fig. 38.2B). Monochromatic laser energy examines tissue to 200–350 μm in depth over a surface area of 0.5–0.10 μm [50,51]. Laser light is directed through a pinhole opening and scans the region of interest in a specific focal plane. Reflected light is directed to a confocal point on a photodetector to generate data in one plane. Serial optical sections are collected by changing the focal depth, and the resulting data generates 3D images. RCM is used in vivo, providing greater resolution than dermoscopy or digital imaging. Thirty-seven features of melanoma/melanocytic lesions have been identified with RCM [52], including six differentiated malignant versus nonmalignant lesions: basaloid cord-like structures, pagetoid cells, aspecific dermal–epidermal junction, ringed dermal–epidermal junction, atypical cells, and ulceration, among nine RCM users [53]. Accuracy, specificity, and sensitivity were 76–89%, 69–91%, and 83–100%, respectively. Higher specificity was associated with greater experience, suggesting that interrater reliability remains a factor. Method refinements are needed to further differentiate features and improve training.

RCM has been used for inflammatory conditions, eg, to demonstrate parakeratosis and the absence of stratum granulosum in psoriasis [54]. Epidermal spongiosis, necrosis, and vesicles were noted for allergic contact dermatitis. Irritant contact dermatitis had parakeratosis, spongiosis, epidermal thickening, necrosis, and SC disruption. Comprehensive studies are warranted to determine the full potential of RCM.

Spectral, Multispectral, and Hyperspectral Imaging

Spectral imaging, including spectrophotometry and diffuse reflectance spectroscopy, uses specific wavelengths of 400–1000 nm [55] to examine skin in vivo [56]. Melanin, oxyhemoglobin, and deoxyhemoglobin can be evaluated separately when skin surface reflection is eliminated [57]. Near-infrared wavelengths are used to map and quantity collagen (papillary dermis). Deeper melanin can be analyzed once more superficial melanin and hemoglobin are removed [56]. Skin erythema from surfactant irritation, histamine release, ultraviolet exposure, and reactive hyperemia have been quantified [58].

Multispectral imaging systems use specific wavelengths within the visible, ultraviolet, and infrared regions to acquire skin information. Further color contrast has been obtained with a multispectral camera and illumination with light-emitting diodes from L*a*b* color space [59]. A similar system quantified erythema from postocclusion hyperemia in darkly pigmented subjects [60]. Multispectral images at 10 nm intervals from 450–950 nm were obtained for 334 subjects with vascular or pigmented lesions [61]. Melanomas absorbed significantly more energy than other pigmented tumors.

Hyperspectral imaging utilizes reflected light and natural fluorescence from tissue components, eg, collagen and elastin, to collect spectral data at wavelengths throughout the ultraviolet, visible, and infrared regions and generate a series of two-dimensional planes for a 3D image (hypercube) [62]. Hyperspectral imaging produced more information from cutaneous wounds than other modalities [63]. Necrotic tissue, granulating tissue, shaved skin, and normal skin with hair could be differentiated, although scar and new epithelium were not different than normal skin.

Optical Coherence Tomography and Multiphoton Tomography

OCT imaging uses two incident light paths to penetrate to depths of up to 2 mm with high (10 μm) resolution [64]. OCT can distinguish tumor borders from normal skin and quantitatively characterize nonmelanoma skin cancer in vivo. Epidermal thickening and edema were observed for irritant dermatitis and psoriasis.

Multiphoton tomography employs rapid (femtoseconds) pulsed-laser energy (titanium:sapphire laser, range 700–1000 nm), simultaneous absorption of two photons in the near-infrared region, and molecular level autofluorescence to achieve penetration and resolution at subcellular levels [65]. The need for: (1) greater resolution than computed tomography and magnetic resonance imaging and (2) improved diagnosis for approximately 10,000 diseases that could not be diagnosed with existing methods motivated its development [66]. Six features discriminated melanoma from melanocytic lesions: poorly defined borders, large intercellular distance, pleomorphic cells, architectural disarray, ascending melanocytes, and dendritic cells. Specificity and sensitivity were 71–95% and 69–97%, respectively, and higher for more experienced investigators [67].

Thermal Imaging

Skin emits infrared radiation as a result of physiological processes, such as metabolism and blood perfusion, thermal conductivity, sympathetic nervous system activity, and response to environmental changes [68–70]. Temperatures of tumors, including melanoma and IHs, are higher than uninvolved skin [71], in part because of differences in perfusion and metabolism [68,72]. Skin surface temperature depends on tumor size, ie, higher with increasing size, and depth, ie, inversely with vasculature distance [68,73,74]. Infrared thermography detects 7.5–13-μm skin surface radiation, ie, steady-state conditions (static mode) [69]. Dynamic thermography applies a stress (eg, cooling, heating) to the skin and eliminates environmental effects [70]. When the stress is removed, tissues respond to restore equilibrium. Changes in the infrared signal over time provide functional information [75–77] and can reveal physiological abnormalities [78]. Tumors respond abnormally, thereby differentiating them from uninvolved tissue [68]. Recovery from cool stress for IHs was more rapid for subjects aged 2 months than those over 5 months, indicative of proliferating tumors [79].

Photoacoustic Imaging

Many imaging methods employ energy in the ultraviolet, visible, and near-infrared regions and are limited in skin penetration depth. Functional photoacoustic microscopy uses a tunable laser (neodymium: yttrium-aluminum-garnet) with fast pulses (6 ns) to perturb the skin and create ultrasonic waves [80]. The energy causes elastic tissue expansion, allows penetration greater than 1 mm and achieves high resolution. Three-dimensional reconstructed images of melanoma and the surrounding vasculature have been obtained with this method. Currently studies are exploring the diagnostic potential for skin cancer and other conditions.

SKIN IMAGE PROCESSING

The purpose of skin imaging and therefore image processing is generally in two categories: (1) diagnosis, ie, presence, absence, and which condition; and (2) extent/severity of the condition over time. Both the method and processing strategy depend on the purpose.

General Strategy

Image capture is the first step, given the inherent variation in skin properties with season. Standardized conditions for positioning, distance, direction, and instrumental settings are essential. Collection of image data for an uninvolved, site (control) is recommended for image processing and analysis.

Image processing involves four major steps: (1) preprocessing, (2) feature extraction, (3) feature selection, and (4) classification [81,82]. For melanoma detection, they are: (1) preprocessing, (2) border detection, (3) feature descriptor extraction, and (4) classification [83]. Preprocessing maximizes the information and removes confounders, such as correction for uneven illumination with filters. An image histogram shows the intensity distribution of pixels from values of 0–255. Contrast is increased by normalization or "stretching" the range. This step is important in cases where border detection is needed, eg, melanoma or solar lentigos. The histogram shape determines whether the image information has distinct regions, eg, normal and abnormal intensity, or is more uniform.

The next step is feature extraction. Segmentation divides the image into parts, ie, segments, to simplify the data and increase the signal-to-noise ratio. Images may be segmented on gray level, color, texture, depth, and motion. For example, the image is converted to grayscale and a threshold is applied to separate the affected regions from normal skin. The threshold value is typically dictated by the specific purpose of the work. The histogram shape may be used to select a threshold value. Texture, ie, repeating elements and intensities within an image, is used when thresholding is not effective.

Feature selection is used to determine the characteristics that differentiate a condition from a reference or control, such as the ABCD rule [83]. *The characteristics are typically derived from clinical (visual) evaluation, histopathology, and mechanistic studies.* Classification, the last step, determines the meaning of the information in the image. In general, processing techniques for diagnosis are based on expert assessment of image sets that are used to generate computer algorithms derived from the training set. Feature selection and classification are discussed in greater detail in the following sections. Image processing examples from four categories are provided for specific imaging methods.

Category 1: Irritant Dermatitis, Erythema, and General Skin Structure

Digital Imaging

Digital images record the clinical examination, typically in red-green-blue (RGB) color space. RBG images are commonly converted to the International Commission on Illumination (CIE) 1976 L*a*b* (also written as *CIELab, Lab*) color space in preprocessing because this colorspace relates to human color perception [84]. CIE L*a*b* is a uniform 3D color space where L* is lightness-darkness (L* = 255, white; L* = 0, black), a* is red-green (positive to negative) and b* is yellow-blue (positive to negative) [85–87]. Hue, saturation, and value colorspace was combined with L*a*b* to discern color shading and quantify treatment effects for port wine stains [88].

Skin irritation from repeated exposure to water, cleansers, and disinfectants manifests as erythema from capillary dilation and increased blood flow [89]. Erythema is difficult to quantify visually because of high skin compromise and heterogeneity (knuckles, fingers). Digital image processing was applied to objectively quantify hand-skin erythema from irritants [90]. RGB images were converted to L*a*b* form and separated into L*, a*, and b* channels (Fig. 38.3). The a* image, ie, red color, was correlated with erythema [91].

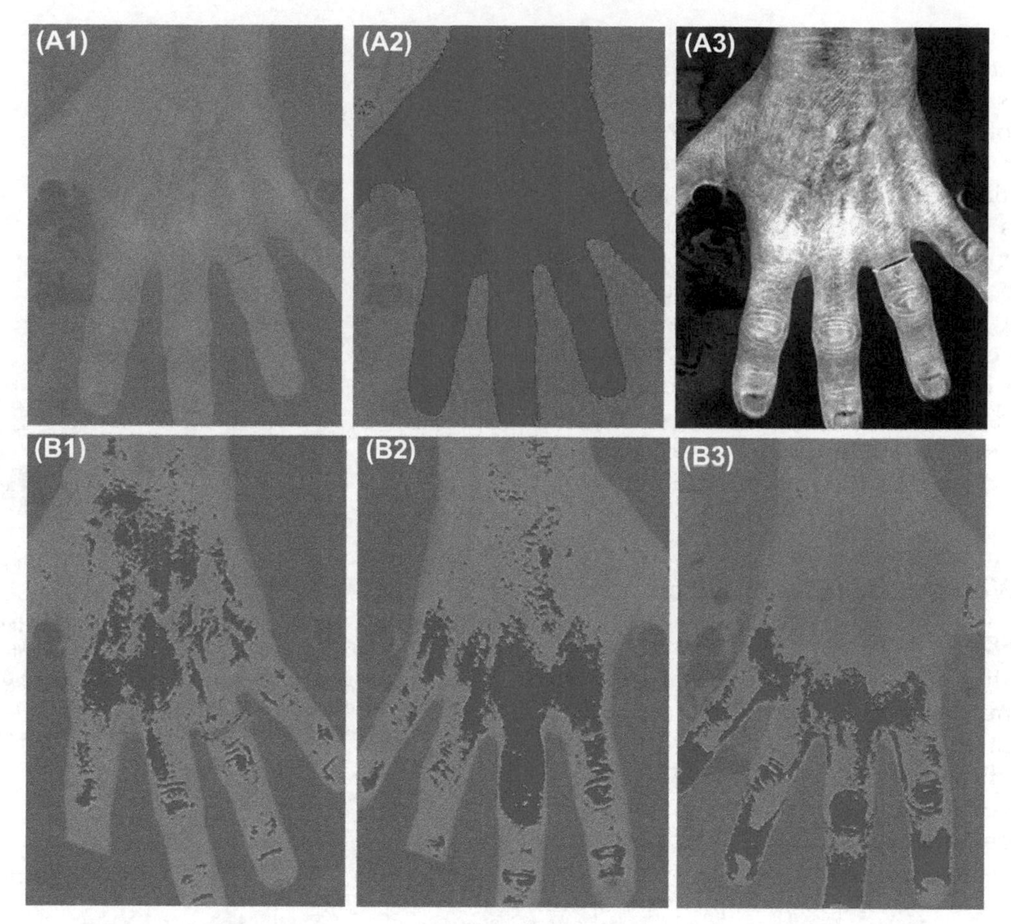

FIGURE 38.3 **Image processing procedures for skin erythema caused by irritant contact dermatitis.** (A and B) Digital imaging and processing were applied to objectively quatify hand-skin erythema from irritant contact dermatitis caused by repeated hand hygiene procedures. Images were collected in RGB, converted to L*a*b* color, and separated into L*, a*, and b* images. Red color was correlated with erythema [91]. (A) The a* image was converted to grayscale (A1). A threshold was applied to extract the features of interest (hands) and separate the background (A2). The grayscale image contrast was increased to reduce noise (A3). The red pixel mean (μ) and standard deviation (σ) values were computed. The pixel values greater than the mean (μ) plus one standard deviation (σ), ie, $\mu + \sigma$ represent the most erythematous regions and were defined as excess erythema [90]. The areas of greatest erythema are visualized by applying the threshold to show $\mu + \sigma$, excess erythema. (B) The highest erythema is on the dorsum and knuckles (B1) and knuckles and fingers in (B2 and B3).

It was converted to grayscale (see Fig. 38.3A1), a threshold applied to extract the features of interest (hands) and separate the background (see Fig. 38.3A2), and the contrast increased (see Fig. 38.3A3). Pixels greater than the red pixel mean (μ) plus one standard deviation ($\mu + \sigma$) represented the most erythematous regions. Excess erythema was defined as the percent of red pixels greater than ($\mu + \sigma$) [90], as shown in Fig. 38.3B. The highest erythema is on the dorsum and knuckles (see Fig. 38.3B1) and the knuckles and fingers (see Fig. 38.3B2 and B3).

Multispectral Imaging

Multispectral imaging quantified occlusion-induced erythema (arterial pressures) in highly pigmented subjects [60]. The images were coregistered and corrected for shading variations. The feature, erythema, was extracted by separation from melanin with five fusion algorithms. Background was removed using a control without erythema and algorithms evaluated with and without contrast enhancement. The highest sensitivity (100%) and specificity (92%) were with contrast and an algorithm based on hemoglobin (550 nm), melanin (650 nm), and water (950 nm for water absorption of 94–970 nm).

Multiphoton Tomography

Processing methods to differentiate the epidermis and dermis were developed from 3D multiphoton tomography images to address image capture limitation, eg, varying signal intensity in x, y, and z sampling intervals and second harmonic generation signals caused by dissimilarities in collagen content [92]. Markers were used for image segmentation for the two-photon excited fluorescence and second harmonic generation signals. Voxels were separated with a watershed algorithm and graph cuts to generate three gray levels (ie, three layers). Epidermal thickness, dermal–epidermal junction area, elastic fiber density, and melanin content were extracted. An evaluation of 0.3% retinol and 0.025% retinoic acid showed greater epidermal thickness, greater dermal–epidermal junction irregularity and lower melanin for both treatments versus controls [93]. Combined RCM and multiphoton microscopy in vivo were effective for identifying subcellular epidermal features [94]. The confocal image identified cell boundaries that were separated from cell matter with a gaussian filter and a watershed algorithm. The intensities of the cytoplasm and nuclei were not particularly distinct, so a global energy level–set formulation and local Lagrange level–set method were used for segmentation.

Category 2: Wounds, Burns, and Ulcers

Digital Imaging

Wound imaging may be used to determine healing, infection, inflammation, and rate of change. Clinical assessment is often the starting point for processing. Wound color indicates granulation tissue (red), necrosis (black), fibrin (yellow), and hyperkeratosis (white) [95]. After preprocessing, edge detection based on abrupt changes in pixel intensity separates the wound from uninvolved tissue. For example, multiple color spaces, ie, RGB; L*a*b*; hue, saturation, and intensity; and L*u*v* were used to extract 111 features from wound images [95]. Cooccurrence matrix algorithms, ie, frequency pixels with given gray intensities over width and length, were applied. The features that best quantified the wound were selected and classified with four algorithms: k-nearest neighbor ($n = 10$), multilayer perceptron, decision tree, and naïve Bayes (k-nearest-neighbor with 5 neighbors, shown in Fig. 38.4). An expert (dermatologist) identified the regions to create a training set for supervised machine learning. The 111 features were evaluated with 12 processes against the training set [96]. Contrast between a* and b* and u* and v*, and entropy and heterogeneity between a* and b* discriminated the wound features.

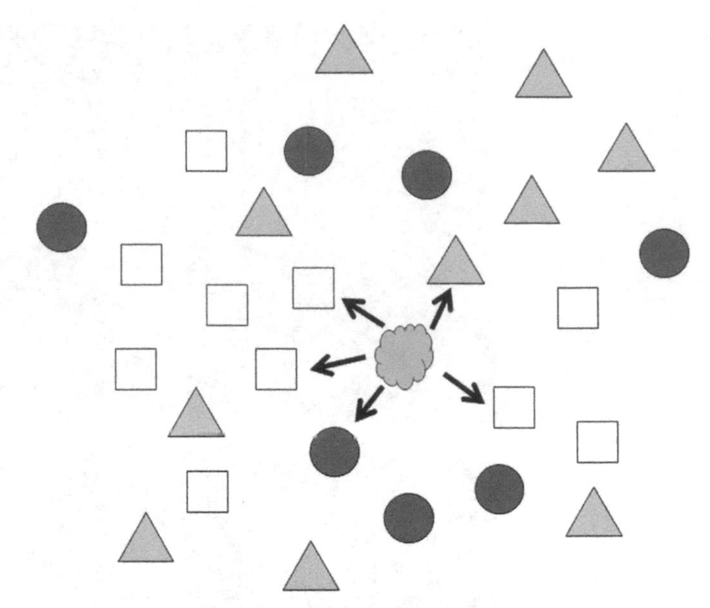

FIGURE 38.4 **Schematic for the k-nearest neighbor classification method.** Three types of image pixels are represented by *white squares, red circles,* and *gold triangles.* To classify the *green pixel,* the five nearest neighbors are examined. There are three *white squares,* one *red circle,* and one *gold triangle.* Therefore the green pixel is classified as a *white square.*

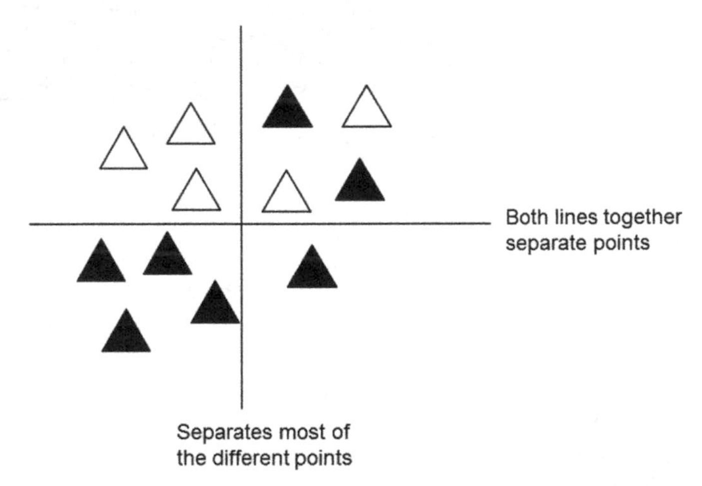

FIGURE 38.5 **Schematic for support vector machine (SVM) classification method.** Two types of image pixels are represented by *white* and *black triangles.* The horizontal boundary separates the white from most of the black. Addition of a second boundary (*perpendicular line*) allows separation of white and black pixels.

Stressing the importance of standardized image collection and preprocessing, a classification system for quantifying granulation tissue, necrosis, and slough (fibrin) was described for teledermatology [27]. Four feature extraction algorithms, ie, J-value segmentation, efficient graph-based image segmentation, mean shift, and color structure code, were evaluated by: (1) each automated system, and (2) an expert group who manually determined wound boundaries. The expert values were combined for the training set and features were classified with clustering and support vector machine (SVM) methods (Fig. 38.5). Classification effectiveness was not impacted by variations in skin color, lighting, or moisture. The J-value segmentation for granulation, necrosis, slough, and healthy skin performed the best and was comparable to the combined expert results.

Mean shift procedures followed by segmentation with a region-growing algorithm were used to identify regions of greatest density in pressure ulcers [97]. The borders between granulation tissue, wound bed, and slough were not sufficiently distinct from the threshold by histogram or watershed methods. To overcome this limitation, images were segmented with adaptive mean shift and region-growing procedures. Principal component analysis was applied to reduce the color and texture features. Wound care experts categorized tissues from segmented images as granulation, slough, necrosis, healing, and healthy skin for a training set. Neural networks (pattern recognition based on a large number of interconnecting processes simulating neuronal transmission) and bayesian classifiers (method to decide whether an image area is in one of two classes based on known information for two classes and comparison to the area) were used to classify wound tissues. Regions were classified as necrosis or nonnecrosis (binary classification) and then nonnecrosis was classified as slough or not slough. Next, the region was classified as healing or healthy skin. The overall accuracy for the bayesian method was 94% (range 87–99%) with lowest values for healing tissue and healthy skin. The overall sensitivity was 84% (range 75–89%) and lowest for slough.

Hyperspectral Imaging

Hyperspectral imaging was investigated for efficacy in differentiating granulation, blood clots, necrosis, and epithelial skin from uninvolved tissue. A mixture-tuned matched filtering algorithm to classify wound tissues was evaluated [63]. Noise was reduced in preprocessing with two principal component data transformations, followed by an indexing procedure to threshold image pixels against pure spectral values. Features were extracted with clustering methods and displayed as a map.

Category 3: Melanoma, Pigmented Lesions, Basal Cell Carcinoma, and Photoaging

Dermoscopy

Image processing for assessing suspected malignancies needs to determine presence or absence and differentiate visually similar lesions, eg, melanoma versus pigmented lesion. Various processing strategies have been employed, as reviewed comprehensively [81]. Preprocessing included removal or minimization of hair, specular reflectance, and other noise [98]. Feature extraction involved: (1) histogram and adaptive thresholding, (2) color feature segmentation with statistical procedures (eg, principal component transformation, sequential forward selection, sequential backward selection), (3) edge detection including radial search and active contours, (4) segmentation into smaller data sets based on regional similarities, and (5) pixel classification. Identification of the extracted features that accurately differentiate one lesion type from another is essential. Sets of 10–15 features provided optimum discrimination, whereas more than 20 features reduced it [99]. The pigmented lesion features used in the literature for the major categories of shape, color and texture are in Table 38.1 [82]. The 2015 Masood et al. review reported the frequencies of classification methods as: (1) statistical procedures (23%), (2) artificial neural network (27%), (3) SVM (16%), (4) decision tree (13%), (5) rule-based (10%), (6) k-nearest neighbor (6%), (7) other (4%), and (8) extreme learning machine (1%) [81].

The image processing methods for pigmented lesions were comprehensively and critically reviewed, including the complexities and lack of standardization [83]. The following methods were cited: SVMs, artificial neural networks, decision trees, k-nearest neighbors, discriminant analysis, regression analysis, bayesian classifiers, fuzzy logic, principal direction divisive partitioning, adaptive wavelet transformation–based tree structures, minimum distance classifiers, hidden Markov models, multiple classifiers, AdaBoost metaclassifier, and attributional calculus. Table 38.2 shows the features for pattern recognition analysis. These authors called for the compilation of a robust database of dermoscopic images as necessary for improving computer-assisted diagnosis of pigmented lesions.

Differentiation of melanoma from dysplastic nevi has been difficult to achieve by both clinical assessment and dermoscopy [100]. A new set of color and texture image processing methods was applied to this distinction problem for 4090 benign, 950 dysplastic, and 90 melanoma lesions [82]. They were: (1) histogram of oriented

TABLE 38.1 Features of Pigmented Lesions [81]

Feature Name
SHAPE
Area, perimeter
Irregularity and compactness ratio
Fractal geometry
Fourier features
COLOR
Color quantization
Color variance and ratio
Color opponent angle
Alternative color space
TEXTURE
Histogram of gradients
Contrast and entropy
Gabor filter
Gray-level cooccurrence matrix
Difference of Gaussian
Wavelet-based descriptor
Local binary pattern
Scale-invariant feature transformation (SIFT) and color SIFT
Harris/Hessian-Laplace

TABLE 38.2 Features of Pigmented Lesions Used in Pattern Recognition [19,83]

Pattern
GLOBAL
Globular
Reticular
Cobblestone
Homogeneous
Parallel
Starburst
Multicomponent
Nonspecific
LOCAL
Pigment network
Dots/globules
Streaks
Blue-whitish veil
Regression structures
Hypopigmentation
Blotches
Vascular structures

gradients and scale-invariant feature transformation, (2) Gabor filter, (3) opponent color space, and (4) local binary pattern. They were used with first color and gray level cooccurrence matrix methods for shape for 90 melanoma and 90 random dysplastic nevi images (Table 38.3). After feature selection, three classification algorithms were evaluated: SVM, gradient boosting to improve performance, and random forest algorithm with decision trees. Combinations of texture, color, and shape produced the highest sensitivity, ie, differentiating melanoma from dysplastic nevi. Results ranged from 56.2% sensitivity for shape and SVM classifier to 98.5% for four texture features with the random forest classifier. Lesion color features did not contribute to the most effective system. Global lesion features provided greater sensitivity than local attributes.

TABLE 38.3 Features Evaluated in Differentiation of Melanoma and Dysplastic Nevi [81]

Description	Type and Number
Thinness ratio	Shape
Border asymmetry	Shape
Distance variance of border points to center	Shape
Statistics of gradient operator on lesion border	Shape
Color variance and response for RGB, HSI, and L*a*b* colorspace	Color
Color histogram in RGB	Color ($n = 42$)
Opponent colorspace angle and hue histogram	Color ($n = 42$)
Completed local binary pattern	Texture
Gray-level cooccurrence matrix	Texture ($n = 32$)
Gabor filter	Texture
Histogram of oriented gradient	Texture
Scale-invariant feature transformation (SIFT)	Texture

HSI, Hue, saturation, and intensity; *L*a*b**, orthogonal color space representing lightness, red/green, and blue/yellow color; *RGB*, red-green-blue.

Reflectance Confocal Microscopy

In vivo RCM, digital dermoscopy, and histology were combined to classify: (1) melanocytic from nonmelanocytic lesions and (2) melanoma from nevi [101]. Four features differed for melanocytic versus nonmelanocytic lesions: pagetoid spread, cobblestone pattern, presence

TABLE 38.4 Features of Aging and Photoaging From Reflectance Confocal Imaging [104]

Location and Depth	Description
Subsurface epidermis (15 ± 5 μm)	Small rhomboidal furrow pattern
	Large rhomboidal furrow pattern
	Disarranged rhomboidal furrow pattern
	Linear furrow pattern
	Hyperkeratotic regions
Middle epidermis (30 ± 5 μm)	Sunburn cells
	Regular honeycomb pattern
	Irregular honeycomb pattern
	Disarray of honeycomb pattern
	Mottled pigmentation
Dermal epidermal junction (60 ± 5 μm)	Regular rings
	Irregular rings
	Polycyclic rings
	Absence of rings
Upper dermis (60 ± 5 μm)	Thin reticulated collagen fibers
	Coarse collagen fibers
	Huddled collagen
	Curled fibers (elastosis)
	Inflammatory cells
	Longitudinal vessels
Image stack through all skin layers (Maximum 100 μm)	Minimal epidermal thickness

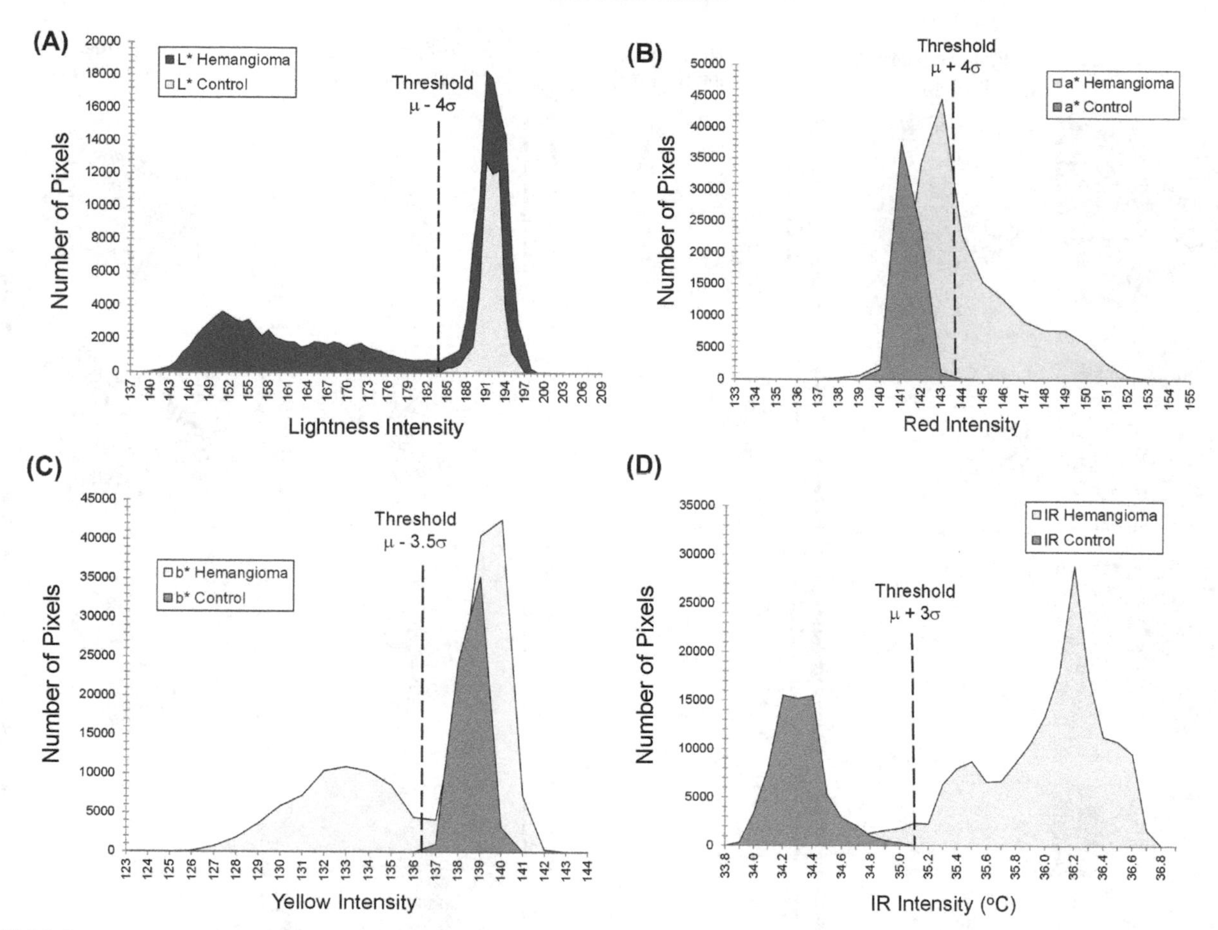

FIGURE 38.6 **Threshold selection for color and thermal image analysis of infantile hemangiomas (IHs).** Color images were separated into L*, a*, and b* images. Histograms of IH and controls were compared for 120 evaluations to determine thresholds. They were: L* (lightness) values less than $\mu - 4\sigma$ (mean: 4 × standard deviation) (A) a* (red color) values higher than $\mu + 4\sigma$ (B) b* (yellow color) values less than $\mu - 3.5\sigma$ (C) and infrared temperature higher than $\mu + 3\sigma$ (D). *IR*, Infrared.

of dermal cell clusters or dermal nests, and meshlike dermal–epidermal junction. Melanoma lesions had round pagetoid cells and atypical dermal nucleated cells, whereas nevi had normal basal cells and dermal papillae. Scores of −2 to +2 were assigned to the image, based on degree of each feature, with −1 as the threshold for melanoma versus nevi decision.

To distinguish pigmented nevi from surface spreading melanoma, serial optical sections were collected with RCM to generate 3D images [102]. An algorithm based on pagetoid spread and dermal–epidermal junction distinguished pigmented nevi from surface spreading melanoma. Preprocessing corrected for motion artifacts and curvature. The depth of the most reflective point on the pigmented surface generated a roughness map. Nevi had low roughness at the dermal–epidermal junction, whereas surface spreading melanoma had high roughness.

RCM with serial sections to 100 μm was used to quantify photodamage in subjects aged 20–30 or 50–60 years [103]. "Photoaging" was a continuum from regularly shaped keratinocytes, increasing irregularity over time, progressing to actinic keratosis and to squamous cell carcinoma. Experts assigned a score (0–10) from hand and forearm images for 18 features (0, none; 3, severe) with the Score of Intrinsic and Extrinsic Skin Aging scale. Two experts evaluated confocal images for 21 features (Table 38.4) [104]. An entropy function was used to reduce fuzziness and images were segmented with erosion methods to remove background. A two-dimensional Fourier transform was applied to determine tissue regularity. The study concluded: (1) in vivo RCM images may augment clinical judgment in early detection of nonmelanoma skin cancer, and (2) objective evaluation is likely to combine supervised (modeling experts) and unsupervised image processing.

Multispectral Imaging

Multispectral images of pigmented lesions (1856 total, 243 melanomas) were acquired in vivo. A threshold–reflectance graph representing four regions within the lesion and surrounding tissue was generated [105]. The agreement between results from the segmentation algorithm and tracings by experts was 97%.

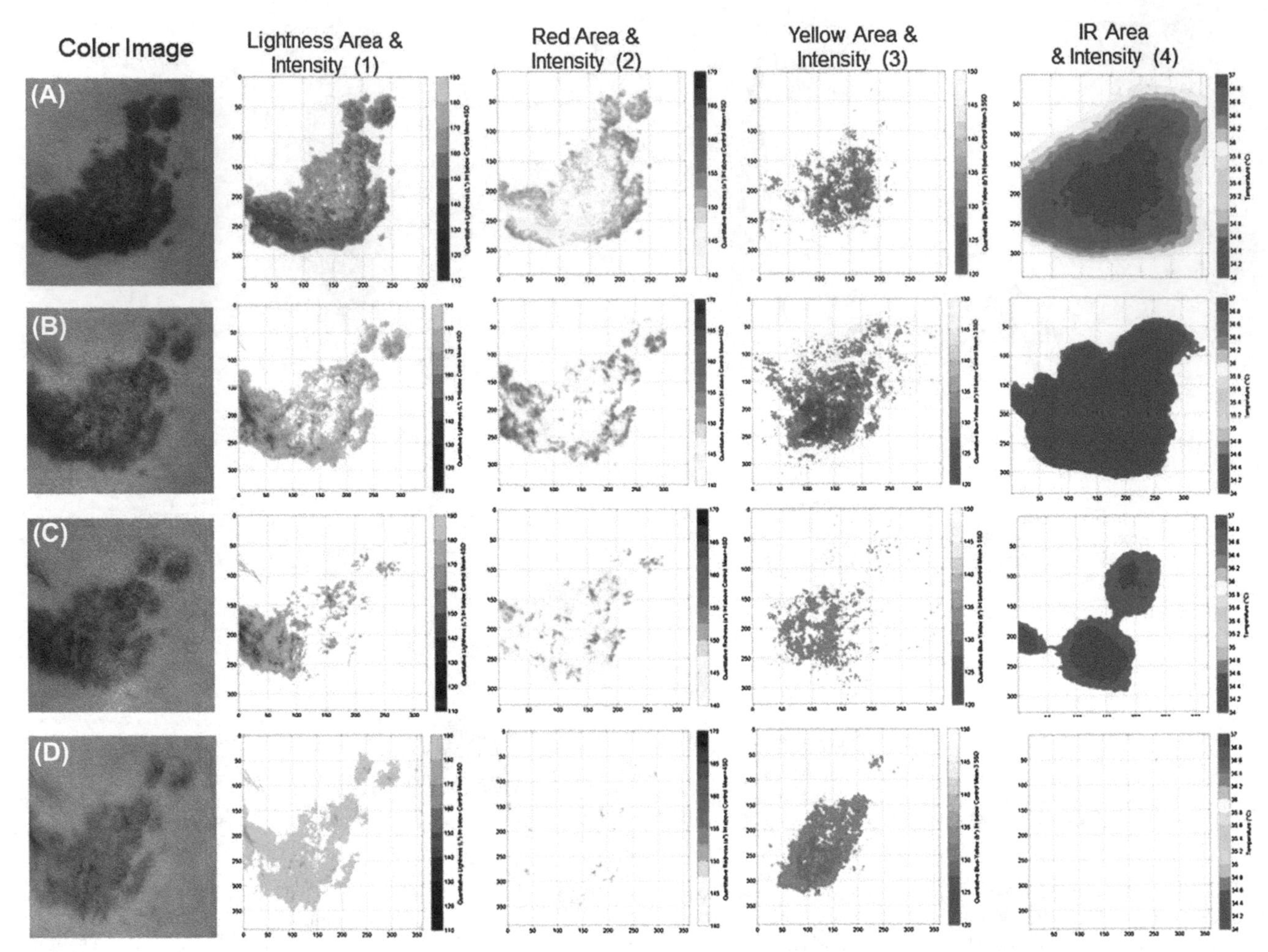

FIGURE 38.7 **Extracted features and intensity distribution maps of a mixed infantile hemangioma (IH).** Segmentation algorithms were used to generate intensity distribution maps (for coregistered images) and quantify the features for a mixed IH at baseline (A) and over 7 months of propranolol [79]. Data for each age are provided with areas (pixels) and intensities as mean pixels above each threshold.

Age	Height (mm)	Volume (mm^3)	Lightness Intensity	Lightness Area	Red Intensity	Red Area	Yellow Intensity	Yellow Area	Infrared Intensity	Infrared Area
3.0	5.3	2.5	−16.6	46,071	4.3	32,697	−1.9	11,590	0.70	82,074
4.3	2.5	1.2	−13.7	36,660	5.1	36,005	−2.5	26,289	0.43	59,163
6.0	1.8	0.9	−10.5	14,839	3.2	21,638	−1.0	9901	0.20	23,608
9.9	1.8	0.6	−5.4	5570	1.8	9693	−1.5	16,876	0.00	0

Red intensity, red area, and temperature increased and yellow intensity decreased after 1 month (B). Height, lightness intensity, lightness area, and thermal area decreased. By month 2, height, lightness intensity, red intensity, infrared intensity and infrared area decreased (B and C). Yellow intensity increased over time (D). At 6 months, infrared intensity and area were zero, ie, no region of the IH above the threshold temperature (D).

Multiphoton Tomography

Multiphoton multispectral tomography (fluorescence lifetime) imaging was conducted in vivo and later ex vivo (excised lesions) for subjects with basal cell carcinoma and compared with normal skin [106]. Serial images were obtained at 10-mm depth intervals throughout the epidermis to the dermis. The regions of interest were specified by manual thresholds to reduce background. A pixel-by-pixel transformation image segmentation algorithm was used to enhance brightness. User-specified thresholds were applied. Red and yellow channel images produced the greatest differentiation in basal cell carcinoma versus normal skin.

Category 4: Vascular Lesions

Thermal Imaging

High-resolution digital images of IHs were collected. Simultaneous infrared thermal and low-resolution color images were taken. Color and thermal images were coregistered with a landmark-based registration algorithm because the region of thermal activity extended beyond the visible tumor borders. Coregistration was accomplished with at least three control points easily identified in both images, eg, anatomical landmarks, tumor features, and/or an interface point (clothing, background). The procedure was infrared to low-resolution color image registration, low-resolution color to high-resolution color registration, control site low- to high-resolution color, and control high-resolution color to IH high-resolution color. Affine transformations were performed at each registration step to maintain collinearity and distance ratios.

Algorithms were applied to the coregistered images to extract physiologically relevant information regarding IH status [79]. Color images were separated into L*, a*, and b* images. Histograms of IH and controls were compared for 120 evaluations to determine thresholds (Fig. 38.6). They were: L* (lightness) values less than $\mu - 4\sigma$ (mean: 4 $\times$ standard deviations), a* (red color) values higher than $\mu + 4\sigma$, b* (yellow color) values less than $\mu - 3.5\sigma$, and infrared temperature higher than $\mu + 3\sigma$. Segmentation algorithms were generated intensity-distribution maps (coregistered images) and quantified IH features (Fig. 38.7).

Photoacoustic Imaging

Photoacoustic imaging (laser excitation at 422 and 530 nm) generated different responses for vascular and pigmented lesions among 15 subjects [107]. Analyses by the method of probability density (assumed to be gaussian) and a bayesian probit regression model were compared. The lesion classification rates were 100% using probability density and 80% with the bayesian model.

CONCLUSIONS

Skin imaging has undergone considerable, impressive growth, resulting in substantial progress by investigators with varying expertise. The techniques have proliferated from the first use of photography to provide a clinical record and measure progress over time.

The evidence strongly suggests that visual examination is inaccurate, expert-intensive, and time consuming. The expert-intensive requisite is worthy of comment because medical care is based upon multiple observers, eg, physicians and nurses who change daily and who may not have sufficient training. Consequently, there is variability in evaluation and conclusions about skin condition status that may lead to delayed or ineffective treatment, delayed healing, infection, pain, and increased costs. The recognized need for objective, reliable measurement of skin condition features has driven the development of skin imaging modalities. During the past 10 years, techniques, such as hyperspectral imaging, multiphoton tomography, OCT, RCM, thermography, and photoacoustic imaging have emerged to increase capability for examining skin structure, functionality, and treatment response.

Selection of image processing methods depends upon four interrelated factors: (1) purpose; (2) skin condition, eg, disease, wound; (3) condition characteristics; and (4) imaging technique.

The purpose is generally: (1) diagnosis, ie, presence or absence or which of several conditions; and/or (2) extent/severity of the abnormal skin condition at baseline and over time to determine treatment effectiveness. Diagnostic accuracy from digital images depends upon image quality [108].

An assortment of image processing methods have been developed, implemented, and evaluated for accuracy, sensitivity, and specificity. Key drivers for the development of new image processing strategies have been: (1) increased capabilities of the imaging techniques, such as RCM, multiphoton tomography, and multispectral imaging to obtain subcellular information; (2) the need for more accurate disease diagnosis that does not rely on years of experience; and (3) in vivo methods to reduce the need for invasive biopsies. These methods require that more sophisticated image processing techniques be developed to handle the complexity of 3D data reconstruction [63].

FUTURE PERSPECTIVE

The future of skin imaging and image processing shows promise for further advancement. New application and opportunities are likely, as are the challenges. The following points emerge from this review of the current status:

- Investigators often cite limitations of their work because of "small sample sizes" for the condition of interest. Given the importance and value to patients, there are needs for accessible databases for use in development, effective forums for information sharing, and education for first-level healthcare providers.
- Implementation of validated techniques and processing methods into general use will require

organized effort. The 2011 technical report by Parsons et al. critically reviews the state of the science, advantages, disadvantages, and indicated next steps [109]. This is a practical view in contrast to the excitement and insight typically found in the research literature. How do we move from research to the clinic? What partners do dermatologists, nurses, surgeons, and healthcare providers need? Change is difficult.

- New technique development is costly and may limit access by the healthcare community where costs for imaging have to be justified. The stage is set for the creation of new partnerships to overcome this limitation and benefit method development.
- Reports using combinations of imaging and analysis techniques are expected to increase to extend the information available to clinicians. Color and texture modalities are being combined with methods to evaluate skin function, including ultrasound, thermography, and laser Doppler perfusion imaging. Image processing methods to coregister the information are essential for the multimodal approach.
- There are opportunities for improved patient outcomes by implementation of objective imaging and processing modalities, such as wound care [110]. Wounds are prevalent, expected to increase as lifespans get longer, and need to be monitored for treatment optimization. Resource limitations regarding expert observers present opportunities for imaging and automatic processing. Costs for chronic wounds are estimated to be $25 billion in the United States and growing [111]. Early detection of skin compromise, before it is visible to an observer, is possible with imaging methods and will most likely be critical for reducing wound care costs.

List of Acronyms and Abbreviations

3D Three-dimensional
ABCD Asymmetry, border sharpness, color variegation, diameter
CIEL*a*b* An orthogonal color space representing lightness, red/green and blue/yellow color
CT Computerized tomography
HSI Hue, saturation, and intensity
HSV Hue, saturation, and value
IH Infantile hemangioma
IR Infrared
L*a*b* An orthogonal color space representing lightness, red/green, and blue/yellow color
LED Light emitting diode
m + s ($\mu + \sigma$) Mean plus standard deviation
mm Micrometer
MRI Magnetic resonance imaging
Nd:YAG Neodymium:yttrium-aluminum-garnet
nm nanometer
ns nanosecond
OCT Optical coherence tomography
PubMed Free information database maintained by the National Center for Biotechnology Information at the National Library of Medicine
RCM Reflectance confocal microscopy
RGB Red, green, blue color space
ROI Region of interest
SC Stratum corneum
SCINEXA Score of intrinsic and extrinsic skin aging
SIFT Scale-invariant feature transformation
SVM Support vector machine
VAS Visual analog scale

Glossary

Doppler effect The change in frequency of energy that occurs with movement toward the source.
Feature An attribute or characteristic of interest within an image, generally descriptive of the condition, eg, red color of irritated skin.
Histogram A plot showing the intensity distribution within an image from values of 0–255.
Hypercube The three-dimensional data set obtained with hyperspectral imaging.
Innate immunity Defense mechanisms that provide immediate protection from antigen, eg, the skin barrier.
Photodamage Skin characteristics that result from cumulative exposure to ultraviolet radiation, eg, pigmented spots, wrinkles.
Pixel The smallest part of a two-dimensional image with equal dimensions for x and y.
Threshold A limit or value used to separate parts of an image, such as noise.
Tomography A method to produce a three-dimensional image by collecting a series of parallel two-dimensional images.
Voxel The smallest part of a three dimensional image with equal dimensions for x, y, and z.

Acknowledgments

Some of the information presented in this chapter is derived from research that was funded by the Society of Pediatric Dermatology, the American Foundation of Pharmaceutical Education Pre-Doctoral Fellowship, the Center for Clinical & Translational Science & Training, and the Imaging Research Center. The project was supported by the National Center for Research Resources and the National Center for Advancing Translational Sciences, National Institutes of Health, through Grant eight UL1 TR000077-05. The content is solely the responsibility of the authors and does not necessarily represent the official views of the NIH.

References

[1] Visscher M, Narendran V. The ontogeny of skin. Adv Wound Care 2014;3(4):291–303.
[2] Visscher MO. Imaging skin: past, present and future perspectives. Giornale italiano di dermatologia e venereologia : organo ufficiale. Societa italiana di dermatologia e sifilografia 2010; 145(1):11–27.
[3] The American heritage science dictionary. Orlando, FL, USA: Houghton Mifflin Company; 2005 [Imaging].
[4] Taylor S, Westerhof W, Im S, Lim J. Noninvasive techniques for the evaluation of skin color. J Am Acad Dermatol 2006;54 (5 Suppl. 2):S282–90.

[5] Serup J. Skin irritation: objective characterization in a clinical perspective. In: Wilhelm KP, Elsner P, Berardesca E, Maibach HI, editors. Bioengineering of the skin: skin surface imaging and analysis. Boca Raton: CRC Press; 1997. p. 261–73.
[6] Galdino GM, Vogel JE, Vander Kolk CA. Standardizing digital photography: it's not all in the eye of the beholder. Plast Reconstr Surg 2001;108(5):1334–44.
[7] Anderson RR, Parrish JA. The optics of human skin. J Invest Dermatol 1981;77(1):13–9.
[8] Arce-Lopera C, Igarashi T, Nakao K, Okajima K. Image statistics on the age perception of human skin. Skin Res Technol 2012.
[9] Nkengne A, Bertin C, Stamatas GN, Giron A, Rossi A, Issachar N, et al. Influence of facial skin attributes on the perceived age of Caucasian women. J Eur Acad Dermatol Venereol 2008;22(8): 982–91.
[10] Wagner JK, Jovel C, Norton HL, Parra EJ, Shriver MD. Comparing quantitative measures of erythema, pigmentation and skin response using reflectometry. Pigment Cell Res 2002; 15(5):379–84.
[11] Jordan WE, Lawson KD, Berg RW, Franxman JJ, Marrer AM. Diaper dermatitis: frequency and severity among a general infant population. Pediatr Dermatol 1986;3(3):198–207.
[12] Hanifin JM, Thurston M, Omoto M, Cherill R, Tofte SJ, Graeber M. The eczema area and severity index (EASI): assessment of reliability in atopic dermatitis. EASI Evaluator Group. Exp Dermatol 2001;10(1):11–8.
[13] Sugarman JL, Fluhr JW, Fowler AJ, Bruckner T, Diepgen TL, Williams ML. The objective severity assessment of atopic dermatitis score: an objective measure using permeability barrier function and stratum corneum hydration with computer-assisted estimates for extent of disease. Arch Dermatol 2003;139(11): 1417–22.
[14] Pressure Ulcer Scale for Healing (PUSH). Push Tool 3.0. National pressure ulcer Advisory Panel. 1998.
[15] Horfelt C, Funk J, Frohm-Nilsson M, Wiegleb Edstrom D, Wennberg AM. Topical methyl aminolaevulinate photodynamic therapy for treatment of facial acne vulgaris: results of a randomized, controlled study. Br J Dermatol 2006;155(3):608–13.
[16] Sullivan T, Smith J, Kermode J, McIver E, Courtemanche DJ. Rating the burn scar. J Burn Care Rehabil 1990;11(3):256–60.
[17] Vierkotter A, Ranft U, Kramer U, Sugiri D, Reimann V, Krutmann J. The SCINEXA: a novel, validated score to simultaneously assess and differentiate between intrinsic and extrinsic skin ageing. J Dermatol Sci 2009;53(3):207–11.
[18] Visscher M. A practical method for rapid measurement of skin condition. Newborn Infant Nurs Rev 2014;14:147–53.
[19] Malvehy J, Puig S, Argenziano G, Marghoob AA, Soyer HP. International Dermoscopy Society Board m. Dermoscopy report: proposal for standardization: results of a consensus meeting of the International Dermoscopy Society. J Am Acad Dermatol 2007; 57(1):84–95.
[20] Bertrand J, McCuaig C, Dubois J, Hatami A, Ondrejchak S, Powell J. Propranolol versus prednisone in the treatment of infantile hemangiomas: a retrospective comparative study. Pediatr Dermatol 2011;28(6):649–54.
[21] Ho NT, Lansang P, Pope E. Topical imiquimod in the treatment of infantile hemangiomas: a retrospective study. J Am Acad Dermatol 2007;56(1):63–8.
[22] Kawai K, Kawai J, Nakagawa M, Kawai K. Effects of detergents. In: Wilhelm K, Elsner P, Berardesca E, Maibach HI, editors. Bioengineering of the skin: skin surface imaging and analysis. Boca Raton: CRC Press; 1997. p. 303–14.
[23] Mattsson U, Jonsson A, Jontell M, Cassuto J. Digital image analysis (DIA) of colour changes in human skin exposed to standardized thermal injury and comparison with laser Doppler measurements. Comput Methods Progr Biomed 1996;50(1):31–42.
[24] Barczak CA, Barnett RI, Childs EJ, Bosley LM. Fourth national pressure ulcer prevalence survey. Adv Wound Care 1997;10(4): 18–26.
[25] Henderson CT, Ayello EA, Sussman C, Leiby DM, Bennett MA, Dungog EF, et al. Draft definition of stage I pressure ulcers: inclusion of persons with darkly pigmented skin. NPUAP Task Force on Stage I Definition and Darkly Pigmented Skin. Adv Wound Care 1997;10(5):16–9.
[26] De Luca D, Zecca E, Zuppa AA, Romagnoli C. The joint use of human and electronic eye: visual assessment of jaundice and transcutaneous bilirubinometry. Turk J Pediatr 2008;50(5): 456–61.
[27] Wannous H, Treuillet S, Luas Y. Robust tissue classification for reproducible wound assessment in telemedicine environments. J Electron Imaging 2010;19(2):023002.
[28] Harvey NT, Budgeon CA, Leecy T, Beer TW, Kattampallil J, Yu L, et al. Interobserver variability in the diagnosis of circumscribed sebaceous neoplasms of the skin. Pathology 2013;45(6):581–6.
[29] Morton CA, Mackie RM. Clinical accuracy of the diagnosis of cutaneous malignant melanoma. Br J Dermatol 1998;138(2): 283–7.
[30] Mukherjee R, Manohar DD, Das DK, Achar A, Mitra A, Chakraborty C. Automated tissue classification framework for reproducible chronic wound assessment. Biomed Res Int 2014; 2014:851582.
[31] Rock I, Gallant J, Kube P. An introduction to vision science. In: Palmer SE, editor. Vision science. Cambridge: MIT Press; 1999; p. 760.
[32] Chardon A, Cretois I, Hourseau C. Skin colour typology and suntanning pathways. Int J Cosmet Sci 1991;13(4):191–208.
[33] Takiwaki H. Measurement of skin color: practical application and theoretical considerations. J Med Invest 1998;44(3–4):121–6.
[34] Aspres N, Egerton IB, Lim AC, Shumack SP. Imaging the skin. Aust J Dermatol 2003;44(1):19–27.
[35] Fogelberg A, Ioffreda M, Helm KF. The utility of digital clinical photographs in dermatopathology. J Cutan Med Surg 2004;8(2): 116–21.
[36] Nystrom J, Geladi P, Lindholm-Sethson B, Rattfelt J, Svensk AC, Franzen L. Objective measurements of radiotherapy-induced erythema. Skin Res Technol 2004;10(4):242–50.
[37] Oduncu H, Hoppe A, Clark M, Williams RJ, Harding KG. Analysis of skin wound images using digital color image processing: a preliminary communication. Int J Low Extrem Wounds 2004; 3(3):151–6.
[38] Setaro M, Sparavigna A. Quantification of erythema using digital camera and computer-based colour image analysis: a multicentre study. Skin Res Technol 2002;8(2):84–8.
[39] Coelho SG, Miller SA, Zmudzka BZ, Beer JZ. Quantification of UV-induced erythema and pigmentation using computer-assisted digital image evaluation. Photochem Photobiol 2006; 82(3):651–5.
[40] Szychta P, Stewart K, Anderson W. Treatment of infantile hemangiomas with propranolol: clinical guidelines. Plast Reconstr Surg 2014;133(4):852–62.
[41] Choi JW, Kim BR, Lee HS, Youn SW. Characteristics of subjective recognition and computer-aided image analysis of facial erythematous skin diseases: a cornerstone of automated diagnosis. Br J Dermatol 2014;171(2):252–8.
[42] Lock-Andersen J, Wulf HC. Seasonal variation of skin pigmentation. Acta Derm Venereol 1997;77(3):219–21.
[43] Quigley EA, Tokay BA, Jewell ST, Marchetti MA, Halpern AC. Technology and technique standards for camera-acquired digital dermatologic images: a systematic review. JAMA Dermatol 2015.
[44] Soendergaard C, Nielsen OH, Skak K, Ropke MA, Seidelin JB, Kvist PH. Objective quantification of immune cell infiltrates and epidermal proliferation in psoriatic skin: a comparison of

digital image analysis and manual counting. Appl Immunohistochem Mol Morphol 2015.

[45] Malvehy J, Puig S, Argenziano G, Marghoob AA, Soyer HP. Dermoscopy report: proposal for standardization: results of a consensus meeting of the International Dermoscopy Society. J Am Acad Dermatol 2007;57(1):84–95.

[46] Dhawan AP, Gordon R, Rangayyan RM. Nevoscopy: three-dimensional computed tomography of nevi and melanomas in situ by transillumination. IEEE Trans Med Imaging 1984; 3(2):54–61.

[47] Demirli R, Otto P, Viswanathan R, Patwardhan S, Larkey J. RBXTM technology overview. Fairfield, NJ, USA: Canfield Imaging Systems; 2007.

[48] Vestergaard ME, Macaskill P, Holt PE, Menzies SW. Dermoscopy compared with naked eye examination for the diagnosis of primary melanoma: a meta-analysis of studies performed in a clinical setting. Br J Dermatol 2008;159(3):669–76.

[49] Kittler H, Pehamberger H, Wolff K, Binder M. Diagnostic accuracy of dermoscopy. The Lancet Oncol 2002;3(3):159–65.

[50] Calzavara-Pinton P, Longo C, Venturini M, Sala R, Pellacani G. Reflectance confocal microscopy for in vivo skin imaging. Photochem Photobiol 2008;84(6):1421–30.

[51] Gonzalez S. Confocal reflectance microscopy in dermatology: promise and reality of non-invasive diagnosis and monitoring. Actas dermo-sifiliograficas 2009;100(Suppl. 2):59–69.

[52] Pellacani G, Guitera P, Longo C, Avramidis M, Seidenari S, Menzies S. The impact of in vivo reflectance confocal microscopy for the diagnostic accuracy of melanoma and equivocal melanocytic lesions. J Invest Dermatol 2007;127(12):2759–65.

[53] Farnetani F, Scope A, Braun RP, Gonzalez S, Guitera P, Malvehy J, et al. Skin cancer diagnosis with reflectance confocal microscopy: reproducibility of feature recognition and accuracy of diagnosis. JAMA Dermatol 2015.

[54] Hoogedoorn L, Peppelman M, van de Kerkhof PC, van Erp PE, Gerritsen MJ. The value of in vivo reflectance confocal microscopy in the diagnosis and monitoring of inflammatory and infectious skin diseases: a systematic review. Br J Dermatol 2015; 172(5):1222–48.

[55] Stamatas GN, Kollias N. Blood stasis contributions to the perception of skin pigmentation. J Biomed Opt 2004;9(2):315–22.

[56] Moncrieff M, Cotton S, Claridge E, Hall P. Spectrophotometric intracutaneous analysis: a new technique for imaging pigmented skin lesions. Br J Dermatol 2002;146(3):448–57.

[57] Stamatas GN, Zmudzka BZ, Kollias N, Beer JZ. In vivo measurement of skin erythema and pigmentation: new means of implementation of diffuse reflectance spectroscopy with a commercial instrument. Br J Dermatol 2008;159(3):683–90.

[58] Kollias N, Seo I, Bargo PR. Interpreting diffuse reflectance for in vivo skin reactions in terms of chromophores. J Biophotonics 2010;3(1–2):15–24.

[59] Wang H, Cuijpers RH, Luo MR, Heynderickx I, Zheng Z. Optimal illumination for local contrast enhancement based on the human visual system. J Biomed Opt 2015;20(1):015005.

[60] Sprigle S, Zhang L, Duckworth M. Detection of skin erythema in darkly pigmented skin using multispectral images. Adv Skin Wound Care 2009;22(4):172–9.

[61] Kuzmina I, Diebele I, Jakovels D, Spigulis J, Valeine L, Kapostinsh J, et al. Towards noncontact skin melanoma selection by multispectral imaging analysis. J Biomed Opt 2011;16(6): 060502.

[62] Lu G, Fei B. Medical hyperspectral imaging: a review. J Biomed Opt 2014;19(1):10901.

[63] Calin MA, Coman T, Parasca SV, Bercaru N, Savastru R, Manea D. Hyperspectral imaging-based wound analysis using mixture-tuned matched filtering classification method. J Biomed Opt 2015;20(4):046004.

[64] Mogensen M, Thrane L, Joergensen TM, Andersen PE, Jemec GB. Optical coherence tomography for imaging of skin and skin diseases. Semin Cutan Med Surg 2009;28(3):196–202.

[65] Tsai TH, Jee SH, Dong CY, Lin SJ. Multiphoton microscopy in dermatological imaging. J Dermatol Sci 2009;56(1):1–8.

[66] Konig K. Clinical multiphoton tomography. J Biophotonics 2008; 1(1):13–23.

[67] Dimitrow E, Ziemer M, Koehler MJ, Norgauer J, Konig K, Elsner P, et al. Sensitivity and specificity of multiphoton laser tomography for in vivo and ex vivo diagnosis of malignant melanoma. J Invest Dermatol 2009;129(7):1752–8.

[68] Wu Z, Liu HH, Lebanowski L, Liu Z, Hor PH. A basic step toward understanding skin surface temperature distributions caused by internal heat sources. Phys Med Biol 2007;52(17): 5379–92.

[69] Jones BF. A reappraisal of the use of infrared thermal image analysis in medicine. IEEE Trans Med Imaging 1998;17(6):1019–27.

[70] Fujimasa I, Chinzei T, Saito I. Converting far infrared image information to other physiological data. IEEE Eng Med Biol Mag 2000;19(3):71–6.

[71] Keyserlingk JR, Ahlgren PD, Yu E, Belliveau N, Yassa M. Functional infrared imaging of the breast. IEEE Eng Med Biol Mag 2000;19(3):30–41.

[72] Tepper M, Gannot I. Parametric study of different contributors to tumor thermal profile. Proc SPIE 2014;8940. 89400P-1-P-12.

[73] Draper JW, Boag JW. Skin temperature distributions over veins and tumours. Phys Med Biol 1971;16(4):645–54.

[74] Deng ZS, Liu J. Mathematical modeling of temperature mapping over skin surface and its implementation in thermal disease diagnostics. Comput Biol Med 2004;34(6):495–521.

[75] Pirtini Çetingül M, Herman C. The assessment of melanoma risk using the dynamic infrared imaging technique. J Therm Sci Eng Appl 2011;3(3):031006.

[76] Leaute-Labreze C, Hoeger P, Mazereeuw-Hautier J, Guibaud L, Baselga E, Posiunas G, et al. A randomized, controlled trial of oral propranolol in infantile hemangioma. N Engl J Med 2015; 372(8):735–46.

[77] Santa Cruz GA, Bertotti J, Marin J, Gonzalez SJ, Gossio S, Alvarez D, et al. Dynamic infrared imaging of cutaneous melanoma and normal skin in patients treated with BNCT. Appl Radiat Isot 2009;67(7–8 Suppl.):S54–8.

[78] Cetingul MP, Herman C. A heat transfer model of skin tissue for the detection of lesions: sensitivity analysis. Phys Med Biol 2010; 55(19):5933–51.

[79] Burkes SA, Adams DM, Hammill AM, Chute C, Eaton KP, Welge JA, et al. Skin imaging modalities quantify progression and stage of infantile hemangiomas. Br J Dermatol 2015.

[80] Zhang HF, Maslov K, Stoica G, Wang LV. Functional photoacoustic microscopy for high-resolution and noninvasive in vivo imaging. Nat Biotechnol 2006;24(7):848–51.

[81] Masood A, Al-Jumaily AA. Computer aided diagnostic support system for skin cancer: a review of techniques and algorithms. Int J Biomed Imaging 2013;2013:323268.

[82] Rastgoo M, Garcia R, Morel O, Marzani F. Automatic differentiation of melanoma from dysplastic nevi. Comput Med Imaging Graphics 2015;43:44–52.

[83] Korotkov K, Garcia R. Computerized analysis of pigmented skin lesions: a review. Artif Intell Med 2012;56(2):69–90.

[84] Lee HC. Color imaging science. Cambridge: Cambridge University Press; 2005.

[85] CIE publications No. 15.2 Colorimetry. 2nd ed. Vienna: Central Bureau of the CIE; 1986.

[86] Wyszecki G, Stiles WS. Color science: concepts and methods, quantitative data and formulae. 2nd ed. New York: Wiley; 1982.

[87] Weatherall IL, Coombs BD. Skin color measurements in terms of CIELAB color space values. J Invest Dermatol 1992;99(4):468–73.

[88] Szychta P, Al-Nakib K, Anderson W, Stewart K, Quaba A. Quantitative method for evaluation of aesthetic results after laser treatment for birthmarks. Lasers Med Sci 2013;28(6):1567–72.
[89] Elsner P. Skin color. In: Berardesca E, Elsner P, Wilhelm KP, Maibach HI, editors. Bioengineering of the skin: methods and instrumentation. Boca Raton: CRC Press; 1995. p. 29–40.
[90] Canning J, Barford B, Sullivan D, Wickett R, Visscher M. Use of digital photography and image analysis techniques to quantify erythema in health care workers. Skin Res Technol 2009;15(1): 24–34.
[91] Westerhof W, van Hasselt BA, Kammeijer A. Quantification of UV-induced erythema with a portable computer controlled chromameter. Photodermatol 1986;3(5):310–4.
[92] Decenciere E, Tancrede-Bohin E, Dokladal P, Koudoro S, Pena AM, Baldeweck T. Automatic 3D segmentation of multiphoton images: a key step for the quantification of human skin. Skin Res Technol 2013;19(2):115–24.
[93] Tancrede-Bohin E, Baldeweck T, Decenciere E, Brizion S, Victorin S, Parent N, et al. Non-invasive short-term assessment of retinoids effects on human skin in vivo using multiphoton microscopy. J Eur Acad Dermatol Venereol 2015;29(4): 673–81.
[94] Chen G, Lui H, Zeng H. Image segmentation for integrated multiphoton microscopy and reflectance confocal microscopy imaging of human skin in vivo. Quant Imaging Med Surg 2015;5(1):17–22.
[95] Pereira SM, Frade MA, Rangayyan RM, Azevedo Marques PM. Classification of color images of dermatological ulcers. IEEE J Biomed Health Inform 2013;17(1):136–42.
[96] Singh B, Kushwaha N, Vyas OP. A feature subset selection technique for high dimensional data using symmetric uncertainty. J Data Anal Inf Process 2014;2:95–105.
[97] Veredas F, Mesa H, Morente L. Binary tissue classification on wound images with neural networks and bayesian classifiers. IEEE Trans Med Imaging 2010;29(2):410–27.
[98] Abbas Q, Fondon I, Rashid M. Unsupervised skin lesions border detection via two-dimensional image analysis. Comput Methods Progr Biomed 2011;104(3):e1–15.
[99] Ganster H, Pinz A, Rohrer R, Wildling E, Binder M, Kittler H. Automated melanoma recognition. IEEE Trans Med Imaging 2001;20(3):233–9.
[100] Burroni M, Sbano P, Cevenini G, Risulo M, Dell'eva G, Barbini P, et al. Dysplastic naevus vs. in situ melanoma: digital dermoscopy analysis. Br J Dermatol 2005;152(4):679–84.
[101] Segura S, Puig S, Carrera C, Palou J, Malvehy J. Development of a two-step method for the diagnosis of melanoma by reflectance confocal microscopy. J Am Acad Dermatol 2009;61(2):216–29.
[102] Gareau D, Hennessy R, Wan E, Pellacani G, Jacques SL. Automated detection of malignant features in confocal microscopy on superficial spreading melanoma versus nevi. J Biomed Opt 2010;15(6):061713.
[103] Raphael AP, Kelf TA, Wurm EM, Zvyagin AV, Soyer HP, Prow TW. Computational characterization of reflectance confocal microscopy features reveals potential for automated photoageing assessment. Exp Dermatol 2013;22(7):458–63.
[104] Wurm EM, Longo C, Curchin C, Soyer HP, Prow TW, Pellacani G. In vivo assessment of chronological ageing and photoageing in forearm skin using reflectance confocal microscopy. Br J Dermatol 2012;167(2):270–9.
[105] Carrara M, Tomatis S, Bono A, Bartoli C, Moglia D, Lualdi M, et al. Automated segmentation of pigmented skin lesions in multispectral imaging. Phys Med Biol 2005;50(22):N345–57.
[106] Patalay R, Talbot C, Alexandrov Y, Lenz MO, Kumar S, Warren S, et al. Multiphoton multispectral fluorescence lifetime tomography for the evaluation of basal cell carcinomas. PLoS One 2012; 7(9):e43460.
[107] Swearingen JA, Holan SH, Feldman MM, Viator JA. Photoacoustic discrimination of vascular and pigmented lesions using classical and Bayesian methods. J Biomed Opt 2010;15(1):016019.
[108] Weingast J, Scheibbock C, Wurm EM, Ranharter E, Porkert S, Dreiseitl S, et al. A prospective study of mobile phones for dermatology in a clinical setting. J Telemed Telecare 2013;19(4): 213–8.
[109] Parsons SK, Chan JA, Yu WW, Obadan N, Raticheck SJ, Lee J, et al. Noninvasive diagnostic techniques for the detection of skin cancers. 2011. Contract No.: Technical Brief No. 11.
[110] Salcido RS. Beyond photography: wound imaging. Adv Skin Wound Care 2011;24(2):56.
[111] Sen CK, Gordillo GM, Roy S, Kirsner R, Lambert L, Hunt TK, et al. Human skin wounds: a major and snowballing threat to public health and the economy. Wound Repair Regen 2009; 17(6):763–71.

Index

Note: 'Page numbers followed by "f" indicate figures, "t" indicate tables', and "b" indicate boxes'.

N

Q

R

S

T

CPSIA information can be obtained
at www.ICGtesting.com
Printed in the USA
LVHW050535040820
662268LV00010B/672